Lecture Notes in Computer Science 8656

Commenced Publication in 1973
Founding and Former Series Editors:
Gerhard Goos, Juris Hartmanis, and Jan van Leeuwen

Barry O'Sullivan (Ed.)

Principles and Practice of Constraint Programming

20th International Conference, CP 2014
Lyon, France, September 8-12, 2014
Proceedings

Springer

Volume Editor

Barry O'Sullivan
Insight Centre for Data Analytics
School of Computer Science and Information Technology
University College Cork
Western Road, Cork, Ireland
E-mail: barry.osullivan@insight-centre.org

ISSN 0302-9743 e-ISSN 1611-3349
ISBN 978-3-319-10427-0 e-ISBN 978-3-319-10428-7
DOI 10.1007/978-3-319-10428-7
Springer Cham Heidelberg New York Dordrecht London

Library of Congress Control Number: 2014946201

LNCS Sublibrary: SL 2 – Programming and Software Engineering

Typesetting: Camera-ready by author, data conversion by Scientific Publishing Services, Chennai, India

Printed on acid-free paper

Springer is part of Springer Science+Business Media (www.springer.com)

Preface

This volume contains the proceedings of the 20th International Conference on the Principles and Practice of Constraint Programming (CP 2014), which was held in Lyon, France, from September 8–12, 2014. A comprehensive web-site about the conference is available at http://cp2014.a4cp.org.

The CP conference is the premier annual international conference on constraint programming. It is concerned with all aspects of computing with constraints, including theory, algorithms, environments, languages, models, systems, and applications such as decision making, resource allocation, scheduling, configuration, planning, etc. The CP community is very keen to ensure it remains open to interdisciplinary research at the intersection between constraint programming and related fields such as search, satisfiability, knowledge representation and reasoning, machine learning, multi-agent systems, and operations research.

The CP 2014 program included presentations of high-quality scientific research papers and applications of constraints technology. In addition, for the first time, the program included a journal presentation track that was designed to provide a forum to discuss important results in the area of constraint programming that appeared recently in relevant journals, but had not been previously presented at CP, CPAIOR, or any other major AI conference.

The review process for CP 2014 relied on a multi-tier approach involving a senior Program Committee, dedicated regular Program Committees for both the main technical and application tracks, along with a set of additional reviewers recruited by Program Committee members. Authors chose to submit either long or short papers to either the main technical track or the application track. Alternatively, authors submitting to the journal presentation track submitted abstracts for review by a dedicated committee. All submissions to the technical track were assigned to a member of the senior Program Committee and three members of the Program Committee. All submissions to the application track were assigned to the chair of that track and three members of its Program Committee, which was the same approach adopted for the journal presentation track. Authors were given an opportunity to respond to reviews before a detailed discussion was undertaken at the level of the Program Committees, overseen by the Program Chair, the senior Program Committee member or track chair, as appropriate.

A meeting of the senior Program Committee was held at University College Cork at the end of May, chaired by the Program Chair, where the reviews, author feedback, and discussions on every paper were discussed in detail. The principle under which these discussions took place was that all papers deemed to be of sufficient quality were accepted into the program. The result of this was that the acceptance rate for the technical track was a little over 50% while the application track accepted 66% of papers. Abstracts submitted to the journal presentation track that satisfied the requirements for that track were accepted. Overall, the quality of submissions to the conference was very high, and the final program, as evidenced by these proceedings, was excellent. We selected a set of prize-winning papers, which are presented later in the front matter of the proceedings, including a best technical track paper, a best application track paper, a best student paper, and a runner-up best student paper.

The conference included four invited talks from distinguished scientists: Maria Fox, Patrick Prosser, Louis-Martin Rousseau, and Vijay Saraswat. Abstracts of these talks are included in the proceedings. We also benefitted from an excellent program of tutorials and workshops; these are also detailed further in the front matter of these proceedings.

Two elements of the conference program that are not reflected in the proceedings are the doctoral program and the 20th anniversary celebration. The doctoral program provided an opportunity for PhD students to meet each other as well as senior researchers in the field. The focus of the program was on mentoring students and providing a forum for them to exchange ideas, get feedback on their research, and benefit from a specially designed tutorial program. To mark the 20th anniversary of the conference, a special celebratory session at the conference was organized.

The task of producing an excellent scientific program for a conference like CP 2014 is a truly international undertaking, involving a large number of people from around the world. I would like to sincerely thank the members of the senior Program Committee, who not only took responsibility for overseeing the reviewing of a number of papers, but also took time out from their busy schedules to attend a weekend meeting in Cork in May. I would like to thank the members of the Program Committee, and the additional reviewers they recruited, for providing high-quality reviews and discussions on each and every paper submitted to the conference. A special word of thanks goes to the authors of all submissions to the conference.

I was very fortunate to work with a great team of people who chaired aspects of the conference: Mark Wallace (Application Track Chair), Justyna Petke and Andrea Rendl (Doctoral Program Chairs), Michela Milano (Workshop and Tutorial Chair), Francesca Rossi (Journal Track Chair), Pascal Van Hentenryck (20th Anniversary Celebration Chair), and Pierre Schaus (Publicity Chair). A very special thanks is deserved by Yves Deville and Christine Solnon, the Conference Chairs, who managed all aspects of the local arrangements, logistics, finances, and sponsorship. They were the very generous hosts of the conference itself, and provided exceptional hospitality to the delegates.

I would like to thank the Association for Constraint Programming, who entrusted the scientific program of the conference to me. It was a huge honour for me, as well as a career highlight. I would like to thank the many sponsors who provided generous financial support for the conference. A complete list of sponsors is provided later in these proceedings. Without the support of these sponsors the conference would not have been financially viable.

Finally, on a personal note, I would like to dedicate my work on this conference and this volume to the memory of my late uncle and godfather, Alan Lee (August 17, 1942 – November 7, 2013).

September 2014

Barry O'Sullivan

Prize-Winning Papers

Following the reviewing process and the senior Program Committee (SPC) meeting, a small committee of (S)PC members was established to assist the program chair in the selection of the best papers from the technical track of the conference. The committee for best technical track paper and best student paper comprised Nicolas Beldiceanu (TASC(CNRS/Inria), Mines Nantes), Peter Jeavons (University of Oxford), and Ian Miguel (University of St. Andrews). The best application track paper was recommended by the application track chair, Mark Wallace (Monash University and Opturion), who worked closely with Helmut Simonis (University College Cork). The best papers for CP 2014 are listed below. A runner-up was also deemed appropriate in the case of best student paper.

Best Technical Track Paper

On Broken Triangles
Martin Cooper, Achref El Mouelhi, Cyril Terrioux, and Bruno Zanuttini

Best Application Track Paper

Using CP in Automatic Test Generation for ABB Robotics' Paint Control System
Morten Mossige, Arnaud Gotlieb, and Hein Meling

Best Student Paper

On Compiling CNF into Decision-DNNF
Umut Oztok and Adnan Darwiche

Runner-Up Best Student Paper

A Complete Solver for Constraint Games
Thi-Van-Anh Nguyen and Arnaud Lallouet

Tutorials and Workshops

A feature of the CP 2014 conference program was a set of tutorials and workshops. Tutorials were expected to give an in-depth presentation of emerging and exciting topics that are relevant to a broad swath of the constraint programming community. On the other hand, the workshops provided an informal venue where participants were given the opportunity to present, discuss, and brainstorm on new ideas, technical topics, exciting new application areas, and cross-fertilization with other domains. The Workshop and Tutorial Chair for CP 2014 was Michela Milano (University of Bologna) who, with the Program Chair, selected the following tutorials and workshops for inclusion in the conference program. Each tutorial and workshop was submitted in response to an open call for proposals, and each was subjected to peer review.

Tutorials

The Past and Future of `csplib.org`: Why and How to Contribute?
Christopher Jefferson

Automated Reformulation of Constraint Models in Savile Row
Peter Nightingale

Social Choice
Francesca Rossi, Kristen Brent Venable, and Toby Walsh

MiniZinc 2.0
Peter J. Stuckey and Guido Tack

Workshops

ModRef 2014 - the 13th International Workshop on Constraint Modelling and Reformulation
Carlos Ansótegui

Constraint Programming Meets Verification 2014
Parosh Aziz Abdulla, Mohamed Faouzi Atig, Pierre Flener, Arnaud Gotlieb, and Justin Pearson

Constraint-Based Methods for Bioinformatics
Simon de Givry and Nicos Angelopoulos

Bridging the Gap Between Theory and Practice in Constraint Solvers
Philippe Jégou, Martin Cooper, Lakhdar Sais, and Bruno Zanuttini

Cloud Computing and Optimization
Jean-Charles Régin and Bertrand Le Cun

Conference Organization

Conference Chairs

Yves Deville	UCLouvain, Belgium
Christine Solnon	LIRIS, INSA Lyon/CNRS, France

Program Chair

Barry O'Sullivan	University College Cork, Ireland

Application Track Chair

Mark Wallace	Monash University and Opturion, Australia

Doctoral Program Chairs

Justyna Petke	University College London, UK
Andrea Rendl	NICTA and Monash University, Australia

Workshop and Tutorial Chair

Michela Milano	University of Bologna, Italy

Journal Track Presentation Chair

Francesca Rossi	University of Padova, Italy

20th Anniversary Celebration Chair

Pascal Van Hentenryck	NICTA and University of Melbourne, Australia

Publicity Chair

Pierre Schaus	UCLouvain, Belgium

Senior Program Committee

J. Christopher Beck	University of Toronto, Canada
Nicolas Beldiceanu	TASC(CNRS/Inria), Mines Nantes, France

Christian Bessiere	CNRS and University of Montpellier, France
Ken Brown	University College Cork, Ireland
Berthe Y. Choueiry	University of Nebraska-Lincoln, USA
David Cohen	Royal Holloway, University of London, UK
Yves Deville	UCLouvain, Belgium
Jimmy Lee	The Chinese University of Hong Kong, SAR China
Ian Miguel	University of St. Andrews, UK
Michela Milano	University of Bologna, Italy
Barry O'Sullivan	University College Cork, Ireland (Chair)
Patrick Prosser	Glasgow University, UK
Jean-Charles Régin	Université de Nice-Sophia Antipolis, France
Francesca Rossi	University of Padova, Italy
Christian Schulte	KTH Royal Institute of Technology, Sweden
Helmut Simonis	University College Cork, Ireland
Christine Solnon	LIRIS, INSA Lyon/CNRS, France
Peter Stuckey	NICTA and the University of Melbourne, Australia
Mark Wallace	Monash University and Opturion, Australia
Toby Walsh	NICTA and UNSW, Australia

Technical Track Program Committee

Carlos Ansótegui	Universitat de Lleida, Spain
Hadrien Cambazard	University of Grenoble Alpes, G-SCOP, France
Hubie Chen	Universidad del País Vasco and Ikerbasque, Spain
Geoffrey Chu	NICTA VRL, University of Melbourne, Australia
Remi Coletta	CNRS and University of Montpellier, France
Martin Cooper	IRIT, University of Toulouse, France
Rina Dechter	University of California at Irvine, USA
Boi Faltings	EPFL, Switzerland
Pierre Flener	Uppsala University, Sweden
Ian Gent	University of St. Andrews, UK
Diarmuid Grimes	University College Cork, Ireland
Emmanuel Hebrard	LAAS-CNRS, France
Brahim Hnich	Izmir University of Economics, Turkey
Peter Jeavons	University of Oxford, UK
Philippe Jégou	Université d'Aix-Marseille, LSIS, France
Peter Jonsson	Linköping University, Sweden
George Katsirelos	INRA, Toulouse, France
Zeynep Kiziltan	University of Bologna, Italy
Javier Larrosa	UPC, Spain

Christophe Lecoutre	Université d'Artois, France
Inês Lynce	INESC-ID, IST, Universidade de Lisboa, Portugal
Felip Manya	IIIA-CSIC, Spain
Radu Marinescu	IBM Research, Ireland
Christopher Mears	Monash University, Australia
Deepak Mehta	University College Cork, Ireland
Amnon Meisels	Ben-Gurion University, Israel
Pedro Meseguer	IIIA-CSIC, Spain
Laurent Michel	University of Connecticut, USA
Nina Narodytska	University of Toronto, Canada
Alexandre Papadopoulos	Université Pierre et Marie Curie (Paris 6), France
Justin Pearson	Uppsala University, Sweden
Gilles Pesant	Ecole Polytechnique de Montreal, Canada
Justyna Petke	University College London, UK
Luis Quesada	University College Cork, Ireland
Claude-Guy Quimper	Université Laval, Canada
Andrea Rendl	NICTA, Australia
Michel Rueher	University of Nice Sophia Antipolis, France
Lakhdar Sais	Université d'Artois, France
Pierre Schaus	UCLouvain, Belgium
Thomas Schiex	INRA Toulouse, France
Kostas Stergiou	University of Western Macedonia, Greece
Guido Tack	NICTA, Monash University, Australia
Johan Thapper	Université Paris-Est, Marne-la-Vallée, France
Michael Trick	Carnegie Mellon University, USA
Willem-Jan van Hoeve	Carnegie Mellon University, USA
Gerard Verfaillie	ONERA, France
Roland Yap	National University of Singapore, Singapore
Yuanlin Zhang	Texas Tech University, USA
Roie Zivan	Ben-Gurion University of the Negev, Israel
Stanislav Zivny	University of Oxford, UK

Application Track Program Committee

Theirry Benoist	Innovation 24, France
Lucas Bordeaux	Microsoft Research, UK
Mats Carlsson	SICS, Sweden
Håkan Kjellerstrand	Malmo, Sweden
Laurent Perron	Google, France
Siddhartha SenGupta	Tata Consultancy Services, India
Paul Shaw	IBM, France
Helmut Simonis	University College Cork, Ireland

Willem-Jan Van Hoeve	Carnegie Mellon University, USA
Mark Wallace	Monash University and Opturion, Australia (Chair)

Journal Presentation Track Program Committee

Christian Bessiere	CNRS and University of Montpellier, France
Jimmy Lee	The Chinese University of Hong Kong, SAR China
Patrick Prosser	Glasgow University, UK
Francesca Rossi	University of Padova, Italy (Chair)
Helmut Simonis	University College Cork, Ireland
K. Brent Venable	Tulane University, USA
Toby Walsh	NICTA and UNSW, Australia

Additional Reviewers

Arbelaez, Alejandro
Beyersdorff, Olaf
Bistarelli, Stefano
Bova, Simone
Carbonnel, Clément
Cire, Andre
Davies, Jessica
Di Gaspero, Luca
Duck, Gregory
Fontaine, Daniel
Gabàs, Joel
Gao, Yong
Gavanelli, Marco
Gay, Steven
Grinshpoun, Tal
Gutierrez, Julian
Gutierrez, Patricia
Hartert, Renaud
Jabbour, Said
Janota, Mikoláš
Kell, Brian
Leo, Kevin
Lhomme, Olivier
Li, Chu-Min
Li, Wei
Likitvivatanavong, Chavalit
Lombardi, Michele
Mairy, Jean-Baptiste
Marques-Silva, Joao
Martins, Ruben
Mauro, Jacopo
Mengel, Stefan
Michel, Claude
Monette, Jean-Noël
Nattaf, Margaux
Neveu, Bertrand
Nightingale, Peter
Okamoto, Steven
Paparrizou, Anastasia
Prestwich, Steve
Pérez, Jorge A.
Rollon, Emma
Roy, Pierre
Saint-Guillain, Michael
Schutt, Andreas
Siala, Mohamed
Slivovsky, Friedrich
Tabary, Sebastien
Terrioux, Cyril
Tjandraatmadja, Christian
Van Gelder, Allen
Vismara, Philippe
Wahbi, Mohamed
Zytnicki, Matthias

Sponsors

CP 2014 is very grateful to the following sponsors for their generous support of the conference.

AIMMS
Artificial Intelligence Journal (Elsevier)
Association for Constraint Programming (ACP)
Association Française pour la Programmation par Contraintes (AFPC)
Cadence
Centre National de la Recherche Scientifique (CNRS), France
Faculté des Sciences et Technologies, Université Lyon 1, France
Graduate School in Computing Science, Belgium (Grascomp)
ICTEAM/UCLouvain, Belgium
INSA Lyon, France
Laboratoire d'Informatique en Image et Systèmes d'information (LIRIS), France
Quintiq

The Association for Constraint Programming

The Association for Constraint Programming (ACP) aims to promote constraint programming in every aspect of the scientific world, by encouraging its theoretical and practical development, its teaching in academic institutions, its adoption in the industrial world, and its use in applications. The ACP is a non-profit association that uses the funds raised from its events to support activities for the CP community. Further information about the ACP, its activities, and membership, is available from its website at `http://www.a4cp.org`

Executive Committee

The current Executive Committee, which was formed on January 1, 2013, has the following membership:

Officers

President – Helmut Simonis (elected 2011)
Treasurer – Thomas Schiex (elected 2011)
Secretary – Willem-Jan van Hoeve (elected 2013)
Conference Coordinator – Pierre Flener (elected 2013)

Other Members

Yves Deville (elected 2011)
Guido Tack (elected 2013)
Roland Yap (elected 2011)

Ex-Officio Members

Past President – Barry O'Sullivan

Table of Contents

Invited Talks

Best Technical Track Paper

Best Application Track Paper

Best Student Paper

Runner-Up Best Student Paper

Technical Track

Application Track

Journal Presentation Track

A Modular Architecture for Hybrid Planning with Theories*

Maria Fox

Dept. Informatics, King's College London, UK
maria.fox@kcl.ac.uk

Planning technology has made huge strides, alongside other combinatorial optimisation solving technologies, over the past decade. Automated planning systems now exist for temporal and metric problems, including management of continuous time and concurrency, continuous numeric resources and action costs [3,1,2,12,7,8,11,9]. There is an increasing interest in combining planners with specialised solvers, such as optimisation alogorithms, to achieve a hybrid form of planning. In this context, the relationship between planning and model-checking, planning and constraint-solving and planning and control are all being clarified.

Synergies between different optimisation modelling and solving paradigms can be exploited to achieve new capabilities and improved performance of solvers. An example of this is recent work exploiting the developments in SAT solving, SAT Modulo Theories, in which atoms can be built from predicates, functions and constants whose interpretations are provided through external theory modules [10,5]. In planning, extension to support external modules allows a much richer expression of preconditions and state variables. A motivation for exploring this idea is that the increased expressiveness can allow planners to work with models of application domains using specialised solvers, necessary for reasoning within those applications, alongside the generic solving cores developed in the planning community. Since this is a common requirement of planning applications, it is important to provide clean and well-understood methods for linking planners to external libraries, choosing heuristics and exchanging constraints.

In this talk we present the Planning Modulo Theories paradigm, first proposed in 2012 [6], describing how the paradigm has been extended to incorporate the latest advances in temporal planning. We discuss how the use of constraint reasoning can provide an additional source of powerful solving capabilities within this framework. In general, constraint solvers prune choices from the search space by inference, while most modern planners focus on heuristic guidance of the search towards good choices. Complex interactions in resource-constrained models can be obscure, making heuristic evaluation of states much more difficult, while at the same time offering more opportunity for leverage from inference [13]. We consider, with reference to two real application domains, how constraint solving can contribute to making planners suitable for deployment in applications with demanding requirements.

One of the important challenges in extending the capabilities of planners is to continue to be able to efficiently validate plans and domain models. We will describe how

* Thanks to my generous collaborators and co-authors who have contributed to this work.

B. O'Sullivan (Ed.): CP 2014, LNCS 8656, pp. 1–2, 2014.

the VAL system [4], developed incrementally over the last 10 years for validation of plans and domains in the mixed discrete-continuous expressiveness of PDDL+, is now being extended to cope with richer behaviours encountered in the PMT framework.

References

1. Coles, A., Coles, A., Fox, M., Long, D.: COLIN: Planning with continuous linear numeric change. Journal of Art. Int. Research 44, 1–96 (2012)
2. Coles, A.I., Fox, M., Long, D., Smith, A.J.: A Hybrid Relaxed Planning Graph-LP Heuristic for Numeric Planning Domains. In: Proc. 18th Int. Conf. on Automated Planning and Scheduling (ICAPS) (2008)
3. Coles, A.J., Coles, A.I., Fox, M., Long, D.: Forward-Chaining Partial-Order Planning. In: Proc. 20th Int. Conf. on Aut. Planning and Scheduling, ICAPS (2010)
4. Fox, M., Howey, R., Long, D.: Validating plans in the context of processes and exogenous events. In: Veloso, M.M., Kambhampati, S. (eds.) Proc. Nat. Conf. on AI, AAAI, pp. 1151–1156. AAAI Press / The MIT Press (2005)
5. Gao, S., Kong, S., Clarke, E.: Satisfiability Modulo ODEs. In: Proc. Formal Methods in Computer-Aided Design, FMCAD (2013)
6. Gregory, P., Long, D., Fox, M., Beck, J.C.: Planning Modulo Theories: Extending the Planning Paradigm. In: Proc. 22nd Int. Conf. on Automated Planning and Scheduling, ICAPS (2012)
7. Ivankovic, F., Haslum, P., Thiebaux, S., Shivashankar, V., Nau, D.: Optimal planning with global numerical state constraints. In: Proceedings of 24th Int. Conf. on Aut. Planning and Scheduling, ICAPS (2014)
8. Lipovetzky, N., Burt, C.N., Pearce, A.R., Stuckey, P.J.: Planning for Mining Operations with Time and Resource Constraints. In: Proceedings of 24th Int. Conf. on Aut. Planning and Scheduling, ICAPS (2014)
9. Löhr, J., Eyerich, P., Winkler, S., Nebel, B.: Domain Predictive Control Under Uncertain Numerical State Information. In: Proc. 23rd Int. Conf. on Automated Planning and Scheduling (ICAPS) (2013)
10. Nieuwenhuis, R., Oliveras, A., Tinelli, C.: Solving SAT and SAT Modulo Theories: From an abstract Davis–Putnam–Logemann–Loveland procedure to DPLL(*T*). J. ACM 53(6), 937–977 (2006)
11. Ono, M., Williams, B.C., Blackmore, L.: Probabilistic Planning for Continuous Dynamic Systems under Bounded Risk. Journal of AI Research (JAIR) 46, 511–577 (2013)
12. Penna, G.D., Intrigila, B., Magazzeni, D., Mercorio, F.: Upmurphi: a tool for universal planning on pddl+ problems. In: Proc. 19th Int. Conf. on Automated Planning and Scheduling (ICAPS), pp. 19–23 (2009)
13. Vidal, V., Geffner, H.: Branching and pruning: An optimal temporal pocl planner based on constraint programming. Artif. Intell. 170(3), 298–335 (2006)

Teaching Constraint Programming

Patrick Prosser

School of Computing Science, University of Glasgow, UK
pat@dcs.gla.ac.uk

How do we do research? We start with a question. Then we read books, journal and conference papers, maybe even speak to people. Then we do our own work, make our own contribution, maybe coming up with an improved technique or a greater insight. We then write up our findings, maybe submit this to a conference, present our work and get feedback, and this results in further research. This is a feedback loop, open to scrutiny by our peers.

And what about teaching? *You* teach yourself and become competent. *You* decide how to teach your subject. *You* then teach and mark students. *You* analyze students' performance and use this to modify what you teach. *You* continue to learn your subject and use this new knowledge to modify your teaching. Again, there is a feedback loop. But it is a closed loop, in the sense that no one really gets to critique what *you* do. If you are teaching Constraint Programming (CP) it is unlikely that there are many teaching colleagues who can actually evaluate what you are doing, other than looking at the final exam marks. So you can wander off topic, away from the target and this can be dangerous.

I am fortunate enough to be allowed to teach CP to final year and masters students at Glasgow University. I have been doing this for about 10 years. What I teach and how I teach has evolved over time. I now recognize some things that I did that were clearly wrong and some things that I did that were really good. I know that I do not teach in a vacuum, that my students take many other courses. So I try and identify *stuff* that I think a Constraint Programmer should know that is not being taught in other courses. Consequently, my CP course contains *stuff* that might be considered unusual. I also expect that there's *stuff* that I should teach but do not.

In my talk I will describe the content of my CP course (the stuff of it), some things I have done wrong and some things that really work well. I will cover lecture material, assessed exercises and even exam questions! In essence, I will open my feedback loop allowing *you* to give *me* feedback on what I teach.

Acknowledgments. I would like to thank my students.

B. O'Sullivan (Ed.): CP 2014, LNCS 8656, p. 3, 2014.

One Problem, Two Structures, Six Solvers, and Ten Years of Personnel Scheduling

Louis-Martin Rousseau

CIRRELT, École Polytechnique de Montréal, Montréal,
C.P. 6128, succ. Centre-ville, Montréal, H3C 3J7,Canada
louis-martin.rousseau@polymtl.ca

The shift-scheduling problem was originally introduced by Edie in 1954 [8] in the context of scheduling highway toll booth operators. It was solved a short time later, by Georges Dantzig [6], using a set covering formulation. However, the Multi-Activity Shift Scheduling (MASSP) version of that problem, where one not only needs to schedule when employees are working or resting, but more precisely, what activity they are performing, still remains a challenge. During this invited lecture, we will recall the turning points of this 60-year journey, focusing particularly on the efforts of the last decade to solve MASSPs.

The first breakthrough came from Constraint Programming (CP), with the introduction of the Regular Language Membership Constraint [13,1], which enabled us to specify shift regulations through Deterministic Finite Automata. Two years later, the Context-free Grammar Constraint [15,18] was introduced, shortly followed by both a decomposed formulation [16] and incremental filtering algorithm [11]. From these constraints it is possible to identify two network structures (paths in a layered directed acyclic graph for *Regular* and hyper-paths in a hyper-graph for *Grammar*).

Using these graph structures, Mixed Integer Programming (MIP) models were initially proposed [3] to address the MASSP. Thanks to Orbital Shrinking [9], a new MIP formulation [4], and hybrid CP-MIP branch and bound [17] were proposed which allowed us to solve these models more efficiently.

Dynamic Programming (DP) algorithms were also developed to optimize (find the shortest paths and hyper-paths) for both *Regular* and *Grammar* given that marginal costs are associated with performing certain activities at a given time. These costs can be estimated manually during a Large Neighbourhood Search (LNS) [14] or obtained through dual values in the context of a Branch-and-Price approach [7,5]. Finally Lazy-Clause Generation (LCG) within CP [10] has shown to produce very good results for a subset of the benchmark originally introduced in [7].

From a practical point of view, the concepts of [5] were implemented into commercial software (*Planora*), while the models using the decomposition of *Regular* were used in case studies involving a major fashion retailer [2] and Hydro Québec's large call center [12].

B. O'Sullivan (Ed.): CP 2014, LNCS 8656, pp. 4–5, 2014.

References

1. Beldiceanu, N., Carlsson, M., Petit, T.: Deriving filtering algorithms from constraint checkers. In: Wallace, M. (ed.) CP 2004. LNCS, vol. 3258, pp. 107–122. Springer, Heidelberg (2004)
2. Chapados, N., Joliveau, M., L'Ecuyer, P., Rousseau, L.M.: Retail Store Scheduling for Profit. European Journal of Operations Research (2014), doi:10.1016/j.ejor.2014.05.033
3. Côté, M.C., Gendron, B., Quimper, C.G., Rousseau, L.M.: Formal languages for integer programming modeling of shift scheduling problems. Constraints 16(1), 54–76 (2011)
4. Côté, M.C., Gendron, B., Rousseau, L.M.: Grammar-based integer programming models for multiactivity shift scheduling. Management Science 57(1), 151–163 (2011)
5. Côté, M.C., Gendron, B., Rousseau, L.M.: Grammar-based column generation for personalized multi-activity shift scheduling. INFORMS Journal on Computing 25(3), 461–474 (2013)
6. Dantzig, G.: A comment on Edie's traffic delay at toll booths. Journal of the Operations Research Society of America 2, 339–341 (1954)
7. Demassey, S., Pesant, G., Rousseau, L.M.: A cost-regular based hybrid column generation approach. Constraints 11(4), 315–333 (2006)
8. Edie, L.: Traffic delays at toll booths. Journal of the Operations Research Society of America 2, 107–138 (1954)
9. Fischetti, M., Liberti, L.: Orbital shrinking. In: Mahjoub, A.R., Markakis, V., Milis, I., Paschos, V.T. (eds.) ISCO 2012. LNCS, vol. 7422, pp. 48–58. Springer, Heidelberg (2012)
10. Gange, G., Stuckey, P.J., Van Hentenryck, P.: Explaining Propagators for Edge-Valued Decision Diagrams. In: Schulte, C. (ed.) CP 2013. LNCS, vol. 8124, pp. 340–355. Springer, Heidelberg (2013)
11. Kadioglu, S., Sellmann, M.: Grammar constraints. Constraints 15(1), 117–144 (2010)
12. Pelleau, M., Rousseau, L.-M., L'Ecuyer, P., Zegal, W., Delorme, L.: Scheduling of Agents from Forecasted Future Call Arrivals at Hydro-Québec' s Call Centers. In: Principles and Practice of Constraint Programming, CP 2013, Springer, Heidelberg (2014)
13. Pesant, G.: A regular language membership constraint for finite sequences of variables. In: Wallace, M. (ed.) CP 2004. LNCS, vol. 3258, pp. 482–495. Springer, Heidelberg (2004)
14. Quimper, C.G., Rousseau, L.M.: A large neighbourhood search approach to the multi-activity shift scheduling problem. Journal of Heuristics 16(3), 373–392 (2010)
15. Quimper, C.G., Walsh, T.: Global grammar constraints. In: Benhamou, F. (ed.) CP 2006. LNCS, vol. 4204, pp. 751–755. Springer, Heidelberg (2006)
16. Quimper, C.G., Walsh, T.: Decomposing global grammar constraints. In: Bessière, C. (ed.) CP 2007. LNCS, vol. 4741, pp. 590–604. Springer, Heidelberg (2007)
17. Salvagnin, D., Walsh, T.: A hybrid MIP/CP approach for multi-activity shift scheduling. In: Milano, M. (ed.) CP 2012. LNCS, vol. 7514, pp. 633–646. Springer, Heidelberg (2012)
18. Sellmann, M.: The theory of grammar constraints. In: Benhamou, F. (ed.) CP 2006. LNCS, vol. 4204, pp. 530–544. Springer, Heidelberg (2006)

Concurrent Constraint Programming Research Programmes – Redux

Vijay Saraswat

IBM TJ Watson Research Center, Yorktown, Heights, NY, USA

At the first PPCP conference in 1995, I was honored to be one of the invited speakers. Twenty conferences later, much has changed in the computational world. We have seen the penetration of the Internet in every aspect of human life; the establishment of the multi-core era; the arrival of petaflop high performance computing; the rise of big data, analytics and machine learning; and the emergence of the planet-wide computer (the "cloud").

With this backdrop, we review the many developments in CCP over the last twenty years, and revisit the core idea behind this framework: the use of constraints for communication and control in concurrent programming languages. Surprisingly, in this age of concurrency and big data, these ideas remain foundational. CCP remains the premier framework for determinate concurrency. By supporting the notion of concurrent composition as intersection of sets of (constraint) stores rather than shuffling of sets of interleaved store sequences, it offers interesting new ideas (declarative debugging [9,1], diagnosis [7]) to deal with the problem of debugging concurrent programs running on tens of thousands of cores. Interestingly, these ideas work even in the presence of global non-monotonic change (and hence support "constraint imperative programming" [8]); this is accomplished by introducing a temporal modality in a principled fashion [15] and using soft constraints [2].

Concretely, we review the goals of the C10 project, being started in collaboration between researchers at IBM TJ Watson and many universities world-wide. C10 is intended for use in the areas of constraint-solving, probabilistic programming, machine learning, and big data analytics. It is a pure, declarative, implicitly concurrent, statically-typed, object-oriented, timed, probabilistic [11,10] realization of the CCP framework. C10 is intended to be compiled to the high-performance, multi-node, concurrent programming language X10 [5], but does not itself have any explicit concurrency or distribution constructs. C10 permits recursive queries against the constraint store (based on [12]), thus subsuming pure (constraint) logic programming. It proposes new indexicals (cf [17]), including set-forming operations that make it easy to write ad hoc queries over large data sets. It exploits random variables ([11,10]) to represent various probabilistic graphical models (Bayesian networks, Markov networks, probabilistic CP nets [4,6]) directly as programs.

We expect C10 to bring into focus several implementation challenges. Besides the traditional challenges of implicit parallelism (statically and dynamically chunking fine-grained parallelism into sizes adequate for efficient exploitation on today's multi-core architectures, [16]), C10 requires the development of efficient, incremental constraint query procedures over the constraint store (cf query compilation challenges of [3,13]). It requires the integration of multiple, efficient, probabilistic inference procedures into

B. O'Sullivan (Ed.): CP 2014, LNCS 8656, pp. 6–8, 2014.

the run-time. Based on the query being asked, the model, and the training data available, the right combination of inference procedures to use may have to be determined dynamically (e.g. using ideas from work on automatic algorithm configuration and selection procedures [14]).

References

1. Ajiro, Y., Ueda, K.: Kima: An Automated Error Correction System for Concurrent Logic Programs. Automated Software Engg. 9(1), 67–94 (2002)
2. Bistarelli, S., Montanari, U., Rossi, F.: Soft Concurrent Constraint Programming. ACM Trans. Comput. Logic 7(3), 563–589 (2006)
3. Boag, S., Chamberlin, D., Fernandez, M., Florescu, D., Robie, J., Simeon, J.: XQuery 1.0: An XML Query Language. Technical report, W3C (2003), http://www.w3.org/TR/xquery
4. Boutilier, C., Brafman, R.I., Domshlak, C., Hoos, H.H., Poole, D.: CP-nets: A Tool for Representing and Reasoning with Conditional Ceteris Paribus Preference Statements. J. Artif. Int. Res. 21(1), 135–191 (2004)
5. Charles, P., Grothoff, C., Saraswat, V., Donawa, C., Kielstra, A., Ebcioglu, K., von Praun, C., Sarkar, V.: X10: an Object-Oriented Approach to Non-uniform Cluster Computing. SIGPLAN Not. 40(10), 519–538 (2005)
6. Cornelio, C., Goldsmith, J., Mattei, N., Rossi, F., Venable, K.B.: Updates and Uncertainity in CP-nets. In: Proceedings of the Australian Conference on Artificial Intelligence, pp. 301–312 (2013)
7. Falaschi, M., Olarte, C., Palamidessi, C., Valencia, F.: Declarative Diagnosis of Temporal Concurrent Constraint Programs. In: Dahl, V., Niemelä, I. (eds.) ICLP 2007. LNCS, vol. 4670, pp. 271–285. Springer, Heidelberg (2007)
8. Freeman-Benson, B.N.: Kaleidoscope: Mixing Objects, Constraints, and Imperative Programming. In: Proceedings of the European Conference on Object-oriented Programming on Object-oriented Programming Systems, Languages, and Applications, OOPSLA/ECOOP 1990, pp. 77–88. ACM, New York (1990)
9. Fromherz, M.P.J.: Towards Declarative Debugging of Concurrent Constraint Programs. In: Fritzson, P.A. (ed.) AADEBUG 1993. LNCS, vol. 749, pp. 88–100. Springer, Heidelberg (1993)
10. Gupta, V., Jagadeesan, R., Panangaden, P.: Stochastic Processes as Concurrent Constraint Programs. In: Proceedings of the 26th ACM SIGPLAN-SIGACT on Principles of Programming Languages, POPL 1999, San Antonio, TX, January 20-22, pp. 189–202. ACM Press, New York (1999)
11. Gupta, V., Jagadeesan, R., Saraswat, V.: Probabilistic Concurrent Constraint Programming. In: Mazurkiewicz, A., Winkowski, J. (eds.) CONCUR 1997. LNCS, vol. 1243, pp. 1–4. Springer, Heidelberg (1997)
12. Jagadeesan, R., Nadathur, G., Saraswat, V.: Testing Concurrent Systems: an Interpretation of Intuitionistic Logic. In: Sarukkai, S., Sen, S. (eds.) FSTTCS 2005. LNCS, vol. 3821, pp. 517–528. Springer, Heidelberg (2005)
13. Khatchadourian, S., Consens, M., Siméon, J.: Chuql: Processing XML with XQuery Using Hadoop. In: Proceedings of the 2011 Conference of the Center for Advanced Studies on Collaborative Research, CASCON 2011, Riverton, NJ, USA, pp. 74–83. IBM Corp. (2011)
14. Kothoff, L., Malitsky, Y., O'Sullivan, B.: Advances in Algorithm Selection and Configuration for Constraint Solving and Satisfiability. In: Tutorial at IJCAI 2013 (2013)

15. Saraswat, V., Jagadeesan, R., Gupta, V.: Timed Default Concurrent Constraint Programming. Journal of Symbolic Computation 22(5-6), 475–520 (1996); Extended abstract appeared in the Proceedings of the 22nd ACM Symposium on Principles of Programming Languages, San Francisco (January 1995)
16. Ueda, K., Morita, M.: Moded Flat GHC and Its Message-oriented Implementation Technique. New Gen. Comput. 13(1), 3–43 (1994)
17. van Hentenryck, P., Deville, Y., Saraswat, V.: Design, Implementation and Evaluation of the Constraint Language cc(FD). Journal of Logic Programming 37(1-3), 139–164 (1998)

On Broken Triangles*

Martin C. Cooper[1], Achref El Mouelhi[2], Cyril Terrioux[2], and Bruno Zanuttini[3]

[1] IRIT, University of Toulouse III, 31062 Toulouse, France
cooper@irit.fr
[2] LSIS, Aix-Marseille University, 13397 Marseille, France
{achref.elmouelhi,cyril.terrioux}@lsis.org
[3] GREYC, University of Caen Basse-Normandie, 14032 Caen, France
bruno.zanuttini@unicaen.fr

Abstract. A binary CSP instance satisfying the broken-triangle property (BTP) can be solved in polynomial time. Unfortunately, in practice, few instances satisfy the BTP. We show that a local version of the BTP allows the merging of domain values in arbitrary instances of binary CSP, thus providing a novel polynomial-time reduction operation. Extensive experimental trials on benchmark instances demonstrate a significant decrease in instance size for certain classes of problems. We show that BTP-merging can be generalised to instances with constraints of arbitrary arity and we investigate the theoretical relationship with resolution in SAT. A directional version of the general-arity BTP then allows us to extend the BTP tractable class previously defined only for binary CSP.

1 Introduction

At first sight one could assume that the discipline of constraint programming has come of age. On the one hand, efficient solvers are regularly used to solve real-world problems in diverse application domains while, on the other hand, a rich theory has been developed concerning, among other things, global constraints, tractable classes, reduction operations and symmetry. However, there often remains a large gap between theory and practice which is perhaps most evident when we look at the large number of deep results concerning tractable classes which have yet to find any practical application. The research reported in this paper is part of a long-term project to bridge the gap between theory and practice. Our aim is not only to develop new tools but also to explain why present tools work so well.

Most research on tractable classes has been based on classes defined by placing restrictions either on the types of constraints or on the constraint hypergraph whose vertices are the variables and whose hyper-edges are the constraint scopes. Another way of defining classes of binary CSP instances consists in imposing conditions on the microstructure, a graph whose vertices are the possible variable-value assignments with an edge linking each pair of compatible assignments [9,12]. If each vertex of the microstructure, corresponding to a

* Supported by ANR Project ANR-10-BLAN-0210.

B. O'Sullivan (Ed.): CP 2014, LNCS 8656, pp. 9–24, 2014.

variable-value assignment $\langle x, a\rangle$, is labelled by the variable x, then this so-called coloured microstructure retains all information from the original instance. The broken-triangle property (BTP) is a simple local condition on the coloured microstructure which defines a tractable class of binary CSP [5]. Inspired by the BTP, investigation of other forbidden patterns in the coloured microstructure has led to the discovery of new tractable classes [1,4,6,8] as well as new reduction operations based on variable elimination [2].

For simplicity of presentation we use two different representations of constraint satisfaction problems. In the binary case, our notation is fairly standard, whereas in the general-arity case we use a notation close to the representation of SAT instances. This is for presentation only, though, and our algorithms do *not* need instances to be represented in this manner.

Definition 1. *A* binary CSP instance *I consists of*

- *a set X of n variables,*
- *a domain $\mathcal{D}(x)$ of possible values for each variable $x \in X$,*
- *a relation $R_{xy} \subseteq \mathcal{D}(x) \times \mathcal{D}(y)$, for each pair of distinct variables $x, y \in X$, which consists of the set of compatible pairs of values (a, b) for variables (x, y).*

A partial solution *to I on $Y = \{y_1, \ldots, y_r\} \subseteq X$ is a set $\{\langle y_1, a_1\rangle, \ldots, \langle y_r, a_r\rangle\}$ such that $\forall i, j \in [1, r]$, $(a_i, a_j) \in R_{y_i y_j}$. A* solution *to I is a partial solution on X.*

For simplicity of presentation, Definition 1 assumes that there is exactly one constraint relation for each pair of variables. The number of constraints e is the number of pairs of variables x, y such that $R_{xy} \neq \mathcal{D}(x) \times \mathcal{D}(y)$. An instance I is *arc consistent* if for each pair of distinct variables $x, y \in X$, for each value $a \in \mathcal{D}(x)$, there is a value $b \in \mathcal{D}(y)$ such that $(a, b) \in R_{xy}$.

In our representation of general-arity CSP instances, we require the notion of *tuple* which is simply a set of variable-value assignments. For example, in the binary case, the tuple $\{\langle x, a\rangle, \langle y, b\rangle\}$ is *compatible* if $(a, b) \in R_{xy}$ and *incompatible* otherwise.

Definition 2. *A* (general-arity) CSP instance *I consists of*

- *a set X of n variables,*
- *a domain $\mathcal{D}(x)$ of possible values for each variable $x \in X$,*
- *a set NoGoods(I) consisting of incompatible tuples.*

A partial solution *to I on $Y = \{y_1, \ldots, y_r\} \subseteq X$ is a tuple $t = \{\langle y_1, a_1\rangle, \ldots, \langle y_r, a_r\rangle\}$ such that no subset of t belongs to NoGoods(I). A* solution *is a partial solution on X.*

2 Value Merging in Binary CSP Based on the BTP

In this section we consider a method, based on the BTP, for reducing domain size while preserving satisfiability. Instead of eliminating a value, as in classic reduction operations such as arc consistency or neighbourhood substitution, we merge two values. We show that the absence of broken-triangles [5] on two values for a variable x in a binary CSP instance allows us to merge these two values in the domain of x while preserving satisfiability. This rule generalises the notion of virtual interchangeability [11] as well as neighbourhood substitution [10].

It is known that if for a given variable x in an arc-consistent binary CSP instance I, the set of (in)compatibilities (known as a broken-triangle) shown in Figure 1 occurs for no two values $a, b \in \mathcal{D}(x)$ and no two assignments to two other variables, then the variable x can be eliminated from I without changing the satisfiability of I [5,2]. In figures, each bullet represents a variable-value assignment, assignments to the same variable are grouped together within the same oval and compatible (incompatible) pairs of assignments are linked by solid (broken) lines. Even when this variable-elimination rule cannot be applied, it may be the case that for a given pair of values $a, b \in \mathcal{D}(x)$, no broken triangle occurs. We will show that if this is the case, then we can perform a domain-reduction operation which consists in merging the values a and b.

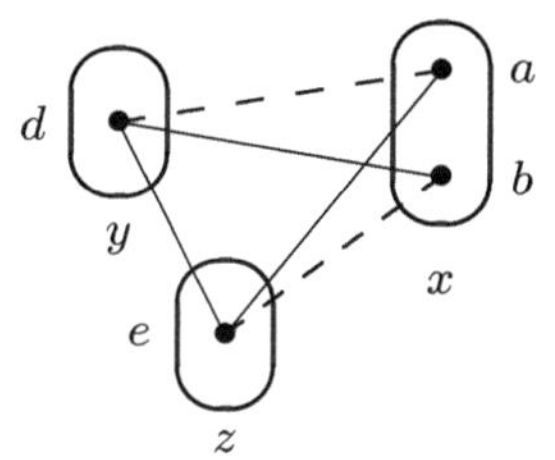

Fig. 1. A broken triangle on two values a, b for a given variable x

Definition 3. Merging *values $a, b \in \mathcal{D}(x)$ in a binary CSP consists in replacing a, b in $\mathcal{D}(x)$ by a new value c which is compatible with all variable-value assignments compatible with at least one of the assignments $\langle x, a\rangle$ or $\langle x, b\rangle$. A* value-merging condition *is a polytime-computable property $P(x, a, b)$ of assignments $\langle x, a\rangle$, $\langle x, b\rangle$ in a binary CSP instance I such that when $P(x, a, b)$ holds, the instance I' obtained from I by merging $a, b \in \mathcal{D}(x)$ is satisfiable if and only if I is satisfiable.*

We now formally define the value-merging condition based on the BTP.

Definition 4. *A* broken triangle *on the pair of variable-value assignments $a, b \in \mathcal{D}(x)$ consists of a pair of assignments $d \in \mathcal{D}(y)$, $e \in \mathcal{D}(z)$ to distinct variables $y, z \in X \setminus \{x\}$ such that $(a, d) \notin R_{xy}$, $(b, d) \in R_{xy}$, $(a, e) \in R_{xz}$, $(b, e) \notin R_{xz}$ and $(d, e) \in R_{yz}$. The pair of values $a, b \in \mathcal{D}(x)$ is* BT-free *if there is no broken triangle on a, b.*

Proposition 1. *In a binary CSP instance, being BT-free is a value-merging condition. Furthermore, given a solution to the instance resulting from the merging of two values, we can find a solution to the original instance in linear time.*

Proof. Let I be the original instance and I' the new instance in which a,b have been merged into a new value c. Clearly, if I is satisfiable then so is I'. It suffices to show that if I' has a solution s which assigns c to x, then I has a solution. Let s_a, s_b be identical to s except that s_a assigns a to x and s_b assigns b to x. Suppose that neither s_a nor s_b are solutions to I. Then, there are variables $y, z \in X \setminus \{x\}$ such that $\langle a, s(y)\rangle \notin R_{xy}$ and $\langle b, s(z)\rangle \notin R_{xz}$. By definition of the merging of a, b to produce c, and since s is a solution to I' containing $\langle x, c\rangle$, we must have $(b, s(y)) \in R_{xy}$ and $(a, s(z)) \in R_{xz}$. Finally, $(s(y), s(z)) \in R_{yz}$ since s is a solution to I'. Hence, $\langle y, s(y)\rangle$, $\langle z, s(z)\rangle$, $\langle x, a\rangle$, $\langle x, b\rangle$ forms a broken-triangle, which contradicts our assumption. Hence, the absence of broken triangles on assignments $\langle x, a\rangle$, $\langle x, b\rangle$ allows us to merge these assignments while preserving satisfiability.

Reconstructing a solution to I from a solution s to I' simply requires checking which of s_a or s_b is a solution to I. □

We can see that the BTP-merging rule, given by Proposition 1, generalises neighbourhood substitution [10]: if b is neighbourhood substitutable by a, then no broken triangle occurs on a, b and merging a and b produces a CSP instance which is identical (except for the renaming of the value a as c) to the instance obtained by simply eliminating b from $\mathcal{D}(x)$. BTP-merging also generalises the merging rule proposed by Likitvivatanavong and Yap [11]. The basic idea behind their rule is that if the two assignments $\langle x, a\rangle$, $\langle x, b\rangle$ have identical compatibilities with all assignments to all other variables except concerning at most one other variable, then we can merge a and b. This is clearly subsumed by BTP-merging.

The BTP-merging operation is not only satisfiability-preserving but, from Proposition 1, we know that we can also reconstruct a solution in polynomial time to the original instance I from a solution to an instance I^m to which we have applied a sequence of merging operations until convergence. It is known that for the weaker operation of neighbourhood substitutability, all solutions to the original instance can be generated in $O(N(de + n^2))$ time, where N is the number of solutions to the original instance, n is the number of variables, d the maximum domain size and e the number of constraints [3]. We now show that a similar result also holds for the more general rule of BTP-merging.

Proposition 2. *Let I be a binary CSP instance and suppose that we are given the set of all solutions to the instance I^m obtained after applying a sequence of BTP-merging operations. All N solutions to I can then be determined in $O(Nn^2d)$ time.*

Proof. Let I' be the CSP instance which results after performing a single BTP-merging operation of values $a, b \in \mathcal{D}(x)$ in I. As we saw in the proof of Proposition 1, given the set of solutions $sol(I')$ to I' we can generate the set of solutions to I by testing for each $s \in Sol(I')$ whether s_a or s_b (or both) are solutions to I. This requires $O(n)$ time per solution to I, since there are at most $n - 1$ constraints to be tested involving the variable x, and at least one of s_a or s_b is a solution to I.

The total number of BTP-merging operations performed to transform I into I^m is at most $n(d-1)$. Therefore, the total time to generate all N solutions to I from the set of solutions to I^m is $O(Nn^2d)$. □

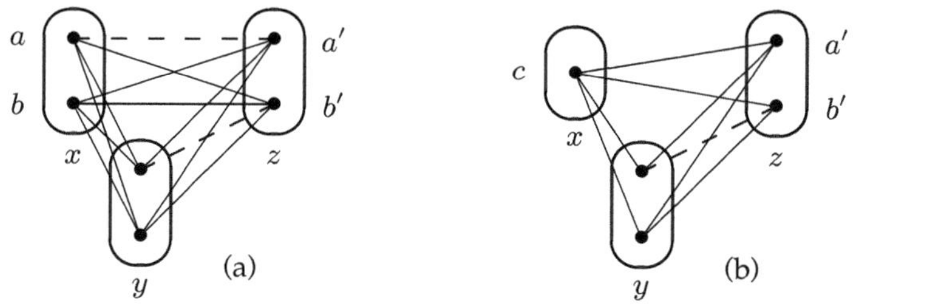

Fig. 2. (a) A broken triangle exists on values a', b' at variable z. (b) After BTP-merging of values a and b in $\mathcal{D}(x)$, this broken triangle has disappeared.

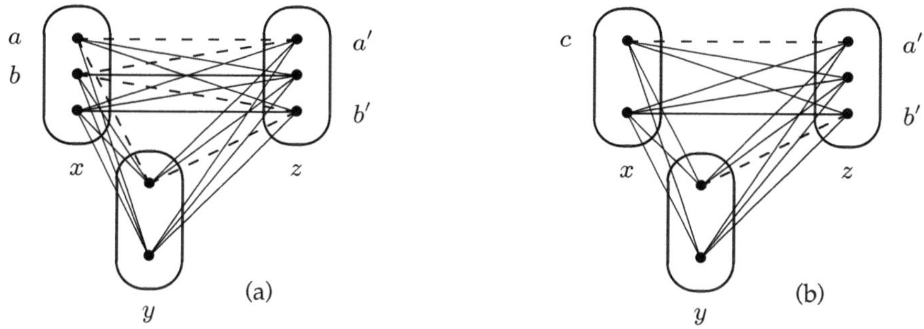

Fig. 3. (a) This instance contains no broken triangle. (b) After BTP-merging of values a and b in $\mathcal{D}(x)$, a broken triangle has appeared on values $a', b' \in \mathcal{D}(z)$.

The weaker operation of neighbourhood substitution has the property that two different convergent sequences of eliminations by neighbourhood substitution necessarily produce isomorphic instances I_1^m, I_2^m [3] . This is not the case for BTP-merging. Firstly, and perhaps rather surprisingly, BTP-merging can have as a side-effect to eliminate broken triangles. This is illustrated in the 3-variable instance shown in Figure 2. The instance in Figure 2(a) contains a broken triangle on values a', b' for variable z, but after BTP-merging of values $a, b \in \mathcal{D}(x)$ into a new value c, as shown in Figure 2(b), there are no broken triangles in the instance. Secondly, BTP-merging of two values in $\mathcal{D}(x)$ can introduce a broken triangle on a variable $z \neq x$, as illustrated in Figure 3. The instance in Figure 3(a) contains no broken triangle, but after the BTP-merging of $a, b \in \mathcal{D}(x)$ into a new value c, a broken triangle has been created on values $a', b' \in \mathcal{D}(z)$.

3 Experimental Trials

To test the utility of BTP-merging we performed extensive experimental trials on benchmark instances from the International CP Competition[1]. For each instance not including global constraints, we performed BTP-mergings until convergence with a time-out of one hour. In total, we obtained results for 2,547 instances out of 3,811 benchmark instances. In the other instances the search for all BTP-mergings did not terminate within a time-out of one hour.

Table 1. Results of experiments on CSP benchmark problems

domain	no. instances	no. values	no. values deleted	%age deleted
BH-4-13	6	7,334	3,201	44%
BH-4-4	10	674	322	48%
BH-4-7	20	2,102	883	42%
ehi-85	98	2,079	891	43%
ehi-90	100	2,205	945	43%
graph-coloring/school	8	4,473	104	2%
graph-coloring/sgb/book	26	1,887	534	28%
jobShop	45	6,033	388	6%
marc	1	6400	6,240	98%
os-taillard-4	30	2,932	1,820	62%
os-taillard-5	28	6,383	2,713	43%
rlfapGraphsMod	5	14,189	5,035	35%
rlfapScens	5	12,727	821	6%
rlfapScensMod	9	9,398	1,927	21%
others	1919	1,396	28	0.02%

All instances from the benchmark-domain `hanoi` satisfy the broken-triangle property and BTP-merging reduced all variable domains to singletons. After establishing arc consistency, 38 instances from diverse benchmark-domains satisfy the BTP, including all instances from the benchmark-domain `domino`. We did not count those instances for which arc consistency detects inconsistency by producing a trivial instance with empty variable domains (and which trivially satisfies the BTP). In all instances from the `pigeons` benchmark-domain with a suffix `-ord`, BTP-merging again reduced all domains to singletons. This is because BTP-merging can eliminate broken triangles, as pointed out in Section 2, and hence can render an instance BTP even though initially it was not BTP. The same phenomenon occurred in a 680-variable instance from the benchmark-domain `rlfapGraphsMod` as well as the 3-variable instance `ogdPuzzle`.

Table 1 gives a summary of the results of the experimental trials. We do not include those instances mentioned above which are entirely solved by

[1] `http://www.cril.univ-artois.fr/CPAI08`

BTP-merging. We give details about those benchmark-domains where BTP-merging was most effective. All other benchmark-domains are grouped together in the last line of the table. The table shows the number of instances in the benchmark-domain, the average number of values (i.e. variable-value assignments) in the instances from this benchmark-domain, the average number of values deleted (i.e. the number of BTP-merging operations performed) and finally this average represented as a percentage of the average number of values.

We can see that for certain types of problem, BTP-merging is very effective, whereas for others (grouped together in the last line of the table) hardly any merging of values occurred.

4 Generalising BTP-Merging to Constraints of Arbitrary Arity

In the remainder of the paper, we assume that the constraints of a general-arity CSP instance I are given in the form described in Definition 2, i.e. as a set of incompatible tuples NoGoods(I), where a tuple is a set of variable-value assignments. For simplicity of presentation, we use the predicate Good(I, t) which is true iff the tuple t is a partial solution, i.e. t does not contain any pair of distinct assignments to the same variable and $\nexists t' \subseteq t$ such that $t' \in$ NoGoods(I). We first generalise the notion of broken triangle and merging to the general-arity case, before showing that absence of broken triangles allows merging.

Definition 5. *A* general-arity broken triangle *(GABT) on values $a, b \in \mathcal{D}(x)$ consists of a pair of tuples t, u (containing no assignments to variable x) satisfying the following conditions:*

1. *Good($I, t \cup u$) $\wedge$ Good($I, t \cup \{\langle x, a\rangle\}$) $\wedge$ Good($I, u \cup \{\langle x, b\rangle\}$)*
2. *$t \cup \{\langle x, b\rangle\} \in$ NoGoods(I) $\wedge$ $u \cup \{\langle x, a\rangle\} \in$ NoGoods(I)*

The pair of values $a, b \in \mathcal{D}(x)$ is GABT-free *if there is no broken triangle on a, b.*

Observe that Good($I, t \cup \{\langle x, a\rangle\}$) entails $t \cup \{\langle x, a\rangle\} \notin$ NoGoods(I). Hence to decide whether there is a GABT on a, b in a CSP instance, one can either explore all pairs $t \cup \{\langle x, b\rangle\}, u \cup \{\langle x, a\rangle\} \in$ NoGoods(I), as suggested by Definition 5, or, equivalently, explore all pairs $t \cup \{\langle x, a\rangle\}, u \cup \{\langle x, b\rangle\}$ of tuples explicitly allowed by the constraints in I. Whatever the representation, a pair t, u can be checked to be a GABT on a, b by evaluating the properties of Definition 5, all of which involve only constraint checks. Hence deciding whether a pair a, b is GABT-free is polytime for constraints given in extension (as the set of satisfying assignments) as well as for those given by nogoods (the set of assignments violating the constraint).

Definition 6. Merging *values $a, b \in \mathcal{D}(x)$ in a general-arity CSP instance I consists in replacing a, b in $\mathcal{D}(x)$ by a new value c which is compatible with all variable-value* *assignments compatible with at least one of the assignments* $\langle x,a\rangle$ *or* $\langle x,b\rangle$*, thus producing an instance* I' *with the new set of nogoods defined as follows:*

$$\begin{aligned}\text{NoGoods}(I') = \{t \in \text{NoGoods}(I) \mid \langle x,a\rangle,\langle x,b\rangle \notin t\}\\ \cup \{t \cup \{\langle x,c\rangle\} \mid t \cup \{\langle x,a\rangle\} \in \text{NoGoods}(I) \wedge\\ \exists t' \in \text{NoGoods}(I) \text{ s.t. } t' \subseteq t \cup \{\langle x,b\rangle\}\}\\ \cup \{t \cup \{\langle x,c\rangle\} \mid t \cup \{\langle x,b\rangle\} \in \text{NoGoods}(I) \wedge\\ \exists t' \in \text{NoGoods}(I) \text{ s.t. } t' \subseteq t \cup \{\langle x,a\rangle\}\}\end{aligned}$$

A value-merging condition *is a polytime-computable property* $P(x,a,b)$ *of assignments* $\langle x,a\rangle$*,* $\langle x,b\rangle$ *in a CSP instance* I *such that when* $P(x,a,b)$ *holds, the instance* I' *is satisfiable if and only if* I *is satisfiable.*

Clearly, this merging operation can be performed in polynomial time whether constraints are represented positively in extension or negatively as nogoods.

Proposition 3. *In a general-arity CSP instance, being GABT-free is a value-merging condition. Furthermore, given a solution to the instance resulting from the merging of two values, we can find a solution to the original instance in linear time.*

Proof. In order to prove that satisfiability is preserved by this merging operation, it suffices to show that if s is a solution to I' containing $\langle x,c\rangle$, then either $s_a = (s \setminus \{\langle x,c\rangle\}) \cup \{\langle x,a\rangle\}$ or $s_b = (s \setminus \{\langle x,c\rangle\}) \cup \{\langle x,b\rangle\}$ is a solution to I. Suppose, for a contradiction that this is not the case. Then there are tuples $t,u \subseteq s \setminus \{\langle x,c\rangle\}$ such that $t \cup \{\langle x,b\rangle\} \in \text{NoGoods}(I)$ and $u \cup \{\langle x,a\rangle\} \in \text{NoGoods}(I)$. Since t,u are subsets of the solution s to I' and t,u contain no assignments to x, we have $\text{Good}(I, t \cup u)$. Since $t \cup \{\langle x,c\rangle\}$ is a subset of the solution s to I', we have $t \cup \{\langle x,c\rangle\} \notin \text{NoGoods}(I')$. By the definition of NoGoods(I') given in Definition 6, and since $t \cup \{\langle x,b\rangle\} \in \text{NoGoods}(I)$, we know that $\nexists t' \in \text{NoGoods}(I)$ such that $t' \subseteq t \cup \{\langle x,a\rangle\}$. But then $\text{Good}(I, t \cup \{\langle x,a\rangle\})$. By a symmetric argument, we can deduce $\text{Good}(I, u \cup \{\langle x,b\rangle\})$. This provides the contradiction we were looking for, since we have shown that a general-arity broken triangle occurs on t, u, $\langle x,a\rangle$, $\langle x,b\rangle$.

Reconstructing a solution to the original instance can be achieved in linear time, since it suffices to verify which (or both) of s_a or s_b is a solution to I. □

Relationship with Resolution in SAT

We now show that in the case of Boolean domains, there is a close relationship between merging two values a,b on which no GABT occurs and a common preprocessing operation used by SAT solvers. Given a propositional CNF formula φ in the form of a set of clauses (each clause C_i being represented as a set of literals) and a variable x occurring in φ, recall that *resolution* is the process of inferring the clause $(C_0 \cup C_1)$ from the two clauses $(\{\bar{x}\} \cup C_0), (\{x\} \cup C_1)$.

Define the formula $Res(x, \varphi)$ to be the result of performing all such resolutions on φ, removing all clauses containing x or $\bar{x}$, and removing subsumed clauses:

$$Res(x, \varphi) = \min_{\subset}(\{C \mid C \in \varphi; x, \bar{x} \notin C\} \cup \{(C_0 \cup C_1) \mid (\{\bar{x}\} \cup C_0), (\{x\} \cup C_1) \in \varphi\})$$

It is a well-known fact that $Res(x, \varphi)$ is satisfiable if and only if φ is.

Eliminating variables in this manner from SAT instances, to get an equisatisfiable formula with less variables, is a common preprocessing step in SAT solving, and is typically performed provided it does not increase the size of the formula [7]. A particular case is when it amounts to simply removing all occurrences of x, which is the case, for instance, if x or $\bar{x}$ is unit or pure in φ, or if all resolutions on x yield a tautological clause.

Definition 7. *A variable x is said to be* erasable *from a CNF φ if*

$$Res(x, \varphi) \subseteq \{C \mid C \in \varphi; x, \bar{x} \notin C\} \cup \{C_0 \mid (\{\bar{x}\} \cup C_0) \in \varphi\} \cup \{C_1 \mid (\{x\} \cup C_1) \in \varphi\}$$

A CNF φ can be seen as the CSP instance I_φ on the set X of variables occurring in φ, with $\mathcal{D}(x) = \{\top, \bot\}$ for all $x \in X$, and NoGoods$(I_\varphi) = \{\overline{C} \mid C \in \varphi\}$, where $\overline{(\{x_1, \cdots x_p, \bar{x}_{p+1}, \cdots, \bar{x}_q\})} = \{\langle x_1, \bot\rangle, \ldots, \langle x_p, \bot\rangle, \langle x_{p+1}, \top\rangle, \ldots, \langle x_q, \top\rangle\}$.

Proposition 4. *Assume that no GABT occurs on values $\bot, \top$ for x in I_φ. Assume moreover that no clause in φ is subsumed by another one*[2]*. Then x is erasable from φ.*

Proof. Rephrasing Definition 5 in terms of clauses, for any two clauses $(\{\bar{x}\} \cup C_0), (\{x\} \cup C_1) \in \varphi$ we have one of (i) $\exists C \in \varphi, C \subseteq (C_0 \cup C_1)$, (ii) $\exists C' \in \varphi, C' \subseteq (C_0 \cup \{x\})$, or (iii) $\exists C'' \in \varphi, C'' \subseteq (C_1 \cup \{\bar{x}\})$. Moreover, in Case (ii) C' must contain x, for otherwise the clause $(\{\bar{x}\} \cup C_0)$ would be subsumed in φ, contradicting our assumption. Similarly, in Case (iii) C'' must contain $\bar{x}$.

In Case (i) the resolvent $(C_0 \cup C_1)$ of $(\{\bar{x}\} \cup C_0), (\{x\} \cup C_1)$ is subsumed by C in $Res(x, \varphi)$, and hence does not occur in it. Similarly, in the second case $(C_0 \cup C_1)$ is subsumed by the resolvent of $(\{\bar{x}\} \cup C_0)$ and C', which is precisely C_0. The third case is dual. We finally have that the only resolvents added are of the form C_0 (resp. C_1) for some clause $(\{\bar{x}\} \cup C_0)$ (resp. $(\{x\} \cup C_1)$) of φ, as required. □

We can show the converse is also true provided that a very reasonable property holds.

Proposition 5. *Assume that φ satisfies: $\forall(\{x\} \cup C) \in \varphi, \nexists C' \subseteq C, (\{\bar{x}\} \cup C') \in \varphi$ and dually $\forall(\{\bar{x}\} \cup C) \in \varphi, \nexists C' \subseteq C, (\{x\} \cup C') \in \varphi$. If x is erasable from φ, then no GABT occurs on values $\bot, \top$ for x in I_φ.*

Proof. Assume for a contradiction that there is a GABT on values $\bot, \top$ for x in I_φ, let t, u be witnesses to this, and write $t \cup \{\langle x, \top\rangle\} = \overline{(\{\bar{x}\} \cup C_0)}$, $u \cup \{\langle x, \bot\rangle\} = \overline{(\{x\} \cup C_1)}$. Then the clause $(C_0 \cup C_1)$ is produced by resolution on x.

[2] This is without loss of generality since such clauses can be removed in polytime and such removal preserves logical equivalence.

Since x is erasable, $(C_0 \cup C_1)$ is equal to or subsumed by a clause $C \in Res(x, \varphi)$, where (applying Definition 7 in reverse) either C, or $(\{x\} \cup C)$, or $(\{\bar{x}\} \cup C)$ is in φ. The first case contradicts Good($I_\varphi, t \cup u$), so assume by symmetry $(\{x\} \cup C) \in \varphi$. From $C \notin \varphi$ and $C \in Res(x, \varphi)$ we get $\exists C' \subseteq C, (\{\bar{x}\} \cup C') \in \varphi$. But then the pair of clauses $(\{x\} \cup C), (\{\bar{x}\} \cup C') \in \varphi$ violates the assumption of the claim. □

5 A Tractable Class of General-Arity CSP

In binary CSP, the broken-triangle property defines an interesting tractable class when broken-triangles are forbidden according to a given variable ordering. Unfortunately, the original definition of BTP was limited to binary CSPs [5]. Section 4 described a general-arity version of the broken-triangle property whose absence on two values allows these values to be merged while preserving satisfiability. An obvious question is whether GABT-freeness can be adapted to define a tractable class. In this section we show that this is possible for a fixed variable ordering, but not if the ordering is unknown.

Definition 5 defined a general-arity broken triangle (GABT). What happens if we forbid GABTs according to a given variable ordering? Absence of GABTs on two values a, b for the last variable x in the variable ordering allows us to merge a and b while preserving satisfiability. It is possible to show that if GABTs are absent on all pairs of values for x, then we can merge all values in the domain $D(x)$ of x to produce a singleton domain. This is because (as we will show later) merging a and b, to produce a merged value c, cannot introduce a GABT on c, d for any other value $d \in \mathcal{D}(x)$. Once the domain $D(x)$ becomes a singleton $\{a\}$, we can clearly eliminate x from the instance, by deleting $\langle x, a \rangle$ from all nogoods, without changing its satisfiability. It is at this moment that GABTs may be introduced on other variables, meaning that forbidding GABTs according to a variable ordering does not define a tractable class.

Nevertheless, we will show that strengthening the general-arity BTP allows us to avoid this problem. The resulting directional general-arity version of BTP (for a known variable ordering) then defines a tractable class which includes the binary BTP tractable class as a special case.

Note that the set of general-arity CSP instances whose dual instance satisfies the BTP also defines a tractable class which can be recognised in polynomial time even if the ordering of the variables in the dual instance is unknown [8]. This DBTP class is incomparable with the class we present in the present paper (which is equivalent to BTP in binary CSP) since DBTP is known to be incomparable with the BTP class already in the special case of binary CSP [8].

5.1 Directional General-Arity BTP

We suppose given a total ordering $<$ of the variables of a CSP instance I. We write $t^{<x}$ to represent the subset of the tuple t consisting of assignments to variables occurring before x in the order $<$, and $Vars(t)$ to denote the set of all variables assigned by t.

Definition 8. *A* directional general-arity (DGA) broken triangle *on assignments a, b to variable x in a CSP instance I is a pair of tuples t, u (containing no assignments to variable x) satisfying the following conditions:*

1. *$t^{<x}$ and $u^{<x}$ are non-empty*
2. *$Good(I, t^{<x} \cup u^{<x}) \wedge Good(I, t^{<x} \cup \{\langle x, a\rangle\}) \wedge Good(I, u^{<x} \cup \{\langle x, b\rangle\})$*
3. *$\exists t'$ s.t. $Vars(t') = Vars(t) \wedge (t')^{<x} = t^{<x} \wedge t' \cup \{\langle x, a\rangle\} \notin NoGoods(I)$*
4. *$\exists u'$ s.t. $Vars(u') = Vars(u) \wedge (u')^{<x} = u^{<x} \wedge u' \cup \{\langle x, b\rangle\} \notin NoGoods(I)$*
5. *$t \cup \{\langle x, b\rangle\} \in NoGoods(I) \wedge u \cup \{\langle x, a\rangle\} \in NoGoods(I)$*

I satisfies the directional general-arity broken-triangle property (DGABTP) *according to the variable ordering $<$ if no directional general-arity broken triangle occurs on any pair of values a, b for any variable x.*

We will show that any instance I satisfying the DGABTP can be solved in polynomial time by repeatedly alternating the following two operations: (i) merge all values in the last remaining variable (according to the order $<$); (ii) eliminate this variable when its domain becomes a singleton. We will give the two operations (merging and variable-elimination) and show that both operations preserve satisfiability and that neither of them can introduce DGA broken triangles. Moreover, as for GABT-freeness, the DGABTP can be tested in polynomial time for a given order whether constraints are given as tables of satisfying assignments or as nogoods. Indeed, in the former case, using items (3) and (4) in Definition 8 we can restrict the search for a DGA broken triangle to pairs of tuples satisfying some constraint (there must be a constraint with scope $Vars(t' \cup \{x\})$ since there is a nogood on these variables by item (5), and similarly for u'). This is sufficient to define a tractable class.

5.2 Merging

Let x be the last variable according to the variable order $<$. When values a, b in the domain of variable x do not belong to any DGA broken triangle, we can replace a, b by a new value c to produce an instance I' with the new set of nogoods given by Definition 6. Since x is the last variable in the ordering $<$, DGA broken triangles on $a, b \in \mathcal{D}(x)$ are GA broken triangles (and vice versa). Thus, from Proposition 3 we can deduce that satisfiability is preserved by this merging operation. What remains to be shown is that merging two values in the domain of the last variable cannot introduce the forbidden pattern.

Lemma 1. *Merging two values a, b into a value c in the domain of the last variable x (according to the variable order $<$) in an instance I cannot introduce a directional general-arity broken triangle (DGABT) in the resulting instance I'.*

Proof. We first claim that this operation cannot introduce a DGABT on a variable $y < x$. Indeed, assume there is a DGABT on $d, e \in \mathcal{D}(y)$ in I', that is, that there are tuples v, w such that

1. $v^{<y}$ and $w^{<y}$ are non-empty
2. $\text{Good}(I', v^{<y} \cup w^{<y}) \wedge \text{Good}(I', v^{<y} \cup \{\langle y, d\rangle\}) \wedge \text{Good}(I', w^{<y} \cup \{\langle y, e\rangle\})$

3. $\exists v'\ Vars(v') = Vars(v) \ \wedge\ (v')^{<y} = v^{<y} \ \wedge\ v' \cup \{\langle y, d\rangle\} \notin$ NoGoods(I')
4. $\exists w'\ Vars(w') = Vars(w) \ \wedge\ (w')^{<y} = w^{<y} \ \wedge\ w' \cup \{\langle y, e\rangle\} \notin$ NoGoods(I')
5. $v \cup \{\langle y, e\rangle\} \in$ NoGoods(I') $\wedge$ $w \cup \{\langle y, d\rangle\} \in$ NoGoods(I')

If v' contains the assignment $\langle x, c\rangle$ then, by construction of NoGoods(I') (Definition 6), $\exists v'' \in \{(v' \setminus \langle x, c\rangle) \cup \{\langle x, a\rangle\}, (v' \setminus \langle x, c\rangle) \cup \{\langle x, b\rangle\}\}$ such that $v'' \cup \{\langle y, d\rangle\}$ $\notin$ NoGoods(I). If v' does not contain $\langle x, c\rangle$ then let $v'' = v'$. Define w'' in a similar way. Now considering the last item, if v contains $\langle x, c\rangle$ then by construction of NoGoods(I') there is v''' assigning a or b to x and otherwise equal to v, such that $v''' \cup \{\langle y, e\rangle\}$ was in NoGoods(I), and if $v \not\ni \langle x, c\rangle$ we let $v''' = v$. We define w''' similarly. Then:

1. $(v''')^{<y} = v^{<y}$ and $(w''')^{<y} = w^{<y}$ are non-empty
2. Good($I, (v''')^{<y} \cup (w''')^{<y}$) $\wedge$ Good($I, (v''')^{<y} \cup \{\langle y, d\rangle\}$) $\wedge$ Good(I, $(w''')^{<y} \cup \{\langle y, e\rangle\}$) (since x is the last variable, $(v''')^{<y} = v^{<y}$ and $(w''')^{<y} = w^{<y}$)
3. $Vars(v'') = Vars(v''') \ \wedge\ (v'')^{<y} = (v''')^{<y} \ \wedge\ v'' \cup \{\langle y, d\rangle\} \notin$ NoGoods(I)
4. $Vars(w'') = Vars(w''') \wedge (w'')^{<y} = (w''')^{<y} \ \wedge\ w'' \cup \{\langle y, e\rangle\} \notin$ NoGoods(I))
5. $v''' \cup \{\langle y, e\rangle\} \in$ NoGoods(I) $\wedge$ $w''' \cup \{\langle y, d\rangle\} \in$ NoGoods(I)

that is, there was a DGABT on d, e in I, contradicting our assumption.

We now show that a broken triangle cannot be introduced on x. Observe that since x is the last variable, for all tuples t not containing an assignment to x, $t^{<x} = t$ holds. We use this tacitly in the rest of the proof. Suppose for a contradiction that I contained no DGABT, but that after merging $a, b \in D(x)$ in I to produce the instance I', in which a, b have been replaced by a new value c, we have a DGABT on c, d. Then there is a pair of non-empty tuples t, u (containing no assignments to variable x) satisfying in particular the following conditions:

(1) Good($I', t \cup u$)
(2) Good($I', t \cup \{\langle x, c\rangle\}$)
(3) Good($I', u \cup \{\langle x, d\rangle\}$)
(4) $t \cup \{\langle x, d\rangle\} \in$ NoGoods(I')
(5) $u \cup \{\langle x, c\rangle\} \in$ NoGoods(I')

We show that there was a DGABT in I either on a, d, on b, d or on a, b.

Since merging only affects tuples containing $\langle x, a\rangle$ or $\langle x, b\rangle$, (1) implies that Good($I, t \cup u$) and hence Good($I, t \cup u'$) for all $u' \subseteq u$. Similarly, (3) implies that Good($I, u \cup \{\langle x, d\rangle\}$) and hence Good($I, u' \cup \{\langle x, d\rangle\}$) for all $u' \subseteq u$. Similarly, (4) implies that $t \cup \{\langle x, d\rangle\}$ $\in$NoGoods(I).
There are three possible cases to consider:

(a) Good($I, t \cup \{\langle x, a\rangle\}$),
(b) Good($I, t \cup \{\langle x, b\rangle\}$),
(c) $\exists t_1, t_2 \subseteq t$ such that $t_1 \cup \{\langle x, a\rangle\}, t_2 \cup \{\langle x, b\rangle\} \in$ NoGoods(I).

case (a): By Definition 6 of the creation of nogoods during merging, (5) implies that $\exists u' \subseteq u$ such that $u' \cup \{\langle x, a\rangle\} \in$ NoGoods(I). We know that u' is non-empty since $u' \cup \{\langle x, a\rangle\} \in$ NoGoods(I) but Good($I, t \cup \{\langle x, a\rangle\}$) (and hence Good($I, \{\langle x, a\rangle\}$)). We have Good($I, t \cup u'$), Good($I, t \cup \{\langle x, a\rangle\}$) (and hence $t \cup \{\langle x, a\rangle\} \notin$ NoGoods(I)), Good($I, u' \cup \{\langle x, d\rangle\}$) (and hence $u' \cup \{\langle x, d\rangle\} \notin$

NoGoods(I)), $t \cup \{\langle x,d\rangle\} \in$ NoGoods(I), $u' \cup \{\langle x,a\rangle\} \in$ NoGoods(I) and hence there was a DGABT on a, d in I.
case (b): Symmetrically to case (a), there was a DGABT on b, d in I.
case (c): We claim that Good($I, t_1 \cup \{\langle x,b\rangle\}$). If not, then we would have $\exists t_3 \subseteq t_1$ such that $t_3 \cup \{\langle x,b\rangle\} \in$ NoGoods(I) which would imply $t_1 \cup \{\langle x,c\rangle\} \in$ NoGoods(I') which is impossible since, by (2) above, we have Good($I', t \cup \{\langle x,c\rangle\}$). By a symmetrical argument, we can deduce Good($I, t_2 \cup \{\langle x,a\rangle\}$). Since Good($I, t \cup u$) and $t_1, t_2 \subseteq t$, we have Good($I, t_1 \cup t_2$). Since $t_1 \cup \{\langle x,a\rangle\} \in$ NoGoods(I) and Good($I, t_2 \cup \{\langle x,a\rangle\}$) (and hence Good($I, \{\langle x,a\rangle\}$)), we must have $t_1 \neq \emptyset$. By a symmetric argument, $t_2 \neq \emptyset$. We therefore have non-empty tuples t_1, t_2 such that Good($I, t_1 \cup t_2$), Good($I, t_1 \cup \{\langle x,b\rangle\}$ (and hence $t_1 \cup \{\langle x,b\rangle\} \notin$ NoGoods(I)), Good($I, t_2 \cup \{\langle x,a\rangle\}$) (and hence $t_2 \cup \{\langle x,a\rangle\} \notin$ NoGoods(I)), $t_1 \cup \{\langle x,a\rangle\} \in$ NoGoods(I), $t_2 \cup \{\langle x,b\rangle\} \in$ NoGoods(I) and hence we have a DGABT in I on a, b.

Since in each of the three possible cases, we produced a contradiction, this completes the proof. □

5.3 Tractability of DGABTP for a Known Variable Ordering

Theorem 1. *A CSP instance I satisfying the DGABTP on a given variable ordering can be solved in polynomial time.*

Proof. Suppose that I satisfies the DGABTP for variable ordering $<$ and that x is the last variable according to this ordering. Lemma 1 tells us that DGA broken triangles cannot be introduced by merging all elements in $\mathcal{D}(x)$ to form a singleton domain $\{a\}$. At this point it may be that $\{\langle x,a\rangle\}$ is a nogood. In this case the instance is clearly unsatisfiable and the algorithm halts returning this result. If not then we simply delete $\langle x,a\rangle$ from all nogoods in which it occurs. This operation of variable elimination clearly preserves satisfiability. It is polynomial time to recursively apply this merging and variable elimination algorithm until a nogood corresponding to a singleton domain is discovered or until all variables have been eliminated (in which case I is satisfiable).

To complete the proof of correction of this algorithm, it only remains to show that elimination of the last variable x cannot introduce a DGA broken triangle on another variable y. For all tuples t, u and all values $c, d \in D(y)$, none of Good($I, t^{<y} \cup u^{<y}$), Good($I, t^{<y} \cup \{\langle y,c\rangle\}$) and Good($I, u^{<y} \cup \{\langle y,d\rangle\}$) can become true due to the variable elimination operation described above. On the other hand it is possible that $t \cup \{\langle y,d\rangle\}$ or $u \cup \{\langle y,c\rangle\}$ becomes a nogood due to variable elimination. Without loss of generality, suppose that $t \cup \{\langle y,d\rangle\}$ becomes a nogood and that $t' \cup \{\langle y,d\rangle\}$ is not a nogood for some t' such that $(t')^{<y} = t^{<y}$. Then by construction there was a nogood $t \cup \{\langle y,d\rangle\} \cup \{\langle x,a\rangle\}$ before the variable x (with singleton domain $\{a\}$) was eliminated, and $t' \cup \{\langle y,d\rangle\} \cup \{\langle x,a\rangle\}$ was not a nogood. But then there was a DGA broken triangle (given by tuples $t \cup \{\langle x,a\rangle\}$, u on values $c, d \in D(y)$) before elimination of x. This contradiction shows that variable elimination cannot introduce DGA broken triangles. □

5.4 Finding a DGABTP Variable Ordering Is NP-Hard

An important question is the tractability of the recognition problem of the class DGABTP when the variable order is not given, i.e. testing the existence of a variable ordering for which a given instance satisfies the DGABTP. In the case of binary CSP, this test can be performed in polynomial time [5]. Unfortunately, as the following theorem shows, the problem becomes NP-complete in the general-arity case.

Theorem 2. *Testing the existence of a variable ordering for which a CSP instance satisfies the DGABTP is NP-complete (even if the arity of constraints is at most 5).*

Proof. The problem is in NP since verifying the DGABTP is polytime for a given order, so it suffices to give a polynomial-time reduction from the well-known NP-complete problem 3SAT. Let I_{3SAT} be an instance of 3SAT with variables $X_1, \ldots, X_N$ and clauses $C_1, \ldots, C_M$. We will create a CSP instance I_{CSP} which has a DGABTP variable-ordering if and only if I_{3SAT} is satisfiable. For each variable X_i of I_{3SAT}, we add two variables x_i, y_i to I_{CSP}. To complete the set of variables in I_{CSP}, we add three special variables v, w, z. We add constraints to I_{CSP} in such a way that each DGABTP ordering of its variables corresponds to a solution to I_{3SAT} (and vice versa). The role of the variable z is critical: a DGABTP ordering $>$ of the variables of I_{CSP} corresponds to a solution to I_{3SAT} in which $X_i = \texttt{true} \Leftrightarrow x_i > z$. The variables y_i are used to code $\overline{X_i}$: $y_i > z$ in a DGABTP ordering if and only if $X_i = \texttt{false}$ in the corresponding solution to I_{3SAT}. The variables v, w are necessary for our construction and will necessarily satisfy $v, w < z$ in a DGABTP ordering. Each clause $C = l_1 \vee l_2 \vee l_3$, where l_1, l_2, l_3 are literals in I_{3SAT}, is imposed in I_{CSP} by adding constraints which force one of $\overline{l_1}, \overline{l_2}, \overline{l_3}$ to be false. To give a concrete example, if $C = X_1 \vee X_2 \vee X_3$, then constraints are added to I_{CSP} to force $y_1 < z$ or $y_2 < z$ or $y_3 < z$ in a DGABTP ordering. If the clause C contains a negated variable $\overline{X_i}$ instead of X_i, it suffices to replace y_i by x_i.

We now give in detail the necessary gadgets in I_{CSP} to enforce each of the following properties in a DGABTP ordering:

1. $v, w < z$
2. $y_i < z \Leftrightarrow x_i > z$
3. $y_i < z$ or $y_j < z$ or $y_k < z$

We introduce broken triangles in order to impose these properties. However, it is important not to inadvertently introduce other broken triangles. This can be avoided by making all pairs of assignments $\langle x, a \rangle$, $\langle x', a' \rangle$ from two different gadgets incompatible (i.e. $\{\langle x, a \rangle, \langle x', a' \rangle\} \in \text{NoGoods}(I_{CSP})$). We also assume that two gadgets which use the same variable x use distinct domain values in $\mathcal{D}(x)$. To avoid creating a trivial instance in which the gadgets disappear after establishing arc consistency, we can also add extra values in each domain which are compatible with all variable-value assignments in the gadgets.

We give the details of the three types of gadget:

1. The gadget to force $v, w < z$ in a DGABTP ordering consists of values $a_0 \in \mathcal{D}(z)$, $b_0, b_1 \in \mathcal{D}(v)$, $c_0, c_1 \in \mathcal{D}(w)$ and three nogoods $\{\langle z, a_0\rangle, \langle v, b_0\rangle\}$, $\{\langle z, a_0\rangle, \langle w, c_0\rangle\}$, $\{\langle v, b_1\rangle, \langle w, c_1\rangle\}$. The only way to satisfy the DGABTP on this triple of variables is to have $v, w < z$ since there are broken triangles on variables v and w.
2. To force $y_i < z \Leftrightarrow x_i > z$ in a DGABTP ordering we use two gadgets, the first to force $y_i > z \vee x_i > z$ and the second to force $y_i < z \vee x_i < z$.
   The first gadget is a broken triangle consisting of values $a_1, a_2 \in \mathcal{D}(z)$, $d_0 \in \mathcal{D}(x_i)$, $e_0 \in \mathcal{D}(y_i)$ and two nogoods $\{\langle z, a_1\rangle, \langle x_i, d_0\rangle\}$, $\{\langle z, a_2\rangle, \langle y_i, e_0\rangle\}$. In a DGABTP ordering we must have $y_i > z \vee x_i > z$.
   The second gadget consists of values $a_3, a_4 \in \mathcal{D}(z)$, $b_2 \in \mathcal{D}(v)$, $c_2 \in \mathcal{D}(w)$, $d_1 \in \mathcal{D}(x_i)$, $e_1 \in \mathcal{D}(y_i)$ and four nogoods $\{\langle z, a_3\rangle, \langle v, b_2\rangle, \langle x_i, d_1\rangle\}$, $\{\langle z, a_4\rangle, \langle v, b_2\rangle, \langle x_i, d_1\rangle\}$, $\{\langle z, a_4\rangle, \langle w, c_2\rangle, \langle y_i, e_1\rangle\}$, $\{\langle z, a_3\rangle, \langle w, c_2\rangle, \langle y_i, e_1\rangle\}$. We assume that we have forced $v, w < z$ using the gadget described in point (1). The tuples $t = \{\langle v, b_2\rangle, \langle x_i, d_1\rangle\}$, $u = \{\langle w, c_2\rangle, \langle y_i, e_1\rangle\}$ then form a DGA broken triangle on assignments $a_3, a_4 \in \mathcal{D}(z)$ if $x_i, y_i > z$. If either $x_i < z$ or $y_i < z$ then there is no DGA broken triangle; for example, if $x_i < z$, then we longer have Good(I_{CSP},$t^{<z} \cup \{\langle z, a_3\rangle\}$) since $t^{<z} \cup \{\langle z, a_3\rangle$ is precisely the nogood $\{\langle z, a_3\rangle, \langle v, b_2\rangle, \langle x_i, d_1\rangle\}$. Thus this gadget forces $y_i < z \vee x_i < z$ in a DGABTP ordering.
3. The gadget to force $y_i < z$ or $y_j < z$ or $y_k < z$ in a DGABTP ordering consists of values $a_5, a_6 \in \mathcal{D}(z)$, $b_3 \in \mathcal{D}(v)$, $c_3 \in \mathcal{D}(w)$, $e_2 \in \mathcal{D}(y_i)$, $e_3 \in \mathcal{D}(y_j)$, $e_4 \in \mathcal{D}(y_k)$ and five nogoods, namely $\{\langle z, a_6\rangle, \langle v, b_3\rangle, \langle y_i, e_2\rangle, \langle y_j, e_3\rangle, \langle y_k, e_4\rangle\}$, $\{\langle z, a_5\rangle, \langle w, c_3\rangle\}$, $\{\langle z, a_5\rangle, \langle y_i, e_2\rangle\}$, $\{\langle z, a_5\rangle, \langle y_j, e_3\rangle\}$, $\{\langle z, a_5\rangle, \langle y_k, e_4\rangle\}$. The tuples $t = \{\langle v, b_3\rangle, \langle y_i, e_2\rangle, \langle y_j, e_3\rangle, \langle y_k, e_4\rangle\}$, $u = \{\langle w, c_3\rangle\}$ form a DGA broken triangle on $a_5, a_6 \in \mathcal{D}(a)$ if $y_i, y_j, y_k > z$. If $y_i < z$ or $y_j < z$ or $y_k < z$, then there is no DGA broken triangle; for example, if $y_i < z$, then we longer have Good(I_{CSP},$t^{<z} \cup \{\langle z, a_5\rangle\}$) since $\{\langle z, a_5\rangle, \langle y_i, e_2\rangle\}$ is a nogood. Thus this gadget forces $y_i < z$ or $y_j < z$ or $y_k < z$ in a DGABTP ordering.

The above gadgets allow us to code I_{3SAT} as the problem of testing the existence of a DGABTP ordering in the corresponding instance I_{CSP}. To complete the proof it suffices to observe that this reduction is clearly polynomial. □

Our proof of Theorem 2 used large domains. The question still remains whether it is possible to detect in polynomial time whether a DGABTP variable ordering exists in the case of domains of bounded size, and in particular in the important case of SAT.

6 Conclusion

This paper described a novel reduction operation for binary CSP, called BTP-merging, which is strictly stronger than neighbourhood substitution. Experimental trials have shown that in several benchmark-domains applying

BTP-merging until convergence can significantly reduce the total number of variable-value assignments. We gave a general-arity version of BTP-merging and demonstrated a theoretical link with resolution in SAT. From a theoretical point of view, we then went on to define a general-arity version of the tractable class defined by the broken-triangle property for a known variable ordering. Further research is required to find optimal algorithms for BTP-merging and to investigate the tractability of applying BTP-merging in instances containing global constraints.

References

1. Cohen, D.A., Cooper, M.C., Creed, P., Marx, D., Salamon, A.Z.: The tractability of CSP classes defined by forbidden patterns. Journal of Artificial Intelligence Research 45, 47–78 (2012)
2. Cohen, D.A., Cooper, M.C.: Guillaume Escamocher and Stanislav Živný, Variable elimination in binary CSP via forbidden patterns. In: Proceedings of IJCAI, pp. 517–523 (2013)
3. Cooper, M.C.: Fundamental properties of neighbourhood substitution in constraint satisfaction problems. Artificial Intelligence 90(1-2), 1–24 (1997)
4. Cooper, M.C., Escamocher, G.: A dichotomy for 2-constraint forbidden CSP patterns. In: Proceedings of AAAI, pp. 464–470 (2012)
5. Cooper, M.C., Jeavons, P.G., Salamon, A.Z.: Generalizing constraint satisfaction on trees: Hybrid tractability and variable elimination. Artificial Intelligence 174(9-10), 570–584 (2010)
6. Cooper, M.C., Živný, S.: Tractable Triangles and Cross-Free Convexity in Discrete Optimisation. Journal of Artificial Intelligence Research 44, 455–490 (2012)
7. Eén, N., Biere, A.: Effective Preprocessing in SAT Through Variable and Clause Elimination. In: Bacchus, F., Walsh, T. (eds.) SAT 2005. LNCS, vol. 3569, pp. 61–75. Springer, Heidelberg (2005)
8. Mouelhi, A.E., Jégou, P., Terrioux, C.: A Hybrid Tractable Class for Non-Binary CSPs. In: Proceedings of ICTAI, pp. 947–954 (2013)
9. Jégou, P.: Decomposition of Domains Based on the Micro-Structure of Finite Constraint-Satisfaction Problems. In: Proceedings of AAAI, pp. 731–736 (1993)
10. Freuder, E.C.: Eliminating interchangeable values in constraint satisfaction problems. In: Proceedings of AAAI, pp. 227–233 (1991)
11. Likitvivatanavong, C., Yap, R.H.C.: Eliminating redundancy in csps through merging and subsumption of domain values. ACM SIGAPP Applied Computing Review 13(2) (2013)
12. Salamon, A.Z., Jeavons, P.G.: Perfect Constraints Are Tractable. In: Stuckey, P.J. (ed.) CP 2008. LNCS, vol. 5202, pp. 524–528. Springer, Heidelberg (2008)

Using CP in Automatic Test Generation for ABB Robotics' Paint Control System

Morten Mossige[1,3], Arnaud Gotlieb[2], and Hein Meling[3]

[1] ABB Robotics, Norway
morten.mossige@no.abb.com
[2] Simula Research Laboratory, Norway
arnaud@simula.no
[3] University of Stavanger, Norway
hein.meling@uis.no

Abstract. Designing industrial robot systems for welding, painting, and assembly, is challenging because they are required to perform with high precision, speed, and endurance. ABB Robotics has specialized in building highly reliable and safe robotized paint systems based on an *integrated process control system*. However, current validation practices are primarily limited to manually designed test scenarios. A tricky part of this validation concerns testing the timing aspects of the control system, which is particularly challenging for paint robots that need to coordinate paint activation with the robot motion control.

To overcome these challenges, we have developed and deployed a cost-effective, automated test generation technique based on Constraint Programming, aimed at validating the timing behavior of the process control system. We designed a constraint optimization model in SICStus Prolog, using arithmetic and logic constraints including use of global constraints. This model has been integrated into a fully automated continuous integration environment, allowing the model to be solved on demand prior to test execution, which allows us to obtain the most optimal and diverse set of test scenarios for the present system configuration.

After three months of daily operational use of the constraint model in our testing process, we have collected data on its performance and bug finding capabilities. We report on these aspects, along with our experiences and the improvements gained by the new testing process.

1 Introduction

Developing reliable software for Complex Industrial Robots (CIRs) is a complex task, because typical robots are comprised of numerous components, including computers, field-programmable gate arrays (FPGAs), and sensor devices. These components typically interact through a range of different interconnection technologies, e.g. Ethernet and dual port RAM, depending on delay and latency requirements on their communication. As the complexity of robot control systems continues to grow, developing and validating software for CIRs is becoming increasingly difficult. For robots performing process-intensive tasks such as

B. O'Sullivan (Ed.): CP 2014, LNCS 8656, pp. 25–41, 2014.

painting, gluing, or sealing, the problem is even worse as their dedicated process control systems is loosely coupled with the robot motion control system. A key feature of robotized painting is the ability to perform precise activation of the process equipment along a robot's programmed path. At ABB Robotics, Norway, they develop and validate Integrated Painting control Systems (IPS) for CIRs and are constantly improving the processes to deliver more reliable products to their customers.

Current practices for validating the IPS software involve designing and executing manual test scenarios. In order to reduce the testing costs and to improve quality assurance, there is a growing trend to automate the generation of test scenarios and multiplying them in the context of continuous testing.

In this paper, we report on our work to use Constraint Programming (CP) over finite domains to *generate* automatically timed-event sequences (i.e., test scenarios) for the IPS and *execute* them within a Continuous Integration (CI) process [1]. Building on initial ideas sketched in a poster [2] one year ago, we have developed a constrained optimization model in SICStus Prolog `clpfd` [3] to help test the IPS under operational conditions. Due to online configurability of the IPS, test scenarios must be reproduced every day, meaning that indispensable trade-offs between optimality and efficiency must be found, to increase the capabilities of the CI process to reveal software defects as early as possible. Using CP to generate model-based test scenario is not a completely new idea [4,5], but, according to our knowledge, this is the first time that a CP model and its solving process been integrated into a CI environment for testing complex distributed systems.

Organization. The rest of the paper is organized as follows: Section 2 presents some background on robotized painting, with an example serving as a basis for describing the mathematical relations involved ; Section 3 describes ABB Robotic's current testing practices of the IPS and the rationale behind our validation choices ; Section 4 presents the CP model with its decision variables, test objectives and optimization principle ; Section 5 explains how the model and its solving process are implemented and included in the CI process ; Finally, Section 6 discusses some lessons learnt and summarizes the impact of using CP in ABB Robotics's industrial context.

Notation. Throughout the paper a constant in the CP model is prefixed with a *, as in $^*SeqLen$. This is typically a value set by a validation engineer or queried from the system under test prior to launching the model.

2 Robotized Painting

This section briefly introduces robotized painting, and highlights some of the challenges faced when testing such systems. A robot system dedicated to painting typically consists of two main parts: the robot controller, responsible for moving the mechanical arm, and the IPS, responsible for controlling the paint process. That is, to control the activation and deactivation of several physical processes such as paint pumps, air flows, air pressures, and to synchronize these with the motion of the robot arm. A *spray pattern* is defined as the combination of the different physical processes. Typically, the physical processes involved in a spray pattern will have different response times. For instance, a pump may have a response time in the range 40-50 ms, while the airflow response time is in the range 100-150 ms. The IPS can adjust for these differences using sophisticated algorithms that have been analyzed and tuned over the years to serve different needs. In this paper, we focus on validating the timing aspects of the IPS.

2.1 Example of Robotized Painting

We now give a concrete example of how a robot controller communicates with the IPS in order to generate a spray pattern along the robot's path. A schematic overview of the example is shown in Figure 1, where the node marked *robot controller* is the CPU interpreting a user program and controlling the servo motors of the robot in order to move it. The example is realistic, but simplified, in order to keep the explanations as simple as possible.

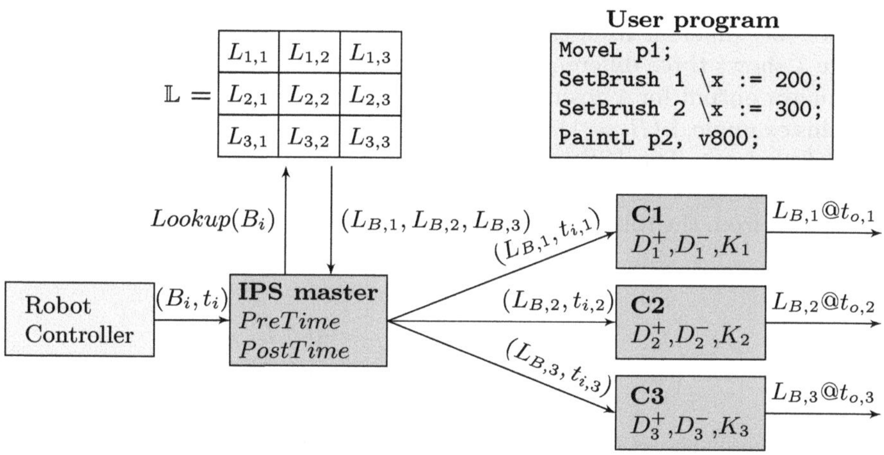

Fig. 1. Logical overview of a robot controller and the IPS

The program listing of Figure 1 shows an example user program. The first instruction `MoveL p1` moves the robot to the Cartesian point `p1`. The next two `SetBrush` instructions tells the robot to apply spray pattern number 1 when the robot reaches $x = 200$ on the x-plane, and to apply spray pattern number 2 when it reaches $x = 300$. Both `SetBrush` instructions tell the IPS to apply a specific behavior when the physical robot arm is at a given position. The last instruction (`PaintL`) starts the movement of the robot from the current position `p1` to `p2` and activates the painting process. The `v800` argument of `PaintL` gives the speed of the movement (i.e., 800 mm/s).

Assuming the path from `p1` to `p2` results in a movement from $x = 0$ to $x = 500$. The robot controller interprets the user program ahead of the actual physical movement of the robot, and can therefore estimate *when* the robot will be at a specific position. Assuming that the movement starts at time $t = 0$, the robot can compute that the two `SetBrush` activations should be triggered at $t_1 = 250$ ms and $t_2 = 375$ ms.

The robot controller now sends the following messages (a.k.a. *events*) to the IPS master: $(B_1 = 1, t_1 = 250), (B_2 = 2, t_2 = 375)$, which means apply spray pattern 1 at 250 ms, and spray pattern 2 at 375 ms. The messages are sent around 200 ms before the actual activation time, or at ≈ 50 ms for spray pattern 1, and at ≈ 175 ms for spray pattern 2. These messages simply convert position into an absolute global activation time. Note also that the IPS receives the second message before the first spray pattern is bound for execution, which means that the IPS must handle a queue of scheduled spray patterns.

IPS Master: When the IPS receives a message from the robot controller, it first determines the physical outputs associated with the logical spray pattern number. Many different spray patterns can be generated based on factors like paint type or equipment in use. In the IPS each spray pattern is translated into 3 to 6 different physical actuator outputs that must be activated at appropriate times, possibly different from each other.

Figure 1 shows three different actuator outputs (**C1, C2, C3**). The value of each actuator output for a given spray pattern is resolved by using a brush table ($\mathbb{L}$). In this example, $\mathbb{L}(B_1 = 1)$ returns $(L_{1,1}, L_{1,2}, L_{1,3})$, while $\mathbb{L}(B_2 = 2)$ results in $(L_{2,1}, L_{2,2}, L_{2,3})$. The IPS master now passes these values to each actuator output along with its activation time, which may be different from the original time received from the robot controller. Possible modifications can be formalized as follows:

$$t_i' = \begin{cases} t_i - PreTime & \text{if } L_{1,B_{i-1}} = 0 \land L_{1,B_i} \neq 0 \\ t_i - PostTime & \text{if } L_{1,B_{i-1}} \neq 0 \land L_{1,B_i} = 0 \end{cases} \tag{1}$$

What equation (1) shows is that the activation time of each actuator output may be adjusted by a constant factor (*PreTime, PostTime*), depending on changes from other actuator outputs. This is done because small adjustments may be necessary when there is a direct link between the timing of different actuator outputs. In our example, the timing on **C2** is influenced by changes on **C1**.

Activation of Actuator Outputs: Referring to Figure 1, we now present how messages are processed when sent from the IPS master to a single actuator output. Let us assume that message (L, t_i) is sent, and the current actuator output is L'. Since painting involves many slow physical processes, the actuator output compensates for this by computing an adjusted activation time t_o, that accounts for the time it takes the physical process to apply the change.

The IPS can adopt two different strategies to compute this time compensation. The first one is to adjust the time with a *constant* factor: D^+ for positive change, and D^- for negative change. The second one is using a *linear* timing function to adjust the change of the physical value.

Equation (2) combines these strategies into a single compensation function, where $^*Min$ (resp. $^*Max$) is the physical minimum (resp. maximum) value possibly handled[1] by some actuator output.

$$t_o = t_i - \begin{cases} D^- \cdot (\frac{L-L'}{^*Max-^*Min})^K & \text{if } L' < L \\ D^+ \cdot (\frac{L'-L}{^*Max-^*Min})^K & \text{if } L' > L \\ 0 & \text{otherwise} \end{cases} \quad (2)$$

Physical Layout of the IPS: Figure 1 only shows the logical connections in a possible IPS configuration. In real applications, each component (**IPS master, C1, C2, C3**) may be located on different embedded controllers, interconnected through an industrial-grade network. As such, the different components may be located at different physical locations on the robot, depending on which physical process it is responsible of.

3 Testing the IPS

Having a distributed control system such as the IPS mounted on a physical robot makes its validation unnecessarily complex, and current testing practices involve a considerable amount of manual work, including setup and collecting observations. If while developing a new version of the IPS software, test scenarios are only run when approaching the release date, then development costs can grow substantially, as correcting software defects late in the development process may require developers to dig into the early stages of development. Even worse, if a software failure is observed during operation (i.e., by the customer), costs become even higher since corrections may need to take place at the customer site.

3.1 Continuous Integration

CI is a software engineering practice aimed at uncovering software defects at the earliest stage of development, by regularly building the system and executing tests automatically [1].

A good engineering practice requires developers to submit only small source code changes frequently, instead of large sets of changes occasionally. Together with this practice, CI has been shown to be a very efficient way of uncovering defects when developers are geographically distributed or large teams are involved. Typically a CI infrastructure includes tools for source control repository, automated build servers, and testing engines.

[1] These values are determined by the physical equipment involved in the paint process (pumps, valves, air, etc.).

3.2 Testing in a CI Environment

We have developed an automated testing framework for the IPS as an integrated part of ABB's CI environment, where we have used CP to generate both the configuration for the IPS, the test sequence, the brush table and the output of each actuator output and finally execute the test as part of a CI cycle.

Compared to traditional software testing, running a test scenario in a CI environment has additional requirements. In particular, as pointed out by Fowler [1], mastering the total *round-trip time* is crucial for a successful CI deployment. Here, round-trip time refers to the time it takes for a developer to submit a change to the source control repository and get feedback from the build and test processes. Thus in order to keep the round-trip time as small as possible, we have identified a few areas where special care must be taken:

- **Test complexity:** In CI, a less accurate but faster test will always be preferred over a slow but accurate test. In practice, a test must satisfy the so-called *good enough criterion*, frequently used in industry [6].
- **Solving time:** Constraint-based optimization is most often a time-consuming task, especially if a global optimum solution is sought [7]. Thus, when used in CI, it becomes imperative that a time-contracted optimization procedure be used. In other words, it is important to have precise control over the time needed to compute the optima, by sacrificing the solution quality.
- **Execution time:** We observe that test execution time is dependent on the length of the test sequence, i.e., the number of test scenarios. This must be accounted for, together with the time needed to generate the test sequence.

In essence, balancing between the length of a test sequence (its execution time) and the time needed to generate the test sequence (its solving time) is a way to find the appropriate trade-off to fully integrate CP into a CI process.

4 CP Model of the IPS

We now present our CP model for the IPS. We emphasize that test models, as proposed in model-based testing [8], are usually limited in their scope. They are not intended to reflect the full behavior of the system they represent. In our case, we confine ourselves to modeling the timing aspects of the IPS in order to build an efficient CP model for generating test scenarios.

4.1 Decision Variables and Domains

While still referring to Figure 1, we now assume that the number of actuator outputs is a constant input parameter C, instead of 3. The decision variables for our problem can be divided into three distinct groups: the variables of the input sequence $\mathbb{I}$, the configuration variables $\mathbb{C}$, and the variables of the brush table $\mathbb{L}$. In principle, a solution of the CP model is formed by an instantiation of these variables, in addition to the so-called test oracle $\mathbb{O}$, which is the expected output computed by the system formed by each actuator output and its corresponding time.

Formally, the test input sequence $\mathbb{I}$ corresponds to $((B_1, t_1), \ldots, (B_N, t_N))$, where $N = {}^*SeqLen$ and each $B_i \in [0, {}^*BTabSize]$ and each $t_i \in [0, {}^*MaxTime]$. The configuration $\mathbb{C}$ contains parameter variables for each actuator output and for the IPS master:

$\mathbb{C} = [PreTime, PostTime, D_1^+, D_1^-, K_1, \ldots, D_{*C}^+, D_{*C}^-, K_{*C}]$

The domain of the variables in $\mathbb{C}$ is given by configurable constants to the CP model. In the brush table $\mathbb{L}$, the number of columns corresponds to the number of actuator outputs, i.e., ${}^*C$, and the number of rows is a constant ${}^*BTabSize$. The domain of each variable in $\mathbb{L}$ is extracted from ${}^*Min$ and ${}^*Max$ for the corresponding actuator output. The test oracle $\mathbb{O}$ corresponds to the physical output of each actuator output with its corresponding time. Each actuator output has output and time corresponding to a single input: $(B_i, t_i) \mapsto \big((L_{1,i}, t_{1,i}), \ldots, (L_{C,i}, t_{C,i})\big)$ for $i \in [1, N]$.

4.2 Test Scenarios

We have identified several distinct test scenarios, and we present three of them here, as shown in Figure 2. Scenarios *overlap* and *kill brush* represent failure conditions, where the IPS is forced into an error state. When generating such scenarios it is our interest to check whether the IPS can respond correctly (i.e., shutdown, error messages, etc.). On the contrary, the scenario *normal* represents acceptable behavior and they are targeted to check whether the IPS behaves as expected. Whenever the CP model is solved, a scenario is given as a test objective to the solver, and the solving process intends to find an assignment of variables that can drive the execution of the IPS in the corresponding scenario status. The *overlap* scenario is used throughout the paper, as it is the hardest to find and therefore, it corresponds to the most difficult objective to solve for the CP model. Let us explain it in more details. As explained in Section 2.1, the IPS can queue up a sequence of actuator output changes. However, a sequence spray pattern number sent to the IPS can cause one or several of the actuator outputs to come out of order with respect to time. This can be due to changes over time *between* spray patterns, or due to usage of $PreTime/PostTime$ configuration or else due to different configuration of the actuator output. In principle, the IPS must handle these issues by sending an appropriate error message to the control system.

4.3 Avoiding Trivial and Enforcing Diversity

An additional objective in test sequence generation for the IPS is to introduce diversity in the test input sequence $\big((B_1, t_1), \ldots, (B_i, t_i)\big)$, in the values of the brush table ($\mathbb{L}$) and in the configuration parameters for each actuator outputs (D^+, D^-, K). By diversity, we mean variations in the test scenarios so that the chances to discover an error-prone scenario are greater.

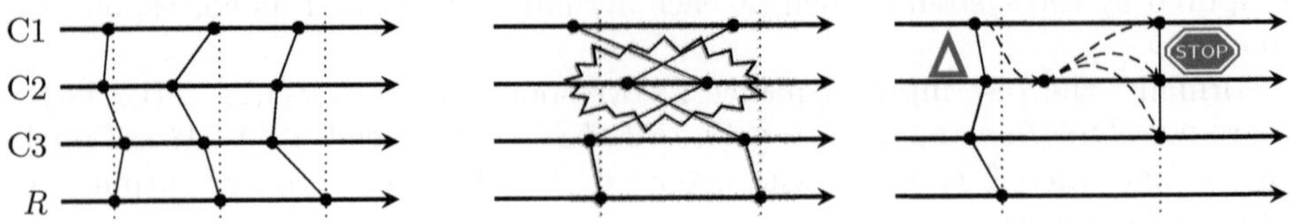

(a) Normal (b) Overlap (c) Kill

Fig. 2. Test scenarios considered as test objectives. Horizontal axis represent time and black dots correspond to output activation. A specific *spray pattern* is a collection of output activations, and is visualized by a line connecting the black dots.

As solving the CP model is a deterministic process, introducing diversity is a way to cope with the possible generation of useless scenarios. Let us consider the setup in Figure 1 where configuration $\mathbb{L} = \begin{smallmatrix}0&0&0\\0&0&0\\0&0&0\end{smallmatrix}$, $D_j^+ = 0$, $D_j^- = 0$ and $K_j = 0$, $j \in 1..3$ is given, we see that no matter what the input sequence $(\ldots,(B_i,t_i),\ldots,)$ is, the actuator output output is always $(0,t_i)$. This is of course a solution, but it has no practical interest, as it does not correspond to a possible behavior of the IPS. Using randomization in the CP model, in order to introduce diversity, has clearly been discarded. In fact, one of our initial requirements is to maintain a reproducible process. When testing the IPS, it is important to document failure cases and to help debug the system with the generated test scenarios.

4.3.1 Variation in $\mathbb{I}$ Values

Let $\mathbb{I} = ((B_1,t_1),\ldots,(B_N,t_N))$ be an input sequence to the IPS. Significant variation on values for t_i is not interesting, as the only requirement is that time must increase monotonically with a minimum step ($^*MinBrushSep$). More interesting is the variation in B_i. Obviously, enforcing a change in two successive brush element selections is important, i.e., $\forall i, B_i \neq B_{i+1}$, but this is not sufficient to guarantee that all indexes in the $\mathbb{L}$ are tested. Diversity on sequence $B = B_1,\ldots,B_N$ could be implemented by using global constraint `nvalue( LenLookupTab, B )`, enforcing that every index is present at least once. But, it will not be sufficient for our testing purposes.

Let us define the notion of *diversity entropy* (DE): given a sequence of integers, DE is the product of the number of occurrences of each value in the sequence. For example $DE([0,1,0,1,0,1,0,1,2,3]) = 4\cdot4\cdot1\cdot1 = 16$, while $DE([0,1,2,1,2,3,1,3,2,3]) = 1\cdot3\cdot3\cdot3 = 27$. With this example, we see that the first solution, respecting both previously mentioned constraints, has a diversity entropy lower than the second solution. We therefore come up with another solution in which we use the `global_cardinality` constraint. By specifying the minimum number of times an index of $\mathbb{L}$ must appear in the input sequence, we increase the diversity entropy of the solution. For example, given $^*SeqLen = 10$, $^*BTabSize = 4$, variation in the input sequence is enforced by using

```
global_cardinality( [B1,...,B10], [1-N1,2-N2, 3-N3, 4-N4] )
N1 #>= Ob, N2 #>= Ob, N3 #>= Ob N4 #>= Ob
```

where `Ob` is a given constant input parameter of the CP model. This implementation is flexible enough to consider solutions with a satisfactory DE.

4.3.2 Variation in $\mathbb{L}$ Values

Variation in $\mathbb{L}$ is equally important. When the validation engineers create these tables manually, they try to enforce that each actuator outputs $^*Min$ and $^*Max$ is part of the $\mathbb{L}$, and that the whole operating area of the actuator output is used. If each entry in $\mathbb{L}$ is regarded as coordinates in an Euclidian space, $\mathbb{R}^C$, an approach could be to maximize the distance between each point, i.e., each entry in $\mathbb{L}$. However, we observed that this approach is too costly to compute in practice, and we prefer a more light-weight approach. For each row in $\mathbb{L}$, $(R_1, \dots, R_{*BTabSize})$, we exploit the global constraints $minimum(^*Min_j, R_j)$ and $maximum(^*Max_j, R_j)$ to enforce usage of extremal values. In addition, the `all_min_dist` [9] constraint is used to make sure the values are spread out.

To introduce variation *between* the entries in $\mathbb{L}$, additional constraints are used to enforce at least one transition where all except one value is changed. For example, from Figure 1, if two entries is $[L_{1,i}, L_{2,i}, L_{3,i}]$ and $[L_{1,j}, L_{2,j}, L_{3,j}]$ then there should exist i and j such that $L_{1,i} < L_{1,j} \wedge L_{2,i} \geq L_{2,j} \wedge L_{3,i} \geq L_{3,j}$, and so on other entries. This is clearly far away from maximizing the Euclidian distance between each entry, but this approach turns out to perform fairly well together with the scenarios presented earlier. Of course, there is room for improvement here in further work.

4.3.3 Variation in $\mathbb{C}$ Values

The generated configuration for a specific test scenario includes both the values for each actuator output (D^+, D^-, K) and the value for IPS master (*PreTime*, *PostTime*). In many setups, validation engineers select these values manually without questioning the error-proneness of a given configuration with respect to another. By adding simple constraints for each actuator output, such that $D^+ \neq D^- \wedge D^- \neq 0 \wedge D^+ \neq 0$, we offer an opportunity for the CP model to introduce diversity in the configuration values as well. By using global constraint `all_different` $(D_1^+, \dots, D_C^+)$,etc, we also enforce diversity between actuator output values. For the $PreTime$ and $PostTime$ values, a similar strategy is employed: $PreTime \neq PostTime \wedge PreTime \neq 0 \wedge PostTime \neq 0$.

It is worth noticing that these variation strategies have served well the *good enough principle*, as introducing diversity is important but not at the cost of losing efficiency.

4.4 Search and Optimization

We now briefly present the optimization function and the search heuristics used in our model. In our framework, finding optimal solutions that respect the set of above-mentioned constraints is the most interesting. Optimal solution here means a sequence of timed events $\mathbb{I} = ((B_1, t_1), \ldots, (B_N, t_N))$ of the *smallest execution time*, i.e., where t_N is minimized. By doing this, we increase the capabilities of the CI process to execute more tests during a limited period. Of course, reaching exactly the global minimum over t_N is interesting from an intellectual perspective, but not really necessary in our industrial setting.

As mentioned in Section 3.2, managing the time needed to generate and execute test sequences when running tests in a CI environment is of crucial importance. Considering a test sequence $\mathbb{I} = ((B_1, t_1), \ldots, (B_N, t_N))$, and the fact that each spray pattern is sent approximately 200 ms before execution, as explained in Section 2.1, we see that the execution time of the test can be roughly estimated to be t_N. This means that the total time used is roughly $t_s + t_N$, where t_s correspond to the solving time of the model.

Knowing that the constrained optimization model tends to minimize t_N, the goal is therefore to control the time needed to find an optimal solution. CP offers means to control the time taken to optimize by using a *branch-and- bound* procedure. That is, we can give a contract of time to this procedure, and it returns the current feasible solution after the contract of time has passed. We found this option very useful to compromise between the time spent on search and solving, and the time spent on execution of the test.

4.5 Search Heuristics

When searching for solutions, many heuristics can be employed or programmed in CP. Observing the absence of evident *structure* in our CP model, we have considered variable orderings as the first element to examine systematically. In order to extract useful information, we considered 72 distinct static variable orderings depending on rearrangements of the decision variables $(\mathbb{I}, \mathbb{L}, \mathbb{C})$. In addition to this systematic exploration, we took as a reference two well-known dynamic variable orderings, namely *first-fail* and *first-fail constraint* [3]. We also tested *up* and *down* which dictates the direction the domain is searched (ascending or descending). This analysis and the experiments revealed two points:

1. Even if first-fail and its variation are efficient for timed sequences containing few events, they quickly become unusable for larger sequences. This can be easily explained by the necessary computation of comparison between domain sizes during the search, which becomes intractable as soon as the number of variables grows.
2. Looking at search heuristics with static variable orderings, we can group the result into three groups: H_1, H_2 and H_3.
   $H_1 = (\mathbb{C}, \mathbb{L}, \mathbb{I}'), (\mathbb{L}, \mathbb{C}, \mathbb{I}')$ $(\mathbb{C}, \mathbb{L}^{\mathbb{T}}, \mathbb{I}')$,$(\mathbb{L}^{\mathbb{T}}, \mathbb{C}, \mathbb{I}')$, $H_2 = (\mathbb{C}, \mathbb{B}, \mathbb{L}, \mathbb{T}), (\mathbb{B}, \mathbb{C}, \mathbb{L}, \mathbb{T})$, $(\mathbb{B}, \mathbb{L}, \mathbb{C}, \mathbb{T})$, $(\mathbb{C}, \mathbb{B}, \mathbb{L}^{\mathbb{T}}, \mathbb{T})$, $(\mathbb{B}, \mathbb{C}, \mathbb{L}^{\mathbb{T}}, \mathbb{T})$, $(\mathbb{B}, \mathbb{L}^{\mathbb{T}}, \mathbb{C}, \mathbb{T})$, H_3 = The other 62 tested combinations, where $\mathbb{I}' = (t_1, B_1, t_2, B_2, \ldots)$, $\mathbb{B} = (B_1, B_2, \ldots)$, $\mathbb{T} = (t_1, t_2, \ldots)$ and $\mathbb{L}^{\mathbb{T}} = transp(\mathbb{L})$. H_1 is the only heuristics able to produce a solution within an acceptable timeframe for small values of $^*BTabSize$ combined with large values of $^*SeqLen$, e.g. $^*BTabSize = 10, ^*SeqLen = 200$.

For configurations of large values of $^*SeqLen$ combined with large values of $^*BTabSize$, e.g. $^*BTabSize = 40, ^*SeqLen = 600$, H_2 is the only heuristic able to generate a solution within reasonable time. The result for H_3 is either no solution at all, or only solution for small $^*BTabSize$ and small $^*SeqLen$. Understanding precisely why H_1 and H_2 performs so well is part of our planned further work.

5 Implementation and Exploitation

This section details our implementation of the CP model [10] with SICStus Prolog and its `clpfd` library [3], and its exploitation in the CI process at ABB Robotics. It also gives some insights on the rationale behind the selection of CP instead of other possible techniques.

5.1 Selection of CP and the CP Solver

The mathematical model of the IPS could have been implemented with other techniques than CP, including SAT- or SMT-solving [11], local search techniques for test data generation [12], or Mixed Integer Programming (MIP) [13]. We briefly review the reasons why these other techniques have been discarded[2]:

1. The selected technique had to be flexible enough to accommodate the many alternatives in the dynamic configuration of the IPS. CP offers a higher degree of flexibility to handle disjunctive constraint systems, by authorizing the usage of *backtracking*, *reification*, or *constructive disjunction* [14] ;
2. Time-constrained optimization was essential in our industrial context in order to accommodate with the CI process. SAT- and SMT-solving are very efficient to handle boolean and theory-based satisfiability problems [11], but they are not tuned to solve optimization problems (i.e., to minimize a cost function in a given contract of time). Even if extensions exist to handle optimization problems, classical off-the-shelf SMT-solvers do not provide implementations of these extensions. On the contrary, CP integrates time-aware optimization methods on discrete combinatorial problems ;
3. As the model is used to predict the expected outputs of the IPS, using exact methods was mandatory. Despite the efficiency of local search techniques for test data generation [12], the absence of guarantee on the satisfiability of the constraints (e.g., no possible detection of unsatisfiability or no guarantee on the determination of satisfiability for complex constraint sets) was sufficient to discard these techniques ;
4. Input formats of the constraint solver had to be easily tunable to accommodate the high-level tuning of IPS parametrization. SAT- and SMT-solvers takes specific formats as inputs (e.g., SMTLIB formats) while CP-solvers are usually hosted by a programming language (e.g., Prolog, Java or C++) which includes high-level programming features such as predicate/method invocation, recursivity, inheritance, and so on ;

[2] Note that no general claim is made, just specific claims to illuminate our choice of CP in the case of validating the IPS.

5. The availability of global constraints to implement diversity in test sequences was a strong advantage, even if, to be honest, we discovered it after our choice was made.

We found that SICStus 4.2.3 in combination with `clpfd` responded well to our industrial requirements and decided to use it as back-end, and Python 2.7 as front-end.

5.2 Overall Implementation

The complete system contains around 2k lines of Prolog code, 300 lines of C code (an interface DLL between Python and SICStus), and finally around 3k lines of Python code. A schematic overview of the implementation and how it is executed can be found in Figure 3.

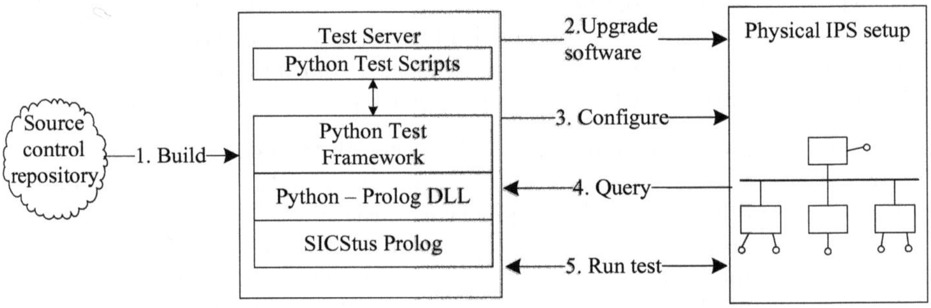

Fig. 3. Integration between the test server and IPS

The modeling part of the project has started early in 2013 ; at the beginning, just by using the user interface of SICStus. In April 2013, a first running model was available on a desktop for testing IPS, running over a single embedded board. In May, the model was integrated into the source control repository and the first automatic test running in a full CI environment was executed. From May to October 2013, the system was further extended to also cover testing over complete distributed systems (i.e., several embedded boards) of the IPS. Today, the model is used in the CI process and solved daily. It generates test sequences for 11 different physical embedded IPS boards. For testing on the full-distributed setting, we currently run the model on one single physical setup, but we run 10 different configurations on this setup. To summarize, the number of measurable activations of physical actuator outputs shows that around 20.000 distinct test scenarios are executed during each individual CI cycle. It means that these test scenarios are executed at least once every 24 hours.

5.3 Execution of the Model

Test execution is typically triggered by a build server upon a successful build of the IPS software. These steps are illustrated in Figure 3 and explained below.

1. **Build:** Building the software is scheduled to run every night, or a developer can trigger a build manually.
2. **Upgrade:** Upgrade all connected embedded controllers with the newly built software. A failure in this step aborts the complete cycle.
3. **Configure:** Configuration fetched from the source control repository is loaded onto the IPS. The configuration describes the interconnections of embedded boards and the properties of this specific paint-setup.
4. **Query and Solve Model:** Data retrieved from the IPS is fed into the CP model. This enables us to keep the generated test in sync with changes in the newly built software or changes in the configuration.
5. **Run Test:** Finally, the actual test is executed by applying the generated test sequence, and comparing the actuator outputs with the model generated *oracle*, $\mathbb{O}$.

5.4 Using the Flexibility of CP

As described in the previous sections, we have designed the model to be flexible enough and to be able to generate realistic test sequences. In particular, introducing diversity by applying global constraints between variables has been a key factor for satisfying our industrial requirements. However, the CP model can also handle specific parameter values, directly given by the validation engineers not having a strong knowledge of CP. This is simply implemented by guarding the posting of each constraint with some groundness conditions. For example, using `(var(X), var(Y)) -> X \#= Y ; true` to guard the posting of `X \#= Y`. Thanks to the Prolog commodity, our Python front-end can give value to *any* variable in the model and avoid posting spurious constraints that would slow down the solving process, or prevent a solution.

5.5 Performance of Model

Recalling that H_1 and H_2 represent two different groups of variable orderings with similar performance, Figure 4a compares the total time for a test execution $(t_s + t_N)$ for H_1 and H_2 with two different sizes of $^*BTabSize$.

When used only to find a solution to the constraints (i.e., without optimization), H_1 gives better results for $^*BTabSize = 10$, while H_2 performs better for $^*BTabSize = 40$. This experiment revealed 1) that H_1 usually produces a relative small value for t_N, by using more time than H_2 and 2) that H_2 usually produces a larger value for t_N but faster than H_1.

We also compared H_1 and H_2 when minimizing the overall time of a test sequence, i.e. $minimize(t_N)$. Figure 4b shows that H_2 provides the first solution faster than H_1, and that the quality of solution is better when more time is

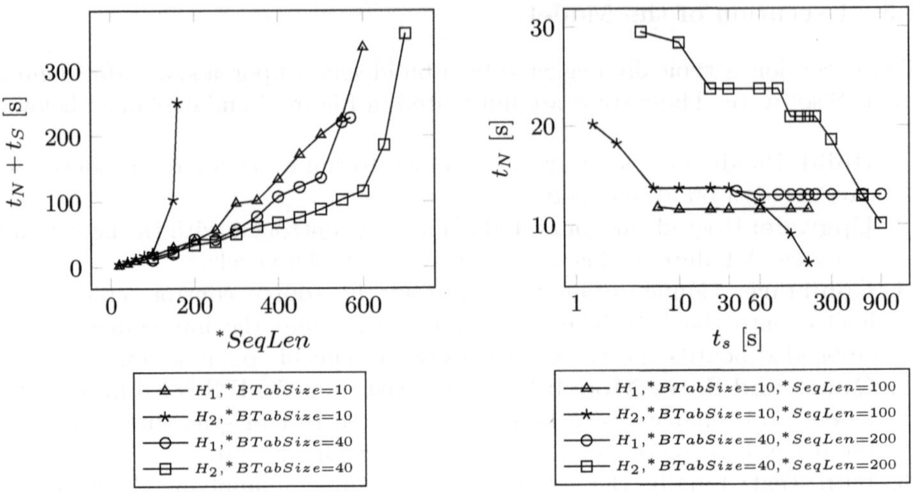

(a) Total time ($t_s + t_n$) (b) Optimization, $minimize(t_N)$

allocated to the search for optimality. For H_1, Figure 4b shows that there is no gain in minimizing t_N.

For the setup $^*SeqLen = 100, ^*BTabSize = 10$ we see that the solver must run for $\approx$ 60 s before H_2 gives a smaller t_N than for H_1. For the setup $^*SeqLen = 200, ^*BTabSize = 40$ the solver must run for $\approx$ 600 s before a break-even occurs.

From the results on Figure 4a and Figure 4b, no strong conclusion can be drawn when it comes to select between H_1 and H_2. If a test sequence is generated for multiple uses, i.e. reusing the same test sequence multiple times, then using H_2 is beneficial at the price of allocating more time to the optimization procedure. On the contrary, if a single usage is targeted, as is in the CI process, then using H_1 should be preferred by considering than the total time $t_s + t_N$ is the actual target of our test generation and execution procedure. Consequently, at ABB Robotics, we decided to keep the choice between these two heuristics as an option in our CP model. From a practical point of view, it permits the validation engineers to tune the test generation process according to their needs.

6 Lessons Learned and Conclusions

This section concludes the paper by presenting some lessons learned at ABB Robotics from our experience with introducing CP in our CI process.

6.1 CP for Validation Engineers

As previously stated, validation of robotized painting involves a fair amount manual, labour-intensive work. Therefore, replacing parts of this validation process with automation is necessary, and is perceived by validation engineers as a means to strengthen the process. However, it also comes with some drawbacks.

Two factors must be distinguished: (1) **The automation through the CI process** including automatic building of software, software upgrade, test execution and results reporting, and (2) **Test generation through use of CP**, which permits validation engineers to focus on validating other parts of the CIRs.

Point (1) does not have any drawbacks except the effort required to set up the CI process. From an industrial perspective, point (2) is the most critical, especially because (a) validation engineers are not yet sufficiently trained in CP, to change the model without help ; (b) validation engineers are usually reluctant to trust any tool that produces results, that are very difficult to compute by hand or with an easily understandable process. It is also recognized [15,16] as a concern that many optimization problems require expert knowledge. In order to reduce the risks, we decided to build a Python front-end to our CP model, so that some details can be hidden from the validation engineers. We also organized basic training in CP with simple and understandable examples in order to facilitate the adoption. Of course, we do not claim that these actions form a recipe for adopting CP in general, but we observe that it worked well in the context of ABB Robotics IPS validation.

6.2 Actual Defects Found with the CP Model

After the model was put into production at ABB Robotics, it immediately detected two new unknown defects related to timing aspects of IPS. These defects were however classified as non-critical, as they correspond to very unlikely scenarios. Digging into the causes of these defects, we saw that they had been present in the IPS for several years without any significant consequences and that they had been spotted by the CP model through enforcing diversity in the selection of test sequences. These defects were corrected and the test sequences used for spotting them were introduced into our non-regression test suite.

For validating the CP model, we also reintroduced five old, historical, defects into the source control repository. These defects were known by the validation engineers to be extremely hard to find. After a round of experiments, the CP model produced test sequences that spotted all the five defects. This was considered as a strong justification for the continued use of the CP model in production.

6.3 Return on Investment with the Use of CP

Computing the ROI for the use of CP for ABB Robotics' IPS validation is not easy. Possibly, one can measure the number of defects found with and without the CP model during the validation of a new IPS release. It is also possible to compare the human effort required in both cases. However, another important factor is the increased confidence of the engineers to the validation process, which is a factor that is very difficult to measure. After the introduction of the CP model in production, we observed a much higher confidence among the engineers to the testing framework and their appetite to perform necessary code re-factoring is now higher. They are more willing to make critical, but needed, changes in the software and they rely on the test framework to detect undesired side-effects. If a side-effect is discovered, they can simply roll back the change.

In the long term, we expect to see the benefits of using CP being recognized as a way to increase the general quality of the testing process, since necessary re-factoring will be performed before the technical depth grows beyond control.

6.4 Further Work

In the previous section, we mentioned at least two main points to dig into, in order to get a better understanding of the benefits of the CP model in ABB Robotics' IPS validation. Firstly, as introducing diversity in the selection of test sequences is crucial in our application, more dedicated global constraints could be built to capture the needs of validation engineers. In particular, constraining the variables of the brush table to take balanced values is highly desirable. Secondly, a deep understanding of the reasons why our heuristics H_1 and H_2 perform significantly better than other variable ordering choices would help us improving the constraint model by refining constraint posting.

Acknowledgements. We thank the anonymous reviewers for their comments. This work has been realized with the partial support of the Research Council of Norway, as part of the Industrial PhD of Morten Mossige, ABB Robotics and Certus SFI - Project 8.

References

1. Fowler, M., Foemmel, M.: Continuous integration (2006) (accessed August 13, 2013)
2. Mossige, M., Gotlieb, A., Meling, H.: Poster: Test generation for robotized paint systems using constraint programming in a continuous integration environment. In: 2013 IEEE Sixth International Conference on Software Testing, Verification and Validation (ICST), pp. 489–490 (2013)
3. Carlsson, M., Ottosson, G., Carlson, B.: An open-ended finite domain constraint solver. In: Glaser, H., Hartel, P., Kuchen, H. (eds.) PLILP 1997. LNCS, vol. 1292, pp. 191–206. Springer, Heidelberg (1997)
4. Di Alesio, S., Nejati, S., Briand, L., Gotlieb, A.: Stress testing of task deadlines: A constraint programming approach. In: 2013 IEEE 24th International Symposium on Software Reliability Engineering (ISSRE), pp. 158–167. IEEE (2013)
5. Balck, K., Grinchtein, O., Pearson, J.: Model-based protocol log generation for testing a telecommunication test harness using CLP. In: Design, Automation and Test in Europe Conference and Exhibition (DATE), pp. 1–4 (2014)
6. Stolberg, S.: Enabling agile testing through continuous integration. In: Agile Conference, AGILE 2009, pp. 369–374. IEEE (2009)
7. Marriott, K., Stuckey, P.J.: Programming with constraints: an introduction. MIT Press (1998)
8. Utting, M., Legeard, B.: Practical Model-Based Testing: A Tools Approach. Morgan Kaufmann Publishers Inc., San Francisco (2007)

9. Régin, J.C.: The global minimum distance constraint. Technical report, Technical report, ILOG (1997)
10. Mossige, M.: Prolog Model of ABB's Paint Control System for test case generation (2014), http://www.ux.uis.no/~mortenm/ips/trigdev_bt.pl
11. de Moura, L., Bjørner, N.S.: Z3: An efficient SMT solver. In: Ramakrishnan, C.R., Rehof, J. (eds.) TACAS 2008. LNCS, vol. 4963, pp. 337–340. Springer, Heidelberg (2008)
12. McMinn, P.: Search-based software test data generation: A survey. Software Testing, Verification and Reliability 14, 105–156 (2004)
13. IBM, ILOG Labs, I.: IBM CPLEX: High-performance software for mathematical programming and optimization (2006), http://www.ilog.com/products/cplex/
14. Rossi, F., Beek, P.V., Walsh, T.: Handbook of Constraint Programming (Foundations of Artificial Intelligence). Elsevier Science Inc., New York (2006)
15. de la Banda, M.G., Stuckey, P.J., Van Hentenryck, P., Wallace, M.: The future of optimization technology. Constraints, 1–13 (2013)
16. Francis, K., Brand, S., Stuckey, P.: Optimisation modelling for software developers. In: Milano, M. (ed.) CP 2012. LNCS, vol. 7514, pp. 274–289. Springer, Heidelberg (2012)

On Compiling CNF into Decision-DNNF

Umut Oztok and Adnan Darwiche

Computer Science Department, University of California,
Los Angeles, CA 90095, USA
{umut,darwiche}@cs.ucla.edu

Abstract. Decision-DNNF is a strict subset of decomposable negation normal form (DNNF) that plays a key role in analyzing the complexity of model counters (the searches performed by these counters have their traces in Decision-DNNF). This paper presents a number of results on Decision-DNNF. First, we introduce a new notion of CNF width and provide an algorithm that compiles CNFs into Decision-DNNFs in time and space that are exponential only in this width. The new width strictly dominates the treewidth of the CNF primal graph: it is no greater and can be bounded when the treewidth of the primal graph is unbounded. This new result leads to a tighter bound on the complexity of model counting. Second, we show that the output of the algorithm can be converted in linear time to a sentential decision diagram (SDD), which leads to a tighter bound on the complexity of compiling CNFs into SDDs.

1 Introduction

Decision-DNNF is a tractable propositional language that is a strict subset of DNNF. One key role of this language is in the complexity analysis of modern model counters. We will therefore start with a motivation of model counters.

Model counting is the problem of counting the number of satisfying assignments of a Boolean formula. It has various applications, such as inference in Bayesian networks [1,5]. Although model counting has been shown to be a hard problem (#P-complete [21]), there are two common approaches that have proven effective in practice.

One approach is based on DPLL [11,10], which is a family of algorithms that were initially developed for SAT: the problem of deciding whether a Boolean formula has a satisfying assignment. In essence, it is a systematic search algorithm that searches the space of truth assignments until finding a satisfying one or identifying that such an assignment does not exist. This search method can easily be extended to compute the number of satisfying assignments of the formula. Simply, by not stopping the search when a single satisfying assignment is found, and exhaustively continuing to look for all other satisfying assignments, one can obtain a naive model counter. To make this approach more practical, various sophisticated techniques were incorporated into the core exhaustive DPLL algorithm, such as component analysis [13] and formula caching [14,1,20].

Another approach for model counting is based on *knowledge compilation*. The basic idea of knowledge compilation is to compile a Boolean formula represented

B. O'Sullivan (Ed.): CP 2014, LNCS 8656, pp. 42–57, 2014.

in a source language into a target language that supports model counting in polytime. Negation normal form (NNF) circuits have been established as the basis of a number of such languages [9]. These circuits have *and* nodes (representing conjunctions) and *or* nodes (representing disjunctions) as internal gates, and literals or constants as inputs (see Fig. 1(c)). In [9], two fundamental properties on NNF circuits are identified to ensure the tractability of model counting: *decomposability* and *determinism*. Decomposability is a property of *and* nodes, requiring that the children of *and* nodes share no variables. Determinism is a property of *or* nodes, requiring that each two children of an *or* node be mutually exclusive (i.e., contradict each other). Determinism and decomposability characterize deterministic-DNNF (d-DNNF), a strict subset of DNNF [6], which includes other languages such as sentential decision diagrams (SDD) [8], free binary decision diagrams (FBDD) [3], and ordered binary decision diagrams (OBDD) [4]. Although d-DNNF is the most general language known that supports efficient model counting, a strict subset, Decision-DNNF, has been used in state-of-the-art model counters based on knowledge compilers [7,15].

Although the approaches described above look conceptually different than each other, a strong connection between them has been established [12]. In particular, the *traces*[1] of the searches performed by state-of-the-art model counters has been shown to be in Decision-DNNF. In other words, model counters based on exhaustive DPLL effectively generates the compilation of the Boolean formula in Decision-DNNF. By this result, Decision-DNNF has the role of bridging model counters and knowledge compilers. More importantly, any new result pertaining to Decision-DNNFs will have a possibly significant further impact on model counters. For instance, the relationship between Decision-DNNFs and FBDDs has been recently studied in [2]. Accordingly, Decision-DNNFs can be converted into FBDDs with only a quasipolynomial increase in the representation size. This result allowed the authors to show new exponential lower bounds on Decision-DNNFs, by leveraging the existing lower bounds on FBDDs, which are immediately applicable to model counters.

In this work, we present new results on Decision-DNNFs. First, we introduce a new notion of width for CNFs, called *decision-width*. We show a compilation algorithm that can compile CNFs into Decision-DNNFs in time and space that are exponential only in decision-width. This new width is no greater than the treewidth of the CNF primal graph, and can be bounded while the latter is unbounded. This result not only improves the existing complexity results on d-DNNF compilation but also the existing results on the complexity of model counting. Second, we show that Decision-DNNFs constructed by our algorithm can be converted to SDDs in linear time. SDD is a recently discovered tractable language that is a strict superset of the influential language OBDD. It comes with many interesting properties, including a polytime `Apply`[2] operation and canonicity, which are two key features underlying the success of OBDDs. Our result leads to a tighter bound on the complexity of compiling CNFs into SDDs.

---

[1] See Section 4 for a detailed discussion of the trace of an exhaustive search algorithm.
[2] `Apply` takes two SDDs and uses a binary operator to combine them.

The rest of the paper is organized as follows. Section 2 starts with technical preliminaries. Section 3 then introduces decision-width and discusses a compilation algorithm that can compile CNFs into Decision-DNNFs in time and space that are exponential only in this width. This is followed by a comparison of decision-width and the treewidth of the CNF primal graph. Next, Section 4 discusses the importance of Decision-DNNFs in model counting by reviewing in detail the strong connection that has been established between Decision-DNNFs and model counters. Section 5 shows that the output of our algorithm for compiling CNFs into Decision-DNNFs can be transformed in linear time to SDDs. We close with a discussion of related work and some concluding remarks. Due to space limitations, some proofs are delegated to the full version of the paper.[3]

2 Technical Preliminaries

Upper case letters (e.g., X) will denote variables and lower case letters (e.g., x) will denote their instantiations. Bold upper case letters (e.g., $\mathbf{X}$) will denote sets of variables and bold lower case letters (e.g., $\mathbf{x}$) will denote their instantiations.

A *Boolean function* f over variables $\mathbf{Z}$ maps each instantiation $\mathbf{z}$ of variables $\mathbf{Z}$ to *true* or *false*. The *conditioning* of f on instantiation $\mathbf{x}$, written $f|\mathbf{x}$, is a *subfunction* that results from setting variables $\mathbf{X}$ to their values in $\mathbf{x}$. A conjunction is *decomposable* if each pair of its conjuncts share no variables. A disjunction is *deterministic* if each pair of its disjuncts are mutually exclusive. A *negation normal form* (NNF) circuit is a rooted DAG whose internal nodes correspond to disjunctions and conjunctions, and whose leaf nodes correspond to literals or the constants $\top$ (*true*) and $\bot$ (*false*). An NNF circuit is decomposable and deterministic (called a d-DNNF) iff each of its conjunctions is decomposable and each of its disjunctions is deterministic; see Fig. 1(c). We will identify an NNF circuit by its root node N, use $Vars(N)$ to denote the set of variables mentioned by circuit N, and $F(N)$ to denote the Boolean function represented by N. The *size* of NNF circuit N, denoted $|N|$, is the total number of edges in the circuit.

The literals of variable X are denoted by x and $\neg x$. A *conjunctive normal form* (CNF) is a set of clauses, where each clause is a disjunction of literals, and the set represents the conjunction of its clauses (e.g., $\{x \vee \neg y \vee \neg z,\ y \vee z,\ \neg x\}$ represents the Boolean formula $(x \vee \neg y \vee \neg z) \wedge (y \vee z) \wedge \neg x$). Conditioning a CNF Δ on literal ℓ, denoted $\Delta|\ell$, amounts to removing literal $\neg\ell$ from all clauses and then dropping all clauses that contain literal ℓ.

3 Compiling CNFs into Decision-DNNFs

The purpose of this section is to show an algorithm that compiles CNFs into Decision-DNNFs with a complexity guarantee. To analyze the complexity of the algorithm, we will also introduce a new notion of width and study its properties.

[3] Available at `http://reasoning.cs.ucla.edu`.

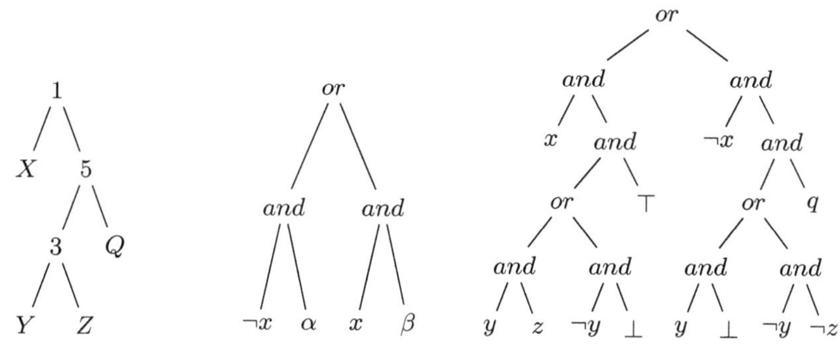
(a) A vtree (b) A decision node (c) A Decision-DNNF

Fig. 1. A vtree, a decision node, and a Decision-DNNF

3.1 Decision-DNNF

A *decision node* is a special form of an *or* node, which is depicted in Fig. 1(b) where X is a variable, and α and β are arbitrary NNF nodes. A d-DNNF is called a Decision-DNNF iff each of its *or* nodes is a decision node; see Fig. 1(c). In this case, determinism is always ensured by the decision nodes.

3.2 Decision Vtrees

Both the width and the compilation algorithm we will present in this section are driven by a tree-structure, which is introduced next. A *vtree* for a set $\mathbf{Z}$ of variables is a rooted, full binary tree whose leaves are in one-to-one correspondence with variables in $\mathbf{Z}$. Figure 1(a) depicts an example vtree. We will use v^l and v^r to refer to the left and right children of an internal vtree node v. We will also use $Vars(v)$ to denote the set of variables at or below a vtree node v. A vtree node is called a *Shannon node* iff its left child is a leaf. A vtree in which every node is a Shannon node will be called *right-linear*. Given a vtree v, we will sometimes refer to v as the root of the vtree.

A vtree for a CNF is a vtree over the CNF variables. Our focus is on a special type of vtrees, defined next.

Definition 1 (decision vtree). *A clause is compatible with an internal vtree node v iff the clause mentions some variables inside v^l and some variables inside v^r. A vtree for CNF Δ is said to be a decision vtree for Δ iff every clause in Δ is compatible with only Shannon nodes.*[4]

Figure 1(a) depicts a decision vtree for the CNF $\{y \vee \neg z,\ \neg x \vee z,\ x \vee \neg y,\ x \vee q\}$. We will later show that one can always construct a decision vtree for any CNF.

[4] A unit clause (one containing a single literal) is not compatible with any vtree node. Hence, a unit clause trivially satisfies the condition of being compatible with only Shannon nodes.

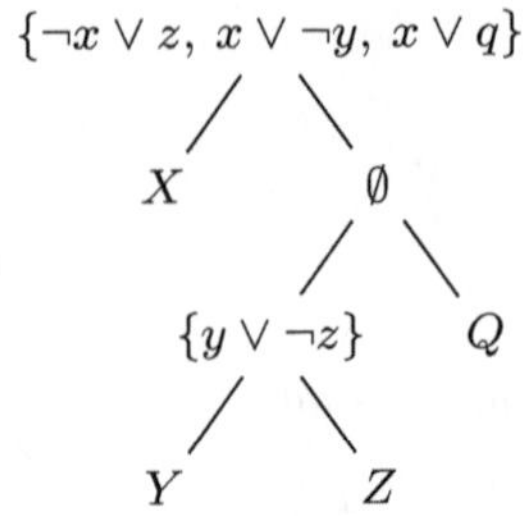

Fig. 2. Distributing the clauses of CNF $\{y \lor \neg z,\ \neg x \lor z,\ x \lor \neg y,\ x \lor q\}$ over a vtree

3.3 A Compilation Algorithm

We will next present an algorithm that compiles a CNF into a Decision-DNNF using a decision vtree for the CNF. This compilation method is given by Algorithm 1, which takes a decision vtree v and an auxiliary CNF S over the variables of vtree v (S is initially empty). The CNF Δ to be compiled is passed with the vtree as follows. Each clause of Δ is assigned to the lowest vtree node that contains the clause variables. Figure 2 depicts an example of how clauses are assigned to vtree nodes. Note that the (non-unit) clauses are assigned only to Shannon nodes as the vtree is a decision vtree. We use $Clauses(v)$ to denote the clauses assigned to a vtree node v. We also use $CNF(v)$ to denote the clauses assigned to all nodes in the vtree rooted at v. A recursive call $\texttt{c2d}(v, S)$ will return a Decision-DNNF for $CNF(v) \cup S$. The algorithm keeps a cache at every vtree node, which is indexed by S.

Consider now a run of Algorithm 1. An *or* node can only be constructed on Line 4. Accordingly, each *or* node created by this algorithm is a decision node. Using this fact with an inductive argument is enough to prove the soundness of the algorithm.

Lemma 1. *Let v be a decision vtree for $CNF(v)$. Let S be a CNF over $Vars(v)$ whose clauses are compatible with only Shannon nodes of v. The call $\texttt{c2d}(v, S)$ to Algorithm 1 returns a Decision-DNNF equivalent to $CNF(v) \cup S$.*

Proof. The proof is by induction on vtree nodes. The base case is when v is a leaf node. This case is trivially satisfied by Line 2. Assume now that v is an internal node. As an induction hypothesis, consider that for each vtree node v' below v, the call $\texttt{c2d}(v', S')$ computes a Decision-DNNF equivalent to $CNF(v') \cup S'$, where S' is a CNF over $Vars(v')$ whose clauses are compatible with only Shannon nodes of v'. During the call to v, we will compute a Decision-DNNF equivalent to $CNF(v) \cup S$ by utilizing the following decomposition of a Boolean function f (known as Shannon decomposition): $f = (x \land f|x) \lor (\neg x \land f|\neg x)$. Note that in our context $f = CNF(v) \cup S$. Assume v is a Shannon node. Then v^l is a leaf node (with variable X). The possible clauses that can be assigned to v^l are $\{x\}$ and $\{\neg x\}$. Lines 4–4 consider all four possible assignments of those two

Algorithm 1. $\texttt{c2d}(v, S)$
$cache(v, \Delta)$ is a hash table that maps v and Δ into a Decision-DNNF.
$terminal(\Delta)$ returns the literal or constant equivalent to Δ.

Input: v : a vtree node, S : a CNF over $Vars(v)$.
Output: A Decision-DNNF for $CNF(v) \cup S$.
1 **if** $cache(v, S) \neq nil$ **then return** $cache(v, S)$ $C \leftarrow Clauses(v)$
2 **if** *v is a leaf* **then return** $terminal(C \cup S)$ **if** *v is a Shannon node* **then**
3 $\quad X \leftarrow$ variable of v^l
4 $\quad$ **if** $\{x\}$ *and* $\{\neg x\}$ *assigned to* v^l **then** $\alpha \leftarrow \bot$ **else if** $\{x\}$ *assigned to* v^l **then**
$\quad \alpha \leftarrow x \wedge \texttt{c2d}(v^r, (C \cup S)|x)$ **else if** $\{\neg x\}$ *assigned to* v^l **then**
$\quad \alpha \leftarrow \neg x \wedge \texttt{c2d}(v^r, (C \cup S)|\neg x)$ **else**
$\quad \alpha \leftarrow \Big(x \wedge \texttt{c2d}(v^r, (C \cup S)|x)\Big) \vee \Big(\neg x \wedge \texttt{c2d}(v^r, (C \cup S)|\neg x)\Big)$
5 **else**
6 $\quad S_1 \leftarrow$ clauses in S that only mention variables in v^l
7 $\quad S_2 \leftarrow$ clauses in S that only mention variables in v^r
8 $\quad \alpha \leftarrow \Big(\texttt{c2d}(v^l, S_1) \wedge \texttt{c2d}(v^r, S_2)\Big)$
9 $cache(v, S) \leftarrow \alpha$
10 **return** α

clauses to v^l: (1) both $\{x\}$ and $\{\neg x\}$ are assigned and $f = \bot$, (2) only $\{x\}$ is assigned and $f = x \wedge f|x$, (3) only $\{\neg x\}$ is assigned and $f = \neg x \wedge f|\neg x$, and (4) no clause is assigned and $f = (x \wedge f|x) \vee (\neg x \wedge f|\neg x)$. Except for the first case, in which f is trivially computed as $\bot$, by the induction hypothesis, $\texttt{c2d}(v^r, (C \cup S)|x)$ and $\texttt{c2d}(v^r, (C \cup S)|\neg x)$ compute Decision-DNNFs for $f|x$ and $f|\neg x$, respectively.[5] Note that we construct an *or* node only in the last case, which is a decision node. So, when v is a Shannon node, we compute a Decision-DNNF equivalent to $CNF(v) \cup S$. Assume now that v is a non-Shannon node. In this case, C must be empty because the vtree is a decision vtree. Thus, $CNF(v) = CNF(v^l) \cup CNF(v^r)$. Also, S cannot contain any clause that mentions variables from both v^l and v^r as no clause in S can be compatible with v. Then, by the induction hypothesis, on Line 8, we compute a Decision-DNNF equivalent to $CNF(v) \cup S$. □

Corollary 1. *Let v be a decision vtree for $CNF(v)$. The call $\texttt{c2d}(v, \{\})$ to Algorithm 1 returns a Decision-DNNF that is equivalent to $CNF(v)$.*

For instance, when the vtree in Fig. 2 is passed to Algorithm 1, it computes the Decision-DNNF in Fig. 1(c). To analyze time and space complexities of the algorithm, we next introduce a new notion of width.

[5] If a clause is not compatible with a vtree node v then every conditioning of the clause will also not be compatible with v.

3.4 Decision-Width

Before defining the new notion of width, we will introduce two concepts.

Definition 2. *Consider a CNF and a corresponding vtree. Let v be an internal vtree node. The context clauses of v are the clauses that mention variables inside v and outside v.*

For example, consider the CNF $\{y \vee \neg z,\ \neg x \vee z,\ x \vee \neg y,\ x \vee q\}$ and the vtree node $v = 5$ in Fig. 1(a). The context clauses of v are $\{\neg x \vee z,\ x \vee \neg y,\ x \vee q\}$.

Definition 3. *Consider a CNF Γ and a set of variables $\mathbf{V}$. We denote by $CNFs(\Gamma, \mathbf{V})$ the set of CNFs that is obtained from conditioning Γ on each instantiation $\mathbf{v}$ of $\mathbf{V}$.*

Consider the CNF $\Gamma = \{x \vee y \vee z,\ x \vee y \vee q,\ \neg x \vee \neg y \vee z,\ x \vee \neg y \vee z\}$ and the set of variables $\mathbf{V} = \{Y\}$. Then,

$$\Gamma|y = \{\neg x \vee z,\ x \vee z\},$$
$$\Gamma|\neg y = \{x \vee z,\ x \vee q\},$$
$$CNFs(\Gamma, \mathbf{V}) = \{\{\neg x \vee z,\ x \vee z\},\ \{x \vee z,\ x \vee q\}\}.$$

We are now ready to introduce the new notion of width.

Definition 4 (decision-width). *Consider a CNF Δ and a corresponding decision vtree. Let v be an internal vtree node with context clauses Γ. Let $\mathbf{Y}$ be variables outside v. Then, the width of v is the ceiling of* $\log(|CNFs(\Gamma, \mathbf{Y})|)$, *where* $\log 0$ *is defined as* 0. *The decision-width of the decision vtree is the largest width of any of its internal nodes minus 1. The decision-width of the CNF is the smallest decision-width attained by any of its decision vtrees.*

For instance, consider the vtree in Fig. 1(a). Assuming that the vtree corresponds to the CNF $\{y \vee \neg z,\ \neg x \vee z,\ x \vee \neg y,\ x \vee q\}$, the width of node $v = 5$ is 1, and the decision-width of the vtree is 0.

Having defined decision-width, we can now establish the complexity of Algorithm 1.

Theorem 1. *If decision vtree v is over n variables and has decision-width w, and if $CNF(v)$ has size m, then the call* **c2d**$(v, \{\})$ *to Algorithm 1 takes time in $O(nm2^w)$ and returns a Decision-DNNF whose size is in $O(n2^w)$.*

Proof (Sketch). Each distinct call to a Shannon node (Lines 2–4) takes time in $O(2m)$: we perform at most two conditionings of $C \cup S$, which has at most m clauses, on a single literal. This process contributes to the size at most three nodes, each having two children. Each distinct call to a non-Shannon node (Lines 5–8) takes time in $O(m)$: we partition the set S, which has at most m clauses, into two subsets. This case contributes to the size one node with two children. Also, due to caching, the number of distinct calls to a vtree node v is at most 2^k where k is the width of v. As there are $O(n)$ nodes in the vtree, Algorithm 1 takes time in $O(nm2^w)$ and returns a Decision-DNNF whose size is in $O(n2^w)$. □

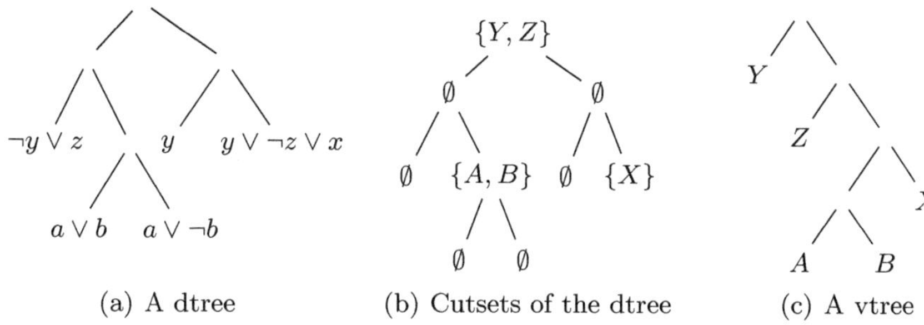

Fig. 3. A dtree, its cutsets, and a vtree for the CNF $\{y,\ \neg y \vee z,\ y \vee \neg z \vee x,\ a \vee b,\ a \vee \neg b\}$

3.5 Relationship to Treewidth

One of the classical parameters for characterizing the structural properties of a CNF is treewidth [19]. This is a property of some graph abstraction of the CNF, such as the primal, dual or incidence graph. We will now compare decision-width with the treewidth of the CNF primal graph, and show that decision-width strictly dominates the treewidth.

The *primal graph* of a CNF is obtained by treating CNF variables as graph nodes, while adding an edge between two variables iff they appear in the same clause. We will use *twp* to denote the treewidth of the primal graph.

We will now compare decision-width with *twp*. First, we will show that decision-width dominates *twp*.

Theorem 2. *Consider a CNF whose primal graph has treewidth w. We can construct a decision vtree of this CNF whose decision-width is no greater than w.*

Proof (Sketch). The primal graph must have a tree decomposition in the form of a dtree [6] that has width w. A *dtree* of a CNF is a rooted, full binary tree whose leaves are in one-to-one correspondence with the CNF clauses; see Fig. 3(a). We define the cutset of an internal dtree node d as the variables that appear in the left child of d and in the right child of d but not in any cutset of the ancestors of d. Also, the cutset of a leaf dtree node d is defined as the variables of d that do not appear in any cutset of the ancestors of d. Figure 3(b) shows the cutsets of a dtree. We construct decision vtrees from dtrees using the following recursive procedure. At a leaf dtree node d, we construct a right-linear vtree from the variables appearing in the cutset of d. At an internal dtree node d, we recursively construct a decision vtree v^l for the dtree d^l, and a decision vtree v^r for the dtree d^r. We then create another vtree node v by assigning v^l as v's left child and v^r as v's right child. Finally, we create and return a right-linear vtree using the variables in the cutset of dtree node d, with vtree node v as the right-most child of this right-linear vtree. Due to cutset properties, the resulting vtree is a decision vtree. Figure 3(c) shows a decision vtree obtained from the dtree in Fig. 3(a) using the method we just described. The full paper shows that the decision-width of the resulting decision vtree is no greater than w. □

The above result shows that decision-width dominates *twp*. It also implies that one can always construct a decision vtree for any CNF. We will next show that decision-width can be bounded when *twp* is unbounded.

Theorem 3. *The CNF $\Delta_n = \{x_1 \vee \ldots \vee x_n\}$, $n \geq 1$, has a primal graph with treewidth $n-1$ and decision-width* 0.

Proof. The primal graph of Δ_n is a complete graph, and complete graphs are known to have unbounded treewidth, which is $n-1$ in this case. Consider the right-linear vtree induced by the variable ordering $X_1, \ldots, X_n$. That is, the left child of the vtree root v is X_1. The left child of v^r is X_2, and so on. Consider a vtree node v' whose left child is X_i. Let Γ be the context clauses of v'. If $i = 1$, then Γ is empty and the width of v' is 0. Otherwise, $\Gamma = \{x_1 \vee \ldots \vee x_n\}$. Let $\mathbf{Y}$ be the variables outside v'. Then, the set of CNFs that are obtained from $\Gamma|\mathbf{y}$ is $\{\{\}, \{x_i \vee \ldots \vee x_n\}\}$. The width of v' is then 1. The decision-width of the vtree is then 0. □

The best known upper bounds for the complexities of compiling CNFs into both Decision-DNNFs and d-DNNFs are exponential in the treewidth of the CNF primal graph [12]. As decision-width strictly dominates treewidth, and also Decision-DNNFs is a strict subset of d-DNNFs, by Theorem 1, we obtain a tighter upper bound both for Decision-DNNF and d-DNNF compilation. Furthermore, as the traces of most model counters are in Decision-DNNF (see next section), this result also provides a tighter bound for model counters based on Decision-DNNFs (under some assumptions to be discussed next).

4 Decision-DNNFs and Model Counters

In this section, we will discuss the close relationship between Decision-DNNFs and model counters based on exhaustive DPLL.

The original DPLL algorithm was developed for SAT. It is a systematic search algorithm that searches the space of truth assignments until finding a satisfying one or identifying that such an assignment does not exist. In particular, given a Boolean formula f, it chooses a variable X of the formula, and then considers two cases recursively, which correspond to $f|x$ and $f|\neg x$. It then decides that the formula f is satisfiable when at least one of $f|x$ and $f|\neg x$ is satisfiable. This method can easily be extended to compute the number of satisfying assignments of the formula by not stopping the search when a single satisfying assignment is found. That is, by exhaustively continuing to look for all other satisfying assignments, one can obtain a naive model counter.

The tree in Fig. 4(a) shows all the paths that are traversed during an exhaustive DPLL on a Boolean formula. Each circled node represents a variable on which two decisions are performed: the variable is set to false (dashed edge) or set to true (solid edge). This way, paths from the root to leaf nodes represent (partial) variable assignments. Each leaf node then represents the result of the search when the variable assignment on the path from the root to the leaf is applied on the Boolean formula, with the label **unsat** being unsatisfiable and the

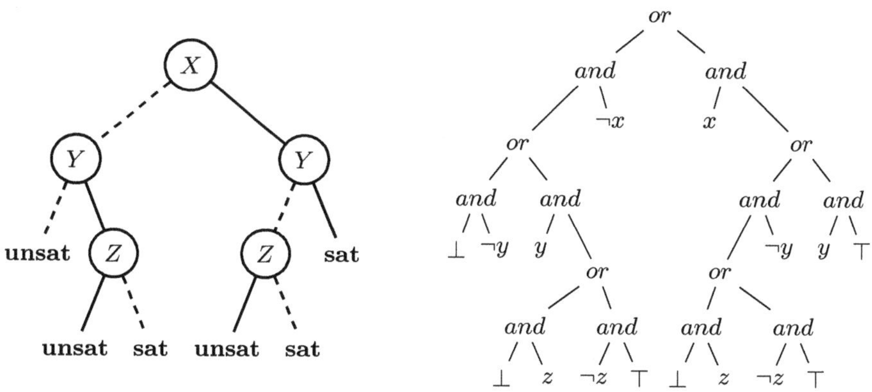

(a) Termination tree (b) Equivalent NNF circuit

Fig. 4. The trace of an exhaustive DPLL

label **sat** being satisfiable. This tree is called the *trace* of the search performed by an exhaustive DPLL [12]. Note that one can think of each circled node of the tree as an *or* node, by utilizing the following conversion:

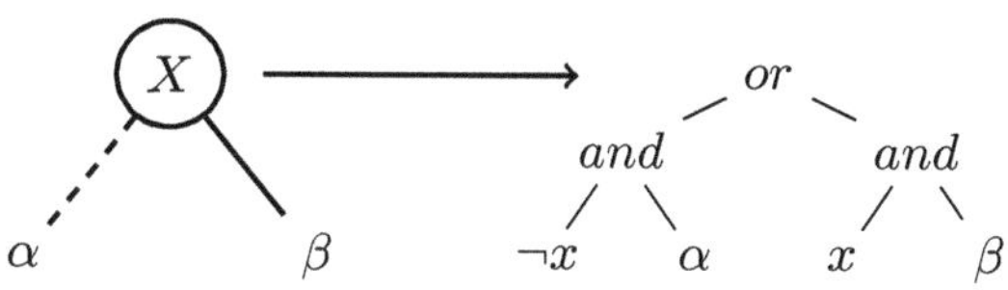

Figure 4(b) is the NNF circuit obtained from Fig. 4(a) using the above conversion, and also replacing each **sat** with ⊤, and each **unsat** with ⊥. One can always interpret the trace of an exhaustive DPLL as an equivalent NNF circuit, which turns out to satisfy both decomposability and determinism. In fact, the traces of the searches performed by such model counters correspond to FBDDs [12]. Moreover, when these model counters are augmented with component analysis, their traces correspond to Decision-DNNFs [12].

This connection has two implications. First, it allows one to translate Decision-DNNF lower bounds immediately into lower bounds on the complexity of model counters. Second, but under some assumptions, it allows one to translate Decision-DNNF upper bounds into ones on the complexity of model counters. For example lower bounds, it was recently shown that Decision-DNNFs can be converted into FBDDs with only a quasipolynomial increase in size [2]. As a result, known lower bounds for FBDDs immediately translate into lower bounds on model counters whose traces are in Decision-DNNF (see [2] for examples).

Translating upper bounds on Decision-DNNF to upper bounds on arbitrary model counters, however, is not as direct. Here, one needs, for example, to assume that the traces of the model counter are optimal, and that the time complexity of the counter is polynomial in the size of the trace. Under these assumptions, an upper bound on Decision-DNNF translates directly into an upper bound on the model counter. Interestingly enough, Algorithm 1 satisfies the second condition. The algorithm does not satisfy the first condition, but we know that its traces (i.e., compilations) are bounded exponentially only by the decision-width. Since this width dominates the primal graph treewidth, we now have a tighter upper bound on model counting in general (realized by Algorithm 1). We also have a tighter upper bound on any model counter that satisfies the previous conditions.

5 From Decision-DNNF to SDD

We finally show a new complexity result on compiling CNFs into SDDs.[6] The existing upper bound on SDDs is exponential in the treewidth of the CNF primal graph. We show a tighter bound that is exponential only in decision-width, which strictly dominates treewidth as shown in Sect. 3.5. We will obtain this result by showing that Decision-DNNFs generated by Algorithm 1 can be converted into compressed and trimmed SDDs in linear time and by at most doubling the size. That is, Algorithm 1 is effectively compiling SDDs.

Note that the output of Algorithm 1 is a special form of Decision-DNNF. In particular, the vtree used in the compilation provides the generated Decision-DNNF with a specific structure. That is, every node N in the Decision-DNNF is associated with some vtree node v in the following way:

- an *and* node is associated with v when $Vars(N^l) \subseteq Vars(v^l)$ and $Vars(N^r) \subseteq Vars(v^r)$, and
- an *or* node, $(x \wedge \alpha) \vee (\neg x \wedge \beta)$, is associated with v when $X \subseteq Vars(v^l)$ and $Vars(\alpha \cup \beta) \subseteq Vars(v^r)$.

For instance, each node of the Decision-DNNF in Fig. 1(c) is associated with a vtree node in Fig. 1(a).

Algorithm 2 shows how to convert a Decision-DNNF into an SDD. It takes a Decision-DNNF that is constructed by Algorithm 1 and computes two compressed and trimmed SDDs that are equivalent to the Boolean functions represented by N and the negation of N. The conversion is done in a bottom-up fashion. Terminal SDDs are obtained from leaf nodes (Line 1). Then, an *or* node (Lines 1–3) or an *and* node (Lines 4–6) is obtained from the results of recursive calls. To prevent redundant calculations, the results are cached (Line 7). The following theorem establishes the soundness and the complexity of the algorithm.

Theorem 4. *If N is a Decision-DNNF generated by Algorithm 1 and has size m, then the call* `d2sdd`(N) *takes time in $O(m)$, and returns two compressed and trimmed SDDs for $F(N)$ and $\neg F(N)$, whose sizes are in $O(m)$.*

Proof. The proof is by induction on NNF nodes. The base case is when N is a leaf node, which is satisfied by Line 1. As an induction hypothesis (IH), assume

[6] The definition of SDD and some of its properties are delegated to Appendix A.

Algorithm 2. d2sdd(N)

cache(N) is a hash table that maps N to an SDD.
terminal(N) returns the terminal SDD equivalent to $F(N)$.
unique(α) removes an element from α if its prime is $\bot$. It then returns s if $\alpha = \{(p_1, s), (p_2, s)\}$ or $\alpha = \{(\top, s)\}$; p_1 if $\alpha = \{(p_1, \top), (p_2, \bot)\}$; else the unique SDD node with elements α.

Input: N : a Decision-DNNF generated by Algorithm 1.
Output: Two compressed and trimmed SDDs equivalent to $F(N)$ and $\neg F(N)$.

1 **if** *cache*(N) $\neq$ *nil* **then return** *cache*(N) **if** *N is a leaf node* **then return** *terminal*(N), *terminal*($\neg N$) **if** $N = (x \wedge N_1) \vee (\neg x \wedge N_2)$ **then**
2 $\quad s_1, \neg s_1 \leftarrow$ d2sdd(N_1), $\quad s_2, \neg s_2 \leftarrow$ d2sdd(N_2)
3 $\quad \alpha \leftarrow unique(\{(x, s_1), (\neg x, s_2)\}), \quad \neg\alpha \leftarrow unique(\{(x, \neg s_1), (\neg x, \neg s_2)\})$
4 **else** // $N = N^l \wedge N^r$
5 $\quad p, \neg p \leftarrow$ d2sdd(N^l), $\quad s, \neg s \leftarrow$ d2sdd(N^r)
6 $\quad \alpha \leftarrow unique(\{(p, s), (\neg p, \bot)\}), \quad \neg\alpha \leftarrow unique(\{(p, \neg s), (\neg p, \top)\})$
7 *cache*(N) $\leftarrow \alpha, \neg\alpha$
8 **return** $\alpha, \neg\alpha$

that for every NNF circuit whose root N' is below N, and whose size $|N'|$ is k, the call d2sdd(N') takes time in $O(k)$, and returns two compressed and trimmed SDDs for $F(N')$ and $\neg F(N')$, whose sizes are in $O(k)$. Suppose that N is an *or* node, $(x \wedge N_1) \vee (\neg x \wedge N_2)$, where N_1 and N_2 are NNF nodes. Let $|N_1| = m_1$ and $|N_2| = m_2$. Then, $m = 6 + m_1 + m_2$. By the IH, the call d2sdd(N_1) (Line 2) takes time in $O(m_1)$, and returns the SDDs for $F(N_1)$ and $\neg F(N_1)$, whose sizes are in $O(m_1)$. Similarly, the call d2sdd(N_2) (Line 2) takes time in $O(m_2)$, and returns the SDDs for $F(N_2)$ and $\neg F(N_2)$, whose sizes are in $O(m_2)$. Then, using the structure of N, we construct the SDD for $F(N)$ (Line 3), whose size is at most $2 + O(m_1) + O(m_2) = O(m)$. To compute the SDD for $\neg F(N)$, we just need to negate the subs of the SDD for $F(N)$, which are already computed by the recursive calls. So, the call to an *or* node takes time in $O(m)$. Assume now that N is an *and* node, $N^l \wedge N^r$, where N^l and N^r are NNF nodes. Let $|N^l| = m_1$ and $|N^r| = m_2$. Then, $m = 2 + m_1 + m_2$. By the IH, the call d2sdd(N^l) (Line 5) takes time in $O(m_1)$, and returns the SDDs for $F(N^l)$ and $\neg F(N^l)$, whose sizes are in $O(m_1)$. Similarly, the call d2sdd(N^r) (Line 5) takes time in $O(m_2)$, and returns the SDDs for $F(N^r)$ and $\neg F(N^r)$, whose sizes are in $O(m_2)$. We then construct the SDD for $F(N)$ (Line 6) by making use of the following: $F(N) = \big(F(N^l) \wedge F(N^r)\big) \vee \big(\neg F(N^l) \wedge \bot\big)$. Thus, the constructed SDD has size at most $2 + O(2m_1) + O(m_2) = O(2m)$. Again, the SDD for $\neg F(N)$ is computed by negating the subs. Thus, the call to an *and* node takes time in $O(m)$. □

For instance, when we pass the Decision-DNNF in Fig. 1(c) to Algorithm 2, the algorithm computes the compressed and trimmed SDDs in Fig. 5.

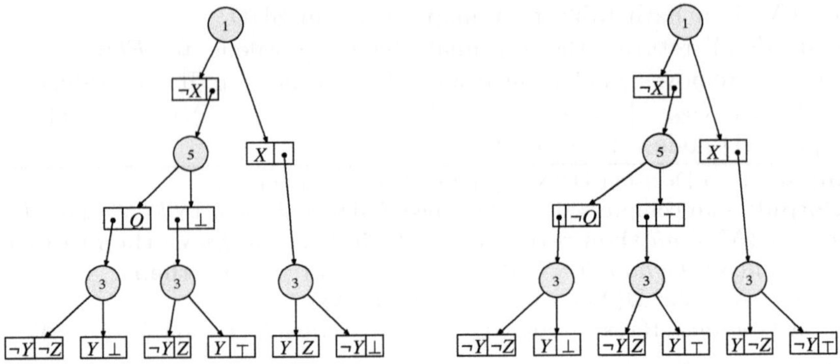

(a) An SDD (b) Negation of the SDD in (a)

Fig. 5. An SDD and its negation (both defined over the vtree in Fig. 1(a))

The result we obtained in this section shows that we can compile CNFs into SDDs in time and space that are exponential only in decision-width. As the best known upper bound for compiling CNFs into SDDs is exponential in the treewidth of the CNF primal graph [8], we obtain a tighter upper bound on the complexity of SDD compilation.

6 Related Work

The notion of decision-width is closely connected to another notion of width that we introduced recently [16], called *CV-width*. In this latter work, an algorithm was provided for compiling CNFs into structured DNNFs, based on vtrees, with space and time guarantees that are exponential only in CV-width. This width was defined over arbitrary vtrees. However, when restricted to decision vtrees, it reduces to the notion of decision-width defined in this paper. In fact, when the compilation algorithm of [16] is passed a decision vtree, it reduces to the compilation algorithm given in this paper. This is an interesting phenomenon, where language fragments are generated by restricting the type of vtrees passed to a compilation algorithm. In fact, in [16], we also showed that when right-linear vtrees are used, the compilation algorithm yields OBDD compilations (with new complexity bounds that improve on what existed before).

When the current work is considered from the viewpoint of [16], it corresponds to identifying a class of vtrees (decision vtree), together with a corresponding notion of width (decision-width) and a corresponding fragment of structured DNNF. These vtrees are particularly important given their role in model counting and sentential decision diagrams.

Interestingly enough, decision vtrees and the method for constructing them have been used twice in the past, yet without realizing the corresponding properties and guarantees that we discussed [18,8]. Moreover, no specific notion of width has been defined before based on this restricted type of vtrees. In [18], these vtrees were used to show an upper bound on the compilation of structured DNNFs. Similarly, in [8], these vtrees were used to show an upper bound on the compilation of SDDs. In both cases, the bounds were in terms of the primal graph treewidth. In this work, we used these vtrees to provide an upper bound on a subset of Decision-DNNF, which is included in both structured DNNF and SDDs. Moreover, our bound (based on decision-width) is tighter than the ones based on the primal graph treewidth. More importantly though, we have identified decision vtrees as a distinct class of vtrees for the first time, explicated their characteristic property (clauses are compatible only with Shannon nodes), equipped them with a corresponding notion of width, and characterized their corresponding compilations (as a subset of Decision-DNNF).

7 Conclusion

In this paper, we presented new results on Decision-DNNFs. We showed a compilation algorithm that compiles CNFs into Decision-DNNFs. To analyze the complexity of the algorithm, we defined a new notion of width, called decision-width, and showed that the algorithm has time and space complexities that are exponential only in decision-width. To better evaluate decision-width, we compared it with the treewidth of the CNF primal graph, and showed that decision-width strictly dominates treewidth. Not only did we obtain a tighter upper bound for compiling CNFs into d-DNNFs (through Decision-DNNFs) but we also obtained a tighter upper bound on model counting, for which the previously best known bounds were both exponential in the treewidth. We finally showed that Decision-DNNFs compiled by the algorithm can be transformed into SDDs in linear time. This led to a tighter upper bound on compiling CNFs into SDDs, for which the previously best known bound was exponential in the treewidth of the CNF primal graph.

Acknowledgements. This work has been partially supported by ONR grant #N00014-12-1-0423 and NSF grant #IIS-1118122.

A Sentential Decision Diagrams

Consider a Boolean function $f(\mathbf{X}, \mathbf{Y})$ with disjoint sets of variables $\mathbf{X}$ and $\mathbf{Y}$. If $f(\mathbf{X}, \mathbf{Y}) = (p_1(\mathbf{X}) \wedge s_1(\mathbf{Y})) \vee \ldots \vee (p_n(\mathbf{X}) \wedge s_n(\mathbf{Y}))$, the set $\{(p_1, s_1), \ldots, (p_n, s_n)\}$ is called an $(\mathbf{X}, \mathbf{Y})$-decomposition of the function f and each pair (p_i, s_i) is called an *element* of the decomposition [17]. The decomposition is further called an $(\mathbf{X}, \mathbf{Y})$-*partition* iff the p_i's form a partition [8]. That is, $p_i \neq \bot$ for all i; and $p_i \wedge p_j = \bot$ for $i \neq j$; and $\bigvee_i p_i = \top$. In this case, each p_i is called a *prime* and each s_i is called a *sub*. An $(\mathbf{X}, \mathbf{Y})$-partition is *compressed* iff its subs are distinct, i.e., $s_i \neq s_j$ for $i \neq j$ [8]. Compression can always be ensured by

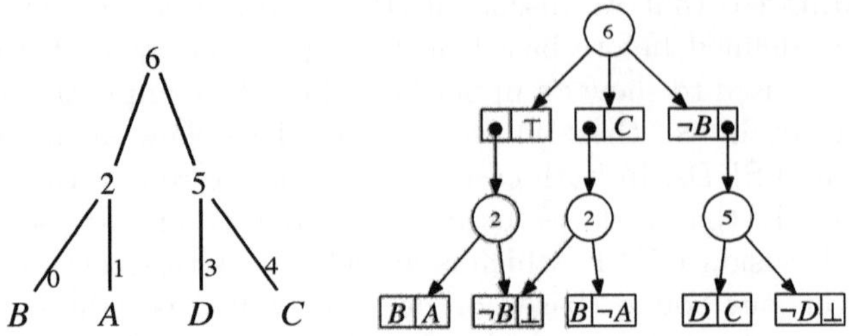

(a) A vtree (b) An SDD

Fig. 6. Function $f = (A \wedge B) \vee (B \wedge C) \vee (C \wedge D)$

repeatedly disjoining the primes of equal subs. Moreover, a function $f(\mathbf{X}, \mathbf{Y})$ has a *unique*, compressed $(\mathbf{X}, \mathbf{Y})$-partition. Finally, the *size* of a decomposition, or partition, is the number of its elements.

Note that $(\mathbf{X}, \mathbf{Y})$-partitions generalize Shannon decompositions, which fall as a special case when $\mathbf{X}$ contains a single variable. OBDDs result from the recursive application of Shannon decompositions, leading to decision nodes that branch on the states of a single variable (i.e., literals). As we show next, SDDs result from the recursive application of $(\mathbf{X}, \mathbf{Y})$-partitions, leading to decision nodes that branch on the state of multiple variables (i.e., arbitrary sentences).

Consider the vtree in Fig. 6(a). Consider also the Boolean function $f = (A \wedge B) \vee (B \wedge C) \vee (C \wedge D)$ over the same variables. Node $v = 6$ is the vtree root. Its left subtree contains variables $\mathbf{X} = \{A, B\}$ and its right subtree contains $\mathbf{Y} = \{C, D\}$. Decomposing function f at node $v = 6$ amounts to generating an $(\mathbf{X}, \mathbf{Y})$-partition of function f. The unique compressed $(\mathbf{X}, \mathbf{Y})$-partition here is

$$\{(\underbrace{A \wedge B}_{\text{prime}}, \underbrace{\top}_{\text{sub}}), (\underbrace{\neg A \wedge B}_{\text{prime}}, \underbrace{C}_{\text{sub}}), (\underbrace{\neg B}_{\text{prime}}, \underbrace{D \wedge C}_{\text{sub}})\}.$$

This partition is represented by the root node of Fig. 6(b). This node, which is a circle, represents a *decision node* with three branches. Each branch corresponds to one element $\boxed{p|s}$ of the above partition. Here, the left box contains a prime when the prime is a literal or a constant; otherwise, it contains a pointer to a prime. Similarly, the right box contains a sub or a pointer to a sub. The three primes are decomposed recursively, but using the vtree rooted at $v = 2$. Similarly, the subs are decomposed recursively, using the vtree rooted at $v = 5$. This recursive decomposition process moves down one level in the vtree with each recursion, terminating when it reaches leaf vtree nodes. The full SDD for this example is depicted in Fig. 6(b).

SDDs obtained from the above process are called *compressed* iff the $(\mathbf{X}, \mathbf{Y})$-partition computed at each step is compressed. These SDDs may contain trivial decision nodes which correspond to $(\mathbf{X}, \mathbf{Y})$-partitions of the form $\{(\top, \alpha)\}$ or $\{(\alpha, \top), (\neg\alpha, \bot)\}$. When these decision nodes are removed (by directing their parents to α), the resulting SDD is called *trimmed*. Compressed and trimmed SDDs are canonical for a given vtree [8].

References

1. Bacchus, F., Dalmao, S., Pitassi, T.: DPLL with Caching: A new algorithm for #SAT and Bayesian Inference. In: Electronic Colloquium on Computational Complexity (ECCC), vol. 10(003) (2003)
2. Beame, P., Li, J., Roy, S., Suciu, D.: Lower Bounds for Exact Model Counting and Applications in Probabilistic Databases. In: Proc. of UAI (2013)
3. Blum, M., Chandra, A.K., Wegman, M.N.: Equivalence of free boolean graphs can be decided probabilistically in polynomial time. Inf. Process. Lett. 10(2), 80–82 (1980)
4. Bryant, R.E.: Graph-Based Algorithms for Boolean Function Manipulation. IEEE Trans. Computers 35(8), 677–691 (1986)
5. Chavira, M., Darwiche, A.: On Probabilistic Inference by Weighted Model Counting. Artif. Intell. 172(6-7), 772–799 (2008)
6. Darwiche, A.: Decomposable Negation Normal Form. J. ACM 48(4), 608–647 (2001)
7. Darwiche, A.: New Advances in Compiling CNF into Decomposable Negation Normal Form. In: Proc. of ECAI, pp. 328–332 (2004)
8. Darwiche, A.: SDD: A New Canonical Representation of Propositional Knowledge Bases. In: Proc. of IJCAI, pp. 819–826 (2011)
9. Darwiche, A., Marquis, P.: A Knowledge Compilation Map. J. Artif. Intell. Res. (JAIR) 17, 229–264 (2002)
10. Davis, M., Logemann, G., Loveland, D.W.: A Machine Program for Theorem-Proving. Commun. ACM 5(7), 394–397 (1962)
11. Davis, M., Putnam, H.: A Computing Procedure for Quantification Theory. J. ACM 7(3), 201–215 (1960)
12. Huang, J., Darwiche, A.: The Language of Search. J. Artif. Intell. Res. (JAIR) 29, 191–219 (2007)
13. Bayardo Jr., R.J., Pehoushek, J.D.: Counting Models Using Connected Components. In: AAAI/IAAI, pp. 157–162 (2000)
14. Majercik, S.M., Littman, M.L.: Using Caching to Solve Larger Probabilistic Planning Problems. In: AAAI/IAAI, pp. 954–959 (1998)
15. Muise, C., McIlraith, S.A., Beck, J.C., Hsu, E.I.: DSHARP: Fast d-DNNF Compilation with sharpSAT. In: Kosseim, L., Inkpen, D. (eds.) Canadian AI 2012. LNCS, vol. 7310, pp. 356–361. Springer, Heidelberg (2012)
16. Oztok, U., Darwiche, A.: CV-width: A New Complexity Parameter for CNFs. In: Proc. of ECAI (to appear, 2014)
17. Pipatsrisawat, K., Darwiche, A.: A Lower Bound on the Size of Decomposable Negation Normal Form. In: Proc. of AAAI (2010)
18. Pipatsrisawat, K., Darwiche, A.: Top-Down Algorithms for Constructing Structured DNNF: Theoretical and Practical Implications. In: Proc. of ECAI. pp. 3–8 (2010)
19. Robertson, N., Seymour, P.D.: Graph minors. III. Planar tree-width. J. Comb. Theory, Ser. B 36(1), 49–64 (1984)
20. Sang, T., Bacchus, F., Beame, P., Kautz, H.A., Pitassi, T.: Combining Component Caching and Clause Learning for Effective Model Counting. In: Proc. of SAT (2004)
21. Valiant, L.G.: The complexity of computing the permanent. Theor. Comput. Sci. 8, 189–201 (1979)

A Complete Solver for Constraint Games

Thi-Van-Anh Nguyen and Arnaud Lallouet

Université de Caen, GREYC, Campus Côte de Nacre,
Boulevard du Maréchal Juin, BP 5186, 14032 Caen Cedex, France
{thi-van-anh.nguyen,arnaud.lallouet}@unicaen.fr

Abstract. Game Theory studies situations in which multiple agents having conflicting objectives have to reach a collective decision. The question of a compact representation language for agents utility function is of crucial importance since the classical representation of a n-players game is given by a n-dimensional matrix of exponential size for each player. In this paper we use the framework of *Constraint Games* in which CSP are used to represent utilities. Constraint Programming –including global constraints– allows to easily give a compact and elegant model to many useful games. Constraint Games come in two flavors: Constraint Satisfaction Games and Constraint Optimization Games, the first one using satisfaction to define boolean utilities. In addition to multimatrix games, it is also possible to model more complex games where hard constraints forbid certain situations. In this paper we study complete search techniques and show that our solver using the compact representation of Constraint Games is faster than the classical game solver Gambit by one to two orders of magnitude.

1 Introduction

Game theory has proven to be highly successful in modeling interaction of selfish agents [21,20]. In a strategic game, each player is given a set of actions and has to choose one to perform. A reward is given to a player by a utility function which depends on the actions taken by all players. One of the best known solution concepts for this type of game is the Pure Nash Equilibrium (PNE), which occurs when no player is able to improve his utility by changing his chosen action to another one. There are many ways do define solution concepts [24] but PNE has the notable advantage of giving a deterministic decision for the players. Indeed, PNE for games are similar to solutions for CSP: not all games own a PNE, and when available, some PNE may be more desirable than others.

The basic representation of games is a multimatrix called *normal form* whose size is exponential in the number of players. The intractability of this representation is a severe limitation to the widespread use of game-based modeling. This key issue has been addressed by several types of compact representations. Some are based on some assumptions on the interactions between players, like graphical games [18] or action-graph games [17] while other are language-based, like Boolean Games [14,9,4] or Constraint Games [22].

We focus our interest here in Constraint Games for which utilities are expressed by Constraint Satisfaction Problems (CSP) or Constraint Optimization Problems (COP).

[1] This work is supported by Microsoft Research grant MRL-2011-046.

B. O'Sullivan (Ed.): CP 2014, LNCS 8656, pp. 58–74, 2014.

Constraint Games provide a rich modeling language which allows a natural formulation of the players goals. In particular, it allows to express in a compact way most classical games like congestion games [26], network games [6], strategic scheduling [29], to name a few. In addition, hard constraints [25] can be provided to limit the joint strategies all players may take. This allow unrealistic equilibria to be ruled out. Note that these constraints are global to all players, unlike the local constraints defined in [3], and they provide crucial expressivity in modeling. Constraint Games are a generic modeling tool for strategic games and can be used to study new algorithms to solve them.

Despite the high interest of games for modeling strategic interaction, it is still difficult to find a PNE game solver. The normal form game solver Gambit [19] is currently considered as state-of-the-art. Also non-compact logic transformations have been studied [8,12]. For Boolean Games, there are techniques limited to specific categories of games like games with acyclic interaction graph [2] or using specific bargaining techniques [10]. For Constraint Games, only a solver based on local search has been proposed [22]. Not surprisingly, large games can be solved but with no guarantee of finding an equilibrium or prove the absence of equilibrium. This is due for a large part to the high complexity of finding a PNE [13,4].

Few other works are related to Games and Constraint Programming. In [5], it has been proposed to compute a mixed equilibrium using continuous constraints. Some other formalisms try to solve a combinatorial problem by multiple agents, either with a predefined assignment of variables to agents like in DCOP [11] or by letting the agents select dynamically their variable like in SAT-Games [30] and Adversarial CSP [7].

In this paper, we prove that Constraint Games are Σ_2^p-complete (like Boolean Games) and then we focus on the problem of finding all PNE for a Constraint Game. Although finding one PNE is a very interesting problem in itself, finding all of them allows more freedom for choosing equilibrium that fulfills some additional requirements. For example the correctness of the computation of Pareto Nash equilibrium relies on the completeness of PNE enumeration. We present in this paper ConGa, a new correct and complete solver for Constraint Games. This solver is based on a fast computation of equilibrium condition that we call Nash consistency and a pruning algorithm for never best responses. We demonstrate the effectiveness of our approach on publicly available benchmarks from the Gamut suite [23] as well as on real-life applications.

2 Constraint Games

Let V be a set of variables and $D = (D_x)_{x \in V}$ be the family of their (finite) domains. For $W \subseteq V$, we denote by D^W the set of tuples on W, namely $\Pi_{x \in W} D_x$. Projection of a tuple (or a set of tuples) on a variable (or a set of variables) is denoted by $|$: for $t \in D^V$, $t|_x = t_x$, $t|_W = (t_x)_{x \in W}$ and for $E \subseteq D^V$, $E|_W = \{t|_W \mid t \in E\}$. For $W, U \subseteq V$, the join of $A \subseteq D^W$ and $B \subseteq D^U$ is $A \bowtie B = \{t \in D^{W \cup U} \mid t|_W \in A \wedge t|_U \in B\}$. When $W \cap U = \emptyset$, we denote the join of tuples $t \in D^W$ and $u \in D^U$ by (t, u). A *constraint* $c = (W, T)$ is a couple composed of a subset $W = \mathit{var}(c) \subseteq V$ of variables and a relation $T = \mathit{sol}(c) \subseteq D^W$ (called *solutions*). A *Constraint Satisfaction Problem* (or CSP) is a set of constraints. We denote by $\mathit{sol}(C) = \bowtie_{c \in C} \mathit{sol}(c)$ its set of solutions. To simplify the exposition, we identify $\mathit{sol}(C)$ with its cylindric extension to all variables of V (i.e. with any combination of values for the variables which do not belong to C).

Let $\mathcal{P}$ be a set of n players and V a finite set of variables. The set of variables is partitioned into *controlled* variables $V_c = \bigcup_{i \in \mathcal{P}} V_i$ where V_i is the subset of variables controlled by Player i, and V_E the set of *uncontrolled* or *existential* variables ($V_E = V \setminus V_c$).

Definition 1 (Constraint Satisfaction Game). *A* Constraint Satisfaction Game *(or CSG) is a 4-tuple $(\mathcal{P}, V, D, G)$ where $\mathcal{P}$ is a finite set of players, V is a finite set of variables composed of a family of disjoint sets (V_i) for each player $i \in \mathcal{P}$ and a set V_E of* existential *variables disjoint of all the players variables, $D = (D_x)_{x \in V}$ is the family of their domains and $G = (G_i)_{i \in \mathcal{P}}$ is a family of CSP on V.*

The CSP G_i is called the *goal* of the player i. The intuition behind CSG is that, while Player i can only control his own subset of variables V_i, his satisfaction will depend also on variables controlled by all the other players (Example 1). The intuition behind existential variables is that they are existentially quantified (but most of the time they will be functionally defined from decision variables).

Example 1. We consider the following CSG: the set of players is $\mathcal{P} = \{X, Y, Z\}$. Each player owns one variable: $V_X = \{x\}, V_Y = \{y\}$ and $V_Z = \{z\}$ with $D_x = D_y = \{1, \ldots, 9\}$ and $D_z = \{1, 2, 3\}$. The goals are $G_X = \{x = yz\}$, $G_Y = \{y = xz\}$ and $G_Z = \{xy \leq z \leq x + y, (x+1)(y+1) \neq 3z\}$.

A *strategy* for player i is an assignment of the variables V_i controlled by player i. A *strategy profile* $s = (s_i)_{i \in \mathcal{P}}$ is the given of a strategy for each player.

Definition 2 (Winning Strategy). *A strategy profile s is winning for i if it satisfies the goal of i: $s \in sol(G_i)$.*

A CSG can be interpreted as a classical game with a boolean payoff function which takes value 1 when the player's CSP is satisfied and 0 when not.

We denote by s_{-i} the projection of s on $V_{-i} = V \setminus V_i$. Given a strategy profile s, a player i has a *beneficial deviation* if $s \notin sol(G_i)$ and $\exists s'_i \in D^{V_i}$ such that $(s'_i, s_{-i}) \in sol(G_i)$. Beneficial deviation represents the fact that a player will try to maximize his satisfaction by changing the assignment of the variables he can control if he is unsatisfied by the current assignment. A tuple s is *best response* for Player i if this player is not able to make any beneficial deviation. Then we define the notion of solution of a CSG by pure Nash equilibrium:

Definition 3 (Pure Nash Equilibrium). *A strategy profile s is a* Pure Nash Equilibrium (or PNE) *of the CSG C if and only if no player has a beneficial deviation, i.e. s is* best response *of all players.*

Theorem 1. *CSG are Σ_2^p-complete.*

Proof. The proof is adapted from the Σ_2^p-completeness of boolean games [4].

Membership comes from the simple algorithm in which one guesses a strategy profile and checks that no player has a beneficial deviation. Each verification consists in proving that a player i has no solution if the strategy profile is not winning for i. This verification is in coNP because for a strategy profile s, proving that there exists a solution for Player i amounts to solve the CSP

G_i, which is in NP. Since the number of players is finite, there is a polynomial number of calls to a coNP oracle (actually one for each player) and thus the problem is in Σ_2^p.

For hardness, we introduce a special case of CSG: the 2-players 0-sum game. In this kind of game, when one player wins, the other player looses. Thus it is enough to represent only the goal of the first player, the other one being deduced by negation. Such a CSG can be represented by $(\mathcal{P} = \{1, 2\}, V, D, C)$ where C is the goal of player 1 (the goal of Player 2 is straightforwardly deduced by negation).

We perform a reduction from a $\exists\forall$-QCSP to a 2-players 0-sum CSG. $\exists\forall$-QCSP are known to be Σ_2^p-complete. This reduction proves that even 2-players 0-sum CSG are at least as hard as solving a $\exists\forall$-QCSP. Together with membership of the Σ_2^p class, it gives the exact complexity for n-players CSG.

The reduction is from the QCSP $Q = \exists X \forall Y C$ where X and Y are disjoint sets of variables to the 2-players 0-sum CSG $G = (\{1, 2\}, X \cup Y \cup \{x, y\}, (D, D_x, D_y), C \vee (x = y))$ where x is a new variable controlled by player 1, y a new variable controlled by player 2 and $D_x = D_y$ are domains composed at least of 2 elements. It is obvious that the conversion can be performed in polynomial time. If Q is valid, then let s_1 be the assignment of variables of X and let s_2 be an assignment of variables of Y. Because Q is valid, $\forall s_2' \in D^Y$, $(s_1, s_2') \in sol(C)$. Thus (s_1, s_2) is a PNE because player 1 is winning and player 2 has no beneficial deviation. Conversely, if Q is not valid then for any assignment $s_1 \in D^X$ of player 1, player 2 can play $s_2' \in D^Y$ such that $(s_1, s_2') \notin sol(C)$. Then if player 1 plays $x = v$ and if $(s_1, s_2) \in sol(C)$, then player 2 can play s_2' and $y = w$ with $w \neq v$. Thus player 2 has a beneficial deviation and (s_1, s_2, v, w) is not an equilibrium. If $(s_1, s_2) \notin sol(C)$ and player 2 plays $y = w$, then player 1 can play $x = w$ and player 1 has a beneficial deviation. Thus (s_1, s_2, w, w) is not an equilibrium. In conclusion, G has a PNE if and only if Q is valid, proving the Σ_2^p-hardness. □

The players' goals could be considered as soft constraints or preferences. It may happen however that some games have rules that forbid some strategy profiles as they model impossible situations. It is natural to reject such profiles by setting *hard constraints* shared by all players [25]. Hard constraints can be easily expressed in the framework of Constraint Games by adding an additional CSP on the whole set of variables in order to constrain the set of possible strategy profiles:

Definition 4 (Constraint Satisfaction Game with Hard Constraints). *A* Constraint Satisfaction Game with Hard Constraints *(or CSG-HC) is a 5-tuple* $(\mathcal{P}, V, D, C, G)$ *where* $(\mathcal{P}, V, D, G)$ *is a CSG and* C *is a CSP on* V.

It is useful to distinguish a strategy profile which does not satisfy any player's goal from a strategy profile which does not satisfy the hard constraints. The former can be a PNE if no player has a beneficial deviation while the latter cannot. Therefore hard constraints provide an increase of modeling expressibility (without however changing the general complexity of CSG).

By adding an optimization condition it is possible to represent classical games. A *Constraint Optimization Game* (or COG) is an extension of CSG in which each player tries to optimize his goal. This is achieved by adding for each player i an optimization condition $\min(x)$ or $\max(x)$ where $x \in V$ is a variable to be optimized by Player i.

Definition 5 (Constraint Optimization Game). *A* Constraint Optimization Game *(or COG) is a 5-tuple $(\mathcal{P}, V, D, G, opt)$ where $(\mathcal{P}, V, D, G)$ is a CSG and $opt = (opt_i)_{i\in\mathcal{P}}$ is a family of optimization conditions for each player of the form* $\min(x)$ *or* $\max(x)$ *where $x \in V$.*

A winning strategy for player i is still a strategy profile which satisfies G_i. However, the notion of beneficial deviation needs to be slightly adapted. We denote by $<_{opt_i}$ the (partial) order on strategy profiles such that $s <_{opt_i} s'$ if $s_{-i} = s'_{-i}$ and $s|_x < s'|_x$ when $opt_i = \min(x)$ (resp. $s|_x > s'|_x$ when $opt_i = \max(x)$). Given a strategy profile s, a player i has a *beneficial deviation* if $\exists s'_i \in D^{V_i}$ such that $s' = (s'_i, s_{-i}) \in sol(G_i)$ and $s' <_{opt_i} s$. Given this, the notion of solution is the same as for CSG. In addition, COG can be extended with hard constraints the same way CSG are, yielding COG-HC.

3 Modeling with Constraint Games

In this section, we show that complex games can be easily expressed using constraint games.

Example 2 (Location Game). In this example inspired by [15], n ice cream vendors from a set $\mathcal{P} = \{1, 2, \ldots, n\}$ want to choose a location numbered from 1 to m for their stand in a street. Each seller i wants to find a location l_i. He already has fixed the price of his ice cream to p_i and we assume there is a customer's house at each location. No two vendors may choose the same location. The customers choose their vendor by minimizing the sum of the distance between their house and the seller plus the price of the ice cream.

In order to build the model, we need the following existential variables (which are functionally determined by the decision variables l_i): cost_{ic} defines the cost customer c has to pay if he chooses the stand of seller i, min_c defines the minimal cost customer c has to pay for an ice cream, choice_{ic} is a boolean variable which equals 1 if customer c chooses seller i and benefit_i defines the number of customers actually buying from seller i. The Location Game (LG) can be easily modeled by a COG-HC in which each seller wants to maximize his profit:

- $\mathcal{P} = \{1, \ldots, n\}$; $\forall i \in \mathcal{P}, V_i = \{l_i\}$ and $\forall i \in \mathcal{P}, D(l_i) = \{1, \ldots, m\}$
- the hard constraints C are the following:
  - no two vendors are located at the same place: *all_different*$(l_1, l_2, \ldots, l_n)$
  - $\forall i \in \mathcal{P}, \forall c \in [1..m], \text{cost}_{ic} = |c - l_i| + p_i$
  - $\forall c \in [1..m], \text{min}_c = \min(\text{cost}_{1c}, \ldots, \text{cost}_{nc})$
  - $\forall c \in [1..m], (\text{min}_c = \text{cost}_{ic}) \leftarrow (\text{choice}_{ic} = 1)$
  - $\forall c \in [1..m], \sum_{i\in\mathcal{P}} \text{choice}_{ic} = 1$
- $\forall i \in \mathcal{P}$, G_i contains the following constraint: $\text{benefit}_i = p_i . \sum_{c=1}^{m} \text{choice}_{ic}$
- $\forall i \in \mathcal{P}$, the optimization condition $opt_i = \max(\text{benefit}_i)$

An interesting feature of this example is that it uses global constraints like *all_different* the same way as in Constraint Programming. It also shows the interest of modeling hard constraints in games since it is perfectly natural to think that no two vendors can settle at the same place. It is possible to transform this problem into a CSG by fixing a minimal profit mp_i for each player i and stating that player i is satisfied if his benefits is over mp_i. It can be done by adding the constraint benefit$_i \geq mp_i$ to G_i instead of the optimization condition. In the Gamut [23] version of the game, vendors do not choose locations but prices, because there is no way to express that sellers should choose different locations in a normal form game like we do here with the *all_different* constraint.

Example 3 (Cloud Resource Allocation Game). Resource allocation is a central issue in cloud computing where clients use and pay computing resources on demand. In order to manage conflicting interests between clients, [16] has proposed the framework of *CRAG* (Cloud Resource Allocation Game) in which resource assignments are defined by game equilibrium.

A cloud computing provider owns a set $\mathcal{M} = \{M_1, \ldots, M_m\}$ of m machines, each machine M_j having a capacity c_j representing the amount of resource available (for example CPU-hour, memory). The cost of using machine j is given by $l_j(x) = x \times u_j$ where x is the number of resources requested and u_j some unit cost. A set of n clients $\mathcal{P} = \{1, 2, \ldots, n\}$ wants to use simultaneously the cloud in order to perform tasks. Client $i \in \mathcal{P}$ has m_i tasks $\{T_{i1}, \ldots, T_{im_i}\}$ to perform, with respective requested capacity of $\{d_{i1}, \ldots, d_{im_i}\}$. Each client $i \in \mathcal{P}$ chooses selfishly an allocation r_{ik} for the task T_{ik} ($k \in [1..m_i]$) and wishes to minimize his cost $cost_i = \sum_{k=1}^{m_i} l_{r_{ik}}(d_{ik})$. We assume that the provider's resources amount is sufficient to accommodate the resources requested by all of the clients: $\sum_{i=1}^{n} \sum_{k=1}^{m_i} d_{ik} \leq \sum_{j=1}^{m} c_j$. This problem can be modeled by the following COG-HC:

- $\mathcal{P} = \{1, \ldots, n\}; \forall i \in \mathcal{P}, V_i = \{r_{i1}, \ldots, r_{im_i}\}$ and $\forall i \in \mathcal{P}, \forall k \in [1..m_i], D(r_{ik}) = \{1, \ldots, m\}$
- C is composed of the following constraints:
  - channeling constraints for boolean variables stating that machine j is requested by task t_{ik}: $(r_{ik} = j) \leftrightarrow (choice_{ijk} = 1)$
  - capacity constraints: $\forall j \in [1..m], \sum_{i=1}^{n} \sum_{k=1}^{m_i} choice_{ijk} \times d_{ik} \leq c_j$
- $\forall i \in \mathcal{P}, G_i = \{cost_i = \sum_{j=1}^{m} \sum_{k=1}^{m_i} choice_{ijk} \times l_j(d_{ik})\}$
- $\forall i \in \mathcal{P}, opt_i = \min(cost_i)$

Other interesting examples can be modeled by Constraint Games like network games [6], strategic scheduling [29], or games from the Gamut suite [23].

4 Pruning Techniques

A natural algorithm is to use generate and test to find an equilibrium. This naive algorithm is however the only known algorithm for finding PNE [28] and from the complexity result, it is unlikely that any fast (polynomial) algorithm could exist. This algorithm is therefore the basis of the implementation of the Gambit solver [19] for PNE enumeration. We first show that this technique is subject to a form of trashing. In order to simplify the exposition, we assume in the remaining of the paper that each player i only controls one variable x_i with domain D_i. The extension to more than one variable per player is not difficult (indeed our solver Conga does not have this limitation since many examples require a player to control several variables).

The *enum1* algorithm (Algorithm 1) consists in enumerating all strategy profiles, testing each of them for each player for deviation and skipping to the next profile when the first deviation is found.

Algorithm 1. enum1

```
function ENUM1(Game CG): setof tuples
    Nash ← ∅
    for s ∈ D^{V_c} do
        if IsNash(s) then
            Nash = Nash ∪ {s}
        end if
    end for
    return Nash
end function
```

```
function ISNASH(tuple s): boolean
    for i ∈ P do
        for v ∈ D_i, v ≠ s_i do
            if (s_{-i}, v) <_{opt_i} s then
                return false
            end if
        end for
    end for
    return true
end function
```

The following example shows that some deviations are performed several times.

Example 4. Let G be the 2-players game defined by the following table:

		y		
		1	2	3
x	a	$(0,1)_\alpha$	(1,0)	(1,0)
	b	$(0,1)_\beta$	(0,0)	(1,0)
	c	(1,0)	(1,1)	(0,0)

We assume that the enumeration starts by Player x. The first tuple to be enumerated is $(a, 1)$ denoted by α. Deviation is checked for Player y and no deviation is found. Then deviation is checked for Player x and a deviation towards $(c, 1)$ is found. Thus this tuple is not a PNE. The next candidate is $(b, 1)$ denoted by β. This tuple is checked for Player y and again no deviation is found. But when checked for Player x, the same deviation towards $(c, 1)$ is found as for tuple α.

This form of trashing is a strong motivation to investigate search and pruning techniques for Constraint Games. To introduce our technique, we first recall [13] where the authors introduce (originally for graphical games) a CSP composed of *Nash constraints* to represent best responses for each player.

Definition 6 (GGS-CSP). *Let* $CG = (\mathcal{P}, V, D, G, opt)$ *be a COG. The* Nash constraint *of player* $i \in \mathcal{P}$ *is* $g_i = (V_c, T)$ *where* $T = \{t \in D^V \mid \nexists t' \in D^V \text{ s.t. } t' <_{opt_i} t\}$. *The* GGS-CSP $\mathcal{G}(CG)$ *of* CG *is the set of Nash constraints for all players.*

This CSP has the important property that it defines the PNE of the game:

Theorem 2. *([13])* t *is a PNE of* $CG \leftrightarrow t \in sol(\mathcal{G}(CG))$

Then it follows that a PNE of a Constraint Game CG has a support in all of its Nash constraints.

Our technique consists to perform a traversal of the search space by assigning the variables of each player in turn according to a predefined ordering on $\mathcal{P}$. For each candidate tuple, we seek for supports by performing an incremental computation of the Nash constraints. Each computed deviation is recorded in a table for each player. By retrieving tuples in Nash constraints, we can avoid computing costly deviations.

However, since we are studying general games, each Nash constraint has the same arity as the whole problem, which is challenging in terms of space requirements. First, note that any tuple deleted from a table does not hinder the correctness of the Nash test. It may only forces a deviation to be computed twice. Hence we are free to limit the size of the tables and trade space with time. In practice, two features limit the size of the tables.

First, deviation checks are performed in reverse player ordering. It means that a tuple checked for the first player must have succeed the deviation test for all other players. In practice for most problems, this limits the number of tuples reaching the upper levels. Second, we can delete a tuple t recorded in a table when we can ensure that no candidate t' will deviate anymore towards t. This property is given by the following independence of subgames theorem. Let $CG = (\mathcal{P}, V, D, G, opt)$ be a constraint game. A game $CG' = (\mathcal{P}, V, D', G, opt)$ is a subgame of CG if $\exists i \in \mathcal{P}, D'_i \subseteq D_i$ and $i \neq j \rightarrow D'_j = D_j$. We denote by $br_i(t) = \{t' \in sol(G_i) \mid t'_{-i} = t_{-i}\}$ the set of best response strategies from t for Player i.

Proposition 1. *Let CG be a constraint game and CG' a subgame of CG such that $D'_i \subseteq D_i$. Let $D'' = D \setminus D'$, $t' \in D'^V$ and $t'' \in D''^V$. Then $\forall j \in \mathcal{P}, j \neq i \rightarrow br_j(t') \cap br_j(t'') = \emptyset$.*

Proof. The proof is by contradiction. Suppose there exists $t' \in D'^V$ and $t'' \in D''^V$ such that $br_j(t') \cap br_j(t'') \neq \emptyset$. Let $s \in br_j(t') \cap br_j(t'')$. Then since $s \in br_j(t')$, $s_i \in D'^{V_i}$. Since $s \in br_j(t'')$, $s_i \in D''^{V_i}$. Hence the contradiction. □

In other words, if we split the search space following Player Z, best responses for Players X and Y are forced to remain in different subspaces.

By applying inductively Proposition 1 on a sequence of assignments of strategies for Player 1 to k with $k < n$, we see that two branches of the search tree will not share best responses for the remaining unassigned players. Hence we can freely remove all tuples from the table of subsequent players once the branch is explored.

The last optimization consists in elimination of *never best responses* (NBR).

Definition 7. *A strategy s_i for Player i is a* never best response *if $\forall t_{-i}, \exists s'_i$ such that $(s'_i, t_{-i}) <_{opt_i} (s_i, t_{-i})$.*

Iterative elimination of NBR is a sound pruning for games [1] which additionally is stronger than elimination of strongly dominated strategies. But unfortunately their detection is very costly in the n-players case since it needs to know that this action will never been chosen by Player i for all strategy profiles of the other players. However, being a NBR in a subgame is a sufficient condition for not being an equilibrium:

Proposition 2. *Let CG be a constraint game and CG' a subgame of CG such that $D'_i \subseteq D_i$. Let $s_j \in D'_j$ be a NBR in CG' with $j \neq i$. Then for all s_{-j}, if $s = (s_j, s_{-j})$ is a PNE, then $s_i \notin D'_i$.*

Proof. The proof is by contradiction. Suppose there exists a PNE $s = (s_j, s_{-j})$ with $(s_{-j})_i \in D'_i$. Because s is a PNE, we have $\forall k \in \mathcal{P}, s_k \in br(s)|_k$. Then because s_j is a NBR for j in CG', there exists s'_j such that $(s'_j, s_{-j}) <_{opt_j} (s_j, s_{-j})$. Thus $s_j \notin br(s)|_j$. □

By applying inductively Proposition 2 on a sequence of assignments of strategies for Player 1 to k with $k < n$, we see that if we detect that a strategy v is NBR for player k in the subgame defined by the sequence of assignments, then this strategy will not participate to a PNE and we can prune it.

5 An Algorithm for Nash Equilibrium Enumeration

We propose a tree-search algorithm for finding all PNE. The general method is based on three ideas:

- all candidates (except those which are detected as NBR) are generated in lexicographic order;
- Best responses for each player are recorded in a table BR;
- whenever a domination check is performed, it first checks this player's recorded best responses.

We assume given an ordering on players from 1 to n. The main algorithm (Algorithm 2) launches the recursive traversal (Algorithm 3) starting by Player 1. We distinguish the original domains of the variables (called D) used to compute deviations from their actual domain explored by the search tree (called A) and subject to pruning by arc-consistency on hard constraints.

Algorithm 2. ConGa

global:
  BR: array[1..n] of tuples ▷ *best responses for all players*
  cnt: array[1..n] of integer ▷ *counters for NBR detection*
  $Nash$: set of tuples ▷ *Nash equilibrium*
  S: global solver

function CONGA(Game CG): **setof** tuples
  Nash $\leftarrow \emptyset$
  Initialize solver S with hard constraints
  $A \leftarrow D$
  enum(A, 1)
  return Nash
end function

Propagation of hard constraints allows to ensure that no forbidden tuple will be explored. If the propagation returns *false*, then at least one domain has been wiped out and there is no solution in this subspace. Otherwise domains A are reduced according to arc-consistency. Values for each player are then recursively enumerated.

Algorithm 3. enum

```
1: procedure ENUM(domains A, int i)
2:     if S.propagate(A) then
3:         if i > n then
4:             checkNash(tuple(A),n)
5:         else
6:             BR[i] ← ∅
7:             cnt[i] ← Π_{j>i}|D_j|
8:             while A_i ≠ ∅ do
9:                 choose v ∈ A_i
10:                B ← A
11:                enum((B_{-i}, (B_i = {v})), i + 1)
12:                A_i ← A_i − {v}
13:                if cnt[i] ≤ 0 then
14:                    checkEndOfTable(A, i)
15:                    break
16:                end if
17:            end while
18:        end if
19:    end if
20: end procedure
```

Algorithm 4. checkNash

```
1: procedure CHECKNASH(tuple t, int i)
2:     if i = 0 then
3:         Nash ← Nash ∪ {t}
4:     else
5:         d ← search_table(t, BR, i)
6:         if d = ∅ then
7:             d ← findBestResponses-C*G(t, i)
8:             if d = ∅ then d ← D_i end if
9:             insert_table(i, BR, d)
10:            cnt[i] --
11:        end if
12:        if t_i ∈ d then checkNash(t, i − 1) end if
13:    end if
14: end procedure
```

When a tuple is reached, it is checked for Nash condition (line 4) by Algorithm 4. Otherwise, at least one domain remains to be explored. Each player i owns a table $BR[i]$ of best responses, initialized empty and a counter $cnt[i]$ initialized with the size of the subspace needed to detect potential never best responses.

Algorithm 5. findBestResponses-CSG

```
function FINDBESTRESPONSES-CSG(tuple t, int i): set of tuples
    d ← ∅
    initialize solver S_i with G_i (and Hard Constraints if required)
    add constraints x_j = t_j for all j ≠ i
    sol ← S_i.getSolution()
    while sol ≠ nil do
        d ← d ∪ {sol}
        sol ← S_i.getSolution()
    end while
    return d
end function
```

After the recursive call of *enum*, we test whether all the subspace after Player i has been checked for deviation. Then all subsequent values are *never best responses*. In this case an exit from the loop causes backjumping to the ancestor node. This backjumping is done after the exploration by *checkEndOfTable* (Algorithm 7) of the potential values which are stored in the table and belong to the unexplored space (lines 13-16).

Algorithm 6. findBestResponses-COG

```
function FINDBESTRESPONSES-COG(tuple t, int i): set of tuples
    d ← ∅
    initialize solver S_i with G_i and opt_i (and Hard Constraints if required)
    add constraints x_j = t_j for all j ≠ i
    save the current state of S_i
    sol ← S_i.getOptimalSolution()
    if sol = nil then
        return d
    else
        get the optimal value opt of Player i from sol
        restore the previous state of S_i
        for opt_i = min/max(X), add constraint X = opt
        sol ← S_i.getSolution()
        while sol ≠ nil do
            d ← d ∪ {sol}
            sol ← S_i.getSolution()
        end while
    end if
    return d
end function
```

The *checkNash* procedure in Algorithm 4 verifies whether a player can make a beneficial deviation from a tuple. Since the exploration of the search tree is done level by level, the verification starts from the deepest level. First the tuple is searched in the table of stored best response for this player (line 5). If not found, a solver for G_i is called in function *findBestResponses-C*G* (line 7, with * standing for S for CSG and O for COG). This function returns the set of deviation for Player i from a tuple t. There can be more than one deviation. In a CSG (Algorithm 5), it means that several assignments satisfy the constraints of G_i. In a COG (Algorithm 6), it means that the optimal value is reached for more than one point.

Algorithm 7. checkEndOfTable

procedure CHECKENDOFTABLE(domain A, int i)
  for all $t \in BR[i]$ such that $t \in \Pi_{i=1..n} A_i$ **do**
    checkNash(t, n)
  end for
end procedure

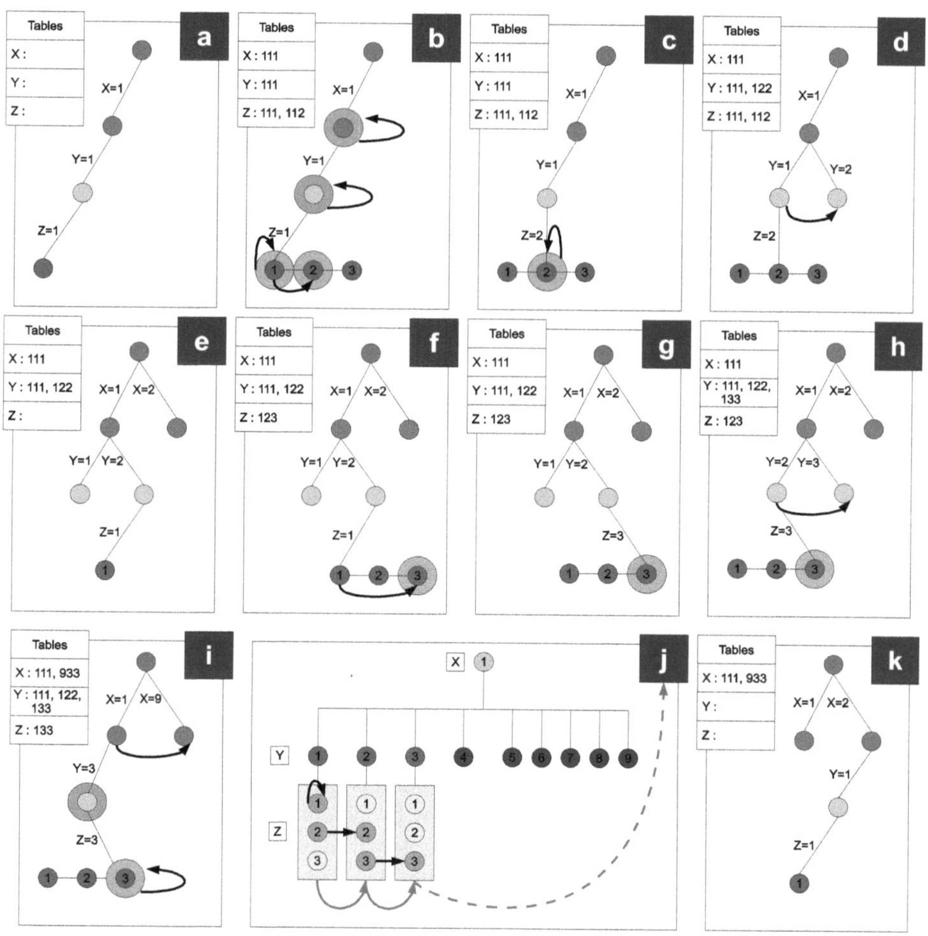

Fig. 1. ConGa algorithm

If d is empty, it means that there is no possible action for Player i which can satisfy the constraints of her goal. Indeed, a tuple can be an equilibrium even if a (or all) player is unsatisfied. In this case we return the whole initial domain as deviation since any value can participate to a PNE.

The procedure *checkEndOfTable* depicted in Algorithm 7 is used when the subset has been explored and we are able to perform backjumping. In this case, all tuple of the table which belong to the unexplored zone are checked for PNE.

In Figure 1, we apply ConGa algorithm on Example 1 of Section 2. The resolution starts in Figure 1.**a** (abbreviated 1.**a**) with tuple 111. During the descent to the leaf, X counter is initialized to $|D_Y| \times |D_Z| = 9 \times 3 = 27$ and Y's counter to $|D_Z| = 3$. Tuple 111 is first checked for deviation for Z. It is found in 1.**b** that best responses for Z are 111 itself and 112, thus we store these two values in Z's table. Then we check Players Y and X for deviation and find that none of them deviate. Thus 111 is PNE, we record it and store it in Y and X's tables. Y's counter is decremented to 2.

Since the possible deviations for Z are explored, we continue by *checkEndOfTable* for Z to 112 as in 1.**c**. This tuple is obviously a best response for Z (as found in Z's table), and is checked for deviation for the other players in reverse order. It is thus checked for Y in 1.**d** and a deviation is found to 122, inserting 122 in Y's table. Y's counter is decremented to 1.

Then backtracking occurs in 1.**e** at Y's level, which resets the table for Z. The next candidate is 121. In 1.**f**, a deviation is found for Z to 123. In 1.**g**, the tuple 123 is considered by *checkEndOfTable* for Z, checked for deviation for Z and found stable by looking up in Z's table. A further check for Y in 1.**h** finds a deviation to 133 and stores this tuple in Y's table. Y's counter is decremented to 0.

In this example, the domain of Player Z is of size 3. Hence $cnt[Y]$ was initialized to 3. All tested tuples are of the form $1yz$ where 1 is the value on X. It happens in this example that only by exploring values 1, 2, 3 for Y yield a complete traversal of the subspace defined by Z (all values from Z's domain have been considered). Thus we know that, from that point, only 133 recorded in Y's table can be a PNE with $X = 1$. The other remaining values of Player Y are NBR. It is sufficient to check tuple 133 in *checkEndOfTable* as depicted in 1.**i**. Deviation is found for Player X to 933, so 133 is not a PNE. Then, backjumping can occur. Figure 1.**j** makes a small summary of Player Y's deviation possibilities. We see that tuples 111, 112, 123 are stable for Z, thus these tuples were lifted to Y level to be checked for Y's deviation. Solid arrows depict the deviations recorded for Player Y on the different values for Z. After having checked 133, we can backjump to the next value of Player X (dotted arrows) (1.**k**).

In general *checkEndOfTable* tests all tuples of $BR[Y]$ which belong to the unexplored part of the search space. This NBR detection is incomplete but comes almost for free because it only takes a counter. Note that by Proposition 1, when exploring $X = 2$, the table for Y can be reset because no other tuple will deviate to a tuple where $X = 1$.

Note that the tables are actually implemented by a tree whose nodes represent all players but i, with i's best responses attached on the leaves. Thus the search for deviation in the table does not depend on the number of recorded tuples but only on the number of players and is performed in $O(|\mathcal{P}|)$.

Proposition 3. *ConGa is correct and complete.*

Proof. Correctness comes from the one of PNE check (Theorem 2). A reported PNE has been checked for deviation for every player. Either the tuple has been recorded in the table as deviation from another one, or had been directly checked by the solver against the player's goal. Completeness is due to the traversal of the whole search space and soundness of never best response pruning. □

6 Experiments

We have performed experiments on classical games of the Gamut suite [23] and some games with hard constraints. Results are summarized in Table 1 in which the name of the game is followed by the number of players and the size of the domain. Gamut [23] games are CG (Congestion Game), GTTA (Guess Two Third Average), LG(GV) (Location Game, Gamut version), MEG (Minimum Effort Game) and TD (Traveller's Dilemma). The other games are LG(HC) (Location Game with Hard Constraints, example 2) and CRAG (Cloud Resource Allocation Game, example 3).

Table 1. Results for Gamut and other games

Name	NF Gen.		Gambit	enum1			ConGa			#PNE
	Time	*Size*	*Time*	*Time*	*#Cand*	*#Dev*	*Time*	*#Cand*	*#Dev*	
CG.7.15	259	5.1	*MO*	75	1.7E+8	1.8E+8	**30**	2.1E+7	1.3E+7	630
CG.8.15	4742	89	*MO*	1037	2.5E+9	2.7E+9	**371**	3.1E+8	1.9E+8	1680
CG.9.15	*TO*	–	–	17314	3.8E+10	4.2E+10	**6035**	4.9E+9	2.9E+9	5040
GTTA.3.101	2	0.1	18	4	1.0E+6	1.4E+6	**1**	1.0E+4	1.0E+4	1
GTTA.4.101	125	1.8	1844	337	1.0E+8	1.3E+8	**12**	1.0E+6	1.0E+6	1
GTTA.5.101	14705	216	*MO*	*TO*	–	–	**816**	1.0E+8	1.1E+8	1
LG(GV).2.1000	2	0.01	131	340	1.0E+6	1.0E+6	**8**	2.0E+3	1.5E+3	0
LG(GV).2.2000	6	0.04	534	1448	4.0E+6	4.0E+6	**33**	4.0E+3	3.5E+3	0
LG(GV).2.3500	18	0.1	2614	6893	1.2E+7	1.2E+7	**96**	7.0E+3	6.0E+3	0
LG(GV).2.5000	35	0.2	7696	19990	2.5E+7	2.5E+7	**205**	1.0E+4	9.0E+3	0
LG(GV).2.20000	551	3.7	*MO*	*TO*	–	–	**3462**	4.0E+4	3.9E+5	0
MEG.3.100	1	0.1	14	1	1.0E+6	1.0E+6	**1**	1.9E+4	1.5E+4	100
MEG.4.100	96	1.9	1555	46	1.0E+8	1.0E+8	**10**	1.9E+6	1.3E+6	100
MEG.5.100	11414	241	*MO*	3844	1.0E+10	1.0E+10	**447**	1.9E+8	1.2E+8	100
MEG.30.2	8998	91	*MO*	**437**	1.1E+9	2.1E+9	1085	5.4E+8	1.1E+9	2
MEG.35.2	*TO*	–	–	**15408**	3.4E+10	6.9E+10	*TO*	–	–	2
TD.3.99	2	0.1	15	1	9.7E+5	9.8E+5	**1**	1.9E+4	1.5E+4	1
TD.4.99	82	1.9	1572	28	9.6E+7	9.7E+7	**10**	1.9E+6	1.3E+6	1
TD.5.99	9301	119	*MO*	3338	9.1E+9	9.6E+9	**488**	1.8E+8	1.2E+8	1
CRAG.7.9	N/A	N/A	N/A	302	4.7E+6	5.3E+6	**58**	1.0E+6	5.9E+5	1
CRAG.8.9	N/A	N/A	N/A	3137	4.2E+7	4.8E+7	**546**	9.5E+6	5.3E+6	1
CRAG.9.9	N/A	N/A	N/A	*TO*	–	–	**4980**	4.3E+7	4.8E+7	1
LG(HC).4.30	N/A	N/A	N/A	27	6.5E+5	8.0E+5	**9**	1.4E+5	4.4E+4	24
LG(HC).5.30	N/A	N/A	N/A	701	1.7E+7	2.1E+7	**231**	4.1E+6	1.2E+5	240
LG(HC).6.30	N/A	N/A	N/A	19040	4.3E+8	5.5E+8	**17263**	1.1E+8	3.2E+7	2160

For each instance, we have compared ConGa which has been built on top of the library Choco [27] to the game solver Gambit [19] and to a base solver called *enuml* (Algorithm 1). This solver works like Gambit by examining each tuple but the only difference is that it uses the compact Constraint Game representation. All experiments have been executed on a server of 48-core AMD Opteron 6174 with 4-processor at 2,2 GHz with 256 GB of RAM. The operating system installed is ubuntu 64bit 12.04 LTS. In Table 1, time taken for each solver is given in seconds and time out is set to 20000 seconds. The number $aE{+}b$ is equal to $a \times 10^b$.

The experiment on Gambit is divided into two steps: we first generate the normal form matrix (column NF gen) and then we launch the command line `gambit-enumpure` on the normal form to find all PNE. We have measured the time needed to generate the normal form and its size (in GB), then the time required for the game to be solved by Gambit. TO stands for *Time Out*, MO for *Memory Out* and "–" means there is no information (for example, if the generation times out, it is not possible to launch the resolution). As expected, the size of the normal form soon becomes intractable and exceeds the capacity of Gambit.

For *enuml* and *ConGa*, we have measured the time needed to solve an instance, the number of candidate profiles and the number of deviation checks performed when checking whether a candidate is a PNE. From a simple reasoning, the number of candidates for *enuml* is simply $|D^{V_c}|$, and the number of checks is comprised between the number of candidates and an upper bound of $|D^{V_c}| \times |\mathcal{P}|$. Not surprisingly, we see that *ConGa* prunes most of the time a large part of the search space, mainly thanks to NBR detection. But interestingly, most of the time, it also saves deviation checks, meaning that the solution is found in the tables before a check is performed. A notable counterexample is the Minimum Effort Game with a domain of size 2 (MEG.30.2 and MEG.35.2) for which neither the tables or the NBR detection are working because the domains are too small. Note that games with hard constraints are not implementable in Gambit, indicated by N/A for *not applicable*. In all other benchmarks, ConGa outperforms both Gambit and *enuml* by one order of magnitude or even more.

A potential problem of ConGa could be that the size of the tables grows too much. It is easy to build an example for which the first player will get a table of exponential size: the game with no constraint for each player. In this game each profile is a PNE and is thus stored in the table of the first player. However, this behavior has not been observed in practice in any of our examples. Tables rather stay of reasonable size, either because they belong to lower level players and they are reset often or because many profiles do not reach high level players.

7 Conclusion

Constraint Games provide a new compact yet natural encoding to games. In this paper, we propose the first complete solver for constraint games based on a fast computation of Nash consistency and pruning of Never Best Responses. We show that these techniques are yet able to outperform the existing state-of-the-art solver Gambit. But they also open directions for further investigations on efficient algorithmic techniques to compute Nash equilibrium. Future work include the use of heuristics, graphical constraint games and other solution concepts like Pareto equilibrium.

References

1. Apt, K.R.: Order independence and rationalizability. In: van der Meyden, R. (ed.) TARK, pp. 22–38. National University of Singapore (2005)
2. Bonzon, E., Lagasquie-Schiex, M.C., Lang, J.: Dependencies between players in boolean games. Int. J. Approx. Reasoning 50(6), 899–914 (2009)
3. Bonzon, E., Lagasquie-Schiex, M.C., Lang, J.: Effectivity functions and efficient coalitions in boolean games. Synthese 187(1), 73–103 (2012)
4. Bonzon, E., Lagasquie-Schiex, M.C., Lang, J., Zanuttini, B.: Boolean games revisited. In: Brewka, G., Coradeschi, S., Perini, A., Traverso, P. (eds.) ECAI. Frontiers in Artificial Intelligence and Applications, vol. 141, pp. 265–269. IOS Press (2006)
5. Bordeaux, L., Pajot, B.: Computing equilibria using interval constraints. In: Faltings, B.V., Petcu, A., Fages, F., Rossi, F. (eds.) CSCLP 2004. LNCS (LNAI), vol. 3419, pp. 157–171. Springer, Heidelberg (2005)
6. Bouhtou, M., Erbs, G., Minoux, M.: Joint optimization of pricing and resource allocation in competitive telecommunications networks. Networks 50(1), 37–49 (2007)
7. Brown, K.N., Little, J., Creed, P.J., Freuder, E.C.: Adversarial constraint satisfaction by game-tree search. In: de Mántaras, R.L., Saitta, L. (eds.) ECAI, pp. 151–155. IOS Press (2004)
8. De Vos, M., Vermeir, D.: Choice logic programs and nash equilibria in strategic games. In: Flum, J., Rodríguez-Artalejo, M. (eds.) CSL 1999. LNCS, vol. 1683, pp. 266–276. Springer, Heidelberg (1999)
9. Dunne, P.E., van der Hoek, W.: Representation and complexity in boolean games. In: Alferes, J.J., Leite, J. (eds.) JELIA 2004. LNCS (LNAI), vol. 3229, pp. 347–359. Springer, Heidelberg (2004)
10. Dunne, P.E., van der Hoek, W., Kraus, S., Wooldridge, M.: Cooperative boolean games. In: Padgham, L., Parkes, D.C., Müller, J.P., Parsons, S. (eds.) AAMAS (2), pp. 1015–1022. IFAAMAS (2008)
11. Faltings, B.: Distributed Constraint Programming, ch. 20. Handbook of Constraint Programming, pp. 699–729. Elsevier (2006)
12. Foo, N.Y., Meyer, T., Brewka, G.: Lpod answer sets and nash equilibria. In: Maher, M.J. (ed.) ASIAN 2004. LNCS, vol. 3321, pp. 343–351. Springer, Heidelberg (2004)
13. Gottlob, G., Greco, G., Scarcello, F.: Pure nash equilibria: Hard and easy games. J. Artif. Intell. Res. (JAIR) 24, 357–406 (2005)
14. Harrenstein, P., van der Hoek, W., Meyer, J.J.C., Witteveen, C.: Boolean Games. In: van Benthem, J. (ed.) TARK. Morgan Kaufmann (2001)
15. Hotelling, H.: Stability in competition. Economic Journal, 41–57 (1929)
16. Jalaparti, V., Nguyen, G., Gupta, I., Caesar, M.: Cloud resource allocation games. Technical report, University of Illinois at Urbana-Champaign (2010), `http://hdl.handle.net/2142/17427`
17. Jiang, A.X., Leyton-Brown, K., Bhat, N.A.R.: Action-graph games. Games and Economic Behavior 71(1), 141–173 (2011)
18. Kearns, M.J., Littman, M.L., Singh, S.P.: Graphical models for game theory. In: Breese, J.S., Koller, D. (eds.) UAI, pp. 253–260. Morgan Kaufmann (2001)
19. McKelvey, R.D., McLennan, A.M., Turocy, T.L.: Gambit: Software tools for game theory (2014), `http://www.gambit-project.org`
20. Nash, J.: Non-cooperative games. Annals of Mathematics 54(2), 286–295 (1951)
21. Neumann, J.V., Morgenstern, O.: Theory of Games and Economic Behavior. Princeton University Press (1944), `http://jmvidal.cse.sc.edu/library/neumann44a.pdf`

22. Nguyen, T.V.A., Lallouet, A., Bordeaux, L.: Constraint games: Framework and local search solver. In: Grégoire, É., Mazure, B. (eds.) ICTAI, pp. 963–970. Special Track on SAT and CSP Technologies, Springer (2013)
23. Nudelman, E., Wortman, J., Shoham, Y., Leyton-Brown, K.: Run the gamut: A comprehensive approach to evaluating game-theoretic algorithms. In: AAMAS, pp. 880–887. IEEE Computer Society (2004), http://gamut.stanford.edu/, http://gamut.stanford.edu/
24. Osborne, M., Rubinstein, A.: A Course in Game Theory. The MIT Press (1994)
25. Rosen, J.B.: Existence and uniqueness of equilibrium points for concave n-person games. Econometrica 33(3), 520–534 (1965)
26. Rosenthal, R.W.: A class of games possessing pure-strategy nash equilibria. International Journal of Game Theory 2(1), 65–67 (1973)
27. The Choco Team: Choco: An Open Source Java Constraint Programming Library. Ecole des Mines de Nantes (2008-2014), http://www.emn.fr/z-info/choco-solver/
28. Turocy, T.L.: Personal communication (2013)
29. Vöcking, B.: Selfish load balancing, ch. 20. Algorithmic game theory, pp. 517–542. Cambridge University Press (2007)
30. Zhao, L., Müller, M.: Game-SAT: A preliminary report. In: Hoos, H., Mitchell, D. (eds.) SAT, pp. 357–362 (2004)

Encoding Linear Constraints into SAT

Ignasi Abío[1] and Peter J. Stuckey[1,2]

[1] NICTA Victoria Laboratory, Australia
[2] Department of Computing and Information Systems, The University of Melbourne, Australia

Abstract. Linear integer constraints are one of the most important constraints in combinatorial problems since they are commonly found in many practical applications. Typically, encoding linear constraints to SAT performs poorly in problems with these constraints in comparison to constraint programming (CP) or mixed integer programming (MIP) solvers. But some problems contain a mix of combinatoric constraints and linear constraints, where encoding to SAT is highly effective. In this paper we define new approaches to encoding linear constraints into SAT, by extending encoding methods for pseudo-Boolean constraints. Experimental results show that these methods are not only better than the state-of-the-art SAT encodings, but also improve on MIP and CP solvers on appropriate problems. Combining the new encoding with lazy decomposition, which during runtime only encodes constraints that are important to the solving process that occurs, gives a robust approach to many highly combinatorial problems involving linear constraints.

1 Introduction

In this paper we study linear integer (LI) constraints, that is, constraints of the form $a_1x_1 + \cdots + a_nx_n \# a_0$, where the a_i are integer given values, the x_i are finite-domain integer variables, and the relation operator $\#$ belongs to $\{<, >, \leqslant, \geqslant, =\}$. We will assume w.l.o.g that $\#$ is $\leqslant$, the a_i are positive and all the domains of the variables are $[0, d_i]$, since other cases can be reduced to this one.[1]

Linear integer constraints appear in many combinatorial problems such as scheduling, planning or software verification; they are also present in MaxSAT problems [8]; or are part of some MaxSAT techniques, as in *Fu & Malik algorithm* [17] (and some other algorithms based on it). Therefore, all approaches to combinatorial optimization have studied how to best handle them, including for MIP solvers, CP solvers [21], SMT solvers [15,11], and SAT solvers [27,6].

In this paper we examine how we can extend the state-of-the-art methods for SAT encoding of pseudo-Boolean (PB) constraints of the form $a_1x_1 + \cdots + a_nx_n \# a_0$ where x_i are Booleans, to the general linear integer case.

The method proposed here roughly consists in encoding the linear integers constraints into Reduced Ordered Multi-Decision Diagram (MDD for short), and

[1] Although replacing an equality by two inequalities substantially reduces propagation strength.

B. O'Sullivan (Ed.): CP 2014, LNCS 8656, pp. 75–91, 2014.

then decomposing the MDD to SAT. There are different reasons for choosing this approach: firstly, most state-of-the-art encoding methods define one auxiliary variable for every different possible value of the partial sum $s_i = a_1x_1 + a_2x_2 + \cdots + a_ix_i$. However, some values of the partial sums may be equivalent in the constraint. For instance, if a_j is even for every $j > i$, there is no difference between $s_i = a_0$ and $s_i = a_0 - 1$. With MDDs, due to the reduction process, we can identify these situations, and encode all these indistinguishable values with a single variable, producing a more compact encoding.

Secondly, BDDs are one of the best methods for encoding pseudo-Boolean constraints into SAT [3], and MDDs seem the natural tool to generalize the pseudo-Boolean encoding. Although the resulting encoding may be exponential; however, in real-world problems we have not found any exponential examples.

The goal of this encoding is not for use in arbitrary problems involving LI constraints. In fact, a specific linear integer (MIP) solver will probably outperform any SAT encoding in problems with more LI constraints than Boolean clauses.

Nevertheless, a fairly common kind of combinatorial problem mainly consist of Boolean variables and clauses, but also a few integer variables and LI constraints. Among these problems, an important class correspond to SAT problems with a linear integer objective function. In these cases, SAT solvers are the optimal tool for solving the problem, but a good encoding for the linear integer constraints is needed to make the optimization effective. Therefore, in these problems the decomposition presented here can make a significant difference.

Note, however, that decomposing the constraint may not always be the best option. In some cases the encoding might produce a large number of variables and clauses, transforming an easy problem for a CP solver into a huge SAT problem. In some other cases, nevertheless, the auxiliary variables may give an exponential reduction of the search space. Lazy decomposition [5,4] is a hybrid approach that has been successfully used to handle this issue for cardinality and pseudo-Boolean constraints. Here, we show that it also can be applied successfully on LI constraints.

The method proposed here uses the order encoding for representing the integer variables. In some cases, however, the domains of the integer variables are too large for order encoding. We also propose a new alternative method for encoding linear integer constraints with the logarithmic encoding.

The contributions of this paper are:

- A new encoding (MDD) for LI constraints using MDDs that outperforms the state-of-the-art encodings.
- An alternative encoding (BDD-Dec) for LI constraints for large constraints or variables with huge domains.
- An improved encoding of PB constraints which share coefficients, by converting these constraints to LI constraints.
- A rigorous and extensive experimental comparison of our methods with respect to other decompositions to SAT and other solvers. A total of 9 methods are compared, over approximately 3000 benchmarks, both industrial and crafted.

2 Preliminaries

2.1 SAT Solving

Let $\mathcal{X} = \{x_1, x_2, \ldots\}$ be a fixed set of propositional *variables*. If $x \in \mathcal{X}$ then x and $\neg x$ are *positive* and *negative literals*, respectively. The *negation* of a literal l, written $\neg l$, denotes $\neg x$ if l is x, and x if l is $\neg x$. A *clause* is a disjunction of literals $\neg x_1 \vee \cdots \vee \neg x_p \vee x_{p+1} \vee \cdots \vee x_n$, sometimes written as $x_1 \wedge \cdots \wedge x_p \rightarrow x_{p+1} \vee \cdots \vee x_n$. A *CNF formula* is a conjunction of clauses.

A (partial) *assignment* A is a set of literals such that $\{x, \neg x\} \nsubseteq A$ for any $x \in \mathcal{X}$, i.e., no contradictory literals appear. A literal l is *true* in A if $l \in A$, is *false* in A if $\neg l \in A$, and is *undefined* in A otherwise. True, false or undefined is the *polarity* of the literal l. A clause C is true in A if at least one of its literals is true in A. A formula F is true in A if all its clauses are true in A. In that case, A is a *model* of F. Systems that decide whether a formula F has any model are called SAT-solvers, and the main inference rule they implement is *unit propagation*: given a CNF F and an assignment A, find a clause in F such that all its literals are false in A except one, say l, which is undefined, add l to A and repeat the process until reaching a fix-point. See e.g. [24] for more details.

2.2 LCG and LD Solvers

Many modern CP solvers, so called Lazy Clause Generation (LCG) solvers, include the ability to explain their propagation and generate nogoods just as in SAT solvers. Propagation of LI constraints is well understood [21] and standard. And adding explanation for LI constraints is also well understood [16], although there are often a number of choices of explanation that result.

More recently, Lazy Decomposition (LD) solvers have been proposed. An LD solver is a LCG solver that, when one complex constraint propagator is very active (that is, is frequently asked to generate explanations), then the solver replaces the propagator by either partially or totally decomposing the constraint into SAT (see [5,4] for more details). The advantage of LD solvers is that the exposure of intermediate variables in the SAT encodings can substantially benefit search, but it avoids the up front cost of encoding all complex constraints, only those that are important in the solving process.

2.3 Order and Logarithmic Encoding

There are different methods for encoding integer variables into SAT (see for instance [28,18]). In this paper we use the order and the logarithmic encoding.

Let y be an integer variable with domain $[0, d]$. The *order encoding* [19,7] (sometimes called the *ladder* or *regular* encoding) introduces Boolean variables y^i for $0 \leqslant i < d$. A variable y^i is true iff $y \leqslant i$. The encoding also introduces the clauses $y^i \rightarrow y^{i+1}$ for $0 \leqslant i < d - 1$.

The *logarithmic encoding* introduces only $\log d$ variables y_b^i which codify the binary representation of the value of y, as $y = \sum_{i=0}^{\lfloor log(d) \rfloor} 2^i y_b^i$. It is a more compact encoding, but it usually gives poor propagation performance.

2.4 Multi Decision Diagrams

A directed acyclic graph is called an *ordered Multi Decision Diagram* if it satisfies the following properties:

- It has two terminal nodes, namely $\mathcal{T}$ (true) and $\mathcal{F}$ (false).
- Each non-terminal node is labeled by an integer variable $\{x_1, x_2, \cdots, x_n\}$. This variable is called *selector variable.*
- Every node labeled by x_i has the same number of outgoing edges, namely $d_i + 1$.
- If an edge connects a node with a selector variable x_i and a node with a selector variable x_j, then $j > i$.

The MDD is *quasi-reduced* if no isomorphic subgraphs exist. It is *reduced* if, moreover, no nodes with only one child exist. A *long edge* is an edge connecting two nodes with selector variables x_i and x_j such that $j > i+1$. In the following we only consider quasi-reduced ordered MDDs without long edges, and we just refer to them as MDDs for simplicity.

An MDD represents a function

$$f : \{0, 1, \ldots, d_1\} \times \{0, 1, \ldots, d_2\} \times \cdots \times \{0, 1, \ldots, d_n\} \to \{0, 1\}$$

in the obvious way. Moreover, given the variable ordering, there is only one MDD representing that function. We refer to [26] for further details about MDDs.

3 Linear Integer Constraints

In this paper we consider linear integer constraints of the form $a_1x_1 + \cdots + a_nx_n \leqslant a_0$, where the a_i are positive integer coefficients and the x_i are integer variables with domains $[0, d_i]$. Other LI constraints can be easily reduced to this one:

$$a_1x_1 + \cdots + a_nx_n = a_0 \quad \Longrightarrow \quad \begin{cases} a_1x_1 + \cdots + a_nx_n \leqslant a_0 \;\wedge \\ a_1x_1 + \cdots + a_nx_n \geqslant a_0 \end{cases}$$

$$a_1x_1 + \cdots + a_nx_n < a_0 \quad \Longrightarrow \quad a_1x_1 + \cdots + a_nx_n \leqslant a_0 - 1$$

$$a_1x_1 + \cdots + a_nx_n \geqslant a_0 \quad \Longrightarrow \quad -a_1x_1 + \cdots + -a_nx_n \leqslant -a_0$$

$$a_1x_1 + \cdots + a_nx_n > a_0 \quad \Longrightarrow \quad -a_1x_1 + \cdots + -a_nx_n \leqslant -a_0 - 1$$

$$\left.\begin{array}{l} a_1x_1 + \cdots + a_ix_i + \cdots \\ +a_nx_n \leqslant a_0 \\ \text{when } x_i \in [l, u], l \neq 0, a_i > 0 \end{array}\right\} \quad \Longrightarrow \quad \begin{cases} a_1x_1 + \cdots + a_ix'_i + \cdots \\ +a_nx_n \leqslant a_0 + a_i \times l \;\wedge \\ x'_i \in [0, u-l] \;\wedge\; x'_i = x_i - l \end{cases}$$

$$\left.\begin{array}{l} a_1x_1 + \cdots + a_ix_i + \cdots \\ +a_nx_n \leqslant a_0 \\ \text{when } a_i < 0 \text{ and } x_i \in [l, u] \end{array}\right\} \quad \Longrightarrow \quad \begin{cases} a_1x_1 + \cdots + -a_ix'_i + \cdots \\ +a_nx_n \leqslant a_0 - a_i \times u \;\wedge \\ x'_i \in [0, u-l] \;\wedge\; x'_i = u - x_i \end{cases}$$

$$\left.\begin{array}{l} x_i \in [l, u],\ l \neq 0 \;\wedge\; x'_i = x_i - l \\ \wedge\; x_i^j \equiv x_i \leqslant j \ \text{ for } l \leqslant j < u \end{array}\right\} \quad \Longrightarrow \quad x_i^{j-l} \equiv x'_i \leqslant j \ \text{ for } 0 \leqslant j < u - l$$

$$\left.\begin{array}{l} x_i \in [l,u] \ \wedge \ x'_i = u - x_i \\ \wedge \ x_i^j \equiv x_i \leqslant j \ \text{ for } l \leqslant j < u \end{array}\right\} \implies \neg x_i^{u-j-1} \equiv x'_i \leqslant j \ \text{ for } 0 \leqslant j < u - l$$

The goal of this paper is to find a SAT encoding for a given LI constraint. That is, given a LI constraint C, construct an equivalent formula F such that any model for F restricted to the variables of C is a model of C. Two extra properties are usually sought:

- *consistency checking by unit propagation* or simply *consistency*: whenever a partial assignment A cannot be extended to a model for C, unit propagation on F and A produces a contradiction (a literal l and its negation $\neg l$);
- *domain consistency* (again by unit propagation): given an assignment A that can be extended to a model of C, but such that $A \cup \{x\}$ cannot, unit propagation on F and A produces $\neg x$.

4 Construction of the MDD

In this section we describe an efficient method for building MDDs. Let us fix a LI constraint $a_1x_1 + \cdots + a_nx_n \leqslant a_0$ and a variable ordering $[x_1, x_2, \ldots, x_n]$. Before explaining the algorithm, we need a preliminary definition.

Let $\mathcal{M}$ be the MDD of the given LI constraint and let ν be a node of $\mathcal{M}$ with selector variable x_i. We define the *interval* of ν as the set of values α such that the MDD rooted at ν represents the LI constraint $a_ix_i + \cdots + a_nx_n \leqslant \alpha$. It is easy to see that this definition corresponds in fact to an interval.

Example 1. Figure 1 contains the MDD of $3x_1+2x_2+5x_3 \leqslant 15$, where $x_1 \in [0,4]$, $x_2 \in [0,2]$ and $x_3 \in [0,3]$. The root interval is $[15,15]$: this means that the root does not correspond to any constraint $3x_1 + 2x_2 + 5x_3 \leqslant \alpha$, apart from $\alpha = 15$. This means that this constraint is not equivalent to $3x_1 + 2x_2 + 5x_3 \leqslant 14$ or $3x_1+2x_2+5x_3 \leqslant 16$. However, the left node with selector variable x_2 has interval $[15,16]$. This means that $2x_2+5x_3 \leqslant 15$ and $2x_2+5x_3 \leqslant 16$ are both represented by the MDD rooted at that node. In particular, that means that $2x_2+5x_3 \leqslant 15$ and $2x_2 + 5x_3 \leqslant 16$ are two equivalent constraints. □

The next proposition shows how to compute the intervals of every node:

Proposition 1. *Let $\mathcal{M}$ be the MDD of a LI constraint $a_1x_1 + \cdots + a_nx_n \leqslant a_0$. Then, the following holds:*

- *The interval of the true node $\mathcal{T}$ is $[0, \infty)$.*
- *The interval of the false node $\mathcal{F}$ is $(-\infty, -1]$.*
- *Let ν be a node with selector variable x_i and children $\{\nu_0, \nu_1, \ldots, \nu_{d_i}\}$. Let $[\beta_j, \gamma_j]$ be the interval of ν_j. Then, the interval of ν is $[\beta, \gamma]$, with*

$$\beta = \max\{\beta_r + ra_i \mid 0 \leqslant r \leqslant d_i\}, \qquad \gamma = \min\{\gamma_r + ra_i \mid 0 \leqslant r \leqslant d_i\}.$$

The proof of this proposition is very similar to the Proposition 7 of [3].

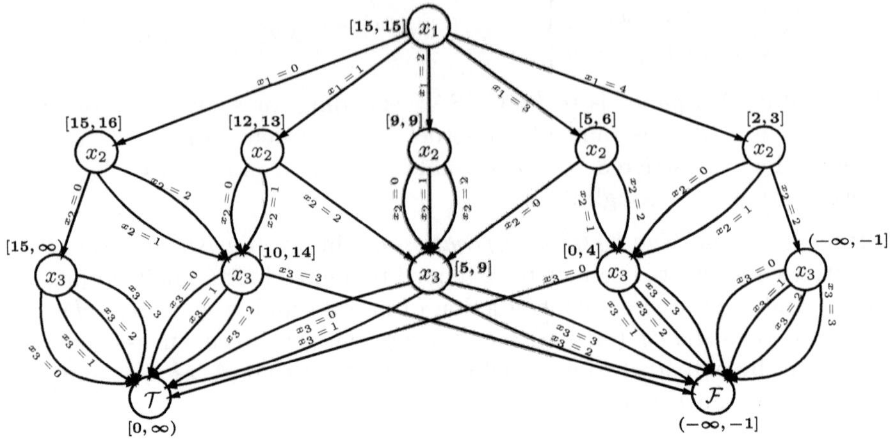

Fig. 1. MDD of $3x_1 + 2x_2 + 5x_3 \leqslant 15$

Example 2. Again, let us consider the constraint $3x_1 + 2x_2 + 5x_3 \leqslant 15$, whose MDD is represented at Figure 1. By the previous Proposition, $\mathcal{T}$ and $\mathcal{F}$ have, respectively, intervals $[0, \infty)$ and $(-\infty, -1]$. Applying again the same proposition, we can compute the intervals of the nodes having x_3 as selector variable. For instance, the interval from the left node is

$$[0, \infty) \cap [5, \infty) \cap [10, \infty) \cap [15, \infty) = [15, \infty),$$

and the interval from the node having selector variable x_3 in the middle is

$$[0, \infty) \cap [5, \infty) \cap (-\infty, 9] \cap (-\infty, 14] = [5, 9].$$

After computing all the intervals from the nodes with selector variable x_3, we can compute the intervals of the nodes with selector variables x_2 in the same way, and, after that, we can compute the interval of the root. □

The key point of the MDDCreate algorithm, detailed in Algorithm 1 and Algorithm 2, is to label each node of the MDD with its interval $[\beta, \gamma]$.

In the following, for every $i \in \{1, 2, \ldots, n+1\}$, we use a set L_i consisting of pairs $([\beta, \gamma], \mathcal{M})$, where $\mathcal{M}$ is the MDD of the constraint $a_i x_i + \cdots + a_n x_n \leqslant a_0'$ for every $a_0' \in [\beta, \gamma]$ (i.e., $[\beta, \gamma]$ is the interval of $\mathcal{M}$). All these sets are kept in a tuple $\mathcal{L} = (L_1, L_2, \ldots, L_{n+1})$.

Note that by definition of the MDD's intervals, if both $([\beta_1, \gamma_1], \mathcal{M}_1)$ and $([\beta_2, \gamma_2], \mathcal{M}_2)$ belong to L_i then either $[\beta_1, \gamma_1] = [\beta_2, \gamma_2]$ or $[\beta_1, \gamma_1] \cap [\beta_2, \gamma_2] = \emptyset$. Moreover, the first case holds if and only if $\mathcal{M}_1 = \mathcal{M}_2$. Therefore, L_i can be represented with a *binary search tree-like* data structure, where insertions and searches can be done in logarithmic time. The function **search**(K, L_i) searches whether there exists a pair $([\beta, \gamma], \mathcal{M}) \in L_i$ with $K \in [\beta, \gamma]$. Such a tuple is returned if it exists, otherwise an empty interval is returned in the first component of the pair. Similarly, we also use function **insert**$(([\beta, \gamma], \mathcal{M}), L_i)$ for

Algorithm 1. Procedure MDDCreate

Require: Constraint $C : a_1x_1 + \cdots + a_nx_n \leqslant a_0$
Ensure: returns $\mathcal{M}$ the MDD of C.
for all i such that $1 \leq i \leq n$ **do**
  $L_i \leftarrow \emptyset$.
end for
$L_{n+1} \leftarrow \Big\{ \big((-\infty, -1], \mathcal{F}\big),\ \big([0, \infty), \mathcal{T}\big) \Big\}$.
$\mathcal{L} \leftarrow (L_1, \ldots, L_{n+1})$.
$([\beta, \gamma], \mathcal{M}) \leftarrow$ **MDDConstruction**$(1, a_1x_1 + \cdots + a_nx_n \leqslant a_0, \mathcal{L})$.
return $\mathcal{M}$.

Algorithm 2. Procedure MDDConstruction

Require: $i \in \{1, 2, \ldots, n+1\}$, constraint $C : a_ix_i + \cdots + a_nx_n \leqslant a'_0$ and tuple $\mathcal{L}$
Ensure: returns $[\beta, \gamma]$ interval of C and $\mathcal{M}$ its MDD
$([\beta, \gamma], \mathcal{M}) \leftarrow$ **search**(a'_0, L_i).
if $[\beta, \gamma] \neq \emptyset$ **then**
  return $([\beta, \gamma], \mathcal{M})$.
else
  for all j such that $0 \leq j \leq d_i$ **do**
    $([\beta_j, \gamma_j], \mathcal{M}_j) \leftarrow$ **MDDConstruction**$(i+1, a_{i+1}x_{i+1} + \cdots + a_nx_n \leqslant a'_0 - ja_i, \mathcal{L})$.
  end for
  $\mathcal{M} \leftarrow$ **mdd**$(x_i, [\mathcal{M}_0, \ldots, \mathcal{M}_{d_i}])$.
  $[\beta, \gamma] \leftarrow [\beta_0, \gamma_0] \cap [\beta_1 + a_1, \gamma_1 + a_1] \cap \cdots \cap [\beta_{d_i} + d_ia_i, \gamma_{d_i} + d_ia_1]$.
  insert$(([\beta, \gamma], \mathcal{M}), L_i)$.
  return $([\beta, \gamma], \mathcal{M})$.
end if

insertions. The size of the MDD in the worst case is $O(na_0)$ (exponential in the size of the rhs coefficient) and algorithm complexity is $O(nw \log w)$ where w is the maximum width of the MDD ($w \leq a_0$).

5 Encoding MDDs into CNF

In this section we generalize the encoding for monotonic BDDs described in [3] to monotonic MDDs. The encoding assumes that the selector variables are encoded with the order encoding.

Let $\mathcal{M}$ be an MDD with the variable ordering $[x_1, \ldots, x_n]$. Let $[0, d_i]$ be the domain of the i-th variable, and let $\{x_i^0, \ldots, x_i^{d_i-1}\}$ be the variables of the order encoding of x_i (i.e., x_i^j is true iff $x_i \leqslant j$). Let μ be the root of $\mathcal{M}$, and let $\mathcal{T}$ and $\mathcal{F}$ be respectively the true and false terminal nodes. In the following, given a non-terminal node ν of $\mathcal{M}$, we define SelVar(ν) as the selector variable of ν, and Child(ν, j) as the j-th child of ν.

The encoding introduces the variables $\{z_\nu \mid \nu \in \mathcal{M}\}$; and the clauses

$$\{z_\mu,\ z_\mathcal{T},\ \neg z_\mathcal{F}\} \cup \Big\{ \neg z_\nu \vee x_i^{j-1} \vee z_{\nu'} \mid \nu \in \mathcal{M} \setminus \{\mathcal{T},\mathcal{F}\}, \\ \mathrm{SelVar}(\nu) = x_i,\ 0 \leqslant j \leqslant d_i,\ \nu' = \mathrm{Child}(\nu, j)\Big\},$$

where x_i^{-1} is a dummy false variable.

Notice that this encoding coincides with the BDD encoding of [3] if the MDD is a BDD.

Theorem 2. *Unit propagation on the encoding described above enforces domain consistency (and hence also consistency).* □

The proof is very similar to the BDD case described in [3].

Example 3. Let us consider the MDD represented in Figure 1. The encoding introduces the variables $z_1, z_2, \ldots, z_{11}, z_\mathcal{T}, z_\mathcal{F}$, one for each node of the MDD; and the following clauses:

z_1,	$z_\mathcal{T}$,	$\neg z_\mathcal{F}$,	$\neg z_1 \vee z_2$,
$\neg z_1 \vee x_1 \leqslant 0 \vee z_3$,	$\neg z_1 \vee x_1 \leqslant 1 \vee z_4$,	$\neg z_1 \vee x_1 \leqslant 2 \vee z_5$,	$\neg z_1 \vee x_1 \leqslant 3 \vee z_6$,
$\neg z_2 \vee z_7$,	$\neg z_2 \vee x_2 \leqslant 0 \vee z_8$,	$\neg z_2 \vee x_2 \leqslant 1 \vee z_8$,	$\neg z_3 \vee z_8$,
$\neg z_3 \vee x_2 \leqslant 0 \vee z_8$,	$\neg z_3 \vee x_2 \leqslant 1 \vee z_9$,	$\neg z_4 \vee z_9$,	$\neg z_4 \vee x_2 \leqslant 0 \vee z_9$,
$\neg z_4 \vee x_2 \leqslant 1 \vee z_9$,	$\neg z_5 \vee z_9$,	$\neg z_5 \vee x_2 \leqslant 0 \vee z_{10}$,	$\neg z_5 \vee x_2 \leqslant 1 \vee z_{10}$,
$\neg z_6 \vee z_{10}$,	$\neg z_6 \vee x_2 \leqslant 0 \vee z_{10}$,	$\neg z_6 \vee x_2 \leqslant 1 \vee z_{11}$,	$\neg z_7 \vee z_\mathcal{T}$,
$\neg z_7 \vee x_3 \leqslant 0 \vee z_\mathcal{T}$,	$\neg z_7 \vee x_3 \leqslant 1 \vee z_\mathcal{T}$,	$\neg z_7 \vee x_3 \leqslant 2 \vee z_\mathcal{T}$,	$\neg z_8 \vee z_\mathcal{T}$,
$\neg z_8 \vee x_3 \leqslant 0 \vee z_\mathcal{T}$,	$\neg z_8 \vee x_3 \leqslant 1 \vee z_\mathcal{T}$,	$\neg z_8 \vee x_3 \leqslant 2 \vee z_\mathcal{F}$,	$\neg z_9 \vee z_\mathcal{T}$,
$\neg z_9 \vee x_3 \leqslant 0 \vee z_\mathcal{T}$,	$\neg z_9 \vee x_3 \leqslant 1 \vee z_\mathcal{F}$,	$\neg z_9 \vee x_3 \leqslant 2 \vee z_\mathcal{F}$,	$\neg z_{10} \vee z_\mathcal{T}$,
$\neg z_{10} \vee x_3 \leqslant 0 \vee z_\mathcal{F}$,	$\neg z_{10} \vee x_3 \leqslant 1 \vee z_\mathcal{F}$,	$\neg z_{10} \vee x_3 \leqslant 2 \vee z_\mathcal{F}$,	$\neg z_{11} \vee z_\mathcal{F}$,
$\neg z_{11} \vee x_3 \leqslant 0 \vee z_\mathcal{F}$,	$\neg z_{11} \vee x_3 \leqslant 1 \vee z_\mathcal{F}$,	$\neg z_{11} \vee x_3 \leqslant 2 \vee z_\mathcal{F}$.	

Notice that some clauses are redundant. This issue is handled in Section 7.2. □

6 Optimization Problems

In this section we describe how to deal with combinatorial problem where we minimize a linear integer optimization function. A similar idea is used in [14], where the authors use BDDs for encoding problems with pseudo-Boolean objectives. Combinatorial optimization problems can be efficiently solved with a branch-and-bound strategy. In this way, all the lemmas learned in the previous steps are reused for finding the next solutions or proving the optimality. For implementing branch-and-bound, we need to be able to create a decomposition of the constraint $a_1x_1 + \cdots + a_nx_n \leqslant a_0'$ from the decomposition of $a_1x_1 + \cdots + a_nx_n \leqslant a_0$ where $a_0' < a_0$.

This is easy for cardinality constraints, since, when we have encoded a constraint $x_1+\cdots+x_n \leqslant a_0$ with a sorting network, we can encode $x_1+\cdots+x_n \leqslant a_0'$ by adding a single clause (see [9]).

Algorithm 3. MDD Construction: Optimization version

Require: Constraint $C : a_1x_1 + \cdots + a_nx_n \leqslant a_0'$ and tuple $\mathcal{L}$.
Ensure: returns $\mathcal{M}$ the MDD of C.
1: $([\beta,\gamma],\mathcal{M}) \leftarrow$ **MDDConstruction**$(1, a_1x_1 + \cdots + a_nx_n \leqslant a_0', \mathcal{L})$.
2: **return** $\mathcal{M}$.

In order to reuse the previous encodings for the MDD encoding of an LI constraint, we have to save the tuple $\mathcal{L}$ used in Algorithm 1. When a new solution of cost $a_0' + 1$ is found, Algorithm 3 is called.

Notice that the encoding creates at most one variable for every element of $L_i \in \mathcal{L}$, $1 \leqslant i \leqslant n$. Therefore, after finding optimality, the encoding has generated at most na_0 variables in total, where a_0 is the cost of the first solution found. The number of clauses generated can be bounded by na_0d, where $d = \max\{d_i\}$.

7 Improvements

In this section we describe some improvements of the method. The first improvement is to reorder the constraint such that $a_1 \geqslant a_2 \geqslant \cdots \geqslant a_n$. The MDD obtained in this way is usually smaller.

7.1 Grouping Identical Coefficients

Let us fix the LI constraint $C : a_1x_1 + \cdots + a_nx_n \leqslant a_0$. Assume that some coefficients are equal; for simplicity, let us assume $a_1 = a_2 = \cdots = a_r$. In this case, we can define the integer variable $s = x_1 + \cdots + x_r$ and decompose the constraint $C' : a_1s + a_{r+1}x_{r+1} + \cdots + a_nx_n \leqslant a_0$ instead of C. The domain of s is $[0, d_s]$ with $d_s = \min\{a_0/a_1, d_1 + \cdots + d_r\}$.

Notice that we do not need to encode the constraint $s = x_1 + \cdots + x_r$ defining the integer variables s, instead we can encode $c \equiv s \geqslant x_1 + \cdots + x_r$ since we are only interested in lower bounds. The encoding of c can be done with cardinality networks [2], which usually gives a more compact encoding than the MDD of c.

In industrial problems where constraints are not randomly generated, the coefficients have some meaning. It may be likely, that a large LI constraint has only a few different coefficients. In this case this technique can be very effective.

7.2 Removing Subsumed Clauses

The encoding explained at Section 5 can easily be improved by removing some unnecessary clauses. We apply the following rule when producing the encoding:

Given a non-terminal node ν with SelVar$(\nu) = x_i$, if Child$(\nu, j) = $ Child$(\nu, j-1)$, then the clause $\neg z_z\nu \vee x_i^{j-1} \vee z_{\nu'}$ is subsumed by the clause $\neg z_\nu \vee x_i^{j-2} \vee z_{\nu'}$; therefore, we can remove it.

Additionally, we also improve the encoding by reinstating long edges (since the dummy nodes used to eliminate long edges do not provide any information); that is, we encode the reduced MDD instead of the quasi-reduced MDD.

7.3 Solution Phase Saving

In decision problems, last phase saving described in [25] has proven to be a very effective strategy. According to this scheme, when the SAT solver makes a decision, the variable is chosen with the same polarity as in the last assignment.

However, in optimization problems this is not the best option. As seen in [1], a better strategy is to take the polarity that minimizes the objective function in the variables which directly appear in the objective function, or the polarity in the last solution in the other variables. That method, called *solution phase saving*, emulates a local search: after finding a solution, the method explores the neighbourhood of the solution in order to find a better solution nearby.

7.4 Lazy Decomposition

Lazy decomposition [5,4] has proved to be very successful for handling cardinality and pseudo-Boolean constraints. Lazy decomposition for LI constraints implements each LI constraint as a propagator initially, and later when we see that a constraint is generating many explanations we replace the propagator fully or partially by a SAT decomposition. We use the approach of [4] which fully replaces a propagator. Our strategy for when to decompose an LI constraint is: when the constraint has generated more explanations than half its encoding size, or generated more than 50,000 explanations; except that we never encode LI constraints whose encoding size is ≥ 50 million clauses.

8 Related Work and Extensions

The simplest decomposition of linear integer constraints to SAT uses binary adders (Adder) [29]. The encoding is very compact, but it has a poor performance in practice since information does not propagate effectively through the encoding.

Decision diagrams methods have been widely used to handle LI constraints. The best current method [10] (BDD) uses a logarithmic encoding of the coefficients to create a BDD of the constraint, sorting the variables in a clever way. The encoding size is reduced to $O(n \log d \sum a_i)$, which is polynomial in the domain size but exponential in the coefficient size. We can improve this encoding if we also decompose the coefficients as is done in [3]: in this way, the encoding size is $O(n^2 \log d \log a_m)$, where a_m is the largest coefficient. We call this BDD-Dec.

Example 4. Consider the LI constraint $3x_1+2x_2+5x_3 \leqslant 15$ from Example 1. After encoding the integer variables with the logarithmic encoding, the constraint becomes the pseudo-Boolean $3x^0_{b,1} + 6x^1_{b,1} + 12x^2_{b,1} + 2x^0_{b,2} + 4x^1_{b,2} + 5x^0_{b,3} + 10x^1_{b,3} \leqslant 15$. Bartzis and Bultan [10] construct the BDD of the pseudo-Boolean $2x^0_{b,2} + 3x^0_{b,1} + 4x^1_{b,2} + 5x^0_{b,3} + 6x^1_{b,1} + 10x^1_{b,3} + 12x^2_{b,1} \leqslant 15$. Our method decomposes the coefficients (i.e., considers $x^0_{b,1} + 2x^0_{b,1}$ instead of $3x^0_{b,1}$) and builds the resulting BDD; so we encode the constraint $x^0_{b,1} + x^0_{b,3} + 2x^0_{b,2} + 2x^0_{b,1} + 2x^1_{b,1} + 2x^1_{b,3} + 4x^1_{b,2} + 4x^0_{b,3} + 4x^1_{b,1} + 4x^2_{b,1} + 8x^1_{b,3} + 8x^2_{b,1} \leqslant 15$. □

Formally, the BDD-Dec method encodes LI constraint $a_1x_1 + \cdots + a_nx_n \leqslant a_0$ with $x_i \in [0, d_i], 1 \leq i \leq n$ by first creating the PB constraint

$$\sum_{i=1}^{n} \sum_{j \in 0..\lfloor \log_2 d_i \rfloor, (d_i \div 2^j) \bmod 2 = 1} \sum_{k \in 0..\lfloor \log_2 a_i \rfloor, (a_i \div 2^k) \bmod 2 = 1} 2^{j+k} \times x_{b,i}^j \leq a_0$$

over the logarithmic encoding variables x_b and encoding this using the state-of-the-art encoding for PB constraints given in [3].

Note however, logarithmic encoding of integers, while compact, is usually a bad option since it significantly hampers propagation of information, leading to poor solving performance. Neither BDD or BDD-Dec enforce consistency.

The most similar encoding to the approach we define is the *support encoding* (Support) [27,6]. While the encodings both effectively define auxiliary variables for the values of the partial sums $s_i = a_1x_1 + \cdots + a_ix_i$, the support encoding fails to identify which values of these partial sums are indistinguishable in the constraint. The result is to create an encoding equivalent to a non-reduced ordered MDD. If the MDD cannot be reduced further (for instance, if all the coefficients are 1), the two encodings would be identical (ignoring the further improvement discussed above). In general, however, the support encoding generates redundant variables and clauses. Another important improvement in our encoding is to group identical coefficients (see Section 7.1).

9 Experimental Results

In this section we compare our encoding with other LI constraints encodings and specific LI solvers. Unfortunately, we have not found in the literature a rigorous comparison of the different approaches for solving SAT+MIP problems. Here, we consider state-of-the-art methods from different areas and a huge set of benchmarks (about 2,900) coming from different industrial and crafted families.

We do not expect to be the best method in all the families of the problems. The main goals of this section are to:

- Detect the type of problems where it is worthwhile to encode an LI constraint instead of using a specific solver.
- Decide, in these problems, which encoding is better.
- Evaluate the lazy decomposition approach with different encodings.

All experiments were performed in a 2x2GHz Intel Quad Core Xeon E5405, with 2x6MB of Cache and 16 GB of RAM. All the benchmarks we used can be found at `www.cs.mu.oz.au/~pjs/encodeli/` in MiniZinc [23] format with scripts to generate CNF format.

We compare our new encodings MDD and BDD-Dec with those from the literature Adder, BDD and Support. We also consider the Gurobi mixed integer programming solver [20] (Gurobi) and the Barcelogic SMT solver [13] (LCG). We also consider the use of lazy decomposition [4] together with the two domain-consistent encoding approaches: using the MDD decomposition explained here (LD-MDD), and using the support encoding (LD-Sup).

Table 1. Multiple knapsack solving average time

	Different values of n						Different values of $a_{\max}$						Different values of d					
	5	10	20	40	80	160	1	2	4	8	16	32	1	2	4	10	25	100
Adder	0.05	9.55	186	276	296	300	57.3	60.4	74.9	114	105	117	0.02	0.15	1.84	32.7	80.3	215
BDD	0.04	7.11	185	272	298	300	21.1	39.2	59.1	111	110	129	**0.01**	0.06	0.58	26.8	77.6	220
BDD-Dec	0.12	4.73	163	269	295	300	10.5	25.6	47.7	90.9	84.8	82.5	**0.01**	0.13	0.46	18.6	56.3	202
MDD	0.05	6.45	175	268	290	300	51.8	57.2	62.9	103	107	119	**0.01**	0.03	0.17	18.9	80.6	258
Support	0.12	16.0	197	278	300	300	53.9	78.6	90.3	142	133	145	0.02	0.07	0.57	32.8	108	272
LD-MDD	**0.01**	3.23	165	264	287	300	44.3	48.2	59.4	90.0	91.5	91.1	0.02	0.01	0.09	16.4	63.1	244
LD-Sup	**0.01**	8.44	179	270	288	300	50.3	68.7	76.6	115	108	110	0.02	0.01	0.19	21.9	82.0	254
LCG	**0.01**	4.87	173	265	288	300	117	97.2	88.0	118	94.0	70.6	0.02	0.01	0.13	22.9	75.3	242
Gurobi	**0.01**	**0.09**	**0.02**	**0.03**	**0.02**	**0.03**	**0.01**	**0.01**	**0.01**	**0.01**	**0.02**	**0.02**	**0.01**	**0.02**	**0.01**	**0.01**	**0.02**	**0.01**

The Barcelogic SAT solver was used for all the SAT-based methods; this ensured that the lazy decomposition approaches were implemented using the same solver. The solution phase saving policy (see Section 7.3) was used in all SAT-based methods. Gurobi used its default settings.

9.1 Multiple Knapsack

First we consider the classic multiple knapsack problem.

$$\begin{array}{llllll}\text{Max} & a_1^0 x_1 & + a_2^0 x_2 & + \cdots & + a_n^0 x_n & \qquad \text{such that} \\ & a_1^1 x_1 & + a_2^1 x_2 & + \cdots & + a_n^1 x_n & \leqslant a_0^1 \\ & \ldots & & & & \\ & a_1^m x_1 & + a_2^m x_2 & + \cdots & + a_n^m x_n & \leqslant a_0^m, \end{array}$$

where x_i are integer variables with domain $[0, d]$ and the coefficients belong to $[0, a_{\max}]$.

Since it only consists of linear integer constraints it is ideal for MIP solvers. We consider this problem since it is easy to modify the parameters of the constraints, and, therefore, we can easily compare the encodings in different situations. More precisely, we have considered different constraint sizes, coefficient sizes and domain sizes. In these problems, n is the number of variables, $m = 20$ is the number of LI constraints, $d+1$ is the domain size of the variables; and $a_{\max}$ is the bound of the coefficients.

Table 1 contains the results on these benchmarks. For each parameter configuration, 100 benchmarks are considered and the average time for solving them is reported. Timeout are considered as 300s response in the average computation. In columns 2-7 $m = 20$, $d = 20$, $a_{\max} = 10$ and different values of n are taken. Columns 8-13 consider different values of $a_{\max}$, with $m = 20$, $n = 15$ and $d = 20$. In columns 14-19 $n = 15$, $m = 20$ and $a_{\max} = 10$, with different values of d. For each group of problems, the best encoding is underlined and the best method is bolded.

As expected Gurobi is by far the best method. SAT-based methods do not compete, however, we can effectively compare the encodings in different situations.

Table 2. Average quality from 600 RCPSP benchmarks

	15s	60s	300s	900s	3600s
Adder	0.905	0.963	<u>0.986</u>	<u>0.992</u>	<u>0.996</u>
BDD	0.784	0.859	0.928	0.957	0.977
BDD-Dec	<u>**0.929**</u>	<u>0.967</u>	0.985	0.99	0.994
MDD	0.727	0.75	0.858	0.889	0.899
Support	0.727	0.774	0.861	0.872	0.876
LD-MDD	0.918	**0.982**	0.992	0.994	0.996
LD-Sup	0.918	0.98	0.991	0.993	0.994
LCG	0.92	0.981	**0.993**	**0.995**	**0.997**
Gurobi	0.598	0.618	0.647	0.671	0.721

In general, BDD-Dec and MDD are the best encodings, MDD is specially efficient if the domains are small and BDD-Dec in large ones. Also, notice that lazy decomposition performs, in general, better than both the decomposition and the propagator approaches.

9.2 RCPSP

Resource-constrained project scheduling problem [12] (RCPSP) is possibly the most studied scheduling problem. It consists of tasks consuming one or more resources, precedences between some tasks, and resources. Here we consider the case of non-preemptive tasks and renewable resources with a constant resource capacity over the planning horizon. A solution is a schedule of all tasks so that all precedences and resource constraints are satisfied.

Usually, the objective of RCPSP is to find a solution minimizing the makespan. Here, however, the objective is to minimize a weighted sum of start times, i.e., minimize $\sum w_i s_i$, where w_i is the weight of the i-th task and s_i is its starting time. These weights represent the importance of the tasks: usually, a company not only needs to finish all the tasks in the minimum time, but also wants to give more importance to some of them. For the examples considered here, only ten tasks have a non-zero weight, but terminal tasks (this is, tasks with no successors with respect the precedence constraints) are never given a zero weight. Here we have considered the 600 RCPSP problems with 120 tasks (ie, the largest ones) from PSPlib [22].

The results are summarized in Table 2. Columns contain the *quality* average after X seconds. Quality is computed by dividing the cost of the best known solution by the cost of the current solution of the method; therefore, quality $= 0$ if no solution has been found, and quality $= 1$ when the best solution has been found. Again, the best method is bolded and the best encoding is underlined.

Since in these benchmarks, the variables' domains are very large (frequently $d > 200$); the logarithmic encodings Adder and BDD-Dec are the best encodings. MDD and Support have a similar performance, they are far from the best methods. However, the best method is LCG. Both lazy decomposition methods perform almost identically to LCG. It is clear (and well known) that MIP is not competitive on RCPSP problems.

Table 3. Average quality from 320 graph coloring benchmarks

	15s	60s	300s	900s	3600s
Adder	0.421	0.468	0.511	0.527	0.546
BDD	0.404	0.444	0.483	0.497	0.513
BDD-Dec	0.417	0.462	0.499	0.512	0.53
MDD	0.615	**0.624**	**0.644**	**0.651**	**0.657**
Support	0.605	0.615	0.636	0.641	0.648
LD-MDD	0.616	0.621	0.64	0.642	0.648
LD-Sup	0.613	0.618	0.635	0.639	0.643
LCG	**0.617**	0.623	0.64	0.643	0.646
Gurobi	0.443	0.45	0.452	0.453	0.454

9.3 Graph Coloring

The classical graph coloring problem consists in, given a graph, assign to each node a color $\{0, 1, \ldots, c-1\}$ such that two nodes connected by an edge have different colors. Usually, the problem consists in finding a solution that minimizes the number of colors (i.e., c). In this section we have considered a variant of this problem. Let us consider a graph that can be colored with c colors: For each node ν of the graph, let us define an integer value a_ν. Now, we want to color the graph with c colors $\{0, 1, \ldots, c-1\}$ minimizing the function $\sum a_\nu x_\nu$, where x_ν is the color of the node ν.

We have considered the 80 graph coloring instances from `http://mat.gsia.cmu.edu/COLOR08/` that have less than 500 nodes. For each graph problem, we have considered 4 different benchmarks: in the i-th one, $1 \leqslant a_\nu \leqslant 3i-2$ for $i = 1, 2, 3, 4$. Results are presented on Table 3 similarly to the previous section.

The best encoding in this problem is clearly MDD. The best methods are LCG and LD-MDD and MDD. Gurobi and logarithmic methods are not a good option in these problems.

9.4 Sport Leagues Scheduling

The last experiment considers scheduling a double round-robin sports league of N teams. All teams meet each other once in the first $N-1$ weeks and again in the second $N-1$ weeks, with exactly one match per team each week. A given pair of teams must play at the home of one team in one half, and at the home of the other in the other half, and such matches must be spaced at least a certain minimal number of weeks apart. Additional constraints include, e.g., that no team ever plays at home (or away) three times in a row, other (public order, sportive, TV revenues) constraints, blocking given matches on given days, etc.

Additionally, the different teams can propose a set of constraints with some importance (low, medium or high). It is desired not only to maximize the number of these constraints satisfied, but also to assure that at least some of the constraints of every team are satisfied. More information can be found at [1].

Table 4. Average quality from 200 sport scheduling league benchmarks

	15s	60s	300s	900s	3600s
Adder	0	0	0.031	0.108	0.167
BDD	0.038	0.061	0.26	0.397	0.537
BDD-Dec	0.037	0.069	0.267	0.441	0.582
MDD	0.034	0.073	**0.292**	**0.46**	0.583
Support	0.035	0.07	0.254	0.404	0.545
LD-MDD	**0.039**	**0.077**	0.277	0.443	**0.592**
LD-Sup	0.035	0.066	0.269	0.417	0.555
LCG	0.023	0.063	0.159	0.287	0.421
Gurobi	0	0	0	0	0

Low-importance constraints are given a weight of 1; medium-importance, 5, and high-importance, 10. For every constraint proposed by a team i, a new Boolean variable $x_{i,j}$ is created. This variable is set to true if the constraint is violated. For every team, a pseudo-Boolean constraint $\sum_j w_{i,j}x_{i,j} \leqslant K_i$ is imposed. The objective function to minimize is $\sum_i \sum_j w_{i,j}x_{i,j}$. The data is based on real-life instances.

Even though this problem only has pseudo-Boolean constraints, linear integer constraints arise from grouping identical coefficients. We have considered 10 different problems with 20 random seeds. In all the problems, the optimal value was found around 30. The results are shown in Table 4.

For sports league scheduling problems MDD is clearly the best encoding, followed by BDD-Dec. MDD and LD-MDD are the best methods. Gurobi is unable to handle these problems well (at least with the best model we could devise).

The BDD encoding for these problems is equivalent to using the BDD encoding of [3] for the original pseudo-Boolean constraints. Comparing BDD and MDD illustrates that by using LI constraint encoding we can improve one of the best known approach for PB constraints, in cases where the PB shares coefficients.

10 Conclusion

We have introduced a new domain-consistent encoding (MDD) for linear integer constraints. For small and medium-sized domains, this decomposition substantially improves the current state-of-the-art SAT encodings for LI constraints. It uniformly beats the only other domain consistent encoding (Support) as execution time increases. Combining this encoding with lazy decomposition, we create a hybrid method for LI constraints which is robust across the benchmark suite and rarely substantially bettered by any encoding or propagation method.

We have also introduced a new method (BDD-Dec) for encoding LI constraints based on the logarithmic decomposition of domains and coefficients, substantially improving on the previous state-of-the-art logarithmic method (BDD). This provides a robust alternative to domain-consistent methods in problems with large domains.

As future work, we want to combine lazy decomposition with logarithmic methods for large-domain problems. We are also designing a lazy decomposition solver which dynamically selects which type of encoding apply to every constraint to decompose.

Acknowledgments. NICTA is funded by the Australian Government through the Department of Communications and the Australian Research Council through the ICT Centre of Excellence Program.

References

1. Abío, I.: Solving hard industrial combinatorial problems with SAT. Ph.D. thesis, Technical University of Catalonia, UPC (2013)
2. Abío, I., Nieuwenhuis, R., Oliveras, A., Rodríguez-Carbonell, E.: A Parametric Approach for Smaller and Better Encodings of Cardinality Constraints. In: Schulte, C. (ed.) CP 2013. LNCS, vol. 8124, pp. 80–96. Springer, Heidelberg (2013)
3. Abío, I., Nieuwenhuis, R., Oliveras, A., Rodríguez-Carbonell, E., Mayer-Eichberger, V.: A New Look at BDDs for Pseudo-Boolean Constraints. Journal of Artificial Intelligence Research (JAIR) 45(1), 443–480 (2012)
4. Abío, I., Nieuwenhuis, R., Oliveras, A., Rodríguez-Carbonell, E., Stuckey, P.J.: To Encode or to Propagate? The Best Choice for Each Constraint in SAT. In: Schulte, C. (ed.) CP 2013. LNCS, vol. 8124, pp. 97–106. Springer, Heidelberg (2013)
5. Abío, I., Stuckey, P.J.: Conflict-Directed Lazy Decomposition. In: Milano, M. (ed.) CP 2012. LNCS, vol. 7514, pp. 70–85. Springer, Heidelberg (2012)
6. Ansótegui, C., Bofill, M., Manyà, F., Villaret, M.: Extending Multiple-Valued Clausal Forms with Linear Integer Arithmetic. In: ISMVL, pp. 230–235. IEEE (2011)
7. Ansótegui, C., Manyà, F.: Mapping problems with finite-domain variables to problems with boolean variables. In: Hoos, H.H., Mitchell, D.G. (eds.) SAT 2004. LNCS, vol. 3542, pp. 1–15. Springer, Heidelberg (2005)
8. Argelich, J., Manyà, F.: Exact Max-SAT solvers for over-constrained problems. Journal of Heuristics 12(4-5), 375–392 (2006)
9. Asín, R., Nieuwenhuis, R., Oliveras, A., Rodríguez-Carbonell, E.: Cardinality Networks: a theoretical and empirical study. Constraints 16(2), 195–221 (2011)
10. Bartzis, C., Bultan, T.: Efficient BDDs for bounded arithmetic constraints. International Journal on Software Tools for Technology Transfer 8(1), 26–36 (2006)
11. Berezin, S., Ganesh, V., Dill, D.L.: An Online Proof-Producing Decision Procedure for Mixed-Integer Linear Arithmetic. In: Garavel, H., Hatcliff, J. (eds.) TACAS 2003. LNCS, vol. 2619, pp. 521–536. Springer, Heidelberg (2003)
12. Blazewicz, J., Lenstra, J., Kan, A.: Scheduling subject to resource constraints: classification and complexity. Discrete Applied Mathematics 5(1), 11–24 (1983)
13. Bofill, M., Nieuwenhuis, R., Oliveras, A., Rodríguez-Carbonell, E., Rubio, A.: The Barcelogic SMT Solver. In: Gupta, A., Malik, S. (eds.) CAV 2008. LNCS, vol. 5123, pp. 294–298. Springer, Heidelberg (2008)
14. Bofill, M., Palahí, M., Suy, J., Villaret, M.: Boosting Weighted CSP Resolution with Shared BDDs. In: 12th International Workshop on Constraint Modelling and Reformulation (ModRef 2013). pp. 57–73 (2013)

15. Dutertre, B., de Moura, L.: A Fast Linear-Arithmetic Solver for DPLL(T). In: Ball, T., Jones, R.B. (eds.) CAV 2006. LNCS, vol. 4144, pp. 81–94. Springer, Heidelberg (2006)
16. Feydy, T., Stuckey, P.J.: Lazy clause generation reengineered. In: Gent, I.P. (ed.) CP 2009. LNCS, vol. 5732, pp. 352–366. Springer, Heidelberg (2009)
17. Fu, Z., Malik, S.: Solving the minimum-cost satisfiability problem using SAT based branch-and-bound search. In: Proceedings of the 2006 IEEE/ACM International Conference on Computer-aided Design, ICCAD 2006, pp. 852–859. ACM, New York (2006)
18. Gent, I.P.: Arc consistency in SAT. In: Proceedings of ECAI 2002, pp. 121–125. IOS Press (2002)
19. Gent, I.P., Nightingale, P.: A new encoding of AllDifferent into SAT. In: 3rd International Workshop on Modelling and reformulating Constraint Satisfaction Problems (CP 2004), pp. 95–110 (2004)
20. Gurobi Optimization, Inc. Gurobi optimizer reference manual (2013), `http://www.gurobi.com`
21. Harvey, W., Stuckey, P.: Improving linear constraint propagation by changing constraint representation. Constraints 8(2), 173–207 (2003)
22. Kolisch, R., Sprecher, A.: PSPLIB – A project scheduling problem library (1996), `http://129.187.106.231/psplib/`
23. Nethercote, N., Stuckey, P.J., Becket, R., Brand, S., Duck, G.J., Tack, G.R.: MiniZinc: Towards a standard CP modelling language. In: Bessière, C. (ed.) CP 2007. LNCS, vol. 4741, pp. 529–543. Springer, Heidelberg (2007)
24. Nieuwenhuis, R., Oliveras, A., Tinelli, C.: Solving SAT and SAT Modulo Theories: From an abstract Davis–Putnam–Logemann–Loveland procedure to DPLL(T). Journal of the ACM, JACM 53(6), 937–977 (2006)
25. Pipatsrisawat, K., Darwiche, A.: On modern clause-learning satisfiability solvers. Journal of Automated Reasoning 44(3), 277–301 (2010)
26. Srinivasan, A., Ham, T., Malik, S., Brayton, R.: Algorithms for discrete function manipulation. In: 1990 IEEE International Conference on Computer-Aided Design, ICCAD-90. Digest of Technical Papers, pp. 92–95 (1990)
27. Tamura, N., Taga, A., Kitagawa, S., Banbara, M.: Compiling finite linear CSP into SAT. Constraints 14(2), 254–272 (2009)
28. Walsh, T.: SAT *v* CSP. In: Dechter, R. (ed.) CP 2000. LNCS, vol. 1894, pp. 441–456. Springer, Heidelberg (2000)
29. Warners, J.P.: A Linear-Time Transformation of Linear Inequalities into Conjunctive Normal Form. Information Processing Letters 68(2), 63–69 (1998)

Efficient Application of Max-SAT Resolution on Inconsistent Subsets

André Abramé and Djamal Habet

Aix Marseille Université, CNRS, ENSAM, Université de Toulon, LSIS UMR 7296, 13397 Marseille, France
{andre.abrame,djamal.habet}@lsis.org

Abstract. Max-SAT resolution is the adaption of the powerful SAT resolution rule for the Max-SAT problem. One of the differences between these two rules is that Max-SAT resolution adds, besides the resolvent, several compensation clauses to keep the formula's equivalency. We address in this paper the problem of reducing both the number and the size of these compensation clauses. We show that the order in which the Max-SAT resolution steps are applied on the inconsistent subsets of clauses has a direct impact on the number and the sizes of the compensation clauses added to the formula. Based on this observation, we present a new algorithm for applying Max-SAT resolution on inconsistent subsets which reduces the number and the sizes of the produced compensation clauses. We demonstrate experimentally the interest of our contribution.

1 Introduction

The Max-SAT problem consists in finding an assignment which maximizes the number of satisfied clauses of a given CNF formula. In Weighted Max-SAT, a positive weight is associated to each clause and the goal is to maximize the sum of the weights of the satisfied clauses. There are other variants of Max-SAT (Partial and Weighted Partial) which are not considered in this paper.

A typical Branch and Bound (BnB) algorithm for the (weighted) Max-SAT problem makes a depth-first exploration of the search space. At each node, it evaluates if it is worth exploring the sub-nodes by comparing the current sum of the falsified clause weights plus an (under-)estimation of the weights of the ones which will become falsified (the lower bound, LB) to the best solution found so far (the upper bound UB). If $LB \geq UB$, then no better solution can be found by exploring the sub-part of the search tree and the algorithm performs a backtrack. The lower bound is generally computed by counting the disjoint inconsistent subsets of the formula (sets of clauses which cannot be all satisfied). In recent years, the definition of new inference rules [3,5,6,7,8,9] has significantly improved the performances of BnB solvers by limiting redundancies in the LB computation. They allow solvers to make more incremental the LB computation by memorizing in the sub-part of the search tree some of the detected inconsistent subsets. It has been shown [4] that almost all the existing inference rules for

B. O'Sullivan (Ed.): CP 2014, LNCS 8656, pp. 92–107, 2014.

Max-SAT are special cases of Max-SAT resolution, the Max-SAT version of the well-known resolution rule.

However, these inference rules catch only a small part of the inconsistent subsets. One of the reasons why the memorization is only applied on few inconsistent subsets is that the Max-SAT resolution rule adds, beside the resolvent, several compensation clauses to maintain the formula's equivalency. Thus, the size of the formula increases quickly which slows down the BnB solver.

We address in this paper the problem of reducing the number and the size of the compensation clauses added to the formula when applying the Max-SAT resolution rule. At each node of the search tree, BnB Max-SAT solvers detect inconsistent subsets by unit propagation. When an empty clause (a falsified clause) is found, it is possible to transform the corresponding inconsistent subset (IS) of clauses by several Max-SAT resolution steps. We study the impact of the order of the Max-SAT resolution steps on the number and the size of compensation clauses. The Max-SAT resolution steps are incremental, ie. the resolvent clause produced by the current resolution step will be used as input clause by a next resolution step. Moreover, the number and the sizes of the compensation clauses produced by a Max-SAT resolution step depend on the sizes of the input clauses. Thus, minimizing the intermediary resolvents will reduce the number and the sizes of compensation clauses added by the next resolution steps. Accordingly, we propose a new order of application of Max-SAT resolution on the clauses belonging to an inconsistent subset. We have implemented this new order in our solver AHMAXSAT. The obtained results confirm the practical interest of our contribution and give interesting insights of the solver's behavior.

This paper is organized as follows. We give some basic definitions and notations in Section 2. In Section 3, we present the Max-SAT resolution rule and describe in Section 4 how it can be used to transform inconsistent subsets. We propose a new order of application of the resolution steps in Section 5. Eventually, we present in Section 6 the results of the experimental study we have performed and we discuss the practical impact of our contribution before concluding in Section 7.

2 Formalism and Definitions

A weighted formula Φ in conjunctive normal form (CNF) defined on a set of propositional variables $X = \{x_1, \ldots, x_n\}$ is a conjunction of weighted clauses. A weighted clause c_j is a weighted disjunction of literals and a literal l is a variable x_i or its negation $\overline{x}_i$. We denote $var(l)$ the variable of a literal l. Alternatively, a weighted formula can be represented as a multiset of weighted clauses $\Phi = \{c_1, \ldots, c_m\}$ and a weighted clause as a tuple $c_j = (\{l_{j_1}, \ldots, l_{j_k}\}, w_j)$ with $\{l_{j_1}, \ldots, l_{j_k}\}$ a set of literals and w_j a strictly positive weight. We denote the number of clauses of a formula Φ by $|\Phi|$ and the number of literals of a clause c_j by $|c_j|$.

An assignment can be represented as a set I of literals which cannot contain both a literal and its negation. If x_i is assigned to $true$ (resp. $false$) then $x_i \in I$

(resp. $\overline{x}_i \in I$). I is a complete assignment if $|I| = n$ and it is partial otherwise. A literal l is said to be satisfied by an assignment I if $l \in I$ and falsified if $\overline{l} \in I$. A variable which does not appear either positively or negatively in I is unassigned. A clause is satisfied by I if at least one of its literals is satisfied, and it is falsified if all its literals are falsified. Empty clauses (denoted by $\square$) are always falsified. A subset ψ of Φ is inconsistent if there is no assignment which satisfies all its clauses. Solving the weighted Max-SAT problem consists in finding, for a formula Φ, a complete assignment which maximizes the sum of the weights of the satisfied clauses of Φ. Two formulas are equivalent for (weighted) Max-SAT iff they have the same sum of falsified clause weights for each partial assignment.

3 Max-SAT Resolution

The Max-SAT version of the resolution rule, first introduced by Heras and Larrosa [3] and further refined by Bonet et al. [1,2], can be defined as follows (on top the premises of the rule and at the bottom the conclusions) :

$$\frac{c_1 = \{x, y_1, \ldots, y_s\},\ c_2 = \{\overline{x}, z_1, \ldots, z_t\}}{cr = \{y_1, \ldots, y_s, z_1, \ldots, z_t\},\ cc_1,\ \ldots,\ cc_t,\ cc_{t+1},\ \ldots,\ cc_{t+s}}$$

with:

$$\begin{aligned}
cc_1 &= \{x, y_1, \ldots, y_s, \overline{z}_1, z_2, \ldots, z_t\} & cc_{t+1} &= \{\overline{x}, z_1, \ldots, z_t, \overline{y}_1, y_2, \ldots, y_s\}\\
cc_2 &= \{x, y_1, \ldots, y_s, \overline{z}_2, \ldots, z_t\} & cc_{t+2} &= \{\overline{x}, z_1, \ldots, z_t, \overline{y}_2, \ldots, y_s\}\\
&\vdots & &\vdots\\
cc_t &= \{x, y_1, \ldots, y_s, \overline{z}_t\} & cc_{t+s} &= \{\overline{x}, z_1, \ldots, z_t, \overline{y}_s\}
\end{aligned}$$

Contrary to SAT resolution, the original clauses c_1 and c_2 are removed from the formula and, besides the resolvent cr, several compensation clauses $cc_1 \ldots cc_{t+s}$ are added to preserve the equivalency of the formula. The size of the resolvent cr and the cardinality of the set of the compensation clauses $CC = \{cc_1, \ldots, cc_{t+s}\}$ can be expressed as follows:

$$\begin{aligned}
|cr| &= |c_1 \cup c_2 \setminus \{x, \overline{x}\}| = |c_1| + |c_2| - |c_1 \cap c_2| - 2\\
|CC| &= |c_1| + |c_2| - 2 * |c_1 \cap c_2| - 2
\end{aligned}$$

The sizes of the compensation clauses $cc_1, \ldots, cc_t$ range from $|c_1| + 1$ to $|c_1| + |c_2| - |c_1 \cap c_2| - 1$ and the ones of $cc_{t+1}, \ldots, cc_{t+s}$ range from $|c_2| + 1$ to $|c_1| + |c_2| - |c_1 \cap c_2| - 1$. The Max-SAT resolution rule can easily be extended to weighted formulas. If m is the minimum weight of c_1 and c_2, then m is subtracted from the weight of the premise clauses and all the clauses of the conclusion take the weight m.

4 Transforming Inconsistent Subsets

Recent BnB Max-SAT solvers apply unit propagation (UP) to detect inconsistent subsets. For each unit clause $\{l\}$, they remove all the occurrences of $\bar{l}$ from the clauses and all the clauses containing l. This process is repeated until an empty clause (a conflict) is found or no more unit clauses remain. The unit clause $\{l\}$ causing the propagation of l is called its predecessor and the clauses which are reduced by l are its successors. BnB Max-SAT solvers memorize the propagation steps by an implication graph, which can be defined as follows (see for instance [10] for a definition in the SAT context).

Definition 1 (Implication Graph). *Let $\Phi = \{c_1, \ldots, c_m\}$ be a (weighted) CNF formula defined on a set of Boolean variables $X = \{x_1, \ldots, x_n\}$ and I a partial assignment (with both decisions and propagations) of the variables of X. We assume that there can be only one falsified clause, ie. UP is stopped when a conflict is discovered. An implication graph is a directed labeled acyclic graph $G = (V, E)$ with:*

$$\begin{aligned} V &= \{l \in I\} \cup \{\Diamond_{c_i} \text{ s.t. } \exists c_i \in \Phi, |c_i| = 1\} \cup \{\Box \text{ if } \exists c_j \in \Phi \text{ falsified by } I\} \\ E &= \{(l, l', c_k) \text{ s.t. } \exists c_k \in \Phi \text{ which is reduced by } l \text{ and propagates } l'\} \cup \\ &\quad \{(\Diamond_{c_p}, l, c_p) \text{ s.t. } \exists c_p = \{l\} \in \Phi\} \cup \\ &\quad \{(l, \Box, c_q) \text{ s.t. } \exists c_q \in \Phi \text{ falsified by } I \text{ and } l \in c_q\} \end{aligned}$$

We use the two special nodes $\Diamond$ and $\Box$ to represent respectively the initial vertices of the unit clauses and the terminal one of the falsified clause. For clarity reason, we hide the nodes $\Diamond$ in the graphical representation of the implication graphs. Each arc is labeled with the clause it comes from.

When an empty clause is found by UP, the corresponding inconsistent subset (IS) of the formula can be built by analyzing the implication graph G. Each clause which is in a path between decisions (or unit clauses) and the empty clause belongs to this inconsistent subset. We consider in the rest of this paper that implication graphs are restricted to the nodes and arcs which lead to the empty clause. Then, the IS can be transformed by eliminating successively by Max-SAT resolution the propagated variables participating in the conflict (ie. which appear in the implication graph). For a Max-SAT resolution step between the predecessor c_j and the successor c_k of a variable $x_i \in X$ we use the following notation: $cr_{<x_i>}$ is the resolvent clause and $CC_{<x_i>}$ the set of the compensation clauses. If the undirected subgraph of G induced by $X' = \{x_{i_1}, \ldots, x_{i_k}\} \subset X$ is connected, these definitions can be extended to any sequence of Max-SAT resolution steps $< x_{i_1}, \ldots, x_{i_k} >$. After a Max-SAT resolution step, the implication graph can be updated by removing the resolved variable and replacing the removed clauses by the new resolvent. This operation is formally defined below.

Definition 2 (Updated Implication Graph). *Let $G = (V, E)$ be an implication graph. After the application of Max-SAT resolution on the predecessor c_j*

and the unique successor c_k *of a variable* $x_i \in X$ *such that* $\{x_i, \overline{x_i}\} \cap V \neq \emptyset$, *G can be transformed into* $G_{<x_i>} = (V_{<x_i>}, E_{<x_i>})$ *with:*

$$
\begin{aligned}
V_{<x_i>} &= (V \setminus \{x_i, \overline{x_i}\}) \setminus \{\lozenge_{c_j} \textit{ if } |c_j| = 1\} \cup \{\lozenge_{cr_{<x_i>}} \textit{ if } |cr_{<x_i>}| = 1\} \\
E_{<x_i>} &= \{(l, l', c_p) \in E \textit{ s.t. } var(l) \neq x_i,\ var(l') \neq x_i,\ c_p \neq c_j \textit{ and } c_p \neq c_k\} \cup \\
&\quad \{(l, l', cr_{<x_i>}) \textit{ s.t. } \bar{l}, l' \in cr_{<x_i>} \textit{ and } l, l' \in V_{<x_i>}\} \cup \\
&\quad \{(\lozenge_{cr_{<x_i>}}, l, cr_{<x_i>}) \textit{ if } cr_{<x_i>} = \{l\}\} \cup \\
&\quad \{(l, \square, cr_{<x_i>}) \textit{ s.t. } \bar{l} \in cr_{<x_i>} \textit{ and } cr_{<x_i>} \textit{ is falsified by } I\}
\end{aligned}
$$

This notation can be extended to all sequences of Max-SAT resolution steps $< x_{i_1}, x_{i_2}, \ldots, x_{i_k} >$: $G_{<x_{i_1}, x_{i_2}, \ldots, x_{i_k}>} = (\ldots((G_{<x_{i_1}>})_{<x_{i_2}>})\ldots)_{<x_{i_k}>}$.

When all the propagated variables of the implication graph have been eliminated, the final resolvent clause contains no propagated variables. It is empty or directly falsified by the decisions. The conflict is memorized and unit propagation is not needed anymore to detect it, neither at this decision level nor in the sub-part of the search tree.

It should be noted that a propagated variable $x_i \in X$ can only have one predecessor but it can reduce more than one clause, and thus it can lead to several propagations. In such a case, Max-SAT resolution cannot be applied on x_i until all its successors have been merged by Max-SAT resolution steps into a single intermediary resolvent. For this reason, the resolution steps are usually applied in reverse propagation order. Algorithm 1 shows how Max-SAT resolution can be applied on an inconsistent subset from the implication graph. The algorithm first applies Max-SAT resolution between the empty clause and the predecessor of the most recently propagated variable. Then it applies iteratively Max-SAT resolution between the last intermediary resolvent obtained and the predecessor of the most recently propagated untreated variable until all the variables have been treated. Eventually, it adds the last resolvent obtained to the formula. The algorithm uses a function `max_sat_resolution(`Φ,c_1,x,c_2`)` which returns the resolvent clause resulting of the application of Max-SAT resolution between

Algorithm 1. Classical transformation of IS by Max-SAT resolution

Data: A CNF formula Φ, an assignment I which falsify a clause $c_f \in \Phi$ and the corresponding implication graph $G = (V, E)$.
Result: A transformed formula Φ.

```
begin
    cr ← c_f;
    while |V| > 0 do
        l ← remove the most recently propagated literal from V;
        c ← predecessor of var(l);
        cr ← max_sat_resolution(Φ,cr,var(l),c);
    Φ = Φ ∪ {cr};
```

c_1 and c_2 on x. It removes c_1 and c_2 and adds the compensation clauses to the formula. The following example illustrates how this algorithm works.

Example 1. Let us consider the formula $\Phi = \{c_1, c_2, \ldots, c_6\}$ defined on a set of Boolean variables $\{x_1, \ldots, x_5\}$ with $c_1 = \{x_1\}$, $c_2 = \{x_2\}$, $c_3 = \{\overline{x}_2, x_3\}$, $c_4 = \{\overline{x}_2, x_4\}$, $c_5 = \{\overline{x}_1, \overline{x}_3, x_5\}$ and $c_6 = \{\overline{x}_4, \overline{x}_5\}$. The application of unit propagation on Φ leads to the assignments $< x_1@c_1, x_2@c_2, \ldots, x_5@c_5 >$ (meaning that x_1 is propagated by clause c_1, then x_2 by c_2, etc.). The clause c_6 is falsified. Fig. 1 shows the corresponding implication graph and Fig. 2 the Max-SAT resolution steps applied on the formula, with the compensation clauses in boxes. The original clauses $c_1, \ldots, c_6$ are removed from the formula and besides the compensation clauses $cc_1, \ldots, cc_8$, the resolvent $cr_{<x_1,\ldots,x_5>} = \Box$ is added to the formula. The intermediary resolvents $cr_{<x_1>}, \ldots, cr_{<x_1,\ldots,x_4>}$ are consumed by the Max-SAT resolution steps. We obtain the formula $\Phi' = \{\Box, cc_1, \ldots, cc_8\}$. Note that Φ' contains, besides the empty clause $\Box$, one clause of size two, three clauses of size three and four clauses of size four.

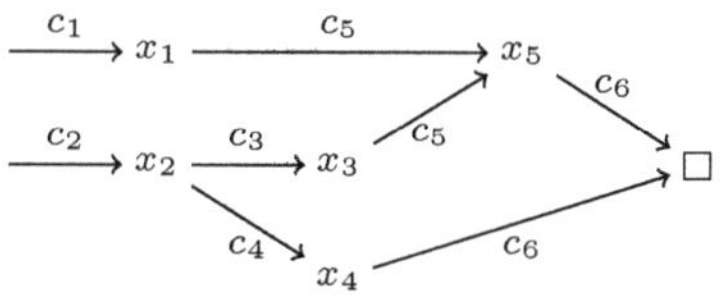

Fig. 1. Implication graph for the formula Φ of Example 1

5 Improved Transformation of Inconsistent Subsets

We have seen in the previous section how inconsistent subsets can be transformed thanks to Max-SAT resolution. These transformations produce compensation clauses, and thus the size of the formula can grow quickly. It is interesting to observe that the number and the sizes of the compensation clauses added depend on the order in which the Max-SAT resolution steps are performed. Let us take the subset $\{c_4, c_5, c_6\}$ from the formula Φ of Example 1. We can apply two sequences of Max-SAT resolution steps: $< x_4, x_5 >$ or $< x_5, x_4 >$. Fig. 3 shows the application of these two sequences. In both cases we obtained the same resolvent $cr_{<x_4,x_5>} = cr_{<x_5,x_4>} = \{\overline{x}_1, \overline{x}_2, \overline{x}_3\}$, but in the second case six compensation clauses are added (two of size three and four of size four) while in the first case only five are added (two of size four and three of size three).

There is a direct relation between the size of the intermediary resolvents and the number and the sizes of the compensation clauses added. As we have seen in Section 3, the bigger the resolved clauses are, the bigger are the size of the resolvent and the number of compensation clauses. Thus, by making first the resolution steps which produce small intermediary resolvents, we reduce the number

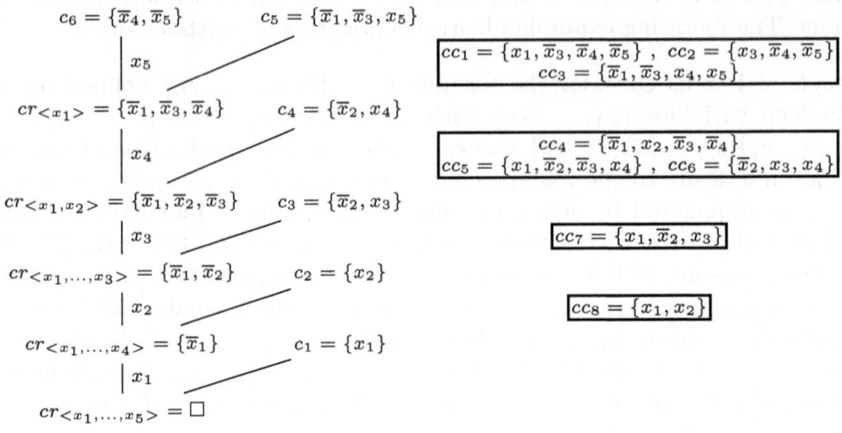

Fig. 2. Max-SAT resolution steps applied on the formula Φ in Example 1

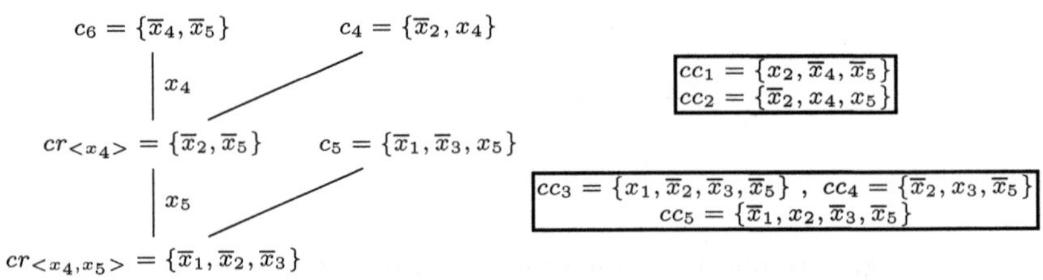

(a) Max-SAT resolution steps $< x_4, x_5 >$

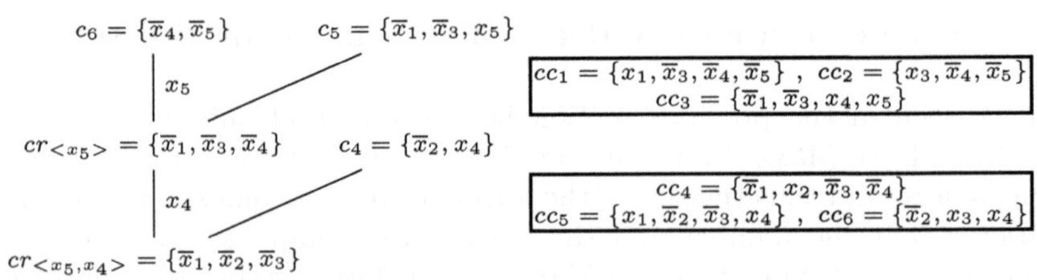

(b) Max-SAT resolution steps $< x_5, x_4 >$

Fig. 3. Application of Max-SAT resolution on $\{c_4, c_5, c_6\}$

of compensation clauses added by the next resolution steps which use these resolvents. To evaluate which resolution step must be made first, we use the notion of score of a variable. This score corresponds to the size of the resolvent obtained if we apply Max-SAT resolution between the predecessor and the successor of a variable. If a variable has more than one successor then, as explained in the previous section, it cannot be transformed immediately and we give it an infinite score. Formally, for a variable $x_i \in X$ propagated by a clause c_j:

$$score(x_i) = \begin{cases} |c_j| + |c_k| - |c_j \cap c_k| - 2 & \text{if } x_i \text{ reduces a single clause } c_k \\ \infty & \text{if } x_i \text{ reduces more than one clause} \end{cases}$$

We can show that for two adjacent variables in an implication graph, applying Max-SAT resolution first on the one of lowest score always produces a smaller number of compensation clauses.

Property 1. Let $G = (V, E)$ be an implication graph. Let us consider two variables $x_i, x_j \in X$ $(i \neq j)$ such that $\exists l, l' \in V$ with $var(l) = x_i$, $var(l') = x_j$ and $\exists (l, l')$ or $(l', l) \in E$. We have the following relation:

$$score(x_i) \leq score(x_j) \text{ iff } |CC_{<x_i,x_j>}| \leq |CC_{<x_j,x_i>}|$$

Proof. We assume without loss of generality that the predecessors and successors of x_i and x_j are respectively c_{k_1}, c_{k_2} and c_{k_2}, c_{k_3}. We first express $|CC_{<x_i,x_j>}|$ and $|CC_{<x_j,x_i>}|$ in size of c_{k_1}, c_{k_2} and c_{k_3}. $|CC_{<x_i,x_j>}|$ is the number of compensation clauses obtained by applying two Max-SAT resolution steps, first between c_{k_1} and c_{k_2} on the variable x_i and second between $c_{<x_i>}$ (the resolvent of the previous resolution step) and c_{k_3} on x_j. These two resolution steps produce respectively the sets of compensation clauses $CC_{<x_i>}$ and $CC_{<x_j>|<x_i>}$ (we denote $CC_{<x_l>|<x_{m_1},\ldots,x_{m_p}>}$ the set of compensation clauses obtained by applying Max-SAT resolution on a variable x_l after the sequence of resolution steps $< x_{m_1}, \ldots, x_{m_p} >$). We have:

$$|CC_{<x_i,x_j>}| = |CC_{<x_i>}| + |CC_{<x_j>|<x_i>}|$$

From the definition of the Max-SAT resolution rule, we know that $|CC_{<x_i>}| = |c_{k_1}| + |c_{k_2}| - 2 * |c_{k_1} \cap c_{k_2}| - 2$ and $|CC_{<x_j>|<x_i>}| = |cr_{<x_i>}| + |c_{k_3}| - 2 * |cr_{<x_i>} \cap c_{k_3}| - 2$. Moreover, again from the definition of Max-SAT resolution, we know that $|cr_{<x_i>}| = |c_{k_1}| + |c_{k_2}| - |c_{k_1} \cap c_{k_2}| - 2$. Since $cr_{<x_i>} = c_{k_1} \cup c_{k_2} \setminus \{x_i, \overline{x}_i\}$ and $c_{k_3} \cap \{x_i, \overline{x}_i\} = \emptyset$ we have $|cr_{<x_i>} \cap c_{k_3}| = |c_{k_1} \cap c_{k_3}| + |c_{k_2} \cap c_{k_3}| - |c_{k_1} \cap c_{k_2} \cap c_{k_3}|$. We obtain:

$$\begin{aligned} |CC_{<x_i,x_j>}| &= 2 * |c_{k_1}| + 2 * |c_{k_2}| + |c_{k_3}| - 3 * |c_{k_1} \cap c_{k_2}| - 2 * |c_{k_1} \cap c_{k_3}| \\ &\quad -2 * |c_{k_2} \cap c_{k_3}| + 2 * |c_{k_1} \cap c_{k_2} \cap c_{k_3}| - 6 \\ &= score(x_i) + k \end{aligned}$$

with $k = |c_{k_1}| + |c_{k_2}| + |c_{k_3}| - 2 * |c_{k_1} \cap c_{k_2}| - 2 * |c_{k_1} \cap c_{k_3}| - 2 * |c_{k_2} \cap c_{k_3}| + 2 * |c_{k_1} \cap c_{k_2} \cap c_{k_3}| - 4$. In the same way, we can obtain:

$$\begin{aligned} |CC_{<x_j,x_i>}| &= |c_{k_1}| + 2 * |c_{k_2}| + 2 * |c_{k_3}| - 2 * |c_{k_1} \cap c_{k_2}| - 2 * |c_{k_1} \cap c_{k_3}| \\ &\quad -3 * |c_{k_2} \cap c_{k_3}| + 2 * |c_{k_1} \cap c_{k_2} \cap c_{k_3}| - 6 \\ &= score(x_j) + k \end{aligned}$$

□

The next property links, for two adjacent variables in the implication graph, the maximum size of the compensation clauses produced to the score of the variables.

Property 2. Let $G = (V, E)$ be an implication graph. Let us consider two variables $x_i, x_j \in X$ $(i \neq j)$ such that $\exists l, l' \in V$ with $var(l) = x_i$, $var(l') = x_j$ and $\exists (l, l')$ or $(l', l) \in E$. We have the following relation: $score(x_i) \leq score(x_j)$ iff the upper bounds of the ranges of the compensation clause sizes obtained with the sequence of max-resolution steps $< x_i, x_j >$ are lower or equal to ones obtained with the sequence $< x_j, x_i >$.

Proof. As for the proof of Property 1, we assume without loss of generality that the predecessors and successors of x_i and x_j are respectively c_{k_1}, c_{k_2} and c_{k_2}, c_{k_3}. We know from the definition of the max-resolution rule (see Section 3) that the compensation clause sizes obtained by resolving c_{k_1} and c_{k_2} on x_i are ranging from $|c_{k_1}| + 1$ to $|c_{k_1}| + |c_{k_2}| - |c_{k_1} \cap c_{k_2}| - 1$ and from $|c_{k_2}| + 1$ to $|c_{k_1}| + |c_{k_2}| - |c_{k_1} \cap c_{k_2}| - 1$. The sizes of the compensation clauses obtained by the second max-resolution step between c_{k_3} and $cr_{<x_i>}$ range from $|cr_{<x_i>}| + 1$ to $|cr_{<x_i>}| + |c_{k_3}| - |cr_{<x_i>} \cap c_{k_3}| - 1$ and from $|c_{k_3}| + 1$ to $|cr_{<x_i>}| + |c_{k_3}| - |cr_{<x_i>} \cap c_{k_3}| - 1$. We denote respectively $ub_{<x_i>}$ and $ub_{<x_i,x_j>}$ the two upper bounds of these ranges, which can be expressed in size of c_{k_1}, c_{k_2} and c_{k_3} as follows:

$$\begin{aligned} ub_{<x_i>} &= |c_{k_1}| + |c_{k_2}| - |c_{k_1} \cap c_{k_2}| - 1 \\ &= score(x_i) + 1 \end{aligned}$$

and

$$\begin{aligned} ub_{<x_i,x_j>} &= |cr_{<x_i>}| + |c_{k_3}| - |cr_{<x_i>} \cap c_{k_3}| - 1 \\ &= score(x_i) + |c_{k_3}| - |cr_{<x_i>} \cap c_{k_3}| - 1 \\ &= score(x_i) + |c_{k_3}| - |c_{k_1} \cap c_{k_3}| - |c_{k_2} \cap c_{k_3}| + |c_{k_1} \cap c_{k_2} \cap c_{k_3}| - 1 \\ &= score(x_i) + score(x_j) - |c_{k_2}| - |c_{k_1} \cap c_{k_3}| + |c_{k_1} \cap c_{k_2} \cap c_{k_3}| + 1 \end{aligned}$$

In the same way, we can express in size of c_{k_1}, c_{k_2} and c_{k_3} the upper bounds obtained by applying max-resolution on x_j then on x_i, respectively denoted $ub_{<x_j>}$ and $ub_{<x_j,x_i>}$:

$$\begin{aligned} ub_{<x_j>} &= |c_{k_2}| + |c_{k_3}| - |c_{k_2} \cap c_{k_3}| - 1 \\ &= score(x_j) + 1 \end{aligned}$$

and

$$\begin{aligned} ub_{<x_j,x_i>} &= |cr_{<x_j>}| + |c_{k_1}| - |cr_{<x_j>} \cap c_{k_1}| - 1 \\ &= score(x_j) + |c_{k_1}| - |cr_{<x_j>} \cap c_{k_1}| - 1 \\ &= score(x_j) + |c_{k_1}| - |c_{k_1} \cap c_{k_2}| - |c_{k_1} \cap c_{k_3}| + |c_{k_1} \cap c_{k_2} \cap c_{k_3}| - 1 \\ &= score(x_j) + score(x_i) - |c_{k_2}| - |c_{k_1} \cap c_{k_3}| + |c_{k_1} \cap c_{k_2} \cap c_{k_3}| + 1 \end{aligned}$$

Thus, if $score(x_i) \leq score(x_j)$ we have $ub_{<x_i>} \leq ub_{<x_j>}$ and conversely. The second upper bounds obtained ($ub_{<x_i,x_j>}$ and $ub_{<x_j,x_i>}$) are equal regardless of x_i and x_j scores. □

The two previous properties show that for two adjacent variables in an implication graph, applying Max-SAT resolution first on the one of lowest score produces less compensation clauses and particularly less long compensation clauses. Algorithm 2 shows how these properties can be put into practice by applying Max-SAT resolution steps in ascendant score order. Firstly, it computes the scores of the variables (line 2). Then, it selects the variable of V of smallest score and it applies Max-SAT resolution between its predecessor and successor (lines 4-7). It replaces in the implication graph the consumed clauses by the intermediary resolvent produced (duplicates arcs are removed) and it updates the scores of the variables (lines 8-9). This process is repeated until the implication graph contains no more variable. The last resolvent produced is added to the formula (line 10). The following example details how this algorithm works.

Algorithm 2. Improved transformation of IS by Max-SAT resolution

Data: A CNF formula Φ, an assignment I which falsify a clause and the corresponding implication graph $G = (V, E)$.
Result: A transformed formula Φ

1 **begin**
2   **for** $x \in V$ **do** compute x score
3   **while** $|V| > 0$ **do**
4         $l \leftarrow$ literal of smallest score of V;
5         $c_1 \leftarrow$ predecessor of $var(l)$;
6         $c_2 \leftarrow$ successor of $var(l)$;
7         $cr \leftarrow$ `max_sat_resolution`(Φ, c_1, x, c_2);
8         $G \leftarrow G_{<var(l)>}$;
9   update the scores of the variables of c_1 and c_2;
10     $\Phi = \Phi \cup \{cr\}$;

Example 2. Let us consider the formula Φ of Example 1. As previously, the application of unit propagation on Φ leads to the assignments $< x_1@c_1, x_2@c_2, \ldots, x_5@c_5 >$ and the clause c_6 is empty. Fig. 4 shows the evolution of the implication graph (with in brackets the variable's score) during the transformation and Fig. 5 shows the Max-SAT resolution steps.

Initially, there are two variables with the lowest score 2, x_1 and x_4. The algorithm applies Max-SAT resolution between the predecessor of x_1, c_1, and its successor c_5. It produces the intermediary resolvent $cr_{<x_1>} = \{\overline{x}_3, x_5\}$, then it removes x_1 from the implication graph, it replaces the tag c_5 of the arc (x_3, x_5) by $cr_{<x_1>}$ and it updates the scores of x_3 and x_5. Fig. 4(b) show the modified implication graph, with in bold the replaced arcs and the updated scores.

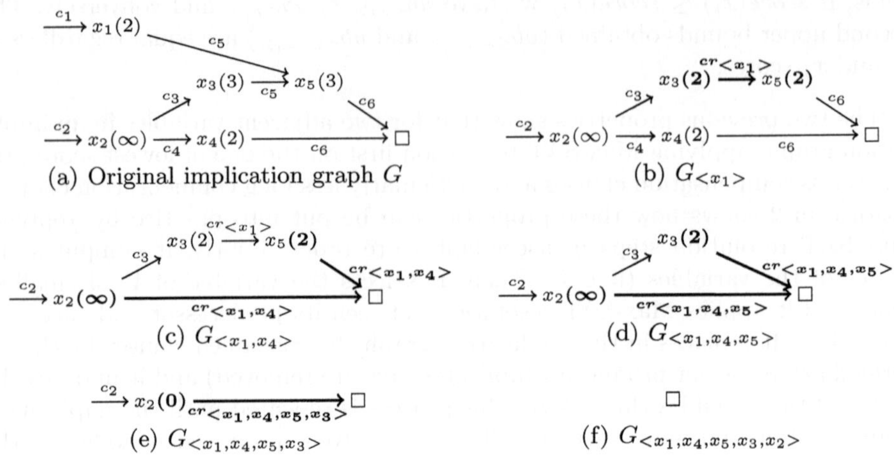

(a) Original implication graph G (b) $G_{<x_1>}$ (c) $G_{<x_1,x_4>}$ (d) $G_{<x_1,x_4,x_5>}$ (e) $G_{<x_1,x_4,x_5,x_3>}$ (f) $G_{<x_1,x_4,x_5,x_3,x_2>}$

Fig. 4. Evolution of the implication graph during the transformation of the formula Φ of Example 2

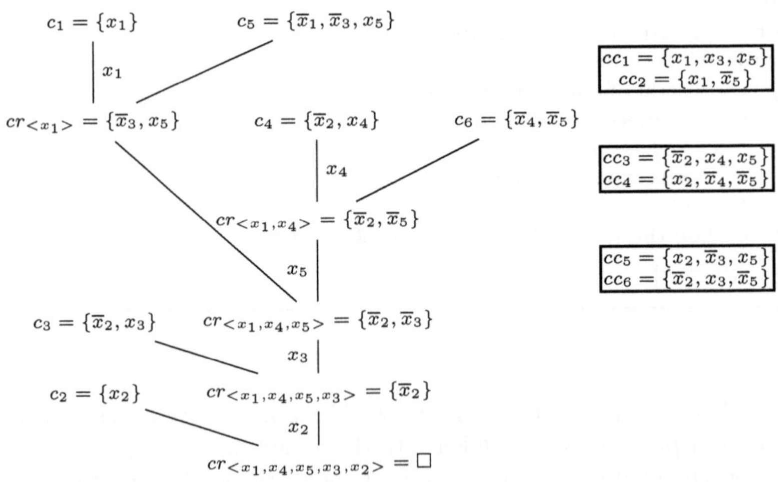

Fig. 5. Max-SAT resolution steps applied on the formula Φ in Example 2

Then, it takes the next variables of score 2, x_4, and it applies Max-SAT resolution between its predecessor and successor (c_4 and c_6 respectively). It updates the implication graph and the variable scores (Fig. 4(c)). It applies the same treatment on x_5 (Fig. 4(d)) then x_3 (Fig. 4(e)). At this point, the two successors of x_2 have been merged and Max-SAT resolution can be applied between its predecessor c_2 and its new successor, the intermediary resolvent $cr_{<x_1,x_4,x_5,x_3>}$. This last resolution step produces an empty clause and, after its updating, the implication graph does not contain any more variable (Fig. 4(f)).

We obtain the transformed formula $\Phi'' = \{\square, cc_1, \ldots, cc_6\}$ which contains, besides the empty clause $\square$, six clauses (five of size three and one of size two). There are two clauses less than in Example 1 and the clause sizes are smaller. It is also interesting to observe that the max-resolution steps are no longer applied in a topological order of the unit propagation steps.

6 Experimental Study

We have implemented Algorithm 2 in our solver AHMAXSAT[1], which is a BnB solver for Max-SAT and weighted Max-SAT. AHMAXSAT detects inconsistent subsets using simulated unit propagation [6] and the failed literals mechanism [7]. Each detected IS is transformed by applying max-resolution steps between its clauses. If an IS matches (partially or totally) a restrictive sets of patterns [8], then the formula transformations are kept (partially or totally) in the sub-part of the search tree. Otherwise, they are simply restored before the next decision. AHMAXSAT uses a Jeroslow-Wang branching heuristic (a slight variation of the one of WMAXSATZ2009 [8]). In the rest of this section, we call AHMAXSAT the variant applying Max-SAT resolution in the classical reverse propagation order and AHMAXSAT$^+$ the variant based on the scores of the variables. We have run the two variants on all the random and crafted instances of the Max-SAT and Weighted Max-SAT categories of the Max-SAT Competition 2013[2]. The experiments are performed on a cluster of servers equipped with Intel Xeon 2.4 Ghz processors and 24 Gb of RAM and running under a GNU/Linux operating system. The cutoff time for each instance is fixed to 1800 seconds.

One can note that we include neither (weighted) Partial Max-SAT instances nor industrial ones in our experiments. Even if the results presented in this paper can naturally be extended to these instance categories, our solver AHMAXSAT does not handle them efficiently. A performing BnB solver for (weighted) Partial Max-SAT must handle both the soft and the hard parts of the instances. Thus, it must include SAT mechanisms such as nogood learning, activity-based branching heuristic or backjumping and our solver currently does not. For the industrial instances, solvers must have a very efficient memory management. To the best of our knowledge, none of the best performing BnB solvers (including ours) handles huge industrial instances efficiently.

Table 1 shows the detailed results of AHMAXSAT and AHMAXSAT$^+$. The columns S, D and T give respectively, for each variant of the solver, the number

[1] An early version of AHMAXSAT has been submitted to the Max-SAT Competition 2013. It was the version 1.16. Since that competition, we have made numerous improvement, such as code optimization and new data structures. The version presented in this paper is numbered 1.54.

[2] Available from `http://maxsat.ia.udl.cat:81`.

Table 1. Compared performances of AHMAXSAT and AHMAXSAT$^+$. The two first columns give the instance classes and the number of instances in each class. For each solver, the columns S, D and T give respectively the number of solved instances, the average number of decisions made and the average solving time. The two last columns give the percentages of variation between the two variants of the averages of the number of decisions and the solving time respectively. The columns tagged with * include only the instances solved by both variants of the solver.

	Instance classes	#	AHMAXSAT S	AHMAXSAT D	AHMAXSAT T*	AHMAXSAT T	AHMAXSAT$^+$ S	AHMAXSAT$^+$ D	AHMAXSAT$^+$ T*	AHMAXSAT$^+$ T	Δ_T*	Δ_D*
unweighted	crafted/bipartite	100	100	40184	113.2	114.3	100	34909	97	97.7	-13%	-14%
	crafted/maxcut	67	55	214561	45.4	45.6	56	177964	44.3	44.6	-17%	-2%
	random/highgirth	82	6	5075697	1337.2	1347	6	4597721	1119.2	1122.9	-9%	-16%
	random/max2sat	100	100	53451	114.6	116.2	100	38841	79.7	80.3	-27%	-30%
	random/max3sat	100	99	477402	329.7	345.3	100	426143	300.3	302.1	-11%	-9%
	random/min2sat	96	96	1115	2.8	2.9	96	979	2.4	2.4	-12%	-16%
weighted	crafted/frb	34	14	220164	40	40.2	14	210245	38.2	38.5	-5%	-5%
	crafted/ramsey	15	4	160192	69.6	70.6	4	157478	57.2	58.2	-2%	-18%
	crafted/wmaxcut	67	62	23073	35.7	35.9	62	19993	31.1	31.8	-13%	-13%
	random/wmax2sat	120	120	4877	68.5	69.3	120	3898	48.9	49.3	-20%	-29%
	random/wmax3sat	40	40	55523	154.3	155.7	40	44804	123.1	123.6	-19%	-20%
	Global results	821	696	153195	119.9	123.2	698	135686	101.2	101.8	-11%	-16%

of solved instances, the average number of decisions and the average execution time. The first observation which can be made is that AHMAXSAT$^+$ solves 3 more instances than AHMAXSAT. The average solving time (measured on the instances solved by both variants of the solver) is significantly reduced (-19% on average). The gain is especially high on random (unweighted and weighted) instances, where the average solving time reductions vary from -16% to -30%. The gain is slightly lower on crafted instances, with reduction varying from -5% to -18%. Fig. 6 compares for each instance the solving time of the two variants. For clarity reason, only the instances solved in more than ten seconds by the two variants are displayed. It confirms the previous observation and shows that the gain is not limited to some instances, it is spread over the majority of the benchmark.

Let us analyze the impact of the score based order of application of the Max-SAT resolution steps on the solver behavior. Tab. 2 shows some detailed statistics. We can first observe that both the number and the size (columns NB and SZ) of the compensation clauses are significantly reduced, respectively by 22% and 20% on average. Since the number and the size of the added compensation clauses is reduced, one can have expected a speedup in the exploration of the search space (measured by the number of decisions per second, column D/s). It is the case on the majority of the instance categories but, to the contrary, on some other categories (e.g. crafted/maxcut or crafted/wmaxcut) the average number of decisions per second is reduced. This behavior can be explained as follows. The reduction of the size of the compensation clauses improves the application of unit propagation, and AHMAXSAT$^+$ makes in average 4,2% more propagations at each decisions (columns P/D). Consequently, it detects and transforms more inconsistent subsets at each decision (in average 4,5% more, columns □/D). Thus

Table 2. Detailed statistics of AHMAXSAT and AHMAXSAT$^+$. For each solver, the columns NB, SZ, D/s, □/D and P/D give respectively the averages of: number of compensation clauses added per IS transformation, size of these compensation clauses, number of decisions per seconds, number on IS detected per decision and number of propagations per decision. The two last columns give the percentages of variation between the two variants of the average of respectively the columns NB and SZ.

	Instance classes	AHMAXSAT					AHMAXSAT$^+$					Δ_{NB}	Δ_{SZ}
		NB	SZ	D/s	□/D	P/D	NB	SZ	D/s	□/D	P/D		
unweighted	crafted/bipartite	15.7	3.46	347.7	71.26	1106.4	**13**	**2.88**	352.8	73.49	1132.1	-17%	-17%
	crafted/maxcut	8.7	2.9	1515.2	15.24	97	**7.7**	**2.54**	1405	18.35	134.9	-12%	-12%
	random/highgirth	34	5.23	278	2.43	40.5	**21.6**	**3.79**	302	2.53	43.5	-37%	-28%
	random/max2sat	18	3.9	472.8	42.79	859.4	**13.5**	**3.03**	489.2	45.93	900.4	-25%	-22%
	random/max3sat	14.6	3.52	1578	18.44	179.2	**11.7**	**2.92**	1570.4	19.48	192.4	-20%	-17%
	random/min2sat	18.7	4.1	83.1	34.35	1053.2	**13.8**	**2.93**	88.9	35.99	1100.3	-26%	-28%
weighted	crafted/frb	8.7	3.22	806.4	4.23	35.8	**7.4**	**2.7**	801.5	4.28	36	-15%	-16%
	crafted/ramsey	11.1	3.49	875.6	4.8	11.9	**8.4**	**2.85**	1003.9	4.82	12.9	-24%	-19%
	crafted/wmaxcut	9.4	2.93	223.8	63.39	192.1	**7.6**	**2.55**	216.4	63.23	193.3	-19%	-13%
	random/wmax2sat	17	4	78.2	254.53	2459.1	**12.8**	**3.1**	88.2	266.32	2563	-25%	-23%
	random/wmax3sat	15.4	3.52	384.5	87.54	495.9	**12**	**2.9**	393.9	94.17	550.9	-22%	-18%
	Global results	15.4	3.64	551.1	79.65	931.6	**11.9**	**2.9**	550.3	83.26	971.5	-22%	-20%

it spends more time on each decision and it makes fewer decisions per second. On the other hand, it detects more IS, thus the quality of the LB is improved and the average number of decisions made by AHMAXSAT$^+$ is 11% smaller than the one of AHMAXSAT (Tab. 1, columns D). This reduction of the number of decisions, combined to the speedup induced by the reduction of the sizes and number of the compensation clauses, explains the gain in execution time.

We have compared the best variant of our solver AHMAXSAT$^+$ with two of the best performing BnB solvers of the Max-SAT Competition 2013: WMAXSATZ2009 and WMAXSATZ2013 [5,7,8]. The results are presented in Fig. 7. AHMAXSAT$^+$ is very competitive. It solves seven more instances than WMAXSATZ2013 and

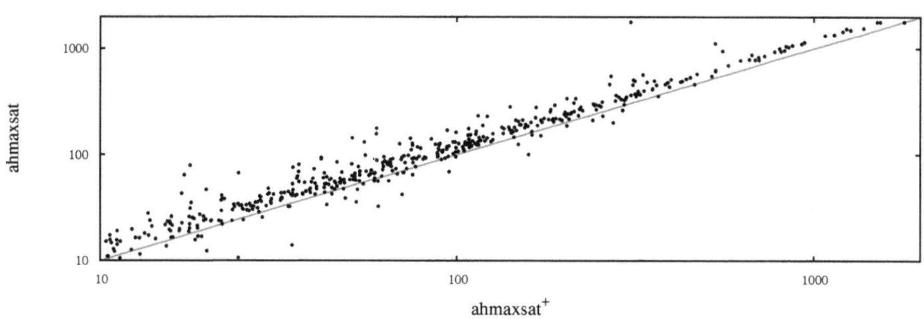

Fig. 6. Comparison of the solving times of AHMAXSAT and AHMAXSAT$^+$. Each point represents an instance and the farther a point is from a solver axes the better for the solver. All axis are in logarithmic scale.

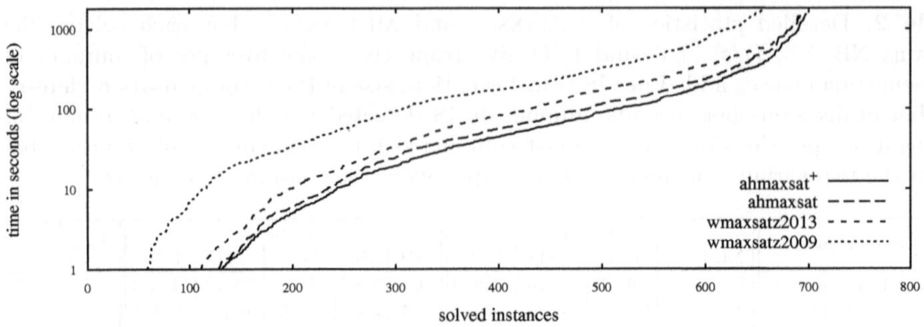

Fig. 7. Comparison of the performances of AHMAXSAT$^+$, WMAXSATZ2009 and WMAXSATZ2013.

41 more than WMAXSATZ2009. AHMAXSAT$^+$ is also significantly faster than WMAXSATZ2013 which was the best performing BnB solver of the Max-SAT Competition 2013 on the considered instances.

7 Conclusion

We have addressed in this paper the problem of reducing the number and the size of the compensation clauses produced by the transformation of inconsistent subsets by Max-SAT resolution. We have presented a new order of application of Max-SAT resolution based on the size of the intermediary resolvent. The experimental study we have performed shows that this order reduces efficiently both the number and the size of the compensation clauses. The performances of our solver are significantly improved.

This work could be a step toward a more general learning scheme for Max-SAT BnB solver. In the future, we will try to overcome the remaining obstacles to this learning scheme. Especially, we will study the impact of the Max-SAT resolution rule on the unit propagation mechanism. We will also extend this work to partial Max-SAT.

References

1. Bonet, M.L., Levy, J., Manyà, F.: A complete calculus for max-sat. In: Biere, A., Gomes, C.P. (eds.) SAT 2006. LNCS, vol. 4121, pp. 240–251. Springer, Heidelberg (2006)
2. Bonet, M.L., Levy, J., Manyá, F.: Resolution for max-sat. Artificial Intelligence 171(8-9), 606–618 (2007)
3. Heras, F., Larrosa, J.: New inference rules for efficient max-sat solving. In: Proceedings of the 21st National Conference on Artificial Intelligence, AAAI 2006, vol. 1, pp. 68–73. AAAI Press (2006)
4. Larrosa, J., Heras, F., de Givry, S.: A logical approach to efficient max-sat solving. Artificial Intelligence 172(2-3), 204–233 (2008)

5. Li, C.M., Manyà, F., Mohamedou, N., Planes, J.: Exploiting cycle structures in max-sat. In: Kullmann, O. (ed.) SAT 2009. LNCS, vol. 5584, pp. 467–480. Springer, Heidelberg (2009)
6. Li, C.M., Manyà, F., Planes, J.: Exploiting unit propagation to compute lower bounds in branch and bound max-sat solvers. In: van Beek, P. (ed.) CP 2005. LNCS, vol. 3709, pp. 403–414. Springer, Heidelberg (2005)
7. Li, C.M., Manyà, F., Planes, J.: Detecting disjoint inconsistent subformulas for computing lower bounds for max-sat. In: Proceedings of the 21st National Conference on Artificial Intelligence, AAAI 2006, pp. 86–91. AAAI Press (2006)
8. Li, C.M., Manyà, F., Planes, J.: New inference rules for max-sat. Journal of Artificial Intelligence Research 30, 321–359 (2007)
9. Li, C., Manyá, F., Mohamedou, N., Planes, J.: Resolution-based lower bounds in maxsat. Constraints 15(4), 456–484 (2010)
10. Marques-Silva, J.P., Sakallah, K.A.: Grasp: A search algorithm for propositional satisfiability. IEEE Transactions on Computers 48(5), 506–521 (1999)

Sequential Time Splitting and Bounds Communication for a Portfolio of Optimization Solvers

Roberto Amadini[1] and Peter J. Stuckey[2]

[1] Dept. of Computer Science and Engineering/Lab. Focus INRIA, University of Bologna, Italy
[2] National ICT Australia
Department of Computing and Information Systems
University of Melbourne, VIC 3010, Australia

Abstract. Scheduling a subset of solvers belonging to a given portfolio has proven to be a good strategy when solving Constraint Satisfaction Problems (CSPs). In this paper, we show that this approach can also be effective for Constraint Optimization Problems (COPs). Unlike CSPs, sequential execution of optimization solvers can communicate information in the form of bounds to improve the performance of the following solvers. We provide a hybrid and flexible portfolio approach that combines static and dynamic time splitting for solving a given COP. Empirical evaluations show the approach is promising and sometimes even able to outperform the best solver of the porfolio.

1 Introduction and Related Work

One of the main uses of Constraint Programming (CP) is to model and solve *Constraint Satisfaction Problems* (CSP) [19]. Solving CSPs is hard, and there are plenty of approaches that can be used to tackle them. One of the more recent trend in this research area—especially in the SAT field—is trying to solve a given problem by using a portfolio approach [12, 23].

An *algorithm portfolio* is a general methodology that exploits a number of different algorithms in order to get an overall better algorithm. A portfolio of CP solvers can therefore be seen as a particular solver, the *portfolio solver*, that exploits a collection of $m > 1$ different constituent solvers $s_1, \ldots, s_m$ in order to obtain a globally better CP solver. When a new unseen instance i arrives, the portfolio solver tries to predict which are the best constituent solvers $s_1, \ldots, s_k$ ($k \leq m$) for solving i and then runs such solver(s) on i. This solver selection process is clearly a fundamental part for the success of the approach and it is usually performed by exploiting Machine Learning techniques.

There has been a significant body of work in using portfolios to leverage and combine a number of different solvers in order to get an overall better solver [15, 18]. A particular case of portfolio approach consists in *scheduling* (even in parallel) a subset of the constituent solvers within a certain time window

B. O'Sullivan (Ed.): CP 2014, LNCS 8656, pp. 108–124, 2014.

(see for instance [3,7,14,16,22,24]. This would seem to be a winning strategy, in particular due to the fact that often the solving time of a satisfaction problem is either relatively short or very long. It hence naturally handles the heavy tailed nature of solving.

Surprisingly, most of the focus of algorithm portfolios has been on constraint satisfaction problems. In practice most of the combinatorial problems of interest are *Constraint Optimization Problems* (COPs), where we are interested in finding a solution which minimizes as much as possible a given objective function. As pointed out also in [25], there is a lack of use of meta-learning for algorithm selection: to the best of our knowledge, COP portfolios are mostly developed just for some specific optimization problems like Knapsack, Most Probable Explanation, Set Partitioning, Travel Salesman Problem [13,15,26] or for example to properly tune the parameters of a single COP solver [17,27].

It is hence natural to ask how to create a scheduling algorithm portfolio for COPs. A crucial difference from CSPs arises because a COP solver may yield sub-optimal *partial solutions* before finding the best one (and possibly proving its optimality). This means that one solver can *transmit* useful bounds information to another if they are scheduled sequentially. This key feature is the basis of our work. Indeed, in a portfolio scenario, a partial solution found by a solver s_1 is a token of knowledge that s_1 can pass to another solver s_2 in order to prune its search space and therefore possibly improve its solving process. In this paper, we thus address the problem of boosting optimization by exploiting the sequential cooperation of different COP solvers. To do so, we will introduce the notion of solving *behaviour* for taking into account the anytime performance of the solvers.

A work related to this paper is [9], in which algorithm control techniques are used to share bounds information between the scheduled solvers without, however, explicitly rely on the solvers behaviours (as in our technical definition). In [20] the authors provide a generic approach to knowledge sharing, based on the communication of learned clauses and cuts information, which is suitable for sequential SAT solvers but is less likely to be useful when solvers are very disparate in nature. Finally, [4] reports an empirical evaluation of different portfolio approaches applied to COPs, without however taking into account the anytime performance of the solvers as well as the possible bounds communication between them.

Paper Structure. In Section 2 we give the technical definitions of solving behaviour and timesplit solver; then, in Section 3, we provide and evaluate `TimeSplit`: an algorithm aimed to determine the best time splitting according to already known behaviours. By exploiting the results of `TimeSplit` in Section 4 we introduce `TPS`, a generic and flexible *portfolio* approach that relies on two steps: when a new unseen problem arrives, a static solver schedule (computed off-line) is run first, while a dynamic schedule is executed then by possibly exploiting the best solution found in the first stage. In Section 5 we describe the methodology and the results achieved by `TPS`, using the SUNNY [3,4] algorithm as a baseline for computing and evaluating different portfolio approaches. Finally, in Section 6 we conclude by providing some possible future directions.

2 Solving Behaviour and Timesplit Solvers

Let us fix a dataset of minimization[1] problems Δ, a universe of COP solvers Σ (which can include a particular portfolio $\Pi \subseteq \Sigma$) and a solving time window $[0, T]$. We wish to determine the best sequence of solvers in Σ to run on p and for how long to run each solver within the interval $[0, T]$ in order obtain the best result for instance p. Ideally we aim to *improve the best solver* of Σ for the instance p.

We define the *(solving) behaviour* of each solver $s \in \Sigma$ applied to a problem $p \in \Delta$ over time $[0, T]$ as a sequence of pairs $\mathcal{B}(s,p) = [(t_1, v_1), \dots, (t_n, v_n)]$ where $t_i \in [0, T]$ is the time when s finds a solution, and v_i is the objective value of such a solution. Note that we can consider the pairs ordered so that $t_1 < \dots < t_n$ while $v_1 > \dots > v_n$ since we assume the solving process is *monotonic* (we can omit the non-monotonic entries if any). For example, consider the behaviours $\mathcal{B}(s_1,p) = [(10,40),(50,25),(100,15)]$ and $\mathcal{B}(s_2,p) = [(800,45),(900,10)]$ illustrated in Figure 1a with a timeout of $T = 1000$ seconds. The best value $v^* = 10$ is found by s_2 after 900 seconds, but it takes 800 seconds to find its first solution ($v = 45$). Meanwhile, s_1 finds a better value ($v = 40$) after just 10 seconds and even better values in just 100 seconds. So, the question is: what happens if we *"inject"* the upper bound 40 from s_1 to s_2? Considering that starting from $v = 45$ the solver s_2 is able to find v^* in 100 seconds (from 800 to 900), hopefully starting from any better (or equal) value $v' \leq 45$ the time needed by s_2 to find v^* is no more than 100 seconds. Note that from a graphical point of view what we would like to do is therefore to *"shift"* the curve of s_2 towards the left from $t = 800$ to 10, by exploiting the fact that after 10 seconds s_1 can suggest to s_2 the upper bound $v = 40$. The cooperation between s_1 and s_2 would thereby reduce by $\Delta t = 790$ seconds the time needed to find v^*, and moreover would allow us to exploit the remaining Δt seconds for finding better solutions or even proving the optimality of v^*. However, note that the *virtual* behaviour may not occur: it may be that s_2 calculates important information in the first 800 seconds required to find the solution $v^* = 10$, and therefore the injection of $v = 40$ could be useless (if not harmful!).

Given a problem $p \in \Delta$ and a schedule $\sigma = [(s_1, t_1), \dots, (s_k, t_k)]$ of k solvers we define the corresponding *timesplit solver* as a particular solver such that: *i)* first, runs s_1 on p for t_1 seconds; *ii)* then, for $i = 1, \dots, k-1$, runs s_{i+1} on p for t_{i+1} seconds possibly exploiting the best solution found by the previous solver s_i in t_i seconds. We will use the notation $\underline{\sigma}$ to indicate the *base* of timesplit solver σ where we omit the last solver in the schedule, i.e. $\underline{\sigma} = [(s_1, t_1), \dots, (s_{k-1}, t_{k-1})]$. Intuitively, if s_k is the last solver of the schedule, $\underline{\sigma}$ is the timesplit solver that ideally contributes to improve s_k.

As an example, in Figure 1a the ideal timesplit solver would be defined by $\sigma = [(s1, 10), (s2, 990)]$, but note that there are cases in which the timesplit solver is actually a single solver, since the best solver is not virtually improvable

[1] We can convert a maximization problem to a minimization problem by simply negating the objective function.

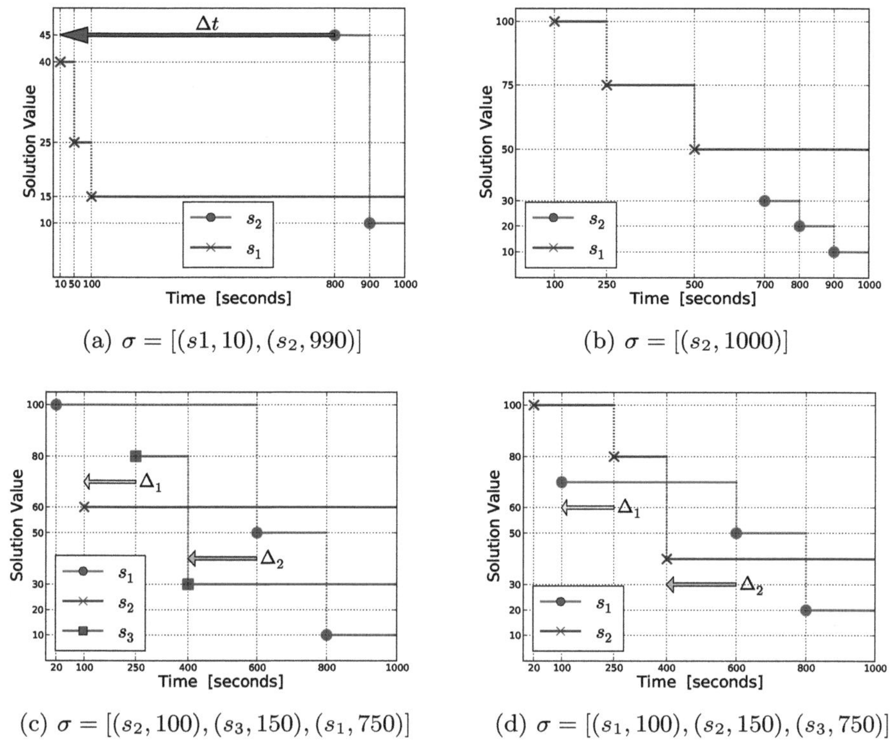

(a) $\sigma = [(s1, 10), (s_2, 990)]$ (b) $\sigma = [(s_2, 1000)]$

(c) $\sigma = [(s_2, 100), (s_3, 150), (s_1, 750)]$ (d) $\sigma = [(s_1, 100), (s_2, 150), (s_3, 750)]$

Fig. 1. Examples of solving behaviours and corresponding time splitting σ

by any other: this happens when every solution found by the best solver is also the best solution found so far (e.g., see Figure 1b). Moreover, there may be also cases in which splitting the time window in more than two slots (even alternating the same solvers) may ideally lead to better performances. Indeed, the "overall" best solver at the time edge T might no longer be the best one at a previous time $t < T$. For example, in Figure 1c the best solver at time $t \geq 800$ is s_1, at time $400 \leq t < 800$ is s_3 while for $t \leq 400$ is s_2; in Figure 1d the best solver is s_1 if $t < 400$ or $t \geq 800$, while for $400 \leq t < 800$ is s_2.

3 Splitting Selection and Evaluation

Once we have informally hypothesized the potential benefits of timesplit solvers, some questions naturally arise. First, which metric(s) is reasonable to formally define the "best solver"? Furthermore, how do we split the time window between solvers for determining the (virtually) best timesplit solver? Finally, to what extent do timesplit solvers act like the virtual timesplit solvers? In order to answer these questions, we fixed some proper metrics, defined a splitting algorithm and empirically evaluated the assumptions previously introduced.

3.1 Evaluation Metrics

In order to evaluate the performances of different COP solvers (and thus formally define the notion of best solver) we examine a number of metrics for grading a solver s on a problem p over a time limit T.

Analogous to the usual metric for CSP solvers, let $\texttt{proven}(s,p) = 1$ if solver s finds and proves the optimal solution (including proving unsatisfiability or unboundedness) for problem p in T seconds, and 0 otherwise. A slightly better metric measures *optimization time*, i.e. the time to find an optimal solution. Let $\texttt{otime}(s,p) = t$ if s finds and proves the optimal solution of p in time t, and $\texttt{otime}(s,p) = T$ if $\texttt{proven}(s,p) = 0$. Unfortunately, both these metrics are rather poor at discriminating: for many optimization problems no solver may be able to prove optimality.

The score function introduced in [4] gives to each solver a score in $[0.25, 0.75]$ proportional to the distance between the best solution it finds and the best known solution. An additional reward ($\texttt{score} = 1$) is given if $\texttt{proven}(s,p) = 1$ while a punishment ($\texttt{score} = 0$) is given if no solution is found without proving unsatisfiability. Let $\texttt{val}(s,p,t) = \min(\{+\infty\} \cup \{v \mid (t',v) \in \mathcal{B}(s,p),\ t' \leq t\})$ be the (possible) best objective value found by solver s for instance p at time t. Let $V_p = \{\texttt{val}(s,p,T) \in \mathbb{Z} \mid s \in \Sigma\}$ be the set of all the objective values found by any solver s at the time limit T. The score of solver s on a problem p (at the time limit T) is a value $\texttt{score}(s,p) \in \{0,1\} \cup [0.25, 0.75]$ such that:

$$\texttt{score}(s,p) = \begin{cases} 0 & \text{if } \texttt{val}(s,p,T) = +\infty;\ \textit{else} \\ 1 & \text{if } \texttt{proven}(s,p) = 1;\ \textit{else} \\ 0.75 & \text{if } \texttt{val}(s,p,T) = \min V_p = \max V_p;\ \textit{else} \\ 0.75 - 0.5 \cdot \dfrac{\texttt{val}(s,p,T) - \min V_p}{\max V_p - \min V_p} & \end{cases}$$

Under this measure, a better solver has a higher score. While more discriminatory than the previous measures, the score measure is still only considers the result at the time limit T, without considering how it was reached (i.e. the behaviour).

In this work we introduce a new metric able to estimate the *anytime* solver performance. Let $W_p = \{\texttt{val}(s,p,t) \in \mathbb{Z} \mid s \in \Sigma,\ t \in [0,T]\}$ be the set of all the solutions found by any solver at any time, so $\min W_p$ is the best solution found for problem p and $\max W_p$ is the worst one. If $\mathcal{B}(s,p) = [(t_1,v_1),\ldots,(t_n,v_n)]$ is the behaviour of solver s on problem p, we define the *(solving) area* of s on p as:

$$\texttt{area}(s,p) = t_1 + \sum_{i=1}^{n} \left(0.25 + 0.5 \cdot \frac{\texttt{val}(s,p,t_i) - \min W_p}{\max W_p - \min W_p}\right)(t_{i+1} - t_i)$$

where $t_{n+1} = \texttt{otime}(s,p)$. As the name implies, area is a normalized measure of the area under a solver behaviour. This metric is similar to the *primal integral* [8] used for measuring impact of heuristics for MIP solvers, but differs since the primal integral assumes the optimal solution is known, while area also differentiates between finding and proving a solution optimal. The area measure folds in a number of measures of the strength of an optimization algorithm: the quality of the best solution found, how quickly any solution is found, whether optimality is proven, and how quickly good solutions are found. Even though the ideal goal is to find the best objective value and hopefully proving its optimality, `area` allows us to discriminate much more between solvers, since we capture the tradeoff between speed and solution quality. Two solvers which eventually reach the same best solution (without proving optimality) are indistinguishable with the other measures, but we would almost certainly prefer the solver that finds the solution(s) faster. Furthermore, consider two solvers that prove optimality at the same instant $t < T$: while both will have $\texttt{otime} = t$, `area` will reward the solver in $[0, t]$ that finds better solutions faster.

Finally, we can now define the *best solver* of Σ for a given problem p as the solver $s \in \Sigma$ which minimizes (w.r.t. the lexicographic ordering) the set of triples $(1 - \texttt{score}(s, p), \texttt{otime}(s, p), \texttt{area}(s, p))$ i.e., the solver that finds the best solution within the time limit T, breaking ties using minimum optimization time first, and then minimum area (i.e., giving priority to the solvers that prove optimality in less time, or at least that quickly find sub-optimal solutions).

3.2 `TimeSplit` Algorithm

Our goal is now to find a suitable timesplit solver for instance p which can improve upon the best solver for p. The algorithm `TimeSplit` described with pseudo-code in Listing 1.1 encodes what was informally explained earlier (see Figure 1). Given as input a problem p, a portfolio $\Pi \subseteq \Sigma$, and the timeout T, the basic idea of `TimeSplit` is to start from the behaviour of the best solver $s_2 \in \Pi$ for p and then examine other solvers behaviours looking for the maximum ideal "left shift" toward another solver $s_1 \in \Pi \setminus \{s_2\}$. Then, starting from s_1, this process is iteratively repeated until no other shift is found. The best solver of Π is assigned to s_2 via function `best_solver` in line 2, while line 3 set the current schedule σ to $[(s_2, T)]$. In line 4 auxiliary variables are initialized: *tot_shift* keeps track of the sum of all the shifts identified, *max_shift* is the current maximum shift that s_2 can perform, *split_time* is the time instant from which s_2 will start its execution, while *split_solver* is the solver that has to be run before s_2 until *split_time* instant. The `while` loop enclosed in lines 5-18 is repeated until no more shifts are possible (i.e., *max_shift* $= 0$). The three nested loops starting at lines 7-9 find two pairs (t_1, v_1) and (t_2, v_2) such that s_2 can virtually shift to another solver s_1, i.e., such that in the current solving window $[0, \textit{split\_time}]$ we have that at time $t_1 < t_2$ solver s_1 finds a value v_1 better than or equal to v_2.

If the actual shift $\Delta t = t_2 - t_1$ is greater than *max_shift*, in lines 11-13 the auxiliary variables are updated accordingly. At the end of each such loop, if at least one shift has been detected (*max_shift* > 0) the current schedule σ needs to be updated. In line 15, the allocated time of the current first solver of σ (i.e., s_2) is decreased by an amount of time *max_shift* + *split_time* (note that $\texttt{first}(\sigma)$ is a reference to the first element of σ, while `snd` returns the second element of a pair, i.e. the allocated time in this case). This is because *split_time* seconds will be allocated to *split_solver* (line 16: `push_front` inserts an element on top of the list) while *max_shift* seconds corresponding to the ideal shift will be later donated to the 'overall' best solver of Π (i.e., the last solver of σ) via *tot_shift* variable. At this stage, the search for a new shift is restricted to the time interval $[0, split\_time]$ in which the new best solver s_2 will be *split_solver* (line 18). Once out of the `while` loop (no more shifts are possible) the total amount of all the shifts found is added to the best solver (line 19: `last`(σ) is a reference to the last element of σ) and the final schedule is finally returned in line 20.

Listing 1.1. `TimeSplit` Algorithm

1 `TimeSplit`(p, Π, T):
2   $s_2 =$ `best_solver`(p, Π, T)
3   $\sigma = [(s_2, T)]$
4   $tot\_shift = 0$; $max\_shift = 1$; $split\_time = T$; $split\_solver = s_2$
5   `while` $max\_shift > 0$:
6     $max\_shift = 0$
7   `for` (t_2, v_2) `in` $\{(t, v) \in \mathcal{B}(s_2, p) \mid t \leq split\_time\}$:
8   `for` s_1 `in` $\Pi \setminus \{s_2\}$:
9   `for` (t_1, v_1) `in` $\{(t, v) \in \mathcal{B}(s_1, p) \mid t < t_2 \wedge v \leq v_2\}$:
10   `if` $t_2 - t_1 > max\_shift$:
11         $max\_shift = t_2 - t_1$
12         $split\_time = t_1$
13         $split\_solver = s_1$
14   `if` $max\_shift > 0$:
15   `first`(σ)`.snd` $-= max\_shift + split\_time$
16   `push_front`$(\sigma, (split\_solver, split\_time))$
17       $tot\_shift \mathrel{+}= max\_shift$
18       $s_2 = split\_solver$
19   `last`(σ)`.snd` $\mathrel{+}= tot\_shift$
20   `return` σ

3.3 `TimeSplit` Evaluation

In order to experimentally verify the correctness of our assumptions of the behaviour of timesplit solvers, we tested `TimeSplit` by considering a portfolio Π constructed from the solvers of the MiniZinc 1.6 suite [21] (i.e., CPX, G12/FD, G12/LazyFD, and G12/MIP) with some additional solvers disparate in their nature, namely: Gecode [11] (CP solver), MinisatID [10] (SAT-based solver), Chuffed (Lazy Clause CP solver), and G12/Gurobi (MIP solver). We retrieved and filtered an initial dataset Δ of 4864 MiniZinc COPs from MiniZinc 1.6 benchmarks and the MiniZinc Challenges 2012/13 and then ran `TimeSplit` using a solving timeout of $T = 1800$ seconds. In particular, we ran and compared two versions of the algorithm: the original one and a variant (denoted `TS-2` in what follows) in which we imposed a maximum size of 2 solvers for each schedule σ. This is because splitting $[0, T]$ in too many slots can be counterproductive in practice: excessive fragmentation of the time window may produce time slots that are too short to be useful. Once executed these algorithms, in order to evaluate their significance we discarded all the "degenerate" instances for which the

Table 1. Average performances

	score	proven	otime	area
VBS	**82.40%**	**34.73%**	1298.67	478.05
TimeSplit	80.49%	33.67%	1263.74	347.91
TS-2	80.60%	33.89%	**1259.98**	**343.97**

Table 2. Pairwise Comparisons

	VBS	TimeSplit	TS-2
VBS	—	222	232
TimeSplit	**373**	—	40
TS-2	364	13	—

potential total shift was minimal (less than 5 seconds). We then ended up with a reduced dataset $\Delta' \subset \Delta$ of 596 instances. We ran timesplit solvers defined by the schedule returned by each algorithm on every instance of Δ'. In addition, we added as a baseline the *Virtual Best Solver* (VBS) i.e. a fictitious portfolio solver that always choose the best solver for every instance according to a given metric. Finally, we evaluated and compared the average performance in terms of the above mentioned metrics: score, proven, otime, area.

Table 1 shows the average results for each approach. As can be seen, the performances are rather close. On average, VBS is still the best solver if we focus on score metric (i.e., considering only the values found at the time limit T). Regarding proven and otime metrics, we can observe a substantial equivalence: VBS is slightly better in terms of percentage of optima proven, while it is worse than TimeSplit and TS-2 if we consider the average time to prove optimality. Conversely, looking at area the situation appears to be more clearly defined: on average, VBS is substantially worse than both TimeSplit and TS-2. This means that, even if the virtual behaviour does not always occur, often the time splitting we propose is able to find good partial solutions more quickly than the best solver of Π. Focusing just on the two versions of TimeSplit, we can also note that these are substantially equivalent: this confirms the hypothesis that limiting the algorithm to schedule only two solvers is a reasonable choice (TS-2 seems slightly better than TimeSplit on average). Indeed, among all the instances of Δ', only for 53 of them TimeSplit has produced a schedule with more than two solvers.

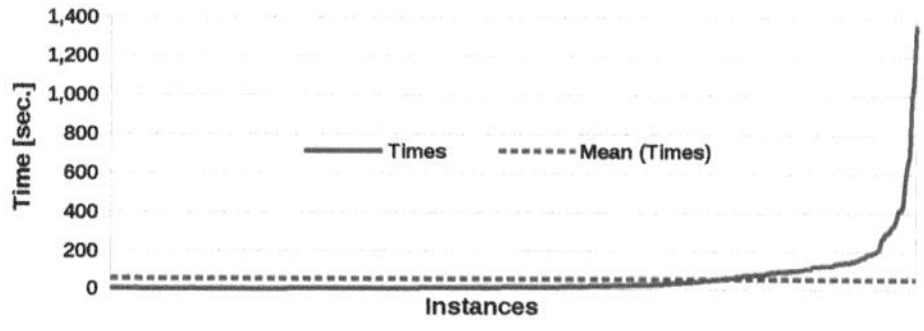

Fig. 2. Times allocated to $\underline{\sigma(p)}$

Table 2 shows instead how many times the approach on the i-th row is better than the one on the j-th column. In this case we can note that TimeSplit and TS-2 perform better than VBS: indeed, in the cases in which the score is the same for both the approaches, often the timesplit solvers take less time to find a (partial) solution. Note that for 375 problems (62.92% of Δ') *at least one* between TimeSplit and TS-2 is better than the VBS. Let Δ^* be the set of such instances, and considering the base $\underline{\sigma(p)}$ of each schedule $\sigma(p)$ returned by the best approach between TimeSplit and TS-2 for each instance of $p \in \Delta^*$, we noticed an interesting fact: the time allocated to $\underline{\sigma(p)}$ is usually pretty low. Figure 2 reports the distribution of such each times. As can be seen, almost all the times are concentrated in the lower part of the graph: even if the maximum value is 1363 seconds, the mean is less than a minute (54.18 seconds to be precise) while the median value is significantly lower (9 seconds).

4 Timesplit Portfolio Solvers

The results of Section 3.3 show that in a non-negligible number of cases the benefits of using a timesplit solver are tangible. Unfortunately, in such experiments for every instance we already knew the corresponding runtimes of each solver of the portfolio. The main motivation of this work is instead to try to predict and run the best timesplit solver for a new unseen instance. Regrettably, given runtime prediction of a solver is a non-trivial task, predicting the detailed solver behaviour on a new test instance is even harder. Indeed, in our case we can not simply limit ourselves to guess the best solver for a new instance, but we should instead predict a suitable timesplit solver $[(s_1, t_1), \ldots, (s_k, t_k)]$. Moreover, even if in most cases the `TimeSplit` algorithm works pretty well, on the others we noticed a considerable number of instances for which this algorithm is ineffective (or even harmful). Therefore, a successful strategy should be able not only to predict a proper timesplit solver, but also to distinguish between the instances for which the timesplit is actually useful and those where it is counterproductive. Furthermore, another interesting observation that has emerged from the results of Section 3.3 is that often for the "significant" timesplit solvers is sufficient to run the base of the schedule for a relatively low number of seconds in order to allow an effective improvement of the best solver.

On the basis of these observations and motivations, what we propose is therefore a generic and hybrid framework that we called *Timesplit Portfolio Solver* (TPS). When a new instance p arrives, we compute and run on p a corresponding timesplit solver $\texttt{TPS}(p) = [(\mathbb{S}, C), (\mathbb{D}(p), T - C)]$, where $[(\mathbb{S}, C)]$ is a *static* timesplit solver pre-computed off-line that will run for $C < T$ seconds, while for the remaining $T - C$ seconds we execute a *dynamic* timesplit solver $[(\mathbb{D}(p), T - C)]$ computed on-line by means of a given prediction algorithm $\mathbb{D}(p)$.

The underlying idea of TPS is to exploit for the first C seconds a fixed schedule calculated *a priori*, whose purpose is to produce as many good sub-optimal solutions as possible. If after C seconds the optimality is still not proven, in the remaining $T - C$ seconds the algorithm $\mathbb{D}(p)$ tries to predict which is the best (timesplit) solver for p, that will be executed taking advantage of any upper bound provided by $\mathbb{S}$. Since TPS is a general model that can be arbitrarily specialized, in the rest of the Section we explain in more detail what choices we made and what algorithms we used to define and evaluate (variants of) TPS.

4.1 Static Splitting

Drawing inspiration from what was done in [16] for SAT problems, we decided to compute a static schedule of solvers according to the outcomes of `TimeSplit` on a given set of training instances. While in [16] the authors solve a Resource Constrained Set Covering Problem (RCSCP) in order to get a schedule that maximizes the number of training instances that can be solved within a time limit of $C = 180$ seconds, in our case the objective is different. What we would like is indeed to compute a schedule that may act as a good base for the solver(s) who will be executed in the remaining $T - C$ seconds. To do this, we first identify by means of `TimeSplit` algorithm the set Δ^* of all the training instances for which a timesplit solver outperforms the VBS. Let $\sigma(p) = [(s_{p,1}, t_{p,1}), \ldots, (s_{p,k}, t_{p,k})]$ be the schedule returned by `TimeSplit` on each $p \in \Delta^*$. We look for a schedule $\mathbb{S}$ that maximizes the number of time slots $t_{p,i} \in \overline{\sigma(p)}$ for $i = 1, \ldots, k-1$ that are *covered*, that is the portfolio solver allocates at least $t_{p,i}$ seconds to solver $s_{p,i}$. Again, note that we consider the base $\overline{\sigma(p)}$ instead of $\sigma(p)$ since at this stage we are not interested in choosing the best solver: we want to determine an effective timesplit solver able to quickly find suitable sub-optimal solutions. However, a nice side-effect of this approach is that it also may be able to solve quickly those instances that are extremely difficult for some solvers but very easy for others.

For each $p \in \Delta^*$, we define $\nabla_p = \{(s_{p,i}, t) \mid (s_{p,i}, t_{p,i}) \in \overline{\sigma(p)},\ t_i \leq t \leq C\}$ as the set of all the pairs $(s_{p,i}, t)$ that *cover* the time slot $t_{p,i}$ within C seconds. Named $\Pi^* = \bigcup_{p \in \Delta^*} \{s \in \Pi : (s,t) \in \nabla_p\}$ the set of the solvers of the portfolio that appear in at least a ∇_p, and fixed $C = T/10$, we solve the following problem:

$$\min \left[(C+1) \sum_{p \in \Delta^*} y_p + \sum_{s \in \Pi^*} \sum_{t \in [0,C]} t\, x_{s,t} \right] \quad s.t.$$

$$y_p + \sum_{(s,t) \in \nabla_p} x_{s,t} \geq 1 \qquad \forall p \in \Delta^*$$

$$\sum_{s \in \Pi^*} \sum_{t \in [0,C]} t\, x_{s,t} \leq C$$

$$y_p,\ x_{s,t} \in \{0,1\} \qquad \forall p \in \Delta^*,\ \forall s \in \Pi^*,\ \forall t \in [0,C]$$

For each pair (s,t) there is a binary variable $x_{s,t}$ that will be equal to one if and only if in $\mathbb{S}$ the solver s will run for t seconds. For each problem p, the binary variable y_p will be one if and only if $\mathbb{S}$ cannot cover any time slot of $\overline{\sigma(p)}$. Constraint $y_p + \sum x_{s,t} \geq 1$ imposes that instance p is covered (possibly setting $y_p = 1$ in the worst case) while $\sum t\, x_{s,t} \leq C$ ensures that $\mathbb{S}$ will not exceed the time limit C. The objective is thus to minimize the number of uncovered instances first (by means of $C+1$ coefficient for each y_p), and the total time of $\mathbb{S}$ then (using t coefficient for each $x_{s,t}$).

Note that the solution of the problem defines an allocation $\xi = \{(s,t) : x_{s,t} = 1\}$ and not actually a schedule: we still have to define the execution order of the solvers. Since the interaction between different solvers is not easily predictable, and neither generalizable, we decided to use a simple and reasonable heuristic: we get the schedule $\mathbb{S}$ by sorting each $(s,t) \in \xi$ by increasing allocated time t.

4.2 Dynamic Splitting

Once defined the static part of TPS, we want to determine an algorithm $\mathbb{D}(p)$ able to predict for a new unseen instance p a proper (timesplit) solver to run for $T-C$ seconds after $[(\mathbb{S}, C)]$. Inspired by the results of [4], we made use of the SUNNY algorithm [3,4]. The reasons behind this choice are essentially two. First, even if originally designed for CSP portfolios [3], the adaption of SUNNY for COPs

Table 3. Pairs (score, otime) of each solver s_i for every problem p_j

	p_1	p_2	p_3	Total
s_1	(1, 150)	(0.25, 1000)	(**0.75**, 1000)	(**2**, *2150*)
s_2	(0, 1000)	(1, 10)	(0, 1000)	(**1**, *2010*)
s_3	(1, 100)	(0.75, 1000)	(0.7, 1000)	(2.45, *2100*)
s_4	(0.75, 1000)	(0.75, 1000)	(0.25, 1000)	(1.75, 3000)

turns out to perform well according to the results of [4]. Second, SUNNY is not limited to predict a single solver but selects instead a schedule of the constituent solvers: in other terms, it implicitly returns a timesplit solver.

SUNNY is a new lazy algorithm portfolio originally tailored for CSPs: given a CSP p and a portfolio Π, it uses a k-NN algorithm to select from a set of training instances a subset $N(p, k)$ of the k instances closer to p according to the Euclidean distance. Then, on-the-fly, it computes a schedule of solvers by considering the smallest sub-portfolio $S \subseteq \Pi$ able to solve the maximum number of instances in the neighborhood $N(p, k)$ and by allocating to each solver of S a time proportional to the number of solved instances in $N(p, k)$. In [4] SUNNY was adapted in order to deal with COPs: this variant selects the sub-portfolio $S \subseteq \Pi$ that maximizes the score in the neighborhood and allocates to each solver a time in $[0, T]$ proportional to its total score in $N(p, k)$. In particular, while in the CSP version SUNNY allocates to a *backup solver*[2] an amount of time proportional to the number of instances not solved in $N(p, k)$, in the COP version it assigns to it a slot of time proportional to $k - h$ where h is the maximum score achieved by the sub-portfolio S. While for CSPs the final schedule is obtained by sorting the solvers by increasing solving time, for COPs the sorting is done by using increasing otime. In a nutshell, the underlying idea behind SUNNY is to minimize the probability of choosing the wrong solvers(s) by exploiting instance similarities in order to get the smallest possible schedule of solvers. Padding the uncovered instances of $N(p, k)$ with the backup solver has the purpose of filling the "gray area" between the best sub-portfolio found and a virtual solver always able to find the optimal solution with the (hopefully) most reliable solver of Π. Of course, this is an arbitrary choice that biases the schedule toward the backup solver. But experimental results have proven the effectiveness of this approach.

Example 1. Let us suppose that $\Pi = \{s_1, s_2, s_3, s_4\}$, the backup solver is s_3, $T = 1000$ seconds, $k = 3$, $N(p, k) = \{p_1, p_2, p_3\}$, and the scores/optimization times are defined as listed in Table 3. The minimum size sub-portfolio that reaches the highest score $h = 1 + 1 + 0.75 = 2.75$ is $\{s_1, s_2\}$. On the basis of the sum of the scores reached by s_1 and s_2 in $N(p, k)$ (resp. 2 and 1, see the last column of Table 3) we then determine a slot size of $t = T/(2+1+(k-h)) = 307.69$ seconds and assign $t_1 = 2 * t = 615.38$ seconds to s_1 and $t_2 = 1 * t = 307.69$ seconds to s_2. The remaining $t_3 = 1000 - (t_1 + t_2) = 76.93$ seconds are finally

[2] A backup solver is a special solver of the portfolio (typically, its *single best solver*) aimed to handle exceptional circumstances (e.g.,premature failures of other solvers).

allocated to the backup solver s_3. The final schedule $[(s_2, t_2), (s_3, t_3), (s_1, t_1)]$ is then obtained by sorting s_1, s_2, s_3 by their total optimization time in $N(p, k)$ (i.e., 2010, 2100, and 2150 respectively: see the last column of Table 3).

5 Empirical Evaluation

In order to measure the performances of the TPS described in Sections 4.1 and 4.2, in the following referred to as `sunny-tps`, we considered a solving timeout of $T = 1800$ seconds, a threshold of $C = 180$ seconds, the portfolio $\Pi = \{$Chuffed, CPX, G12/FD, G12/LazyFD, G12/Gurobi, G12/MIP, Gecode, MinisatID$\}$ and the dataset Δ of 4864 MiniZinc instances introduced in Section 3.3.

We evaluated `sunny-tps` using a 10-fold cross validation [6]: Δ was randomly partitioned in 10 disjoint folds $\Delta_1, \ldots, \Delta_{10}$ treating in turn one fold Δ_i as the test set TS_i $(i = 1, \ldots, 10)$ and the union of the remaining folds $\bigcup_{j \neq i} \Delta_j$ as the training set TR_i. For each training set TR_i we then computed a corresponding static schedule $[(\mathbb{S}_i, 180)]$ as explained in Section 4.1, and for every instance $p \in \mathrm{TS}_i$ we computed and executed the timesplit solver $[(\mathbb{S}_i, 180), (\mathbb{D}_i(p), 1620)]$ where $\mathbb{D}_i(p)$ is the schedule returned by SUNNY algorithm for problem p using a reduced solving window of $T - C = 1620$ seconds.

Note that, for computing $\mathbb{D}_i(p)$, SUNNY has to retrieve the k instances of TR_i closest to p. In order to do so, a proper set of *features* has to be extracted from p (and each instance of TR_i). Instead of using the whole set of 155 features extracted by the `mzn2feat` tool described in [2,5] (as in [4]) we decided to select a proper subset of them by exploiting the new extractor `mzn2feat-1.0`.[3] This tool is a new version of `mzn2feat` designed to be more portable, light-weight, flexible, and independent from the particular machine on which it is run as well as from the specific global redefinitions of a given solver. Indeed, `mzn2feat-1.0` does not compute features based on graph measures (since this process could be very time/space consuming), solver specific features (like global constraint redefinitions) and dynamic features (to decouple the extractor from a particular solver and from the given machine on which it is executed). In more detail, `mzn2feat-1.0` extracts in total 95 features: the variables (27), domains (18), constraints (27), and solving (11) features are exactly the same of `mzn2feat`; the objective features (8) are the 12 objective features of [5] except the 4 features that involve graph measures; finally, the global constraints features are just 4 and no longer bound to the Gecode solver, namely: the number of global constraints n, the number of different global constraints m, the ratio m/n and the ratio n/c where c is the total number of constraints of the problem. We finally removed all the constant features and scaled them in [-1, 1], obtaining thus a reduced set of 88 features.

As in Section 2, we evaluated the average performance of `sunny-tps` in terms of `score`, `proven`, `otime`, `area` by varying the neighbourhood size k in $\{10, 15, 20\}$. Finally, we compared `sunny-tps` vs. the following approaches:

[3] Available at `http://www.cs.unibo.it/~amadini/mzn2feat-1.0.tar.bz2`

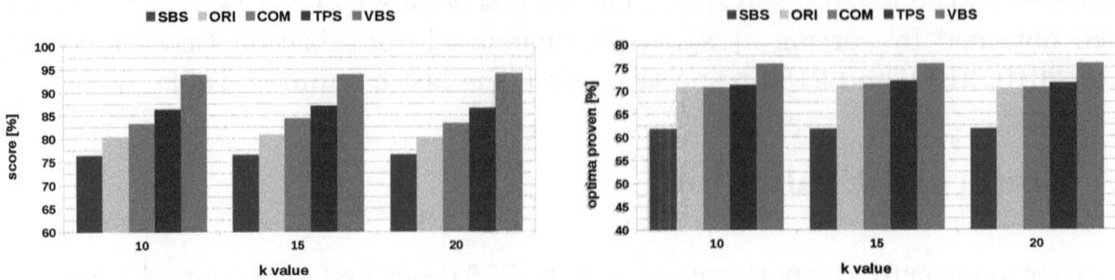

(a) `score` results (in percent) (b) `proven` results (in percent)

- `SBS`: is the overall Single Best Solver of Π according to the given metric; [4]
- `VBS`: is the Virtual Best Solver of Π defined as in Section 3.3;
- `sunny-ori`: is the original SUNNY algorithm evaluated in [4], that is a portfolio solver in which the selected solvers are executed independently (i.e., without any bounds communication) in the time window $[0, T]$ without the "warm start" provided by $\mathbb{S}$ for the first C seconds;
- `sunny-com`: is a portfolio solver that acts basically as `sunny-ori`, with the only exception that solvers execution is not independent: the best value found by a solver within its time slot is subsequently exploited by the following solver of the schedule.

The universe of all the (timesplit) solvers we used for our empirical evaluation was therefore $\Sigma = \Pi \cup \{$`SBS`, `VBS`, `sunny-com`, `sunny-ori`, `sunny-tps`$\}$. It is worth nothing that, while in [1,4] the evaluation was based on *simulations* of the portfolio approaches according to the already computed behaviours of every solver of Π on every instance of Δ, in this work all the approaches have been *actually* run and evaluated. Indeed, as shown also in Section 3.3, in this case we can not make use of simulations since the side effects of bounds communication are unpredictable in advance.

5.1 Test Results

The average `score` results (in percent) by varying the k parameter are reported in Figure 3a. The plot clearly shows a common pattern: `SBS`, `sunny-ori`, `sunny-com`, `sunny-tps`, and `VBS` are respectively sorted by increasing score for every value of k. In general, we can see a rather sharp separation between the various approaches: this witnesses the effectiveness of bounds communication for reaching a better score or, in other terms, for improving the objective value (possibly proving its optimality). For example, the percentage difference between `sunny-ori` and `sunny-com` ranges between 2.83% and 3.45%. Furthermore, running the static schedule for the first 180 seconds (and therefore shrinking the

[4] Regarding the `score`, `otime`, and `area` metrics the single best solver of Π turned out to be CPX, while for the `proven` metric it is Chuffed. According to these results, we elected CPX as the backup solver of Π.

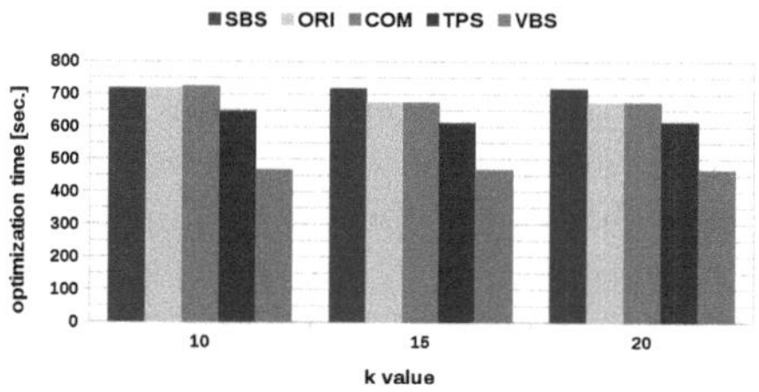

(c) `otime` results (in seconds)

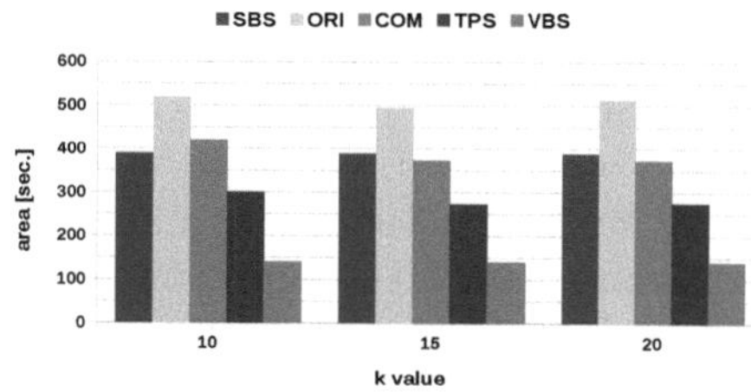

(d) `area` results (in seconds)

dynamic schedule of `sunny-com` in the remaining 1620 seconds) seems to be advantageous: `sunny-tps` is always better than `SBS`, `sunny-ori`, and `sunny-com`. The peak performance (86.91%) is reached with $k = 15$, but the difference with $k = 10$ and 20 is minimal (0.73% and 0.59% respectively). Considering $k = 15$, `sunny-tps` has an average score higher than `SBS` by 10.55%, and lower than `VBS` by 6.9%. Moreover, in 82 cases (1.69% of Δ) it scores better than `VBS`.

When considering the `proven` metric (Figure 3b) the performance difference between the different SUNNY approaches is not so pronounced. Indeed, `sunny-ori`, `sunny-com`, and `sunny-tps` are pretty close: for every k, the percentage difference between the worst and the best SUNNY approach ranges between 0.45% and 1.13%. In this case we can say that the remarkable difference in performance between the portfolio solvers and the `SBS` is mainly due to the SUNNY algorithm rather than the bounds communication. In other words, passing the bound is not so effective if we just focus on proving optimality. A possible explanation is that communicating an upper bound can be useful to find a better solution (see Figure 3a) but ineffective when it comes to prove optimality. In these cases probably the time needed by a solver to compute information for completing the search process can not be offset by the mere knowledge of an objective bound. Nonetheless, the plot shows how the "warm start" provided by the static schedule is helpful: in fact, the performance of `sunny-tps` is always better than the other approaches. The peak performance ($k = 15$) is 72.06%, about 10.36% more than `SBS` and only 3.74% less than `VBS`. For 27 instances (0.56% of Δ) `sunny-tps` is able to prove the optimality while `VBS` is not.

Let us now focus on optimization time. In Figure 3c we see, in contrast to all the `score` and `proven` results that appear to be pretty robust by varying k, a slight discrepancy between $k = 10$ and $k > 10$. This delay time in proving optimality is due to the scheduling order of the constituent solvers. However, for $k = 15$ the results improve and for $k = 20$ are substantially the same. The peak performance is achieved with $k = 15$ (272.61 seconds), 105.07 less than `SBS` and 145.9 more than `VBS`; in 53 cases (1.09% of Δ) `sunny-tps` is able to prove the optimality in less time than `VBS`.

The `area` results depicted in Figure 3d clearly show the benefits of bounds communication. First, note that `sunny-ori` is always worse than `SBS`: this is because each solver scheduled by `sunny-ori` is executed independently, and therefore for every solver the search is always (re-)started from scratch without exploiting previously found solutions. `sunny-com` significantly improves `sunny-ori`, even if its average area is very close to SBS (even worse for $k = 10$). On the other hand, the fixed schedule run by `sunny-tps` often allows one to quickly find partial solutions and thus to noticeably outperform both `sunny-ori` and `sunny-com`. Like the `otime` metric, the average area is not so close to VBS (the peak performance, with $k = 15$, is 272.61 seconds: 132.77 seconds more than VBS, and 114.9 less than SBS), but `sunny-tps` outperforms VBS in 110 cases (2.26% of Δ).

6 Conclusions and Future Work

In this work we addressed the problem of boosting optimization by exploiting the sequential cooperation of different COP solvers. Exploiting the fact that finding good solutions early can significantly improve optimization solvers, we first provided a proper `TimeSplit` algorithm that relies on the behaviour of different solvers on an instance for determining a good timesplit solver for this instance (i.e., ideally able to outperform the best solver of a portfolio). Our results show that on average the actual timesplit solver does perform similarly to (and sometimes even better than) the Virtual Best Solver of the portfolio. We therefore exploited the results of `TimeSplit` in order to define the Timesplit Portfolio Solver (TPS), a generic and hybrid framework that combines a static schedule (computed off-line and run for a limited time) as well as a dynamic schedule (computed on-line by means of a proper prediction algorithm and run in the remaining time) for solving a new unseen instance by exploiting the bounds communication between the scheduled solvers. In particular, on the one hand, we determined the static schedule by solving a Set Covering problem according to the results of `TimeSplit` on a set of training instances, and, on the other, we defined the dynamic selection by exploiting the SUNNY algorithm [3, 4]. Empirical results shows that this idea can be beneficial and sometimes even able to outperform the Virtual Best Solver according to different metrics that we introduced (namely, `score`, `proven`, `otime`, and `area`) in order to evaluate the performance of different (portfolio) solvers.

We see this work a cornerstone for portfolio approaches to solving Constraint Optimization Problems. Clearly bounds communication should be taken into account in this case. It is natural to think of extensions to this work, for example, one may try to maximize the ideal shift by considering all the constituent solvers instead of focusing just on improving the best one. Moreover, the nature of TPS naturally allows one to instantiate its generic schema with new algorithms and techniques (perhaps by simply adapting the most successful portfolio approaches of the SAT and CSP fields). For instance, one might study how to select the set of features to improve solver selection. Note that, contrary to [1,4], for example, significant experimentation is required since it is clearly not predictable in a deterministic way what the side effects of transmitting bounds from a solver to another are. Finally, other interesting directions concerns the study of parallel timesplit solvers as well as the communication of not only the objective bounds but also other additional information (such as for instance cuts which is common in SAT portfolios).

Acknowledgments. NICTA is funded by the Australian Government through the Department of Communications and the Australian Research Council through the ICT Centre of Excellence Program. This work was partially supported by Asian Office of Aerospace Research and Development grant 12-4056.

References

1. Amadini, R., Gabbrielli, M., Mauro, J.: An Empirical Evaluation of Portfolios Approaches for Solving CSPs. In: Gomes, C., Sellmann, M. (eds.) CPAIOR 2013. LNCS, vol. 7874, pp. 316–324. Springer, Heidelberg (2013)
2. Amadini, R., Gabbrielli, M., Mauro, J.: Features for Building CSP Portfolio Solvers. CoRR, abs/1308.0227 (2013)
3. Amadini, R., Gabbrielli, M., Mauro, J.: SUNNY: a Lazy Portfolio Approach for Constraint Solving. In: ICLP (2014), `http://www.cs.unibo.it/~amadini/iclp_2014.pdf`
4. Amadini, R., Gabbrielli, M., Mauro, J.: Portfolio Approaches for Constraint Optimization Problems. In: LION (2014), `http://www.cs.unibo.it/~amadini/lion_2014.pdf`
5. Amadini, R., Gabbrielli, M., Mauro, J.: An Enhanced Features Extractor for a Portfolio of Constraint Solvers. In: SAC (2014), `http://www.cs.unibo.it/~amadini/sac_2014.pdf`
6. Arlot, S., Celisse, A.: A survey of cross-validation procedures for model selection. Statistics Surveys 4, 40–79 (2010)
7. Audemard, G., Hoessen, B., Jabbour, S., Lagniez, J.-M., Piette, C.: PeneLoPe, a Parallel Clause-Freezer Solver. In: SAT Challenge 2012 (2012)
8. Berthold, T.: Measuring the impact of primal heuristics. Operations Research Letters 41(6), 611–614 (2013)
9. Carchrae, T., Beck, J.C.: Low-knowledge algorithm control. In: AAAI, pp. 49–54 (2004)
10. DeCat, B.: KRR Software: MinisatID (2013), `http://dtai.cs.kuleuven.be/krr/software/minisatid`
11. GECODE - An open, free, efficient constraint solving toolkit, `http://www.gecode.org`
12. Gomes, C.P., Selman, B.: Algorithm portfolios. Artif. Intell (2001)
13. Guo, H., Hsu, W.H.: A machine learning approach to algorithm selection for NP-hard optimization problems: a case study on the MPE problem. Annals OR 156(1), 61–82 (2007)
14. Hamadi, Y., Jabbour, S., Sais, L.: ManySAT: A Parallel SAT Solver. JSAT 6(4), 245–262 (2009)
15. Hutter, F., Xu, L., Hoos, H.H., Leyton-Brown, K.: Algorithm Runtime Prediction: The State of the Art. CoRR, abs/1211.0906 (2012)
16. Kadioglu, S., Malitsky, Y., Sabharwal, A., Samulowitz, H., Sellmann, M.: Algorithm Selection and Scheduling. In: Lee, J. (ed.) CP 2011. LNCS, vol. 6876, pp. 454–469. Springer, Heidelberg (2011)

17. Kadioglu, S., Malitsky, Y., Sellmann, M., Tierney, K.: ISAC - Instance-Specific Algorithm Configuration. In: ECAI 2010 (2010)
18. Kotthoff, L.: Algorithm Selection for Combinatorial Search Problems: A Survey. CoRR, abs/1210.7959 (2012)
19. Mackworth, A.K.: Consistency in Networks of Relations. Artif. Intell. (1977)
20. Malitsky, Y., Sabharwal, A., Samulowitz, H., Sellmann, M.: Boosting sequential solver portfolios: Knowledge sharing and accuracy prediction. In: Nicosia, G., Pardalos, P. (eds.) LION 7. LNCS, vol. 7997, pp. 153–167. Springer, Heidelberg (2013)
21. Minizinc version 1.6, `http://www.minizinc.org/download.html`
22. O' Mahony, E., Hebrard, E., Holland, A., Nugent, C., O' Sullivan, B.: Using case-based reasoning in an algorithm portfolio for constraint solving. In: AICS 2008 (2009)
23. Rice, J.R.: The Algorithm Selection Problem. Advances in Computers (1976)
24. Roussel, O.: ppfolio, `http://www.cril.univ-artois.fr/~roussel/ppfolio/`
25. Smith-Miles, K.: Cross-disciplinary perspectives on meta-learning for algorithm selection. ACM Comput. Surv. 41(1) (2008)
26. Telelis, O., Stamatopoulos, P.: Combinatorial Optimization through Statistical Instance-Based Learning. In: ICTAI (2001)
27. Xu, L., Hutter, F., Hoos, H.H., Leyton-Brown, K.: Hydra-MIP: Automated Algorithm Configuration and Selection for Mixed Integer Programming. In: RCRA workshop on Experimental Evaluation of Algorithms for Solving Problems with Combinatorial Explosion (2011)

Scoring-Based Neighborhood Dominance for the Subgraph Isomorphism Problem

Gilles Audemard[1], Christophe Lecoutre[1], Mouny Samy-Modeliar[1], Gilles Goncalves[2], and Daniel Porumbel[2]

[1] CRIL-CNRS, UMR 8188, University of Artois, Lens, France
[2] LGI2A, University of Artois, Bethune, France
{audemard,lecoutre,modeliar}@cril.fr,
{gilles.goncalves,daniel.porumbel}@univ-artois.fr

Abstract. This paper presents an original filtering approach, called SND (Scoring-based Neighborhood Dominance), for the subgraph isomorphism problem. By reasoning on vertex dominance properties based on various scoring and neighborhood functions, SND appears to be a filtering mechanism of strong inference potential. For example, the recently proposed method LAD is a particular case of SND. We study a specialization of SND by considering the number of k-length paths in graphs and three ways of relating sets of vertices. With this specialization, we prove that SND is stronger than LAD and incomparable to SAC (Singleton Arc Consistency). Our experimental results show that SND achieves most of the time the same filtering performances as SAC (while being several orders of magnitude faster), which allows one to find subisomorphism functions far more efficiently than MAC, while slightly outperforming LAD.

1 Introduction

The scientific literature is replete with graph-related objects (e.g., trees, combinatorial maps, attributed or weighted graphs, etc.) that are routinely used to describe different inter-linked or relational structures, e.g, networks, hierarchies and patterns. Many applications involving such concepts require finding a "matching" between two graph objects. To deal with such applications, a variety of (sub-)graph isomorphism problems have been proposed over the years, e.g., graph matching, error-correcting isomorphism, induced subgraph isomorphism, maximum common sub-graph and epimorphism [2,3].

Most of these matching problems are NP-hard, except perhaps the pure Graph Isomorphism (GI) that has an undefined status. From a theoretical perspective, an important effort has been done to determine the complexity class of (sub-)graph isomorphism problems, e.g., see the Fulkerson prize work on bounded degree graphs [14]. Complexity results and (polynomial) algorithms for new isomorphism problems represent an active area of research [7,8]. Interestingly enough, many practical applications have been proposed over the years.

Historically, the first practical (sub-)graph isomorphism algorithms date back to the 1970s [6,23]. Generally speaking, these early methods recursively construct a match matrix by backtracking, using "look-ahead" techniques to anticipate isomorphism violations and prune branches. But important progress has been done over the years in this

B. O'Sullivan (Ed.): CP 2014, LNCS 8656, pp. 125–141, 2014.

area. For example, VF2 [5] is shown to exhibit an exponential speed-up compared to implementations of other similar approaches (i.e., re-implementations of the algorithms of Ulmann Schmidt-Druffel). One can also mention the methods based on the computation of graph automorphism groups for pure isomorphism, and the "color-refining" techniques (e.g., the Weisfeiler-Lehman algorithm [24]) that can be used to classify vertices according to different invariants (e.g., degrees).

Incomplete and heuristic approaches are also appropriate to deal with NP-hardness. They can actually be found in the very vast graph matching (GM) literature, e.g., hundreds of references are given in [2,3]. Certain families of algorithms can be particularized to pure (sub-)graph isomorphism. For instance, the heuristic from [17] tries to minimize a similarity measure between two graphs: reaching a value of 0 is equivalent to finding an isomorphism.

With respect to complete methods, a work direction resides in using the rich filtering and propagation tools from the Constraint Programming (CP) literature (see [22,26,16] for recent developments). Indeed, the (sub-)graph isomorphism problem can be easily cast as an instance of the Constraint Satisfaction Problem (CSP). In such view, one can exploit long-acknowledged constraint propagation techniques based on properties like, e.g., Arc Consistency (AC) or Singleton Arc Consistency (SAC) (see other properties in [13]). To reinforce the inference process, several isomorphism-specific filtering techniques have also been proposed; for instance, ILF [26] and LAD [22] uses degree information, multi-sets and neighborhood to classify vertices and filter domains.

We propose a general approach to reason about vertex dominance following two main directions: first, by associating a set of scores with each pair of vertices in any graph, and second, by considering different ways of defining the neighborhood of vertices. We call this approach SND for Scoring-based Neighborhood Dominance. We show that LAD is a particular case of SND and we study a specialization of SND by considering the number of k-length paths in graphs and three ways of relating sets of vertices. With this specialization, we prove that SND is stronger than LAD and incomparable to SAC (Singleton Arc Consistency). Our experimental results show the effectiveness of our approach.

This paper is organized as follows. Section 2 presents constraint approaches for the subgraph isomorphism problem. Section 3 gives a detailed description of Scoring-based Neighborhood Dominance. Section 4 presents a theoretical comparison of SND with respect to other filtering algorithms. In Section 5, an algorithm for a weak version of SND is introduced, and experimental results are presented in Section 6. Then, we conclude.

2 CP for the Subgraph Isomorphism Problem

Many combinatorial problems can be modeled and solved using tools of Constraint Programming (CP). This is the case for the subgraph isomorphism problem, whose instances can be straightforwardly modeled as constraint networks.

2.1 Technical Background

A (discrete) *constraint network* (CN) P is composed of a finite set of variables, denoted by vars(P), and a finite set of constraints, denoted by cons(P). Each *variable* x has a *domain*, which is the finite set of values that can be feasibly assigned to x. The *initial* domain of a variable x is denoted by $dom^{init}(x)$ whereas the *current* domain of x is denoted by $dom(x)$; we always have $dom(x) \subseteq dom^{init}(x)$. Each *constraint* c involves an ordered set of variables, called the *scope* of c and denoted by $scp(c)$. An *instantiation* I of an ordered set of variables $X = \{x_1, \ldots, x_r\}$ is a set $\{(x_1, a_1), \ldots, (x_r, a_r)\}$ such that $\forall i \in 1..r, a_i \in dom^{init}(x_i)$; each a_i is denoted by $I[x_i]$. Importantly, each constraint c is semantically defined by a *relation*, which contains the set of *allowed* instantiations of the scope of c ; these instantiations are also said to *satisfy* c. A *solution* of a CN P is an instantiation I of vars(P) such that every constraint $c \in$ cons(P) is satisfied by I (restricted to the variables in $scp(c)$). The set of solutions of a CN P is denoted by $sols(P)$.

The initial search space of a CN is exactly the Cartesian product of the initial domains for the variables of the CN. Testing the existence of a solution in this space is a NP-complete task (assuming that checking whether a constraint is satisfied or not can be made in polynomial time). Fortunately, it is possible to reduce the search space by removing certain inconsistent values from domains[1], while preserving the set of solutions, by executing so-called *filtering* algorithms. Interestingly enough, filtering algorithms can be called in sequence, until a fixed point is reached, leading to a process known as *constraint propagation*.

The level of filtering reached by constraint propagation is typically determined by a target property called *consistency*. For example, Arc Consistency (AC), called Generalized Arc Consistency (GAC) when constraints are non-binary, is the most famous one. We define it now. An instantiation I is *valid* iff $\forall(x, a) \in I, a \in dom(x)$. An instantiation I of $scp(c)$ is a *support* on c iff I is a valid instantiation satisfying c. If I is a support on c such that $x \in scp(c)$ and $I[x] = a$, we say that I is a *support for* (x, a) on c. A constraint c is *generalized arc consistent* (GAC) iff $\forall x \in scp(c), \forall a \in dom(x)$, there exists at least one support for (x, a) on c. A CN P is GAC iff every constraint of P is GAC.

(G)AC is the workhorse of general-purpose constraint solvers. It corresponds to the maximum level of filtering when constraints are considered independently. If P is a CN, then GAC(P) denotes the CN obtained from P by removing iteratively all values detected without any support on a constraint of P. Of course, it is possible to filter further by extending the reach of local reasoning. One such consistency, stronger than GAC, is called Singleton Arc Consistency (SAC) [9]. If P is a CN, and (x, a) a pair such that $x \in scp(c)$ and $a \in dom(x)$, called *value of* P, then $P|_{x=a}$ denotes the CN obtained from P by restricting the domain of x to a. A value (x, a) of P is *singleton arc consistent* (SAC) iff GAC$(P|_{x=a}) \neq \bot$, meaning that no domain has become empty by propagation. A CN P is SAC iff every value of P is SAC.

Using the terminology from [10], we say that a domain filtering process ϕ is *stronger* than another one ψ iff, on any CN, ϕ always deletes at least the values deleted by ψ, and *strictly stronger* if there is one CN where ϕ deletes more values than ψ. Accordingly, ϕ is *incomparable* to ψ iff neither is stronger than the other.

[1] Other inference techniques that do not directly filter domains exist, but they are beyond the scope of this paper.

2.2 Isomorphism Model and Filtering Procedures

Before introducing the subgraph isomorphism problem, and different ways of addressing it using CP, we recall some definitions on graphs. A simple *graph* G is a pair (V, E) where V is a set of *vertices* and E is a set of *edges*, which are subsets of V composed of exactly two vertices. An edge corresponds to an unordered pair of vertices. In a *simple* graph, the vertices in each edge are distinct, and there is not more than one edge between each pair of distinct vertices. Hereafter, graph alone means a simple graph. For any vertex $v \in V$, $\Gamma(v)$ denotes the set of nodes *adjacent* to v, i.e., the set of vertices sharing an edge with v.

An instance of the *subgraph isomorphism problem* is defined by a pattern graph $G_p = (V_p, E_p)$ and a target graph $G_t = (V_t, E_t)$: the objective is to determine whether G_p is isomorphic to some subgraph(s) in G_t. Finding a solution to such a problem instance means then finding a *subisomorphism function*, that is an injective mapping $f : V_p \rightarrow V_t$ such that all edges of G_p are preserved: $\forall(v, v') \in E_p, (f(v_p), f(v'_p)) \in E_t$. Note that we refer in this paper to the partial, and not the induced subgraph isomorphism problem. Our problem is also referred to as (subgraph) monomorphism, see [3, p. 268].

We introduce a natural way of representing an instance of this problem by means of a CN noted P. First, a variable x_{v_p} is introduced in vars(P) for each vertex v_p of the pattern graph such that the domain of x_{v_p} is exactly the set V_t : if x_{v_p} is set to the value v_t, it means that the pattern vertex v_p is mapped to the target vertex v_t. Second, a global constraint allDifferent(vars(P)) is introduced in cons(P), which guarantees injection, i.e., the fact that pattern vertices are mapped to distinct target vertices. Finally, a binary extensional constraint $\{x_{v_p}, x_{v'_p}\} \in E_t$ is introduced for each edge $\{v_p, v'_p\} \in E_p$, which guarantees that every edge in the pattern graph is associated to an edge in the target graph, i.e., the mapping does not need to be edge-preserving in both directions. Clearly, the set of solutions to the CN P is exactly the set of possible sub-isomorphism functions.

There are other possible CP models for the subgraph isomorphism problem, but the one introduced above is certainly the simplest one, and was used basically in many previous pieces of work (e.g., see [23,15,19,12,26,22]). In the literature, attempts to solve efficiently instances of this model have been made with various levels of filtering. For example, a partial form of arc consistency, called Forward Checking (FC) was employed in [23,15]. Later, J.-C. Régin [19] used GAC on the constraint allDifferent and AC on the binary extensional constraints. More recently, there have been a few proposals to combine constraints in order to filter out more values. In [12], in addition to GAC, it is proposed to reason from the cardinality of the sets of neighbors of each vertex, while considering the current domains of variables; we shall denote LV2002 this level of filtering (as in [22]). A so-called iterated labeling filtering (ILF) has been introduced in [26]. The idea is to exploit the structure of both pattern and target graphs in order to associate a global label with each vertex. A compatibility relationship defined over the set of labels can then be used to remove from the domain of a variable x_{v_p} every target vertex v_t such that the label associated with v_t is not compatible with the label associated with v_p. ILF is an iterative procedure that starts from an initial labeling. For example, one can choose vertex degrees; we shall denote ILF^{deg} the filtering method obtained from this choice. At each iteration, the new label for a vertex is a multiset composed of the labels adjacent to the vertex. Reasoning from sorted multisets allows us to detect and remove inconsistent values. ILF^{dom} uses another iterative labeling that integrates current domains in the compatibility relation [26]. Note that this idea of using multisets (of degrees) was also used outside the constraint programming literature [24,21]. Finally, LAD is a filtering procedure [22] that is equivalent to adding $n_p \times n_t$ constraints to the CN, and to enforcing GAC. For each pattern vertex v_p and each target vertex v_t, there is thus an implicit constraint, that we denote by $\texttt{LAD}(v_p, v_t)$, added to cons(P) such that its scope is $\{x_{v_p}\} \cup \{x_{v'_p} \mid v'_p \in \Gamma(v_p)\}$, and its semantics is given by:

$$\begin{array}{r} x_{v_p} = v_t \Rightarrow \\ allDifferent(\{x_{v'_p} \mid v'_p \in \Gamma(v_p)\}) \wedge x_{v'_p} \in \Gamma(v_t), \forall v'_p \in \Gamma(v_p) \end{array} \quad (1)$$

This constraint states that if v_p is mapped to v_t then all pattern vertices adjacent to v_p must be mapped to distinct target vertices adjacent to v_t. In practice, such "neighborhood" constraints are not added to the CN, but a specific filtering algorithm is used to reach the level of filtering equivalent to GAC. Note that the filtering algorithm introduced in [12] corresponds to enforcing a partial form of GAC on LAD constraints: it basically ensures that the number of vertices adjacent to v_p is smaller than or equal to the number of target vertices that are both adjacent to v_t and belong to at least one domain of a variable in $\{x_{v'_p} \mid v'_p \in \Gamma(v_p)\}$.

For simplicity in our presentation, we say that GAC is enforced on LAD constraints. This is not exactly true in practice because for each constraint $\texttt{LAD}(v_p, v_t)$, we only focus on the value (x_{v_p}, v_t): if (x_{v_p}, v_t) has no support on $\texttt{LAD}(v_p, v_t)$, we delete this value. In other words, we use the converse of (1) to make a single possible inference. In case the domain of x_{v_p} is reduced to $\{v_t\}$, we simply use the other constraints of the CN to make the inferences that would be possible with the constraint $\texttt{LAD}(v_p, v_t)$. So, the result is the same: at the end of constraint propagation, the CN is GAC.

3 Scoring-Based Neighborhood Dominance

3.1 Principle and Correctness

We propose to generalize the reasoning behind LAD according two dimensions: first, by associating a set of scores with any pair of vertices in a graph, and second, by considering different ways of defining the neighborhood of vertices in a graph (contrary to LAD, these sets do not necessarily correspond to the vertices that are adjacent to a specified vertex). We call this approach SND for Scoring-based Neighborhood Dominance.

Hence, SND can be seen as a general approach that is parametrized by two elements denoted by $\mathcal{S}$ (for scoring) and $\mathcal{N}$ (for neighborhood). A specialization SND$_{\mathcal{S},\mathcal{N}}$ of scoring-based neighborhood dominance is then defined by a set of scoring functions $\mathcal{S}$ such that each $S \in \mathcal{S}$ gives for each pair of vertices (v, v') of a graph a score $S(v, v')$ (defined as an integer or real number computed by any relevant means) and a set of neighborhood functions $\mathcal{N}$ such that each $N \in \mathcal{N}$ gives for each vertex v of a graph a set $N(v)$ of vertices in the same graph.

As for LAD, for each pair (v_p, v_t) of pattern-target vertices, we implicitly consider a constraint. It is denoted by $\text{SND}_{\mathcal{S},\mathcal{N}}(v_p, v_t)$, its scope is $\{x_{v_p}\} \bigcup_{N \in \mathcal{N}} \{x_{v'_p} \mid v'_p \in N(v_p)\}$, and its semantics is given by:

$$x_{v_p} = v_t \Rightarrow$$
$$\bigwedge_{N \in \mathcal{N}} \Big(allDifferent(\{x_{v'_p} \mid v'_p \in N(v_p)\}) \wedge x_{v'_p} \in N(v_t), \forall v'_p \in N(v_p) \tag{2}$$
$$\wedge\ S(v_p, v'_p) \le S(v_t, x_{v'_p}), \forall v'_p \in N(v_p), \forall S \in \mathcal{S} \Big)$$

When the last condition of Equation 2 holds (in the context of a neighborhood function N), i.e., $S(v_p, v'_p) \le S(v_t, v'_t)$ where $v'_t = x_{v'_p}$, we say that (v_p, v'_p) is *dominated* by (v_t, v'_t). If there is no way of finding an instantiation of $\{x_{v'_p} \mid v'_p \in N(v_p)\}$ such that all pairs (v_p, v'_p) are dominated by pairs $(v_t, x_{v'_p})$, then we can conclude that $x_{v_p} \neq v_t$. Of course, the way the scores are computed and the way the neighborhoods are defined must be coherent with respect to the CN. In other words, any constraint $\text{SND}_{\mathcal{S},\mathcal{N}}(v_p, v_t)$, added to cons$(P)$ must be a consequence of P. We shall say that a constraint of the form $\text{SND}_{\mathcal{S},\mathcal{N}}(v_p, v_t)$ is *implied* (w.r.t. P) iff $sols(P) = sols(P \oplus \text{SND}_{\mathcal{S},\mathcal{N}}(v_p, v_t))$ where $P \oplus c$ is the CN P with the additional constraint c.

In this paper, we propose some original combinations of scoring and neighborhood functions. On the one hand, we propose to use numbers of paths in graphs as a basis of our scoring functions. Indeed, it is well-known that if M_G is the adjacency matrix associated with a graph G, then $M_G^k[v][v']$ indicates the number of k-length paths in G going from vertex v to vertex v'. The idea is that it is not possible to map a pair of pattern vertices (v_p, v'_p) to a pair (v_t, v'_t) of target vertices if for a certain k, we have $M_{G_p}^k[v_p][v'_p] > M_{G_t}^k[v_t][v'_t]$. Consequently, such an observation can be used for filtering. For simplicity and by abuse of notation, M^k will denote the scoring function S such that for any pair of vertices (v, v') in a graph G, we have $S(v, v') = M_G^k[v][v']$. Hereafter, we shall only consider such functions, so-called *length-scoring* functions. In this context, $\mathcal{M}$ will denote the set of all length-scoring functions, i.e., $\mathcal{M} = \{M^1, M^2, \dots\}$.

On the other hand, we propose three different neighborhood functions defined as follows on any graph $G = (V, E)$:

- `Id` is the identity function defined as $\texttt{Id}(v) = \{v\}, \forall v \in V$. Combined with $\mathcal{M}$, it can be used to filter domains by considering the number of paths going from one vertex to itself.
- Γ is the function (defined earlier) that returns the set of adjacent vertices. This is the kind of neighborhood used by LAD.
- `all` is the function that returns the full set of vertices except the input; $\texttt{all}(v) = V \setminus \{v\}, \forall v \in V$. This function is complementary to `Id`.

After the observation that LAD is a particular case of SND, we prove the correctness of our approach with length-scoring functions and neighborhood functions `Id`, Γ, and `all`.

Remark 1. LAD is equivalent to $\text{SND}_{\mathcal{S},\mathcal{N}}$, with $\mathcal{S} = \emptyset$ and $\mathcal{N} = \{\Gamma\}$.

Proposition 1. *Any constraint* $SND_{\mathcal{S},\mathcal{N}}(v_p, v_t)$*, with* $\mathcal{S} \subseteq \mathcal{M}$ *and* $\mathcal{N} \subseteq \{\texttt{Id}, \Gamma, \texttt{all}\}$ *is implied.*

Proof. Suppose first that $\mathcal{N} = \{\Gamma\}$. We know from [22] that any constraint LAD is implied. The only difference $\texttt{SND}_{\mathcal{S},\{\Gamma\}}$ makes, compared to LAD, is that, if $M^k \in \mathcal{S}$ then it is safe to prevent from matching a node v_p' adjacent to v_p to a node v_t' adjacent to v_t when the number of k-length paths going from v_p to v_p' is strictly greater than the number of paths of length k going from v_t to v_t'. Clearly, if $M^k_{G_p}[v_p][v_p'] > M^k_{G_t}[v_t][v_t']$, there is no subisomorphism function that maps v_p to v_t while v_p' is mapped to v_t'. Hence, the additional restriction imposed by M^k guarantees that the constraint is implied. Suppose now that $\mathcal{N} = \{\texttt{Id}\}$. With this neighborhood, assuming that $M^k \in \mathcal{S}$, v_p is prevented from being matched to v_t when the number of k-length paths going from v_p to itself is strictly greater than the number of paths of length k going from v_t to itself. If $M^k_{G_p}[v_p][v_p] > M^k_{G_t}[v_t][v_t]$, there is no subisomorphism function that maps v_p to v_t. So the constraint is implied. The proof is similar for $\mathcal{N} = \{\texttt{all}\}$, and can be extended to any subset of $\{\texttt{Id}, \Gamma, \texttt{all}\}$. □

In our experimentation, described later, we shall use the neighborhood functions defined previously. However, note that there are many other possibilities. For example, one could reason with neighborhoods defined as sets of vertices at distance 1 or 2. Also, instead of using length-scoring functions, we might exploit other scoring functions based, for instance, on numbers of cliques or cycles. Studying these alternatives is beyond the scope of the paper, and left as a perspective of our work.

3.2 Filtering SND Constraints

Filtering SND constraints in order to reach GAC is similar to what has been proposed for the constraints allDifferent [18] and LAD. This is based on the concept of covering matching in a bipartite graph. A *matching* in a graph is a subset of its edges, no two of which share a vertex. A matching *covers* a set of vertices iff each of its vertices appears in an edge of the matching. A *bipartite* graph is a graph $G = (V, E)$ where V is partitioned into two sets V_1 and V_2 such that each edge E is incident to a vertex in V_1 and a vertex in V_2; to emphasize the partition, we note $V = V_1|V_2$.

We can associate one bipartite graph G_N per constraint $\texttt{SND}_{\mathcal{S},\mathcal{N}}(v_p, v_t)$ and neighborhood function $N \in \mathcal{N}$. The bipartite graph $G_N = (V, E)$, associated with this constraint and N, is defined as follows:

- $V = \text{N}(v_p)|\text{N}(v_t)$;
- $E = \{(v_p', v_t') \in \text{N}(v_p) \times \text{N}(v_t) \mid v_t' \in dom(x_{v_p'}) \wedge S(v_p, v_p') \leq S(v_t, v_t'), \forall S \in \mathcal{S}\}$

If there does not exist a matching of the bipartite graph that covers $\text{N}(v_p)$, then the right-hand side of Equation 2 (after symbol $\Rightarrow$) evaluates to $\texttt{false}$, and then (x_{v_p}, v_t) has no support on the constraint. This value can be safely deleted (provided that the constraint is implied).

Example 1. Figure 1 shows an instance of the subgraph isomorphism problem. The adjacency matrices M_{G_p} and M_{G_t} as well as $M^2_{G_p}$ and $M^2_{G_t}$, are:

$$\underset{M_{G_p}}{\begin{pmatrix} 0&1&1&1\\ 1&0&1&1\\ 1&1&0&1\\ 1&1&1&0 \end{pmatrix}} \qquad \underset{M_{G_t}}{\begin{pmatrix} 0&1&0&1&1\\ 1&0&1&0&1\\ 0&1&0&1&1\\ 1&0&1&0&1\\ 1&1&1&1&0 \end{pmatrix}} \qquad \underset{M^2_{G_t}}{\begin{pmatrix} 3&2&2&2\\ 2&3&2&2\\ 2&2&3&2\\ 2&2&2&3 \end{pmatrix}} \qquad \underset{M^2_{G_t}}{\begin{pmatrix} 3&1&3&1&2\\ 1&3&1&3&2\\ 3&1&3&1&2\\ 1&3&1&3&2\\ 2&2&2&2&4 \end{pmatrix}}$$

Let us consider the constraints $\text{SND}_{\{M^1\},\{\Gamma\}}(1, A)$ and $\text{SND}_{\{M^2\},\{\Gamma\}}(1, A)$ (for simplicity, we only consider a unique length-scoring function and a unique neighborhood function). The bipartite graphs associated with those constraints (and Γ) are depicted in Figure 2, where $\Gamma(1) = \{2, 3, 4\}$ and $\Gamma(A) = \{B, D, E\}$ (assuming that the current domains of x_2, x_3 and x_4 contain all initial values). Note that the scoring is shown next to vertices. Clearly, with M^1, we cannot deduce anything on $(1, A)$ since there is a matching that covers $\Gamma(1)$ (shown in boldface in Figure 2(a)). However, with M^2, we can deduce that $x_1 \neq A$. Indeed, when we consider the scoring based on M^2, all dotted edges can be discarded. For example, the score of x_2 is $M^2_p[x_1][x_2] = 2$ while the score of B is only $M^2_t[A][B] = 1$. This means that 2 cannot be mapped to B, and so the edge is deleted. Finally, when only considering the remaining edges (non dotted ones), clearly there is no covering matching of $\Gamma(1)$.

3.3 Simplifying the Target Graph

When domains are reduced during propagation, it is sometimes possible to refine the scoring functions we use. Indeed, an edge can be deleted from the target graph when we have the guarantee that no pair of pattern vertices can be mapped anymore to the target vertices corresponding to this edge. More precisely, we say that an edge $(v_t, v'_t) \in E_t$ is *unreachable* iff for every edge $(v_p, v'_p) \in E_p$, $(v_t, v'_t) \notin dom(x_{v_p}) \times dom(x_{v'_p})$ and $(v_t, v'_t) \notin dom(x_{v'_p}) \times dom(x_{v_p})$. So, after any filtering process, we can simplify the target graph by removing all its unreachable edges. Scoring functions can possibly be updated, as for example, length-scoring functions of $\mathcal{M}$ since as soon as an edge is removed, the corresponding adjacency matrix is modified. Consequently, we can start again reasoning with the SND constraints.

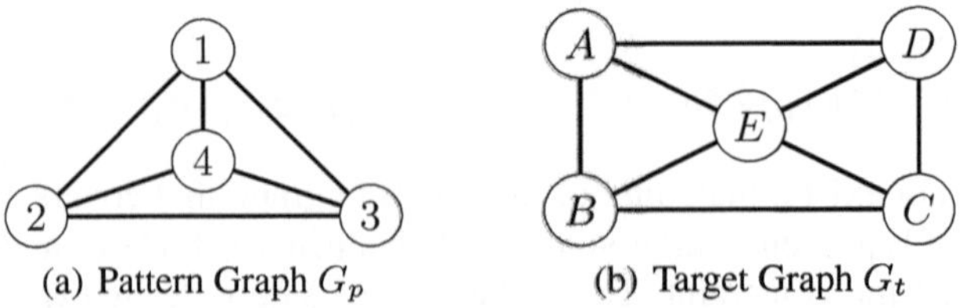

(a) Pattern Graph G_p (b) Target Graph G_t

Fig. 1. An instance of the subgraph isomorphism problem

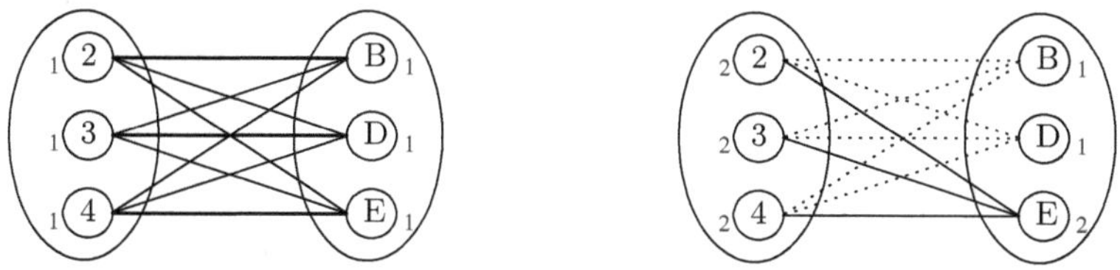

(a) $\texttt{SND}_{\{M^1\},\{\Gamma\}}(1, A)$ (b) $\texttt{SND}_{\{M^2\},\{\Gamma\}}(1, A)$

Fig. 2. Existence of a covering matching in the bipartite graph for $\texttt{SND}_{\{M^1\},\{\Gamma\}}(1, A)$, and absence of covering matching in the bipartite graph for $\texttt{SND}_{\{M^2\},\{\Gamma\}}(1, A)$

4 Theoretical Filtering Comparisons

From now on, $\text{SND}_{\mathcal{S}}$ will be used as an abbreviation for $\text{SND}_{\mathcal{S},\{\texttt{Id},\Gamma,\texttt{all}\}}$, and will stand for the filtering level corresponding to enforce GAC on the initial CN supplemented with all constraints $\text{SND}_{\mathcal{S},\{\texttt{Id},\Gamma,\texttt{all}\}}(v_p, v_t)$. In our study, $\mathcal{S}$ will be either $\mathcal{M}$ (the set of all length-scoring functions) or a unique function M^k.

Our first results indicate that it can be useful to reason with both M^k and M^{k+1}.

Proposition 2. *For some $k \geq 1$, $SND_{\{M^k\}}$ is not stronger than $SND_{\{M^{k+1}\}}$.*

Proof. Example 1 proves that $\text{SND}_{\{M^1\}}$ cannot be stronger than $\text{SND}_{\{M^2\}}$, since $\text{SND}_{\{M^2\}}$ is able to infer $x_1 \neq A$, contrary to $\text{SND}_{\{M^1\}}$. □

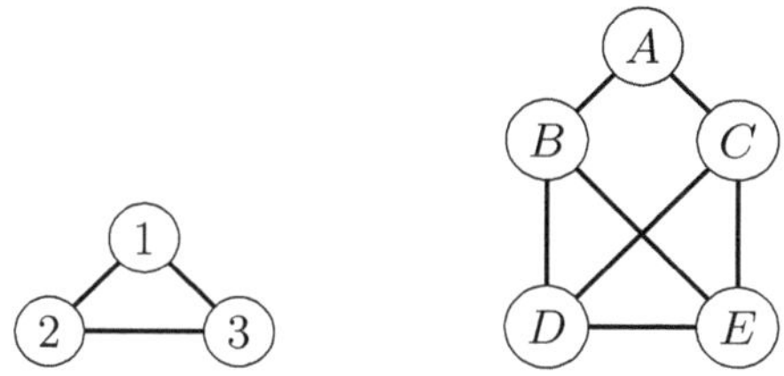

(a) G_p (b) G_t

Fig. 3. An Instance showing that $\text{SND}_{\{M^6\}}$ is not stronger than $\text{SND}_{\{M^5\}}$

Proposition 3. *For some $k \geq 1$, $SND_{\{M^{k+1}\}}$ is not stronger than $SND_{\{M^k\}}$.*

Proof. Let us consider the problem defined by Figure 3. The matrices $M^5_{G_p}$, $M^5_{G_t}$, $M^6_{G_p}$ and $M^6_{G_t}$ are as follows:

$$\underset{M^5_{G_p}}{\begin{pmatrix} 10 & 11 & 11 \\ 11 & 10 & 11 \\ 11 & 11 & 10 \end{pmatrix}} \quad \underset{M^5_{G_t}}{\begin{pmatrix} 8 & 40 & 40 & 22 & 22 \\ 40 & 26 & 26 & 51 & 51 \\ 40 & 26 & 26 & 51 & 51 \\ 22 & 51 & 51 & 40 & 41 \\ 22 & 51 & 51 & 41 & 40 \end{pmatrix}} \quad \underset{M^6_{G_p}}{\begin{pmatrix} 22 & 21 & 21 \\ 21 & 22 & 21 \\ 21 & 21 & 22 \end{pmatrix}} \quad \underset{M^6_{G_t}}{\begin{pmatrix} 80 & 52 & 52 & 102 & 102 \\ 52 & 142 & 142 & 103 & 103 \\ 52 & 142 & 142 & 103 & 103 \\ 102 & 103 & 103 & 143 & 142 \\ 102 & 103 & 103 & 142 & 143 \end{pmatrix}}$$

On this example, $\text{SND}_{\{M^6\}}$ cannot make any inference (observe that all values in $M^6_{G_t}$ are systematically greater than those in $M^6_{G_p}$, which guarantees a covering matching for each neighborhood function). On the other hand, as $M^5_{G_p}[1][1] = 10 > M^5_{G_t}[A][A] = 8$, with `Id`, we can infer that $x_1 \neq A$. □

Proposition 4. *$\text{SND}_{\mathcal{M}}$ is incomparable to SAC*

Proof. On the one hand, Example 1 shows that $\text{SND}_{\mathcal{M}}$ can filter more values than SAC. Indeed, $\text{SND}_{\mathcal{M}}$ is able to infer $x_1 \neq A$ whereas no singleton test performed by SAC on this instance enables us to identify an inconsistent value. For example, enforcing GAC after the assignment $x_1 = A$ just reduces the domains of x_2, x_3 and x_4 to $\{B, E, D\}$; the value A being removed by the allDifferent constraint and the value C being removed by the binary constraints. As there is no domain wipeout, we cannot deduce that $x_1 \neq A$.

On the other hand, Figure 4 shows an instance where SAC filters more values than $\text{SND}_{\mathcal{M}}$. Indeed, let us consider the singleton test for (x_3, A). By using the allDifferent constraint, A is first removed from the domain of all other variables. By using the binary constraints involving x_3, values E, F and G are removed from the domain of x_2, x_4 and x_5, which gives at this point $dom(x_2) = dom(x_4) = dom(x_5) = \{B, C, D\}$. We have here a Hall set, three values for three variables, so the allDifferent constraint removes B, C and D from the domain of x_1, giving $dom(x_1) = \{E, F, G\}$. Finally, the binary constraint between x_1 and x_2 removes B and C from $dom(x_2)$, which entails that D can be removed from $dom(x_4)$, leading to an inconsistency when considering the constraint between x_4 and x_1. So, SAC infers that $x_3 \neq A$. Now, consider M_{G_p}, M_{G_t}, $M^2_{G_p}$ and $M^2_{G_t}$:

$$M_{G_p} = \begin{pmatrix} 0&1&0&1&0 \\ 1&0&1&0&0 \\ 0&1&0&1&1 \\ 1&0&1&0&1 \\ 0&0&1&1&0 \end{pmatrix} \quad M_{G_t} = \begin{pmatrix} 0&1&1&1&0&0&0 \\ 1&0&1&1&0&0&0 \\ 1&1&0&1&0&0&0 \\ 1&1&1&0&1&1&1 \\ 0&0&0&1&0&1&1 \\ 0&0&0&1&1&0&1 \\ 0&0&0&1&1&1&0 \end{pmatrix} \quad M^2_{G_p} = \begin{pmatrix} 2&0&2&0&1 \\ 0&2&0&2&1 \\ 2&0&3&1&1 \\ 0&2&1&3&1 \\ 1&1&1&1&2 \end{pmatrix} \quad M^2_{G_t} = \begin{pmatrix} 3&2&2&2&1&1&1 \\ 2&3&2&2&1&1&1 \\ 2&2&3&2&1&1&1 \\ 2&2&2&6&2&2&2 \\ 1&1&1&2&3&2&2 \\ 1&1&1&2&2&3&2 \\ 1&1&1&2&2&2&3 \end{pmatrix}$$

One can verify that reasoning with $\text{SND}_{\{M^1, M^2\}}$ does not allow to identify any inconsistent value. There is always a possibility of finding a covering matching. This is also true with M^3 (not shown here), and because of the structure of both graphs, the situation renders scoring useless from level 4: the greatest value in $M^4_{G_p}$ is less than the smallest value in $M^4_{G_t}$ (and this will be also true at levels greater than 4). In other words, SND does not identify any inconsistent value. □

Proposition 5. *$\text{SND}_{\mathcal{M}}$ is strictly stronger than LAD*

Proof. LAD only involves Γ as neighborhood function and no scoring function at all. Besides, on Example 1, LAD behaves as $\text{SND}_{\{M^1\}}$, which is strictly weaker than $\text{SND}_{\mathcal{M}}$. □

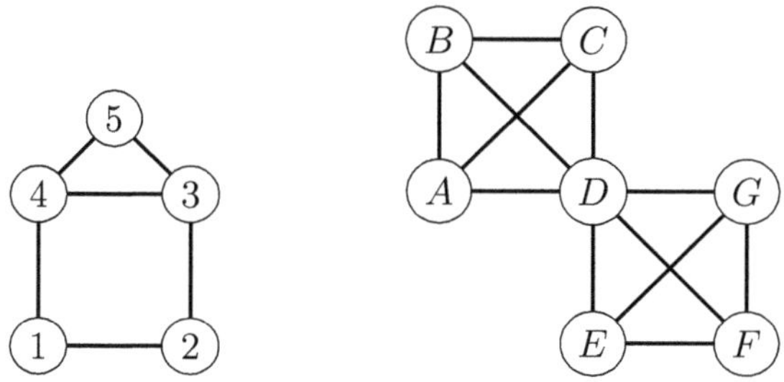

(a) G_p (b) G_t

Fig. 4. An instance showing that $\text{SND}_{\mathcal{M}}$ is not stronger than SAC

A summary of the relationships existing between the domain filtering processes introduced in the literature for the subgraph isomorphism problem, and $\text{SND}_{\mathcal{M}}$ introduced in this paper is given by Figure 5.

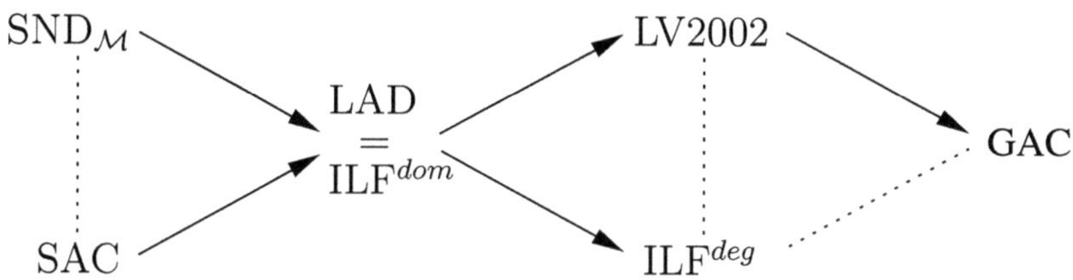

Fig. 5. Summary of the relationships between domain filtering processes. An arrow from ϕ to ψ means that ϕ is strictly stronger than ψ. A dotted line means incomparability.

5 A Weak SND Algorithm

This section describes an algorithm called SND^w for applying scoring-based neighborhood dominance. In fact, this algorithm enforces a level of filtering that is weaker than the full $\text{SND}_{\mathcal{M},\{\texttt{Id},\Gamma,\texttt{all}\}}$; more details are given at the end of the section.

The main SND^w steps are specified in Algorithm 1, the input of which is a pattern graph G_p and a target graph G_t. To reduce the memory usage, this algorithm avoids recording too many matrices and it exploits scoring functions M^k independently. This means that we reason first with M^1, then with M^2 and so on, through calls to Function filterUsingScoring at line 6. The constant `MIN_ITERATIONS` is set to a small value (e.g., 4) in order to guarantee exploiting at least all length-scoring functions M^i with $1 \leq i \leq$ `MIN_ITERATIONS`. The inner **repeat-until** loop (Lines 5-8) is stopped when the filtering performed using M^k does not allow any additional inference (Function filterUsingScoring returns `false`) and, as indicated just above, enough steps have been done. After this, we attempt to remove some unreachable edges from the target graph, as

Algorithm 1: SND$^w(G_p = (V_p, E_p)$: Pattern, $G_t = (V_t, E_t)$: Target)

1 `finished` $\leftarrow$ *false* ;
2 **while** $\neg$`finished` **do**
3   $M^0_{G_p} \leftarrow I_{n_p}$; $M^0_{G_t} \leftarrow I_{n_t}$;   `// Initialize to identity matrices`
4   $k \leftarrow 1$;
5   **repeat**
6   `modified` $\leftarrow$ filterUsingScoring(k) ;
7     $k \leftarrow k + 1$;
8   **until** $k >$ `MIN_ITERATIONS` $\wedge$ $\neg$**`modified`**;
9   **if** $\neg$removeTargetEdges() **then**
10   `finished` $\leftarrow$ *true* ;

indicated in Section 3.3, by using a function called removeTargetEdges (not described here). If this operation is effective, then the entire process starting at line 2 is run again with the target matrix modified according to these removals.

Function filterUsingScoring(k: integer): Boolean

1 $M^k_{G_p} \leftarrow M^{k-1}_{G_p} \times M_{G_p}$;
2 $M^k_{G_t} \leftarrow M^{k-1}_{t} \times M_{G_t}$;
3 `modified` $\leftarrow$ *false* ;
4 **foreach** *variable* $x_{v_p} \in$ vars(P) **do**
5   **foreach** *value* $v_t \in dom(x_{v_p})$ **do**
6   **if** $\neg$isCoherent$((v_p, v_t), \mathtt{Id})$
7      $\vee$ $\neg$isCoherent$((v_p, v_t), \Gamma)$
8      $\vee$ $\neg$isCoherent$((v_p, v_t), \mathtt{all})$ **then**
9   remove v_t from $dom(x_{v_p})$;
10   **if** $dom(x_{v_p}) = \emptyset$ **then**
11   throw exception INCONSISTENT ;
12   `modified` $\leftarrow$ *true* ;
13 **return** `modified` ;

Function `filterUsingScoring` starts with the computation of $M^k_{G_p}$ and $M^k_{G_t}$ from $M^{k-1}_{G_p}$ and $M^{k-1}_{G_t}$. At any time, we just need storage for matrices at two different levels, which allows us to control the memory space required by our approach. Then, for each pair (v_p, v_t) such that $v_t \in dom(x_{v_p})$, the conditions executed at lines 6–8 correspond to the evaluation of the right-hand side of Equation 2, with M^k as unique length-scoring function and $\{\mathtt{Id}, \Gamma, \mathtt{all}\}$ as neighborhood functions. When one of these conditions evaluates to `false` (according to Function isCoherent), the value v_t is removed from $dom(x_{v_p})$. If the deleted value was the last value in the domain, an exception is thrown to indicate that no subisomorphism function exists. Note that

Function isCoherent((v_p,v_t): vertices, N : Neighborhood): Boolean

$E \leftarrow \emptyset$;
foreach $v'_p \in N(v_p)$ **do**
  foreach $v'_t \in N(v_t) \cap dom(x_{v'_p})$ **do**
    if $M^k_{G_p}[v_p][v'_p] \leq M^k_{G_t}[v_t][v'_t]$ **then**
      $E \leftarrow E \cup \{(v'_p, v'_t)\}$;
return findCoveringMatching($(N(v_p)|N(v_t), E)$) ;

Function filterUsingScoring returns `true` when at least one value has been deleted. Finally, Function isCoherent builds first the bipartite graph associated with the pair (v_p, v_t) of vertices and the scoring function M^k, and once it is built, a covering matching is sought by a call to findCoveringMatching (not described here).

Our algorithm SND^w is clearly weaker than $\text{SND}_{\mathcal{M},\{\texttt{Id},\Gamma,\texttt{all}\}}$ for several reasons. First, as already indicated, we exploit scoring functions independently (for simplicity and for controlling memory usage). Second, in our implementation, we have used an incomplete procedure to determine whether a covering matching exists. Roughly speaking, it is similar to identifying every Hall set of size 1 (only one possible target vertex dominating a pattern vertex) and some Hall sets of greater size by a simple reasoning on the cardinality of vertex sets used for dominance. This implementation choice was guided by our main objective of having a fast algorithm (although we believe that there is room for getting a fast incremental full covering matching implementation; this is one perspective of our work). Third, because domains can be permanently reduced, we could have envisaged considering again the scoring functions previously exploited (i.e., in our current approach, we only execute once the outer loop in Algorithm 1). However, we do believe that the overhead would be very significant.

In terms of time complexity, a matrix multiplication can be carried out in $O(n^3)$, where n is the matrix order, and faster algorithms do exist, i.e., the algorithm from [25] requires $O(n^{2.3729})$. However, when the (adjacency) matrices are sparse, the complexity tends to $O(n^2)$. In our implementation, we highly exploit this feature through dedicated data structures.

6 Experimental Results

In order to show the practical interest of our approach, we have performed several experiments using a cluster of Xeon 3.0GHz with 13GB of RAM. For our experimentation, we have considered the subgraph isomorphism instances used in [22] and classified as follows:

- the series `lv` is composed of 793 instances with different properties, based on the Stanford GraphBase [11,12]
- the series `si2`, `si4`, and `si6` are composed of 390 instances each, based on the Vflib database [4,20]

– the series sf is composed of 100 scale-free networks that have been randomly generated using a power law distribution of degrees [26]

We have first compared the cpu time (expressed in seconds) and the filtering level, given by the number of deleted values (del), of GAC, SAC[2], LAD and SND^w. So, we just applied the filtering algorithms stand-alone (during what can be considered as a preprocessing step). For GAC, SAC and SND^w, we used our platform AbsCon (written in Java), whereas for LAD we used the software (written in C) available at the author's page [22]. Table 1 shows the average results obtained on the five series introduced above. For each series, # represents the number of instances, and D the average total number of values (D is the average of $D_P = \sum_{x \in vars(P)} |dom^{init}(x)|$ over all CNs P of the series). A first observation is that SAC and SND^w are close to each other in term of deleted values: on some instances/series, SAC is more powerful, and on some other instances/series, this is SND^w. However, when turning our attention to cpu time, SAC appears to be slower than SND^w by around two orders of magnitude. In order to have a fair comparison between SAC and SND^w, we only kept for those two methods the instances where SAC finished within $1,200$ seconds. This can explain why LAD seems to filter more on the series sf. To summarize, SND^w is significantly stronger than LAD (and close to SAC) while being more expensive to a reasonable extent (remember too that different programming languages are used).

In our second set of experiments, we searched to solve each problem instance, i.e., to find a solution or prove that none exists, within $1,200$ seconds. We used MAC, the classical algorithm that maintains GAC during a backtrack search, LAD, and also MSND^w the algorithm that maintains SND^w during search (however, we only executed one iteration of the main loop of Algorithm 1, disabling the call to removeTargetEdges). We also used MAC after either SAC executed at preprocessing (pSAC) or SND^w (pSND^w). Table 2 shows that, in term of the number of solved instances (sol), MSND^w is clearly the best approach, except for the series sf where LAD is very efficient.

Table 3 presents the results on a few selected instances. One can observe that maintaining SND^w can be very effective. Indeed, for some instances, the number of nodes

Table 1. *Preprocessing.* For each filtering technique, cpu represents the average time (in seconds) and del represents the average number of removed values (K stands for thousands).

Series			**GAC**		**LAD**		**SAC**		**SND^w**	
Name	#	D	cpu	del	cpu	del	cpu	del	cpu	del
lv	793	5,877	0.5	188	0.01	857	24	1,237	0.6	**1,502**
si2	390	78K	0.6	3	0.29	50K	168	**52K**	1.1	51K
si4	390	156K	0.75	8	1.05	107K	177	**109K**	1.4	107K
si6	390	235K	0.8	12	2.19	164K	193	165K	2.1	**184K**
sf	100	284K	1	0	0.32	**276K**	567	234K	2.5	244K

[2] We tried both algorithms SAC1 and SAC3 [1], but for this problem, the results were rather similar on average. So the results are given here for SAC1.

Table 2. *Finding one solution.* For each filtering technique, `cpu` represents the average time (in seconds) needed to solved instances and `sol` represents the number of solved instances (timeout set to 1, 200 seconds).

Series		**MAC**		**LAD**		**pSAC**		**pSND^w**		**MSND^w**	
Name	#	cpu	sol	cpu	sol	cpu	sol	cpu	sol	cpu	sol
lv	793	24	683	8.3	726	89	695	7.6	724	9.6	**729**
si2	390	38	352	13	345	153	236	25	356	28	**357**
si4	390	42	357	19	358	142	212	24	359	22	**366**
si6	390	46	346	20	372	191	204	22	367	24	**375**
sf	100	120	58	2.6	**100**	557	62	4.3	99	18	99

can be highly reduced. But this is not always sufficient, as can be seen for instance si6-m4D-m1296-00, where pSND^w remains slightly faster than MSND^w although the number of nodes is five times lower. This comforts us toward studying in the near future optimized incremental variants of SND^w.

Table 3. *Selected instances.* For each technique, `cpu` represents the time needed to solve it (finding one solution), `nodes` the number of nodes of the search tree and `del` the number of values removed during the preprocessing step. The rightmost column is for both pSND^w and MSND^w.

Instances	**LAD**			**pSND^w**		**MSND^w**		
Name	cpu	nodes	del	cpu	nodes	cpu	nodes	del
val35-49	Time-OUT		0	169	29,579	**54**	2,985	4,558
si2-m4D-m625-01	**1.10**	384	30,994	56	2,222	18	870	33,138
si4-r005-m400-09	456	1,553	10	684	18,934	**99**	1,017	524
si6-m4D-m1296-00	324	470,007	456,936	**220**	4,135	224	848	468,970
E-14	**5.29**	5	44,126	7.02	35	14.2	8	44,578

7 Conclusion

In this paper, we have proposed a general approach for solving instances of the subgraph isomorphism problem. This general approach called SND (Scoring-based Neighborhood Dominance) is parametrized by two sets of scoring and neighborhood functions. We have shown both the theoretical and practical interest of SND on a selection of a few functions. A natural continuation of this work consists in optimizing our current implementation, e.g., to ensure full covering matching incrementally. A promising perspective concerns the design of new scoring functions and the integration of other neighborhood functions from the literature.

Acknowledgments. This work has been supported by a project BQR (Bonus Qualité Recherche) from the University of Artois. The three first authors also benefit from a support by both CNRS and OSEO within the ISI project 'Pajero'.

References

1. Bessiere, C., Cardon, S., Debruyne, R., Lecoutre, C.: Efficient algorithms for singleton arc consistency. Constraints 16(1), 25–53 (2011)
2. Bunke, H.: Recent developments in graph matching. In: Proceeding of ICPR 2000, vol. 2, pp. 117–124 (2000)
3. Conte, D., Foggia, P., Sansone, C., Vento, M.: Thirty years of graph matching in pattern recognition. International Journal of Pattern Recognition and Artificial Intelligence 18(3), 265–298 (2004)
4. Cordella, L.P., Foggia, P., Sansone, C., Vento, M.: Performance evaluation of the VF graph matching algorithm. In: Proceedings of ICIAP 1999, pp. 1172–1177 (1999)
5. Cordella, L.P., Foggia, P., Sansone, C., Vento, M.: A (sub)graph isomorphism algorithm for matching large graphs. IEEE Transactions on Pattern Analysis & Machine Intelligence 26(10), 1367–1372 (2004)
6. Corneil, D.G., Gotlieb, C.C.: An efficient algorithm for graph isomorphism. Journal of the ACM 17(1), 51–64 (1970)
7. Damiand, G., Solnon, C., de la Higuera, C., Janodet, J.-C., Samuel, E.: Polynomial algorithms for subisomorphism of nD open combinatorial maps. Computer Vision and Image Understanding 115(7), 996–1010 (2011)
8. de la Higuera, C., Janodet, J.-C., Samuel, E., Damiand, G., Solnon, C.: Polynomial algorithms for open plane graph and subgraph isomorphisms. Theoretical Computer Science 498, 76–99 (2013)
9. Debruyne, R., Bessiere, C.: Some practical filtering techniques for the constraint satisfaction problem. In: Proceedings of IJCAI 1997, pp. 412–417 (1997)
10. Debruyne, R., Bessiere, C.: Domain filtering consistencies. Journal of Artificial Intelligence Research 14, 205–230 (2001)
11. Knuth, D.E.: The Stanford GraphBase: A Platform for Combinatorial Computing. ACM Press (1993)
12. Larrosa, J., Valiente, G.: Constraint satisfaction algorithms for graph pattern matching. Mathematical Structures in Computer Science 12(4), 403–422 (2002)
13. Lecoutre, C.: Constraint networks: techniques and algorithms. ISTE/Wiley (2009)
14. Luks, E.M.: Isomorphism of graphs of bounded valence can be tested in polynomial time. Journal of Computer and System Sciences 25(1) (1982)
15. McGregor, J.J.: Relational consistency algorithms and their application in finding subgraph and graph isomorphisms. Information Sciences 19, 229–250 (1979)
16. Ndiaye, S.N., Solnon, C.: CP models for maximum common subgraph problems. In: Lee, J. (ed.) CP 2011. LNCS, vol. 6876, pp. 637–644. Springer, Heidelberg (2011)
17. Porumbel, D.C.: Isomorphism testing via polynomial-time graph extensions. Journal of Mathematical Modelling and Algorithms 10(2), 119–143 (2011)
18. Régin, J.C.: A filtering algorithm for constraints of difference in CSPs. In: Proceedings of AAAI 1994, pp. 362–367 (1994)
19. Régin, J.C.: Développement d'outils algorithmiques pour l'intelligance artificielle. Application à la chimie organique. PhD thesis, Université Montpellier II (1995)
20. De Santo, M., Foggia, P., Sansone, C., Vento, M.: A large database of graphs and its use for benchmarking graph isomorphism algorithms. Pattern Recognition Letters 24(8), 1067–1079 (2003)
21. Shervashidze, N., Schweitzer, P., Jan van Leeuwen, E., Mehlhorn, K., Borgwardt, K.M.: Weisfeiler-lehman graph kernels. Journal of Machine Learning Research 12, 2539–2561 (2011)

22. Solnon, C.: Alldifferent-based filtering for subgraph isomorphism. Artificial Intelligence 174(12-13), 850–864 (2010)
23. Ullmann, J.R.: An algorithm for subgraph isomorphism. Journal of the ACM 23(1), 31–42 (1976)
24. Weisfeiler, B., Lehman, A.: A reduction of a graph to a canonical form and an algebra arising during this reduction. Nauchno-Technicheskaya Informatsia 2(9) (1968)
25. Williams, V.V.: Multiplying matrices faster than coppersmith-winograd. In: Proceedings of STOC 2012, pp. 887–898 (2012)
26. Zampelli, S., Deville, Y., Solnon, C.: Solving subgraph isomorphism problems with constraint programming. Constraints 15(3), 327–353 (2010)

Linking Prefixes and Suffixes for Constraints Encoded Using Automata with Accumulators

Nicolas Beldiceanu[1], Mats Carlsson[2], Pierre Flener[3],
María Andreína Francisco Rodríguez[3], and Justin Pearson[3]

[1] TASC Team (CNRS/INRIA), Mines de Nantes, 44307 Nantes, France
Nicolas.Beldiceanu@mines-nantes.fr
[2] SICS, P.O. Box 1263, 164 29 Kista, Sweden
Mats.Carlsson@sics.se
[3] Uppsala University, Dept. of Information Technology, 751 05 Uppsala, Sweden
{FirstName.LastName}@it.uu.se

Abstract. Consider a constraint on a sequence of variables functionally determining a result variable that is unchanged under reversal of the sequence. Most such constraints have a compact encoding via an automaton augmented with accumulators, but it is unknown how to maintain domain consistency efficiently for most of them. Using such an automaton for such a constraint, we derive an implied constraint between the result variables for a sequence, a prefix thereof, and the corresponding suffix. We show the usefulness of this implied constraint in constraint solving, both by local search and by propagation-based systematic search.

1 Introduction

Deterministic finite automata augmented with accumulators [4] were motivated by the need to encode a constraint C on a sequence X of variables using an automaton whose size does not depend on the size of X: accumulators are initialised at the start state and evolve through the transitions; upon acceptance, the accumulators are often mapped to a result variable R of C. The *Global Constraint Catalogue* [1] gives very compact automata with accumulators for 56 constraints (and some will be given shortly), but it is unknown how to maintain domain consistency efficiently for most of them, so implied constraints can help improve the propagation. In this paper, we consider such constraints $C(X, R)$ where R is the same for both X and its reverse X^{rev}; this covers 45 of those 56 constraints. Such constraints have proved very useful, for instance in production sequencing and staff rostering. Given a partition of X into a prefix P and a suffix T, we derive an implied constraint, shown to exist and be unique, between R, $\overrightarrow{R}$, and $\overleftarrow{R}$ when $C(X, R)$, $C(P, \overrightarrow{R})$, and $C(T^{\text{rev}}, \overleftarrow{R})$ hold. We show the usefulness of this implied constraint in constraint solving, whether by local search or by propagation-based systematic search.

We now define the GROUP constraint, to which we will be referring heavily throughout. We then give a motivating example, introducing our terminology and serving as running example throughout the rest of the paper.

B. O'Sullivan (Ed.): CP 2014, LNCS 8656, pp. 142–157, 2014.

Definition 1 ([1]). *In a sequence, a* group *is a maximal contiguous subsequence with values from a given set. The constraint* GROUP(X, W, G, V, H, L) *holds if there are G groups of a total of V values from the given set W in the possibly empty sequence X of variables, the highest and lowest group sizes being H and L respectively, with $H = 0 = L$ if $G = 0$. (W.l.o.g., we omit two parameters.)*

The instance GROUP$([\mathrm{d}, \mathrm{a}, \mathrm{c}, \mathrm{b}, \mathrm{e}, \mathrm{a}, \mathrm{b}], \{\mathrm{a}, \mathrm{e}\}, 2, 3, 2, 1)$ holds since there are $G = 2$ groups of a total of $V = 3$ occurrences of 'a' and 'e' in the sequence $[\mathrm{d}, \mathrm{a}, \mathrm{c}, \mathrm{b}, \mathrm{e}, \mathrm{a}, \mathrm{b}]$, namely the groups $[\mathrm{a}]$ and $[\mathrm{e}, \mathrm{a}]$, the highest group size being $H = 2$ and the lowest group size being $L = 1$. GROUP has no known propagator. Its decomposition [1] into the conjunction GROUPG$(X, W, G) \wedge$ GROUPV$(X, W, V) \wedge$ GROUPH$(X, W, H) \wedge$ GROUPL(X, W, L) can be encoded using four AUTOMATON constraints on automata with accumulators [4]: see Figure 1 and Section 2 for details. These constraints are very useful, for instance in staff rostering, where multiple counting constraints on the same sequence (the shift assignments of an employee over a planning horizon) are quite frequent.

Example 1. Consider the instance GROUP$([X_1, X_2, X_3], \{\mathrm{a}\}, G, V, H, L)$, where $\mathrm{dom}(X_i) = \{\mathrm{a}, \mathrm{b}\}$, $\mathrm{dom}(G) = \{0, 1, 2\} = \mathrm{dom}(V)$, $\mathrm{dom}(H) = \{2, 3\} = \mathrm{dom}(L)$, with $\mathrm{dom}(\alpha)$ denoting the current domain of variable α. Using the encoding mentioned above, *no* domain pruning is achieved on this instance.

However, it *is* possible to achieve some propagation on this instance, whose solutions are GROUP$([\mathrm{a}, \mathrm{a}, \mathrm{b}], \{\mathrm{a}\}, 1, 2, 2, 2)$ and GROUP$([\mathrm{b}, \mathrm{a}, \mathrm{a}], \{\mathrm{a}\}, 1, 2, 2, 2)$: among others, there cannot be $G = 2$ groups (as that would require a sequence of at least five elements, as groups must have at least two elements), the groups cannot have a total of $V = 0$ values (as the largest group must have at least two elements), and X_2 cannot be 'b' (as X_2 must participate in a group of 'a').

We now discuss three schemes for achieving more propagation than with just the encoding by four AUTOMATON constraints.

Scheme 1. The GROUP constraint has so-called graph invariants [5], which can be seen as implied constraints. For instance, consider the following bounds on V:

$$\max(G - 1, 0) \cdot L + H \leq V \leq \max(G - 1, 0) \cdot H + L \tag{1}$$

Intuitively, the lower bound corresponds to having one group of H elements while all the other groups are as small as possible, that is, they have L elements. The upper bound is justified in a similar way. Consider again the instance above: if the implied constraint (1) is added to the four AUTOMATON constraints, then 2 is pruned from $\mathrm{dom}(G)$, but all the other domains remain unchanged. There are 90 graph invariants in [5] for the GROUP constraint: the pruning upon adding *all* the corresponding implied constraints is evaluated in Section 5.

Scheme 2. Note that GROUP$([X_1, \ldots, X_n], W, G, V, H, L)$ holds if and only if GROUP$([X_n, \ldots, X_1], W, G, V, H, L)$ holds with the same set and the same integer variables for the reverse sequence: in Section 3.1, we will say that GROUP is its own *reverse constraint*. Let us focus on the variable V, representing the total number of group values. If we split a sequence $[X_1, \ldots, X_n]$ with $n \geq 2$ elements into a non-empty prefix $[X_1, \ldots, X_i]$ and a non-empty suffix $[X_{i+1}, \ldots, X_n]$, with

$1 \leq i < n$, then observe that the numbers V, $\overrightarrow{V}$, and $\overleftarrow{V}$ of group values respectively in the entire sequence, the prefix, and the reverse suffix are related by the constraint $V = \overrightarrow{V} + \overleftarrow{V}$: in Section 3.2, we will say that this constraint implied by the conjunction of $\text{GROUPV}([X_1, \ldots, X_n], W, V)$, $\text{GROUPV}([X_1, \ldots, X_i], W, \overrightarrow{V})$, and $\text{GROUPV}([X_n, \ldots, X_{i+1}], W, \overleftarrow{V})$ is a *glue constraint*. Glue constraints for all the integer variables of GROUP are given in Figure 2 and will be explained in Section 3.2. While $\text{GROUPV}([X_n, \ldots, X_{i+1}], W, \overleftarrow{V})$ could be replaced above by $\text{GROUPV}([X_{i+1}, \ldots, X_n], W, \overleftarrow{V})$, this will be seen to be impossible in general with our approach, where the third implying constraint must be on the *reverse suffix*, not on the suffix itself. Now consider again the instance above: if we add the glue constraints in Figure 2 for *every* possible split of the sequence (with $1 \leq i < n$), but not the implied constraint (1), then 'b' is pruned from $\text{dom}(X_2)$ and 0 is pruned from $\text{dom}(V)$, but all the other domains remain unchanged.

Note that the extra pruning achieved by Scheme 1 is incomparable with that achieved by Scheme 2.

Scheme 3. The idea of sequence splitting (underlying the glue constraints) can be applied also to the implied constraints stemming from graph invariants: for instance, instead of adding (1) on the integer variables for the *entire* sequence, we can add (1) on the integer variables for the prefix and reverse suffix of *every* possible split of the sequence. On the instance under consideration, applying Scheme 1+2+3 achieves the same pruning as applying Scheme 1+2, but, in general, more pruning is possible, as we will show in Section 5. In Section 4, we formalise Scheme 3. □

Multiple constraints on a sequence can originate from sources other than the decomposition of a constraint, unlike the previous example. For instance, a conjunction of about 20 constraints on the same sequence (of energy produced by a plant every half an hour for two consecutive days) is the pattern learned in the context of the EDF model seeker [7]. Also, in staff rostering, one has a matrix indexed in the rows by the employees and in the columns by the days of a planning horizon: each matrix cell is to be assigned an identifier (or a special off-duty value) giving the shift assigned to the corresponding employee on the corresponding day. The constraints on the columns are usually cardinality constraints, stemming from a performance contract, and there are often multiple constraints on each row, stemming from employee preferences as well as labour union and legislative restrictions. In [2], we have addressed the lack of interaction between such row and column constraints; in this paper, we address the lack of interaction between the constraints on a given row.

The contributions and the organisation of the rest of this paper are as follows, after first recalling (and slightly extending) in Section 2 the concept of automaton with accumulators [4], which can be used for compactly encoding a constraint on a sequence of variables using the AUTOMATON constraint [4]:

– In Section 3, our main result, after introducing the notion of reverse of a constraint, we define a glue constraint as an implied constraint, shown to exist and be unique, linking the result variable R under a constraint $C(X, R)$

on a sequence X of variables to the result under C on a prefix and the result under the reverse of C on the corresponding reverse suffix of X, where both C and its reverse are encoded using automata with accumulators. We show how to derive glue constraints automatically for a useful class of such constraints.
- In Section 4, we formalise Scheme 3 of Example 1 so as to cover any conjunction of constraints $C_j(X, R_j)$ on the same sequence X, whose results R_j do not vary independently but are linked by an implied constraint.
- In Section 5, we evaluate the effectiveness and efficiency of the three schemes.
- In Section 6, we use the glue constraint to compute, in constant time, the violation cost of C when probing an assignment move in local search.

We conclude and discuss related and future work in Section 7.

2 Background: Automata with Accumulators

Recall that a *deterministic finite automaton* (DFA) is a tuple $\langle Q, \Sigma, \delta, \phi, A\rangle$, where Q is the set of *states*, Σ the *alphabet*, $\delta\colon Q \times \Sigma \to Q$ the *transition function*, $\phi \in Q$ the *start state*, and $A \subseteq Q$ the set of *accepting states*. When $\delta(q, \sigma) = q'$, there is a transition from state q to state q' upon reading alphabet symbol σ in the word given to the DFA. Let Σ^* denote the infinite set of words built from Σ, including the empty word, denoted ϵ. The *extended transition function* $\widehat{\delta}\colon Q \times \Sigma^* \to Q$ for words (instead of symbols) is recursively defined by $\widehat{\delta}(q, \epsilon) = q$ and $\widehat{\delta}(q, w\sigma) = \delta(\widehat{\delta}(q, w), \sigma)$ for a word w and symbol σ. An example will be given shortly, but first we augment DFAs with a memory, in the spirit of [4], in order to encode more compactly many constraints.

We here define a *memory-DFA* (mDFA) with a memory of $k \geq 0$ accumulators as a tuple $\langle Q, \Sigma, \delta, \phi, I, A, \alpha\rangle$, where Q, Σ, ϕ, and A are as in a DFA, while the transition function δ has signature $(Q \times \mathbb{Z}^k) \times \Sigma \to Q \times \mathbb{Z}^k$, and similarly for its extended version $\widehat{\delta}$. Further, I is the k-tuple of initial values of the variables in the memory. Finally, $\alpha\colon A \times \mathbb{Z}^k \to \mathbb{Z}$ is called the *acceptance function* and transforms the memory of an accepting state into an integer. Given a word w, the mDFA returns $\alpha(\widehat{\delta}(\langle \phi, I\rangle, w))$ if w is accepted. Note that δ, $\widehat{\delta}$, and α are total functions. This definition can be generalised, but suffices for the purpose of the examples in this paper.

Example 2. Consider the mDFA $\mathcal{H}$ depicted in Figure 1c. It returns the highest size of all groups of ones within a given word of zeros and ones. It uses two accumulators: at any moment, c stores the size of the current group, while h stores the highest size of all the groups seen so far. The state set Q is $\{s\}$. The alphabet Σ is $\{0, 1\}$. The start state ϕ is s, and is indicated by an arrow coming from nowhere, annotated within braces by the initialisation to zero of h and c, hence $I = \langle 0, 0\rangle$. A transition $\delta(\langle q, \langle h, c\rangle\rangle, \sigma) = \langle q', \langle h', c'\rangle\rangle$, where h' and c' are functional expressions in terms of h and c, is depicted by an arrow going from state q to state q', annotated by symbol σ and, within braces, the memory update $\langle h, c\rangle := \langle h', c'\rangle$. For instance, the lower self-loop depicts that $\delta(\langle s, \langle h, c\rangle\rangle, 1) = \langle s, \langle \max(h, c + 1), c + 1\rangle\rangle$ for all h and c. If a memory update

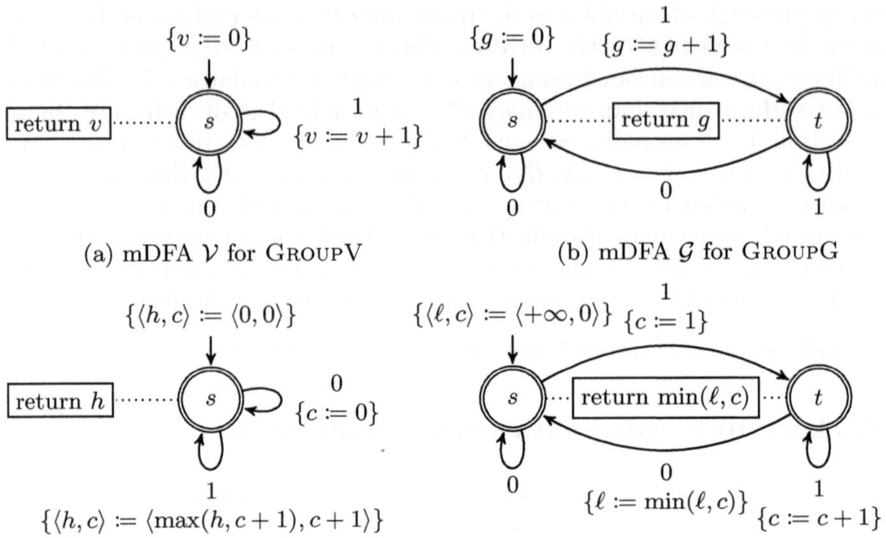

(a) mDFA $\mathcal{V}$ for GroupV (b) mDFA $\mathcal{G}$ for GroupG

(c) mDFA $\mathcal{H}$ for GroupH (d) mDFA $\mathcal{L}$ for GroupL

Fig. 1. Memory-DFAs for the constraints of the decomposition of Group

corresponds to the identity function, then we omit it; for instance, the right self-loop has the memory update $c := 0$ as an abbreviation for $\langle h, c\rangle := \langle h, 0\rangle$. All states are accepting, hence $A = Q$, an accepting state being marked by a double circle. The acceptance function α transforms a memory $\langle h, c\rangle$ at state s into h, and is depicted by a box linked to s by a dotted line. □

Automata are useful for encoding a constraint on a sequence X of variables: the Regular($\mathcal{D}, X$) constraint [14] takes a constraint encoded by a DFA $\mathcal{D}$ and holds if and only if the word represented by X is accepted by $\mathcal{D}$; similarly, the Automaton($\mathcal{M}, X$) constraint [4] takes a constraint encoded by an mDFA $\mathcal{M}$ and specialises to Regular for a memory of $k = 0$ accumulators. We define Automaton($\mathcal{M}, X, R$) for an mDFA with a memory of $k > 0$ accumulators: this constraint is equivalent to the conjunction of Automaton($\mathcal{M}, X$) and the *acceptance constraint* that variable R be equal to the integer returned by $\mathcal{M}$, that is $R = \alpha(\widehat{\delta}(\langle \phi, I\rangle, X))$. Note that R functionally depends on X, as α and $\widehat{\delta}$ are total functions. Automaton does not maintain domain consistency if $k > 0$.

Further, a constraint C on a sequence X of variables can sometimes be encoded with the help of an (m)DFA that operates not on X, but on a sequence S of *signature variables*, each depending via a *signature constraint* [4] under a total function on a sliding window of a consecutive variables within X. The constant $a \geq 1$ is called the *arity* of the signature constraints, and is linked to the lengths n of X and m of S by $m = n+1-a$. The arity gives a precondition on C, namely $n \geq a-1$, as the signature constraints fail otherwise. The sliding windows within X for two consecutive signature variables overlap by $a - 1$ variables.

Example 3. Consider the GROUPH($[X_1, \ldots, X_n], W, H$) constraint with $n \geq 0$. We constrain a sequence $[S_1, \ldots, S_m]$ of $m = n$ signature 0/1 variables S_i with the signature constraints $(X_i \in W) \Leftrightarrow (S_i = 1)$ for all $1 \leq i \leq n$: we have $a = 1$ since each signature constraint is on a single X_i. Using the mDFA $\mathcal{H}$ of Example 2 and Figure 1c, we encode GROUPH($[X_1, \ldots, X_n], W, H$) by AUTOMATON($\mathcal{H}, [S_1, \ldots, S_n], H$) and these signature constraints. For the constraint instance GROUPH($[\mathrm{d}, \mathrm{a}, \mathrm{c}, \mathrm{b}, \mathrm{e}, \mathrm{a}, \mathrm{b}], \{\mathrm{a}, \mathrm{e}\}, 2$), the mDFA $\mathcal{H}$ indeed returns $h = 2$ on the sequence $S = [0, 1, 0, 0, 1, 1, 0]$ of signature values. Similarly, the other constraints of the given decomposition of the GROUP constraint can be encoded with the help of the other three mDFAs in Figure 1. □

In the absence of signature variables and constraints, we consider $S = X$ to be a signature constraint, as this simplifies the discussion.

3 Reverse Constraints and Glue Constraints

After defining the concept of reverse of a constraint in Section 3.1, we define in Section 3.2 a glue constraint as an implied constraint for a constraint with a reverse constraint, both encoded using an automaton with accumulators, and we show that the glue constraint is unique and always exists. Finally, we show in Section 3.3 how to derive, mechanically and efficiently, the glue constraint for a useful large class of constraints.

3.1 The Reverse of a Constraint

A constraint $C(V_1, \ldots, V_n)$ is a *total-function constraint* [3] if its variables V_i can be partitioned into two non-empty sets, D and R, such that for any assignment to the variables of D there is a *unique* assignment to the variables of R that satisfies C. For example, the constraints GROUP(X, W, G, V, H, L), GROUPG(X, W, G), GROUPV(X, W, V), GROUPH(X, W, H), GROUPL(X, W, L) are total-function constraints, where X and W uniquely determine G, V, H, and L. Also, signature constraints (see Section 2) are total-function constraints.

We write a constraint $C(D, R)$ as $C(D \rightarrow R)$ when the variables D functionally determine the variables R. We denote the reverse of a word or variable sequence w by w^{rev}. We now define our first core concept.

Definition 2. *The* reverse *of a total-function constraint $C(D \rightarrow R)$, where D is a sequence of variables, is a total-function constraint $C'(D' \rightarrow R')$, where D' is a sequence of variables, such that, for any sequence X of variables, both X and its reverse functionally determine the same result variables Y under C and C' respectively, that is both $C(X \rightarrow Y)$ and $C'(X^{rev} \rightarrow Y)$ hold.*

Example 4. The constraints GROUP, GROUPG, GROUPV, GROUPH, GROUPL are their own reverses. The constraint LENGTHFIRSTSEQUENCE(X, L), which holds if L is the size of the *first* group of identical values within the sequence X of variables [1], does not have itself as reverse, but LENGTHLASTSEQUENCE(X, L), which holds if L is the size of the *last* group of identical values within X [1]. □

If a total-function constraint $C(X, R)$ is encoded by AUTOMATON$(\mathcal{M}, S, R)$ and signature constraints linking the sequences X and S of variables, using a memory-DFA $\mathcal{M} = \langle Q, \Sigma, \delta, \phi, I, A, \alpha \rangle$ with k accumulators, then $\mathcal{M}$ necessarily only has accepting states, that is $A = Q$, because the total function α is only defined on accepting states, so that the value returned by $\mathcal{M}$ on the word represented by S might be undefined if there were some non-accepting states.

If a total-function constraint C has a memory-DFA $\mathcal{M}$, then a memory-DFA $\mathcal{M}'$ for its reverse constraint C' can in some cases be derived automatically from $\mathcal{M}$. Indeed, $\mathcal{M}'$ is then the *reverse* of $\mathcal{M}$, in the sense that $\mathcal{M}'$ recognises the reverse of every word recognised by $\mathcal{M}$, and both return the same value. For instance, if $\mathcal{M}$ has a single accumulator (hence $k = 1$), which is initialised to zero (hence $I = 0$) at the start state ϕ and increased by a non-negative quantity at each transition, and if the acceptance function α returns that accumulator increased by a non-negative quantity, then $\mathcal{M}$ is a *weighted DFA* [12] over the tropical semiring over the integers, and the algorithms implemented in [13] can be used for reversal. Among the 45 of 56 constraints of the catalogue covered by this paper, there are 16 with weighted DFAs, such as GROUPG and GROUPV, but not GROUPH and GROUPL, whose accumulator updates use the max and min operators. (We revisit this useful class of constraints in Section 3.3.)

3.2 Glue Constraints

We need the AUTOMATON constraint to be implemented, in extension to how it is done in [4], by a decomposition that introduces variables representing not only the accumulators but also the state of the argument mDFA $\mathcal{M} = \langle Q, \Sigma, \delta, \phi, I, A, \alpha \rangle$ after reading each symbol of an argument sequence S of variables. Upon reading the entire S, we then have that $\widehat{\delta}(\langle \phi, I \rangle, S)$ is a tuple $\langle q, V \rangle$, where variable q represents the reached state of $\mathcal{M}$, and V is an array of k variables representing the k obtained accumulator values of $\mathcal{M}$.

We now show that the result of $\mathcal{M}$ on a word w can be computed from only these state and accumulator variables, as they encode all information on w. We will then exploit this insight by constructing a function g, which is unique and correctly computes the result of $\mathcal{M}$ on a word $w = pt$ by combining the state and accumulator variables reached by its prefix p and the reverse of its suffix t.

Theorem 1. *Consider an mDFA $\mathcal{M} = \langle Q, \Sigma, \delta, \phi, I, A, \alpha \rangle$ and its reverse $\mathcal{M}' = \langle Q', \Sigma, \delta', \phi', I', A', \alpha' \rangle$, over the same alphabet Σ. Consider four words p_1, p_2, t_1, and t_2. Assume p_1 and p_2 reach the same tuple in $\mathcal{M}$, that is $\widehat{\delta}(\langle \phi, I \rangle, p_1) = \langle q_p, V_p \rangle = \widehat{\delta}(\langle \phi, I \rangle, p_2)$. Assume the reverses of t_1 and t_2 reach the same tuple in $\mathcal{M}'$, that is $\widehat{\delta}'(\langle \phi', I' \rangle, {t_1}^{rev}) = \langle q'_t, V'_t \rangle = \widehat{\delta}'(\langle \phi', I' \rangle, {t_2}^{rev})$. We then have $\alpha(\widehat{\delta}(\langle \phi, I \rangle, p_1 t_1)) = \alpha(\widehat{\delta}(\langle \phi, I \rangle, p_2 t_2))$, so that the result on a word is independent of its prefixes p_i and corresponding suffixes t_i transiting through the same tuples.*

Proof. We have $\alpha(\widehat{\delta}(\langle \phi, I \rangle, p_1 t_1)) = \alpha(\widehat{\delta}(\langle q_p, V_p \rangle, t_1)) = \alpha(\widehat{\delta}(\langle \phi, I \rangle, p_2 t_1))$. Similarly, $\alpha'(\widehat{\delta}'(\langle \phi', I' \rangle, (p_2 t_1)^{\mathrm{rev}})) = \alpha'(\widehat{\delta}'(\langle q'_t, V'_t \rangle, {p_2}^{\mathrm{rev}})) = \alpha'(\widehat{\delta}'(\langle \phi', I' \rangle, (p_2 t_2)^{\mathrm{rev}}))$.

As $\mathcal{M}'$ is the reverse of $\mathcal{M}$, we have $\alpha(\widehat{\delta}(\langle\phi, I\rangle, p_2t_1)) = \alpha'(\widehat{\delta}'(\langle\phi', I'\rangle, (p_2t_1)^{\text{rev}})) = \alpha'(\widehat{\delta}'(\langle\phi', I'\rangle, (p_2t_2)^{\text{rev}})) = \alpha(\widehat{\delta}(\langle\phi, I\rangle, p_2t_2))$, and hence $\alpha(\widehat{\delta}(\langle\phi, I\rangle, p_1t_1)) = \alpha(\widehat{\delta}(\langle\phi, I\rangle, p_2t_2))$. □

Hence there exists a unique total function g that takes two tuples $\langle q, V\rangle$ and $\langle q', V'\rangle$ of $\mathcal{M}$ and $\mathcal{M}'$ respectively, such that

$$g(\langle q, V\rangle, \langle q', V'\rangle) = \alpha(\widehat{\delta}(\langle\phi, I\rangle, pt)) \tag{2}$$

for *any* prefix p that reaches the tuple $\langle q, V\rangle$ in $\mathcal{M}$ and *any* suffix t such that t^{rev} reaches the tuple $\langle q', V'\rangle$ in $\mathcal{M}'$. It follows from Theorem 1 that this function is well-defined, as it is independent of the prefix p and suffix t picked.

Consider a total-function constraint $C(X \to R)$, for which an mDFA $\mathcal{M}$ reads a sequence S of signature variables channelled with the sequence X by signature constraints, such that the variable R must be the result returned by $\mathcal{M}$ on S. Hence $C(X, R)$ can be encoded by $\text{Automaton}(\mathcal{M}, S, R)$ and the signature constraints. Consider a split of S into the concatenation of a possibly empty prefix P and a possibly empty suffix T, that is $S = PT$, with $R = \alpha(\widehat{\delta}(\langle\phi, I\rangle, PT))$. We now define our second core concept:

Definition 3. *Suppose, in addition to* $\text{Automaton}(\mathcal{M}, PT, R)$*, we post the constraint* $\text{Automaton}(\mathcal{M}, P, \overrightarrow{R})$ *on the prefix* P*, as well as the constraint* $\text{Automaton}(\mathcal{M}', T^{rev}, \overleftarrow{R})$ *on the reverse suffix* T^{rev}*, where* $\mathcal{M}'$ *is the reverse of* $\mathcal{M}$*. Let* $\widehat{\delta}(\langle\phi, I\rangle, P) = \langle\overrightarrow{q}, \overrightarrow{V}\rangle$ *and* $\widehat{\delta}'(\langle\phi', I'\rangle, T^{rev}) = \langle\overleftarrow{q}, \overleftarrow{V}\rangle$*. The function* g *of (2) gives rise to an implied constraint, called the* glue constraint*:*

$$R = g(\langle\overrightarrow{q}, \overrightarrow{V}\rangle, \langle\overleftarrow{q}, \overleftarrow{V}\rangle) \tag{3}$$

Example 5. The glue constraints for all four numeric variables of the Group constraint are given in Figure 2. They are organised as matrices, called *glue matrices*. The glue matrix in Figure 2d represents the following glue constraint:

$$L = \begin{cases} \min(\overrightarrow{\ell}, \overrightarrow{c} + \overleftarrow{c}, \overleftarrow{\ell}) & \text{if } \overrightarrow{q} = s \wedge \overleftarrow{q} = s \\ \min(\overrightarrow{\ell}, \overleftarrow{c}, \overleftarrow{\ell}) & \text{if } \overrightarrow{q} = s \wedge \overleftarrow{q} = t \\ \min(\overrightarrow{\ell}, \overrightarrow{c}, \overleftarrow{\ell}) & \text{if } \overrightarrow{q} = t \wedge \overleftarrow{q} = s \\ \min(\overrightarrow{\ell}, \overrightarrow{c} + \overleftarrow{c}, \overleftarrow{\ell}) & \text{if } \overrightarrow{q} = t \wedge \overleftarrow{q} = t \end{cases}$$

Figure 3 illustrates its use on an instance of GroupL. □

We post the three Automaton constraints of Definition 3 and their implied glue constraint (3) for every split of the sequence S of signature variables into a possibly empty prefix P and a possibly empty suffix T.

We have no general calculus (yet) for deriving the glue function g mechanically from an mDFA, but we now illustrate the typical reasoning on an example. In Section 3.3, we then show how to derive g mechanically and efficiently for the useful class of constraints mentioned at the end of Section 3.1.

	s
s	$V = \overrightarrow{v} + \overleftarrow{v}$

(a) Glue constraint for GroupV

	s	t
s	$G = \overrightarrow{g} + \overleftarrow{g}$	$G = \overrightarrow{g} + \overleftarrow{g}$
t	$G = \overrightarrow{g} + \overleftarrow{g}$	$G = \overrightarrow{g} - 1 + \overleftarrow{g}$

(b) Glue constraint for GroupG

	s
s	$H = \max(\overrightarrow{h}, \overrightarrow{c} + \overleftarrow{c}, \overleftarrow{h})$

(c) Glue constraint for GroupH

	s	t
s	$L = \min(\overrightarrow{\ell}, \overrightarrow{c} + \overleftarrow{c}, \overleftarrow{\ell})$	$L = \min(\overrightarrow{\ell}, \overleftarrow{c}, \overleftarrow{\ell})$
t	$L = \min(\overrightarrow{\ell}, \overrightarrow{c}, \overleftarrow{\ell})$	$L = \min(\overrightarrow{\ell}, \overrightarrow{c} + \overleftarrow{c}, \overleftarrow{\ell})$

(d) Glue constraint for GroupL

Fig. 2. Glue constraints for the constraints of the decomposition of Group: a row index refers to the state of the corresponding mDFA in Figure 1 reached by the prefix, and a column index refers to the state reached by the corresponding reverse suffix; recall that each of the four mDFAs is its own reverse.

Example 6. The mDFA $\mathcal{H}$ in Figure 1c has a single state, whose semantics is as follows, given current accumulator values c and h: a word matching the regular expression $\pi 1^c$ has been read so far, where $\pi = \varepsilon \mid \Sigma^* 0$, with $c \geq 0$ (let $1^0 = \varepsilon$) and $\max(\theta(h, \pi), c) = h$, where $\theta(a, w)$ denotes the value of accumulator a upon reading word w. Observe that $\theta(h, \pi)$ is *not* accessible any more in any accumulator after reading $\pi 1^c$ when $c > 0$. Note that $\theta(h, w) = \theta(h, w^{\text{rev}})$ for any word w, because $\mathcal{H}$ and GroupH are their own reverses.

When GroupH(X, W, H) is encoded by Automaton$(\mathcal{H}, S, H)$ and the signature constraints of Example 3 between X and S, let us split S into the concatenation of a possibly empty prefix P and a possibly empty suffix T.

Upon feeding the prefix P to $\mathcal{H}$, we end up with acceptance at state $\overrightarrow{q} = s$, since $\mathcal{H}$ has only that state. Let us call $\overrightarrow{c}$ and $\overrightarrow{h}$ the obtained accumulator values. From the semantics above of state s, we know that $\overrightarrow{c} \geq 0$ and $P = \pi 1^{\overrightarrow{c}}$ for a possibly empty prefix π of P, with:

$$\max(\theta(h, \pi), \overrightarrow{c}) = \overrightarrow{h} \tag{4}$$

Similarly, upon feeding the reverse suffix T^{rev} to $\mathcal{H}$, we get $\overleftarrow{q} = s$, $\overleftarrow{c} \geq 0$, and $T^{\text{rev}} = \tau^{\text{rev}} 1^{\overleftarrow{c}}$ for a possibly empty prefix τ^{rev} of T^{rev}, that is $T = 1^{\overleftarrow{c}} \tau$, with:

$$\max(\theta(h, \tau^{\text{rev}}), \overleftarrow{c}) = \overleftarrow{h} \tag{5}$$

Overall, we have $S = PT = \pi 1^{\overrightarrow{c}} 1^{\overleftarrow{c}} \tau = \pi 1^{\overrightarrow{c} + \overleftarrow{c}} \tau$, hence:

$$\begin{aligned}
\theta(h, S) &= \max(\theta(h, \pi), \overrightarrow{c} + \overleftarrow{c}, \theta(h, \tau)) && \text{by the semantics of } \mathcal{H} \\
&= \max(\theta(h, \pi), \overrightarrow{c} + \overleftarrow{c}, \theta(h, \tau^{\text{rev}})) && \text{by } \mathcal{H} \text{ being its own reverse} \\
&= \max(\overrightarrow{h}, \overrightarrow{c} + \overleftarrow{c}, \theta(h, \tau^{\text{rev}})) && \text{by } \overleftarrow{c} \geq 0 \text{ and (4)} \\
&= \max(\overrightarrow{h}, \overrightarrow{c} + \overleftarrow{c}, \overleftarrow{h}) && \text{by } \overrightarrow{c} \geq 0 \text{ and (5)}
\end{aligned} \tag{6}$$

GROUPL([b, a, a, a, b, b, a, a, b, a, a, a, a], {a}, 2)

Signature variables: ⟨0 1 1 1 0 0 1⟩ ⟨1 0 1 1 1 1⟩

i	0	1	2	3	4	5	6	7	⋯
$\overrightarrow{q_i}$	s	s	t	t	t	s	s	t	⋯
$\overrightarrow{\ell_i}$	$+\infty$	$+\infty$	$+\infty$	$+\infty$	$+\infty$	3	3	**3**	⋯
$\overrightarrow{c_i}$	0	0	1	2	3	3	3	**1**	⋯

⋯	6	5	4	3	2	1	0	i
⋯	t	s	t	t	t	t	s	$\overleftarrow{q_i}$
⋯	**4**	4	$+\infty$	$+\infty$	$+\infty$	$+\infty$	$+\infty$	$\overleftarrow{\ell_i}$
⋯	**1**	4	4	3	2	1	0	$\overleftarrow{c_i}$

GROUPL ([b, a, a, a, b, b, a], {a}, **1**) $\wedge$ GROUPL ([a, a, a, a, b, a], {a}, **1**)

glue matrix entry associated with $\langle t,t\rangle$: $L = \min\left(\overrightarrow{\ell_7}, \overrightarrow{c_7} + \overleftarrow{c_6}, \overleftarrow{\ell_6}\right) = \min(\mathbf{3}, \mathbf{1}+\mathbf{1}, \mathbf{4}) = 2$

Fig. 3. Use of the entry for the state pair $\langle t,t\rangle$ in the glue matrix of Figure 2d for linking the result variable L with the state and accumulator variables after reading the prefix [b, a, a, a, b, b, a] and corresponding suffix [a, b, a, a, a, a] of a sequence. The left (resp. right) table shows the initialisation (for $i = 0$) and evolution of the state of the mDFA $\mathcal{L}$ in Figure 1d and its accumulators ℓ and c upon reading the symbol at index i of the sequence (resp. its reverse).

The law underlying the last two equalities is that $\max(x, y) = h$ implies that $\max(x, y + z) = \max(h, y + z)$ when $z \geq 0$. Note that $\theta(h, S)$ is now entirely defined in terms of the accumulator values after processing P and T^{rev}.

Posting AUTOMATON($\mathcal{H}, P, \overrightarrow{H}$) gives access to $\overrightarrow{q}$, $\overrightarrow{c}$, $\overrightarrow{h}$ as variables, with $\overrightarrow{H} = \overrightarrow{h}$. Similarly, posting AUTOMATON($\mathcal{H}, T^{\text{rev}}, \overleftarrow{H}$) gives access to $\overleftarrow{q}$, $\overleftarrow{c}$, $\overleftarrow{h}$ as variables, with $\overleftarrow{H} = \overleftarrow{h}$. Since the acceptance function α computes $H = \theta(h, S)$, the implied glue constraint is $H = \max(\overrightarrow{h}, \overrightarrow{c} + \overleftarrow{c}, \overleftarrow{h})$, due to (6). □

3.3 Deriving the Glue Constraint

Reconsider the useful class of constraints mentioned at the end of Section 3.1, namely those that can be encoded using an mDFA $\mathcal{M} = \langle Q, \Sigma, \delta, \phi, I, A, \alpha\rangle$ with a single accumulator (hence $k = 1$), which is initialised to zero (hence $I = 0$) at the start state ϕ and increased by a non-negative quantity at each transition as well as by the acceptance function α. We denote $\gamma(q, \sigma)$ the accumulator increase on the transition from state q upon reading symbol σ. Similarly, we denote $\widehat{\gamma}(q, w)$ the *total* accumulator increase on the path from state q upon reading a possibly empty word w. Let $\mathcal{M}' = \langle Q', \Sigma, \delta', \phi', I', A', \alpha'\rangle$ be the reverse of $\mathcal{M}$ (computed as seen in Section 3.1), on the *same* alphabet. The glue constraint then always takes a particular form, as described after defining a needed concept.

Definition 4. *Let PT be a word such that possibly empty prefix P leads to state $\overrightarrow{q}$ of mDFA $\mathcal{M}$, and the reverse of possibly empty suffix T leads to state $\overleftarrow{q}$ of the reverse mDFA $\mathcal{M}'$ of $\mathcal{M}$. The* correction term *for $\overrightarrow{q}$ and $\overleftarrow{q}$ is:*

$$\Delta(\overrightarrow{q}, \overleftarrow{q}) = \widehat{\gamma}(\phi, PT) - (\widehat{\gamma}(\phi, P) + \widehat{\gamma}'(\phi', T^{rev}))$$

Suppose, in addition to $\text{AUTOMATON}(\mathcal{M}, PT, R)$, we post the constraints $\text{AUTOMATON}(\mathcal{M}, P, \overrightarrow{R})$ and $\text{AUTOMATON}(\mathcal{M}', T^{\text{rev}}, \overleftarrow{R})$. Let $\widehat{\delta}(\langle \phi, I\rangle, P) = \langle \overrightarrow{q}, \overrightarrow{v}\rangle$ and $\widehat{\delta}(\langle \phi', I'\rangle, T^{\text{rev}}) = \langle \overleftarrow{q}, \overleftarrow{v}\rangle$. The glue constraint for states $\overrightarrow{q}$ and $\overleftarrow{q}$ is then always of the form $R = \alpha(\overrightarrow{v} + \Delta(\overrightarrow{q}, \overleftarrow{q}) + \overleftarrow{v})$, because $\overrightarrow{v} = \widehat{\gamma}(\phi, P)$, $\overleftarrow{v} = \widehat{\gamma}'(\phi', T^{\text{rev}})$, and $R = \alpha(\widehat{\delta}(\langle \phi, I\rangle, PT)) = \alpha(\widehat{\gamma}(\phi, PT))$. Therefore, we can abbreviate the glue matrix into the matrix of its correction terms.

Example 7. The glue matrices in Figures 2a and 2b can be abbreviated into (0) and $\left(\begin{smallmatrix} 0 & 0 \\ 0 & -1 \end{smallmatrix}\right)$, respectively, which contain constants independent of P and T. □

We now show that the matrix of constant correction terms can be computed efficiently, as it obeys a recurrence relation. By Definition 4, we have the base cases $\Delta(\phi, q') = 0$ for every state q' of $\mathcal{M}'$, and $\Delta(q, \phi') = 0$ for every state q of $\mathcal{M}$. For the step case of two non-start states $\overrightarrow{q}$ and $\overleftarrow{q}$, let $P\sigma T$ be a word such that non-empty prefix P leads to state $\overrightarrow{q}$ of $\mathcal{M}$, with successor $\overrightarrow{r}$ upon reading symbol σ, and the reverse of possibly empty suffix T leads to state $\overleftarrow{r}$ of $\mathcal{M}'$, with successor $\overleftarrow{q}$ upon reading symbol σ. Using Definition 4 with σT as suffix, we get the following recurrence relation:

$$\begin{aligned} \Delta(\overrightarrow{q}, \overleftarrow{q}) &= (\widehat{\gamma}(\phi, P) + \widehat{\gamma}(\overrightarrow{q}, \sigma T)) - (\widehat{\gamma}(\phi, P) + \widehat{\gamma}'(\phi', T^{\text{rev}}\sigma)) \\ &= \widehat{\gamma}(\overrightarrow{q}, \sigma T) - \widehat{\gamma}'(\phi', T^{\text{rev}}\sigma) \\ &= (\gamma(\overrightarrow{q}, \sigma) + \widehat{\gamma}(\overrightarrow{r}, T)) - (\widehat{\gamma}'(\phi', T^{\text{rev}}) + \gamma'(\overleftarrow{r}, \sigma)) \\ &= \Delta(\overrightarrow{r}, \overleftarrow{r}) + \gamma(\overrightarrow{q}, \sigma) - \gamma'(\overleftarrow{r}, \sigma) \end{aligned} \tag{7}$$

The matrix of correction terms for all state pairs can then be computed by dynamic programming. It suffices to initialise to zero the row of ϕ and the column of ϕ'. Then let name(q) denote a shortest word reaching state q from the start state of its mDFA. The dynamic program uses the recurrence relation (7) for every pair $\langle \overrightarrow{q}, \overleftarrow{q}\rangle$ of non-start states with $\text{name}(\overleftarrow{q}) = \text{name}(\overleftarrow{r})\sigma$. Note that both σ and $\overleftarrow{r}$ are uniquely determined by $\text{name}(\overleftarrow{q})$, and that $\overrightarrow{r}$ is uniquely defined as *the* successor of $\overrightarrow{q}$ under σ, since an mDFA is deterministic. Proceeding by increasing lexicographic order of $\langle |\text{name}(\overrightarrow{q})| + |\text{name}(\overleftarrow{q})|, -|\text{name}(\overrightarrow{q})|\rangle$, where $|w|$ denotes the length of word w, ensures that the right-hand side of (7) is always already determined.

Example 8. The glue matrix $\left(\begin{smallmatrix} 0 & 0 \\ 0 & -1 \end{smallmatrix}\right)$ of Example 7 is for the GROUPG constraint, whose mDFA $\mathcal{G}$ in Figure 1b is its own reverse. There are zeros in the left column and top row, by the base cases. The -1 in the lower-right cell follows from the following application of (7): we have $\overrightarrow{q} = t = \overleftarrow{q}$, with $\text{name}(\overleftarrow{q}) = \text{“1”} = \text{name}(\overleftarrow{r})\sigma$, hence $\sigma = 1$ and $\text{name}(\overleftarrow{r}) = \epsilon$, so $\overleftarrow{r} = s$; since $\overrightarrow{r}$ is the successor of $\overrightarrow{q}$ under σ, we have $\overrightarrow{r} = t$; hence (7) gives $\Delta(t,t) = \Delta(t,s) + \gamma(t,1) - \gamma'(s,1) = 0 + 0 - 1 = -1$. Indeed, if the last symbol of prefix P and the first symbol of suffix T are in a group, then the sum of the numbers of groups of P and T^{rev} would be one unit too high and has to be downward adjusted by -1. □

Filling the matrix of $|Q| \cdot |Q'|$ correction terms takes $\Theta(|Q| \cdot |Q'| + |\Sigma|)$ time. Indeed, each correction term is computed in constant time, and the state names can be computed in $\Theta(|Q| + |Q'| + |\Sigma|)$ time.

4 Implied Constraints on Prefixes and Suffixes

In Scheme 3 of Example 1, we showed how to improve pruning in the presence of an implied constraint on the result variables of multiple total-function constraints on the same sequence of variables. We now formalise this idea.

Consider a conjunction of p total-function constraints $C_j(X \to R_j)$ on the *same* sequence X of n variables. For each constraint C_j, assume we have an mDFA $\mathcal{M}_j = \langle Q_j, \Sigma_j, \delta_j, \phi_j, I_j, A_j, \alpha_j \rangle$ that reads a sequence S^j of m_j signature variables channelled with X by signature constraints of arity $a_j \geq 1$ (hence $m_j = n + 1 - a_j$), such that the variable R_j is constrained to be equal to the result returned by $\mathcal{M}_j$ on S^j. (We write S^j rather than S_j so that, in line with the rest of the paper, we can write S^j_i to refer to the element at index i of S^j.) Hence each $C_j(X \to R_j)$ is encoded by AUTOMATON$(\mathcal{M}_j, S^j, R_j)$ and its signature constraints. Let $\mathcal{M}'_j = \langle Q'_j, \Sigma_j, \delta'_j, \phi'_j, I'_j, A'_j, \alpha'_j \rangle$ be the reverse of $\mathcal{M}_j$, over the *same* alphabet Σ_j: it is used for encoding the reverse of C_j by AUTOMATON$(\mathcal{M}'_j, (S^j)^{\text{rev}}, R_j)$, using the *same* signature constraints and variables. Note that all the R_j are only defined when X is sufficiently long, namely $n \geq \overline{a} - 1$, where $\overline{a} = \max(a_1, \dots, a_p)$.

Consider that the p result variables R_j are not independent and that we have an implied constraint $\Im(R_1, \dots, R_p)$, called a *graph invariant* in [5], on them.

The idea is to improve the propagation on the conjunction of the C_j with $\Im$ (which constrains the *overall* results R_j under the C_j on the *entire* sequence X) by adding also $\Im$ on the *partial* results under the C_j for every sufficiently long *prefix* of X, as well as adding $\Im$ on the *partial* results under the reverses of the C_j for the reverse of every sufficiently long *suffix* of X.

We thus post the implied constraint $\Im$ on the results R^i_j for every not necessarily strict prefix $[X_1, \dots, X_{i+a_j-1}]$ of X, each prefix being long enough for all the R^i_j to be defined. We also post $\Im$ on the results R'^i_j for the reverse of every not necessarily strict suffix $[X_n, \dots, X_{n-a_j-i+2}]$ of X, each suffix being long enough for all the R'^i_j to be defined. We get:

$$\forall j : \forall 1 \leq i \leq n - a_j + 1 : \text{AUTOMATON}(\mathcal{M}_j, [X_1, \dots, X_{i+a_j-1}], R^i_j)$$
$$\forall j : \forall 1 \leq i \leq n - a_j + 1 : \text{AUTOMATON}(\mathcal{M}'_j, [X_n, \dots, X_{n-a_j-i+2}], R'^i_j)$$
$$\forall \overline{a} \leq i \leq n : \Im(R_1^{i-a_1+1}, \dots, R_p^{i-a_p+1}) \wedge \Im(R'^{i-a_1+1}_1, \dots, R'^{i-a_p+1}_p)$$

Rather than posting $2 \cdot (n - a_j + 1)$ AUTOMATON constraints for a given $\mathcal{M}_j$, we only post two AUTOMATON constraints over the sequence X and its reverse, where our implementation of AUTOMATON provides access to the internal variables R^i_j and R'^i_j.

Note that the glue constraints of Section 3 make each implied constraint on a prefix communicate, through shared variables, with the implied constraint on the reverse of the corresponding suffix: we evaluate this experimentally in Section 5.

Example 9. In Scheme 3 of Example 1, we had $p = 4$, $a_1 = a_2 = a_3 = a_4 = 1 = \overline{a}$, and $\Im$ as the implied constraint (1). Further experiments are in Section 5. □

5 Experiments

We have implemented as a generic framework for our method a Prolog predicate taking as arguments (i) a shared sequence of variables, (ii) a list of p mDFAs with their corresponding signature and glue constraints, (iii) a list of p numeric variables R_j to be computed, (iv) a conjunction of implied constraints over the R_j, and (v) some options for controlling the encoding scheme (see below). We have also extended the *Global Constraint Catalogue* [1] with glue constraints for most of its mDFAs and with implied constraints relating arguments of different constraints (e.g. those of [5] and the relation between the number of valleys and peaks [7] of a sequence).[1]

To evaluate our methods, we ran three experiments on a conjunction of two GROUP constraints over a shared sequence of a/b variables, one with $W = \{a\}$ and one with $W = \{b\}$. Thus, a total of eight numeric values are computed from the sequence. Each experiment was run on two sets of 1000 randomly generated unique instances. An instance consists of initial domains for the sequence and numeric variables. In each experiment, four encoding schemes were compared: **B** the baseline, i.e. as $p = 8$ AUTOMATON constraints; **BI** as **B** plus 90 implied constraints linking the eight numeric variables, provided by [1, Section 4.3], see Scheme 1 of Example 1; **BG** as **B** plus the appropriate glue constraints posted at every prefix-suffix junction, see Scheme 2; and **BGI** as **BG** plus the same 90 implied constraints posted on the full sequence as well as on every nonempty prefix and suffix, i.e. using Scheme 1+2+3.

All experiments were run in SICStus Prolog 4.3 [9] on a quad core 2.8 GHz Intel Core i7-860 machine with 8MB cache per core, running Ubuntu Linux.

Special attention was devoted to generating meaningful instances, since the eight numeric variables are all but independent, and a truly random choice of their initial domains leads to an unsatisfiable instance in the vast majority of cases. We came up with two instance sets:

Sloppy (11% satisfiable) is generated as follows: First, each a/b variable is assigned 'a' with 10% probability, 'b' with 10% probability, and left unassigned with 80% probability. Then, each numeric variable is given a random subinterval of its feasible interval. If posting the 90 implied constraints on the obtained candidate instance detects failure without search, then the candidate is rejected. Otherwise, it is included in the set.

Strict (96% satisfiable) is generated like the **Sloppy** set, but also, if posting the full **BGI** scheme on the candidate detects failure without search, then the candidate is rejected. Otherwise, it is included in the set.

The results of the three experiments are shown in Figure 4. In the first experiment (left column), sequences of length 20 were used. In the two plots, each point (x, y) denotes that y instances reached a total domain size of at most x

[1] All code and data for the experiments as well as the extended version of the catalogue can be found at `http://www.sics.se/~matsc/research/reversible`.

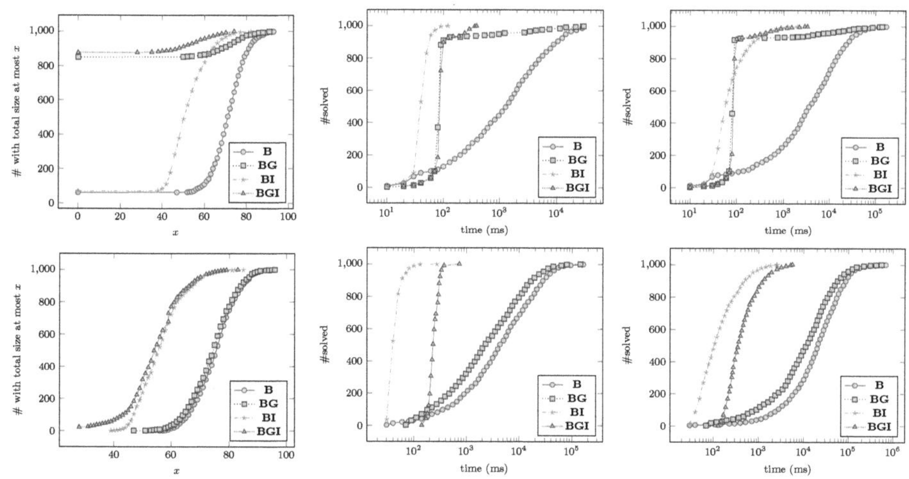

Fig. 4. Experimental results. Comparison of the amount of pruning after posting (left), the time to find the first solution or detect unsatisfiability (centre), and the time to find all solutions (right). Top row: **Sloppy**. Bottom row: **Strict**.

after initial pruning, for the given scheme. We find that **BGI** is the most effective in pruning and **B** the least effective. For **Sloppy**, we see that **BG** detects unsatisfiability far more effectively than **BI**. For **Strict**, we see that **BI** gives much more pruning than **BG**.

For the remaining experiments (centre and right column), we plot the number of instances that can be solved by a given deadline. Sequences of length 10 were used here, to avoid having to impose a time limit, so that we can compare all methods on all instances. We find that the schemes involving implied constraints outperform the schemes that do not.

The **BG** curves are similar throughout the **Sloppy** row, confirming that the glue method more effectively detects unsatisfiability. For the **Strict** instances, **BI** outperformed **BGI**. Partial benchmark results for length 20 and for combining GROUP and CHANGECONTINUITY [1,5] paint an almost identical picture, except there is some indication that **BGI** is more effective on longer sequences.

Evaluating our schemes on the model inferred by the EDF model seeker [7] is ongoing work, and requires more work and analysis since the model is nontrivial and contains 20 interacting constraints.

6 Constant-Time Move Probing in Local Search

In the context of constraint-based local search [15], consider a total-function constraint $C([X_1, \ldots, X_n] \to R)$ that is encoded using an mDFA $\mathcal{M}$. The violation cost of C under the current assignment β is $|\beta(R) - r|$, where r is the result returned by $\mathcal{M}$ on $[\beta(X_1), \ldots, \beta(X_n)]$. We now show, on an example, how to use the glue constraint of C in order to compute, in *constant* time, the violation cost of C when probing a move $X_i := v$, which changes the current assignment.

For example, let us start from the ground instance of Figure 3, namely GroupL([b, a, a, a, b, b, a, a, b, a, a, a, a], {a}, 2). It is satisfied, hence its violation cost under the current assignment is 0. Assume now we want to probe changing variable X_7 from 'a' into 'b', that is changing signature variable S_7 from 1 into 0. The violation cost under the resulting new assignment, and its increase compared to the current assignment, are computed in *constant* time as follows:

1. Starting from the prefix column for $i = 6$, we compute the new column for $i = 7$, upon the mDFA $\mathcal{L}$ in Figure 1d reading a 0 instead of a 1: we get $\overrightarrow{q}_7 = s$ and $\overrightarrow{\ell}_7 = 3 = \overrightarrow{c}_7$. The suffix column for $i = 6$ remains unchanged.
2. Using the glue matrix entry in Figure 2d for the state pair $\langle \overrightarrow{q}_7, \overleftarrow{q}_6 \rangle = \langle s, t \rangle$, we know that the new sequence has $\min(\overrightarrow{\ell}_7, \overleftarrow{c}_6, \overleftarrow{\ell}_6) = \min(3, 1, 4) = 1$ as the lowest group size (that is not 2 anymore).
3. Hence the violation cost under the new assignment is $|2 - 1| = 1$ (as we still have $R = 2$), and has thus increased by $1 - 0 = 1$.

Thus, with one matrix of states and accumulator values for the mDFA and one for the reverse mDFA, as in Figure 3, probing can be done in constant time, thereby beating the linear time achieved for the Automaton constraint in [11].

Commitment to a move actually selected by the search (meta-)heuristic follows the same steps as above, namely once for updating the values of the left table (in Figure 3) from the concerned column until the last column, and once for updating the values in the right table from the concerned column until the first column. This takes time linear in the length of X.

7 Conclusion

For a total-function constraint on a sequence of variables whose result is invariant under sequence reversal, we have shown how to derive, from a compact encoding of the constraint via an automaton with accumulators, an implied constraint between the result variables for a sequence of variables, a prefix thereof, and the corresponding suffix. Such total-function constraints have proved very useful, for instance in production sequencing and staff rostering. We have shown that the glue constraint is unique and always exists. We have also shown the usefulness of the derived implied constraint in constraint solving, both by local search, where the implied constraint enables constant-time move probing, and by propagation-based systematic search, where the implied constraint improves propagation: our concept is thus not oriented toward a specific solver technology.

Other constraints than Automaton could be used for handling memory-DFAs and deriving glue constraints: recall that our method supports multiple accumulators, and needs access as variables to the sequence of values for each accumulator, as well as access as variables to the sequence of automaton states. For instance, CostRegular [10] is currently limited to the class of single-accumulator mDFAs discussed in Section 3.3, and is compared in detail with Automaton in [6]. Encodings based on Slide [8] could also be investigated.

Acknowledgements. The first author is supported by the *Gaspard Monge Program for Optimisation and operations research.* The last three authors are supported by grants 2011-6133 and 2012-4908 of the Swedish Research Council (VR). We thank Pascal Van Hentenryck, with whom we discovered a few years ago the unpublished result of Section 3.3. We thank the anonymous reviewers for their valuable comments.

References

1. Beldiceanu, N., Carlsson, M., Demassey, S., Petit, T.: Global constraint catalogue: Past, present, and future. Constraints 12(1), 21–62 (2007)
2. Beldiceanu, N., Carlsson, M., Flener, P., Pearson, J.: On matrices, automata, and double counting in constraint programming. Constraints 18(1), 108–140 (2013)
3. Beldiceanu, N., Carlsson, M., Flener, P., Pearson, J.: On the reification of global constraints. Constraints 18(1), 1–6 (2013)
4. Beldiceanu, N., Carlsson, M., Petit, T.: Deriving filtering algorithms from constraint checkers. In: Wallace, M. (ed.) CP 2004. LNCS, vol. 3258, pp. 107–122. Springer, Heidelberg (2004)
5. Beldiceanu, N., Carlsson, M., Rampon, J.-X., Truchet, C.: Graph invariants as necessary conditions for global constraints. In: van Beek, P. (ed.) CP 2005. LNCS, vol. 3709, pp. 92–106. Springer, Heidelberg (2005)
6. Beldiceanu, N., Flener, P., Pearson, J., Van Hentenryck, P.: Propagating regular counting constraints. In: Brodley, C.E., Stone, P. (eds.) AAAI 2014. AAAI Press (2014)
7. Beldiceanu, N., Ifrim, G., Lenoir, A., Simonis, H.: Describing and generating solutions for the EDF unit commitment problem with the ModelSeeker. In: Schulte, C. (ed.) CP 2013. LNCS, vol. 8124, pp. 733–748. Springer, Heidelberg (2013)
8. Bessière, C., Hebrard, E., Hnich, B., Kiziltan, Z., Walsh, T.: SLIDE: A useful special case of the CARDPATH constraint. In: Ghallab, M., et al. (eds.) ECAI 2008, pp. 475–479. IOS Press (2008)
9. Carlsson, M., Ottosson, G., Carlson, B.: An open-ended finite domain constraint solver. In: Glaser, H., Hartel, P., Kuchen, H. (eds.) PLILP 1997. LNCS, vol. 1292, pp. 191–206. Springer, Heidelberg (1997), `http://sicstus.sics.se/`
10. Demassey, S., Pesant, G., Rousseau, L.M.: A `Cost-Regular` based hybrid column generation approach. Constraints 11(4), 315–333 (2006)
11. He, J., Flener, P., Pearson, J.: An *automaton* constraint for local search. Fundamenta Informaticae 107(2-3), 223–248 (2011)
12. Mohri, M.: Weighted automata algorithms. In: Droste, M., Kuich, W., Vogler, H. (eds.) Handbook of Weighted Automata. Monographs in Theoretical Computer Science, pp. 213–254. Springer (2009)
13. Mohri, M., Pereira, F.C.N., Riley, M.: The design principles of a weighted finite-state transducer library. Theoretical Computer Science 231(1), 17–32 (2000), The OpenFst Library is available `http://www.openfst.org/`
14. Pesant, G.: A regular language membership constraint for finite sequences of variables. In: Wallace, M. (ed.) CP 2004. LNCS, vol. 3258, pp. 482–495. Springer, Heidelberg (2004)
15. Van Hentenryck, P., Michel, L.: Constraint-Based Local Search. MIT Press (2005)

The Propagation Depth of Local Consistency

Christoph Berkholz

RWTH Aachen University, Aachen, Germany

Abstract. We establish optimal bounds on the number of nested propagation steps in k-consistency tests. It is known that local consistency algorithms such as arc-, path- and k-consistency are not efficiently parallelizable. Their inherent sequential nature is caused by long chains of nested propagation steps, which cannot be executed in parallel. This motivates the question "What is the minimum number of nested propagation steps that have to be performed by k-consistency algorithms on (binary) constraint networks with n variables and domain size d?"

It was known before that 2-consistency requires $\Omega(nd)$ and 3-consistency requires $\Omega(n^2)$ sequential propagation steps. We answer the question exhaustively for every $k \geq 2$: there are binary constraint networks where any k-consistency procedure has to perform $\Omega(n^{k-1}d^{k-1})$ nested propagation steps before local inconsistencies were detected. This bound is tight, because the overall number of propagation steps performed by k-consistency is at most $n^{k-1}d^{k-1}$.

1 Introduction

A constraint network (X, D, C) consists of a set X of n variables over a domain D of size d and a set of constraints C that restrict possible assignments of the variables. The *constraint satisfaction problem* (CSP) is to find an assignment of the variables with values from D such that all constraints are satisfied. The constraint satisfaction problem can be solved in exponential time by exhaustive search over all possible assignments. *Constraint propagation* is a technique to speed up the exhaustive search by restricting the search space in advance. This is done by iteratively propagating new constraints that follow from previous ones. Most notably, in local consistency algorithms the overall goal is to propagate new constraints to achieve some kind of consistency on small parts of the constraint network. Additionally, if local inconsistencies were detected, it follows that the constraint network is also globally inconsistent and hence unsatisfiable.

The k-consistency test [8] is a well-known local consistency technique, which enforces that every satisfying $(k-1)$-partial assignment can be extended to a satisfying k-partial assignment. At the beginning, all partial assignments that violate a constraint were marked as inconsistent. Then the following inference rule is applied iteratively:

> If h is a consistent ℓ-partial assignment ($\ell < k$) for which there exists a variable $x \in X$ such that $h \cup \{x \mapsto a\}$ is inconsistent for all $a \in D$, then mark h and all its extensions as inconsistent.

B. O'Sullivan (Ed.): CP 2014, LNCS 8656, pp. 158–173, 2014.

After at most $n^{k-1}d^{k-1}$ propagation steps this procedure stops. If the empty assignment becomes inconsistent, we say that (strong) k-consistency *cannot be established*. In this case we know that the constraint network is globally inconsistent. Otherwise, if k-consistency *can be established*, we can use the propagated constraints to restrict the search space for a subsequent exhaustive search. There are several different k-consistency algorithms in the literature, especially for $k = 2$ (arc consistency) and $k = 3$ (path consistency), which all follow this propagation scheme. The main difference between these algorithms are the underlying data structure and the order in which they apply the propagation rule. It seems plausible to apply the propagation rule in parallel in order to detect local inconsistencies in different parts of the constraint network at the same time. Indeed, this intuition has been used to design parallel arc and path consistency algorithms [15,16]. On the other hand, the k-consistency test is known to be PTIME-complete [10,11] and hence not efficiently parallelizable (unless NC=PTIME). The main bottleneck for parallel approaches are the sequential dependencies in the propagation rule: some assignments will be marked as inconsistent after some other assignments became inconsistent.

For 2-consistency the occurrence of long chains of sequential dependencies has been observed very early [6] and was recently studied in depth in [4]. There are simple constraint networks for which 2-consistency requires $\Omega(nd)$ nested propagation steps. Ladkin and Maddux [14] used algebraic techniques to show that 3-consistency requires $\Omega(n^2)$ nested propagation steps on binary constraint networks with constant domain. We extend these previous results and obtain a complete picture of the propagation depth of k-consistency. Our main result (Theorem 1) states that for every constant $k \geq 2$ and given integers n, d there is a constraint network with n variables and domain size d such that every k-consistency algorithm has to perform $\Omega(n^{k-1}d^{k-1})$ nested propagation steps. This lower bound is optimal as it is matched by the trivial upper bound $n^{k-1}d^{k-1}$ on the overall number of propagation steps. It follows that every parallel propagation algorithm for k-consistency has a worst case time complexity of $\Omega(n^{k-1}d^{k-1})$. Since the best-known running time of a sequential algorithm for k-consistency is $O(n^k d^k)$ [5] it follows that no significant improvement over the sequential algorithm is possible.

2 Preliminaries

As first pointed out by Feder and Vardi [7] the CSP is equivalent to the structure homomorphism problem where two finite relational structures A and B are given as input. The universe $V(\mathsf{A})$ of structure A corresponds to the set of variables X and the universe $V(\mathsf{B})$ of structure B corresponds to the domain D. The constraints are encoded into relations such that every homomorphism from A to B corresponds to a solution of the CSP. For the rest of this paper we mainly stick to this definition as it is more convenient to us. In fact, our main result benefits to a large extend from the fruitful connection between these two viewpoints.

In the introduction we have presented k-consistency as a propagation procedure on constraint networks. Below we restate the definition in terms of a formal inference system (which is inspired by the proof system in [1] and is a generalization of [4]). This view allows us to gain insight into the structure of the propagation process and to formally state our main theorem afterwards. At the end of this section we provide a third characterization of k-consistency in terms of the existential pebble game, which is the tool of our choice in the proof of the main theorem.

2.1 CSP-Refutations

Given two σ-structures A and B, every line of our derivation system is a partial mapping from $V(\mathsf{A})$ to $V(\mathsf{B})$. The axioms are all partial mappings $p\colon V(\mathsf{A}) \to V(\mathsf{B})$ that are not partial homomorphisms. We have the following derivation rule to derive a new inconsistent assignment p. For all partial mappings $p'_i \subseteq p$, $x \in V(\mathsf{A})$ and $V(\mathsf{B}) = \{a_1, \dots, a_n\}$:

$$\frac{p'_1 \cup \{x \mapsto a_1\} \quad \cdots \quad p'_n \cup \{x \mapsto a_n\}}{p} \tag{1}$$

A *CSP-derivation* of p is a sequence $(p_1, \dots, p_\ell = p)$ such that every p_i is either an axiom or derived from lines p_j, $j < i$, via the derivation rule (1). A *CSP-refutation* is a CSP-derivation of $\emptyset$. Every derivation of p can naturally be seen as a directed acyclic graph (dag) where the nodes are labeled with lines from the derivation, one node of in-degree 0 is labeled with p and all nodes of out-degree 0 are labeled with axioms. If p_i is derived from $p_{j_1}, \dots, p_{j_n}$ using (1), then there is an arc from p_i to each $p_{j_1}, \dots, p_{j_n}$.

Given a CSP-derivation P, we let $\mathrm{Prop}(P)$ be the set of propagated mappings $p \in P$, i.e. all lines in the derivation that are not axioms. We define the *width* of a derivation P to be $\mathrm{width}(P) = \max_{p \in \mathrm{Prop}(P)} |p|$.[1] Furthermore, $\mathrm{depth}(P)$ denotes the *depth* of P which is the number of edges on the longest path in the dag associated with P. This measure characterizes the maximum number of nested propagation steps in P. Since CSP-derivations model the propagation process mentioned in the introduction, there is a CSP-refutation of width $k-1$ if and only if k-consistency cannot be established.

Furthermore, every propagation algorithm produces some CSP-derivation P. The total number of propagation steps performed by this algorithm is $|\mathrm{Prop}(P)|$ and the maximum number of nested propagation steps is $\mathrm{depth}(P)$. Let A and B be two relational structures such that k-consistency cannot be established. We define the *propagation depth* $\mathrm{depth}^k(\mathsf{A}, \mathsf{B}) := \min_P \mathrm{depth}(P)$ where the minimum is taken over all CSP-refutations P of width at most $k-1$. Hence, the $\mathrm{depth}^k(\mathsf{A}, \mathsf{B}) \le |V(\mathsf{A})|^{k-1}|V(\mathsf{B})|^{k-1}$ is the number of sequential propagation steps that have to be performed by any sequential or parallel propagation algorithm for k-consistency.

[1] Note that this implies $|p| \le \mathrm{width}(P) + 1$ for all axioms p used in the derivation P. However, the size of the axioms can always be bounded by the maximum arity of the relations in A and B.

2.2 Results and Related Work

Our main theorem is a tight lower bound on the propagation depth.

Theorem 1. *For every integer $k \geq 2$ there exists a constant $\varepsilon > 0$ and two positive integers n_0, m_0 such that for every $n \geq n_0$ and $m \geq m_0$ there exist two binary structures A_n and B_m with $|V(\mathsf{A}_n)| = n$ and $|V(\mathsf{B}_m)| = m$ such that* $\mathrm{depth}^k(\mathsf{A}_n, \mathsf{B}_m) \geq \varepsilon n^{k-1} m^{k-1}$.

We are aware of two particular cases that have been discovered earlier. First, for the case $k = 2$ (arc consistency) the theorem can be shown by rather simple examples that occurred very early in the AI-community. The structure of this exceptional case is discussed in deep in a joint work of Oleg Verbitsky and the author of this paper [4]. Second, for $k = 3$ Ladkin and Maddux [14] showed that there is a fixed finite binary structure B and an infinite sequence of binary structures A_i such that $\mathrm{depth}^3(\mathsf{A}_i, \mathsf{B}) = \Omega(|V(\mathsf{A}_i)|^2)$. They used this result to argue that every parallel propagation algorithm for path consistency needs at least a quadratic number of steps. This is tight only for fixed structures B, Theorem 1 extends their result to the case when B is also given as input.

Other related results investigate the decision complexity of the k-consistency test. To address this more general question one analyzes the computational complexity of the following decision problem.

k-Cons
Input: Two binary relational structures A and B.
Question: Can k-consistency be established for A and B?

Kasif [10] showed that 2-Cons is complete for PTIME under LOGSPACE reductions. Kolaitis and Panttaja [11] extended this result to every fixed $k \geq 2$. Moreover, they established that the problem is complete for EXPTIME if k is part of the input. In [3] the author showed that k-Cons cannot be decided in $O(n^{\frac{k-3}{12}})$ on deterministic multi-tape Turing machines, where n is the overall input size. Hence, any algorithm solving k-Cons (regardless of whether it performs constraint propagation or not) cannot be much faster than the standard propagation approach. It also follows from this result that, parameterized by the number of pebbles k, k-Cons is is complete for the parameterized complexity class XP. It is also worth noting that Gaspers and Szeider [9] investigated the parameterized complexity of other parameterized problems related to k-consistency.

2.3 The Existential Pebble Game

In this paragraph we introduce a third view on the k-consistency heuristic in terms of a combinatorial pebble game. The *existential k-pebble game* [12] is played by two players *Spoiler* and *Duplicator* on two relational structures A and B. There are k pairs of pebbles $(p_1, q_1), \ldots, (p_k, q_k)$ and during the game Spoiler moves the pebbles $p_1, \ldots, p_k$ to elements of $V(\mathsf{A})$ and Duplicator moves the pebbles $q_1, \ldots, q_k$ to elements of $V(\mathsf{B})$. At the beginning of the game, Spoiler places pebbles $p_1, \ldots, p_k$ on elements of $V(\mathsf{A})$ and Duplicator answers by putting pebbles $q_1, \ldots, q_k$ on elements of $V(\mathsf{B})$. In each further round Spoiler picks up a pebble pair (p_i, q_i) and places p_i on some element in $V(\mathsf{A})$. Duplicator answers by moving the corresponding pebble q_i to one element in $V(\mathsf{B})$. Spoiler wins the game if he can reach a position where the mapping defined by $p_i \mapsto q_i$ is not a partial homomorphism from A to B.

The connection between the existential k-pebble game and the k-consistency heuristic was made by Kolaitis and Vardi [13]. They showed that one can establish k-consistency by computing a winning strategy for Duplicator. Going a different way, the next lemma states that there is also a tight correspondence between Spoiler's strategy and CSP-refutations. The proof is a straightforward induction over the depth and included in the full version of the paper [2].

Lemma 2. *Let* A *and* B *be two relational structures. There is a CSP-refutation for* A *and* B *of width $k-1$ and depth d if and only if Spoiler has a strategy to win the existential k-pebble game on* A *and* B *within d rounds.*

Using this lemma it suffices to prove lower bounds on the number of rounds in the existential pebble game in order to prove Theorem 1. To argue about strategies in the existential pebble game we use the framework developed in [3]. We start with a formal definition of strategies for Duplicator.

Definition 3. *A* critical strategy *for Duplicator in the existential k-pebble game on structures* A *and* B *is a nonempty family $\mathcal{H}$ of partial homomorphisms from* A *to* B *together with a set* $\mathrm{crit}(\mathcal{H}) \subseteq \mathcal{H}$ *of critical positions satisfying the following properties:*

1. *All critical positions are $(k-1)$-partial homomorphisms.*
2. *If $h \in \mathcal{H}$ and $g \subset h$, then $g \in \mathcal{H}$.*
3. *For every $g \in \mathcal{H} \setminus \mathrm{crit}(\mathcal{H})$, $|g| < k$, and every $x \in V(\mathsf{A})$ there is an $a \in V(\mathsf{B})$ such that $g \cup \{x \mapsto a\} \in \mathcal{H}$.*

If $\mathrm{crit}(\mathcal{H}) = \emptyset$, *then $\mathcal{H}$ is a* winning strategy.

The set $\mathcal{H}$ is the set of good positions for Duplicator (therefore they are all partial homomorphisms). Non-emptiness and the closure property (2.) ensure that $\mathcal{H}$ contains the start position $\emptyset$. Furthermore, the closure property guarantees that the current position remains a good position for Duplicator when Spoiler picks up pebbles. The extension property (3.) ensures that, from every non-critical position, Duplicator has an appropriate answer if Spoiler puts a free pebble on x. It follows that if there are no critical positions, then Duplicator can always answer accordingly and thus wins the game. Otherwise, if Spoiler reaches a critical position, then Duplicator may not have an appropriate answer and the game reaches a critical state. In the next lemma we describe how to use critical strategies to prove lower bounds on the number of rounds.

Lemma 4. *If $\mathcal{H}_1, \ldots, \mathcal{H}_l$ is a sequence of critical strategies on the same pair of structures and for all $i < l$ and all $p \in \mathrm{crit}(\mathcal{H}_i)$ it holds that $p \in \mathcal{H}_j \setminus \mathrm{crit}(\mathcal{H}_j)$ for some $j \leq i+1$, then Duplicator wins the l-round existential k-pebble game.*

Proof. Starting with $i = 1$, Duplicator answers according to the extension property of $\mathcal{H}_i$, if the current position p is non-critical in $\mathcal{H}_i$. Otherwise, p is non-critical in $\mathcal{H}_j$ for some $j \leq i+1$ and Duplicator answers according to the extension property of $\mathcal{H}_j$. This allows Duplicator to survive for at least l rounds. □

The two structures A and B we construct are vertex colored graphs. They are built out of smaller graphs, called *gadgets*. Every gadget Q consists of two graphs Q_S and Q_D for Spoiler's and Duplicator's side, respectively. Hence, Q_S and Q_D will be subgraphs of A and B in the end. The gadgets contain *boundary vertices*, which are the vertices shared with other gadgets. To combine two strategies on two connected gadgets we need to ensure that the strategies agree on the boundary of the gadgets. Formally, let a *boundary function* of a strategy $\mathcal{H}$ on a gadget Q be a mapping β from the boundary of Q_S to the boundary of Q_D such that $\beta(z) = h(z)$ for all $h \in \mathcal{H}$ and all z in the domain of β and h. We say that two strategies $\mathcal{G}$ and $\mathcal{H}$ on gadgets Q and Q' are *connectable*, if their boundary functions agree on the common boundary vertices of Q and Q'. If $\mathcal{G}$ and $\mathcal{H}$ are two connectable critical strategies on gadgets $Q = (Q_S, Q_D)$ and $Q' = (Q'_S, Q'_D)$ it is not hard to see that the *composition*

$$\mathcal{G} \uplus \mathcal{H} = \{g \cup h \mid g \in \mathcal{G}, h \in \mathcal{H}\}$$

is a critical strategy on $Q_S \cup Q'_S$ and $Q_D \cup Q'_D$ with $\mathrm{crit}(\mathcal{G} \uplus \mathcal{H}) = \mathrm{crit}(\mathcal{G}) \cup \mathrm{crit}(\mathcal{H})$. Intuitively, playing according to the strategy $\mathcal{G} \uplus \mathcal{H}$ on Q and Q' means that Duplicator uses strategy $\mathcal{G}$ on Q and strategy $\mathcal{H}$ on Q'.

3 The Construction

3.1 Overview of the Construction

In this section we prove Theorem 1 for $k \geq 3$. We let $\mathsf{k} := k-1 \geq 2$ and construct two vertex colored graphs A_n and B_m with $O(n)$ and $O(m)$ vertices such that Spoiler needs $\Omega(n^{\mathsf{k}} m^{\mathsf{k}})$ rounds to win the existential $(\mathsf{k}+1)$-pebble game. We color the vertices of both graphs such that the colors partition the vertex set into independent sets, i. e. every vertex gets one color and there is no edge between vertices of the same color. The basic building blocks in our construction are sets of vertices which allow to store $n^{\mathsf{k}} m^{\mathsf{k}}$ partial homomorphisms with k pebbles.

Fig. 1. Basic vertex blocks. Two vertices x_j^i and $x_{j'}^{i'}$ get the same color iff $i = i'$.

We introduce vertices x_j^i ($i \in [\mathsf{k}]$, $j \in [n]$) in A_n and vertices x_j^i ($i \in [\mathsf{k}]$, $j \in [m] \cup \{0\}$) in B_m. For every $i \in [\mathsf{k}]$ the vertices x_j^i form a *block* and are colored with the same color (say P_{x^i}), which is different from any other color in the entire construction. The vertices x_0^i in structure B_m play a special role in our construction and are visualized by ∘ instead of • in the pictures. However, they are colored with the same color P_{x^i} as the other vertices x_j^i. Because of the coloring, Duplicator has to answer with some $x_{j'}^i$ whenever Spoiler pebbles a vertex x_j^i. Since there are nm positions for one pebble pair on • vertices in one block, we get $n^{\mathsf{k}}m^{\mathsf{k}}$ positions if every block has exactly one pebble pair on • vertices. The ∘ vertices are used by Duplicator whenever Spoiler does not play the intended way. That is, if Spoiler pebbles a vertex in block i that he is not supposed to pebble now, then Duplicator answers with x_0^i. The construction will have the property that this is always a good situation for Duplicator.

To describe pebble positions on such vertex blocks, we define mappings $\mathfrak{a}\colon [\mathsf{k}] \to [n]$ and $\mathfrak{b}\colon [\mathsf{k}] \to [m]$ and call the pebble position $\{(x^i_{\mathfrak{a}(i)}, x^i_{\mathfrak{b}(i)}) \mid i \in [\mathsf{k}]\}$ *valid.* If such valid position is on the board, then Duplicator answers with $x^i_{\mathfrak{b}(i)}$ if Spoiler pebbles $x^i_{\mathfrak{a}(i)}$ and with x_0^i if Spoiler pebbles x_j^i for some $j \neq \mathfrak{a}(i)$. We also need to name positions where Duplicator answers with x_0^i for every vertex in block i and let T be the set of blocks where this happens. For $\mathfrak{a}\colon [\mathsf{k}] \to [n]$, $\mathfrak{b}\colon [\mathsf{k}] \to [m]$ and $T \subseteq [\mathsf{k}]$ we call $\mathfrak{q} = (\mathfrak{a}, \mathfrak{b}, T)$ a *configuration.* The configuration $\mathfrak{q}$ is *valid* if $T = \emptyset$ and *invalid* otherwise. For every configuration $\mathfrak{q}$ and a set of x_j^i vertices as in Figure 1 we define the following homomorphism that describes Duplicator's behavior:

$$h^x_{\mathfrak{q}}(x^i_j) = \begin{cases} x^i_{\mathfrak{b}(i)}, \text{ if } j = \mathfrak{a}(i) \text{ and } i \notin T, \\ x^i_0, \text{ otherwise.} \end{cases}$$

By $h^x_{\mathfrak{0}}$ we denote the homomorphism $h^x_{\mathfrak{0}}(x^i_j) := x^i_0$ for all $i \in [\mathsf{k}], j \in [n]$. We say that a position of (at most $\mathsf{k}+1$) pebble pairs on these vertices is *invalid* if it is a subset of $h^x_{\mathfrak{q}}$ for some invalid configuration $\mathfrak{q}$. For valid configurations $\mathfrak{q} = (\mathfrak{a}, \mathfrak{b}, \emptyset)$ we say "$\mathfrak{q}$ on x" to name the valid pebble position $\{(x^i_{\mathfrak{a}(i)}, x^i_{\mathfrak{b}(i)}) \mid i \in [\mathsf{k}]\}$. Note that valid pebble positions are not invalid.[2]

In the entire construction there is one unique copy of the x_j^i-vertices, which are denoted by x_j^i. Our goal is to force Spoiler to pebble every valid position on x before he wins the game. He is supposed to do so in a specific predefined order. To fix this order we define a bijection α between valid configurations $(\mathfrak{a}, \mathfrak{b}, \emptyset)$ and the numbers $0, \dots, n^{\mathsf{k}}m^{\mathsf{k}} - 1$:

$$\alpha(\mathfrak{q}) := m^{\mathsf{k}} \sum_{i=1}^{\mathsf{k}} (\mathfrak{a}(i) - 1) n^{\mathsf{k}-i} + \sum_{i=1}^{\mathsf{k}} (\mathfrak{b}(i) - 1) m^{\mathsf{k}-i}.$$

Thus, $\alpha(\mathfrak{q})$ is the rank of the tuple $(\mathfrak{a}(1), \dots, \mathfrak{a}(\mathsf{k}), \mathfrak{b}(1), \dots, \mathfrak{b}(\mathsf{k}))$ in lexicographical order. If $\alpha(\mathfrak{q}) < n^{\mathsf{k}}m^{\mathsf{k}} - 1$, we define the successor $\mathfrak{q}^+ = (\mathfrak{a}^+, \mathfrak{b}^+, \emptyset)$ to be the unique valid configuration satisfying $\alpha(\mathfrak{q}^+) = \alpha(\mathfrak{q}) + 1$. In the sequel we introduce gadgets to make sure that:

[2] There are pebble positions on the x_j^i vertices that are neither valid nor invalid. However, such positions will not occur in our strategies.

- Spoiler can reach the position $\alpha^{-1}(0)$ on x from $\emptyset$,
- Spoiler can reach $\alpha^{-1}(i+1)$ on x from $\alpha^{-1}(i)$ on x and
- Spoiler wins from $\alpha^{-1}(n^{\mathsf{k}}m^{\mathsf{k}}-1)$ on x.

If we have these properties, we know that Spoiler has a winning strategy in the $(\mathsf{k}+1)$-pebble game. To show that Spoiler needs at least $n^{\mathsf{k}}m^{\mathsf{k}}$ rounds we argue that this is essentially the only way for Spoiler to win the game.

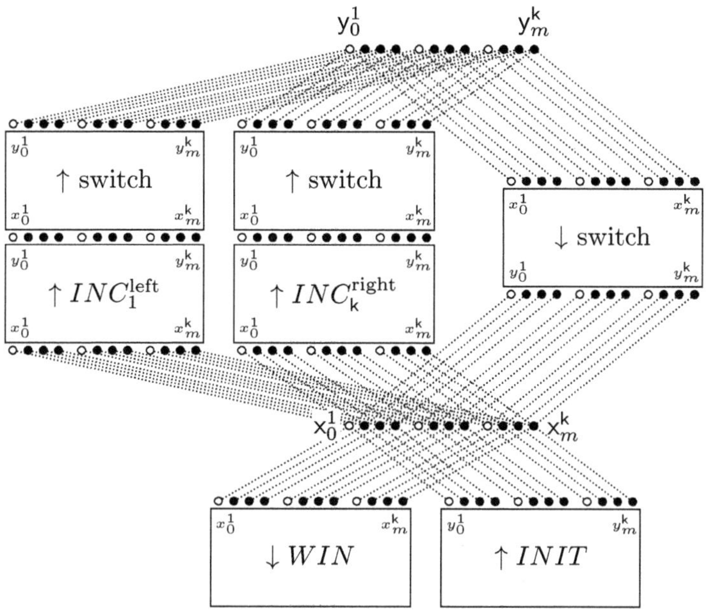

Fig. 2. The graph B_m. The boundaries of the gadgets are connected as indicated by the dotted lines (which need to be contracted). The arrows point from the input to the output vertices of the gadgets.

We start with an overview of the gadgets and how they are glued together to form the structures A_n and B_m. The boundary of our gadgets consists of *input* vertices and *output* vertices. For every gadget the set of input (output) vertices is a copy of the vertex set in Figure 1 and we write x_j^i (y_j^i) to name them. This enables us to glue together the gadgets at their input and output vertices. The overall construction for the graph B_m is shown in Figure 2. The schema for A_n is similar, it contains Spoiler's side of the corresponding gadgets which are glued together the same way as in B_m (just replace m by n and drop the $\circ$ vertices). There are four types of gadgets: the initialization gadget, the winning gadget, several increment gadgets and the switch.

The *initialization gadget* ensures that Spoiler can reach $\alpha^{-1}(0)$ on x, i. e. the pebble position $\{(\mathsf{x}_1^1, \mathsf{x}_1^1), \ldots, (\mathsf{x}_1^k, \mathsf{x}_1^k)\}$. This gadget has only output boundary vertices and is used by Spoiler at the beginning of the game. There are *increment gadgets* $\text{INC}_i^{\text{left}}$ and $\text{INC}_i^{\text{right}}$ for all $i \in [\mathsf{k}]$. The input vertices of every increment gadget are identified with the x vertices as depicted in Figure 2. The increment gadgets (all together) ensure that Spoiler can increment a configuration. More precisely, for every valid configuration $\mathfrak{q}$ with $\alpha(\mathfrak{q}) < n^{\mathsf{k}}m^{\mathsf{k}} - 1$, there is one increment gadget INC such that Spoiler can reach $\mathfrak{q}^+$ on the output of INC from $\mathfrak{q}$ on the input. Every increment gadget is followed by a copy of the *switch*. The input of $2\mathsf{k}$ switches is identified with the output of the $2\mathsf{k}$ increment gadgets and the output of these switches is identified with a unique block of y-vertices and the input of one additional *single switch* (see Figure 2). The output of this switch is in turn identified with the unique block of x-vertices. The switches are used to perform the transition in the game from $\alpha^{-1}(i)$ on x to $\alpha^{-1}(i+1)$ on x. Spoiler can pebble a valid position through one switch: from $\mathfrak{q}$ on the input of a switch Spoiler can reach $\mathfrak{q}$ on the output of that switch. Hence, Spoiler can simply pebble the incremented position $\alpha^{-1}(i+1)$ from the output of an increment gadget through two switches to the x-block.

Finally, the *winning gadget* ensures that from $\alpha^{-1}(n^{\mathsf{k}}m^{\mathsf{k}}-1)$ on x Spoiler wins the game. The winning gadget has only input vertices, which are identified with the x-vertices. From $\alpha^{-1}(n^{\mathsf{k}}m^{\mathsf{k}} - 1)$ on the input, Spoiler can win the game by playing on this gadget. On the other hand, the gadget ensures that Spoiler can *only* win from $\alpha^{-1}(n^{\mathsf{k}}m^{\mathsf{k}} - 1)$ on x and Duplicator does not lose from any other configuration on x.

3.2 The Gadgets

We now describe the winning gadget and the increment gadgets in detail and provide strategies for Spoiler and Duplicator on them. Afterwards we briefly discuss the switch and the initialization gadget. In the next section we combine the partial strategies on the gadgets to prove Theorem 1.

The *winning gadget* is shown in Figure 3. On Spoiler's side there is just one additional vertex a, which is connected to x_n^i for all $i \in [\mathsf{k}]$. On Duplicator's side there are k additional vertices $a^i, i \in [\mathsf{k}]$. Every a^i is connected to all input vertices except x_m^i. We use one new vertex color to color the vertex a and all vertices a_i. From the position $\{(x_n^1, x_m^1), \dots, (x_n^k, x_m^k)\}$ "$\alpha^{-1}(n^{\mathsf{k}}m^{\mathsf{k}} - 1)$ on x" Spoiler wins the game by placing the $(\mathsf{k}+1)$st pebble on a. Duplicator has to answer with some a_i (because of the coloring). Since there is an edge between x_n^i and a in WIN_S but none between x_m^i and a_i in WIN_D, Spoiler wins immediately. It is also not hard to see that for any other position where at least one pebble pair (x_n^j, x_m^j) is missing Duplicator can survive by choosing a_j.

The *increment gadgets* enable Spoiler to reach the successor $\mathfrak{q}^+$ from $\mathfrak{q}$. Recall that we identify every valid configuration $\mathfrak{q} = (\mathfrak{a}, \mathfrak{b}, \emptyset)$ with the tuple $(\mathfrak{a}(1), \dots, \mathfrak{a}(\mathsf{k}), \mathfrak{b}(1), \dots, \mathfrak{b}(\mathsf{k})) \in [n]^k \times [m]^k$ and define $\alpha(\mathfrak{q})$ to be the rank (from 0 to $n^k m^{\mathsf{k}} - 1$) of this tuple in lexicographical order. Let $\mathfrak{q}$ be a valid configuration with $\alpha(\mathfrak{q}) < n^{\mathsf{k}}m^{\mathsf{k}} - 1$ and successor $\mathfrak{q}^+ = (\mathfrak{a}^+(1), \dots, \mathfrak{a}^+(\mathsf{k}), \mathfrak{b}^+(1), \dots, \mathfrak{b}^+(\mathsf{k}))$. We use two types of increment gadgets, *left* and *right*, depending on whether the left-hand side of the tuple changes after incrementation or not. There are k increment gadgets of each type. Spoiler uses them depending on which position

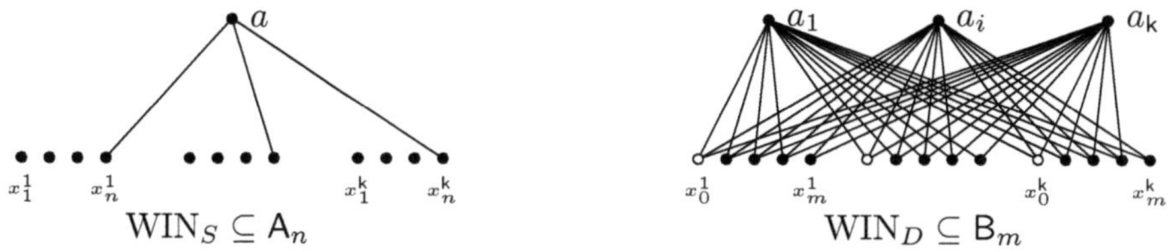

Fig. 3. The winning gadget

the last carryover occurs. If

$$\begin{aligned} \mathfrak{q} &= (\mathfrak{a}(1), \ldots, \mathfrak{a}(\mathsf{k}), \quad \mathfrak{b}(1), \ldots, \mathfrak{b}(\ell-1), \quad \mathfrak{b}(\ell) < m, \quad m, \ldots, m) \text{ and hence} \\ \mathfrak{q}^+ &= (\mathfrak{a}(1), \ldots, \mathfrak{a}(\mathsf{k}), \quad \mathfrak{b}(1), \ldots, \mathfrak{b}(\ell-1), \quad \mathfrak{b}(\ell)+1, \quad 1, \ldots, 1), \end{aligned}$$

then Spoiler uses the increment gadget $\mathrm{INC}_\ell^{\mathrm{right}}$ to reach $\mathfrak{q}^+$ on the output from $\mathfrak{q}$ on the input. If

$$\begin{aligned} \mathfrak{q} &= (\mathfrak{a}(1), \ldots, \mathfrak{a}(\ell-1), \quad \mathfrak{a}(\ell) < n, \quad n, \ldots, n, \quad m, \ldots, m) \text{ and hence} \\ \mathfrak{q}^+ &= (\mathfrak{a}(1), \ldots, \mathfrak{a}(\ell-1), \quad \mathfrak{a}(\ell)+1, \quad 1, \ldots, 1, \quad 1, \ldots, 1), \end{aligned}$$

then Spoiler uses $\mathrm{INC}_\ell^{\mathrm{left}}$. Thus, for every valid configuration $\mathfrak{q}$ with $\alpha(\mathfrak{q}) < n^{\mathsf{k}}m^{\mathsf{k}}-1$ there is exactly one *applicable* increment gadget. The increment gadgets

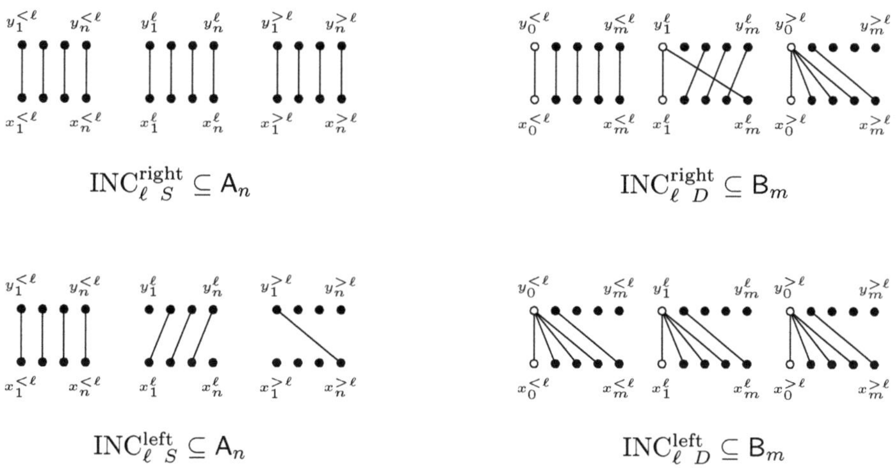

Fig. 4. The increment gadgets

are shown in Figure 4. All input vertices x^i_j have at most one output vertex $y^i_{j'}$ as neighbor. Furthermore, if the gadget is applicable to a valid configuration $\mathfrak{q} = (\mathfrak{a}, \mathfrak{b}, \emptyset)$, then the unique neighbor of $x^i_{\mathfrak{a}(i)}$ is $y^i_{\mathfrak{a}^+(i)}$ and the unique neighbor of $x^i_{\mathfrak{b}(i)}$ is $y^i_{\mathfrak{b}^+(i)}$. This enables Spoiler to reach $\mathfrak{q}^+$ on the output from $\mathfrak{q}$ on the input by the following procedure. First, Spoiler places the remaining pebble on $y^1_{\mathfrak{a}^+(1)}$. Since this vertex is adjacent to $x^1_{\mathfrak{a}(1)}$, Duplicator has to answer with $y^1_{\mathfrak{b}^+(1)}$, the only vertex that is adjacent to $x^1_{\mathfrak{b}(1)}$. Afterwards, Spoiler picks up the pebble pair from $(x^1_{\mathfrak{a}(1)}, x^1_{\mathfrak{b}(1)})$. On the second block Spoiler proceeds the same way: he pebbles $y^2_{\mathfrak{a}^+(2)}$, forces the position $(y^2_{\mathfrak{a}^+(2)}, y^2_{\mathfrak{b}^+(2)})$ and picks up the pebbles from $(x^2_{\mathfrak{a}(2)}, x^2_{\mathfrak{b}(2)})$. By iterating this procedure Spoiler reaches $\mathfrak{q}^+$ on the output.

If Spoiler tries to move a configuration through one increment gadget that is *not* applicable, then Duplicator can answer with an invalid configuration on the output as follows. On the one hand, if the gadget is not applicable because some $\mathfrak{b}(i)$ does not have the specified value, then $x^i_{\mathfrak{b}(i)}$ is adjacent to y^i_0. On the other hand, if some $\mathfrak{a}(i)$ has the wrong value, then $x^i_{\mathfrak{a}(i)}$ is not adjacent to an output vertex. In both cases Duplicator can safely pebble y^i_0 if Spoiler queries some y^i_j and hence maintain an invalid output position. The next lemma summarizes the strategies on the increment gadget.

Lemma 5. *Let* $\mathfrak{q} = (\mathfrak{a}, \mathfrak{b}, T)$ *be a configuration and* INC *an increment gadget.*

1. *If* INC *is applicable to* $\mathfrak{q}$*, then Spoiler can reach* $\mathfrak{q}^+$ *on the output from* $\mathfrak{q}$ *on the input.*
2. *If* INC *is applicable to* $\mathfrak{q}$*, then there is a winning strategy for Duplicator with boundary function* $h^x_{\mathfrak{q}}$ *on the input and* $h^y_{\mathfrak{q}^+}$ *on the output.*
3. *If* INC *is not applicable to* $\mathfrak{q}$*, then there is a winning strategy for Duplicator with boundary function* $h^x_{\mathfrak{q}}$ *on the input and* $h^y_{\mathfrak{q}_{\text{inv}}}$ *on the output for an invalid configuration* $\mathfrak{q}_{\text{inv}}$.

The *switch* is an extension of the "multiple input one-way switch" defined in [3] (which in turn is a generalization of [11]). The difference is that the old switch can only be used for the case $n = 1$. It requires some work to adjust the old switch to make it work for the more general setting. But since these modifications require a deeper inspection into this technical construct (and are not the main contribution of this paper), we refer to the full version of the paper [2] and use the switch as black box at this point.

We briefly explain the strategies on the switch and provide them in Lemma 6. As mentioned earlier, Spoiler can simply move a valid position from the input to the output of the switch (Lemma 6(i)). Duplicator has a winning strategy called *output strategy,* where any position is on the output and $h^x_{\mathbf{0}}$ is on the input (Lemma 6(ii)). This ensures that Spoiler cannot move backwards to reach $\mathfrak{q}$ on the input from $\mathfrak{q}$ on the output. Hence, this strategy forces Spoiler to play through the switches in the intended direction (as indicated by arrows Figure 2). Furthermore, for every invalid $\mathfrak{q}_{\text{inv}}$ Duplicator has a winning strategy where

$h^x_{\mathfrak{q}_{\mathrm{inv}}}$ is on the input and $h^y_{\mathbf{0}}$ is on the output (Lemma 6(iii)), which ensures that Spoiler cannot move invalid positions through the switch. This strategy is used by Duplicator whenever Spoiler plays on an increment gadget that is not applicable. By Lemma 5, Duplicator can force an invalid configuration on the output of that increment gadget and hence on the input of the subsequent switch.

To ensure that Spoiler picks up all pebbles when reaching $\mathfrak{q}$ on the output from $\mathfrak{q}$ on the input, Duplicator has a critical *input strategy* with $\mathfrak{q}$ on the input and $h^y_{\mathbf{0}}$ on the output (Lemma 6(iv)). The critical positions are either contained in an output strategy, where $\mathfrak{q}$ is on the output, or (for technical reasons) in a *restart strategy.* If Duplicator plays according to this input strategy, the only way for Spoiler to bring $\mathfrak{q}$ from the input to the output is to pebble an output critical position inside the switch (using all the pebbles) and force Duplicator to switch to the corresponding output strategy.

Lemma 6. *For every configuration* $\mathfrak{q} = (\mathfrak{a}, \mathfrak{b}, T)$*, the following statements hold in the existential* $(\mathsf{k}+1)$*-pebble game on the switch:*

(i) If $\mathfrak{q}$ *is valid, then Spoiler can reach* $\mathfrak{q}$ *on the output from* $\mathfrak{q}$ *on the input.*
(ii) Duplicator has a winning strategy $\mathcal{H}^{out}_{\mathfrak{q}}$ *with boundary function* $h^x_{\mathbf{0}} \cup h^y_{\mathfrak{q}}$*.*
(iii) If $\mathfrak{q}$ *is invalid, then Duplicator has a winning strategy* $\mathcal{H}^{restart}_{\mathfrak{q}}$ *with boundary function* $h^x_{\mathfrak{q}} \cup h^y_{\mathbf{0}}$*.*
(iv) If $\mathfrak{q}$ *is valid, then Duplicator has a critical strategy* $\mathcal{H}^{in}_{\mathfrak{q}}$ *with boundary function* $h^x_{\mathfrak{q}} \cup h^y_{\mathbf{0}}$ *and sets of restart critical positions* $\mathcal{C}^{restart\text{-}crit}_{\mathfrak{q},t}$ *(for* $t \in [\mathsf{k}]$*) and output critical positions* $\mathcal{C}^{out\text{-}crit}_{\mathfrak{q}}$ *such that:*
  (a) $\mathrm{crit}(\mathcal{H}^{in}_{\mathfrak{q}}) = \bigcup_{t\in[\mathsf{k}]} \mathcal{C}^{restart\text{-}crit}_{\mathfrak{q},t} \cup \mathcal{C}^{out\text{-}crit}_{\mathfrak{q}}$*,*
  (b) $\mathcal{C}^{restart\text{-}crit}_{\mathfrak{q},t} \subseteq \mathcal{H}^{restart}_{(\mathfrak{a},\mathfrak{b},\{t\})}$ *and*
  (c) $\mathcal{C}^{out\text{-}crit}_{\mathfrak{q}} \subseteq \mathcal{H}^{out}_{\mathfrak{q}}$*.*

At the beginning of the game we want that Spoiler can reach the start configuration $\alpha^{-1}(0)$ on x, which is the pebble position $\{(\mathsf{x}^1_1, \mathsf{x}^1_1), \dots, (\mathsf{x}^k_1, \mathsf{x}^k_1)\}$. To ensure this, we use the *initialization gadget* and identify its output vertices y^i_j with the block of x^i_j vertices. As for the switch, this gadget is an extension of the initialization gadget presented in [3] and we use it as a black box here. The strategies on this gadget are provided in Lemma 7, the proof of Lemma 7 is given in the full version of the paper [2]. The main property of the gadget is that Spoiler can reach the start position $\mathfrak{q}$ at the boundary (i) and Duplicator has a corresponding counter strategy (ii) in this situation. Furthermore, if an arbitrary position occurs at the boundary during the game, Duplicator has a strategy to survive (iii). This is only a critical strategy, but Duplicator can switch to the initial strategy (hence "restart" the game) if Spoiler moves to one of the critical positions.

Lemma 7. *Let* $\mathfrak{q} = \alpha^{-1}(0)$*. The following holds in the existential* $(\mathsf{k}+1)$*-pebble game on* INIT*:*

(i) Spoiler can reach $\mathfrak{q}$ *on the output.*

(ii) *There is a winning strategy* $\mathcal{I}^{init}$ *for Duplicator with boundary function* $h^{\mathsf{y}}_{\mathsf{q}}$.
(iii) *For every (valid or invalid) configuration* q' *there is a critical strategy* $\mathcal{I}^{init}_{\mathsf{q}'}$ *with boundary function* $h^{\mathsf{y}}_{\mathsf{q}'}$ *and* $\mathrm{crit}(\mathcal{I}^{init}_{\mathsf{q}'}) \subseteq \mathcal{I}^{init}$.

3.3 Proof of Theorem 1

The size of the vertex set in every gadget is linear in n on Spoiler's side and linear in m on Duplicator's side. Since the overall construction uses a constant number of gadgets it follows that $|V(\mathsf{A}_n)| = O(n)$ and $|V(\mathsf{B}_m)| = O(m)$. To prove the lower bound on the number of rounds Spoiler needs to win the existential $(\mathsf{k}+1)$-pebble game we provide a sequence of critical strategies in Lemma 8 satisfying the properties stated in Lemma 4. For a critical strategy $\mathcal{S}$ we let $\widehat{\mathcal{S}} := \mathcal{S} \setminus \mathrm{crit}(\mathcal{S})$.

Lemma 8. *Spoiler has a winning strategy in the existential* $(\mathsf{k}+1)$*-pebble game on* A_n *and* B_m. *Furthermore, there is a sequence of critical strategies for Duplicator* $\mathcal{G}^{\mathsf{start}}, \mathcal{F}_1, \mathcal{G}_1, \mathcal{F}_2, \mathcal{G}_2, \ldots, \mathcal{G}_{n^{\mathsf{k}}m^{\mathsf{k}}-2}, \mathcal{F}_{n^{\mathsf{k}}m^{\mathsf{k}}-1}$ *such that*

$$\begin{aligned} \mathrm{crit}(\mathcal{G}^{\mathsf{start}}) &\subseteq \widehat{\mathcal{F}}_1, \\ \mathrm{crit}(\mathcal{G}_i) &\subseteq \widehat{\mathcal{F}}_{i+1} \cup \widehat{\mathcal{G}}^{\mathsf{start}}, && 1 \le i \le n^{\mathsf{k}}m^{\mathsf{k}} - 2, \\ \mathrm{crit}(\mathcal{F}_i) &\subseteq \widehat{\mathcal{G}}_i \cup \widehat{\mathcal{G}}^{\mathsf{start}}, && 1 \le i \le n^{\mathsf{k}}m^{\mathsf{k}} - 2. \end{aligned}$$

Proof (Proof of Theorem 1). For $k = 2$ the theorem follows from [4]. For $k \ge 3$ consider the structures A_n and B_m (for $\mathsf{k} = k - 1$) defined above. By Lemma 8 Spoiler wins the existential k-pebble game on A_n and B_m. Furthermore, it follows via Lemma 4 that Spoiler needs at least $\Omega(n^{k-1}m^{k-1})$ rounds to win the game. To get structures with exactly n and m vertices we take the largest n', m' such that $|V(\mathsf{A}_{n'})| \le n$, $|V(\mathsf{B}_{m'})| \le m$ and fill up the structures with an appropriate number of isolated vertices. □

Proof (Proof of Lemma 8). To show that Spoiler has a winning strategy it suffices to prove the following three statements:

(1) Spoiler can reach the position $\alpha^{-1}(0)$ on x from $\emptyset$,
(2) Spoiler can reach $\alpha^{-1}(i+1)$ on x from $\alpha^{-1}(i)$ on x (for $i < n^{\mathsf{k}}m^{\mathsf{k}} - 1$) and
(3) Spoiler wins from $\alpha^{-1}(n^{\mathsf{k}}m^{\mathsf{k}} - 1)$ on x.

Assertion (1) follows from Lemma 7 and (3) is ensured by the winning gadget. For (2), Spoiler starts with the position $\mathsf{q} = \alpha^{-1}(i)$ on x. Since $i < n^{\mathsf{k}}m^{\mathsf{k}} - 1$ there is exactly one increment gadget applicable to q. Spoiler uses Lemma 5 to reach $\mathsf{q}^+ = \alpha^{-1}(i+1)$ on the output of that gadget. By applying Lemma 6.(i) twice, Spoiler can pebble q^+ through the two switches to the x vertices.

To define the sequence of global critical strategies we combine the partial critical strategies on the gadgets using the $\uplus$-operator. There are three types of strategies: $\mathcal{G}^{\mathsf{start}}$, $\mathcal{F}_i$ and $\mathcal{G}_i$. To define $\mathcal{G}_i$ we let $\mathsf{q} = \alpha^{-1}(i)$. Duplicator plays according to $h^{\mathsf{x}}_{\mathsf{q}}$ on x and according to $h^{\mathsf{y}}_{\mathsf{0}}$ on y. She plays according to this strategy in the case when Spoiler reaches "$\mathfrak{q}$ on x". The critical strategy $\mathcal{G}_i$ is the combination of the following (pairwise connectable) strategies on the gadgets:

- The critical strategy $\mathcal{I}_{\mathfrak{q}}^{\text{init}}$ on the initialization gadget (Lemma 7).
- The winning strategy with boundary $h_{\mathfrak{q}}^{x}$ and $h_{\mathfrak{q}+}^{y}$ on the increment gadget applicable to $\mathfrak{q}$ (Lemma 5).
- The critical input strategy $\mathcal{H}_{\mathfrak{q}+}^{\text{in}}$ on the switch following the applicable increment gadget (Lemma 6).
- The winning strategy with boundary $h_{\mathfrak{q}}^{x}$ and $h_{\mathfrak{q}_{\text{inv}}}^{y}$ on the other increment gadgets not applicable to $\mathfrak{q}$ (Lemma 5).
- The winning strategy $\mathcal{H}_{\mathfrak{q}_{\text{inv}}}^{\text{restart}}$ on the switches following the inapplicable increment gadgets (Lemma 6). Here, $\mathfrak{q}_{\text{inv}}$ is the invalid configuration on the output of the corresponding increment gadget.
- The output winning strategy $\mathcal{H}_{\mathfrak{q}}^{\text{out}}$ on the single switch (Lemma 6).

If in the above setting Spoiler increments $\mathfrak{q}$ through the applicable increment gadget and moves $\mathfrak{q}^{+} = \alpha^{-1}(i+1)$ through the subsequent switch, then Duplicator switches to the strategy $\mathcal{F}_{i+1}$. To define $\mathcal{F}_i$ we fix $\mathfrak{q} = \alpha^{-1}(i)$. In this strategy, Duplicator plays according to $h_{\mathfrak{0}}^{\mathsf{x}}$ on x and according to $h_{\mathfrak{q}}^{\mathsf{y}}$ on y. This critical strategy is the combination of the following strategies on the gadgets.

- The critical strategy $\mathcal{I}_{\mathfrak{0}}^{\text{init}}$ on the initialization gadget.
- The winning strategy with boundary $h_{\mathfrak{0}}^{x}$ and $h_{\mathfrak{0}}^{y}$ on the increment gadgets.
- The output strategy $\mathcal{H}_{\mathfrak{q}}^{\text{out}}$ on the switches following the increment gadgets.
- The critical input strategy $\mathcal{H}_{\mathfrak{q}}^{\text{in}}$ on the single switch.

The critical positions in the strategies $\mathcal{G}_i$ and $\mathcal{F}_i$ are inside the switches and the initialization gadget. Recall that by Lemma 6.(iv) the critical positions on the switch can be divided into restart critical positions and output critical positions. Furthermore, all output critical positions of $\mathcal{G}_i$, which are inside the switch following the applicable increment gadget, are contained as non-critical positions in $\mathcal{F}_{i+1}$. All output critical position in $\mathcal{F}_i$, which are inside the single switch, are contained as non-critical positions in $\mathcal{G}_i$. Now we define $\mathcal{G}^{\text{start}}$, which contains all other critical positions of $\mathcal{G}_i$ and $\mathcal{F}_i$. The critical strategy $\mathcal{G}^{\text{start}}$ is the union of several other global strategies. The first one is $\mathcal{G}^{\text{init}}$, which is defined as $\mathcal{G}_0$ except that it contains the winning strategy $\mathcal{I}^{\text{init}}$ on the initialization gadget. Thus, by Lemma 7, it contains every critical position on the initialization gadget as non-critical position. Note that the output critical positions of $\mathcal{G}^{\text{init}}$ are contained as non-critical positions in $\mathcal{F}_1$. Since $\mathcal{G}^{\text{init}}$ handles the critical positions on the initialization gadget and we discussed the output critical positions on the switches, it remains to consider the restart critical positions of the strategies. For this we construct a strategy $\mathcal{G}_i^{\text{restart}}$ to handle the restart critical positions of $\mathcal{G}_i$ (for $i \geq 1$) and of $\mathcal{G}^{\text{init}}$ (for $i = 0$). Furthermore, we define for every $i \geq 1$ a strategy $\mathcal{F}_i^{\text{restart}}$ to handle the restart critical positions of $\mathcal{F}_i$.

For $0 \leq i \leq n^{\mathsf{k}}m^{\mathsf{k}} - 2$ and $t \in [\mathsf{k}]$ we let $\mathfrak{q} = \alpha^{-1}(i) = (\mathfrak{a}, \mathfrak{b}, \emptyset)$ and $\mathfrak{q}_t$ be the invalid configuration $(\mathfrak{a}, \mathfrak{b}, \{t\})$. The global strategy $\mathcal{G}_{i,t}^{\text{restart}}$ is the combination of the following strategies on the gadgets.

- The critical strategy $\mathcal{I}_{\mathfrak{q}_t}^{\text{init}}$ on the initialization gadget.
- The winning strategy with boundary $h_{\mathfrak{q}_t}^x$ and $h_{\mathfrak{q}_{\text{inv}}}^y$ on the increment gadgets. Note that, since $\mathfrak{q}_t$ is invalid, no increment gadget is applicable to $\mathfrak{q}_t$.
- The winning strategy $\mathcal{H}_{\mathfrak{q}_{\text{inv}}}^{\text{restart}}$ on the switches following the increment gadgets. Again, $\mathfrak{q}_{\text{inv}}$ is the invalid configuration at the output of the preceding increment gadget.
- The output winning strategy $\mathcal{H}_{\mathfrak{q}_t}^{\text{out}}$ on the single switch.

Finally, we let $\mathcal{G}_i^{\text{restart}} := \bigcup_{i\in[\mathsf{k}]} \mathcal{G}_{i,t}^{\text{restart}}$. Note that by Lemma 6.(iv) every restart critical position of $\mathcal{G}_i$ is contained in $\mathcal{G}_i^{\text{restart}}$ and every restart critical position of $\mathcal{G}^{\text{init}}$ is contained in $\mathcal{G}_0^{\text{restart}}$. Now we define for $1 \leq i \leq n^{\mathsf{k}}m^{\mathsf{k}} - 2$, $t \in [\mathsf{k}]$, $\mathfrak{q} = \alpha^{-1}(i) = (\mathfrak{a}, \mathfrak{b}, \emptyset)$ and $\mathfrak{q}_t := (\mathfrak{a}, \mathfrak{b}, \{t\})$ the strategy $\mathcal{F}_{i,t}^{\text{restart}}$ analogously. It consists of the following partial strategies.

- The critical strategy $\mathcal{I}_{\mathfrak{0}}^{\text{init}}$ on the initialization gadget.
- The winning strategy with boundary $h_{\mathfrak{0}}^x$ and $h_{\mathfrak{0}}^y$ on the increment gadgets.
- The winning strategy $\mathcal{H}_{\mathfrak{0}}^{\text{restart}}$ on the switches after the increment gadgets.
- The winning strategy $\mathcal{H}_{\mathfrak{q}_t}^{\text{restart}}$ on the single switch.

In the end we let $\mathcal{F}_i^{\text{restart}}$ be the union of all $\mathcal{F}_{i,t}^{\text{restart}}$. Note that every restart critical position of $\mathcal{F}_i$ is contained as non-critical position in $\mathcal{F}_i^{\text{restart}}$. Finally, let

$$\mathcal{G}^{\text{start}} := \mathcal{G}^{\text{init}} \cup \bigcup_{0\leq i\leq n^{\mathsf{k}}m^{\mathsf{k}}-2} \mathcal{G}_i^{\text{restart}} \cup \bigcup_{1\leq i\leq n^{\mathsf{k}}m^{\mathsf{k}}-2} \mathcal{F}_i^{\text{restart}}.$$

To conclude the proof note that the critical positions of $\mathcal{G}_i^{\text{restart}}$ and $\mathcal{F}_i^{\text{restart}}$ are inside the initialization gadget and hence contained in $\widehat{\mathcal{G}}^{\text{init}}$. Thus they are not critical positions of $\mathcal{G}^{\text{start}}$. Hence, $\text{crit}(\mathcal{G}^{\text{start}}) = \text{crit}(\mathcal{G}^{\text{init}}) \subseteq \widehat{\mathcal{F}}_1$. □

4 Conclusion

We have proven an optimal lower bound of $\Omega(n^{k-1}d^{k-1})$ on the number of nested propagation steps in the k-consistency procedure on constraint networks with n variables and domain size d. It follows that every parallel propagation algorithm has to perform at least $\Omega(n^{k-1}d^{k-1})$ sequential steps. Using $(n+d)^{O(k)}$ processors (one for every instance of the inference rule), k-consistency can be computed in $O(n^{k-1}d^{k-1})$ parallel time, which is optimal for propagation algorithms. In addition, the best sequential algorithm runs in $O(n^k d^k)$. The overhead compared to the parallel approach is mainly caused by the time needed to search for the next inconsistent assignment that might be propagated – and this seems to be the only task that can be parallelized.

Although we have proven an optimal lower bound in the general setting, it might be interesting to investigate the propagation depth of k-consistency on restricted classes of structures. Especially, if in such cases the propagation depth is bounded by $O(\log(n+d))$, we know that k-consistency is in NC and hence parallelizable.

References

1. Atserias, A., Kolaitis, P.G., Vardi, M.Y.: Constraint propagation as a proof system. In: Wallace, M. (ed.) CP 2004. LNCS, vol. 3258, pp. 77–91. Springer, Heidelberg (2004)
2. Berkholz, C.: The Propagation Depth of Local Consistency. ArXiv e-prints (2014), `http://arxiv.org/abs/1406.4679`
3. Berkholz, C.: Lower bounds for existential pebble games and k-consistency tests. Logical Methods in Computer Science 9(4) (2013), `http://arxiv.org/abs/1205.0679`
4. Berkholz, C., Verbitsky, O.: On the speed of constraint propagation and the time complexity of arc consistency testing. In: Chatterjee, K., Sgall, J. (eds.) MFCS 2013. LNCS, vol. 8087, pp. 159–170. Springer, Heidelberg (2013)
5. Cooper, M.C.: An optimal k-consistency algorithm. Artificial Intelligence 41(1), 89–95 (1989)
6. Dechter, R., Pearl, J.: A problem simplification approach that generates heuristics for constraint-satisfaction problems. Tech. rep., Cognitive Systems Laboratory, Computer Science Department, University of California, Los Angeles (1985)
7. Feder, T., Vardi, M.Y.: The computational structure of monotone monadic snp and constraint satisfaction: A study through datalog and group theory. SIAM Journal on Computing 28(1), 57–104 (1998)
8. Freuder, E.C.: Synthesizing constraint expressions. Commun. ACM 21, 958–966 (1978)
9. Gaspers, S., Szeider, S.: The parameterized complexity of local consistency. In: Lee, J. (ed.) CP 2011. LNCS, vol. 6876, pp. 302–316. Springer, Heidelberg (2011)
10. Kasif, S.: On the parallel complexity of discrete relaxation in constraint satisfaction networks. Artificial Intelligence 45(3), 275–286 (1990)
11. Kolaitis, P.G., Panttaja, J.: On the complexity of existential pebble games. In: Baaz, M., Makowsky, J.A. (eds.) CSL 2003. LNCS, vol. 2803, pp. 314–329. Springer, Heidelberg (2003)
12. Kolaitis, P.G., Vardi, M.Y.: On the expressive power of datalog: Tools and a case study. J. Comput. Syst. Sci. 51(1), 110–134 (1995)
13. Kolaitis, P.G., Vardi, M.Y.: A game-theoretic approach to constraint satisfaction. In: Proc AAAI/IAAI 2000, pp. 175–181 (2000)
14. Ladkin, P.B., Maddux, R.D.: On binary constraint problems. J. ACM 41(3), 435–469 (1994), `http://doi.acm.org/10.1145/176584.176585`
15. Samal, A., Henderson, T.: Parallel consistent labeling algorithms. International Journal of Parallel Programming 16, 341–364 (1987)
16. Susswein, S., Henderson, T., Zachary, J., Hansen, C., Hinker, P., Marsden, G.: Parallel path consistency. International Journal of Parallel Programming 20(6), 453–473 (1991), `http://dx.doi.org/10.1007/BF01547895`

The Balance Constraint Family

Christian Bessiere[1], Emmanuel Hebrard[2], George Katsirelos[3], Zeynep Kiziltan[4], Émilie Picard-Cantin[5], Claude-Guy Quimper[5], and Toby Walsh[6]

[1] CNRS, University of Montpellier, France
bessiere@lirmm.fr
[2] LAAS-CNRS, Toulouse, France
hebrard@laas.fr
[3] INRA, Toulouse, France
george.katsirelos@toulouse.inra.fr
[4] DISI, University of Bologna, Italy
zeynep@cs.unibo.it
[5] Université Laval, Canada
epicardcantin@petalmd.com, claude-guy.quimper@ift.ulaval.ca
[6] NICTA and University of New South Wales, Australia
toby.walsh@nicta.com.au

Abstract. The BALANCE constraint introduced by Beldiceanu ensures solutions are balanced. This is useful when, for example, there is a requirement for solutions to be fair. BALANCE bounds the difference B between the minimum and maximum number of occurrences of the values assigned to the variables. We show that achieving domain consistency on BALANCE is NP-hard. We therefore introduce a variant, ALLBALANCE with a similar semantics that is only polynomial to propagate. We consider various forms of ALLBALANCE and focus on ATMOSTALLBALANCE which achieves what is usually the main goal, namely constraining the upper bound on B. We provide a specialized propagation algorithm, and a powerful decomposition both of which run in low polynomial time. Experimental results demonstrate the promise of these new filtering methods.

1 Introduction

In many scheduling, rostering and related problems, we want to share tasks out as equally as possible. For example, in the nurse rostering problems in [17,5], we wish for all nurses to have a similar workload. As a second example, in the Balanced Academic Curriculum Problem (prob030 in CSPLib.org), we want to assign time periods to courses in a way which balances the academic load across periods. As a third example, when scheduling viewing times on a satellite, we might want agents to be assigned a similar number of observations. The BALANCE constraint introduced by Beldiceanu in the Global Constraint Catalog[1] [3] can be used to model situations like this where we need to minimize the difference B in the number of times different values, which typically represent different resources, are used.

Beldiceanu proposes an automaton based filtering algorithm for the BALANCE constraint that uses a counter for each value. This requires exponential space and time

[1] http://www.emn.fr/z-info/sdemasse/gccat

B. O'Sullivan (Ed.): CP 2014, LNCS 8656, pp. 174–189, 2014.

to work. Alternatively, Beldiceanu proposes a decomposition that reorders the variables and then computes the difference between the longest and the smallest sequences of consecutive values. As we show, such a decomposition can hinder propagation. We therefore revisit this global constraint. We prove that propagating the BALANCE constraint completely is intractable in general. We then introduce ALLBALANCE with a similar semantics that is only polynomial to propagate. We consider various forms of ALLBALANCE, focusing on ATMOSTALLBALANCE which constrains the upper bound of B. This can be used when we desire solutions to be as balanced as possible and thus want to minimize B. We present a flow-based algorithm to maintain domain consistency on ATMOSTALLBALANCE and compare it empirically to a decomposition augmented with implied constraints. The results show that the implied constraints significantly improve the performance of the decomposition, whilst the filtering algorithm in turn further improves performance.

2 Background

A *Constraint Network* consists of a set of variables $\mathcal{X}$, a domain $\mathcal{D}$ mapping each variable $X \in \mathcal{X}$ to a finite set of values $\mathrm{D}(X)$, and a set of constraints $\mathcal{C}$. An *assignment* σ is a mapping from variables in $\mathcal{X}$ to values in their domains: for all $X_i \in \mathcal{X}$ we have $\sigma(X_i) \in \mathrm{D}(X_i)$. We denote $\max(\mathrm{D}(X))$ by $\max(X)$ and $\min(\mathrm{D}(X))$ by $\min(X)$. When σ is implied from the context, we write $X_i = v$ instead of $\sigma(X_i) = v$ and X_i instead of $\sigma(X_i)$. A *constraint* C is a relation on a set of variables. An assignment *satisfies* C iff it is a tuple of this relation. We use capitals for variables and lower case for values. Constraint solvers typically use backtracking search to explore the space of partial assignments. At each assignment, propagation algorithms prune the search space by enforcing local consistency properties like domain consistency. A constraint C on $\mathcal{X}$ is *domain consistent* (DC) if and only if, for every $X_i \in \mathcal{X}$ and for every $v \in \mathrm{D}(X_i)$, there is an assignment σ satisfying C such that $\sigma(X_i) = v$. Such an assignment is a *support*. A CSP is DC iff all its constraints are DC. A constraint is *disentailed* iff there is no possible support.

A *decomposition* of a constraint C is a reformulation of C into a conjunction of constraints that is logically equivalent to C, potentially including extra variables. A decomposition N_1 is *stronger* than N_2 if and only if propagation on N_1 detects a superset of the inconsistent values detected by N_2 [4].

The domain of a variable X_i is an *interval* iff $|\mathrm{D}(X_i)| = \max(X_i) - \min(X_i) + 1$. Let $\{X_1, \dots, X_n\}$ be a set of variables. We call $\mathrm{occ}(v) = |\{i \mid X_i = v\}|$ the number of occurrences of the value v in this set. The constraint GCC is defined over the variables $[X_1, \dots, X_n]$ and is parameterized by two sets of integers $\{l_1, \dots, l_m\}$ and $\{u_1, \dots, u_m\}$. It ensures that we have $\forall j \in [1, \dots, m]$, $l_j \leq \mathrm{occ}(v_j) \leq u_j$. Achieving DC on GCC is polynomial [13]. If the lower and upper bounds on the occurrences are given by variables $[O_1, \dots, O_m]$, DC on the variables X_i can be achieved with the same computational complexity provided that the domains of the occurrences are intervals.

3 The Balance Constraint Family

BALANCE bounds the difference in the number of occurrences of values.

Definition 1 (BALANCE).

$$\text{BALANCE}([X_1,\ldots,X_n],B) \iff$$
$$B = \max_{v\in\{X_1,\ldots,X_n\}} occ(v) - \min_{v\in\{X_1,\ldots,X_n\}} occ(v)$$

Notice that only values occurring at least once are considered. Depending on the application, this may or may not be desirable. For instance, if we want to select a subset of resources and distribute tasks among them in a balanced way, then BALANCE is suited. However, if resources are already selected, then such a solution might be imbalanced as some resources may receive no tasks. Moreover, it is hard to know if a value will be used for sure until all variables are set. As a consequence, filtering is weak. We therefore consider a variant, ALLBALANCE in which all values in a set $\mathcal{V}$ are considered (without loss of generality, we shall assume that $\mathcal{V} = \{1,\ldots,m\}$). We shall see that achieving DC on BALANCE is NP-hard, while it is polynomial for ALLBALANCE.

Definition 2 (ALLBALANCE).

$$\text{ALLBALANCE}(\mathcal{V},[X_1,\ldots,X_n],B) \iff$$
$$B = \max_{v\in\mathcal{V}} occ(v) - \min_{v\in\mathcal{V}} occ(v) \;\wedge\; \forall i\; X_i \in \mathcal{V}$$

We also consider the variants of BALANCE and ALLBALANCE where B is only a lower or an upper bound. By replacing "=" in Definition 1 and 2 by "$\geq$" and "$\leq$", we define the constraints ATMOSTBALANCE, ATLEASTBALANCE, ATMOSTALLBALANCE and ATLEASTALLBALANCE. While ATMOSTBALANCE and ATMOSTALLBALANCE ensure that a solution is "balanced enough", ATLEASTBALANCE and ATLEASTALLBALANCE ensure that the solution is "somewhat unbalanced". The first two constraints are useful when we seek balanced solutions and want to minimize B, whilst the last two are useful when we cannot make B lower than a certain value or desire some level of imbalance.

Theorem 1. *Enforcing DC on* BALANCE *takes polynomial time if the number of values m is bounded.*

Proof. We construct a REGULAR constraint [9] with states containing counters for every value, i.e. a state is labeled with a tuple $\langle c_{v_1},\ldots,c_{v_m}\rangle$ where c_{v_i} is the number of times the value $v_i \in \mathcal{V}$ was encountered. The unfolded automaton therefore has $O(n^m)$ states. Enforcing DC then takes $O(n^{m+1}m)$ time. One can reduce the number of counters by 1. Choose a value and delete its counter. After parsing the string, the missing counter should have value n - sum of the other counters. Total complexity is then $O(n^m m)$. □

Theorem 2. *Enforcing DC on* BALANCE *is NP-hard.*

Proof. Reduction from 3-SAT to the problem of finding a support of BALANCE. Given a formula φ with n atoms $1, \ldots, n$ and m clauses, we build a BALANCE constraint finding the balance B over a set of $(n+1)(m+1)$ variables. Let l_j^k be the k-th literal of the j-th clause of φ. We define the variables:

$$B = 0 \tag{3.1}$$
$$\mathrm{D}(X_{1,j}) = \{0\} \qquad \forall j \in 1..m+1 \tag{3.2}$$
$$\mathrm{D}(X_{2,i}) = \{\bar{i}, i\} \qquad \forall i \in 1..n \tag{3.3}$$
$$\mathrm{D}(X_{3,j}) = \{l_j^1, l_j^2, l_j^3\} \qquad \forall j \in 1..m \tag{3.4}$$
$$\mathrm{D}(X_{4,l}) = \{\bar{n}, \ldots, \bar{1}, 1, \ldots, n\} \qquad \forall l \in 1..(n-1)m \tag{3.5}$$

The domains of variables (3.1) and (3.2) force every value to occur 0 or $m+1$ times. The existence of a model of φ implies that there is a solution of BALANCE. Indeed atom and clause variables (3.3,3.4) can be assigned using only the n literals appearing in the model. The total number of occurrences of these n values on the variables $X_{2,i}$ and $X_{3,j}$ is $n+m$. Thanks to the $(n-1)m$ filler variables (3.5), we can ensure that each value of these n values occurs exactly $m+1$ times. Thus, 0 for B is consistent.

Now consider a solution of BALANCE. Since $B = 0$, every value occurs $m+1$ times or never. Since each $X_{2,i}$ must take a value, either the value i or $\bar{i}$ must occur at least once, hence $m+1$ times. There are $(n+1)(m+1)$ variables in total and we identified $n+1$ values (counting the value 0) that must occur $m+1$ times each. Therefore, the other values must not occur at all. The model which contains the literal i iff the value i occurs $m+1$ times, and $\bar{i}$ otherwise, is a model of φ. Indeed, for every clause $c_j \in \varphi$, since $X_{3,j}$ (3.4) must be assigned a value, it follows that the model above has at least one literal from c_j. □

This proof also shows that ATMOSTBALANCE is NP-hard to propagate as it only requires an upper bound on B. By comparison, it is easier to reason with ALLBALANCE, ATLEASTALLBALANCE, ATLEASTBALANCE, and ATMOSTALLBALANCE. For the first three constraints, we give complexity results in the form of an algorithm intended only to prove polynomiality. For ATMOSTALLBALANCE, we will present a practical filtering algorithm.

Theorem 3. *Enforcing DC on* ALLBALANCE, ATLEASTALLBALANCE *and* ATLEASTBALANCE *takes polynomial time.*

Proof. Consider a restricted case of ALLBALANCE where the least (most) occurring value is required to be v_{least} (v_{most}) and must occur exactly c times ($c+b$ times). DC can be enforced using GCC with $\mathrm{D}(O_{v_{least}}) = \{c\}$, $\mathrm{D}(O_{v_{most}}) = \{c+b\}$ and $\mathrm{D}(O_v) = [c, c+b]$ for all values $v \notin \{v_{least}, v_{most}\}$. To filter ALLBALANCE, one can test whether a value $v \in \mathrm{D}(X_i)$ has a support in one of the restricted cases where $b \in \mathrm{D}(B)$, $c \in [0, n-b]$, $v_{least} \in [1, m]$, and $v_{most} \in [1, m]$. Since there are $O(|\mathrm{D}(B)|nm^2)$ such cases, the filtering can be done in polynomial time.

For ATLEASTBALANCE, we set the domain $\mathrm{D}(O_{v_{least}}) = [1, c]$ to ensure that v_{least} occurs at least once. We set $\mathrm{D}(O_{v_{most}}) = [c + \min(B), n]$ so that the balance is at least $\min(B)$ and we set the domains of the other occurrence variables to $O_i \in [0, n]$ for $i \notin$

$\{v_{least}, v_{most}\}$. There are $O(nm^2)$ restricted cases to test since $v_{least} \in [1, m], v_{most} \in [1, m]$, and $c \in [0, n-\min(B)]$. The restricted cases apply for ATLEASTALLBALANCE except that the domain $D(O_{v_{least}})$ is $[0, c]$ to allow the value v_{least} to not occur. □

4 Decompositions

We focus mainly on BALANCE and ALLBALANCE, as their decompositions can be used also for the others by suitably constraining only the lower or upper bound of B. The Global Constraints Catalog [3] proposes a decomposition of BALANCE that uses the constraint SORTEDNESS($[X_1, \ldots, X_n], [Y_1, \ldots, Y_n]$) to count the minimum and maximum length of stretches of equal value in the sequence $[Y_1, \ldots, Y_n]$. Then, it makes sure that the difference between the length of the maximum and minimum stretch is equal to B. We propose another decomposition of BALANCE using GCC. Let $\bigcup_{i=1}^{n} D(X_i) = \{1, \ldots, m\}$:

$$
\begin{aligned}
&\text{GCC}([X_1, \ldots, X_n], [O_1, \ldots, O_m]) \;\; \& \\
&P = \max(\{O_1, \ldots, O_m\}) \;\; \& \\
&Q = \min(\{O_1, \ldots, O_m\} \setminus \{0\}) \;\; \& \\
&B = P - Q
\end{aligned}
$$

As DC on BALANCE is NP-hard, it is no surprise that neither decomposition enforces DC. However, Example 1 shows that, even if we assume perfect communication between the variables Y_j's and B in the decomposition using SORTEDNESS[2], then this decomposition is not stronger than the new decomposition using the GCC constraint.

Example 1. $m = 6, X_1 \in \{1, 6\}, X_2 \in \{2, 5\}, X_3 \in \{3, 4\}, Y_1 \in \{1, 2, 3, 4\}, Y_2 \in \{2, 3, 4, 5\}, Y_3 \in \{3, 4, 5, 6\}, B = 1$. The domains of occurrences variables O_v are set to $\{0, 1\}$ by GCC, hence the variables P and Q are both set to 1, and thus the constraint is found inconsistent. However, the SORTEDNESS decomposition allows stretches greater than 1. It is therefore consistent, irrespective of the reasoning used on Y_i and B.

Similar to BALANCE, the ALLBALANCE constraint can also be decomposed using the GCC constraint. Let $\mathcal{V} = \{1, \ldots, m\}$.

$$
\begin{aligned}
&\forall i \in \{1, \ldots, n\},\; X_i \in \mathcal{V} \\
&\text{GCC}([X_1, \ldots, X_n], [O_1, \ldots, O_m]) \;\; \& \\
&P = \max(\{O_1, \ldots, O_m\}) \;\; \& \\
&Q = \min(\{O_1, \ldots, O_m\}) \;\; \& \\
&B = P - Q \\
&D(P) = [\lceil \tfrac{n}{m} \rceil, n] \\
&D(Q) = [0, \lfloor \tfrac{n}{m} \rfloor]
\end{aligned}
$$

[2] Such filtering can be obtained, for instance, through a REGULAR constraint.

The domains of the variables P and Q are based on the observation that the average of the occurrence variables is exactly $\frac{n}{m}$. Consequently, the greatest occurrence should be no smaller than the average and the smallest occurrence should be no greater than the average. Example 2 shows that this decomposition does not maintain DC, even on ATMOSTALLBALANCE.

Example 2. $m = 4$, $B \in [0, 2]$, $X_1 = X_2 = 1$, $X_3 \in \{1, 2, 3\}$, $X_4 \in \{1, 3, 4\}$, $X_5 \in \{1, 3, 4\}$. After propagation, the domains of these variables remain the same and we get:

$$O_1 \in [2, 3], \quad O_2 \in [0, 1], \quad O_3 \in [0, 3], \quad O_4 \in [0, 2], \quad P \in [2, 3], \quad Q \in [0, 1]$$

However, the only way for the occurrence variables to sum to 5 and have a balance of at most 2 is to take their values in the multiset $\{2, 2, 1, 0\}$ or $\{2, 1, 1, 1\}$. In other words, a value cannot occur three times and the value 1 should be removed from the domains of X_3, X_4, and X_5.

In order to investigate the limits of a decomposition of ALLBALANCE based on the GCC constraint, we consider another decomposition using the constraints together with an automaton defined on the occurrence variables. Observe that we have perfect communication from the X_i's domains to the O_j's bounds (through GCC) and perfect communication between the O_j's and B (through REGULAR). However, we shall see that this is still not sufficient to achieve DC on ALLBALANCE.

$$\begin{array}{l}\text{GCC}([X_1, \dots, X_n], [O_1, \dots, O_m]) \ \& \\ \text{REGULAR}([O_1, \dots, O_m, B], \mathcal{A})\end{array}$$

The automaton $\mathcal{A}$ has $O(n^3 m)$ states. Non-final states are tuples $\langle i, S, q, p \rangle$ which respectively encode the current variable, the current sum, the minimum value encountered, and the maximum value encountered. The final state is denoted f. The starting state is $\langle 1, 0, n, 0 \rangle$. The transition function is:

$$\delta(\langle i, S, q, p \rangle, x) = \begin{cases} \langle i + 1, S + x, \min(q, x), \max(p, x) \rangle & \text{if } i \leq m \\ f & \text{if } i = m + 1 \text{ and } x = p - q \end{cases}$$

Since this automaton is acyclic, unfolding does not alter the number of states, hence the total complexity to propagate this REGULAR constraint is $O(mn^4)$. This decomposition is costly yet is still insufficient to maintain DC. Consider the following example.

Example 3. $m = 4$, $B \in [0, 2]$, $X_1, X_2 = 1$, $X_3 \in \{1, 2, 3\}$, and $X_4, X_5, X_6 \in \{1, 3, 4\}$. After filtering the REGULAR constraint, we obtain the domains $O_1 \in [2, 3]$, $O_2 \in [0, 1]$, $O_3 \in [1, 2]$, and $O_4 \in [1, 2]$ that do not allow the GCC to filter 1 from the domain of X_3.

In the rest of the paper, we will focus on the GCC decomposition of ALLBALANCE. In addition to being cheaper than the REGULAR decomposition, it can be strengthened by adding implied constraints, as we will show next.

4.1 Constraints Implied by ALLBALANCE

We can strengthen the GCC decomposition of ALLBALANCE thanks to the following inequality: $P + (m-1)Q \leq n$. This is true because at least one value will occur P times, and at most $m-1$ values will occur Q times, where n is the total number of occurrences. We have $Q = P - B$, hence $P + (m-1)(P-B) \leq n$, that is:

$$mP - (m-1)B \leq n \tag{4.1}$$

In other words, we have an upper bound $P \leq \lfloor \frac{n-B}{m} \rfloor + B$. Consider again Example 2 which shows that the GCC decomposition of ALLBALANCE does not maintain DC. Due to (4.1), we discover that $P \leq \left\lfloor \frac{n+(m-1)B}{m} \right\rfloor \leq \lfloor \frac{5+3\times 2}{4} \rfloor = 2$ and the upper bound of P, O_1, and O_3 is reduced to 2. Therefore, the constraint GCC removes 1 from the domains of X_3, X_4, and X_5.

We can make a similar argument to obtain a lower bound on Q. We have: $Q + (m-1)P \geq n$ which is equivalent to:

$$mQ + (m-1)B \geq n \tag{4.2}$$

Again in Example 2, thanks to (4.2), we discover that $Q \geq \left\lceil \frac{n-(m-1)B}{m} \right\rceil \geq \lceil \frac{5-3\times 0}{4} \rceil = 1$ and the lower bound of Q, O_2, and O_3 are increased to 1, and the variable X_3 is set to 2.

It is possible to add implied constraints providing even stronger level of filtering. The following constraints are implied by the decomposition whilst being stronger than constraints (4.1) and (4.2).

$$(\sum_{j=1}^{m} \max(P-B, O_j)) \leq n \leq (\sum_{j=1}^{m} \min(P, O_j)) \tag{4.3}$$

$$(\sum_{j=1}^{m} \min(Q+B, O_j)) \geq n \geq (\sum_{j=1}^{m} \max(Q, O_j)) \tag{4.4}$$

Indeed, consider the following example.

Example 4. $m = 7, X_1, X_2 \in \{1\}, X_3, X_4 \in \{2\}, X_5, X_6 \in \{3\}, X_7, X_8, X_9 \in \{4,5,6,7\}, B \in \{1,2\}$. The domains of occurrences variables O_1, O_2, O_3 are set to 2 and O_4, O_5, O_6, O_7 to $[0,3]$ by GCC, hence the variables P and Q are set respectively to $[2,3]$ and $[0,1]$, and thus the GCC decomposition as well as constraints (4.1) and (4.2) are DC. However, $P = 3$ is not consistent with the constraint (4.3) and $Q = 1$ is not consistent with the constraint (4.4). Therefore we can deduce $B = 2$.

Notice that these two extra constraints require a dedicated, albeit rather straightforward, filtering algorithm because using SUM and MIN/MAX constraints would hinder propagation. The algorithm proceeds by *shaving* the bounds of the variables P and Q. For instance, after having temporarily fixed P to its upper bound, we find a support for the relation $(\sum_{j=1}^{m} \max(P - B, O_j)) \leq n$ by using the maximum value for B and the minimum value for each O_j. If this is not sufficient to keep the sum below n, then we can deduce that $P = \max(P)$ is inconsistent. In Example 4, assuming $P = 3$, we have $P - \max(B) = 1$, and $\sum_{j=1}^{m} \max(P - B, O_j) = 2+2+2+1+1+1+1 > 9$.

Last, we can add another cheap implied constraint. If the number of values (m) does not divide the number of variables (n), then B cannot be equal to 0. Conversely, if $n = mk$, then B cannot be equal to 1. Indeed, suppose that the balance is 1. Furthermore, suppose that a value occurs $k - 1$ times or less. Then since $n = mk$, at least one other value occurs $k + 1$ times or more, hence the balance is greater or equal to 2. The same contradiction arises if we suppose that a value occurs $k+1$ times or more. Therefore, the value of B cannot be equal to 1. These two rules can be combined together as follows:

$$1 + \left\lfloor \frac{n}{m} \right\rfloor - \left\lceil \frac{n}{m} \right\rceil \neq B \tag{4.5}$$

As we will show later in the empirical results, the implied constraints presented in this section turn out to be very effective in propagating ALLBALANCE.

4.2 Special Cases of ALLBALANCE

There exist some special cases of the ALLBALANCE where we have a simple encoding that does not hurt DC propagation. For instance, if $B = n$ then all variables must be equal. We can thus post: $X_i = X_{i+1}$ for $1 \leq i < n$. Another case is when $m = 2$. In this case, the implied constraints (4.1) and (4.2) reveal that: $P \leq \frac{n+B}{2}$, $Q \geq \frac{n-B}{2}$. Since there are two values, one occurs P times, and the other Q times, and $P + Q = n$. Therefore, we have $P = \frac{n+B}{2}$ and $Q = \frac{n-B}{2}$. It follows that the value of B must be even if and only if n is even. Moreover, we can safely assume that the two values are 0 and 1 since any binary domain can be mapped to these values. Therefore, the expression $\sum_{i=1}^{n} X_i$ gives either P or Q. Thus, we post:

$$B \bmod 2 = n \bmod 2 \;\wedge\; \sum_{i=1}^{n} X_i \in \left\{ \frac{n-B}{2}, \frac{n+B}{2} \right\}$$

There are other cases where the decomposition with the implied constraints is sufficient.

Proposition 1. *The* GCC *decomposition with the implied constraints (4.1),(4.2) and (4.5) achieves DC on* ALLBALANCE *if* $B \leq 1$.

Proof. When $B = 0$, the implied constraints (4.1) and (4.2) entail that: $P \leq \frac{n}{m}$, $Q \geq \frac{n}{m}$. This enforces the occurrence variables O_v for all $v \in \mathcal{V}$ of the GCC decomposition to be set to $\frac{n}{m}$. Since P, Q and B are fixed, the constraint is now equivalent to GCC. When $B = 1$, the implied bounds are:

$$P \leq \left\lfloor \frac{n-1}{m} \right\rfloor + 1 \leq \left\lceil \frac{n}{m} \right\rceil, \qquad Q \geq \left\lceil \frac{n+1}{m} \right\rceil - 1 \geq \left\lfloor \frac{n}{m} \right\rfloor$$

This will enforce $\mathrm{D}(O_v) = \left[\left\lfloor \frac{n}{m} \right\rfloor, \left\lceil \frac{n}{m} \right\rceil\right]$ for all $v \in \mathcal{V}$ of the GCC decomposition. We know that either m does not divide n or $B = 1$ is inconsistent. Since the latter case has already been treated, we check the former. In this case, constraint (4.5) implies that a balance of 0 is not consistent, hence $D(B) = \{1\}$. However, any assignment consistent with GCC with the bounds given by P and Q will have a balance of 1. Therefore, in either case, we achieve DC on ALLBALANCE. □

Another special case is as follows.

Proposition 2. *The* GCC *decomposition with the implied constraints (4.1),(4.2) and (4.5) achieves DC on* ATMOSTALLBALANCE *if* $m \leq 2$.

Proof. Let $b = \max(B)$. The implied constraints (4.1) and (4.2) put an upper bound on P of $\lfloor \frac{n-b}{m} \rfloor + b$, and a lower bound on Q of $\lceil \frac{n+b}{m} \rceil - b$. Hence:

$$\max(P) - \min(Q) \leq 2b - \left\lceil \frac{n+b}{m} \right\rceil + \left\lfloor \frac{n-b}{m} \right\rfloor$$
$$\leq 2b - \frac{n+b}{m} + \frac{n-b}{m} = 2b - \frac{2b}{m}$$

Now, suppose $m \leq 2$. Then $\max(P) - \min(Q) \leq b$ and for any $1 \leq j \leq m$, $\max(O_j) - \min(O_j) \leq b$. Thus, all solutions of the GCC satisfy ATMOSTALLBALANCE. □

In summary, we have shown that our decomposition with the introduced implied constraints achieves DC on ALLBALANCE if $B \leq 1$, and on ATMOSTALLBALANCE if $m \leq 2$. Moreover, there exists a decomposition achieving DC on ALLBALANCE if $m \leq 2$. In general, the decomposition does not achieve DC, even given perfect communication between the variables O_j, P, Q and B (Example 3).

5 A Filtering Algorithm for ATMOSTALLBALANCE

We present now a filtering algorithm that achieves DC on ATMOSTALLBALANCE. The algorithm (see Algorithm 1) proceeds in two steps. First, it finds a support by iteratively reducing the balance of a support for GCC until it is minimal. Second, it computes the union of the supports over each possible window of width $\max(B)$ for the values' occurrences. The resulting union can be computed efficiently and corresponds to the domain consistent values.

5.1 Finding a Support

To find a support, the algorithm computes a flow in a graph similar to the one used for the GCC. There is one node X_i per variable, one node v per value, a source s, and a sink t. Each edge has a capacity $[a, b]$, i.e. a lower capacity a and an upper capacity b. There is an edge of capacity $[0, 1]$ between s and each variable node X_i. There is an edge of capacity $[0, 1]$ between each node X_i and value v for $v \in D(X_i)$. Finally, there is an edge of capacity $[0, n]$ between each value v and t. Let $f(a, b)$ be the amount of flow that circulates from node a to b. A maximum flow [1] from s to t corresponds to an assignment of the variables, i.e. $X_i = v \iff f(X_i, v) = 1$. The value v occurs exactly $f(v, t)$ times in the assignment. To modify the assignment so that it satisfies

Algorithm 1. FilterAtMostAllBalance($[\mathcal{V}, [X_1, \ldots, X_n], B)$

$\overline{b} \leftarrow \max(B)$;
$\mathrm{D}(X'_i) \leftarrow \mathrm{D}(X_i)$ for all $i = 1..n$;
1 Find a support σ for ATMOSTALLBALANCE whose balance is minimal and let q be the occurrence of the least occurring value;
Set $\min(B)$ to be the balance of the support σ;
$\mathrm{D}(O_i) \leftarrow [q, q + \overline{b}]$ for all $i = 1..m$;
2 filter $\mathrm{GCC}([\mathrm{D}(X_1), \ldots, \mathrm{D}(X_n)], [O_1, \ldots, O_m])$;
if *no filtering occurred* **then return**;
if *filtering occurred because of a Hall set* **then** $k \leftarrow 1$;
else $k \leftarrow -1$;
$\mathrm{D}(O_i) \leftarrow [q + k, q + \overline{b} + k]$ for all $i = 1..m$;
filter $\mathrm{GCC}([\mathrm{D}(X'_1), \ldots, \mathrm{D}(X'_n)], [O_1, \ldots, O_m])$;
$\mathrm{D}(X_i) \leftarrow \mathrm{D}(X_i) \cup \mathrm{D}(X'_i)$ for all $i = 1..n$;

the ATMOSTALLBALANCE constraint, the algorithm finds a path in the residual graph from the most occurring (or least occurring) value v to any value v' such that $f(v', t) \leq f(v, t) - 2$ (or such that $f(v', t) \geq f(v, t) + 2$). The algorithm pushes a unit of flow along this path to modify the assignment. The algorithm repeats this operation until no such path exists. If no such path exists and if the balance of the current assignment is strictly greater than $\max(B)$, then no support exists. To prove correctness, we show that if there is a solution of ATMOSTALLBALANCE, then there is a sequence of such paths leading to it from any maximum flow. In other words, if no such path exists, then the gap between the maximum and minimum flow going through an edge for a value node to the sink node is a lower bound of B.

Lemma 1. *If v is the most occurring value and there is no value v' such that $f(v', t) \leq f(v, t) - 2$ and that v can reach v' in the residual graph and if w is the least occurring value and there is no value w' such that $f(w', t) \geq f(w, t) + 2$ and that w can reach w' in the residual graph, then the balance of the current assignment is minimal.*

Proof. We prove the contraposition. Suppose there is a flow f^* whose corresponding assignment has a smaller balance than the assignment given by f. Let the most occurring and least occuring values in each flow be:

$$v_{most} = \operatorname*{argmax}_{v \in \mathcal{V}} f(v, t), \qquad v_{least} = \operatorname*{argmin}_{v \in \mathcal{V}} f(v, t),$$
$$v^*_{most} = \operatorname*{argmax}_{v \in \mathcal{V}} f^*(v, t), \qquad v^*_{least} = \operatorname*{argmin}_{v \in \mathcal{V}} f^*(v, t).$$

Necessarily, we have $f(v_{most}, t) > f^*(v^*_{most}, t) \vee f(v_{least}, t) < f^*(v^*_{least}, t)$. Suppose that $f(v_{most}, t) > f^*(v^*_{most}, t)$, we have $f(v_{most}, t) > f^*(v^*_{most}, t) \geq f^*(v_{most}, t)$. Since both flows have the same flow value, the difference of the vectors $f^* - f$ describes a circulation, i.e. a collection of cycles on which the flow circulates. Since $f^*(v_{most}, t) - f(v_{most}, t) < 0$, the flow circulates from t to v_{most} in the circulation which means that there is a value v' for which the flow circulates from v' to t which implies $f^*(v', t) - f(v', t) > 0$. We conclude that $f(v_{most}, t) > f^*(v^*_{most}, t) \geq$

$f^*(v',t) > f(v',t)$ thus $f(v_{most},t) \geq f(v',t) + 2$. Finally, the edges (v',t) and (t,v) lie on the same cycle in the circulation. Hence there is a path that connects v to v' in the residual graph. The case $f(v_{least},t) < f^*(v^*_{least},t)$ is symmetric. □

5.2 Filtering the Domains

First, we filter the lower bound of B to the balance value of the support found in the first phase. The balance of this solution is, by Lemma 1, the maximum lower bound on B. Next, we set $q = \min_v f(v,t)$ to be the frequency of the least occurring value. Let $\overline{b} = \max(B)$. We then run the filtering algorithm of the GCC with the domains of the occurrence variables set to $[q, q + \overline{b}]$. If this does no filtering, then each value in the domains belongs to a support where the occurrences of the values lie between q and $q+\overline{b}$. All these supports satisfy the ATMOSTALLBALANCE and we are done. However, if the filtering algorithm detects that the assignment $X_i = v$ is inconsistent for the GCC, it is not necessarily inconsistent for ATMOSTALLBALANCE. The assignment $X_i = v$ can occur in a support where the maximum and minimum number of occurrences do not belong to $[q, q+\overline{b}]$. Therefore, we need to test for a support with different domains such as $[q-1, q+\overline{b}-1]$ and $[q+1, q+\overline{b}+1]$. Fortunately, we do not need to test all possible intervals of size $\overline{b}$. A *Hall set* is a set of values H for which exactly $(q + \overline{b}) \times |H|$ variable domains are subset of H, since $q + \overline{b}$ is the maximum allowed occurrences for any value in H. Conversely, an *unstable set* is a set of values U for which exactly $q \times |U|$ variable domains intersect U. From [11], an assignment $X_i = v$ is filtered either because v belongs to a Hall set or the domain of X_i intersects with an unstable set. The following lemmas restrict search to two windows.

Lemma 2. *If H is a Hall set for a* GCC *with occurrences bounded by q and $q + \overline{b}$ and k a positive integer, then the bounds $q - k$ and $q + \overline{b} - k$ are inconsistent.*

Proof. Since H is a Hall set when the upper bound is equal to $q + \overline{b}$, then there are $(q + \overline{b}) \times |H|$ variables whose domains are included in H. Therefore the total number occurrences of values in H is at least $(q + \overline{b}) \times |H|$. Therefore at least one value must occur at least $q + \overline{b}$ times. This is a contradiction with the upper bound $q + \overline{b} - k$. □

The dual result for unstable sets can be obtained in a similar way (proof omitted):

Lemma 3. *If U is an unstable set for a* GCC *with occurrences bounded by q and $q + \overline{b}$ and k a positive integer, then the bounds $q + k$ and $q + \overline{1} + k$ are inconsistent.*

These two lemmas imply that we only need to check the window $[q + 1, q + \overline{b} + 1]$ if the pruning was due to Hall sets only, the window $[q - 1, q + \overline{b} - 1]$ if it was due to unstable sets only, and no other window otherwise. This leads to the following theorem.

Theorem 4. *Enforcing DC on* ATMOSTALLBALANCE *takes $O(n^2m)$ time.*

Proof. First, the pruning on B is correct by Lemma 1. Since B is only an upper bound, its own upper bound is never pruned. It follows that the pruning on B is complete and a support is found if and only if the constraint is not disentailed. Now, consider Algorithm 1. Let $\overline{b} = \max(B)$ and q be the minimum occurrence of any value found in the support (Line 1). We compute all consistent values for a GCC constraint where occurrence variables are bounded by $[q, q + \bar{b}]$. Suppose first that no pruning occurs. Then every value is consistent for GCC. In each support the difference between maximum and minimum occurrence is at most $\bar{b}$. Hence it is a support for ATMOSTALLBALANCE.

Suppose now that there is at least one Hall set. Then by Lemma 2, we know that a GCC on the same variables but with all occurrence variables bounded by $[q - k, q + \bar{b} - k]$ would be inconsistent. In other words, these values have no support for ATMOSTALLBALANCE on lower windows. By Lemma 2, since the GCC is consistent for the window $[q, q + \bar{b}]$, there will not be any Hall set on higher windows. It follows that values inconsistent for GCC on the window $[q, q + \bar{b}]$ are inconsistent for ATMOSTALLBALANCE only if they are pruned because of an unstable set on $[q+1, q+\bar{b}+1]$ and all higher windows. However, by Lemma 3, if there is an unstable set on the window $[q + 1, q + \bar{b} + 1]$, then all higher windows will be inconsistent. It follows that values pruned by GCC on the windows $[q, q + \bar{b}]$ are inconsistent if and only if they are also pruned on the window $[q + 1, q + \bar{b} + 1]$. The second case is when there is at least one unstable set when setting the occurrence variables to the window $[q, q + \bar{b}]$. Symmetrically, values are inconsistent if and only if they would be pruned also for the window $[q - 1, q + \bar{b} - 1]$. Finally, if the window $[q, q + \bar{b}]$ has both a Hall and an unstable set, all other windows are inconsistent.

The running time is bounded by the time to find a support. This requires finding $O(n - \max(B))$ augmenting paths, each with a depth-first search (DFS) in $O(nm)$ time. The two calls to the filtering algorithm of GCC take $O(n^{3/2}m)$ time. Finding what caused the filtering uses a DFS in the transposed residual graph that marks nodes that can reach the sink t. If the value v was filtered out of the domain of X_i and that v cannot reach the sink, then the filtering occurred because of a Hall set. Otherwise, it occurred because of an unstable set. The total complexity is thus $O(n^2m)$. □

In practice, the complexity can be considerably reduced. For instance, the support σ can remain valid for multiple consecutive calls to the filtering algorithm. Then, the running time is equivalent to executing the filtering algorithm of GCC twice. If the support is no longer valid, it can usualy be updated rather than computed from scratch. This involves finding much fewer than $n - \max(B)$ augmenting paths.

6 Related Work

The problem of ensuring a certain balance in the assignments of $[X_1, \ldots, X_n]$ has been previously studied with the SPREAD [10,15,14] and DEVIATION [16,14] constraints. Both constraints look at the deviation from the mean $m = \frac{1}{n}\sum_{i=1}^{n} X_i$ with balancing criteria $D = \sum_{i=1}^{n}(X_i - m)^2$ and $D = \sum_{i=1}^{n}|X_i - m|$, respectively. The constraints in the BALANCE family, however, cannot be expressed using SPREAD and DEVIATION. In particular, SPREAD and DEVIATION consider the values taken by X_is, as opposed to the number of occurrences of each value. For instance, for DEVIATION, an assignment $[1, 2, 2, 2, 2, 2, 2, 2, 3]$ is better than $[1, 1, 1, 2, 2, 2, 3, 3, 3]$ as the deviation from the mean is lower in the first, but the latter has balance $B = 0$ (3 occurences for each value), so is preferred by ATMOSTALLBALANCE. This criterion is important in applications where we want to balance the occurrence of values where each occurrence of a value represents the use of a resource such as an employee or machine. The balance criterion B is also different from DEVIATION on the occurrences of values. The criteria coincide when $B = D = 0$, but not otherwise. Assume that the occurrences of some 4 values in some 16 variables can be $[2, 2, 6, 6]$ or $[2, 3, 4, 7]$. While in the first vector $B = 4$ and $D = 8$, in the second $B = 5$ and $D = 6$.

Similar criteria have been studied in graph theory in problems involving generating balanced cuts, such as judicious partitioning [7] and graph conductance [2]. These are all NP-hard but can be approximated in polynomial time.

7 Experimental Results

We evaluate our DC algorithm for ATMOSTALLBALANCE and its decompositions on the Balanced Academic Curriculum Problem and a shift scheduling problem. All experiments use Choco (version 2.1.5) under Linux on a 3Ghz CPU with 12 GB of RAM.

7.1 Balanced Academic Curriculum Problem (BACP)

In BACP (prob030 in CSPLib.org), a set of n courses must be assigned to m time periods, such that (1) a lower and an upper bound on the number of courses per period must be respected; (2) some courses are prerequisite for others; (3) the *load* (i.e., the sum of the credits of the assigned courses) of each period should be balanced.

The "standard" model [6] has a variable X_i for each course i, whose value is the period allocated to this course. A variable O_j for each period j gives the load on this period. In order to channel these two sets, $\{0, 1\}$ variables $Y_{i,j}$ are introduced and constrained such that $X_i = j \Leftrightarrow Y_{i,j} = 1$ and $O_j = \sum_{i=1}^{n} c(i) \cdot Y_{i,j}$ where $c(i)$ stands for the the number of credits of a course i. Constraint (1) is modeled by a GCC constraint on $\{X_1, \ldots, X_n\}$, constraints (2) are simple precedences between the corresponding X_i and X_j. For the objective (3), a criterion is minimized representing how balanced the set $\{O_1, \ldots, O_m\}$ is. In [6], the criterion is the maximum load. We here consider the gap between the minimum and maximum load, denoted $L(\infty)$ in [16]. We thus have three variables P, Q and B with the constraints $P = \max(\{O_1, \ldots, O_m\})$, $Q = \min(\{O_1, \ldots, O_m\})$ and $B = P - Q$, and we minimize B.

We propose an alternative model which respects $L(\infty)$ using our ALLBALANCE constraint. It uses the same variables $\{X_1, \ldots, X_n\}$ and same constraints for (1) and (2). However, we do not need $Y_{i,j}$ and O_j. We can directly post ALLBALANCE using $\mathcal{V} = \{1, \ldots, m\}$ and B on the multiset of variables containing $c(i)$ times each X_i.

We report the results obtained by running 7 models on 3 real instances involving 8, 10 and 12 periods. `Standard` is the first model described above. The rest correspond to the alternative model using ALLBALANCE. The first uses the basic GCC decomposition for ALLBALANCE (`Decomp.`), the second uses the decomposition with implied constraints (4.1) and (4.2) (`Implied`), the third uses the implied constraints (4.3) and (4.4) (`Implied`$^+$), and the fourth uses the DC algorithm (`AllBalance`). Finally, in the two last models we balance the load using the DEVIATION constraint [16] on the load variables O_j. The way that DEVIATION is used differs in the two models. In the former (`Deviat.`), following [16], O_j variables are channelled to the original X_i variables via the $Y_{i,j}$ variables as in the standard model. In the latter model

Table 1. Balanced Academic Curriculum Problem

	`Standard`			`Decomp.`			`Implied`			`Implied`$^+$			`AllBalance`			`Deviat.`			`Gcc+Deviat.`		
	B	#	Time	B	#	Time	B	#	Time	B	#	Time	B	#	Time	B	#	Time	B	#	Time
`08`	16.3		-	**1.0**	**20**	116	**1.0**	**20**	**112**	**1.0**	**20**	120	**1.0**	**20**	174	1.3	19	388	**1.0**	**20**	579
`10`	15.2		-	**1.0**	**20**	**2626**	**1.0**	**20**	2665	**1.0**	**20**	21261	**1.0**	**20**	17509	1.1	19	63100	1.6	18	9850
`12`	31.6		-	2.1	15	19104	2.0	15	21984	1.4	19	28999	**0.0**	**20**	45503	0.1	19	2832	0.2	18	15023

(`Gcc+Deviat.`), the O_j variables are channelled to the duplicated X_i variables using a GCC. That is, we use a GCC and a DEVIATION constraint together to balance the load. We then report the $L(\infty)$ value of the best solutions provided by these two models. Notice that (almost) perfectly balanced solutions exist for the instances we used, thus the $L(\infty)$ and deviation criteria are very close to each other.

As branching strategy, we use Impact Based Search [12] in all cases. Each instance is run 20 times with a 900s cutoff by each method after randomly shuffling the variables so that initial ties are broken randomly. We report in Table 1 the average observed balance ($B = L(\infty)$) over the 20 runs, and the number of runs where optimality was proven (#). Moreover, when possible, we report the average run time in milliseconds (Time) over all completed runs (i.e., in which optimality was proven). In other words, CPU times can be compared only when the same set of instances have been solved by all methods. As also shown in [6], the standard model rarely solves the instances to optimality which renders difficult the computation of averages for which optimality is proven by all methods. We use **bold** to highlight the best results. When multiple methods solve the same instances, we also highlight the best average CPU time.

We observe that the standard CP model has extremely poor performance, the solutions found are all suboptimal. Notice that the criterion optimized by the DEVIATION models is different. Several symmetric solutions for the $L(\infty)$ criterion have a different deviation. Indeed, we can see that the `Deviat.` model, which is the same as the standard model where the simple objective function is replaced by DEVIATION , greatly outperforms the standard model. Neither of the DEVIATION models is able to find an optimal solution and prove it in every case. However, adding the implied constraint as suggested in [8], significantly improves the model `Deviat.`: it then finds optimal solutions in all but one of the instances, making it nearly as good as the `AllBalance` model. The models using the decompositions of the ALLBALANCE constraint is very efficient on instances `08` and `10`, however, only the filtering algorithm is able to find an optimal solution and prove it within the cutoff time in all cases.

7.2 Shift Scheduling

In order to better assess the advantages of the propagator over the different decompositions, we ran another series of tests (under the same conditions). We consider a task assignment problem. We have m tasks per day. Each task requires a separate worker so we have m workers. Over the n days of the schedule, we want each worker to receive

Table 2. Shift Scheduling

m	n	Decomp. #	Decomp. B	Decomp. Time	Decomp. Bkt	Implied #	Implied B	Implied Time	Implied Bkt	Implied$^+$ #	Implied$^+$ B	Implied$^+$ Time	Implied$^+$ Bkt	AllBalance #	AllBalance B	AllBalance Time	AllBalance Bkt
6	16	8	1.92	6634	87398	**25**	**1.88**	37	472	**25**	**1.88**	35	423	**25**	**1.88**	**33**	**260**
6	17	11	**2.16**	60637	1073765	**25**	**2.16**	78	1123	**25**	**2.16**	63	877	**25**	**2.16**	**36**	**249**
6	18	16	3.2	8869	166146	**25**	**1.84**	127	1903	**25**	**1.84**	114	1617	**25**	**1.84**	**36**	**279**
6	19	8	3.24	106003	1352600	**25**	**2.64**	607	6983	**25**	**2.64**	504	6923	**25**	**2.64**	**61**	**408**
6	20	7	3.04	2302	27839	**25**	**2.80**	910	10027	**25**	**2.80**	734	8221	**25**	**2.80**	**169**	**1085**
7	16	6	**1.44**	32540	476847	**25**	**1.44**	2361	29767	**25**	**1.44**	2112	28382	**25**	**1.44**	**1828**	**12383**
7	17	9	2.04	159790	1542016	**25**	**1.96**	8416	90680	**25**	**1.96**	6697	72236	**25**	**1.96**	**1576**	**9378**
7	18	3	2.36	135580	1439674	22	1.76	19432	236671	22	1.76	14300	183069	**24**	**1.68**	13920	90665
7	19	4	2.04	80636	804503	22	1.88	21981	230327	22	1.88	13840	151262	**23**	**1.76**	6378	36822
7	20	2	2.72	25779	290430	23	1.56	46267	600434	**24**	**1.52**	55260	715789	23	1.68	18772	116406
8	16	8	2.12	128618	2109236	22	0.92	17420	231594	23	0.72	34216	462257	**25**	**0.44**	3797	14999
8	17	3	1.84	154183	1271700	21	1.68	55193	716866	21	1.68	49859	689151	**25**	**1.28**	12900	68059
8	18	1	1.76	4033	35971	15	1.56	56785	542326	**16**	**1.52**	84438	745177	**16**	1.56	5264	15636
8	19	2	2.12	176092	1675776	**24**	**1.40**	64074	665200	**24**	**1.40**	51899	544990	**24**	**1.40**	**31092**	**201842**
8	20	2	5.84	242901	2082063	11	2.76	51041	468643	11	2.68	35712	316148	**15**	**2.32**	12654	52741

an assignment as balanced as possible. We have one variable $X_{i,j}$ per worker i and per day j. We make sure that on any day j, all tasks are performed by distinct workers through an ALL-DIFFERENT constraint over $[X_{1,j}, \ldots, X_{m,j}]$. We bound the balance of the tasks assigned to each worker i by a shared variable B which we minimize with ATMOSTALLBALANCE over $[X_{i,1}, \ldots, X_{i,n}]$. To make problems hard, we ensure that not all workers are available for every task every day. Given a ratio $0 \leq \alpha < 1$, we randomly forbid $\lceil \alpha n^2 m \rceil$ triples ⟨i, j, k⟩ for which we remove the value k from the variable $X_{i,j}$ (so that worker i cannot do task k on day j). We make sure that $i \in X_{i,j}$ for all i and j, to ensure a feasible solution exists. We randomly generated instances for 6 to 8 workers/tasks (m) and 16 to 20 days (n). For each pair (m, n), we generated 25 instances with α ranging from 0.1 to 0.58 by increments of 0.02.

We compare the basic GCC decomposition (`Decomp.`), the decompositions with implied constraints (`Implied`) and (`Implied`$^+$), and the DC algorithm (`AllBalance`). We report the same statistics as for BACP, but compute averages over the values of α instead of over random runs. A static variable and value ordering was used so that the decrease in number of backtracks is only due to stronger propagation. We also report the average number of backtracks over instances solved to optimality within the cutoff.

We can clearly see that the implied constraints have a huge impact for a very low overhead. On the smaller instances, while the filtering algorithm saves backtracks, it is almost twice as slow (in terms of backtracks per second) as either decomposition with implied constraints. It is nevertheless almost always faster, but only by a small margin. As the instances get larger, the benefits of the algorithm over the decompositions, and of the stronger decompositions over the weaker ones, become more evident. Indeed, the algorithm allows to prove optimality in 84% of the cases for $m = 8$, whereas the decompositions (`Decomp.`, `Implied`, `Implied`$^+$) can only do it in 13%, 74% and 76% of the cases, respectively. Moreover, still for $m = 8$ the objective value is decreased 48%, 16% and 12% in average with respect to these three decompositions.

8 Conclusions

We have studied constraints for ensuring solutions are balanced. We first proved that enforcing domain consistency on the ATMOSTBALANCE and therefore on the BALANCE constraint is NP-hard. This is due to the disjunctive choice in the semantics of BALANCE that ignores a value which does not occur. We therefore introduced a variant, ALLBALANCE with a similar semantics in which all values are considered. We provided a propagation algorithm, and a powerful decomposition, which both work in low polynomial time. Experimental results demonstrated the promise of these new filtering methods.

References

1. Ahuja, R.K., Magnanti, T.L., Orlin, J.B.: Networks Flows, Theory, Algorithms, and Applications. Prentice Hall (1993)
2. Arora, S., Rao, S., Vazirani, U.: Expander flows, geometric embeddings and graph partitioning. Journal of the ACM (JACM) 56(2), 5 (2009)
3. Beldiceanu, N., Carlsson, M., Demassey, S., Petit, T.: Global Constraint Catalogue: Past, Present and Future. Constraints 12(1), 21–62 (2007)
4. Bessiere, C.: Constraint propagation. In: Rossi, F., van Beek, P., Walsh, T. (eds.) Handbook of Constraint Programming. Elsevier (2006)
5. Cattafi, M., Herrero, R., Gavanelli, M., Nonato, M., Malucelli, F.: Improving Quality and Efficiency in Home Health Care: an application of Constraint Logic Programming for the Ferrara NHS unit. In: ICLP, pp. 415–424 (2012)
6. Hnich, B., Kiziltan, Z., Walsh, T.: Modelling a Balanced Academic Curriculum Problem. In: CPAIOR, pp. 121–131 (2002)
7. Lee, C., Loh, P.-S., Sudakov, B.: Bisections of graphs. Journal of Combinatorial Theory, Series B 103(5), 599–629 (2013)
8. Monette, J.-N., Schaus, P., Zampelli, S., Deville, Y., Dupont, P.: A CP Approach to the Balanced Academic Curriculum Problem. In: The Seventh International Workshop on Symmetry and Constraint Satisfaction Problems, Symcon 2007 (2007)
9. Pesant, G.: A regular language membership constraint for finite sequences of variables. In: Wallace, M. (ed.) CP 2004. LNCS, vol. 3258, pp. 482–495. Springer, Heidelberg (2004)
10. Pesant, G., Régin, J.-C.: SPREAD: A Balancing Constraint Based on Statistics. In: van Beek, P. (ed.) CP 2005. LNCS, vol. 3709, pp. 460–474. Springer, Heidelberg (2005)
11. Quimper, C.-G., Golynski, A., López-Ortiz, A., van Beek, P.: An Efficient Bounds Consistency Algorithm for the Global Cardinality Constraint. Constraints 10, 115–135 (2005)
12. Refalo, P.: Impact-Based Search Strategies for Constraint Programming. In: Wallace, M. (ed.) CP 2004. LNCS, vol. 3258, pp. 557–571. Springer, Heidelberg (2004)
13. Régin, J.-C.: Generalized Arc Consistency for Global Cardinality Constraint. In: IAAI, pp. 209–215 (1996)
14. Schaus, P.: Solving Balancing and Bin-Packing problems with Constraint Programming. PhD thesis, Universite Catholique de Louvain (2009)
15. Schaus, P., Deville, Y., Dupont, P., Régin, J.-C.: Simplification and Extension of the SPREAD Constraint. In: Proc. of the 3rd Int'l Workshop on Constraint Propagation and Implementation, held alongside CP-06, pp. 77–91 (2006)
16. Schaus, P., Deville, Y., Dupont, P.E., Régin, J.-C.: The Deviation Constraint. In: Van Hentenryck, P., Wolsey, L.A. (eds.) CPAIOR 2007. LNCS, vol. 4510, pp. 260–274. Springer, Heidelberg (2007)
17. Schaus, P., Van Hentenryck, P., Régin, J.-C.: Scalable Load Balancing in Nurse to Patient Assignment Problems. In: van Hoeve, W.-J., Hooker, J.N. (eds.) CPAIOR 2009. LNCS, vol. 5547, pp. 248–262. Springer, Heidelberg (2009)

Experimental Comparison of BTD and Intelligent Backtracking: Towards an Automatic Per-instance Algorithm Selector

Loïc Blet[1,3], Samba Ndojh Ndiaye[1,2], and Christine Solnon[1,3]

[1] Université de Lyon, LIRIS, UMR5205, France
[2] Université Lyon 1, 69622 France
[3] INSA-Lyon, 69621, France
{loic.blet,samba-ndojh.ndiaye,christine.solnon}@liris.cnrs.fr

Abstract. We consider a generic binary CSP solver parameterized by high-level design choices, i.e., backtracking mechanisms, constraint propagation levels, and variable ordering heuristics. We experimentally compare 24 different configurations of this generic solver on a benchmark of around a thousand instances. This allows us to understand the complementarity of the different search mechanisms, with an emphasis on Backtracking with Tree Decomposition (BTD). Then, we use a per-instance algorithm selector to automatically select a good solver for each new instance to be solved. We introduce a new strategy for selecting the solvers of the portfolio, which aims at maximizing the number of instances for which the portfolio contains a good solver, independently from a time limit.

1 Introduction

Backtracking approaches solve constraint satisfaction problems by building a search tree (or graph). In Chronological BackTracking (CBT) [1], this tree is explored in a depth-first way: When a failure occurs, the search backtracks to the last choice point. CBT is known to explore redundant subtrees when a failure is not due to the last decision (trashing). To overcome trashing, intelligent backtrackings have been proposed, such as Conflict-directed BackJumping (CBJ) [2], Dynamic BackTracking (DBT) [3] and Decision Repair (DR) [4]: They dynamically exploit the structure of the problem to directly backjump to failure causes thus avoiding trashing. Backtracking with Tree Decomposition (BTD) [5] uses a different idea to avoid trashing: It captures the static problem structure by identifying independent subproblems which are solved separately.

CBJ, DBT and CBT have already been experimentally compared (e.g., [6]). BTD has also been experimentally compared to CBT and CBJ (e.g., [5]). However, BTD has never been compared to CBJ and DBT on a wide benchmark, and it has never been compared to DR. It is interesting to compare them as they all exploit structure to guide the search: CBJ, DBT and DR exploit a dynamic structure thanks to explanations, whereas BTD exploits a static structure thanks to decompositions. Furthermore, these backtracking mechanisms may be combined with different constraint propagation mechanisms, such as Forward-Checking (FC) and Maintaining Arc Consistency (MAC), and with different variable ordering heuristics. In particular, [7] proposes to exploit information

B. O'Sullivan (Ed.): CP 2014, LNCS 8656, pp. 190–206, 2014.

about previous states of the search when selecting the next variable to be assigned. In some sense, this heuristic also exploits the structure of the instance to guide the search.

In this paper, we describe a generic CSP solver which has three parameters: (i) the search strategy, which may be instantiated to CBT, CBJ (with or without variable re-ordering), DBT, DR, or BTD; (ii) the constraint propagation mechanism, which may be instantiated to FC or MAC; and (iii) the variable ordering heuristic, which may be instantiated to minDomain over dynamic degree or over weighted degree.

A first contribution of the paper is to experimentally compare the 24 configurations of this generic solver on a wide benchmark of around a thousand instances. In particular, we compare BTD-based variants with other variants based on intelligent backtracking frameworks. This extensive experimental study shows us that, even though one configuration has better global success rates than all others, some configurations (such as BTD-based ones) which have low global success rates are very good on a large number of instances. In particular, we identify a minimal subset of 13 complementary configurations such that, for every instance of our benchmark, there is always at least one of these 13 configurations which is good for it, i.e., which is not significantly outperformed by any other configuration on this instance.

The next step is to exploit the complementarity of these configurations to improve success rates. This may be done by hybridizing mechanisms. In particular, we have proposed to combine BTD with approaches which dynamically exploit the structure (CBJ and DR) in [8]. Such hybrid approaches are able to solve more efficiently some instances but they are outperformed by some other configurations on other instances. Recent works on portfolios and per-instance algorithm selectors (e.g., [9,10,11,12,13]) have shown us that we may much more significantly improve success rates by learning selection models, which are able to choose a good solver for each new instance to be solved. Therefore, we combine our generic solver with a per-instance algorithm selector. Like other recent approaches, we extract features from instances, and we use machine learning techniques to learn a selection model. A key point is to choose a subset of solvers that may be selected by the selector: The goal is to keep a subset S of solvers with complementary performances so that S contains a solver which performs well on every instance of the training set. We compare two different strategies for achieving this task, called *Solved* and *Good*. The *Solved* strategy maximizes the number of instances solved at a given CPU time limit (ties are broken by minimizing CPU time), as proposed in [12]. The *Good* strategy maximizes the number of instances for which S contains a good solver, and uses statistical tests to decide whether a solver is good for an instance. We experimentally show that this new strategy outperforms *Solved*.

The paper is organized as follows. In Section 2 we describe our generic framework for solving CSPs. In Section 3, we experimentally compare different configurations of this framework. In Section 4, we describe the per-instance algorithm selector and the two selection strategies. In Section 5, we experimentally compare the two selection strategies. We conclude in Section 6 with ideas for some further works.

2 Generic Framework for Binary CSPs

Background. A CSP instance is defined by a triplet (X, D, C). X is a finite set of variables. D associates a finite set of values $D(x_i)$ with every variable $x_i \in X$. C is a set of constraints. Each constraint is defined over a subset of variables and defines tuples of values that can be assigned simultaneously to its variables. In this paper, we consider binary CSPs, which only contain binary constraints defined over 2 variables. A solution is an assignment of all variables satisfying all constraints.

We focus on backtracking approaches which structure the assignment space in a search tree (or graph for DBT and DR) whose nodes correspond to variable/value assignments. We introduce a generic algorithm which is parameterized by the backtracking mechanism, the constraint propagation mechanism and the variable ordering heuristic. This generic algorithm allows us to compare in a unified framework state-of-the-art backtracking approaches for binary CSPs. It basically extends the generic algorithm of [6] by adding 3 new backtracking mechanisms (CBJR, DR and BTD).

Backtracking. In *Chronological Backtracking* (CBT) [1], the tree is explored with a depth-first search. When a failure occurs, the search backtracks to the last choice point. When the cause of the failure is not due to the last decision, but to an earlier one, CBT explores redundant subtrees. To overcome this issue, *Conflict directed BackJumping* (CBJ) [2] backtracks immediately to the last assigned variable involved in the failure, and unassigns all variables assigned after it. In this study, we use improvements proposed in [14,15] to get a version of CBJ similar to the one in [6]. It maintains for each value unsuccessfully tried the set of assigned variables involved in this failure. Moreover, if this set is empty, the value is permanently removed from the problem.

Dynamic BackTracking (DBT) [3] does not unassign variables between the current node and the cause of the failure, but simply backjumps over them since they are not involved in the current failure. Due to the poor performance of DBT combined with FC and a good variable ordering [16], [17] proposes a new version of CBJ combined with a retroactive ordering of already assigned variable (CBJR). After each new assignment, the variable ordering heuristic is used to try to replace the assigned variable higher in the search tree in light of the current state of the problem (for example if many values are removed from the domain of a variable, the ordering heuristic may move up this variable in the tree). Yet, to ensure completeness, the assigned variable cannot be moved before a variable involved in the filtering or failure of a value in its domain.

Decision Repair (DR) [18] is a generic framework that generalizes [4] with several parameters. Each instantiation of this framework corresponds to a different hybridization between tree search, local search and constraint propagation. We consider the *DR(mindestroy, uvar)* instantiation of this framework, as proposed in [18]. It performs a depth first search combined with Forward-Checking (FC). When a failure occurs, the current assignment is repaired by unassigning one variable among those involved in the failure (not necessarily the last) and removing all explanations involving this variable. To ease inconsistency proofs, a variable minimizing the number of removed explanations is chosen randomly. We have extended DR to allow its combination with arc consistency (MAC) instead of FC. Note that DR does not guarantee a complete exploration of the search space.

Finally, *Backtracking with Tree Decomposition* (BTD) [5] uses a tree-decomposition of the constraint graph which captures the problem structure by identifying independent subproblems. BTD computes the order in which the subproblems must be solved, resulting in a partial order on the variables. Moreover, it records *goods* associated with subproblem solutions, and *nogoods* associated with subproblem failures. This information is exploited to avoid solving the same subproblem more than once. In this study, the tree decomposition is computed using the minimum-fill heuristic [19] to triangulate the constraint graph.

Constraint propagation. At each node of the search tree (or graph), constraints are propagated in order to filter domains and detect local inconsistencies. In this study, we consider two well-known filtering mechanisms [1]: *Forward Checking* (FC), which removes values which are not arc-consistent with the last variable/value assignment; and *Maintaining Arc Consistency* (MAC), which ensures arc consistency of all constraints. For CBJ, DBT and DR, we maintain arc consistency with AC3 [20], whereas for CBT and BTD we use AC2001 [21]. AC2001 considers an ordering of the values in the domains and records for each value its first compatible value in the other domains. If this compatible value is removed, AC2001 searches for a new compatible value starting from the position of the removed one.

Variable ordering heuristics. At each node of the tree (or graph), the search chooses the next variable to be assigned among the set of all non assigned variables. It uses a variable ordering heuristic to guide this choice. A classical variable ordering heuristic is minDomain, which chooses a variable which has the smallest domain. In this study, we consider two well-known improvements of this heuristic [22,7]: *minDomain over dynamic degree* (d), which chooses a variable x which minimizes the ratio between the size of $D(x)$ and the number of unassigned variables sharing a constraint with x; and *minDomain over weighted degree* (w), which chooses a variable x which minimizes the ratio between the size of $D(x)$ and the sum of weights of constraints which involve x with another unassigned variable (where the weight of a constraint is the number of failures it has generated since the beginning of the search).

Note that when the backtracking mechanism is BTD, the variable ordering heuristic is used to choose the next variable within the current cluster of the tree decomposition, and not within the set of all unassigned variables (see [23]).

Generic framework. [6] defines a first generic framework that encompasses several state-of-the-art backtracking algorithms. In this study, we have extended this framework with three new backtracking mechanisms, namely CBJR, DR and BTD. From this generic framework, we can obtain configurations denoted by triplets (b, c, o) where $b \in$ {CBT, CBJ, DBT, CBJR, DR, BTD} defines the backtracking mechanism, $c \in$ {FC, MAC} the constraint propagation mechanism, and $o \in$ {d,w} the variable ordering heuristic.

For all configurations, we first decompose the constraint graph into its set of connected components to obtain independent subproblems which are solved independently and consecutively. Also, each subproblem is made arc consistent before starting the solving process.

All configurations are non deterministic: When choosing variables, ties are randomly broken; furthermore, we do not consider any value ordering heuristic and values are randomly chosen.

Table 1. Classes of the benchmark. For each class, the table displays its name, number of instances, number of variables, domain sizes, number of constraints and constraint tightness (ratio of forbidden tuples over number of possible tuples): minimum, average and maximum values.

Class	#Instances	#Variables			#Values			#Constraints			Constraint tightness		
		min	avg	max	min	avg	max	min	avg	max	min	avg	max
ACAD	75	10	116	500	2	146	2187	45	691	4950	0.001	0.692	0.998
PATT	238	16	263	1916	3	66	378	48	4492	65390	0.002	0.795	0.996
QRND	80	50	220	315	7	11	20	451	2968	4388	0.122	0.578	0.823
RAND	206	23	37	59	8	36	180	84	282	753	0.095	0.613	0.984
REAL	193	200	628	1000	2	152	802	1235	6394	17447	0.0	0.519	1.0
STRUCT	300	150	257	500	20	23	25	617	1641	3592	0.544	0.647	0.753

3 Experimental Comparison

Benchmark. Our benchmark is composed of 1092 instances grouped into 6 classes described in Table 1. The first 5 classes come from the CSP'08 competition. We have only considered the binary instances. If classes contained too many similar instances we only took the first 10 instances. We have removed from the benchmark every instance which has not been solved by any of our 24 configurations within a time limit of 30 minutes among 15 runs for each instance. The last class (STRUCT) contains structured instances which are randomly generated as described in [24]. These instances have a structure similar to RLFAP instances which are real-world instances. This structure is defined by a tree of variable clusters, and the level of structure depends on the density of constraints in clusters and the sizes of the clusters. The class contains subclasses of instances with different levels of structure, sizes and constraint tightness.

Experimental results. Table 2 compares the success rates of the 24 configurations at different CPU-time limits on an Intel(R) Xeon(R) CPU E5-2670 0 @ 2.60GHz, 20480 KB cache size, 3GB RAM. As all configurations are non deterministic, we have performed 15 runs for each instance and each configuration. Table 2 also gives success rates of Gecode (with the model proposed in [25]) with 3 different propagation levels: ICL_VAL, ICL_DOM and ICL_DEF. It shows us that our implementation is competitive with Gecode. Of course, our implementation is dedicated to binary CSPs, whereas Gecode is a generic solver which has not be tailored for solving binary CSPs.

(CBT,MAC,w) is the best configuration when considering global success rates. This result is not surprising and has already been observed, for example, in [6]. Without surprise, we also note that configurations which use weighted degrees for ordering variables outperform configurations which consider dynamic degrees (as already observed in [7]). However the gain depends on the considered backtracking mechanism as pointed out in [26]. In particular, using weighted degrees greatly improves the solution process for CBT, DBT and DR whereas the improvement is not so high for CBJ. Also, configurations using DBT or DR as backtracking mechanism and FC for propagating constraints perform poorly. It was already hinted that DBT could lead to poor performances in [16].

Table 2. Comparison by means of global success rates. Each line successively displays the parameters of the solver and the percentage of successful runs at different CPU time limits (in seconds, over 15 runs on 1092 instances). For each backtracking mechanism we highlight in bold the best configuration. We highlight in blue the best over all configurations.

			1	5	10	50	100	500	1000	1800
CBT	FC	d	37.0	45.2	47.6	52.7	55.3	60.0	61.1	61.7
		w	41.8	51.7	56.8	65.9	69.4	77.8	81.5	83.2
	MAC	d	43.0	51.7	56.7	65.5	69.1	75.3	76.5	77.7
		w	**47.1**	**61.5**	**68.3**	**80.5**	**85.2**	**92.3**	**94.3**	**95.4**
CBJ	FC	d	**41.3**	50.4	55.2	66.9	70.5	81.9	85.6	88.0
		w	39.6	51.0	55.0	67.8	72.6	84.0	88.1	91.0
	MAC	d	38.0	50.2	54.3	68.2	74.2	85.3	88.8	90.4
		w	39.7	**53.1**	**57.6**	**73.7**	**79.6**	**90.7**	**93.5**	**95.1**
CBJR	FC	d	**39.9**	49.4	53.3	63.3	66.9	75.8	78.1	79.5
		w	39.1	**50.5**	**55.0**	**67.5**	**72.6**	**84.2**	**88.3**	**90.9**
	MAC	d	29.2	37.3	41.1	46.4	48.5	53.9	55.4	56.5
		w	31.6	40.1	44.9	55.0	58.9	67.7	69.4	70.7
DBT	FC	d	33.8	38.0	38.8	40.8	41.5	43.9	45.8	46.7
		w	37.7	47.4	50.5	61.9	66.5	77.0	80.3	83.7
	MAC	d	35.8	46.2	49.4	56.6	60.0	66.6	68.0	69.3
		w	**37.9**	**49.5**	**54.1**	**68.6**	**74.5**	**85.7**	**89.5**	**91.8**
DR	FC	d	32.7	37.5	39.2	41.9	42.6	44.0	44.6	45.0
		w	**35.1**	**44.4**	48.1	55.4	59.8	71.9	76.3	79.4
	MAC	d	32.5	41.6	45.1	51.9	53.8	59.0	60.8	62.3
		w	34.4	44.3	**48.7**	**57.8**	**62.5**	**75.2**	**80.3**	**84.5**
BTD	FC	d	31.4	45.1	52.2	65.1	69.2	76.1	77.3	78.0
		w	33.5	48.0	55.5	70.2	74.9	82.1	83.9	84.5
	MAC	d	32.8	45.1	53.9	70.1	75.8	84.6	86.0	87.1
		w	**37.4**	**51.3**	**61.9**	**77.6**	**83.3**	**91.9**	**93.6**	**94.2**
Gecode ICL_DEF			29.7	34.9	38.1	48.9	55.4	**66.7**	69.3	71.9
Gecode ICL_VAL			27.7	32.9	35.2	45.3	51.3	63.8	67.2	70.4
Gecode ICL_DOM			**29.9**	**35.8**	**38.9**	**50.9**	**56.6**	**66.7**	**70.5**	**73.4**

These global success rates on the 1092 instances of our benchmark hide very different results when we look at each instance separately. In particular, some configurations which have rather low success rates on the whole benchmark are the best performing ones on some instances. We apply a simple rule to decide if a configuration is the *best* for an instance i: we first compare the number of successful runs within a 30 minute CPU time limit, and we break ties by comparing the CPU time of the successful runs.

Line (b) of Table 3 displays the percentage of instances for which a configuration is the best among the whole set of configurations. It shows us that, even though (CBT,FC,d) only solves 61.7% instances of the whole benchmark after 30 minutes of CPU time, it is the best configuration for 27.7% of the 1092 instances. Of course, it is well known that simple configurations like (CBT,FC,d) outperform more complicated configurations

Table 3. For each configuration c, line (b) gives the percentage of instances for which c is the best configuration; line (b/h) (resp. (g/h)) gives the percentage of hard instances for which c is the best configuration (resp. c is a good configuration); line (sb/h) (resp. (sb2/h)) gives the percentage of hard instances for which c is the best configuration and all other configurations are significantly worse than c (resp. all other configurations except (CBJ,FC,w), (CBJR,FC,d), (CBJR,MAC,*), (DBT,*,*), (DR,FC,d) and (DR,MAC,*) are significantly worse than c)

	CBT				CBJ				CBJR				DBT				DR				BTD			
	FC		MAC		FC		MAC		FC		MAC		FC		MAC		FC		MAC		FC		MAC	
	d	w	d	w	d	w	d	w	d	w	d	w	d	w	d	w	d	w	d	w	d	w	d	w
(b)	**27.7**	4.5	11.3	9.2	2.0	1.2	1.0	0.7	0.7	1.1	1.6	0.9	1.4	0.4	0.7	0.1	1.0	3.4	0.5	0.2	14.2	12.3	0.7	3.1
(b/h)	17.8	2.3	8.8	11.1	1.1	0.3	0.3	1.1	0.8	1.6	0.0	0.8	0.8	0.2	0.5	0.0	0.5	2.7	0.5	0.2	**23.3**	18.8	1.3	5.1
(g/h)	27.7	9.8	12.9	19.9	4.0	3.4	2.7	7.1	2.3	4.8	2.3	2.3	1.3	5.5	2.7	2.7	2.7	7.1	2.9	2.9	**42.1**	26.5	5.3	10.5
(sb/h)	6.4	0.3	5.5	5.8	0	0	0.2	0.5	0	1.1	0	0	0	0	0	0	0	1.9	0	0	**14.0**	7.2	0.2	1.3
(sb2/h)	6.4	0.6	6.1	7.1	1.6	-	0.2	0.8	-	1.4	-	-	-	-	-	-	-	2.3	-	-	**14.6**	7.9	0.3	1.3

on very simple instances, for which there is no need for intelligent but expensive mechanisms, whereas they usually have very poor performance on harder instances. In line (b/h) of Table 3, we have removed easy instances from the benchmark: we consider that an instance is *easy* if it has been solved in less than one second by (CBT,MAC,w), for each of the 15 runs. With this definition, 470 instances of our benchmark are easy, and 622 are more difficult. When focusing on these harder instances, line (b/h) of Table 3 shows us that some configurations (such as those using DBT) only have very few instances for which they are the best.

As several configurations may have close results for a given instance, we also study the number of instances for which a configuration performs well: We consider that a configuration is *good* for an instance i either if it is the best one, or if there is no statistical difference between its 15 runs and the 15 runs of the best configuration for i. We used the Student's t-test with $p = 0.01$ to decide whether a configuration is not significantly different from another one on a given instance. Line (g/h) of Table 3 displays the percentage of hard instances for which a configuration is good. Again, we note that configurations which are good for many instances do not always have high global success rates on the whole benchmark. In particular, the two configurations which are good for the largest numbers of instances (i.e., (BTD,FC,d) and (CBT,FC,d)) are far from having the highest success rates in Table 2.

All configurations are good for at least one instance of the benchmark. However, it may happen that some configurations are good only for instances for which other configurations are also good, i.e., some configurations are dominated by other ones. To study this, line (sb/h) of Table 3 displays the percentage of hard instances for which a configuration is the best and all other configurations are significantly worse than it. It shows us that 12 configurations are dominated by the other configurations: (CBJ,FC,*), (CBJR,FC,d), (CBJR,MAC,*), (DBT,*,*), (DR,FC,d) and (DR,MAC,*). Hence, we have removed these configurations from our study, except (CBJ,FC,d): there are 10 instances for which all good configurations belong to the set of 12 dominated configurations; as for these 10 instances (CBJ,FC,d) is good (and the only configuration to be good on those 10 instances), we keep (CBJ,FC,d). Finally, line (sb2/h) of Table 3 gives the percentage of hard instances for which a configuration is the best and all other configurations except

Table 4. Description of the 3 sets of instances. For each set, the table displays the number of hard instances in each benchmark class, and the average tree width and separator size (in percentage of the number of variables) of the tree decomposition.

	Number of hard instances							Sep size	Tree width
	ACAD	PATT	QRND	RAND	REAL	STRUCT	Total	(avg)	(avg)
Decompose	7	8	1	14	5	177	212	4,7%	17,1%
Don't decompose	5	50	12	23	77	63	230	25,3%	31,8%
Don't know	9	35	5	99	3	29	180	32,9%	54,5%

(CBJ,FC,*), (CBJR,FC,d), (CBJR,MAC,*), (DBT,*,*), (DR,FC,d) and (DR,MAC,*) are significantly worse than it. The set composed of the 13 remaining configurations contains a good configuration for every instance. This set is minimal as each of these 13 solvers is the only one to be good for at least one instance.

BTD is very effective on many instances. In particular, (BTD,FC,d) is good on more than 42% of the hard instances, and it is significantly better than all other configurations on more than 14% of the hard instances. However, on some other instances it also performs poorly so that its global success rate is rather low compared to other approaches. In order to give an insight into which instances are better solved by BTD, we partitioned the 622 hard instances in 3 sets:

- The *Decompose* set contains all hard instances which are best solved by one of the 4 BTD-based configurations (i.e., (BTD,*,*)), and for which none of the 9 non BTD-based configuration (i.e., (CBT,*,*), (CBJ,FC,d), (CBJ,MAC,*), (CBJR,FC,w)), and (DR,FC,w)) is good;
- The *Don't decompose* set contains all hard instances which are best solved by one of the 9 non BTD-based configurations and for which none of the 4 BTD-based configuration is good;
- The *Don't know* set contains all other instances.

Table 4 shows us how the instances of the benchmark are distributed into these sets. Many instances of the *Decompose* set come from the STRUCT class, which contains structured instances. This is not a surprise that BTD-based approaches perform better than other approaches on these instances (see, e.g., [5]). As BTD has never been compared with intelligent backtracking approaches, it is interesting to note that BTD outperforms them on many of these structured instances. Only 35 instances of the 2008 competition belong to the *Decompose* set: Many instances of this benchmark do not exhibit static structures that can be exploited by BTD. When looking at parameters of the tree decomposition, we note that instances of the *Decompose* set have a smaller tree width (half the size of the *Don't decompose* set) and a smaller separator size (one fifth the size of the *Don't decompose* set). Instances of the *Don't know* set have large tree width. Actually, when the tree width is close to the number of variables, BTD behaves like CBT as nearly all the variables belong to the same cluster.

On some instances, most notably some *rlfap* instances, the decomposition is good but the instance is in *Don't decompose*. These instances are easy (the solution is found very quickly by (CBT,FC,d)) but huge (up to 900 variables). Restricting the search to clusters can be detrimental as we forbid the search from going directly to the solution.

4 Per-instance Algorithm Selector

Experimental results reported in the previous section have shown us that the best performing configurations on some instances may have very bad performance on other instances so that they are far from having the best average success rates on the whole benchmark. This illustration of the well-known no-free-lunch theorem motivates our study on a per-instance algorithm selector which aims at selecting a good configuration for each new instance to be solved.

In this study, we do not aim at improving the state-of-the-art of per-instance algorithm selectors such as, e.g., CPHydra [9], ISAC [11] or EISAC [27], but we focus on a key point of these approaches, *i.e.*, the selection of the solvers to be included in the portfolio. Indeed, [12] shows us that better performance may be obtained with smaller portfolios. However, it is also important that the portfolio contains a large enough number of solvers so that there is a good solver for every instance. Experimental results reported in the previous section may be used to definitely remove some solvers from the portfolio: The 11 configurations which are dominated by the 13 other configurations can be removed from the portfolio without significantly changing the performance of a *Virtual Best Selector* (VBS), which always selects the best solver in the portfolio. In this section, we describe and compare two different strategies for selecting a subset of these 13 solvers in the portfolio: the strategy used in [12], and a new strategy. Before describing these two strategies, we describe the basic framework of our per-instance algorithm selector.

4.1 Basic Framework of the Selector

We consider a classical framework similar to the "off-the-shelf" framework of [12]. The idea is to train a supervised classifier by giving a training set of labeled CSP instances to it: Each instance of the training set is described by an input vector of features and is associated with an output label, corresponding to the best solver for this instance. This training phase allows the classifier to learn a selection model. Then, this model is used to select solvers for new instances to be solved, given their input feature vectors.

Features. Each CSP instance is described by a vector of features. We consider classical features, similar to those used in [9,12], for example. The main difference is that we also extract features from the tree decomposition, as Table 4 has shown us that the performance of BTD depends on tree widths and separator sizes.

More precisely, we extract the following static features from each instance: Number of variables, number of constraints, size of domains (average and standard deviation), constraint tightness, i.e., ratio of forbidden tuples with respect to all possible tuples in the relation (average and standard deviation), and variable degree in the constraint graph (average and standard deviation). As the constraint graphs of some instances are not connected, we also extract the following features: Number of connected components in the constraint graph, number of variables in a connected component (average and standard deviation), and number of constraints in a connected component (average and standard deviation). Finally, we also extract features from a tree decomposition which is computed using the greedy algorithm *minFill* of [19] to triangulate the constraint graph: Number of clusters, maximum separator and cluster size, and density of constraints in a cluster (average and standard deviation).

In order to gather more information on the instance to be solved, we also perform a short run on it and extract dynamic features from this run. We have limited the time of this run to 1 second. As (CBT,MAC,w) is the best configuration within this time limit, we have chosen to run (CBT,MAC,w). Furthermore, this configuration allows us to gather information on variable weights (used by the variable ordering heuristic) and the number of values filtered by MAC. We collect the following dynamic features: Number of nodes in the search tree, maximum depth of a node in the search tree, number of failed nodes, number of values removed by MAC (average and standard deviation), and weight of a variable (average and standard deviation). In order to gather insights into the dynamics of the run, we collect these features for 3 time limits, i.e., 0.25, 0.5 and 1 second.

Training. Given a portfolio of solvers and a training set I of instances such that each instance $i \in I$ is described by a vector of features and is associated with the solver of the portfolio which is the best for i, the goal is to train a classifier to associate instances with solvers. This is a classical supervised classification problem and there exist different well-known approaches to solve this problem [28]. In this study, we have used the Weka library [29,30] to perform this task. We have compared the different supervised classifiers which are implemented in Weka. The best classification results are obtained with *ClassificationViaRegression* with default parameters [31] so that we have used this classifier in our experiments.

Once the classifier has been trained on the training set of instances, we can use it to dynamically choose the best configuration of our generic solver for each new instance to be solved. More precisely, to solve a new instance i we proceed as follows: We first run (CBT,MAC,w) on i with a CPU-time limit of 1 second; if i is not solved within this time limit, we extract static features from i, and dynamic features from the run of (CBT,MAC,w); we give these features to the classifier which returns a configuration and we run this configuration on i. Note that the time spent to extract the features and classify i is very short (less than 0.1 seconds on average).

4.2 Selection of a Subset of Solvers

A key point of per-instance algorithm selection is to select the solvers to be included in the portfolio. The goal is to select solvers with complementary behaviors so that the portfolio contains a good solver for every instance. We may include in our portfolio the 13 non dominated solvers identified in Section 3. However, the larger the portfolio, the harder the learning task. Therefore, better results may be obtained with fewer configurations, as observed in [12].

We compare two strategies for selecting a subset S_k of k solvers (where $k \in [2;13]$ is a parameter to be fixed). The first strategy, called *Solved*, is the one used in [12]. It selects in S_k the k solvers which maximize the number of instances solved by a VBS at the CPU time limit. Further ties are broken by minimizing the solving time of the VBS. The second strategy, called *Good*, selects in S_k the k solvers which maximize the number of instances for which S_k contains a good solver (i.e., a solver which is not statistically different from the best solver for this instance). Further ties are broken by maximizing the number of instances solved by a VBS at the CPU time limit.

For both strategies, finding the optimal subset S_k is NP-hard: It is a set covering problem between solvers and instances, where a solver s covers an instance i if s is able to solve i (for *Solved*) or if s is good for i (for *Good*). In this study, we approximately solve it in a greedy way: Starting from the subset S_1, which contains the solver which covers the largest number of instances, we define S_i from S_{i-1} by adding to S_{i-1} the solver which most increases the number of covered instances.

When considering the 15 runs of our 13 solvers on the 622 hard instances, we obtain the following orders:

- Order of selection of solvers with the *Solved* strategy:
  1-(CBT,MAC,w), 2-(BTD,FC,w), 3-(CBJR,FC,w), 4-(DR,FC,w), 5-(CBJ,MAC,w), 6-(CBT,MAC,d), 7-(BTD,FC,d), 8-(BTD,MAC,w), 9-(CBT,FC,d), 10-(CBJ,FC,d), 11-(CBJ,MAC,d), 12-(BTD,MAC,d), 13-(CBT,FC,w)
- Order of selection of solvers with the *Good* strategy:
  1-(BTD,FC,d), 2-(CBT,MAC,w), 3-(BTD,FC,w), 4-(CBT,FC,d), 5-(CBT,MAC,d), 6-(DR,FC,w), 7-(CBJ,MAC,w), 8-(BTD,MAC,w), 9-(CBJ,FC,d), 10-(CBJR,FC,w), 11-(CBT,FC,w), 12-(BTD,MAC,d), 13-(CBJ,MAC,d)

Of course, this order strongly depends on the composition of the benchmark. For example, if we remove half of the STRUCT instances, the *Good* strategy selects (CBT,FC,d) in third position and (BTD,FC,w) in fourth position, while all other positions are unchanged. However, if we remove all STRUCT instances, the order becomes very different and the best BTD-based approach is (BTD,FC,w) and it is selected in fourth position.

Let us note S_k^s the subset which contains every solver whose rank is lower than or equal to k for the strategy $s \in \{Solved, Good\}$, and VBS(S_k^s) the VBS associated with S_k^s. This VBS selects the best solver of S_k^s for each instance to be solved so that any selector built upon S_k^s cannot outperform VBS(S_k^s).

Table 5 compares the two strategies by means of VBS success rates. Let us first note that VBS(S_k^{Solved})=VBS(S_k^{Good}) when $k = 10$ or $k = 13$ as $S_k^{Solved} = S_k^{Good}$ in these two cases. Also, VBS(S_k^{Solved}) outperforms VBS(S_k^{Good}) at the time limit of 1800s, when $k \leq 9$, and both approaches are equivalent when $k \geq 10$. This comes from the fact that the *Solved* strategy maximizes the number of solved instances at the time limit. As a counterpart, VBS(S_k^{Good}) outperforms VBS(S_k^{Solved}) for lower time limits or smaller values of k. This comes from the fact that the *Good* strategy maximizes the number of instances for which the portfolio contains a good solver, independently from the time limit.

For example, when $k = 4$, the difference between the success rates of VBS(S_4^{Good}) and VBS(S_4^{Solved}) is equal to 2.2, 3.5, 2.4, 1.3, and 0.2 when the time limit is equal to 1, 5, 10, 50, and 100s, respectively, whereas it becomes negative after 100s. However, the difference is less important (-0.4, -0.7, and -0.7 at 500, 1000 and 1800s, respectively). Actually, with S_3^{Solved}={(CBT,MAC,w), (BTD,FC,w), (CBJR,FC,w)}, a VBS is able to solve 99.4% of the runs, but S_3^{Solved} contains a good solver for only 294 of the

Table 5. Comparison of *Solved* and *Good*. Each line successively displays: the number k of solvers selected in S_k and, for different time limits in seconds, percentages of successful runs of virtual best selectors built upon the sets defined with *Solved* and *Good* (over 15 runs on the 1092 instances). For each (S_k,time) couple, we highlight the strategy with the highest success rate.

	1		5		10		50		100		500		1000		1800	
	Solved	*Good*	*Solved*	*Good*	*Solved*	*Good*	*Solved*	*Good*	*Solved*	*Good*	*Solved*	*Good*	*Solved*	*Good*	*Solved*	*Good*
2	52.0	52.7	68.9	70.6	75.4	77.2	87.5	88.7	91.8	92.3	96.7	96.9	98.2	98.2	98.7	98.6
3	52.4	53.2	69.2	71.6	75.9	77.8	87.8	89.0	92.5	92.8	97.6	97.3	99.0	98.4	99.4	98.8
4	52.5	54.7	69.4	72.9	76.1	78.5	88.2	89.5	92.8	93.0	97.8	97.4	99.1	98.4	99.5	98.8
5	52.7	55.8	69.4	73.1	76.1	78.7	88.3	89.7	93.0	93.0	97.8	97.4	99.2	98.4	99.6	98.8
6	54.5	55.9	69.7	73.3	76.4	79.0	88.8	90.3	93.1	93.6	97.8	97.8	99.2	98.8	99.7	99.3
7	55.6	56.0	72.3	73.3	78.7	79.0	90.2	90.3	94.0	93.9	98.3	98.1	99.3	99.2	99.7	99.6
8	55.7	56.1	72.3	73.3	78.8	79.2	90.4	90.5	94.0	93.9	98.3	98.1	99.4	99.2	99.7	99.6
9	56.2	56.4	73.4	73.5	79.3	79.3	90.5	90.6	94.1	94.1	98.4	98.2	99.4	99.2	99.7	99.6
10	56.5	56.5	73.5	73.5	79.5	79.5	90.7	90.7	94.2	94.2	98.4	98.4	99.4	99.4	99.7	99.7
11	56.5	56.5	73.5	73.7	79.5	79.6	90.8	90.7	94.3	94.2	98.4	98.4	99.4	99.4	99.7	99.7
12	56.5	56.5	73.5	73.7	79.5	79.6	90.8	90.7	94.3	94.2	98.4	98.4	99.4	99.4	99.7	99.7
13	56.6	56.6	73.7	73.7	79.6	79.6	90.8	90.8	94.3	94.3	98.4	98.4	99.4	99.4	99.7	99.7

622 hard instances. When adding new solvers to S_3^{Solved}, we only very slightly increase the success rate of the VBS. The solver which most increases the number of solved instances is (DR,FC,w) and it allows us to solve 26 more runs (among 622*15 runs). However, (DR,FC,w) is a good solver for a rather small number of instances and adding it to S_3^{Solved} increases the number of hard instances for which we have a good solver by 24. As a comparison, adding (BTD,FC,d) to S_3^{Solved} would allow us to solve 12 more runs (instead of 26, among 622*15) but it would increase the number of hard instances for which we have a good solver by 168 (instead of 24, among 622 instances).

5 Experimental Evaluation

Experimental Setting. We consider the 1092 instances of the benchmark described in Section 3, and the training set is composed of the 622 hard instances of this benchmark. We use a leave-one-out scheme: for each instance i of the benchmark, if i is a hard instance which belongs to the training set, then we remove i from it and we train the classifier on all hard instances but i; finally we ask the classifier to select a solver for i.

Comparison of classification rates obtained with different sets S_k. The learnt solver of an instance i is the solver returned by the classifier, and we say that i is *well-classified* if its learnt solver is the best solver for i among the set S_k^s of candidate solvers (or if it is not statistically different from the best solver for i in S_k^s). The second and third columns of the left part of Table 6 gives the percentage of well-classified hard instances for *Solved* and *Good*, respectively. It shows us that this percentage decreases when the number of configurations in S_k^s increases, both for *Solved* and *Good*: It decreases from 81.7% and 84.6% with 2 configurations to 65.4% with 13 configurations. However, the

Table 6. Ranking and goodness of learnt solvers. For each set S_k and each rank $j \in \{1, \ldots, k\}$, the left table displays the percentage of hard instances whose learnt configuration is the jth best among the k configurations in S_k. For each set S_k, the right table gives the percentage of hard instances for which the learnt solver is a good solver.

Ranking of the learnt solvers

	1		2		3		4		5		6		≥7	
	Solved	*Good*	*Solved*	*Good*	*Solved*	*Good*	*Solved*	*Good*	*Solved*	*Good*	*Solved*	*Good*	*Solved*	*Good*
2	81.7	84.6	18.3	15.4										
3	78.6	77.3	12.4	16.2	9.0	6.4								
4	78.6	75.7	10.6	15.8	6.3	6.8	4.5	1.8						
5	77.0	75.1	7.2	14.6	6.6	5.9	4.7	3.2	4.5	1.1				
6	73.2	71.5	6.6	13.8	6.6	7.1	3.9	5.0	5.8	2.3	4.0	0.3		
7	66.1	71.2	13.8	11.9	5.5	5.5	4.3	3.9	5.5	4.7	2.4	2.6	2.4	0.3
8	63.8	70.7	11.6	10.5	5.9	5.8	5.6	5.8	5.9	3.1	3.1	1.9	4.0	2.3
9	65.6	67.7	11.4	9.8	6.1	5.6	4.3	7.1	4.0	3.9	3.4	2.4	5.1	3.5
10	66.1	66.1	10.9	10.9	5.9	5.9	4.0	4.0	3.2	3.2	3.5	3.5	6.3	6.3
11	66.2	66.7	10.0	10.5	6.3	6.1	3.5	3.9	3.4	2.7	1.9	2.9	8.6	7.2
12	64.8	65.8	9.8	10.5	6.1	6.8	3.9	3.7	4.3	3.2	1.9	2.1	9.1	8.1
13	65.4	65.4	9.8	9.8	6.8	6.8	3.4	3.4	3.9	3.9	1.8	1.8	8.9	8.9

good solvers

	Solved	*Good*
2	36.7	54.5
3	36.5	56.4
4	37.3	61.4
5	38.4	67.2
6	41.3	66.4
7	59.8	67.8
8	58.7	69.0
9	64.8	67.2
10	65.4	65.4
11	65.8	66.6
12	64.1	65.6
13	65.4	65.4

left part of Table 6 also shows us that the learnt solvers of instances which are not well classified often correspond to solvers which perform well: Given a set S_k of solvers, and given an instance i, we rank each solver of S_k from 1 to k according to its performance on i (the solver ranked 1 being the best one for i, and the solver ranked k being the worst one). For example, let us look at the results for S_{13}: For 65.4% of the instances, the learnt solver is the best one; for 9.8% of the instances, it is the second best one; for 6.8% it is the third best one; ...; and finally, for 8.9% of the instances it is the seventh best one, or it is worse than the seventh best one.

The fact that the learnt solver is well-classified for an instance i does not necessarily imply that it is good for i (except when $k = 13$): This depends on whether S_k contains a good solver for i or not. The right part of Table 6 displays the percentage of hard instances for which the learnt solver is good (i.e., it is the best among the 13 solvers, or it is not statistically different from the best on this instance). For the *Solved* strategy, this percentage increases from 36.7% with S_2^{Solved} to 65.8% with S_{11}^{Solved}, whereas for the *Good* strategy it increases from 54.5% with S_2^{Good} to 69% with S_8^{Good}.

Comparison of success rates. Table 7 displays the percentage of instances solved at different time limits for the best solver, (CBT,MAC,w), and for the per-instance algorithm selector with different portfolios S_k^s with $k \in [2; 13]$ and $s \in \{Solved, Good\}$, on average over 15 runs. We have used the Student's t-test with $p = 0.01$ to decide whether the 15 success rates at a given time t and a given size k are significantly different for the two strategies and we highlight in blue the best strategy when the test is positive. At one second, all variants of the selector have the same success rate as (CBT,MAC,w) because the selector runs (CBT,MAC,w) during one second before starting the selection process. However, after 5 seconds, all variants of the selector have better success rates

Table 7. Each line displays the size k of the portfolio, followed by success rates of per-instance solver selectors built upon S_k^{Solved} and S_k^{Good} at different time limits (for 15 runs on the 1092 instances). For each time limit and each size k, we highlight in blue the cell with the best result if it is significantly better. For each time limit and each strategy $s \in \{Solved, Good\}$, we highlight in bold the highest success rate whatever the size k. The last line of the table recalls the success rates of (CBT,MAC,w).

Success rates of per instance solver selectors (average on 15 runs):

	1		5		10		50		100		500		1000		1800	
k	*Solved*	*Good*	*Solved*	*Good*	*Solved*	*Good*	*Solved*	*Good*	*Solved*	*Good*	*Solved*	*Good*	*Solved*	*Good*	*Solved*	*Good*
2	47.1	47.1	64.2	66.7	72.1	73.9	84.7	86.5	88.8	**90.2**	94.4	95.1	**96.0**	96.3	**96.6**	96.8
3	47.1	47.1	64.5	66.5	71.9	73.8	84.4	86.0	88.6	90.1	94.3	95.3	95.6	96.2	96.1	96.6
4	47.1	47.1	64.8	67.8	72.1	74.6	84.7	86.3	88.7	89.9	94.4	95.4	95.7	96.3	96.2	96.8
5	47.1	47.1	64.9	68.4	72.0	**75.1**	84.5	86.3	88.6	89.9	94.4	**95.6**	95.8	**96.5**	96.3	**96.9**
6	47.1	47.1	64.3	**68.7**	71.5	75.0	83.5	**86.6**	87.7	89.7	93.2	95.1	94.6	95.9	95.2	96.4
7	47.1	47.1	66.7	68.2	73.1	74.5	85.0	86.1	88.4	89.4	94.0	94.8	95.0	95.9	95.7	96.5
8	47.1	47.1	66.0	68.2	72.7	74.7	84.8	86.2	88.0	89.7	93.7	95.0	94.8	95.7	95.6	96.3
9	47.1	47.1	67.8	68.2	74.0	74.6	85.1	86.0	88.4	89.4	94.0	94.7	94.9	95.5	95.5	96.1
10	47.1	47.1	67.9	67.9	73.9	73.9	85.3	85.3	88.9	88.9	94.3	94.3	95.2	95.2	95.7	95.7
11	47.1	47.1	67.9	68.2	73.9	74.3	85.3	85.5	89.1	88.8	94.4	94.5	95.3	95.3	95.7	95.7
12	47.1	47.1	67.8	68.1	73.7	74.4	85.3	85.8	88.6	89.2	94.5	94.7	95.4	95.7	95.8	96.2
13	47.1	47.1	**68.5**	68.5	**74.5**	74.5	**85.8**	85.8	**89.2**	89.2	**94.6**	94.6	95.7	95.7	96.3	96.3

Success rates of (CBT,MAC,w) (average on 15 runs):

47.1	61.5	68.3	80.5	85.2	92.3	94.3	95.4

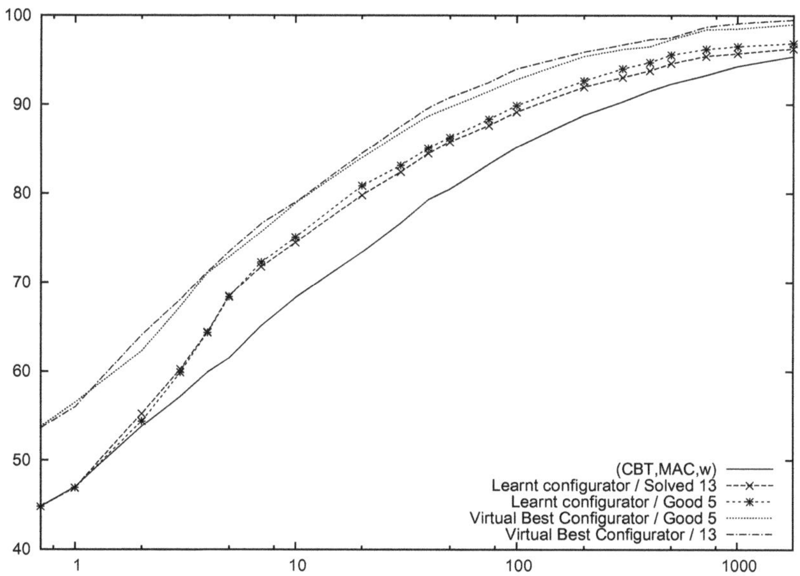

Fig. 1. Evolution of the percentage of solved instances with respect to CPU time (in seconds)

than (CBT,MAC,w). The *Good* strategy is significantly better than the *Solved* one when $k \leq 9$, at all time limits. When $k \geq 10$, the two strategies often have results which are not significantly different.

For the *Solved* strategy, the best results are obtained with the largest portfolio, S_{13}, up to 500 seconds. After that, the best results are obtained with S_2. For the *Good* strategy, the best results are often obtained with a portfolio of 5 or 6 solvers. Figure 1 plots the evolution of the percentage of solved instances with respect to CPU time for the best solver (CBT,MAC,w) and for the selector with S_5^{Good} and S_{13}^{Solved}. It also plots results of VBS(S_5^{Good}) and VBS(S_{13}^{Solved}).

6 Conclusion

We have extended the generic framework of [6] by adding three new backtracking mechanisms (CBJR, DR and BTD), thus defining a unified framework for comparing 24 different configurations corresponding to state-of-the-art approaches. As far as we know, this is the first time that approaches based on tree decomposition (BTD) are extensively compared with other search mechanisms such as CBJ, DBT, and DR, when combined with two different constraint propagation techniques (MAC and FC) and with two different variable ordering heuristics (d and w). Experiments have shown us that although BTD has lower global success rates than the best approaches, it also performs significantly better than them on many instances.

We have used a per-instance algorithm selector to choose a good configuration for each new instance to be solved. This selector is parametrized by the size k of the portfolio and we have introduced a new strategy for selecting the k solvers. This strategy is independent from the CPU time limit and aims at maximizing the number of instances for which the portfolio contains a good solver. We compare this strategy with the one used in [12], which aims at maximizing the number of instances solved within a given CPU time limit. We experimentally show that our new strategy allows the selector to solve more instances.

In this first study, we have extracted rather simple features to characterize instances and we plan to study (i) the usefulness of these different features for the classification task and (ii) the possibility of adding new features such as other dynamic features gathered when running other algorithms (e.g., greedy search or local search). We also plan to extend this work to build runtime prediction models by using linear regression techniques, as done for example in SATzilla. This kind of prediction model could then be used to schedule configurations in a portfolio approach, as done for example in CPHydra. Further work will also concern the extension of our generic solver to n-ary constraints and to impact-based or activity-based variable ordering heuristics [32]. Finally, our generic framework allows us to change dynamically the configuration during the solving process. Therefore, we plan to extend our work to dynamic configuration as proposed, for example, in [33] or [34].

Acknowledgements. Many thanks to Pierre Flener and Justin Pearson for enriching discussions on this work.

References

1. Rossi, F., van Beek, P., Walsh, T.: Handbook of Constraint Programming (Foundations of Artificial Intelligence). Elsevier Science Inc., New York (2006)
2. Prosser, P.: Hybrid algorithms for the constraint satisfaction problem. Computational Intelligence 9, 268–299 (1993)
3. Ginsberg, M.: Dynamic backtracking. Journal of Artificial Intelligence Research 1, 25–46 (1993)
4. Jussien, N., Lhomme, O.: Local search with constraint propagation and conflict-based heuristics. Artif. Intell. 139(1), 21–45 (2002)
5. Jégou, P., Terrioux, C.: Hybrid backtracking bounded by tree-decomposition of constraint networks. Artif. Intell. 146, 43–75 (2003)
6. Lecoutre, C., Boussemart, F., Hemery, F.: Backjump-based techniques versus conflict-directed heuristics. In: 16th IEEE International Conference on Tools with Artificial Intelligence, ICTAI 2004, pp. 549–557. IEEE (2004)
7. Boussemart, F., Hemery, F., Lecoutre, C., Sais, L.: Boosting systematic search by weighting constraints. ECAI 16, 146 (2004)
8. Blet, L., Ndiaye, S.N., Solnon, C.: A generic framework for solving csps integrating decomposition methods. In: CP Doctoral Program, Quebec, Canada (2012)
9. O'Mahony, E., Hebrard, E., Holland, A., Nugent, C., O'Sullivan, B.: Using case-based reasoning in an algorithm portfolio for constraint solving. In: Irish Conference on Artificial Intelligence and Cognitive Science (2008)
10. Xu, L., Hoos, H., Leyton-Brown, K.: Hydra: Automatically configuring algorithms for portfolio-based selection. In: AAAI, vol. 10, pp. 210–216 (2010)
11. Kadioglu, S., Malitsky, Y., Sellmann, M., Tierney, K.: Isac-instance-specific algorithm configuration. In: ECAI, vol. 215, pp. 751–756 (2010)
12. Amadini, R., Gabbrielli, M., Mauro, J.: An empirical evaluation of portfolios approaches for solving CSPs. In: Gomes, C., Sellmann, M. (eds.) CPAIOR 2013. LNCS, vol. 7874, pp. 316–324. Springer, Heidelberg (2013)
13. Geschwender, D.J., Karakashian, S., Woodward, R.J., Choueiry, B.Y., Scott, S.D.: Selecting the appropriate consistency algorithm for csps using machine learning classifiers. In: Twenty-Seventh AAAI Conference on Artificial Intelligence (2013)
14. Bacchus, F.: Extending forward checking. In: Dechter, R. (ed.) CP 2000. LNCS, vol. 1894, pp. 35–51. Springer, Heidelberg (2000)
15. Jussien, N., Debruyne, R., Boizumault, P.: Maintaining arc-consistency within dynamic backtracking. In: Dechter, R. (ed.) CP 2000. LNCS, vol. 1894, pp. 249–261. Springer, Heidelberg (2000)
16. Baker, A.B.: The hazards of fancy backtracking. In: AAAI, pp. 288–293 (1994)
17. Zivan, R., Shapen, U., Zazone, M., Meisels, A.: Retroactive ordering for dynamic backtracking. In: CP, pp. 766–771 (2006)
18. Pralet, C., Verfaillie, G.: Travelling in the world of local searches in the space of partial assignments. In: Régin, J.-C., Rueher, M. (eds.) CPAIOR 2004. LNCS, vol. 3011, pp. 240–255. Springer, Heidelberg (2004)
19. Kjaerulff, U.: Triangulation of graphs: Algorithms giving small total state space. Technical report, University of Aalborg (1990)
20. Mackworth, A.K.: Consistency in networks of relations. Artificial intelligence 8(1), 99–118 (1977)
21. Bessière, C., Régin, J.-C.: Refining the basic constraint propagation algorithm. In: IJCAI, vol. 1, pp. 309–315 (2001)

22. Bessiere, C., Régin, J.-C.: Mac and combined heuristics: Two reasons to forsake fc (and cbj?) on hard problems. In: Freuder, E.C. (ed.) CP 1996. LNCS, vol. 1118, pp. 61–75. Springer, Heidelberg (1996)
23. Jégou, P., Ndiaye, S., Terrioux, C.: Dynamic heuristics for backtrack search on tree-decomposition of CSPs. In: IJCAI, pp. 112–117 (2007)
24. Jégou, P., Ndiaye, S.N., Terrioux, C.: Strategies and Heuristics for Exploiting Tree-decompositions of Constraint Networks. In: Inference methods based on graphical structures of knowledge (WIGSK 2006), ECAI Workshop, pp. 13–18 (2006)
25. Morara, M., Mauro, J., Gabbrielli, M.: Solving xcsp problems by using gecode. In: 26th Italian Conference on Computational Logic (CILC). CEUR Workshop Proceedings, vol. 810, pp. 401–405. CEUR-WS.org (2011)
26. Chen, X., Beek, P.v.: Conflict-directed backjumping revisited. Journal of Artificial Intelligence Research 14, 53–81 (2001)
27. Malitsky, Y., Mehta, D., O'Sullivan, B.: Evolving instance specific algorithm configuration. In: Symposium on Combinatorial Search, SOCS (2013)
28. Battiti, R., Brunato, M.: The LION Way: Machine Learning plus Intelligent Optimization. Lionsolver Inc. (2013)
29. Holmes, G., Donkin, A., Witten, I.H.: Weka: A machine learning workbench. In: Proceedings of the 1994 Second Australian and New Zealand Conference on Intelligent Information Systems, pp. 357–361. IEEE (1994)
30. Hall, M., Frank, E., Holmes, G., Pfahringer, B., Reutemann, P., Witten, I.H.: The weka data mining software: an update. ACM SIGKDD Explorations Newsletter 11(1), 10–18 (2009)
31. Frank, E., Wang, Y., Inglis, S., Holmes, G., Witten, I.H.: Using model trees for classification. Machine Learning 32(1), 63–76 (1998)
32. Kadioglu, S., O'Mahony, E., Refalo, P., Sellmann, M.: Incorporating variance in impact-based search. In: Lee, J. (ed.) CP 2011. LNCS, vol. 6876, pp. 470–477. Springer, Heidelberg (2011)
33. Sakkout, H.E., Wallace, M.G., Richards, E.B.: An instance of adaptive constraint propagation. In: Freuder, E.C. (ed.) CP 1996. LNCS, vol. 1118, pp. 164–178. Springer, Heidelberg (1996)
34. Liberto, G.D., Kadioglu, S., Leo, K., Malitsky, Y.: Dash: Dynamic approach for switching heuristics. CoRR, abs/1307.4689 (2013)

Solving Intensional Weighted CSPs by Incremental Optimization with BDDs

Miquel Bofill*, Miquel Palahí*,**, Josep Suy*, and Mateu Villaret*

Departament d'Informàtica, Matemàtica Aplicada i Estadística
Universitat de Girona, Spain
{mbofill,mpalahi,suy,villaret}@imae.udg.edu

Abstract. We present a method for solving weighted Constraint Satisfaction Problems, based on translation into a Constraint Optimization Problem and iterative calls to an SMT solver, with successively tighter bounds of the objective function. The novelty of the method herewith described lies in representing the bound constraint as a shared Binary Decision Diagram, which in turn is translated into SAT. This offers two benefits: first, BDDs built for previous bounds can be used to build the BDDs for new (tighter) bounds, considerably reducing the BDD construction time; second, as a by-product, many clauses asserted to the solver in previous iterations can be reused.

The reported experimentation on the `WSimply` system shows that this technique has better performance in general than other methods implemented in the system. Moreover, with the new technique `WSimply` outperforms some state-of-the-art solvers in most of the studied instances.

1 Introduction

A Constraint Satisfaction Problem (CSP) is a decision problem where the goal is to determine whether an assignment of values to a set of variables exists which satisfies a given set of constraints. It is common to find CSPs where, additionally to determine if there exists a solution for the problem, the possible solution has to minimize or maximize some objective function. These kinds of CSP are known as Constraint Optimization Problems (COP).

Occasionally, some real-world CSP instances have no solution. In such situations, we can relax the CSP by allowing the violation of a subset of the constraints, and try to maximize the number of satisfied constraints. This CSP variant is known as Maximum CSP (MaxCSP) [17]. Furthermore, there can exist preferences over which constraints to violate. A convenient way of expressing these preferences is by giving a weight to each constraint, denoting its violation cost. The constraints that can be violated (the ones with a non-infinite weight) are usually called *soft*, while those constraints that must be satisfied

* Supported by the Spanish Ministry of Science and Innovation (project TIN2012-33042).
** Supported by UdG grant (BR 2010).

B. O'Sullivan (Ed.): CP 2014, LNCS 8656, pp. 207–223, 2014.

are called *hard*. Then, the objective is to find an assignment which satisfies all hard constraints and minimizes the aggregated cost of the violated soft constraints [22]. These problems are known as Weighted CSP (WCSP) [20] or, alternatively, as Cost Function Networks (CFN) [13].

`WSimply` [3,5] is a pioneering language and system for solving intensionally represented WCSPs by reformulation into Satisfiability Modulo Theories (SMT) [7], namely, into SAT modulo Linear Integer Arithmetic (LIA). An SMT formula can be seen as a generalization of propositional Boolean formula, where some predicates have predefined interpretations from background theories, and any satisfying assignment has to be compatible with those theories. Leveraging the advances made in SAT solvers in the last decade, SMT solvers have proved to be competitive with classical decision methods in many areas, and in particular in CSP solving [4,9]. Most modern SMT solvers integrate a SAT solver with decision procedures (theory solvers) for sets of literals belonging to each theory. For example, variations of the simplex method are used for dealing with LIA predicates. This way, one can hopefully get the best of both worlds: in particular, the efficiency of the SAT solver for the Boolean reasoning and the efficiency of special-purpose algorithms for the theory reasoning. As shown in [9], SMT outperforms other methods on instances with a significant Boolean component, i.e., instances with Boolean decision variables and disjunctions of (arithmetic) constraints.

`WSimply` benefits from the expressiveness of the SMT language and the performance of current SMT solvers. However, SMT solvers are decision procedures, and they rarely support optimization. A few solvers support (weighted partial) MaxSMT [15,14,12], and there is a recent attempt to introduce optimization into SMT by means of a theory of costs [11]. In `WSimply`, optimization is implemented by means of successive calls to the decision procedure in several (user choosable) ways: performing sequential or binary search, or using algorithms based on unsatisfiable cores like WPM1 [6].

In this paper we extend the WCSP solving capability of `WSimply` by introducing a new optimization method, based on representing the bound constraint on the objective function (generated from the violation cost of soft constraints) as a BDD [2]. This allows us to encode the objective as a pure propositional formula, following the generalized arc-consistent encoding proposed in [1]. This way, on the one hand we increase the Boolean component of the instances and, on the other hand, we tighten the link between optimization and the logical structure of the problem, with the hope of benefiting from crucial capabilities of the underlying solver, such as conflict driven learning [21]. Since the BDD for the objective can be really big, and so the number of clauses to represent it, an interesting aspect of our approach is the reutilization of BDDs in successive calls to the decision procedure. Although changing the bound of the objective function implies building a new BDD, some parts can be easily reused. We create a Shared BDD [19], also known in the literature as Multi-Rooted BDD, keeping all the generated BDDs. This allows not only to improve the performance of the

BDD construction algorithm, but also to keep a number of learned clauses from the solver, as we reuse the clauses representing the previous BDDs.

Since the size of BDDs strongly depends on the number of variables involved, and we use them to represent bound constraints on objective functions encoding the violation cost of soft constraints, our method is especially well suited for WCSP instances involving a *small* number of soft constraints. We provide an experimentation section where we show that the new BDD-based solving method outperforms previous `WSimply` solving methods in the majority of the problems. Moreover, we also provide comparisons with other state-of-the-art optimization and WCSP solvers, showing that our method is, in general, the most robust one.

The paper is structured as follows. In Section 2 we introduce the required background. In Section 3 we introduce Pseudo-Boolean constraints and (Reduced Ordered) Binary Decision Diagrams (ROBDD). In Section 4 we present our incremental method of solving WCSPs using shared ROBDDs. In Section 5 we study the performance of the new method and we compare it with other methods implemented in `WSimply` and other state-of-the-art systems. In the same section, we report on some experiments showing how the use of BDDs to represent the objective, instead of using a linear equation, has a positive impact in the learning of the SMT solver. Finally, in Section 6 we conclude and propose some future work.

2 Preliminaries

In this section we recall some basic definitions related to our research. First, we introduce the kind of WCSPs we are interested to solve. Second, we briefly explain what is SMT. Finally, we briefly review how the `WSimply` system faces the optimization process by iterative calls to an SMT solver.

2.1 WCSPs and COPs

A *Constraint Satisfaction Problem (CSP)* instance is defined as a triple $\langle X, D, C\rangle$, where $X = \{x_1, \ldots, x_n\}$ is a set of variables, $D = \{d(x_1), \ldots, d(x_n)\}$ is a set of domains containing the values the variables may take, and $C = \{C_1, \ldots, C_m\}$ is a set of constraints defining relations between subsets of variables. In this paper we assume to deal with intensional Weighted CSPs, where weighted constraints have the form $(c, w(c))$, being c a constraint as defined for a CSP and $w(c)$ the cost corresponding to its falsification, which can be a natural number or infinity. We call those constraints whose associated cost is infinity *hard*, if otherwise *soft*. A solution to a WCSP is an assignment that satisfies all hard constraints and minimizes the sum of falsified soft-constraints costs.

A *Constraint Optimization Problem (COP)* instance consists of an optimization variable O, matched to an objective function to be minimized (maximized) subject to the constraints of a CSP instance $\langle X, D, C\rangle$, where $O \in X$. A solution to a COP instance is a solution to the CSP instance that minimizes (maximizes) the value of the optimization variable O. A WCSP can be seen as a COP where the objective function to minimize is $\sum_{i=1}^{m} w_i * o_i$, being w_i the cost of the soft-constraint i of the WCSP and o_i a pseudo-Boolean variable representing if the soft-constraint i is violated.

2.2 SMT and Weighted SMT

A *Satisfiability Modulo Theories (SMT)* formula is a generalization of a Boolean formula in which some propositional variables have been replaced by predicates with predefined interpretations from background theories such as, e.g., linear integer arithmetic. For example, a formula can contain clauses like $p \vee q \vee (x + 2 \leq y) \vee (x > y + z)$, where p and q are Boolean variables and x, y and z are integer variables. A *solution* to an SMT instance is an assignment that satisfies the formula, provided that predicates over non-Boolean variables, such as linear integer inequalities, are evaluated according to the rules of a background theory [7]. As in the CSP case, we can extend SMT to *weighted SMT (WSMT)* as follows.

A *weighted SMT clause* is a pair (C, w), where C is an SMT clause[1] and w is a natural number or infinity (indicating the penalty for violating C). A *weighted SMT formula* is a multiset of weighted SMT clauses

$$\{(C_1, w_1), \ldots, (C_m, w_m), (C_{m+1}, \infty), \ldots, (C_{m+m'}, \infty)\}$$

where the first m clauses are soft and the last m' clauses are hard. The optimal cost of a weighted SMT formula is the minimal cost of all its assignments. An optimal assignment is an assignment with optimal cost. The *WSMT problem*[2] for a WSMT formula is the problem of finding an optimal assignment for that formula.

2.3 Solving WCSP with (Weighted) SMT

`WSimply` [5] is a tool with its own language for WCSP and COP modelling that solves the corresponding instances translating them into (weighted) SMT. First of all, the hard-constraints of the instance are translated into an SMT formula. Second, depending on the user-specified way to solve the instance, the soft-constraints are reformulated either into an objective function or into weighted SMT formulas.[3] Finally, the generated (weighted) SMT formula is solved using one of three solving methods: `yices`, `core` or `dico`, where the two former are used to solve WCSP instances, while the latter is used to solve COP instances. In order to give the reader some idea of these methods we add a short description of them:

[1] In fact these can be general SMT formulas, not necessarily disjunctions of literals.

[2] In the literature the weighted SMT problem is also referred to as weighted MaxSMT, same as in the SAT formalism. We prefer to talk about WSMT because it is closer to WCSP.

[3] Note that in case it is wanted to solve a COP instance using WSMT, the objective function can be easily translated into WSMT clauses.

- The `yices` method uses an algorithm that performs a sequence of satisfiability checks until the optimum is found. It is the default Yices [15] algorithm for solving WSMT (`WSimply` is built on top of Yices). This algorithm is not exact since Yices defines a maximum number of iterations for the search.[4] We are not aware of any document describing the procedures used there.
- The `core` method is an implementation, introduced in [5], of the core based WPM1 algorithm [6] for MaxSAT.
- In the `dico` method, the system first translates the constraints of the COP into SMT formulae, and then iteratively calls the SMT solver, bounding the optimization variable O by adding the unit clause $O \leq K$, where K is an integer constant determined by the system using binary search.
  It is worth noting that in the case we are bounding the objective function of a COP encoding a WCSP instance, this results in a pseudo-Boolean constraint (see Subsection 2.1).

For deeper details about `WSimply` and its solving techniques we refer the reader to [5].

3 Binary Decision Diagrams

A typical data structure to represent Boolean functions is a *Binary Decision Diagram (BDD)*, which consists of a rooted, directed, acyclic graph, where each non-terminal (decision) node corresponds to a Boolean variable x and has two child nodes with edges representing a *true* and a *false* assignment to x, respectively. We talk about the *true child* (resp. *false child*) to refer to the child node linked by the *true* (resp. *false*) edge. Terminal nodes are called 0-terminal and 1-terminal, representing the truth value of the formula for the assignment leading to them. A BDD is called *ordered* if different variables appear in the same order on all paths from the root. A BDD is said to be *reduced* if the following two rules have been applied to its graph until fixpoint:

- Merge any isomorphic subgraphs.
- Eliminate any node whose two children are isomorphic.

A *Reduced Ordered Binary Decision Diagram (ROBDD)* is canonical (unique) for a particular function and variable order. Figure 1 shows an example of a ROBDD.

An interesting property of ordered BDDs is that when multiple ordered BDDs contain isomorphic subgraphs, they can be joined in a single *shared BDD (SBDD)* [19], providing a more compact representation of the Boolean functions.

3.1 SAT Encodings of Pseudo-Boolean Constraints Using BDDs

Pseudo-Boolean (PB) constraints [10] are constraints of the form $a_1x_1 + \cdots + a_nx_n \;\#\; K$, where the a_i and K are integer coefficients, the x_i are pseudo-Boolean (0/1) variables, and the relation operator $\#$ belongs to $\{<, >, \leq, \geq, =\}$.

[4] `http://yices.csl.sri.com/language.shtml`

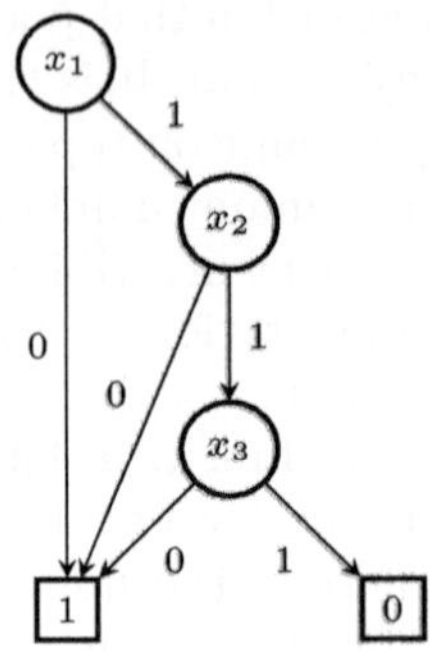

Fig. 1. ROBDD for the Boolean function $\overline{x_1} + x_1 \cdot \overline{x_2} + x_1 \cdot x_2 \cdot \overline{x_3}$

For our purposes we will assume that $\#$ is $\leq$ and that the a_i and K are positive. Under these assumptions, these constraints are monotonic (decreasing) Boolean functions $C : \{0,1\}^n \rightarrow \{0,1\}$, i.e., any solution for C remains a solution after flipping input values from 1 to 0.

It is quite common to use BDDs to represent PB-constraints. For example, the ROBDD of Figure 1 also corresponds to the PB constraint $2x_1 + 3x_2 + 4x_3 \leq 7$.

An interval of a PB constraint C is the set of values of K for which C has identical solutions. More formally, given a constraint C of the form $a_1x_1 + \cdots + a_nx_n \leq K$, the *interval of* C is the set of all integers M such that the constraint $a_1x_1 + \cdots + a_nx_n \leq M$, seen as a Boolean function, is equivalent to C (i.e., that the corresponding Boolean functions have the same truth table). For instance, the interval of $2x_1 + 3x_2 + 4x_3 \leq 7$ is $[7, 8]$ since, as no combination of coefficients adds to 8, we have that the constraint $2x_1 + 3x_2 + 4x_3 \leq 7$ is equivalent to $2x_1 + 3x_2 + 4x_3 \leq 8$.

There exist several BDD-based approaches for reformulating PB constraints into propositional clauses [16]. We focus on the recent work of [1], that proposes a simple and efficient algorithm to construct ROBDDs for monotonic Boolean functions, and a corresponding generalized arc-consistent SAT encoding. The algorithm proposed in [1] is a dynamic, bottom up BDD construction algorithm, which runs in polynomial time with respect to the ROBDD size and the number of variables. The key point is that it keeps the intervals of the PB constraints built for the already visited nodes: for a given variable ordering, say $x_1, x_2, \ldots, x_n$, a list of *layers* $\mathcal{L} = L_1, \ldots, L_{n+1}$ is maintained, where each layer L_i is a set of pairs of the form $([\beta, \gamma], \mathcal{B})$, being $\mathcal{B}$ the ROBDD of the constraint $a_ix_i + \cdots + a_nx_n \leq K$, for every K in the interval $[\beta, \gamma]$.

These intervals are used to detect if some needed ROBDD has already been constructed. That is, if for some node at level i, the ROBDD for the constraint $a_ix_i + \cdots + a_nx_n \leq K$ is needed for a given K, and K belongs to some interval already computed in layer L_i, then the same ROBDD can be used for this node. Otherwise, a new ROBDD is constructed and a new pair (the ROBDD and its respective interval) is added to the layer. It is important to recall here that ROBDDs are unique for a given function and variable ordering.

The encoding of the ROBDD to SAT that we borrow from [1] is generalized arc-consistent, and works as follows. For each node with a selector variable x we create, a new auxiliary variable n, which represents the state[5] of the node, and two clauses:

$$\bar{f} \rightarrow \bar{n} \qquad \bar{t} \wedge x \rightarrow \bar{n}$$

being f the state variable of its false child and t the state variable of its true child. Finally, we add a unit clause with the state variable of the 1-terminal node, another clause with the negation of the variable of the 0-terminal node. To make the encoding generalized arc-consistent, a unit clause forcing the state variable of the root node to be *true* must be added.

We refer the reader to [1] for additional details on the BDD construction algorithm and the SAT encoding.

4 Solving WCSPs by Incremental Optimization Using Shared ROBDDs

The WCSP solving method presented here consists in reformulating the WCSP into a COP, and solving the resulting optimization problem by iteratively calling an SMT solver with the problem instance, together with successively tighter bounds for the objective function.

The novelty of the method lies in the way the objective function is treated. Inspired by the idea of intervals in the BDD construction algorithm of [1], our aim is to represent the pseudo-Boolean objective function resulting from the WCSP (see Subsection 2.1) as a BDD, and take profit of BDD reuse in successive iterations of the optimization process. That is, instead of creating a (reduced ordered) BDD from scratch at every iteration, we build a shared BDD. Since the PB constraint encoded at each iteration is almost the same, with only the bound constant K changing, this will hopefully lead to high node reuse.

We claim that using shared BDDs has two important benefits. The first one, fairly evident, is that node reuse considerably reduces the BDD construction time. The second one, which is not so evident, is that, as a by-product, we will be reusing many literals and clauses resulting from the SAT encoding of BDDs from previous iterations (in addition to clauses learned at previous steps). In Section 5 we present some experiments to support these claims.

4.1 Incremental Optimization Algorithm

Algorithm 1 describes our WCSP solving method. The input of the algorithm is a WCSP instance divided into a set φ_s of soft constraints and a set φ_h of hard constraints, and its output is the optimal cost of $\varphi_s \cup \varphi_h$ if φ_h is satisfiable, and UNSAT otherwise.

[5] That is, if the PB constraint corresponding to (the interval of) the node is satisfied or not.

Algorithm 1 Solving a WCSP by incremental optimization using a shared ROBDD

Input: $\varphi_s = \{(C_1, w_1), \ldots, (C_m, w_m)\}$, $\varphi_h = \{C_{m+1}, \ldots, C_{m+m'})\}$
Output: Optimal cost of $\varphi_s \cup \varphi_h$ or *UNSAT*

```
φ ← φh ∪ reif_soft(φs)
(st, M) ← SMT_algorithm(φ)
if st = UNSAT then
    return UNSAT
else
    ub ← sum({wi | (Ci, wi) ∈ φs and Ci is falsified in M})
end if
lb ← −1
ℒ ← init_layers(φs)
while ub > lb + 1 do
    K ← ⌊(ub + lb)/2⌋
    ([β, γ], ℬ) ← ROBDD(φs, K, ℒ)
    (root, φ) ← BDD2SAT(ℬ, φ)
    φ ← φ ∪ {root}
    (st, M) ← SMT_algorithm(φ)
    if st = UNSAT then
        lb ← γ
        φ ← (φ \ {root}) ∪ {¬root}
    else
        ub ← min(β, sum({wi | (Ci, wi) ∈ φs and Ci is falsified in M}))
    end if
end while
return ub
```

The first step is to reify the soft constraints (**reif_soft**), and create a formula φ with the reified soft constraints together with the hard constraints. By reifying a soft constraint C_i we mean adding a hard constraint $x_i \leftrightarrow \overline{C_i}$, where x_i is a new pseudo-Boolean variable. Note that in fact we are reifying the negation of C_i, since the cost w_i is associated to its violation (see Subsection 2.1).

Then, the satisfiability of φ is checked by an SMT solver. **SMT_algorithm** returns a tuple with the satisfiability (st) of φ and, if satisfiable, an assignment (model) M of φ. Note that if φ is unsatisfiable, this implies that φ_h is unsatisfiable, and hence we return UNSAT. Otherwise, we use the solution found to compute an upper bound ub of the objective function by aggregating the weights of the violated soft constraints, and set the lower bound lb to -1.

Before starting a binary search procedure to determine the optimal cost, in $\mathcal{L} \leftarrow$ **init_layers**(φ_s) we initialize each layer i of $\mathcal{L}$ (for i in $1..m+1$) with the two pairs of intervals and (trivial) ROBDDs $\{((-\infty, -1], \mathbf{0}), ([\sum_{j=i}^{m} w_j, \infty), \mathbf{1})\}$, meaning that the PB constraint $w_i x_i + \cdots + w_m x_m \leq K$ is trivially false for $K \in (-\infty, -1]$ and trivially true for $K \in [\sum_{j=i}^{m} w_j, \infty)$, where $x_i, \ldots, x_m$ denote the reification variables of the soft constraints $C_i, \ldots, C_m$ as described above (see Subsection 3.1 for the definition of layer and interval).

In the first step of the `while` statement, we determine a new tentative bound K for the objective function. Then, we call the **ROBDD** construction algorithm of [1] (briefly described in Subsection 3.1) with the set of soft clauses φ_s, the new bound K and the list of layers $\mathcal{L}$, being this last an input/output parameter. This way, $\mathcal{L}$ will contain the shared ROBDD with all the computed ROBDDs, and may be used in the following iterations of the search, significantly reducing the construction time and avoiding the addition of repeated clauses. This procedure returns the ROBDD $\mathcal{B}$ representing the objective function for the specific K in the current iteration.

In the next step we call the **BDD2SAT** procedure, which generates the propositional clauses from $\mathcal{B}$, as explained in Subsection 3.1, but only for the new nodes. The procedure inserts these clauses into the formula φ, and the new formula is returned together with the auxiliary variable *root* associated to the root node of $\mathcal{B}$. This variable is inserted into φ as a unit clause to effectively force the objective function to be less or equal than K.

At this point we call the SMT solver to check the satisfiability of the new φ. We remark that, since we are using the SMT solver through its API, we only need to feed it with the new clauses. If φ is satisfiable we can keep all the learned clauses. Otherwise, we need to remove the unit clause for the root node. This way, we will (only) remove the learned clauses related to this unit clause. In addition, we add a unit clause with the negation of the root node variable, stating that the objective function value must be greater than K.

Finally, we update either the lower or upper bound according to the interval $[\beta, \gamma]$ of the ROBDD $\mathcal{B}$ and the computed assignment M for φ: if φ is unsatisfiable, then the lower bound is set to γ; otherwise, the upper bound is set to $min(\beta, sum(\{w_i \mid (C_i, w_i) \in \varphi_s \text{ and } C_i \text{ is falsified in } M\}))$. From the invariant that the lower bound always corresponds to an unsatisfiable case, while the upper bound corresponds to a satisfiable case, when $ub = lb + 1$ this value corresponds to the optimum.

Note that, thanks to the intervals, in fact we are checking the satisfiability of the PB constraints for several values at the same time and hence, sometimes, this can allow us to obtain better lower/upper bound updates.

Example 1. Figure 2 and Figure 3 show, respectively, the evolution of the shared ROBDD and the propositional clauses added to the SMT formula for the objective $2x_1 + 3x_2 + 4x_3 \leq K$, with successive values $K = 7$, $K = 3$, $K = 5$ and $K = 4$.

5 Benchmarking

In this section we first compare the performance of the presented solving method with that of other methods implemented in `WSimply`. Second, we compare the performance of `WSimply`, using this new method, with that of several state-of-the-art CSP, WCSP and ILP solvers. Third, we study the benefits, such as learning, obtained from using a shared BDD for the objective instead of a linear

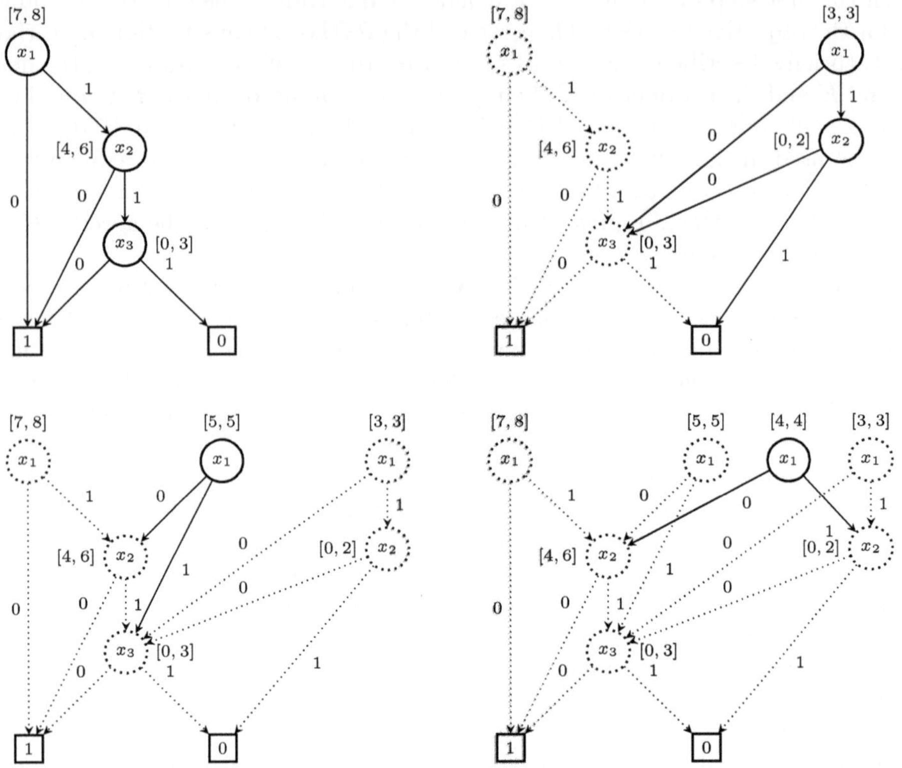

Fig. 2. Shared ROBDDs and intervals for objective $2x_1+3x_2+4x_3 \leq K$, with successive values $K = 7$ (top left), $K = 3$ (top right), $K = 5$ (bottom left) and $K = 4$ (bottom right), illustrating the reuse of previous ROBDDs

equation, and the amount of BDD node reutilization between successive iterations in the optimization process.

For the comparison we use (variants of) six problems:[6] five variants of the Soft BACP (SBACP) [5] (a softened version of the well-known Balanced Academic Curriculum Problem), three variants of the Still Life Problem and one variant of the Talent Scheduling Problem from the MiniZinc distribution[7] (all of them reformulated as a WCSP) and three classical WCSPs: CELAR, SPOT5 and Combinatorial Auctions.

Since the size of the generated BDDs strongly depends on the number of variables and the number of distinct coefficients in the objective function, we briefly describe these features for the studied problems in Table 1.

The experiments were run on a cluster of Intel® Xeon™CPU@3.1GHz machines, with 8GB of RAM, under 64-bit CentOS release 6.3 (kernel 2.6.32).

---

[6] All instances available in `http://ima.udg.edu/Recerca/lap/simply`

[7] `http://www.minizinc.org`

$$K = 7: \ \varphi_1 = \varphi \cup \{\overline{t} \to \overline{n_{1,1}}, \overline{n_{2,1}} \wedge x_1 \to \overline{n_{1,1}}, \overline{t} \to \overline{n_{2,1}}, \overline{n_{3,1}} \wedge x_2 \to \overline{n_{2,1}}, \overline{t} \to \overline{n_{3,1}},$$
$$\overline{f} \wedge x_3 \to \overline{n_{3,1}}, t, \overline{f}\} \cup \{n_{1,1}\}$$
$$K = 3: \ \varphi_2 = \varphi_1 \cup \{\overline{n_{3,1}} \to \overline{n_{1,2}}, \overline{n_{2,2}} \wedge x_1 \to \overline{n_{1,2}}, \overline{n_{3,1}} \to \overline{n_{2,2}}, \overline{f} \wedge x_2 \to \overline{n_{2,2}}\} \cup \{n_{1,2}\}$$
$$K = 5: \ \varphi_3 = (\varphi_2 \setminus \{n_{1,2}\}) \cup \{\overline{n_{1,2}}\} \cup \{\overline{n_{2,1}} \to \overline{n_{1,3}}, \overline{n_{3,1}} \wedge x_1 \to \overline{n_{1,3}}\} \cup \{n_{1,3}\}$$
(supposing that φ_2 has been found to be unsatisfiable in iteration 2)
$$K = 4: \ \varphi_4 = \varphi_3 \cup \{\overline{n_{2,1}} \to \overline{n_{1,4}}, \overline{n_{2,2}} \wedge x_1 \to \overline{n_{1,4}}\} \cup \{n_{1,4}\}$$

Fig. 3. Formula update (clauses added and removed), for objective $2x_1+3x_2+4x_3 \leq K$ with successive values $K = 7$, $K = 3$, $K = 5$ and $K = 4$, according to the shared ROBDDs of Figure 2. φ_j denotes the SMT formula at hand when checking satisfiability in iteration number j of Algorithm 1. Each atom $n_{i,j}$ is the state variable of the node with selector variable x_i, that is added in iteration j, and t and f are the variables of the 1-terminal and 0-terminal nodes, respectively.

Table 1. Average number of variables, and weights, of the soft constraints in each pack of instances. Weights are denoted as $(min - max)$ for ranges and as $(val_1, val_2, \ldots)$ for packs with a small number of distinct values in weights.

Problem	#Variables	Weights	Problem	#Variables	Weights
sbacp	67	1	*talent*	80	(4,5,10,20,40)
sbacp_h1	67	1			
sbacp_h2	67	1	*auction*	138	~(100-1000)
sbacp_h2_ml2	312	(1,246)			
sbacp_h2_ml3	332	(1,21,5166)	*spot5*	385	(1-5,1000,2000)
s.l.	30	1			(1,10,100,1000)
s.l. free	30	1	*celar*	1482	(1,100,10000,100000)
s.l. no border	30	1			(1,2,3,4)

WSimply was run on top of the Yices 1.0.33 [15] SMT solver. It is worth noting that by calling Yices through its API, we are able to keep learned clauses from previous calls that are still valid.

5.1 WSimply Solving Methods Comparison

Table 2 shows the aggregated time (sum of the times to produce an optimal solution and prove its optimality, for all instances) for the solved instances. We consider the yices, core and dico solving methods described in Subsection 2.1, plus the new method sbdd$\leq$ using shared ROBDDs, where variables in the BDD are ordered from small (root) to big (leaves) coefficients. We also studied the performance of sbdd$\geq$, where variables are ordered from big (root) to small (leaves) coefficients, but the performance of sbdd$\leq$ was slightly better.

The new solving method with shared ROBDDs is clearly the best in the majority of problems. The performance of the sbdd$\leq$ method is better on instances with a small number of distinct coefficients (especially when all coefficients are 1), namely, in the *sbacp* and *still life* problems. In these cases, the BDDs are

Table 2. Aggregated time in seconds for the solved instances in each set, with a cutoff of 600s per instance. The column # indicates the number of instances per set. The indication (n^1) refers to the number of instances for which the solver ran out of time, (n^2) refers to the number of instances for which the solver returned *unknown*, and (n^3) refers to the number of instances for which the solver ran out of memory.

Problem	#	dico	yices	core	sbdd≤
sbacp	28	70.53	40.17	1342.16 (13^1)	**9.83**
sbacp_h1	28	**6.68**	12.02	363.78 (12^1)	6.93
sbacp_h2	28	26.04	26.65	748.81 (13^1)	**9.97**
sbacp_h2_ml2	28	280.68	109.42	75.82 (19^1)	**78.36**
sbacp_h2_ml3	28	804.74	516.51	547.55 (23^1)	**92.37**
s.l.	10	118.05 (1^1)	102.97 (1^2)	3.35 (4^1)	**191.56**
s.l. free	10	434.05 (2^1)	2.69 (3^2)	386.91 (3^1)	**98.75**
s.l. no border	10	187.94 (1^1)	166.91 (1^2)	81.24 (3^1)	**105.30**
talent	11	289.90	60.72	94.82 (8^1)	**38.25**
auction	170	9084,03 (93^1)	**1083.44** $(\mathbf{85^2})$	0 $(80^1\ 90^3)$	12028.50 $(78^1\ 11^3)$
celar	16	582,64 (14^1)	750.32 $(10^1\ 2^2)$	94.82 $(8^1\ 6^3)$	**1417.19** $(\mathbf{6^1})$
spot5	20	205,2 (18^1)	9.32 $(5^1\ 13^2)$	94.82 $(14^1\ 3^3)$	**1990.59** $(\mathbf{7^1})$

relatively small, having the bigger ones only thousands of nodes. On the other side, the timeouts occurring in the *celar* and *spot5* problems are mostly in the instances with higher number of variables in the objective function (e.g., from approximately 1200 to 5300 variables in *celar*), generating BDDs of hundreds of thousands nodes (a similar situation occurs in the *auction* instances running out of time, where there are not so many variables but a lot of distinct variable coefficients). In spite of this, `sbdd≤` is still the best method on the *celar* and *spot5* problems.

It is worth to notice that there are some *unknowns* in the `yices` method, due to the fact that the Yices MaxSMT solver is non exact and incomplete. The `core` method has really bad performance on the crafted instances considered in this experiment, probably due to the bad quality of the unsatisfiable cores found during the solving process.

5.2 SBDD-Based versus State-of-the-Art CSP and WCSP Solvers

For the sake of completeness we tested the performance of the MATHSAT5-MAX and LL_{WPM} MaxSMT solvers of [12] on the weighted SMT instances generated by `WSimply` in the previous experiment. However we do not report these results, since they are comparable to those of the `WSimply core` method.

The second part of our experiments consists in reformulating the WCSPs into (MiniZinc) COPs, and compare the performance of `WSimply` using the `sbdd≤` method with that of IBM ILOG CPLEX 12.6 and some state-of-the-art CSP and WCSP solvers, namely Toulbar2 0.9.5.0, G12-CPX 1.6.0 and Opturion 1.0.2, all of them capable of solving MiniZinc instances. We used Numberjack [18] as a platform for solving MiniZinc COPs through the Toulbar2 API. We also used the Toulbar2 binary to solve the original WCSP instances of *auction*, *celar* and

Table 3. Aggregated time in seconds for the solved instances in each set, with a cutoff of 600s per instance. The column # indicates the number of instances per set. The indication (n^1) refers to the number of instances for which the solver ran out of time, (n^3) refers to the number of instances for which the solver ran out of memory, and (n^4) refers to the number of instances for which the solver returned another type of error.

Problem	#	TB2	G12-CPX	Opturion	`sbdd`≤	CPLEX
sbacp	28	662.58 (24^1)	76.31	83.47	**9.83**	36.81
sbacp_h1	28	707.78	**4.00**	13.25	6.93	21.78
sbacp_h2	28	483.98	19.53	30.51	**9.97**	24.89
sbacp h2_ml2	28	3849.38 (6^1)	166.69	91.55	78.36	**64.98**
sbacp h2_ml3	28	2408.66 (12^1)	336.25	99.77	92.37	**72.95**
s.l.	10	166.08 (2^1)	232.1 (1^1)	43.81 (2^1)	**191.56**	234.71
s.l. free	10	69.85 (3^1)	267.13 (2^1)	394.28 (2^1)	98.75	**29.91**
s.l. no border	10	122.05 (2^1)	354.41 (1^1)	102.29 (2^1)	105.3	**68.37**
talent	11	2.38 (9^4)	1.81 (8^1)	2.07 (8^1)	**38.25**	1269.13 (2^1)
auction	170	888.99	7288.73 (95^1)	7053.43 (100^1)	12028.5 (78^1 11^3)	**220.12**
auction (wcsp)		5038.49 (3^1)				
celar	16	0 (16^1)	0 (16^4)	0 (16^1)	1417.19 (6^1)	0 (16^1)
celar (wcsp)		**311.23 (4^1)**				
spot5	20	114.66 (8^1 8^4)	0 (20^1)	0 (20^1)	**1990.59 (7^1)**	297.59 (7^1 10^4)
spot5 (wcsp)		76.28 (16^1)				

spot5, indicated in Table 3 as *auction (wcsp)*, *celar (wcsp)* and *spot5 (wcsp)*, respectively. In order to test the performance of CPLEX on the considered problems, we used `WSimply` to translate the instances to pseudo-Boolean constraints as in [8]. For the experiments with Opturion we used a slightly different computer (Intel® Core™CPU@2.8GHz, with 12GB of RAM, under 64-bit Ubuntu 12.04.3, kernel 3.2.0) due to some library dependence problems.

Table 3 shows the results of this second experiment. We can observe that, in general, `sbdd`≤ is the most robust method, considering the number of instances solved and their aggregated solving time. The *sbacp* and *still life* problems seem to be reasonably well suited for SMT solvers (in particular, the latter consists of disjunctions of arithmetic constraints). We highlight the *talent scheduling* problem, where `sbdd`≤ clearly outperforms the other solvers, being the only solver capable to solve all the instances. In fact, this is probably the best suited problem for SMT solvers, as it consists of binary clauses of *difference logic* constraints. Unfortunately, in most of the instances of this problem, Toulbar2 reported an error saying that the model decomposition was too big. In the *auction* problem, CPLEX is by far the best solver, solving all 170 instances in only 220.12 seconds. This is clearly one of the worst kind of problems for SMT solvers, as it simply consists of a conjunction of arithmetic (0/1) constraints, i.e., it has a trivial Boolean structure and all deductions need to be performed by the theory solver. For *celar*, only `sbdd`≤ and Toulbar2 (in the WCSP version) were able to solve some instances (10 and 12 out of 16, respectively). This problem also has a balanced combination of arithmetic constraints and clauses. G12-CPX reported an error (not finding the `int_plus` constraint), and Toulbar2 (in the COP version), Opturion and CPLEX ran out of time on all instances. Finally, for *spot5*, sbdd$\leq$ was able to solve some instances (13 out of 20), Toulbar2 only 4, CPLEX only 3, and Opturion and G12-CPX ran out of time on all instances. This problem is also well suited for SMT solvers since it basically consists of clauses with equality constraint literals.

We remark that we also tested the G12-Lazy, G12-FD and Gecode 4.2.1 solvers, but since they presented slightly worst performance than G12-CPX we have not included them in Table 3.

5.3 SBDD Incrementality

In this section we study the benefits of using a shared BDD for the objective instead of a linear equation, in particular, the effect that this has on the learning capability of the SMT solver. Note that with this technique we are increasing the Boolean component of the formula at hand and, hence, we should expect some improvement in learning. Finally, we quantify the amount of BDD node reutilization through iterations.

We compare the performance of four solving approaches: using either SBDDs or arithmetic expressions to bound the objective function, with and without using the learning capabilities of the solver. From now on we will denote by SBDD+L, SBDD-L, LIA+L and LIA-L these four possibilities, where LIA stands for linear integer arithmetic expressions and +L/-L indicates the use or not of learning. Note that `WSimply` uses Yices as a back-end solver, and it does not provide statistics about learned clauses. Therefore, we used the Yices commands *push* and *pop* to define backtrack points and to backtrack to that points, respectively, in order to force the solver to forget learned clauses.

Table 4 summarizes the solving times for the four options. They do not include neither BDD construction, clause generation nor assertion times. Since solving times are very sensitive to the bounds on the objective function, first of all we solved all the instances with the SBDD+L approach and stored the bounds at each iteration. Then these bounds were passed to the other three methods to obtain their solving times under the similar conditions.

If we first compare SBDD-L and LIA-L we can appreciate that using BDDs considerably reduces the number of timeouts in *still life*, *auction*, *celar* and *spot5*, and the solving time, almost 5 times in *sbacp* and 10 times in *talent scheduling*.

Furthermore, comparing SBDD+L and SBDD-L, we can easily appreciate that learning reduces even more the solving times. In *sbacp* the number of timeouts is reduced to 0 and the sum of solving times is reduced almost 4 times; in *talent scheduling* and *still life* the solving time is reduced almost to the half; and in *celar* and *spot5* the number of timeouts is reduced in 2 and 3 instances respectively. In *auction*, the SBDD+L method has the drawback of increasing the number of memory outs.

Table 4. Aggregated time in seconds for the solved instances of each problem, with a cutoff of 600s per instance, and indication of the number of unsolved instances due to time out (TO) and memory out (MO)

		SBDD+L		**SBDD-L**		**LIA+L**		**LIA-L**	
Problem	# Inst.	Solving	# TO/MO	Solving	# TO/MO	Solving	# TO	Solving	# TO
sbacp	140	58.03	0	221.30	1	853.56	0	924.83	2
s.l.	30	393.22	0	715.17	0	795.62	4	895.28	4
talent	11	35.36	0	55.57	0	363.43	0	472.09	0
auction	170	9631.19	73/12	5673.22	75/7	8540.76	92	9919.79	90
celar	16	1199.99	6	1299.99	8	10.94	15	8.29	15
spot5	20	1675.90	7	943.96	10	165.48	18	59.99	18

If we compare SBDD-L and LIA-L with respect to SBDD+L and LIA+L, we can see that the improvement in terms of number of timeouts and solving time is higher in the SBDD approach than in the LIA approach.

Finally, to quantify the contribution of the shared BDD to the amount of BDD node reutilization, we computed the percentage of reused nodes at each iteration of the optimization process. Our experiments shown an average 50% of node reuse when the solving process was about the 40.73% of the search, and a 80% of node reuse when it was about the 48.72% of the search. This is especially relevant because the BDDs attained hundreds of thousands of nodes.

6 Conclusions and Future Work

We have presented a new WCSP solving method, implemented in the `WSimply` system, based on using shared ROBDDs to generate propositional clauses representing the objective. We think that it opens a promising research line, taking into account that the presented method clearly outperforms not only the previously implemented solving methods in `WSimply`, but also some state-of-the art solvers, on several problems. We have also shown how to boost the generation of ROBDDs for objective functions using previously generated ROBDDs, more precisely constructing a shared ROBDD.

As future work we want to study more deeply the efficiency of our method on other weighted CSPs. Also, as a well-known challenge, a crucial aspect to study is how to find a good variable ordering for the objective function. Although the problem of finding the optimal variable ordering in order to generate a minimal BDD is known to be NP-hard, we are interested in finding a variable ordering that maximizes the node reuse through iterations. Another aspect that could be interesting to explore is to extend the new method to deal with objective functions with (finite domain) integer variables, using Multi-valued Decision Diagrams (MDDs) to represent them. Finally, we also want to test if making visible the intermediate literals of the arithmetic representation of the objective function could benefit the `dico` solving method.

Acknowledgments. The authors would like to thank Simon de Givry for providing the CELAR, SPOT5 and Combinatorial Auctions WCSP instances, and Roberto Sebastiani for providing the MATHSAT5-MAX and LL_{WPM} MaxSMT solvers for the experiments. The authors also thank the anonymous referees for suggestions to improve the paper.

References

1. Abío, I., Nieuwenhuis, R., Oliveras, A., Rodríguez-Carbonell, E., Mayer-Eichberger, V.: A New Look at BDDs for Pseudo-Boolean Constraints. Journal of Artificial Intelligence Research (JAIR) 45, 443–480 (2012)
2. Akers, S.B.: Binary Decision Diagrams. IEEE Transactions on Computers 27(6), 509–516 (1978)
3. Ansótegui, C., Bofill, M., Palahí, M., Suy, J., Villaret, M.: A Proposal for Solving Weighted CSPs with SMT. In: Proceedings of the 10th International Workshop on Constraint Modelling and Reformulation (ModRef 2011), pp. 5–19 (2011)
4. Ansótegui, C., Bofill, M., Palahí, M., Suy, J., Villaret, M.: Satisfiability Modulo Theories: an Efficient Approach for the Resource-Constrained Project Scheduling Problem. In: Proceedings of the 9th Symposium on Abstraction, Reformulation and Approximation (SARA 2011), pp. 2–9 (2011)
5. Ansótegui, C., Bofill, M., Palahí, M., Suy, J., Villaret, M.: Solving weighted CSPs with meta-constraints by reformulation into Satisfiability Modulo Theories. Constraints 18(2), 236–268 (2013)
6. Ansótegui, C., Bonet, M.L., Levy, J.: Solving (Weighted) Partial MaxSAT through Satisfiability Testing. In: Kullmann, O. (ed.) SAT 2009. LNCS, vol. 5584, pp. 427–440. Springer, Heidelberg (2009)
7. Barrett, C.W., Sebastiani, R., Seshia, S.A., Tinelli, C.: Satisfiability Modulo Theories. In: Handbook of Satisfiability. Frontiers in Artificial Intelligence and Applications, vol. 185, pp. 825–885. IOS Press (2009)
8. Bofill, M., Espasa, J., Palahí, M., Villaret, M.: An extension to Simply for solving Weighted Constraint Satisfaction Problems with Pseudo-Boolean Constraints. In: XII Spanish Conference on Programming and Computer Languages (PROLE 2012), Almería, Spain, pp. 141–155 (September 2012)
9. Bofill, M., Palahí, M., Suy, J., Villaret, M.: Solving constraint satisfaction problems with SAT modulo theories. Constraints 17(3), 273–303 (2012)
10. Boros, E., Hammer, P.L.: Pseudo-Boolean optimization. Discrete Applied Mathematics 123(1-3), 155–225 (2002)
11. Cimatti, A., Franzén, A., Griggio, A., Sebastiani, R., Stenico, C.: Satisfiability Modulo the Theory of Costs: Foundations and Applications. In: Esparza, J., Majumdar, R. (eds.) TACAS 2010. LNCS, vol. 6015, pp. 99–113. Springer, Heidelberg (2010)
12. Cimatti, A., Griggio, A., Schaafsma, B.J., Sebastiani, R.: A Modular Approach to MaxSAT Modulo Theories. In: Järvisalo, M., Van Gelder, A. (eds.) SAT 2013. LNCS, vol. 7962, pp. 150–165. Springer, Heidelberg (2013)
13. de Givry, S., Zytnicki, M., Heras, F., Larrosa, J.: Existential arc consistency: getting closer to full arc consistency in weighted CSPs. In: Proceedings of the 19th International Joint Conference on Artificial Intelligence (IJCAI 2005), pp. 84–89 (2005)

14. de Moura, L., Bjørner, N.: Z3: An Efficient SMT Solver. In: Ramakrishnan, C.R., Rehof, J. (eds.) TACAS 2008. LNCS, vol. 4963, pp. 337–340. Springer, Heidelberg (2008)
15. Dutertre, B., de Moura, L.: The Yices SMT solver (August 2006), Tool paper available at `http://yices.csl.sri.com/tool-paper.pdf`
16. Eén, N., Sörensson, N.: Translating Pseudo-Boolean Constraints into SAT. Journal on Satisfiability, Boolean Modeling and Computation (JSAT) 2(1-4), 1–26 (2006)
17. Freuder, E.C., Wallace, R.J.: Partial constraint satisfaction. Artificial Intelligence 58(1-3), 21–70 (1992)
18. Hebrard, E., O'Mahony, E., O'Sullivan, B.: Constraint Programming and Combinatorial Optimisation in Numberjack. In: Lodi, A., Milano, M., Toth, P. (eds.) CPAIOR 2010. LNCS, vol. 6140, pp. 181–185. Springer, Heidelberg (2010)
19. Minato, S.-I., Ishiura, N., Yajima, S.: Shared Binary Decision Diagram with Attributed Edges for Efficient Boolean function Manipulation. In: Proceedings of the 27th ACM/IEEE Conference on Design Automation (DAC 1990), pp. 52–57 (1990)
20. Larrosa, J., Schiex, T.: Solving Weighted CSP by Maintaining Arc-Consistency. Artificial Intelligence 159(1-2), 1–26 (2004)
21. Marques-Silva, J., Lynce, I., Malik, S.: Conflict-driven clause learning SAT solvers. In: SAT Handbook, pp. 131–154 (2009)
22. Meseguer, P., Rossi, F., Schiex, T.: Soft constraints. In: Rossi, F., van Beek, P., Walsh, T. (eds.) Handbook of Constraint Programming, vol. 9. Elsevier (2006)

On Backdoors to Tractable Constraint Languages*

Clément Carbonnel[1,3], Martin C. Cooper[2], and Emmanuel Hebrard[1]

[1] CNRS, LAAS, 7 avenue du colonel Roche, 31400 Toulouse, France
{carbonnel,hebrard}@laas.fr
[2] IRIT, University of Toulouse III, 31062 Toulouse, France
cooper@irit.fr
[3] University of Toulouse, INP Toulouse, LAAS, 31400 Toulouse, France

Abstract. In the context of CSPs, a strong backdoor is a subset of variables such that every complete assignment yields a residual instance guaranteed to have a specified property. If the property allows efficient solving, then a small strong backdoor provides a reasonable decomposition of the original instance into easy instances. An important challenge is the design of algorithms that can find quickly a small strong backdoor if one exists. We present a systematic study of the parameterized complexity of backdoor detection when the target property is a restricted type of constraint language defined by means of a family of polymorphisms. In particular, we show that under the weak assumption that the polymorphisms are idempotent, the problem is unlikely to be FPT when the parameter is either r (the constraint arity) or k (the size of the backdoor) unless P = NP or FPT = W[2]. When the parameter is $k+r$, however, we are able to identify large classes of languages for which the problem of finding a small backdoor is FPT.

1 Introduction

Unless P=NP, the constraint satisfaction problem (CSP) is in general intractable. However, one can empirically observe that solution methods scale well beyond what a worst-case complexity analysis would suggest.

In order to explain this gap, Williams, Gomes and Selman introduced the notion of *backdoor* [1]. A strong backdoor is a set of variables whose complete assignments all yield an easy residual problem. When it is small, it therefore corresponds to a weak spot of the problem through which it can be attacked. Indeed, by branching first on the variables of a backdoor of size k, we ensure that the depth of the search tree is bounded by k. There exists a similar notion of weak backdoor, ensuring that at least one assignment yields an easy problem, however, we shall focus on strong backdoors and omit the adjective "strong".

Finding small backdoors is then extremely valuable in order to efficiently solve constraint problems, however, it is very likely to be itself intractable. In order

* Supported by ANR Project ANR-10-BLAN-0210.

B. O'Sullivan (Ed.): CP 2014, LNCS 8656, pp. 224–239, 2014.

to study the computational complexity of this problem, we usually consider backdoors with respect to a given tractable class T, i.e., such that all residual problems fall into the class T. In Boolean Satisfiability (SAT), it was shown that finding a minimum backdoor with respect to HornSAT, 2-SAT and their disjunction is fixed-parameter tractable with respect to the backdoor size [2][3]. It is significantly harder, however, to do so with respect to bounded treewidth formulas [4]. In this paper we study the computational complexity of finding a strong backdoor to a *semantic* tractable class of CSP, in the same spirit as a very recent work by Gaspers et al. [3]. Assuming that P $\neq$ NP, semantic tractable classes are characterized by unions and intersections of languages of constraints closed by some operations. We make the following three main contributions:

- We first consider the case where these operations are *idempotent*, and show that computing a k-backdoor with respect to such a class is NP-hard even on bounded arity CSPs, and W[2]-hard for the parameter k if the arity is not bounded. Observe that the scope of this result is extremely wide, as most tractable classes of interest are idempotent.
- Then, we characterize another large category of tractable classes, that we call *Helly*, and for which finding k-backdoors is fixed-parameter tractable in $k + r$ where r is the maximum arity of a constraint.
- Lastly, we show that finding k-backdoors with respect to many semantic tractable classes that are not Helly is W[2]-hard for $k + r$ (and remains W[2]-hard for k if r is fixed). However, we do not prove a strict dichotomy since a few other conditions must be met besides not being Helly.

The paper is organized as follows: After introducing the necessary technical background in Section 2, we study idempotent tractable classes in Section 3 and Helly classes as well as a family of non-Helly classes in Section 4.

2 Preliminaries

Constraint Satisfaction Problems. A constraint satisfaction problem (CSP) is a triplet (X, D, C) where X is a set of variables, D is a domain of values, and C is a set of constraints. For simplicity, we assume D to be a finite subset of $\mathbb{N}$. A constraint is a pair (S, R) where $S \subseteq X$ is the scope of the constraint and R is an $|S|$-ary relation on D, i.e. a subset of $D^{|S|}$ representing the possible assignments to S. A solution is an assignment $X \to D$ that satisfies every constraint. The goal is to decide whether a solution exists.

A *constraint language* is a set of relations. The domain of a constraint language Γ is denoted by $D(\Gamma)$ and contains all the values that appear in the tuples of the relations in Γ. Given a constraint language Γ, CSP(Γ) is the restriction of the generic CSP to instances whose constraints are relations from Γ. The *Dichotomy Conjecture* by Feder and Vardi says that for every finite Γ, CSP(Γ) is either in P or NP-complete [5]. Since the conjecture is still open, the complexity of constraint languages is a very active research area (see e.g. [6][7][8]).

It is known that the complexity of a language is determined by its set of closure operations [9]. Specifically, an operation $f : D(\Gamma)^a \rightarrow D(\Gamma)$ of arity a is a *polymorphism* of Γ if for every $R \in \Gamma$ of arity r and $t_1, \ldots, t_a \in R$, $f(t_1, \ldots, t_a) = (f(t_1[1], \ldots, t_a[1]), \ldots, f(t_1[r], \ldots, t_a[r])) \in R$. A polymorphism f is *idempotent* if $\forall x \in D, f(x, x, \ldots, x) = x$. We denote by $\text{Pol}(\Gamma)$ (resp. $\text{IdPol}(\Gamma)$) the set of all polymorphisms (resp. idempotent polymorphisms) of Γ. Given two languages Γ_1, Γ_2 with $D(\Gamma_2) \subseteq D(\Gamma_1)$, we write $\text{Pol}(\Gamma_1) \subseteq \text{Pol}(\Gamma_2)$ if the restriction to $D(\Gamma_2)$ of every $f \in \text{Pol}(\Gamma_1)$ is in $\text{Pol}(\Gamma_2)$. A wide range of operations have been shown to induce polynomial-time solvability of any language they preserve: these include near-unanimity operations [10], edges [11], semilattices [9], 2-semilattices [12] and totally symmetric operations of all arities [13].

Composite Classes. A semantic class is a set of languages. A semantic class T is tractable if $\text{CSP}(\Gamma) \in P$ for every $\Gamma \in T$, and recognizable in polynomial time if the membership problem 'Does $\Gamma \in T$?' is in P. We say that a semantic class T is *atomic* if there exists an operation $f : \mathbb{N}^a \rightarrow \mathbb{N}$ such that $\Gamma \in T$ if and only if $f_{|D(\Gamma)} \in \text{Pol}(\Gamma)$, where $f_{|D(\Gamma)}$ denotes the restriction of f to $D(\Gamma)$. We sometimes denote such a class by T_f and say that f *induces* T_f. We call a semantic class T *simple* if there exists a set $\mathcal{T}$ of atomic classes such that $T = \cap_{T_f \in \mathcal{T}} T_f$. Finally, a semantic class T is *composite* if there exists a set $\mathcal{T}$ of simple classes such that $T = \cup_{T_s \in \mathcal{T}} T_s$. In both cases, the set $\mathcal{T}$ is allowed to be infinite. Using the distributivity of intersection over union, it is easy to see that any class derived from atomic classes through any combination of intersections and unions is composite. We say that an atomic class T_f is *idempotent* if f is idempotent. By extension, a composite class is idempotent if can be obtained by intersections and unions of idempotent atomic classes.

Example 1. Consider the class of max-closed constraints, introduced in [14]. This class is tractable as any CSP instance over a max-closed constraint language can be solved by establishing (generalised) arc-consistency. Using our terminology, this class is exactly the atomic class induced by the operation $\max(.,.)$, and thus it is composite. Max-closed constraints have been generalized to any language that admits a semilattice polymorphism, i.e. a binary operation f such that $f(x, x) = x$, $f(x, y) = f(y, x)$ and $f(f(x, y), z) = f(x, f(y, z))$ for any $x, y, z \in D$ [9]. If we denote by Sml the set of all possible semilattice operations on $\mathbb{N}$, this larger class corresponds to $\cup_{f \in \text{Sml}} T_f$, which is composite but not atomic.

Example 2. For a given language Γ, let $\overline{\Gamma}$ be the language obtained from Γ by adding all possible unary relations over $D(\Gamma)$ with a single tuple. Consider the very large class T_{BW} of languages Γ such that $\text{CSP}(\overline{\Gamma})$ (and thus $\text{CSP}(\Gamma)$) can be solved by achieving k-consistency for some k that only depends on Γ. This property is equivalent to the existence of two idempotent polymorphisms f and g such that for every $x, y \in D(\Gamma)$ [15][16],

$$\begin{aligned}
(i) \quad & g(y, x, x, x) = g(x, y, x, x) = g(x, x, y, x) = g(x, x, x, y) \\
(ii) \quad & f(y, x, x) = f(x, y, x) = f(x, x, y) \\
(iii) \quad & f(x, x, y) = g(x, x, x, y)
\end{aligned}$$

If we denote by FGBW the set of all pairs of operations (f, g) on $\mathbb{N}$ satisfying these three conditions, the class T_{BW} is composite and idempotent since it can be written as $T_{BW} = \cup_{(f,g)\in\text{FGBW}}(T_f \cap T_g)$.

The choice to study composite classes shows multiple advantages. First, they are general enough to capture most natural semantic tractable classes defined in the literature, and they also allow us to group together tractable languages that are solved by the same algorithm (such as arc consistency or Gaussian elimination). Second, membership in these classes is hereditary: if $\Gamma \in T$ and $\text{Pol}(\Gamma) \subseteq \text{Pol}(\Gamma')$, then $\Gamma' \in T$. In particular, any sublanguage of a language in T is in T, and every composite class contains the empty language.

Strong Backdoors. Given an instance (X, D, C) of CSP(Γ), assigning a variable $x \in X$ to a value $d \in D$ is done by removing the tuples inconsistent with $x \leftarrow d$ from the constraints whose scope include x, and then removing the variable x from the instance (thus effectively reducing the arity of the neighbouring constraints by one). A *strong backdoor* to a semantic class T is a subset $S \subseteq X$ such that every complete assignment of the variables from S yields an instance whose language is in T. Note that assigning a variable involves no further inference (e.g., arc consistency); indeed doing so has been shown to make backdoors potentially much harder to detect [17]. There exist alternative forms of backdoors, such as weak backdoors [1] and partition backdoors [18], but we only consider strong backdoors throughout this paper so we may omit the word "strong" in proofs. The goal of this work is to study how the properties of the target semantic class T affect the (parameterized) complexity of the following problem.

Strong T-Backdoor: Given a CSP instance I and an integer k, does I have a strong backdoor to T of size at most k?

Parameterized Complexity. A problem is *parameterized* if each instance x is coupled with a nonnegative integer k called the *parameter*. A parameterized problem is *fixed-parameter tractable* (FPT) if it can be solved in time $O(f(k)|x|^{O(1)})$, where f is any computable function. For instance, Vertex Cover parameterized with the size k of the cover is FPT as it can be solved in time $O(1.2738^k + kn)$ [19], where n is the number of vertices of the input graph. The class XP contains the parameterized problems that can be solved in time $O(f(k)|x|^{g(k)})$ for some computable functions (f, g). FPT is known to be a proper subset of XP [20]. Between these extremes lies the *Weft Hierarchy*:

$$\text{FPT} = \text{W}[0] \subseteq \text{W}[1] \subseteq \text{W}[2] \subseteq \ldots \subseteq \text{XP}$$

where for every t, W[t+1] is believed to be strictly larger than W[t]. These classes are closed under FPT-reductions, which map an instance (x, k) of a problem L_1 to an instance (x', k') of a problem L_2 such that:

- (x', k') can be built in time $O(f(k)|x|^{O(1)})$ for some computable function f
- (x', k') is a yes-instance if and only if (x, k) is
- $k' \leq g(k)$ for some computable function g

For instance, k-CLIQUE is W[1]-complete [20] when the parameter is k. Note that if considering the parameter as a constant yields an NP-hard problem, the parameterized version is not in XP (and thus not FPT) unless P = NP.

3 General Hardness

We consider two parameters: k, the size of the backdoor, and r, the maximum arity of the constraint network. Under the very weak assumption that T is composite and idempotent, we show that STRONG T-BACKDOOR is unlikely to be FPT for either of the parameters taken separately, assuming that P $\neq$ NP and FPT $\neq$ W[2], as we shall do througout the paper. In both cases, we show that our results extend to the class of Boolean CSPs with minor modifications.

Our hardness results will be obtained by reductions from various forms of the p-HITTING SET problem: given a universe U, a collection $S = \{S_i \mid i = 1..n\}$ of subsets of U with $|S_i| = p$ and an integer k, does there exist a subset $H \subseteq U$ such that $|H| \leq k$ and $\forall i$, $H \cap S_i \neq \emptyset$? This problem is NP-complete for every fixed $p \geq 2$ [21], and W[2]-complete when the parameter is k and p is unbounded [20]. The special case $p = 2$ is called VERTEX COVER, and the input is typically given in the form of a graph $G = (U, S)$ and an integer k.

We will make use of two elementary properties of idempotent composite classes. First, any relation with a single tuple is closed by every idempotent operation. Thus, adding such a relation to a language does not affect its membership in idempotent classes. The second property is slightly more general. Given a relation R of arity r, let M_R be the matrix whose rows are the tuples of R sorted by lexicographic order (so that M_R is unique). We say that a relation R is an *extension* of a relation R' if M_R has all the columns of $M_{R'}$, plus extra columns that are either constant (i.e. every value in that column is the same) or copies of some columns of $M_{R'}$. In that case, since $\text{IdPol}(\{R\}) = \text{IdPol}(\{R'\})$, $\{R\} \in T$ if and only if $\{R'\} \in T$, for every idempotent composite class T.

For the rest of the document, we represent relations as lists of tuples delimited by square brackets (e.g. $R = [t_1, \ldots, t_n]$), while tuples are delimited by parentheses (e.g. $t_1 = (d_1, \ldots, d_r)$).

3.1 Hardness on Bounded Arity CSPs

Theorem 1. STRONG T-BACKDOOR *is NP-hard for every idempotent composite tractable class T, even for binary* CSP*s.*

Proof. We reduce from VERTEX COVER. Let $I = (G, k)$ be an instance of VERTEX COVER. We consider two cases. First, suppose that $\Gamma = \{[(1), (2)], [(2), (3)], [(1), (3)]\} \in T$. We create a CSP with one variable per vertex in G, and if two variables correspond to adjacent vertices we add the constraint $\neq_{1,2,3}$ (inequality over the 3-element domain) between them. Since $\text{CSP}(\{\neq_{1,2,3}\})$ is NP-hard and T is tractable, a valid backdoor of size at most k must correspond to a vertex cover on G. Conversely, the variables corresponding to a vertex cover form a backdoor: after every complete assignment to these variables, the language of the reduced instance is a subset of Γ, which is in T since T is composite and hence hereditary. Now, suppose that $\Gamma \notin T$. We duplicate the column of each relation in Γ to obtain the binary language $\Gamma' = \{R_1, R_2, R_3\}$ with $R_1 = [(1,1),(2,2)]$, $R_2 = [(2,2),(3,3)]$ and $R_3 = [(1,1),(3,3)]$. Since Γ' is an extension of Γ and T is idempotent, Γ' is not in T. Then, we follow the same reduction as in the first case, except that we add the three constraints R_1, R_2, R_3 instead of $\neq_{1,2,3}$ between two variables associated with adjacent vertices. By construction, a backdoor must be a vertex cover. Conversely, if we have a vertex cover, after any assignment of the corresponding variables we are left with at most one tuple per constraint, and the resulting language is in T by idempotency. □

In the case of Boolean CSPs, Theorem 1 cannot apply verbatim. This is due to the fact that every binary Boolean language is a special case of 2-SAT and is therefore tractable. Thus, a binary Boolean CSP has always a backdoor of size 0 to any class that is large enough to contain 2-SAT, and the minimum backdoor problem is trivial. The next proposition shows that this is the only case for which STRONG T-BACKDOOR is not NP-hard under the idempotency condition. Note that looking for a strong backdoor in a binary Boolean CSP has no practical interest; however this case is considered for completeness.

Proposition 1. *On Boolean* CSP*s with arity at most r,* STRONG T-BACKDOOR *is NP-hard for every idempotent composite tractable class T if $r \geq 3$. For $r = 2$,* STRONG T-BACKDOOR *is either trivial (if every binary Boolean language is in T) or NP-hard.*

We omit the proof as it is similar to that of Theorem 1 (with more cases). All omitted proofs are available in an extended version [22].

3.2 Hardness When the Parameter Is the Size of the Backdoor

In general, a large strong backdoor is not of great computational interest as the associated decomposition of the original instance is very impractical. Thus, it makes sense to design algorithms that are FPT when the parameter is the size of the backdoor. In this section we show that in the case of idempotent composite classes such algorithms cannot exist unless FPT = W[2]. Furthermore, we establish this result under the very restrictive condition that the input CSP has a single constraint, which highlights the fact that STRONG T-BACKDOOR is more than a simple pseudo-HITTING SET on the constraints outside T.

For any natural numbers m, e ($m \geq 3$), we denote by $R_3^m(e)$ the relation obtained by duplicating the last column of $[(e+1,e,e),(e,e+1,e),(e,e,e+1)]$ until the total arity becomes m. It is straightforward to see that CSP($\{R_3^m(e)\}$) is NP-hard for every m, e by a reduction from 1-in-3-SAT. In a similar fashion, we define $R_2^m(e)$ as an extension of $[(e+1,e),(e,e+1)]$ of arity m.

Theorem 2. STRONG T-BACKDOOR *is W[2]-hard for every idempotent composite tractable class T when the parameter is the size of the backdoor, even if the* CSP *has a single constraint.*

$$U = (u_1, \ldots, u_7)$$
$$S_1 = (u_3, u_4, u_5)$$
$$S_2 = (u_2, u_5, u_6)$$
$$S_3 = (u_1, u_3, u_7)$$

(a) 3-HITTING SET

$$\{R_2^2(2)\} \in T$$
$$\{R_2^2(4)\} \notin T$$
$$\{R_2^2(6)\} \notin T$$

(b) class T

u_1	u_2	u_3	u_4	u_5	u_6	u_7
2	2	3	2	2	2	2
2	2	2	3	2	2	2
2	2	2	2	3	2	2
4	5	4	4	4	4	4
4	4	4	4	5	5	4
7	6	6	6	6	6	6
6	6	7	6	6	6	7

(c) Constraint C

Fig. 1. Example of reduction from a 3-HITTING SET instance to the problem of finding a backdoor to the class T. The reduction produces a single constraint C.

Proof. The proof is an FPT-reduction from p-HITTING SET parameterized with solution size k. Let (p, U, S) be an instance of p-HITTING SET, where U is the universe ($|U| = n$) and $S = \{S_i \mid i = 1..s\}$ is the collection of p-sets. We assume without loss of generality that $p \geq 3$ (if this is not the case we pad each set with unique elements). We build an n-ary relation R, where each column is associated with a value from U, as follows. For every $S_i \in S$, we consider two cases. If $\{R_2^2(2i)\} \notin T$, we add two tuples t_1, t_2 to R such that the restriction of $[t_1, t_2]$ to the columns corresponding to the values appearing in S_i form the relation $R_2^p(2i)$, and the other columns are constant with value $2i$. If $\{R_2^2(2i)\} \in T$, we add 3 tuples t_1, t_2, t_3 such that the restriction of t_1, t_2, t_3 to the columns corresponding to S_i form $R_3^p(2i)$, and the remaining columns are constant with value $2i$. Once the relation is complete, we apply it to n variables to obtain an instance of our backdoor problem. See Figure 1 for an example of the construction.

Suppose we have a backdoor of size at most k, and suppose there exists a set S_i such that none of the corresponding variables belong to the backdoor. Then, if we assign every variable in the backdoor to $2i$, the reduced constraint must belong to T. By idempotency, we can further assign every remaining variable outside of S_i to the value $2i$ and the resulting constraint must still be in T. The reduced constraint becomes either $R_2^p(2i)$ if $\{R_2^2(2i)\} \notin T$, or $R_3^p(2i)$, which is not in T since T is tractable and we assume P $\neq$ NP. In both cases, this constraint does not belong to T, and we have a contradiction. Therefore, if there is a backdoor of size at most k, we also have a hitting set of size at most k.

Conversely, suppose we have a hitting set of size at most k. We prove that the associated set of variables form a backdoor. Observe that two blocks (i.e. pairs/triples) of tuples of the constraint C associated with different sets do not share any common value; hence, after assigning the variables corresponding to the hitting set to any values, the resulting constraint is either empty or a subrelation of a single block associated with the set S_i. The latter case yields two possibilities. If T does not contain $\{R_2^2(2i)\}$, then the block i must have been reduced to a single tuple, since the two initial tuples t_1, t_2 satisfy $t_1[x_j] \neq t_2[x_j]$ for all x_j associated with a value in S_i. Thus, by idempotency, the resulting constraint is in T. Now, if T contains $\{R_2^2(2i)\}$, the resulting constraint has at most two tuples (same argument as above), which can only happen if all the variables are assigned to the value $2i$. If we are in this situation, the new constraint must be an extension of $R_2^2(2i)$ and hence is in T. Therefore, our hitting set provides a strong backdoor in our CSP instance, which concludes the reduction. □

This theorem still holds on Boolean CSPs if we allow multiple constraints in the target instance, even if these constraints are all the same relation. We omit the proof, as it follows the same principles as that of Theorem 2.

Proposition 2. *On Boolean* CSPs*,* Strong T-Backdoor *is W[2]-hard for every idempotent composite tractable class T when the parameter is the size of the backdoor, even if the* CSP *has a single type of constraint.*

Remark 1. Partition Backdoors is an alternative form of backdoors recently introduced by Bessiere et al [18]. Such backdoors are especially interesting in the case of conservative classes (conservativity is more restrictive than idempotency, since each polymorphism is required to satisfy $f(x_1, \ldots, x_a) \in \{x_1, \ldots, x_a\}$). The authors argue that, given a partition of the constraints $C = \{C_1, C_2\}$ such that the language of C_1 is in a conservative class T, the vertex cover of the primal graph of C_2 is a strong backdoor to T. The minimum-size *partition backdoor* is then the best such backdoor over every possible partition of the constraints. Computing the minimum-size partition backdoor is FPT in the parameter $k+l$, where l is the size of the constraint language; our results show that computing the actual minimum strong backdoor is a much harder problem as it is still W[2]-hard for the larger parameters $k+m$ (Theorem 2) and $d+k+l$ (Proposition 2), where m is the number of constraints and d the size of the domain.

4 Combined Parameters: Helly Classes and Limits

We have shown in sections 3.1 and 3.2 that considering independently the maximum constraint arity r and the size of the backdoor k as parameters is unlikely to yield FPT tractability. We now consider the combined parameter $k + r$ and show that FPT tractability ensues for numerous tractable composite classes.

In order to design an algorithm for Strong T-Backdoor that is FPT for $k+r$, it is important to have a procedure to *check* whether a subset of variables of size at most k is a strong backdoor to T. The natural algorithm for this task runs in time $O(mrtd^k P(\Gamma))$ (where m is the number of constraints, t the maximum number of tuples and $P(\Gamma)$ the complexity of the membership problem of a language Γ in T) by checking independently each of the d^k possible assignments of B. In our case this approach is not satisfactory: since d is not a parameter, the term d^k is problematic for the prospect of an algorithm FPT in $k + r$. The next lemma presents an alternative algorithm for the "backdoor check" problem that is only exponential in the number of constraints m. Although it may seem impractical at first sight (as m is typically much larger than k), we will show that it can be exploited for many tractable classes.

Lemma 1. *Let T be a composite class recognizable in time $P(\Gamma)$. Let $I = (X, D, C)$ be a* CSP *instance with m constraints of arity at most r and containing at most t tuples, and $B \subseteq X$. It is possible to decide whether B is a strong backdoor to T in time $O(mrt^2 + m^2 r(2t)^m P(\Gamma))$.*

Proof. We first focus on a single constraint (S, R). Let $B_S = B \cap S$. Observe that at most t different assignments of B_S can leave R nonempty, since the subrelations of R obtained with each assignment are pairwise disjoint and their union is R. To compute these assignments in polynomial time, one can explore a search tree. Starting from a node labelled R, we pick a nonfixed variable $v \in B_S$ and for every $d \in D(v)$ such that the subrelation $R_{v=d}$ is not empty we create a child node labelled with $R_{v=d}$. Applying this rule recursively, we obtain a tree of depth at most r and with no more than t leaves, so it has at most rt nodes. The time spent at each node is $O(t)$, so computing all leaves can be done in time $O(rt^2)$. Now, suppose that for each constraint $c = (R, S) \in C$ we have computed this set ϕ_c of all the locally consistent assignments of B_S and stored the resulting subrelation. For every $\phi \in \prod_{c \in C} \phi_c$ and every possible subset C' of the constraints, we check if the restriction ϕ_s of ϕ to the constraints of C' is a consistent assignment (i.e., no variable is assigned multiple values). If ϕ_s is consistent, we temporarily remove from the instance the constraints outside C', we apply the assignment ϕ_s and we check whether the language of the resulting instance is in T. The algorithm returns that B is a backdoor if and only if each membership test in T is successful. To prove the correctness of the algorithm, suppose that ψ is an assignment of B such that the resulting language is not in T. Then, at least one subset of the constraints have degraded into non-empty subrelations. For each of these constraints, the restricted assignment ψ_R is consistent with the others, so the algorithm must have checked membership of the resulting language in T and concluded that B is not a strong backdoor. Conversely, if B is a strong backdoor, every complete assignment of B yields an instance in T. In particular, if we consider only a subset of the constraints after each assignment, the language obtained is also in T since T is composite (and hence hereditary). Thus, none of the membership tests performed by the algorithm will fail. The complexity of the algorithm is $O(mrt^2 + m^2rt^m2^mP(\Gamma))$. □

We say that a composite class T is h-Helly if it holds that for any language Γ, if every $\Gamma_h \subseteq \Gamma$ of size at most h is in T then Γ is in T. This property is analogous to the well-studied Helly properties for set systems. We call Helly number of T the minimum positive integer h such that T is h-Helly. Being characteristic of a class defined exclusively in terms of polymorphisms over $\mathbb{N}$, the Helly number is independant from parameters like the domain size or the arity of the languages. The next theorem is the motivation for the study of such classes, and is the main result of this section.

Theorem 3. *For every fixed composite class T recognizable in polynomial time, if T has a finite Helly number then* STRONG T-BACKDOOR *is FPT when the parameter is $k + r$, where k is the size of the backdoor and r is the maximum constraint arity.*

Proof. Let h denote the Helly number of T. The algorithm is a bounded search tree that proceeds as follows. Each node is labelled by a subset of variables B. The root of the tree is labelled with the empty set. At each node, we examine every possible combination of h constraints and check if B is a strong backdoor for the subset in time $O(hrt^2+h^2r(2t)^hP(\Gamma))$ (where $P(\Gamma)$ is the polynomial complexity of deciding the membership of a language Γ in T) using Lemma 1. Suppose that B is a strong backdoor for every h-subset. Then, for any possible assignment of B, each h-subset of the constraints of the resulting instance must be in T: otherwise, B would not be a strong backdoor for the h original constraints that generated them. Since T is h-Helly, we can conclude that B is a valid strong backdoor for the whole instance. Now suppose that we have found a h-subset for which B is not a strong backdoor. For every variable x in the union of the scopes of the constraints in this subset that is not already in B (there are at most rh such variables x), we create a child node labelled with $B \cup \{x\}$. At each step we are guaranteed to add at least one variable to B, so we stop creating child nodes when we reach depth k. The algorithm returns 'YES' at the first node visited that corresponds to a strong backdoor, and 'NO' if no such node is found.

If no strong backdoor of size at most k exists, it is clear that the algorithm correctly returns 'NO'. Now suppose that a strong backdoor $\mathbf{B}$, $|\mathbf{B}| \leq k$, exists. Observe that if a node is labelled with $B \subset \mathbf{B}$ and B is not a backdoor for some h-subset of constraints, then $\mathbf{B}$ contains at least one more variable within this subset. Since the algorithm creates one child per variable that can be added and the root is labelled with a subset of $\mathbf{B}$, by induction there must be a path from the root to a node labelled with $\mathbf{B}$ and the algorithm returns 'YES'.

The complexity of the procedure is $O\left((rh)^k m^h (hrt^2 + h^2 r(2t)^h P(\Gamma))\right) = O\left(f(k+r)m^h(ht^2 + h^2(2t)^h P(\Gamma))\right)$. □

In contrast to the previous hardness results, the target tractable class is not required to be idempotent. However, the target class must have a finite Helly number, which may seem restrictive. The following series of results aims to identify composite classes with this particular property.

Lemma 2. *A composite class T is simple if and only if it is 1-Helly.*

Proof. Let T be a simple class, i.e. an intersection of atomic classes $T = \cap_{f \in \mathcal{F}} T_f$. Let Γ be a constraint language such that each $\{R\} \subseteq \Gamma$ is in T. Then, every $f \in \mathcal{F}$ preserves every relation in Γ and thus preserves Γ, so $\Gamma \in T$ and T is 1-Helly. Conversely, let T be a 1-Helly composite class. Let $\mathcal{F} = \{f \mid f$ preserves every $\{R\} \in T\}$. Every $\Gamma \in T$ admits as polymorphism every $f \in \mathcal{F}$ (as each $\{R\} \subseteq \Gamma$ is in T and thus is preserved by f), so $T \subseteq \cap_{f \in \mathcal{F}} T_f$. The other way round, a language Γ in $\cap_{f \in \mathcal{F}} T_f$ is preserved by every $f \in \mathcal{F}$ and thus must be a sublanguage of $\Gamma_\infty = \cup_{\{R\} \in T}\{R\}$, which is in T since T is 1-Helly, so $\Gamma \in T$ and $\cap_{f \in \mathcal{F}} T_f \subseteq T$. Finally, $T = \cap_{f \in \mathcal{F}} T_f$ and so T is simple. □

Proposition 3. *Let h be a positive integer and $\mathcal{T}$ be a set of simple classes. Then, $T = \{\Gamma \mid \Gamma$ belongs to every $T_i \in \mathcal{T}$ except at most $h\}$ is a $(h+1)$-Helly composite class.*

Proof. T is composite since it is the union of every possible intersection of all but h classes from $\mathcal{T}$ and any class derived from atomic classes through any combination of intersections and unions is composite. We write $\mathcal{T} = \{T_i \mid i \in I\}$. Let Γ be a language such that every sublanguage of size at most $h+1$ is in T. For each $R \in \Gamma$ we define $S(R) = \{T_i \mid \{R\} \notin T_i\}$. By Lemma 2, simple classes are 1-Helly so $\Gamma \notin T_i \Leftrightarrow (\exists R \in \Gamma$ such that $\{R\} \notin T_i) \Leftrightarrow T_i \in \cup_{R\in\Gamma} S(R)$. So $\Gamma \in T$ if and only if $|\cup_{R\in\Gamma} S(R)| \leq h$. We discard from Γ every relation R such that $|S(R)| = 0$ as they have no influence on the membership of Γ in T. If that process leaves Γ empty, then it belongs to T. Otherwise, let s_j denote the maximum size of $\cup_{R\in\Gamma_j} S(R)$ over all size-j subsets Γ_j of Γ. Since each sublanguage Γ_j of size $j \leq h+1$ is in T, from the argument above we have $1 \leq s_1 \leq \ldots \leq s_{h+1} \leq h$, thus there exists $j < h+1$ such that $s_j = s_{j+1}$. Let $\Gamma_j \subseteq \Gamma$ denote a set of j relations such that $|\cup_{R\in\Gamma_j} S(R)| = s_j$. Suppose there exists $R_0 \in \Gamma$ such that $S(R_0) \nsubseteq \cup_{R\in\Gamma_j} S(R)$. Then, $|\cup_{R\in\Gamma_j\cup\{R_0\}} S(R)| > s_j = s_{j+1}$, and we get a contradiction. So $\cup_{R\in\Gamma} S(R) \subseteq \cup_{R\in\Gamma_j} S(R)$, hence $|\cup_{R\in\Gamma} S(R)| \leq h$ and Γ is in T. Therefore, T is $(h+1)$-Helly. □

In the particular case where $\mathcal{T}$ is finite and $h = |\mathcal{T}| - 1$, we get the following nice corollary. Recall that a composite class is any union of simple classes.

Corollary 1. *Any union of h simple classes is h-Helly.*

Example 3. Let $T = \{\Gamma \mid \Gamma$ is either min-closed, max-closed or 0/1/all$\}$. T is the union of 3 well-known tractable semantic classes. By definition, min-closed and max-closed constraints are respectively the languages that admit min(.,.) and max(.,.) as polymorphisms. Likewise, 0/1/all constraints have been shown to be exactly the languages that admit as polymorphism the majority operation [9]

$$f(x,y,z) = \begin{cases} y \text{ if } y = z \\ x \text{ otherwise} \end{cases}$$

Thus, T is the union of 3 atomic classes and hence is 3-Helly by Corollary 1. Since T is also recognizable in polynomial time, by Theorem 3 STRONG T-BACKDOOR is FPT when parameterized by backdoor size and maximum arity.

In the light of these results, it would be very interesting to show a dichotomy. Is STRONG T-BACKDOOR with parameter $k + r$ at least W[1]-hard for every tractable composite class T that does not have a finite Helly number? While we leave most of this question unanswered, we have identified generic sufficient conditions for W[2]-hardness when r is fixed and the parameter is k.

Given a bijection $\phi : D_1 \to D_2$, we denote by R_ϕ the relation $[(d, \phi(d)), d \in D_1]$. Given a language Γ, a subdomain D_1 of $D(\Gamma)$ is said to be *conservative* if every $f \in \text{Pol}(\Gamma)$ satisfies $f(x_1, \ldots, x_m) \in D_1$ whenever $\{x_1, \ldots, x_m\} \subseteq D_1$. For instance, $D(\Gamma')$ is conservative for every $\Gamma' \subseteq \Gamma$, and for every column of some $R \in \Gamma$ the set of values that appear in that column is conservative. Then, we say that a class T is *value-renamable* if for every $\Gamma \in T$ and $\phi : D_1 \to D_2$, where D_1 is a conservative subdomain of Γ and $D_2 \cap D(\Gamma) = \emptyset$, $\Gamma \cup \{R_\phi\}$ is in T. For instance, the class of 0/1/all constraints introduced in Example 3 is value-renamable, but max-closed constraints are not (as they rely on a fixed order on

$\mathbb{N}$). We also say that a composite class T is *domain-decomposable* if for each pair of languages $\Gamma_1 \in T$ and $\Gamma_2 \in T$, $D(\Gamma_1) \cap D(\Gamma_2) = \emptyset$ implies $\Gamma_1 \cup \Gamma_2 \in T$. Value-renamability and domain-decomposability are natural properties of any class that is large enough to be invariant under minor (from the algorithmic viewpoint) modifications of the constraint languages.

Given a language Γ and a bijection $\phi : D(\Gamma) \to D'$, we denote by $\phi(\Gamma)$ the language over D' obtained by replacing every tuple $t = (d_1, \ldots, d_r)$ in every relation in Γ by $\phi(t) = (\phi(d_1), \ldots, \phi(d_r))$.

Lemma 3. *Let $\Gamma = \Gamma_1 \cup \Gamma_2 \cup \{R_\phi\}$ where ϕ is a bijection from $D(\Gamma_2)$ to some domain D_1. Then, Pol(Γ) $\subseteq$ Pol($\Gamma_1 \cup \phi(\Gamma_2)$).*

Proof. Let $f \in \text{Pol}(\Gamma)$ of arity a. We only need to show that f preserves $\phi(\Gamma_2)$, as f already preserves Γ_1. Since f preserves R_ϕ, for each $(d_1, \phi(d_1)), \ldots, (d_a, \phi(d_a)) \in R_\phi$ we have $(f(d_1, \ldots, d_a), f(\phi(d_1), \ldots, \phi(d_a))) \in R_\phi$, so $f(\phi(d_1), \ldots, \phi(d_a)) = \phi(f(d_1, \ldots, d_a))$ for every $d_1, \ldots, d_a \in D(\Gamma_2)$. Then, given a tuples $\phi(t_1), \ldots, \phi(t_a)$ of $\phi(\Gamma_2)$, $f(\phi(t_1), \ldots, \phi(t_a)) = \phi(f(t_1, \ldots, t_a)) \in \phi(\Gamma_2)$ since $f(t_1, \ldots, t_a) \in \Gamma_2$. Therefore, f is a polymorphism of $\phi(\Gamma_2)$ and $\text{Pol}(\Gamma) \subseteq \text{Pol}(\Gamma_1 \cup \phi(\Gamma_2))$.

Theorem 4. *On* CSP*s with arity at most r, if T is a composite class that is*

- *idempotent*
- *not 1-Helly for constraints of arity at most r*
- *value-renamable*
- *domain-decomposable*

then Strong T-Backdoor *is W[2]-hard when the parameter is k.*

Proof. Since T is not 1-Helly for constraints of arity at most r, there exists a language $\Gamma_m = \{R_i \mid i \in 1..l_m\}$ (of arity $r_m \leq r$ and over a domain $D_m, |D_m| = d_m$) such that $l_m > 1$ and every sublanguage of Γ_m is in T but Γ_m is not. Since T is fixed, we shall consider that Γ_m is fixed as well and hence has constant size. We assume for simplicity of presentation that every $R \in \Gamma_m$ has arity r_m.

We perform an FPT-reduction from p-Hitting Set parameterized with solution size as follows. Let (p, U, S) be an instance of p-Hitting Set, with $S = \{S_1, \ldots, S_s\}$ and $U = \{u_1, \ldots, u_n\}$. For every $u_i \in U$, we associate a unique variable x_i. For every set $S_j = (u_{\sigma_j(1)}, \ldots, u_{\sigma_j(p)})$, we add $2r_m$ new variables $y_j^1, \ldots, y_j^{r_m}, z_j^1, \ldots, z_j^{r_m}$ and we create $p+2$ new disjoint domains D_j^i, $i \in [0 \ldots p{+}1]$ of size d_m. Then, we pick a chain of $p{+}1$ bijections $\psi_j^i : D_j^i \to D_j^{i+1}$, $i \in [0 \ldots p]$ and we add a chain of constraints $R_{\psi_j^i}$ between the $p+2$ variables $(y_j^{r_m}, x_{\sigma_j(1)}, \ldots, x_{\sigma_j(p)}, z_j^1)$. Afterwards, we pick a bijection $\phi_j : D_m \to D_j^0$ and we apply $\phi_j(R_1)$ to $y_j^1, \ldots, y_j^{r_m}$. In the same fashion, if we denote by ψ_j the bijection from D_j^0 to D_j^{p+1} obtained by composition of all the ψ_j^i, we apply every constraint in $(\psi_j \circ \phi_j)(\Gamma_m \backslash \{R_1\})$ to the variables $z_j^1, \ldots, z_j^{r_m}$. The main idea behind the construction is that both $\Gamma_m \backslash \{R_1\}$ and $\{R_1\}$ are in T but Γ_m is not: by adding $\phi_j(R_1)$ on the variables y, $\psi_j \circ \phi_j(\Gamma_m \backslash \{R_1\})$ on the variables z and the chain of bijections $R_{\psi_j^i}$ of the x variables, we have a language that is not in

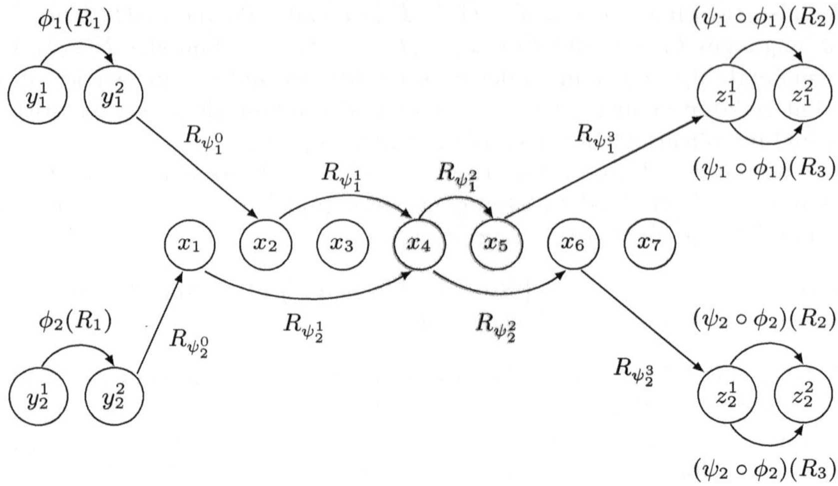

Fig. 2. Example of the construction for $U = (u_1, \ldots, u_7)$, two sets $S_1 = (u_2, u_4, u_5)$, $S_2 = (u_1, u_4, u_6)$, $\Gamma_m = \{R_1, R_2, R_3\}$ and $r_m = 2$. Each arrow is a (binary) constraint. The upper part of the instance is constructed from S_1 and the lower part from S_2.

T but assigning any value to x yields a residual language in T (the proof can be found below). We use this property to encode a HITTING SET instance. See Figure 2 for an example of the reduction.

Suppose we have a backdoor to T of size at most k. Then, for each set S_j, at least one variable from $(y_j^1, \ldots, y_j^{r_m}, x_{\sigma_j(1)}, \ldots, x_{\sigma_j(p)}, z_j^1, \ldots, z_j^{r_m})$ must belong to the backdoor. Suppose this is not the case. Then, the language Γ of any reduced instance would contain the relations of $\{\phi_j(R_1), (\psi_j \circ \phi_j)(\Gamma_m \backslash \{R_1\})\}$ plus the relations $R_{\psi_j^i}$. Applying Lemma 3 $p+1$ times, we get $\text{Pol}(\Gamma) \subseteq \text{Pol}((\psi_j \circ \phi_j)(R_1) \cup (\psi_j \circ \phi_j)(\Gamma_m \backslash \{R_1\})) = \text{Pol}((\psi_j \circ \phi_j)(\Gamma_m))$. Thus, if Γ is in T, then so is $(\psi_j \circ \phi_j)(\Gamma_m)$. Then, by value-renamability $\{(\psi_j \circ \phi_j)(\Gamma_m) \cup R_{(\psi_j \circ \phi_j)^{-1}}\}$ is also in T and using Lemma 3 again, Γ_m is in T, which is a contradiction. Therefore, a hitting set of size at most k can be constructed by including every value u_i such that x_i is in the backdoor, and if any variable from $y_j^1, \ldots, y_j^{r_m}, z_j^1, \ldots, z_j^{r_m}$ belongs to the backdoor for some j, we also include $u_{\sigma_j(1)}$.

Conversely, a hitting set forms a backdoor. After every complete assignment of the variables from the hitting set, the set of constraints associated with any set S_j can be partitioned into sublanguages whose domains have an empty intersection(see Figure 2). The sublanguages are either:

- $\phi_j(R_1)$ together with some constraints $R_{\psi_j^i}$ and a residual unary constraint with a single tuple. This language is in T by Lemma 3, value-renamability and idempotency.
- $(\psi_j \circ \phi_j)(\Gamma_m \backslash \{R_1\})$ together with some constraints $R_{\psi_j^i}$ and a residual unary constraint with a single tuple. This case is symmetric.

– A (possibly empty) chain of constraints $R_{\psi_j^i}$ plus unary constraints with a single tuple, which is again in T since T is idempotent, value-renamable and contains the language $\{\emptyset\}$.

Furthermore, the sublanguages associated with different sets S_j also have an empty domain intersection. Since T is domain-decomposable, the resulting language is in T. □

Note that this result does not conflict with Theorem 3, since any class T that is domain-decomposable, value-renamable and not 1-Helly cannot have a finite Helly number (part of the proof of Theorem 4 amounts to showing that one can build in polynomial time arbitrarily large languages Γ such that every sublanguage is in T but Γ is not). While the proof may seem technical, the theorem is actually easy to use and applies almost immediately to many known tractable classes: to prove W[2]-hardness of STRONG T-BACKDOOR on CSPs of arity bounded by r, one only has to prove value-renamability, domain-decomposability (which is usually straightforward) and exhibit a language Γ such that each $\{R\} \subset \Gamma$ is in T but Γ is not.

Example 4. An idempotent operation f is totally symmetric (TSI) if it satisfies $f(x_1, \ldots, x_a) = f(y_1, \ldots, y_a)$ whenever $\{x_1, \ldots, x_a\} = \{y_1, \ldots, y_a\}$. Using the same notations as in Example 2, it has been shown in [13] that CSP($\overline{\Gamma}$) is solved by arc-consistency if and only if Γ has TSI polymorphisms of all arities. We show that this class of languages (which we denote by T_{TSI}) falls in the scope of Theorem 4 even for binary relations. First, this class is composite and idempotent: If we denote by TS(a) the set of all possible TSI operations on $\mathbb{N}$ of arity a and ATS $= \prod_{a \in \mathbb{N}^*} TS(a)$, we have $T_{\text{TSI}} = \cup_{\mathcal{F} \in \text{ATS}} (\cap_{f \in \mathcal{F}} T_f)$. To prove domain-decomposability and value-renamability, we will use the equivalent and more convenient characterization that Γ is in T_{TSI} if and only if Γ has a TSI of arity $|D(\Gamma)|$. Without loss of generality, we consider TSI polymorphisms as set functions and write $f(x_1, \ldots, x_a) = f(\{x_1, \ldots, x_a\})$.

– Domain-decomposability: Let $\Gamma_1, \Gamma_2 \in T$ be constraint languages with respective TSI polymorphisms f_1, f_2 (of respective arities $|D(\Gamma_1)|, |D(\Gamma_2)|$), and $D(\Gamma_1) \cap D(\Gamma_2) = \emptyset$. Let f be the operation on $D(\Gamma_1) \cup D(\Gamma_2)$ of arity $|D(\Gamma_1) \cup D(\Gamma_2)|$ defined as follows:

$$f(x_1, \ldots, x_m) = \begin{cases} f_1(\{x_1, \ldots, x_m\}) \text{ if } \{x_1, \ldots, x_m\} \subseteq D(\Gamma_1) \\ f_2(\{x_1, \ldots, x_m\}) \text{ if } \{x_1, \ldots, x_m\} \subseteq D(\Gamma_2) \\ \max(x_1, \ldots, x_m) \text{ otherwise} \end{cases}$$

f is totally symmetric and preserves both Γ_1 and Γ_2, so f is a polymorphism of $\Gamma_1 \cup \Gamma_2$ and $\Gamma_1 \cup \Gamma_2 \in T_{\text{TSI}}$. Therefore, T_{TSI} is domain-decomposable.

– Value-renamability: Let $\Gamma \in T_{\text{TSI}}$ be a language with a TSI polymorphism f_1 of arity $|D(\Gamma)|$. Let $\phi : D_1 \to D_2$ be a bijection, where D_1 is a conservative subdomain of $D(\Gamma)$ and $D_2 \cap D(\Gamma) = \emptyset$. Then, the operation of arity

$|D(\Gamma) \cup D_2|$ defined as

$$f(x_1, \ldots, x_m) = \begin{cases} f_1(\{x_1, \ldots, x_m\}) & \text{if } \{x_1, \ldots, x_m\} \subseteq D(\Gamma) \\ \phi(f_2(\{\phi^{-1}(x_1), \ldots, \phi^{-1}(x_m)\})) & \text{if } \{x_1, \ldots, x_m\} \subseteq D_2 \\ \max(x_1, \ldots, x_m) & \text{otherwise} \end{cases}$$

is a TSI and preserves both Γ and R_ϕ (the proof is straightforward using the fact that D_1 is a conservative subdomain), so $\Gamma \cup \{R_\phi\} \in T_{\text{TSI}}$ and T_{TSI} is value-renamable.

- Not 1-Helly: Let $R_1 = [(0,0),(0,1),(1,0)]$ and $R_2 = [(1,1),(0,1),(1,0)]$. Both $\{R_1\}$ and $\{R_2\}$ are in T_{TSI} (as they are respectively closed by min and max, which are 2-ary TSIs), but $\{R_1, R_2\}$ is not (R_1 forces $f(0,1) = 0$ and R_2 forces $f(0,1) = 1$ for every TSI polymorphism f), so T_{TSI} is not 1-Helly.

Finally, we conclude that Theorem 4 applies to T_{TSI} even for binary constraints. The same reasoning also applies to many other tractable classes, such as languages preserved by a near-unanimity ($f(y,x,\ldots,x) = f(x,y,x,\ldots,x) = \ldots = f(x,\ldots,x,y) = x$) or a Mal'tsev ($f(x,x,y) = f(y,x,x) = y$) polymorphism.

5 Related Work

A very recent paper by Gaspers et al. [3] has independently investigated the same topic (parameterized complexity of strong backdoor detection for tractable semantic classes) and some of their results seem close to ours. In particular, one of their theorems (Theorem 5) is similar to our Proposition 2, but is less general as they assume the target class to be an union of atomic classes. They also study the case where r is bounded and k is the parameter, as we do, but their result (Theorem 6) is more specific and can be shown to be implied by our Theorem 4.

6 Conclusion

We have shown that finding small strong backdoors to tractable constraint languages is often hard. In particular, if the tractable class is a set of languages closed by an idempotent operation, or can be defined by arbitrary conjunctions and disjunctions of such languages, then finding a backdoor to this class is NP-hard even when all constraints have a fixed arity. Moreover, it is W[2]-hard with respect to the backdoor size k.

When considering the larger parameter $k + r$, however, we have shown that strong backdoor detection is FPT provided that the target class is h-Helly for a constant h, that is, membership in this class can be decided by checking all h-tuples of relations. We then give a complete characterization of 1-Helly classes, and we use this result to show that any finite union of 1-Helly classes induces a backdoor problem FPT in $k + r$. Finally, we characterize another large family of tractable classes for which backdoor detection is W[2]-hard for the parameter k even if r is fixed. This result can be used to derive hardness of backdoor detection for many known large tractable classes, provided they have certain natural properties (which we call value-renamability and domain-decomposability).

References

1. Williams, R., Gomes, C.P., Selman, B.: Backdoors to typical case complexity. In: Proceedings of the 18th International Joint Conference on Artificial Intelligence, IJCAI 2003, pp. 1173–1178. Morgan Kaufmann Publishers Inc., San Francisco (2003)
2. Nishimura, N., Ragde, P., Szeider, S.: Detecting Backdoor Sets with Respect to Horn and Binary Clauses. In: SAT (2004)
3. Gaspers, S., Misra, N., Ordyniak, S., Szeider, S., Živný, S.: Backdoors into heterogeneous classes of sat and csp. In: Proceedings of the 28th AAAI Conference on Artificial Intelligence, AAAI 2014 (2014)
4. Gaspers, S., Szeider, S.: Strong Backdoors to Bounded Treewidth SAT. In: FOCS, pp. 489–498 (2013)
5. Feder, T., Vardi, M.Y.: The computational structure of monotone monadic snp and constraint satisfaction: A study through datalog and group theory. SIAM Journal on Computing 28(1), 57–104 (1998)
6. Bulatov, A.A.: Complexity of conservative constraint satisfaction problems. ACM Trans. Comput. Logic 12(4), 24:1–24:66 (2011)
7. Barto, L., Bulin, J.: Csp dichotomy for special polyads. IJAC 23(5), 1151–1174 (2013)
8. Barto, L., Kozik, M.: Constraint satisfaction problems solvable by local consistency methods. J. ACM 61(1), 3:1–3:19 (2014)
9. Jeavons, P., Cohen, D., Gyssens, M.: Closure properties of constraints. J. ACM 44(4), 527–548 (1997)
10. Jeavons, P., Cohen, D., Cooper, M.: Constraints, consistency, and closure. Artificial Intelligence 101, 101–1 (1998)
11. Idziak, P., Markovic, P., McKenzie, R., Valeriote, M., Willard, R.: Tractability and learnability arising from algebras with few subpowers. SIAM Journal on Computing 39(7), 3023–3037 (2010)
12. Bulatov, A.A.: Combinatorial problems raised from 2-semilattices. Journal of Algebra 298(2), 321–339 (2006)
13. Dalmau, V., Pearson, J.: Closure functions and width 1 problems. In: Jaffar, J. (ed.) CP 1999. LNCS, vol. 1713, pp. 159–173. Springer, Heidelberg (1999)
14. Jeavons, P., Cooper, M.: Tractable constraints on ordered domains. Artificial Intelligence 79, 327–339 (1995)
15. Barto, L., Kozik, M.: Constraint satisfaction problems solvable by local consistency methods. Journal of the ACM (JACM) 61(1), 3 (2014)
16. Kozik, M., Krokhin, A., Valeriote, M., Willard, R.: Characterizations of several maltsev conditions (2013) (preprint)
17. Dilkina, B., Gomes, C.P., Sabharwal, A.: Tradeoffs in the complexity of backdoors to satisfiability: dynamic sub-solvers and learning during search. Annals of Mathematics and Artificial Intelligence, 1–33 (2014)
18. Bessiere, C., Carbonnel, C., Hebrard, E., Katsirelos, G., Walsh, T.: Detecting and exploiting subproblem tractability. In: International Joint Conference on Artificial Intelligence (IJCAI), Beijing, China (August 2013)
19. Chen, J., Kanj, I.A., Xia, G.: Improved upper bounds for vertex cover. Theor. Comput. Sci. 411(40-42), 3736–3756 (2010)
20. Downey, R.G., Fellows, M.R.: Parameterized Complexity, 530 p. Springer (1999)
21. Karp, R.M.: Reducibility among combinatorial problems. In: Complexity of Computer Computations, pp. 85–103 (1972)
22. Carbonnel, C., Cooper, M.C., Hebrard, E.: On backdoors to tractable constraint languages (extended paper), `http://arxiv.org/abs/1404.3675`

Nested Constraint Programs

Geoffrey Chu and Peter J. Stuckey

National ICT Australia, Victoria Laboratory,
Department of Computing and Information Systems,
University of Melbourne, Australia
{gchu,pjs}@cis.unimelb.edu.au

Abstract. Many real world discrete optimization problems are expressible as nested problems where we solve one optimization or satisfaction problem as a subproblem of a larger meta problem. Nested problems include many important problem classes such as: stochastic constraint satisfaction/optimization, quantified constraint satisfaction/optimization and minimax problems. In this paper we define a new class of problems called nested constraint programs (NCP) which include the previously mentioned problem classes as special cases, and describe a search-based CP solver for solving NCP's. We briefly discuss how nogood learning can be used to significantly speedup such an NCP solver. We show that the new solver can be significantly faster than existing solvers for the special cases of stochastic/quantified CSP/COP's, and that it can solve new types of problems which cannot be solved with existing solvers.

1 Introduction

An *aggregator constraint* takes the form: $y = agg([f(x_1, \ldots, x_n, z_1, \ldots, z_m) \mid z_1, \ldots, z_m \text{ where } C(x_1, \ldots, x_n, z_1, \ldots, z_m)])$ where *agg* is an aggregator function such as **sum, max, min, and, or**, f is a function which we will call the local function, c is a (set of) local constraint(s), x_i are some input variables, y is an output variable, and z_i are some local variables.

Aggregator constraints are an extremely flexible and powerful modelling construct, especially when we allow them to be nested inside each other. Problems such as constraint satisfaction/optimization problems, stochastic constraint satisfaction/optimization problems, quantified constraint satisfaction/optimization problems, bi-level and multi-level programming, and many others, can be expressed using aggregator constraints. Unfortunately, most of these problem classes, and the solvers designed for them, only support a very restricted subset of aggregator constraints. For example CP solvers typically cannot handle aggregator constraints natively at all, and rely on some sort of *unrolling* procedure to convert them into primitive constraints first.

An aggregator constraint can be unrolled by eliminating the local variables and local constraints in the aggregator. If we can statically find the set of local solutions S to the constraint $C(x_1, \ldots, x_n, z_1, \ldots, z_m)$ (either independent of x_i or if the x_i are fixed), then for each $\theta \in S$, we create a variable $a[\theta]$ and post the constraints $a[\theta] = f(\theta)$, then post the constraint $y = agg([a[\theta] \mid \theta \in S])$. This completely eliminates the need to handle those local variables and constraints

B. O'Sullivan (Ed.): CP 2014, LNCS 8656, pp. 240–255, 2014.

during solving, but at a potential cost of creating exponentially many variables and constraints.

Example 1. Suppose we have variable arrays p and q. Suppose we have aggregator constraint $y = \mathbf{max}([p[z] + z \times q[z] \mid z \textit{ where } z \in \{1, 2, 3\})$. We can unroll this to: $y = max([a[1], a[2], a[3]])$, $a[1] = p[1] + q[1]$, $a[2] = p[2] + 2 \times q[2]$, $a[3] = p[3] + 3 \times q[3]$. □

Such an approach can handle many common CSP problems, and can also be used to convert stochastic CSP's and quantified CSP's into normal CSP's which can be solved using standard CP solvers. For example, scenario-based methods for solving stochastic CSP's [1] eliminate the stochastic variables in order to convert the problem into a CSP.

However, there are significant problems with this approach. Firstly, in general, it may not be possible or efficient to calculate the set of local solutions S statically. E.g., if the input variables x_i are not fixed at compile time, then it may not be possible at all, or if we have complex constraints in c (like other aggregator constraints), it may take exponential time just to check the satisfiability of an assignment. Secondly, if there are many nested aggregator constraints, then such unrolling could create an exponential number of variables and constraints, causing the solver to run out of memory. For example, in a k-stage stochastic CSP, we have k nested **max** and weighted **sum** aggregators. If each stage had $O(S)$ scenarios and we were to unroll all the weighted sum aggregator constraints by eliminating their local variables, then we end up with $O(S^k)$ variables and constraints which could easily cause the solver to run out of memory. Thirdly, not all the terms in the aggregation are necessarily relevant, especially with aggregators like **max**, **min**, **and**, **or**, etc, where evaluating one term may mean that other terms do not need to be evaluated or do not need to be evaluated fully.

An alternative method for handling aggregator constraints is to keep the local variables and constraints, and dynamically calculate the local solutions to the aggregator constraint during solving, rather than statically at compile time. Such search-based approaches have been used in stochastic CSP/COP solvers [2], in quantified CSP/COP solvers [3], and in quantified Boolean formulae solvers, e.g., [4,5]. In this paper we go further than these works by defining a much more general class of problems which we will call *nested constraint programs* (NCP's). Rather than only allowing linear aggregation structures where the output of one aggregator is immediately used as the function for the next as in stochastic CSP/COP's and quantified CSP/COP's, we represent the output of aggregators with a variable and allow these variables to be used in an arbitrary manner in the parent context. This means that they can be used in constraints, or as part of some complex expression for the parent's local function. It also means that we can have multiple aggregator constraints in the local constraints of another aggregator constraint, and thus we can have tree-like quantification structures. NCP's include stochastic CSP/COP, quantified CSP/COP/COP+ and many more as special cases, but can model problems which do not fit in any of these subclasses. We describe a new CP-based solver for solving NCP's, and briefly describe how to apply nogood learning in such a solver.

The contributions of this paper are:

- An expressive framework for nested constraint programs (Section 3).
- A propagation based solver architecture that supports this class of problems (Section 4).
- Experiments showing that the resulting system is highly competitive with existing solvers for specialized subclasses of NCP (Section 5).

2 Preliminaries

A *valuation*, θ, is a mapping of variables to values, denoted $\{x_1 \mapsto d_1, \ldots, x_n \mapsto d_n\}$. Let $vars(\theta) = \{x_1, \ldots, x_n\}$. We can apply a valuation to a variable $\theta(x_i)$ to return the value d_i, and extend application of valuations θ to arbitrary expressions involving $vars(\theta)$ in the obvious way.

A *constraint*, c, is a set of valuations over a set of variables $vars(c)$. A valuation θ is a *solution* of c if $\{x \mapsto \theta(x) \mid x \in vars(c)\} \in c$. A valuation θ is a solution of a set of constraints C if it is a solution for each $c \in C$. We write $c_1 \models c_2$ if every solution of c_1 is a solution of c_2.

A *literal* is a unary constraint (we can restrict to the forms $x = d, x \neq d, x \geq d, x \leq d$), or *false*. A *domain* D is a conjunction of literals over $vars(D)$. We use notation $D(x) = \{\theta(x) \mid \theta$ is a solution of $D\}$. We use *range notation* $[\, l \,..\, u\,] = \{d \mid l \leq d \leq u\}$. A *singleton domain* is one where $|D(x)| = 1, x \in vars(D)$, and we let $\theta_D = \{x \mapsto d_x \mid x \in vars(D), D(x) = \{d_x\}\}$ in this case.

A *propagator* $p(c)$ for constraint c is an inference algorithm, it maps a domain D to a conjunction of literals $p(c)(D)$, where $D \wedge c \models p(c)(D)$. We shall sometimes treat this conjunction as a set. We assume each propagator is *checking*, that is if $\forall x \in vars(c).|D(x)| = 1$ then $p(c)(D) = \emptyset$ if θ_D is a solution of c and $\{false\}$ otherwise.

In *lazy clause generation (LCG)* solvers [6,7] propagators are also required to return explanations for each new consequence $l \in p(c)(D)$, that is an explanation clause $e \equiv l_1 \wedge \cdots l_n \rightarrow l$ where $\forall 1 \leq i \leq n, D \models l_i$ and $c \models e$. LCG solvers, like SAT solvers, create an implication graph, where every new consequence is attached to a reason. On failure this used to create a *nogood* by repeatedly replacing literals in the explanation of failure until only one literal that became true after the last decision remains. This nogood is guaranteed to generate new propagation information. See [5] for more details.

3 Aggregators and Nested Constraint Programs

An *aggregator function* is a function which maps a multiset (list) of values to a single value by performing some sort of aggregation over them, e.g., by summing over them, or taking the maximum, etc. Aggregators are functions on multisets, they cannot make use of the order of elements in the list they operate on. They may be partial functions.

An *aggregator constraint* is of the form: $y = agg([f(x_1, \ldots, x_n, z_1, \ldots, z_m) \mid z_1, \ldots, z_m \textit{ where } C(x_1, \ldots, x_n, z_1, \ldots, z_m)])$ where agg is an aggregator function, f is the *local function* of the aggregator constraint, and C is a set (or conjunction) of *local constraints* of this aggregator constraint. We assume the local function f is total in its inputs, if not we can add constraints to C to ensure it is total for all possible local solutions, thus implementing the relational semantics [8]. The scope of this aggregator constraint is $vars(a) = \{y, x_1, \ldots, x_n\}$. Given aggregator constraint a, let $ovar(a) = y$ be the *output variable*, $ivars(a) = \{x_1, \ldots, x_n\}$ be the *input variables*, $lvars(a) = \{z_1, \ldots, z_m\}$ be the *local variables*, and $lcons(a) = C$.

The solutions of an aggregator constraint $c \equiv y = agg([f(x_1, \ldots, x_n, z_1, \ldots, z_m) \mid z_1, \ldots, z_m \text{ where } C(x_1, \ldots, x_n, z_1, \ldots, z_m)])$ are defined inductively on the depth of nesting. Let Θ be the solutions of the constraint C, then the solutions of c are

$$\{x_1 \mapsto d_1, \ldots, x_n \mapsto d_n, y \mapsto agg([f(d_1, \ldots, d_n, \theta(z_1), \ldots, \theta(z_m) \mid \theta \in \Theta, \theta(x_i) = d_i])\}$$

for all $(d_1, \ldots, d_n)$ where $\exists \theta \in \Theta . \forall 1 \leq i \leq n . \theta(x_i) = d_i$. If C includes no aggregator constraints then this definition is self-contained, otherwise we can determine the solutions of C by induction (since the depth of nesting is 1 less) and use them to define the solutions of c. Note that an aggregator constraint may have no solutions if the aggregation function is not defined on the resulting multiset of solutions, e.g. $y = \textbf{average}([])$.

Example 2. Suppose we have $y = \textbf{sum}([z_1 \mid z_1, z_2 \text{ where } z_1, z_2 \in [1..3], z_1 + z_2 \leq x])$. Then given a particular value of x, we find all solutions to z_1, z_2 which satisfy $z_1, z_2 \in [1..3], z_1 + z_2 \leq x$, and sum over the z_1 values of those solutions in order to calculate y. So for example, this constraint would allow the tuples: $(x, y) \in \{\ldots, (0,0), (1,0), (2,1), (3,4), (4,10), (5,15), (6,18), (7,18), \ldots\}$. □

Some commonly used aggregators are **sum**, **product**, **min**, **max**, **and**, **or**, **average**, **stddev**, **variance**, whose definitions are already well known. Note that we have not restricted the types of the variables or the input or output arguments to the aggregator functions. This means for example, we can define aggregator functions which take in a list of tuples as arguments, or which return a tuple as the output, etc. For example, weighted average can be defined on a list of pairs where the first element is the weight and the second element is the value and it returns a single value as output. Similarly, we can extend **min** and **max** to take a list of tuples as argument and use the lexico-graphical ordering to return the smallest or largest tuple as the return value.

A *nested constraint program* (NCP) consists of a single aggregator constraint with no input variables: e.g. $y = agg([f \mid z_1, \ldots, z_m \text{ where } C(z_1, \ldots, z_m)])$ The goal of an NCP is to determine the value of the output variable of the top-level aggregator constraint. The power of NCPs arise from the fact that the local constraints of one aggregator constraint can contain other aggregator constraints. Thus in general, we can have a nested structure where we have a tree of aggregator constraints, each with its own local variables and constraints, and where each aggregator constraints is a local constraint of its parent aggregator constraint.

Example 3. In the simple production planning problem studied in [2], in each stage, we can choose to produce 0 or more books. After production, there is a stochastic demand for books between 100 and 105 with equal probabilities for each. There are soft constraints enforcing that the available stock be sufficient to satisfy the demand. The problem is to find a policy whose expected satisfiability is above a certain threshold α. We can model a 3 stage instance as follows:

$$\begin{aligned}
&r = \mathbf{or}([m_1 \geq \alpha \mid m_1 \textit{ where} \\
&\quad m_1 = \mathbf{max}([a_1 \mid s_1, p_1, a_1 \textit{ where } s_1 = 0 \wedge \\
&\quad\quad a_1 = \mathbf{average}([\texttt{bool2int}(s_1 + p_1 \geq d_1) \times m_2 \mid d_1 \in [100..105], m_2 \textit{ where} \\
&\quad\quad\quad m_2 = \mathbf{max}([a_2 \mid s_2, p_2, a_2 \textit{ where } s_2 = s_1 + p_1 - d_1 \wedge \\
&\quad\quad\quad\quad a_2 = \mathbf{average}([\texttt{bool2int}(s_2 + p_2 \geq d_2) \times m_3 \mid d_2 \in [100..105], m_3 \textit{ where} \\
&\quad\quad\quad\quad\quad m_3 = \mathbf{max}([a_3 \mid s_3, p_3, a_3 \textit{ where } s_3 = s_2 + p_2 - d_2 \wedge \\
&\quad\quad\quad\quad\quad\quad a_3 = \mathbf{average}([\texttt{bool2int}(s_3 + p_3 \geq d_3) \mid d_3 \in [100..105]])])])])])])
\end{aligned}$$

Example 4. Consider the 2-player Nim-Fibonacci game [9]. The game starts with n matches. The first player may take between 1 to $n-1$ of the matches. Thereafter, the turns alternate and the current player may take between 1 to $2k$ of the matches where k is the number of matches taken by the previous player. The player who takes the final match wins. The problem is to find out for each n whether the first player has a winning strategy. It has the interesting property that the first player has a winning strategy iff n is not a Fibonacci number. The problem can be modelled as follows. Given turn i, let r_i be the number of matches left, l_i be the maximum number of matches that can be taken during that turn, t_i the actual number of matches taken, and w_i whether there is a winning strategy from that position.

$$\begin{aligned}
&w_1 = \mathbf{or}([l_1 \geq r_1 \vee \neg w_2 \mid w_2, l_1, r_1, t_1 \textit{ where} \\
&\quad l_1 = n - 1 \wedge r_2 = n \wedge 1 \leq t_1 \leq l_1 \wedge \\
&\quad w_2 = \mathbf{or}([l_2 \geq r_2 \vee \neg w_3 \mid w_3, l_2, r_2, t_2 \textit{ where} \\
&\quad\quad l_2 = 2 \times t_1 \wedge r_2 = r_1 - t_1 \wedge 1 \leq t_2 \leq l_2 \wedge \\
&\quad\quad w_3 = \mathbf{or}([l_3 \geq r_3 \vee \neg w_4 \mid w_4, l_3, r_3, t_3 \textit{ where} \\
&\quad\quad\quad l_3 = 2 \times t_2 \wedge r_3 = r_2 - t_2 \wedge 1 \leq t_3 \leq l_3 \wedge \\
&\quad\quad\quad \ldots \\
&\quad\quad\quad w_n = \textit{true}])\ldots])
\end{aligned}$$

The ability to model problems using tree-like quantification structures rather than the linear quantification structure used in stochastic CSP and quantified CSP allows certain kinds of optimisations.

Example 5. Consider a stochastic scheduling problem with precedence and non-overlap constraints C on n tasks where we fix the (array of) start times $\bar{s}$ within the makespan $[0..m]$, written as $\bar{s} \in \overline{[0..m]}$, but then each task duration d_i can then independently be one of three values $L_i = \{f_i, r_i, w_i\}$ fast, regular or slow, and we need to pay recourse $\textit{recourse}_c(s_{c_1}, s_{c_2}, d_{c_1}, d_{c_2})$ for the violation of each constraint $c \in C$ involving at most two tasks numbered c_1 and c_2. Given n tasks there are 3^n scenarios. The natural stochastic model is

$$\begin{aligned}
u = \mathbf{min}([\mathbf{average}([\mathbf{sum}([\textit{recourse}_c(s_{c_1}, s_{c_2}, d_{c_1}, d_{c_2}) \mid c \in C]) \\
\mid \bar{d} \in \bar{L}]) \mid \bar{s} \in \overline{[0..m]}])
\end{aligned}$$

where we find a schedule (start times) $\bar{s}$ which minimizes the expected recourse cost, over all possible durations $\bar{d} \in \bar{L}$ scenarios The problem is there are 3^n scenarios and hence evaluating a schedule is $O(|C|3^n)$. But, since the recourse for each constraint $c \in C$ is dependent on at most two stochastic durations d_{c_1} and d_{c_2}, we can model this instead as

$$u = \mathbf{min}([\mathbf{sum}([\mathbf{average}([recourse_c(s_{c_1}, s_{c_2}, d_{c_1}, d_{c_2}) \mid d_{c_1} \in L_{c_1}, d_{c_2} \in L_{c_2}]) \mid c \in C])) \mid \bar{s} \in \overline{[0..m]}])$$

There are at most 9 scenarios for each constraints recourse calculation, hence to evaluate a schedule is $O(|C|)$. Note that such a quantification structure is not supported by stochastic CSP solvers but can be solved with our more generic NCP solver. □

Aggregator functions involving tuples allows us to model things like bi-level programs.

Example 6. The Network Links Pricing problem [3] can be described as follows. The problem is to set the tariffs on the network links in order to maximise the profit of the owner of the links. The $i \in I$ customer (or data movement) will route their d_i data from src_i to snk_i using the smallest possible cost path. Each path has to cross a tolled arc $j \in J$ with cost per unit data t_j. We assume the cost per unit data from src_i to snk_i via j is $c_{i,j}$ for the rest of the network. Hence the cost to the customer of a path through arc j is $(c_{i,j} + t_j) \times d_i$. The income to the network is $t_j \times d_i$. The customer can choose another independent network provider with cost u_i instead of using this network. The problem is determine the toll t_j from some set of possibilities T_j ($\bar{t} \in \bar{T}$) for each arc j to maximise revenue. We assume customers will pick the cheapest link for them, but if there are ties, they will pick the one most profitable for the operator, as the network operator can simply adjust their tariffs by some small ϵ to make that choice the cheapest for that customer. The interesting thing here is that in the subproblem, we need to minimize one quantity, i.e., the cost the customer, but return another value as the result, i.e., the profit to the operator. Such problems cannot be modelled as normal QCOPs, but can be modelled as QCOP+s or as NCPs by using tuples.

$$y = \mathbf{max}([\mathbf{sum}(p * d_i \mid c, p \ where \ (c, p) = \mathbf{max}([(-(c_{i,j} + t_j), t_j) \mid j \in J \ where \ (c_{i,j} + t_j) \times d_i \leq u_i])) \mid i \in I]) \mid \bar{t} \in \bar{T}])$$

Note that the inner (lexico-graphical) **max** on the pair picks the link with the lowest cost, and among those the one with highest profit, and returns that pair as the return value. □

The greater expressivity of NCP allow us to model all kinds of meta problems where the results or properties of subproblems can be used in constraints or the objective function of the parent problem, or the output of one subproblem can be used as the input to another, etc.

Example 7. Consider a Sudoku problem, given a set of possible clues $\{x_{i_l j_l} = d_l \mid 1 \leq l \leq n\}$, find the smallest subset of these clues such that the resulting Sudoku problem has a unique solution. We can model this as a NCP, using a

Boolean variable b_l to indicate whether a clue is used:

$$\begin{aligned} c = \mathbf{min}([\mathbf{sum}([b_l \mid l \in [1..n]]) \mid \overline{b \in [0..1]} \ \textit{where} \\ 1 = \mathbf{sum}([1 \mid \overline{\bar{x} \in [1..9]} \ \textit{where} \\ \mathbf{and}([b_l \rightarrow x_{i_l j_l} = d_l \mid l \in [1..n]]) \wedge \textit{sudoku}(\bar{x})])]) \end{aligned}$$

where $\textit{sudoku}(\bar{x})$ are constraints enforcing that $\bar{x}$ is a solution to a Sudoku problem. □

CSP/COP, stochastic CSP/COP [2], QCSP [10], QCOP, QCOP+ [3], stochastic SAT, MAXSAT, QBF [11], influence diagrams, finite horizon markov decision processes, MPE and MAP queries over stochastic graphical models such as Bayesian nets, and probably many more, are all expressible as NCP. In addition to this, there are many problems expressible as NCP which cannot easily be expressed as any of the previously mentioned problem classes. Thus NCP is a very expressive and generic problem class. In this paper, we are interested in solving NCPs using a modified CP solver. Thus we restrict our attention to *finite NCPs*, i.e., where every local variable is constrained to have a fixed finite initial domain given by explicit local constraints, e.g. $z \in [1..100]$.

4 Solving NCP's

In this section, we describe how to solve NCPs. The main idea is to augment a standard CP solver with a new class of propagators which can encapsulate the subproblem modelled by an aggregator constraint. Such a propagator constrains the input and output variables of the aggregator constraint and performs propagation on these variables. The major difference with normal CP propagators is that unlike a normal CP propagator which run a self-contained algorithm to perform propagation, these new propagators may take control of the search engine itself to do some search in order to perform their propagation. However, at the end of their propagation algorithm, they must return the search engine to its former state so that it is as if nothing has happened. Thus these new propagators simply change the domains of variables, just like all other standard CP propagtors, and there is no need to treat them specially in any way. The fact that the set of assignments allowed by the aggregator constraint is defined via aggregating the solutions to a CP problem rather than extensionally or intensionally as in normal CP constraints is irrelevant. As far as the parent subproblem is concerned, the aggregator constraint is just a normal constraint which is satisfied by a particular set of assignments, and we have a propagator which is capable of enforcing the constraint, and thus the parent subproblem can be treated as a completely normal (non-nested) CP problem. This allows us to solve NCPs using CP solvers by simply adding a new kind of propagator.

Recall that there are two main ways to deal with an aggregator constraint. We can either unroll it by eliminating the local variables and local constraints as described in Section 1, or we can post a propagator which can handle it natively as described above. The first suffers from a potential combinatorial explosion in the number of variable/constraints required, while the second tends to have weaker propagation strength. The key here is to try to find the best tradeoff between propagation strength and time/memory usage. For simple aggregator constraints, we would often get a better tradeoff simply by unrolling them. Whereas for more complicated ones with non-trivial local constraints, the search based approach may give a better tradeoff. In general, we use the following policy: given an aggregator constraint a, if the local solutions of a can be computed at compile time and there are no more than L of them where L is some user defined parameter, and there are no nested aggregator constraints within a, then we unroll a. Otherwise, we leave it as is and post a propagator that can handle it natively. Such a policy ensures that we avoid any sort of exponential blowup in problem size that can occur when we unroll aggregator constraints.

After unrolling, we create a domain object for each variable and a propagator object for each constraint and aggregator. Unlike a normal CP solver where all variables exist in the one existential context, variables in a NCP may belong to different contexts. Each variable is either a local variable to one aggregator constraint or is the root variable. It can also be an input variable of zero or more descendant aggregator constraints. Each local constraint can contain local variables from the same aggregator constraint and also input variables which are local to the ancestor aggregator constraints.

First, we consider a non-aggregator constraint c. For each such constraints c, we create a modified propagator $p(c)$ which is identical to the standard CP propagator, except that it is only allowed to prune values from the domains of the *local variables* of the parent aggregator constraint $y = agg([...])$ to which it belongs. It is *incorrect* in general to prune values from the input variables. The reason for this is that when a local constraint has no solutions, this does not mean that the parent aggregator constraint has no solutions, rather it means that the value of y is given by $agg([])$ since the local problem has no solutions. This does not constrain the input variables, so pruning their domains is incorrect!

Example 8. Consider the following problem: $y = \mathbf{min}([x_2 \mid x_1, x_2 \ \textit{where} \ x_1 \in [0..3] \wedge x_2 \in [-4..4] \wedge x_2 = \mathbf{sum}([x_3 \mid x_3 \ \textit{where} \ x_3 \in [1..5] \wedge x_3 \leq x_1])])$. The local constraint $x_2 = \mathbf{sum}([x_3 \mid x_3 \ \textit{where} \ x_3 \in [1..5] \wedge x_3 \leq x_1])$ has the solutions $(x_1, x_2) \in \{\ldots, (-1, 0), (0, 0), (1, 1), (2, 3), (3, 6), (4, 10), (5, 15), (6, 15), \ldots\}$. This means that the local solutions of the **min** aggregator are $(x_1, x_2) \in \{(0, 0), (1, 1), (2, 3)\}$, giving $y = 0$. Suppose however, we allowed the constraint $x_3 \leq x_1$ to propagate on its input variable x_1. Then at the root, since we have $x_3 \in [1..5]$ we can immediately propagate $x_1 \geq 1$, and we have completely pruned off a totally valid local solution ($(x_1, x_2) = (0, 0)$) of the **min** aggregator, leading to an answer of $y = 1$ which is simply wrong. □

For each aggregator constraint a, we create a propagator $p(a)$. This can use the semantics of its aggregator function to propagate domain changes to the output variable based on the domains of its input and local variables. For example, consider an aggregator constraint $a \equiv y = \mathbf{min}([f(z, x) \mid z \ \textit{where} \ c(z, x)])$. Suppose that a has not yet taken control of the search (i.e., x is not yet fixed), but propagation of the local constraints c has already forced a lower bound l on $f(z, x)$. Then we can immediately propagate $y \geq l$, because no matter what x ends up being set to, any local solution must have local objective value greater than or equal to l.

Secondly, we maintain a copy of each aggregator constraint in A which will propagate in a more complex way. When all the input variables of a are fixed, then a can take control of the search engine and perform search on its local variables in order to calculate the value of the output variable. Different aggregator constraints can use different search strategies. The search strategy can either be defined in the model, or some sort of autonomous or default search can be used. They will have different conditions for when they can yield control of the search engine. For example, an **or** (resp. **and**) aggregator can yield control as soon as a *true* (resp. *false*) solution is found. A **min** aggregator will perform local branch and bound in order to find its output value. The first branching decision that it will make will be of the form $o < k$ where o is the objective variable, and k is either the value of the best solution found so far, or $\max D(y) + 1$ if it has just taken control.[1] If it finds a solution with objective value k, it can propagate $y \leq k$. If the branch and bound decision $o < k$ produces failure, then it can immediately propagate $y \geq k$. It can yield control if it either proves $y \leq \min D(y) - 1$, $y \geq \max D(y) + 1$ or it finds the optimal solution and proves optimality. Similarly, a **sum** aggregator would perform a search to find all of its local solutions to calculate the sum of the local function values. If it proves that $y \leq \min D(y) - 1$ or $y \geq \max D(y) + 1$, it can terminate early.

Pseudo-code for the algorithm is given in Figure 1. We set up an initial domain D for all variables, and set P to be the propagators $p(c)$ for each $c \in C \cup A$. Initially, the root aggregator *root* is in control of the search engine with the call $\mathsf{agg}(root, D, P, A)$. The aggregator constraint sets the variables V as its input and local variables, and invokes search.

Then propagation is performed to fixed point or failure by $\mathsf{propagate}$. Propagation repeatedly chooses a propagator p from the queue Q, calculates a new domain D' then adds all the propagators in P that may need to be recomputed due to changes in the domain computed by $\mathsf{new}(P, D, D', a)$, repeating until the queue is empty. There is a subtle difference with regular CP propagation. If a variable x gets an empty domain, this does not necessarily mean that the last decision made was infeasible. If $x \notin V$ is not a local variable then it simply means that the aggregator a' that introduces x (which must be a descendent of a in the aggregator tree) has no local solution given the decisions of its ancestor aggregators. This means that a' needs to be woken up so that it can propagate $y = agg([])$ where y is its output variable and agg is its aggregator function. Note that a' has to be a descendant of a, because decisions made by a can only cause domain changes to variables belonging to descendants of a.

If we reach propagation fixed point, then we need to check whether any other aggregator constraints become eligible for taking over control of the search. An aggregator a is *eligible* if:

- its not currently suspended and either all its input variables are fixed or one of its local variables has an empty domain, and
- its output variable has not already been fixed by a when it executed earlier on the same fixed inputs or empty domain, and
- its output variable appears in at least one constraint which is not already satisfied.

[1] Assuming y is integer for simplicity of explanation.

```
agg(a, D, P, A)
   let a = agg([o|lvars(a) where lcons(a)])
   search(D, ivars(a) ∪ lvars(a), P, A, {p(c) | c ∈ lcons(a)}, a)
   let Θ be the set of solutions processed
   D := D ∧ ovar(a) = agg([θ(o)|θ ∈ Θ])
   return D

search(D, V, P, A, Q, a)
   D := propagate(D, V, P, A, Q, a)
   if (∃x ∈ V.D(x) = ∅) return false
   if (∀x ∈ V.|D(x)| = 1)
      let θ = {x ↦ d_x | x ∈ V, D(x) = {d_x}}
      return process_solution(θ, a, D)
   else
      {c_1, ..., c_m} := branch(a, D)
      for i ∈ 1..m
         if (search(D, P ∪ Q, A, {p(c_i)}, a))
            return true
      return false

propagate(D, V, P, A, Q, a)
   P := P ∪ Q
   repeat
      while (∀x ∈ V.D(x) ≠ ∅ ∧ ∃p' ∈ Q)
         Q := Q − {p'}
         D' := D ∧ p'(D)
         Q := Q ∪ new(P, D, D', a)
         D := D'
      if (∃a ∈ A.eligible(a))
         A := A − {a}
         D' := agg(a, D, P, A)
         Q := Q ∪ new(P, D, D', a)
         D := D'
   until Q = ∅
   return D
```

Fig. 1. Pseudo-code for evaluating NCPs

In the last case, no other constraint cares about the value so we do not need to calculate it. If an aggregator is eligible, it will immediately take over control of the search engine and the aggregator constraint which was previously in control will be suspended until this one returns. If multiple aggregators become eligible at the same time, then we choose one of the aggregators closest to the root in the aggregator tree.

After propagation quiesces there are three cases. If a *local* variable has no solution this indicates failure, hence search backtracks. If all the local variables are fixed then we have discovered a new solution θ. Then the aggregator constraint will do whatever sort of bookkeeping it needs to do to calculate its aggregate value using process_solution. If it determines it now has enough information to determine the final aggregate value or fail, process_solution returns *true* and the search finishes, otherwise it backtracks and continues the search. If propagation neither fails nor succeeds, the aggregator constraint a in control of the search makes a branching decision $c_1 \vee \cdots \vee c_n$ using its branching heuristic $\mathsf{branch}(a, D)$, and searches each resulting subproblem. Search continues until either the entire subtree for this subproblem is explored or process_solution detects early termination.

When search finishes the aggregator calculates the result on its output variable and updates the domain accordingly, then yields control to its parent. The algorithm terminates when the root aggregator constraint yields control, at which point, we have calculated the value of its output variable and solved the NCP.

Example 9. Consider the problem of Example 8. We create an initial domain $D(x_1) = [0..3]$, $D(x_2) = [-4..4]$, $D(x_3) = [1..5]$, $D(y) = [-\infty..\infty]$. We create propagators for the constraints $x_3 \leq x_1$ and the two aggregator constraints. Execution begins by calling the search on the root **min** aggregator. Propagation uses $x_3 \leq x_1$ to set $D(x_3) = [1..3]$ and queisces. Assume the **min** aggregator makes a branching decision on the x_1 variable $x_1 = 0 \vee x_1 \geq 1$ (since the branch and bound decision $x_2 < +\infty \vee x_2 \geq +\infty$ is not useful). Searching on the left branch sets $D(x_1) = \{0\}$ which causes $D(x_3) = \emptyset$ which wakes the **sum** aggregator (with an empty domain for the local variable x_3) which immediately returns setting $D(x_2) = \{0\}$. The **min** aggregator processes the solution $(x_1, x_2) = (0, 0)$, (we will at this stage propagate that $D(y) = [-\infty..0]$). Search then tries the right branch where propagation returns the domain $D(x_1) = [1..3]$, $D(x_2) = [-4..4]$, $D(x_3) = [1..3]$ by propagating the constraint $x_1 \geq 1$. Once again the **min** aggregator makes a branching decision $x_1 = 1 \vee x_1 \geq 2$, and taking the left branch sets $D(x_1) = \{1\}$ and $D(x_3) = \{1\}$ waking the **sum** aggregator since its input variable x_1 is fixed. This aggregator finds a single solution $(x_1, x_3) = (1, 1)$ and then sets $D(x_2) = \{1\}$. The **min** aggregator processes the solution $(x_1, x_2) = (1, 1)$ by just throwing it away. Search continues with the right branch where eventually the **min** aggregator finds the remaining solution $(x_1, x_2) = (2, 3)$, and returns $y = 0$.

Note that if we use a cleverer propagator for the **sum** aggregator then at the first propagation step it will set $D(x_2) = [0..4]$ since the sum of any number of positive values ($x_3 \in [1..3]$) is at least 0. The **min** propagator can also add the bounding constraint $x_2 < 0$ after it finds the first solution. These two together would cause search to terminate immediately once the first solution was found. □

In this paper we only consider computing the result of the NCP, in practice we will may want to know the "policy" of decisions that lead to this result. It is easy enough to expose the values of the local variables of the root aggregator, thus giving the "first-stage" decisions. To record the entire policy of decisions we would need to store a shorthand form of the entire search tree (including nested search trees) analogous to the approach used in QCOP+ [3].

4.1 Complexity

The time and space complexity of the algorithm depends on many things, such as the time and space complexity of the propagators and aggregators, and the search strategy. However, a very large subclass of finite NCP is PSPACE-complete. In particular, consider the subclass where all non-aggregator constraints have a polynomial space propagator (this is true for all commonly used CP constraints) and all aggregator functions require only a polynomial space to compute their output value when the elements of its list argument are fed in one by one (this is true for **sum**, **product**, **min**, **max**, **and**, **or**, **average**, **stddev**, **variance**, but not for aggregators like **median**). This subclass includes QBFs and quantified CSPs and thus is PSPACE-HARD. For such problems, the algorithm is PSPACE-complete.

4.2 Learning for NCPs

Nogood learning [5,6] significantly improves both SAT and CP solving performance. Similarly, it dramatically improve the efficiency of an NCP solver. Unfortunately we do not have sufficient space to adequately describe how to add nogood learning to an NCOP solver. Instead we will briefly discuss the uses of nogood learning, and some of the issues that arise in implementing it. There are generally two different kinds of nogoods that we can learn: ones which explain a local failure, and ones which explain the return value of a subproblem.

- Nogoods learned within one execution of an aggregator constraint a become new local constraints for a. Just like other local constraints they are only allowed to prune local variables. When we reexecute a with different input variable values, much of the search in a may be repeated, and these local nogoods can substantially reduce this repeated search, similar to the case for inter-problem nogood learning [14].
- We can cache the return value of an execution of an aggregator constraint a using a nogood, e.g. if we run a with input variables fixed to $x_1 = d_1, \ldots, x_n = d_n$ and find that output $y = d$ we can cache this as $x_1 = d_1 \wedge \cdots x_n = d_n \rightarrow y = d$. This nogood prevents us from having to run the aggregator again on the same input values.

  Nogood learning can be better than this however, since it may determine that only some of the input constraints are required to give the result of aggregator a leading to a much more general nogood. Consider for example the aggregator of Example 2, setting $x = 0$ gives $y = 0$, but nogood learning will learn that $x \leq 1 \rightarrow y = 0$. Similarly, $x = 7$ gives $y = 18$, but nogood learning will learn universally that $y \leq 18$.

The challenge for implementing nogood learning in an NCOP solver is to extend the propagators for aggregator constraints to explain their propagation behaviour. To do so we must be able to determine what parts of the input constraints contributed to the result of the aggregator. This requires combining the usual uses of nogoods, to explain why some part of the search *failed*, with explaining why some part of the search *succeeded with a certain value of the local function*. The success explanation part is entirely novel, and is a generalisation of solution analysis in QBF, (see e.g. [11]) where all constraints are clauses and they use specialized heuristics to pick a satisfying literal for each clause.

5 Experiments

Due to the very large range of problem classes that can be modelled as NCPs, it is difficult to compare against the current state of the art in all those problem classes. Instead, we concentrate on the problem classes where CP-based solvers have had some success, namely in stochastic CSP/COP problems and quantified CSP/COP problems. We implemented a NCP solver in Chuffed, a state-of-the-art CP solver that supports nogood learning. Experiments are run on Intel Xeon 2.40GHz processors, with a 1800s timeout. Times are given in seconds. We use an unrolling limit (L in Section 4) of 100. We use input order search and try the

Table 1. Comparison of the search-based NCP solver with learning (learn) and without learning (no-learn) with QeCode (qecode) on the Nim-Fibonacci Problem

Size	no-learn		learn		qecode	
	fails	*time*	*fails*	*time*	*fails*	*time*
5	2	0.01	2	0.01	23	0.01
10	20	0.01	13	0.01	1310	0.02
15	174	0.01	45	0.01	11116	0.26
20	1438	0.01	101	0.01	56560	1.61
30	62313	0.36	272	0.01	483346	16.24
40	4773553	30.13	727	0.01	—	TO
50	—	TO	1502	0.0	—	TO
100	—	TO	8461	0.21	—	TO
200	—	TO	45227	2.12	—	TO
500	—	TO	414152	55.29	—	TO

smallest value first. We compare against QeCode 2.0 [3], which is a state of the art QCOP+ solver, on the problems that it supports. QeCode does not support floating point variables or weighted sum aggregators. As a result, it is unable to solve stochastic COP's, so we only compare against it on integer and Boolean problems. We also compare against the published results of other systems that are not publicly available.

The Nim-Fibonacci problem is described in Example 4. It is a QCSP and can potentially be unrolled into a CSP by eliminating the universal variables and solved using a CP solver. However, this is not really practical as this produces $O(n^{n/2})$ variables and constraints and causes the solver to run out of memory on all but the smallest instances. In this experiment, we compare the new search-based NCP solver with and without learning with QeCode. It can be seen from Table 1 that our search-based NCP solver does significantly better than QeCode on this problem. Even without learning, we are much faster, due to the fact that we have variables representing the output values of subproblems and it is possible to propagate domain changes on them. Such variables do not exist in Qecode. Nogood learning provides a massive benefit due to its ability to explain successes. It is able to learn that if there is a winning strategy with a particular number of matches where you take k matches next, then in any other branch where you had the same number of matches but the limit on the number of matches you can take is greater than or equal to k, its a won game. In fact, the asymptotic complexity of the NCP solver is polynomial when nogood learning is used, as opposed to $O(n^n)$ if no learning is used.

The Network Links Pricing problem [3] is described in Example 6. The results are shown in Table 2. The no-learn solver is substantially faster than QeCode, which appears to be because Qecode does not use branch and bound during solving this problem. Nogood learning is only slightly beneficial for this problem, giving a constant factor reduction in node count and run times. This is not surprising, as the problem is very shallow (only 3 layers), and there is only a single inequality constraint in the final layer, so there is not much propagation going on, and thus not much opportunity for learning.

Table 2. Comparison of the search-based NCP solver without learning (no-learn) and with learning (learn) with QeCode (qecode) on the network link pricing problem

Stages	no-learn		learn		qecode	
	fails	*time*	*fails*	*time*	*fails*	*time*
6	37399	2.63	5037	1.80	376234	7.10
7	342823	23.06	52301	18.54	2218653	44.48
8	985622	68.24	121815	56.82	12148442	255.66
9	17566514	1225.80	2253269	903.01	—	TO

Table 3. Comparison on a simple production planning problem of the search-based NCP with solver learning (learn) with published results (from [1], run on a different machine) for the search-based stochastic CSP solver of [2] (search) and the scenario-based solver of [1] (scen), also showing how learn scales to larger numbers

Stages	no-learn		learn		search		scen	
	fails	*time*	*fails*	*time*	*fails*	*time*	*fails*	*time*
1	7	0.01	8	0.01	10	0.01	4	0.00
2	178	0.01	16	0.01	148	0.03	8	0.02
3	4574	0.36	24	0.01	3604	0.76	24	0.16
4	116305	9.24	32	0.01	95570	19.07	42	1.53
5	2955371	233.05	40	0.01	2616858	509.95	218	18.52
6	—	TO	48	0.01	—	TO	1260	474.47

Stages	learn	
	fails	*time*
10	80	0.03
20	160	0.11
30	240	0.27
40	320	0.53
50	400	0.93
100	800	7.40

Consider the simple production planning problem of Example 3 which was examined in [2] using search-based approaches and [1] using unrolling or scenario generation. From Table 3, it is clear there is an asymptotic difference in complexity. This occurs since each subproblem only involves a single input parameter, i.e., how much stock we currently have. If the stock is over the maximum demand for this period, then increasing it does not help at all, since we can just produce it during the next period instead. Nogood learning is able to derive a nogood that expresses this. Similarly, if the current stock is under the minimum demand for this period, then it fails in all scenarios, and again, nogood learning is able to derive a nogood that expresses this. Thus at each stage, the solver only has to try $O(S)$ values for the production, where S is the number of different possible demands, and we end up needing to examine only $O(Sk)$ nodes where k is the number of stages. If we do not use nogood learning, then our solver has virtually identical behaviour to the search based approach of [1] (search) and has an exponential complexity. Although our learning solver only requires a linear number of nodes, the run time appears to be growing as $O(k^3)$ due to the fact that as k increases, we have more variables and constraints to propagate at each node.

6 Related Work

NCPs are a very general form of optimization problem. They include stochastic CSP/COP, quantified CSP/COP, QCOP+ [3], QBF, bi-level and multi-level programming. The evaluation approach we define for NCP is a generalization of the search-based approaches used for many of these problems, although not many focus on propagation, and only QBF solvers also consider learning. NCPs are more general than these other formalisms principally because aggregator terms can appear arbitrarily nested in both function terms and constraints.

Probably the closest work to NCPs is Quantified Constraint Optimization (QCOP+) [3]. QCOP+ are based on extending quantified CSPs to include local objective functions. They are limited compared to NCPs since only a single chain of nesting is allowed, meaning they cannot for example use the efficient form of the problem in Example 5. QCOP+ is implemented in a system QeCode which we compare with in the experiments section. They do not consider learning.

Another closely related work is the Plausibility-Feasibility-Utility (PFU) framework [15]. However, PFU does not allow tuple types to be the result of aggregator constraints, which means it cannot express bilevel problems such as the Network Link Pricing problem of Example 6. The tree search algorithms for evaluating PFUs is similar to that for NCPs, but they do not consider shortcircuit evaluation or learning. The PFU framework is studied theoretically in [15] and does not appear to have an implementation.

QBF is the form of NCP with only **and** and **or** aggregator constraints and clauses, and there is a significant body of work about how propagation and clause learning can be used in this context. The learning used in QBF is considerably simpler than for NCP, which tackles finite domain and interval variables and constraints and a much larger range of aggregators and more complicated nesting. SAT modulo theory solvers (e.g. Z3 [16]) are extended to handled quantified formula but principally through instantiation [17] akin to unrolling, which does not require (partial) search trees to be explained and only considers the **and** and **or** aggregators.

7 Conclusion

In summary, NCP's are a highly expressive formalism that unifies CSP/COP's, stochastic CSP/COP's, quantified CSP/COP's, bi-level and multi-level programming in the one framework, and allows many other kinds of nested problems to be expressed. We have demonstrated an effective search-based CP solver for evaluating them, which is significantly improved by the use of nogood learning to avoid repeating similar search. The resulting solver is competitive with or significantly better than state of the art CP-based approaches for many of the problems in these problem classes and brings us much closer to a universal CP solver that can "solve them all". Interesting directions for further investigation are: lazy or partial unrolling of aggregator constraints, model analysis and transformation for NCPs, improving propagation using the structure of the aggregation tree, approximation methods, and hybrid methods where each subproblem is solved using the technology most suited for it, e.g., using an LP or MIP propagator.

Acknowledgments. NICTA is funded by the Australian Government through the Department of Communications and the Australian Research Council through the ICT Centre of Excellence Program. This work was partially supported by Asian Office of Aerospace Research and Development grant 12-4056.

References

1. Tarim, A., Manandhar, S., Walsh, T.: Stochastic Constraint Programming: A Scenario-Based Approach. Constraints 11, 53–80 (2006)
2. Walsh, T.: Stochastic Constraint Programming.. In: van Harmelen, F. (ed.) ECAI, pp. 111–115. IOS Press (2002)
3. Benedetti, M., Lallouet, A., Vautard, J.: Quantified constraint optimization. In: Stuckey, P.J. (ed.) CP 2008. LNCS, vol. 5202, pp. 463–477. Springer, Heidelberg (2008)
4. Giunchiglia, E., Narizzano, M., Tacchella, A.: Clause/term resolution and learning in the evaluation of quantified Boolean formulas. J. Artif. Intell. Res. (JAIR) 26, 371–416 (2006)
5. Zhang, L., Malik, S.: Conflict driven learning in a quantified Boolean satisfiability solver. In: Proceedings of the 2002 IEEE/ACM International Conference on Computer-aided Design, pp. 442–449. ACM (2002)
6. Ohrimenko, O., Stuckey, P., Codish, M.: Propagation via lazy clause generation. Constraints 14, 357–391 (2009)
7. Feydy, T., Stuckey, P.J.: Lazy Clause Generation Reengineered. In: Gent, I.P. (ed.) CP 2009. LNCS, vol. 5732, pp. 352–366. Springer, Heidelberg (2009)
8. Marriott, K., Nethercote, N., Rafeh, R., Stuckey, P., Garcia de la Banda, M., Wallace, M.: The design of the Zinc modelling language. Constraints 13, 229–267 (2008)
9. Schwenk, A.J.: Take-away games. Fibonacci Quarterly 8, 225–234 (1970)
10. Chen, H.: The Computational Complexity of Quantified Constraint Satisfaction. PhD thesis, Cornell University (2004)
11. Samulowitz, H.: Solving Quantified Boolean Formulas. PhD thesis, University of Toronto (2007)
12. Benedetti, M., Lallouet, A., Vautard, J.: Reusing CSP propagators for qCSPs. In: Azevedo, F., Barahona, P., Fages, F., Rossi, F. (eds.) CSCLP. LNCS (LNAI), vol. 4651, pp. 63–77. Springer, Heidelberg (2007)
13. Schulte, C., Stuckey, P.J.: Effcient constraint propagation engines. ACM Transactions on Programming Languages and Systems (TOPLAS) 31, 2 (2008)
14. Chu, G., Stuckey, P.J.: Inter-instance nogood learning in constraint programming. In: Milano, M. (ed.) CP 2012. LNCS, vol. 7514, pp. 238–247. Springer, Heidelberg (2012)
15. Pralet, C., Verfailles, G., Schiex, T.: An algebraic graphical model for decision with uncertainties, feasibilities, and utilities. Journal of Artificial Intelligence Research 29, 421–489 (2007)
16. de Moura, L., Bjørner, N.S.: Z3: An efficient SMT solver. In: Ramakrishnan, C.R., Rehof, J. (eds.) TACAS 2008. LNCS, vol. 4963, pp. 337–340. Springer, Heidelberg (2008)
17. Ge, Y., Barrett, C., Tinelli, C.: Solving quantified verification conditions using satisfiability modulo theories. In: Pfenning, F. (ed.) CADE 2007. LNCS (LNAI), vol. 4603, pp. 167–182. Springer, Heidelberg (2007)

Beyond Consistency and Substitutability*

Martin C. Cooper

IRIT, University of Toulouse III, 31062 Toulouse, France
cooper@irit.fr

Abstract. Elimination of inconsistent values in instances of the constraint satisfaction problem (CSP) conserves all solutions. Elimination of substitutable values conserves at least one solution. We show that certain values which are neither inconsistent nor substitutable can also be deleted while conserving at least one solution. This allows us to state novel rules for the elimination of values in a binary CSP. From a practical point of view, we show that one such rule can be applied in the same asymptotic time complexity as neighbourhood substitution but is strictly stronger.

An alternative to the elimination of values from domains is the elimination of variables. We give novel satisfiability-preserving variable elimination operations. In each case we show that if the instance is satisfiable, then a solution to the original instance can always be recovered in low-order polynomial time from a solution to the reduced instance.

1 Introduction

Operations to reduce the worst-case exponential time complexity of exhaustive search are essential for the efficient resolution of large-scale constraint satisfaction problems. Reduction operations are most effective at reducing search space size when applied during search, but if this is too computationally expensive they can still be usefully applied just once during a preprocessing phase. Most previous research in this domain has concentrated on domain-filtering operations based on various forms of consistency: a value is removed from a domain if an algorithm running in low-order polynomial time demonstrates that this assignment cannot be part of a solution. Other reduction operations include the elimination of values by interchangeability or substitutability [7,11], the merging of domain values [10], the elimination of variables [9,5,3] and the introduction of symmetry-breaking constraints [1,8].

This paper studies local (and hence polytime-detectable) properties of binary CSP instances which allow value elimination or variable elimination while preserving satisfiability. We show that allowing arbitrary quantification over variables and values as well as arbitrary conditions on the compatibilities of pairs of assignments provides a rich and largely unexplored source of reduction operations.

* Supported by ANR Project ANR-10-BLAN-0210 and EPSRC grant EP/L021226/1.

B. O'Sullivan (Ed.): CP 2014, LNCS 8656, pp. 256–271, 2014.

Definition 1. *A binary CSP instance I consists of*

- *a set X of n variables,*
- *a domain $\mathcal{D}(x)$ of possible values for each variable $x \in X$,*
- *a relation $R_{xy} \subseteq \mathcal{D}(x) \times \mathcal{D}(y)$, for each pair of distinct variables $x, y \in X$, which consists of the set of pairs of values (a, b) which can simultaneously be assigned to variables (x, y).*

A partial solution *to I on $Y \subseteq X$ is a mapping $s : Y \rightarrow D$ where, for all $x \neq y \in Y$ we have $(s(x), s(y)) \in R_{xy}$. A* solution *to I is a partial solution on X.*

For simplicity of presentation, Definition 1 assumes that there is exactly one constraint relation for each pair of variables $\{x, y\}$. If $R_{xy} \neq \mathcal{D}(x) \times \mathcal{D}(y)$, then we say that variable x *constrains* variable y. If $(a, b) \in R_{xy}$, then the assignments $\langle x, a \rangle$ and $\langle y, b \rangle$ are *compatible*, otherwise *incompatible*.

In previous work we showed that there are exactly four variable-elimination rules based on so-called irreducible existential patterns [3]. In the present paper we give strict generalisations of all these rules. We also give value-elimination rules which are strict generalisations of neighbourhood substitution [7]. The paper is organised as follows: Section 2 and Section 3 present rules for, respectively, value elimination and variable elimination, Section 4 gives a particular value-elimination rule which generalises neighbourhood substitution but can be applied in the same time complexity, Section 5 gives the complexity of recovering all solutions after applying our value or variable elimination rules, while Section 6 discusses the difficulty of characterising all value or variable elimination rules based on local properties.

2 Value Elimination

For each rule which tells us when a value can be eliminated from a domain, there is a corresponding property which holds if and only if no value eliminations can be performed by this rule. Following the tradition of consistency properties, we state our rules in the form of positive properties which are satisfied if and only if no eliminations are possible.

We begin by recalling the notions of arc consistency and neighbourhood substitution, illustrated in Figure 1. In figures, each bullet represents a variable-value assignment, assignments to the same variable are grouped together within the same oval and compatible (incompatible) pairs of assignments are linked by solid (broken) lines.

Definition 2. *A value $b \in \mathcal{D}(x)$ is* AC-supported *if $\forall y \in X \setminus \{x\}$, $\exists c \in \mathcal{D}(y)$ such that $(b, c) \in R_{xy}$. We say that c is an* AC support *for $\langle x, b \rangle$ at y.*

Any assignment value $b \in \mathcal{D}(x)$ which is not AC-supported can be eliminated from $\mathcal{D}(x)$ without losing any solutions since the assignment $\langle x, b \rangle$ cannot be part of any solution.

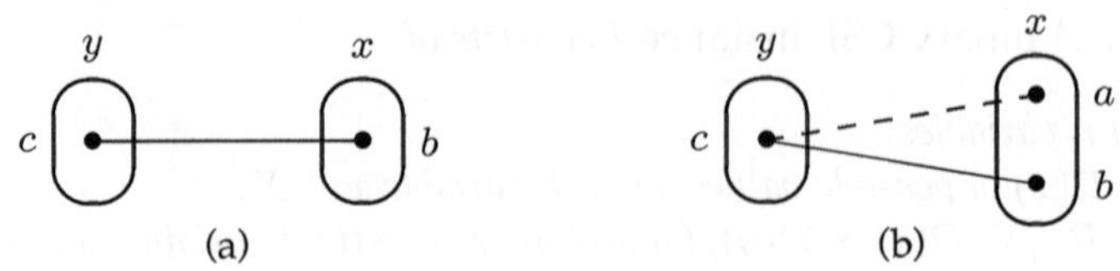

Fig. 1. Illustration of the notions of (a) arc consistency, (b) neighbourhood substitution

In a binary CSP instance we can eliminate a value $b \in \mathcal{D}(x)$ by neighbourhood substitution if $\exists a \in \mathcal{D}(x) \setminus \{b\}$ such that $\forall y \in X \setminus \{x\}$, $\nexists c \in \mathcal{D}(y)$ such that $(a, c) \notin R_{xy}$ and $(b, c) \in R_{xy}$ [7]. This is because in any solution the assignment $\langle x, b\rangle$ can be replaced by the assignment $\langle x, a\rangle$. The corresponding positive property can be defined as follows.

Definition 3. *A value $b \in \mathcal{D}(x)$ is* neighbour-supported *if $\forall a \in \mathcal{D}(x) \setminus \{b\}$, $\exists y \in X \setminus \{x\}$, $\exists c \in \mathcal{D}(y)$ such that $(a, c) \notin R_{xy}$ and $(b, c) \in R_{xy}$. We say that $\langle y, c\rangle$ is a* neighbour-support *of (x, b, a)*

Clearly, an elimination by arc consistency is possible if and only if some variable-value assignment has no AC-support and a neighbourhood substitution elimination is possible if and only if some variable-value assignment is not neighbour-supported. We require the following definition in order to give more general rules for value elimination than arc consistency and neighbourhood substitution.

Definition 4. *A* value-elimination condition *(or simply a* val-elim condition*) is a polytime-computable property $P(x, b)$ of an assignment $\langle x, b\rangle$ in a CSP instance I such that when $P(x, b)$ holds, the instance I' obtained from I by eliminating b from $\mathcal{D}(x)$ is satisfiable if and only if I is satisfiable.*

A val-elim condition allows us to eliminate values from domains while conserving at least one solution (if one exists). In binary CSP, two val-elim conditions on assignment $\langle x, b\rangle$ are: (1) $b \in \mathcal{D}(x)$ is not AC-supported, (2) $b \in \mathcal{D}(x)$ is not neighbour-supported. We now introduce two other notions of support which if not satisfied allow us to eliminate a value from a domain. The first of these is illustrated in Figure 2.

Given a binary CSP instance I, let $I[\langle y, c\rangle]$ denote the instance which results by assigning c to y and by eliminating all values e from other domains $\mathcal{D}(w)$ ($w \in X \setminus \{y\}$) such that $(c, e) \notin R_{yw}$. Suppose that for all possible assignments c to y in I, b is neighbourhood substitutable by some value (not necessarily the same for each value c) in $I[\langle y, c\rangle]$. Then b can be deleted from $\mathcal{D}(x)$ in I without changing the satisfiability of I. This idea is captured by the following positive property of conditional neighbour (CN) support.

Definition 5. *A value $b \in \mathcal{D}(x)$ is* CN-supported *if $\forall y \in X \setminus \{x\}$, $\exists c \in \mathcal{D}(y)$ such that: (1) $(b, c) \in R_{xy}$ and (2) $\forall a \in \mathcal{D}(x) \setminus \{b\}$ with $(a, c) \in R_{xy}$, $\exists z \in X \setminus \{x, y\}$, $\exists d \in \mathcal{D}(z)$ such that $(c, d) \in R_{yz}$, $(b, d) \in R_{xz}$ and $(a, d) \notin R_{xz}$.*

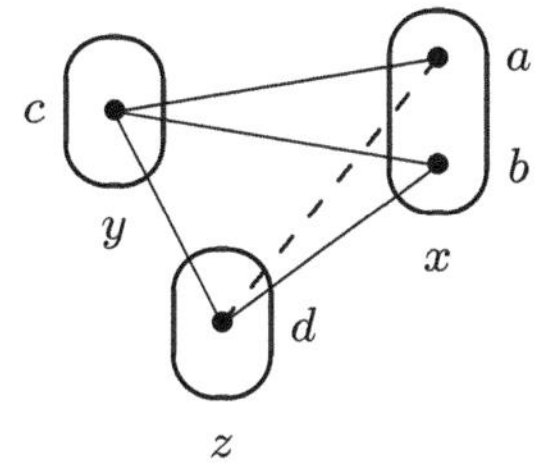

Fig. 2. Illustration of the definition that b is CN-supported

In other words, $b \in \mathcal{D}(x)$ is *CN-supported* if $\forall y \in X \setminus \{x\}$, $\langle x, b \rangle$ has an AC support c at y such that $\forall a \in \mathcal{D}(x) \setminus \{b\}$ with $(a, c) \in R_{xy}$, $\exists z \in X \setminus \{x, y\}$, $\exists d \in \mathcal{D}(z)$ such that $(c, d) \in R_{yz}$ and $\langle z, d \rangle$ is a neighbour-support of (x, b, a). It follows immediately from this definition that a CN-supported assignment is also AC-supported, and (almost immediately) that it is neighbour-supported.

When $\langle y, c \rangle$ is a neighbour-support of (x, b, a), as illustrated in Figure 1(b), it may still be possible to replace b by a in all solutions provided we also replace c by another value d. As we will see in the proof of Proposition 1, below, this is the motivation behind the following notion of extended-neighbour (EN) support, illustrated in Figure 3.

Definition 6. *A value $b \in \mathcal{D}(x)$ is* EN-supported *if $\forall a \in \mathcal{D}(x) \setminus \{b\}$, $\exists y \in X \setminus \{x\}$, $\exists c \in \mathcal{D}(y)$ such that: (1) $(a, c) \notin R_{xy}$, $(b, c) \in R_{xy}$ and (2) $\forall d \in \mathcal{D}(y)$ with $(a, d) \in R_{xy}$, $\exists z \in X \setminus \{x, y\}$, $\exists e, f \in \mathcal{D}(z)$ such that $(a, e) \in R_{xz}$, $(d, e) \notin R_{yz}$, $(c, f) \in R_{yz}$ and $(d, f) \notin R_{yz}$.*

In other words, $b \in \mathcal{D}(x)$ is EN-supported if $\forall a \in \mathcal{D}(x) \setminus \{b\}$, there is a neighbour-support $\langle y, c \rangle$ of (x, b, a) such that condition (2) of Definition 6 holds. It follows that a EN-supported assignment is also neighbour-supported.

Example 1. Suppose that in a binary CSP instance I, there is a subset S of the variables such that each domain $\mathcal{D}(x)$ ($x \in S$) contains a default value 0, where assigning 0 to a variable in S is only possible if we assign 0 to all variables in S, and this partial solution ($s(x) = 0$ for $x \in S$) is compatible with all possible

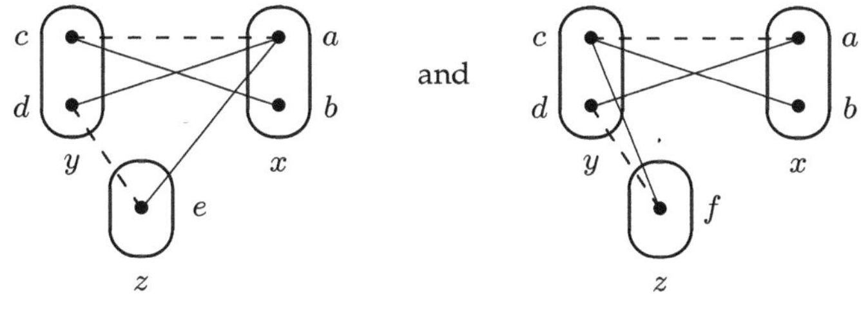

Fig. 3. Illustration of the definition that b is EN-supported

assignments to all variables in $X \setminus S$, i.e. $\forall x, y \in S$, $(0,c) \in R_{xy} \Leftrightarrow c = 0$, and $\forall x \in S, \forall y \notin S$, $(0,c) \in R_{xy}$ for all $c \in \mathcal{D}(y)$. Let b be any value in $\mathcal{D}(x) \setminus \{0\}$, for some $x \in S$. We will show that b is *not* EN-supported. Setting $a = 0$, if $y \in X \setminus \{x\}$ and $c \in \mathcal{D}(y)$ are such that $(a,c) \notin R_{xy}$ and $(b,c) \in R_{xy}$, then we necessarily have $y \in S$. So $\exists d = 0 \in \mathcal{D}(y)$ with $(a,d) = (0,0) \in R_{xy}$, such that

- $\forall z \in S \setminus \{x,y\}$, $\forall e \in \mathcal{D}(z)$ with $(a,e) \in R_{xz}$, we have $e = 0$ and hence $(d,e) = (0,0) \in R_{yz}$, and
- $\forall z \in X \setminus S, \forall f \in \mathcal{D}(z)$, we have $(d,f) = (0,f) \in R_{yz}$

It follows from Definition 6 that no $b \in \mathcal{D}(x) \setminus \{0\}$ is EN-supported, and hence, anticipating Proposition 1, we can reduce the domains of all variables $x \in S$ to a singleton $\{0\}$ by eliminating all such values b.

We now show that the notions of support given in Definitions 5 and 6 allow us to define val-elim conditions: Proposition 1, below, tells us that assignments with no CN-support or no EN-support can be eliminated while conserving satisfiability. If $b \in \mathcal{D}(x)$ is not neighbour-supported, then it is neither CN-supported nor EN-supported. Thus eliminating values that are not CN-supported or not EN-supported implies eliminating a superset of those values that can be eliminated by neighbourhood substitution.

Proposition 1. *Both of the following conditions on assignment $\langle x,b\rangle$ are val-elim conditions in binary CSP instances:*

1. *$b \in \mathcal{D}(x)$ is not CN-supported.*
2. *$b \in \mathcal{D}(x)$ is not EN-supported.*

Proof. Let I be a binary CSP instance and suppose that s is a solution to I such that $s(x) = b$. In both cases, we will show that I has a solution s' in which $s'(x) \neq b$.

1. Since $b \in \mathcal{D}(x)$ is not CN-supported, $\exists y \in X \setminus \{x\}$ such that $\forall c \in \mathcal{D}(y)$ with $(b,c) \in R_{xy}$, $\exists a(y,c) \in \mathcal{D}(x) \setminus \{b\}$ with $(a(y,c),c) \in R_{xy}$ such that $\forall z \in X \setminus \{x,y\}$, $\forall d \in \mathcal{D}(z)$ with $(c,d) \in R_{yz}$ and $(b,d) \in R_{xz}$, we have $(a(y,c),d) \in R_{xz}$. Define s' to be identical to s, except that $s'(x) = a(y,s(y))$. Now the assignment $\langle x, a(y,s(y))\rangle$ is compatible with $\langle y, s(y)\rangle$ (by definition of $a(y,s(y))$). Furthermore (again by definition of $a(y,s(y))$) the assignment $\langle x, a(y,s(y))\rangle$ is compatible with all assignments which are compatible with both of $\langle x,b\rangle$ and $\langle y,s(y)\rangle$ and hence with all assignments $\langle z,s(z)\rangle$ ($z \in X \setminus \{x,y\}$). It follows that s' is a solution.

   It is worth pointing out that this proof is valid even in the special case in which $X = \{x,y\}$.
2. Since $b \in \mathcal{D}(x)$ is not EN-supported, $\exists a \in \mathcal{D}(x) \setminus \{b\}$ such that $\forall y \in X \setminus \{x\}$, $\forall c \in \mathcal{D}(y)$ with $(a,c) \notin R_{xy}$ and $(b,c) \in R_{xy}$, $\exists d(y,c) \in \mathcal{D}(y)$ with $(a,d(y,c)) \in R_{xy}$, such that $\forall z \in X \setminus \{x,y\}$, either
   (a) $\forall e \in \mathcal{D}(z)$ with $(a,e) \in R_{xz}$, we have $(d(y,c),e) \in R_{yz}$, or
   (b) $\forall f \in \mathcal{D}(z)$ with $(c,f) \in R_{yz}$, we have $(d(y,c),f) \in R_{yz}$

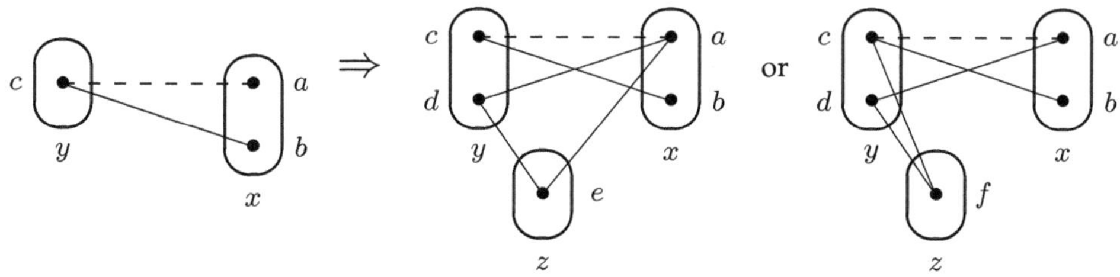

Fig. 4. $b \in \mathcal{D}(x)$ is *not* EN-supported

This is illustrated in Figure 4. Recall that s is a solution such that $s(x) = b$. Let $Y := \{y \in X \mid (a, s(y)) \in R_{xy}\}$ and $\overline{Y} := X \setminus (Y \cup \{x\})$. Define s' as follows

$$s'(v) = \begin{cases} a & \text{if } v = x, \\ s(v) & \text{if } v \in Y, \\ d(v, s(v)) & \text{otherwise.} \end{cases}$$

The assignments $\langle v, s(y)\rangle$ ($v \in Y$) are all compatible with $\langle x, a\rangle$ (by definition of Y) and with each other (since they are all part of the solution s). The assignments $\langle v, d(v, s(v))\rangle$ ($v \in \overline{Y}$) are all compatible with $\langle x, a\rangle$ (by definition of $d(v, s(v))$). Furthermore (again by the definition of $d(v, s(v))$, whether it is (a) or (b) that holds) the assignments $\langle v, d(v, s(v))\rangle$ ($v \in \overline{Y}$) are all compatible with all assignments which are compatible both with $\langle x, a\rangle$ and $\langle v, s(v)\rangle$ and hence with all assignments $\langle w, s(w)\rangle$ for all $w \in Y$ (which are compatible with $\langle x, a\rangle$ by definition of Y and with $\langle v, s(v)\rangle$ since s is a solution). To complete the proof that s' is a solution, it suffices to prove that $\forall v \neq w \in \overline{Y}$, $(d(v, s(v)), d(w, s(w))) \in R_{vw}$. Suppose, for a contradiction, that $(d(v, s(v)), d(w, s(w))) \notin R_{vw}$. Then, when $y = v$, $c = s(v)$ and $z = w$, we necessarily fall into case (b), since setting $e = d(w, s(w))$ contradicts case (a). But then setting $f = d(w, s(w))$ in case (b) implies that $(s(v), d(w, s(w))) \notin R_{vw}$. By a symmetrical argument, exchanging v and w, we immediately have that $(d(v, s(v)), s(w)) \notin R_{vw}$. But applying case (b) to $y = v$, $c = s(v)$, $z = w$ and $f = s(w)$ implies that $(d(v, s(v)), s(w)) \in R_{vw}$. From this contradiction we can deduce that s' is a solution.

It is worth pointing out that this proof is valid even in the special case in which I has only two variables x, y. In this case, either b can be eliminated by neighbourhood substitution, or there is some solution $(a, d(y, c))$ to I which does not require the assignment $\langle x, b\rangle$.

The following two examples show that the two rules given in Proposition 1 allows us to eliminate certain values which are neither arc-inconsistent nor neighbourhood substitutable.

Example 2. Consider the arc-consistent CSP instance I_4 with $X = \{w, x, y, z\}$, $\mathcal{D}(w) = \mathcal{D}(x) = \mathcal{D}(y) = \mathcal{D}(z) = \{1, 2, 3\}$ and five constraints $x \neq y$, $y = z$, $x \neq z$,

$w \neq y$, $w = z$. No eliminations are possible by neighbourhood substitution, but any $b \in \mathcal{D}(x)$ can be eliminated since it is not CN-supported: $\forall c \in \mathcal{D}(y)$, $b \in \mathcal{D}(x)$ is neighbourhood substitutable in $I_4[\langle y, c\rangle]$.

Example 3. Consider the arc-consistent CSP instance with $X = \{x, y, z\}, \mathcal{D}(x) = \mathcal{D}(y) = \mathcal{D}(z) = \{1, 2, 3\}$ and three constraints $x \neq y, x \neq z, (y, z) \notin \{(1, 3), (3, 1)\}$. No eliminations are possible by neighbourhood substitution, but the assignment $b = 2 \in \mathcal{D}(x)$ can be eliminated since it is not EN-supported: this follows from the symmetry between variables y,z and the fact that the value $a = 1 \in \mathcal{D}(x)$ is such that $\forall c \in \mathcal{D}(y)$ with $(a, c) \notin R_{xy}$ and $(b, c) \in R_{xy}$ (i.e. $c = 1$), $\exists d = 2 \in \mathcal{D}(y)$ with $(a, d) \in R_{xy}$, such that $\forall e \in \mathcal{D}(z)$, $(d, e) \in R_{yz}$.

Our two value-elimination rules are complementary since in Example 2, all variable-value assignments are EN-supported and in Example 3, all variable-value assignments are CN-supported.

3 Variable Elimination

In this section we present conditions under which a variable x can be eliminated from a binary CSP instance while preserving satisfiability. A simple example of such a condition is that $\exists a \in \mathcal{D}(x)$ which is compatible with all assignments to all other variables. Another simple example is that the variable x has a singleton domain $\{a\}$. This second example demonstrates that when eliminating the variable x we need to retain the projections onto $X \setminus x$ of all constraints whose scope includes x, since in this example we must first eliminate from all domains $\mathcal{D}(y)$ $(y \neq x)$ those values that are not compatible with $\langle x, a\rangle$. Thus, the instance I' obtained by *eliminating a variable* x from a binary CSP instance I is identical to I except that (1) $\forall y \neq x$, we have deleted from $\mathcal{D}(y)$ all values b such that $\langle y, b\rangle$ has no AC-support at x in I, and (2) we have deleted the variable x and all constraints with x in their scope.

As another example, consider the case when an assignment $\langle x, a\rangle$ is such that all other values in the domain of x can be removed one by one, by elimination thanks to one of the val-elim conditions given in Section 2. The variable x can again be eliminated while preserving satisfiability. We can relate this to previous variable-elimination rules as follows. By the above discussion, it is possible to eliminate a variable x when all values $b \in \mathcal{D}(x)$, except for an assignment $\langle x, a\rangle$, are not EN-supported at $\langle x, a\rangle$ (in the sense that there is no neighbour-support $\langle y, c\rangle$ of (x, b, a) which satisfies Condition (2) of Definition 6). This rule strictly subsumes two previously published variable-elimination rules (corresponding to the absence of the existential patterns ∃snake or ∃invsubBTP in arc-consistent binary CSP instances) [3].

We require the following formal definition in order to give further variable-elimination rules.

Definition 7. *A* satisfiability-preserving variable-elimination condition *(or a* var-elim condition*) is a polytime-computable property $P(x)$ of a variable x in a binary CSP instance I such that when $P(x)$ holds the instance I' obtained from I by* *eliminating x from I is satisfiable if and only if I is satisfiable. Such a property $P(x)$ is a* solution-preserving variable-elimination condition (sol-var-elim condition) *if it is possible to construct a solution to I from any solution s' to I' in polynomial time.*

A sol-var-elim condition not only allows us to eliminate variables while preserving satisfiability but also allows the polynomial-time recovery of at least one solution to the original instance I from a solution to the reduced instance I'. All the var-elim properties given in this paper are also sol-var-elim properties.

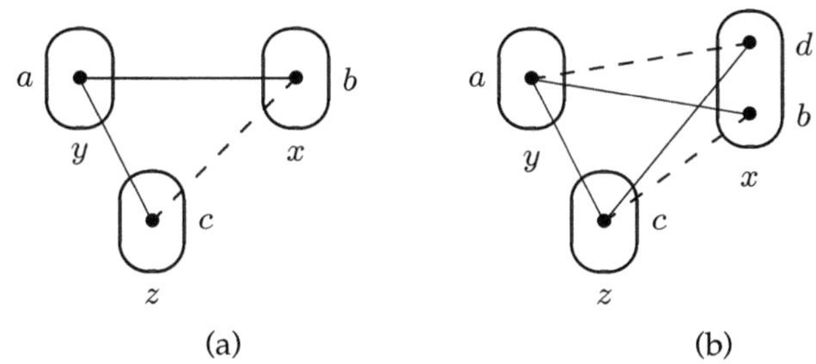

Fig. 5. Illustration of the definition that (a) a variable x is Triangle-supported, (b) a variable x is $\exists\forall$BTP-supported

The following notion of support is illustrated in Figure 5(a). It says that it is not the case that $\exists y \neq x$ such that for all $a \in \mathcal{D}(y)$ to y, in $I[\langle y, a\rangle]$ (the reduced instance consisting of the set of assignments compatible with $\langle y, a\rangle$) there is an assignment $\langle x, b\rangle$ compatible with all assignments to all variables $z \in X \setminus \{x, y\}$.

Definition 8. *A variable x is* Triangle-supported *if $\forall y \in X \setminus \{x\}$, $\exists a \in \mathcal{D}(y)$ such that $\forall b \in \mathcal{D}(x)$ with $(b, a) \in R_{xy}$, $\exists z \in X \setminus \{x, y\}$, $\exists c \in \mathcal{D}(z)$ such that $(a, c) \in R_{yz}$ and $(b, c) \notin R_{xz}$.*

It is known that if for a given variable x in an arc-consistent binary CSP instance I, the set of (in)compatibilities (known as a broken triangle) shown in Figure 5(b) occurs for no two values $b, d \in \mathcal{D}(x)$ and no two assignments a, c to two other variables y, z, then the variable x can be eliminated from I without changing the satisfiability of I [5,3]. The following notion of support, based on the same broken triangle shown in Figure 5(b), leads to a strict generalisation of the broken-triangle property (BTP) variable-elimination rule [5]. We can observe that, unlike BTP, this new rule does not require arc consistency. The corresponding positive property is given by the following definition.

Definition 9. *A variable x is* $\exists\forall$BTP-supported *if $\exists y \in X \setminus \{x\}$, $\exists a \in \mathcal{D}(y)$ such that $\forall b \in \mathcal{D}(x)$ with $(b, a) \in R_{xy}$, $\exists z \in X \setminus \{x, y\}$, $\exists c \in \mathcal{D}(z)$ with $(a, c) \in R_{yz}$ and $(b, c) \notin R_{xz}$, such that $\exists d \in \mathcal{D}(x)$ with $(d, c) \in R_{xz}$ and $(d, a) \notin R_{xy}$.*

The following notion of support is illustrated in Figure 6.

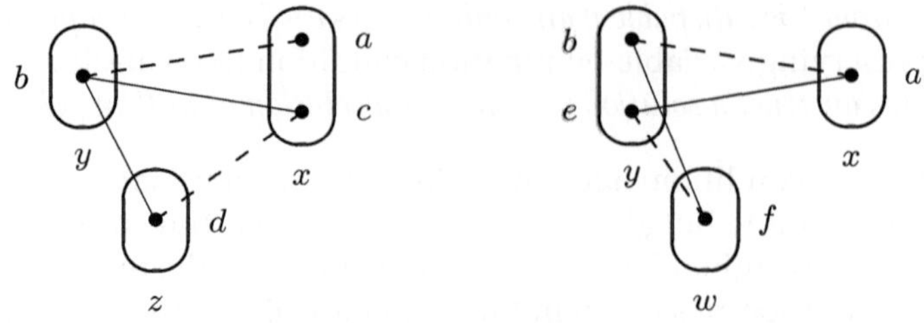

Fig. 6. Illustration of cases (a) and (b) of Definition 10 that variable x is crab-supported

Definition 10. *A variable x is* crab-supported *if* $\forall a \in \mathcal{D}(x)$, $\exists y \in X \setminus \{x\}$, $\exists b \in \mathcal{D}(y)$ *with* $(a, b) \notin R_{xy}$ *such that (1)* $\forall c \in \mathcal{D}(x)$ *with* $(c, b) \in R_{xy}$, $\exists z \in X \setminus \{x, y\}$, $\exists d \in \mathcal{D}(z)$ *with* $(b, d) \in R_{yz}$ *and* $(c, d) \notin R_{xz}$, *and (2)* $\forall e \in \mathcal{D}(y)$ *with* $(a, e) \in R_{xy}$, $\exists w \in X \setminus \{x, y\}$, $\exists f \in \mathcal{D}(w)$ *with* $(b, f) \in R_{yw}$ *and* $(e, f) \notin R_{yw}$.

Proposition 2. *Each of the following properties of variable x are sol-var-elim conditions in binary CSP instances:*

1. *x is not Triangle-supported.*
2. *x is not $\exists\forall$BTP-supported and $\mathcal{D}(x) \neq \emptyset$.*
3. *x is not crab-supported.*

Proof. Let I be a binary CSP instance and suppose that s' is a solution to I', the instance obtained by eliminating variable x from I. In each of the three cases, we will show that I has a solution s. In each case our proof is constructive and there is an obvious polynomial-time algorithm to produce s from s'.

1. Since x is not Triangle-supported, $\exists y \in X \setminus \{x\}$ such that $\forall a \in \mathcal{D}(y)$, $\exists b(a) \in \mathcal{D}(x)$ with $(b(a), a) \in R_{xy}$ such that $\forall z \in X \setminus \{x, y\}$, $\forall c \in \mathcal{D}(z)$ with $(a, c) \in R_{yz}$, we have $(b(a), c) \in R_{xz}$. Define s as follows: $s(v) = s'(v)$ $(v \in X \setminus \{x\})$ and $s(x) = b(s'(y))$. The assignment $\langle x, b(s'(y))\rangle$ is compatible with $\langle y, s'(y)\rangle$ (by definition of $b(s'(y))$) and is compatible with all of the assignments $\langle v, s'(v)\rangle$ $(v \in X \setminus \{x\})$ again by definition of $b(s'(y))$ since $(s'(y), s'(v)) \in R_{yv}$. Hence s is a solution to I.
   It is easily verified that this proof is valid even in the special case $X = \{x, y\}$.
2. If $X = \{x\}$, then since $\mathcal{D}(x) \neq \emptyset$, I has a solution. So from now on we assume that $|X| \geq 2$. Since x is not $\exists\forall$BTP-supported, $\forall y \in X \setminus \{x\}$, $\forall a \in \mathcal{D}(y)$, $\exists b(y, a) \in \mathcal{D}(x)$ with $(b(y, a), a) \in R_{xy}$ such that $\forall z \in X \setminus \{x, y\}$, $\forall c \in \mathcal{D}(z)$ with $(a, c) \in R_{yz}$ and $(b(y, a), c) \notin R_{xz}$, $\forall d \in \mathcal{D}(x)$ with $(d, c) \in R_{xz}$, we have $(d, a) \in R_{xy}$.
   For $v \in X \setminus \{x\}$, let $\mathrm{Im}(v) := \{d \in \mathcal{D}(x) \mid (d, s'(v)) \in R_{xv}\}$. If $y, z \in X \setminus \{x\}$ are such that $(b(y, s'(y)), s'(z)) \notin R_{xz}$, then setting $a = s'(y)$, $c = s'(z)$, we can deduce that $(d, s'(z)) \in R_{xz} \Rightarrow (d, s'(y)) \in R_{xy}$ and hence that $\mathrm{Im}(z) \subseteq \mathrm{Im}(y)$. Indeed, we have $\mathrm{Im}(z) \subset \mathrm{Im}(y)$ since $b(y, s'(y)) \in \mathrm{Im}(y) \setminus \mathrm{Im}(z)$. Now choose some $y \in X \setminus \{x\}$ such that $\mathrm{Im}(y)$ is minimal for inclusion among the

sets $\text{Im}(v)$ ($v \in X \setminus \{x\}$). Then the assignment $\langle x, b(y, s'(y))\rangle$ is compatible with all the assignments $s'(z)$ ($z \in X \setminus \{x, y\}$) (otherwise we would have $\text{Im}(z) \subset \text{Im}(y)$ which would contradict the minimality of $\text{Im}(y)$). Therefore, s is a solution to I, where $s(v) = s'(v)$ ($v \in X \setminus \{x\}$) and $s(x) = b(y, s'(y))$.

3. Since x is not crab-supported, $\exists a \in \mathcal{D}(x)$ such that $\forall y \in X \setminus \{x\}$, $\forall b \in \mathcal{D}(y)$ with $(a, b) \notin R_{xy}$, at least one of the following two conditions holds:
   (a) $\exists c(y, b) \in \mathcal{D}(x)$ with $(c(y, b), b) \in R_{xy}$ such that $\forall z \in X \setminus \{x, y\}$, $\forall d \in \mathcal{D}(z)$ with $(b, d) \in R_{yz}$, we have $(c(y, b), d) \in R_{xz}$.
   (b) $\exists e(y, b) \in \mathcal{D}(y)$ with $(a, e(y, b)) \in R_{xy}$ such that $\forall w \in X \setminus \{x, y\}$, $\forall f \in \mathcal{D}(w)$ with $(b, f) \in R_{yw}$, we have $(e(y, b), f) \in R_{yw}$.

   If $X = \{x\}$, then since $a \in \mathcal{D}(x)$, I has a solution. If $X = \{x, y\}$ (with $y \neq x$), then I has a solution, either of the form $(c(y, b), b)$ or of the form $(a, e(y, b))$. So, from now on we assume that $|X| \geq 3$.

   Let $b \in \mathcal{D}(y)$ be such that $(a, b) \notin R_{xy}$. Let I_1 be identical to I except that we have made $\langle x, a\rangle$ compatible with the assignment $\langle y, b\rangle$. We will show that I_1 is satisfiable iff I is satisfiable. Furthermore, it follows directly from Definition 10 that I_1 is also not crab-supported. It will then follow, by a simple inductive argument, that we can make $\langle x, a\rangle$ compatible with all assignments to all other variables in I without changing its satisfiability. But then we can eliminate x from I since there is an assignment to x which is compatible with all assignments to all other variables.

   Suppose first that $\langle y, b\rangle$ satisfies condition (a), i.e. $\exists c(y, b) \in \mathcal{D}(x)$ with $(c(y, b), b) \in R_{xy}$ such that $\forall z \in X \setminus \{x, y\}$, $\forall d \in \mathcal{D}(z)$ with $(b, d) \in R_{yz}$, we have $(c(y, b), d) \in R_{xz}$. Let I_1 be identical to I except that $(a, b) \in R_{xy}$ in I_1. Suppose that I_1 has a solution s_1 such that $s_1(x) = a$ and $s_1(y) = b$. To show that I and I_1 have the same satisfiability, it suffices to show that I also has a solution. Consider any $z \in X \setminus \{x, y\}$ and let $d = s_1(z)$. Since s_1 is a solution to I_1, $(b, d) \in R_{yz}$. Thus, by condition (a), $(c(y, b), d) \in R_{xz}$. Furthermore, $(c(y, b), b) \in R_{xy}$. Define s by $s(v) = s_1(v)$ ($v \in X \setminus \{x\}$) and $s(x) = c(y, b)$. Then s is a solution to I, since we have just shown that $\langle x, c(y, b)\rangle$ is compatible with $\langle v, s_1(v)\rangle$ for all $v \in X \setminus \{x\}$.

   Suppose now that $\langle y, b\rangle$ satisfies condition (b), i.e. $\exists e(y, b) \in \mathcal{D}(y)$ with $(a, e(y, b)) \in R_{xy}$ such that $\forall w \in X \setminus \{x, y\}$, $\forall f \in \mathcal{D}(w)$ with $(b, f) \in R_{yw}$, we have $(e(y, b), f) \in R_{yw}$. Again, let I_1 be identical to I except that $(a, b) \in R_{xy}$ in I_1. Suppose that I_1 has a solution s_1 such that $s_1(x) = a$ and $s_1(y) = b$. To show that I and I_1 have the same satisfiability, it suffices to show that I also has a solution. Consider any $w \in X \setminus \{x, y\}$ and let $f = s_1(w)$. Since s_1 is a solution to I_1, $(b, f) \in R_{yw}$. Thus by condition (b), $(e(y, b), f) \in R_{yw}$. Furthermore, $(a, e(y, b)) \in R_{xy}$. Define s by $s(v) = s_1(v)$ ($v \in X \setminus \{y\}$) and $s(y) = e(y, b)$. Then s is a solution to I, since we have just shown that $\langle y, e(y, b)\rangle$ is compatible with $\langle v, s_1(v)\rangle$ for all $v \in X \setminus \{y\}$.

The var-elim rule given by Proposition 2(2) subsumes the BTP var-elim rule [5]. Examples of the BTP var-elim rule include a variable x which is only constrained by one other variable in an arc-consistent instance or a Boolean variable x in a path-consistent instance. However, eliminating a variable with no

$\exists\forall$BTP support is strictly stronger then the BTP var-elim rule. This is demonstrated by the fact that it also subsumes the rule that allows us to eliminate a variable x when an assignment to x is compatible with all assignments to all other variables. Another generic example is when all occurrences of the BTP pattern shown in Figure 5(b) on variable x occur on pairs of values $b, d \in S \subset \mathcal{D}(x)$ and each assignment a to each other variable $y \neq x$ has an AC-support at x in $\mathcal{D}(x) \setminus S$.

The var-elim rule given by Proposition 2(3) is a strict generalisation of two previously published var-elim rules (corresponding to the absence of the existential patterns $\exists$subBTP or $\exists$snake in arc-consistent binary CSP instances) [3].

4 Practical Considerations

In binary CSP instances with a large number of variables and/or with large domains, applying the value and variable elimination rules given in this paper may not be practical. Thus, to demonstrate the practical utility of our approach, we now give a weaker version of the notion of EN-support which is nevertheless strictly stronger than the notion of neighbour-support. It leads to a val-elim rule that is strictly stronger than neighbourhood substitution but that can be applied in the same worst-case time complexity [6].

Definition 11. *A value $b \in \mathcal{D}(x)$ is* snake-supported *if $\forall a \in \mathcal{D}(x) \setminus \{b\}$, $\exists y \in X \setminus \{x\}$, $\exists c \in \mathcal{D}(y)$ such that: (1) $(a, c) \notin R_{xy}$, $(b, c) \in R_{xy}$ and (2) $\forall d \in \mathcal{D}(y)$ with $(a, d) \in R_{xy}$, $\exists z \in X \setminus \{x, y\}$, $\exists f \in \mathcal{D}(z)$ such that $(c, f) \in R_{yz}$ and $(d, f) \notin R_{yz}$.*

In other words, $b \in \mathcal{D}(x)$ is snake-supported if $\forall a \in \mathcal{D}(x) \setminus \{b\}$, there is a neighbour-support $\langle y, c\rangle$ of (x, b, a) such that $\forall d \in \mathcal{D}(y)$ with $(a, d) \in R_{xy}$, (y, c, d) has a neighbour-support $\langle z, f\rangle$ for some $z \in X \setminus \{x, y\}$. This is illustrated by the right-hand side of Figure 3. An assignment which is not snake-supported is not EN-supported and hence, by Proposition 1, can be eliminated.

In order to establish and maintain the property that all assignments are snake-supported, we use the following data structures: AC-supps(x,s,y) (for all $x, y \in X$ such that y constrains x and for all $s \in \mathcal{D}(x)$), neighbour-supps(y,p,q), neighbour-supp-vars(y,p,q), diamond-supps(y,p,q), snake-supps(y,p,q) (for all $y \in X$ and for all $p, q \in \mathcal{D}(y)$), neighbour-supps-at(y,p,q,z) (for all $y, z \in X$ such that y constrains z and for all $p, q \in \mathcal{D}(y)$), and hinge-supps(y,p,x,s) (for all $x, y \in X$ such that y constrains x and for all $p \in \mathcal{D}(y)$, $s \in \mathcal{D}(x)$), where

- AC-supps(x,s,y) $= \{q \in \mathcal{D}(y) \mid (s, q) \in R_{xy}\}$
- neighbour-supps(y,p,q) $= \{\langle z, r\rangle \mid r \in \mathcal{D}(z) \wedge (p, r) \in R_{yz} \wedge (q, r) \notin R_{yz}\}$
- neighbour-supps-at(y,p,q,z) $= \{r \in \mathcal{D}(z) \mid \langle z, r\rangle \in$ neighbour-supps(y,p,q)$\}$
- neighbour-supp-vars(y,p,q) $= \{z \in X \mid$ neighbour-supps-at(y,p,q,z) $\neq \emptyset\}$
- diamond-supps(y,p,q) $= \{\langle x, s\rangle \in$ neighbour-supps(y,q,p) $\mid$
  $\exists\langle z, r\rangle \in$ neighbour-supps(y,p,q) with $z \neq x$ $\}$
- hinge-supps(y,p,x,s) $= \{q \in \mathcal{D}(y) \setminus \{p\} \mid \langle x, s\rangle \in$ diamond-supps(y,p,q)$\}$
- snake-supps(x,t,s) $= \{\langle y, p\rangle \in$ neighbour-supps(x,t,s) $\mid$
  $|$hinge-supps(y,p,x,s)$| = |$AC-supps(x,s,y)$|$ $\}$.

These different notions of support are illustrated in Figure 7: in Figure 7(a), $\langle z, r \rangle \in$ neighbour-supps(y,p,q); in Figure 7(b), $\langle x, s \rangle \in$ diamond-supps(y,p,q) and $q \in$ hinge-supps(y,p,x,s); in Figure 7(c), $\langle y, p \rangle \in$ snake-supps(x,t,s) if $\forall q \in \mathcal{D}(y)$, $q \in$ AC-supps(x,s,y) $\Rightarrow q \in$ hinge-supps(y,p,x,s).

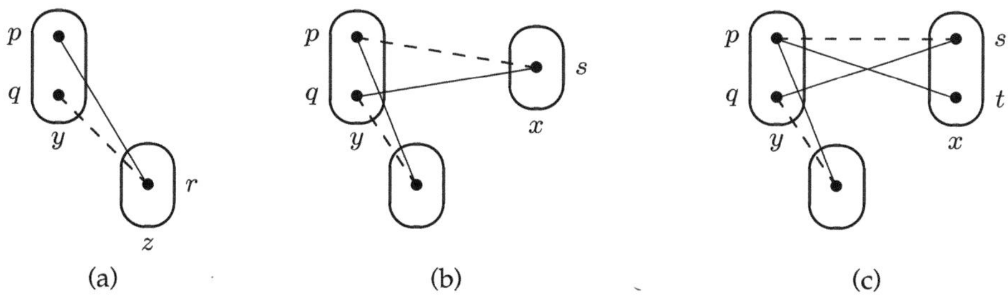

Fig. 7. Illustration of (a) neighbour, (b) diamond and hinge, and (c) snake supports

We can see from Figure 7 and Definition 11, that $t \in \mathcal{D}(x)$ is snake-supported if and only if $\forall s \in \mathcal{D}(x) \setminus \{t\}$, snake-supps($x$,$t$,$s$) $\neq \emptyset$. A value t is therefore deleted from $\mathcal{D}(x)$ when snake-supps(x,t,s) $= \emptyset$ for some $s \in \mathcal{D}(x) \setminus \{t\}$.

Let e be the number of pairs of variables which constrain each other, and let d be the maximum domain size. We can store subsets of a finite set S (such as the set of all variable-value assignments) in the form of a doubly linked list (whose length is also stored) and an array indexed by elements of S and containing pointers to this list. This allows the basic operations of addition, deletion and test of membership and of size to be performed in $O(1)$ time. The six data structures, given above, require $O(ed^3)$ space when stored in this way. We calculate and maintain diamond-supps(y,p,q) using the fact that

$$\text{diamond-supps}(y,p,q) = \text{neighbour-supps}(y,q,p) \quad \text{if } |\text{neighbour-supp-vars}(y,p,q)| > 1,$$

$$\text{diamond-supps}(y,p,q) = \{\langle z, r \rangle \in \text{neighbour-supps}(y,q,p) \mid z \neq x\} \quad \text{if neighbour-supp-vars}(y,p,q) = \{x\}.$$

The above six data structures can be calculated in $O(ed^3)$ from their definitions. Then values $t \in \mathcal{D}(x)$ which are not snake-supported can be eliminated, which may provoke new eliminations. Maintaining the above data structures until convergence (i.e. to the point at which all assignments are snake-supported) can be achieved in $O(ed^3)$ time since assignments can only be deleted and never added to the data structures.

5 Recovering All Solutions

In some applications, it is important to return all solutions to a CSP instance. We therefore study in this section whether it is possible to efficiently recover all solutions to a binary CSP instance after elimination of variables and/or values by our rules.

Proposition 3. *Let I be a binary CSP instance and let S be the set of all solutions to the instance I' obtained after applying a sequence σ of operations given by the elimination of values that are not CN-supported or the elimination of variables that are not Triangle-supported, or not ∃∀BTP-supported, or not crab-supported. Then the set of all solutions to I can be found from (S, σ) in $O(|S_I|ed + 1)$ time, where S_I is the set of solutions to I.*

Proof. In the trivial case in which $|S_I| = 0$, we necessarily have as input $S = \emptyset$ which can clearly be tested for in $O(1)$ time.

First consider the elimination of a single variable x from an instance I by one of the three variable-elimination rules. As observed in the proof of Proposition 2, each solution of the reduced instance can be extended to a solution of I. This implies that the number of solutions cannot decrease when we reinstate the variable x. Clearly each solution of I is an extension of a solution of the reduced instance. So testing all possible extensions of each solution of the reduced instance will produce all solutions of I in time $O(|S_I|e_x d)$, where e_x is the number of binary constraints with x in their scope.

Now consider the elimination of a value b from the domain of a variable x due to the fact that b is not CN-supported. As observed in the proof of Proposition 1, s is a solution to I with $s(x) = b$ implies that there is a solution s' to the reduced instance I' such that $s'(x) \neq b$ and $s'(v) = s(v)$ for $v \neq x$. To determine all solutions of I including the assignment $\langle x, b \rangle$ from the set of all solutions of the reduced instance thus requires only $O(|S_I|e_x)$ time.

Summing over all variables x and, in the case of value-eliminations, over all assignments to x, we obtain a total time complexity of $O(|S_I|ed+1)$, as claimed.

On the other hand, the following proposition indicates that eliminating values with no snake-support or no EN-support is not useful if we require all solutions. Since a value which is not snake-supported is not EN-supported, we only need to consider the former.

Proposition 4. *Let I be a binary CSP instance and let I' be the instance obtained from I after eliminating all values that are not snake-supported or not arc consistent. Even if we are given the set of all solutions to I', determining whether I has more than one solution is NP-complete.*

Proof. This problem is clearly in NP. It therefore suffices to give a polynomial reduction from the known NP-complete problem binary CSP. Let J be an arbitrary instance of binary CSP on variables X where, without loss of generality, we assume $\forall x \in X, 0 \notin \mathcal{D}(x)$ in J. We build an instance I on variables $X \cup \{x_0\}$ where $x_0 \notin X$ and the domain of variable x_0 in I is $\{0, 1\}$. We add an extra value 0 to each domain $\mathcal{D}(x)$ $(x \in X)$. In I, for all variables $y \in X$, the assignment $\langle x_0, 0 \rangle$ is compatible only with the assignment $\langle y, 0 \rangle$, whereas the assignment $\langle x_0, 1 \rangle$ is compatible with all the assignments $\langle y, a \rangle$ for $a \neq 0$; furthermore for each $y, z \in X$, the assignment $\langle y, 0 \rangle$ is compatible with all assignments to z.

In I, the value $1 \in \mathcal{D}(x_0)$ is not snake-supported, and hence can be eliminated from the domain of x_0. After establishing arc consistency, all domains are reduced to the singleton $\{0\}$. Hence the reduced instance has exactly one solution. In the instance I, the assignment $\langle x_0, 0\rangle$ only belongs to the solution assigning 0 to each variable, whereas the assignment $\langle x_0, 1\rangle$ is compatible with exactly the set of solutions to the instance J. Thus, determining the existence of a second solution to I is equivalent to determining the satisfiability of J.

6 Theoretical Discussion

We now look into the question of whether there are other rules for the elimination of values or variables (which are not subsumed by known rules or the rules we have given in this paper). To avoid confusion, we use the specific terms CSP-value and CSP-variable to refer to names of values and variables to be quantified. We consider very general rules of the form $Q(A_{var} \cup A_{val})f(E(A))[v]$, where A is a set of variable-value assignments $\langle x, a\rangle$ in which each CSP-value a occurs exactly once, A_{var} (A_{val}) is the set of CSP-variables (CSP-values) occurring in A, $Q(A_{var} \cup A_{val})$ is a sequence of quantifications on $A_{var} \cup A_{val}$, $E(A)$ is the list of the compatibilities of all pairs of assignments from A to two distinct CSP-variables (i.e. the list of truth values of $(a, b) \in R_{xy}$ for each $(\langle x, a\rangle, \langle y, b\rangle) \in A^2$ with $x \neq y$), $f : \{0,1\}^m \to \{0,1\}$ is any Boolean function (where $m = |E(A)|$), and v is the CSP-variable or CSP-value which can be eliminated whenever $Q(A_{var} \cup A_{val})f(E(A))$ holds.

For $Q(A_{var} \cup A_{val})f(E(A))[v]$ to be *well-formed* we require that

1. Each CSP-value in A_{val} and each CSP-variable in A_{var} occurs exactly once in $Q(A_{var} \cup A_{val})$,
2. In $Q(A_{var} \cup A_{val})$ each CSP-variable $x \in A_{var}$ is quantified $\exists x \in X \setminus Y$ or $\forall x \in X \setminus Y$ where Y is the set of CSP-variables which has already been quantified (i.e. those CSP-variables appearing to the left of x in $Q(A_{var} \cup A_{val})$),
3. In $Q(A_{var} \cup A_{val})$ each CSP-value a is quantified $\forall a \in \mathcal{D}(x)$ or $\exists a \in \mathcal{D}(x) \setminus H_x$ where x is a CSP-variable which has already been quantified, and H_x is the set of CSP-values which have already been quantified over $\mathcal{D}(x)$,
4. v is a CSP-variable in A_{var} or a CSP-value in A_{val},
5. f is not identically equal to FALSE.

We have chosen to impose that universal quantification of CSP-values be over all values in a domain whereas existential quantification of CSP-values be over all unused values, since all the rules given in this paper can be expressed using this convention. Note that since the various kinds of support (such as neighbour-support, CN-support, etc.) are the negation of the corresponding elimination rule, in the definition of each kind of *support*, existential quantification of CSP-values is over all values in a domain and universal quantification of CSP-values is over all unused values.

Unfortunately, exhaustive search even concerning rules on a small number of CSP-variables and CSP-values rapidly becomes impossible since the number of

Boolean functions on m arguments is 2^{2^m}. Previously, we have studied different forms of forbidden patterns [2,3,4]. Forbidding a *flat pattern* on assignments A corresponds to a rule $Q(A_{var} \cup A_{val})f(E(A))$ where all quantifiers in Q are $\forall$ and the function f is a clause. *Quantified* (respectively, *existential*) *patterns* are of the same form except that the sequence of quantifications Q begins $\exists x \in X$ (respectively, $\exists x \in X, \exists a_1 \in \mathcal{D}(x), \ldots, \exists a_r \in \mathcal{D}(x)$) [3]. The rules we consider in this paper are thus much more general in that we allow any (well-formed) sequence of quantifications but also because we allow any Boolean function of the compatibilities.

A valid rule is interesting if it is not too expensive to apply and there is no other rule which both strictly subsumes it and is no more expensive to apply. Not only is the number of cases to consider very large, but the number of interesting var-elim or val-elim rules $Q(A_{var} \cup A_{val})f(E(A))[v]$ could possibly turn out to be very large. It should also be pointed out that certain reduction operations, such as singleton arc consistency, cannot be expressed as local properties.

7 Conclusion

This paper describes several novel reduction operations for binary CSP which are neither based on consistency nor on substitutability. They reduce search space size either by elimination of variables or by the elimination of values. We showed that one of these operations can be applied in the same time complexity as neighbourhood substitution but is strictly stronger. From a practical point of view, further research is required to determine the utility of the rules given in this paper, for example, as preprocessing operations on large-scale real-world instances, or to identify tractable problem domains in which all variables can be eliminated by our variable-elimination rules. From a theoretical point of view, the most interesting challenge is the characterisation of all such rules.

Acknowledgements. I would like to thank David Cohen, Guillaume Escamocher, Peter Jeavons and Stanislav Živný for their insightful comments during discussions concerning this work.

References

1. Cohen, D.A., Jeavons, P., Jefferson, C., Petrie, K.E., Smith, B.M.: Symmetry definitions for constraint satisfaction problems. Constraints 11(2-3), 115–137 (2006)
2. Cohen, D.A., Cooper, M.C., Creed, P., Marx, D., Salamon, A.Z.: The tractability of CSP classes defined by forbidden patterns. J. Artif. Intell. Res. (JAIR) 45, 47–78 (2012)
3. Cohen, D.A., Cooper, M.C., Escamocher, G., Zivny, S.: Variable elimination in binary CSP via forbidden patterns. In: Rossi, F. (ed.) IJCAI, pp. 517–523. AAAI Press, Menlo Park (2013)
4. Cooper, M.C., Escamocher, G.: A dichotomy for 2-constraint forbidden CSP patterns. In: Hoffmann, J., Selman, B. (eds.) AAAI, pp. 464–470. AAAI Press (2012)

5. Cooper, M.C., Jeavons, P.G., Salamon, A.Z.: Generalizing constraint satisfaction on trees: Hybrid tractability and variable elimination. Artif. Intell. 174(9-10), 570–584 (2010)
6. Cooper, M.C.: Fundamental properties of neighbourhood substitution in constraint satisfaction problems. Artif. Intell. 90(1-2), 1–24 (1997)
7. Freuder, E.C.: Eliminating interchangeable values in constraint satisfaction problems. In: Proceedings of AAAI-91, pp. 227–233 (1991)
8. Gent, I.P., Petrie, K.E., Puget, J.-F.: Symmetry in constraint programming. In: van Beek, P., Walsh, T., Rossi, F. (eds.) Handbook of Constraint Programming, pp. 327–374. Elsevier (2006)
9. Larrosa, J., Dechter, R.: Boosting search with variable elimination in constraint optimization and constraint satisfaction problems. Constraints 8(3), 303–326 (2003)
10. Likitvivatanavong, C., Yap, R.H.C.: Eliminating redundancy in csps through merging and subsumption of domain values. ACM SIGAPP Applied Computing Review 13(2) (2013)
11. Prestwich, S.D.: Full Dynamic Substitutability by SAT Encoding. In: Wallace, M. (ed.) CP 2004. LNCS, vol. 3258, pp. 512–526. Springer, Heidelberg (2004)

Subexponential Time Complexity of CSP with Global Constraints*

Ronald de Haan[1], Iyad Kanj[2], and Stefan Szeider[1]

[1] Institute of Information Systems, Vienna University of Technology, Vienna, Austria
[2] School of Computing, DePaul University, Chicago, USA

Abstract. Not all NP-complete problems share the same practical hardness with respect to exact computation. Whereas some NP-complete problems are amenable to efficient computational methods, others are yet to show any such sign. It becomes a major challenge to develop a theoretical framework that is more fine-grained than the theory of NP-completeness, and that can explain the distinction between the exact complexities of various NP-complete problems. This distinction is highly relevant for constraint satisfaction problems under natural restrictions, where various shades of hardness can be observed in practice.

Acknowledging the NP-hardness of such problems, one has to look beyond polynomial time computation. The theory of subexponential time complexity provides such a framework, and has been enjoying increasing popularity in complexity theory. Recently, a first analysis of the subexponential time complexity of classical CSPs (where all constraints are given extensionally as tables) was given.

In this paper, we extend this analysis to CSPs in which constraints are given intensionally in the form of global constraints. In particular, we consider CSPs that use the fundamental global constraints *AllDifferent*, *AtLeastNValue*, *AtMost-NValue*, and constraints that are specified by compressed tuples (*cTable*). We provide tight characterizations of the subexponential time complexity of the aforementioned problems with respect to several natural structural parameters.

1 Introduction

It has been observed in various practical contexts that some NP-hard problems are accessible to efficient exact computational methods, whereas for others such methods are futile. In particular, there seem to be various grades of "empirical hardness" among several NP-complete variants of the constraint satisfaction problem (CSP). It is a central challenge for theoreticians to develop a framework, that is more fine graded than the theory of NP-completeness, and that can explain the distinction between the exact complexities of NP-hard problems. Subexponential time complexity is a framework of complexity theory that provides such a distinction [27]. It is based on the observation that for some NP-complete problems, one can improve the exponent in the exponential term of the upper bound on their running time indefinitely—such problems admit subexponential time algorithms—whereas for others this is apparently not possible under commonly-believed hypotheses in complexity theory. In particular, subexponential time

* Supported by the European Research Council (ERC), project 239962 (COMPLEX REASON), and the Austrian Science Fund (FWF), project P26200 (Parameterized Compilation).

B. O'Sullivan (Ed.): CP 2014, LNCS 8656, pp. 272–288, 2014.

algorithms were developed for many graph problems, including INDEPENDENT SET and DOMINATING SET, under natural structural restrictions (see [8,12]). The benchmark problem for subexponential time computation is the satisfiability problem for CNF formulas, where each clause contains at most three literals, denoted 3-CNF-SAT. The *Exponential Time Hypothesis* (ETH), proposed by Impagliazzo and Paturi [21], states that 3-CNF-SAT with n variables is not decidable in subexponential time, i.e., not decidable in time $2^{o(n)}$ (omitting polynomial factors).

In a recent paper, Kanj and Szeider [23] provided a first analysis of the subexponential time complexity of the classical CSP, where all constraints are given extensionally in the form of tables. In this paper, we extend this line of research by considering CSPs where constraints are specified intensionally using *global constraints*. This extension is highly relevant since it is central for the modeling and the solving of real-world problems, to use various global constraints that come along with efficient propagation and filtering techniques [33,36].

In particular, we consider CSPs in which the global constraints are all either *AllDifferent* constraints, *NValue* constraints, *AtLeastNValue* constraints, *AtMostNValue* constraints, or constraints that are specified by tables with compressed tuples (*cTable*). We provide tight characterizations of the subexponential time complexity of the aforementioned CSPs with respect to several natural parameters of the problem instance. For example, we show that the CSP with *AllDifferent* constraints is solvable in subexponential time if the domain size is $\omega(1)$ (that is, lower bounded by any nondecreasing unbounded function of the number of variables), whereas, unless the ETH fails, the problem is not solvable in subexponential time for any constant domain size that is at least 3. For the CSP with *AtLeastNValue* constraints, we show that the problem is solvable in subexponential time if the number of constraints is constant and the domain size is $\omega(1)$, and unless the ETH fails, the problem is not solvable in subexponential time if the number of constraints is linear and the domain size is constant. The results in this paper shed some light on which instances of the aforementioned CSPs with global constraints are feasible with respect to exact computation.

2 Preliminaries

2.1 CSP

An instance I of the CONSTRAINT SATISFACTION PROBLEM (or CSP, for short) is a triple $(V, D, \mathcal{C})$, where V is a finite set of *variables*, D is a mapping that assigns each variable $v \in C$ a finite set $D(v)$ of *domain values*, and $\mathcal{C}$ is a finite set of *constraints*. We write $D = \bigcup_{v \in V} D(x)$.

Each constraint in $\mathcal{C}$ is a pair (S, R), where S, the *constraint scope*, is a non-empty sequence of distinct variables of V, and R, the *constraint relation*, is a relation over D whose arity matches the length of S; a relation is considered as a set of tuples. Therefore we also call such a constraint a *table constraint*. The *size* of a CSP instance $I = (V, D, \mathcal{C})$ is the sum $\sum_{(S,R) \in \mathcal{C}} |S| \cdot |R|$. We write $var(C)$ for the set of variables that occur in the scope of constraint C.

An *assignment* or *instantiation* is a mapping from the set V of variables to the domain D. An assignment τ *satisfies* a constraint $C = ((x_1, \ldots, x_n), R)$ if

$(\tau(x_1), \ldots, \tau(x_n)) \in R$, and τ satisfies the CSP instance if it satisfies all its constraints. An instance I is *consistent* or *satisfiable* if it is satisfied by some assignment. CSP is the problem of deciding whether a given instance of CSP is consistent.

Bounding the *treewidth* is a classical method for restricting the structure of CSP instances. The method dates back to Freuder [15]. Treewidth is a graph parameter that can be applied to CSP in terms of *primal graphs* or *incidence graphs* giving rise to the CSP parameters *primal treewidth* (also called *induced width* [11]) and *incidence treewidth*, respectively [35]. For self-containment we give the definitions. The primal graph of a CSP instance I has as vertices the variables of I, and two variables are joined by an edge if and only if the variables occur together in some constraint of I. The incidence graph is a bipartite graph, one side of which consists of the variables and the other side consists of the constraints; a variable and a constraint are joined by an edge if the variable occurs in the constraint. A tree decomposition of a graph $G = (V, E)$ is a pair (T, χ) consisting of a tree T and a mapping χ that assigns to each node t of T a subset $\chi(t) \subseteq V$ such that the following conditions are satisfied: (i) for every edge $\{u, v\} \in E$ there is a node t of T such that $u, v \in \chi(t)$; and (ii) for any three nodes t_1, t_2, t_3 of T we have $\chi(t_2) \subseteq \chi(t_1) \cap \chi(t_3)$ if t_2 lies on a path between t_1 and t_3. The width of (T, χ) is the size of a largest set $\chi(t)$ minus 1. The treewidth of G is the smallest width over all its tree decompositions.

For an instance $I = (V, D, \mathcal{C})$ of CSP we define the following basic parameters:

- vars: the number $|V|$ of variables, usually denoted by n;
- dom: the number $|D|$ of values;
- cons: the number $|\mathcal{C}|$ of constraints;
- arity: the maximum size of a constraint scope;
- tw: the treewidth of the primal graph of I;
- tw*: the treewidth of the incidence graph of I.

BOOLEAN CSP denotes the CSP with the *Boolean domain* $\{0, 1\}$. CNF-SAT is the satisfiability problem for propositional formulas in conjunctive normal form (CNF). k-CNF-SAT denotes CNF-SAT restricted to formulas where each clause is of width at most k, i.e., contains at most k literals.

2.2 Global Constraints

It is often preferred to represent a constraint more succinctly than by listing all the tuples of the constraint relation. Such an intensionally represented constraint is called a *global constraint* [33,36]. The *Global Constraints Catalogue* [1] lists several hundred of global constraints. In this paper we focus on the following global constraints.

- The *AllDifferent* global constraint is probably the best-known, most influential, and most studied global constraint in constraint programming [36]. It admits efficient matching based filtering algorithms [31]. An *AllDifferent* constraint over a set S of variables is satisfied if each variable in S is assigned a different value.
- The global constraints *NValue* [29], *AtLeastNValue* [32], and *AtMostNValue* [2] are widely used in constraint programming [1]. Each such constraint C is associated with an integer $n_C \in \mathbb{N}$. The *NValue* constraint C over a set S of variables is

satisfied if the number of distinct values assigned to the variables in S is exactly n_C. The *AtLeastNValue* and *AtMostNValue* constraints are satisfied if the number of distinct values is $\leq n_C$ or $\geq n_C$, respectively. The special case of an *NValue* or *AtLeastNValue* constraint C where n_C equals the arity of C is equivalent to an *AllDifferent* constraint.

- The global constraint *cTable* is a *table constraint with compressed tuples.* This global constraint admits a potentially exponential reduction in the space compared to an extensional table constraint and can be propagated using a variant of the GAC-schema algorithm [24]. *cTable* constraints have also been studied under the name *generalized DNF constraints* [6]. A *cTable* constraint is a pair (S, U) where $S = (v_1, \ldots, v_r)$ is a non-empty sequence of distinct variables, and U is a set of *compressed tuples*, which are sequences of the form $(V_1, \ldots, V_r)$, where $V_i \subseteq D(v_i)$, $1 \leq i \leq r$. One compressed tuple $(V_1, \ldots, V_r)$ represents all the tuples $(d_1, \ldots, d_r)$ with $d_i \in V_i$. Thus, by "decompression" one can compute from (S, U) a (unique) equivalent table constraint (S, R) where R contains all the tuples that are represented by the compressed tuples in U.

The CSP where all constraints are *AllDifferent* constraints is denoted CSP$^{\neq}$. This variant of the CSP was studied by Fellows et al. [13] who called it MAD-CSP (multiple all different CSP). The CSP where all constraints are *NValue*, *AtLeastNValue*, or *AtMostNValue* constraints, is denoted CSP$^{=}$, CSP$^{\geq}$, and CSP$^{\leq}$, respectively. The CSP where all constraints are *cTable* constraints is denoted CSP^{c}.

We note that all the problems CSP$^{\neq}$, CSP$^{=}$, CSP$^{\geq}$, CSP$^{\leq}$, CSP^{c}, are NP-complete. In fact, CSP$^{\neq}$ (and therefore the more general problems CSP$^{\geq}$, CSP$^{\leq}$) is even NP-hard for instances consisting of two constraints only [26], and CSP$^{\leq}$ is even NP-hard for instances consisting of a single constraint [3]. CSP^{c} is clearly NP-hard as it contains the classical CSP (with table constraints) as a special case. Hence all the considered problems admit the representation of NP-hard combinatorial problems.

Consider a CSP instance that models some real-world problem and uses, among others, some of the global constraints considered above, say the *AllDifferent* constraint. Then, we can combine all the *AllDifferent* constraints in the instance into a new global constraint, a multi-AllDifferent constraint. Filtering this combined constraint is polynomial time equivalent to solving one instance of CSP$^{\neq}$. Such a combination of several global constraints into a new one has been considered for several different global constraints (see, e.g., [20,34]).

Guarantees and limits for polynomial-time preprocessing for single *NValue*, *AtLeastNValue*, and *AtMostNValue* constraints have been given by Gaspers and Szeider [18].

The Boolean versions of the above global constraints problems, and the parameters vars, dom, cons, arity, tw, and tw*, are defined as in the CSP.

2.3 Subexponential Time Complexity

The time complexity functions used in this paper are assumed to be proper complexity functions that are unbounded and nondecreasing.

It is clear that CSP and CNF-SAT are solvable in time $\mathsf{dom}^n|I|^{O(1)}$ and $2^n|I|^{O(1)}$, respectively, where I is the input instance and n is the number of variables in I. We say that the CSP (resp. CNF-SAT) problem is solvable in (uniform) *subexponential time* if there exists an algorithm that solves the problem in time $\mathsf{dom}^{o(n)}|I|^{O(1)}$ (resp. $2^{o(n)}|I|^{O(1)}$). Using the results of [9,14], the above definition is equivalent to the following: The CSP (resp. CNF-SAT) problem is solvable in *subexponential time* if there exists an algorithm that for all $\varepsilon = 1/\ell$, where ℓ is a positive integer, solves the problem in time $\mathsf{dom}^{\varepsilon n}|I|^{O(1)}$ (resp. $2^{\varepsilon n}|I|^{O(1)}$). This means that we can improve the exponent in the exponential-term of the running time of the algorithm indefinitely.

Let Q and Q' be two problems, and let μ and μ' be two parameter functions defined on instances of Q and Q', respectively. In the case of CSP and CNF-SAT, μ and μ' will be the number of variables in the instances of these problems. A *subexponential time Turing reduction family* (SERF-reduction) [14,22] is an algorithm A with an oracle to Q' such that there are computable functions $f, g : \mathbb{N} \longrightarrow \mathbb{N}$ satisfying: (1) given a pair (I, ε) where $I \in Q$ and $\varepsilon = 1/\ell$ (ℓ is a positive integer), A decides I in time $f(1/\varepsilon)\mathsf{dom}^{\varepsilon\mu(I)}|I|^{O(1)}$ (for CNF-SAT $\mathsf{dom} = 2$); and (2) for all oracle queries of the form "$I' \in Q'$" posed by A on input (I, ε), we have $\mu'(I') \leq g(1/\varepsilon)(\mu(I) + \log|I|)$.

Since we focus on the super-polynomial factor in the running time, we will often use the O^* notation, which suppresses the polynomial factor in the input length $|I|$.

The optimization class SNP consists of all search problems expressible by second-order existential formulas whose first-order part is universal [30]. [22] introduced the notion of *completeness* for the class SNP under serf-reductions, and identified a class of problems which are complete for SNP under serf-reductions, such that the subexponential time solvability for any of these problems implies the subexponential time solvability of all problems in SNP. Many well-known NP-hard problems are proved to be complete for SNP under the serf-reduction, including 3-SAT, VERTEX COVER, and INDEPENDENT SET, for which extensive efforts have been made in the last three decades to develop subexponential time algorithms with no success. This fact has led to the *exponential-time hypothesis*, ETH, which is equivalent to the statement that not all SNP problems are solvable in subexponential time:

Exponential-Time Hypothesis (ETH): The problem k-CNF-SAT, for any $k \geq 3$, cannot be solved in time $2^{o(n)}$, where n is the number of variables in the formula. Therefore, there exists $c > 0$ such that k-CNF-SAT cannot be solved in time 2^{cn}.

The following result is implied, using the standard technique of renaming variables, from [22, Corollary 1] and from the proof of the Sparsification Lemma [22], [14, Lemma 16.17].

Lemma 1. k-CNF-SAT *($k \geq 3$) is solvable in $2^{o(n)}$ time if and only if* k-CNF-SAT *with a linear number of clauses and in which the number of occurrences of each variable is at most 3 is solvable in time $2^{o(n)}$, where n is the number of variables in the formula (note that the size of an instance of* k-CNF-SAT *is polynomial in n). In particular, choosing $k = 3$ we get:* 3-CNF-SAT *in which every variable occurs at most 3 times, denoted* 3-3-SAT, *is not solvable in $2^{o(n)}$ time unless the ETH fails.*

The ETH has become a standard hypothesis in complexity theory [27].

Remark 1. In this paper, when we consider the CSP with global constraints restricted to instances in which a certain parameter is $\Omega(g(n))$ (resp. $\omega(g(n))$, $O(g(n))$, $o(g(n))$), for some proper complexity function $g(n)$ of the number of variables n in the instance, we mean the CSP restricted to all the instances in which the parameter is *upper bounded* by a (prespecified) function that is $\Omega(g(n))$ (resp. $\omega(g(n))$, $O(g(n))$, $o(g(n))$).

3 The Problem CSP$^{\neq}$

Let $\mathcal{I}$ be an instance of CSP$^{\neq}$ with constraints $C_1, \ldots, C_c$ for some integer $c > 0$, over the set of variables $\{x_1, \ldots, x_n\}$. Denote by D_i, $i = 1, \ldots, n$, the domain of x_i.

Proposition 1. CSP$^{\neq}$ *can be solved in time* $O^*(2^n)$.

Proof. We reduce the instance $\mathcal{I}$ to an instance of LIST COLORING. Construct the graph G whose vertices are $x_1, \ldots, x_n$ (without loss of generality, we label the vertices in G with their corresponding variables' names in $\mathcal{I}$) and such that there is an edge between two vertices x_i and x_j, $1 \leq i < j \leq n$ if and only if x_i and x_j appear together in some constraint in $\mathcal{I}$. For each vertex x_i in G, associate with it a list of colors $L_i = D_i$. It is not difficult to see that $\mathcal{I}$ is a yes-instance of CSP$^{\neq}$ if and only if the graph G has a proper list coloring. It is known that LIST COLORING is solvable in time $O^*(2^n)$ [4], and hence so is CSP$^{\neq}$. □

Corollary 1. *Let* $d(n) = \omega(1)$ *be a proper complexity function. The* CSP$^{\neq}$ *restricted to instances in which* $\mathsf{dom} \geq d(n)$ *is solvable in subexponential time.*

Proof. Let $d(n) = \omega(1)$ be a proper complexity function, and consider the CSP$^{\neq}$ restricted to instances in which $\mathsf{dom} \geq d(n)$. By Proposition 1, CSP$^{\neq}$ is solvable in time $O^*(2^n) = O^*(d(n)^{n/\log(d(n))}) \subseteq O^*(\mathsf{dom}^{o(n)})$. □

By Corollary 1, we can focus our investigation of the subexponential time complexity of the problem CSP$^{\neq}$ on instances in which $\mathsf{dom} = O(1) = d$, for some constant d. Note that dom is an upper bound on arity because each constraint must have arity at most dom (otherwise it cannot be satisfied). If $d \leq 2$, then each constraint can have arity at most 2, and CSP$^{\neq}$ in this case reduces to 2-CNF-SAT, which is in P. Therefore, we can assume in the remainder of this section that $d \geq 3$.

Proposition 2. *Unless the ETH fails,* CSP$^{\neq}$ *restricted to instances in which* $\mathsf{dom} = d \geq 3$ *and* $\mathsf{cons} = \Omega(n)$ *is not solvable in subexponential time.*

Proof. It suffices to prove the result for $\mathsf{cons} = s(n)$, where $s(n)$ is any *specific* function such that $s(n) = \Theta(n)$, as the result would extend using a padding argument to any function that is linear in n (we can add new "dummy" variables and new "dummy" constraints on those new variables to make the relation between the constraints and the variables satisfy the desired function $s()$).

By Lemma 1, 3-3-SAT is not solvable in subexponential time unless ETH fails. The standard polynomial-time reduction from 3-SAT to 3-COLORABILITY (see [10]), establishing the NP-hardness of 3-COLORABILITY, reduces an instance of 3-SAT on n variables and m clauses to an instance of 3-COLORABILITY with $O(n+m)$ vertices and $O(n+m)$ edges. Therefore, if we use the same reduction but start from 3-3-SAT instead of 3-SAT, we end up with an instance of 3-COLORABILITY in which the number of vertices is $O(n)$ and the number of edges is $O(n)$ as well. Let LINEAR-3-COLORABILITY be the restriction of 3-COLORABILITY to instances in which the number of edges is linear in the number of vertices. The previous argument shows that if LINEAR-3-COLORABILITY is solvable in subexponential time then so is 3-3-SAT, and then the ETH would fail. Now if we use the standard reduction from 3-COLORABILITY to $\mathrm{CSP}^{\neq}$ (in which each vertex becomes a variable, each edge becomes a constraint of arity 2, and the domain is the set of 3 colors), but instead we start from an instance of LINEAR-3-COLORABILITY, we obtain an instance of $\mathrm{CSP}^{\neq}$ on n variables (the same as the number of vertices in the graph), linear number of constraints, and domain size $\mathsf{dom} = 3$. Therefore, the previous reduction is a SERF-reduction from LINEAR-3-COLORABILITY to the restriction of $\mathrm{CSP}^{\neq}$ to instances in which the number of constraints is linear, and $\mathsf{dom} = 3$. Combining the above sequence of arguments proves the proposition. □

Remark 2. We do not consider the restriction of $\mathrm{CSP}^{\neq}$ to instances in which $\mathsf{cons} = o(n)$ and $\mathsf{dom} = O(1)$. This is because each constraint must have $\mathsf{arity} \leq \mathsf{dom}$, and hence, if $\mathsf{cons} = o(n)$ then it would follow that the total number of variables is $o(n)$. It follows that Proposition 2 and Corollary 1 provide tight characterizations of the subexponential time complexity of $\mathrm{CSP}^{\neq}$ with respect to each of cons and dom.

The following proposition provides a tight characterization of the subexponential time complexity of $\mathrm{CSP}^{\neq}$ with respect to the treewidth of the primal graph:

Proposition 3. $\mathrm{CSP}^{\neq}$ *is solvable in subexponential time for instances in which* $\mathsf{tw} = o(n)$*, and unless the ETH fails,* $\mathrm{CSP}^{\neq}$ *is not solvable in subexponential time for instances in which* $\mathsf{tw} = \Omega(n)$.

Proof. Let $\mathcal{I}$ be an instance of $\mathrm{CSP}^{\neq}$ such that the treewidth of its primal graph is $o(n)$. Since the arity of each constraint in $\mathcal{I}$ is at most d and the domain size is d, in polynomial time we can reduce $\mathcal{I}$ to an instance of CSP on the same set of variables, and with the same domain, constraints, and primal treewidth. It is well known [16] that CSP is solvable in time $O^*(d^{\mathsf{tw}}) \subseteq O^*(d^{o(n)})$, and hence $\mathcal{I}$ can be decided in subexponential time.

The hardness result follows from a general observation about the primal treewidth of the CSP. First note that the number of variables n is an upper bound on the primal treewidth; that is, $\mathsf{tw} \leq n$. Therefore, for any upper bound $s(n) = \Omega(n)$ on tw, using a padding argument (adding a linear number of dummy new variables and singleton constraints that do not increase the primal treewidth) we can reduce a general instance of $\mathrm{CSP}^{\neq}$ to an instance in which $\mathsf{tw} \leq s(n)$ at the cost of a linear increase in the number of variables and the instance size. This provides a SERF-reduction from a general instance of $\mathrm{CSP}^{\neq}$ to an instance in which $\mathsf{tw} \leq s(n) = \Omega(n)$. The result now follows from the same result for $\mathrm{CSP}^{\neq}$ on general instances (implied, e.g., from Proposition 2).[1] □

It is well-known that (see [25]) $\mathsf{tw} \leq \mathsf{arity} \cdot (\mathsf{tw}^* - 1)$ and $\mathsf{tw}^* \leq \mathsf{tw} + 1$. If $\mathsf{arity} = O(1)$, then tw and tw^* are within a multiplicative constant from one another. Therefore, from Proposition 3 we can infer the following tight result:

[1] This padding argument applies as well to the other variants of the CSP with global constraints considered in this paper, and will prove useful for the hardness results on their subexponential time complexity when $\mathsf{tw} \leq s(n) = \Omega(n)$.

Proposition 4. $\mathrm{CSP}^{\neq}$ *is solvable in subexponential time for instances in which* $\mathsf{tw}^* = o(n)$*, and unless the ETH fails,* $\mathrm{CSP}^{\neq}$ *is not solvable in subexponential time for instances in which* $\mathsf{tw}^* = \Omega(n)$.

Remark 3. There are several width parameters for CSP that are even more general than tw^* in the sense that any instances for which tw^* is small, also the other width parameter is small; but there are instances for which the other width parameter is small but tw^* can be arbitrarily large. Prominent examples for such with parameters are *hypertree width* [19] and *submodular width* [28]. The lower bound statement of Proposition 4 clearly carries over to the more general width parameters. The same holds true for the lower bound statements in Proposition 7 and Theorem 3.

4 The Problems $\mathrm{CSP}^{=}$, $\mathrm{CSP}^{\geq}$, and $\mathrm{CSP}^{\leq}$

We start by presenting an exact algorithm for $\mathrm{CSP}^{\geq}$; we do so by reducing $\mathrm{CSP}^{\geq}$ to $\mathrm{CSP}^{\neq}$. We use the example illustrated in Figure 1 as a running example to explain the idea behind this reduction. In this example, the instance $\mathcal{I}$ of $\mathrm{CSP}^{\geq}$ consists of three constraints C_1, C_2, C_3, where the variables in C_1 are x_1, x_2, x_3, x_4, the variables in C_2 are x_4, x_5, and the variables in C_3 are x_1, x_5, x_6, x_7. The domain of x_1 is $\{a, b\}$, the domain of both x_2 and x_3 is $\{b\}$, the domain of x_4 is $\{b, c\}$, the domain of x_5 is $\{a\}$, and the domain of both x_6 and x_7 is $\{d, e\}$. The number of distinct values that need to be assigned to the variables of C_1 is at least 3, to the variables of C_2 is at least 2, and to the variables of C_3 is at least 3.

In a solution $\mathcal{S}$ (i.e., an assignment of variables to domain values) to an instance $\mathcal{I}$ of $\mathrm{CSP}^{\geq}$, and for a constraint C in $\mathcal{I}$, it is possible for several variables in C to be assigned the same value by the solution $\mathcal{S}$ (in the running example we are forced to assign both x_2 and x_3 the value b). Therefore, if we attempt a straightforward reduction from $\mathrm{CSP}^{\geq}$ to $\mathrm{CSP}^{\neq}$ that produces the same instance $\mathcal{I}$, the solution $\mathcal{S}$ to $\mathcal{I}$ as an instance of $\mathrm{CSP}^{\geq}$ may not be a solution to $\mathcal{I}$ as an instance of $\mathrm{CSP}^{\neq}$. It is possible that the above happens due to the fact that there are variables in $\mathcal{I}$ that can be removed without affecting the satisfiability of $\mathcal{I}$, because there is a solution to $\mathcal{I}$ in which each constraint will still be satisfied without considering the values assigned to those variables.

The algorithm starts by trying each subset of the variables as a subset for which there exists a solution in which each of those variables is "essential" for this solution; the algorithm then removes all the other (nonessential) variables, updates the instance, and works toward finding a solution under this assumption in the resulting instance. (In the running example, we remove x_3 from C_1; see the Venn diagram on the left in Figure 1.) Even with the above assumption, it is still possible that in a solution to the resulting instance, two variables in a constraint C are assigned the same value. One cannot simply ignore (remove) one of these variables on the basis that removing it will not affect the satisfiability of C, because the removed variable may contribute to the satisfiability of a constraint other than C, in which this variable appears as well. (In the running example, we are forced to assign both x_1 and x_5 the same value, which would violate constraint C_3 of $\mathrm{CSP}^{\neq}$.) Therefore, the resulting instance, even though it may be a satisfiable instance of $\mathrm{CSP}^{\geq}$, it may not be a satisfiable instance of $\mathrm{CSP}^{\neq}$. However, as it will be shown in Lemma 2, it is possible in such an instance to "reassign" each variable to a subset of the constraints that it appears in, so that after this reassignment/repartitioning each variable contributes to the satisfiability of each constraint that it appears in. After such a reassignment, the resulting instance of $\mathrm{CSP}^{\geq}$ becomes an equivalent instance of $\mathrm{CSP}^{\neq}$. (In the running example, variable x_5 is not contributing to C_3, and can be safely reassigned to C_2; see the Venn diagram on the right in Figure 1.) We now proceed to the formal proofs.

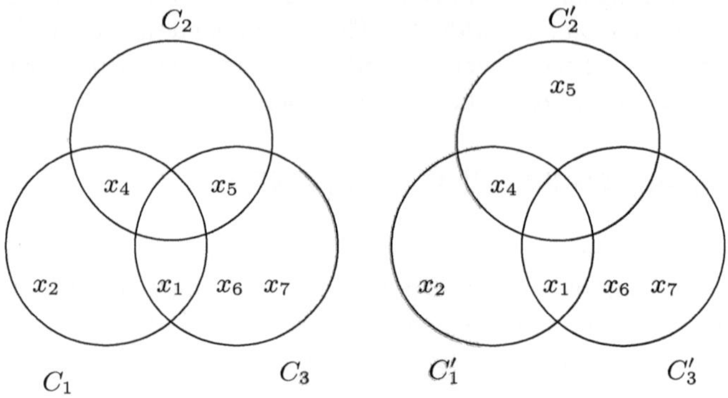

Fig. 1. Illustration of the example of the reduction from $\mathrm{CSP}^{\geq}$ to $\mathrm{CSP}^{\neq}$

Let $\mathcal{I}$ be an instance of $\mathrm{CSP}^{\geq}$ with constraints $C_1, \ldots, C_c$ for some integer value $c > 0$, over the variables $x_1, \ldots, x_n$. Let n_i, $i = 1, \ldots, c$, be the nonnegative integer associated with constraint C_i. Denote by D_i, $i = 1, \ldots, n$, the domain of variable x_i, and let $D = \bigcup_{i=1}^{n} D_i$. Set $k = |D|$. If we consider each C_i, $i = 1, \ldots, c$, as a set consisting of all the variables in C_i, and we draw the Venn diagram for the C_i's, then this Venn diagram consists of at most $s \leq 2^c$ many nonempty *regions*, where each region R_j, $j = 1, \ldots, s$, is defined as the intersection of all the sets containing the variables that lie in R_j in the Venn diagram. For a solution $\mathcal{S}$ to the instance $\mathcal{I}$, we call a variable x_i *essential* (to $\mathcal{S}$) if discounting the value assigned to x_i by $\mathcal{S}$ violates at least one of the constraints (containing x_i), and hence no longer gives a solution to $\mathcal{I}$. It is clear that by enumerating every subset of the variables in $\mathcal{I}$, which takes $O(2^n)$ time, we can work under the assumption that we are looking for a solution such that every variable is essential to $\mathcal{S}$. Since we are working on an instance of $\mathrm{CSP}^{\geq}$, adding the nonessential variables to the solution afterwards (and assigning them values from their domains) will not hurt the solution. Therefore, without loss of generality, we will assume that each of the variables $x_1, \ldots, x_n$ is essential to the solution sought (if any exists). We start with the following lemma.

Lemma 2 (The Repartitioning Lemma). *Let $\mathcal{I}$ be an instance of* $\mathrm{CSP}^{\geq}$. *There is a solution to $\mathcal{I}$ if and only if there is an instance $\mathcal{I}'$ on the same set of variables as $\mathcal{I}$, and whose constraints are $C_1', \ldots, C_c'$, such that:*

(1) the variables in C_i' are a subset of those in C_i, for $i = 1, \ldots, c$;
(2) the numbers $n_1, \ldots, n_c$ are the same in both $\mathcal{I}$ and $\mathcal{I}'$; and

(3) *there is a solution to $\mathcal{I}'$ satisfying that for every value v, and for any two distinct variables x_i, x_j that are assigned the value v in the solution for $\mathcal{I}'$, the set of constraints that x_i belongs to in $\mathcal{I}'$ is disjoint from that that x_j belongs to in $\mathcal{I}'$.*

Proof. Suppose that $\mathcal{I}$ has a solution $\mathcal{S}$; by the discussion preceding this lemma, we can assume that every variable is essential to $\mathcal{S}$. We define the instance $\mathcal{I}'$ on the same set of variables as $\mathcal{I}$ as follows. The constants $n_1, \ldots, n_c$ remain the same in $\mathcal{I}'$. We define the constraints in $\mathcal{I}'$ by a sequence of changes performed to the constraints in $\mathcal{I}$; initially the constraints of $\mathcal{I}'$ are identical to those of $\mathcal{I}$. For every value $v \in D$ assigned to some variable by the solution $\mathcal{S}$, let $x_v^1, \ldots, x_v^\ell$ be the variables assigned the value v by $\mathcal{S}$. For each x_v^j, $j = 1, \ldots, \ell - 1$, considered in the listed order, let $\mathcal{C}_v^j$ be the set of constraints containing x_v^j in $\mathcal{I}'$, and let $\mathcal{C}_{v,\cup}^j$ be the union of all constraints containing any of the variables $x_v^{j+1}, \ldots, x_v^\ell$. Remove x_v^j from each constraint in $\mathcal{C}_v^j \cap \mathcal{C}_{v,\cup}^j$.

We claim that the same solution to $\mathcal{I}$ is a solution to $\mathcal{I}'$ that satisfies all the conditions in the statement of the lemma. First, from the construction of the constraints in $\mathcal{I}'$, for any value v in the solution, the set of constraints containing each variable assigned the value v are mutually disjoint because each variable x_v^i ($i < \ell$) assigned a value v is removed from each constraint that some subsequent variable in $x_v^{i+1}, \ldots, x_v^\ell$ is contained in. Moreover, because each constraint C_i' is obtained from C_i only by (possibly) removing variables from C_i, we have $C_i' \subseteq C_i$, for $i = 1, \ldots, c$. Finally, when a variable x_v^i that is assigned a value v is removed from a constraint C_j', this removal will not affect the number of different values assigned to the variables in C_j' by $\mathcal{S}$; this is because we know for sure that there will be a subsequent variable x_v^p, $p \in \{i+1, \ldots, \ell\}$, that is assigned value v and that will remain in C_j', namely the variable x_v^p with the maximum index p that appears in C_j'.

Conversely, because each C_i' is a subset of C_i, for $i = 1, \ldots, c$, it is easy to see that any solution to $\mathcal{I}'$ is also a solution to $\mathcal{I}$. □

Theorem 1. $\mathrm{CSP}^{\geq}$ *can be solved in time* $O^*((2^{(\mathsf{cons}+1)} + 1)^n)$.

Proof. Let $\mathcal{I}$ be an instance of $\mathrm{CSP}^{\geq}$ with constraints $C_1, \ldots, C_c$ for some integer $c > 0$, over the variables $x_1, \ldots, x_n$. Let n_i, $i = 1, \ldots, c$, be the nonnegative integer associated with constraint C_i.

We first enumerate each subset of the variables $\{x_1, \ldots, x_n\}$ as the subset of essential variables for the solution $\mathcal{S}$ sought. Fix such an enumerated subset X, remove the other variables from $\mathcal{I}$, and update the instance accordingly (i.e., update the constraints); without loss of generality, we will still refer to the resulting instance as $\mathcal{I}$.

By Lemma 2, there is a solution to $\mathcal{I}$ if and only if there is an instance $\mathcal{I}'$ on the same set of variables as $\mathcal{I}$, and whose constraints are $C_1', \ldots, C_c'$, such that: (1) the variables in C_i' form a subset of those in C_i, for $i = 1, \ldots, c$, (2) the numbers $n_1, \ldots, n_c$ are the same in both $\mathcal{I}$ and $\mathcal{I}'$, and (3) there is a solution to $\mathcal{I}'$ satisfying that for every value v, and for any two distinct variables x_i, x_j that are assigned the value v in the solution for $\mathcal{I}'$, the set of constraints that x_i belongs to in $\mathcal{I}'$ is disjoint from that that x_j belongs to in $\mathcal{I}'$.

To find the instance $\mathcal{I}'$, we will try every possible partitioning of the variables in X into c constraints to determine the new constraints $C_1', \ldots, C_c'$ in $\mathcal{I}'$. For each such partitioning π in which $C_i' \subseteq C_i$ and at least n_i variables are in C_i', for $i = 1, \ldots, c$, we form the instance of CSP$^{\neq}$ on the set of variables X and the set of constraints $C'_1, \ldots, C'_c$, and invoke the algorithm for CSP$^{\neq}$ described in Proposition 1 on this instance; if the algorithm returns a solution then we return the same solution as a solution to $\mathcal{I}$. If for each enumerated subset X and each enumerated partitioning π the algorithm for CSP$^{\neq}$ rejects, then we reject the instance $\mathcal{I}$.

It is easy to see the correctness of the above algorithm. Clearly, if there is a solution to the CSP$^{\neq}$ instance then there is a solution to $\mathcal{I}'$, and hence to $\mathcal{I}$. This is because each constraint contains at least n_i variables, which must receive n_i distinct values in the solution to the CSP$^{\neq}$ instance, hence satisfying each constraint C_i and satisfying $\mathcal{I}$. On the other hand, if $\mathcal{I}$ has a solution, then there is an enumerated partitioning of the variables in X that will correspond to the constraints in $\mathcal{I}'$. Now because there is a solution to $\mathcal{I}'$ that satisfies properties (1)-(3) in Lemma 2, no two variables in the same constraint of $\mathcal{I}'$ receive the same value v in this solution (by property (3)). Therefore, this solution will also be a solution to the constructed instance of CSP$^{\neq}$. This shows the correctness of the above algorithm.

The running time of the algorithm is the time taken to enumerate all subsets of the variables, and for each subset X, the time to enumerate all partitions of X into c constraints, and finally for each such partition the time taken to invoke the CSP$^{\neq}$ algorithm on the resulting instance. The number of subsets of variables of $\{x_1, \ldots, x_n\}$ is $\sum_{i=0}^{n} \binom{n}{i}$. For each subset of cardinality i, there are at most 2^{ci} many ways of partitioning it into c constraints. Finally, for each instance on i variables, the CSP$^{\neq}$ algorithm takes $O^*(2^i)$ time. Putting everything together, the overall running time of the algorithm is a polynomial factor multiplied by:

$$\sum_{i=0}^{n} \binom{n}{i} \cdot 2^{ci} \cdot 2^i = \sum_{i=0}^{n} \binom{n}{i} \cdot 2^{(c+1)i} = (2^{(c+1)} + 1)^n.$$

Therefore, the running time of the algorithm is $O^*((2^{(\mathsf{cons}+1)} + 1)^n)$ as claimed. □

Corollary 2. CSP$^{\geq}$ *restricted to instances in which* $\mathsf{cons} = O(1)$ *is solvable in* $O^*(2^{O(n)})$ *time.*

Corollary 3. CSP$^{\geq}$ *restricted to instances in which* $\mathsf{cons} = o(\log \mathsf{dom})$ *is solvable in subexponential time.*

Proof. The result follows from Theorem 1 after noticing that if $\mathsf{cons} = o(\log \mathsf{dom})$ then $2^{\mathsf{cons}} = \mathsf{dom}^{o(1)}$. □

Proposition 5. *Let* $d(n) = \omega(1)$ *be a proper complexity function. Then* CSP$^{\geq}$ *restricted to instances in which* $\mathsf{cons} = O(1)$ *and* $\mathsf{dom} \geq d(n)$ *is solvable in subexponential time, and unless the ETH fails,* CSP$^{\geq}$ *restricted to instances in which* $\mathsf{cons} = \Omega(n)$ *(even when* $\mathsf{dom} = O(1)$*) is not solvable in subexponential time.*

Proof. The positive result follows from Corollary 3. The hardness result follows from the hardness result for CSP$^{\neq}$ in Proposition 2 (CSP$^{\neq}$ is a special case of CSP$^{\geq}$). □

Theorem 2. CSP$^{\leq}$ *restricted to instances where* $\mathsf{dom} = O(1)$ *and* $\mathsf{cons} = \Omega(n)$ *is not solvable in subexponential time, unless the ETH fails.*

Proof. We give a SERF-reduction from 3-3-SAT to CSP$^{\leq}$; the result will then follow by Lemma 1. Take an instance φ of 3-3-SAT with n variables. We construct in polynomial time an instance of CSP$^{\leq}$, with $\mathsf{cons} = O(n)$ and $\mathsf{dom} = O(1)$ that is a yes-instance if and only if $\varphi \in$ 3-3-SAT. We proceed in two steps: firstly, we modify the well-known polynomial-time reduction from 3-SAT to VERTEX COVER [17] to a reduction from 3-3-SAT to CSP$^{\leq}$, resulting in an instance with $\mathsf{cons} = O(n)$ and $\mathsf{dom} = O(n)$; secondly, we transform this instance of CSP$^{\leq}$ to an equivalent instance of CSP$^{\leq}$ with $\mathsf{cons} = O(n)$ and $\mathsf{dom} = O(1)$.

We start with the first step. Let φ consist of the clauses $c_1, \dots, c_m$, where $c_i = l_1^i \vee l_2^i \vee l_3^i$ for each $1 \leq i \leq m$. The well-known reduction to VERTEX COVER produces a graph $G = (V, E)$, containing vertices $v_x, v_{\overline{x}}$ for each variable x occurring in φ, and a vertex v_j^i for each literal occurrence, where $1 \leq i \leq m$ and $1 \leq j \leq 3$. The variables v_x and $v_{\overline{x}}$ are adjacent, for each variable x, and the vertices v_1^i, v_2^i, v_3^i form a triangle, for each $1 \leq i \leq m$. Moreover, there is an edge between v_j^i and v_l, where $l = l_j^i$. Then φ is satisfiable if and only if G has a vertex cover consisting of $n + 2m$ vertices. More specifically, φ is satisfiable if and only if G has a vertex cover containing exactly one vertex from $v_x, v_{\overline{x}}$ for each variable x and exactly two vertices from v_1^i, v_2^i, v_3^i for each $1 \leq i \leq m$. We now construct an instance of CSP$^{\leq}$ as follows. For each edge $e = \{v_1, v_2\} \in E$, we introduce a variable u_e with domain $\{v_1, v_2\}$. Then, for each clause c_i, we define the set E_{c_i} to consist of all edges between v_1^i, v_2^i, v_3^i, between v_j^i and $v_{l_j^i}$ and between $v_{l_j^i}$ and $v_{\overline{l_j^i}}$, for each $1 \leq j \leq 3$. Then, we add a constraint ensuring that the variables u_e for all nine $e \in E_{c_i}$ take at most 5 different values. The assignments to the variables u_e that satisfy all these constraints exactly correspond to the vertex covers of G containing exactly one vertex from $v_x, v_{\overline{x}}$ for each variable x and exactly two vertices from v_1^i, v_2^i, v_3^i for each $1 \leq i \leq m$. These particular vertex covers, in turn, correspond exactly to truth assignments (which set one of $x, \overline{x}$ to true, for each variable x) satisfying φ. The construction of such a constraint is illustrated in Figure 2.

In the second step, we transform the instance of CSP$^{\leq}$ in such a way that $\mathsf{dom} = O(1)$. In order to do so, we will use the following observation. Whenever two vertices $v_1, v_2 \in V$ have the property that there is no constraint both containing a

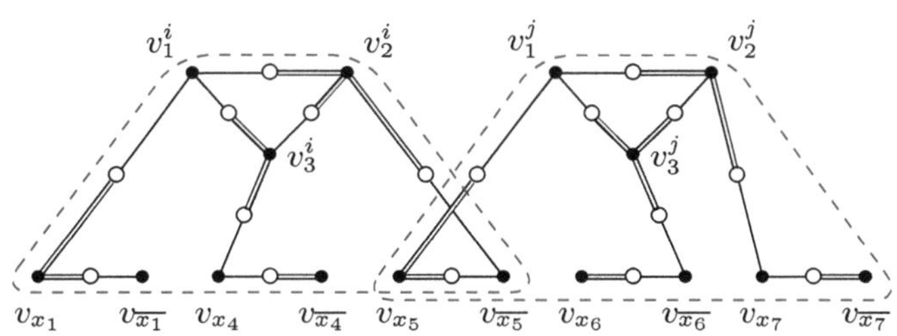

Fig. 2. The CSP$^{\leq}$ constraints corresponding to example clauses $c_i = (x_1 \vee x_4 \vee \overline{x_5})$ and $c_j = (x_5 \vee \overline{x_6} \vee x_7)$. Variables are denoted by $\circ$, and values by $\bullet$. The constraints are indicated by dashed lines. The nine variables in each constraint must be assigned to at most 5 different values. The double lines indicate an assignment to the variables satisfying the constraint that corresponds to the truth assignment $\{x_1 \mapsto \top, x_4 \mapsto \bot, x_5 \mapsto \top, x_6 \mapsto \top, x_7 \mapsto \bot\}$.

variable u_{e_1} for some edge e_1 incident with v_1 and a variable u_{e_2} for some edge e_2 incident with v_2, then we can safely identify the domain values v_1 and v_2 in the instance of $\mathrm{CSP}^{\leq}$. Consequently, we can identify all m many domain values $v_1^1, \dots, v_1^m$ into a single value, and similarly identify all domain values $v_2^1, \dots, v_2^m$ and $v_3^1, \dots, v_3^m$. Next, to reduce dom even more, we will identify a number of domain values v_x with each other (and similarly identify their complementary values $v_{\overline{x}}$ with each other). Consider the primal graph of φ, i.e., the graph G_φ^p containing as vertices the variables of φ where two vertices x, x' are adjacent if and only if x and x' occur together in a clause (positively or negatively). Since each variable occurs at most 3 times in φ, we know that the maximum degree of G_φ^p is bounded above by 8. Then, by Brooks' Theorem [5], we know that there exists a proper coloring of G_φ^p by at most 9 colors, and that such a coloring can be computed in linear time. Take such a proper coloring c of G_φ^p. Now, for each color b used by the coloring c, we let $X_b \subseteq \mathrm{Var}(\varphi)$ be the set of variables x such that $c(x) = b$. Then, since c is a proper coloring of the primal graph G_φ^p of φ, we know that for any color b no two variables $x, x' \in X_b$ occur together in any clause of φ. Therefore, for each color $1 \leq b \leq 3$ we can safely identify all domain values v_x for $x \in X_b$ with each other in the instance of $\mathrm{CSP}^{\leq}$, and similarly we can safely identify all domain values $v_{\overline{x}}$ for $x \in X_b$ with each other. This results in an equivalent instance of $\mathrm{CSP}^{\leq}$ with $\mathsf{cons} = O(n)$ and $\mathsf{dom} = O(1)$. □

We next consider the subexponential time complexity of the $\mathrm{CSP}^{=}$, $\mathrm{CSP}^{\geq}$, and $\mathrm{CSP}^{\leq}$ with respect of the primal treewidth. We have the following tight result:

Proposition 6. $\mathrm{CSP}^{=}$, $\mathrm{CSP}^{\geq}$, *and* $\mathrm{CSP}^{\leq}$ *restricted to instances in which* $\mathsf{tw} = o(n)$ *are solvable in subexponential time, and unless the ETH fails,* $\mathrm{CSP}^{=}$, $\mathrm{CSP}^{\geq}$, *and* $\mathrm{CSP}^{\leq}$ *restricted to instances in which* $\mathsf{tw} = \Omega(n)$ *are not solvable in subexponential time.*

Proof. The proof of this proposition for each of the $\mathrm{CSP}^{=}$, $\mathrm{CSP}^{\geq}$, and $\mathrm{CSP}^{\leq}$ is exactly the same as the proof of Proposition 3. □

Finally, the following hardness result for $\mathrm{CSP}^{=}$ and $\mathrm{CSP}^{\geq}$ with respect to tw^* follows from Proposition 4 since $\mathrm{CSP}^{\neq}$ is a special case of each of $\mathrm{CSP}^{=}$ and $\mathrm{CSP}^{\geq}$:

Proposition 7. *Unless the ETH fails,* $\mathrm{CSP}^{=}$ *and* $\mathrm{CSP}^{\geq}$ *are not solvable in subexponential time for instances in which* $\mathsf{tw}^* = \Omega(n)$.

5 The Problem CSP^{c}

We start by providing strong evidence that BOOLEAN CSP^{c} is not solvable in subexponential time. By SAT[3] we denote the satisfiability of normalized propositional formulas of depth 3 (see [14]), that is, propositional formulas that are the conjunction-of-disjunction-of-conjunction of literals. It is well known that if SAT[3] is solvable in time $O^*(2^{o(n)})$ then the W-hierarchy in parameterized complexity collapses at the second level [7], that is, $W[2] = \mathrm{FPT}$, which is a consequence that is deemed very unlikely and would imply that the ETH fails [14]. We have the following result:

Proposition 8. *Unless* $W[2] = \text{FPT}$, BOOLEAN CSP^c *is not solvable in time* $O^*(2^{o(n)})$.

Proof. It is easy to see that an instance of SAT[3] is polynomial-time reducible to an instance of BOOLEAN CSP^c on the same set of variables. In this reduction, every disjunction of conjunction of literals in the Boolean formula is associated with a *cTable* constraint, where each compressed tuple $(V_1, \ldots, V_r)$ of this constraint represents a conjunction of literals: a positive literal x_i is represented by $V_i = \{1\}$, a negative literal $\neg x_i$ is represented by $V_i = \{0\}$, and if a variable x_i does not occur in the conjunction, it is represented by $V_i = \{0, 1\}$. Therefore, there is a SERF-reduction from SAT[3] to BOOLEAN CSP^c. The statement now follows from the result in [7]. □

Next, we consider the subexponential time complexity of CSP^c with respect to the number of constraints cons. We have the following proposition:

Proposition 9. CSP^c *restricted to instances in which* $\mathsf{cons} = O(1)$ *is solvable in subexponential time (even in* P*), and unless the ETH fails,* CSP^c *restricted to instances in which* $\mathsf{cons} = \omega(1)$ *is not solvable in subexponential time.*

Proof. If the number of constraints in an instance is $O(1)$, then in polynomial time we can enumerate each subset of tuples T such that T contains exactly one compressed tuple from each constraint in the instance (because the size of T is $O(1)$). We can then verify consistency, and deduce an instantiation of the set of variables if it exists in polynomial time. The hardness result follows from the same hardness result for CSP [23] since CSP is a special case of CSP^c. □

The following theorem provides a tight characterization of the subexponential time complexity of CSP^c with respect to the primal and incidence treewidth.

Theorem 3. *The following statements are true:*

(i) CSP^c *restricted to instances in which* $\mathsf{tw} = o(n)$ *is solvable in subexponential time, and unless the ETH fails,* CSP^c *restricted to instances in which* $\mathsf{tw} = \Omega(n)$ *is not solvable in subexponential time.*
(ii) CSP^c *restricted to instances in which* $\mathsf{tw}^* = O(1)$ *is solvable in subexponential time (even in* P*), and unless the ETH fails,* CSP^c *restricted to instances in which* $\mathsf{tw}^* = \omega(1)$ *is not solvable in subexponential time.*

Proof. (i) Note that an upper bound on the primal treewidth implies the same upper bound on the arity. Let $\mathcal{I}$ be an instance of CSP^c whose $\mathsf{tw} = o(n)$. Since $\mathsf{arity} = o(n)$, each constraint contains at most $d(n)^{o(n)}$ many satisfying tuples. By decompressing compressed tuples, i.e., by enumerating all the satisfying tuples in each constraint in time $O^*(d(n)^{o(n)})$ we can reduce the instance $\mathcal{I}$ to an instance of CSP on the same set of variables, domain, and primal tree width. It is well known [16] that CSP is solvable in time $O^*(d(n)^{\mathsf{tw}}) \subseteq O^*(d(n)^{o(n)})$, and hence $\mathcal{I}$ can be decided in subexponential time. The hardness result follows from the same hardness result for the CSP [23].

(ii) The hardness result is a direct consequence of the hardness result in Proposition 9, since cons is an upper bound on tw^*. Establishing the first statement requires some work. Consider an instance $\mathcal{I}$ of CSP^c whose incidence treewidth is a constant w.

We apply a construction of [35] to transform $\mathcal{I}$ into an equivalent instance $\mathcal{I}'$ of CSP^c whose incidence treewidth is at most $w+1$ and where each variable appears in the scope of at most 3 constraints. The construction keeps all constraints of $\mathcal{I}$ and adds binary equality constraints and copies of variables. The equality constraints enforce that a variable and all its copies get assigned the same value. The construction in [35] is stated for table constraints but clearly works also for *cTable*, since the constraints of $\mathcal{I}$ are not changed at all, and the newly introduced constraints are binary.

Consider the *dual graph* G^d of $\mathcal{I}'$ which has as vertices the constraints of $\mathcal{I}'$, and where two constraints are joined by an edge if and only if they share at least one variable. Because each variable appears in the scope of at most 3 constraints, a further result of [35, Lemma 2(5)] applies, which is based on a construction due to Kolaitis and Vardi [25], and from which it follows that the treewidth of G^d is at most $2w+2$.

Next we obtain the CSP instance $\mathcal{I}''$ which is "dual" to the instance $\mathcal{I}'$. This construction is a straightforward generalization of a known construction for CSP with table constraints (see, e.g., [11, Definition 2.1]). Each constraint $C=(S,U)$ of $\mathcal{I}'$ gives rise to a variable $x[C]$ of $\mathcal{I}''$; the domain $D(x[C])$ is U, a set of compressed tuples. Between any two variables $x[C_1], x[C_2]$ of $\mathcal{I}''$ corresponding to constraints $C_1=(S_1,U_1)$ and $C_2=(S_2,U_2)$, respectively, of $\mathcal{I}'$ that share at least one variable we add a binary table constraint $((x[C_1],x[C_2]),R)$. Here, the relation R contains all pairs $(t_1,t_2)\in U_1\times U_2$ that are consistent in the sense that for all variables x that appear in the scopes of C_1 and C_2, the coordinate V_i^1 of t_1 corresponding to x and the coordinate V_j^2 of t_2 corresponding x have a nonempty intersection. It is straightforward to see that $\mathcal{I}'$ and $\mathcal{I}''$ are equivalent. It remains to observe that G^d is isomorphic to the primal graph of $\mathcal{I}''$, and hence the primal treewidth of $\mathcal{I}''$ is $2w+2$, a constant. Hence we can solve $\mathcal{I}''$ in polynomial time [16]. □

As it turns out, both CSP and CSP^c exhibit the same subexponential time complexity behavior with respect to the same restrictions on the structural parameters considered above. On the other hand, the negative result proved in Proposition 8 for the BOOLEAN CSP^c is stronger than that known for BOOLEAN CSP [23], the latter of which states that a (nonuniform) subexponential time algorithm for CSP implies a (nonuniform) subexponential time algorithm for CNF-SAT.

6 Conclusion

We have provided a first analysis of the subexponential time complexity of CSP with global constraints, focusing on instances that are composed of the fundamental global constraints *AllDifferent*, *AtLeastNValue*, *AtMostNValue*, and *cTable*, respectively. Our results show a detailed complexity landscape for these problems under various natural structural restrictions. In most cases, we were able to obtain tight bounds that exactly determine the borderline between the classes of instances that can be solved in subexponential time, and those for which the existence of subexponential time algorithms is unlikely. There are several ways for extending the current work such as considering other global constraints, the combination of different global constraints, and other structural restrictions on the primal or incidence graphs.

References

1. Beldiceanu, N., Carlsson, M., Rampon, J.-X.: Global constraint catalog. Technical Report T2005:08, SICS, SE-16 429 Kista, Sweden (August 2006), `http://www.emn.fr/x-info/sdemasse/gccat/`
2. Bessiere, C., Hebrard, E., Hnich, B., Kiziltan, Z., Walsh, T.: Filtering algorithms for the NValue constraint. Constraints 11(4), 271–293 (2006)
3. Bessière, C., Hebrard, E., Hnich, B., Walsh, T.: The complexity of global constraints. In: McGuinness, D.L., Ferguson, G. (eds.) Proceedings of the Nineteenth National Conference on Artificial Intelligence, San Jose, California, USA, July 25-29, pp. 112–117. AAAI Press / The MIT Press (2004)
4. Björklund, A., Husfeldt, T., Koivisto, M.: Set partitioning via inclusion-exclusion. SIAM J. Comput. 39(2), 546–563 (2009)
5. Brooks, R.L.: On colouring the nodes of a network. Mathematical Proceedings of the Cambridge Philosophical Society 37, 194–197 (1941)
6. Chen, H., Grohe, M.: Constraint satisfaction with succinctly specified relations. J. of Computer and System Sciences 76(8), 847–860 (2010)
7. Chen, J., Huang, X., Kanj, I.A., Xia, G.: Strong computational lower bounds via parameterized complexity. J. of Computer and System Sciences 72(8), 1346–1367 (2006)
8. Chen, J., Kanj, I., Perkovic, L., Sedgwick, E., Xia, G.: Genus characterizes the complexity of certain graph problems: Some tight results. Journal of Computer and System Sciences 73(6), 892–907 (2007)
9. Chen, J., Kanj, I.A., Xia, G.: On parameterized exponential time complexity. Theoretical Computer Science 410(27-29), 2641–2648 (2009)
10. Cormen, T.H., Leiserson, C.E., Rivest, R.L., Stein, C.: Introduction to Algorithms, 3rd edn. The MIT Press, Cambridge (2009)
11. Dechter, R.: Constraint Processing. Morgan Kaufmann (2003)
12. Demaine, E., Fomin, F., Hajiaghayi, M., Thilikos, D.: Subexponential parameterized algorithms on bounded-genus graphs and H-minor-free graphs. J. ACM 52, 866–893 (2005)
13. Fellows, M.R., Friedrich, T., Hermelin, D., Narodytska, N., Rosamond, F.A.: Constraint satisfaction problems: Convexity makes alldifferent constraints tractable. In: Walsh, T. (ed.) IJCAI 2011, Proceedings of the 22nd International Joint Conference on Artificial Intelligence, Barcelona, Catalonia, Spain, July 16-22, pp. 522–527. IJCAI/AAAI (2011)
14. Flum, J., Grohe, M.: Parameterized Complexity Theory. Texts in Theoretical Computer Science. An EATCS Series, vol. XIV. Springer, Berlin (2006)
15. Freuder, E.C.: A sufficient condition for backtrack-bounded search. J. of the ACM 29(1), 24–32 (1982)
16. Freuder, E.C.: Complexity of k-tree structured constraint satisfaction problems. In: Shrobe, H.E., Dietterich, T.G., Swartout, W.R. (eds.) Proceedings of the 8th National Conference on Artificial Intelligence, Boston, Massachusetts, July 29-August 3, 2 vols., pp. 4–9. AAAI Press / The MIT Press (1990)
17. Garey, M.R., Johnson, D.R.: Computers and Intractability. W. H. Freeman and Company, New York (1979)
18. Gaspers, S., Szeider, S.: Kernels for global constraints. In: Walsh, T. (ed.) Proceedings of the 22nd International Joint Conference on Artificial Intelligence, IJCAI 2011, pp. 540–545. AAAI Press/IJCAI (2011)
19. Gottlob, G., Leone, N., Scarcello, F.: Hypertree decompositions and tractable queries. J. of Computer and System Sciences 64(3), 579–627 (2002)
20. Hnich, B., Kiziltan, Z., Walsh, T.: Combining symmetry breaking with other constraints: Lexicographic ordering with sums. In: AI&M 1-2004, Eighth International Symposium on Artificial Intelligence and Mathematics, Fort Lauderdale, Florida, USA, January 4-6 (2004)

21. Impagliazzo, R., Paturi, R.: On the complexity of k-SAT. J. of Computer and System Sciences 62(2), 367–375 (2001)
22. Impagliazzo, R., Paturi, R., Zane, F.: Which problems have strongly exponential complexity? J. of Computer and System Sciences 63(4), 512–530 (2001)
23. Kanj, I., Szeider, S.: On the subexponential time complexity of CSP. In: Proceedings of the Twenty-Seventh AAAI Conference on Artificial Intelligence. AAAI Press (2013)
24. Katsirelos, G., Walsh, T.: A compression algorithm for large arity extensional constraints. In: Bessière, C. (ed.) CP 2007. LNCS, vol. 4741, pp. 379–393. Springer, Heidelberg (2007)
25. Kolaitis, P.G., Vardi, M.Y.: Conjunctive-query containment and constraint satisfaction. J. of Computer and System Sciences 61(2), 302–332 (2000); Special issue on the Seventeenth ACM SIGACT-SIGMOD-SIGART Symposium on Principles of Database Systems, Seattle, WA (1998)
26. Kutz, M., Elbassioni, K., Katriel, I., Mahajan, M.: Simultaneous matchings: hardness and approximation. J. of Computer and System Sciences 74(5), 884–897 (2008)
27. Lokshtanov, D., Marx, D., Saurabh, S.: Lower bounds based on the exponential time hypothesis. Bulletin of the European Association for Theoretical Computer Science 105, 41–72 (2011)
28. Marx, D.: Tractable hypergraph properties for constraint satisfaction and conjunctive queries. J. of the ACM 60(6), Art. 42, 51 (2013)
29. Pachet, F., Roy, P.: Automatic generation of music programs. In: Jaffar, J. (ed.) CP 1999. LNCS, vol. 1713, pp. 331–345. Springer, Heidelberg (1999)
30. Papadimitriou, C.H., Yannakakis, M.: Optimization, approximation, and complexity classes. J. of Computer and System Sciences 43(3), 425–440 (1991)
31. Régin, J.-C.: A filtering algorithm for constraints of difference in CSPs. In: Hayes-Roth, B., Korf, R.E. (eds.) Proceedings of the 12th National Conference on Artificial Intelligence, Seattle, WA, USA, July 31-August 4, vol. 1, pp. 362–367. AAAI Press / The MIT Press (1994)
32. Régin, J.-C.: Développement d'outils algorithmiques pour l'Intelligence Artificielle. PhD thesis, Montpellier II (1995) (in French)
33. Régin, J.-C.: Global constraints: A survey. In: van Hentenryck, P., Milano, M. (eds.) Hybrid Optimization: The Ten Years of CPAIOR. Optimization and Its Applications, vol. 45, ch. 3, pp. 63–134. Springer (2011)
34. Régin, J.-C., Rueher, M.: A global constraint combining a sum constraint and difference constraints. In: Dechter, R. (ed.) CP 2000. LNCS, vol. 1894, pp. 384–395. Springer, Heidelberg (2000)
35. Samer, M., Szeider, S.: Constraint satisfaction with bounded treewidth revisited. J. of Computer and System Sciences 76(2), 103–114 (2010)
36. van Hoeve, W.-J., Katriel, I.: Global constraints. In: Rossi, F., van Beek, P., Walsh, T. (eds.) Handbook of Constraint Programming, ch. 6. Elsevier (2006)

A New Characterization of Relevant Intervals for Energetic Reasoning

Alban Derrien and Thierry Petit

TASC (Mines Nantes, LINA, CNRS, INRIA),
4, Rue Alfred Kastler, 44307 Nantes Cedex 3, France
{alban.derrien,thierry.petit}@mines-nantes.fr

Abstract. Energetic Reasoning (ER) is a powerful filtering algorithm for the Cumulative constraint. Unfortunately, ER is generally too costly to be used in practice. One reason of its bad behavior is that many intervals are considered as relevant, although most of them should be ignored. In the literature, heuristic approaches have been developed in order to reduce the number of intervals to consider, leading to a loss of filtering. In this paper, we provide a sharp characterization that allows to reduce the number of intervals by a factor seven without loss of filtering.

1 Introduction

Due to its relevance in many industrial contexts, the NP-Hard Cumulative Scheduling Problem (CuSP) has been widely studied in Constraint Programming (CP). This problem is defined on a set of activities $\mathcal{A}$ consuming a resource of capacity C. Each activity $a \in \mathcal{A}$ is defined by four variables: its starting time s_a, its processing time p_a, its ending time e_a and its height h_a, which represents the amount of resource consumed by the activity when it is processed. We use the notation $a = \{s_a, p_a, e_a, h_a\}$. Usually, variables p_a and h_a are fixed integers, as well as C. In this paper, we make such assumptions. A solution to a CuSP is a schedule that satisfies the following constraints:

$$\forall a \in \mathcal{A} : s_a + p_a = e_a \quad \wedge \quad \forall t \in \mathbb{N} : \textstyle\sum_{t \in [s_a, e_a[, a \in \mathcal{A}} h_a \leq C$$

In CP, this problem is generally represented by the global constraint *Cumulative* [1]. The Energetic Reasoning of Baptiste et al. (ER) is one of the most powerful filtering algorithms for *Cumulative* [2]. This algorithm uses a characterization of relevant intervals, that is, intervals that are sufficient to check in order to ensure that all the undergoing rules used for filtering domains are satisfied. Unfortunately, ER is often too costly to be used in practice. First, its time complexity is $O(n^3)$. Moreover, the hidden constant in that time complexity is huge, as many intervals are characterized to be relevant although most of them should be ignored. In the literature, only heuristic approaches have been proposed for reducing the number of checked intervals [3].

This article provides a sharper characterization of relevant intervals. We reduce the number of intervals by a factor seven without loss of reasoning. From this theoretical work, we improve the ER checker and we introduce a new ER propagator. Compared with state-of-the-art ER techniques for *Cumulative*, our experiments show a significant reduction in the running time of both the ER checker and the ER propagator.

B. O'Sullivan (Ed.): CP 2014, LNCS 8656, pp. 289–297, 2014.

2 Background

Given a variable x, let $\underline{x}$ be the minimum value in its domain and $\overline{x}$ the maximum value. The principle of ER is to compare the available energy within a given time interval (length of that interval × capacity) with the energy necessarily taken by activities that should partially or totally overlap this interval. The minimum energy for an activity can be found either when the activity is left shifted or right shifted.

We define the part of a left shifted activity a in intersection with an interval $[t_1, t_2[$ as $LS(a, t_1, t_2) = \max(0, \min(\underline{e_a}, t_2) - \max(\underline{s_a}, t_1))$. Similarly, for the right shifted intersection we define $RS(a, t_1, t_2) = \max(0, \min(\overline{e_a}, t_2) - \max(\overline{s_a}, t_1))$. Then the *minimal intersection* of activity a with an interval $[t_1, t_2[$ is:

$$MI(a, t_1, t_2) = \min(LS(a, t_1, t_2), RS(a, t_1, t_2))$$

Proposition 1 (ER checker [5]). *If the condition*

$$\forall t_1, t_2 \in \mathbb{N}^2, t_1 < t_2 \quad C \times (t_2 - t_1) \geq \sum_{i \in \mathcal{A}} h_i \times MI(i, t_1, t_2) \tag{1}$$

is violated then the problem represented by Cumulative *is unfeasible.*

One issue is then to find the smallest sufficient set of intervals $[t_1, t_2[$ that should be checked to detect the unfeasibility.

Proposition 2 (Baptiste et al. characterization). *In order to ensure that the condition of Proposition 1 holds, it is sufficient to consider all pairs of activities (i, j) and check intervals $[t_1, t_2[$ from the set $O_B = \bigcup_{(i,j) \in \mathcal{A}^2} O_B(i, j)$, with:*

$$O_B(i,j) = \begin{cases} (t_1, t_2), t_1 \in O_1(i) < t_2 \in O_2(j) \\ (t_1, t_2), t_1 \in O_1(i) < t_2 \in O_{t_1}(j) \\ (t_1, t_2), t_2 \in O_2(j) > t_1 \in O_{t_2}(i) \end{cases}$$

and $O_1(i) = \{\underline{s_i}, \overline{s_i}, \underline{e_i}\}$, $O_2(i) = \{\overline{s_i}, \underline{e_i}, \overline{e_i}\}$, $O_t(i) = \{\underline{s_i} + \overline{e_i} - t\}$.

Proposition 1 can also be used to adjust bounds of starting and ending time variables. We examine if scheduling an activity a at its minimum schedule does not lead to a failure of condition (1). We first define the available energy for a over interval $[t_1, t_2[$ as the capacity of the interval minus the minimum intersection of all other activities:

$$\text{Avail}(a, t_1, t_2) = C \times (t_2 - t_1) - \sum_{i \in \mathcal{A} \setminus \{a\}} h_i \times MI(i, t_1, t_2)$$

Proposition 3. *For any activity a if there exists an interval $[t_1, t_2[$ such that $Avail(a, t_1, t_2) < h_a \times LS(a, t_1, t_2)$ then the left shift placement of a is not valid and the activity can not start before $t_2 - \frac{1}{h_a} \times Avail(a, t_1, t_2)$.*

Proposition 4. *For any activity a there exists an interval $[t_1, t_2[$ such that $Avail(a, t_1, t_2) < h_a \times RS(a, t_1, t_2)$ then the right shift placement of activity a is not valid and a can not end after $t_1 + \frac{1}{h_a} \times Avail(a, t_1, t_2)$.*

Definition 1 (Complete ER propagation). *The* Complete ER Propagation *is obtained when no activity can be adjusted using Proposition 3 or 4.*

The characterization of Proposition 2 is proved to be sufficient in [2] (Proposition 19) for the ER checker. Two open questions remain. The first one is related to the checker: The set of relevant intervals O_B is proved to be sufficient but could it be reduced? The second one is related to the propagator: Is O_B also sufficient to perform a complete ER propagation? In the next section, we demonstrate that one can respond affirmatively to those two questions.

3 The Energetic Reasoning Checker Revisited

Baptiste et al. showned that $f_1 : (t_1, t_2) \to C \times (t_2 - t_1) - \sum_{i \in \mathcal{A}} h_i \times MI(i, t_1, t_2)$ is continuous and piecewise linear, and that any piece can be bounded by points defined in their characterization. As extrema of a continuous and piecewise linear function can only be found on bounds of the pieces their characterization is sufficient. Out of the scope of Constraint Programming, Schwindt proposed in [9] a study of f_1 limited to local minima in order to compute a lower bound of the makespan. We propose a study adapted to the computation of relevant intervals for the Energetic Reasoning checker.

Lemma 1. *f_1 is locally minimum in (t_1, t_2) only if there exist two activities i and j such that the two following conditions are satisfied.*

$$\frac{\partial^- MI(i, t_1, t_2)}{\partial t_1} > \frac{\partial^+ MI(i, t_1, t_2)}{\partial t_1} \quad (2)$$

$$\frac{\partial^- MI(j, t_1, t_2)}{\partial t_2} > \frac{\partial^+ MI(j, t_1, t_2)}{\partial t_2} \quad (3)$$

Proof. By contradiction, let (t_1, t_2) such that for all activities in $\mathcal{A}$ condition (2) is not satisfied. Then $\sum_{i \in \mathcal{A}} h_i \times MI(i, t_1, t_2)$ has its left derivative lower than or equals to it's right derivative and f_1 has its left derivative greater than or equal to its right. By the second derivative test, minimal value of a function can only be found at points where its left derivative is lower than its right derivative. (t_1, t_2) can not be a local minimum. Proof is similar for condition (3). This proves the lemma. □

The set of intervals O_B characterizes for any couple of activity (i, j) a total number of 15 intervals. This number can be reduced thanks to Lemma 1: We can deduce necessary conditions for determining the subset of intervals that are really relevant. We first characterize the condition for which the end of an interval may be relevant.

Lemma 2. *For any activity j and any interval starting time t_1 there exists at most one interval $[t_1, t_2[$ such that $\frac{\partial^- MI(j,t_1,t_2)}{\partial t_2} > \frac{\partial^+ MI(j,t_1,t_2)}{\partial t_2}$:*

1. *if $t_1 \leq \underline{s_j}$ then only $[t_1, \overline{e_j}[$ has to be considered*
2. *if $t_1 > \underline{s_j} \wedge t_1 \geq \underline{e_j}$ then no interval has to be considered*
3. *if $t_1 > \underline{s_j} \wedge t_1 < \underline{e_j} \wedge t_1 < \overline{s_j}$ then only $[t_1, \underline{s_j} + \overline{e_j} - t_1[$ has to be considered*
4. *if $t_1 > \underline{s_j} \wedge t_1 < \underline{e_j} \wedge t_1 \geq \overline{s_j}$ then only $[t_1, \underline{e_j}[$ has to be considered*

Proof. Let us study the variation of the function $f_2^j : t_2 \rightarrow MI(j, t_1, t_2)$ when t_2 varies. As an example that illustrates the case of the first item, Figure 1 is a representation of the evolution of the minimal intersection of an activity with the following data: $j = \{s_j \in [2,4],\ p_j = 4,\ e_j \in [6,8], h_j\}$. We can distinguish three cases.

- If $t_2 \leq \overline{s_j}$ then
  $MI(j, t_1, t_2) = 0$.
- If $\overline{s_j} \leq t_2 \leq \overline{e_j}$ then
  $MI(j, t_1, t_2) = t_2 - \overline{s_j}$.
- And finally if $\overline{e_j} \leq t_2$ then
  $MI(j, t_1, t_2) = p_j$.

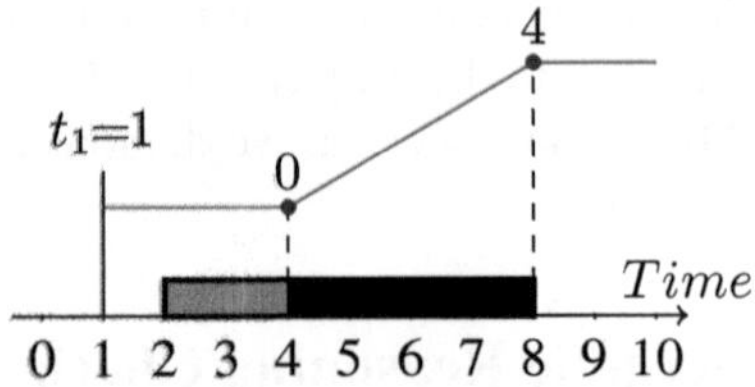

Fig. 1. A graphical exemple

The only interval for which $\frac{\partial^- MI(j,t_1,t_2)}{\partial t_2} > \frac{\partial^+ MI(j,t_1,t_2)}{\partial t_2}$ is then $[t_1, \overline{e_j}[$; $[1, 8[$ in the example. Similar case-based proofs apply for other items [4]. □

Lemma 3. *f_1 is locally minimum in (t_1, t_2) only if there exist two activities i and j such that $(t_1, t_2) \in O_C(i, j)$ with*

$$O_C(i,j) = \begin{cases} [\underline{s_i},\ \overline{e_j}[& \text{if } \underline{s_i} \leq \underline{s_j} \wedge \overline{e_j} \geq \overline{e_i} \\ [\underline{s_i},\ \underline{s_j} + \overline{e_j} - \underline{s_i}[& \text{if } \underline{s_i} > \underline{s_j} \wedge \underline{s_i} < \underline{e_j} \wedge \underline{s_i} < \overline{s_j} \wedge \underline{s_j} + \overline{e_j} - \underline{s_i} \geq \overline{e_i} \\ [\underline{s_i},\ \underline{e_j}] & \text{if } \underline{s_i} > \underline{s_j} \wedge \underline{s_i} < \underline{e_j} \wedge \underline{s_i} \geq \overline{s_j} \wedge \underline{e_j} \geq \overline{e_i} \\ [\overline{s_i}, \overline{e_j}] & \text{if } \overline{s_i} \leq \underline{s_j} \wedge \overline{e_j} < \overline{e_i} \wedge \overline{e_j} > \overline{s_i} \wedge \overline{e_j} \leq \underline{e_j} \\ [\overline{s_i},\ \underline{s_j} + \overline{e_j} - \overline{s_i}] & \text{if } \overline{s_i} > \underline{s_j} \wedge \overline{s_i} < \underline{e_j} \wedge \overline{s_i} < \overline{s_j} \wedge \\ & \quad \overline{s_i} < \underline{s_j} + \overline{e_j} - \overline{s_i} \leq \underline{e_i} \wedge \underline{s_j} + \overline{e_j} - \overline{s_i} < \overline{e_i} \\ [\overline{s_i},\ \underline{e_j}] & \text{if } \overline{s_i} > \underline{s_j} \wedge \overline{s_i} < \underline{e_j} \wedge \overline{s_i} \geq \overline{s_j} \wedge \\ & \quad \underline{e_j} < \overline{e_i} \wedge \underline{e_j} > \overline{s_i} \wedge \underline{e_j} \leq \underline{e_i} \\ [\underline{s_i} + \overline{e_i} - \overline{e_j},\ \overline{e_j}] & \text{if } \overline{e_j} < \overline{e_i} \wedge \overline{e_j} > \overline{s_i} \wedge \overline{e_j} > \underline{e_i} \wedge \underline{s_i} + \overline{e_i} - \overline{e_j} \leq \overline{s_j} \\ [\underline{s_i} + \overline{e_i} - \underline{e_j},\ \underline{e_j}] & \text{if } \underline{e_j} < \overline{e_i} \wedge \underline{e_j} > \overline{s_i} \wedge \underline{e_j} > \underline{e_i} \wedge \\ & \quad \overline{s_j} \leq \underline{s_i} + \overline{e_i} - \underline{e_j} < \underline{e_j} \wedge \underline{s_j} < \underline{s_i} + \overline{e_i} - \underline{e_j} \end{cases}$$

Proof. Suppose $\nexists(i, j)$ such that $(t_1, t_2) \in O_C(i, j)$ then by Lemma 2 and its symmetric both condition $\frac{\partial^- MI(j,t_1,t_2)}{\partial t_2} > \frac{\partial^+ MI(j,t_1,t_2)}{\partial t_2}$ and $\frac{\partial^- MI(j,t_1,t_2)}{\partial t_1} > \frac{\partial^+ MI(j,t_1,t_2)}{\partial t_1}$ can not be satisfied; by Lemma 1 f_1 can not be minimal. This proves the Lemma. □

Theorem 1. *In order to ensure ER checker property holds (condition (1)), it is enough to check intervals of the form $O_C(\mathcal{A}) = \bigcup_{(i,j) \in \mathcal{A}^2} O_C(i, j)$.*

Proof. Suppose $\exists [t_1, t_2[$ such that $\sum_{i \in \mathcal{A}} h_i \times MI(i, t_1, t_2) - C \times (t_2 - t_1) < 0$. By Lemma 3, $\exists [t_1^*, t_2^*[\in O_C(\mathcal{A})$ such that $\sum_{i \in \mathcal{A}} h_i \times MI(i, t_1^*, t_2^*) - C \times (t_2^* - t_1^*) \leq \sum_{i \in \mathcal{A}} h_i \times MI(i, t_1, t_2) - C \times (t_2 - t_1)$. f_1 is negative in (t_1^*, t_2^*), thus checking $[t_1^*, t_2^*[$ leads to a failure. The characterization is sufficient. □

This precise characterization reduces the number of relevant intervals for any pair of activities. Our characterization leads to 2 intervals for any pair of activities, as no more than two conditions can be simultaneously valid. We have thus reduced the number of intervals by a factor 7 compared with Baptiste et al. characterization. Moreover, no intervals start by $\underline{e_i}$ or end by $\overline{s_j}$.

4 Characterization of Intervals for the Propagator

Similarly to the checker, we aim to find minimal values of the induced function $f_3^a : (t_1, t_2) \rightarrow \text{Avail}(a, t_1, t_2) - h_a \times LS(a, t_1, t_2)$. If f_3^a takes a negative value, the lower bound of activity a can be adjusted (thanks to Proposition 3).

Lemma 4. *f_3^a is locally minimum in (t_1, t_2) only if one of the four conditions is satisfied:*

$$\exists (i,j), \frac{\partial^- MI(i,t_1,t_2)}{\partial t_1} > \frac{\partial^+ MI(i,t_1,t_2)}{\partial t_1} \wedge \frac{\partial^- MI(j,t_1,t_2)}{\partial t_2} > \frac{\partial^+ MI(j,t_1,t_2)}{\partial t_2} \quad (4)$$

$$\exists i, \frac{\partial^- MI(i,t_1,t_2)}{\partial t_1} > \frac{\partial^+ MI(i,t_1,t_2)}{\partial t_1} \wedge \frac{\partial^- LS(a,t_1,t_2)}{\partial t_2} > \frac{\partial^+ LS(a,t_1,t_2)}{\partial t_2} \quad (5)$$

$$\exists j, \frac{\partial^- LS(a,t_1,t_2)}{\partial t_1} > \frac{\partial^+ LS(a,t_1,t_2)}{\partial t_1} \wedge \frac{\partial^- MI(j,t_1,t_2)}{\partial t_2} > \frac{\partial^+ MI(j,t_1,t_2)}{\partial t_2} \quad (6)$$

$$\frac{\partial^- LS(a,t_1,t_2)}{\partial t_1} > \frac{\partial^+ LS(a,t_1,t_2)}{\partial t_1} \wedge \frac{\partial^- LS(a,t_1,t_2)}{\partial t_2} > \frac{\partial^+ LS(a,t_1,t_2)}{\partial t_2} \quad (7)$$

Proof. Similar to proof of Lemma 1. □

We can build from Lemma 4 the set of relevant intervals for a couple of activities from the four conditions. Intervals satisfying condition (4) have already been defined: $O_C(\mathcal{A} \backslash a)$. From conditions (5), (6) and (7) we can similarly build the set L^a studying the conditions from the left shift placement function $f_4^a : (t_1, t_2) \rightarrow LS(a, t_1, t_2)$.

Lemma 5. *For any activity a and any interval starting time t_1 there exists at most one interval $[t_1, t_2[$ such that $\frac{\partial^- LS(a,t_1,t_2)}{\partial t_2} > \frac{\partial^+ LS(a,t_1,t_2)}{\partial t_2}$:*

- *If $t_1 < \underline{e_a}$ then only $[t_1, \underline{e_a}[$ has to be considered.*
- *If $t_1 \geq \underline{e_a}$ then no intervals have to be considered.*

Proof. We consider 3 different cases :

1. $t_1 < \underline{s_a}$:
   Then $LS(a, t_1, t_2) = \max(0, \min(\underline{e_a}, t_2) - \underline{s_a})$
   (a) if $t_2 \leq \underline{s_a}$ then $LS(a, t_1, t_2) = 0$.
   (b) if $\underline{s_a} \leq t_2 \leq \underline{e_a}$ then $LS(a, t_1, t_2) = t_2 - \underline{s_a}$.
   (c) if $\underline{e_a} \leq t_2$ then $LS(a, t_1, t_2) = p_a$.
   The only interval for which $\frac{\partial^- LS(a,t_1,t_2)}{\partial t_2} > \frac{\partial^+ LS(a,t_1,t_2)}{\partial t_2}$ is then $[t_1, \underline{e_a}[$.
2. $\underline{s_a} \leq t_1 < \underline{e_a}$:
   Then $LS(a, t_1, t_2) = \max(0, \min(\underline{e_a}, t_2) - t_1)$
   (a) if $t_2 \leq \underline{e_a}$ then $LS(a, t_1, t_2) = t_2 - t_1$.
   (b) if $\underline{e_a} \leq t_2$ then $LS(a, t_1, t_2) = \underline{e_a} - t_1$.
   The only interval for which $\frac{\partial^- LS(a,t_1,t_2)}{\partial t_2} > \frac{\partial^+ LS(a,t_1,t_2)}{\partial t_2}$ is then $[t_1, \underline{e_a}[$.
3. $\underline{e_a} \leq t_1$: Then $LS(a, t_1, t_2) = 0$ and no interval satisfies the condition.

Combination of cases 1, 2 and 3 proves the lemma. □

We now precisely characterize relevant intervals for the left shift placement of activity a, from the conditions 5 , 6 and 7 : $L^a = \bigcup_{i \in \mathcal{A} \setminus a} L_1^a(i) \bigcup_{j \in \mathcal{A} \setminus a} L_2^a(j) \bigcup L_3^a$.

From Lemma 5 and the symmetric of Lemma 2, we can characterize for any i the interval that satisfy condition (5).

$$L_1^a(i) = \begin{cases} [\underline{s_i},\, \underline{e_a}[& \text{if } \underline{s_i} < \overline{s_a} \wedge \overline{e_i} < \underline{e_a} \\ [\underline{s_i} + \overline{e_i} - \underline{e_a},\, \underline{e_a}[& \text{if } \underline{s_i} + \overline{e_i} - \underline{e_a} < \underline{e_a} \wedge \underline{s_i} + \overline{e_i} - \underline{e_a} < \overline{s_a} \wedge \\ & \quad \underline{e_a} < \overline{e_i} \wedge \underline{e_a} > \overline{s_i} \wedge \underline{e_a} > \underline{e_i} \\ [\overline{s_i},\, \underline{e_a}[& \text{if } \overline{s_i} < \overline{s_a} \wedge \underline{e_a} < \overline{e_i} \wedge \underline{e_a} < \overline{s_i} \wedge \underline{e_a} \leq \underline{e_i} \end{cases}$$

From the symmetric of Lemma 5 and Lemma 2 we can characterize for any j the interval that satisfy condition (6).

$$L_2^a(j) = \begin{cases} [\underline{s_a},\, \overline{e_j}[& \text{if } \underline{s_a} \leq \underline{s_j} \wedge \overline{e_j} < \overline{e_a} \\ [\underline{s_a},\, \underline{s_j} + \overline{e_j} - \underline{s_a}[& \text{if } \underline{s_a} > \underline{s_j} \wedge \underline{s_a} < \underline{e_j} \wedge \underline{s_a} < \overline{s_j} \wedge \\ & \quad \underline{s_j} + \overline{e_j} - \underline{e_a} > \underline{s_a} \wedge \underline{s_j} + \overline{e_j} - \underline{e_a} < \overline{e_a} \\ [\underline{s_a},\, \underline{e_j}[& \text{if } \underline{s_a} > \underline{s_j} \wedge \underline{s_a} < \underline{e_j} \wedge \underline{s_a} \geq \overline{s_j} \wedge \overline{e_j} < \overline{e_a} \end{cases}$$

From Lemma 5 and its symmetric we can build the interval that satisfy condition (7).

$$L_3^a = \{\, [\underline{s_a},\, \underline{e_a}[\}$$

Lemma 6. *f_3^a is locally minimum only in $(t_1, t_2) \in O_L^a$ with $O_L^a = O_C(\mathcal{A} \setminus a) \cup L^a$.*

Proof. Same proof as Lemma 3. □

The same reasoning leads to the characterization of relevant intervals for the right shift placement $R^a = \bigcup_{j \in \mathcal{A} \setminus a} R_1^a(j) \bigcup_{i \in \mathcal{A} \setminus a} R_2^a(i) \bigcup R_3^a$.The precise characterization is symmetrical to the left shift placement characterization.

The number of relevant intervals for any activity a is then $|O_C(\mathcal{A} \setminus a) \cup L^a \cup R^a|$. By construction, $|O_C(\mathcal{A} \setminus a)| = 2(n-1)^2$ and $|L^a| = |R^a| = 2.n + 1$. Compared with Baptiste et al. characterization, our characterization reduces by a factor 7 the number of relevant intervals.

Theorem 2. *In order to ensure a complete ER propagation (Definition 1) it is sufficient to check intervals $[t_1, t_2[$ in $O_P = O_C(\mathcal{A}) \bigcup_{a \in \mathcal{A}} L^a \bigcup_{a \in \mathcal{A}} R^a$.*

Proof. Same proof as Theorem 1. □

We can thus respond affirmatively to the second open question:

Property 1. Baptiste et al. characterization of relevant intervals O_B is sufficient to ensure a complete ER propagation.

Proof. By Theorem 2, O_P is sufficient and $O_P \subset O_B$. □

5 Algorithms and Experiments

5.1 Checker

Baptiste et al. proposed an $O(n^2)$ checker algorithm based on their characterization. Their algorithm loops over set $O_1 = \bigcup_{a \in \mathcal{A}} \{\underline{s_a}, \overline{s_a}, \underline{e_a}\}$ to compute all relevant intervals starting by a value in O_1. We have shown that $\underline{e_a}$ is not relevant as a starting value. We propose a version of the algorithm adapted to our characterization, reducing the relevant starting values. We replace O_1 by $O'_1 = \bigcup_{a \in \mathcal{A}} \{\underline{s_a}, \overline{s_a}\}$ and apply the same algorithm.

5.2 Propagator

The same adaptation could be made to Baptiste et al's propagator using the reduced set O'_B, removing $\underline{e_a}$ from $O_1(a)$ and $\overline{s_a}$ from $O_2(a)$. This adaptation is simple but it deals with a superset of the relevant intervals obtained with our sharp characterization. Therefore, we propose a new ER algorithm. As the characterization given in Theorem 2, the algorithm is in 3 parts. First, we apply Baptiste et al's algorithm reduced to the set of relevant intervals $O_C(\mathcal{A})$ (lines 1 to 9). Then, for all activities we check its left and right shifted placements with sets L^a (lines 11 to 15) and R^a (lines 16 to 20).

Algorithm 1. ERpropagator()

1 **foreach** $(t_1, t_2) \in O_C(\mathcal{A})$ **do**
2     $W := \sum_{a \in \mathcal{A}} h_a \times MI(a, t_1, t_2)$;
3   **if** $W > C \times (t_2 - t_1)$ **then** fail;
4   **else foreach** $a \in \mathcal{A}$ **do**
5         $avail := C \times (t_2 - t_1) - W + h_a \times MI(a, t_1, t_2)$;
6   **if** $avail < h_a.LS(a, t_1, t_2)$ **then**
7             $\underline{s_a} := \max(\underline{s_a},\ t_2 - \frac{1}{h_a} \times avail)$;
8   **if** $avail < h_a.RS(a, t_1, t_2)$ **then**
9             $\overline{e_a} := \min(\overline{e_a},\ t_1 + \frac{1}{h_a} \times avail)$;

10 **foreach** $a \in \mathcal{A}$ **do**
11   **foreach** $(t_1, t_2) \in L^a$ **do**
12         $avail := C \times (t_2 - t_1) - \sum_{i \in \mathcal{A} \setminus a} h_a \times MI(i, t_1, t_2)$;
13   **if** $avail < h_a.MI(a, t_1, t_2)$ **then** fail;
14   **else if** $avail < h_a.LS(a, t_1, t_2)$ **then**
15             $\underline{s_a} := \max(\underline{s_a},\ t_2 - \frac{1}{h_a} \times avail)$;

16   **foreach** $(t_1, t_2) \in R^a$ **do**
17         $avail := C \times (t_2 - t_1) - \sum_{i \in \mathcal{A} \setminus a} h_a \times MI(i, t_1, t_2)$;
18   **if** $avail < h_a.MI(a, t_1, t_2)$ **then** fail;
19   **else if** $avail < h_a.RS(a, t_1, t_2)$ **then**
20             $\overline{e_a} := \min(\overline{e_a},\ t_1 + \frac{1}{h_a} \times avail)$;

5.3 Experiments

Experiments were run on a 2.9 GHz Intel Core i7, in Choco [10] version 3 (release 13.03). In order to check the gain obtained with the new characterization we have considered 100 random instances and the instances from the PSPLIB [7]. Random instances have either 10 or 20 activities. Their processing times were chosen within $[1, 10]$, their heights within $[1, 5]$. We used the *first fail* [6] search strategy (the current default strategy of Choco) and compared our algorithms with the corresponding state of the art algorithms [2], both combined with the Time-Table (TT) filtering algorithm of Letort et al. [8]. The number of nodes is identical for all proved instances, as expected. Table 1 shows a running time improvement of 20 to 36% using the new checker (measured in $\mu s/node$). Table 2 shows a time improvement of 49 to 72% using the new propagator.

Table 1. Comparison of average running of ER checkers

Instances	New checker ($\mu s/node$)	Baptiste et al ($\mu s/node$)	Gain in %
Random10	16	25	36
Random20	44	56	21
PspLib 30	451	619	27
PspLib 120	1 339	1 683	20

Table 2. Comparison of average running of ER propagators

Instances	Algorithm 1 ($\mu s/node$)	Baptiste et al ($\mu s/node$)	Gain in %
Random10	91	244	62
Random20	327	641	49
PspLib 30	4 372	8 809	50
PspLib 120	41 418	151 390	72

We also compared those combinations with the state-of-the-art filtering combination: TT + Time-Table Edge-Finding (TTEF) [11]. We tried to prove optimality. On the random10 instances, TT associated with our new ER propagator proved 63 out of 100 instances in the given time limit of five minutes. TT+TTEF was only able to prove 8 instances, mainly due to the fact that TTEF does not include an energetic checker whereas our ER propagator does; The combination TT+TTEF+ our ER Checker proved 72 instances. This shows the interest of an energetic checker as a standard feature of *Cumulative* in existing solvers. Regarding the ER propagator, a promising perspective of our work is to exploit the theoretical characterization to design a light version, with a lower time complexity than the current propagator but still filtering more values than TTEF.

6 Discussion and Conclusion

We have proposed a new characterization of relevant intervals for the energetic reasoning. Our characterization reduces by a factor seven the number of relevant intervals for the checker and for filtering any activity. We answered to an open question: Baptiste et al. characterization is sufficient to ensure a complete bounds adjustment. Compared with state-of-the-art ER techniques for Cumulative, our experiments show a significant reduction in the running time of both the ER checker and the ER propagator. Our sharpened characterization opens the new possibility to analyze the impact, in terms of filtering, of each type of relevant interval. This may help to design heuristics for ignoring some intervals without decreasing too much the pruning power of ER.

References

1. Aggoun, A., Beldiceanu, N.: Extending chip in order to solve complex scheduling and placement problems. Math. Comput. Model. 17(7), 57–73 (1993)
2. Baptiste, P., Pape, C.L., Nuijten, W.: Constraint-Based Scheduling: Applying Constraint Programming to Scheduling Problems. In: International Series in Operations Research and Management Science. Kluwer (2001)
3. Berthold, T., Heinz, S., Schulz, J.: An Approximative Criterion for the Potential of Energetic Reasoning. In: Marchetti-Spaccamela, A., Segal, M. (eds.) TAPAS 2011. LNCS, vol. 6595, pp. 229–239. Springer, Heidelberg (2011)
4. Derrien, A., Petit, T.: The Energetic Reasoning Checker Revisited. In: CP Doctoral Program 2013, Uppsala, Sweden, pp. 55–60 (September 2013)
5. Erschler, J., Lopez, P., Thuriot, C.: Scheduling under time and resource constraints. In: Proc. of Workshop on Manufacturing Scheduling, 11th IJCAI, Detroit, USA (1989)
6. Haralick, R.M., Elliot, G.L.: Increasing tree search efficiency for constraint satisfaction problems. Artificial Intelligence 14(3), 263–313 (1980)
7. Kolisch, R., Sprecher, A.: Psplib – a project scheduling problem library. European Journal of Operational Research 96, 205–216 (1996)
8. Letort, A., Beldiceanu, N., Carlsson, M.: A scalable sweep algorithm for the *cumulative* constraint. In: Milano, M. (ed.) CP 2012. LNCS, vol. 7514, pp. 439–454. Springer, Heidelberg (2012)
9. Schwindt, C.: Verfahren zur Lösung des ressourcenbeschränkten Projektdauerminimierungsproblems mit planungsabhängigen Zeitfenstern. PhD thesis, Fakultät für wirtschaftswissenschaften der Universität Fridericiana zu Karlsruhe (1998) (in German)
10. CHOCO Team. Choco: an open source Java CP library. Research report 10-02-INFO, Ecole des Mines de Nantes (2010)
11. Vilím, P.: Timetable edge finding filtering algorithm for discrete cumulative resources. In: Achterberg, T., Beck, J.C. (eds.) CPAIOR 2011. LNCS, vol. 6697, pp. 230–245. Springer, Heidelberg (2011)

A Declarative Paradigm for Robust Cumulative Scheduling

Alban Derrien, Thierry Petit, and Stéphane Zampelli

TASC (Mines Nantes, LINA, CNRS, INRIA),
4, Rue Alfred Kastler, 44307 Nantes Cedex 3, France
{alban.derrien,thierry.petit}@mines-nantes.fr,
szampelli@gmail.com

Abstract. This paper investigates cumulative scheduling in uncertain environments, using constraint programming. We present a new declarative characterization of robustness, which preserves solution quality. We highlight the significance of our framework on a crane assignment problem with business constraints.

1 Introduction

Scheduling consists in assigning activities over time. When a solution is executed in a real-world environment, activities may take longer to execute than expected. In many practical cases, solutions cannot be re-computed at anytime when disruptions occur. For instance, in Crane Assignment [17], planners need a fixed schedule which guarantees that the vessel processing will be completed ahead of schedule. The solution should meet the deadline while being able to absorb activity delays during its execution. We wish a tradeoff between robustness and performance.

We aim to address this issue for the Cumulative Scheduling Problem (CuSP), with possibly additional constraints. In a CuSP, each activity $a \in \mathcal{A}$ has a starting time variable s_a and an ending time variable e_a. Its duration p_a (processing time) and resource consumption h_a are usually strictly positive integers. We use the notation $a = \langle s_a, p_a, e_a, h_a \rangle$. Given an integer capacity C, a solution to a CuSP satisfies the following constraints: $\forall a \in \mathcal{A}, s_a + p_a = e_a$ and $\forall t \in \mathbb{N}, \sum_{t \in [s_a, e_a[, a \in \mathcal{A}} h_a \leq C$. In Constraint Programming, the *Cumulative*$(\mathcal{A}, C)$ constraint [3] represents a CuSP. A usual objective is to minimize the *makespan*, i.e., the latest end among all activities.

In this paper, we integrate the notion of robustness directly into the problem definition. We define a new generic problem, such that any activity can be delayed up to a certain time without being forced to re-schedule the other activities in its neighborhood. We introduce *FlexC*, a constraint dedicated to this problem. Our paradigm deals with the three following aspects at the same time. 1. *Declarative framework*: We solve problems such that solutions cannot be recomputed at anytime. Maximum allowed delays of activities are a data and may vary from one activity to another. 2. *Specialized definition*: In order to fit the practical needs, we use a robustness definition based on the semantics of the core problem. Notably, all the variables have not the same status. 3. *Modular approach*: We design a framework such that the model should not be totally re-written each time we add a new business constraint.

Crane Assignment is a real-world example with these three requirements. In order to validate our approach, we experiment on this problem.

B. O'Sullivan (Ed.): CP 2014, LNCS 8656, pp. 298–306, 2014.

2 Robust Cumulative Scheduling

Related Work. In the literature, some frameworks deal with the three aspects (*declarative approach*, *specialized definition* and *modularity*), but not all at the same time for the CuSP. In a super-solution [9; 10], the loss of the values of at most a variables can be repaired by changing the values of these variables and at most b other variables. This notion is generic. All variables have the same status. It has been applied to job-shop benchmarks [9]. A low-performance technique for obtaining robust schedules is to augment the duration of each activity. To improve it, some slack-based techniques incorporate the reasoning about uncertainty in the solving process [7]. This approach does not deal with CuSP and its modularity was not investigated but it has some links with our work: Schedules absorb some level of unexpected events without rescheduling. Other Operations Research techniques for robustness in scheduling problems different from the CuSP can be found in [5; 16]. A Mixed-Integer Linear Programming formulation of robust RCPSP (i.e., CuSP with precedences) has been proposed in [14]. This formulation requires an exponential number of variables and constraints.

A New Framework for Robust CuSP. We use the following notation for i-order maximum heights of activities: Given $\mathcal{A}^{\downarrow}$ the collection of activities in a set $\mathcal{A}$ sorted by decreasing heights, $max^i_{a \in \mathcal{A}}(h_a)$ is the height of the ith activity in $\mathcal{A}^{\downarrow}$.

We propose a new definition of cumulative problems, where each activity a can be delayed up to k_a points in time, without modifying the position of any other activity in its neighborhood. This can be viewed as a specialization of the notion of super-solutions that takes into account the semantics of the CuSP. Formally, given an integer $r \geq 1$, we define the Robust Cumulative Problem of order r (RCuSP^r).

Definition 1 (RCuSP^r). *Given a set of activities $\mathcal{A}$, let $\mathcal{K}$ be a set of positive integers slacks associated with activities, such that to each $a \in \mathcal{A}$ corresponds $k_a \in \mathcal{K}$. Let r be an integer, $r \geq 1$. A solution to a* RCuSP^r *satisfies the following constraints:*

$$\forall a \in \mathcal{A}, s_a + p_a = e_a \quad \wedge \quad \forall t \in \mathbb{N}, \sum_{\substack{a \in \mathcal{A}, \\ t \in [s_a, e_a[}} h_a + \sum_{i=1}^{i=r} \max^i_{a \in \{b \in \mathcal{A}, t \in [e_b, e_b + k_b[\}} (h_a) \leq C$$

Definition 1 considers a set $\mathcal{K}$ of positive integers for slacks. The material presented in this paper is consistent with the case where $\mathcal{K}$ is a set of variables, provided that values in their domains are greater than or equal to 0 and for each activity a we consider k_a as the minimum valid value for the variable in $\mathcal{K}$ mapped with a. Using variables, constraints can be defined on slacks, e.g., dependencies on starting time of activities.

We now focus on the problem RCuSP (RCuSP^r with $r = 1$). We express a RCuSP with a constraint, *FlexC*($\mathcal{A}, C, \mathcal{K}$), as it is done for the CuSP.

Property 1. *FlexC*($\mathcal{A}, C, \mathcal{K}$) $\Rightarrow$ *Cumulative*($\mathcal{A}, C$).

Proof. By Definition 1 and Definition of the CuSP. □

Property 2. Assume activities in $\mathcal{A}$ are fixed. Let $a = \langle s_a, p_a, e_a, h_a \rangle$ be an activity in $\mathcal{A}$, and k an integer, $0 \leq k \leq k_a$. Given $a' = \langle s'_a = (s_a + k), p_a, e'_a = (e_a + k), h'_a = h_a \rangle$ and $\mathcal{A}' = \mathcal{A} \cup \{a'\} \setminus \{a\}$, we have: *FlexC*($\mathcal{A}, C, \mathcal{K}$) $\Rightarrow$ *Cumulative*($\mathcal{A}', C$).

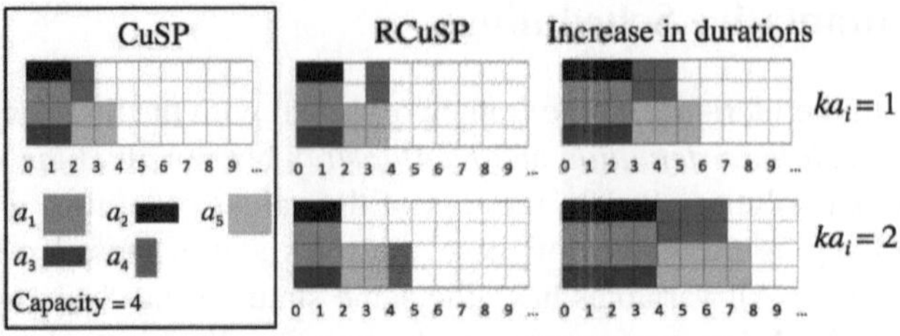

Fig. 1. Solutions minimizing the makespan. Grey rectangles are activities, the horizontal axis is time and the vertical one is consumption. On the left, a CuSP without robustness. In the middle, optimal solutions of the RCuSP with values in $\mathcal{K}$ respectively all equal to 1 (top) and 2 (bottom). On the right, optimal solutions with all durations increased respectively by 1 and 2.

Proof. Assume $\neg$ *Cumulative*$(\mathcal{A}', C)$ $\wedge$ *FlexC*$(\mathcal{A}, C, \mathcal{K})$ (hypothesis). As $\neg$ *Cumulative*$(\mathcal{A}', C)$, $\exists t_{fail} \in \mathbb{N}$, $\sum_{b \in \mathcal{A}', t_{fail} \in [s_b, e_b[} h_b > C$ (1). From Property 1, *FlexC*$(\mathcal{A}, C, \mathcal{K}) \Rightarrow$ *Cumulative*$(\mathcal{A}, C)$. Then, $\forall t \in \mathbb{N}, \sum_{b \in \mathcal{A}, t \in [s_b, e_b[} h_b \leq C$. As $\mathcal{A}$ and $\mathcal{A}'$ differ only wrt. a and a', $t_{fail} \in [\max(e_a, s'_a), e'_a[$. As *FlexC*$(\mathcal{A}, C, \mathcal{K})$ is satisfied, $(\sum_{b \in \mathcal{A}, t_{fail} \in [s_b, e_b[} h_b) + (\max_{b \in \mathcal{A}, t_{fail} \in [e_b, e_b + k_b[} h_b) \leq C$ (2). (2) minus (1) leads to $(\max_{b \in \mathcal{A}, t_{fail} \in [e_b, e_b + k_b[} h_b) - h'_a < 0$. As $t_{fail} \in [\max(e_a, s'_a), e'_a[$, $h_a \leq (\max_{b \in \mathcal{A}, t_{fail} \in [e_b, e_b + k_b[} h_b)$. Thus, $h_a - h'_a < 0$, absurd by definition of $\mathcal{A}'$. □

Delaying the starting time of an activity or increasing its duration are equivalent in the context of the RCuSP, provided we do not both delay and enlarge the activity. In case of enlargement ($a' = \langle s_a, p'_a = (p_a + k), e'_a = (e_a + k), h'_a = h_a \rangle$), the proof of Property 2 is the same. Given any set of activities, comparing RCuSP^r when r varies, we can say that RCuSP^r has an optimum makespan less than or equal to the minimum makespan of RCuSP^{r+1}. When r is too high (e.g., with sum_r the sum of the r minimum heights of distinct activities, $sum_r \geq C$) we obtain the naive approach consisting in adding k_a to the duration of each activity. This naive method is the least performant and the most robust, as Fig. 1 shows. Conversely, the RCuSP (when r=1) is the most performant tradeoff. Nevertheless, we claim that the level of robustness in solutions of a RCuSP will be widely satisfactory in most of cases. Indeed, the RCuSP allows to delay several activities in a solution provided they are scheduled in disjoint intervals in time even when they are delayed. This property is the key of the practical significance of the RCuSP. Furthermore, in some particular solutions of *FlexC*, more than one activity (e.g., a_2 and a_3 in the medium picture of Fig. 1, with all values in $\mathcal{K}$ equal to 1) can be delayed in the same time window without violating *Cumulative*. Regarding modularity, as we define a constraint, our declarative paradigm can be combined with other constraints. To guarantee the robustness, additional constraints may also have to be modified. We demonstrate this possibility in Sect. 4.

3 Filtering Technique

This Sect. presents a Time-Table filtering for *FlexC*. We selected this algorithm because it is the best one in terms of scaling (number of activities) in the CuSP case [13]. This algorithm does not directly depends on the selected time unit. Given a variable x, $\underline{x}$ (resp. $\overline{x}$) denotes the minimum value (resp. the maximum value) in its domain.

Time-Table Failure Detection. We first study the failure condition of the Time-Table filtering Algorithm for *Cumulative*, such as Letort et al.'s algorithm [13]. It is based on the profile of compulsory parts [11]. The compulsory part of an activity $a \in \mathcal{A}$ is the interval in time where a has to be processed. This interval is $[\overline{s_a}, \underline{e_a}[$ (empty if $\overline{s_a} \geq \underline{e_a}$). The *profile* is the cumulated sum of heights of compulsory parts for each point in time t, which should never exceed the capacity C. From [2], we have:

Proposition 1 (Time-Table failure check for *Cumulative***).**
If $\exists t \in \mathbb{N}, (\sum_{a\in\mathcal{A},t\in[\overline{s_a},\underline{e_a}[} h_a) > C$, *Cumulative*$(\mathcal{A}, C)$ *has no solution.*

To provide a similar failure condition for *FlexC*, we have to add the necessary height to preserve the robustness (array $\mathcal{K}$) to the cumulated sum of heights of activities having a non-empty compulsory part. To do so, we introduce $\mathcal{K}$-compulsory parts. Given the compulsory part I_a of an activity a computed with the hypothesis that duration of a is $p_a + k_a$, the $\mathcal{K}$-compulsory part of a is the sub-interval of I_a that is not intersecting the initial compulsory part of a, $[\overline{s_a}, \underline{e_a}[$, if such a sub-interval exists (it can be empty).

Definition 2 ($\mathcal{K}$-compulsory part). *Let* $a \in \mathcal{A}$ *be an activity and* $k_a \in \mathcal{K}$. *The* $\mathcal{K}$-compulsory part *of* a, *denoted* KCP_a, *is the interval* $[\max(\overline{s_a}, \underline{e_a}), \underline{e_a} + k_a]$.

The Time-Table failure condition of *FlexC* integrates in the profile, at any time t, the maximum height among activities having a $\mathcal{K}$-compulsory part intersecting t.

Proposition 2 (Time-Table failure check for *FlexC***).**
If $\exists t\in\mathbb{N}, (\sum_{a\in\mathcal{A},t\in[\overline{s_a},\underline{e_a}[} h_a)+(\max_{a\in\mathcal{A},t\in KCP_a} h_a)>C$ *then FlexC*$(\mathcal{A}, C, \mathcal{K})$ *fails.*

Proof. Assume $\exists t, \sum_{a\in\mathcal{A},t\in[\overline{s_a},\underline{e_a}[} h_a + \max_{a\in\mathcal{A},t\in KCP_a} h_a > C$. Let b be an activity such that $t \in KCP_b$ and $h_b = \max_{a\in\mathcal{A},t\in KCP_a} h_a$. Consider $\mathcal{A}' = \mathcal{A} \setminus \{b\} \cup \{b' = \langle s'_b = s_b, p'_b = p_b + k_b, e'_b = e_b + k_b, h_b\rangle\}$. By construction $t \in [\overline{s_{b'}}, \underline{e_{b'}}[$. $\sum_{a\in\mathcal{A}',t\in[\overline{s_a},\underline{e_a}[} h_a > C$. *Cumulative*$(\mathcal{A}', C)$ is violated. By Property 2, *FlexC*$(\mathcal{A}, C, \mathcal{K})$ is violated. □

Pruning Characterization. We assume now that, for each activity $a \in \mathcal{A}$, the solver maintains Bounds-Consistency [4] (BC) on the constraint $s_a + p_a = e_a$, independently from our propagator. A special case of *FlexC*$(\mathcal{A}, C, \mathcal{K})$ is the case where all values in K are equal to 0. In this case, from Definition 1, *FlexC*$\Leftrightarrow$ *Cumulative*. As enforcing BC for *Cumulative* is NP-Hard [1], it is NP-Hard for *FlexC*. Therefore, we consider a weaker form of BC. Our goal is that the achieved consistency corresponds to the filtering enforced by Time-Table in the case of *Cumulative*.

Definition 3. *Given a scheduling constraint, a propagator is* Time-Table *if* $\forall a \in \mathcal{A}$, *fixing* s_a *at time* $\underline{s_a}$ *(respectively,* e_a *at time* $\overline{e_a}$*) does not lead to a contradiction if we apply the Time-Table Failure check of the constraint.*

Fix-Point Property. The following property holds when Letort et al.'s *sweep_min* algorithm reaches its fixpoint (Property 1 in [13]) on lower bounds of start variables.

Property 3. Given *Cumulative*$(\mathcal{A}, C)$, *sweep_min* ensures that: $\forall b\in\mathcal{A}, \forall t \in [\underline{s_b}, \underline{e_b}[, h_b + \sum_{a\in\mathcal{A}\setminus\{b\},t\in[\overline{s_a},\underline{e_a}[} h_a \leq C$.

To adapt the fixpoint Property 3 to the case of *FlexC*, we have to ensure that any activity $b \in \mathcal{A}$ could be able to be scheduled at its earliest time $\underline{s_b}$ without leading directly to a fail when we apply Prop. 2.

$$\forall t \in [\underline{s_b}, \underline{e_b}[, (h_b + \sum_{\substack{a \in \mathcal{A} \setminus \{b\}, \\ t \in [\overline{s_a}, \underline{e_a}[}} h_a) + (\max_{\substack{a \in \mathcal{A}, \\ t \in KCP_a}} h_a) \leq C$$

This condition guarantees that when all variables are instantiated we have a solution of *FlexC*. The obtained filtering is weaker than Time-Table. For instance, consider a capacity $C = 1$ and two activities a_1 and a_2, such that a_2 is fixed to $\langle s_{a_2} = 4, p_{a_2} = 2, e_{a_2} = 6, h_{a_2} = 1 \rangle$, with $k_{a_2} = 0$. Assume $s_{a_1} = [0, 1000]$, $p_{a_1} = 2$, $h_{a_1} = 1$ and $k_{a_1} = 3$. The lower bound $\underline{s_{a_1}} = 0$ satisfies the previous condition. However, scheduling a_1 at $s_{a_1} = 0$ leads to a fail using Prop. 2. The condition ensures the consistency of each activity a all along its duration if scheduled at $\underline{s_a}$, but it does not guarantee that the space required after a to make it robust does not induce an inconsistency (because some activities may end after $\underline{e_a}$). The complete Time-Table fixpoint conditions are the following. Any activity which would lead to a Time-Table fail if fixed at its earliest (resp. latest) date violates one of the conditions, and reciprocally.

Property 4 (FlexC (lower bounds)). Given *FlexC*($\mathcal{A}, C, \mathcal{K}$), the propagator should ensure $\forall b \in \mathcal{A}$:

$$\forall t \in [\underline{s_b}, \underline{e_b}[, (h_b + \sum_{\substack{a \in \mathcal{A} \setminus \{b\}, \\ t \in [\overline{s_a}, \underline{e_a}[}} h_a) + (\max_{\substack{a \in \mathcal{A}, \\ t \in KCP_a}} h_a) \leq C \quad (1)$$

$$\wedge \forall t \in [\underline{e_b}, \underline{e_b} + k_b[, (\sum_{\substack{a \in \mathcal{A}, \\ t \in [\overline{s_a}, \underline{e_a}[}} h_a) + h_b \leq C \quad (2)$$

Property 5 (FlexC (upper bounds)). Given *FlexC*($\mathcal{A}, C, \mathcal{K}$), the propagator should ensure the same conditions as Property 4 with intervals $[\overline{s_b}, \overline{e_b}[$ (condition (1)) and $[\overline{e_b}, \overline{e_b} + k_b[$ (condition (2)).

We can obtain this filtering using either a decomposition or a dedicated algorithm.

Decomposition. Let $\mathcal{A} = \{a_1, a_2, \ldots, a_n\}$ be a set of activities. For each $a_i = \langle s_{a_i}, p_{a_i}, e_{a_i}, h_{a_i} \rangle$ in $\mathcal{A}$ we define $a'_i = \langle s_{a_i}, p_{a'_i} = (p_{a_i} + k_{a_i}), e_{a'_i} = (e_{a_i} + k_{a_i}), h_{a_i} \rangle$ and $\mathcal{A}_i = \mathcal{A} \cup \{a'_i\} \setminus \{a_i\}$. The set of solutions of *FlexC*($\mathcal{A}, C, \mathcal{K}$) is the set obtained by projecting on variables in $\mathcal{A}$ all solutions of the following constraint network $\mathcal{CN}$: *Cumulative*($\mathcal{A}_1, C, \mathcal{K}$)$\wedge$*Cumulative*($\mathcal{A}_2, C, \mathcal{K}$)$\wedge \ldots \wedge$*Cumulative*($\mathcal{A}_n, C, \mathcal{K}$). Representing a global constraint with n global constraints may be costly. With respect to R$_r$CuSP, $\binom{n}{r}$ *Cumulative* constraints are required. However, using Time-Table for each *Cumulative* of the decompositon prunes the same values as Time-Table for *FlexC*.

Dynamic Sweep Time-Table algorithm. We have adapted [8] the Time-Table Letort et al.'s dynamic Sweep algorithm [13; 12] for *Cumulative*, in order to design a propagator for *FlexC*. This algorithm is in two steps: Filtering of lower bounds of starting time variables (*Sweep_min*) and upper bounds of ending-time variables (*Sweep_max*). *Sweep_min* for *FlexC* is in $O(n^2)$ time, as for *Cumulative* [12, p. 55]. Conversely to

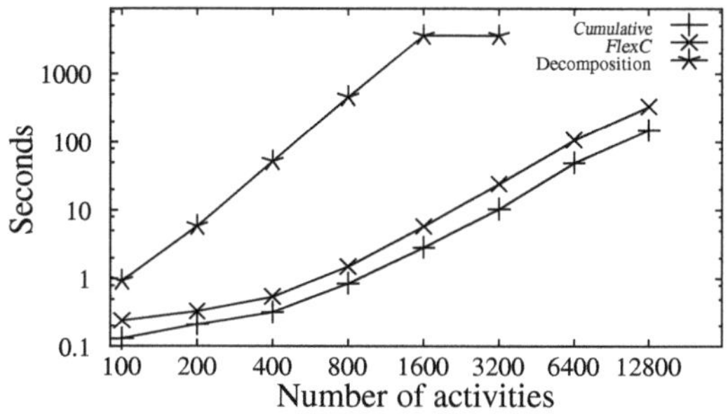

Fig. 2. Scaling of Dynamic Sweep for *FlexC*

the case of *Cumulative*, the filtering of *FlexC* is not symmetrical. In *Sweep_max*, our implementation adds new events in the sweep process to handle $\mathcal{K}$-compulsory parts, leading to a $O(n^2 \times \max_{a \in \mathcal{A}}(k_a)))$ time algorithm. As there is some differences with Sweep for *Cumulative*, we have experimented the limits of our algorithm with respect to problems size. We used Choco [6] with a 2.9 Ghz Intel i7 and 8GB of RAM. Following experiments provided in [12], we generated large random instances with p_a from 5 to 10, h_a from 1 to 5, $C = 30$. Values in $\mathcal{K}$ are not null, with an average equal to 4. Similar results are obtained with fixed k_a. Figure 2 shows that our filtering algorithm scales on problems up 12800 activities for a first solution. The decomposition reaches the time limit of 1h:00m with 1600 activities and leads to a memory crash with 6400 *Cumulative*.

4 Experiments with Side Constraints

FlexC can be used in a closed world, but external constraints can also be defined. It may be necessary to make them robust. For instance, precedence constraints of the form $e_{a_i} + \Delta_{ij} \leq s_{a_j}$. As the constraint $\leq$ is monotone, using the natural ordering of integers, augmenting s_{a_j} by k_{a_j} does not reduce the set of solutions of $e_{a_i} + \Delta_{ij} \leq s_{a_j}$. Conversely, to ensure that solutions are robust to the increase of e_{a_i}, the precedences should be strengthened: $e_{a_i} + k_{a_i} + \Delta_{ij} \leq s_{a_j}$. The principle can be extended to more complex constraints, e.g., business constraints of the Crane Assignment Problem (CAP).

We now present experiments on this problem. Our goal is to show how the model with business side constraints (such as precedence constraints, transition times, machine assignments) can be transformed into a robust one, and to measure the impact of the robust model on the objective function compared to a naive model where all durations are extended. The CAP is a specialization of the berth and crane problem [17], where we focus on the detailed scheduling of a single-container cargo's discharge. A cargo vessel is made of bays. Bays are transverse sections storing containers. Each bay is split into above deck and below deck parts. Below and above bays hold containers. A fixed number of cranes is assigned to the vessel. Cranes are operated on a single rail, they cannot cross each other. The goal is to minimize the makespan: The terminal has to pay a fee proportional to the leaving time of the cargo. Crane productivity depends on the wind and sea conditions, on the driver, and on the position of the containers in the cargo. To avoid fees, customers wish a fixed "worst case" schedule which guarantees to meet

```
1  range A=1..40//Range of acts
2  int nbc=4; //nbr of cranes
3  int pos[A]=...; //act position
4  int tt[A,A]=...;//acts transition time
5  bool preced[A,A]=...;//act precedence
6  Solver m();
7  IntVar s[A](...);//start
8  IntVar p[A](m,rand[5,maxd]));//duration
9  IntVar e[A](...);//end
10 IntVar h[A](m,1);//resource
11 IntVar c[A](m,[0,nbc−1]);//crane
12 IntVar k[i∈A](m,rand([0,.25])*p[i]);
```

Fig. 3. CAP Input Data

```
1  //1.cumu cstr, nbc resources
2  m.post(Cumulative(s,p,e,h,nbc))
3  //2.precedence constraints
4  for (i,j) s.t. preced[i,j]==1:
5     m.post(e[i]<s[j]);
6  //3.crane alloc, transition times
7  for (i,j) i!=j∧pos[i]<pos[j]:
8     m.post( ((s[i]<e[j]+tt[i,j])
9         ∧ (s[j]<e[i]+tt[i,j]))
10               => c[i]<c[j] );
11 //4.no intersec for nearby acts
12 for (i,j) i<j∧|pos[i]−pos[j]|<=2:
13    m.post(s[i]>e[j] ∨ e[i]<s[j]);
14 minimize obj = min({e[i]} i∈ A)
```

Fig. 4. CAP model

```
1  //1.flex cumu cstr, nbc resources
2  m.post(FlexC(s,p,e,h,k,nbc))
3  //2.precedence constraints
4  for (i,j) s.t. preced[i,j]==1:
5     m.post(e[i]+k[i] <s[j]);
6  //3.crane alloc, transition times
7  for (i,j) i!=j∧pos[i]<pos[j]:
8     m.post( ((s[i]<e[j]+tt[i,j]+k[j])
9         ∧ (s[j]<e[i]+tt[i,j]+k[i]))
10               => c[i]<c[j] );
11 //4.no intersec for nearby acts
12 for (i,j) i<j∧|pos[i]−pos[j]|<=2:
13    m.post(s[i]>e[j]+k[i] ∨ e[i]+k[i]<s[j]);
14 minimize obj = min({e[i]+k[i]} i∈ A)
```

Fig. 5. Robust CAP model

a deadline, given a precise robustness definition, as our framework does. Simulation is not relevant: Uncertainty has to be taken into account *a priori* in the problem definition. The maximum allowed slack k_a for each activity a is a data. Generating durations a posteriori is not relevant, as they would have to match the input robustness criteria.

Data. Figure 3 provides the pseudo code for the input data and decision variables. Line 1 is the range of activities. The cargo has 20 bays with one activity below and above deck. Line 2 sets the number of resources to 4, the typical number of cranes for such a cargo of 20 bays. Lines 3-5 declare the bay position of each activity, the transition time and the presence of a precedence constraint. Transition times are computed based on the distance between two positions, multiplied by a factor. For each bay, we have a precedence constraint between below and above deck activities. An additional 5% of precedences are randomly set to reflect discharge balance constraints on the cargo. Lines 7-10 declare the start, duration, end, and resource variables for each activity. Durations (in minutes) are randomly chosen in $[5, \text{maxd} = 800]$ to capture many scenarios. Value 800 is the maximum duration for discharging a large bay. In lines 11-12, additional crane and robustness variables are created for each activity. The robust factor k, fixed, is randomly chosen in $[0, 25\%]$ multiplied by the duration of the activity. This is a bad case for our approach (performance improves as the ratio k_a/p_a increases).

Constraints. Figure 4 shows the constraints of the CAP model without robustness. Following [17], we model this application as a cumulative scheduling problem. Lines 2-5 post the cumulative constraint and precedence constraints. Lines 8-10 post the crane allocation constraints. Given two different activities i and j with i being on the left of j, if those activities intersect in time, we ensure that crane assignment follows their position, assuming crane 0 is on the left of crane 1. Those constraints ensure that any set of activities intersecting with a given point in time has a feasible crane assignment. The activities should not be assigned to the same crane if they intersect in time while being extended by the transition time, because a crane would not have the required time to travel from activity i to activity j. Line 13 ensures that, for security reasons, cranes should not work on nearby bays. If two activities are two bays away from each other, they should have no intersection in time.

Robust Model. In Fig. 5, *FlexC* is used with the k variables in line 2. The precedences constraints are modified in line 5 to accomodate the robust factor. The left side of the constraint in lines 8-10 is an intersection in time condition. The solution should ensure that if an activity is pushed or extended and intersects after this change with another activity, the schedule is still valid. Line 13 posts the negation of an intersection in time condition, and we can add $k[i]$ to $e[i]$.

Heuristic. For the three models, first the starting variables are assigned based on min domain and min value, then on crane variables. The search strategy uses propagation-guided LNS [15] and fixes randomly 70% of the starting time of the activities.

Table 1. Lower bound distance in %

#	CAP	*FlexC*	Naive	α	γ
1	10.2 (0.08)	20.7 (2.20)	36.2 (3.28)	0.40	2.47
2	8.6 (0)	19.4 (2.13)	32.5 (1.62)	0.45	2.21
3	5.7 (0)	19.2 (3.08)	28.5 (0.60)	0.59	1.68
4	9.3 (0)	17.7 (0.47)	30.9 (3.66)	0.38	2.57
5	6.3 (0)	20.2 (2.10)	35.3 (0)	0.47	2.08
6	6.4 (0)	17.8 (1.37)	30.7 (0.15)	0.46	2.13
7	4 (0)	15.5 (1.08)	35.8 (2.40)	0.36	2.76
8	1.8 (0)	19.8 (2.35)	27.1 (1.42)	0.71	1.40
9	13.2 (0)	15.5 (2.61)	22 (0.31)	0.26	3.82
10	7.2 (0)	19.4 (1.63)	31.6 (0.07)	0.5	2.0

Performance. We measure the distance of the objective value to a lower bound computed by adding the durations and dividing by the number of cranes. This lower bound ignores side constraints. We compare the CAP, robust CAP, and a naive model where all durations are extended. To make a fair comparison with the naive approach, using *FlexC*($\mathcal{A}, \mathcal{K}$) we minimize $\max_{a \in \mathcal{A}}(e_a + k_a)$. Table 1 shows ten instances with a time-out of 5 minutes. Entries in the first columns are the distance in percentage with the lower bound. The standard deviation on 5 runs is in parenthesis. Column α is the relative position of the robust approach compared with the initial and the naive model. Column γ is the ratio between the naive and *FlexC* results. A value of $\gamma = 2$ means the naive model doubles the distance with respect to the CAP model, compared with *FlexC*. Our approach produces a robust solution adding on average 10% to the makespan, while the naive model adds 20% on average. A fully loaded cargo would take a lower bound of (20*2*800)/4=8000 minutes to discharge, that is, 5d:13h:20m. The naive model adds 26h:40m. The robust approach adds 13h:20m, a good worst case compromise.

5 Conclusion

This paper has introduced a new declarative paradigm in order to deal with robustness in cumulative problems. We have defined a new constraint and adapted the Time-Table dynamic sweep algorithm. The experiments showed that our approach is modular, as solution performance can be preserved for a problem with many business constraints. Future work includes the adaption of other solving techniques and the use of a similar approach for other classes of optimization problems.

References

1. Baptiste, P., Le Pape, C., Nuijten, W.: Satisfiability tests and time-bound adjustments for cumulative scheduling problems. Annals of Operations Research 92, 305–333 (1999)
2. Beldiceanu, N., Carlsson, M.: A new multi-resource *cumulatives* constraint with negative heights. In: Van Hentenryck, P. (ed.) CP 2002. LNCS, vol. 2470, pp. 63–79. Springer, Heidelberg (2002)
3. Beldiceanu, N., Contejean, E.: Introducing global constraints in CHIP. Journal of Mathematical and Computer Modelling 20(12), 97–123 (1994)
4. Bessiere, C.: Constraint propagation. Research report 06020. In: Rossi, F., van Beek, P., Walsh, T. (eds.) Handbook of Constraint Programming, LIRMM, ch. 3, Elsevier (2006)
5. Billaut, J.-C., Moukrim, A., Sanlaville, E. (eds.): Flexibility and Robustness in Scheduling. Wiley (2010)
6. Choco. 3.1.0 (2013), http://choco.sourceforge.net/
7. Davenport, A.-J., Jefflot, C., Beck, J.-C.: Slack-based techniques for robust schedules. In: Proc. European Conference on Planning, pp. 7–18 (2001)
8. Derrien, A., Petit, T., Zampelli, S.: Dynamic sweep filtering algorithm for FlexC. Research report RR14/1/INFO, Mines Nantes (2014)
9. Hebrard, E.: Super solutions in constraint programming. In: Sattler, U. (ed.) IJCAR Doctoral Programme. CEUR Workshop Proceedings, vol. 106. CEUR-WS.org (2004)
10. Hebrard, E., Hnich, B., Walsh, T.: Super solutions in constraint programming. In: Régin, J.-C., Rueher, M. (eds.) CPAIOR 2004. LNCS, vol. 3011, pp. 157–172. Springer, Heidelberg (2004)
11. Lahrichi, A.: The notions of Hump, Compulsory Part and their use in Cumulative Problems. C.R. Acad. Sc. 294, 20–211 (1982)
12. Letort, A.: Passage à l'échelle pour les contraintes d'ordonnancement multi-ressources. Ph.D dissertation (2013)
13. Letort, A., Beldiceanu, N., Carlsson, M.: A scalable sweep algorithm for the *cumulative* constraint. In: Milano, M. (ed.) CP 2012. LNCS, vol. 7514, pp. 439–454. Springer, Heidelberg (2012)
14. Leus, R., Artigues, C., Talla Nobibon, F.: Robust optimization for resource-constrained project scheduling with uncertain activity durations. In: Proc. IEEM, pp. 101–105 (2011)
15. Perron, L., Shaw, P., Furnon, V.: Propagation guided large neighborhood search. In: Wallace, M. (ed.) CP 2004. LNCS, vol. 3258, pp. 468–481. Springer, Heidelberg (2004)
16. Wu, C.W., Brown, K.N., Beck, J.C.: Scheduling with uncertain durations: Modeling beta-robust scheduling with constraints. Computers & OR 36(8), 2348–2356 (2009)
17. Zampelli, S., Vergados, Y., Van Schaeren, R., Dullaert, W., Raa, B.: The berth allocation and quay crane assignment problem using a CP approach. In: Schulte, C. (ed.) CP 2013. LNCS, vol. 8124, pp. 880–896. Springer, Heidelberg (2013)

Improving DPOP with Branch Consistency for Solving Distributed Constraint Optimization Problems

Ferdinando Fioretto[1,2], Tiep Le[1], William Yeoh[1], Enrico Pontelli[1], and Tran Cao Son[1]

[1] Department of Computer Science, New Mexico State University, USA
[2] Department of Mathematics and Computer Science, University of Udine, Italy
{ffiorett,tile,epontell,wyeoh,tson}@cs.nmsu.edu

Abstract. The DCOP model has gained momentum in recent years thanks to its ability to capture problems that are naturally distributed and cannot be realistically addressed in a centralized manner. Dynamic programming based techniques have been recognized to be among the most effective techniques for building complete DCOP solvers (e.g., DPOP). Unfortunately, they also suffer from a widely recognized drawback: their messages are exponential in size. Another limitation is that most current DCOP algorithms do not actively exploit hard constraints, which are common in many real problems. This paper addresses these two limitations by introducing an algorithm, called BrC-DPOP, that exploits arc consistency and a form of consistency that applies to paths in pseudo-trees to reduce the size of the messages. Experimental results shows that BrC-DPOP uses messages that are up to one order of magnitude smaller than DPOP, and that it can scale up well, being able to solve problems that its counterpart can not.

1 Introduction

Distributed Constraint Optimization Problems (DCOPs) are constraint optimization problems where variables and constraints are distributed among a group of agents, and where each agent can only interact with agents that share a common constraint [20,24,30]. As a result, agents need to coordinate their value assignments to maximize the overall sum of resulting constraint utilities and lead to an optimal solution of the optimization problem. DCOPs provide an elegant and effective modeling of problems that have a distributed nature, and where a collective is trying to achieve a globally optimal solution within the confines of the localized communication. Researchers have used DCOPs to model various distributed optimization problems, such as meeting scheduling [19,34], resource allocation [8,33], and power network management problems [16].

In recent years, we have witnessed a growing interest towards DCOPs, with the development of a number of complete and incomplete distributed algorithms. A number of implementations have been proposed and are publicly available [18,28,7]. The majority of the existing DCOP algorithms can be placed in one of three classes. *Search-based* algorithms perform a distributed search over the space of solutions to determine optimum [20,9,32]. *Inference-based* algorithms, on the other hand, make use

B. O'Sullivan (Ed.): CP 2014, LNCS 8656, pp. 307–323, 2014.

of techniques from dynamic programming to propagate aggregate information among agents [24,8]. Finally, *sampling-based* algorithms rely on sampling applied to the overall search space [23,22]. Of these methods, the *Distributed Pseudo-tree Optimization Procedure*[1] (DPOP) [24] is one of the most efficient DCOP solvers; DPOP has also been extended in several ways to enhance its performance and capabilities (e.g., O-DPOP and MB-DPOP trade off memory requirement for longer runtimes [26,27], A-DPOP trades off solution optimality for shorter runtimes [25], SS-DPOP trades off runtime for increased privacy [10], H-DPOP exploits hard constraints for smaller runtimes [17], and DPOP with function filtering exploits utility bounds for smaller runtimes [3]).

This paper proposes a novel variant of DPOP, called *Branch-Consistency DPOP* (BrC-DPOP), that takes advantage of hard constraints present in the problem to prune the search space. BrC-DPOP introduces a new form of consistency, called *branch consistency*, which can be viewed as a weaker version of path consistency [21] tailored to variables ordered in a pseudo-tree, and where each agent can only communicate with neighboring agents. The effect of enforcing this consistency in DPOP is the ability to actively use hard constraints (either explicitly provided in the problem specification or implicitly described in utility tables) to prune the search space and to reduce the size of the utility tables exchanged among agents.

2 Background

2.1 Distributed Constraint Optimization Problems (DCOPs)

A *DCOP* [20,24,30] is defined by a tuple $\langle \mathcal{X}, \mathcal{D}, \mathcal{F}, \mathcal{A}, \alpha \rangle$, where $\mathcal{X} = \{x_1, \ldots, x_n\}$ is a set of *variables*; $\mathcal{D} = \{D_1, \ldots, D_n\}$ is a set of finite *domains*, where D_i is the domain of variable x_i; $\mathcal{F} = \{f_1, \ldots, f_e\}$ is a set of *utility functions* (also called *constraints*), $f_i : \times_{x_j \in scope(f_i)} D_j \mapsto \mathbb{N} \cup \{0, -\infty\}$, specifying the utility of each combination of values to the variables in the *scope* of the constraint (where $scope(f_i) \subseteq \mathcal{X}$); $\mathcal{A} = \{a_1, \ldots, a_p\}$ is a set of *agents*; $\alpha : \mathcal{X} \rightarrow \mathcal{A}$ maps each variable to one agent. In this paper, we will focus on *unary* and *binary* constraints. For simplicity, we will refer to unary constraints as f_{ii} and binary constraints as f_{ij} to denote the fact that their scope is $\{x_i\} \subseteq \mathcal{X}$ and $\{x_i, x_j\} \subseteq \mathcal{X}$, respectively. We also assume that each agent has exactly one variable mapped to it. Thus, we will use the terms "variable" and "agent" interchangeably. This is a common assumption in the DCOP literature as there exist pre-processing techniques that transform a general DCOP into our more restrictive DCOP [31,4]. A solution is a value assignment for a subset of variables. Its utility is the evaluation of all utility functions on it. A solution is *complete* if it is a value assignment for all variables in $\mathcal{X}$. The goal is to find a utility-maximal complete solution.

Each constraint in $\mathcal{F}$ can be either *hard*, indicating that some value combinations result in a utility of $-\infty$ and must be avoided, or *soft*, indicating that all value combinations result in a finite utility and need not be avoided. We use $\mathbf{H}_i$ and $\mathbf{S}_i$ to denote the set of hard and soft constraints, respectively, whose scope includes x_i. With a slight abuse of notation, we will also often view a unary constraint f_{ii} as a subset of D_i, defined as $f_{ii} = \{(u) \in D_i \mid f_{ii}(u) \neq -\infty\}$, and a binary constraint f_{ij} as a subset of $D_i \times D_j$, defined as $f_{ij} = \{(u, v) \in D_i \times D_j \mid f_{ij}(u, v) \neq -\infty\}$.

[1] This algorithm has also been referred to as Dynamic Programming Optimization Protocol.

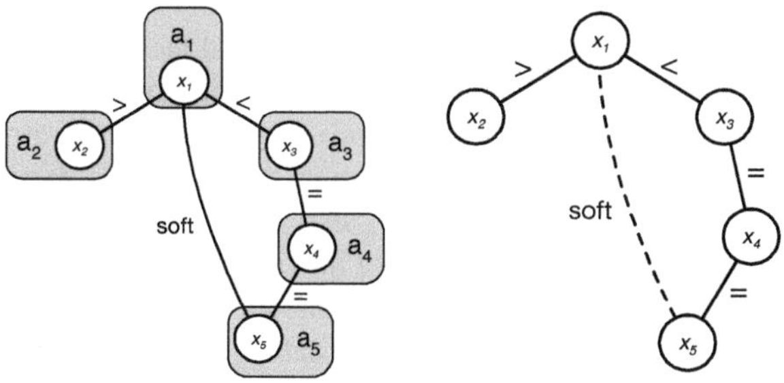

x_1	x_5	Utilities
0	0	20
0	1	8
0	2	10
0	3	3
...		
3	3	2

(a) Constraint Graph (b) Pseudo-tree (c) Constraint Table

Fig. 1. Example DCOP

Table 1. Example UTIL Phase Computations of a_5

x_1	x_4	Utilities
0	0	max(20+0, 8-∞, 10-∞, 3-∞) = 20
0	1	max(20-∞, 8+0, 10-∞, 3-∞) = 8
0	2	max(20-∞, 8-∞, 10+0, 3-∞) = 10
0	3	max(20-∞, 8-∞, 10-∞, 3+0) = 3
...		...

A *constraint graph* visualizes a DCOP, where nodes correspond to the variables and the edges connect pairs of variables in the scope of the same utility function. A *DFS pseudo-tree* arrangement has the same nodes and edges as the constraint graph and satisfies two conditions: *(i)* there is a subset of edges (*tree edges*) that form a rooted tree, and *(ii)* two variables in the scope of the same utility function appear in the same branch of the tree. The other edges are called *backedges*. Tree edges connect parent-child nodes, while backedges connect a node with its pseudo-parents and pseudo-children. We also use the following notation: C_i, PC_i, P_i, and PP_i refer to the set of children, pseudo-children, parent, and pseudo-parents of agent a_i, respectively; and $sep(a_i)$ refers to the *separator* of agent a_i, which is the set of ancestor agents that are constrained with agent a_i or one of its descendant agents.

Figure 1(a) shows the constraint graph of a simple DCOP with five agents, a_i, with $i = 1, \ldots, 5$, each owning exactly one variable x_i. The domain of each variable is the set $\{0, 1, 2, 3\}$. Figure 1(b) shows one possible pseudo-tree for the problem, where the agent a_1 has one pseudo-child, a_5 (the dotted line is a backedge). Figure 1(c) describes few value combinations of the utility function associated with the constraint f_{15}.

2.2 Distributed Pseudo-Tree Optimization Procedure (DPOP)

DPOP [24] has the following three phases:

(1) The first phase is the pseudo-tree generation phase, realized through an existing distributed pseudo-tree construction algorithm, like Distributed DFS [15].

(2) The second phase is the UTIL propagation phase, where each agent, starting from the leaves of the pseudo-tree, computes the optimal sum of utilities in its subtree for

each value combination of variables in its separator. The agent does so by summing the utilities of its constraints with the variables in its separator and the utilities in the UTIL messages received from its children agents, and then projecting out its own variables by optimizing over them. In our example problem, agent a_5 computes the optimal utility for each value combination of variables x_1 and x_4, as shown in Table 1, and sends the utilities to its parent agent a_4 in a UTIL message. Such a table consists of $4^3 = 64$ utilities before projecting out its variable x_5, and $4^2 = 16$ utilities after the projection. The value 0 ($-\infty$) represents the utility for the hard constraint f_{45} for values of x_4, x_5 that satisfy (do not satisfy) it. When the root agent a_1 receives the UTIL message from each of its children, it computes the maximum utility of the entire problem.

(3) The third phase is the VALUE propagation phase, where each agent, starting from the root of the pseudo-tree, determines the optimal value for its variable. The root agent does so by choosing the value of its variable from its UTIL computations—selecting the value with the maximum utility. It sends the selected value to its children in a VALUE message. Each agent receiving a VALUE message will determine the value for its variable producing the maximum utility given the variable assignments (of the agents in its separator) indicated in the VALUE message. Once determined, such assignment is further propagated to the children via VALUE messages.

3 Branch-Consistent DPOP (BrC-DPOP)

3.1 Preliminaries

Definition 1. *The* consistency graph *of a DCOP* $\langle \mathcal{X}, \mathcal{D}, \mathcal{F}, \mathcal{A}, \alpha \rangle$ *is* $\mathbf{G} = (\mathbf{V}, \mathbf{E})$ *where* $\mathbf{V} = \{(i,k) \mid x_i \in \mathcal{X}, k \in D_i\}$ *and* $\mathbf{E} = \{\langle (i,r),(j,c) \rangle \mid r \in D_i, c \in D_j, f_{ij} \in \mathcal{F}, (r,c) \in f_{ij}\}$.

Definition 2. *Given a pseudo-tree associated with a DCOP problem instance, we define a* linear ordering $\prec$ *on its variables:* $x_i \prec x_j$ *iff* $x_j \in P_i$. *Similarly,* $x_i \succ x_j$ *iff* $x_j \in C_i$. *We denote with* $\preceq$ *(and* $\succeq$*) the reflexive closure of* $\prec$ *(and* $\succ$*), and with* $\overset{*}{\prec}$ *(and* $\overset{*}{\succ}, \overset{*}{\preceq}, \overset{*}{\succeq}$*) the transitive closure of* $\prec$ *(and* $\succ, \preceq, \succeq$ *).*

Definition 3. *A pair of values* $(r,c) \in D_i \times D_j$ *of two variables* x_i, x_j *that share a constraint* f_{ij} *is* branch consistent (BrC) *iff for any sequence of variables* $(x_i = x_{k_1}, \dots, x_{k_m} = x_j)$*, such that* $f_{k_p k_q} \in \mathcal{F}$*, where* $p \leq q \leq p+1$*, and* $x_{k_1} \preceq \cdots \preceq x_{k_m}$*, there exists a tuple of values* $(r = v_{k_1}, \dots, v_{k_m} = c)$ *such that* $v_{k_q} \in D_{k_q}$ *and* $(v_{k_p}, v_{k_q}) \in f_{k_p k_q}$*, for each* $1 \leq q \leq m$ *and* $p \leq q \leq p+1$.

Definition 4. *A DCOP is* branch consistent (BrC) *iff for any pair of variables* (x_i, x_j) *with* $x_i \preceq x_j$ *and any* $(u,v) \in f_{ij}$*,* (u,v) *is branch consistent.*

Definition 5. *Given a DCOP, the* Value Reachability Matrix (VRM) M_{ij} *of two variables* x_i *and* x_j *of* $\mathcal{X}$*, with* $x_i \overset{*}{\preceq} x_j$*, is a binary matrix of size* $D_i{\times}D_j$*, where* $M_{ij}[r,c]{=}1$ *iff there exists at least one sequence of variables* $(x_i = x_{k_1}, \dots, x_{k_m} = x_j)$*, such that* $x_{k_1} \preceq \cdots \preceq x_{k_m}$*, and a tuple of values* $(r = v_{k_1}, v_{k_2}, \dots, v_{k_m} = c)$ *such that* $v_{k_p} \in D_{k_p}$ *and* $(v_{k_p}, v_{k_q}) \in f_{k_p k_q}$*, for each* $1 \leq q \leq m$ *and* $p \leq q \leq p+1$.

Proposition 1. *For each variable* x_i, x_j, *and* x_k, *the* regular product *of two VRMs* M_{ik} *and* M_{kj} *is a VRM* $M_{ij} = M_{ik} \times M_{kj}$, *where each entry* (r, c) *of* M_{ij} *is given by*

$$M_{ij}[r, c] = \min \left\{ 1, \sum_{l=1}^{|D_k|} M_{ik}[r, l] \cdot M_{kj}[l, c] \right\}$$

Proposition 2. *For each variable* x_i *and* x_j, *the* entrywise product *of two VRMs* M_{ij} *and* $\hat{M}_{ij}$ *is a VRM* $M'_{ij} = M_{ij} \circ \hat{M}_{ij}$, *where each entry* (r, c) *of* M'_{ij} *is given by*

$$M'_{ik}[r, c] = M_{ij}[r, c] \cdot \hat{M}_{ij}[r, c]$$

Definition 6. *Given a VRM* M_{ij}, *a pair of values (r,c) is a* valid *pair iff* $M_{ij}[r, c] = 1$.

Definition 7. *If* $f_{ij} \in \mathcal{F}$, *then* M_{ij} *is* branch consistent (BrC) *iff all its valid pairs are branch consistent. If* $f_{ij} \notin \mathcal{F}$, *then* M_{ij} *is* branch consistent *iff it is a regular product of branch consistent VRMs.*

3.2 High-Level Algorithm Description

Let us now illustrate the high-level structure of BrC-DPOP on the example DCOP shown in Figure 1. BrC-DPOP consists of the following phases:

- **Pseudo-tree Generation Phase:** This phase is identical to that of DPOP.
- **Path Construction Phase:** In this phase, each agent builds the VRMs associated with the constraints involving its variables along with the structures describing the paths between pseudo-parents and pseudo-children. Figure 2(a) shows the VRMs (in a consistency graph representation); we do not show the soft constraint between variables x_1 and x_5 as it allows every value combination of the two variables.
- **Arc Consistency Enforcement Phase:** In this phase, the agents enforce arc consistency in a distributed manner. At the end of this phase, each agent has the updated VRMs shown in Figure 2(b). Arc consistency causes the removal of exactly two values from the domain of each variable of the DCOP: values 0 and 3 from D_1, 0 and 1 from D_2, and 2 and 3 from D_3, D_4, and D_5.
- **Branch Consistency Enforcement Phase:** In this phase, the agents enforce branch consistency in a distributed manner. In our example, branch consistency needs to be enforced for the pairs of values of variables x_1 and x_5 only. The values for all other pairs of variables are already branch consistent. Agent a_1 starts this process by sending a message containing VRM M_{11} to its child a_3 (since a_5 is in the subtree rooted at a_3). Once agent a_3 receives the message, it computes the VRM M_{31} by multiplying its VRM M_{31} with the VRM M_{11} just received, and sends a message containing this VRM to its child a_4. Agent a_4 repeats this process by multiplying its VRM M_{43} with the VRM M_{31}, resulting in VRM M_{41}, which it sends to its child a_5. This process repeats until agent a_5 computes the VRM M_{51}, after which it knows its set of reachable values in x_5 for each value in x_1. Figure 2(c) shows the VRMs.
- **UTIL and VALUE Propagation Phases:** This phase is identical to the corresponding UTIL and VALUE propagation phases of DPOP, except that each agent constructs

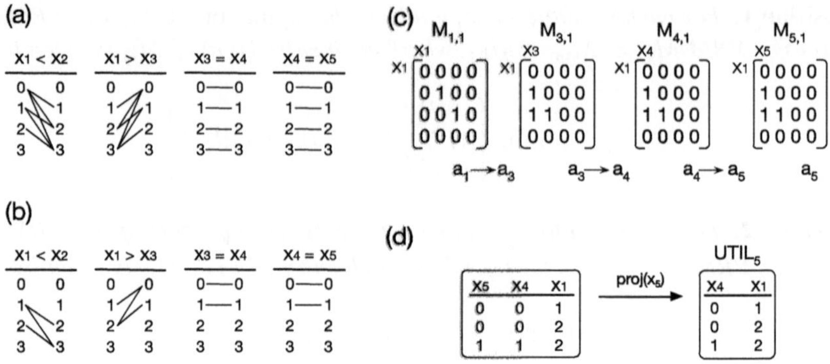

Fig. 2. Example Trace

a UTIL table that contains utilities for each combination of *unpruned* values of variables in its VRMs. In our example, agent a_5 is able to project out its variable x_5 and construct its UTIL table, shown in Figure 2(d). Note that the UTIL table consists of only 3 utilities, both before and after projection. In contrast, DPOP's UTIL table consists of $4^3 = 64$ utilities before projection and $4^2 = 16$ utilities after projection.

3.3 Messages and Data Structures

During the execution of BrC-DCOP, each agent a_i maintains the following data structures, where the first three are used in the arc consistency phase and the last two are used in the branch consistency phase.

- The set of hard constraints $\hat{\mathbf{H}}_i = \{f_{ij} \in \mathbf{H}_i \mid a_i \overset{*}{\preceq} a_j\}$ to check for consistency.
- The set of VRMs $\hat{\mathbf{M}}_i = \{\hat{M}_{ij} \mid f_{ij} \in \mathcal{F}, a_j \overset{*}{\preceq} a_i\}$, which includes the VRMs for each parent and pseudo-parent a_j.
- The flag *fixed*$_i$ for each agent a_i, which is initialized to true. It indicates if agent a_i has reached a fixed point in the arc consistency phase.
- The set of VRMs $\mathbf{M}_i = \{M_{ij} \mid a_j \in sep(a_i)\}$, which includes the VRMs for each separator agent a_j.
- The set of paths $\text{PATHS}_i = \{(a_s \overset{a_j}{\rightsquigarrow} a_d) \mid a_j \in C_i,\ a_s \overset{*}{\succeq} a_i \succ a_j \overset{*}{\succeq} a_d\}$, which the agent uses to send updated VRMs in the branch consistency phase. Each path $(a_s \overset{a_j}{\rightsquigarrow} a_d)$ indicates that there is a branch in the pseudo-tree from a_s to a_d that passes through a_i and its child a_j. This data is needed by agent a_i to know which child it should send its updated VRM to, if the VRM originated from agent a_s. For example, in our example trace, agent a_1 knows to send its VRM to its child a_3 and not a_2. To preserve privacy, the information about the destination agent a_d can be omitted from each path. Each agent thus maintains only $(a_s \overset{a_j}{\rightsquigarrow} ?)$, which is sufficient to ensure correctness.

In addition to the UTIL and VALUE messages used in the UTIL and VALUE propagation phases, each agent a_i uses the following types of messages, where the first

Algorithm 1. BRC-DPOP

1 PSEUDO-TREE-GENERATION-PHASE()
2 PATH-CONSTRUCTION-PHASE()
3 AC-PROPAGATION-PHASE()
4 BRC-PROPAGATION-PHASE()
5 UTIL-AND-VALUE-PHASES()

Procedure Path-Construction-Phase()

6 **foreach** $a_p \in PP_i$ **do**
7 $\quad$ $\text{PATHS}_i \leftarrow \text{PATHS}_i \cup (a_p \overset{NULL}{\rightsquigarrow} ?)$
8 **if** $C_i \neq \emptyset$ **then**
9 $\quad$ **while** *not received all* $PATH_c^{\uparrow}(\cdot)$ *from each* $a_c \in C_i$ **do**
10 $\quad\quad$ **if** *receive* $PATH_c^{\uparrow}(a_s)$ *from* $a_c \in C_i$ **then**
11 $\quad\quad\quad$ $\text{PATHS}_i \leftarrow \text{PATHS}_i \cup (a_s \overset{a_c}{\rightsquigarrow} ?)$
12 **foreach** $a_s \neq a_i$ *such that* $(a_s \overset{a_c}{\rightsquigarrow} ?) \in \text{PATHS}_i$ **do**
13 $\quad$ Send $PATH_i^{\uparrow}(a_s)$ to P_i
14 **if** $PATH_i^{\uparrow}(\cdot)$ *has not been sent to* P_i **then**
15 $\quad$ Send $PATH_i^{\uparrow}(NULL)$ to P_i

two are used in the arc consistency phase, while the last two in the branch consistency phase:[2]

- $AC_i^{\uparrow}(D'_j, \mathit{fixed}_i)$, which is sent from an agent a_i to an agent $a_j \overset{*}{\succ} a_i$ such that $f_{ij} \in \mathbf{H}_i$. It contains a copy of the domain of the variable x_j, D'_j, updated with the changes caused by the propagation of the constraints in $\hat{\mathbf{H}}_i$, and a flag, fixed_i, which denotes whether changes have occurred in the domain of some variable in the subtree rooted at a_i during the last iteration of the $AC^{\uparrow}$ messages.
- $AC_i^{\downarrow}(D_i)$, which is sent from an agent a_i to the agents $a_j \overset{*}{\prec} a_i$ such that $f_{ij} \in \mathbf{H}_i$. It contains a copy of the domain of the variable x_i, D_i, updated with the changes caused by the propagation of the constraints in $\hat{\mathbf{H}}_i$.
- $PATH_i^{\uparrow}(a_s)$, which is sent from an agent a_i to its parent P_i to inform it that it is part of a tree path in the pseudo-tree between agents a_s and some pseudo-child of a_s.
- $BrC_i^{\downarrow}(M_{is})$, which is used to determine the branch consistent value pairs of x_s and x_i.

3.4 Algorithm Description

Algorithm 1 shows the pseudo-code of BrC-DPOP. It can be visualized as a process composed of 5 phases:

- **Phase 1 - Pseudo-tree Generation Phase:** This phase is identical to that of DPOP, where a pseudo-tree is generated (line 1).

[2] We use the superscript $^{\uparrow}$ to denote the messages being propagated from the leaves of the pseudo-tree to the root, and $^{\downarrow}$ to denote the ones propagated from the root to the leaves.

Procedure AC-Propagation-Phase()

16 $iteration \leftarrow 0$
17 **repeat**
18   **if** $C_i \neq \emptyset$ **then**
19   Wait until received $AC_c^{\uparrow}(D'_i, fixed_c)$ from each $a_c \in C_i \cup PC_i$ in this iteration
20   **foreach** $AC_c^{\uparrow}(D'_i, fixed_c)$ *received* **do**
21     $D_i \leftarrow D_i \cap D'_i$
22   $\langle \hat{\mathbf{M}}_i, D_i \rangle \leftarrow$ ENFORCEAC$(\hat{\mathbf{H}}_i, \hat{\mathbf{M}}_i, D_i)$
23   $fixed_i \leftarrow \neg$CHANGED$(D_i) \wedge \bigwedge_{a_c \in C_i} fixed_c$
24   Send $AC_i^{\uparrow}(\hat{M}_{ij|j}, fixed_i)$ to each $a_j \in P_i \cup PP_i$
25   **if** $P_i \neq NULL$ **then**
26   Wait until received $AC_p^{\downarrow}(D_p)$ from each $a_p \in P_i \cup PP_i$ in this iteration or received $BrC_p^{\downarrow}(\cdot)$ from parent a_p
27   **if** *received* $BrC_p^{\downarrow}(\cdot)$ *from parent* a_p **then break**
28   **if** $\neg fixed_i$ **then**
29   **foreach** $AC_p^{\downarrow}(D_p)$ *received* **do**
30   update $\hat{M}_{ip}$ with D_p
31   **if** $P_i \neq NULL$ **then**
32       $\langle \hat{\mathbf{M}}_i, D_i \rangle \leftarrow$ ENFORCEAC$(\hat{\mathbf{H}}_i, \hat{\mathbf{M}}_i, D_i)$
33   Send $AC_i^{\downarrow}(D_i)$ to each $a_c \in C_i \cup PC_i$
34     $iteration \leftarrow iteration + 1$
35 **until** $P_i = NULL$ **and** $fixed_i$

- **Phase 2 - Path Construction Phase:** The phase is used to construct the direct paths from each agent to its parent and pseudo-parents. At the start of this phase (line 2), each agent, starting from the leaves of the pseudo-tree, recursively populates its PATHS$_i$ as follows: It saves a path information ($a_p \overset{NULL}{\rightsquigarrow} ?$) for each of its pseudo-parents a_p (lines 6-7) and sends a $PATH_i^{\uparrow}(a_p)$ message to its parent. When the parent a_i receives a $PATH_c^{\uparrow}$ message from each of its child a_c, it stores the path information in the message in its PATHS$_i$ data structure (lines 9-11). For each path in PATHS$_i$, if it is not the destination agent, then it sends a $PATH_j^{\uparrow}$ message that contains that path to its parent (lines 12-13). If it does not send a $PATH_j^{\uparrow}$ message to its parent, then it sends an empty $PATH_j^{\uparrow}$ message (lines 14-15). These path information will be used in the branch consistency propagation phase later. When the root processes and stores the path information from each of its children, it ends this phase and starts the next AC propagation phase.
- **Phase 3 - Arc Consistency (AC) Propagation Phase:** In this phase, the agents enforce arc consistency in a distributed manner, by interleaving the direction of the AC message flows: from the leaves to the root (lines 18-24) and from the root to the leaves (lines 25-34), until a fixed point is detected at the root (line 35).

  In the first part of this phase (lines 18-24), each agent, starting from the leaves up to the root, recursively enforces the consistency of its hard constraints in $\hat{\mathbf{H}}_i$ (line 22) via the ENFORCEAC procedure, which we implemented using the AC-2001 algorithm [2]. In this process, the agent also updates the VRMs $\hat{\mathbf{M}}_i$ associated with all its constraints $f_{ij} \in \hat{\mathbf{H}}_i$ and its domain D_i to prune all unsupported values. If

Procedure BrC-Propagation-Phase()

```
36 if P_i ≠ NULL then
37     Wait until received a BrC_p^↓(M_ps) for each path (a_s ⇝^{a_c} ?) ∈ PATHS_i from parent a_p
38 foreach (a_s ⇝^{a_c} ?) ∈ PATHS_i do
39     if a_s = a_i then M_is ← M̂_ii
40     else M_is ← M̂_ip × M_ps
41     M_is ← M̂_is ∘ M_is
42     if a_c ≠ NULL then
43         Send BrC_i^↓(M_is) to a_c
44 foreach a_c ∈ C_i that has not been sent a BrC_i^↓ message do
45     Send BrC_i^↓(NULL) to a_c
```

any of its values are pruned, indicating that it has not reached a fixed point, it sets its *fixed*$_i$ flag to false (line 23). It then sends an $AC_i^{\uparrow}$ message to each of its parent and pseudo-parent a_j, which contains its *fixed*$_i$ flag as well as a copy of their domains D'_j[3] to notify them about which unsupported values were pruned (line 24). The domain of each agent is updated before enforcing the arc consistency, as soon as it receives all the $AC_i^{\uparrow}$ messages from each of its children and pseudo-children (lines 20-21).

Once the root enforces the consistency of its hard constraints, it checks if it has reached a fixed point (line 28). If it has not, then it starts the next part of this phase, which is similar to the previous one except for the direction of the recursion and the AC message flow (lines 29-34). This phase is carried from the root down to the leaves of the pseudo-tree, and it ends when all the leaves have enforced the consistency of their hard constraints. Then the procedure repeats the first part where the recursion and the AC message flow starts from the leaves again and continues up to the root. This process repeats until a fixed point is reached at the root (line 35), which ends this phase, and starts the next BrC propagation phase.

- **Phase 4 - Branch Consistency (BrC) Propagation Phase:** In this phase, the agents enforce branch consistency in a distributed manner, that is, every pair of values of an agent and its pseudo-parents are mutually reachable throughout every tree path connecting them in the pseudo-tree.

  At the start of this phase, each agent, starting from the root down to the leaves, recursively enforces branch consistency for all tree paths from the root to that agent and sends a $BrC_i^{\downarrow}$ message to each of its children. This message includes the VRM for each path through that child. Once an agent a_i receives all the VRM messages from its parent (lines 36-37), for each path that goes through it (line 38), it creates a new VRM M_{is}. If it is the start of the path, then it sets its VRM $\hat{M}_{ii}$ (line 39), which is arc consistent, as the new VRM M_{is}. Otherwise, it performs the regular product of its VRM $\hat{M}_{ip}$ for the constraint between itself and its parent a_p and the VRM received from the parent M_{ps} and sets it to M_{is} (line 40). Then, to ensure that the VRM M_{is} is branch consistent, it performs the *entrywise product* with the VRM $\hat{M}_{is}$ of its pseudo-parent a_s (line 41). If the agent is the destination of the path, then

[3] In the pseudo-code, we use the notation $\hat{M}_{ij|j}$ to indicate D'_j.

it will use the resulting VRM in the construction of the UTIL messages in the UTIL phase. Otherwise, it will send the VRM to its child agent that is in that path in a $BrC_i^{\downarrow}$ message (lines 42-43). Finally, it will send an empty $BrC_i^{\downarrow}$ to all remaining child agents to ensure that the propagation reaches all the leaves (lines 44-45).

- **Phase 5 - DPOP's UTIL and VALUE Phases:** This phase is identical to the corresponding UTIL and VALUE propagation phases of DPOP, except that each agent constructs a UTIL table that contains utilities for each combination of *unpruned* values of variables in its VRMs.

4 Theoretical Analysis

In this section, we use n, e, and d to denote $|\mathcal{A}|$, $|\mathcal{F}|$, and $\max_{x_i \in \mathcal{X}} |D_i|$, respectively.

Theorem 1. *The AC propagation phase requires $O(nde)$ messages, each of size $O(d)$.*

Proof Sketch: In the worst case, each AC iteration removes exactly one value from one domain. Thus, there are only $O(nd)$ iterations, as there are only $O(nd)$ values among all variables. In each iteration, each agent sends exactly one $AC^{\uparrow}$ message to each parent and pseudo-parent and one $AC^{\downarrow}$ message to each child and pseudo-child. Thus, there are at most $O(e)$ messages sent in each iteration. Each message contains at most the full domain of a variable and the fixed flag, which is $O(d)$. ■

Theorem 2. *The BrC propagation phase requires $O(e)$ messages, each of size $O(d^2)$.*

Proof Sketch: In the BrC propagation phase, each agent sends exactly one $BrC^{\downarrow}$ message to each child, and the phase ends after all the leaves in the pseudo-tree receives a $BrC^{\downarrow}$ message. Each message contains at most a VRM, which is $O(d^2)$. ■

Theorem 3. *The DCOP is arc consistent after the AC propagation phase.*

Proof Sketch: We prove this result by contradiction. Assume that there are $a_i, a_j \in \mathcal{A}$ and $a \in D_i$ such that $\forall b \in D_j$, $(a, b) \notin f_{ij}$. Let $b_1, \ldots, b_m$ be all the (pruned) values in D_j supporting a. We have the following two cases:

- $a_i \in P_j \cup PP_j$. If agent a_j pruned all its values b_r $(1 \leq r \leq m)$ from D_j, then the value a is pruned from the copy of the domain D_i held at a_j ($\hat{\mathbf{M}}_{ji|i}$ will not include the value a) (line 22). When a_i receives an $AC^{\uparrow}$ message from each $a_k \in C_i \cup PC_i$ (including a_j), it updates its own domain with the copy received from each agent (lines 20-21) removing a from D_i and resulting in a contradiction.
- $a_i \in C_j \cup PC_j$. Agent a_j can prune all its values b_r $(1 \leq r \leq m)$ from D_j in the following two ways. In case 1, agent a_i prunes all the values b_r from a copy of D_j during its AC consistency enforcement (line 22), sends up an $AC^{\uparrow}$ message to a_j, and a_j prunes all its values b_r from its D_j. However, in this case, agent a_i would have also pruned value a from its domain, resulting in a contradiction. In case 2, some other agent a_k that shares a constraint f_{kj} with agent a_j prunes all the values b_r from the copy of D_j during its AC consistency enforcement, sends up an $AC^{\uparrow}$ message to

a_j, and a_j prunes all its values b_r from its D_j. In this case, a_j will eventually send an $AC^{\downarrow}$ message to a_i that contains its updated domain without the values b_r. Then, agent a_i will prune value a from its domain in its AC consistency enforcement (line 22), resulting in a contradiction. ∎

Theorem 4. *The DCOP is branch consistent after the BrC propagation phase.*

Proof Sketch: We prove by induction on the number of variables in the paths $x_i = x_{k_1}, \ldots, x_{k_m} = x_j$, such that $x_{k_1} \succ \ldots \succ x_{k_m}$.
Base Case ($m = 2$)**:** We know that $x_j \in C_i$ and there is only one path from x_i to x_j via the constraint f_{ij}. Additionally, this constraint is arc consistent because the BrC propagation phase runs after the AC propagation phase. Thus, all the remaining pairs of values in both variables are by definition branch consistent (Definition 3). The VRM M_{ji} is thus branch consistent.
Induction Assumption: Assume that for any $2 \leq q \leq r$ and paths $x_i = x_{k_1}, \ldots, x_{k_q} = x_j$ with $x_{k_1} \succ \ldots \succ x_{k_q}$, there is a VRM M_{ji} that is branch consistent.
Induction Case ($m = r + 1$)**:** We know that the paths from $x_i = x_{k_1}$ to x_{k_r} is branch consistent from the induction assumption. Thus, the VRM $M_{k_r k_1}$ received by $x_{k_{r+1}}$ is branch consistent. Additionally, all the constraints between any x_{k_p} $(1 \leq p \leq r)$ and $x_{k_{r+1}}$ are arc consistent because the BrC propagation phase runs after the AC propagation phase. Thus, the VRMs $\hat{M}_{k_{r+1} k_p}$ are also branch consistent.
We now show that the algorithm removes values of $x_{k_{r+1}}$ that are not branch consistent with values of its ancestors in the following two cases:

- For paths that include the constraint between x_r and x_{r+1}, BrC-DPOP takes the regular product (line 40), which removes all inconsistent values.
- For paths that do not include the constraint between x_r and x_{r+1} and, thus, must include the constraint between x_{k_1} and $x_{k_{r+1}}$, BrC-DPOP performs the entrywise product (line 41), which removes all inconsistent values. ∎

Theorem 5. *BrC-DPOP is complete and correct.*

Proof Sketch: The completeness and correctness of BrC-DPOP follows from the correctness and completeness of DPOP [24] and the correctness and completeness of the AC and BrC propagation phases (Theorems 1, 2, 3, and 4). ∎

Property 1. Both the UTIL and the VALUE phases require $O(n)$ number of messages.

Property 2. The memory requirement of BrC-DPOP is in the worst case exponential in the induced width of the problem for each agent.

Both properties follow trivially from the properties of DPOP since no values are pruned from the AC and BrC propagation phases in the worst case.

5 Related Work

We characterize the approaches that prune values of variables in DCOPs along two general types. Algorithms in the first category *propagates exclusively hard constraints* (BrC-DPOP falls into this category). To the best of our knowledge, the only existing work that falls into this category is H-DPOP [17], which, like BrC-DPOP, is also an extension of DPOP. The main difference between H-DPOP and BrC-DPOP is that instead of VRMs, each agent a_i in H-DPOP uses *constraint decision diagrams* (CDDs) to represent the space of possible value assignments of variables in its separator set $sep(a_i)$. A CDD is a rooted directed acyclic graph structured by levels, one for each variable in $sep(a_i)$. In each level, a non-terminal node represents a possible value assignment for the associated variable. Each non-terminal node v has a list of successors: one for each value u in the next variable for which the pair (u, v) is satisfied by the constraint between the two variables. As a result of using CDDs, H-DPOP suffers from two limitations: (1) H-DPOP can be slower than DPOP because maintaining and performing join and projection operations on CDD are computationally expensive. In contrast, maintaining and performing operations on VRMs can be faster, which we will demonstrate in the experimental results section later. (2) H-DPOP cannot fully exploit information of hard constraints to reduce the size of UTIL messages. Consider the DCOP instance of Figure 2, where the domains for the variables x_1, x_3, x_4, and x_5 are represented by the set $\{1, \ldots, 100\}$, while the domain for variable x_2 is the set $\{1, 2\}$. In H-DPOP, a_5 is not aware of the constraints $x_1 < x_2$ and $x_1 < x_3$—neither x_2 nor x_3 are in $sep(a_5)$, thus no pruning will be enforced. Its UTIL table will hence contain $100^2 = 10,000$ utilities for each combination of values of x_4 and x_1. This is the same table that DPOP would construct. In contrast, in BrC-DPOP, the domains of x_1 and x_2 will be pruned to $\{1\}$ and $\{2\}$, respectively, and the domains of x_3, x_4, and x_5 to $\{2, \ldots, 100\}$. Therefore, the UTIL table that a_5 sends to a_4 contains $99 \times 1 = 99$ utilities. Aside from these two limitations, a more critical limitation of H-DPOP is its assumption that each agent has knowledge of all the constraints whose scope is a subset of its separator set. This assumption is stronger than the assumptions made by most DCOP algorithms and might cause privacy concerns in some applications. In contrast, BrC-DPOP does not make such assumptions.

Algorithms in the second category *propagates lower and upper bounds*. Researchers have extended search-based DCOP algorithms (e.g., BnB-ADOPT and its enhanced versions [29,12,14]) to maintain soft-arc consistency in a distributed manner [1,13,11]. Such techniques are typically very effective in search-based algorithms as their runtime depends on the accuracy of its lower and upper bounds.

Finally, it is important to note the differences between branch consistency and path consistency [21]. One can view branch consistency as a weaker version of path consistency, where all the variables in a path must be ordered according to the relation $\prec$, and only a subset of all possible paths have to be examined for consistency. Thus, one can view branch consistency as a form of consistency tailored to pseudo-trees, where each agent can only communicate with neighboring agents.

6 Experimental Results

We implemented a variant of BrC-DPOP, called AC-DPOP, that enforces arc consistency only in order to assess the impact of the branch consistency phase in BrC-DPOP. Moreover, in order to be as comprehensive as possible in our evaluations, we also implemented a variant of H-DPOP called PH-DPOP, which stands for Privacy-based H-DPOP, that restricts the amount of information that each agent can access to the amount

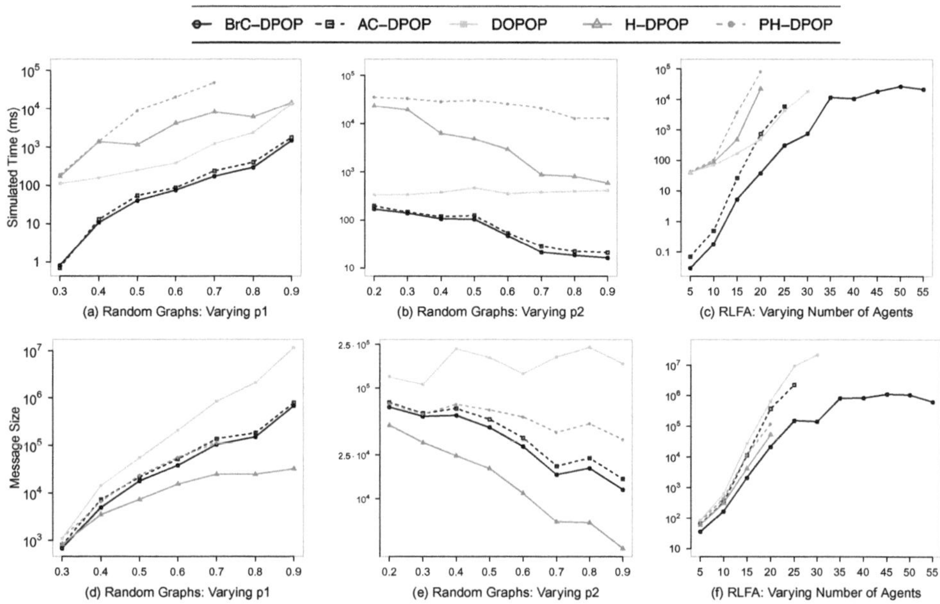

Fig. 3. Runtimes and Message Sizes

common in most DCOP algorithms including BrC-DPOP. Specifically, agents in PH-DPOP can only access their own constraints and, unlike H-DPOP, cannot access their neighboring agents' constraints.

In our experiments,[4] we compare AC-DPOP and BrC-DPOP against DPOP [24], H-DPOP [17], and PH-DPOP. We use a publicly-available implementation of DPOP available in the FRODO framework [18] and an implementation of H-DPOP provided by the authors. We ensure that all algorithms use the same pseudo-tree for fair comparisons. All experiments are performed on an Intel i7 Quadcore 3.4GHz machine with 16GB of RAM with a 300-second timeout. If an algorithm fails to solve a problem, it is due to either memory limitations or timeout. We conduct our experiments on random graphs [6], where we systematically vary the constraint density p_1 and constraint tightness p_2,[5] and distributed Radio Link Frequency Assignment (RLFA) problems [5], where we vary the number of agents $|\mathcal{A}|$ in the problem. We generated 50 instances for each experimental setting, and we report the average runtime, measured using the simulated runtime metric [28], and the average total message size, measured in the number of utility values in the UTIL tables. For the distributed RLFA problems, we also report the percentage of satisfiable instances solved to show the scalability of the algorithms.

Random Graphs: In our experiments, we set $|\mathcal{A}| = 10$, $|\mathcal{X}| = 10$, $|D_i| = 8$ for all variables. We vary p_1 (while setting $p_2 = 0.6$) and vary the p_2 (while setting $p_1 = 0.6$). We did not bound the tree-width, which is determined based on the underlying graph

[4] Available at `http://www.cs.nmsu.edu/klap/brc-dpop_cp14/`

[5] p_1 and p_2 are defined as the ratio between the number of binary constraints in the problem and the maximum possible number of binary constraints in the problem and the ratio between the number of hard constraints and the number of constraints in the problem, respectively.

and randomly generated. We used hard constraints that are either the "less than" or "different" constraints. We also assign a unary constraint to each variable that gives it a utility corresponding to each its value assignments.

Figures 3(a-b) show the runtimes of the algorithms and Figures 3(d-e) show the message sizes. We omit results of an algorithm for a specific parameter if it fails to solve 50% of the satisfiable instances for that parameter. We make the following observations:

- On message sizes, BrC-DPOP uses smaller messages than AC-DPOP because BrC-DPOP prunes more values due to its BrC propagation enforcement. H-DPOP uses smaller messages than BrC-DPOP and AC-DPOP because agents in H-DPOP utilize more information about the neighbors' constraints to prune values. In contrast, agents in BrC-DPOP and AC-DPOP only utilize information on their own constraints to prune values. BrC-DPOP and AC-DPOP use smaller messages than PH-DPOP at large p_2 values, highlighting the strength of the AC and BrC propagation phases compared to the pruning techniques in PH-DPOP. Finally, they all use smaller messages than DPOP because they all prune values while DPOP does not.
- On runtimes, BrC-DPOP is slightly faster than AC-DPOP because BrC-DPOP prunes more values than AC-DPOP. Additionally, these results also indicate that the overhead of the BrC propagation phase is relatively small. BrC-DPOP and AC-DPOP are faster than DPOP because they do not need to perform operations on the pruned values. This also indicates that the overhead of the AC propagation phase is small. In our experiments, the number of *AC* messages exchanged during the AC propagation phase never exceeds $3|\mathcal{F}|$ and is, on average, as small as $|\mathcal{F}|$. DPOP is faster than H-DPOP and PH-DPOP because they maintain and perform operations on CDDs, which are computationally very expensive. In contrast, BrC-DPOP maintains and performs operations on matrices, which are more computationally efficient.

Distributed RLFA Problem: In these problems, we are given a set of links $\{L_1, \ldots, L_r\}$, each consisting of a transmitter and a receiver. Each link must be assigned a frequency from a given set F. At the same time the total interference at the receivers must be reduced below an acceptable level using as few frequencies as possible. In our model, each transmitter corresponds to an agent (and a variable). The domain of each variable consists of the frequencies that can be assigned to the corresponding transmitter. The interference between transmitters are modeled as constraints of the form $|x_i - x_j| > s$, where x_i and x_j are variables and $s \geq 0$ is a randomly generated required frequency separation. We also assign a utility value to each frequency taken by each agent, represented as a unary soft constraint, which represents a preference for an agent to transmit at a given frequency.

For generating the constraint graphs, we vary $|\mathcal{A}|$ and fix the other parameters: $|D_i| = 6, p_2 = 0.55, s \in \{2, 3\}$. We also set the maximum number of neighbors for each agent to 3 in order to generate more satisfiable instances. Figure 3(c) shows the runtimes and Figure 3(f) shows the message sizes. We omit results of an algorithm for a specific parameter if it fails to solve 50% of the satisfiable instances for that parameter.

We observe trends that are similar to those in the earlier random graphs except that the message size of H-DPOP is slightly larger than of those of BrC-DPOP. Therefore, as we have described in Section 5, it is possible for H-DPOP to prune fewer values despite using more information. Additionally, both H-DPOP and PH-DPOP can only solve

Table 2. Percentage of Satisfiable Instances Solved

$\lvert\mathcal{A}\rvert$	5	10	15	20	25	30	35	40	45	50	55
BrC-DPOP	**1.00**	**1.00**	**1.00**	**1.00**	**1.00**	**0.97**	**0.52**	**0.78**	**0.73**	**0.70**	**0.51**
AC-DPOP	**1.00**	**1.00**	**1.00**	**1.00**	**1.00**	0.39	0.11	0.30	0.15	0.15	0.19
H-DPOP	**1.00**	**1.00**	**1.00**	**1.00**	0.46	0.12	0.00	0.00	0.00	0.00	0.00
PH-DPOP	**1.00**	**1.00**	**1.00**	**1.00**	0.21	0.09	0.00	0.00	0.00	0.00	0.00
DPOP	**1.00**	**1.00**	**1.00**	**1.00**	**1.00**	0.67	0.23	0.35	0.23	0.29	0.19

small problems and failed to solve some problems that DPOP successfully solved. Table 2 tabulates the percentage of satisfiable problem instances solved by each algorithm (the largest percentage in each parameter setting is shown in bold), where it is clear that BrC-DPOP is more scalable than all its counterparts.

7 Conclusions and Future Work

To the best of our knowledge, H-DPOP is the only existing DCOP algorithm that propagates exclusively hard constraints. Unfortunately, it suffers from high computational requirements as well as its overly strong assumption on the knowledge of each agent. In this paper, we alleviate these limitations by introducing the concept of branch consistency as well as the BrC-DPOP algorithm, a DPOP extension that enforces arc consistency and branch consistency. We experimentally show that BrC-DPOP can prune as much as a version of H-DPOP that limits its knowledge to the same amount as BrC-DPOP in a much smaller amount of time. We also show that it can scale to larger problems than DPOP and H-DPOP. Therefore, these results confirm the strengths of this approach, leading to enhanced efficiency and scalability.

For future work, we plan to extend BrC-DPOP to handle higher arity constraints, which can be done by substituting the VRM structures with either consistency graphs or higher dimension VRMs. We suspect that there will be a tradeoff between runtime and memory requirement between the two approaches, where using higher dimension VRMs is faster but uses more memory. We also plan to extend BrC-DPOP to memory-bounded versions similar to MB-DPOP [27] in order to scale to even larger problems. Finally, we plan to explore propagation of soft constraints similar to the versions of BnB-ADOPT with soft AC enforcement [1,13,11].

Acknowledgment. We would like to thank Akshat Kumar for providing us his implementation of H-DPOP. This research is partially supported by the National Science Foundation under grant number HRD-1345232. The views and conclusions contained in this document are those of the authors and should not be interpreted as representing the official policies, either expressed or implied, of the sponsoring organizations, agencies, or the U.S. government.

References

1. Bessiere, C., Gutierrez, P., Meseguer, P.: Including Soft Global Constraints in DCOPs. In: Milano, M. (ed.) CP 2012. LNCS, vol. 7514, pp. 175–190. Springer, Heidelberg (2012)
2. Bessiere, C., Regin, J.: Refining the Basic Constraint Propagation Algorithm. In: Proc. of IJCAI, pp. 309–315 (2001)
3. Brito, I., Meseguer, P.: Improving DPOP with function filtering. In: Proc. of AAMAS, pp. 141–158 (2010)
4. Burke, D., Brown, K.: Efficiently Handling Complex Local Problems in Distributed Constraint Optimisation. In: Proc. of ECAI, pp. 701–702 (2006)
5. Cabon, B., De Givry, S., Lobjois, L., Schiex, T., Warners, J.P.: Radio Link Frequency Assignment. Constraints 4(1), 79–89 (1999)
6. Erdös, P., Rényi, A.: On Random Graphs I. Publicationes Mathematicae Debrecen 6, 290 (1959)
7. Ezzahir, R., Bessiere, C., Belaissaoui, M., Bouyakhf, E.: DisChoco: A Platform for Distributed Constraint Programming. In: Proc. of the Distributed Constraint Reasoning Workshop, pp. 16–27 (2007)
8. Farinelli, A., Rogers, A., Petcu, A., Jennings, N.: Decentralised Coordination of Low-Power Embedded Devices Using the Max-Sum Algorithm. In: Proc. of AAMAS, pp. 639–646 (2008)
9. Gershman, A., Meisels, A., Zivan, R.: Asynchronous Forward-Bounding for Distributed COPs. Journal of Artificial Intelligence Research 34, 61–88 (2009)
10. Greenstadt, R., Grosz, B., Smith, M.: SSDPOP: Improving the Privacy of DCOP with Secret Sharing. In: Proc. of AAMAS, pp. 1098–1100 (2007)
11. Gutierrez, P., Lee, J.H.M., Lei, K.M., Mak, T.W.K., Meseguer, P.: Maintaining Soft Arc Consistencies in BnB-ADOPT$^+$ during Search. In: Schulte, C. (ed.) CP 2013. LNCS, vol. 8124, pp. 365–380. Springer, Heidelberg (2013)
12. Gutierrez, P., Meseguer, P.: Saving redundant messages in BnB-ADOPT. In: Proc. of AAAI, pp. 1259–1260 (2010)
13. Gutierrez, P., Meseguer, P.: Improving BnB-ADOPT$^+$-AC. In: Proc. of AAMAS, pp. 273–280 (2012)
14. Gutierrez, P., Meseguer, P., Yeoh, W.: Generalizing ADOPT and BnB-ADOPT. In: Proc. of IJCAI, pp. 554–559 (2011)
15. Hamadi, Y., Bessière, C., Quinqueton, J.: Distributed Intelligent Backtracking. In: Proc. of ECAI, pp. 219–223 (1998)
16. Kumar, A., Faltings, B., Petcu, A.: Distributed Constraint Optimization with Structured Resource Constraints. In: Proc. of AAMAS, pp. 923–930 (2009)
17. Kumar, A., Petcu, A., Faltings, B.: H-DPOP: Using Hard Constraints for Search Space Pruning in DCOP. In: Proc. of AAAI, pp. 325–330 (2008)
18. Léauté, T., Ottens, B., Szymanek, R.: FRODO 2.0: An Open-Source Framework for Distributed Constraint Optimization. In: Proc. of the Distributed Constraint Reasoning Workshop, pp. 160–164 (2009)
19. Maheswaran, R., Tambe, M., Bowring, E., Pearce, J., Varakantham, P.: Taking DCOP to the Real World: Efficient Complete Solutions for Distributed Event Scheduling. In: Proc. of AAMAS, pp. 310–317 (2004)
20. Modi, P., Shen, W.-M., Tambe, M., Yokoo, M.: ADOPT: Asynchronous Distributed Constraint Optimization with Quality Guarantees. Artificial Intelligence 161(1-2), 149–180 (2005)
21. Mohr, R., Henderson, T.C.: Arc and Path Consistency Revisited. Artificial Intelligence 28(2), 225–233 (1986)

22. Nguyen, D.T., Yeoh, W., Lau, H.C.: Distributed Gibbs: A Memory-Bounded Sampling-Based DCOP Algorithm. In: Proc. of AAMAS, pp. 167–174 (2013)
23. Ottens, B., Dimitrakakis, C., Faltings, B.: DUCT: An Upper Confidence Bound Approach to Distributed Constraint Optimization Problems. In: Proc. of AAAI, pp. 528–534 (2012)
24. Petcu, A., Faltings, B.: A Scalable Method for Multiagent Constraint Optimization. In: Proc. of IJCAI, pp. 1413–1420 (2005)
25. Petcu, A., Faltings, B.V.: Approximations in Distributed Optimization. In: van Beek, P. (ed.) CP 2005. LNCS, vol. 3709, pp. 802–806. Springer, Heidelberg (2005)
26. Petcu, A., Faltings, B.: ODPOP: An algorithm for open/distributed constraint optimization. In: Proc. of AAAI, pp. 703–708 (2006)
27. Petcu, A., Faltings, B.: MB-DPOP: A New Memory-Bounded Algorithm for Distributed Optimization. In: Proc. of IJCAI, pp. 1452–1457 (2007)
28. Sultanik, E., Lass, R., Regli, W.: DCOPolis: a Framework for Simulating and Deploying Distributed Constraint Reasoning Algorithms. In: Proc. of the Distributed Constraint Reasoning Workshop (2007)
29. Yeoh, W., Felner, A., Koenig, S.: BnB-ADOPT: An Asynchronous Branch-and-Bound DCOP Algorithm. Journal of Artificial Intelligence Research 38, 85–133 (2010)
30. Yeoh, W., Yokoo, M.: Distributed Problem Solving. AI Magazine 33(3), 53–65 (2012)
31. Yokoo, M. (ed.): Distributed Constraint Satisfaction: Foundation of Cooperation in Multi-agent Systems. Springer (2001)
32. Zhang, W., Wang, G., Xing, Z., Wittenberg, L.: Distributed Stochastic Search and Distributed Breakout: Properties, Comparison and Applications to Constraint Optimization Problems in Sensor Networks. Artificial Intelligence 161(1-2), 55–87 (2005)
33. Zivan, R., Glinton, R., Sycara, K.: Distributed Constraint Optimization for Large Teams of Mobile Sensing Agents. In: Proc. of IAT, pp. 347–354 (2009)
34. Zivan, R., Okamoto, S., Peled, H.: Explorative anytime local search for distributed constraint optimization. Artificial Intelligence 212, 1–26 (2014)

Constraint-Based Lagrangian Relaxation

Daniel Fontaine[1], Laurent Michel[1], and Pascal Van Hentenryck[2]

[1] University of Connecticut, Storrs, CT 06269-2155, USA
[2] NICTA and Australian National University, Australia

Abstract. This paper studies how to generalize Lagrangian relaxation to high-level optimization models, including constraint-programming and local search models. It exploits the concepts of constraint violation (typically used in constraint programming and local search) and constraint satisfiability (typically exploited in mathematical programming). The paper considers dual and primal methods, studies their properties, and discusses how they can be implemented in terms of high-level model combinators and algorithmic templates. Experimental results suggest the potential benefits of Lagrangian methods for improving high-level constraint programming and local search models.

1 Introduction

Lagrangian relaxation (e.g., [1]) is a significant optimization paradigm that typically applies to models that feature both easy and hard constraints. Its idea is to relax the hard constraints into the objective, using Lagrangian multipliers to adjust the coefficients of these relaxed constraints. Consider, for instance, the application of Lagrangian relaxation to a mixed integer program:

$$Z = \texttt{min } c \cdot x \quad \text{s.t.} \begin{cases} A_h x \geq b_h \\ A_e x \geq b_e \\ x \in \{0,1\} \end{cases} \longrightarrow \quad Z_{LR}(\lambda) = \texttt{min } c \cdot x + \lambda^T \cdot (b_h - A_h x) \quad \text{s.t.} \begin{cases} A_e x \geq b_e \\ x \in \{0,1\} \end{cases}$$

The Lagrangian relaxation can be used to find both dual and primal bounds. The value $Z_{LR}(\lambda)$ is a lower bound of Z for any $\lambda \geq 0$ and primal bounds can be obtained by searching for saddle points to the Lagrangian relaxation (in the x and λ space).

The idea of Lagrangian relaxation is largely independent of how the model is expressed (and solved). It is heavily used in continuous optimization (e.g., [8]) and in mixed integer programming (e.g., [5]). It had attracted some attention in constraint satisfaction problems and SAT in the late 1990s (e.g., [17,3]) but has not been a topic of much research since then.

This research originated from an attempt to build model combinators for Lagrangian relaxation that would apply to arbitrary optimization models. The design of these combinators required a systematic investigation of the semantics of

B. O'Sullivan (Ed.): CP 2014, LNCS 8656, pp. 324–339, 2014.

Lagrangian relaxation over these models, which revealed some interesting modeling, computational, and implementation issues. It also raised the question about the potential benefits of Lagrangian relaxation for constraint programming, hybrid methods, and constraint-based local search, a topic which has been largely neglected in the constraint-programming community.

This paper reports some findings in this direction. It starts by introducing the concept of *satisfiability degree* that provides an alternative to the notion of constraint violation used in constraint programming (e.g., [2]) and constraint-based local search [4,9,14]. The paper then provides natural generalizations of traditional Lagrangian relaxation concepts, including the Lagrangian duals, subgradient optimization [5], surrogate subgradient optimization [18], and primal Lagrangian methods (e.g., [17,3]). These generalizations then makes it possible to design model combinators for Lagrangian relaxation that apply to arbitrary models and algorithmic templates for Lagrangian methods that are independent of the solving technology. The paper also presents some preliminary experimental results indicating the potential of Lagrangian methods for constraint programming and local search, as well as synergies between mathematical programming, constraint programming, and large-neighborhood search.

The remainder of the paper is organized as follows. Section 2 introduces the generalized Lagrangian relaxation idea. Sections 3 and 4 discuss dual and primal Lagrangian methods respectively. Section 5 sketches the implementation in OJECTIVE-CP. Section 6 presents the experimental results techniques. Section 7 discusses related work and Section 8 concludes the paper.

2 Generalized Lagrangian Relaxation

This section explores how the traditional formulation of Lagrangian relaxations and Lagrangian duals can be systematically generalized.

2.1 Violation and Satisfiability Degrees

Generalized Lagrangian relaxations are based on the concepts of violation and satisfiability degrees. The violation degree of a constraint is a key concept in constraint programming (e.g., [2]) and constraint-based local search (e.g., [9,14,4]). The violation degree is constraint-dependent and exploits the constraint structure. Intuitively, the violation degree denotes how violated the constraint is.

Definition 1 (Violation Degree). *The violation degree of constraint* $c : \Re^n \rightarrow$ *Bool is a function* $\nu_c : \Re^n \rightarrow \Re^+$ *such that*

$$c(v_1, \ldots, v_n) \equiv \nu_c(v_1, \ldots, v_n) = 0.$$

In contrast, the satisfiability degree of a constraint captures both how much the constraint is violated and the constraint slackness when it is satisfied. It generalizes the Lagrangian relaxation typically used in mathematical programming. Intuitively, when the satisfiability degree is strictly positive, it denotes how much the constraint is violated; when it is negative, the constraint is satisfied and the satisfiability degree denotes the *slack* of the constraint. When the satisfiability degree is zero, the constraint is satisfied but tight.

Definition 2 (Satisfiability Degree). *The satisfiability degree of constraint* $c : \Re^n \rightarrow Bool$ *is a function* $\sigma_c : \Re^n \rightarrow \Re$ *such that*

$$c(v_1, \ldots, v_n) \equiv \sigma_c(v_1, \ldots, v_n) \leq 0.$$

Example 1 ($A \cdot x \geq b$). Consider the constraint $c(x)$ defined by $A \cdot x \geq b$. Its satisfiability degree is given by the function $\sigma_c(x) = b - A \cdot x$. Observe that, when $\sigma_c(v) \leq 0$, $A \cdot v \geq b$ and $c(v)$ is satisfied. When $\sigma_c(v) > 0$, $A \cdot v < b$ and $\sigma_c(v)$ represents how much the constraint is violated. When $\sigma_c(v) = 0$, the constraint is satisfied at equality. When $\sigma_c(v) < 0$, $\sigma_c(v)$ captures the slackness of the inequality.

Example 2 ($\texttt{disjunctive}(s_1, d_1, s_2, d_2)$). Constraint $\texttt{disjunctive}(s_1, d_1, s_2, d_2)$ holds if $s_1 + d_1 \leq s_2 \vee s_2 + d_2 \leq s_1$. Its degree of satisfiability is given by

$$\sigma_d(s_1, d_1, s_2, d_2) = \min(s_1 + d_1, s_2 + d_2) - \max(s_1, s_2).$$

When $\max(s_1, s_2) < \min(s_1 + d_1, s_2 + d_2)$, the two intervals $[s_1, s_1 + d_1[$ and $[s_2, s_2 + d_2[$ overlap, the disjunctive constraint is violated and $\sigma_d(s_1, d_1, s_2, d_2) > 0$. Similarly, if $\max(s_1, s_2) \geq \min(s_1 + d_1, s_2 + d_2)$, the two intervals are temporally separated and $|\sigma_d(s_1, d_1, s_2, d_2)|$ is the temporal slack separating the end of the first activity from the start of the second activity.

The following example illustrates that, for some constraints, the satisfiability degree reduces to the violation degree.

Example 3 ($\texttt{permutation}(x_1, \cdots, x_n)$). Constraint $\texttt{permutation}(x_1, \cdots, x_n)$ holds if $x_1, \cdots, x_n$ is a permutation of the values in interval $1..n$. Its degree of satisfiability is given by

$$\sigma_p(v_1, \cdots, v_n) = \sum_{j=1}^{n} \max\left(0, \left(\sum_{i=1}^{n} (x_i = j)\right) - 1\right).$$

When $\sigma_p(v_1, \cdots, v_n) > 0$, the constraint is violated. When $\sigma_p(v_1, \cdots, v_n) = 0$, each value is selected exactly once and the constraint is satisfied. However, $\sigma_p(v_1, \cdots, v_n)$ is never negative as all permutations are equally good: None satisfies the constraint more than the others.

Observe that the violation degree ν_c of a constraint c can always be defined in terms of its satisfiability degree σ_c by stating

$$\nu_c(x_1, \ldots, x_n) = \max(0, \sigma_c(x_1, \ldots, x_n)).$$

We make this assumption in the rest of this paper, when comparing relaxations based on ν_c and σ_c.

2.2 Generalized Lagrangian Relaxations

This section considers Lagrangian relaxations based on violation and satisfiability degrees. It first defines the concepts of constraint softening and constraint relaxation.

Definition 3 (Constraint Softening). *The softening of a constraint c over $\Re^n$ is a constraint soft(c) over $\Re^{n+1}$ defined as*

$$soft(c)(x_1, \cdots, x_n, y) \equiv y = \nu_c(x_1, \cdots, x_n).$$

Definition 4 (Constraint Relaxation). *The relaxation of a constraint c over $\Re^n$ is a constraint relax(c) over $\Re^{n+1}$ defined as*

$$relax(c)(x_1, \cdots, x_n, y) \equiv y = \sigma_c(x_1, \cdots, x_n).$$

We are now in a position to define generalized and soft Lagrangian relaxations.

Definition 5 (Generalized and Soft Lagrangian Relaxations). *Consider the optimization problem*

$$\begin{array}{l} P = \min_x f(x) \\ subject\ to \begin{cases} c_h(x)\ (h \in H) \\ c_e(x)\ (e \in E) \end{cases} \end{array}$$

where H and E denote, respectively, the index sets of hard and easy constraints. The generalized Lagrangian relaxation of P for a set of Lagrangian multipliers $\lambda_h \geq 0$ is given by

$$\begin{array}{ll} GLR(\lambda) & = \min_x f(x) + \sum_{h \in H} \lambda_h \cdot \sigma_h \\ subject\ to & \begin{cases} c_e(x) & (e \in E) \\ relax(c_h)(x, \sigma_h) & (h \in H) \end{cases} \end{array}$$

The soft Lagrangian relaxation of P for a set of Lagrangian multipliers $\lambda_h \geq 0$ is given by

$$\begin{array}{ll} SLR(\lambda) & = \min_x f(x) + \sum_{h \in H} \lambda_h \cdot \nu_h \\ subject\ to & \begin{cases} c_e(x) & (e \in E) \\ soft(c_h)(x, \nu_h) & (h \in H) \end{cases} \end{array}$$

The definitions of the generalized and soft Lagrangian relaxations are compositional and constraint-driven. This makes it possible to define model combinators that systematically obtain a Lagrangian relaxation from a high-level model as discussed in Section 5. The following (direct) lemma establishes the soundness of the approach.

Lemma 1 (Relaxations). *Consider the optimization problem P defined above, an optimal solution x^* of P, and the generalized and soft relaxations $GLR(\lambda)$ and $SLR(\lambda)$ for a vector $\lambda \geq 0 \in \mathbb{R}^{|H|}$. Then, $GLR(\lambda)$ and $SLR(\lambda)$ are relaxations of P, i.e., $GLR(\lambda) \leq f(x^*)$ and $SLR(\lambda) \leq f(x^*)$.*

Proof. Each feasible solution of P satisfies $\nu_h = 0$ and $\sigma_h \leq 0$ for all $h \in H$. Hence, in $SLR(\lambda)$ (resp. $GLR(\lambda)$), the objective value of a feasible solution is the same as (resp. no greater than) the objective value of a feasible solution in P. The results follows since the λ_h are nonnegative. □

The following (also direct) lemma shows that the soft relaxation is at least as strong as the generalized relaxation.

Lemma 2 (GLR versus SLR). *For any $\lambda \geq 0$, we have $GLR(\lambda) \leq SLR(\lambda)$.*

This lemma seems to suggest the use of $SLR(\lambda)$ instead of $GLR(\lambda)$ (which generalizes the traditional mathematical approach), since it is a stronger relaxation. Indeed, $SLR(\lambda)$ could be defined as

$$\begin{array}{ll} SLR(\lambda) & = \min_x f(x) + \sum_{h \in H} \lambda_h \cdot \nu_h \\ \text{subject to} & \begin{cases} c_e(x) & (e \in E) \\ relax(c_h)(x, \sigma_h) & (h \in H) \\ \nu_h \geq 0 & (h \in H) \\ \nu_h \geq \sigma_h & (h \in H) \end{cases} \end{array}$$

which does not change the theoretical complexity of the relaxation if $GLR(\lambda)$ is a linear program or a mixed integer program. The experimental results in Section 6 will shed some light on this issue.

3 Generalized Lagrangian Duals

Lagrangian relaxation is often used to find tight dual bounds to optimization problems. The aim is thus to determine the set of Lagrangian multipliers λ that gives the strongest dual bound. This section focuses on the generalized Lagrangian relaxation but the results apply to the soft Lagrangian relaxation as well. The generalized Lagrangian dual can then be defined as

$$GLR^* = \max_{\lambda \geq 0} GLR(\lambda).$$

The generalized Lagrangian dual satisfies the following property.

Theorem 1 (Optimality Test). *Let $\hat{x}$ be an optimal solution to $GLR(\lambda)$ for some $\lambda \geq 0$ such that*

1. *$c_h(\hat{x})$ holds for all $h \in H$.*
2. *$\lambda_h \cdot \sigma_h = 0$ for all $h \in H$.*

Then, $\hat{x}$ is an optimal solution to P and $GLR^* = GLR(\lambda)$.

Proof. By condition (1) and the definition of $GLR(\lambda)$, $\hat{x}$ is a feasible solution. Moreover, by condition (2), $GLR(\lambda) = f(\hat{x}) + \sum_{h \in H} \lambda_h \cdot \sigma_h = f(\hat{x})$. Since $f(\hat{x})$ is both a lower and an upper bound, the result follows. □

Note that, for SLR, condition (1) implies condition (2).

```
1  function SubgradientSolve(GLR(λ), Z_UB)
2       π = 2
3       k = 0
4       λ^0 = 0
5       Z_best = −∞
6       noImproveCount = 0
7       do
8          x^{k+1} = solve(GLR(λ^k))
9          Z^{k+1} = f(x^{k+1}) + Σ_{h∈H} λ_h^k · σ_h(x^{k+1})
10         Δ^{k+1} = π(Z_UB − Z^{k+1})/||σ(x^{k+1})||^2
11         forall(h ∈ H) λ_h^{k+1} = max(0, λ_h^k + Δ^{k+1} ∗ σ_h(x^{k+1}))
12         if Z^{k+1} > Z_best
13            Z_best = Z^{k+1}
14            noImproveCount = 0
15         else noImproveCount = noImproveCount + 1
16         if noImproveCount > 30
17            π = π/2
18            noImproveCount = 0
19         k = k + 1
20      while the termination criterion is not met;
21      return Z_best
```

Fig. 1. The *subgradient* Algorithm Template

Subgradient Optimization. The generalized Lagrangian dual can be rewritten explicitly as

$$\begin{aligned}&\max_{\lambda \geq 0} w\\ &\text{subject to}\\ &\quad w \leq f(x) + \lambda^T \cdot \sigma_h(x) \quad \forall x, e \in E : c_e(x).\end{aligned}$$

This formulation has exponentially many constraints but it can be solved by a subgradient method which iterates two steps

$$\begin{aligned}x^{k+1} &= solve(GLR(\lambda^k))\\ \lambda_h^{k+1} &= \max\left(0, \lambda_h^k + \Delta^{k+1}\ \sigma_h(x^{k+1})\right)\ (h \in H)\end{aligned}$$

where Δ^k is the step size at iteration k. What remains to determine is the initial value of the multipliers and the step size at each iteration. The algorithmic schema for subgradient optimization is depicted in Figure 1 and is independent of the model and the solving technology. Its input is a parametric Lagrangian model $GLR(\lambda)$ and an initial upper bound to the original problem P. The algorithmic template also uses an agility parameter π used to compute the step sizes. The subgradient optimization sets the initial multipliers λ^0 to 0. Lines 8–19 repeatedly solves the parametric model with the current multipliers λ^k and store its solution in x^{k+1} and its objective value in Z^{k+1} (lines 8–9). Lines 10 computes the step function Δ^{k+1} used on line 11 to compute the next multipliers λ^{k+1}. Lines 12–19 record the current best objective and update the agility parameter π when there is no improvement over some time. Observe that the template does not prescribe any technology for solving $GLR(\lambda^k)$.

Generalized Surrogate Optimization. The surrogate gradient method was introduced in [18] and refined in [13] to solve a Lagrangian dual featuring loosely coupled subproblems. By relaxing the coupling constraints, the subproblems can then be optimized independently.

Consider the following simple IP minimization problem:

$$Z = \texttt{min} \ \sum_{i=1}^{4} x_i$$
$$\texttt{s.t.} \begin{cases} x_1 + x_2 & \geq b_1 \\ & x_3 + x_4 \geq b_2 \\ x_1 + & x_3 + x_4 \geq b_3 \end{cases}$$

Relaxing the coupling constraint produces the Lagrangian dual:

$$Z_{LD} = \texttt{min} \ \sum_{i=1}^{4} x_i + \lambda(b_3 - (x_1 + x_3 + x_4))$$
$$\texttt{s.t.} \begin{cases} x_1 + x_2 & \geq b_1 \\ & x_3 + x_4 \geq b_2 \end{cases}$$

The objective function of Z_{LD} can be simplified algebraically in order to obtain two separable subproblems:

$$min[x_1(1-\lambda) + x_2] + [(x_3 + x_4)(1-\lambda)]$$

Such rewritings are not always possible when using arbitrary models featuring global constraints. But the surrogate subgradient algorithm can be generalized to arbitrary models by using ideas from large neighborhood search. At each iteration, a subproblem can be chosen and all the variables not appearing in this subproblem are fixed to their values in the incumbent solution. A subproblem $GLR(\lambda, \hat{x}, V)$, where $\hat{x}$ is an incumbent solution and V is the set of variables associated with one of the subproblems, can be defined as

$$GLR(\lambda, \hat{x}, V) = \min_x f(x) + \sum_{h \in H} \lambda_h \ \sigma_h(x)$$
$$\text{subject to} \quad \begin{cases} c_e(x) \quad (e \in E) \\ x_i = \hat{x}_i \ (i \notin V) \end{cases}$$

With this idea in mind, the generalized surrogate gradient template is presented in Figure 2. It receives as inputs the parametric model and the set of variables appearing in each subproblem. Observe that line 6 solves the initial model entirely before starting the subproblem optimization. Once again, the template does not prescribe any technology for solving $GLR(\lambda^k, \hat{x}, V)$.

4 Generalized Lagrangian Primal Methods

Primal Lagrangian methods are ubiquitous in continuous optimization. In the late 1990s, some of their main concepts were elegantly transferred to discrete optimization [12,17]. Focusing on violation degrees, the resulting Lagrangian primal methods (SPLR) can be viewed as the iteration of two steps:

$$x^{k+1} = \text{argmin}_{x \in \mathcal{N}(x^k)} SLR(\lambda^k, x^k)$$
$$\lambda^{k+1} = \lambda^k + \nu(x^{k+1})$$

```
function SurrogateSolve(GLR(λ), Z_UB, {V_1, ..., V_k})
  k = 0
  λ^0 = 0
  Z_best = -∞
  noImproveCount = 0
  x^0 = Solve(GLR(λ^0))
  do
      Z^k = f(x^k) + Σ_{h∈H} λ_h^k σ_h(x^k)
      Δ^k = (Z_UB - Z^k)/||σ(x^k)||^2
      forall(h ∈ H) λ_h^{k+1} = max(0, λ_h^k + Δ^k * σ_h(x^k))
      if Z^k > Z_best
          Z_best = Z^k
          noImproveCount = 0
      else noImproveCount = noImproveCount + 1
      select  i ∈ 1..k
      y = Solve(GLR(λ^{k+1}, x^k, V_i))
      obj = f(y) + Σ_{h∈H} λ_h^{k+1} σ_h(y)
      if obj < z^k
          x^{k+1} = y
      else x^{k+1} = x^k
      k = k + 1
  while the termination criterion is not met;
  return Z_best
```

Fig. 2. The *Surrogate Subgradient* Algorithm Template

where $\mathcal{N}(x)$ is the neighborhood around x, i.e., a set of points satisfying the easy constraint and including x, and $SLR(\lambda, x) = f(x) + \lambda\nu(x)$. Such primal Lagrangian methods thus descend in the x-space and ascend in the λ-space. Such primal Lagrangian methods were applied to SAT [12] and constraint satisfaction problems [3], with neighborhoods changing the value of one variable. However, they have not attracted much attention in the constraint-programming community since then. It is useful to state the main theoretical results from [17], since they shed some light on the search algorithm.

Definition 6 (Discrete Saddle Point). *A pair (λ^*, x^*) is a discrete saddle point of SLR if $SLR(\lambda, x^*) \leq SLR(\lambda^*, x^*) \leq SLR(\lambda^*, x)$ for all λ and $x \in \mathcal{N}(x^*)$.*

The left condition in the definition can be shown to be equivalent to $\nu(x^*) = 0$. The following theorem is a direct adaptation to our setting of the main results in [17].

Theorem 2 (Saddle Point Theorem). *Point x^* is a local minimum to the original problem P if and only if there exists $\lambda^* \geq 0$ such that (λ^*, x^*) is a discrete saddle point. Moreover, (λ^*, x^*) is a saddle point if and only if $x^* = argmin_{x \in \mathcal{N}(x^*)} SLR(\lambda^*, x^*)$ and $\nu(x^*) = 0$.*

Observe also that if (λ^*, x^*) is a saddle point, so is (λ, x^*) for $\lambda \geq \lambda^*$ [17]. Hence, in theory, there is no need to decrease the Lagrangian multipliers when searching for a saddle point. It is thus unclear whether the satisfiability degree is useful in primal Lagrangian methods.

In contrast to earlier work, this paper studies whether primal Lagrangian methods can provide a simple, systematic, and principled way of boosting existing search methods, such as tabu search or large neighborhood search, when applied to high-level models. In other words, the neighborhood $\mathcal{N}$ in these primal Lagrangian methods is very large and defined by a neighborhood search technique over a high-level model.

5 Practical Implementation

The earlier sections defined a general framework for applying Lagrangian relaxation to high-level models. This section describes how this generality is supported in OBJECTIVE-CP [15]. Intuitively, the implementation starts with a high-level model which is then relaxed by replacing the hard constraints with their relaxation and adding a new term in the objective function to capture the weighted sum of violations or satisfiability degrees. The hard constraints are identified either by users or automatically by a partitioning algorithms. The resulting Lagrangian model is then concretized into an optimization program, which can be a MIP solver, a constraint-programming solver, or a constraint-based local search. The concrete optimization program is then embedded in an algorithmic template (a runnable in OBJECTIVE-CP's terminology [6], e.g., a surrogate dual or a primal Lagrangian methods. We now illustrate this methodology on a few code snippets. Consider the excerpt

```
1 id<ORModel>  P = [ORFactory createModel];
2 ...
3 id<ORIdArray>  H  = ... // array of hard constraints in P
4 id<ORModel>    L  = [ORFactory lagrangianRelax: P relaxingConstraints: H];
5 id<ORProgram>  O  = [ORFactory createMIPProgram: L];
6 id<ORRunnable> r  = [ORFactory subgradient: O];
7 [r run];
```

The code fragment starts by declaring a model P on line 1. Line 3 stores the set of constraints deemed hard in P in array H. Line 4 creates a parametric model L representing $GLR_P(\lambda)$. Line 5 concretizes $GLR_P(\lambda)$ into a parametric MIP program O, which is solved using a subgradient template in Lines 6–7. To switch to a CP solver, it suffices to change line 5 into

```
1 id<ORProgram> O = [ORFactory createCPProgram: L];
```

Similarly, to use violation degrees rather than satisfiability degrees, it is sufficient to edit line 4 to read

```
1 id<ORModel>     L = [ORFactory lagrangianRelax: P softeningConstraints: H];
```

Observe that, following [6], OBJECTIVE-CP stores the fact that L is a relaxation of P and the runnable produces several products in agreement with a relaxation specification, including a stream of lower bounds. It can thus be composed naturally with a primal algorithm.

Consider now the application of a surrogate optimization scheme.

```
1 id<ORModel>      P = [ORFactory createModel];
2 ...
3 id<ORIdArray>    H = ... // array of hard constraints in P
4 id<ORPartition> Vs = [ORFactory autoPartition: P          accordingTo: H];
5 id<ORModel>      S = [ORFactory lagrangianRelax: P relaxingConstraints: H];
6 id<ORProgram>    O = [ORFactory createMIPProgram: S];
7 id<ORRunnable>   r = [ORFactory surrogate: O splitWith: Vs];
8 [r run];
```

Line 4 computes a partition of the variables in P from the hard constraints. Line 5 creates the Lagrangian relaxation of P with respect to H and line 6 creates a MIP program that is then used by a surrogate runnable in line 7. The partition in line 4 is the argument $\{V_1, \cdots, V_k\}$ appearing in the template in Figure 2.

Models with a natural decomposable or "block" structure are often difficult to decompose by hand, particularly for larger problems. Hence, it is useful to have the ability to automatically partition a problem based on sets of *coupling* constraints. OBJECTIVE-CP makes use of a hyper-graph clustering algorithm [7] to provide an automatic decomposition. The variables of a model become nodes and each constraint defines an hyper-edge connecting it variables. The algorithm recursively clusters variables into disjoint sets until the maximal decomposition is achieved.

6 Empirical Results

This section reports some experimental results highlighting the concepts described in this paper. The goal is not to present state-of-the-art results on specific problems but to make the case that Lagrangian relaxation could play a larger role in the constraint-programming community. In addition, the experiments present some interesting perspectives on some design choices in Lagrangian methods. All experimental results are obtained using OBJECTIVE-CP [15] unless specified otherwise. Mixed-Integer programs are solved using Gurobi 5.6.

6.1 Graph Coloring

Graph coloring aims at minimizing the number of colors necessary to color a graph so that no two adjacent vertices have the same color. The following is a typical CP formulation of the problem:

$$Z_{CP} = \texttt{min}\ m$$
$$\text{subject to} \begin{cases} v_i \leq m, & i \in 1..|V| \\ v_i \neq v_j, & (i,j) \in E \\ v_i \in 1..|V|, \ i \in 1..|V| \\ m \in 1..|V| \end{cases}$$

Here, V is the set of vertices, E the set of edges, m a decision variable for the number of colors used and $\{v_i\}_{i \in 1..|V|}$ are decision variables representing the color assigned to the i-th vertex of V. In the Lagrangian relaxation, E is partitioned into a 'hard' and 'easy' edge set $E = E_e \cup E_h$, relaxing E_h. The experiments also use a MIP formulation automatically obtained from the above model

Table 1. Experimental Results on *Graph Coloring*

Instances	Dual											Primal				
	GLR(MIP)				SLR(MIP)				SLR(CP)			MIP			CP	
	time	lb	ub	itr	time	lb	ub	itr	time	lb	itr	time	lb	up	time	ub
20-3-39	0.08	11*	11*	1	0.07	11*	11*	1	0.44	11*	10.95	0.06	11*	11*	**0.01**	11*
120-25-188	300	9	9	173	3.4	9*	9*	2	30.25	9*	8.35	**1.52**	9*	9*	1.98	9*
160-2-846	300	55	160	1	300	55	160	1	300	105	104.5	300	55	160	**0.07**	108*
160-30-187	300	9	9	103	6.77	9*	9*	2	0.11	9*	4.35	2.5	9*	9*	**0.03**	9*
80-10-176	300	13	13	266	2.15	13*	13*	2	8.25	13*	13.3	0.90	13*	13*	**0.02**	13*
200-10-281	300	30	30	30	23.48	30*	30*	2	12.14	30*	30.0	12.03	30*	30*	**11.03**	30*
120-3-465	300	53	53	33	300	53*	53*	34	37.25	53*	51.15	10.0	53*	53*	**0.05**	53*
160-4-498	300	62	62	16	40.4	62*	62*	2	54.04	62*	61.0	20.01	62*	62*	**6.92**	62*
120-5-938	300	34	34	49	300	34	34	54	300	34	33.5	**9.48**	35*	35*	25.94	35*
200-20-201	300	16	16	47	12.8	16*	16*	2	3.58	16*	16.15	5.77	16*	16*	**0.03**	16*
180-5-873	300	51	51	11	176.85	51*	51*	2	300	51*	49.5	**23.29**	51*	51*	37.04	51*
100-2-910	12.5	71*	71*	1	12.40	71*	71*	1	300	69	67.5	12.52	71*	71*	**0.03**	71*
150-5-1803	-	-	-	-	-	-	-	-	300	41	39.5	26.8	46*	46*	**11.11**	46*
140-12-1137	-	-	-	-	-	-	-	-	300	19	10.0	**12.04**	20*	20*	55.7	20*

by a linearization transformation. The OBJECTIVE-CP linearization performs a *binarization* of the variables $\{v_i\}_{i \in 1..|V|}$ over their domains and transforms the (non-linear) disequality constraints into sets of inequalities.

The experiments consider three dual methods (GLR(MIP), SLR(MIP), and SLR(CP)), as well as two primal methods (MIP, CP). The methods are evaluated on randomly generated instances[1] which are built around collections of vertex cliques connected via *coupling edges*. More precisely, the vertex set is first partitioned into randomly sized cliques. Then a subset of vertices are chosen at random and *coupling edges* between these vertices are added. These instances are generated based on three parameters: *number of vertices* (nbv), *number of cliques* (nbc), *number of coupled vertices* (nbcv). In Figure 1, instances are referred to in the following format: 'nbv-nbc-nbcv'. The *relaxed edges* E_h used in GLR(MIP), SLR(MIP), and SLR(CP) are a subset of the *coupling edges*. The problem is first decomposed into independent sets (cliques in this case) using a standard hyper-graph partitioning algorithm. Edges which do not have a vertex in the maximal clique are relaxed. Initial upper bounds provided to the dual problem were about twice the optimal value.

Table 1 describes the results on 15 instances and it reports the runtime and the bounds produced by the various algorithms. Dual algorithms only report a lower bound *lb* while the MIP produces both lower and an upper bounds, and the CP program produces an upper bound only. Dual algorithms may terminate because of a timeout or because the step size is too small in which case the dual solution is typically primal-infeasible. Theorem 1 specifies when the Lagrangian dual produces an optimal solution. The table also reports the number of Lagrangian iterations. Bold entries correspond to the fastest implementation, while *starred*

[1] Python script for generating instances: `http://bit.ly/1jDCgJq`

Table 2. Experiments on *Set Covering* problems

Instances	GLR			SLR 5s			SLR 10s			SLR		
	time	itr	bnd	time	itr	bnd	time	itr	bnd	time	itr	bnd
inst 1	243.8	154	155.8	900	13	164.0	900	15	163.2	900	7	161.6
inst 2	900	109	149.2	900	12	153.1	900	12	155.3	900	9	153.5

entries indicate whether an optimal solution was found and proved optimal. A timeout of 300 seconds is used throughout.

The results bring some interesting conclusions. The dual MIP approaches perform poorly on these benchmarks and are strongly dominated by the primal MIP. **The dual CP is better than the dual MIP approaches but is typically bettered by the Primal CP formulation. The primal CP approach is the most effective approach on a number of benchmarks but is sometimes dominated by the primal MIP. These results seem to indicate that it would be valuable to investigate combinations of Lagrangian relaxation and constraint programming systematically on more applications. Note that the absence of lower-bounds and the large number of symmetries is detrimental to SLR(CP) which explores alternative selections of violated colorings.

6.2 GLR versus SLR

Lemma 2 indicated that SLR is a stronger relaxation than GLR, although GLR is the traditional Lagrangian relaxation in mathematical programming. Experimental results on graph coloring indicated that $SLR(MIP)$ systematically outperforms $GLR(MIP)$ on these instances and performs significantly fewer iterations. This section aims at confirming these results on set-covering instances. The results are based on random instances generated in separable blocks of random sizes which are extended with *coupling constraints*. The instances consider 1000 elements and 400 sets partitioned into 10 separable blocks and 250 coupling constraints (which are relaxed). The results for two representative instances are presented in Figure 3 and Table 2. The results again indicate that GLR and SLR behave very differently. SLR tends to have longer iterations but make larger jumps, while GLR features relatively rapid iterations but makes much less progress per iteration. It is also possible to speed-up SLR substantially by using a time limit. SLR then returns its best lower bound at the time limit when an improved lower bound has been found; otherwise, it continues until such a lower bound is found or the search is complete. Figure 3 shows the bound quality of GLR, SLR, and the time-limited SLR with limits of 5 seconds (SLR 5s) and 10 seconds (SLR 10s). A 15 minute (900s) time out is used.

The results on these set-covering instances shed some light on the respective strengths of GLR and SLR. On the first instance, GLR terminates because the step size has become too small (due to lack of bound improvement). Once again, SLR approaches dominate GLR on these instances. Results on instance 2 is

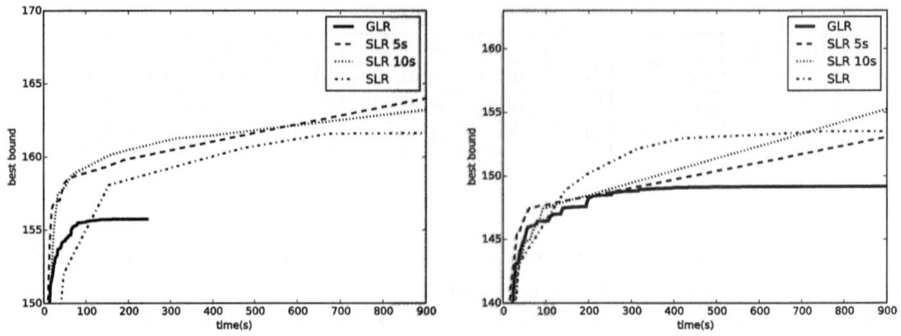

Fig. 3. Bound quality over time, GLR vs SLR

also revealing in that the smaller time limit gives better bounds early on but eventually falls behind and produces poorer bounds. Once again, these results seem to indicate that concepts from constraint programming, i.e., the degree of violations, could bring benefits into traditional Lagrangian dual methods.

6.3 Primal Lagrangian Tabu Search

This section describes the application of Lagrangian primal methods to boost the performance of a tabu-search algorithm. The tabu search is used to explore the neighborhood in the SPLR method; Upon completion, the Lagrangian multipliers are updated using the violation degrees and the tabu-search algorithm is restarted. The experiments are performed on the hardest instances of the progressive party problem using the model and the tabu search presented in [14] but with no restarting component and no manual tuning of the constraint weights. The progressive party problem features a variety of global constraints (e.g., *alldifferent* and *packing* constraints). It is thus fundamentally different from the benchmarks typically used to demonstrate the weighting schemes which uses SAT or binary CSPs, or binary constraints (e.g., [12,3,11]).

Figure 4 reports the experimental results. SPLR is a primal Lagrangian method using the tabu search (with no restart) to explore the neighborhood $\mathcal{N}$. SPLR updates the weights after n^2 iterations of the tabu search, where n is the number of variables. TABU is the tabu search with restarts, i.e., the control case. W-TABU is the COMET implementation of the tabu search with hand-chosen constraint weights and restarts (running on a slightly faster processor). All algorithms have a limit of 1,000,000 iterations and were executed 50 times on each instance. The table reports the configuration C, the number of periods P, the success percentage $\%S$ and the minimum, maximum, average, and standard deviation for the number of iterations and CPU time.

The results shows that SPLR significantly outperforms the tabu search in speed and success rate on these instances, indicating that Lagrangian primal

Algorithm	C	P	$\%S$	$m(I)$	$M(I)$	$\mu(I)$	$\sigma(I)$	$m(T)$	$M(T)$	$\mu(T)$	$\sigma(T)$
SPLR	3	9	98	63019	1000000	373515.82	233297.49	4.68	73.95	26.36	16.65
	4	9	94	73319	1000000	386478.46	246292.98	5.14	73.01	27.59	17.90
	6	7	94	40010	1000000	312445.44	280601.45	3.24	76.61	23.26	21.11
TABU	3	9	30	22729	1000000	816440.38	311171.57	1.855	86.542	59.46	23.15
	4	9	48	17692	1000000	767466.30	313399,19	1.40	82.36	55.59	23.15
	6	7	64	130117	1000000	733028.38	285338.94	8.699	66.11	47.55	18.56
W-TABU	3	9	96	23645	1000000	245331.96	249549.41	4.41	164.44	40.10	40.14
	4	9	94	47962	1000000	363842.50	273432.48	8.35	166.90	59.55	44.44
	6	7	88	19314	1000000	379072.73	328393.13	3.24	152.37	56.92	48.91

Fig. 4. Experimental Results for SPLR on the Progressive Party Problem

methods may provide a simple, systematic, and principled way to boost the performance of meta-heuristics and complex search procedures. Moreover, SPLR exhibits a performance similar to W-TABU on instances (3,9) and (4,9) and outperforms it slightly on instance (6,7). This indicates that primal Lagrangian methods may find proper multipliers quickly (the theory indicates that such multipliers exist but not how fast they can be identified).

It is also interesting to report some additional insights on SPLR. Indeed, experimental results show that using the satisfiability is counter-productive in this setting, the search rarely converging to a feasible solution. Moreover, using a restarting component is also not productive, which is a surprise given the importance on restarts for tabu search on this benchmark. The Lagrangian multipliers are very effective in driving the search out of local minima on this problems.

7 Related Work

This section reviews related work and positions this research, complementing the citations already given in the paper. To the best of our knowledge, this paper presents the first implementation of the Lagrangian dual with constraint programming, showing the benefits of Lagrangian relaxation to improve dual bounds in constraint programming and speed up optimality proofs. The paper also shows how to generalize the surrogate method for solving the Lagrangian dual, using ideas from large neighborhood search. **This is the first** systematic, compositional, and technology-independent implementation of the surrogate method. The paper also suggests that it may be valuable to consider violation degrees instead of satisfiability degrees in a variety of applications. Primal Lagrangian methods were generalized from continuous to discrete optimization in [12,17] and applied to SAT and CSPs in [17,3]. The focus in this paper was to suggest that Lagrangian methods can boost the performance of existing local or large neighborhood search systematically and compositionally. In that sense, this resulting algorithmic template is close to guided local search [16]. There are some interesting differences however, including the fact that the updates of Lagrangian multipliers use the violation degrees and that the underlying search can be arbitrary. Lagrangian relaxation was used to boost the coupling, communication, and propagation capabilities of two global propagators for optimization constraints in in [10]. The key insight is to solve a sequence of Lagrangian relaxations for the two propagators, using dual values at optimality of one propagator to seed the multipliers for the second optimization. This use of Lagrangian relaxation is rather different from this research where Lagrangian relaxation is used at the model level and existing methodologies are generalized to become compositional, to apply to high-level models, and to leverage multiple solution technologies.

8 Conclusion

The key contribution of this paper is to generalize Lagrangian relaxation to arbitrary high-level models. Lagrangian relaxations of such models can then be concretized into a variety of optimization technology (e.g., constraint programming, local search, or MIP). The resulting concrete optimization programs can be entrusted to algorithmic templates to solve Lagrangian duals or use Lagrangian primal methods. The paper also showed that Lagrangian relaxations can be built around the notion of satisfiability degrees (typical in mathematical programming) or violation degrees (typically in constraint programming and local search). Finally, the paper also indicated how to apply surrogate optimization systematically in a generic algorithmic template that optimizes independent problems separately. The experimental results show the versatility of Lagrangian relaxation for a variety of solver technologies and models. In particular, they show that

- The Lagrangian dual coupled with constraint programming is an effective method for some classes of graph coloring problems.
- The concept of violation degree is valuable to improve the quality and performance of the Lagrangian dual when solved with MIP solvers. It is not clear however whether satisfiability degrees can be valuable for constraint programming or local search.
- Primal Lagrangian methods may systematically boost the performance and solution quality of meta-heuristics in a principled way.

Overall, these results tend to indicate that Lagrangian methods could play a much more significant role in constraint programming and large neighborhood search and that further synergies between constraint programming and mathematical programming should be explored.

References

1. Ahuja, R.K., Magnanti, T.L., Orlin, J.B.: Network Flows: Theory, Algorithms, and Applications. Prentice Hall, Englewood Cliffs (1993)
2. Beldiceanu, N., Petit, T.: Cost evaluation of soft global constraints. In: Régin, J.-C., Rueher, M. (eds.) CPAIOR 2004. LNCS, vol. 3011, pp. 80–95. Springer, Heidelberg (2004)

3. Choi, K.M.F., Lee, J.H.-M., Stuckey, P.J.: A lagrangian reconstruction of genet. Artif. Intell. 123(1-2), 1–39 (2000)
4. Codognet, C., Diaz, D.: Yet Another Local Search Method for Constraint Solving. In: AAAI Fall Symposium on Using Uncertainty within Computation, Cape Cod, MA (2001)
5. Fisher, M.: The Lagrangian Relaxation Method for Solving Integer Programming Problems. Management Science 27, 1–18 (1981)
6. Fontaine, D., Michel, L., Van Hentenryck, P.: Model combinators for hybrid optimization. In: Schulte, C. (ed.) CP 2013. LNCS, vol. 8124, pp. 299–314. Springer, Heidelberg (2013)
7. Klimmek, R., Wagner, F.: A simple hypergraph min cut algorithm (1996)
8. Luenberger, D.G.: Introduction to Linear and Nonlinear Programming. Addison-Wesley, Reading (1973)
9. Nareyek, A.: Using global constraints for local search. In: Freuder, E.C., Wallace, R.J. (eds.) Constraint Programming and Large Scale Discrete Optimization. American Mathematical Society Publications, vol. 57, pp. 9–28. DIMACS (2001)
10. Sellmann, M., Fahle, T.: Constraint programming based lagrangian relaxation for the automatic recording problem. Annals of Operations Research 118(1-4), 17–33 (2003)
11. Selman, B., Kautz, H., Cohen, B.: Local search strategies for satisfiability testing. In: DIMACS Series in Discrete Mathematics and Theoretical Computer Science. American Mathematical Society Publications, vol. 26. DIMACS (1996)
12. Shang, Y., Wah, B.: A Discrete Lagrangian-Based Global-Search Method for Solving Satisfiability Problems. Journal of Global Optimization 12, 61–99 (1998)
13. Sun, T., Zhao, Q.C., Luh, P.B.: On the Surrogate Gradient Algorithm for Lagrangian Relaxation. Journal of Optimization Theory and Applications 133(3), 413–416 (2007)
14. Van Hentenryck, P.: Constraint-Based Local Search. The MIT Press, Cambridge (2005)
15. Van Hentenryck, P., Michel, L.: The Objective-CP Optimization System. In: Schulte, C. (ed.) CP 2013. LNCS, vol. 8124, pp. 8–29. Springer, Heidelberg (2013)
16. Voudouris, C., Tsang, E.: Partial constraint satisfaction problems and guided local search. In: Proc., Practical Application of Constraint Technology (PACT 1996), pp. 337–356 (1996)
17. Wah, B.W., Wu, Z.: The theory of discrete lagrange multipliers for nonlinear discrete optimization. In: Jaffar, J. (ed.) CP 1999. LNCS, vol. 1713, pp. 28–42. Springer, Heidelberg (1999)
18. Zhao, X., Luh, P.B., Wang, J.: The surrogate gradient algorithm for Lagrangian relaxation method. In: Proceedings of the 36th IEEE Conference on Decision and Control, pp. 305–310 (1997)

Loop Untangling

Kathryn Francis and Peter J. Stuckey

National ICT Australia, Victoria Research Laboratory,
The University of Melbourne, Victoria 3010, Australia
{kathryn.francis,peter.stuckey}@nicta.com.au

Abstract. An effective translation from procedural code into equivalent constraints is necessary in order to facilitate automated reasoning about the behaviour of programs. We consider the translation of bounded loops, proposing a new form of loop unwinding called *loop untangling*. In comparison to standard loop unwinding the constraints representing each iteration of the loop are greatly simplified. This is achieved by decoupling the execution order from the representation of each individual iteration. We illustrate this new technique using two different examples and provide experimental results verifying that the technique produces simpler models which result in much better solver performance.

1 Introduction

A translation from procedural code into equivalent constraints is a prerequisite for various applications based on automatic reasoning about program behaviour, such as program testing [15], test generation [19] and program verification [10,14]. This paper is concerned specifically with the treatment of loops (**for** loops and **while** loops) during this translation.

We focus on bounded loops, where a limit on the number of iterations is assumed or can be computed. Bounded loops arise in bounded model checking (e.g.[6]), simulation optimization (e.g. [11,5]), and other forms of symbolic execution. The typical approach to handling bounded loops is *loop unwinding*, which involves flattening the loop by creating a copy of the body for each potential iteration. This is used in e.g. [11,12,5,6,3,4].

The key insight of this paper is that the iterations of a loop do not necessarily need to be identified by the order of execution. That is, when creating copies of the loop body we do not have to label them as the iteration reached by execution first, second, and third, as is done in standard loop unwinding. Instead we can choose a different way of identifying each potential iteration, and then link them together using a separate representation of the execution order.

We describe here a new technique called *loop untangling* which does just that. Instead of execution order, iterations are identified by the value taken by a key expression within the loop body. This can vastly simplify the constraints for each copy of the loop body as the value of this key expression is known. As shown in Section 4, the result is greatly improved solver performance.

B. O'Sullivan (Ed.): CP 2014, LNCS 8656, pp. 340–355, 2014.

2 Motivating Examples

We give here two example programs where standard loop unwinding produces a particularly inefficient model, and sketch how loop untangling can provide a better translation. We will later show how this can be achieved automatically.

Our motivating examples come from a tool which allows combinatorial optimisation problems to be defined procedurally [11,12]. A programmer with no modelling experience can define an optimisation problem by writing a Java method which uses provided non-deterministic library methods to build a random solution to the problem, and then evaluates that solution, returning a measure of its quality or throwing an exception if it is invalid. The tool automatically finds the values to be returned by the library functions in order to produce the best return value. This is achieved by translating the code into equivalent constraints and passing the resulting model to a constraint solver. More details can be found in [11] but are not important for this work.

Our first example (below) is a routing problem which was one of the original benchmarks from [11]. Given a set of jobs, each of which has a pickup stop and a delivery stop, the problem is to choose the shortest Hamiltonian route visiting all stops, with no delivery stop visited before the corresponding pickup. Note that the ChoiceMaker argument provides the non-deterministic decision making methods. In this case the method chooseOrder is used, which returns a permutation of the given list.

```
int buildRoute(ChoiceMaker chooser) {
  List<Stop> route = chooser.chooseOrder(allStops);
  // compute arrival times
  int currentLocation = startLocation;
  int currentTime = 0;
  for(Stop stop: route) {
    int nextLocation = stop.getLocation();
    currentTime += travTime(currentLocation, nextLocation);
    stop.arrivalTime = currentTime;
    currentLocation = nextLocation;
  }
  tripFinishTime = currentTime + travTime(currentLocation, startLocation);
  // check no pickup is after the corresponding delivery
  for(Job j : jobs) {
    if(j.pickupStop.arrivalTime > j.deliveryStop.arrivalTime)
      throw new Exception();
  }
  return tripFinishTime;
}
```

Fig. 1. Java code defining a routing problem

Consider the first loop, which computes the arrival time for each stop. Let us assume the list allStops contains three stops [A,B,C], which means these three stops also occur exactly once in the route list, but in an unknown order. Using standard loop unwinding we would create a copy of the loop body for the first, second, and third iteration. For each of these, the value of stop may be A, B, or C. All of the other variables depend on stop, so their values are also unknown. Furthermore, when we later look up the arrival time for the pickup and delivery stop for each job, the value retrieved could be the currentTime value computed in any of the three iterations.

Figure 2(a) shows the (idealized) MiniZinc [16] produced by loop unwinding. The decisions are the permutation of the stops, enforced by `alldifferent`. Expressions computed within the loop body are represented using arrays indexed by iteration time `Ite` $= 1..n$ (where n is the number of iterations), or `Ite0` $= 0..n$ for those having a version before the loop. Constraints simulate the calculation within the loop, using the `locations` and `dist` arrays to look up parameter values referenced within the `getLocation` and `travTime` methods. To constrain the final arrival time for each stop s we need to determine which iteration was the last where we changed the arrivalTime field of the stop s, encoded using `which`. We then can lookup the currentTime in that iteration to give the arrivalTime.

```
array[Ite] of var Stop: route;
constraint alldifferent(route);

array[Ite] of var Location: nextL;
array[Ite0] of var Location: currL;
array[Ite] of var int: travT;
array[Ite0] of var int: currT;
constraint currL[0] = startL;
constraint currT[0] = 0;
constraint forall (i in Ite) (
  nextL[i] = locations[route[i]] ∧
  travT[i] = dist[currL[i-1],nextL[i]] ∧
  currT[i] = currT[i-1] + travT[i] ∧
  currL[i] = nextL[i]
);
array[Stop] of var Ite: which;
constraint forall(s in Stop) (
  which[s] = max(i in Ite)
        (i*bool2int(route[i] = s)));

constraint forall (j in Job) (
  currT[which[pickup[j]]] <=
  currT[which[delivery[j]]]
);
```

(a)

```
array[Stop] of var Stop0: prevS;
constraint path(prevS,_,0);

array[Stop] of var Location: nextL;
array[Stop0] of var Location: currL;
array[Stop] of var int: travT;
array[Stop0] of var int: currT;
constraint currT[0] = 0;
constraint currL[0] = startL;
constraint forall (s in Stop) (
  nextL[s] = locations[s] ∧
  travT[s] = dist[currL[prevS[s]],nextL[s]] ∧
  currT[s] = currT[prevS[s]] + travT[s] ∧
  currL[s] = nextL[s]
);

constraint forall (j in Job) (
  currT[pickup[j]] <= currT[delivery[j]]
);
```

(b)

Fig. 2. (Idealized) Constraints generated using (a) unwinding, and (b) untangling

The observation we make in this paper is that instead of creating copies of the loop body for each iteration in order of execution, it would be much better to create a copy for *the iteration where* stop *equals A, B, and C.* We know that each of these iterations will be executed, the only uncertainty is the order in which this will happen. With a known value for stop, the value of nextLocation is fixed, and crucially the stop whose arrival time we set in each iteration is also known, so when we later look up the arrival time for each stop we know which version of currentTime is relevant in each case. Obviously we still need to link the iterations to each other, as expressions used within the loop depend on the previous iteration. Actually the value of e.g. currentLocation for a given iteration is exactly the value of nextLocation from the previous iteration. So this linking can be achieved with a `path` [13] or `DomReachability` constraint [18].

The (again idealized) MiniZinc produced using loop untangling is shown in Figure 2(b). Since we are identifying iterations by the value of stop, the arrays for expressions calculated within the loop are indexed by Stop or `Stop0` (which includes an artificial initial stop 0). The decisions are `prevS`, that is for each iteration/stop, what is the previous iteration/stop (or 0 for the first iteration). This is used to look up values that depend on the previous loop iteration (or initialization), while a `path` constraint ensures that these predecessor variables correspond to a Hamiltonian path starting anywhere and ending at the artificial stop 0. The arrivalTime for a stop is now simply equal to the currentTime computed in the iteration corresponding to that stop.

Our second example is a pizza ordering problem which was the running example from [12], part of which is shown in Figure 3. The task is to find the cheapest pizza order which will satisfy a group of discriminating pizza eaters. The code computes the acceptable pizzas for each person, then chooses from these for each slice up to the number the person requires. Once the slices are chosen the cost of the order is calculated taking into account a discount for ordering whole pizzas.

```
1  int buildOrder() {
2    order = new Order(menu);
3    for(Person person : people) {
4      // Find acceptable pizzas
5      pizzas.clear();
6      for(OrderItem item : order.items)
7        if(person.willEat(item))
8          pizzas.add(item);
9      // Choose type for each slice
10     for(int i=0; i<person.slices; i++) {
11       OrderItem pizza =
12         chooser.chooseOne(pizzas);
13       pizza.addSlice();
14   }}
15   return order.totalCost();
16 }

17 class OrderItem
18 {
19   int fullPizzas = 0;
20   int numSlices = 0;
21
22   void addSlice() {
23     numSlices = numSlices + 1;
24     if(numSlices == slicesPerPizza) {
25       numSlices = 0;
26       fullPizzas = fullPizzas + 1;
27   }}
28
29   ...
30 }
```

Fig. 3. Extract from a Java simulation of a pizza ordering optimisation problem

The loop we consider this time is the one on lines 10–14, within which we make the decisions and tally up the number of slices and pizzas for each pizza type. Here the order of iterations is actually irrelevant to the final result (the cost of the order). It is only the number of times each type of pizza is chosen which matters. Unwinding the loop introduces symmetries and also creates a lot of added uncertainty as the pizza type whose numSlices and numPizzas field is changed in each iteration is unknown. It would be much better to create a variable giving the number of times each type of pizza is chosen, and then constrain the final value of numSlices and numPizzas for each pizza type to be a function of this variable.

Loop untangling achieves this by labelling the iterations by the return value of chooseOne (assigned to the pizza variable on line 11/12). Note that in this case the label is not unique, so we will need a copy of the body for e.g. the first time Vegetarian is chosen, and the second time, up to the maximum times possible. In each of these iterations the value of numSlices and numPizzas will be fixed. Furthermore, when we later look up these values for a particular pizza type, we know that the result will be the value computed in one of the iterations corresponding to that pizza type. Which iteration will depend only on the number of times that pizza type is chosen. The resulting constraint system is far simpler and propagates much more efficiently.

We describe in the following sections our loop untangling technique which can be applied to any loop, and which when applied to the examples above results in much better performance than standard loop unwinding. It is not necessary to detect specifically that in the first example the value of currentLocation is exactly the value of nextLocation from the previous iteration, nor to detect in the second example that the order of iterations is irrelevant. Provided the appropriate choice of labelling scheme for iterations our generalised implementation automatically produces a model which is equivalent to the better model in both cases.

3 General Loop Untangling Technique

This section explains the process of converting code into equivalent constraints using loop untangling rather than loop unwinding. The underlying translation technique is the query based approach described in [12]. The key feature of this technique is that rather than modelling the current state of the program at each execution step, we simply constrain the value of each state query to correspond correctly to the preceding state changes. This is a necessary prerequisite for loop untangling because it allows the execution order of state changes to be viewed as a decision.

The translation is broken into two phases. First the code is flattened into a list of *basic steps*: state changes, state queries, and path control points. Then the result of each state query in this list is constrained to correspond correctly to the changes and control points. The difference between the technique we describe here and that used in [12] is a new approach to making copies of loop bodies while flattening, and a different representation for the constraints defining which state changes occur before which state queries.

3.1 Programs as Ordered State Changes and State Queries

We consider a Java program to consist of a sequence of *basic steps*, each of which is a *state change*, *state query*, or *path control point*. At the lowest level all state changes are assignments and all state queries are variable references. However, since our application of interest (defining combinatorial optimisation problems using imperative code) tends to make heavy use of collections (sets, lists and maps), we treat the core collection operations as atomic state changes (e.g. add item to list) and state queries (e.g. length of list). Path control points are points in the code where execution branches or merges. That is, **break**, **continue** and **return** statements, plus the beginning and end of **then** blocks, **else** blocks and loop bodies, and the end of methods (if there are multiple **return** statements).

3.2 Flattening

The first step in our translation is to convert the code into a list of basic steps. For example the code in lines 5-12 of the routing example (Figure 1) is flattened as shown in Figure 4. To save space we have not separated compound queries into individual parts. For example stop.getLocation() is actually a query for the value of the stop variable, and then a query for the result of the getLocation method called on that stop variable, which is itself a query for the location field of the stop.

Note that this list is not really *flat* yet, as items inside loop bodies may occur more than once in an execution of the program. To solve this we need to create copies of the loop body in such a way that each copy is executed at most once.

c_1:	currentTime := 0	(5)
c_2:	i := 0	(6)
q_1:	i < route.size()	(6)
p_1:	*start loop* (q_1)	(6)
p_2:	*start loop body*	(6)
q_2:	route.get(i)	(6)
c_3:	stop := q_2	(6)
q_3:	stop.getLocation()	(7)
c_4:	nextLocation := q_3	(7)
q_4:	currentTime+travTime(..)	(8)
c_5:	currentTime := q_4	(8)
q_5:	currentTime	(9)
q_6:	stop	(9)
c_6:	q_6.arrivalTime := q_5	(9)
q_7:	nextLocation	(10)
c_7:	currentLocation := q_7	(10)
q_8:	i+1	(11)
c_8:	i := q_8	(11)
q_9:	i < stopsInOrder.size()	(11)
p_3:	*end loop body* (q_9)	(11)
p_4:	*end loop*	(11)
q_{10}:	currentTime+travTime(..)	(12)
c_9:	tripFinishTime := q_{10}	(12)

Fig. 4. Flattened loop

3.3 Creating Iterations

When standard loop unwinding is used (as in [12]), we create a copy of the loop body for each potential iteration and label them as the first, second, third etc. The execution order is fixed, but each individual iteration can have a large amount of uncertainty. The idea behind loop untangling is to instead create and label our iterations in a way that reduces the uncertainty within each individual iteration.

The first step is to choose a state query inside the body to be used as the *label query*. The label query is how we will refer to the loop iteration, and ideally knowing the value of the label query will make the loop body much easier to model. Currently this choice is specified via annotation, although it seems clear that some simple static analysis should give us good choices. In our illustrative examples, we choose the iteration argument stop (q_2 in Figure 4) and the choice of pizza type assigned to pizza (line 11 in Figure 3).

Given a label query q_L, we determine the maximum number of times the loop body may be executed (n), and for each iteration $i \in 1..n$ we compute the set of possible values D_i which could be taken by q_L. This is exactly the same calculation as would be done as part of standard loop unwinding.

We then create copies of the loop body as follows. For each value v in the union of the domains D_i computed above, we create k copies of the loop body, where k is the number of iterations in which D_i contains v. For each copy, we add a constraint that if this iteration is reached by execution then the value of q_L is v. This means that we can assume a fixed value for each iteration. If execution reaches the iteration then we know its value will be v, and if execution does not reach this iteration then the value of any query contained in it is irrelevant. When multiple copies are created for value v we also impose a fixed execution order on these to eliminate symmetry, and number them accordingly.

Note that the added constraint setting q_L to take value v does not replace the constraints ordinarily used to define the result of the query based on the preceding state changes. These are still needed but they will now impose a constraint on the (no longer fixed) execution order rather than the query result.

Note also that we may create more than n copies of the loop body. A good choice of label query will remove a lot of uncertainty from individual iterations without introducing too many extra iterations. If for the chosen label query every computed domain D_i contains only a single value, then no uncertainty can be removed and loop untangling is equivalent to loop unwinding.

In the routing example we create a single copy of the loop body for each stop in the allStops list, as we know that each occurs exactly once in stopsInOrder and therefore will occur in exactly one iteration. The new list of basic steps will have three copies of the body (assuming there are 3 stops A,B,C). We will add subscripts a, b, c to the listed step ids in Figure 4 to refer to them.

In the pizza example, we need multiple copies for each pizza type. The number of copies for each is the number of iterations in which that pizza type may be contained in the pizzas list, which is calculated by unwinding as described above. For nested loops such as this one, we create copies of the bodies separately. That is, copies of the inner loop body are not associated with a particular outer iteration. However, there will be multiple copies of the start and end loop nodes for the inner loop, and each of these will belong to a particular iteration of the outer loop. Assuming people = [Ant,Bee], if Ant wants one slice of Veg or Capriciosa, and Bee wants two slices which could be Margherita or Veg, then there are 3 copies of the inner loop for Veg, two copies for Mar, and one for Cap.

3.4 Modelling State Queries

A state query is a function of the state changes occurring before it, while the path control points determine which state changes occur before which state queries. To achieve a correct translation from code to constraints we need to constrain each state query in our flattened list to correspond correctly to the state changes (including artificial state changes added at the beginning to set up the initial program state) and path control points. The constraints described below are the same as those used in [12].

Most types of state query (including variable references) are what we call *lookup queries*, which means they return a value which is a function of only the most recent matching state change. What is meant by *matching* depends on the specific query type. For lookup queries we create a variable *changeID* to represent the ID of the most recent matching state change, and then constrain this ID and the retrieved value appropriately. For example, field references are constrained as shown below. Note that only assignments to the queried field (not other state changes) are relevant.

query qstep: var ref qobj.field
changes: step_1: obj_1.field := expr_1
...
step_n: obj_n.field := expr_n
variables: var 1..n: changeID; var int: changestep; var int: qresult;
constraints: $[\text{obj}_1, ..., \text{obj}_n][\text{changeID}] = \text{qobj} \wedge$
$\text{qresult} = [\text{expr}_1, ..., \text{expr}_n][\text{changeID}] \wedge$
$\text{changestep} = [\text{step}_1, ..., \text{step}_n][\text{changeID}] \wedge$
$\mathsf{before}(\text{changestep}, \text{qstep}) \wedge$
forall (i in 1..n) (
$(\text{obj}_i = \text{qobj} \wedge \mathsf{before}(\text{step}_i, \text{qstep})) \rightarrow \text{not } \mathsf{before}(\text{changestep}, \text{step}_i)$);

The first constraint requires the chosen assignment to use the same object as the query, which is the definition of *matching* for field references. The next constraint sets the result of the query to the value from the chosen assignment. The before constraint ensures that the change occurs before the query. A change which is skipped by the execution path or which occurs after the query cannot be chosen. We discuss the implementation of before in the next section. The final constraint is used to ensure that we choose the *latest* matching assignment by requiring that no other matching change overwrites our chosen one.

Other queries return a value which is a function of all matching state changes occurring before the query. We call these *aggregate queries*. For these we use before to constrain which changes should be included in the aggregate. For example, list length is constrained as follows (the length of the list is the number of matching add item changes before the query).

query: qstep: qlist.length()
changes: step_1: list_1.add(item_1)
...
step_n: list_n.add(item_n)
variables: var int: qresult;
constraints: $\text{qresult} = \text{sum (i in 1..n) (bool2int}(\text{list}_i = \text{qlist} \wedge \mathsf{before}(\text{step}_i, \text{qstep})))$;

Sometimes a lookup query behaves like an aggregate query. This happens when the most recent relevant change itself depends on the previous changes, either because of its arguments or because earlier changes affect whether or not execution reaches the later changes. In these cases it is still possible to use the standard representation for lookup queries, but much better performance can be achieved using a specialised translation. In [12] this was called *special cases.* When untangling rather than unwinding loops we can use the same specialised translations described in [12], as in each special case discussed there the query result is not affected by the order in which the changes occur.

3.5 Modelling the Execution Path

When standard loop unwinding is used, path control points can only cause execution to skip state changes. The relative order is known. So in [12], before was implemented by calculating a Boolean expression $cond_a$ for the conditions under which execution reaches step a, and then defining before as follows.

$$\mathsf{before}(step_a, step_b) = (a < b) \wedge cond_a$$

When iterations are identified by something other than execution order, the relative execution order of iterations, and therefore of the basic steps contained in them, is unknown. This means we need a new implementation of before. Our new implementation is based on a graph of the possible execution paths. This graph is very similar to a control flow graph, but it contains a node for every copy of each basic step, rather than a single node for each basic block. The edges are constructed as follows.

- A state query or state change has a single outgoing edge leading to the following step.
- The control point at the beginning of a **then** or **else** block has an edge leading to the first step in the block, and another edge leading directly to the end of the block.
- A **continue**, **break**, or **return** control point has a single outgoing edge to the end of the associated loop body, loop, or method respectively.
- A start loop control point has an edge to its associated end loop control point, and to every start body control point for its loop.
- An end loop body control point has an edge to the start body point for each other iteration of that loop, and to every version of the end loop control point.
- An exception step has no outgoing edges.

Any valid solution must correspond to a path through this graph (between the fixed start and end steps). Note that as required by our application this prevents exception points from being reached.

As we will be looking backwards from queries to the changes affecting them, we assign for each basic step s a predecessor $prev[s]$ which is the basic step

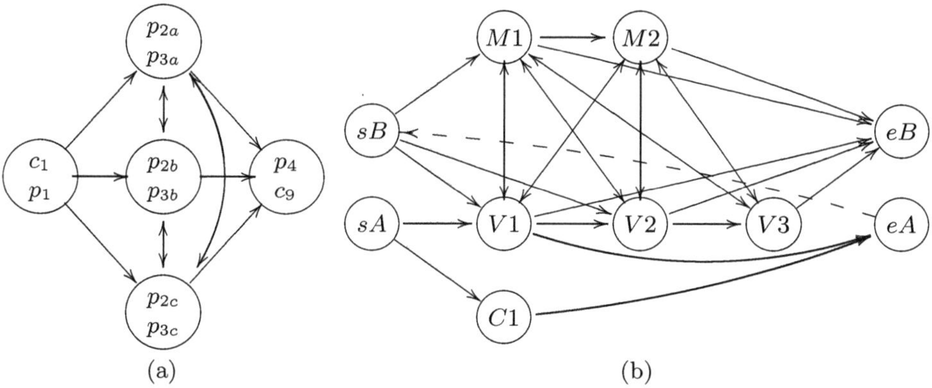

Fig. 5. Path control graphs: (a) lines 5-12 of Figure 1, (b) lines 10-14 of Figure 3

executed immediately before s. Clearly for most steps this is simply the unique predecessor. The *prev* array can be constrained by a `subpath` constraint [13].

Not all paths through the graph represent valid execution paths. The use of certain edges (where execution branches) is conditional on the result of a Boolean state query referred to in that step. The edge leading into a then or else block can only be used if the **if** condition is *true* or *false* respectively. An edge leading into a loop body is only valid if a query for the loop entry condition returns *true*. In Figure 4 the query used to control the use of edges is shown in brackets next to the source node. For loops (both start loop and end loop body control points) if the query shown is *false* then the edge to the end loop control point must be used.

Figure 5(a) shows a portion of the execution graph for our routing example with basic blocks collapsed. The start can reach each loop iteration for stop = A, B or C, and these can each reach each other and the end of the loop. As an example of the conditions on edges, consider the edges leaving step p_{3a}. Setting $prev[p_4] = p_{3a}$ requires $q_{9a} = false$, as this edge represents exiting the loop, while $prev[p_{2b}] = p_{3a}$ (or $prev[p_{2c}] = p_{3a}$) requires that $q_{9a} = true$, as these edges represent re-entering the loop.

For the pizza example (Figure 5(b)), edges which can be discounted upfront due to false edge conditions or constraints on the label query are not shown. Since Ant runs first it can reach only the first instance of Veg or Cap, and each of these can reach its end since Ant only picks one slice. For Bee the start can reach the first Mar or the first or second Veg. Each of these nodes can reach only the next of the same category or any of the other category, and the end of Bee's loop. Outside this part of the graph is a mandatory path from the end of Ant's loop to the start of Bee's.

We need further constraints on the edges for nested loops, to ensure that we do not enter the inner loop from one outer iteration and leave to a different outer iteration. For example we cannot enter node $V1$ from sA and leave to eB.

This is prevented by adding a constraint on the start and end loop body control points (s_i and e_i for i in the iteration set I) for each loop.

$$\forall i, j \in I, \neg(\mathsf{before}(s_i, e_j) \wedge \mathsf{before}(e_j, e_i))$$

This ensures that no other end loop body step from the outer loop can come between a pair of associated start and end body steps for that loop.

As mentioned previously, if we have created multiple iterations for a given value of the label query, we also impose a fixed order on those (using before) to eliminate symmetry. This means that (as shown in Figure 5(b)) edges leading from the start loop control point to the second or later copy of the body for each value of the label query are excluded immediately, and between iterations for the same value we only keep edges leading between successive copies.

3.6 Redefining before

The purpose of constructing the graph described in the previous section is to provide a new definition of the before relation used in our constraints. A simple implementation of before can be achieved by creating a time variable for each node in the graph of possible execution paths, and adding a constraint for each edge to say that if that edge is used then the time of the destination is one greater than the time of the source. Then before can be defined as follows.

$$\mathsf{before}(a, b) = \mathsf{time}[a] < \mathsf{time}[b]$$

In order to prevent changes which are not included on the execution path from affecting queries which are, we require that any step not on the path has a time greater than the number of steps.

While this implementation is correct, it does not provide very strong propagation. Consider the reference to currentTime which forms part of q_4 in the routing example (Figure 4). The options for the *latest matching change* are c_1 and c_5 (which has a version for each iteration of the loop). Imagine we are determining q_{4b} and we have already decided that the a iteration is followed by the b iteration $prev[p_{2b}] = p_{3a}$. Ideally we should know that q_{4b} refers directly to c_{5a}. But after this decision we have that $\mathsf{time}[q_{4b}] \in \{26, 42\}$, $\mathsf{time}[c_{5a}] = \{11, 27\}$, $\mathsf{time}[c_{5c}] = \{11, 43\}$, $\mathsf{time}[c_1] = 1$. So according to the above definition of before each of $\{c_1, c_{5a}, c_{5c}\}$ could be the latest matching state change.

Even harder to handle is when we decide that iteration b does not follow a. We know that all iterations have a matching change for q_{4b}. So if a is not immediately before b, then there must be another iteration in between with a matching change, even though no specific change is known to be between them.

More generally, the logic we would like to have is that whenever all paths between change c and query q go through another change which is known to match q, then change c cannot be chosen for q. Note that although this is related to dominance it is not pure dominance as we do not require that the same change overwrites c on all paths.

Ideally this would be achieved using a global constraint. Such a constraint does not exist, but we have found that including the logic it would use in our simplification phase is sufficient to provide good performance compared with loop unwinding. We intend to implement the global constraint and expect that even better performance should be possible.

3.7 Optimisations and Simplifications

In certain cases it is possible to create a simple expression $closest_c$ defining the conditions which must hold for change c to be the closest matching change to a given lookup query q. When this is possible, we can change the constraints used for q to the simpler form shown below.

query: qstep: var ref qobj.field
changes: $step_1$: obj_1.field := $expr_1$
...
$step_n$: obj_n.field := $expr_n$
variables: var 1..n: changeID; var int: qresult;
constraints: [obj_1, ..., obj_n][changeID] = qobj $\wedge$
[$closest_1$, ..., $closest_n$][changeID] $\wedge$
qresult = [$expr_1$, ..., $expr_n$][changeID];

If all changes potentially matching a lookup query are initialisation changes (added at the beginning to set up the initial program state), then only one can match the query, so we can use *true* as the *closest* expression for all of them.

If for a query q there exists a node in the execution graph n such that every edge leading in to n would create a fixed path between a matching change c for q and q, then we can use the edges into n to define the *closest* conditions. For example, consider the reference to currentTime discussed previously (part of q_4 in Figure 4). The query q_{4b} has a matching change in each iteration of the loop (c_{5a}, c_{5b}, c_{5c}), and one before the loop (c_1). All edges into the node p_{2b} (see Figure 5) create a fixed path between one of these changes and q_{4b}. So we can constrain q_{4b} as shown below. Note that since none of the edges leading in to p_{2b} correspond to the change c_{5b} we can discount this change immediately.

query: q_{4b}: currentTime
changes: c_1: currentTime := 0
c_{5a}: currentTime := q_{4a}
c_{5c}: currentTime := q_{4c}
variables: var 1..3: changeID; var int: qresult;
constraints: [prev[p_{2b}]=p_1, prev[p_{2b}]=p_{3a}, prev[p_{2b}]=p_{3c}][changeID] $\wedge$
qresult = [0, q_{4a}, q_{4c}][changeID];

The same can be done for query q_{10} outside the loop using the edges into p_4. This provides both the positive and negative reasoning discussed in the previous section. If we decide to use an edge then the *closest* condition for that change will become *true* and all others will become *false*. If we decide not to use an edge then that *closest* condition will become *false*, excluding the corresponding change. Although these constraints are more verbose than the idealised MiniZinc shown in Section 2, they provide the same propagation strength.

We can also take into account the known relationship between iterations with the same value for the label query. If all changes relevant to a query belong to iterations with the same label value, then their relative order is known (as we have fixed this to avoid symmetry), so we can use the original definition of before.

If in addition all possibly matching changes are known to actually match and the path through each iteration with a matching change is fixed, then we can do even better. In this case it is not possible for execution to skip the change in iteration i without also skipping those in later iterations. So the closest matching change is the one from the last iteration to be executed before the query. Ordering the changes by iteration version, for each change c_i except the last:

$$closest_{c_i} = \mathsf{before}(c_i, q) \wedge \neg\mathsf{before}(c_{i+1}, q)$$

The change from the last iteration is the closest whenever it is before the query. If the query is outside the loop, then we know that all iterations not skipped by execution will occur before the query, so we can simplify the condition further:

$$closest_{c_i} = in[c_i] \wedge \neg in[c_{i+1}] \qquad in[step_i] = (prev[step_i] \neq step_i)$$

where $in[step_i]$ means step i is included on the execution path. This can be defined as shown above since subpath sets unused nodes to point at themselves.

Finally, for queries which are also contained in an iteration with the same label value, we can assume that all changes in earlier iterations for this value are included on the execution path. If not, then the query will not be reached either so its value does not matter. Therefore for these queries the closest matching change is set to the change from the latest iteration before the query iteration.

The above simplifications are used in the translation of the pizza example. Consider the reference to numSlices on line 23 of the pizza code (Figure 3), in the first Veg iteration. Let us call this query q_{nv1}. The relevant changes are the initialisation assignment for Veg which sets numSlices to 0, and the assignments on lines 23 and 25. All versions of these assignments from non-Veg iterations are known not to match q_{nv1} as they refer to a different pizza object, and all versions from later Veg iterations are known to be after this query. The versions from the current iteration are also after this query, so actually only the initialisation assignment can be chosen. Therefore the value of q_{nv1} can be fixed to 0. This in turn will fix the value assigned to numSlices on line 23 to 1, and the condition on the following line (24) to *false* (assuming slicesPerPizza is fixed to say 2), which means that the assignment on line 25 is not reached by execution.

Now consider the reference to numSlices in the next Veg iteration, query q_{nv2}. As explained above, since the path through the earlier Veg iteration is fixed and the query is also inside a Veg iteration, we can simply use the change from the closest iteration to this one ($V1$). The value of q_{nv2} is therefore 1. The value assigned on line 23 will be 2, and the test on line 24 will succeed. Again the path through this iteration is fixed, but it goes through the assignment on line 25, setting numSlices back to zero. When we consider q_{nv3} applying the same simplification again gives a value of 0.

For the query to Veg.numSlices after the loop (inside the totalCost method), we can use the definition of *closest* described above for queries outside the loop, because all matching changes belong to Veg iterations (except for the init change) and are known to match, and the path through every Veg iteration is fixed. The constraint for this query is therefore:

query: q: var ref Veg.numSlices
changes: c_0: Veg.numSlices := 0
c_{1v1}: Veg.numSlices := 1
c_{2v2}: Veg.numSlices := 0
c_{1v3}: Veg.numSlices := 1
variables: var 1..4: changeID, var int: qresult
constraints: $[\neg in[c_{1v1}], in[c_{1v1}] \wedge \neg in[c_{2v2}], in[c_{2v2}] \wedge \neg in[c_{1v3}], in[c_{1v3}]]$[changeID] $\wedge$ qresult = [0,1,0,1][changeID];

This is the constraint described in Section 2 which links the final value of Veg.numSlices to the number of times Veg is chosen (which is changeID−1).

4 Experimental Results

We have implemented the loop untangling technique described above, and show here experimental results for the two examples discussed. The constraint models for unwinding and untangling are produced fully automatically from the input Java code, the only additional information given to untangling was the choice of label query for each loop. Table 1 compares untangling to unwinding (the version called new+ in [12]) and to a hand written CP model for the same problem (also from [12]). Each figure is the average for 30 instances of the stated size. The times shown include instances which reached the timeout of 10 minutes, while the failures figures (shown in thousands) exclude them. All models were solved using G12 CPX on a 3.40GHz Intel i5-4670K with 16GB RAM.

The results clearly show the benefit of untangling. For these problems, we are clearly better off using a simple model for each iteration and deciding the order through them, rather than deciding what happens in the ith loop iteration where we know the order. We expect that with a specialized global propagator for managing the before constraint this could be further substantially improved.

Table 1. Comparative performance of unwinding and untangling

Problem		Solving Time						Failures (000s)		
		unwind		untangle		hand		unwind	untangle	hand
pizza	4	94.0s	(2)	1.3s		0.1s		127.8	17.2	0.8
	5	320.7s	(13)	5.8s		0.5s		250.5	49.0	5.7
	6	470.1s	(22)	149.7s	(6)	20.9s	(1)	240.1	480.2	12.1
routing	5	12.9s		1.5s		0.4s		24.5	4.4	2.1
	6	102.8s		8.1s		2.2s		112.4	18.4	9.9
	7	569.3s	(20)	40.8s		14.6s		343.4	67.4	45.7

5 Related and Further Work

Loop untangling is related to other forms of program analysis that reason about loops. For example, automatic parallelisation of code needs to reason about when iterations can be reordered [17]. We could improve loop untangling by co-opting methods from this area to detect cases where the execution order of iterations can be fixed arbitrarily. The technique described in [2] for detecting commutativity could be a good starting point as a similar query-based viewpoint is taken when considering whether or not reordering iterations changes the outcome.

Loop untangling could also be improved by employing more general forms of program analysis. Typically optimisations performed by compilers are designed to simplify the remaining code, which would in turn simplify our translation to the constraint model. For example, loop untangling implicitly requires reaching definitions, and can also be simplified by constant propagation. While our tool does a basic form of reaching definition analysis and constant propagation these could be improved by full program analysis techniques (e.g. [1]).

An interesting direction for future work is developing a program analysis which would automatically select the label query. By examining the reaching definitions graph and understanding what data in the program is dependent on decisions and what is not we can choose a label query that, when fixed, fixes much of the computation of the loop body. But we need to trade this off against the number of iterations it will create.

Finally our method, and indeed most methods based on symbolic execution, currently only handles bounded loops. Unbounded loops can be approximated by putting an artificial limit on the number of iterations, but otherwise they require techniques to generate loop invariants including interpolation [8] or abstract interpretation [7]. Loop untangling could possibly be extended to handle unbounded loops using an approach similar to that in [9]. There constraints were added for each iteration of the loop lazily as needed when it became known that the previous iteration was entered. We could do something similar, lazily adding nodes to our execution path graph.

6 Conclusion

Standard loop unwinding unnecessarily ties the actual execution order of iterations with the way an individual iteration is identified. The idea behind loop untangling is to decouple these two things by modelling the execution path explicitly. The labelling scheme can be used to reduce the uncertainty in each copy of the loop body. Although it may be necessary to create more copies of the loop body than through standard loop unwinding, with a good choice of labelling scheme this is far outweighed by the relative simplicity of the constraints required for each copy. The final result is a model which is much easier to solve.

Acknowledgments. NICTA is funded by the Australian Government through the Department of Communications and the Australian Research Council through the ICT Centre of Excellence Program.

References

1. Aho, A.V., Sethi, R., Ullman, J.D., Lam, M.S.: Compilers: Principles, Techniques, and Tools, 2nd edn. Addison-Wesley (2006)
2. Aleen, F., Clark, N.: Commutativity analysis for software parallelization: letting program transformations see the big picture. ACM Sigplan Notices 44(3), 241–252 (2009)
3. Armando, A., Mantovani, J., Platania, L.: Bounded model checking of software using SMT solvers instead of SAT solvers. International Journal on Software Tools for Technology Transfer 11(1), 69–83 (2009)
4. Brandwijk, P.: Verifying software with SMT and random testing using a single property specification. Master's thesis, University of Amsterdam (2012)
5. Brodsky, A., Nash, H.: CoJava: optimization modeling by nondeterministic simulation. In: Benhamou, F. (ed.) CP 2006. LNCS, vol. 4204, pp. 91–106. Springer, Heidelberg (2006)
6. Collavizza, H., Rueher, M., Van Hentenryck, P.: CPBPV: a constraint-programming framework for bounded program verification. Constraints 15(2), 238–264 (2010)
7. Cousot, P., Cousot, R.: Abstract interpretation: A unified lattice model for static analysis of programs by construction or approximation of fixpoints. In: Proceedings of the Fourth ACM Symposium on Principles of Programming Languages, pp. 238–252 (1977)
8. Craig, W.: Linear reasoning: A new form of the Herbrand-Gentzen theorem. Journal of Symbolic Logic 22(3), 250–268 (1957)
9. Denmat, T., Gotlieb, A., Ducassé, M.: An abstract interpretation based combinator for modelling while loops in constraint programming. In: Bessière, C. (ed.) CP 2007. LNCS, vol. 4741, pp. 241–255. Springer, Heidelberg (2007)
10. Floyd, R.W.: Assigning meanings to programs. In: Proceedings of the American Mathematical Society Symposia on Applied Mathematics, vol. 19, pp. 19–31 (1967)
11. Francis, K., Brand, S., Stuckey, P.J.: Optimization modelling for software developers. In: Milano, M. (ed.) CP 2012. LNCS, vol. 7514, pp. 274–289. Springer, Heidelberg (2012)
12. Francis, K., Navas, J., Stuckey, P.J.: Modelling destructive assignments. In: Schulte, C. (ed.) CP 2013. LNCS, vol. 8124, pp. 315–330. Springer, Heidelberg (2013)
13. Francis, K., Stuckey, P.J.: Explaining circuit propagation. Constraints 19(1), 1–29 (2014)
14. Hoare, C.A.R.: An axiomatic basis for computer programming. Communications of the ACM 12(10), 576–580 (1969)
15. King, J.: Symbolic Execution and Program Testing. Com. ACM, 385–394 (1976)
16. Nethercote, N., Stuckey, P.J., Becket, R., Brand, S., Duck, G.J., Tack, G.R.: Minizinc: Towards a standard CP modelling language. In: Bessière, C. (ed.) CP 2007. LNCS, vol. 4741, pp. 529–543. Springer, Heidelberg (2007)
17. Pottenger, W.: The role of associativity and commutativity in the detection and transformation of loop-level parallelism. In: Proceedings of the 12th International Conference on Supercomputing, pp. 188–195. ACM (1998)
18. Quesada, L., Van Roy, P., Deville, Y., Collet, R.: Using dominators for solving constrained path problems. In: Van Hentenryck, P. (ed.) PADL 2006. LNCS, vol. 3819, pp. 73–87. Springer, Heidelberg (2005)
19. Sen, K., Marinov, D., Agha, G.: CUTE: a concolic unit testing engine for C. In: Proceedings of the 10th European Software Engineering Conference, pp. 263–272. ACM (2005)

Discriminating Instance Generation for Automated Constraint Model Selection

Ian P. Gent[1], Bilal Syed Hussain[1], Christopher Jefferson[1], Lars Kotthoff[2], Ian Miguel[1], Glenna F. Nightingale[1], and Peter Nightingale[1]

[1] School of Computer Science, University of St. Andrews, UK
{ian.gent,bh246,caj21,ijm,gfe2,pwn1}@st-andrews.ac.uk
[2] INSIGHT Centre for Data Analytics, Ireland
larsko@4c.ucc.ie

Abstract. One approach to automated constraint modelling is to generate, and then select from, a set of candidate models. This method is used by the automated modelling system CONJURE. To select a preferred model or set of models for a problem class from the candidates CONJURE produces, we use a set of training instances drawn from the target class. It is important that the training instances are discriminating. If all models solve a given instance in a trivial amount of time, or if no models solve it in the time available, then the instance is not useful for model selection. This paper addresses the task of generating small sets of discriminating training instances automatically. The instance space is determined by the parameters of the associated problem class. We develop a number of methods of finding parameter configurations that give discriminating training instances, some of them leveraging existing parameter-tuning techniques. Our experimental results confirm the success of our approach in reducing a large set of input models to a small set that we can expect to perform well for the given problem class.

1 Introduction and Background

Numerous approaches have been taken to automating aspects of constraint modelling, including: learning models from examples [8,4,5,17,3]; automated transformation of medium-level solver-independent constraint models [23,21,24,20]; theorem proving [7]; case-based reasoning [18]; and refinement of abstract constraint specifications [10] in languages such as ESRA [9], ESSENCE [11], $\mathcal{F}$ [14] or Zinc [19,16]. We focus on the refinement approach, where a user writes a constraint specification describing a problem above the level of abstraction at which modelling decisions are made. Constraint specification languages support abstract decision variables with types such as set, multiset, relation and function, as well as *nested* types, such as set of sets and multiset of relations. Therefore, problems can typically be specified very concisely. However, existing constraint solvers do not support these abstract decision variables directly, so abstract constraint specifications must be refined into concrete constraint models.

We use ESSENCE [11]. An ESSENCE specification (see Fig. 1) identifies: the parameters of the problem class (`given`), whose values define an instance; the combinatorial objects to be found (`find`); and the constraints the objects must satisfy (`such that`).

B. O'Sullivan (Ed.): CP 2014, LNCS 8656, pp. 356–365, 2014.

An objective function may also be specified (`min/maximising`) and identifiers declared (`letting`). Our CONJURE system[1] [2] employs refinement rules to convert an ESSENCE specification into the solver-independent constraint modelling language ESSENCE′ [23]. We use SAVILEROW[2] [22] to translate an ESSENCE′ model into input for a particular constraint solver while performing solver-specific model optimisations.

By following alternative refinement paths CONJURE typically produces a large set of models for a given ESSENCE specification. In our previous work [1] we developed a racing process to select among these candidate models, in which a set of training instances drawn from the problem class being modelled is used to gauge relative model performance. For this process to be effective, it is important that the instances chosen are *discriminating*: if all models solve a given instance in a trivial amount of time, or if no model solves it in the time available, then the instance is not useful for model selection. Our previous work assumed that such instances were given. In this paper we address the task of generating discriminating training instances automatically.

2 Racing for Automated Model Selection

Our approach to model selection follows that we reported in [1], but with an improved method of producing a set of winning models described below. It takes as input a set of instances drawn from the target problem class. Our performance measure of a model with respect to an instance is the time taken for SAVILEROW to instantiate the model and translate for input to the MINION constraint solver [12] plus the time taken for MINION to solve the instance. We include the time taken by SAVILEROW since it adds desirable instance-specific optimisations, such as common subexpression elimination [13].

We conduct a *race* [6] for each provided instance. Given a parameter $\rho \geq 1$, a model is *ρ-dominated* on an instance by another model if the measure for the second model is at least ρ times faster than the first. The 'winners' of an instance race are the models not ρ-dominated by any other model. So that trivial instances do not discriminate we do not consider any model that solves within 1s to be dominated. All models enter each race, but for efficiency a model is terminated as soon as it is $\rho-$dominated by some other model. Furthermore, the order in which the models are executed is influenced by their performance in previous races: well-performing models are executed first to establish a good ρ bound early. In order to guide our exploration of the instance space we assign a discriminatory quality value to an instance with respect to the results of the race run. This is the fraction of models that are ρ-dominated.

A set of instances is *ρ-fractured* if every model is ρ-dominated on at least one instance. In the presence of fracturing, care must be taken in defining the set of winning models over a race sequence. We do so as follows. We first find a minimum hitting set of winning models $\{a_1, a_2, a_3, ...\}$ which covers all instance races. We then define the set A_i as the set of models that won *every* race that a_i won. The set of sets $\{A_1, A_2, A_3, ...\}$ then gives a summary of the winning models over all fractured parts of the instance space. Note that each $A_i \cap A_j = \emptyset$ (where $i \neq j$) as otherwise we could find a smaller hitting set. Also note that in an unfractured instance space the unique A_1 is simply

[1] `http://bitbucket.org/stacs_cp/conjure-public`
[2] `http://savilerow.cs.st-andrews.ac.uk`

the set of models which won all races. However, for fractured spaces the set $\{A_i\}$ is not uniquely defined as it depends on the hitting set found: nevertheless it gives us a representation of one particular fracturing of the instance space.

3 Methods for Generating Discriminating Instances

The instance space is defined by the parameters of a problem class. Consider the ESSENCE specifications in Figure 1. Langford's Problem has two independent integer parameters and hence a two-dimensional instance space. The Knapsack Problem is an example of a more complex instance space, consisting of two integers and two functions. The first integer, n, governs the number of items and also the domain of the two given functions, which define the weights and values of those items. Our three methods for generating discriminating instances in these spaces are described below. All run a sequence of races and combine the results following the method described in Section 2.

`Undirected`: For each race in a sequence, `undirected` simply draws a sample from the instance space and runs a race. Section 4 describes our sampling method.

`Markov`: This method is loosely based on the Markov chain Monte Carlo methods used, e.g., to estimate the value of a multi-dimensional integral. We assume that discriminating instances are likely to be found near (by some proximity measure) other discriminating instances, and non-discriminating instances near other non-discriminating instances. This naturally leads to a Markov chain that walks the instance space, is attracted towards known discriminating instances, and is repelled from known non-discriminating instances. Our measure of proximity per parameter type:

Integer. The distance is simply the absolute difference between the two values.

Total Function. Given functions f and g we compute the distance between $f(i)$ and $g(i)$ for each i where both functions are defined, and aggregate using Euclidean distance. When f and g have different domains of definition, some mappings in f and/or g will be discarded. Suppose we had `given weights : function (total) int(1..n) --> int(1..100)` (as in the Knapsack Problem) with `n` a parameter. If the domain of definition `int(1..n)` differs between f and g, then `n` must differ and this will count towards the instance distance.

Set. Given sets S and T, the distance is $\sqrt{|S \setminus T| + |T \setminus S|}$, also Euclidean.

Relation. Treating a relation as a set of tuples, we use the distance measure for sets.

To obtain the instance distance, we combine the distance measure for each parameter again using the Euclidean distance. This combines elegantly with the Euclidean distances computed per parameter. An initial instance is sampled using the method described in Section 4. A race is run using this instance and a record taken of its discriminatory quality. Each subsequent instance (sampled using the same method) is accepted or rejected according to the scheme below. If an instance is accepted a race is run with that instance, otherwise another instance is generated, and so on until the required race sequence is complete. We use the following acceptance function, where x_{i-1} is the previous accepted instance and x'_i is the proposed instance.

$$A(x_{i-1}, x'_i) = \frac{G'(x'_i)}{G(x_{i-1})}$$

LANGFORD'S PROBLEM (CSPLIB 24)
```
given k, n : int(1..)
letting seqLength be k * n
letting seqIndex be domain int(1..seqLength)
find seq : function (total, surjective) seqIndex --> int(1..n)
such that forAll i,j : seqIndex , i < j .
        seq(i) = seq(j) -> seq(i) = j - i - 1
```

THE KNAPSACK PROBLEM
```
given n, totalWeight : int(1..)
given weights, values : function (total) int(1..n) --> int(1..)
find picked: set(maxSize n, minSize 1) of int(1..n)
maximising (sum i in picked . values(i) )
such that (sum i in picked . weights(i) ) <= totalWeight
```

THE PROGRESSIVE PARTY PROBLEM (CSPLIB 14)
```
given n_upper, n_boats, n_periods : int(1..)
letting Boat be domain int(1..n_boats)
given capacity, crew : function (total) Boat --> int(1..n_upper)
where forAll i : Boat . crew(i) <= capacity(i),
find hosts : set of Boat,
     sched : set (size n_periods) of function (total) Boat --> Boat
minimising |hosts|
such that  forAll p in sched . range(p) subsetEq hosts,
   forAll p in sched . forAll h in hosts . p(h) = h,
   forAll p in sched . forAll h in hosts .
     (sum b in preImage(p,h) . crew(b))<= capacity(h),
   forAll b1,b2 : Boat , b1 != b2 .
     (sum p in sched . (p(b1) = p(b2))) <= 1
```

THE WAREHOUSE LOCATION PROBLEM. (CSPLIB 34)
```
given n_upper, n_stores, n_warehouses : int(1..30)
letting Store be domain int(1..n_stores),
        WHouse be domain int(1..n_warehouses)
given capacity : function (total) WHouse --> int(1..n_upper),
      opencost : function (total) WHouse --> int(1..n_upper),
      cost : function (total) tuple (Store, WHouse) --> int(1..n_upper)
find open : function (total) Store --> WHouse
minimising (sum r in range(open). opencost(r)) + sum s : Store . cost((s,open(s)))
such that forAll w : WHouse . |preImage(open,w)| <= capacity(w)
```

Fig. 1. Four sample ESSENCE specifications

G' estimates the discriminatory quality of the instance in the interval $[0, 1]$ using the quality values of previously accepted instances found by racing. We define a radius of influence r, which is 10% of the greatest possible distance between any two instances, using the distance measure above. $G'(x'_i)$ finds the set of all previous accepted instances within distance r of x'_i. If this set is non-empty, $G'(x'_i)$ returns the mean of the true quality values for the set. Otherwise, $G'(x'_i) = 0.5$. $G(x_{i-1})$ gives the true quality of x_{i-1}. Finally, a pseudorandom number a is generated within $[0, 1]$, and x'_i is accepted if $A(x_{i-1}, x'_i) \geq a$ then. Hence, the proposed instance is always accepted if $G'(x'_i)$ is greater than $G(x_{i-1})$.

`Smac`: Our final method is based on SMAC [15], an automatic algorithm configuration system. Given an algorithm, a description of its parameters, and a set of instances, it finds the set of parameters for the algorithm that delivers the best performance on the set of instances. Finding discriminatory instances is a very similar setting – the algorithm is the problem class specification, its parameters instantiate particular instances of the class and the set of problem "instances" is the set of models. We want to find the set of problem class parameters – the set of problem instances – that has the optimum discriminatory power with respect to the models.

We encode the problem class parameters into SMAC's input format, which uses integer and categorical variables. Integer `givens` are converted to integer parameters with the range specified in the problem definition. We model structured `givens` such as functions using multiple SMAC parameters. When one `given` depends on another, such as the size of the domain of the function depending on n in the Knapsack Problem, we must be conservative: sufficient SMAC parameters are used to accommodate the maximum size of the structured `given`. The extraneous parameters are ignored when racing the instances so produced. Although SMAC has demonstrated that it is able to handle large parameter spaces, this conservative encoding may hinder its ability to cover the space effectively, since many of the values it is producing may be ignored. Furthermore, SMAC does not support the complex constraints between parameters that we sometimes require (`where`), so we use CONJURE to validate the instances that SMAC generates and discard those that do not satisfy the constraints.

4 Uniform Versus Non-Uniform Sampling

Except for SMAC, which has its own sampling method, our generation approaches require the ability to sample from the instance space associated with a problem class. Since the instance spaces of the problem classes we consider are typically infinite, our method requires some sensible bounds on the parameters involved in order to circumscribe a finite sub-space of interest. For example in Langford's Problem, as discussed in Section 5, we limit the two integer parameters to the ranges 2..10 and 1..50 respectively.

When the parameters defining the space are independent (e.g. the pair of integer parameters to Langford's problem), uniform sampling is straightforward: simply generate a value uniformly and independently for each parameter. When the parameters are not independent, uniform sampling is more difficult. Reconsider the Knapsack Problem from Figure 1. An approach to sampling from this space is first to generate n uniformly then uniformly and independently generate a mapping for each element of the domain $(1..n)$ of the two functions `weights` and `values`. However, this introduces bias: there are many more possible functions for large values of n than there are for small values of n. Hence, if we generate n uniformly then a particular function for small n is far more likely to be selected than a particular function for large n.

A solution to this problem is to enumerate all of the valid instances of the instance sub-space and sample uniformly from this set. This enumeration problem is naturally cast as a (simple) constraint problem. An ESSENCE specification E^* for the enumeration problem can be obtained automatically from an original ESSENCE specification E simply by replacing in E the `given` (parameter) statements with `find` (decision variable) statements and discarding the rest of E. As a very simple example, performing this process for Langford's Problem produces:

```
find k : int(2..10)
find n : int(1..50)
```

Care must be taken that this transformation produces valid ESSENCE. For the Knapsack Problem (assuming sensible parameter limits) it produces:

```
find n : int(1..100)
find totalWeight : int (1..1000)
find weights : function (total) int(1..n) --> int(1..100)
find values : function (total) int(1..n) --> int(1..100)
```

This is invalid because a decision variable is used to define the size of the domain of the two function variables. The solution we adopt is to leave n as a parameter, solve the problem for each value of n and take the union of the results.

We obtain a single model for E^* automatically using CONJURE and the Compact heuristic [1] — this is sufficient because the enumeration problem is typically easy. MINION is then used to find all solutions to the model. Our `uniform` sampling method is to select uniformly from this set of solutions.

The drawback of `uniform` is that it limits the size of space that we can consider. An alternative approach, which we call `solver-random`, is to sample by requesting a single solution from MINION, employing a random variable and value order. This introduces bias for much the same reason as described above: the distribution of solutions to the model may not be uniform. However, `solver-random` is much more scalable.

In order to compare `uniform` with `solver-random` we performed an experiment on two problem classes: Warehouse Location and the Progressive Party Problem (see Figure 1). So that `uniform` sampling was feasible, the instance space for each problem class was restricted to around a billion instances by limiting the upper bounds of each of their parameters. For each problem class we ran three sequences of thirty races using our `Markov` and `Undirected` generation approaches. In this experiment there was no substantive difference between `Markov` and `Undirected` — probably because of the (lack of) discriminatory quality of the instance sampled as we discuss below. For Warehouse Location the performance of the two sampling methods is identical, reducing 128 input models to 64 non-dominated models. For the Progressive Party Problem `solver-random` actually performs better than `uniform`, reducing 256 input models to 16 non-dominated models, whereas `uniform` produces between 63 and 77 non-dominated models. The bias inherent to `solver-uniform` appears to have been beneficial in this case, guiding the methods to discriminating instances.

These experiments provide some evidence that `solver-random` is a reasonable approach to sampling the instance space. It is also worth noting that the restrictions on the size of the instance space that we explore (to accommodate `uniform`) restricts the discriminatory quality of the instances that we find — producing 64 non-dominated models for Warehouse Location is a relatively weak result. `solver-random` scales easily to more challenging, hence more discriminating, instances as we will see in the following section where we use `solver-random` exclusively.

5 Experimental Results

In this section we report an experimental evaluation of our instance generation methods. Following the outcome of our experiment in Section 4, we use `solver-random` sampling throughout. We experiment on six problem classes described below with results summarised in Table 1. For each class we run three independent sequences of races. For `Markov` and `Undirected` we run 30 races, each with a time budget of 6 hours. This time budget is divided by the number of models to obtain the maximum time allowed for a particular model to solve an instance. Rather than a maximum number of runs, SMAC requires a total time budget. We specify the maximum time either of the other methods took to complete the race sequence.

Given the volume of experiments we use heterogeneous compute resources, while ensuring that all experiments for each problem class are run on the same resource. In all cases, a model is given 6.5GB of RAM and run on an AMD-based architecture. The Social Golfer Problem was run on a 32-core 2.1GHz machine, Warehouse Location on a 64-core 2.1GHz machine, and the remainder on 2.1GHz processors on Microsoft Azure.

Our results are summarised in Table 1. For each problem class we record: the number of models refined by CONJURE from the associated ESSENCE specification; the size of the output set(s) of models as described in Section 2; the number of discovered fractured parts of the instance space; and, for `Markov` and `Undirected`, the number of races until convergence. This last measure indicates how many of the 30 races are necessary to achieve the final result. Detailed discussion follows.

The Knapsack Problem. The parameter space is limited as described in Section 4. All of our methods perform well on this problem class, identifying that the instance space is fractured into two parts. The parts correspond to satisfiable versus unsatisfiable instances. For the satisfiable part, all three methods returned a single winner model (from the 64 input), which employs a 0/1 model of the problem. For the unsatisfiable part, all three methods also returned a single winning model based on an explicit representation of the set of items in the knapsack. Both `Markov` and `Undirected` show some variability in how quickly they converge on this result.

Langford's Problem. The parameters to this problem are a pair of independent integers (limited as described in Section 4). Therefore, `solver-random` sampling is unbiased for this problem. None of our methods found the instance space to be fractured, and each was able to reduce the input set of models drastically to a small set of non-dominated models. Over the three independent runs, `Markov` shows a slight edge in performance returning a single model in one case, whereas `Undirected` on one occasion returns six models. Both show some variability in steps to convergence.

The Social Golfer Problem. The parameters to this problem are a triple of independent integers, hence `solver-random` sampling is again unbiased. Each parameter is limited between 1 and 100. Hence, a large fraction of this instance space is unfeasibly difficult so we would expect our more informed approaches to perform better in this case study. In fact, SMAC struggles with this class, in two cases finding no discriminating instances at all. `Undirected`'s performance is variable, sometimes outputting a single model, but on one occasion also finding no discriminating instances. By contrast

Table 1. Results for the six problem classes over three independent runs

		Markov			SMAC		Undirected		
Problem	#Models	Output Sizes	Frac.	Steps to Conv.	Output Sizes	Frac.	Output Sizes	Frac.	Steps to Conv.
Knapsack-1	64	1,1	2	20	1,1	2	1,1	2	7
Knapsack-2	64	1,1	2	13	1,1	2	1,1	2	30
Knapsack-3	64	1,1	2	2	1,1	2	1,1	2	3
Langford-1	154	4	1	28	4	1	6	1	1
Langford-2	154	1	1	14	3	1	4	1	14
Langford-3	154	4	1	2	4	1	3	1	29
SGP-1	24	1,4	2	26	1	1	1	1	7
SGP-2	24	4	1	13	24	1	1	1	15
SGP-3	24	1	1	7	24	1	24	1	30
PPP-1	256	32,8	2	2	8	1	53,8	2	19
PPP-2	256	32,8	2	12	16	1	27,8	2	11
PPP-3	256	8	1	9	15	1	17	1	7
Warehouse-1	128	32	1	17	32	1	45	1	16
Warehouse-2	128	24	1	13	49	1	72	1	26
Warehouse-3	128	8	1	16	27	1	54	1	26
BACP-1	48	1,1	2	24	26	1	1	1	8
BACP-2	48	1	1	1	8	1	1	1	25
BACP-3	48	1	1	2	25	1	1	1	7

`Markov` performs well, reducing the 24 input models to between 1 and 4 on each run and on one occasion identifying a fracture in the instance space.

The Progressive Party Problem. This class has the most (256) input models. Perhaps unsurprisingly, therefore, the instance space is fractured — but this is only identified by `Markov` and `Undirected`. When fracturing is detected, `Markov` is more consistent than `Undirected` in the size of the returned set for the first fractured part.

The Warehouse Location Problem. All three methods are able to reduce the input set of 128 models significantly. `Markov` has the best performance, followed by SMAC and then `Undirected`. No fracturing was found.

Balanced Academic Curriculum Problem. Only `Markov` is able to detect a fracture in the instance space for this class, and only on one of its runs. Both `Markov` and `Undirected` reduce the input model set drastically from 48 to a single winning model. SMAC performs much less well on this class, perhaps hampered by the necessary encoding compromises (see Section 3) on what is a more complex instance space defined by 7 integers, two functions and a relation.

6 Conclusions

We have developed and investigated three methods for generating discriminating instances for the purpose of automated constraint model selection. Our experimental evaluation shows that all of these methods are capable of reducing a large number of possible models to a much smaller set. The methods are able to detect fracturing if it occurs and successfully determine the best models for each fraction. Overall, our novel `Markov` approach has the best performance on the problem classes in our experiments.

Acknowledgements. This research is supported by UK EPSRC EP/K015745/1. Bilal Syed Hussain is supported by an EPSRC scholarship, and a Google Europe Scholarship. Chris Jefferson is a University Research Fellow funded by the Royal Society. Lars Kotthoff is supported by EU FP7 grant 284715. The Microsoft Azure processors were provided by a Microsoft Azure for Research grant.

References

1. Akgun, O., Frisch, A.M., Gent, I.P., Hussain, B.S., Jefferson, C., Kotthoff, L., Miguel, I., Nightingale, P.: Automated symmetry breaking and model selection in CONJURE. In: Schulte, C. (ed.) CP 2013. LNCS, vol. 8124, pp. 107–116. Springer, Heidelberg (2013)
2. Akgun, O., Miguel, I., Jefferson, C., Frisch, A.M., Hnich, B.: Extensible automated constraint modelling. In: 25th Conference on Artificial Intelligence (AAAI) (2011)
3. Beldiceanu, N., Simonis, H.: A model seeker: Extracting global constraint models from positive examples. In: Milano, M. (ed.) CP 2012. LNCS, vol. 7514, pp. 141–157. Springer, Heidelberg (2012)
4. Bessière, C., Coletta, R., Freuder, E.C., O'Sullivan, B.: Leveraging the learning power of examples in automated constraint acquisition. In: Wallace, M. (ed.) CP 2004. LNCS, vol. 3258, pp. 123–137. Springer, Heidelberg (2004)
5. Bessiere, C., Coletta, R., Koriche, F., O'Sullivan, B.: Acquiring constraint networks using a SAT-based version space algorithm. In: 21st Conference on Artificial Intelligence (AAAI), pp. 1565–1568 (2006)
6. Birattari, M., Stützle, T., Paquete, L., Varrentrapp, K.: A racing algorithm for configuring metaheuristics. In: The Genetic and Evolutionary Computation Conference (GECCO), vol. 2, pp. 11–18 (2002)
7. Charnley, J., Colton, S., Miguel, I.: Automatic generation of implied constraints. In: 17th European Conference on Artificial Intelligence (ECAI), pp. 73–77 (2006)
8. Coletta, R., Bessière, C., O'Sullivan, B., Freuder, E.C., O'Connell, S., Quinqueton, J.: Semi-automatic modeling by constraint acquisition. In: Rossi, F. (ed.) CP 2003. LNCS, vol. 2833, pp. 812–816. Springer, Heidelberg (2003)
9. Flener, P., Pearson, J., Ågren, M.: Introducing ESRA, a relational language for modelling combinatorial problems. In: Rossi, F. (ed.) CP 2003. LNCS, vol. 2833, pp. 971–971. Springer, Heidelberg (2003)
10. Frisch, A.M., Jefferson, C., Hernandez, B.M., Miguel, I.: The rules of constraint modelling. In: 19th International Joint Conference on Artificial Intelligence (IJCAI), pp. 109–116 (2005)
11. Frisch, A.M., Harvey, W., Jefferson, C., Martínez-Hernández, B., Miguel, I.: Essence: A constraint language for specifying combinatorial problems. Constraints 13(3), 268–306 (2008)
12. Gent, I.P., Jefferson, C., Miguel, I.: Minion: A fast scalable constraint solver. In: 17th European Conference on Artificial Intelligence (ECAI), vol. 141, pp. 98–102 (2006)

13. Gent, I.P., Miguel, I., Rendl, A.: Common subexpression elimination in automated constraint modelling. In: Workshop on Modeling and Solving Problems with Constraints, pp. 24–30 (2008)
14. Hnich, B.: Function variables for constraint programming. AI Communications 16(2), 131–132 (2003)
15. Hutter, F., Hoos, H.H., Leyton-Brown, K.: Sequential model-based optimization for general algorithm configuration. In: Coello, C.A.C. (ed.) LION 2011. LNCS, vol. 6683, pp. 507–523. Springer, Heidelberg (2011)
16. Koninck, L.D., Brand, S., Stuckey, P.J.: Data independent type reduction for Zinc. In: 9th International Workshop on Constraint Modelling and Reformulation (2010)
17. Lallouet, A., Lopez, M., Martin, L., Vrain, C.: On learning constraint problems. In: 22nd IEEE International Conference on Tools with Artificial Intelligence (ICTAI), vol. 1, pp. 45–52 (2010)
18. Little, J.J., Gebruers, C., Bridge, D.G., Freuder, E.C.: Using case-based reasoning to write constraint programs. In: Rossi, F. (ed.) CP 2003. LNCS, vol. 2833, p. 983. Springer, Heidelberg (2003)
19. Marriott, K., Nethercote, N., Rafeh, R., Stuckey, P.J., de la Banda, M.G., Wallace, M.: The design of the Zinc modelling language. Constraints 13(3), 229–267 (2008)
20. Mills, P., Tsang, E., Williams, R., Ford, J., Borrett, J.: EaCL 1.5: An easy abstract constraint optimisation programming language. Tech. rep., University of Essex, Colchester, UK (December 1999)
21. Nethercote, N., Stuckey, P.J., Becket, R., Brand, S., Duck, G.J., Tack, G.R.: Minizinc: Towards a standard CP modelling language. In: Bessière, C. (ed.) CP 2007. LNCS, vol. 4741, pp. 529–543. Springer, Heidelberg (2007)
22. Nightingale, P., Akgün, Ö., Gent, I.P., Jefferson, C., Miguel, I.: Automatically improving constraint models in savile row through associative-commutative common subexpression elimination. In: O'Sullivan, B. (ed.) CP 2014. LNCS, vol. 8656, pp. 590–605. Springer, Heidelberg (2014)
23. Rendl, A.: Effective Compilation of Constraint Models. Ph.D. thesis, University of St Andrews (2010)
24. Van Hentenryck, P.: The OPL Optimization Programming Language. MIT Press, Cambridge (1999)

Aggregating CP-nets with Unfeasible Outcomes

Umberto Grandi[1], Hang Luo[2,3], Nicolas Maudet[2], and Francesca Rossi[1]

[1] University of Padova, Italy
umberto.uni@gmail.com, frossi@math.unipd.it
[2] Université Pierre et Marie Curie, France
{nicolas.maudet,hang.luo}@lip6.fr
[3] Tsinghua University, China

Abstract. We consider settings where a collection of agents express preferences over a set of candidates with a combinatorial structure via the use of CP-nets, and we need to exploit the information contained in the CP-nets to choose one of the candidates. Moreover, there is a set of constraints which defines the unfeasible candidates, which cannot be the result of the preference aggregation. We propose a method to achieve this which is based on voting, and considers one variable at a time in a sequence. This method has been studied in the literature to aggregate non-constrained CP-nets. Here we generalise it to work with constrained CP-nets, and we study its properties. The constraints are used to leave in the variable domains only the admissible values. This allows the voting steps to return only feasible values. We find conditions of coherence between the preference expressed in the CP nets and the constraints, in order to guarantee that the classical sequential aggregation method always returns a feasible candidate. Even when such conditions are not met, but the constraints defining the unfeasible candidates have a tree structure (or a structure with bounded tree-width), and the collection of CP-nets is O-legal (that is, the dependency graphs of the CP-nets are compatible), we show that our more general voting procedure can be used, and that it is polynomial in the number of features describing the candidates and in the number of voters.

1 Introduction

Preferences are ubiquitous in real life. Often we need to express our preferences over a large set of objects, which has a combinatorial structure and can be described as the Cartesian product of the domains of a set of decision variables. Consider for example a situation where we give our opinion about cars. A car may be described by several features, like the model, the colour, the engine, the shape, and the maker. Each feature may have several possible choices. We may have various models, colours, engine types, shapes, and makers. If the features are modelled by variables and the choices as variable domains, formally a car can be modelled by an assignment of values to such variables.

However, it may be that not all cars are available in the market. For example, while there are red cars and diesel cars, there could be no red diesel car in the current list of available cars. So we are expressing our preferences over all possible cars, but some of them are actually not available.

Often we take group decisions, together with other individuals. For example, a husband and wife could go and look for a car together, each having his/her own preferences

B. O'Sullivan (Ed.): CP 2014, LNCS 8656, pp. 366–381, 2014.

over all cars. To decide on what car to buy, they need to consider the preferences of both and also the possible feasibility constraints, in order to select a car that is on the market and that satisfies their preferences as well as possible.

In this paper we are interested in modeling and studying these scenarios. To model individual's preferences, we use CP-nets [2], a qualitative formalism to express conditional preferential statements. In the car example, using CP-nets we can say that, if the maker is Citroen, we prefer a gasoline engine to a diesel engine. In particular, we study acyclic CP-nets, which have an acyclic dependency structure, since we believe they are expressive enough to model most rational preference orderings. In acyclic CP-nets, it is computationally easy to determine the most preferred object.

We then use constraints to model the feasibility restrictions. Thus we pair CP-nets with a set of constraints, which define the feasible objects, while the CP-net discriminate among the feasible objects via the preferences. Constrained CP-nets have been studied in the literature, and procedures to obtain a feasible undominated outcome have been proposed [3].

A set of constraints [8] compactly models a set of feasible variable assignments. Each constraint usually involves few variables, and the feasible objects are those that satisfy them all. While the task of finding a feasible variable assignment in a generic constraint set is NP-hard, there are classes of constraint problems where this task is computationally easy. One of these classes consists of those constraint sets whose graph (where nodes model variables and arcs model constraints) is a tree, or a graph with bounded tree-width.

We then consider collections of several acyclic CP-nets, as many as the individuals, plus a set of constraints defining the feasible objects. To aggregate such preferences and make sure we return a feasible object, we use voting rules [1] and we define a sequential voting procedure that considers one variable at a time and uses one of the voting rules to decide the "winning value" for that variable. Sequential voting has been already defined for CP-nets with no feasibility constraints [5]. We show that the classical sequential voting procedure can be used also in this more general setting when CP-nets and constraints are coherent according to three consistency notions that we introduce. In these cases, this procedure returns a feasible outcome. When consistency does not hold, we define a generalization of the classical sequential voting procedure which takes the constraints into account and still returns a feasible outcome. This is computed in time polynomial in the size of the profile, if all the used voting rules are polynomial for winner determination and if the constraints are tractable. This means that the presence of feasibility does not make sequential preference aggregation over CP-nets more difficult, provided that the constraints belong to a tractable class.

2 Background

2.1 CP-nets

CP-nets [2] are a graphical model for compactly representing conditional and qualitative preference relations. CP-nets are sets of *ceteris paribus (cp)* preference statements. For instance, the statement *"I prefer red wine to white wine if meat is served."* asserts that, given two meals that differ *only* in the kind of wine served *and* both containing meat, the meal with red wine is preferable to the meal with white wine.

A CP-net has a set of features, modelled by variables $F = \{x_1, \ldots, x_m\}$, with finite domains $\mathcal{D}(x_1), \ldots, \mathcal{D}(x_m)$. For each feature x_i, it is given a set of *parent* features $Pa(x_i)$ that can affect the preferences over the values of x_i. This defines a *dependency graph* in which each node x_i has $Pa(x_i)$ as its immediate predecessors. Given this structural information, the agent explicitly specifies her preference over the values of x_i for *each complete assignment* on $Pa(x_i)$. This preference is assumed to take the form of total or partial order over $\mathcal{D}(x_i)$. An *acyclic* CP-net is one in which this dependency graph is acyclic. A *separable* CP-net is one in which there is no preferential dependency (that is, the dependency graph has no edges).

Consider a CP-net whose features are A, B, C, and D, with binary domains containing f and $\overline{f}$ if F is the name of the feature, and with preference statements as follows: $a \succ \overline{a}$, $b \succ \overline{b}$, $(a \wedge b) \vee (\overline{a} \wedge \overline{b}) : c \succ \overline{c}$, $(a \wedge \overline{b}) \vee (\overline{a} \wedge b) : \overline{c} \succ c$, $c : d \succ \overline{d}$, $\overline{c} : \overline{d} \succ d$. Here, statement $a \succ \overline{a}$ represents the unconditional preference for $A = a$ over $A = \overline{a}$, while statement $c : d \succ \overline{d}$ states that $D = d$ is preferred to $D = \overline{d}$, given that $C = c$.

A *worsening flip* is a change in the value of a variable to a value which is less preferred by the CP-statement for that variable. For example, in the CP-net above, passing from $abcd$ to $ab\overline{c}d$ is a worsening flip since c is better than $\overline{c}$ given a and b. One outcome α is *better* than another outcome β (written $\alpha \succ \beta$) iff there is a chain of worsening flips from α to β. This definition induces a preorder over the outcomes.

In general, finding the optimal outcome of a CP-net is NP-hard. However, in acyclic CP-nets, there is only one optimal outcome and this can be found in linear time. We simply sweep through the CP-net, following the arrows in the dependency graph and assigning at each step the most preferred value in the preference table. For instance, in the CP-net above, we would choose $A = a$ and $B = b$, then $C = c$ and then $D = d$. The optimal outcome is therefore $abcd$. Determining if one outcome is better than another (a dominance query) is NP-hard even for acyclic CP-nets. Whilst tractable special cases exist, there are also acyclic CP-nets in which there are exponentially long chains of worsening flips between two outcomes. In the CP-net of the example, $\overline{a}b\overline{c}\overline{d}$ is worse than $abcd$.

2.2 Constraints

In this paper we refer to the usual notation and terminology for constraint satisfaction problems (CSPs), that can be found for example in [8]. In a constraint problem (CSP) (V, D, C), where $V = (v_1, \ldots, v_n)$ is a set of variables, with domains $D = (D_1, \ldots, D_n)$, and C is a set of constraints, a solution is an assignment of values to all its variables that satisfies all constraints. In the following, we say that a partial assignment to a subset of the variables in V is *feasible* if it can be extended to a solution.

We will also use classical results about tractability of constraint problems and the relationship between local and global consistency. In particular, we will exploit the fact that CSPs whose constraint graph has a bounded tree-width, such as trees, are tractable. For trees, a standard polynomial algorithm to solve them involves deciding on an ordering over which to instantiate the variables to construct a solution, achieving directional arc-consistency on such an order, and then instantiating the variables with no backtracking.

Another tractability result that we will use in this paper is the fact that, in CSPs with Boolean variables and binary constraints, strong 3-consistency (that is, arc- and path-consistency) is sufficient to assure global consistency. This means that it is polynomial to solve such CSPs. In fact, even if we do not have strong 3- consistency, we can achieve it in polynomial time while remaining in the class of binary constraints with Boolean variables, and then we can apply this result.

We will also use a graph which is obtained by the constraint graph of a CSP by first achieving path-consistency and then taking only those edges that correspond to non-trivial constraints (that is, constraints that are a proper subset of the Cartesian domain of their variables). We will call this the *path-closure* graph of the CSP.

2.3 Constrained CP-nets

A constrained CP-net is just a CP-net N with the addition of a set of constraints C over the same variables. An outcome is feasible if it satisfies all constraints in C. An optimal outcome for a constrained CP-net (N, C) is a feasible outcome which is not dominated by any other feasible outcome in the CP-net preference ordering.

While for acyclic CP-nets finding an optimal outcome is computationally easy, for acyclic constrained CP-nets it is as difficult as solving (possibly several times) the constraint set C. In [3] an algorithm (Search-CP) is defined to find an optimal outcome in a constrained CP-net.

Therefore, when the constraint set is tractable, for example it has a tree structure, then this problem is computationally easy. In this paper we consider constrained CP-nets where the CP-net is acyclic, and for some results we will also exploit the tractability of the constraint set. However, we will not compute the optimal outcome of any single constrained CP-net, but rather use the constraints to make sure we obtain a feasible outcome as the result of the aggregation of several CP-nets.

Even if the literature has focussed on finding optimal outcomes of constrained CP-nets, it is also an interesting question to know whether preferences expressed by the CP-net comply (and to what extent) with the constraints. In Section 3 we discuss several possible ways to define such a compliance.

2.4 Voting Theory

A voting rule allows a set of voters to choose one among a set of candidates. Voters need to submit their vote, that is, their preference ordering over the set of candidates (or part of it), and the voting rule aggregates such votes to yield a final result, usually called the winner. In the classical setting [1], given a set of candidates, a *profile* is a collection of total orderings over the set of candidates, one for each voter. Given a profile, a *voting rule* maps it into a single winning candidate.

Some examples of widely used voting rules are:

- *Plurality*, where each voter states who the preferred candidate is, and the candidate who is preferred by the largest number of voters wins;

- *Borda*, where, given m candidates, each voter gives a ranking of all candidates and the ith ranked candidate scores $m - i$, and the candidate with the greatest sum of scores wins;
- *Approval*, where each voter approves between 1 and $m - 1$ candidates on m total candidates, and the candidate with most approvals wins;
- *Copeland*, where the winner is the candidate that wins the most pairwise competitions against all the other candidates.

When there are ties, a unique winner is chosen according to some tie-breaking rule.

Voting theory has considered many desirable properties of voting rules. Some examples are anonymity, neutrality, consistency, monotonicity, Pareto efficiency, independence of irrelevant alternatives (IIA), non-dictatoriality, strategy proofness, and participation [1]. All the above rules are anonymous, neutral, non-dictatorial, monotone, and Pareto efficient, while only Approval is IIA. All but Copeland are consistent and participative.

2.5 Aggregating CP-nets

There is extensive literature that has considered the task of aggregating preferences modelled via CP-nets [5]. In this setting, n agents express their preferences over a set of candidates with a combinatorial structure: there are m features, and each candidate is an assignment of values to all features. Agents' preferences over the candidates are usually modeled via acyclic CP-nets. Moreover, the dependency graphs of such CP-nets are usually assumed to be compatible with a linear order O over the features: for each voter, the preference over a feature is independent of features following it in O.[1] This implies that the n CP-nets $N_1, \ldots, N_n$ are such that the union of their dependency graphs, that we call $Dep(N_1, \ldots, N_n)$, does not contain cycles. Notice that CP-nets with this property may have different dependency graphs.

Given n agents, m binary features, and a linear ordering O over the features, a profile is thus a collection of n acyclic CP-nets over the m features which are compatible with O.

To aggregate the preferences expressed via CP-nets, what is often done is to employ a sequential approach, with as many steps as the number of features. The algorithm uses as many voting rules as the number of features, say $\langle r_1, \ldots, r_m \rangle$.

Taken the features in the order O, say $\langle x_1, \ldots, x_m \rangle$, the aggregation considers one feature at a time in this order and returns a "winner", that is, a variable assignment for such variables, say $\langle d_1, \ldots, d_n \rangle$. It starts from variable x_1, which is for sure an independent variable, and applies voting rule r_1 to the profile obtained by the preference orderings given by all CP-nets over the domain of x_1, returning a winner value d_1. If we have already considered variables $x_1, \ldots, x_k$, obtaining values $d_1, \ldots, d_k$, we then consider variable x_{k+1} and apply voting rule r_{k+1} to the profile obtained by considering the preference ordering over the domain of x_{k+1} in all CP-nets. If x_{k+1} is a dependent variable, we choose the preference ordering which corresponds to the assignment of the previous variables. Notice that all parents of x_{k+1} must have been assigned already at

[1] This coincides with the notion of O-legality in [5].

this point, because of the way the ordering O is defined. This procedure is sometimes called sequential voting [5], or also Level Aggregation (LA) [7].

Example. Consider three agents, each expressing their preferences over candidates defined by 3 binary features. So we have 3 CP-nets N_1, N_2, and N_3, with features A, B, and C, where each feature X has values x and $\bar{x}$. N_1 contains the preferential statements $a \succ \bar{a}$, $b \succ \bar{b}$, $(a \wedge b) \vee (\bar{a} \wedge \bar{b}) : c \succ \bar{c}$, $(a \wedge \bar{b}) \vee (\bar{a} \wedge b) : \bar{c} \succ c$. We recall that $a \succ \bar{a}$ represents the unconditional preference for $A = a$ over $A = \bar{a}$, while $(a \wedge \bar{b}) \vee (\bar{a} \wedge b) : \bar{c} \succ c$ states that $C = c$ is preferred to $C = \bar{c}$, when $A = a$ and $B = \bar{b}$ and also when $A = \bar{a}$ and $B = b$. Thus, in N_1, A and B are independent variables, while C depends on both A and B. N_2 contains instead the following preferential statements: $a \succ \bar{a}$, $a : b \succ \bar{b}$, $\bar{a} : \bar{b} \succ b$, $b : c \succ \bar{c}$, $\bar{b} : \bar{c} \succ c$. Thus, in N_2, A is an independent variable, while B depends on A and C depends on B. N_3 is defined by: $a \succ \bar{a}$, $b \succ \bar{b}$, $c \succ \bar{c}$. Thus, in N_3, all variables are independent. Figure 1 shows this profile.

Consider now ordering $O = \langle A, B, C \rangle$, and let us apply the sequential voting procedure with voting rule Majority for all three variables. For variable A, we have the preferences $a \succ \bar{a}$ from all agents, thus we select $A = a$. Then, given this choice, we pass on to variable B, getting preferences $b \succ \bar{b}$ from all agents. Thus we choose $B = b$. Notice that, while feature B is independent from A in agent 1 and 3, in agent 2 it depends on A. Thus the preferences on the values of B in such an agent are those corresponding to the value of A chosen in the previous step. Passing on to C, we get preferences $c \succ \bar{c}$ from all agents, thus we choose $C = c$. Thus the sequential procedure chooses the variable assignment $\langle A = a, B = b, C = c \rangle$.

Sequential voting provides an efficient way to determine the winning candidate when preferences are expressed compactly with CP-nets or other preference formalisms.

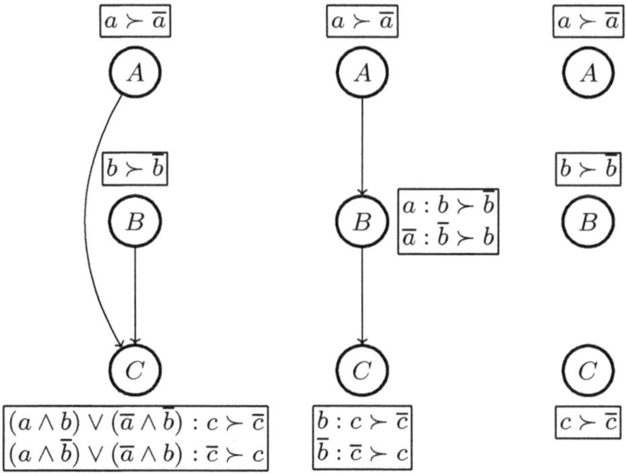

Fig. 1. A profile of CP-nets

Moreover, it maintains many of the desirable properties of the local voting rules used at every step [5].

3 Consistency in Constrained CP-nets

Given a constrained CP-net (N, C), we now study some notions of consistency between the preference structure expressed by N and the set of constraints C. The reason we are interested in these consistency notions is that in some cases they make the aggregation simpler, when preferences are expressed by a collection of constrained CP-nets, as we will see in Section 5.

3.1 Consistency Notions

The first notion of consistency relates the optimal outcome of the CP-net to the constraints.

Definition 1. *A constrained CP-net (N, C) is* top-consistent *if the optimal outcome of N satisfies the constraints in C.*

For example, the CP-net of agent 1 in Figure 1 is top-consistent with the set of constraints $\{A = B\}$.

The next notion of consistency acts at the variable level and makes sure that feasibility is maintained when passing from the parents of the variable to its most preferred value.

Definition 2. *A constrained CP-net (N, C) is* locally-consistent *if there is no line in the CP-tables of N of the form $o : b > \bar{b}$ such that o is feasible but ob is not.*

Since o and ob can be partial outcomes, that is, assigning values to only some of the variables, we recall that a partial outcome is feasible if it can be extended to a solution.

The third notion of consistency we define is a structural property, that related the dependency graph of the CP-net to the path-closure graph of the constraints.

Definition 3. *A constrained CP-net (N, C) is* dependency-consistent *if the path-closure graph of C is a subset of the undirected version of the dependency graph of N.*

Dependency consistency can be natural in several settings. For example, if constraints are known in advance, the process of specifying a CP-net will exploit preferential dependencies among variables connected by a constraint to express qualitative preferences over the partial outcomes over such variables.

Theorem 1. *If a constrained CP-net is both locally and dependency-consistent, then it is also top-consistent.*

Proof. If we have both local and dependency consistency, the optimal outcome is feasible (that is, we have top consistency). In fact, let us compute the optimal outcome by instantiating one variable at a time, in an order which is compatible with the dependency graph of the CP-net (that is, parents come before their children). We start from the independent variables and we give them their most preferred value. This is a feasible partial assignment since, by dependency consistency, there are no constraints among independent variables. At any step, we instantiate a new variable to its most preferred value given the chosen instantiation of its parents. If the partial assignment before this step was feasible, also the new partial assignment is feasible because of local and dependency consistency. Thus all partial assignments built during the procedure, included the last one which is the optimal outcome, are all feasible. A feasible complete assignment is, by definition, a solution. □

It is easy to see that neither local nor dependency consistency alone imply top consistency, and viceversa.

Example. Consider the CP-nets in Figure 1 and the constraints $c_{AB} = \{(A = a, B = b), (A = \overline{a}, B = \overline{b})\}$ *and* $c_{BC} = \{B = b, C = \overline{c}), (B = \overline{b}, C = c)\}$. *None of the CP-nets are top consistent. Moreover,* N_1 *is not locally consistent because of the CP-table for feature* C*:* ab *is a partially feasible assignment but* abc *is not.* N_2 *is not locally-consistent either, again because of the CP-table for* C*:* b *is feasible but* bc *is not. On the other hand,* n_3 *is locally consistent. Only* N_2 *is dependency consistent.*

3.2 Checking the Consistency Notions

We now study the computational complexity of checking the above three notions of consistency in a constrained CP-net. In what follows we assume CP-nets to be acyclic.

Theorem 2. *Given a constrained CP-net* (N, C)*, it is polynomial to check whether it is top-consistent or dependency consistent.*

Proof. For top consistency, it is sufficient to find the optimal outcome of N and check whether it satisfies the constraints in C. Since N is acyclic, this is computationally easy.

For dependency consistency, we just need to compare the dependency graph and the path-closure graph of the constraints. Once we have the two graphs, this is linear in their size. The path-closure graph can be obtained by achieving 3-consistency on the constraints, which is polynomial. □

Theorem 3. *Given a constrained CP-net* (N, C) *with Boolean variables, with* C *a set of binary constraints, it is polynomial to check whether it is locally consistent.*

Proof. We need to check that ob is feasible, for each row in the CP-tables of the form o : $b > \overline{b}$ such that o is feasible. Since constraints are binary, for each row in a CP-table, this can be done in polynomial time. In fact, checking that a partial outcome is feasible with a set of binary constraints is computationally easy if variables are Boolean (it amounts to solving a 2SAT problem). The number of rows in a CP-net may be exponential in the number of issues, but not in the size of the CP-net which is given in the input. Thus the overall complexity is polynomial. □

Observe that local consistency in general cannot be checked in polynomial time if constraints are not binary, even if variables are Boolean, since it would require solving a SAT problem, while in the binary case it is 2-SAT.

3.3 Achieving Top and Local Consistency in Constrained CP-nets

Assume that (N, C) is a constrained CP-net which is not top-consistent or not locally-consistent. This can happen in scenarios in which we have our own preferences over the outcomes expressed via a CP-net, and somebody gives us the constraints describing the feasible outcomes, and the two things together do not have the desired notion of consistency. We would like to modify our CP-net as little as possible in order to obtain either top consistency or local consistency.

Top consistency. To achieve top consistency, we may adopt the following procedure. Let us start from any independent variable (there must be one since N is acyclic) and have one step for each variable, in an order which is compatible with the dependency graph of the CP-net (parents come before their children), computing the optimal outcome. If at any step j, the partial outcome o obtained so far is not feasible, then we modify the row of the CP-table of variable x_j corresponding to the parents' assignment in o doing a switching of the ordering. This assures that the new partial outcome is feasible. This algorithm will produce in polynomial time a CP-net which is top-consistent if the constraints are binary. However, it does not assure that the resulting CP-net is minimally distant from the given one, if the distance is the number of different orderings in the CP-tables. However we conjecture it would be computationally difficult to find the one which is minimally distant.

Local consistency. Instead, to obtain a CP-net which is locally-consistent, it is sufficient to check each row in the CP-tables for the condition of local consistency, again following an order of the variables which is compatible with the dependency graph. If one of the rows fails the consistency check, then the preference expressed in this row needs to be inverted. Since we are moving forward following the dependency structure of N, we are guaranteed that one of the two possible orders in a row of the CP-table must be consistent. Notice that this algorithm is different from the previous one since we need to check all rows of the CP-tables and not just those involved in the computation of the optimal outcome. Again, the assumption of binary constraints is crucial for this algorithm to be polynomial. Unlike the previous one, this algorithm guarantees that the resulting CP-net is minimally distant from the given one, if the distance is the number of different orderings in the CP-tables.

4 Constrained Profiles

A constrained profile models the scenario in which we have several individuals who express their preferences over a common set of outcomes by using CP-nets, and the constraints model the set of feasible outcomes. Only those outcomes that satisfy all constraints can be returned as the result of the aggregation of the preferences of the individuals.

Formally, a *constrained profile* is a collection of CP-nets $\{N_1, \dots, N_n\}$ plus a set of constraints C. This can also be seen as a collection of constrained CP-nets $\{(N_1, C), \dots, (N_n, C)\}$, all having the same constraints.

Notice that all CP-nets share the same set of feasible (and thus unfeasible) candidates, which are those defined by C. Moreover, the CP-nets of all agents share also the variables and the variable domains. So, what can be different in two agents is the dependency graph of their CP-nets, as well as CP-tables of the CP-nets.

In this paper, we restrict our attention to constrained profiles which are O-legal, that is, there is an ordering O of the variables such that, for each CP-net, the preference over a feature is independent of features following it in O.

Notice that O-legality implies that all CP-nets in the constrained profile are acyclic.

Example. As an example of a constrained profile, let us consider the CP-nets in Figure 1, with the addition of the set of constraints $c_{AB} = \{(A = a, B = b), (A = \overline{a}, B = \overline{b})\}$, $C_{BC} = \{B = b, C = \overline{c}), (B = \overline{b}, C = c)\}$. *It is easy to see that this profile is O-legal: there is an ordering O of the variables which is compatible with all dependency links, namely* $O = (A, B, C)$. *Observe that the top outcome is* abc *for all three agents, and this would be the result of sequential majority over the three CP-nets. However, this outcome is not feasible (only* $ab\overline{c}$ *and* $\overline{a}\overline{b}c$ *are).*

5 Aggregating Preferences in Constrained Profiles

The goal is to take a constrained profile and return a feasible outcome, which should satisfy the preferences of the individual CP-nets as much as possible. As we know, when we have no constraints on the feasible outcomes, sequential voting is used to perform such an aggregation. We will now see that sometimes sequential voting is all we need also in presence of constraints. In general, however, we need to take constraints into account. This can be done by adapting the sequential voting procedure, while maintaining a polynomial time complexity if constraints are tractable.

5.1 Top, Local, and Dependency Consistency

Under assumptions the consistency notions introduced in Section 3, sequential aggregation using the majority rule outputs a feasible outcome. The first result applies when we have top consistency, but requires CP-nets to be separable, that is, to have no dependency structure.

Theorem 4. *If* $\langle (N_1, \ldots, N_n), C\rangle$ *is a constrained profile such that all* N_i *are top-consistent and separable, and* C *is a set of binary constraints, then the winner determined by sequential voting with the majority rule is feasible.*

Proof. Since all N_i are separable (and variables are binary), then the order followed by sequential majority is irrelevant, and the problem is equivalent to binary aggregation in which all individuals submit their top outcome and issue-by-issue majority voting is used. We can therefore use the following result from the binary aggregation literature: issue-by-issue majority outputs a feasible outcome given feasible input (that is, it is collectively rational) if and only if the constraints are equivalent to a conjunction of disjunctions of size 2 [4,6]. First we observe that, by top consistency, each individual top outcome satisfies the constraints. Second, since we assume constraints in C to be binary, each constraint can be written as a conjunction of disjunctions of size 2, thus the whole set of constraints can also be written in this way. Therefore this result applies here. If the constraints are not binary, it is possible to find examples in which the outcome of sequential majority is not feasible. □

When the CP-nets have a non-empty dependency structure, we can still apply standard sequential voting to get a feasible outcome if they are both locally and dependency consistent (and thus also top consistent). So we need a stronger property on the CP-nets when we have preferential dependencies.

Theorem 5. *If $\langle (N_1, \ldots, N_n), C \rangle$ is a constrained profile such that all N_i are locally-consistent and dependency-consistent, and C is a set of binary constraints, then the winner determined by sequential voting with the majority rule is feasible.*

Proof. We will prove by induction on the number of variables that, at each step i between 1 and m, the partial assignment generated until step i is feasible. For step 1, it is trivially true since the CP-nets are locally consistent, so the most preferred value in an independent variable must be feasible. This means that all CP-nets vote for a feasible value for the first variable, and thus majority chooses a feasible value.

Let us assume that the statement is true until step i, and let us consider step $i + 1$. We have a feasible partial assignment $\langle v_1, \ldots, v_i \rangle$ obtained so far. For variable $i + 1$, assume that there is a majority in favor of b, i.e., at least a majority of the individual CP-nets prefer b to $\bar{b}$ given the partial assignment obtained so far. This means that if individual j is part of this majority, then N_j contains the row $o_j : b > \bar{b}$, where o_j is the assignment of the parent variables of variable i in CP-net N_j which occurs in the current feasible assignment. By dependency consistency we know that the parent variables of variable i in each individual CP-net include all variables k that are related with i by a constraint. By local consistency, we also know that $o_j b$ is feasible for each j between 1 and n. Thus also $\langle v_1, \ldots, v_i, b \rangle$ is feasible.

Therefore all partial assignments generated during the sequential voting procedure are feasible, including the last one, which is a complete assignment and thus a solution of all the constraints. □

5.2 Aggregation in Non-consistent Profiles

When none of the sufficient conditions mentioned above hold, we can obtain a feasible outcome by modifying the sequential voting procedure to take the constraints into account. Starting from the LA procedure already defined in the literature to aggregate CP-nets, we define the procedure CLA, for Constrained LA procedure. CLA is very similar to LA, except that it will work on possibly reduced variable domains, because of the constraints. As each step, the constraints will tell us what domain values to consider, in order to get a feasible outcome.

The first thing we need to do is to preprocess the constraints in C so to bring to the variable domains the information about the feasible candidates. In fact, since LA is a sequential voting procedure which considers one variable at a time, it is important to leave in the domain of each variable only the values that belong to feasible candidates.

Algorithm 1. CLA

Input: A constrained profile $\langle (N_1, \ldots, N_m), C \rangle$, n voting rules $r_1, \ldots, r_n$, an ordering $O = \langle x_1, \ldots, x_n \rangle$
Output: a variable assignment $\langle x_1 = v_1, \ldots, x_n = v_n \rangle$
for $i = 1$ *to* n **do**
  T_i = the constraint graph of C (a tree), rooted at x_i;
  C' = DAC(T_i);
  D_i = the domain of x_i in C';
  if $D_i = \emptyset$ **then**
    return *No feasible candidate*
  for $j = 1$ *to* m **do**
    o_j = the ordering over D_i given by the CP-table in N_j for $x_1 = v_1, \ldots, x_{i-1} = v_{i-1}$;
    $o'_j = o_j$ restricted to D'_i;
  $v_i = r_i(o'_1, \ldots, o'_n)$;
  Add the constraint $x_i = v_i$ to C;
return $\langle x_1 = v_1, \ldots, x_n = v_n \rangle$

As in the classical sequential voting procedure, we have a collection of m voting rules $\langle r_1, \ldots, r_m \rangle$ that will be used in the m steps of the procedure, one step for each variable. If variables are Boolean, of course all r_i will be the majority voting rule. Assume for now that the constraint set has a bounded tree-width, so it belongs to a tractable class. For sake of easiness of presentation, let us consider a tree- like shape. However, the CLA procedure works also for bounded tree-width constraint sets.

Since the constraints have a tree shape, it is indeed possible to leave in the domain of each variable only those values that appear in some feasible candidate. We just need to consider the variable ordering O, take the first variable x_1, use it as the root of the tree, and achieve directional arc-consistency to this tree. At the end, the new domain of x_1, say D'_1, will contain only those values that appear in some feasible candidate. We can now apply the voting rule r_1 to the profile over variable x_1, where however the domain of x_1 has been reduced to D'_1. This will choose a value for x_1, say v_1, which is feasible (that is, it can be extended to a solution).

Let us now pass to the second variable x_2. Given the value v_1 chosen for x_1, we set $x_1 = v_1$ in C and in all the CP-nets and we apply again DAC bottom-up, now by using x_2 as the root of the tree. This will generate a new domain for x_2, say D'_2, which will contain only those values that appear in some feasible candidate. We can now apply the voting rule r_2 to the profile over variable x_2 (given $x_1 = v_1$), where however the domain of x_2 has been reduced to D'_2. This will choose a value for x_2, say v_2. Now we have the partial assignment $\langle x_1 = v_1, x_2 = v_2 \rangle$, which is feasible.

We then continue like this until all variables have been assigned. The winning candidate is then $\{x_1 = v_1, \ldots, x_n = v_n\}$.

Since C is a tree, the first application of DAC will tell us if there are feasible candidates. If no variable domain is empty after the first DAC, then we know there is at least one feasible candidate, and the later applications of the DAC procedure will never generate any empty variable domain.

Example. *As an example of the application of the CLA algorithm to a constrained profile, consider again the constrained profile in Figure 1, with ordering* $O = \{A, B, C\}$ *and constraints* $c_{AB} = \{(A = a, B = b), (A = \overline{a}, B = \overline{b})\}$, $c_{BC} = \{B = b, C = \overline{c}), (B = \overline{b}, C = c)\}$. *Since we do not have top consistency, nor local and dependency consistency, for all CP-nets, we cannot use classical sequential voting to select a feasible outcome. We will thus use the CLA procedure. Since the variables are binary, we use majority voting at each step. CLA first achieves DAC to the constraint set, which is a tree, rooted at A. This removes the value* $\overline{b}$ *from the domain of* B, *because of the constraint* c_{BC}, *and it also removes the value* $\overline{a}$ *from the domain of A, because of constraint* c_{AB}. *We now apply majority voting to the profile related to variable A, getting* $A = a$. *We then add the constraint* $A = a$ *to the initial set of constraints and we pass on to the second variable, B. We achieve DAC to the tree rooted at B, which does not remove anything from any domain. We apply majority to the profile for B, getting* $B = b$, *and we add this as a new constraint. Finally, we achieve DAC on the tree rooted at C, leaving only* $\overline{c}$ *in the domain of C, and, by majority voting, we get* $C = \overline{c}$. *Thus the result of CLA is* $(A = a, B = b, C = \overline{c})$. *This variable assignment satisfies all constraints. Notice that the outcome of a sequential voting procedure over the same profiles, without the constraints, would be* $(A = a, B = b, C = c)$, *which does not satisfy the constraints.*

On the other hand, if C is not tree-shaped, achieving DAC could leave in the variable domains also values that do not appear in any feasible candidate. Thus, once a value for a variable is chosen, it could be that there is no value for the next variable which is compatible with it. This means that the CLA procedure should backtrack its previous choices (for example the last one made) and replace it with another value. It could also be the case that no feasible candidate exists, and this will result in backtracking over the choices until no more alternative choice is available. Thus the CLA procedure needs to perform search if achieving DAC (or adaptive consistency) does not leave the domains *minimal*, that is, containing only the values that participate in at least a solution.

5.3 Properties of CLA

The most important property to prove is that, in the setting we are considering, CLA always returns feasible outcomes, in time polynomial in the size of the input. We recall that our setting assumes that we have a constrained profile $\langle (N_1, \ldots, N_n), C \rangle$, where C has a tree-like constraint graph, m voting rules $r_1, \ldots, r_m$, and an ordering $O = \langle x_1, \ldots, x_m \rangle$ which makes the profile O-legal.

Theorem 6. *The variable assignment* $\langle x_1 = v_1, \ldots, x_m = v_m \rangle$ *returned by CLA satisfies all constraints in* C.

Proof: Consider the output of CLA, say $\langle v_1, \ldots, v_m \rangle$. Take any constraint in C, say c, between variables x_i and x_j. We need to prove that $\langle x_i = v_i, x_j = v_j \rangle$ satisfies c. At step i, CLA applied DAC to the tree rooted at x_i, restricting the domain of x_i. So, by definition of DAC, v_i is a value for x_i such that there is a value in x_j (and in any other variable) which satisfies c. After doing that, CLA has added the constraint $x_i = v_i$ to

C. Then, at step j, CLA applied DAC again, to the tree rooted at x_j, thereby reducing the domain of x_j to only those values that have support in the domains of all variables, thus also in the domain of x_i, which is now containing just the value x_i. Since v_j is in the domain of x_j, this means that $\langle x_i = v_i, x_j = v_j \rangle$ satisfies c. □

Theorem 7. *CLA works in time $O(n \times (md^2 + n + t))$, where m is the number of variables, d is the size of the largest domain among $D_i, \ldots, D_m$, n is the number of agents, and $t = f(n, d)$ is the time complexity for winner determination in the most computational expensive of the voting rules $r_1, \ldots, r_m$.*

Proof: CLA performs at most m steps. At each step, it achieves DAC on a tree with m variables with domain size at most d. This takes $O(md^2)$ time. It then reduces the n orderings over the current variable domain to the new domains computed by DAC. Finally, it applies the voting rule for that step to such orderings. □

It is worth noting that, in the case of just one voter, we have a single constrained CP-net, and CLA returns a feasible outcome which is undominated in the CP-net preference ordering. This is equivalent to what is done in [3]. However, since we consider tree-shaped constraint sets, we can get this outcome in polynomial time. We therefore get this useful result out of our aggregation procedure. Observe moreover that if we start from acyclic CP-nets, an order O that makes the profile O-legal (or the conclusion that there is no such order) can be found in time polynomial in the number of variables. It is indeed sufficient to take the union of all dependency graphs and take any linearisation of it, if there is one. Any such ordering would give the same result of sequential majority, since preferential dependencies are all taken care of in the union graph.

Theorem 8. *Given a constrained CP-net $\langle N, C \rangle$, where N is acyclic and the constraint graph of C is a tree, finding an undominated feasible outcome is in $\mathcal{P}$.*

Proof: By Theorem 6, the variable assignment returned by CLA is feasible, that is, it satisfies all constraints in C.

We have just one CP-net N and a set of constraints C. Thus, every time we use a voting rule r_i, this voting rule acts as the identity, thus returning the top choice in the preference ordering it gets in input. We will prove by induction on i that, after step i, $\langle x_1 = v_1, \ldots, x_i = v_i \rangle$ is undominated by feasible outcome (in the outcome ordering of N restricted to $x_1, \ldots, x_i$) and it satisfies all constraints in C. It is trivial for step 1, since v_1 is the top choice in the restricted domain of x_1, obtained after applying DAC to the tree C rooted at x_1. Assume the statement is true at step i and let us now consider step $i+1$. CLA returns v_{i+1}, which is the top element of the preference ordering in the restricted domain of x_{i+1}. Since $\langle x_1 = v_1, \ldots, x_i = v_i \rangle$ is an undominated outcome over variables $x_1, \ldots, x_i$, and since x_{i+1} is the top element in the feasible domain of x_{i+1}, there is no other extension of $\langle x_1 = v_1, \ldots, x_i = v_i \rangle$ to variable x_{i+1} which can be more preferred to $\langle x_1 = v_1, \ldots, x_i = v_i, x_{i+1} = v_{i+1} \rangle$. □

Sequential voting with CP-nets has been studied also from the point of view of the properties which can, or cannot, be transferred from the "local" voting rules $r_1, \ldots, r_n$ to the "global" sequential voting rule LA [5]. It is easy to see that, if a property transfers from local to global in the LA procedure, then it also transfers for CLA. In fact, CLA is just LA but on possibly smaller variable domains. So we are performing a domain restriction on the set of possible profiles. If a property is true of a sequential voting procedure when considering a larger set of profiles, it will remain true when we consider a smaller set. This is true for properties like anonymity, consistency, strong monotonicity, and monotonicity (of r_m), as shown in [5].

If instead a property does not transfer from local to global, then by passing from LA to CLA the same examples showing this still hold. Examples of properties that do not transfer from local to global LA are neutrality, efficiency, and participation (see again [5]). However, it could be that the domain restriction imposed by the constraints removes those profiles which are problematic for that property. We plan to study specific constraint classes that could allow some properties to transfer from local to global for CLA, even though they do not do so for LA.

Theorem 9. *If all voting rules $r_1, \ldots, r_m$ are anonymous (resp., consistent, strong monotone), then so is CLA. If r_m is monotone, then so is CLA. If $r_1, \ldots, r_m$ satisfy neutrality (resp., efficiency, participation), it could be that CLA does not satisfy it.*

Proof: It follows directly from the analogous results in [5] for LA. □

6 Conclusions

Often we need to aggregate preferences of several agents over a set of candidates which satisfy certain requirements. Agents may express preferences over a superset of the feasible candidates, with a combinatorial structure, thus being able to use compact preference modeling frameworks such as CP-nets. But the result of preference aggregation can only be a feasible candidate.

We modeled these settings by defining constrained profiles consisting of a collection of CP-nets and a set of constraints, and by proposing a procedure to return a feasible candidate by interleaving constraint solving and voting. If the CP-nets are acyclic and the constraint set has a tree shape, the procedure, called CLA, requires polynomial time if all the voting rules work in polynomial time. Other properties, but not all, transfer from the voting rules to the CLA procedure.

We also identified conditions on the compliance of the CP-nets to the constraints that allow us to forget about the constraints when aggregating the preferences, and thus using standard sequential voting to get a feasible outcome.

We plan to study several other properties of the CLA procedure, among which the complexity of manipulation, bribery, and control. We also plan to test it experimentally on real life profiles from the preflib.org website, where we will add some constraints.

Acknowledgements. The work of Umberto Grandi and Francesca Rossi was partially supported by the strategic project "KIDNEY - Incorporating patients' preferences in kidney transplant decision protocols" funded by the University of Padova.

References

1. Arrow, K.J., Sen, A.K., Suzumura, K.: Handbook of Social Choice and Welfare. North-Holland, Elsevier (2002)
2. Boutilier, C., Brafman, R.I., Domshlak, C., Hoos, H.H., Poole, D.: CP-nets: A tool for representing and reasoning with conditional ceteris paribus preference statements. JAIR 21, 135–191 (2004)
3. Boutilier, C., Brafman, R.I., Domshlak, C., Hoos, H.H., Poole, D.: Preference-based constrained optimization with cp-nets. Computational Intelligence 20, 137–157 (2004)
4. Grandi, U., Endriss, U.: Lifting integrity constraints in binary aggregation. Artificial Intelligence 200, 45–66 (2013)
5. Lang, J., Xia, L.: Sequential composition of voting rules in multi-issue domains. Mathematical Social Sciences 57, 304–324 (2009)
6. List, C., Puppe, C.: Judgment aggregation: A survey. In: Handbook of Rational and Social Choice. Oxford University Press (2009)
7. Maudet, N., Pini, M.S., Rossi, F., Venable, K.B.: Influence and aggregation of preferences over combinatorial domains. In: Proc. AAMAS 2012, pp. 1313–1314 (2012)
8. Rossi, F., Beek, P.V., Walsh, T.: Handbook of Constraint Programming. Elsevier (2006)

The StockingCost Constraint

Vinasétan Ratheil Houndji, Pierre Schaus, Laurence Wolsey, and Yves Deville

Université Catholique de Louvain, Louvain-la-Neuve, Belgium
{vinasetan.houndji,pierre.schaus,laurence.wolsey,
yves.deville}@uclouvain.be

Abstract. Many production planning problems call for the minimization of stocking/storage costs. This paper introduces a new global constraint `StockingCost`$([X_1, \ldots, X_n], [d_1, \ldots, d_n], H, c)$ that holds when each item X_i is produced on or before its due date d_i, the capacity c of the machine is respected, and H is an upper bound on the stocking cost. We propose a linear time algorithm to achieve bound consistency on the `StockingCost` constraint. On a version of the Discrete Lot Sizing Problem, we demonstrate experimentally the pruning and time efficiency of our algorithm compared to other state-of-the-art approaches.

Keywords: Production Planning, Discrete Lot Sizing, Constraint Programming, Global Constraint.

1 Introduction

Production planning problems, such as Lot Sizing and Scheduling Problems, require one to determine a minimum cost production schedule to satisfy the demands for single or multiple items without exceeding machine capacities while satisfying demands. Reviews of those problems and the corresponding Mixed Integer Programming (MIP) formulations are presented in [8,2,5,12]. In many Lot Sizing and Scheduling problems, in particular when the planning horizon is discrete and finite, there are stocking costs to minimize. These costs depend on the time spent between the production of an item and its delivery (due date).

To handle such Lot Sizing Problems in Constraint Programming, we propose an efficient bound consistency filtering algorithm for the `StockingCost`$([X_1, \ldots, X_n], [d_1, \ldots, d_n], H, c)$ constraint that requires each item X_i to be produced on or before its due date d_i and the capacity c of the machine to be respected.

First, we define the `StockingCost` constraint and how one can achieve pruning with the state-of-the-art approaches. After, we present some algorithms to achieve Bound Consistency for the total stocking costs H and for the items $X_i, i \in [1..n]$. Then, we propose a complete $O(n)$ filtering algorithm to achieve Bound Consistency for all variables. Finally, we present some experimental results on a Lot Sizing Problem and conclude.

B. O'Sullivan (Ed.): CP 2014, LNCS 8656, pp. 382–397, 2014.

2 The StockingCost Constraint

The StockingCost constraint has the following form:

$$\texttt{StockingCost}([X_1, \ldots, X_n], [d_1, \ldots, d_n], H, c)$$

where

- the variable X_i is the date of production of item i on the machine,
- the integer d_i is the due-date for item i,
- the integer c is the maximum number of items the machine can produce during one time slot (capacity),
- if an item is produced before its due date, then it must be stocked. The variable H is an upper bound on the total number of slots all the items are need in stock.

The StockingCost constraint holds when each item is produced before its due date ($X_i \leq d_i$), the capacity of the machine is respected (i.e. no more than c variables X_i have the same value), and H is an upper bound on the total stocking cost ($\sum_i (d_i - X_i) \leq H$).

Definition 1. *Each variable has a finite domain. We denote by $X_i^{\min}$ and $H^{\min}$ (resp. $X_i^{\max}$ and $H^{\max}$) the minimal (resp. maximal) value in the domain of variable X_i and H. We also denote $t_{\max} = (\max_i(X_i^{\max}) - \min_i(X_i^{\min}))$.*

Note that `StockingCost` can be viewed as a soft-constraint [13] that would impose every item to be produced exactly at the deadline. The stocking cost variable H is the violation of these deadlines. We use inequality instead of equality (as in van-Hoeve's definition of soft-constraints [13]) because `StockingCost` is an optimization constraint with H typically minimized. Observe that it differs from a standard inequality constraint mainly because H^{max} (the value representing the current best solution) will change during the search for a solution [13]. In particular, it implies that we can filter the domains of variables X_i with respect to $H^{\max}$, and potentially increase $H^{\min}$ with respect to X_i's.

The objective of a filtering algorithm is to remove values that do not participate in any solution of the constraint. In this paper, we are interested in achieving bound-consistency for the `StockingCost` constraint. This consistency level generally offers a good trade-off between speed and filtering power. In a bound consistent constraint, every variable bound (maximum or minimum) occurs in a solution of the constraint. More formally, the bound-consistency definitions for the `StockingCost` constraint are:

Definition 2. *Given a domain $\mathcal{D}$ of variables X_i and H, the constraint $\texttt{StockingCost}([X_1, \ldots, X_n], [d_1, \ldots, d_n], H, c)$ is* bound consistent *with respect to $\mathcal{D}$ iff*

- **BC**$(X_i^{\min})$ $(1 \leq i \leq n)$ *Let $x_i = X_i^{\min}$; there exist $x_j \in [X_j^{\min}..X_j^{\max}]$ $(1 \leq j \leq n, i \neq j)$ and $h = H^{\max}$ such that $\texttt{StockingCost}([x_1, \ldots, x_n], [d_1, \ldots, d_n], h, c)$ holds*

- **BC**(X_i^{max}) ($1 \leq i \leq n$) *Let* $x_i = X_i^{max}$*; there exist* $x_j \in [X_j^{min}..X_j^{max}]$ ($1 \leq j \leq n, i \neq j$) *and* $h = H^{max}$ *such that* *StockingCost*$([x_1, \ldots, x_n], [d_1, \ldots, d_n], h, c)$ *holds*
- **BC**(H^{min}) *Let* $h = H^{min}$*; there exist* $x_i \in [X_i^{min}..X_i^{max}](1 \leq i \leq n)$ *such that* *StockingCost*$([x_1, \ldots, x_n], [d_1, \ldots, d_n], h, c)$ *holds*

Decomposing the Constraint

It is classical to decompose a global constraint into a conjunction of simpler constraints, and applying the filtering algorithms available on the simpler constraints. This raises two questions. First, does the filtering on the decomposition achieves bound consistency? Second, if it achieves the same filtering, what is the complexity of this filtering?

A first decomposition of the constraint StockingCost$([x_1, \ldots, x_n], [d_1, \ldots, d_n], h, c)$ is the following:

$$X_i \leq d_i, \forall i \tag{1}$$

$$\sum_i (X_i = t) \leq c, \forall t \tag{2}$$

$$\sum_i (d_i - X_i) \leq H \tag{3}$$

Assuming that the filtering algorithms for each of the separate constraints achieve bound consistency, the above decomposition does not achieve bound consistency of the StockingCost constraint, as illustrated in the following example.

Example 1. Consider the following instance StockingCost$([X_1 \in [1..2], X_2 \in [1..2]], [d_1 = 2, d_2 = 2], H \in [0..2], c = 1)$. The naive decomposition is not able to increase the lower bound on H because the computation of H gives $(2 - X_1) + (2 - X_2) = [0..1] + [0..1] = [0..2]$. The problem is that it implicitly assumes that both items can be placed at the due date but this is not possible because of the capacity 1 of the machine. The lower bound of H should be set to 1. It corresponds to one item produced in period 1 and the other in period 2.

Other decompositions can be proposed to improve the filtering of the naive decomposition.

A first improvement is to use the global cardinality constraint (gcc) to model the capacity requirement of the machine imposing that no value should occur more than c times. The *gcc* constraint can efficiently replace t_{max} constraints of equation 2 in the basic decomposition. Bound consistency on the gcc constraint can be obtained in $O(n)$ plus the time for sorting the n variables [9]. However, together with equation 3, they do not achieve bound consistency of *StockingCost* constraint.

A second possible improvement is to use a cost based global cardinality constraint (cost-gcc) [10]. In the *cost* − *gcc*, the cost of the arc (X_i, v) is equal to $+\infty$ if $v > d_i$ and $d_i - v$ otherwise. The *cost − gcc* provides more pruning than equations 2 and 3 in the basic decomposition. Enforcing arc-consistency for cost-gcc requires a time complexity of $O(n \cdot S(m, n + d, \gamma))$ to check consistency where n is the number of variables, d is the size of the domains, m is the number arcs and $S(m, n + d, \gamma)$ is the complexity of the search for shortest paths from a node to every node in a graph with m arcs and $n + d$ nodes with a maximal cost γ [10]. For the `StockingCost`, there can be up to $n \cdot t_{\max}$ arcs. Hence[1] the final complexity to obtain arc-consistency[2] on the cost-gcc used to model `StockingCost` can be up to $O(t_{\max}^3)$. To the best of our knowledge the arc-consistent cost-gcc constraint has never been implemented in a solver. Note that for $c = 1$, one can use a minimum assignment constraint with a filtering based on reduced costs [4]. The consistency check for this constraint is achieved in $O(t_{\max}^3)$ (time complexity needed to solve a minimum assignment problem with the Hungarian algorithm). The advantage of the minimum assignment is that a minimum cost assignment can be recomputed in $O((t_{\max})^2)$ for one value removal. It is not possible to clearly characterize the filtering level achieved for the minimum assignment constraint based on reduced-costs.

Without loss of generality, in the rest of paper, we assume that $X_i^{\max} \leq d_i, \forall i$.

3 Pruning the Cost Variable

Given an assignment/solution $\bar{X}$ on variables $X = [X_1, \ldots, X_n]$, we denote by $H(\bar{X})$ the value $\sum_i (d_i - \bar{X}_i)$.

Observation 1. *For two assignments $\bar{X}$ and $\hat{X}$ satisfying $|\{X_i : X_i = t\}| \leq c$, if the sorted sequences of values in these solutions are the same, then $H(\bar{X}) = H(\hat{X})$.*

Let $\mathcal{P}$ denote the problem of computing the optimal lower-bound for H:

$$H^{opt}(\mathcal{P}) = \min \sum_i (d_i - X_i) \; s.t.$$

$$X_i^{\min} \leq X_i \leq X_i^{\max}, \forall i$$

$$|\{X_i : X_i = t\}| \leq c, \forall t$$

Algorithm 1 computes the optimal value $H^{opt}(\mathcal{P})$ in $O(n \cdot log(n))$ and detects infeasibility if the problem not feasible. This algorithm greedily schedules the productions from the latest to the first one. A current time line t is decreased and at each step, all the items such that $X_i^{\max} = t$ are stored into a priority queue (heap) to be scheduled next. Note that each item is added/removed exactly once in the heap and the heap is popped at each iteration (line 11). The items with largest $X_i^{\min}$ must be scheduled first until no more items can be scheduled in time t or the maximum capacity c is reached.

[1] Using Fibonacci heap to implement Dijkstra algorithm for shortest path computation.
[2] Without considering incremental aspects.

Algorithm 1. Filtering of lower bound on H - $\mathbf{BC}(H^{\min})$

```
Input: X = [X_1, ..., X_n] such that X_i ≤ d_i and sorted (X_i^max > X_{i+1}^max)
H^opt ← 0
// total minimum stocking cost
t ← X_1^max
// current time slot
slack ← c
// current slack at this time slot
i ← 1
heap ← {}
// priority queue sorting items in decreasing X_i^min
while i ≤ n do
    while i ≤ n ∧ X_i^max = t do
        heap ← heap ∪ {i}
        i ← i + 1
    while heap.size > 0 do
        // we virtually produce unit j in t
        j ← heap.popFirst
        slack ← slack − 1
        H^opt ← H^opt + (d_j − t)
        if t < X_i^min then
            the constraint is not feasible
        if slack = 0 then
            t ← t − 1
            while i ≤ n ∧ X_i^max = t do
                heap ← heap ∪ {i}
                i ← i + 1
            slack ← c
    if i ≤ n then
        t ← X_i^max
H^min ← max(H^min, H^opt)
```

Let $\mathcal{P}^r$ denote the same problem with relaxed lower bounds of X_i:

$$H^{opt}(\mathcal{P}^r) = \min \sum_i (d_i - X_i) \ s.t.$$

$$X_i \leq X_i^{\max}, \forall i$$
$$|\{X_i : X_i = t\}| \leq c, \forall t$$

Observation 2. *If problem $\mathcal{P}$ is feasible (i.e. the gcc constraint is feasible), then $H^{opt}(\mathcal{P}) = H^{opt}(\mathcal{P}^r)$.*

Proof. If we use a simple queue instead of a priority queue in Algorithm 1, one may virtually assign items to times $t < X_i^{\min}$ and the feasibility test is not valid anymore, but the algorithm terminates with the same ordered sequence of time slots used in the final solution. By Observation 1, the objective values of optimal solutions are the same. The complexity of the algorithm without priority queue is $O(n)$ instead of $O(n \cdot log(n))$. □

The greedy Algorithm 1 is able to compute the best lower bound $H^{opt}(\mathcal{P}^r)$ (in the following we drop problem argument since optimal values are the same) and filters the lower bound of H if possible.

4 Pruning the Item Variable

From now on, since we assume the *gcc* constraint is already bound-consistent and thus feasible, only the cost argument may cause a filtering of lower-bounds $X_i^{\min}$. Therefore, in the rest of the article, we implicitly assumed relaxed domains $[-\infty..X_i^{\max}] \leq d_i$.

Definition 3. *Let $H^{opt}_{X_i \leftarrow v}$ denote the optimal lower bound in a situation where X_i is forced to take the value $v \leq X_i^{\max}$.*

Clearly, v must be removed from the domain of X_i if $H^{opt}_{X_i \leftarrow v} > H^{\max}$. An interesting question is: What is the minimum value v for X_i such that $H^{opt}_{X_i \leftarrow v} = H^{opt}$?

Definition 4. *Let v_i^{opt} denote the minimum value such that $H^{opt}_{X_i \leftarrow v} = H^{opt}$. We have $v_i^{opt} = \min\{v \leq X_i^{max} \; : \; H^{opt}_{X_i \leftarrow v} = H^{opt}\}$.*

The following observation gives a lower bound on the evolution on H^{opt} when a variable X_i is forced to take a value $v < v_i^{opt}$.

Observation 3. *For $v < v_i^{opt}$, we have $H^{opt}_{X_i \leftarrow v} \geq H^{opt} + (v_i^{opt} - v)$*

After the propagation of $H^{\min}$, one may still have some slack between the upper and the lower bound $H^{\max} - H^{\min}$. Since v_i^{opt} is the minimum value such that $H^{opt}_{X_i \leftarrow v} = H^{opt}$, we can use the lower bound of Observation 3 to filter X_i as follows:

$$X_i^{\min} \leftarrow \max(X_i^{\min}, v_i^{opt} - (H^{\max} - H^{\min}))$$

In the following we show that the lower-bound of Observation 3 can be improved and that we can actually predict the exact evolution of $H^{opt}_{X_i \leftarrow v}$ for an arbitrary value $v < v_i^{opt}$. A valuable information to this end is the number of items scheduled at a given time slot t in an optimal solution:

Definition 5. *In an optimal solution $\bar{X}$ (i.e. $H(\bar{X}) = H^{opt}$), let*

$$count[t] = |\{i \; : \; \bar{X}_i = t\}|.$$

Algorithm 2 computes $v_i^{opt}, \forall i$ and $count[t], \forall t$ in linear time $O(t_{\max})$. The first step of the algorithm is to initialize $count[t]$ as the number of variables with upper bound equal to t. This can be done in linear time assuming the time horizon of size $(\max_i\{X_i^{\max}\} - \min_i\{X_i^{\min}\})$ is in $O(n)$. We can initialize an array *count* of the size of the horizon and increment the entry $count[X_i^{\max}]$ of the array in $O(1)$ for each variable X_i.

The idea of the Algorithm 2 is to use a Disjoint-Set T (also called union-find) data structure [1] making it possible to have efficient operations for $T.Union(S_1, S_2)$, grouping two disjoint sets into a same set, and $T.Find(v)$ returning a "representative" of the set containing v. It is easy to extend a disjoint-set data structure with operations $T.min(v)/T.max(v)$ returning the minimum/maximum value of the set containing value v. As detailed in the invariant of the algorithm, time slots are grouped into a set S such that if $X_i^{\max} \in S$ then $v_i^{opt} = \min S$.

Algorithm 2. Compute v_i^{opt} for all i

Initialize *count* as an array such that $count[t] = |\{X_i : X_i^{\max} = t\}|$
Create a disjoint set data structure T with the integers $t \in [\min_i\{X_i^{\min}\}, \max_i\{X_i^{\max}\}]$
$t \leftarrow \max_i\{X_i^{\max}\}$
repeat
    while $count[t] > c$ **do**
        $count[t-1] \leftarrow count[t-1] + count[t] - c$
        $count[t] \leftarrow c$
        $T.Union(t-1, t)$
    `// invariant:` $v_i^{opt} = t, \forall i \in \{i \;:\; t \leq X_i^{\max} \leq T.max(T.find(t))\}$
    $t \leftarrow t - 1$
until $t \leq \min_i\{X_i^{\min}\}$
// if $count[\min_i\{X_i^{\min}] > c$ then the constraint is infeasible
$\forall i : v_i^{opt} = T.min(T.find(X_i^{\max}))$

Example 2. Consider the following instance `StockingCost`($[X_1 \in [1..3], X_2 \in [1..6], X_3 \in [1..7], X_4 \in [1..7], X_5 \in [1..8]], [d_1 = 3, d_2 = 6, d_3 = 7, d_4 = 7, d_5 = 8], H \in [0..4], c = 1$). At the beginning of the algorithm, $count = [0, 0, 1, 0, 0, 1, 2, 1]$ and $T = \{\{1\}, \{2\}, \{3\}, \{4\}, \{5\}, \{6\}, \{7\}, \{8\}\}$. After the loop, $count = [0, 0, 1, 0, 1, 1, 1, 1]$ and $T = \{\{1\}, \{2\}, \{3\}, \{4\}, \{5, 6, 7\}, \{8\}\}$. Thus $v_{X_1}^{opt} = 3$, $v_{X_2}^{opt} = v_{X_3}^{opt} = v_{X_4}^{opt} = 5$ and $v_{X_5}^{opt} = 8$. Next figure shows the different steps of the computation of T. Note that from step 3 to step 8, there is no change since for $t \in [1..6]$, $count[t]$ always ≤ 1 in the loop.

Observation 3 gives a lower bound on the evolution of the optimal stocking cost when assigning variable X_i to v. Unfortunately, this lower bound is not

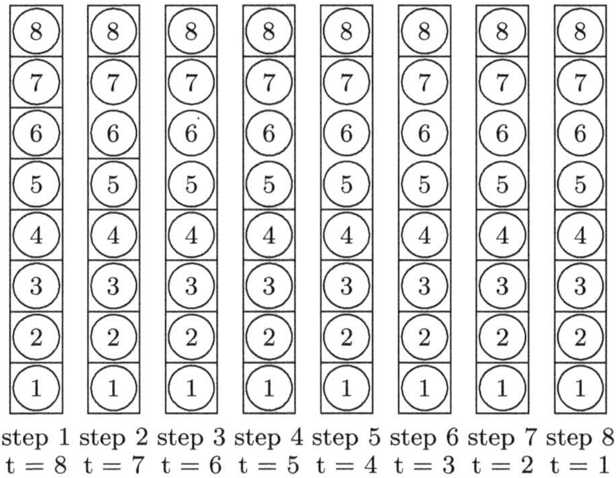

optimal. One can be convinced easily for instance that with $c = 1$, if $v < v_i^{opt}$ is assigned X_i, it virtually imposes to move to left at least all variables X_j such that $\{X_j : X_j^{max} = v\}$. This suggests for c=1, the following improved lower bound for $H_{X_i \leftarrow v}^{opt}$:

$$H_{X_i \leftarrow v}^{opt} \geq H^{opt} + (v_i^{opt} - v) + |\{X_j : X_j^{max} = v\}| \tag{4}$$

Next example illustrates that this lower bound is still not optimal. It is not sufficient just only consider the set $\{X_j : X_j^{max} = v\}$ since more variables could be impacted.

Example 3. Consider the following instance `StockingCost`($[X_1 \in [1..5], X_2 \in [1..4], X_3 \in [1..4]], [d_1 = 5, d_2 = 4, d_3 = 4], H \in [0..10], c = 1$) with $H^{opt} = 1$ and $v_{X_1}^{opt} = 5$. For $v = 4$, $H_{X_1 \leftarrow 4}^{opt} \geq H^{opt} + (v_{X_1}^{opt} - v) + |\{X_j : X_j^{max} = v\}| = 1 + (5 - 4) + 2 = 4$. Here, $H_{X_1 \leftarrow 4}^{opt}$ is really 4. For $v = 3$, $H_{X_1 \leftarrow 3}^{opt} \geq H^{opt} + (v_{X_1}^{opt} - v) + |\{X_j : X_j^{max} = v\}| = 1 + (5 - 3) + 0 = 3$ but here $H_{X_1 \leftarrow 3}^{opt} = 4$.

Definition 6. *A slot t is full if it is using all its capacity $count[t] = c$.*

Observation 4. *There is at most $\lfloor \frac{n}{c} \rfloor$ full time slots.*

Definition 7. *$minfull[t]$ is largest time slot $\leq t$ which is not full. More exactly $minfull[t] = \max\{t' \leq t : count[t'] < c\}$.*

Definition 8. *$maxfull[t]$ is the smallest time slot $\geq t$ which is not full. More exactly $maxfull[t] = \min\{t' \geq t : count[t'] < c\}$.*

Next observation gives the exact evolution of $H_{X_i \leftarrow v}^{opt}$ that will allow the BC filtering of $X_i^{\min}$.

Observation 5. $H_{X_i \leftarrow v}^{opt} = H^{opt} + (v_i^{opt} - v) + (v - minfull[v]), \ \forall v < v_i^{opt}$

To understand the previous observation one can realize that the number of variables affected (that would need to be shifted by one to the left) by assigning $X_i \leftarrow v$ is equivalent to the impact caused by insertion of an artificial item with the domain $[-\infty..v]$. So, the exact impact of $X_i \leftarrow v$ is the number of variables affected by the move plus $(v - v_{X_i}^{opt})$. The Algorithm 3 computes $minfull[t], maxfull[t], \forall t$. The time complexity is thus $O(t_{\max})$.

Algorithm 3. Computation of *minfull*[t],*maxfull*[t] $\forall t$

Create a disjoint set data structure F with the integers $t \in [\min_i\{X_i^{\min}\} - 1, \max_i\{X_i^{\max}\}]$
$t \leftarrow \max_i\{X_i^{\max}\}$
repeat
  if $count[t] = c$ **then**
    $F.Union(t-1, t)$
  $t \leftarrow t - 1$
until $t < \min_i\{X_i^{\min}\}$
$\forall t : minfull(t) = F.min(F.find(t))$
$\forall t : |F.find(t)| > 1 : maxfull(t) = F.max(F.find(t)) + 1$
$\forall t : |F.find(t)| = 1 : maxfull(t) = F.max(F.find(t))$

Observation 6. $H_{X_i \leftarrow t}^{opt} \geq H_{X_i \leftarrow t'}^{opt} \forall t < t' \leq v_i^{opt}$

Observation 7. *If a slot t is full ($count[t] = c$) then $\forall i$:*

$$H_{X_i \leftarrow t}^{opt} = H_{X_i \leftarrow t'}^{opt}, \ \forall t' \in [minfull[t]..maxfull[t]) \text{ such that } t' < v_i^{opt}$$

Proof. Suppose that a slot t is full. We know that $\forall t' \in [minfull[t]..maxfull[t])$, $minfull[t'] = minfull[t]$. Thus, $\forall t' \in [minfull[t]..maxfull[t])$ *such that* $t' < v_i^{opt}$, $H_{X_i \leftarrow t'}^{opt} = H^{opt} + (v_i^{opt} - t') + (t' - minfull[t']) = H^{opt} + v_i^{opt} - minfull[t] = H^{opt} + (v_i^{opt} - t) + (t - minfull[t]) = H_{X_i \leftarrow t}^{opt}$. □

The above observation is very important because if the new minimum for X_i falls on a full time slot, we can increase the lower bound further. The bound consistent filtering rule is given in Algorithm 4.

Algorithm 4. Bound Consistent Filtering of $X_i^{\min}$ - **BC**($X_i^{\min}$)

$newmin \leftarrow v_i^{opt} - (H^{\max} - H^{\min})$
if $count[newmin] = c$ **then**
  $newmin \leftarrow \min\{v_i^{opt}, maxfull[newmin]\}$
$X_i^{\min} \leftarrow \max(X_i^{\min}, newmin))$

Example 4. Considering the following instance StockingCost($[X_1 \in [1..3], X_2 \in [1..6], X_3 \in [1..7], X_4 \in [1..7], X_5 \in [1..8]], [d_1 = 3, d_2 = 6, d_3 = 7, d_4 = 7, d_5 = 8], H \in [0..4], c = 1$). We know that $v^{opt}_{X_1} = 3$, $v^{opt}_{X_2} = v^{opt}_{X_3} = v^{opt}_{X_4} = 5$, $v^{opt}_{X_5} = 8$ and $count = [0, 0, 1, 0, 1, 1, 1, 1]$. After running the algorithm 1 we have $H^{opt} = 2$ and thus $H \in [2..4]$. Algorithm 3 gives $F = \{\{0\}, \{1\}, \{2, 3\}, \{4, 5, 6, 7, 8\}\}$, $minfull = [1, 2, 2, 4, 4, 4, 4, 4]$ and $maxfull = [1, 2, 4, 4, 9, 9, 9, 9]$. Algorithm 4 gives for:
- X_1 : $newmin = 3 - 2 = 1$. $count[1] = 0$ and $X^{min}_1 = \max\{1, 1\} = 1$;
- X_2, X_3, X_4 : $newmin = 5 - 2 = 3$. $count[3] = 1$, $newmin = \min\{4, 5\} = 4$ and $X^{min}_{j\in\{2,3,4\}} = \max\{1, 4\} = 4$. Next figure shows the evolution of $H^{opt}_{X_3 \leftarrow t}$. Note that for $t \in [1..3]$, $H^{opt}_{X_3 \leftarrow t} > H^{max} = 4$.

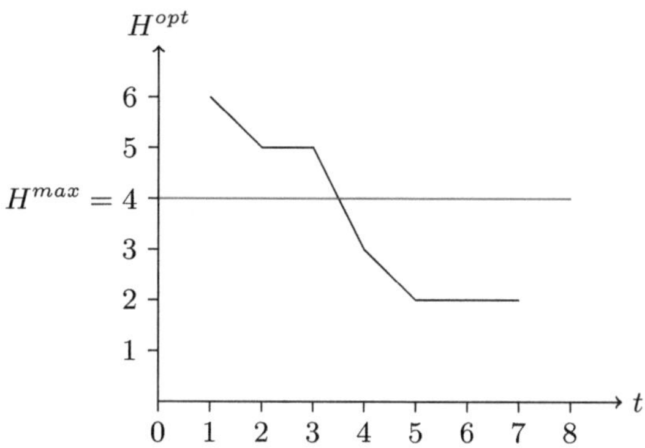

- X_5 : $newmin = 8 - 2 = 6$. $count[6] = 1$, $newmin = \min\{8, 9\} = 8$ and $X^{min}_5 = \max\{1, 8\} = 8$.
Thus $X_1 \in [1..3]$, $X_2 \in [4..6], X_3 \in [4..7], X_4 \in [4..7]$ and $X_5 \in \{8\}$.

5 A Complete Filtering Algorithm in $O(n)$

The Algorithms 2 and 3 for computing $v^{opt}_i, \forall i$ and $maxfull(t), \forall t$ presented so far have a complexity of $O(t_{\max})$. Although for some problems $t_{\max} \approx n$, in practice it can be larger than n if there is some sparsity on the deadlines. Algorithm 5 describes a complete self-contained version of the filtering for StockingCost running in $O(n)$ given a sorted version of the variables. This algorithm keeps tracks of the items in the same set (same v^{opt}) by maintaining two indexes j, k with the following properties:

- After line 10, items in $\{j, \ldots, i\}$ are the *open items* ($X_i : X^{\max}_i \geq t$) that still need to be placed into some slots in an optimal solution.
- After line 10, all the items in $\{k, \ldots, i\}$ have the same v^{opt}. This value v^{opt} is only known when all the current remaining open items can be placed into the current slot. That is when the condition at line 13 is true.

Variable u keeps track of the $maxfull(t)$ potential value with $maxfull(t)$ implemented as a map with constant time insertion. Only time slots t with $maxfull(t) > t$ are added to the map. Each time a full slot t is discovered (at lines 19 and 26), one entry is added to the map. By observation 4 the number of entries added into the map is at most n.

Lines 29 to 34 are just applying the filtering rules from Algorithm 4.

Implementation Details. Although the Algorithm 5 is in $O(n)$, it requires the variables to be sorted. Since the filtering algorithms are called multiple times during the search process and only a few number of variables are modified between each call, simple sorting algorithms such as insertion or bubble sort are generally more efficient than classical sorting algorithms $O(n \cdot log(n))$.

The *map* can be a simple Hashmap but a simple implementation with two arrays of size $t_{\max}$ and a *magic number* incremented at each call can be used to avoid computing hash functions and the map object creation/initialization at each call to the algorithm. One array contains the value for each key index in the map, and the other array contains magic numbers containing the value of the magic number at the insertion. An entry is present only if the value at corresponding index in the magic array is equal to the current magic number. Incrementing the magic number thus amounts at emptying the map in $O(1)$. The cost $O(t_{\max})$ at the map creation has to be paid only once an is thus amortized.

6 Experimental Results

Experiments were conducted on instances $MI - DLS - CC - SC$ (Multi Item - Discrete Lot Sizing - Constant Capacity - Setup Cost) problems described in [8].

Description of the $MI - DLS - CC - SC$ Problem

The Discrete Lot Sizing problem considered here is a multi-item, single machine problem with capacity of production limited to one per period. There are storage costs and sequence-dependent changeover costs, respecting the triangle inequality. Each order consisting of one unit of a particular item has a due date and must be produced at latest by its due date. The stocking (inventory) cost of an order is proportional to the number of periods between the due date and the production period. The changeover cost $q^{i,j}$ is induced when passing from the production of item i to another one j with $q^{i,i} = 0\ \forall i$. Backlogging is not allowed. The objective is to assign a production period for each order respecting its due date and the machine capacity constraint so as to minimize the sum of stocking costs and changeover costs.

Next example shows a tiny instance of the problem.

Example 5. Consider the problem with the following input data: number of items type $nbItems = 2$; number of periods $nbPeriods = 5$; stocking cost $h = 2$; demand times for items of type 1 $d^1_{t\in\{1,\ldots,5\}} = (0, 1, 0, 0, 1)$ and for items of type

Algorithm 5. Complete filtering algorithm in $O(n)$

Input:
$X = [X_1, \dots, X_n, X_{n+1}]$ such that $X_i \leq d_i$ and sorted ($X_i^{\max} > X_{i+1}^{\max}$)
$X_{n+1}^{\max} = -\infty$ `// artificial variable`

$H^{opt} \leftarrow 0$
$t \leftarrow X_1^{\max}$
$i \leftarrow 1$
$j \leftarrow 1$ `// open items` $\{j, \dots, i\}$ `must be placed in some slots`
$k \leftarrow 1$ `// items` $\{k, \dots, i\}$ `have same` v^{opt}
$u \leftarrow t + 1$
$maxfull \leftarrow map()$ `// a map from int to int`
while $i \leq n \vee j < i$ **do**
  while $i \leq n \wedge X_i^{\max} = t$ **do**
    $i \leftarrow i + 1$
  `// place at most` c `items into slot` t
  for $i' \in [j..\min(i-1, j+c-1)]$ **do**
    $H^{opt} \leftarrow H^{opt} + (d_{i'} - t)$
  if $i - j \leq c$ **then** `// all the open items can be placed in` t
    $full \leftarrow i - j = c$ `// true if` t `is fill up completely`
    $v_l^{opt} \leftarrow t, \forall l \in [k..i)$
    $j \leftarrow i$
    $k \leftarrow i$
    if $full$ **then**
      `// invariant` $\forall t' \in [t..u-1], count[t] = c$
      $maxfull(t) \leftarrow u$
      if $X_i^{\max} < t - 1$ **then**
        $u \leftarrow X_i^{\max} + 1$
    else
      $u \leftarrow X_i^{\max} + 1$
    $t \leftarrow X_i^{\max}$
  else `// all open items can not be placed in` t
    `// invariant` $\forall t' \in [t..u-1], count[t] = c$
    $maxfull(t) \leftarrow u$
    $j \leftarrow j + c$ `// place` c `items into slot` t
    $t \leftarrow t - 1$
$H^{\min} \leftarrow \max(H^{\min}, H^{opt})$
for $i \in [1..n]$ **do**
  $newmin \leftarrow v_i^{opt} - (H^{\max} - H^{\min})$
  if $maxfull(t).hasKey(newmin)$ **then**
    $newmin \leftarrow \min\{v_i^{opt}, maxfull(newmin)\}$
  $X_i^{\min} \leftarrow \max(X_i^{\min}, newmin))$

$2\ d^2_{t\in\{1,\ldots,5\}} = (1,0,0,0,1)$; $q^{1,2} = 5$, $q^{2,1} = 3$. A feasible solution of this problem is $productionPlan = (2,1,2,0,1)$ which means that item 2 will be produced in period 1; item 1 in period 2; item 2 in period 3 and item 1 in period 5. Note that there is no production in period 4, it is an idle period. The cost associated to this solution is $q^{2,1}+q^{1,2}+q^{2,1}+2*h = 15$ but it is not the optimal cost. The optimal solution is $productionPlan = (2,1,0,1,2)$ with the cost $q^{2,1}+q^{1,2}+h = 10$.

A Constraint Programming Model

We uniquely identify each order. The aim is to associate to each of these orders a period that respects the due date of the order. Let $date(p) \in [1..nbPeriods]$, $\forall p \in [1..nbDemands]$, represents the period in which the order p is satisfied. This corresponds to period in which the order p is produced/satisfied. Let $dueDate(p)$ be the deadline for order p, that is the period in which p is due.

If $objStorage$ is an upper bound on the total number of periods in which orders have to be held in stock, the stocking part can be modeled by the constraint:

$$\texttt{StockingCost}(date, dueDate, objStorage, 1)$$

Observation 8. *There is no difference between two orders of the same item except for their due dates. Therefore given a feasible production schedule, if it is possible to swap the production periods of two orders involving the same item same item* $(date(p_1), date(p_2)$ *such that* $item(p_1) = item(p_2))$*, we obtain an identical solution with the same stocking cost..*

Based on observation 8, we remove such symmetries by adding precedence constraints on $date$ variables involving by the same item:

$$date(p_1) < date(p_2), \forall (p_1, p_2) \in [1..nbDemands] \times [1..nbDemands]\ such\ that$$

$$dueDate(p_1) < dueDate(p_2) \wedge item(p_1) = item(p_2)$$

Now, the second part of the objective $objChangeover$ concerning changeover costs has to be introduced in the model. This part is similar to a *successor* CP model for the Traveling Salesman Problem (TSP) in which the cities to be visited represent the orders and the distances between them are the corresponding changeover costs. Let $successor(p)$, $\forall p \in [1..nbDemands]$, define the order produced on the machine immediately after producing order p. We additionally create a dummy order $nbDemands + 1$ to be produced after all the other orders. In the first step, a Hamiltonian circuit successor variable is imposed. This is achieved by using the classical *circuit* [7] constraint on successor variables for dynamic subtour filtering. The *date* and *successor* variables are linked with the element constraint by imposing that the production date of p is before the production date of its successors:

$$\forall p \in [1..nbDemands] : date(p) < date(successor(p))$$

As announced, the artificial production is scheduled at the end:

$$date(nbDemands + 1) = nbPeriods + 1$$

Note that as with *date* variables, some symmetries can be broken. For two nodes $n_1, n_2 \in [1..nbDemands]$ such that $dueDate(n_1) < dueDate(n_2)$ and $item(n_1) = item(n_2)$, we force that n_1 cannot be the successor of n_2 with $successor(n_2) \neq n_1$. Finally, a *minAsssignment* constraint [3] is used on the *successor* variables and the changeover part of the objective *objChangeover*.

The objective to minimize is simply the sum of stocking costs and changeover costs : $(objStorage * h) + objChangeover$, where h is the unit stocking cost.

Experimental Results

By assuming that the basic filtering $date(p) \leq dueDate(p)$ $,\forall p \in [1..nbDemands]$ is imposed a priori, we compare the performance of the filtering algorithm due to `StockingCost`$(date, deadline, objStorage, 1)$ constraint with that achieved by the following three sets of constraints:

- the basic decomposition :

$$\sum_p (date(p) = t) \leq 1, \forall t \in [1..nbPeriods]$$

$$\sum_p (dueDate(p) - date(p)) \leq objStorage$$

- t_{max} constraints of the previous decomposition are replaced by the global bound consistency constraint *allDifferentBC* [6], that is the special case of *gcc* constraint when the capacity $c = 1$:

$$allDifferentBC(date)$$

$$\sum_p (dueDate(p) - date(p)) \leq objStorage$$

- the global constraint *minAssignment* [3] on *date* and *objStorage* variables is added to the previous decomposition.

The `StockingCost` filtering algorithm and the $MI - DLS - CC - SC$ model have been implemented in the OscaR open-source solver [11]. They will be available in OscaR from release 1.1.0. As search heuristic, we used a classical static binary search on *date* and *successor* variables in order to reduce the impact of the search on model comparisons. Table 1 shows the results for some randomly generated instances of $MI - DLS - CC - SC$ [3]. We present, for each group of constraints, the number of nodes visited and the time (in seconds) used to complete the search.

[3] Instances available at `http://becool.info.ucl.ac.be/resources/discrete-lot-sizing-problem`

Table 1. Results for 15 MI-DLS-CC-SC instances. The format of instance is the following: *InstanceNumber*(*nbPeriods_nbItems_nbDemands*). "—" means that the model did not complete the search after 3600 seconds.

Instance	StockingCost Nodes	StockingCost Time	Minassignment Nodes	Minassignment Time	AllDifferent Nodes	AllDifferent Time	Basic decomp Nodes	Basic decomp Time
1(15_5_13)	**0.36** 10^6	**26**	0.41 10^6	36	1.25 10^6	87	1.25 10^6	99
2(15_5_14)	**0.98** 10^6	**79**	1.26 10^6	112	3.16 10^6	255	3.17 10^6	283
3(15_8_13)	**1.10** 10^6	**64**	2.34 10^6	156	8.05 10^6	515	8.07 10^6	625
4(15_10_12)	**0.22** 10^6	**12**	0.72 10^6	32	8.02 10^6	385	8.10 10^6	478
5(15_10_14)	**0.32** 10^6	**16**	1.41 10^6	79	18.7 10^6	1350	18.8 10^6	1552
6(20_5_17)	**1.14** 10^6	**135**	1.40 10^6	213	3.33 10^6	353	4.07 10^6	548
7(20_10_18)	**6.90** 10^6	**534**	8.02 10^6	805	9.68 10^6	906	—	—
8(20_10_19)	**1.32** 10^6	**95**	1.34 10^6	120	9.68 10^6	616	—	—
9(30_5_12)	**2.87** 10^6	**124**	3.00 10^6	223	3.00 10^6	127	3.00 10^6	188
10(30_10_11)	**5.51** 10^6	**244**	6.68 10^6	530	7.73 10^6	342	7.73 10^6	494
11(30_10_16)	**2.41** 10^6	**156**	4.64 10^6	439	—	—	—	—
12(100_10_11)	**1.49** 10^6	**60**	1.50 10^6	271	2.30 10^6	110	2.30 10^6	153
13(100_10_18)	**0.11** 10^6	**10**	0.15 10^6	63	0.36 10^6	23	2.96 10^6	331
14(100_15_17)	**2.79** 10^6	**143**	6.51 10^6	1132	22.2 10^6	1305	22.3 10^6	1712
15(200_15_22)	**19.3** 10^6	**854**	—	—	24.6 10^6	1187	24.6 10^6	2024

These results suggest that our `StockingCost` version offers a stronger and faster filtering than other decompositions. In particular, the last four instances suggest that the time complexity of our filtering algorithm scales better than the *minAssignment* decomposition when the number of time slots increases. This is not surprising since filtering algorithm for `StockingCost` is in $O(nbDemands)$ and not a function of the size of horizon as is the case for the *minAssignment* decomposition.

7 Conclusion

In this paper, we have introduced a new global constraint `StockingCost` to handle the stocking aspect of Lot Sizing Problems when using Constraint Programming. We have described an advanced filtering algorithm achieving bound consistency with a time complexity linear in the number of variables. The experimental results show the pruning and time efficiency of the `StockingCost` constraint on a version of the Discrete Lot Sizing Problem compared to various decompositions of the constraint.

References

1. Cormen, T.H., Leiserson, C.E., Rivest, R.L., Stein, C.: Data structures for Disjoint Sets. In: Introduction to Algorithms, ch. 21, 2nd edn., MIT Press, Cambridge (2001)

2. Drexl, A., Kimms, A.: Lot sizing and scheduling - survey and extensions. European Journal of Operational Research, 221–235 (1997)
3. Focacci, F., Lodi, A., Milano, M.: Cost-based domain filtering. In: Jaffar, J. (ed.) CP 1999. LNCS, vol. 1713, pp. 189–203. Springer, Heidelberg (1999)
4. Focacci, F., Lodi, A., Milano, M., Vigo, D.: Solving tsp through the integration of or and cp techniques. Electronic Notes in Discrete Mathematics 1, 13–25 (1999)
5. Jans, R., Degraeve, Z.: Modeling industrial lot sizing problems: A review. International Journal of Production Research (2006)
6. López-Ortiz, A., Quimper, C.-G., Tromp, J., van Beek, P.: A fast and simple algorithm for bounds consistency of the alldifferent constraint. In: International Joint Conference on Artificial Intelligence, IJCAI 2003 (2003)
7. Pesant, G., Gendreau, M., Potvin, J.-Y., Rousseau, J.-M.: An exact constraint logic programming algorithm for the traveling salesman problem with time windows. Transportation Science 32(1), 12–29 (1998)
8. Pochet, Y., Wolsey, L.: Production Planning by Mixed Integer Programming. Springer (2005)
9. Quimper, C.-G., van Beek, P., López-Ortiz, A., Golynski, A., Sadjad, S.B.S.: An efficient bounds consistency algorithm for the global cardinality constraint. In: Rossi, F. (ed.) CP 2003. LNCS, vol. 2833, pp. 600–614. Springer, Heidelberg (2003)
10. Régin, J.-C.: Cost-based arc consistency for global cardinality constraints. Constraints 7(3-4), 387–405 (2002)
11. OscaR Team. Oscar: Scala in or (2014), `https://bitbucket.org/oscarlib/oscar`
12. Ullah, H., Parveen, S.: A literature review on inventory lot sizing problems. Global Journal of Researches in Engineering 10, 21–36 (2010)
13. van Hoeve, W.-J.: Over-constrained problems. In: Hybrid Optimization, pp. 191–225. Springer (2011)

Scalable Parallel Numerical CSP Solver

Daisuke Ishii[1], Kazuki Yoshizoe[1,2] and Toyotaro Suzumura[2,3]

[1] Tokyo Institute of Technology, Tokyo, Japan
[2] Japan Science and Technology Agency, Japan
[3] IBM Research, Dublin, Ireland
{dsksh,yoshizoe,suzumura}@acm.org

Abstract. We present a parallel solver for numerical constraint satisfaction problems (NCSPs) that can scale on a number of cores. Our proposed method runs worker solvers on the available cores and simultaneously the workers cooperate for the search space distribution and balancing. In the experiments, we attained up to 119-fold speedup using 256 cores of a parallel computer.

1 Introduction

Numerical constraint satisfaction problems (NCSPs, Section 2) have been successfully applied to problems described in the domain of reals[3,6]. Given a NCSP with search space represented as a *box* (i.e., interval vector), the *branch and prune algorithm* efficiently computes a *paving*, a set of boxes that encloses the solution set, yet its exponential computational complexity limits the tractable instances. Although the solving process exhibits a parallelism, no parallel NCSP solver has been made available to date because of the difficulty in partitioning the search space equally.

In this research, we parallelize a NCSP solver to scale its solving process on both shared memory and distributed memory parallel computers (Section 3). Our parallel method consists of parallel worker solvers that solve a portion of search space on CPU cores and interact with neighbor workers via message passing for dynamic load balancing. We also propose a preprocess that accelerates the initial search space distribution by sending sets of boxes via static routing between the workers. We have implemented the method by extending the Realpaver solver using the X10 language to realize a process-level parallelization over a number of cores. Section 4 reports experimental results when our method was deployed on two hardware environments.

Related Work. There have been several work regarding parallel solving of CSPs with either discrete or continuous domains. Parallel solving of generic CSPs on massive computer clusters and supercomputers has been explored in [12,2,18]. This work focuses on a massive parallel solver for NCSPs that has a different characteristics compared to generic CSPs. In the survey [7], existing work is classified into (i) search-space splitting methods[16,12,2,5,15,18], (ii) cooperative methods for heterogeneous solver agents (cf. portfolios)[2], and (iii)

B. O'Sullivan (Ed.): CP 2014, LNCS 8656, pp. 398–406, 2014.

parallelization of constraint propagation. Our method belongs to the first category. A few works have used approaches (ii)[11] and (iii)[8] for parallelization of NCSP solving. However, to the best of our knowledge, a massive parallelization method that uses the typical approach (i) has not yet been proposed.

Substantial work regarding the parallelization of the *branch and bound* algorithm with search-space splitting exists [9,13,14]. This approach has also been applied to CSP solvers [16,5] and SAT solvers [15]. This work explores an efficient parallel method for solving NCSPs with similar approach to [9,13].

2 Numerical Constraint Satisfaction Problems

A *numerical constraint satisfaction problem* (NCSP) is defined as a triple (v, v_0, c) that consists of a vector of *variables* $v = (v_1, \ldots, v_n)$, an *initial domain* in the form of a box $v_0 \in \mathbb{IR}^n$ ($\mathbb{IR}$ denotes the set of closed real intervals), and a *constraint* $c(v) \equiv f(v) = 0 \wedge g(v) \geq 0$, where $f : \mathbb{R}^n \to \mathbb{R}^e$ and $g : \mathbb{R}^n \to \mathbb{R}^i$, i.e., a conjunction of e equations and i inequalities. A *solution* of a NCSP is an assignment of its variables $v \in v_0$ that satisfies its constraints. The *solution set* Σ of a NCSP is the region within its initial domain that satisfies its constraints, i.e., $\Sigma(v_0) := \{\tilde{v} \in v_0 \mid c(\tilde{v})\}$. The target of this paper is *under-constrained* NCSPs such that $n > e$. In general, an under-constrained problem may have a continuous set of infinitely many solutions.

Branch and Prune Algorithm. The *branch and prune algorithm* [17] is the standard solving method for NCSPs. It takes a NCSP and a precision ϵ as an input and outputs a set of boxes (or *paving*) S that approximates the solution set with precision ϵ. Examples of S are illustrated in Figure 1.

An intermediate state of the algorithm is represented as a pair of sets of boxes (L, S). The solver receives an initial state $(\{v_0\}, \emptyset)$and iteratively applies the *step computation* (illustrated in Figure 2) until it reaches a final state $(\emptyset, S)$. In the step computation, first, it takes the first element of the queue L of boxes and applies the Prune procedure, which is a filtering procedure that shaves boundary portions of the considered box. In this work, we use an implementation proposed in [6] which provides a verification process based on an interval Newton method combined with a relatively simple filtering process based on the Hull consistency[1]. As a result, a box becomes either empty, precise enough (its width is smaller than ϵ), verified as an *inner* box of the solution set Σ, or undecided. Precise and inner boxes are appended to S and undecided boxes are passed to Branch. Second, the Branch procedure bisects the box at the midpoint along a component corresponding to one of the variables and the sub-boxes will be put back in the queue. In this work, we assume Branch selects variables in an order which makes the search to behave in a breadth-first manner and thus the solving process gradually improves the precision of the overall pavings (Figure 1).

The computation of Prune is expensive and is the bottleneck of the solving process. Under certain conditions, application of Prune contracts a large portion of the search space into a tight box (cf. quadratic convergence of the interval

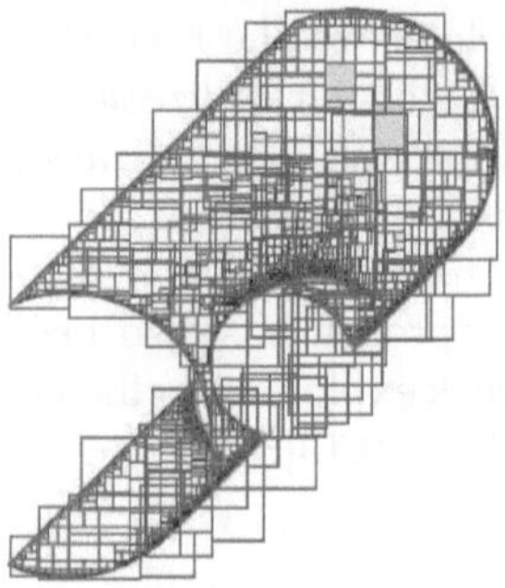

Fig. 1. Overlay of two solution box sets (pavings) with $\epsilon = 0.01, 0.1$

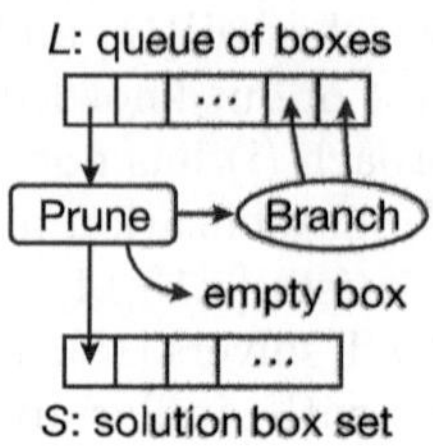

Fig. 2. A step computation of the branch and prune algorithm

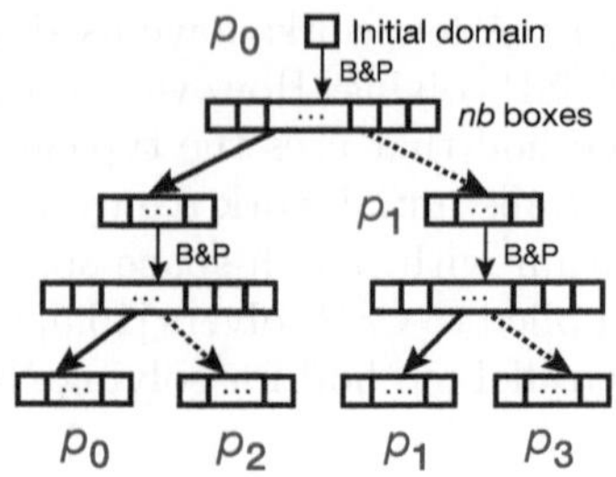

Fig. 3. Distribution of the box queue L in the preprocess

Newton methods). Prune can also filter out the whole box if the considered box is verified as an inner or totally inconsistent region. These characteristics of Prune result in the unbalanced nature of search trees. Therefore, a straightforward parallel method does not work efficiently. It is crucial for efficient NCSP solving to execute Prune on each step of traversing the search tree which makes it more difficult to distribute a search path among processors. These properties will be discussed in Section 4.1.

3 Parallel Branch and Prune

We propose a parallel method that runs several workers on the available CPU cores $p_0, \ldots, p_{\#p-1}$ (we assume a single worker runs on a core and identify both a worker and a core with p_i). Workers homogeneously interleave the following three procedures and cooperate in a decentralized fashion: (i) breadth-first branch and prune search, (ii) distribution and workload balancing of search space in a sender-initiated manner, and (iii) termination detection. Sub-trees in the search space of the branch and prune becomes unbalanced but can be searched independently: there are no confluence of multiple branches, and we have no shared information between branches. Distribution of the search space among workers is done in preprocess and postprocess as described in the following subsections. The preprocess distributes the portions of the search space to the other workers, and the postprocess balances the load of each worker during search. Termination is detected by circulating a termination token via idling workers based on Dijkstra's method (see Section 11.4.4 of [9]).

Preprocess: Search-Space Splitting and Distribution. A solving process is started by a worker p_0 that possesses the initial domain (i.e., a box which contains the whole search space) in the queue L. To distribute the subsequent search space (i.e., a queue of boxes) equally to each core, the preprocess invokes a partition of the queue in two (or more) and then sends a portion to another worker. Figure 3 illustrates in a downward direction some initial transitions of the box queue L distributed among three workers. In our implementation, the distribution routing is formed as a binary tree whose height is $\lceil \log_2 \#p \rceil$. In each node of the tree, the branch and prune process runs until the number of boxes in the queue reaches nb Next, the queue is sorted by the volume of the boxes and the half of the content (i.e., $nb/2$ boxes) is sent to the other core via the right branch.

Postprocess: Dynamic Load Balancing. During search, each worker normalizes the loads within a predefined *neighborhood* which consists of a small number of *neighbor* workers. Because there are sufficiently large number of boxes, we simply regard the number of boxes in the queue L as the amount of load. Assume $\#p$ workers are running and each worker p_i possesses l_i boxes in its queue. We also assume for each worker p_i that N_i is a set of $|N|$ neighbor workers, N_i^{-1} is a set of workers where $p_j \in N_i^{-1} \Leftrightarrow p_i \in N_j$, L_i is a set of loads of the neighbor workers, and Δ is a predefined load margin. The load balancing procedure of a worker p_i performs the following steps once every ns branch and prune steps.

1. For each worker p_j in N_i^{-1}, inform the load l_i and put in the list L_j.
2. Calculate the mean μ of the loads in L_i.
3. If $\mu < \Delta$, for each worker p_j in N_i, send at most $\mu - l_j$ boxes to p_j (to be efficient, a certain number of boxes should be kept locally).

Neighbor workers can be identified e.g. as adjacent nodes in the $|N|$-dimensional mesh of workers. The routing between neighbor workers is fixed during a solving process and thus it may happen that a worker possesses an excessive load than others. However, this load imbalance will be resolved by the subsequent load balancing processes.

4 Experimental Results

We have implemented the proposed method and measured the speedup of the solving process of under-constrained NCSPs. The experiments were performed with an exhaustive set of parameter combinations to explore the optimal settings.

Implementation. We have implemented the proposed method with C++ (gcc ver. 4.4.7 and 4.3.4) and X10 (ver. 2.3.1)[4], a high productivity language for parallel computing. In the following, we use the term *place*, which is a notion of X10 that in our setting represents a CPU core. Libraries Realpaver (ver. 1.1)[10] for sequential NCSP solving and Gaol (ver. 4.0.1)[1] for basic interval computation were used to facilitate the implementation. The `Prune` procedure is realized by calling the sequential implementation in Realpaver. Each run of `Prune` takes around 0.2–1ms and the overall execution becomes the bottleneck of the branch and prune algorithm (occupies greater than 95% of running time in sequential solving). The procedures for search space distribution and load balancing are implemented with X10. Communication of boxes and loads between places are implemented as `async` tasks and performed in parallel to the search process so

[1] `http://sourceforge.net/projects/gaol/`

that the overhead will be hidden. In the experiments, timings t_1 for sequential runs on single core were measured using the C++ implementation described in [6], which works identically and faster than our X10 version.

Experiment Environments. Two sets of experiments were operated using (1) a shared-memory machine equipped with 40 cores (four of 10-core Intel Xeon E7-4860 2.26GHz) and 256GB of local memory and (2) up to 256 cores of SGI UV1000, a pseudo-shared memory machine equipped with 2,048 cores (8-core Xeon E7-8837 2.67GHz) and 16TB of memory. UV1000 works as a single shared memory machine by emulating memory accesses using communication based on a high speed NUMAlink5 network which has a bandwidth of 120Gbps. We used the MPI backend of X10 with options `X10_NTHREADS` = 6 and `GC_NPROCS` = 2.

Experiments on a Shared Memory Machine. We solved the problems shown in [6,3] using 40 cores of the machine (1). We report the results for two representative problems. Parameters in the load balancing method were set as either combination of the following values: $nb = 32, |N| \in \{2, 4\}, ns \in \{10, 100, 1000\}$, and $\Delta = 10$. We also computed with and without the preprocess (when the preprocess is not used, the postprocess is executed from the beginning). For each problem, we solved two instances with two multiplicative precisions. The specification of each instance and the computational results are presented in Table 1. In the table, the columns "problem", "size", "ϵ", "pp", and "$|N|$" represent the name of the problem, size (i.e., the number of projection/parameter variables), the precision, usage of the preprocess, and the number of neighbor workers, respectively. The rest of the columns represents the results. t_1 and $\#\mathrm{br}_1$ represent the running time and the number of branches on single core. $t_{ns,i}$ and $\#\mathrm{br}_{ns,i}$ represent the running time and the largest number of branches performed by a worker when computed with the interval ns and i X10 places (best timings are underlined).Figure 4 illustrates the speedup of the solving process.

Experiments on a Cluster with High-Speed Interconnection. We solved the problem "*3rpr*" using up to 256 cores of the machine (2), UV1000. Parameters in the load balancing method were set as either combination of the following values: $nb = 8, |N| \in \{2, 4\}, ns = 1000$, and $\Delta = 10$. The results are presented in Table 2. Each column of the table represents the same information as presented in Table 1 except that the column "$\#\mathrm{sends}_{1000,256}$" represents the number of loads sent by the load balancer in the solving process with $ns = 1000$ and 256 X10 places. Figure 5 illustrates the speedup of the solving process.

4.1 Discussions

In the experiments, our method scaled up to 256 cores with the optimal configurations. We achieved speedups up to 32.3 fold using 40 cores of the shared memory machine and up to 119 fold using 256 cores of the cluster machine.

The best speedup of 119 fold was obtained with the preprocess. The preprocess facilitates and accelerates the workload distribution in the early stage of

Table 1. Experimental result on the shared memory machine

problem	size	ϵ	pp	$\lvert N \rvert$	t_1	$t_{10,40}$	$t_{100,40}$	$t_{1000,40}$	$\#\mathrm{br}_1$	$\frac{\#\mathrm{br}_1}{\#\mathrm{br}_{100,40}}$
4D sphere	2+4	0.02	yes	2	254	9.52	7.94	10.1	238 319	37.0
and plane				4		11.0	8.73	9.45		37.6
(*sp2-4*)			no	2		9.34	8.13	16.5		35.4
				4		10.8	8.67	10.5		37.3
		0.01	yes	2	721	38.6	22.8	23.9	669 601	38.0
				4		38.1	25.1	24.0		38.7
			no	2		35.7	22.7	30.0		37.6
				4		36.8	24.7	24.7		38.7
3-RPR robot	3+3	0.2	yes	2	1 100	199	73.9	34.1	1 936 939	33.7
(*3rpr*)				4		296	68.5	36.4		38.0
			no	2		185	64.0	34.0		33.5
				4		244	58.3	36.1		38.0
		0.1	yes	2	4 080	1 010	714	282	7 186 845	30.2
				4		2 820	1 070	257		36.0
			no	2		971	678	244		28.5
				4		2 630	901	231		36.0

Table 2. Experimental result on the UV1000 cluster

problem	size	ϵ	pp	$\lvert N \rvert$	t_1	$t_{1000,32}$	$t_{1000,256}$	$\frac{\#\mathrm{br}_1}{\#\mathrm{br}_{1000,256}}$	$\#\mathrm{sends}_{1000,256}$
3-RPR robot	3+3	0.2	yes	2	850	37.4	14.0	131	18 648
(*3rpr*)				4		54.6	33.6	157	85 904
			no	2		39.8	20.4	43.8	10 856
				4		53.4	32.0	123	66 212
		0.1	yes	2	3 040	341	25.6	192	39 608
				4		371	87.8	176	128 084
			no	2		325	54.7	105	34 892
				4		339	52.1	184	129 512

the search process. In some of the experiments without using the preprocess, the speedup ratio became saturated when using many cores (e.g., *sp2-4* with $ns = 1000$ and the experiments on the cluster). This was because the load balancing process was too infrequent for the given number of workers and the work load diffusion became too slow. When comparing the right-hand graphs for the instance *sp2-4*, $ns = 1000$, in Figure 4, we can notice that the point of saturation shifts according to the search space size. On the other hand, in some other experiments, the results got worse with the preprocess (e.g., results with $ns = 10$ on the machine (1)). It occasionally happens that the preprocess mostly solves the problem. However, the preprocess can result in highly unbalanced search trees because of the Prune process, and in such cases the postprocess will not have enough time for load balancing.

Regarding the neighborhood sizes $|N| = 2, 4$, there was a trade off between the workloads balance and the amount of communications required. For the

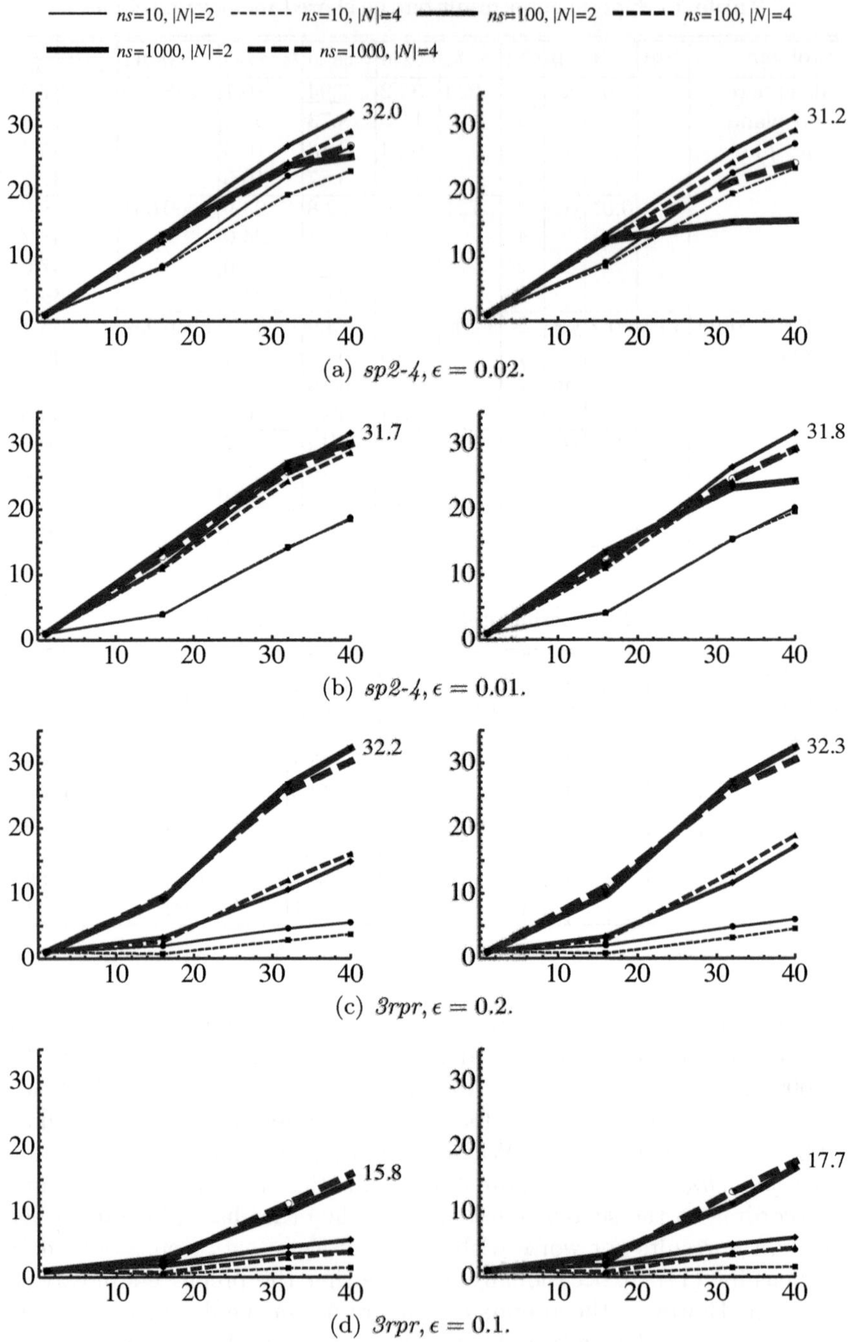

Fig. 4. Speedups on the shared memory machine with 40 cores. Left- and right-hand side graphs correspond to computations with and without the preprocess, respectively.

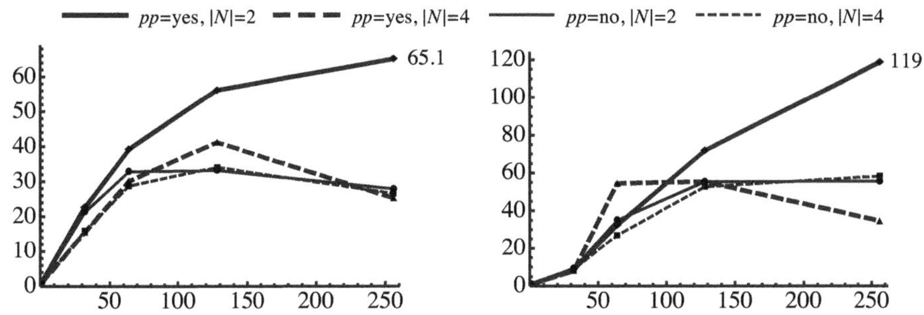

(a) $3rpr, \epsilon = 0.2$. (b) $3rpr, \epsilon = 0.1$.

Fig. 5. Speedups using 256 cores of UV1000

shared memory machine, it was unclear which size had the advantage. However, for UV1000, the solver was notably slower for $|N| = 4$ than $|N| = 2$. It is understandable because larger number of neighbors significantly increased the number of communications (see "#sends$_{1000,256}$" in Table 2) and communications between places were much more costly compared to normal shared memory machines despite the high speed network of UV1000.

Three intervals $ns = 10, 100, 1000$ were used for load balancing which determined the speed of workload distribution. When the distribution was too slow, the speedup ratio did not scale well (e.g., *3rpr*, $\epsilon = 0.2$, with $ns = 1000$ on the machine (1)). Conversely, small intervals required greater amount of communications and therefore we used $ns = 1000$ to draw better performance on the cluster where communications were more costly.

There was a large overhead caused by the workers sending a large number of boxes for load balancing when the number of workers was not sufficient against the problem size. Speedups for *3rpr*, $\epsilon = 0.1$, using 40 workers or less shows an example of such overheads (Figure 4(d)). Resolving this overhead by suppressing redundant box sends is a part of the future work.

5 Conclusions

In this paper, we proposed a parallel branch and prune algorithm, based on, non-portfolio, search-space splitting approach. In the experiments, using 256 X10 places (i.e., cores), we achieved speedup factors of as much as 119. We expect that our parallelized solver will be applied to large practical problems, e.g., the robotics problems in [3].

Acknowledgments. This work was partially funded by JSPS (KAKENHI 25880008 and 25700038). Computing resources for the experiments in this paper were provided by Prof. Kazunori Ueda (Waseda University, Tokyo) and a Compute Canada RAC award, for environments (1) and (2), respectively.

References

1. Benhamou, F., Goualard, F., Granvilliers, L., Puget, J.F.: Revising Hull and Box Consistency. In: Proc. of ICLP, pp. 230–244 (1999)
2. Bordeaux, L., Hamadi, Y., Samulowitz, H.: Experiments with Massively Parallel Constraint Solving. In: Proc. of IJCAI, pp. 443–448 (2006)
3. Caro, S., Chablat, D., Goldsztejn, A., Ishii, D., Jermann, C.: A branch and prune algorithm for the computation of generalized aspects of parallel robots. Artificial Intelligence 211, 34–50 (2014)
4. Charles, P., Grothoff, C., Saraswat, V.: X10: an object-oriented approach to non-uniform cluster computing. In: Proc. of OOPSLA, pp. 519–538 (2005)
5. Chu, G., Schulte, C., Stuckey, P.J.: Confidence-based work stealing in parallel constraint programming. In: Gent, I.P. (ed.) CP 2009. LNCS, vol. 5732, pp. 226–241. Springer, Heidelberg (2009)
6. Ishii, D., Goldsztejn, A., Jermann, C.: Interval-based projection method for under-constrained numerical systems. Constraints Journal 17(4), 432–460 (2012)
7. Gent, I.P., Jefferson, C., Miguel, I., Moore, N.C.A., Nightingale, P., Prosser, P., Unsworth, C.: A Preliminary Review of Literature on Parallel Constraint Solving. In: Proc. of Workshop on Parallel Methods for Constraint Solving, pp. 7–19 (2011)
8. Goldsztejn, A., Goualard, F.: Box consistency through adaptive shaving. In: Proc. of SAC, pp. 2049–2054 (2010)
9. Grama, A., Gupta, A., Karypis, G., Kumar, V.: Introduction to Parallel Computing. Addison Wesley (2003)
10. Granvilliers, L., Benhamou, F.: Algorithm 852: RealPaver: an interval solver using constraint satisfaction techniques. ACM Transactions on Mathematical Software 32(1), 138–156 (2006)
11. Granvilliers, L., Hains, G.: A conservative scheme for parallel interval narrowing. Information Processing Letters 74(3-4), 141–146 (2000)
12. Jaffar, J., Santosa, A., Yap, R., Zhu, K.: Scalable distributed depth-first search with greedy work stealing. In: Proc. of ICTAI, pp. 98–103 (2004)
13. Lüling, R., Monien, B., Reinefeld, A., Tschöke, S.: Mapping Tree-Structured Combinatorial Optimization Problems onto Parallel Computers. In: Ferreira, A., Pardalos, P. (eds.) SCOOP 1995. LNCS, vol. 1054, pp. 115–144. Springer, Heidelberg (1996)
14. Otten, L., Dechter, R.: Towards Parallel Search for Optimization in Graphical Models. In: Proc. of ISAIM (2010)
15. Schubert, T., Lewis, M., Becker, B.: PaMiraXT: Parallel SAT Solving with Threads and Message Passing. JSAT 6, 203–222 (2009)
16. Schulte, C.: Parallel search made simple. In: Proc. of TRICS, pp. 41–57 (2000)
17. Van Hentenryck, P., McAllester, D., Kapur, D.: Solving Polynomial Systems Using a Branch and Prune Approach. SIAM Journal on Numerical Analysis 34(2), 797–827 (1997)
18. Xie, F., Davenport, A.: Massively Parallel Constraint Programming for Supercomputers: Challenges and Initial Results. In: Lodi, A., Milano, M., Toth, P. (eds.) CPAIOR 2010. LNCS, vol. 6140, pp. 334–338. Springer, Heidelberg (2010)

Tree-Decompositions with Connected Clusters for Solving Constraint Networks*

Philippe Jégou and Cyril Terrioux

Aix-Marseille Université, LSIS UMR 7296
13397 Marseille, France
{philippe.jegou,cyril.terrioux}@lsis.org

Abstract. From a theoretical viewpoint, the (tree-)decomposition methods offer a good approach for solving Constraint Satisfaction Problems (CSPs) when their (tree)-width is small. In this case, they have often shown their practical interest. So, the literature (coming from Mathematics or AI) has concentrated its efforts on the minimization of a single parameter, the tree-width. Nevertheless, experimental studies have shown that this parameter is not always the most relevant to consider for solving CSPs. In this paper, we experimentally show that the decomposition algorithms of the state of the art produce clusters (a tree-decomposition is a tree of clusters) having several connected components. Then we highlight that such clusters create a real problem for the efficiency of solving methods. To avoid this kind of problem, we consider here a new kind of graph decomposition called *Bag-Connected Tree-Decomposition*, which considers only tree-decompositions such that each cluster is connected. We propose a first polynomial time algorithm to find such decompositions. Finally, we show experimentally that using these bag-connected tree-decompositions improves significantly the solving of CSPs by decomposition methods.

1 Introduction

Constraint Satisfaction Problems (CSPs, see [1] for a state of the art) provide an efficient way of formulating problems in computer science, especially in Artificial Intelligence.

Formally, a *constraint satisfaction problem* is a triple (X, D, C), where $X = \{x_1, \ldots, x_n\}$ is a set of n variables, $D = (D_{x_1}, \ldots, D_{x_n})$ is a list of finite domains of values, one per variable, and $C = \{C_1, \ldots, C_e\}$ is a finite set of e constraints. Each constraint C_i is a pair $(S(C_i), R(C_i))$, where $S(C_i) = \{x_{i_1}, \ldots, x_{i_k}\} \subseteq X$ is the *scope* of C_i, and $R(C_i) \subseteq D_{x_{i_1}} \times \cdots \times D_{x_{i_k}}$ is its *compatibility relation*. The *arity* of C_i is $|S(C_i)|$. A CSP is called *binary* if all constraints are of arity 2. The structure of a constraint network is represented by a hypergraph (which is a graph in the binary case), called the constraint (hyper)graph, whose vertices correspond to variables and edges to the constraint scopes. In this paper, for sake of simplicity, we only deal with the case of binary CSPs but this work can easily be extended to non-binary CSP by exploiting the 2-section [2] of the constraint hypergraph (also called primal graph), as it will be

* This work was supported by the French National Research Agency under grant TUPLES (ANR-2010-BLAN-0210).

B. O'Sullivan (Ed.): CP 2014, LNCS 8656, pp. 407–423, 2014.

done for our experiments since we will consider binary and non-binary CSPs. Moreover, without loss of generality, we assume that the network is connected. To simplify the notations, in the sequel, we denote the graph $(X, \{S(C_1), \dots S(C_e)\})$ by (X, C). An assignment on a subset of X is said to be *consistent* if it does not violate any constraint. Determining whether a CSP has a *solution* (i.e. a consistent assignment on all the variables) is known to be NP-complete. So, many works have been realized to make the solving of instances more efficient in practice, by using optimized backtracking algorithms which may exploit heuristics, constraint learning, non-chronological backtracking, filtering techniques based on constraint propagation, etc. The time complexity of these backtracking methods is naturally exponential, at least in $O(e.d^n)$ with n the number of variables, d the maximum size of domains and e the number of constraints.

Another way is related to the study of tractable classes defined by properties of constraint networks. E.g., it has been shown that if the structure of this network is acyclic, it can be solved in linear time [3]. Using and generalizing these theoretical results, some methods to solve CSPs have been defined, such as Tree-Clustering [4]. This kind of methods is based on the notion of tree-decomposition of graphs [5]. Their advantage is related to their theoretical complexity, that is d^{w+1} where w is the tree-width of the constraint graph. When this graph has nice topological properties and thus when w is small, these methods allow to solve large instances, e.g. radio link frequency assignment problems [6]. Note that in practice, the time complexity is more related to $d^{w^+ +1}$ where $w^+ \geq w$ is actually an approximation of the tree-width because computing an optimal tree-decomposition (of width w) is an NP-hard problem.

However, the practical implementation of such methods, even though it often shows its interest, has proved that the minimization of the parameter w^+ is not necessarily the most appropriate. Besides the difficulty of computing the optimal value of w^+, i.e. w, it sometimes leads to handle optimal decompositions, but whose properties are not always adapted to a solving that would be the most efficient. This has led to propose graph decomposition methods that make the solving of CSPs more efficient in practice, but for which the value of w^+ can even be really greater than w [7].

In this paper, we show that a reason to this lack of efficiency for solving CSPs using decomposition can be found in the nature of the decompositions for which w^+ is close to w. Indeed, minimizing w^+ can lead to decompositions such that some clusters have several connected components. Unfortunately, this lack of connectedness may lead the solving method to spend large amount of efforts to solve the subproblems related to these disconnected clusters, by passing many times from a connected component to another. To avoid this problem, we consider here a new kind of graph decomposition called *Bag-Connected Tree-Decomposition*[1] and its associated parameter called *Bag-Connected Tree-Width* [8]. This parameter is equal to the minimal width over all the tree-decompositions for which each cluster has a single connected component. So, the Bag-Connected Tree-Width will be the minimum width for all Connected Tree-Decompositions. The notion of Bag-Connected Tree-Width has been introduced very recently and to date, only studied in [8] from a mathematical viewpoint. Here we analyze this concept in terms of its algorithmic properties. So, we firstly prove that its com-

[1] We use the term "bag" rather than "cluster" because it is more compatible with the terminology of Graph Theory.

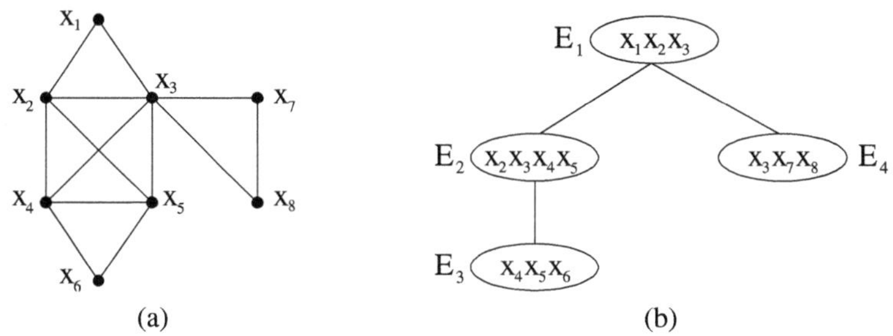

Fig. 1. A constraint graph for 8 variables (a) and an optimal tree-decomposition (b)

putation is NP-hard. So, we propose a first polynomial time algorithm (in $O(n(n+e)))$ in order to approximate this parameter, and the associated decompositions. The experiments we present show the relevance of this parameter, since it allows to significantly improve the solving of CSPs by decomposition.

Note that the present work is applied to tree-decompositions, but it can also be adapted to most decompositions (e.g. [9,10]). Indeed, in most CSP solving methods based on a decomposition approach, the decompositions are computed by algorithms which aim to approximate at best a graphical parameter (width) without taking into account the connectedness of produced clusters, neither the solving step. So, the problems observed here for tree-decomposition can also occur for other decompositions.

Section 2 introduces notations and the principles of tree-decomposition methods for solving CSPs. Section 3 points to some problems due to the computing of "good" tree-decompositions while section 4 presents the notion of bag-connected tree-decomposition, proposing a first algorithm to achieve one. Before concluding, we empirically show the interest of the use of this graph parameter for the practical solving of CSPs in section 5.

2 Solving CSPs Using Graph Decomposition

Tree-Clustering (denoted TC [4]) is the reference method for solving binary CSPs by exploiting the structure of their constraint graph. It is based on the notion of tree-decomposition of graphs [5].

Definition 1. *Given a graph $G = (X, C)$, a tree-decomposition of G is a pair (E, T) with $T = (I, F)$ a tree and $E = \{E_i : i \in I\}$ a family of subsets of X, such that each subset (called cluster or bag in Graph Theory) E_i is a node of T and satisfies (i) $\cup_{i \in I} E_i = X$, (ii) for each edge $\{x, y\} \in C$, there exists $i \in I$ with $\{x, y\} \subseteq E_i$, and (iii) for all $i, j, k \in I$, if k is in a path from i to j in T, then $E_i \cap E_j \subseteq E_k$.*

The width of a tree-decomposition (E, T) is equal to $max_{i \in I}|E_i| - 1$. The tree-width w of G is the minimal width over all the tree-decompositions of G.

Figure 1(b) presents a tree whose nodes correspond to the maximal cliques of the graph depicted in Figure 1(a). It is a possible tree-decomposition for this graph.

So, we get $E_1 = \{x_1, x_2, x_3\}$, $E_2 = \{x_2, x_3, x_4, x_5\}$, $E_3 = \{x_4, x_5, x_6\}$, and $E_4 = \{x_3, x_7, x_8\}$. The tree-width of this graph is 3 as the maximum size of clusters is 4.

The first version of TC [4], begins by computing a tree-decomposition (using the algorithm MCS [11]). In the second step, the clusters are solved independently, considering each cluster as a subproblem, and then, enumerating all its solutions. After this, a global solution of the CSP, if one exists, can be found efficiently exploiting the tree structure of the decomposition. Time and space complexities of this first version is $O(n.d^{w^+ +1})$ where $w^+ + 1$ is the size of the largest cluster ($w+1 \leq w^+ + 1 \leq n$). Note that this first approach has been improved to reach a space complexity in $O(n.s.d^s)$ [12,13] where s is the size of the largest intersection (*separator*) between two clusters ($s \leq w^+$). Unfortunately, this kind of approach which solves completely each cluster is not efficient in practice. So, later, the Backtracking on Tree-Decomposition method (denoted BTD) [14] has been proposed and shown to be really more efficient from a practical viewpoint and appears in the state of the art as a reference method for this type of approach [15]. In contrast to TC, BTD does not need to solve completely each cluster to find a solution. A backtrack search is realized, exploiting a variable ordering induced by a depth first traversal of the tree-decomposition. While this approach has shown its practical interest, from a theoretical viewpoint, in the worst case, it has the same complexities as the improved version of TC, that is $O(n.d^{w^+ +1})$ for time complexity, and $O(n.s.d^s)$ for space complexity. So, to make structural methods efficient, we must a priori minimize the values of w^+ and s when computing the tree-decomposition. Unfortunately, computing an optimal tree-decomposition (i.e. a tree-decomposition of width w) is NP-hard [16]. So, many works deal with this problem. They often exploit an algorithmic approach related to *triangulated* graphs (see [17] for an introduction to triangulated graphs). We can distinguish different classes of approaches. On the one hand, the methods looking for optimal decompositions or their approximations have not shown their practical interest, due to a too expensive runtime w.r.t. the weak improvement of the value w^+. On the other hand, the methods with no guarantee of optimality (like ones based on *heuristic triangulations*) are commonly used. They run in polynomial time (between $O(n + e)$ and $O(n^3)$), are easy to implement and their advantage seems justified. Indeed, these heuristics appear to obtain triangulations reasonably close to the optimum [18]. In practice, the most used methods to find tree-decompositions are based on MCS [11] and Min-Fill [19] which give good approximations of w^+. Moreover, in [7], experiments have shown that the efficiency for solving CSPs is not only related to the value of w^+, but also to the value of s. Nevertheless, to our knowledge, these studies were only focused on the values of w^+ and s, not on the structure of clusters which seems to be a more relevant parameter. This question is studied in the next section, showing that topological properties of clusters constitute also a crucial parameter for solving CSPs.

Before that, we recall how compute a tree-decomposition with the Min-Fill heuristic. The first step, which corresponds to Min-Fill, is to calculate a triangulation of the graph. For a given graph $G = (X, C)$, a set of edges C' will be added so that the resulting graph $G' = (X, C \cup C')$ is triangulated. Min-Fill will order the vertices from 1 to n. At each step, a vertex is numbered by choosing a unnumbered vertex x that minimizes the number of edges to be added in G' to make a clique with the set of unnumbered neighboring vertices of x. Once a vertex is numbered, it will be eliminated. After this processing, the vertices have been numbered from 1 to n, and it is ensured that for a given vertex x with number i, its neighboring vertices in G' with a higher number $j > i$, form a clique. The order defined by these numbers is called a *perfect elimination order*. The cost of this first step is $O(n^3)$. The second step is to compute the maximal cliques of G'. Since G' is triangulated and we have a perfect elimination order, it can be achieved in linear time, i.e. in $O(n + e')$ where $e' = |C \cup C'|$ [20,17]. Each maximal clique corresponds to a cluster of the associated tree decomposition. The third step computes the tree structure of the decomposition. Several approaches exist. A simple way consists in computing a maximum spanning tree (the constraint graph is assumed to be connected) of a graph whose vertices correspond to the maximal cliques (i.e. clusters E_i), and edges link two maximal cliques sharing at least one vertex and are labeled with the size of these intersections. This treatment can be achieved in $O(n^3)$ (e.g. by Prim's algorithm). Overall, the cumulative cost of these three steps is in $O(n^3)$.

3 Disconnected Clusters and Their Impact on the Efficiency of Decomposition Methods

The study of the tree-decompositions shows they can frequently possess clusters that have several connected components. For example, consider a cycle without chord (that is without edge joining two non-consecutive vertices in the cycle) of n vertices (with $n \geq 4$). Any optimal tree-decomposition has exactly $n-2$ clusters of size 3, and among them, $n - 4$ clusters have two connected components.

This phenomenon is also observed for real instances, when we consider tree-decompositions of good quality. For example, the well known RLFAP instance *Scen-06* appearing in the CSP 2008 Competition[2] is defined on 200 variables and its network admit good tree-decompositions which can be found quite easily (e.g. Min-Fill finds one with $w^+ = 20$). Unfortunately, a detailed analysis of these tree-decompositions shows that they have several disconnected clusters. More generally, it turns out that about 32% of the 7,272 instances of the CSP 2008 Competition have a tree-decomposition with at least one disconnected cluster when MCS or Min-Fill are used, what is generally the case of most tree-decomposition methods for solving CSPs. Among these instances for which MCS or Min-Fill produce tree-decompositions with disconnected clusters, we can notably find most of the RLFAP or FAPP instances which are often exploited as benchmarks for decomposition methods for both decision and optimization problems. Moreover, sometimes, the percentage of disconnected clusters in one instance may be very large up to 99% and about 35% in average. For the FAPP instances, the average is about 48% for tree-decompositions produced by Min-Fill, and a greater average using MCS. This observation will be even more striking for algorithms that find decompositions with smaller widths, as suggested by the example of the cycle without chord.

The presence of disconnected clusters in the considered tree-decomposition can have a negative impact on the practical efficiency of decomposition methods which can be penalized by a large amount of time or memory to solve the instance. Firstly, it is well

[2] See `http://www.cril.univ-artois.fr/CPAI08` for more details.

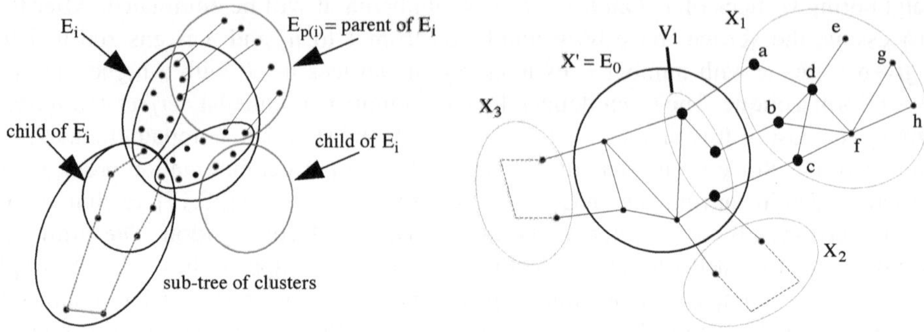

(a) (b)

Fig. 2. (a) Disconnected cluster in a Tree-Decomposition, (b) First pass in the loop for *Bag-Connected-TD*

known that if a constraint network is not connected, this can have important consequences on the effectiveness of its solving. For example, if one of its connected components has no solution, and if the solving first addresses a connected component that has solutions, all of them should be listed before proving the inconsistency of the whole CSP. In the case of decomposition methods, the existence of disconnected clusters is perhaps even more pernicious. In the case of TC, let us consider a disconnected cluster. On the one hand, the phenomenon already encountered in the case of disconnected networks may arise. But it is also possible that this cluster has solutions. All these solutions will be calculated and stored before processing another cluster. Their number can be significant as it is the product of the number of solutions of each of its connected components. Note that for some benchmarks coming from the FAPP instances, the number of connected components in one cluster can be greater than 100. However, many local solutions of this cluster may be globally incompatible, because these connected components may be linked by some constraints which appear in other clusters. Figure 2(a) shows an example of decomposition for which two connected components of a cluster E_i are connected by a sequence of constraints that appear in the subproblem rooted in this cluster. Thus, the overall inconsistency of local solutions of E_i can only be detected when all these clusters have been solved, during the composition of global solutions produced by TC in its last step. This leads TC to a large consumption of time and memory, making this approach unrealistic in practice.

Other methods were proposed to avoid this kind of phenomenon where clusters are initially solved independently. This is notably the case of BTD which is one of the most effective approaches based on decompositions. Although BTD has shown its practical interest, unfortunately, the observed phenomenon still exists, even if it will generally be attenuated. To well understand this, we must remind that BTD solves an instance by solving successively the subproblems rooted in every cluster of the tree-decomposition. But unlike TC which first calculates all the solutions of a cluster, when accessing a cluster, BTD only computes one solution. Roughly speaking, the subproblem rooted in a cluster E_i corresponds to the subproblem involving all the variables of the descendants of E_i in the tree-decomposition (see [14] for more details). In practice, BTD starts

its backtrack search by assigning consistently the variables of the root cluster before exploring a child cluster. When exploring a new cluster E_i, it only assigns the variables which appear in the cluster E_i but not in its parent cluster $E_{p(i)}$, that is all the variables of the cluster E_i except the variables of the separator $E_i \cap E_{p(i)}$[3]. For instance, let us consider the constraint graph of Figure 1 and its associated tree-decomposition. If we assume that E_1 is the root cluster, BTD first tries to assign consistently the variables of E_1. If so, it keeps on the search with one of its child clusters (i.e. E_2 or E_4). If BTD chooses to explore first E_2, it will have to assign consistently the variables of $E_2 \backslash (E_1 \cap E_2)$ (i.e x_4 and x_5).
Now and more generally, let us consider a disconnected cluster E_i. We have two cases:

- if $G[E_i \backslash (E_i \cap E_{p(i)})]$[4] is disconnected: BTD has to consistently assign variables which are distributed in several connected components. If the subproblem rooted in E_i is trivially consistent (for instance it admits a large number of solutions), BTD will find a solution by doing at most a few backtracks and keep on the search on the next cluster. So, in such a case, the non-connectivity of E_i does not entail any problem. In contrast, if this subproblem has few solutions or none, we have a significant probability that BTD passes many times from a connected component of $G[E_i \backslash (E_i \cap E_{p(i)})]$ to another when it solves this cluster. Roughly speaking, BTD may have to explore all the consistent assignments of each connected component by interleaving eventually the variables of the different connected components. Indeed, if BTD exploits filtering techniques, the assignment of a value to a variable x of $E_i \backslash (E_i \cap E_{p(i)})$ has mainly impact on the variables of the connected component of $G[E_i \backslash (E_i \cap E_{p(i)})]$ which contains x. In contrast, the filtering does not modify or slightly the domain of any variable in another connected component. This entails that inconsistencies are often detected later and not necessarily in E_i but in one of its descendant cluster (as illustrated previously by Figure 2(a)). If so, BTD may require a large amount of time or memory (due to (no)good recording) to solve the subproblem rooted in E_i, especially if the variables have large domains. For example, this negative phenomenon has been empirically observed on some FAPP instances (e.g the fapp05-0350-10 instance) with a BTD version using MAC [21].
- if $G[E_i \backslash (E_i \cap E_{p(i)})]$ is connected: it follows that E_i is a disconnected cluster because its separator with its parent cluster is disconnected. As the variables of this separator are already assigned, the non-connectivity of E_i does not cause any problem.

This negative impact of disconnected clusters is compatible with empirical results reported in the literature. We have observed that sometimes, the percentage of disconnected clusters for Min-Fill differs significantly from one for MCS, which may explain some differences of efficiency observed by different authors. Indeed, even if the width is the same, decompositions computed by Min-Fill offer best results for solving than the ones obtained by MCS [7] and is considered as the best heuristic of the state of the art now. Moreover, the analysis of tree-decompositions shows also that the connection

[3] We assume that $E_i \cap E_{p(i)} = \emptyset$ if E_i is the root cluster.

[4] For any $Y \subseteq X$, the subgraph $G[Y]$ of $G = (X, C)$ induced by Y is the graph (Y, C_Y) where $C_Y = \{\{x, y\} \in C | x, y \in Y\}$.

between connected components of some clusters is frequently observed in the leaves (clusters) of the decomposition, further increasing more the negative effects observed. To avoid this kind of phenomenon, we study classes of tree-decompositions for which all the clusters are connected.

4 A New Parameter for Graph Decomposition of CSPS

4.1 Bag-Connected Tree-Decomposition

The facts presented above lead us naturally to consider only tree-decompositions for which all the clusters are connected. This concept has been recently introduced in the context of Graph Theory [8]. It has been studied for some of its combinatorial properties. However, the algorithmic issues related to its computation have not been studied yet, neither in terms of complexity, nor to propose algorithms to find them. [8] provides a central theorem indicating an upper bound of Bag-Connected Tree-Width[5] as a function of the tree-width. We present now the notion of Bag-Connected Tree-Decomposition, which corresponds to tree-decomposition for which each cluster E_i is connected (i.e. the subgraph $G[E_i]$ of G induced by E_i is a connected graph).

Definition 2. *Given a graph $G = (X, C)$, a **Bag-Connected Tree-Decomposition** of G is a tree-decomposition (E, T) of G such that for all $E_i \in E$, the subgraph $G[E_i]$ is a connected graph. The width of a Bag-Connected Tree-Decomposition (E, T) is equal to $max_{i \in I}|E_i| - 1$. The **Bag-Connected Tree-Width** w_c is the minimal width over all the bag-connected tree-decompositions of G.*

Given a graph $G = (X, C)$ of tree-width w, necessarily $w \leq w_c$. The central theorem of [8] provides an upper bound of the Bag-Connected Tree-Width as a function of the tree-width and k which is the maximum length of its geodesic cycles[6]. More precisely, we have $w_c \leq w + \binom{w+1}{2}.(k.w - 1)$ ($k = 1$ if G has no cycle). Note that $w_c = \lceil \frac{n}{2} \rceil$ for graphs defined by cycles of length n and without chord. Nevertheless, if G is a triangulated graph, $w = w_c$.

Furthermore, the fact that $w \leq w_c$, independently of the complexity of achieving a Bag-Connected Tree-Decomposition, indicates that the decomposition methods based on it, necessarily appear below Tree-Decomposition methods in the hierarchy introduced in [9]. But this remark has no real interest here because our contribution mainly concerns practical efficiency of such methods. Nevertheless, the difference between w and w_c can naturally have incidences on the efficiency of solving in practice. Indeed, if we consider the example of the cycle of length n given in section 3 (a geodesic cycle), optimal decompositions give $w = 2$ and $w_c = \lceil \frac{n}{2} \rceil$. But, in such a case, even if the bag-connected tree-width is arbitrarily greater than the tree-width, applying BTD based on

[5] Note that we use the term of Bag-Connected Tree-Width rather than one of Connected Tree-Width exploited in [8] because the term of Connected Tree-Width has been introduced before in [22] but corresponds to a quite different concept.

[6] A cycle is said *geodesic* if for any pair of vertices x and y belonging to the cycle, the distance between x and y in the graph is equal to the length of the shortest path between x and y in the cycle.

MAC is always as effective since as soon as the first variable is assigned, BTD detects the inconsistency or directly finds a solution, due to the arc-consistency propagation which will be realized along the connected paths in the clusters.

The natural question now is related to the computation of optimal Bag-Connected Tree-Decompositions, that is Bag-Connected Tree-Decompositions of width w_c. We show that this problem, as for Tree-Decompositions, is NP-hard.

Theorem 1. *Computing an optimal Bag-Connected Tree-Decomposition is NP-hard.*

Proof. We propose a polynomial reduction from the problem of computing an optimal tree-decomposition to this one. Consider a graph $G = (X, C)$ of tree-width w, the associated tree-decomposition of G being (E, T). Now, consider the graph G' obtained by adding to G an universal vertex x, that is a vertex which is connected to all the vertices in G. Note that from (E, T), we can obtain a tree-decomposition for G' by adding in each cluster $E_i \in E$ the vertex x. It is a bag-connected tree-decomposition since each cluster is necessarily connected (by paths containing x) and its width is $w + 1$. To show that this addition defines a reduction, it is sufficient to show that w is the tree-width of G iff the bag-connected tree-width w_c of G' is $w + 1$.

($\Rightarrow$) We know that at most, the width of the considered tree-decomposition of G' is $w + 1$ since this tree-decomposition is connected and its width is $w + 1$. Thus, $w_c \leq w + 1$. Assume that $w_c \leq w$. So, there is a bag-connected tree-decomposition of G' of width at most w. Using this tree-decomposition of G', we can define the same tree, but deleting the vertex x, to obtain a tree-decomposition of G of width $w - 1$, which contradicts the hypothesis.

($\Leftarrow$) With the same kind of argument as before, we know that the tree-width w of G is at most $w_c - 1$. And by construction, it cannot be strictly less than $w_c - 1$. So, it is exactly $w_c - 1$.

Moreover, achieving G' is possible in linear time. □

We have seen that for solving CSPs, it is not necessary to find an optimal tree-decomposition, and this is even often desirable. Also, we now propose an algorithm which computes a bag-connected tree-decomposition in polynomial time, of course without any guarantee about its optimality. The algorithm *Bag-Connected-TD* described below finds a bag-connected tree-decomposition of a given graph $G = (X, C)$.

4.2 Computing a Bag-Connected Tree-Decomposition

The first step of Algorithm 1 finds a first cluster, denoted E_0, which is a subset of vertices which are connected. X' is the set of already treated vertices. It is initialized to E_0. This first step can be done easily, using an heuristic. Then, let $X_1, X_2, \ldots X_k$ be the connected components of the subgraph $G[X \backslash E_0]$ induced by the deletion of the vertices of E_0 in G. Each one of these sets is inserted in a queue F. For each element X_i removed from the queue F, let $V_i \subseteq X$ be the set of vertices in X' which are adjacent to at least one vertex in X_i. Note that V_i (which can be connected or not) is a separator of the graph G since the deletion of V_i in G makes G disconnected (X_i being disconnected from the rest of G). A new cluster E_i is then initialized by this set V_i. So, we consider the subgraph of G induced by V_i and X_i, that is $G[V_i \cup X_i]$. We choose a first vertex

Algorithm 1. *Bag-Connected-TD*

Input: A graph $G = (X, C)$
Output: A set of clusters $E_0, \ldots E_m$ of a bag-connected tree-decomposition of G
Choose a first connected cluster E_0 in G;
$X' \leftarrow E_0$;
Let $X_1, \ldots X_k$ be the connected components of $G[X \backslash E_0]$;
$F \leftarrow \{X_1, \ldots X_k\}$;
while $F \neq \emptyset$ **do** /* find a new cluster E_i */
  Remove X_i from F;
  Let $V_i \subseteq X'$ be the neighborhood of X_i in G;
  $E_i \leftarrow V_i$;
  Search in $G[V_i \cup X_i]$ starting from $V_i \cup \{x\}$ with $x \in X_i$. Each time a new vertex x is found, it is added to E_i. The process stops once the subgraph $G[E_i]$ is connected;
  if V_i *belongs to the set of clusters already found* **then** Delete the cluster V_i (because $V_i \subsetneq E_i$) $X' \leftarrow X' \cup E_i$;
  Let $X_{i_1}, X_{i_2}, \ldots X_{i_{k_i}}$ be the connected components of $G[X_i \backslash E_i]$;
  $F \leftarrow F \cup \{X_{i_1}, X_{i_2}, \ldots X_{i_{k_i}}\}$;

$x \in X_i$ that is connected to at least one vertex of E_i (so one vertex of V_i). This vertex is added to E_i. If $G[E_i]$ is connected, we stop the process because we are sure that E_i will be a new connected cluster. Otherwise, we continue, taking another vertex of X_i.

Figure 2(b) shows the computation of E_1, the second cluster (after E_0), at the first pass in the loop. After the addition of vertices a, b and c, the subgraph $G[V_1 \cup \{a, b, c\}]$ is not connected. If the next reached vertex is d, it is added to E_1, and thus, $E_1 = V_1 \cup \{a, b, c, d\}$ is a new connected cluster, breaking the search in $G[V_1 \cup X_1]$.

When this process is finished, we add the vertices of E_i to X' and we compute $X_{i_1}, \ldots X_{i_{k_i}}$ the connected components of the subgraph $G[X_i \backslash E_i]$. Each one is then inserted in the queue F. In the example of Figure 2(b), two connected components will be computed, $\{e\}$ and $\{f, g, h\}$. This process continues while the queue is not empty. In the example, in the right part of the graph, the algorithm will compute 3 connected clusters: $\{d, e\}$, $\{b, c, d, f\}$ and $\{f, g, h\}$.

Note that the line 10 is only useful when the set V_i computed at line 7 is a previously built cluster. In such a case, the cluster V_i can be removed. Indeed, as $V_i \subsetneq E_i$, V_i becomes useless in the tree-decomposition.

We now establish the validity of the algorithm and we evaluate its time complexity.

Theorem 2. *The algorithm Bag-Connected-TD computes the clusters of a bag-connected tree-decomposition of a graph G.*

Proof. We need only to prove the lines 5-13 of the algorithm. We first prove the termination of the algorithm. At each pass through the loop, at least one vertex will be added to the set X' and this vertex will not appear later in a new element of the queue because they are defined by the connected components of $G[X_i \backslash E_i]$, a subgraph that contains strictly fewer vertices than was contained in X_i. So, after a finite number of steps, the set $X_i \backslash E_i$ will be an empty set, and therefore no new addition in F will be possible.

We now show that the set of clusters $E_0, E_1, \ldots E_m$ induces a bag-connected tree-decomposition. By construction each new cluster is connected. So, we have only to prove that they induce a tree-decomposition. We prove this by induction on the added clusters, showing that all these added clusters will induce a tree-decomposition of the graph $G(X')$.

Initially, the first cluster E_0 induces a tree-decomposition of the graph $G[E_0] = G[X']$.

For the induction, our hypothesis is that the set of already added clusters $E_0, E_1, \ldots E_{i-1}$ induces a tree-decomposition of the graph $G[E_0 \cup E_1 \cup \cdots \cup E_{i-1}]$. Consider now the addition of E_i. We show that by construction, $E_0, E_1, \ldots E_{i-1}$ and E_i induces a tree-decomposition of the graph $G[X']$ by showing that the three conditions (i), (ii) and (iii) of the definition of tree-decompositions are satisfied.

(i) Each new vertex added in X' belongs to E_i
(ii) Each new edge in $G[X']$ is inside the cluster E_i.
(iii) We can consider two different cases for a vertex $x \in E_i$, knowing that for other vertices, the property is already satisfied by the induction hypothesis:
   (a) $x \in E_i \backslash V_i$: in this case, x does not appear in another cluster than E_i and then, the property holds.
   (b) $x \in V_i$: in this case, by the induction hypothesis, the property was already verified.

Finally, it is easy to see that if the line 10 is applied, we obtain a tree-decomposition of the graph $G[X']$. □

Theorem 3. *The time complexity of the algorithm Bag-Connected-TD is $O(n(n+e))$.*

Proof. The lines 1-4 are feasible in linear time, that is $O(n+e)$, since the cost of computing the connected components of $G[X \backslash E_0]$ is bounded by $O(n+e)$. Nevertheless, we can note that the line 1 can be done by a more expensive heuristic to get a more relevant first cluster, but at most in $O(n(n+e))$ in order not to exceed the time complexity of the most expensive step of the algorithm. We analyze now the cost of the loop (line 5). Firstly, note that there is less than n insertions in the queue F. Now, we analyze the cost of each treatment associated to the addition of a new cluster, and we give for each one, its global complexity.

- Line 6: obtaining the first element X_i of F is bounded by $O(n)$, thus globally $O(n^2)$.
- Line 7: obtaining the neighborhood $V_i \subseteq X'$ of X_i in G is bounded by $O(n+e)$, thus globally by $O(n(n+e))$.
- Line 8: this step is feasible in $O(n)$, thus globally $O(n^2)$.
- Line 9: the cost of the search in $G[V_i \cup X_i]$ starting with vertices of V_i and $x \in X_i$ is bounded by $O(n+e)$. Since the while loop runs at most n times, the global cost of the search in these subgraphs is bounded by $O(n(n+e))$. Moreover, for each new added vertex x, the connectivity of $G[E_i]$ is tested with an additional cost bounded by $O(n+e)$. Note since such a vertex is added at most one time, globally, the cost of this test is bounded by $O(n(n+e))$. So, the cost of the line 9 is globally bounded by $O(n(n+e))$.

- Line 10: using an efficient data structure, this step can be realized in $O(n)$, thus globally $O(n^2)$.
- Line 11: the cost of finding the connected components of $G[X_i \backslash E_i]$ is bounded by $O(n + e)$. So, globally, the cost of this step is $O(n(n + e))$.
- Line 12: the insertion of a set X_{i_j} in F is feasible in $O(n)$, thus globally $O(n^2)$ since there is less than n insertions in F.

Finally, the time complexity of the algorithm *Bag-Connected-TD* is $O(n(n + e))$. □

From a practical viewpoint, it can be assumed that the choice of the first cluster E_0 can be crucial for the quality of the decomposition which is being computed. Similarly, the choice of vertex x, selected in line 9 may be of considerable importance. For these two choices, heuristics can of course be used. This is discussed in the next section. However, a particular choice of these heuristics makes it possible, without any change of the complexity, to compute optimal tree-decompositions for the case of triangulated graphs. Assume that the first cluster E_0 is a maximal clique. This can be done efficiently using a greedy approach. Now, for the choice of the vertex x in line 9, we consider the vertex which has the maximum number of neighbors in the set V_i. As in a triangulated graph, all the clusters of an optimal tree-decomposition are cliques, necessarily, V_i being a clique, x will be connected to all the vertices of V_i and thus, E_i will be a clique. Progressively, each maximal clique will be found and the tree-decomposition will be optimal. Line 10 will be used for the case of maximal cliques including more than one vertex x of a new connected component. In any case, the practical interest of this type of decomposition is based on both the efficiency of its computation, but also on the significance which it may have for solving CSPs. This is discussed in section 5.

5 Experiments

In this section, we mainly compare the solving efficiency and the structural parameters for BTD using tree-decompositions produced by Min-Fill with ones for BTD exploiting bag-connected tree-decompositions. These latter are computed thanks to the *Bag-Connected-TD* algorithm. Regarding the choice of the first cluster in *Bag-Connected-TD*, it consists in computing greedily a maximal clique of the constraint network[7]. To choose the next vertex, we have considered 6 heuristics. We present here the best ones:

- NV1: the next vertex is a vertex in the neighborhood of previously chosen vertices,
- NV2: the vertices are processed in the decreasing degree order,
- NV3: the vertices are processed according to the order they are visited by a breadth-first traversal of the graph from the vertices of V_i,
- NV4: we choose as next vertex the vertex which has the maximum number of neighbors in the set V_i.

The solving is achieved by BTD based on MAC, by using the dom/wdeg variable heuristic [23]. We choose as root cluster the cluster E_i which maximizes the ratio $\frac{e_i}{|E_i|-1}$ with e_i the number of constraints of the cluster E_i. This choice provides better results than

[7] Remind that we use the 2-section for non-binary instances.

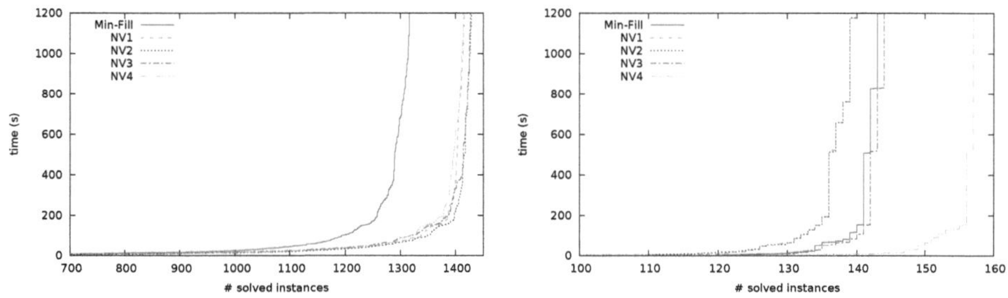

(a) (b)

Fig. 3. The cumulative number of solved instances for each considered tree-decomposition for instances for which Min-Fill produces some disconnected clusters (a), for instances for which Min-Fill produces a bag-connected tree-decomposition (b)

ones of [24]. The decomposition runtimes for Min-Fill and *Bag-Connected-TD* are similar and are included in the runtime of BTD. All the implementations are written in C++. The experimentations are performed on a linux-based PC with an Intel Pentium IV 3.2 GHz and 1 GB of memory.

5.1 Instances for Which Min-Fill Produces Some Disconnected Clusters

In this subsection, we compare the bag-connected tree-decompositions with disconnected ones from the viewpoint of the solving efficiency. With this aim in view, we consider 1,597 instances (of arbitrary arity) among the 2,310 instances of the CSP 2008 Competition for which Min-Fill produces a tree-decomposition with at least one disconnected cluster. The excluded instances are instances which cannot be solved without exceeding the time limit (namely 1,200 s) or which contain global constraints (not implemented yet in our solver). Among the considered instances, we can notably find instances from families rlfap, fapp, modifiedRenault, graphColoring, bqwh or travellingSalesman.

Figure 3(a) presents the cumulative number of solved instances for each considered tree-decomposition. First, we can observe that, by using any bag-connected tree-decomposition, BTD solves more instances than by using the disconnected tree-decompositions produced by Min-Fill. Note that this observation remains true if we use any connected decomposition based on the non presented heuristics. The best number of solved instances is reached thanks to the tree-decomposition based on the heuristic NV2 and NV3. These decomposition allow us to solve respectively 114 and 111 additional instances w.r.t. Min-Fill. Those based on NV1 and NV4 are close to each other. Moreover, for any decomposition, most instances are solved in less than 60 s.

Now, in order to fairly compare the runtimes, we only consider the instances which are solved by BTD for all the considered tree-decompositions, including Min-Fill. The runtime for solving the 1,230 instances by using the decompositions based on Min-Fill is 50,669 s while by using the connected decompositions based on NV1, it requires only 32,372. The connected decomposition based on NV2 is relatively close to NV1 (namely

33,202 s). Those based on NV3 and NV4 are slightly slower (respectively 36,420 s and 36,087 s). Note that the two other heuristics (not presented here) also outperform the Min-Fill decomposition.

If we focus on the 329 instances having a suitable structure (i.e. instances having a ratio n/w^+ greater than 2), again, we observe the same trend, namely that BTD with bag-connected tree-decomposition performs better than BTD with Min-Fill. The best cumulative runtime is achieved by BTD using NV1 in 5,698 s while the worst one is obtained by BTD using Min-Fill in 13,641 s. BTD using NV2 and NV4 are close to each other with respectively 6,137 s and 6,010 s while BTD with NV3 needs 8,483 s.

Finally, if we compare these results with ones obtained by a classical enumerative algorithm like MAC, we can note that some instances solved by BTD with some NV_i are not solved by MAC and conversely. We also observe that MAC performs sometimes better, sometimes worse than BTD using connected decompositions. However, globally, they have similar results. This is explained by the fact that most of the 1,597 instances we consider are far from having a suitable structure. In contrast, when the structure has interesting features, BTD outperforms MAC. For instance, BTD with decomposition based on NV3 requires 856 s for solving 10 instances over the 12 instances of rlfap-Scens11 family while MAC only solves 8 instances in 1,595 s. Moreover, for solving these 8 instances, BTD requires only 63 s, that is more than 25 times faster than MAC.

5.2 Instances for Which Min-Fill Produces a Bag-Connected Tree-Decomposition

This subsection briefly deals with the behavior of BTD when solving instances for which Min-Fill produces a bag-connected tree-decomposition. Of course, for such instances, Min-Fill and our *Bag-Connected-TD* algorithm do not necessarily produce the same tree-decompositions. By lack of place, we focus directly our study on the more relevant instances (i.e. the 191 instances having a suitable structure for a decomposition approach).

As shown in Figure 3(b), BTD using Min-Fill succeeds in solving more instances than BTD using NV1 or NV2 (143 instances against 140) but less than BTD using NV3 and NV4 which solve respectively 144 and 157 instances. If we focus our study on the 132 instances which are solved by BTD for all the considered tree-decompositions, including Min-Fill, BTD using NV3, NV4 or Min-Fill obtain the best cumulative runtime (respectively in 1,283 s, 1,298 s and 1,280 s) while BTD using NV1 or NV2 are slower (respectively 2,226 s and 2,265 s).

5.3 Comparisons of the Structural Parameters

Table 1 presents the value of the structural parameters for some instances. Not surprisingly, Min-Fill produces tree-decompositions with smaller widths and larger numbers of clusters than ones produced by *Bag-Connected-TD*. However, if in some cases, the width obtained by *Bag-Connected-TD* is significantly larger than one provided by Min-Fill (e.g. the width produced by NV3 for instance squares-23-23), in other cases, it remains relatively close (even sometimes equal) to one obtained by Min-Fill. This notably occurs for instance renault-mod-33_ext but also for instances for which Min-Fill

Table 1. Value of the structural parameters for some instances for which Min-Fill produces some disconnected clusters (a), for which Min-Fill produces a bag-connected tree-decomposition (b)

	Instances	n	e	Min-Fill		NV1		NV2		NV3		NV4	
				w^+	s	w_c^+	s	w_c^+	s	w_c^+	s	w_c^+	s
(a)	2-insertions-4-3	149	541	38	34	66	54	95	14	101	66	58	57
	ewddr2-10-by-5-9	50	265	16	15	22	17	21	20	26	23	45	37
	renault-mod-33_ext	111	133	11	11	12	11	14	11	17	15	16	13
	scen7	400	2,865	33	29	90	48	319	9	116	94	81	34
	squares-23-23	1,058	1,268	45	4	45	5	45	5	235	88	45	26
	fapp06-0500-1	500	3,478	221	210	286	284	286	284	314	314	313	248
	js-taillard-15-100-4	225	1785	86	70	114	102	121	97	129	102	210	197
(b)	mps-red-qnet1	5,380	621	970	773	1,272	1,265	1,272	1,265	978	954	998	974
	anna-9	138	493	12	12	14	14	14	14	16	15	14	13
	haystacks-10	100	459	9	1	9	1	9	1	9	1	9	1
	renault-mod-8_ext	111	126	11	11	11	11	12	11	13	12	11	11
	qwh-15-106-9_ext	225	2324	99	99	102	102	102	102	103	103	173	168

produces a bag-connected tree-decomposition (see part (b) of Table 1). We also observe that the quality of the width obtained thanks to *Bag-Connected-TD* may significantly vary depending on the considered instances. If NV1 often presents the best width among ones computed by *Bag-Connected-TD* algorithm, it is sometimes outperformed by NV3 or NV4 (e.g. for instance mps-red-qnet1).

Regarding the parameter s, the observed trends are similar to ones for the width.

6 Conclusion

In this paper, we have introduced the concept of Bag-Connected Tree-Decomposition in the field of constraint network decomposition. After having shown the interest of this new parameter and proposed a first polynomial time algorithm which computes Bag-Connected Tree-Decompositions, we have experimentally demonstrated the relevance of this approach since it allows to significantly improve the solving of CSP using decomposition methods. Indeed, by using bag-connected tree-decompositions, BTD succeeds in solving many more instances. Moreover, the improvement of the runtime is approximately 63% w.r.t. BTD using tree-decompositions with disconnected clusters produced by Min-Fill. This benefit mainly comes from the connectedness of clusters since when Min-Fill produces bag-connected tree-decompositions, the results of the different versions of BTD are close. Finally, thanks to these bag-connected decompositions, BTD also can significantly outperform MAC for well structured instances (e.g. for the rlfapScens11 family, BTD can be 25 times faster than MAC).

The first extension of this work is related to the study of bag-connected tree-decompositions in the more general field of Graphical Models in AI. This concerns the study of this notion for other classes of methods as Hypertree-Decomposition, And/Or Search, Bucket Elimination, etc. This approach is particularly justified by the fact that, even if some of these approaches are based on other parameters (e.g. Hypertree-Decomposition), their efficient implementations use generally algorithms coming from Tree-Decompositions (e.g. Min-Fill for Hypertree-Decomposition [25]). Another promising study is related to the use of bag-connected tree-decompositions in the field of optimization and counting problems. The second extension of this work is related to a theoretical study of this new graph parameter from a mathematical viewpoint. For example, what are the fundamental properties of this parameter? For what classes of graphs, this parameter is easy to compute, or is close to the tree-width? Or, are there problems which are hard when the tree-width is bounded by a constant, and which are easy when the bag-connected tree-width is bounded by a constant?

References

1. Rossi, F., van Beek, P., Walsh, T.: Handbook of Constraint Programming. Elsevier (2006)
2. Berge, C.: Graphs and Hypergraphs. Elsevier (1973)
3. Freuder, E.: A Sufficient Condition for Backtrack-Free Search. JACM 29(1), 24–32 (1982)
4. Dechter, R., Pearl, J.: Tree-Clustering for Constraint Networks. Artificial Intelligence 38, 353–366 (1989)
5. Robertson, N., Seymour, P.D.: Graph minors II: Algorithmic aspects of treewidth. Algorithms 7, 309–322 (1986)
6. Cabon, C., de Givry, S., Lobjois, L., Schiex, T., Warners, J.P.: Radio Link Frequency Assignment. Constraints 4, 79–89 (1999)
7. Jégou, P., Ndiaye, S.N., Terrioux, C.: Computing and exploiting tree-decompositions for solving constraint networks. In: van Beek, P. (ed.) CP 2005. LNCS, vol. 3709, pp. 777–781. Springer, Heidelberg (2005)
8. Müller, M.: Connected tree-width. CoRR, abs/1211.7353 (2012)
9. Gottlob, G., Leone, N., Scarcello, F.: A Comparison of Structural CSP Decomposition Methods. Artificial Intelligence 124, 243–282 (2000)
10. Gyssens, M., Jeavons, P., Cohen, D.: Decomposing constraint satisfaction problems using database techniques. Artificial Intelligence 66, 57–89 (1994)
11. Tarjan, R., Yannakakis, M.: Simple linear-time algorithms to test chordality of graphs, test acyclicity of hypergraphs, and selectively reduce acyclic hypergraphs. SIAM Journal on Computing 13(3), 566–579 (1984)
12. Dechter, R., El Fattah, Y.: Topological Parameters for Time-Space Tradeoff. Artificial Intelligence 125, 93–118 (2001)
13. Dechter, R.: Constraint processing. Morgan Kaufmann Publishers (2003)
14. Jégou, P., Terrioux, C.: Hybrid backtracking bounded by tree-decomposition of constraint networks. Artificial Intelligence 146, 43–75 (2003)
15. Petke, J.: On the bridge between Constraint Satisfaction and Boolean Satisfiability. PhD thesis, University of Oxford (2012)
16. Arnborg, S., Corneil, D., Proskuroswki, A.: Complexity of finding embeddings in a k-tree. SIAM Journal of Discrete Mathematics 8, 277–284 (1987)
17. Golumbic, M.: Algorithmic Graph Theory and Perfect Graphs. Academic Press, New York (1980)
18. Kjaerulff, U.: Triangulation of Graphs - Algorithms Giving Small Total State Space. Technical report, Judex R.R. Aalborg., Denmark (1990)
19. Rose, D.J.: A graph theoretic study of the numerical solution of sparse positive denite systems of linear equations. In: Read, R.C. (ed.) Graph Theory and Computing, pp. 183–217. Academic Press, New York (1973)

20. Gavril, F.: Algorithms for minimum coloring, maximum clique, minimum covering by cliques, and maximum independent set of a chordal graph. SIAM Journal on Computing 1(2), 180–187 (1972)
21. Sabin, D., Freuder, E.: Contradicting Conventional Wisdom in Constraint Satisfaction. In: Borning, A. (ed.) PPCP 1994. LNCS, vol. 874, pp. 125–129. Springer, Heidelberg (1994)
22. Fraigniaud, P., Nisse, N.: Connected treewidth and connected graph searching. In: Correa, J.R., Hevia, A., Kiwi, M. (eds.) LATIN 2006. LNCS, vol. 3887, pp. 479–490. Springer, Heidelberg (2006)
23. Boussemart, F., Hemery, F., Lecoutre, C., Sais, L.: Boosting systematic search by weighting constraints. In: Proceedings of ECAI, pp. 146–150 (2004)
24. Jégou, P., Ndiaye, S.N., Terrioux, C.: An extension of complexity bounds and dynamic heuristics for tree-decompositions of CSP. In: Benhamou, F. (ed.) CP 2006. LNCS, vol. 4204, pp. 741–745. Springer, Heidelberg (2006)
25. Dermaku, A., Ganzow, T., Gottlob, G., McMahan, B.J., Musliu, N., Samer, M.: Heuristic methods for hypertree decomposition. In: Gelbukh, A., Morales, E.F. (eds.) MICAI 2008. LNCS (LNAI), vol. 5317, pp. 1–11. Springer, Heidelberg (2008)

CIP and MIQP Models for the Load Balancing Nurse-to-Patient Assignment Problem

Wen-Yang Ku, Thiago Pinheiro, and J. Christopher Beck

Department of Mechanical & Industrial Engineering,
University of Toronto, Toronto, Ontario M5S 3G8, Canada
{wku,jcb}@mie.utoronto.ca, thiagopj@gmail.com

Abstract. The load balancing nurse-to-patient assignment problem requires the assignment of nurses to patients to minimize the variance of the nurses' workload. This challenging benchmark is currently best solved exactly with constraint programming (CP) using the SPREAD constraint and a problem-specific heuristic. We show that while the problem is naturally modelled as a mixed integer quadratic programming (MIQP) problem, the MIQP does not match the performance of CP. We then develop several constraint integer programming (CIP) models that include bounds propagation, linear relaxations, and cutting planes associated with the QUADRATIC, GCC, and SPREAD global constraints. While the QUADRATIC and GCC techniques are known, our additions to the SPREAD constraint are novel. Our empirical results demonstrate that the CIP approach substantially out-performs the MIQP model, but still lags behind CP. Finally, we propose a simple problem-specific variable ordering heuristic which greatly improves the CIP models, achieving performance about an order of magnitude faster than CP and establishing a new state of the art.

1 Introduction

The load balancing nurse-to-patient assignment problem assigns nurses to a hospital zone (e.g., a nursery room) and to patients within the zone to minimize the variance in the total acuity of patients across nurses. A constraint programming approach, combining the SPREAD constraint [1] with problem-specific search heuristics, is currently the state of the art for solving this problem exactly [2,3]. In this paper, we address the problem with two approaches that, to our knowledge, have not yet been applied: mixed integer quadratic programming (MIQP) and constraint integer programming (CIP) [4,5].

Mixed integer linear programming (MILP) techniques have been extended to reason about convex quadratic constraints [6], allowing MIQPs to be solved by commercial MILP solvers. Since variance minimization can be modelled as an MIQP and underlying MILP technology has seen substantial improvement over the past few years [7], we develop and apply an MIQP model to the problem. The model is natural, given the direct representation of the quadratic constraints, however we show that model performs substantially worse using CPLEX than the existing CP model implemented in COMET.

B. O'Sullivan (Ed.): CP 2014, LNCS 8656, pp. 424–439, 2014.

We then improve the MIQP model in two orthogonal directions. First, in the constraint integer programming (CIP) framework, we add global constraints that embed both inference techniques in the form of bounds propagation and relaxation-based reasoning via constraint-specific linear relaxations and cutting planes. These techniques are functional extensions to global constraints (beyond filtering) that can be implemented in any solver that allows linear relaxations and the dynamic addition of cutting planes. In particular, we augment the MIQP to include QUADRATIC [8], GCC, and SPREAD global constraint. For the first two constraints, the propagation, relaxation, and cutting plane techniques are taken from the literature. For the SPREAD constraint, we implement the bound consistency algorithm due to Schaus et al. [9] and develop novel relaxation and cutting plane techniques. Our augmented global constraints can be soundly used in any models in which the un-augmented global constraints appear. The second direction of improvements is to modify the default branching heuristics. Given the structure of the nurse-to-patient assignment problem, we develop two simple, static variable ordering heuristics that influence the default, general branching rules of the CIP solver, SCIP [4,10].

Our empirical results show that the CIP models without problem-specific branching heuristics significantly outperform the MIQP model in CPLEX, achieving about a three times speed-up. However, the CP model is still more than twice as fast. The primary benefit arises from the use of the QUADRATIC constraint with worse overall run-time performance from the additional inclusion of the GCC and SPREAD constraints.

When problem-specific branching priorities are used, all three CIP models outperform the state-of-the-art CP model. The best model includes all three global constraints and results in an order of magnitude improvement over CP. Compared to the default heuristics, the problem-specific heuristics find and prove optimal solutions with 30 to 40 times less effort in terms of both search tree size and run-time. Subsequent analysis shows that the improved performance arises due to the rapid improvement in both the primal (upper) bound and the dual (lower) bound early in the search.

The paper is organized as follows. In Section 2 we present the problem and previous results. We present the CP, MIQP, and CIP models in Section 3. Section 4 describes the global constraints used in the CIP models while Section 5 defines the branching heuristics. Sections 6 and 7 provide computational results and discussions. We conclude in Section 8.

2 Background

The load balancing nurse-to-patient assignment problem requires the assignment of nurses to new-born infant patients across different zones in a hospital [11]. Each infant is hospitalized in one of the rooms, or zones, in the nursery and requires a certain amount of care depending on the acuity of his/her condition. The problem has two main decisions. First, each nurse has to be assigned to a zone in which he/she will work for the entire shift. Second, each nurse has to be assigned to a set of patients in the same zone. The objective is to minimize the variance of the total acuity assigned to each nurse. Such an assignment will avoid overloading the nurses, which can result in stress and poor quality of the care.

The problem was originally modeled and solved as an MILP with a linear objective function that minimizes the sum of the differences between the maximum and minimum assigned workload in each zone [11]. However, although the objective function ensures the workload of each zone is evenly distributed, the workload difference between nurses in different zones may be large.

Schaus et al. [2,3] proposed a CP model that directly minimizes the standard deviation of the nurses' workloads using the SPREAD constraint [1]. Results showed significant improvements in both solution quality and computational efficiency compared to the MILP model. The CP approach is able to solve problems with two zones exactly, but not very efficiently without using an approximation of the number of nurses assigned to each zone and further decomposition. While the approximation and decomposition techniques solve the problem quickly, optimality is not guaranteed. In this paper, we are interested in exact solutions.

3 Mathematical Models

Formally, the load balancing nurse-to-patient assignment problem is defined by a finite set of m patients, a finite set of n nurses, and a finite set of p zones. In addition, let P_k denote the set of patients in zone k, thus $\{P_1, \ldots, P_p\}$ forms a partition of $\{1, \ldots, m\}$. For each patient, i, a non-negative integer, A_i, represents his/her acuity. The mean of the nurses' workload is therefore computed as $\mu = \sum_{i=1}^{m} A_i/n$. The sum of the acuity levels of the patients assigned to a nurse cannot exceed $MaxAcuity$ and the total number of assigned patients cannot exceed $MaxPatients$. Each patient can only be assigned to one nurse, and each nurse can only be assigned to one zone. The objective is to minimize the variance or, equivalently the standard deviation, of the nurses' workload, measured as the total acuity assigned to the nurses.

3.1 The CP Model

In the state-of-the-art exact CP model [2], the decision variables are defined as follows:

- N_i denotes the nurse assigned to patient i.
- W_j denotes nurse j's workload.
- σ denotes the standard deviation of the variables $W_1, \ldots, W_n$.

The CP model is shown in Fig. 1. Constraint (2) is the SPREAD constraint, which relates a set of variables to their mean and standard deviation. Constraint (3) is a global packing constraint that governs the packing of items, corresponding to patients with "size" equal to their acuity levels, into bins corresponding to the workload of each nurse. Constraint (4) is the GCC placing the

$$
\begin{aligned}
\min \quad & \sigma && (1)\\
\text{s.t.} \quad & \texttt{spread}(\{W_1, \ldots, W_n\}, \mu, \sigma), && (2)\\
& \texttt{multiknapsack}(\{N_1, \ldots, N_m\}, \{A_1, ..., A_m\}, \{W_1, \ldots, W_n\}), && (3)\\
& \texttt{cardinality}(\{N_1, \ldots, N_m\}, \{1, \ldots, n\}, \{1, \ldots, MaxPatients\}), && (4)\\
& \texttt{pairwiseDisjoint}(\{Z_1, \ldots, Z_p\}), && (5)\\
& W_j \in \{min\{A_i\}, \ldots, MaxAcuity\}, \qquad j = 1, \ldots, n && (6)\\
& N_i \in \{1, \ldots, n\}. \qquad i = 1, \ldots, m && (7)
\end{aligned}
$$

Fig. 1. The CP model of Schaus et al. [2]

limit on each nurse in terms of number of assigned patients. Constraint (5) expresses that a nurse can only work in one zone during the shift, where Z_k is the set of nurses assigned to zone k, i.e., $Z_k = \bigcup_{i \in P_k} N_i$. The `pairwiseDisjoint` constraint enforces pairwise empty intersections among variables representing the set of nurses working in each zone. Constraint (6) expresses bounds on the workload of each nurse. Since there are always more patients than nurses, each nurse will be assigned to at least one patient and, therefore, the W_j variables have a lower bound equal to the minimum acuity among all patients.

A customized search heuristic is an important aspect of the success of the CP model. First, the symmetry arising from the interchangeability of the nurses is dynamically broken during search by exploiting the equivalence among all nurses who have not yet been assigned a patient. Second, problem-specific variable and value ordering heuristics are implemented: the unassigned patient with the highest acuity is selected and assigned to the nurse with the current smallest workload.

3.2 The MIQP Model

We propose a mixed integer quadratic programming (MIQP) model for the problem that is mathematically equivalent to the CP model. In addition to μ, σ, and the W_j variables, we define two additional decision variables as follows:

- x_{ij} is equal to 1 if patient i is assigned to nurse j.
- z_{jk} is equal to 1 if nurse j is assigned to work in zone k.

The MIQP model is given in Fig. 2. The objective function is stated in (8). Constraint (9) specifies the quadratic relationship between the standard deviation, σ, and the workload variables, W_j. Constraint (10) ensures that each patient is assigned to exactly one nurse. Constraint (11) ensures that each nurse is assigned to exactly one zone. Constraint (12) calculates the workload of each nurse by summing over all patients. Constraint (13) ensures that the maximum number of patients assigned per nurse does not exceed the capacity of the nurse and that a nurse is only assigned patients from his/her assigned zone. Constraint (14) is the symmetry breaking constraint.

$$\min \quad \sigma \tag{8}$$

$$\text{s.t.} \quad \sigma \geq \sqrt{\frac{\sum_{j=1}^{n}(W_j - \mu)^2}{n}}, \tag{9}$$

$$\sum_{j=1}^{n} x_{ij} = 1, \qquad i = 1, \ldots, m \tag{10}$$

$$\sum_{k=1}^{p} z_{jk} = 1, \qquad j = 1, \ldots, n \tag{11}$$

$$W_j = \sum_{i=1}^{m} x_{ij} A_i, \qquad j = 1, \ldots, n \tag{12}$$

$$\sum_{i \in P_k} x_{ij} \leq z_{jk} MaxPatients, \qquad j = 1, \ldots, n, \quad k = 1, \ldots, p \tag{13}$$

$$W_j \leq W_{j+1}, \qquad j = 1, \ldots, n-1 \tag{14}$$

$$x_{ij} \in \{0, 1\}, \qquad i = 1, \ldots, m, \quad j = 1, \ldots, n \tag{15}$$

$$z_{jk} \in \{0, 1\}, \qquad j = 1, \ldots, n, \quad k = 1, \ldots, p \tag{16}$$

$$W_j \in [min\{A_i\}, MaxAcuity]. \qquad j = 1, \ldots, n \tag{17}$$

Fig. 2. The MIQP model

3.3 The CIP Model

The CIP model adds global constraints to the MIQP model. To do this, we include the N_i variable with the same meaning as in the CP model: the index of the nurse assigned to patient i. We link N_i to x_{ij} as follows:

$$N_i = \sum_{j=1}^{n} x_{ij} j. \qquad i = 1, \ldots, m \tag{18}$$

The global constraints added to the MIQP model to form the CIP model are Constraints (2) and (4) from the CP model (Fig. 1). The complete CIP model is the MIQP model in Fig. 2 plus Constraints (18), (2), and (4).

Note that there is no need for an explicit QUADRATIC global constraint in the CIP model. The SCIP solver used for the CIP models recognizes the quadratic nature of Constraint (9) and automatically includes the QUADRATIC constraint.

4 Global Constraints

Three global constraints are included in the CIP model: QUADRATIC, SPREAD, and GCC. For each constraint, bounds propagation is used as the underlying representation in the solver used, SCIP, employs an interval representation of variable domains. As the functionality of global constraints in CIP is more general than in CP, we discuss each constraint in this section.

4.1 The Quadratic Constraint

The QUADRATIC constraint [8] is used to reason about constraints with quadratic terms and consists of a set of n variables $\{W_1, \ldots, W_n\}$, a $n \times n$ symmetric matrix $\boldsymbol{A}$, and n-dimensional vectors $\boldsymbol{b}$, $\boldsymbol{l}$ and $\boldsymbol{u}$. The representation is given as follows:

$$\texttt{quad}(\{W_1, \ldots, W_n\}, \boldsymbol{A}, \boldsymbol{b}, \boldsymbol{l}, \boldsymbol{u}),$$

where $\boldsymbol{A} \in \mathbb{Q}^{n \times n}$, $\boldsymbol{b} \in \mathbb{Q}^n$, $\boldsymbol{l} \in \mathbb{Q}^n$ and $\boldsymbol{u} \in \mathbb{Q}^n$. The constraint ensures the following condition:

$$\boldsymbol{l} \leq \sum_{i=1}^{n} \sum_{j=1}^{n} A_{i,j} W_i W_j + \sum_{i=1}^{n} b_i \leq \boldsymbol{u}.$$

We use the existing QUADRATIC constraint in SCIP [8] with its default parameter values. It implements a number of problem solving techniques including bounds propagation, addition of linear relaxations, cutting plane generation, problem reformulation, and primal heuristics.

4.2 The Global Cardinality Constraint

The GCC constraint consists of a set of m variables $\{N_1, \ldots, N_m\}$, a set of n values $\{v_1, \ldots, v_n\}$, and a set of n pairs of values $[l_j, u_j]$, for each v_j. The constraint is satisfied if and only if each v_j is assigned at least l_j times and at most u_j times to the N_i variables. The representation is given as follows:

$$\texttt{cardinality}(\{N_1, ..., N_m\}, \{v_1, ..., v_n\}, \{[l_1, u_1], ..., [l_1, u_n]\}).$$

We use the bound consistency filtering algorithm due to Quimper et al. [12].

Relaxation. Linear relaxations have a central role in mixed integer programming. Given an MILP, which is, in general, NP-hard, the standard relaxation arises from ignoring the integrality constraints on the integer variables resulting in a polytime solvable linear program (LP). The LP plays numerous roles in search including providing a dual bound on a problem that is compared to the current best solution to prune search sub-trees in which no improving solution exists and in providing a basis for search heuristics.

A substantial body of work has developed linear relaxations for global constraints (e.g., [13,14,15]). We implement an existing relaxation based on the MILP representation of GCC [15]. Using the notation for GCC introduced above, if we define an additional binary variable $x_{ij} = 1$ if $N_i = v_j$, an exact formulation of GCC is presented in Fig. 3. This formulation is used by the solver to form a linear relaxation of the GCC constraint by ignoring the integrality constraint on x_{ij} (i.e., Constraint (22)) and replacing it with $x_{ij} \in [0, 1]$.

Our CIP model already contains linear constraints that are identical to some of those in Fig. 3. Specifically, Constraint (10), (15), and (18) in the CIP model are identical Constraints (19), (22), and (23), respectively. As Constraint (21) does not fix any x_{ij} variables in our problem, the only constraint we add to the CIP model is Constraint (20).

$$\sum_{j=1}^{n} x_{ij} = 1, \qquad i = 1, \ldots, m \tag{19}$$

$$l_j \leq \sum_{i=1}^{m} x_{ij} \leq u_j, \qquad j = 1, \ldots, n \tag{20}$$

$$x_{ij} = 0, \qquad \forall i, j \notin D_{x_i} \tag{21}$$

$$x_{ij} \in \{0, 1\}, \qquad \forall i, j \tag{22}$$

$$N_i = \sum_{j=1}^{n} v_j x_{ij}. \qquad i = 1, \ldots, m \tag{23}$$

Fig. 3. A MILP formulation of GCC [15]

Cutting Planes. With the importance of the linear relaxation to MILP solving, techniques for improving it via the addition of cutting planes (i.e., implied linear constraints) have been developed [16]. We use the cutting planes for GCC due to Hooker [15].

For the CIP model we generate one inequality constraint using all x_i variables:

$$\sum_{i=1}^{n} p_i v_i \leq \sum_{j=1}^{m} N_j \leq \sum_{i=1}^{n} q_i v_i, \tag{24}$$

where

$$p_i = \min\left\{u_i\ m - \sum_{j=1}^{i-1} p_j - \sum_{j=i+1}^{n} l_j\right\}, \quad i = 1, \ldots, n \tag{25}$$

$$q_i = \min\left\{u_i\ m - \sum_{j=i+1}^{n} q_j - \sum_{j=1}^{i-1} l_j\right\}. \quad i = n, \ldots, 1. \tag{26}$$

Note that p_i and q_i are the maximum number of times that v_i can be selected when, respectively, minimizing and maximizing the sum of the x_i variables, assuming, without loss of generality, that the v_is are ordered from smallest to largest. Similar inequalities including subsets of the x_i variables ranging in cardinality from 2 to $m - 1$ are also valid and could result in a smaller search space. However, our preliminary results indicated that the large number of the constraints added to the problem results in a very large model, reducing search efficiency. Please see Hooker [15] for more details.

4.3 The Spread Constraint

The SPREAD constraint was first proposed by Pesant [1] to achieve the balancing of a set of values based on statistical concepts. A bound consistency algorithm was proposed in the same paper and a simplified and extended filtering algorithm was presented in Schaus et al. [9].

The SPREAD constraint enforces a given mean μ and maximum standard deviation σ among a set of n variables $\{W_1, \ldots, W_n\}$. It can be defined as follows:

$$\texttt{spread}(\{W_1, \ldots, W_n\}, \mu, \sigma).$$

The filtering algorithm reduces the domains of the W_j variables based on μ and σ in $O(n^2)$ time-complexity. Another algorithm filters values in the domain of σ based on domains of the variables W_j in quadratic time [9]. Both algorithms are implemented in the SPREAD constraint in this paper. The complete SPREAD constraint also propagates from W_j and σ to μ. However, since in our problem the total acuity load and number of nurses are fixed and known, the mean, μ, is always fixed to a single value.

Relaxation. A simple relaxation of the SPREAD constraint is a single linear equality that enforces the mean among all the variables in the SPREAD constraint. The constraint guarantees that the sum of all variables W_j is equal to $\mu \times n$, where n is the number of variables.

$$\sum_{j=1}^{n} W_j = \mu \times n. \tag{27}$$

Cutting Planes. When the objective is to minimize the standard deviation, the quadratic objective function can be formulated as follows:

$$\min \sigma = \sqrt{\frac{\sum_{j=1}^{m}(W_j - \mu)^2}{n}}, \tag{28}$$

which can be re-written in the following form:

$$\min \left\| \boldsymbol{\mu} - \boldsymbol{I}\boldsymbol{W} \right\|_2^2, \tag{29}$$

where $\boldsymbol{I} \in \mathbb{Z}^{n \times n}$ is the identity matrix, $\boldsymbol{\mu} = \mu \boldsymbol{e}$, $\boldsymbol{\mu} \in \mathbb{R}^n$ and $\boldsymbol{W} \in \mathbb{R}^n$ are n-dimensional vectors.

Suppose the optimal integer solution $\boldsymbol{W}^*$ to the above problem satisfies the following bound

$$\left\| \boldsymbol{\mu} - \boldsymbol{I}\boldsymbol{W}^* \right\|_2^2 < \beta, \tag{30}$$

where β is a constant. Geometrically, Equation (30) is a hyper-ellipsoid with center $\boldsymbol{\mu}$.

Chang & Golub [17] proposed an efficient way to compute the smallest hyper-rectangle whose edges are parallel to the axes of the coordinate system and that includes the hyper-ellipsoid. Let $\mathcal{B}$ be the smallest hyper-rectangle including a hyper-ellipsoid of the general form $\left\| \boldsymbol{y} - \boldsymbol{A}\boldsymbol{x} \right\|_2^2 \leq \beta$, the lower bound $\boldsymbol{l}$ and the upper bound $\boldsymbol{u}$ can be computed as follows:

$$u_j = \left\lfloor \sqrt{\beta} \left\| \boldsymbol{A}^{-T} \boldsymbol{e}_j \right\|_2 + \boldsymbol{e}_j^T \boldsymbol{A}^{-1} \boldsymbol{y} \right\rfloor,$$

$$l_j = \left\lceil -\sqrt{\beta} \left\| \boldsymbol{A}^{-T} \boldsymbol{e}_j \right\|_2 + \boldsymbol{e}_j^T \boldsymbol{A}^{-1} \boldsymbol{y} \right\rceil .$$

Since in our problem, $\boldsymbol{A} = \boldsymbol{I}$ and $\boldsymbol{y} = \boldsymbol{\mu}$, Equation (30) is actually a hypersphere with center $\boldsymbol{\mu}$ with the same lower bounds and upper bounds for all the variables. Thus, the computation of the lower bounds and the upper bounds can be simplified as follows:

$$u_j = \left\lfloor \sqrt{\beta} + \mu \right\rfloor , \tag{31}$$

$$l_j = \left\lceil -\sqrt{\beta} + \mu \right\rceil . \tag{32}$$

An upper bound of the standard deviation σ_{max} can be used to compute β through the following relationship, based on Equation (28) and (29):

$$\beta = \sqrt{n\sigma_{max}^2}. \tag{33}$$

Therefore, the cutting planes $W_j \geq l_j$ and $W_j \leq u_j$ can be used to update the bounds on the W_j variables. In our implementation, these constraints are computed every time the upper bound of σ is improved. Since the constraints are globally feasible, the global bounds of the variables are updated if the new bounds are tighter than the current global bounds of the variables. We have previously applied a similar approach to solving binary quadratic programming problems within CIP [18].

We also place an initial lower bound on σ. Considering only the standard deviation minimization aspect of the original problem, the resulting relaxed problem can be defined as follows:

$$\min \quad \sigma_{relaxed}, \tag{34}$$

$$\text{s.t.} \quad \sigma_{relaxed} \geq \sqrt{\frac{\sum_{j=1}^{n}(W_j - \mu)^2}{n}}, \tag{35}$$

$$\sum_{j=1}^{n} W_j = n\mu, \tag{36}$$

$$W_j \in \mathbb{Z}, \qquad j = 1, \ldots, n, \tag{37}$$

where the optimal objective value $\sigma_{relaxed}$ is therefore a lower bound on σ, i.e., $\sigma \geq \sigma_{relaxed}$. The optimal solution to this relaxed problem can be obtained trivially by assigning W_js to either $\lceil \mu \rceil$ or $\lfloor \mu \rfloor$ that leads to the smallest standard deviation, while satisfying Constraint (36) and (37). For example, one can start with assigning all W_js to $\lceil \mu \rfloor$, and then assign the last $\lceil n(|\mu - \lceil \mu \rfloor|) \rfloor$ W_js to $\lceil \mu \rfloor + 1$ if $\lceil \mu \rfloor - \mu < 0$ ($\lceil \mu \rfloor - 1$ if $\lceil \mu \rfloor - \mu > 0$).

5 Branching Heuristics

It is well recognized in CP and MILP that the use of search heuristics can have substantial impact on problem solving performance [19,20,10,21]. One simple way to influence search without implementing new heuristics is to modify the heuristic priority of variables. In SCIP, the branching priority of variables can be modified, allowing problem-specific knowledge to be incorporated into the default heuristics. Increasing the branching priority of a set of variables means that they will be branched on earlier in the search.

We investigate two manipulations: 1) increasing the priority of the z_{jk} variables that assign nurses to zones and 2) increasing the branching priority of *both* the z_{jk} variables and the workload variables, W_j. In both conditions, we assign maximum priority to all corresponding variables.

Increasing z_{jk} Priority. We choose to increase the branching priority of the z_{jk} variables based on the intuition that they have a significant effect on many other variables. Pryor & Chinneck [21] have shown that to quickly find a feasible solution in MILP, it is often desirable to branch on variables whose assignment results in substantial change to the linear relaxation. When a z_{jk} variable is set to 1, Constraint (11) immediately results in the $p - 1$ other z_{ik} variables with $i = j$ being assigned to 0. Furthermore, through Constraint (13), fixing a z_{jk} variable to 0 leads to the assignment of m x_{ij} variables to 0. Therefore, whether branching up or down on the z_{jk} variables, we expect to see a substantial change in the linear relaxation.

Increasing W_j Priority. We choose to also increase the branching priority of the W_j variables due to their expected impact on the dual bound. A second intuition for heuristics in a MILP is to branch to quickly increase the dual bound (i.e., the lower bound in a minimization problem). It is often observed that a considerable amount of the effort in solving a problem is not in finding a solution but in proving its optimality. As the primary method for such a proof is through pruning the sub-trees when the dual bound meets or exceeds the incumbent solution value, it is often useful to base branching heuristics on increasing the dual bound (e.g., [4]).

Given Constraint (9), branching on variables other than W_j will tend to lead to relaxed W_j values that are close μ, resulting in a dual bound that is close to 0. Branching on the W_j variables, in contrast, can more quickly increase the dual bound as the upper and lower bounds on other W_j variables must be changed due to Constraint (27).

Our initial experiments showed that only increasing the branching priority of the W_j variables without the z_{jk} led to poor performance due to difficulty in finding feasible solutions. We will return to this point in Section 7.

6 Computational Results

We compare the performance of three solvers and the following 11 solver-model-heuristic combinations.

- The CP model (Fig. 1) is implemented in COMET. We indicate this model by *CP*.

- The MIQP model (Fig. 2) is implemented in CPLEX using default parameters. This model is referred to as *MIQP*.
- The basic CIP is declaratively identical to *MIQP* but, when solved using default SCIP, incorporates the QUADRATIC constraint. We experiment with three models corresponding to the branching priorities: CIP (default), CIP_z (increased z_{jk} priority), and $CIP_{z,w}$ (increased priority for z_{jk} and W_j).
- The CIP model augmented to include only the constraint propagation of the GCC and SPREAD constraints is indicated by a superscript p (for propagation). There are, again, three models: CIP^p, CIP^p_z, and $CIP^p_{z,w}$ corresponding to the branching priorities.
- The full CIP model, indicated by superscript f includes constraint propagation, relaxation, and cutting planes of the QUADRATIC, GCC, and SPREAD constraints: CIP^f, CIP^f_z, and $CIP^f_{z,w}$.

6.1 Experimental Setup

All experiments were performed on a Intel(R) Xeon(R) CPU E5-1650 v2 3.50GHz machine (in 64 bit mode) with 16GB memory running MAC OS X 10.9.2 with one thread. The software is CPLEX v12.5, SCIP v3.0.2, and Comet v2.1-1. The CPU time limit for each run on each problem instance is 7200 seconds.

6.2 Test Sets

Following the methodology of Schaus et al. [2], we generated 24 problem instances to closely resemble the original real world instances [11]. The maximum acuity for a nurse is set to $MaxAcuity = 105$ and the maximum number of patients per nurse is $MaxPatients = 3$. We generate problem instances with number of nurses, $n \in [9, 12]$, number of patients, $m \in [21, 30]$, and number of zones, $p = 2$.

6.3 Results

An overview of the results is given in Table 1. We report the arithmetic mean CPU time "arith", and the shifted geometric mean CPU time "geo" on the 24 instances.[1] The arithmetic mean on the number nodes "Nodes" (number of backtracks "Bts" for CP) and the optimality gap "Opt gap" are also presented. We finally report the number of instances for which an optimal solution is found and proved "# opt" and the number of optimal solutions found "# opt found" without necessarily being proved.

The results of the MIQP model demonstrate that while CPLEX is able to find feasible solutions to all problems, it can only find and prove optimality in 7 of 24 with a substantially larger run-time as compared to the CP model.

[1] The shifted geometric mean time is computed as follows: $\prod(t_i + s)^{1/n} - s$, where t_i is the actual CPU time, n is the number of instances, and s is chosen as 10. Using geometric mean can decrease the influence of outliers [4].

Table 1. Comparison of CP, MIQP, and 9 different CIP variations. For CP, the superscript indicates the number of instances for which no feasible solution was found. The optimality gap "Opt gap" is computed only for the problem instances where feasible solutions are found.

Solver	Model	Time to opt		Nodes or Bts	Opt gap (%)	# opt	# opt found
		arith	geo				
COMET	CP	749.87	137.10	42292088	0^1	23	23
CPLEX	MIQP	5649.15	4247.26	9587569	31.0	7	18
SCIP	CIP	1877.63	565.02	4096764	1.0	20	23
	CIP^p	2795.80	746.89	4683757	3.0	15	18
	CIP^f	2193.95	542.27	4040087	7.0	19	20
	CIP_z	1605.64	267.11	3694590	19.0	20	23
	CIP_z^p	1723.08	331.85	3221449	0	20	24
	CIP_z^f	1459.78	267.16	2549701	6.0	21	21
	$CIP_{z,w}$	100.74	32.26	210485	0	24	24
	$CIP_{z,w}^p$	94.49	32.29	132342	0	24	24
	$CIP_{z,w}^f$	74.05	27.90	130012	0	24	24

The first set of CIP models (CIP, CIP^p, and CIP^f) show a significant gain compared to MIQP: more than twice as many problems solved to proved optimality with about half the number of nodes. We attribute this strong performance to the QUADRATIC constraint. The inclusion of constraint propagation in the GCC and SPREAD constraints *degrades* performance (CIP^p vs. CIP) both in terms of run-time and search tree size. The 15% larger search tree in particular is interesting and deserves more study. We speculate, given results below, that we may be observing a negative interaction between the constraint propagation and relaxation-based MILP-style search, indicating that we cannot necessarily expect improved performance from simply adding global constraint propagation to a MILP-style search. Comparing CIP^f to CIP shows similar run-times and tree sizes but worse solution quality for the full model. We believe that the gains from the global constraints are not large enough to out-weigh the increased computation they incur.

In the second set of CIP results, with increased branching priority for the z_{jk} variables, we see a substantial performance improvement in all models compared to the default heuristic. The inclusion of global constraints, whether just the propagation or the full model, results in smaller search trees by about 13% for CIP_z^p and 30% for $CIP_{z,w}^f$ and better solution quality. However, the run-times are either about the same or are actually worse than the CIP_z model.

Finally, the most substantial gains arise from increasing the branching priority of both the z_{jk} variables and the W_j variables. About an order of magnitude improvement is seen compared to the previous CIP models. The inclusion of global constraint propagation and propagation plus relaxation and cutting planes leads to clear gains with search tree sizes of almost 40% smaller than $CIP_{z,w}$. Furthermore, the three CIP models solve all problems to proved optimality from

7 to 10 times faster (in arithmetic mean) than the CP model. We believe these are the first models of any type that have been able to improve on the performance of the CP state of the art.

7 Discussion

Our experimental results have demonstrated that the hybridization of CP and MIQP techniques within the framework of constraint integer programming results in a new state of the art for the load balanced nurse-to-patient assignment problem.

Primal Bounds and Dual Bounds. Analysis of the solving behavior of the MIQP and default CIP model points to the importance of the dual bound in achieving strong performance. The MIQP approach using CPLEX is able to find high quality solutions early in the search but then only improves the dual bound very slowly, often timing-out without proving optimality. The MIQP model is able to find the optimal solution for 18 of the 24 problems but can only prove optimality in 7. The primary advantage of the default CIP model over MIQP appears to be due to its rapid improvements in the dual bound.

A more complicated pattern emerges from the comparison of the full CIP model with and without the branching heuristics (CIP^f vs. CIP^f_z vs. $CIP^f_{z,w}$). Fig. 4 plots the evolution of the mean primal and dual bounds over time for the three CIP models. For each instance, the primal, p, and dual, d, gaps are computed as follows: $p = (UB(\sigma) - \sigma^*)/\sigma^*$ and $d = (LB(\sigma) - \sigma^*)/\sigma^*$, where σ^* is the known optimal solution cost and $UB(\sigma)$ and $LB(\sigma)$ respectively represent the best upper and lower bounds at a given point in the search.

Consistent with our intuitions in Section 5, $CIP^f_{z,w}$ delivers tight primal and dual bounds, though it is more clearly dominant in the latter. However, the results of CIP^f_z contradict our expectations that increasing the priority of the z_{jk} variables alone would lead to strong primal bounds. We see the opposite, as CIP^f_z dominates CIP^f in terms of the dual bound while performing much worse on the primal bound. Interestingly, experiments that only increasing the priority of the W_j variables (not included here) do match our intuitions: the model performs poorly, timing out without finding *feasible* solutions to 5 problem instances. More detailed experimentation is needed to understand the reasons behind the impact of the search heuristics and, in particular, why $CIP^f_{z,w}$ performs so well.

The Impact of Global Constraints. Global constraints play a primary role in CP, forming the central object in both modeling and solving. Building on this role and the substantial work over the past 15 years on the hybridization of CP and MIP solving techniques [22], we are interested in exploring the concept of a global constraint as a richer object in the search process, exploiting its structure to do more than domain pruning [23]. Several works have shown the success on the pursuing of this direction. For example, global constraints can provide heuristic information [24], generate SAT clauses [25] and cutting planes

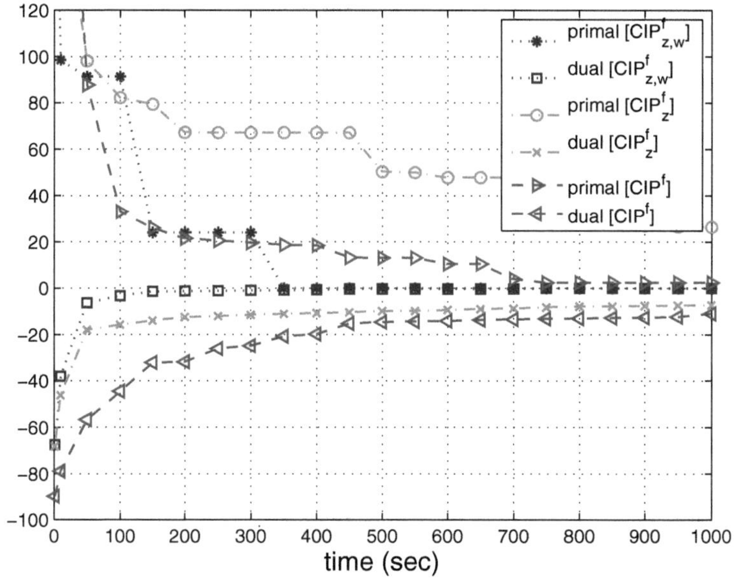

Fig. 4. Comparison of the primal and dual gaps of CIP^f, CIP^f_z and $CIP^f_{z,w}$

[26], detect independent sub-problems [27], and decompose constraints and fix or remove variables [28].

Though we have developed a new state-of-the-art hybrid model for the nurse-to-patient assignment problem and demonstrated the reduction in search effort (in both time and nodes) from the integration of constraint propagation, linear relaxation, and cutting planes within global constraints (i.e., compare CIP_z with CIP^f_z and $CIP_{z,w}$ with $CIP^f_{z,w}$), the importance of augmented global constraints to our results should not be overstated. Without the use of the branching priorities on z_{jk} and W_j variables, none of the CIP models are able to match the performance of the CP model.[2] Therefore, it appears that the primary reason for the strong performance of the new state-of-the-art is the branching priorities and, at best, the *interaction* of the branching priorities with the augmented global constraints. Furthermore, as the comparison of the number of nodes of CIP and CIP^p indicate, the simple addition of global constraint propagation to a MILP-style search does not necessarily result in improved performance: a more nuanced understanding of the interactions is needed.

8 Conclusions

In this paper, we developed a series of mixed integer quadratic programming and constraint integer programming models for the load balanced nurse-to-patient

[2] The CP model also uses a problem-specific heuristic and so we believe the direct comparison of it with the $CIP^f_{z,w}$ is justified.

assignment problem. This problem has been addressed with mixed integer linear programming and constraint programming, with the latter representing the state of the art.

Our approach focused on the integration of augmented global constraints into the CIP model. In addition to constraint propagation, the global constraints implemented constraint-specific linear relaxations and cutting plane generation. Building on the existing work on the QUADRATIC [8], GCC [12,15], and SPREAD [9] constraints, we introduced a linear relaxation and cutting planes for the latter. Our empirical results demonstrate that the CIP approach substantially outperforms the MIQP model, supporting the effectiveness of the extended global constraints, but still does not compete with the CP model. To facilitate the search process, we propose problem-specific branching priorities, which greatly improve the CIP models, resulting in performance about one order of magnitude faster than CP.

References

1. Pesant, G., Régin, J.-C.: Spread: A balancing constraint based on statistics. In: van Beek, P. (ed.) CP 2005. LNCS, vol. 3709, pp. 460–474. Springer, Heidelberg (2005)
2. Schaus, P., Van Hentenryck, P., Régin, J.C.: Scalable load balancing in nurse to patient assignment problems. In: van Hoeve, W.-J., Hooker, J.N. (eds.) CPAIOR 2009. LNCS, vol. 5547, pp. 248–262. Springer, Heidelberg (2009)
3. Schaus, P., Régin, J.C.: Bound-consistent spread constraint. EURO Journal on Computational Optimization, 1–24 (2013)
4. Achterberg, T.: Constraint Integer Programming. PhD thesis, Technische Universität Berlin (2007)
5. Achterberg, T.: SCIP: solving constraint integer programs. Mathematical Programming Computation 1, 1–41 (2009)
6. Bussieck, M.R., Vigerske, S.: Minlp solver software. Wiley Encyclopedia of Operations Research and Management Science. Wiley, Chichester (2010)
7. Koch, T., Achterberg, T., Andersen, E., Bastert, O., Berthold, T., Bixby, R., Danna, E., Gamrath, G., Gleixner, A., Heinz, S., Lodi, A., Mittelmann, H., Ralphs, T., Salvagnin, D., Steffy, D., Wolter, K.: MIPLIB 2010. In: Mathematical Programming Computation, pp. 1–61 (2011)
8. Berthold, T., Heinz, S., Vigerske, S.: Extending a CIP framework to solve MIQCPs. In: Mixed-Integer Nonlinear Programming. The IMA Volumes in Mathematics and its Applications, vol. 154, pp. 427–445. Springer (2012)
9. Schaus, P., Deville, Y., Dupont, P., Régin, J.C.: Simplification and extension of the spread constraint. In: Third International Workshop on Constraint Propagation and Implementation, pp. 77–91 (2006)
10. Achterberg, T., Berthold, T.: Hybrid branching. In: van Hoeve, W.-J., Hooker, J.N. (eds.) CPAIOR 2009. LNCS, vol. 5547, pp. 309–311. Springer, Heidelberg (2009)
11. Mullinax, C., Lawley, M.: Assigning patients to nurses in neonatal intensive care. Journal of the Operational Research Society 53, 25–35 (2002)
12. Quimper, C.G., Van Beek, P., López-Ortiz, A., Golynski, A., Sadjad, S.B.: An efficient bounds consistency algorithm for the global cardinality constraint. In: Rossi, F. (ed.) CP 2003. LNCS, vol. 2833, pp. 600–614. Springer, Heidelberg (2003)

13. Refalo, P.: Tight cooperation and its application in piecewise linear optimization. In: Jaffar, J. (ed.) CP 1999. LNCS, vol. 1713, pp. 375–389. Springer, Heidelberg (1999)
14. Milano, M., Ottosson, G., Refalo, P., Thorsteinsson, E.S.: The role of integer programming techniques in constraint programming's global constraints. INFORMS Journal on Computing 14, 387–402 (2002)
15. Hooker, J.: Integrated Methods for Optimization, 2nd edn. Springer (2012)
16. Marchand, H., Martin, A., Weismantel, R., Wolsey, L.: Cutting planes in integer and mixed integer programming. Discrete Applied Mathematics 123, 397–446 (2002)
17. Chang, X.W., Golub, G.H.: Solving ellipsoid-constrained integer least squares problems. SIAM Journal on Matrix Analysis and Applications 31, 1071–1089 (2009)
18. Ku, W.Y., Beck, J.C.: Combining discrete ellipsoid-based search and branch-and-cut for binary quadratic programming problems. In: Simonis, H. (ed.) CPAIOR 2014. LNCS, vol. 8451, pp. 334–350. Springer, Heidelberg (2014)
19. Haralick, R.M., Elliott, G.L.: Increasing tree search efficiency for constraint satisfaction problems. Artificial Intelligence 14, 263–314 (1980)
20. Beck, J.C., Prosser, P., Wallace, R.J.: Trying again to fail first. In: Faltings, B.V., Petcu, A., Fages, F., Rossi, F. (eds.) CSCLP 2004. LNCS (LNAI), vol. 3419, pp. 41–55. Springer, Heidelberg (2005)
21. Pryor, J., Chinneck, J.W.: Faster integer-feasibility in mixed-integer linear programs by branching to force change. Computers and Operations Research 38, 1143–1152 (2011)
22. Van Hentenryck, P., Milano, M. (eds.): Hybrid Optimization: Ten Years of CPAIOR. Springer (2011)
23. Beck, J.C.: Modeling, global constraints, and decomposition. In: Tenth Symposium of Abstraction, Reformulation, and Approximation (2013)
24. Pesant, G., Quimper, C.G., Zanarini, A.: Counting-based search: Branching heuristics for constraint satisfaction problems. Journal of Artificial Intelligence Research 43, 173–210 (2012)
25. Schutt, A., Feydy, T., Stuckey, P.J., Wallace, M.G.: Explaining the cumulative propagator. Constraints 16, 250–282 (2011)
26. Bergman, D., Hooker, J.N.: Graph coloring facets from all-different systems. In: Beldiceanu, N., Jussien, N., Pinson, É. (eds.) CPAIOR 2012. LNCS, vol. 7298, pp. 50–65. Springer, Heidelberg (2012)
27. Heinz, S., Ku, W.Y., Beck, J.C.: Recent improvements using constraint integer programming for resource allocation and scheduling. In: Gomes, C., Sellmann, M. (eds.) CPAIOR 2013. LNCS, vol. 7874, pp. 12–27. Springer, Heidelberg (2013)
28. Heinz, S., Schulz, J., Beck, J.C.: Using dual presolving reductions to reformulate cumulative constraints. Constraints 18, 166–201 (2013)

On the Erdős Discrepancy Problem

Ronan Le Bras, Carla P. Gomes, and Bart Selman

Computer Science Department,
Cornell University, Ithaca, NY, 14850, USA

Abstract. According to the Erdős discrepancy conjecture, for any infinite ± 1 sequence, there exists a homogeneous arithmetic progression of unbounded discrepancy. In other words, for any ± 1 sequence $(x_1, x_2, ...)$ and a discrepancy C, there exist integers m and d such that $|\sum_{i=1}^{m} x_{i \cdot d}| > C$. This is an 80-year-old open problem and recent development proved that this conjecture is true for discrepancies up to 2. Paul Erdős also conjectured that this property of unbounded discrepancy even holds for the restricted case of completely multiplicative sequences, namely sequences $(x_1, x_2, ...)$ where $x_{a \cdot b} = x_a \cdot x_b$ for any $a, b \geq 1$. The longest such sequence of discrepancy 2 has been proven to be of size 246. In this paper, we prove that any completely multiplicative sequence of size $127,646$ or more has discrepancy at least 4, proving the Erdős discrepancy conjecture for discrepancy up to 3. In addition, we prove that this bound is tight and increases the size of the longest known sequence of discrepancy 3 from $17,000$ to $127,645$. Finally, we provide inductive construction rules as well as streamlining methods to improve the lower bounds for sequences of higher discrepancies.

Introduction

Much like Ramsey theory studies how order must emerge in combinatorial objects as their size increases, discrepancy theory investigates how deviations from uniformity necessarily occur. Namely, discrepancy theory addresses the problem of distributing points uniformly over some geometric object, and studies how irregularities ineluctably appear in these distributions. For example, this subfield of combinatorics aims to answer the following question: for a given set U of n elements, and a finite family $\mathcal{S} = \{S_1, S_2, \ldots, S_m\}$ of subsets of U, is it possible to color the elements of U in red or blue, such that the difference between the number of blue elements and red elements in any subset S_i is small?

Important contributions in discrepancy theory include the Beck-Fiala theorem [1] and Spencer's Theorem [2]. The Beck-Fiala theorem guarantees that if each element appears at most t times in the sets of $\mathcal{S}$, the elements can be colored so that the imbalance, or discrepancy, is no more than $2t - 1$. According to the Spencer's theorem, the discrepancy of $\mathcal{S}$ grows at most as $\Omega(\sqrt{n \log(2m/n)})$.

Nevertheless, some important questions remain open. According to Paul Erdős himself, two of his oldest conjectures relate to the discrepancy of homogeneous arithmetic progressions (HAPs) [3]. Namely, a HAP of length k and of common difference d corresponds to the sequence $(d, 2d, \ldots, kd)$. The first conjecture can be formulated as follows:

B. O'Sullivan (Ed.): CP 2014, LNCS 8656, pp. 440–448, 2014.

Conjecture 1. Let $(x_1, x_2, ...)$ be an arbitrary ± 1 sequence. The discrepancy of x w.r.t. HAPs must be unbounded, i.e. for any integer C there is an integer m and an integer d such that $|\sum_{i=1}^{m} x_{i\cdot d}| > C$.

This problem has been open for over eighty years, as is the weaker form according to which one can restrict oneself to completely multiplicative functions. Namely, f is a completely multiplicative function if $f(a \cdot b) = f(a) \cdot f(b)$ for any a, b. The second conjecture translates to:

Conjecture 2. Let $(x_1, x_2, ...)$ be an arbitrary completely multiplicative ± 1 sequence. The discrepancy of x w.r.t. HAPs must be unbounded, i.e. for any integer C there is a m and a d such that $|\sum_{i=1}^{m} x_{i\cdot d}| > C$.

Hereinafter, when non-ambiguous, we refer to the discrepancy of a sequence as its discrepancy with respect to homogeneous arithmetic progressions. Formally, we denote $disc(x) = max_{m,d}|\sum_{i=1}^{m} x_{i\cdot d}|$. We denote $\mathcal{E}_1(C)$ the minimum length for which any sequence has discrepancy at least $C+1$, or equivalently, one plus the maximum length of a sequence of discrepancy C. Similarly, we define $\mathcal{E}_2(C)$ the minimum length for which any *completely multiplicative* sequence has discrepancy at least $C+1$. [1]

Table 1. Results for the Erdős discrepancy problem. $\mathcal{E}_1(C)$ (resp. $\mathcal{E}_2(C)$) corresponds to the minimum length for which any sequence (resp. completely multiplicative sequence) has discrepancy $C+1$. Bold indicates contribution of the current work.

C	*1*	*2*	*3*
$\mathcal{E}_1(C)$	12	1,161	**≥127,646**
$\mathcal{E}_2(C)$	10	247	**127,646**

A proof or disproof of these conjectures would constitute a major advancement in combinatorial number theory [4]. To date, both conjectures have been proven to hold for the case $C \leq 2$. As illustrated in Table 1, the values of $\mathcal{E}_1(1), \mathcal{E}_2(1)$, and $\mathcal{E}_2(2)$ have been long proven to be 12, 10, and 247 respectively, while recent development proved $\mathcal{E}_1(2) = 1161$ [5]. Konev and Lisitsa [5] also provide a new lower bound for $\mathcal{E}_1(3)$. After 3 days of computation, a SAT solver was able to find a satisfying assignment for a sequence of length $13,000$. Yet, it would fail to find a solution of size $14,000$ in over 2 weeks of computation. They also report a solution of length $17,000$, the longest known sequence of discrepancy 3.

In this work, we explore streamlining for this problem, an effective combinatorial search strategy that exploits regularities in some problem solutions, beyond the structure of the combinatorial problem itself. Streamlining provides and exploits structural information about the problem, and we believe that a sine qua non condition for tackling huge sequences requires deep insights into the structure of the problem. In addition, streamlining provides a vision for a broader strategy for solving problems. Overall, we

[1] Note that, if Conjecture 1 (resp. Conjecture 2) were to be rejected, $\mathcal{E}_1(C)$ (resp. $\mathcal{E}_2(C)$) would correspond to infinity.

substantially increase the size of the longest sequence of discrepancy 3, from $17,000$ to $127,645$. In addition, we prove that $\mathcal{E}_2(3) = 127,646$, making this bound tight, as `Plingeling` was able to prove unsat and `Lingeling` generated an UNSAT proof in DRUP format [6].

This paper is organized as follows. The next section formally defines the Erdős discrepancy problems (for the general case and the multiplicative case) and presents SAT encodings for both problems. We then investigate streamlined search techniques to boost the search for lower bounds of these two problems, and to characterize additional structures that appear in a subset of the solutions. Furthermore, in a subsequent section, we provide construction rules that are based on these streamliners and allow to generate larger sequences of limited discrepancy from smaller ones. The last section presents the results of these approaches.

Problem Formulation

In this section, we first formally define the two conjectures as decision problems and then propose encodings for these problems.

Definition 1 (EDP$_1$). *Given two integers n and C, does there exist a ± 1 sequence $(x_1, \ldots, x_n)$ such that $|\sum_{i=1}^{m} x_{i \cdot d}| \leq C$ for any $1 \leq d \leq n, m \leq n/d$.*

Konev and Lisitsa [5] provide a SAT encoding for this problem that uses an automaton accepting any sub-sequence of discrepancy exceeding C. A state s_j of the automaton corresponds to the sum of the input sequence, while the accepting state s_B captures whether the sequence has exceeded the discrepancy C. A proposition $s_j^{(m,d)}$ is true whenever the automaton is in state $\sum_{i=1}^{m-1} x_{i \cdot d}$ after reading the sequence $(x_d, \ldots, x_{(m-1)d})$. Let p_i be the proposition corresponding to $x_i = +1$. A proposition that tracks the state of the automaton for an input sequence $(x_d, x_{2d}, \ldots, x_{\lfloor n/d \rfloor d})$ can be formulated as:

$$\begin{aligned} \phi(n, C, d) = s_0^{(1,d)} \bigwedge_{m=1}^{n/d} \Big(& \bigwedge_{-C \leq j < C} (s_j^{(m,d)} \wedge p_{id} \rightarrow s_{j+1}^{(m+1,d)}) \wedge \\ & \bigwedge_{-C < j \leq C} (s_j^{(m,d)} \wedge \overline{p_{id}} \rightarrow s_{j+1}^{(m+1,d)}) \wedge \\ & (s_C^{(m,d)} \wedge p_{id} \rightarrow s_B) \wedge \\ & (s_{-C}^{(m,d)} \wedge \overline{p_{id}} \rightarrow s_B) \Big) \end{aligned} \tag{1}$$

In addition, we need to encode that the automaton is in exactly one state at any point in time. Formally, we define this proposition as:

$$\chi(n, C) = \bigwedge_{1 \leq d \leq n/C, 1 \leq m \leq n/d} \Big(\bigvee_{-C \leq j \leq C} s_j^{(i,d)} \wedge \bigwedge_{-C \leq j_1, j_2 \leq C} (\overline{s}_{j1}^{(i,d)} \vee \overline{s}_{j2}^{(i,d)}) \Big) \tag{2}$$

Finally, we can encode the Erdős Discrepancy Problem as follows:

$$\mathbf{EDP}_1(n, C) : \overline{s}_B \wedge \chi(n, C) \wedge \bigwedge_{d=1}^{n} \phi(n, C, d) \tag{3}$$

Furthermore, as the authors of [5], the actual states $s_j^{(m,d)}$ of the automaton do not require $2C + 1$ binary variables to represent the $2C + 1$ values of the states. Instead, one can modify this formulation and use $\lceil log_2(2C + 1) \rceil$ binary variables to encode the automaton states.

For the completely multiplicative case, we introduce additional constraints to capture the multiplicative property of any element of the sequence, i.e. $x_{id} = x_i x_d$ for any $1 \leq d \leq n, 1 \leq i \leq n/d$. With respect to the boolean variables p_i, p_d and p_{id}, such a constraint acts as XNOR gate of input p_i and p_d and of output p_{id}. Formally, we denote this proposition $\mathcal{M}(i, d)$ and define:

$$\mathcal{M}(i, d) = (p_i \vee p_d \vee p_{id}) \wedge (\overline{p_i} \vee \overline{p_d} \vee p_{id}) \wedge (p_i \vee \overline{p_d} \vee \overline{p_{id}}) \wedge (\overline{p_i} \vee p_d \vee \overline{p_{id}}) \tag{4}$$

Importantly, for completely multiplicative sequences, the discrepancy of the subsequence $(x_d, ..., x_{md})$ of length m and common difference d will be the same as the discrepancy of the subsequence $(x_1, ..., x_m)$. Indeed , we have $|\sum_{i=1}^{m} x_{i \cdot d}| = |\sum_{i=1}^{m} x_i x_d| = |x_d| \cdot |\sum_{i=1}^{m} x_i| = |\sum_{i=1}^{m} x_i|$. Therefore, one needs only check that the partial sums $\sum_{i=1}^{m} x_i, 1 \leq m \leq n$ never exceed C nor go below $-C$. Furthermore, note that a completely multiplicative sequence is entirely characterized by the values it takes at prime positions, i.e. $\{x_p | p \text{ is prime}\}$. In addition, if there exists a completely multiplicative sequence $(x_1, ..., x_{p-1})$ of discrepancy C with p prime, then the sequence $(x_1, ..., x_{p-1}, (-1)^{\mathbb{1}_{\sum_{i=1}^{m} x_i \geq 0}})$ will also be a CMS of discrepancy C. As a result, $\mathcal{E}^2(C)$ cannot be a prime number.

Overall, for the completely multiplicative case, we obtain:

$$\mathbf{EDP}_2(n, C) : \overline{s}_B \wedge \chi(n, C) \wedge \phi(n, C, 1) \bigwedge_{1 \leq d \leq n, 1 \leq i \leq n/d} \mathcal{M}(i, d) \tag{5}$$

Streamlined Search

The encoding of $\mathbf{EDP}_1$ given in the previous section has successfully led to prove a tight bound for the case $C = 2$ [5]. On an Intel Core i5-2500K CPU, it takes about 800 seconds for `Plingeling` [7] to find a satisfying assignment for $\mathbf{EDP}_1(1160, 2)$ and less than 6 hours for `Glucose` [8] to generate a proof of $\mathcal{E}_1(2) = 1,161$. Nevertheless, for the case $C = 3$, it requires more than 3 days of computation for `Plingeling` to find a sequence of size $n = 13,000$, and fails to find a sequence of size $14,000$ in over two weeks of computation.

In this section, in order to improve this lower bound and acquire a better understanding of the solution space, we explore streamlining techniques that identifies additional structure occurring in a subset of the solutions. Among the solutions of a combinatorial problem, there might be solutions that possess regularities beyond the structure of the combinatorial problem itself. Streamlining [9] is an effective combinatorial search strategy that exploits these additional regularities. By intentionally imposing additional structure to a combinatorial problem, it focuses the search on a highly structured subspace and triggers increased constraint reasoning and propagation. This search technique is sometimes referred to as "tunneling" [10]. In other words, a streamlined search consists in adding specific desired or observed regularities, such as a partial pattern that appears in a solution, to the combinatorial solver. These additional regularities boost the solver that may find more effectively larger solutions that contain these regularities. If no solution is found, the observed regularities were likely accidental. Otherwise, one can analyze these new solutions and suggests new regularities. This methodology has been successfully applied to find efficient constructions for different combinatorial objects, such as spatially-balanced Latin squares [11], or graceful double-wheel graphs [12].

When analyzing solutions of $\mathbf{EDP}_1(n, 2)$ for $n \in [1, 1160]$, there is a feature that visually stands out of the solutions, as illustrated in Figure 1. When looking at a solution as a $2D$-matrix with entries in $\{-, +\}$ and changing the dimensions of the matrix (see Fig. 1, Left) , there seems to be clear preferred matrix dimensions (say m-by-p) such that the m rows are mostly identical for the columns 1 to $p - 1$, suggesting that $x_i = x_{i \bmod p}$ for $1 \leq i \leq p - 1$. We denote $period(x, p, t)$ the streamliner that enforces this observation and define:

$$period(x, p, t) : x_i = x_{i \bmod p} \; \forall 1 \leq i \leq t, i \not\equiv 0 \bmod p \tag{6}$$

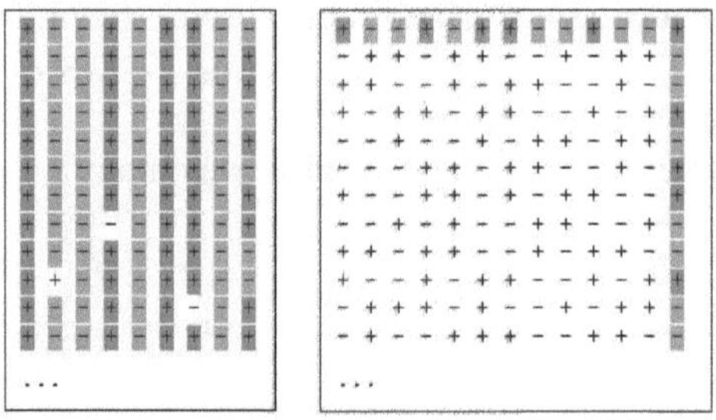

Fig. 1. First elements of a sequence of length 1160 and of discrepancy 2, illustrating the *period* (Left) and *mult* (Right) streamliners

First, while this observation by itself did not allow to improve the current best lower bound for $\mathcal{E}_1(3)$, it led to the formulation of the construction of the next section. Second, it also led to the re-discovery of the so-called 'improved Walters sequence' [13], defined as follows:

$$\mu_3(i) = \begin{cases} +1, & \text{if } i \text{ is } 1 \bmod 3 \\ -1, & \text{if } i \text{ is } 2 \bmod 3 \\ -\mu_3(i/3), & \text{otherwise.} \end{cases} \tag{7}$$

In the following, we denote $walters(x, w)$ the streamliner imposing that the first w elements of a sequence x follow the improved Walters sequence, i.e.:

$$walters(x, w) : x_i = \mu_3(i)\ \forall 1 \leq i \leq w \tag{8}$$

One can easily see that the improved Walters sequence is a special case of the periodic sequence defined previously. Namely, for any sequence x where $walters(x, w)$ holds true, then we have $period(x, 9, w)$.

Finally, another striking feature of the solutions of $\mathbf{EDP}_1(n, 2)$ is that they tend to follow a multiplicative sequence. Interestingly, $\mathbf{EDP}_2$ restricts $\mathbf{EDP}_1$ to the special case of multiplicative functions and we observe for the case $C = 2$ that this restriction substantially impacts the value of the best bound possible (i.e. $\mathcal{E}_1(2) = 1,161$ whereas $\mathcal{E}_2(2) = 247$). Nevertheless, the solutions of $\mathbf{EDP}_1(n, 2)$ exhibit a partial multiplicative property (see Fig. 1, Right) and we define:

$$mult(x, m, l) : x_{i \cdot d} = x_i x_d\ \forall 2 \leq d \leq m, 1 \leq i \leq n/d, i \leq l \tag{9}$$

In the experimental section, we show the speed-ups that are triggered using these streamliners, and how the best lower bound for $\mathbf{EDP}_1(n, 2)$ gets greatly improved.

Construction Rule

In this section, we show how we used insights from the $period(x, p, t)$ streamliner in order to generate an inductive construction rule for sequences of discrepancy C from sequences of lower discrepancy.

Consider a sequence x that is periodic of period p, as defined in the previous section, i.e. $period(x, p, |x|)$ holds true, and is of length $n = p * k$. Then, the sequence x can be written as:

$$\begin{aligned} x = (&y_1, y_2, \ldots, y_{p-2}, y_{p-1}, z_1 \\ &y_1, y_2, \ldots, y_{p-2}, y_{p-1}, z_2 \\ &\ldots \\ &y_1, y_2, \ldots, y_{p-2}, y_{p-1}, z_k) \end{aligned} \tag{10}$$

Let C be the discrepancy of $z = (z_1, z_2, ..., z_k)$ and C' the discrepancy of $(y_1, ..., y_{p-1})$. Given that $\sum_{i=1}^{m} x_{ip} = \sum_{i=1}^{m} z_i$ for any $1 \leq m \leq k$, we have $disc(x) \geq C$. Note that if x was completely multiplicative, then it would hold $disc(x) = C$. We study the general case where x is not necessarily multiplicative, and investigate the conditions under which $disc(x)$ is guaranteed to be less or equal to $C + C'$.

For a given common difference d and length m, we consider the subsequence $(x_d, x_{2d}, ..., x_{md})$. Let $q = \frac{p}{gcd(d,p)}$. Given the definition 10 of x, we have:

$$(x_d, x_{2d}, ..., x_{md}) = (y_{d \bmod p}, y_{2d \bmod p}, ..., y_{(q-1)d \bmod p}, z_q, \tag{11}$$

$$y_{d \bmod p}, y_{2d \bmod p}, ..., y_{(q-1)d \bmod p}, z_{2q}, \tag{12}$$

$$y_{d \bmod p}, ...) \tag{13}$$

Note that if p divides d or d divides p, this subsequence becomes $(z_q, z_{2q}, ..., z_{qm})$ and is of discrepancy at most C. As a result, a sufficient condition for x to be of discrepancy at most $C + C'$ is to have $y_{d \bmod p}, y_{2d \bmod p}, ..., y_{(q-1)d \bmod p}$ of discrepancy C' and summing to 0. We say that such a sequence has a discrepancy mod p of C'. Formally, we define the problem of finding such sequences as follows:

Definition 2 (Discrepancy mod p**).** *Given two integers p and C', does there exist a ± 1 sequence $(y_1, \ldots, y_{p-1})$ such that:*

$$|\sum_{i=1}^{m} y_{i \cdot d \bmod p}| \leq C', \ \forall 1 \leq d \leq n, m < \frac{p}{gcd(d,p)} \tag{14}$$

$$\sum_{i=1}^{\frac{p}{gcd(d,p)}-1} y_{i \cdot d \bmod p} = 0, \ \forall 1 \leq d \leq n \tag{15}$$

Notice that, given the equation 15, p should be odd for such a sequence to exist.

We encode this problem as a Constraint Satisfaction Problem (CSP) in a natural way from the problem definition. We provide the experimental results in the next section.

Results

All experiments were run on a Linux (version 2.6.18) cluster where each node has an Intel Xeon Processor X5670, with dual-CPU, hex-core @2.93GHz, 12M Cache, 48GB RAM. Unless otherwise noted, the results were obtained using the parallel SAT solver `Plingeling`, version `ats1` for the SAT encodings, and using `IBM ILOG CPLEX CP Optimizer`, release `12.5.1` for the CP encodings.

First, we evaluate the proposed streamliners for the two problems. Table 2 reports the length of the sequences that were successfully generated, as well as the computation time. The first clear observation is that, for EDP_1, the streamlined search based on the partial multiplicative property significantly boosts the search and allows to generate solutions that appear to be out of reach of the standard search approach. For example, while it takes about 10 days to find a solution of length $13,900$ without streamliners, the streamlined search generates a substantially-large satisfying assignment of size $31,500$ in about 15 hours. Next, we study streamliners that were used for EDP_2, i.e. partially imposing the walters sequence. The results clearly show the speed up triggered by the combination of the new encoding for EDP_2 with the *walters* streamliners. Interestingly, the longest *walters* sequence of discrepancy 3 is of size 819. Nevertheless, one can successfully impose the first 800 elements of the walters sequence and still expand it to a sequence of length $108,000$. Furthermore, when imposing $walters(730)$, it takes less than 1 hour and an half to find a satisfying assignment for a sequence of size $127,645$. Moreover, without additional streamliners, it takes about 60 hours to prove unsat for the case $127,646$ and allows us to prove that this bound is tight. Indeed, the solver generates a DRUP proof of approximately 29GB and `DRAT-trim`, an independent satisfiability proof checker [6,14], verifies the 88 million lemmas of the proof in about 45 hours.

In terms of the inductive construction described in the previous section, we can generate sequences whose discrepancy mod p is 1, for p in $1, 3, 5, 7$, and 9, while it also

Table 2. Solution runtimes of searches with and without streamliners. The streamlined search leads to new lower bounds for the 2 EDP problems.

Encoding	*Streamliners*	*Size of sequence*	*Runtime (in sec)*
EDP_1	-	13,000	286,247
	-	13,500	560,663
	-	13,900	770,122
	mult(120,2000)	15,600	4,535
	mult(150,2000)	18,800	8,744
	mult(200,1000)	23,900	12,608
	mult(700,10000)	27,000	45,773
	mult(700,20000)	31,500	51,144
EDP_2	walters(800)	81,000	1,364
	walters(800)	108,000	4,333
	walters(700)	112,000	5,459
	walters(730)	127,645	4,501

generates sequences of discrepancy $\bmod\ p$ equal to 2 for p in $11, 13, 15, 17, 25, 27, 45$, and 81. Overall, this proves that one can take any sequence x of length $|x|$ and discrepancy C and generate one of length $9|x|$ and of discrepancy $C+1$, or of length $81|x|$ and of discrepancy $C+2$. As a result, this provides a new bound for the case of discrepancy 4, and proves $\mathcal{E}_1(4) > 9 * 127645 = 1,148,805$. Interestingly, such a long sequence suggests that the proof of the Erdos conjecture for $C > 3$ may require additional insights and analytical proof, beyond the approach proposed in this work.

Conclusions

In this paper, we address the Erdős discrepancy problem for general sequences as well as for completely multiplicative sequences. We adapt a SAT encoding previously proposed and investigate streamlining methods to speed up the solving time and understand additional structures that occur in some solutions. Overall, we substantially improve the best known lower bound for discrepancy 3 from $17,001$ to $127,646$. In addition, we prove that this bound is tight, as evidenced by the unsat proof generated by `Lingeling` and confirmed by the proof checker `DRAT-trim`. Finally, we propose construction rules to inductively generate longer sequences of limited discrepancy.

Acknowledgments. This work was supported by the National Science Foundation (NSF IIS award, grant 1344201). The experiments were run on an infrastructure supported by the NSF Computing research infrastructure for Computational Sustainability grant (grant 1059284). Note that, while this work was submitted on April 14, 2014 to the 20th International Conference on Principles and Practice of Constraint Programming, results [15] were independently submitted on May 13, 2014, confirming the value of $\mathcal{E}_2(3)$.

References

1. Beck, J., Fiala, T.: Integer-Making theorems. Discrete Applied Mathematics 3, 1–8 (1981)
2. Spencer, J.: Six standard deviations suffice. Transactions of the American Mathematical Society 289, 679–706 (1985)
3. Erdös, P.: Some of my favourite problems which recently have been solved. North-Holland Mathematics Studies 74, 59–79 (1982)
4. Nikolov, A., Talwar, K.: On the hereditary discrepancy of homogeneous arithmetic progressions. arXiv preprint arXiv:1309.6034 (2013)
5. Konev, B., Lisitsa, A.: A sat attack on the erdos discrepancy conjecture (2014)
6. Heule, M.J., Hunt, W.A., Wetzler, N.: Trimming while checking clausal proofs. In: Formal Methods in Computer-Aided Design (FMCAD), pp. 181–188. IEEE (2013)
7. Biere, A.: Lingeling, plingeling and treengeling entering the sat competition 2013. In: Proceedings of SAT Competition (2013); Solver and (2013) 51
8. Audemard, G., Simon, L.: Glucose 2.3 in the sat 2013 competition. In: Proceedings of SAT Competition (2013); Solver and (2013) 42
9. Gomes, C., Sellmann, M.: Streamlined constraint reasoning. In: Wallace, M. (ed.) CP 2004. LNCS, vol. 3258, pp. 274–289. Springer, Heidelberg (2004)
10. Kouril, M., Franco, J.: Resolution tunnels for improved sat solver performance. In: Bacchus, F., Walsh, T. (eds.) SAT 2005. LNCS, vol. 3569, pp. 143–157. Springer, Heidelberg (2005)
11. Le Bras, R., Gomes, C.P., Selman, B.: From streamlined combinatorial search to efficient constructive procedures. In: Proceedings of the 15th International Conference on Artificial Intelligence, AAAI 2012 (2012)
12. Le Bras, R., Gomes, C.P., Selman, B.: Double-wheel graphs are graceful. In: Proceedings of the Twenty-Third International Joint Conference on Artificial Intelligence, IJCAI 2013, pp. 587–593. AAAI Press (2013)
13. Polymath1: Matryoshka Sequences, http://michaelnielsen.org/polymath1/index.php?title=Matryoshka_Sequences (accessed April 11, 2014)
14. Wetzler, N., Heule, M.J., Hunt Jr., W.A.: (Drat-trim: Efficient checking and trimming using expressive clausal proofs)
15. Konev, B., Lisitsa, A.: Computer-aided proof of erdos discrepancy properties. arXiv preprint arXiv:1405.3097 (2014)

Towards Practical Infinite Stream Constraint Programming: Applications and Implementation

Jasper C.H. Lee[1] and Jimmy H.M. Lee[2]

[1] Churchill College and Computer Laboratory, University of Cambridge, UK
`chjl2@cam.ac.uk`

[2] Department of Computer Science and Engineering,
The Chinese University of Hong Kong, Shatin, N.T., Hong Kong
`jlee@cse.cuhk.edu.hk`

Abstract. Siu et al. propose stream CSPs (St-CSPs) as a generalisation of finite domain CSPs to cater for constraints on infinite streams, and a solving algorithm that produces a deterministic Büchi automaton recognising the solution language. As a novel application, we demonstrate how St-CSPs can model mathematically and generate a PID controller for driving a self-balancing tray and an inverted pendulum in real-time. We propose and give formally the correctness of an improvement to the implementation that eliminates numerous unnecessary states in the solution automaton for St-CSPs involving the `first` temporal operator, thereby reducing solving time. We give two St-CSP examples that can benefit from our new implementation techniques. Our approach always generates a solution automaton not bigger than, but potentially exponentially smaller than, that produced by the original implementation. Experimental results show substantial improvements.

1 Introduction

Streams of data are ubiquitous. They can either be discrete sequences on their own (e.g. stock market data), or discrete samples of continuous signals (e.g. positional data with respect to time). The evolution of such sequences is typically governed by some physical laws or mathematical equations. However, standard finite domain constraint satisfaction problems (CSPs) do not model such problems very well, because they can only model a finite segment of an otherwise infinite problem. To model such discrete time constraint problems more naturally, Siu et al. [8,11] introduce stream constraint satisfaction problems (St-CSPs) by adapting temporal operators in Wadge and Ashcroft's [12] Lucid programming language. They give the definition of St-CSPs, and a solving algorithm that produces a deterministic Büchi automaton recognising the solutions of an St-CSP. The termination, soundness and completeness of their algorithm are proven. They also suggest practical applications for St-CSPs, such as generating harmonic accompaniment to a melody, and the game engine for the once popular Digi Invaders[1] game in early Casio calculators in the 1970s.

[1] `http://www.youtube.com/watch?v=1YafgAcmov4`

B. O'Sullivan (Ed.): CP 2014, LNCS 8656, pp. 449–464, 2014.

This paper is about practical stream constraint programming. The goal is to push the limit of this relatively new member of the CP family, and take the first step towards putting the theoretical framework into practice. Since there are currently no common modelling idioms, and we know little about implementation technology and applications, we approach this idea from two angles. First, we demonstrate that St-CSPs can be used for solving interesting real life problems. Continuing the work on game engine generation [8,11], we model real-time hardware controllers as St-CSPs. Even though discretisation and approximations have to be applied, we find that the approach produces stable control on our hardware. Second, we propose an improvement for the search algorithm to reduce solving time and the size of the solution automaton. Our improvement is restricted to a certain class of St-CSPs, and we give practical usages of this class on two applications. We also state formally the correctness of our technique. To demonstrate the efficiency of our proposal, we give experimental results to compare our new search algorithm against the original, showing orders of magnitude improvement in terms of runtime and solution automaton size.

2 Background

This section introduces the background for stream constraint solving. We first state the definition of St-CSPs and related notions, followed by the constraint specification language. The solving algorithm of Siu et al. [8,11] is summarised.

2.1 Infinite Strings and Stream Constraint Satisfaction Problems

An *infinite string* α over an *alphabet* Σ is a function $\mathbb{N}_0 \rightarrow \Sigma$. Given i, $\alpha(i)$ is an individual *daton* of α at time point i. The set of all such strings with alphabet Σ is denoted Σ^ω. Infinite strings are also referred to as *streams.*

The notation $\alpha' = \alpha(i, \infty)$ is used for the *string suffix* $\alpha'(j) = \alpha(j+i)$. For a language L, $L(i,\infty) = \{\alpha(i,\infty) \mid \alpha \in L\}$. As for a *finite prefix* of a string, infinite or not, the notation $\alpha' = \alpha[0:i]$ is used to denote the string $\alpha'(j) = \alpha(j)$ if $0 \leq j \leq i$ and undefined otherwise. When $i < 0$, $\alpha[0:i]$ is the empty string.

A *stream constraint satisfaction problem* (St-CSP) is a tuple $P = (X, D, C)$ [8,11], where $X = \{x_1, \ldots, x_n\}$ is a finite set of *variables*, $D(x) = (\Sigma(x))^\omega$ is a function that maps a variable to its *domain* which is the set of all infinite strings with alphabet $\Sigma(x)$, C is a finite set of *constraints*. A constraint $c \in C$ is a relation R defined on an ordered subset *Scope*(c) of variables. The relation gives all the valid simultaneous assignments of values to variables in *Scope*(c). Every constraint $c \in C$ must also be a deterministic ω-regular language [2].

An *assignment* $A(x_i) \in D(x_i)$ is a function mapping a variable to an element in its domain. A *satisfies* a constraint c if and only if $(A(x_{i_1}), A(x_{i_2}), \ldots, A(x_{i_k})) \in c$, where *Scope*$(c) = (x_{i_1}, \ldots, x_{i_k})$. The notion can be generalised to say that the string β of tuples $\beta(i) = (A(x_1)(i), \ldots, A(x_n)(i))$ *satisfies* the constraint c where $X = \{x_1, \ldots, x_n\}$. An St-CSP is *satisfied* by a variable assignment A or a string of tuples β if and only if all constraints are satisfied.

As a corollary of the closure properties of deterministic ω-regular languages, the solution set $sol(P) = \{t = (a_1, a_2, \ldots, a_n) \in \prod_i D(x_i) \mid \forall c \in C. t \text{ satisfies } c\}$ of an St-CSP P is also a deterministic ω-regular language.

In addition, two St-CSPs P and P' are said to be *equivalent*, written as $P \equiv P'$ as usual, if and only if $sol(P) = sol(P')$.

Given a set of constraints C and a point i, the *shifted view* (previously known in the literature as the *limited view* [8,11]) of C is defined as $C(i, \infty) = \{c_k(i, \infty) \mid c_k \in C\}$. Similarly, given an St-CSP $P = (X, D, C)$ and a point i, the *shifted view* of P is defined as $\hat{P}(i) = (X, D, C(i, \infty))$.

2.2 The Stream Constraint Language

In this paper, we are only concerned with St-CSPs whose variable alphabets are integer intervals, i.e. $[m, n]^\omega$ for some $m \leq n \in \mathbb{Z}$.

To specify constraints, there are primitives such as variable streams, which are the variables in the St-CSP, and constant streams. For example, the stream 2 denotes the stream a where $a(n) = 2$.

Three *temporal* operators, in the style of the Lucid programming language [12], `first`, `next` and `fby`, are defined on streams. Suppose α and β are streams. We have `first` α being the constant stream of $\alpha(0)$, and `next` α being the "tail" of α, i.e. `next` $\alpha = \alpha(1, \infty)$. In addition, α `fby` $\beta = \gamma$ is the concatenation of the head of α and β, i.e. $\gamma(0) = \alpha(0)$ and $\gamma(i) = \beta(i-1)$ for $i \geq 1$.

Furthermore, there are *pointwise* operators, such as integer arithmetic operators {`+`, `-`, `*`, `/`, `%`}. They combine two streams point by point using the corresponding arithmetic operator. Integer arithmetic relational operators are {`lt`, `le`, `eq`, `ge`, `gt`, `ne`}. They compare the two argument streams pointwisely and return a *pseudo-Boolean stream*, that is a stream in $[0, 1]^\omega$, where 0 denotes *false* and 1 denotes *true*. Pointwise Boolean operators {`and`, `or`} act on any two pseudo-Boolean streams γ and η. The final pointwise operator supported is `if-then-else`. Suppose γ is pseudo-Boolean, and α, β are streams in general, then (`if` γ `then` α `else` β)(i) is $\alpha(i)$ if $\gamma(i) = 1$ and $\beta(i)$ if $\gamma(i) = 0$.

Given these stream operators, we can now use the following relations to express stream constraints. For integer arithmetic comparisons $\circ \in$ {`<`, `<=`, `==`, `>=`, `>`, `!=`}, the constraint $\alpha \circ \beta$ is *satisfied* if and only if the arithmetic comparison $\circ$ is true at every point in the streams. Therefore, a constraint is *violated* if and only if there exists a time point at which the arithmetic comparison is false.

Care should be taken to distinguish between constraints and the relational operators. Relational operators take two streams and give a pseudo-Boolean stream as output. Constraints, on the other hand, are relations on streams.

2.3 Normalising Constraints

Siu [11] defines an St-CSP to be in *normal form* if it contains only *primitive constraints*. Primitive constraints are in one of the following three forms, assuming x_i are stream variables.

- Primitive first constraints: `first` x_i `==` `first` x_j
- Primitive next constraints: x_i `==` `next` x_j
- Primitive pointwise constraints: constraints with no `first`, `next` or `fby`.

Reducing all occurrences of `first` operators to the primitive form is beneficial, since primitive first constraints can be enforced like a primitive pointwise constraint, but can be deleted after the first time point.

All St-CSPs are reduced to an equivalent normal form before being submitted to the solver. Siu [11] also gives a simple recursive translation of an St-CSP into this normal form. Only the appearance of either "`first` *expr*", "`next` *expr*" or "$expr_1$ `fby` $expr_2$" may violate the normal form property. The cases are translated separately. By adopting notations from programming language semantics theory [13], we write $c[-]$ for *constraint contexts*, i.e. constraints with placeholders for syntactic substitution. For example, if $c[-] = [- \texttt{ + 3 >= 4}]$, then $c[\texttt{first } \alpha] = [(\texttt{first } \alpha) \texttt{ + 3 >= 4}]$. We also write a constraint rewriting transition as $(C_0, C_1) \rightsquigarrow (C_0', C_1')$, where C_0, C_1, C_0' and C_1' are sets of constraints. C_0 is the set of constraints that has to be further normalised, and C_1 is the set that is guaranteed to be in normal form already. Hence, the initial constraint pair for the St-CSP (X, D, C) is $(C, \{\})$. Rules are applied until none are applicable.

1. $(C_0 \cup \{c[\texttt{first } expr]\}, C_1) \rightsquigarrow (C_0 \cup \{c[v_1], v_2 \texttt{ == } expr\}, C_1 \cup \{\texttt{first } v_1 \texttt{ == first } v_2, v_1 \texttt{ == next } v_1\})$ where v_1 and v_2 are auxiliary variables not in any of $c[-]$, C_0 and C_1.
2. $(C_0 \cup \{c[\texttt{next } expr]\}, C_1) \rightsquigarrow (C_0 \cup \{c[v_1], v_2 \texttt{ == } expr\}, C_1 \cup \{v_1 \texttt{ == next } v_2\})$ where v_1 and v_2 are auxiliary variables not in any of $c[-]$, C_0 and C_1.
3. $(C_0 \cup \{c[expr_1 \texttt{ fby } expr_2]\}, C_1) \rightsquigarrow (C_0 \cup \{c[v_1], v_2 \texttt{ == } expr_1, v_3 \texttt{ == } expr_2\}, C_1 \cup \{\texttt{first } v_1 \texttt{ == first } v_2, \texttt{next } v_1 \texttt{ == } v_3\})$ where v_1, v_2 and v_3 are auxiliary variables not in any of $c[-]$, C_0 and C_1.

2.4 Search Trees

A *search tree* for an St-CSP P is a tree with potentially infinite height. Its nodes are St-CSPs, and the root node is P itself. The *level* of a node N is recursively defined as 0 for the root node, and $1 + \ell$ for non-root nodes N where ℓ is the level of the parent of N. A child node $Q' = (X, D, C \cup \{c'\})$ is constructed from a parent node $P' = (X, D, C)$ at level k and an *instantaneous assignment* $\tau(x) \in \Sigma(x)$, where τ takes a stream variable x and returns a daton value for it. In other words, τ gives a scalar assignment to the daton of streams at time point k. The constraint c' specifies that for all $x \in X$, $x(k) = \tau(x)$ and for all $i \neq k$, $x(i)$ is unconstrained. We write $P' \xrightarrow{\tau} Q'$ for such a parent to child construction, and label the edge on the tree between the two nodes with τ. Fig. 1 shows an example search tree.

We can identify a search node Q at level k with the shifted view $\hat{Q}(k)$. Taking this view, if $\hat{P}(k) = (X, D, C)$ is the parent node of $\hat{Q}(k+1)$, then $\hat{Q}(k+1) = (X, D, C \cup \{c'\})(1) = (X, D, (C \cup \{c'\})(1, \infty))$ where c' is the same constraint as defined above.

Recall that a constraint violation requires only a single time point at which the pointwise constraint is false. Therefore, we can generalise the definition of constraint violation such that a finite prefix of an assignment can violate a constraint. A sequence of instantaneous assignments from the root to a node is isomorphic to a finite prefix of an assignment, and so the definition again generalises. Suppose $F = (X, D, C)$ is a node at level k such that $\{\tau_i\}$ is the sequence of instantaneous assignments that constructs F from the root node, i.e. $P \xrightarrow{\tau_0} \dots \xrightarrow{\tau_k} F$. We say node F is a *failure* if and only if $\{\tau_i\}$ violates a constraint $c \in C$.

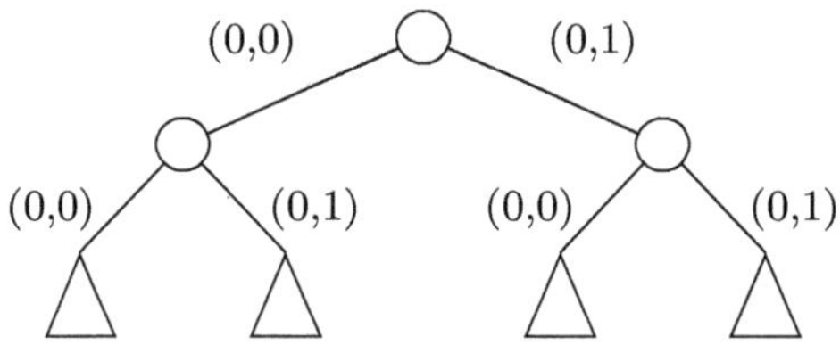

Fig. 1. Example Search Tree

2.5 Solving St-CSPs

Given an St-CSP P, its search tree is explored using depth first search. Backtracking happens when the current search node is a failure. A search node A at level k *dominates* [8,11] another search node B at level k', written as $B \prec A$, if and only if their shifted views are equivalent ($\hat{A}(k) \equiv \hat{B}(k')$) and A is visited before B during the search. When the algorithm visits a search node N that is dominated by a previously seen node M, the edge pointing to N is redirected to M instead. The resulting structure is isomorphic to a deterministic Büchi automaton, which accepts all and only the solutions of P. Siu et al. [8,11] prove the termination, soundness and completeness of the algorithm.

3 Application on Real-Time PID Control

A *proportional-integral-derivative (PID)* controller [10] is a loop feedback control mechanism. A process receives an input signal $u(t)$ and gives an output $y(t)$, which has an error $e(t) = y(t) - r(t)$ from a reference signal $r(t)$. The PID controller produces the input signal $u(t)$ by adding a weighted sum of the following three components. The proportional component is simply the error signal $e(t)$. The integral component is $\int_0^t e(\tau)\,d\tau$. The derivative component is $e'(t)$. Fig. 2 shows the described structure. K_p, K_i and K_d are the corresponding coefficients for the components in the weighted sum.

We can model a PID controller as an St-CSP. The first step is to discretise and scale the domain of the error signal, such that the signal can be represented as an integer stream e. For example, the error signal might have a real interval

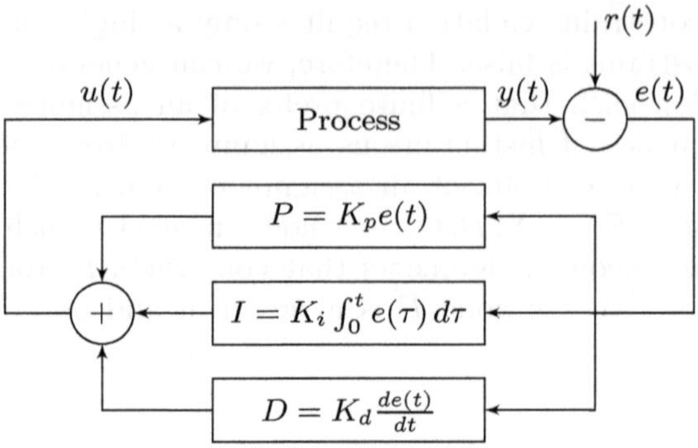

Fig. 2. PID Controller Schematics

$[-15, 15]$ as the domain representing an angle deviation. A possible discretisation is to map the interval to the integer interval $[-60, 60]$ by multiplication with 4 and rounding. The stream e (the error and the proportional component) is unconstrained and acts as an input to the automaton. At each state, the edge with the correct error value is selected in order to proceed to the next state.

There is a tradeoff between having greater precision in the error stream and limiting the size of the solution automaton. The standard approaches to determining the PID coefficients are by experimentation or analysis of a mathematical model of the process. A good discretisation of the error stream can therefore be similarly determined using either of the approaches.

Given the discretisation, we can model the derivative component of the PID control signal. For a stream of discrete time signals α, an analog of the derivative is the finite difference $\alpha(i+1) - \alpha(i)$. Any linear scaling factor required for a better approximation can simply be absorbed into the K_d coefficient for the weighted sum. In order to compute finite differences, a stream l is introduced with constraints l == 0 `fby` e. The derivative stream d is therefore constrained by d == e - l. From this, we deduce that the bounds for d is $[-2c, 2c]$ if the bounds for e is $[-c, c]$.

An analog of the integral for discrete signals is the finite sum $\sum_0^n \alpha(i)$. In an ideal PID controller, the integral component can be unbounded. In practice, it is either restricted by the real number representation of the machine or artificial bounds are introduced. In our case, a bound b is also needed in order to have a finite alphabet for the integral stream i. With the value of b decided, the integral stream can be computed using the constraint i == 0 `fby` (`if` (i+e `gt` b) `then` b `else` (`if` (i+e `lt` -b) `then` -b `else` i+e)). There is an alternative approximation for the integral stream if we know the scaling factor applied to it is close to 0. In this case, instead of summing the error discrete signals, we sum the sign of the error signals. That is, we introduce another stream *tempI* with constraint *tempI* == 0 `fby` (i + `if` (e `gt` 0) `then` 1 `else` (`if` (e `lt` 0) `then` -1 `else` 0)). Instead of using the previous constraint for computing i, we use i == `if` (*tempI* `gt` b) `then` b `else` (`if` (*tempI* `lt` -b) `then` -b `else` *tempI*). It is also possible to inline the definition of *tempI* into i, but we present this as it is here for clarity. This alternative approximation is useful for keeping the alphabet of the stream i small.

As discrete time controllers process input and output streams, St-CSPs are ideal for modelling them. The typical way of implementing controllers is to program the controller equations in an imperative language. Programming with destructive assignments and various control flow commands can be error prone. Bentley [1] gives experimental results that only 10% of professional programmers write correct code for an algorithm as simple as the binary search. Using the St-CSP approach, the imperative code required in a program is only for traversing a solution automaton according to the sensor error input stream and producing control signals to the output streams. This code has to be engineered only once and is largely reusable. The St-CSP specification language is declarative in nature without any side effects in its semantics. Hence, it inherits the advantages of declarative programming over imperative programming, including readability, conciseness, compositionality and referential transparency. Correctness and elegance are therefore more easily achievable than using a conventional programming language like C.

A PID controller for a self-balancing tray[2] was synthesised. The platform has a tray holding a pingpong ball, two motors that allows it to rotate in 3D space and an accelerometer that measures the orientation. The purpose of the controller is to maintain the horizontal position of the tray as the platform rotates, such that the pingpong ball does not fall out. We also applied the technique to control a self-balancing inverted pendulum[3]. It has a vertical body, with wheels at the bottom to allow movement for balancing the body as it tilts sideways. The controller actually uses a variant of PID control with a second derivative component in addition to the original three. Also, a complementary filter and a Kalman filter were applied to the gyroscope sensor input to eliminate noise. The filters however are not part of our St-CSP model.

The traditional controllers of the above hardware happen to be simple and small, even when implemented in C. We anticipate the advantages of our approach to become more apparent when the controllers are more complex. The purpose of the current exercise is really to demonstrate that CP can have applications in real-time hardware control.

4 Improved Handling of the `first` Operator

Our new approach focuses on the handling of streams constructed using the `first` operator. Fig. 3 contains two St-CSPs that show some uses of `first` would increase the number of states in the solution automaton, and some other uses would not. Problem 1 imposes that the first daton of x has to be less than the first daton of y, whilst Problem 2 requires all datons of x to be less than the first daton of y. Therefore, the constraint in Problem 1 only concerns the first time point, whereas the effect of the constraint in Problem 2 persists indefinitely.

[2] A video demonstration of the self-balancing tray in operation can be found at `http://www.youtube.com/watch?v=dT56qAZt8hI`

[3] A video demonstration of the inverted pendulum can be found at `http://www.youtube.com/watch?v=5GvbG3pN0vY`

Streams x, y with alphabet $[0, 2]$

`first` x < `first` y

(a) Problem 1

Streams x, y with alphabet $[0, 2]$

x < `first` y

(b) Problem 2

Fig. 3. Example St-CSPs

The optimal solution automaton (Fig. 4a), in the sense of having the fewest states, for the St-CSP in Fig. 3a has two states, whilst the optimal solution automaton (Fig. 4b) for the St-CSP in Fig. 3b is a three-state automaton. In fact, for the St-CSP in Fig. 3b, as the size of the alphabet of y increases, the number of states in the optimal solution scales linearly. This is because, for each value that `first` y takes, there is a different upper bound on x. Therefore, a different state is needed for each value of `first` y.

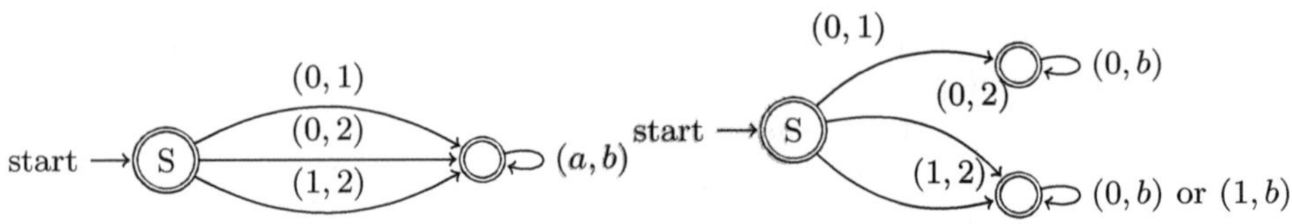

(a) Problem 1 (b) Problem 2

a, b are universally quantified over the integer interval $[0, 2]$

Fig. 4. Optimal Solution Automata for Example Problems

The difference between the two St-CSPs is that Problem 1 has a constraint that involves only the first time point, whereas Problem 2 has a constraint that involves streams with `first` operators and also other constructions of variable streams. These two examples demonstrate that constraints of the former kind do not increase the solution automaton size in general, whilst constraints of the latter kind can potentially multiply the size by a linear factor in the size of the stream alphabet.

However, the original solving approach [8,11] produces an automaton (Fig. 5) of linear size even for Problem 1, as a result of their normalisation rules. Stream expressions of the form "`first` *expr*" are normalised with the introduction of primitive next constraints (x_i == `next` x_j), which increase the size of the solution automaton because the daton values taken by x_i has to be taken by the daton of x_j at the next time point. Therefore, different states are needed to distinguish between the different values, effectively acting as memory for the automaton. Figure 6 shows an example of how states act as memory, where the alphabet of x_i is $[0, 1]$ for simplicity. Each state in Fig. 6 is annotated with the last daton value of the stream x_j that it represents.

Our proposed approach therefore is designed to avoid introduction of primitive next constraints for normalising streams with `first` operators, by improving the normalisation and search procedure. Even though the proposal applies only to a

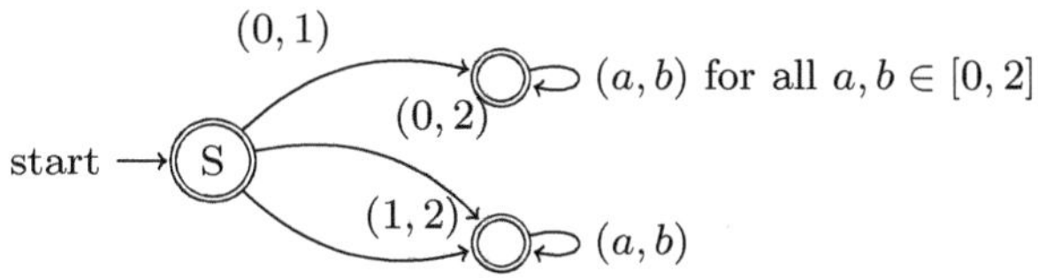

Fig. 5. Siu et al. [8,11]: Solution Automaton for `first` x `<` `first` y

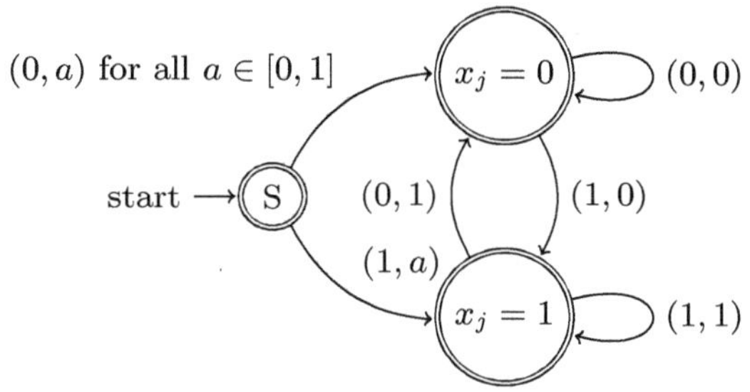

Fig. 6. Solution Automaton for x_i `==` `next` x_j

certain class of St-CSPs, we identify two practical uses for this class, which are presented in Sect. 5.

4.1 Constraint Normalisation

We propose to relax Siu's normal form [11]. An St-CSP is in *normal form* if it contains only constraints of the following two forms.

- Primitive next constraints: x_i `==` `next` x_j
- Primitive pointwise constraints with no `next` or `fby`.

Note that, in our approach, constraints involving only the `first` temporal operator are also considered as pointwise constraints.

The normalisation of `next` and `fby` streams is largely unchanged from Siu's algorithm [11]. The following is our new normalisation algorithm concerning `first` streams.

1. $(C_0 \cup \{c[\texttt{next first } expr]\}, C_1) \rightsquigarrow (C_0 \cup \{c[\texttt{first } expr]\}, C_1)$
2. $(C_0 \cup \{c[\texttt{first first } expr]\}, C_1) \rightsquigarrow (C_0 \cup \{c[\texttt{first } expr]\}, C_1)$
3. $(C_0 \cup \{c[\texttt{first } (expr_1 \texttt{ fby } expr_2)]\}, C_1) \rightsquigarrow (C_0 \cup \{c[\texttt{first } expr_1]\}, C_1)$
4. $(C_0 \cup \{c[\texttt{first } const]\}, C_1) \rightsquigarrow (C_0 \cup \{c[const]\}, C_1)$ where *const* is a constant stream.
5. $(C_0 \cup \{c[\texttt{first next } expr]\}, C_1) \rightsquigarrow (C_0 \cup \{c[\texttt{first } v]\}, C_1 \cup \{v \texttt{ == next } expr\})$ where *expr* is not of the form `first` $expr_1$, `next` $expr_1$ or $expr_1$ `fby` $expr_2$, and v is an auxiliary variable not in any of $c[-]$, C_0 and C_1.

To calculate the alphabets of auxiliary variables, we use interval arithmetic to construct bounds of the expression represented by the variable.

Given our new definition of normal form, the search algorithm has to be adapted.

4.2 Search Algorithm

Constraint specifications are now assumed to be in the normal form defined in the last section. The search algorithm we propose is again similar to the original approach [8,11], but the constraints in our approach can change during search.

We now describe how we construct the set of constraints C' of a child node $\hat{Q}(k+1)$ from the set of constraints C of the parent node $\hat{P}(k)$ and instantaneous assignment τ_k if $\hat{P}(k) \xrightarrow{\tau_k} \hat{Q}(k+1)$. The construction of τ_k should have been such that it satisfies all the primitive next constraints imposed by τ_{k-1}, i.e. $\tau_k(x_j) = \tau_{k-1}(x_i)$ for all primitive next constraints x_i `==` `next` x_j, and also all primitive pointwise constraints in C. In order to construct C', a direct copying from C is not correct. We observe that primitive next constraints are invariant in all shifted views of an St-CSP. It is only the primitive pointwise constraints that may change. In particular, streams with `first` operators are no longer the same when we take the shifted view of the child node $\hat{Q}(k+1)$. Such streams have to be *evaluated*, meaning that all variable streams inside a `first` operator have to be substituted by their assigned values from τ_k. The resulting stream expressions thus contain only constant streams, pointwise operators and `first` operators. Since `first` operators have no effect on constant streams, the expressions can always be reduced to a constant stream by evaluating pointwise operations and deleting `first` operators. We say we *evaluate* a constraint if and only if we evaluate all the stream expressions with `first` operators in the constraint. After evaluating a constraint with τ_k, we test whether it is a tautology. Since arithmetic is not decidable in general, we only consider tautologies of the zeroth order, i.e. those not involving universally quantified variables. If the constraint is a tautology, it is removed from the constraint set C.

Example 1. An example is given here to illustrate the construction process. Consider the St-CSP in Fig. 7a, which is normalised to the one in Fig. 7b. We construct the child node $\hat{Q}(1)$ from the root node and the instantaneous assignment $\tau_0 = \{(x = 0), (y = 0), (v = 2)\}$. Observe that given this τ_0, any τ_1 must obey $\tau_1(y) = 2$ due to the primitive next constraint. The final result of C' is $\{x$ `==` 0, x `<` v, v `==` `next` y$\}$:

- `first` x `==` `first` $(x$ `+` $y)$ is first substituted with values of τ_0 and becomes `first` 0 `==` `first` $(0$ `+` $0)$. The constraint is then evaluated into 0 `==` 0, which is a tautology not involving any variables. Therefore, we remove the constraint from C'.
- x `==` `first` y is evaluated into x `==` 0. Since x is still present, the constraint is not removed from C'.
- v `==` `next` y and x `<` v are unchanged.

Dominance between search nodes is detected in the same way as the original approach [8,11]. We say x is a *signature stream* if and only if x appears on the L.H.S. of a primitive next constraint x `==` `next` y for some stream y. Suppose $\hat{Q}_1(k_1)$ is a search node constructed from the instantaneous assignment τ_{k_1} and

Streams x, y with alphabet [0, 2]

```
first x == first (x + y)
x == first y
x < next y
```

(a) Example St-CSP

Streams x, y, v with alphabet [0, 2]

```
first x == first (x + y)
x == first y
x < v, v == next y
```

(b) Normalised Example St-CSP

Fig. 7. Example St-CSPs Illustrating the Search Algorithm

$\hat{Q}_2(k_2)$ is another search node constructed from τ_{k_2}. Let their corresponding sets of constraints be C_1 and C_2. We say that the two search nodes are *equivalent* if and only if C_1 is syntactically equivalent to C_2 and $\tau_{k_1}(x) = \tau_{k_2}(x)$ for all signature streams x. The proofs of Siu et al. [8,11] can be easily adapted to show that the detection is sound.

The previous example also demonstrates how we remove all information about the values taken for constraints involving only the first time point. They are always evaluated into a tautology that we recognise. Therefore, it is impossible for the solution automaton to have any "memory" on what values were taken, meaning there are no distinct states distinguishing between the different values. This achieves the reduction in automaton size we seek.

Our improvement is applicable whenever there exists a stream expression of the form `first` *expr* that only appears in constraints involving the first time point, and can take multiple values. The size of the solution automaton can be reduced by an exponential factor from the original approach [8,11]. For example, for an St-CSP with $2n$ streams with alphabet $[0, 1]$ and constraints `first` x_{2i} `==` `first` x_{2i+1} for $0 \leq i \leq n-1$, our approach produces a two state automaton since all constraints are removed after the first time point. In contrast, the original approach [8,11] normalises the problem and produces $2n$ streams constrained by primitive next constraints. There are a total of 2^n valid combinations of values taken by the streams with `first` operators. Therefore, considering also the start state, an automaton of size $2^n + 1$ is produced.

Due to space limitation, we state without proof the soundness of our constraint construction algorithm in the following. Recall from Sect. 2.4 that, if a parent node P has the set of constraints C, then the constraints of a child node Q are $(C \cup \{c'\})$ where c' is the constraint stating $x(0) = \tau_k(x)$ for all streams x and $x(i)$ is unconstrained for all $i > 0$. Since our construction algorithm takes shifted views into account, Theorem 1 gives the equivalence of C' and $(C \cup \{c'\})(1, \infty)$ where C' is the result of our construction.

Theorem 1. *Suppose the constraint set C' of the shifted view of child node $\hat{Q}(k+1)$ is constructed from the constraint set C of the parent node $\hat{P}(k)$ and the instantaneous assignment τ_k. Then $C' = (C \cup \{c'\})(1, \infty)$ where c' is the constraint stating $x(0) = \tau_k(x)$ for all streams x and $x(i)$ is unconstrained for all $i > 0$.*

5 Benefitting from the New Implementation

Our improvement applies only to certain usages of the `first` operator in an St-CSP. In this section, we show possible uses of the `first` operator in applications that can benefit from our new search algorithm.

5.1 Symmetry Breaking

The `first` operator can be used for breaking solution symmetry [4] in St-CSPs to reduce search, in the same way symmetry breaking helps with solving standard CSPs.

Symmetry breaking is the avoidance of visiting symmetric counterparts of visited search space. Suppose a CSP has *value symmetry* [9] σ and *variable symmetry* [9] σ'. If $\{x_0 = d_0, x_1 = d_1, \ldots, x_n = d_n\}$ is a solution of P, then $\{x_0 = \sigma(d_0), \ldots, x_n = \sigma(d_n)\}$ and $\{x_{\sigma'(0)} = d_0, \ldots, x_{\sigma'(n)} = d_n\}$ are also solutions respectively.

One technique for breaking value symmetry is by preassignment [7]. An analogous technique for stream constraint solving is to preassign the first daton of streams by constraints of the form "`first` x `==` *const*". This is a constraint that only concerns the first time point. However, since the stream with the `first` operator can only take one value, preassignment constraints do not increase the sizes of solution automata produced by even the original approach [8,11].

To break variable symmetry, a lexicographical ordering of assigned values can be imposed [5]. That is, suppose there is a fixed ordering on the set of variables. Extra constraints are added to enforce that if $x_1 < x_2$, then $d_1 \leq d_2$ where d_1 and d_2 are the values assigned to x_1 and x_2 respectively. An analogous treatment with streams is to enforce such ordering at the first time point by adding constraints such as `first` x_1 `<=` `first` x_2 and `first` x_2 `<=` `first` x_3.

Observe that streams with `first` operators in these lexicographical ordering constraints can take multiple values in general. Therefore, our improvement applies and produces a smaller solution automaton than the original approach [8,11]. Given n streams with the same alphabet of size $|\Sigma|$, our solution automaton is smaller[4] by a multiplicative factor $\binom{|\Sigma|+n-1}{n}$, which is the number of different valid assignments for the first datons of each stream. Section 6 includes experimental results to show the improvement.

5.2 Sequential Planning

Ghallab et al. [6] give a framework for encoding planning problems in traditional CSPs. The framework first states a planning problem in the *state variable representation*, consisting of *state variables* which are descriptions of the world that can change over time, *actions* with *preconditions* and *effects* which cause changes to the world, and *rigid relations* which describe the invariants in the

[4] This calculation excludes the start state, i.e. the size of the original automaton is $\binom{|\Sigma|+n-1}{n} \times (|S| - 1) + 1$ where $|S|$ is the size of our automaton.

world. For example, `at(cat)` is a state variable that holds the location of the `cat` object. The `move(o,a,b)` action has the precondition that `at(o)` = `a` and the effect that `at(o)` = `b`. `adjacent` = {`(desk,wall)`, `(desk,bed)`} is a rigid relation.

The state variable representation can then be encoded [6] as a CSP that expresses a plan of length t. Each time point has an associated $\texttt{action}^t$ variable, denoting the action taken at time t. Each state variable is encoded as a constraint variable for every time point, for example $\texttt{at(cat)}^0, \ldots, \texttt{at(cat)}^3$ for a plan of length 3. Precondition constraints are used to enforce the preconditions of actions. They are of the form ($\texttt{action}^t = \texttt{a}$) $\Rightarrow$ (preconditions of `a` at time $t-1$), such as $(\texttt{action}^t = \texttt{move(o,a,b)}) \Rightarrow (\texttt{at(o)}^{t-1} = \texttt{a})$. Similarly, there are effect constraints of the form ($\texttt{action}^t = \texttt{a}$) $\Rightarrow$ (effects of `a` at time t). Finally, frame constraints enforce that actions do not change anything other than their effects. For example, the constraint $\{(\texttt{action}^t = \texttt{move(o,a,b)}, \texttt{has(balloon)}^t = \texttt{c}, \texttt{has(balloon)}^{t+1} = \texttt{c} \mid \texttt{c} \text{ is an object}\}$.

With a St-CSP formulation, however, a variable for each time point is no longer needed. Only one St-CSP variable is required for each state variable, and another St-CSP variable for the action stream as St-CSP variables inherently span all time points. Precondition and effect constraints do not need to be specified per time point either. Observe that implication can be specified by inequality of pseudo-Boolean streams. Hence the precondition constraints $(\texttt{action}^t = \texttt{move(o,a,b)}) \Rightarrow (\texttt{at(o)}^{t-1} = \texttt{a})$ can be specified in an St-CSP as a single constraint `next` (*action* `eq` *move*(o, a, b)) `<=` *at*(o) `eq` a. It is a coincidence that the inequality appears typographically in the reverse direction of the implication symbol.

Subsequently, to ensure that the goal is achieved within t time points, a goal constraint is added: `first next next next ... next` *goal* `==` 1 where there are t `next` operators and *goal* is the pseudo-Boolean stream expression denoting whether the goal has been achieved or not.

With traditional CSPs, the size of the specification for a t step planning problem scales linearly with t. With St-CSPs, the size stays constant. Therefore, the St-CSP approach achieves representational simplicity that is not possible with the standard CSP approach. Another advantage is that, even though there may be potentially infinitely many solutions to the St-CSP, the solution set can be represented by a finite description, namely the solution automaton.

Example 2. The following is a simple example demonstrating the modelling technique. Suppose there is a unique, physical document to be circulated to n individuals. Only one individual may hold the document at any single time point. We use the state variables `seen(i)` to denote whether individual `i` has seen the document yet. The action `giveTo(i)` has no preconditions and the effect that `seen(i)` becomes true (or 1). Using the formalism above, we get the simple St-CSP in Fig. 8. In this example, we do not specify the length of the plan.

We can also use the `first` operator to specify initial conditions. For example, it can be the case that the first individual who gets the document must be one of senior rank, which is defined by the individual having a number smaller

Stream *goal* with alphabet [0, 1]
Stream *giveTo* with alphabet [0, $n-1$]
Streams $seen_0, \ldots, seen_{n-1}$ with alphabet [0, 1]

Constraints:
For each i,
`first` $seen_i$ `==` 0
`next` $seen_i$ `>=` $seen_i$ `or` ($giveTo$ `eq` i)

`first` $goal = 0$
$goal$ `>=` ($seen_0$ `and` … `and` $seen_{n-1}$)
`next` $goal$ `>=` $goal$;

Fig. 8. Document Circulation Planning St-CSP

than $(n-1)/2$. We introduce a constraint `first` $giveTo$ `<` $(n-1)/2$. Initial conditions produce constraints that only involve the first time point. When an initial condition is not strict, such as the one above, the streams with `first` operators can take multiple values. In this case, our improvement applies again, resulting in a smaller solution automaton and faster search. This shows that our new implementation has relevance to planning problems as well.

6 Experimental Results

We compare our implementation with the original implementation [8,11] using both the runtime and the size of the solution automaton as metrics. Experiments are conducted on an Intel Core i7 (4 × 2.2GHz) machine with 16GB RAM. Both solvers are set to timeout in 1 hour. Columns t and s in the results table corresponding to the runtime in seconds and the number of states in the solution automaton respectively. A "-" means the solver failed to solve the test case within the time limit. We also highlight the best results in bold per test case in the tables. *Note that*, our implementation also includes improvements achieved by using better data structures than the original.

6.1 Juggling Patterns

Siu et al. [8,11] give juggling patterns generation as an application of stream constraint solving. The St-CSP model describes the possible patterns of juggling moves obeying physical laws, parametrised by the number b of balls being juggled and the maximum number f of upward force units that can be applied to the balls. For physical reasons, $b \leq f$. This problem has variable symmetries. The test cases therefore contain symmetry breaking constraints, and can demonstrate the efficiency of our improvement.

Table 1a gives the experimental results. The notation (b, f) is used to denote the b and f values for a test case. Our implementation performs much better than the original solver in both metrics. The reduction in time can be as much as 96% for the case (4, 6), which also achieves a 90% reduction in automaton size.

Table 1. Experimental Results

(a) Juggling Test Cases

	Original		New	
(b, f)	t	s	t	s
(4, 4)	0.00	**5**	**0.00**	**5**
(4, 5)	2.10	481	**0.12**	**121**
(4, 6)	36.28	3601	**1.27**	**361**
(5, 5)	0.10	**6**	**0.01**	**6**
(5, 6)	238.41	3601	**11.33**	**721**
(6, 6)	2.10	**7**	**0.23**	**7**

(b) Document Circulation Test Cases

	Original		New	
n	t	s	t	s
10	5.07	4093	**1.81**	**1920**
11	24.23	10236	**4.57**	**3968**
12	102.43	20476	**11.50**	**7936**
13	1132.53	49147	**28.38**	**16128**
14	-	-	**66.59**	**32256**

6.2 Document Circulation Planning Problem

We use the document circulation example in Sect. 5.2 as a class of test cases. The initial condition described is also included in order to demonstrate our improvement. The number n of individuals is varied to give multiple test cases.

Table 1b gives the experimental results. It shows a significant reduction in both the solving times and the sizes of the solution automata in all test cases. The reduction in search time is at least 64% when $n = 10$ and can be as much as 97% when $n = 13$. As for the sizes of the solution automata, the reduction is at least 53% when $n = 10$ and can achieve 67% when $n = 13$. The results for $n = 14$ are incomparable since the original implementation [8,11] failed to solve the test case within 1 hour.

7 Concluding Remarks

Our contributions in this paper are four-fold. First, we present a novel application of St-CSPs in real-time PID control. We believe this is the first application of CSPs in real-time control on real hardware. The technique was applied to two different pieces of hardware and achieved stable performance as demonstrated in our video recordings. Second, we propose an improvement on the solving technique, which can lead to orders of magnitude reduction in both the search time and the size of the solution automaton. Third, we identify symmetry breaking and the specification of initial conditions in sequential planning problems as good uses of our improvement. Last but not least, we provide empirical evidence of the proposed improvement. All these takes us one step closer to deploying infinite stream constraint programming in practice.

There is ample room for future work. Here are a few possibilities. We can investigate the relationships among model checking [3], μ-calculus [3] and St-CSPs, as they are all related to Büchi automata. We can also extend the constraint language with more temporal operators, for example `asa` (as soon as), `whenever` and `upon` from the Lucid language [12] to increase our framework's expressiveness. In addition, the link between standard CSPs and St-CSPs can be explored. Standard CSP solving techniques may also help with solving St-CSPs.

Acknowledgements. We are grateful to the kind comments and suggestions by the anonymous referees. We thank Simon Wong for building the self-balancing tray and inverted pendulum used for our demonstrations, and Kin-Hong Wong for his help with PID control and robotics. Last but not least, the second author is indebted to Bill Wadge for teaching him dataflow programming and the Lucid language 25 years ago.

References

1. Bentley, J.: Programming Pearls. Addison-Wesley (2000)
2. Büchi, J.: On a decision method in restricted second order arithmetic. In: Mac Lane, S., Siefkes, D. (eds.) The Collected Works of J. Richard Büchi, pp. 425–435. Springer, New York (1990)
3. Clarke, J. E.M., Grumberg, O., Peled, D.A.: Model Checking. MIT Press, Cambridge (1999)
4. Cohen, D., Jeavons, P., Jefferson, C., Petrie, K.E., Smith, B.M.: Constraint symmetry and solution symmetry. In: Proc. AAAI 2006, pp. 1589–1592 (2006)
5. Frisch, A.M., Hnich, B., Kiziltan, Z., Miguel, I., Walsh, T.: Global constraints for lexicographic orderings. In: Van Hentenryck, P. (ed.) CP 2002. LNCS, vol. 2470, pp. 93–108. Springer, Heidelberg (2002)
6. Ghallab, M., Nau, D., Traverso, P.: Automated Planning: Theory & Practice. Morgan Kaufmann Publishers Inc. (2004)
7. Harvey, W.: Symmetry breaking and the social golfer problem. In: Proc. SymCon 2001 (2001)
8. Lallouet, A., Law, Y.C., Lee, J.H.M., Siu, C.F.K.: Constraint programming on infinite data streams. In: Proc. IJCAI 2011, pp. 597–604 (2011)
9. Law, Y.C., Lee, J.H.M.: Symmetry breaking constraints for value symmetries in constraint satisfaction. CONSTRAINTS 11(2-3), 221–267 (2006)
10. Minorsky, N.: Directional stability of automatically steered bodies. Journal of the American Society for Naval Engineers 34(2), 280–309 (1922)
11. Siu, C.F.K.: Constraint Programming on Infinite Data Streams. Ph.D. thesis, Department of Computer Science and Engineering, The Chinese University of Hong Kong (2012)
12. Wadge, W.W., Ashcroft, E.A.: LUCID, the Dataflow Programming Language. Academic Press Professional, Inc. (1985)
13. Winskel, G.: The Formal Semantics of Programming Languages: An Introduction. MIT Press, Cambridge (1993)

An Increasing-Nogoods Global Constraint for Symmetry Breaking During Search*

Jimmy H.M. Lee and Zichen Zhu

Department of Computer Science and Engineering,
The Chinese University of Hong Kong,
Shatin, N.T., Hong Kong
{jlee,zzhu}@cse.cuhk.edu.hk

Abstract. Symmetry Breaking During Search (SBDS) adds conditional symmetry breaking constraints (which are nogoods) dynamically upon backtracking to avoid exploring symmetrically equivalents of visited search space. The constraint store is proliferated with numerous such individual nogoods which are weak in constraint propagation. We introduce the notion of *increasing nogoods*, and give a global constraint of a sequence of increasing nogoods, incNGs. Reasoning globally with increasing nogoods allows extra prunings. We prove formally that nogoods accumulated for a given symmetry at a search node in SBDS and its variants are increasing. Thus we can maintain only one increasing-nogoods global constraint for each given symmetry during search. We give a polynomial time filtering algorithm for incNGs and also an efficient incremental counterpart which is stronger than GAC on each individual nogood. We demonstrate with extensive experimentation how incNGs can increase propagation and speed up search against SBDS, its variants, SBDD and carefully tailored static methods.

1 Introduction

Symmetries are common in many constraint problems. They can be broken statically [18,1,4,11] or dynamically [3,8,19]. While there are pros and cons for each approach, the focus of the paper is on SBDS (symmetry breaking during search) [8,6] and its variants, which add conditional symmetry breaking constraints dynamically during search. ReSBDS [12] is adapted from SBDS that tries to break extra symmetry compositions with a small overhead when only a subset of symmetries is given.

An overhead for SBDS and ReSBDS is the addition of a large number of constraints with weak pruning power. We observe that the symmetry breaking constraints added for each symmetry g at a search node are nogoods that are semantically related. We propose the notion of increasing nogoods. A global constraint (incNGs), which is logically equivalent to a set of increasing nogoods, is derived. Reasoning globally with increasing nogoods allows extra prunings. Thus we can maintain only one incNGs for each given symmetry. *Light ReSBDS* adds only a subset or implied ones of the nogoods added by ReSBDS but has a smaller overhead both in time and space. We give a polynomial time filtering algorithm for incNGs and its incremental version which is stronger

* We thank the anonymous referees and Toby Walsh for their kind and insightful comments and patience.

B. O'Sullivan (Ed.): CP 2014, LNCS 8656, pp. 465–480, 2014.

than GAC on each individual. Extensive experimentations are performed to demonstrate how incNGs can increase propagation and speed up search against SBDS, its variants, GAP-SBDD [7] and carefully tailored static methods.

2 Background

A *constraint satisfaction problem* (CSP) P is a tuple (X, D, C) where X is a finite set of variables, D is a finite set of domains such that each $x \in X$ has a $D(x)$ and C is a set of constraints, each is a subset of the cross product $\bigotimes_{i \in X} D(i)$. A constraint is *generalized arc consistent (GAC)* iff when a variable in the scope of a constraint is assigned any value in its domain, there exist compatible values (called *supports*) in the domains of all the other variables in the scope of the constraint. A CSP is *GAC* iff every constraint is GAC. An *assignment* $x = v$ assigns value v to variable x. A *full assignment* is a set of assignments, one for each variable in X. A *partial assignment* is a subset of a full assignment. A *solution* to P is a full assignment that satisfies every member of C.

A *nogood* is the negation of a partial assignment which cannot be contained in any solution. Nogoods can also be expressed in an equivalent implication form. A *directed nogood* ng ruling out value v_k from the initial domain of variable x_k is an implication of the form $(x_{s_1} = v_{s_1}) \wedge \cdots \wedge (x_{s_m} = v_{s_m}) \Rightarrow (x_k \neq v_k)$, meaning that the assignment $x_k = v_k$ is inconsistent with $(x_{s_1} = v_{s_1}) \wedge \cdots \wedge (x_{s_m} = v_{s_m})$. When a nogood, ng, is represented as an implication, the *left hand side* (LHS) $(lhs(ng) \equiv (x_{s_1} = v_{s_1}) \wedge \cdots \wedge (x_{s_m} = v_{s_m}))$ and the *right hand side* (RHS) $(rhs(ng) \equiv (x_k \neq v_k))$ are defined with respect to the position of $\Rightarrow$. If $lhs(ng)$ is empty, ng is *unconditional*. From now on, we call directed nogoods simply as nogoods when the context is clear.

In this paper, we consider search trees with binary branching, in which every non-leaf node has exactly two children. If a node P_0 is in a subtree under node P_1, P_0 is the *descendant node* of P_1 and P_1 is the *ancestor node* of P_0.

We assume that *the CSP associated with a search tree node is always made GAC* using an AC3-like [13] *constraint filtering algorithm* except our global constraint.

Here we consider symmetry as a property of the set of solutions. A *solution symmetry* [20] is a solution-preserving permutation on assignments.

Symmetry breaking method m_1 is *stronger in nodes (resp. solutions) pruning* than method m_2, denoted by $m_1 \succeq_n$ *(resp.* $\succeq_s$*)* m_2, when all the nodes (resp. solutions) pruned by m_2 would also be pruned by m_1. Symmetry breaking method m_1 is *strictly stronger in nodes (resp. solutions) pruning* than method m_2, denoted by $m_1 \succ_n$ *(resp.* $\succ_s$*)* m_2, when $m_1 \succeq_n$ *(resp.* $\succeq_s$*)* m_2 and $m_2 \not\succeq_n$ *(resp.* $\not\succeq_s$*)* m_1. Note that $\succeq_n$ and $\succ_n$ imply $\succeq_s$ and $\succ_s$ respectively.

Symmetry breaking during search (SBDS) [8,6] adds constraints to a problem during search so that after backtracking from a search node, the added constraints ensure that no symmetric equivalent of that node is ever allowed in subsequent search. An advantage is that this method can break symmetries of arbitrary kind. Partial SBDS (ParSBDS) [4,16] is SBDS but deals with only a given subset of all symmetries. LDSB [14] is a further development of shortcut SBDS [8] which handles only active symmetries and their compositions. Recursive SBDS [12] (ReSBDS) extends ParSBDS by breaking not only the given symmetries but also some symmetry compositions.

SBDD [3,7] is another widely used dynamic symmetry breaking method by checking whether the current state is dominated by recorded nogoods.

3 A Global Constraint for Increasing Nogoods

A set of directed nogoods is *increasing* if the nogoods can form a sequence

$$\begin{aligned} ng_0 &\equiv & A_0 &\Rightarrow x_{k_0} \neq v_{k_0} \\ ng_1 &\equiv & A_1 &\Rightarrow x_{k_1} \neq v_{k_1} \\ & & &\vdots \\ ng_t &\equiv & A_t &\Rightarrow x_{k_t} \neq v_{k_t} \end{aligned} \tag{1}$$

such that (i) for any $i \in [1, t]$, $A_{i-1} \subseteq A_i$ and (ii) no nogoods are implied by another. We consider the nogoods as a set or a sequence according to the context.

Every sequence of increasing nogoods has the following form:

$$\begin{aligned} ng_0 &\equiv & x_{s_{00}} = v_{s_{00}} \wedge \cdots \wedge x_{s_{0r_0}} = v_{s_{0r_0}} &\Rightarrow x_{k_0} \neq v_{k_0} \\ ng_1 &\equiv & lhs(ng_0) \wedge x_{s_{10}} = v_{s_{10}} \wedge \cdots \wedge x_{s_{1r_1}} = v_{s_{1r_1}} &\Rightarrow x_{k_1} \neq v_{k_1} \\ &\vdots & \vdots \\ ng_t &\equiv & lhs(ng_0) \wedge \cdots \wedge lhs(ng_{t-1}) \wedge x_{s_{t0}} = v_{s_{t0}} \wedge \cdots \wedge x_{s_{tr_t}} = v_{s_{tr_t}} &\Rightarrow x_{k_t} \neq v_{k_t}. \end{aligned} \tag{2}$$

A sequence of increasing nogoods can be encoded compactly, using 3 integer lists: I (index), E (equal) and N (not equal).

$$\begin{aligned} I = \langle\, & s_{00}, \ldots, s_{0r_0}, k_0, & E = \langle\, & v_{s_{00}}, \ldots, v_{s_{0r_0}}, \bot, & N = \langle\, & \bot, \ldots, \bot, v_{k_0}, \\ & s_{10}, \ldots, s_{1r_1}, k_1, & & v_{s_{10}}, \ldots, v_{s_{1r_1}}, \bot, & & \bot, \ldots, \bot, v_{k_1}, \\ & \vdots & & \vdots & & \vdots \\ & s_{t0}, \ldots, s_{tr_t}, k_t\rangle & & v_{s_{t0}}, \ldots, v_{s_{tr_t}}, \bot\rangle & & \bot, \ldots, \bot, v_{k_t}\rangle \end{aligned} \tag{3}$$

We encode *in order* every equality on the LHS and every disequality on the RHS of every nogood. The lists have the same length. Consider the ith tuple (I_i, E_i, N_i) from the 3 lists. If $N_i = \bot$, then the tuple is encoding $x_{I_i} = E_i$ on the LHS of a nogood. If $E_i = \bot$, then it is encoding $x_{I_i} \neq N_i$ on the RHS of a nogood.

Suppose we have the four nogoods

$$\begin{aligned} ng_0 &\equiv & x_1 &\neq 2, \\ ng_1 &\equiv & x_2 = 1 \Rightarrow x_3 &\neq 1, \\ ng_2 &\equiv & x_2 = 1 \wedge x_4 = 1 \wedge x_5 = 1 \Rightarrow x_3 &\neq 2, \\ ng_3 &\equiv & x_2 = 1 \wedge x_4 = 1 \wedge x_5 = 1 \wedge x_6 = 2 \Rightarrow x_1 &\neq 1. \end{aligned} \tag{4}$$

These nogoods are increasing because $\emptyset \subseteq \{x_2 = 1\} \subseteq \{x_2 = 1 \wedge x_4 = 1 \wedge x_5 = 1\} \subseteq \{x_2 = 1 \wedge x_4 = 1 \wedge x_5 = 1 \wedge x_6 = 2\}$ and none is implied by another. The 3 lists are derived as follows.

$$\begin{array}{rl} & \quad\ 0 \quad 1 \quad 2 \quad 3 \quad 4 \quad 5 \quad 6 \quad 7 \\ I = & \langle\ 1\ ,\ 2\ ,\ 3\ ,\ 4\ ,\ 5\ ,\ 3\ ,\ 6\ ,\ 1\ \rangle, \\ E = & \langle \perp, 1\ , \perp, 1\ ,\ 1\ , \perp, 2\ , \perp \rangle, \\ N = & \langle\ 2\ , \perp, 1\ , \perp, \perp, 2\ , \perp, 1\ \rangle. \end{array} \tag{5}$$

The 3 lists derived in (3) have the following feature.

Lemma 1. *Each (I_i, E_i, N_i) in I, E, N corresponds to an equality $x_{I_i} = E_i$ or disequality $x_{I_i} \neq N_i$, which can be considered as a variable-value pair. All pairs are distinct.*

An immediate consequence is to ensure the size of these 3 lists has an upper bound.

Theorem 1. *Suppose $P = (X, D, C)$ is a CSP with $|X| = n$, and I, E and N are constructed from a sequence of increasing nogoods. The maximum size of these 3 lists is $\sum_{i=0}^{n-1} D(x_i)$.*

Next we propose a global constraint that is equivalent to these nogoods but has stronger pruning power than each individual nogood. Suppose $P = (X, D, C)$ is a CSP, I, E and N are 3 lists with the same size m in the form of (3). An *increasing-nogoods global constraint* incNGs(I,E,N)(X) specifies

$$\begin{array}{rl} \forall i \in [0, m-1], N_i = \perp \vee (& (E_0 = \perp \vee (x_{I_0} = E_0)) \\ & \wedge (E_1 = \perp \vee (x_{I_1} = E_1)) \\ & \wedge \quad \ldots \\ & \wedge (E_{i-1} = \perp \vee (x_{I_{i-1}} = E_{i-1})) \Rightarrow x_{I_i} \neq N_i) \end{array} \tag{6}$$

meaning that if N_i is a non-$\perp$ value and all variables with indices before i in I are assigned to the corresponding value in E when there is a non-$\perp$ value, value N_i will be pruned from $D(x_{I_i})$. Note that incNGs(I,E,N)(X) is a family of global constraints parameterized by I, E and N.

Due to space limitation, we state without proof that the global constraint constructed from a sequence of increasing nogoods is logically equivalent to the conjunction of the increasing nogoods.

Theorem 2. *Suppose $\langle ng_0, \ldots, ng_t \rangle$ are increasing nogoods. We construct I, E and N as in* (3). *Then incNGs(I,E,N)(X) is logically equivalent to $ng_0 \wedge \cdots \wedge ng_t$.*

4 Deriving incNGs(I,E,N)(X) from SBDS and Its Variants

In this section, we first introduce an adaptation of ReSBDS with a smaller overhead. Next, we prove that constraints added by SBDS or its variants accumulated from the root node to a search node for the same symmetry forms a set of increasing nogoods.

4.1 Light ReSBDS

Domain filtering prune values by an AC3-like [13] constraint filtering algorithm. If a value v is pruned during the propagation of a constraint c, we say this pruning is *effected* by constraint c.

Recursive SBDS [12] (ReSBDS) uses a backtrackable set T to record all the assignments whose violations can indicate that a symmetry breaking constraint is already satisfied. Extra constraints would be added according to these violations. Suppose $x_j = a$ is recorded in T since the constraint $A^g \Rightarrow (x_i \neq v)^g$ is added at node P_0. Suppose further this assignment is violated at a descendant node P_1, i.e. a is pruned from $D(x_j)$. The pruning indicates that $A^g \Rightarrow (x_i \neq v)^g$ is already satisfied. This pruning is effected either by a problem constraint or a symmetry breaking constraint. Considering the latter case only, we propose a light version of the ReSBDS method without T as follows.

> **[Light ReSBDS (LReSBDS)]** Suppose G is a set of symmetries. LReSBDS always adds constraints added by ParSBDS. Once a value v is pruned from $D(x_i)$ *effected by a symmetry breaking constraint* at node P_0 with a partial assignment A, symmetry breaking constraint $A^g \Rightarrow (x_i \neq v)^g$ for all $g \in G$ is added.

Here the recursive addition of constraints is done by the propagation mechanism which stops propagation only when every variable domain does not change anymore.

Due to space limitation, we state without proof the comparison between ReSBDS and LReSBDS.

Theorem 3. *ReSBDS $\succ_n$ LReSBDS and ReSBDS $\succeq_s$ LReSBDS when given the same set of symmetries and both use the same static variable and value orderings.*

We conjecture that ReSBDS $\succ_s$ LReSBDS, but we have not found an example yet.

4.2 Deriving incNGs(I,E,N)(X)

In the following, by variants of SBDS, we mean ReSBDS and LReSBDS.

All the constraints added by SBDS or its variants down a search path for the same symmetry g forms a set of increasing nogoods.

Theorem 4. *Suppose G is a set of symmetries. Suppose further at a search node P_1, the constraints added by SBDS (or its variants) accumulated from root node to P_1 for symmetry g is R, where $g \in G$. R is a set of increasing nogoods.*

During search, all children nodes inherit I, E and N from their parent node. Every time a nogood is introduced by SBDS or its variants, the 3 lists are extended to record the nogood. Supposingly, each node should post a new increasing-nogoods constraint when a new nogood is added. We show in the following, however, that an increasing-nogoods constraint for a symmetry g posted in a parent node will always be subsumed by the corresponding one posted in its child node. Thus, each search node only has to deal with one increasing-nogoods constraint for each given symmetry.

All symmetry breaking constraints added to the parent node would also be added to the child node. It is straightforward to have the following theorem.

Theorem 5. *Suppose g is a symmetry and P_0 and P_1 are search nodes, where P_1 is a descendant of P_0. Suppose P_0 constructs I, E, N and P_1 constructs I', E', N' correspondingly by SBDS (or its variants). We have incNGs(I',E',N')(X) $\Rightarrow$ incNGs(I,E,N)(X).*

To implement LReSBDS, we need to know whether the pruning is effected by problem constraints or symmetry breaking constraints. We therefore need to get access to the filtering algorithm of the symmetry breaking constraints. This can be easily implemented in the constraint filtering algorithm of incNGs for LReSBDS.

5 A Filtering Algorithm

Suppose $\langle ng_0, \ldots, ng_t \rangle$ is a sequence of increasing nogoods. A nogood ng_i is *lower than* nogood ng_j iff $i < j$, and ng_j is *higher than* ng_i.

Suppose Λ is a sequence of increasing nogoods with the current domain D. A nogood ng is *generated* by Λ iff (i) $\exists ng'$ s.t. Λ generates ng' and $\Lambda \cup \{ng'\}$ generates ng, or (ii) we can find an $x \in X$ and a subsequence $\langle ng_{s_1}, \ldots, ng_{s_p} \rangle$ of Λ where $p = |D(x)|$, such that $\forall v \in D(x), \exists j \in [1, p], x \neq v \equiv rhs(ng_{s_j})$ and $ng \equiv \neg lhs(ng_{s_p})$ with $rhs(ng)$ being the rightmost assignment in $lhs(ng_{s_p})$.

A sequence Λ of increasing nogoods is in *reduced form* iff Λ can generate no nogoods. In the rest of the paper, we *assume that Λ is always a sequence of increasing nogoods of the form $\langle ng_0, \ldots, ng_t \rangle$.*

Theorem 6. *Λ is either in reduced form or has an equivalent sequence of increasing nogoods which is in reduced form. The size of the equivalent sequence never increases.*

The reduction procedure repeatedly checks for condition (ii) to generate a new nogood as appropriate. Assume a new nogood ng is generated from Λ according to condition (ii) and ng_k is the lowest nogood in Λ such that $\neg ng \subseteq lhs(ng_k)$. Clausal resolution ensures that $\langle ng_0, \ldots, ng_{k-1}, ng, ng_k, \ldots, ng_t \rangle$ is equivalent to $\langle ng_0, \ldots, ng_t \rangle$. As ng implies $\langle ng_k, \ldots, ng_t \rangle$, $\langle ng_0, \ldots, ng_{k-1}, ng \rangle$ is equivalent to $\langle ng_0, \ldots, ng_t \rangle$. If $\langle ng_0, \ldots, ng_{k-1}, ng \rangle$ cannot generate a new nogood, it is in reduced form. Otherwise, we continue to generate new nogood from $\langle ng_0, \ldots, ng_{k-1}, ng \rangle$.

Consider the nogoods given in (4) with $D(x_1) = D(x_2) = D(x_3) = D(x_6) = \{1, 2\}$ and $D(x_4) = D(x_5) = \{1\}$. Consider ng_1 and ng_2. Between $x_3 \neq 1$ and $x_3 \neq 2$, one of them must be false since x_3 must take a value. Thus we can generate the nogood $(\neg lhs(ng_1) \vee \neg lhs(ng_2)) \Leftrightarrow \neg lhs(ng_2)$, which is $\neg(x_2 = 1 \wedge x_4 = 1 \wedge x_5 = 1)$ and can be expressed as a directed nogood $ng_4 \equiv x_2 = 1 \wedge x_4 = 1 \Rightarrow x_5 \neq 1$ by putting the rightmost assignment of $lhs(ng_2)$ to the right. Now $\langle ng_0, ng_1, ng_4 \rangle$ is a new sequence of increasing nogoods equivalent to $\langle ng_0, ng_1, ng_2, ng_3 \rangle$. Consider ng_4. Domain of x_5 is a singleton. Directed nogood $ng_5 \equiv x_2 = 1 \Rightarrow x_4 \neq 1$ is generated. Now $\langle ng_0, ng_5 \rangle$ is a new sequence of increasing nogoods equivalent to $\langle ng_0, ng_1, ng_4 \rangle$. Domain of x_4 is a singleton. Directed nogood $ng_6 \equiv x_2 \neq 1$ is generated. Now $\langle ng_0, ng_6 \rangle$ is a new sequence of increasing nogoods equivalent to $\langle ng_0, ng_5 \rangle$. No new nogood can be generated and $\langle ng_0, ng_6 \rangle$ is the reduced form of $\langle ng_0, ng_1, ng_2, ng_3 \rangle$.

It might happen that more than one subsequence of Λ satisfies condition (ii). We give the **shortest nogood rule**: whenever more than one nogood can be generated from Λ according to condition (ii), we choose the shortest one to generate. In other words, we generate using a subsequence, the highest nogood of which is the lowest in Λ.

An assignment $x = v$ satisfies the *covering condition* in Λ iff $x = v \in lhs(ng_k) \wedge x = v \notin lhs(ng_{k-1}) \wedge (\forall v' \in D(x) - \{v\}, \exists j \in [0, k-1], rhs(ng_j) \equiv x \neq v')$. In other words, when an assignment $x = v$ appears for the first time in the sequence in the LHS of a nogood (say, ng), $x \neq v'$ would have appeared on the RHSs of nogoods lower than ng for all $v' \in D(x)$ except v.

For a sequence of increasing nogoods Λ in reduced form, we consider only a **single pruning condition**: when the LHS of a nogood in Λ is true, its RHS is enforced to effect value pruning.

The reduction procedure as described seems to call for scanning the sequence of increasing nogoods repeatedly, which is inefficient. In the following, we explain how the covering condition allows us to scan the nogood sequence only once to transform it into reduced form.

Consider the nogoods in (4) again. If, instead, $D(x_5)=\{1,2\}$, ng_4 in $\langle ng_0, ng_1, ng_4\rangle$ cannot generate ng_5 since condition (ii) is not met. This is because $x_5 = 2$ does not satisfy the covering condition. We state without proof the following theorem.

Theorem 7. *Assume Γ is a new and equivalent sequence of increasing nogoods transformed from Λ according to the* **shortest nogood rule** *with the current domain. Γ can generate further new nogoods only if the assignment $\neg rhs(ng)$ satisfies the covering condition in Λ, where ng is the highest nogood in Γ.*

Consider the nogoods in (4) again with the variable domains on page 470: $D(x_1) = D(x_2) = D(x_3) = D(x_6) = \{1,2\}$ and $D(x_4) = D(x_5) = \{1\}$. We scan from ng_0 and up. Assignment $x_2 = 1$ does not satisfy the covering condition since $rhs(ng_0) \not\equiv x_2 \neq 2$. In ng_2, assignments $x_4 = 1$ and $x_5 = 1$ satisfy the covering condition since they have singleton domains. Now the new nogood $ng_4 \equiv \neg lhs(ng_2)$ is generated whose RHS is $x_5 \neq 1$. Another new nogood $ng_5 \equiv \neg lhs(ng_4)$ can be generated immediately without rescanning from the first nogood. Since $\neg rhs(ng_5) \equiv x_4 \neq 1$, another new nogood $ng_6 \equiv \neg lhs(ng_5)$ can be generated immediately. Now $\neg rhs(ng_6) \equiv x_2 \neq 1$, which does not satisfy the covering condition. We can stop and $\langle ng_0, ng_6\rangle$ is equivalent to the original sequence and in reduced form.

As a result, we only have to scan the sequence from ng_0 and up. For every nogood, we check if condition (ii) is met for nogoods from ng_0 up to here. In addition, we also check and record whether new assignments on the LHS satisfies the covering condition or not. Once the first reduction step is launched, there is no further need to check for condition (ii). We stop once $\neg rhs(ng)$ does not satisfy the covering condition, where ng is the last generated nogood.

In our filtering algorithm, the first step is to effect prunings using nogoods whose LHSs are true under the current domain. These nogoods can then be thrown away. The remaining nogoods still form an increasing sequence, which can be turned into reduced form. The single pruning condition for reduced forms is still expensive to check. It turns out that if the leftmost assignment in the LHS of the first nogood in a sequence is not true under the current domain, we can detect pruning in the reduced form much more efficiently. This form is easy to get.

Lemma 2. *Suppose $x = v$ in $lhs(ng_0)$ is the leftmost assignment which is not true under the current domain and Δ is the set of assignments before $x = v$ in $lhs(ng_0)$. Suppose further $\Gamma \equiv \langle ng_{s_0}, \ldots, ng_{s_t}\rangle$ is a sequence of increasing nogoods such that*

$\forall i \in [0,t], lhs(ng_{s_i}) = lhs(ng_i) - \Delta, rhs(ng_{s_i}) = rhs(ng_i)$. Γ is equivalent to Λ and is the simplified version of Λ.

Suppose we have an increasing sequence with the form as given in the last lemma. By repeated applications of the **single pruning condition**, we get the following result which gives a much simplified pruning condition.

Theorem 8. *Suppose Γ is the reduced form of Λ where the first assignment in $lhs(ng_0)$ is not true under the current domain D. A value $v \in D(x)$ is pruned in Γ if $\Gamma \equiv \langle x \neq v\rangle$, where $x \neq v$ is an unconditional nogood.*

In the following, we give an efficient filtering algorithm based on the following two major steps: effect prunings for nogoods whose LHSs are true and transform the simplified version of the remaining sequence of increasing nogoods into reduced form. We explain our algorithm using the I, E and N encodings of Λ. Pointer α is used to index into the 3 lists to effect prunings for nogoods whose LHSs are true.

- Pointer α is set to the *largest* index such that $(\forall i \in [0,\alpha), E_i = \perp \vee x_{I_i} = E_i)$. Note that α points at an unsatisfied equality in $lhs(ng_p) - lhs(ng_{p-1})$ where $p \in [0,t]$ and ng_p is the lowest nogood whose LHS is not true. For all the nogoods lower than ng_p, their RHSs can be enforced to prune values.

Therefore, we examine the LHSs of remaining nogoods starting from α. To transform the simplified version $\langle ng'_p, \ldots, ng'_t\rangle$ of remaining increasing nogoods $\langle ng_p, \ldots, ng_t\rangle$ into reduced form, two pointers β and γ are used to index into the 3 lists with the following conditions.

- Pointer β is used to find the shortest nogood generated by Λ according to condition (ii). If a new nogood is generated, γ is then used to check whether there are extra nogoods that can be generated according to Theorem 7 and find the last generated nogood. If extra nogoods can be generated, β is set to γ. Initially, β is set to the *largest* index such that $\forall i \in [0,\beta), E_i \in D(x_{I_i}) \vee (E_i = \perp \wedge D(x_{I_i}) \neq S_{I_i})$, where $S_{I_i} = \{N_j | I_j = I_i, N_j \neq \perp, N_j \in D(x_{I_i}), j \in [0,i]\}$. The key concept is S_{I_i}, which is the collection of all values N_j still in $D(x_{I_i})$ such that $x_{I_i} \neq N_j$ is a RHS disequality that appears before or at the nogood encoded at i. Note that β points to either
  (a). an equality in $lhs(ng_q) - lhs(ng_{q-1})$ where $q \in [0,t]$, in which case $E_\beta \neq \perp$ and $E_\beta \notin D(x_{I_\beta})$, i.e. ng_q is the lowest nogood whose LHS is false. All nogoods ng_i where $t \geq i \geq q$ are satisfied. We only need to enforce $\langle ng'_p, \ldots, ng_{q-1}\rangle$. No new nogoods can be generated and this increasing nogoods is in reduced form. We say β satisfies condition (a); or
  (b). the disequality in $rhs(ng_q)$ where $q \in [0,t]$, in which case $D(x_{I_\beta}) = S_{I_\beta}$, i.e. ng_q with $rhs(ng_q) \equiv x_{k_q} \neq v_{k_q}$ is the lowest nogood such that all values in the domain of $x_{k_q} (= x_{I_\beta})$ have appeared in $rhs(ng_j)$ for all $j \in [0,q]$. Now new nogood $ng \equiv \neg lhs(ng_q)$ is generated and is the shortest one. Increasing nogoods $\langle ng'_p, \ldots, ng\rangle$ are formed. We say β satisfies condition (b).

- If β satisfies condition (b), the first reduction step is launched since the new nogood ng is generated. Pointer γ is set in such a way that all equalities $x_{I_i} = E_i$ for $i \in [\gamma, s]$, where $\neg rhs(ng) \equiv x_{I_s} = E_s$, satisfy the covering condition. Pointer γ is set to the *smallest* index such that $(\forall i \in (\gamma, \beta), (E_i = \perp) \vee (D(x_{I_i}) = S_{I_i} \cup \{E_i\})) \wedge D(x_{I_\gamma}) = S_{I_\gamma} \cup \{E_\gamma\}$. Note that γ points to the disequality in $rhs(ng')$ where $\neg lhs(ng')$ is the highest nogood of increasing nogoods in reduced form (e.g. for nogoods in (4) with variable domains on page 470, ng' is ng_5 and $\neg lhs(ng')$ is ng_6). If γ takes a value, new nogoods can be generated and β is set to γ. Otherwise, we do not need to update β. Now β still satisfies condition (b).

After updating β according to the above, we have found the increasing nogoods in reduced form. If β still satisfies condition (b), i.e. new nogoods have been generated, we need to check whether values can be pruned according to Theorem 8.

Consider the nogoods given in (4) with variable domains on page 470. If incNGs(I,E,N)(X) has m as the size of the 3 lists, we do propagation in the following ways.

1. Pointer $\alpha = 0$, $\beta = m = 3$ and $\gamma = \perp$.
2. To find α, we scan with i from 0 to $m - 1$.
   - $i = 0$: prune N_0 from $D(x_1)$ $(= D(x_{I_0}))$.
   - $i = 1$: $E_1 \neq \perp$ and E_1 has not been assigned to $x_2 (= x_{I_1})$. Stop scanning and set $\alpha = 1$.
3. To find β, we scan with i from α to $m - 1$.
   - $i = 1$: $E_1 \in D(x_2)$ $(= D(x_{I_1}))$.
   - $i = 2$: $E_2 = \perp$ but value 1 in domain $D(x_3) (= D(x_{I_2}))$ is not in N_j for all $j \leq i$ and $I_j = I_2$.
   - $i = 3$: $E_3 \in D(x_4)$ $(= D(x_{I_3}))$.
   - $i = 4$: $E_4 \in D(x_5)$ $(= D(x_{I_4}))$.
   - $i = 5$: $D(x_{I_5}) = D(x_3) \subseteq \{1, 2\} = \{N_2, N_5\} = S_3$ since $I_2 = I_5 = 3$. Stop scanning and set $\beta = 5$. Now β satisfies condition (b).
4. Pointer β satisfies condition (b), the first reduction step is launched. To find γ, we scan with i from α to $\beta - 1$.
   - $i = 1$: $E_1 \in D(x_2)$ $(= D(x_{I_1}))$ but $S_2 \cup \{E_1\} \neq D(x_2)$.
   - $i = 2$: $E_2 = \perp$.
   - $i = 3$: $E_3 \in D(x_4)$ $(= D(x_{I_3}))$ and $S_4 \cup \{E_3\} = D(x_4)$, set $\gamma = 3$.
   - $i = 4$: $E_4 \in D(x_5)$ $(= D(x_{I_4}))$ and $S_5 \cup \{E_4\} = D(x_5)$, γ is still 3.
5. Pointer β satisfies condition (b) and $\gamma \neq \perp$, set $\beta = \gamma$. Now β points to the disequality in RHS of the newly generated nogood $ng' \equiv x_2 = 1 \Rightarrow x_4 \neq 1$ and $ng \equiv \neg lhs(ng')$ is the final generated nogood. Since $lhs(ng)$ is empty, value 1 is pruned from $D(x_2)$. The propagation is done.

After the propagation, $D(x_1) = \{1\}, D(x_2) = \{2\}, D(x_3) = \{1, 2\}$, $D(x_4) = D(x_5) = \{1\}$, $D(x_6) = \{1, 2\}$. GAC on individual nogoods can only prune value 2 from $D(x_1)$. Our filtering prunes also 1 from $D(x_2)$. The step to find β and γ can be done at the same time. We only need to check whether β should be set to γ or not.

Suppose $P = (X, D, C)$ is a CSP where the size of X is n. We give the filtering algorithm for an incNGs(I,E,N)(X) as follows. For the moment, please ignore highlighted codes in frame boxes, which are reserved for the incremental version of the algorithm.

Algorithm 1. *InGEnforce*()

Require:
X, D, I, E, N
m: the size of I, E and N
$\alpha = 0$
$\beta = m$ | ▲$\beta = \sum_{i=0}^{n-1} D(x_i)$ |
$\gamma = \perp$
$p = 0$: reason of why β is updated

1: **if** $m = 0$ **then**
2: return **ENTAILED**;
3: **end if**
4: *UpdateAlpha*();
5: *UpdateBeta*();
6: **if** $\alpha = \beta$ **then**
7: **if** $p = 1$ **then**
8: return **ENTAILED**;
9: **end if**
10: **if** $p = 2$ **then**
11: return **FAILED**;
12: **end if**
13: **end if**
14: **if** $p = 2$ **then**
15: return *CheckE*();
16: **end if**

Algorithm 1 is the top level of the filtering algorithm. This algorithm is called whenever the domain of a variable in X is modified or I, E and N are extended. The pointer α is initialized to 0, β is initialized to the size of the 3 lists and γ is set to $\perp$. Integer variable p tells the reason of why β is updated, i.e. β satisfies which condition. If the 3 lists are empty (Lines 1-3), the constraint is automatically **ENTAILED**, which means the constraint can be disposed. Line 4 calls the function *UpdateAlpha*() to update the pointer α. Line 5 calls the function *UpdateBeta*() to update the pointer β according to shortest nogood rule and Theorem 7. Lines 6-13 check whether $\alpha = \beta$ or not. If it is true and β is updated as a result of condition (a), this constraint is **ENTAILED** since future pruning can take place only between α and $\beta - 1$. If the two pointers are equal and β is updated because of the condition (b), this constraint is **FAILED** since LHS of nogood at α $(= \beta)$ is satisfied but the negation of the LHS of this nogood is a nogood. Thus the current node should fail. Lines 14-16 calls the function *CheckE*() to check whether the last generated nogood satisfies the condition in Theorem 8.

In the following, *Prune*(v, x) prunes value v from $D(x)$. While x.*assigned*(v) checks whether x is assigned with value v, x.*in*(v) checks whether $v \in D(x)$ and x.*size*() returns $|D(x)|$. We assume the last three functions have constant time complexity.

Algorithm 2. *UpdateAlpha*()

1: int $i = 0$; | ▲int $i = \alpha$; |
2: **while** $i < m$ | ♦and $i < \beta$ | **do**
3: **if** $E_i = \perp$ **then**
4: *Prune*(N_i, x_{I_i});
5: **else**
6: **if** $\neg x_{I_i}$.*assigned*(E_i) **then**
7: break;
8: **end if**
9: **end if**
10: $i = i + 1$;
11: **end while**
12: $\alpha = i$;

Algorithm 2 updates α. It starts scanning from index $i = 0$, and stops only when x_{I_i} is not assigned with non-$\perp$ value E_i (lines 6-8), in which case α is set to i (line 12). During scanning, if $E_i = \perp$, since the LHS of the nogood at this point is true, N_i can be pruned from $D(x_{I_i})$ (lines 3-4).

Algorithm 3. *UpdateBeta*()

```
1: int i = α;
2: int S[n];
3: for each j ∈ [0, n − 1] do
4:     S[j] = 0;
5: end for
6: while i < m ♦and i < β do
7:     if x_{I_i}.in then(E_i)
8:         if γ ≠⊥ ∧S[I_i] ≠ x_{I_i}.size()-1 then
9:             γ =⊥;
10:        else
11:            if γ =⊥ ∧S[I_i] = x_{I_i}.size()-1 then
12:                γ = i;
13:            end if
14:        end if
15:        continue;
16:    end if
17:    if E_i ≠ ⊥ then
18:        if ¬x_{I_i}.in(E_i) then
19:            p = 1;
20:            break;
21:        end if
22:    else
23:        if x_{I_i}.in(N_i) then
24:            S[I_i] + +;
25:        end if
26:        if S[I_i] = x_{I_i}.size() then
27:            p = 2;
28:            break;
29:        end if
30:    end if
31:    i = i + 1;
32: end while
33: if i ≠ m then
34:     if p = 2 ∧ γ ≠⊥ then
35:         β = γ;
36:     else
37:         β = i;
38:     end if
39: end if
```

Algorithm 3 updates β. As $\beta \geq \alpha$, the scan starts from α. We use an array $S[]$, so that $S[i]$ records the number of encountered values during scanning for each variable x_i in the disequalities on the RHS of nogoods. $S[i]$ does not count values already pruned from the domain of x_i (lines 23-25). Lines 7-16 updates γ. Lines 8 and 9 reset γ to $\perp$ if the covering condition is not satisfied. Lines 11-13 set γ to i if γ is $\perp$ and the covering condition is satisfied. Lines 17-21 check if scanning should stop due to condition (a) and set the reason p for updating β, before updating β in lines 33-39. Lines 26-29 check if scanning should stop due to condition (b) and set the reason p for updating β, before updating β in lines 33-39. If the scanning is stopped due to condition (b) and γ is a non-$\perp$ value (line 34), β is set to γ (line 35). Or else, β is set to the interrupted i.

Algorithm 4. *CheckE*()

```
1: int i = β − 1;
2: while i > α do
3:     if E_i ≠ ⊥ then
4:         break;
5:     end if
6:     i = i − 1;
7: end while
8: if i = α then
9:     Prune(E_α,x_{I_α});
10:    return ENTAILED;
11: end if
12: return CONSISTENT;
```

If β is updated because of condition (b), Algorithm 4 is called. Lines 2-7 first check whether there exists non-$\perp$ value in E from $\alpha + 1$ to $\beta - 1$. If yes (line 12), $x_{I_\alpha} = E_\alpha$ must be in the LHS of the last generated nogood which does not satisfy the condition in

Theorem 8. This constraint is **CONSISTENT** means that the domain filtering is done. Now the nogoods between α and β consist of the increasing nogoods in reduced form. If not (lines 8-11), value E_α is pruned from its corresponding variable's domain since $x_{I_\alpha} \neq E_\alpha$ is the only nogood in the increasing nogoods in reduced form. Now this constraint is **ENTAILED** as $\beta = \alpha$.

We do not have an exact characterization on Algorithm 1's consistency level yet, but it is stronger than GAC on individual nogoods and has a polynomial time complexity.

Theorem 9. *Algorithm 1 terminates and enforces a consistency on incNGs(I,E,N)(X) that is strictly stronger than GAC on each individual nogood.*

Theorem 10. *Algorithm 1 runs in $O(db|X|)$ for constraint incNGs(I,E,N)(X), where d is the largest domain size and b is the cost of pruning a value from a variable domain.*

6 Incremental Filtering Algorithm

Though polynomial, Algorithm 1 is expensive to execute from scratch at every invocation of the global constraint during (a) constraint propagation within an AC3-like algorithm and (b) adding new increasing nogoods during search (advancing to child nodes during search and generating extra nogoods during propagation in recursive methods). We can make Algorithm 1 incremental using the following theorems.

Theorem 11. *For a global constraint incNGs(I,E,N)(X), whenever Algorithm 1 is invoked during AC3-like constraint filtering algorithm, the two pointers α and β can be carried over from the last invocation.*

Theorem 12. *Suppose I, E and N are constructed from increasing nogoods Λ. Suppose further that I', E' and N' are constructed from increasing nogoods $\langle ng_0, \ldots, ng_t, ng_{t+1}\rangle$. The two pointers α' and β' for enforcing incNGs(I',E',N')(X) can be initialized to the values of α and β respectively after domain filtering of incNGs(I,E,N)(X) with Algorithm 1.*

The incremental filtering algorithm can be obtained by *adding* the highlighted ones marked by ♦ and *substituting* codes by ones marked by ▲ to the right. Note that at the start of Algorithm 1, the initialization for α, β, γ and p is only for the root node. In subsequent nodes, these four are initialized from the ones after the latest propagation or from the ones of the previous global constraint after its domain filtering.

7 Experiments

This section gives four experiments to demonstrate empirically how globalized SBDS, ParSBDS and ReSBDS can improve the runtime substantially over their original versions. We also implemented the global version of LReSBDS but not the one using decomposed nogooods. When available, we compare our results also against state of the art static methods. All experiments are conducted using Gecode Solver 4.2.0 on Xeon E5620 2.4GHz processors.

SBC uses the static method by Puget [17] to break all variable symmetries and the value symmetries in all-different problems. **Doublelex** [4] lexicographically orders the rows and columns in increasing order. **SBDS** uses SBDS to break *all* symmetries. **ParSBDS** and **ReSBDS** handle the given symmetries by ParSBDS and ReSBDS respectively.

For matrix problems, **ReSBDS**[c] is given as symmetries that adjacent rows (columns) are interchangeable and also cartesian-product of adjacent row symmetries and adjacent column symmetries. The globalized version of **SBDS**, **ParSBDS**, **ReSBDS** and **ReSBDS**[c] are denoted by **[incNGs]**$_S$, **[incNGs]**$_P$, **[incNGs]**$_R$ and **[incNGs]**$_R$[c] respectively. By using the globalized version of LReSBDS and give the same symmetries as **ReSBDS** and **ReSBDS**[c] respectively, we have **[incNGs]**$_{LR}$ and **[incNGs]**$_{LR}$[c]. For decomposed nogood implementation, we use clause constraint whose propagation uses two-watched literals [15,21]. For GAP-SBDD and GAP-SBDS, we do not have their implementation in Gecode, and provide an indirect but machine-independent comparison using results in the literature. LDSB is discarded in the comparison here since ReSBDS is substantially more efficient [12]. *Unless otherwise specified*, we search with input variable order and minimum value order.

In all experiments, we show only the runtime to find all solutions. Runtime is limited to 1 hour. The number of backtracks and number of solutions are in line with the theoretical predictions. We did not report them only because of lack of space. All results are shown in graphical form for easy visualization. The horizontal axis shows instances, and the vertical axis shows the runtime in seconds. N-Queens instances are sorted by size. In the other three experiments, instances are sorted by the runtime of static methods. The last two experiments use *log* graph for better visualization. *Dashed lines* give the results of methods using decomposed nogoods and *solid lines* are for methods using global constraints. *Solid lines with '+'* shows the results for static methods.

7.1 N-Queens

We model the N-Queens problem the standard way using one variable per column. All 8 geometric symmetries are given to **SBDS**. **ParSBDS** and **ReSBDS** are only given the two generators rx (reflection on the vertical axis) and $d1$ (reflection on the diagonal), which can generate all 8 geometric symmetries.

Fig. 1 shows the results. For complete methods, **[incNGs]**$_S$ is up to 1.82 times faster than **SBDS**, and **[incNGs]**$_S$ has up to 434825 less failures than **SBDS**. For partial symmetry breaking methods, only two symmetries are given and these two symmetries

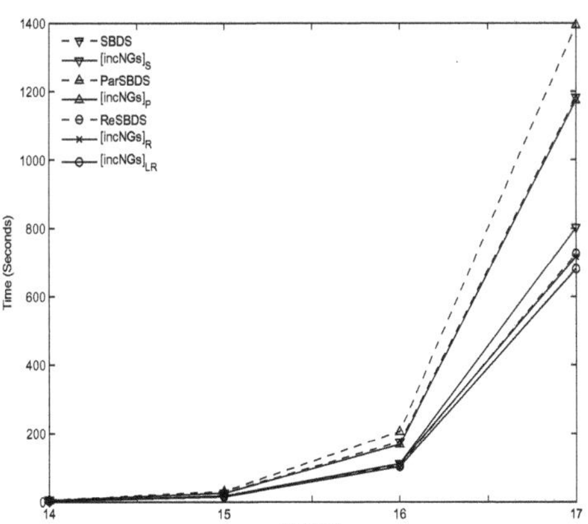

Fig. 1. N-Queens problem

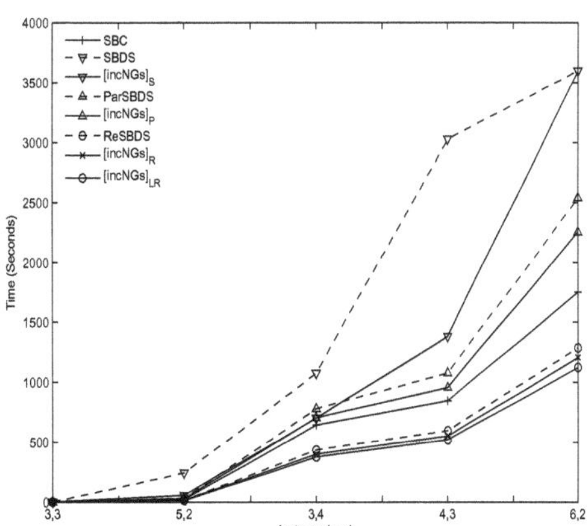

Fig. 2. The Graceful Graph problem

would be broken high up in the search tree. For partial SBDS, **[incNGs]**$_P$ is up to 1.23 times faster than **ParSBDS**. For ReSBDS, **[incNGs]**$_R$ improves only a little over **ReSBDS** due to the overhead to get to know when I, E and N are updated. And **[incNGs]**$_{LR}$ is the most efficient and faster than **ReSBDS**. Note that **[incNGs]**$_S$ is complete and comparable with **ReSBDS** and **[incNGs]**$_{LR}$. This shows how the global constraint can help to prune all symmetric solutions in a competitive manner.

7.2 Graceful Graph

The graceful graph problem is an all-different problem [16]. A $K_n \times P_m$ graph has intra-clique permutations, inter-clique permutations, complement symmetry, and their combinations. **ParSBDS** is given $n * (n-1)/2$ symmetries to describe any two nodes in each clique being permutable simultaneously and two more symmetries to describe inter-clique permutation and complement symmetry. **ReSBDS** is given $(n-1)$ symmetries to describe simultaneous permutation of adjacent nodes in each clique and also one inter-clique permutation and one complement symmetry.

Fig. 2 shows the results. For complete methods, **[incNGs]**$_S$ runs up to 4.25 times faster than **SBDS**, and **[incNGs]**$_S$ has up to 169286 less failures than **SBDS**. This shows the global constraint improves our complete method dramatically. For ParSBDS and ReSBDS, **[incNGs]**$_P$ and **[incNGs]**$_R$ are up to 1.13 and 1.09 times faster than **ParSBDS** and **ReSBDS** respectively as only a small subset of symmetries are given. Using LReSBDS, **[incNGs]**$_{LR}$ is up to 1.16 times faster than **ReSBDS** and is even up to 1.71 times faster than **SBC**. Literature results [17] show that SBC is up to 15 times faster than GAP-SBDD and GAP-SBDS. This demonstrates LReSBDS with global constraints can beat GAP-SBDD, GAP-SBDS and carefully tailored static methods.

7.3 Balanced Incomplete Block Design

A BIBD instance can be determined by its parameters (v, k, λ). We use the 0/1 model [5], which has row and column symmetries since we can permute any rows or columns freely without affecting any of the constraints. **ParSBDS** is given the symmetry that any two rows (columns) are interchangeable. **ReSBDS** is given interchangeability of adjacent rows (columns). All are solved with the maximum value heuristic.

Fig. 3 shows the results for BIBD. For partial SBDS, **[incNGs]**$_P$ runs up to 2.26 times faster than **ParSBDS**. For ReSBDS, the global constraint cannot help much due to the overhead to get to know when I, E and N are updated. Note that **[incNGs]**$_R$[c] is even 1.58 times slower than **ReSBDS**[c]. For light ReSBDS, however, **[incNGs]**$_{LR}$ and **[incNGs]**$_{LR}$[c] are up to 1.46 and 1.32 times faster than **ReSBDS** and **ReSBDS**[c] respectively. The light version improves a lot over ReSBDS. Note that **[incNGs]**$_{LR}$[c] is even 3.03 times faster than **DoubleLex**. The gains come from the advantage of LReSBDS by posting more symmetries and the efficiency of the global constraint. Literature results [7,6] show that GAP-SBDD is about 4 times faster than GAP-SBDS and DoubleLex is at least 10 and up to 38 times faster than GAP-SBDS, we can conclude indirectly that **[incNGs]**$_{LR}$[c] can beat GAP-SBDD and GAP-SBDS dramatically.

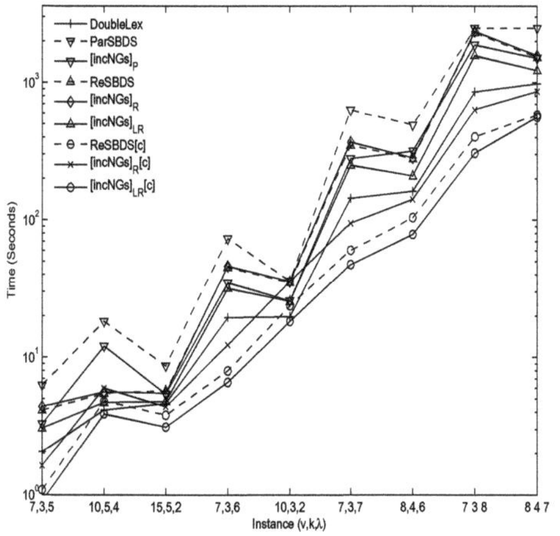

Fig. 3. The BIBD problem

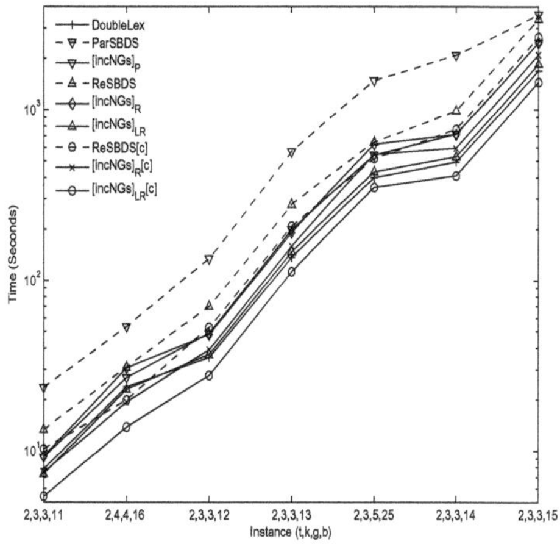

Fig. 4. The CA problem

7.4 Cover Array Problem (CA)

The Cover Array Problem CA(t, k, g, b) is prob045 in CSPLib [9]. We use the integrated model [10], which channels an original model and a compound model. **ParSBDS** and **ReSBDS** are given the same set of symmetries as in BIBD.

Fig. 4 shows the results. For dynamic methods, **[incNGs]**$_P$, **[incNGs]**$_R$ and **[incNGs]**$_R$[c] run up to 2.91, 1.46 and 1.37 times faster than the decomposed ones, and have up to 3014, 3938 and 33383 less failures than the decomposed ones. While **[incNGs]**$_{LR}$ and **[incNGs]**$_{LR}$[c] are up to 1.92 and 1.91 times faster than **ReSBDS** and **ReSBDS**[c] respectively. Note that for the case CA$(2, 4, 4, 16)$, **ReSBDS**[c], **[incNGs]**$_R$[c] and **[incNGs]**$_{LR}$[c] leave 2250, 2076 and 2100 solutions respectively. This demonstrates that with global constraint, ReSBDS prunes more solutions than LReSBDS, and our domain filtering on global constraint can prune more solutions than GAC on each individual nogood. When compared with static methods, the best one **[incNGs]**$_{LR}$[c] runs up to 1.72 times faster than **DoubleLex**. This shows how LReSBDS with global constraint is competitive against static methods.

8 Conclusion and Future Work

Our contributions are five-fold. First, based on the special semantics and structures of increasing nogoods, we propose a global constraint with equivalent meaning but stronger pruning power. Second, we demonstrate that nogoods added by SBDS and its variants are increasng so that the methods can be adapted with the global constraints. Third, benefitting from the global constraint, we devise a light version of ReSBDS with smaller space and time overheads. Fourth, we give a polytime filtering algorithm for the increasing-nogoods constraint, which also has an efficient and simple incremental version. Fifth, extensive experimentations confirm the efficiency of our proposals.

ReSBDS has the advantage that a substantial number of symmetries can be broken with only a small given subset of them. The increasing-nogoods global constraint reduce the overhead of SBDS and its variants dramatically, making it possible to handle larger set of given symmetries which in turn can prune more search space.

Nogood learning is a general technique for improving backtracking search [2]. We envision that the increasing-nogoods constraint is applicable to other scenarios in CP, in addition to symmetry breaking.

References

1. Crawford, J., Ginsberg, M., Luks, E., Roy, A.: Symmetry breaking predicates for search problems. In: KR 1996, pp. 148–159 (1996)
2. Dechter, R.: Enhancement schemes for constraint processing: Backjumping, learning, and cutset decomposition. Artificial Intelligence, 273–312 (1990)
3. Fahle, T., Schamberger, S., Sellmann, M.: Symmetry breaking. In: Walsh, T. (ed.) CP 2001. LNCS, vol. 2239, pp. 93–107. Springer, Heidelberg (2001)
4. Flener, P., Frisch, A.M., Hnich, B., Kiziltan, Z., Miguel, I., Pearson, J., Walsh, T.: Breaking row and column symmetries in matrix models. In: Van Hentenryck, P. (ed.) CP 2002. LNCS, vol. 2470, pp. 187–192. Springer, Heidelberg (2002)
5. Flener, P., Frisch, A., Hnich, B., Kiziltan, Z., Miguel, I., Walsh, T.: Matrix modelling. In: ModRef 2001 (2001)
6. Gent, I.P., Harvey, W., Kelsey, T.: Groups and constraints: Symmetry breaking during search. In: Van Hentenryck, P. (ed.) CP 2002. LNCS, vol. 2470, pp. 415–430. Springer, Heidelberg (2002)
7. Gent, I.P., Harvey, W., Kelsey, T., Linton, S.: Generic SBDD using computational group theory. In: Rossi, F. (ed.) CP 2003. LNCS, vol. 2833, pp. 333–347. Springer, Heidelberg (2003)
8. Gent, I., Smith, B.: Symmetry breaking in constraint programming. In: ECAI 2000, pp. 599–603 (2000)
9. Gent, I.P., Walsh, T.: CSPlib: A benchmark library for constraints. In: Jaffar, J. (ed.) CP 1999. LNCS, vol. 1713, pp. 480–481. Springer, Heidelberg (1999)
10. Hnich, B., Prestwich, S.D., Selensky, E., Smith, B.M.: Constraint models for the covering test problem. Constraints, 199–219 (2006)
11. Law, Y.C., Lee, J.: Global constraints for integer and set value precedence. In: Wallace, M. (ed.) CP 2004. LNCS, vol. 3258, pp. 362–376. Springer, Heidelberg (2004)
12. Lee, J., Zhu, Z.: Boosting SBDS for partial symmetry breaking in constraint programming. In: AAAI 2014 (to appear, 2014)
13. Mackworth, A.: Consistency in networks of relations. Artificial Intelligence, 99–118 (1977)
14. Mears, C., de la Banda, M.G., Demoen, B., Wallace, M.: Lightweight dynamic symmetry breaking. Constraints, 1–48 (2013)
15. Moskewicz, M.W., Madigan, C.F., Zhao, Y., Zhang, L., Malik, S.: Chaff: Engineering an efficient sat solver. In: DAC 2001, pp. 530–535 (2001)
16. Petrie, K.E., Smith, B.M.: Symmetry breaking in graceful graphs. In: Rossi, F. (ed.) CP 2003. LNCS, vol. 2833, pp. 930–934. Springer, Heidelberg (2003)
17. Puget, J.: Breaking symmetries in all different problems. In: IJCAI 2005, pp. 272–277 (2005)
18. Puget, J.F.: On the satisfiability of symmetrical constrained satisfaction problems. In: Komorowski, J., Raś, Z.W. (eds.) ISMIS 1993. LNCS, vol. 689, pp. 350–361. Springer, Heidelberg (1993)
19. Roney-Dougal, C.M., Gent, I.P., Kelsey, T., Linton, S.: Tractable symmetry breaking using restricted search trees. In: ECAI 2004, pp. 211–215 (2004)
20. Rossi, F., Van Beek, P., Walsh, T.: Handbook of constraint programming. Elsevier (2006)
21. Zhang, L., Malik, S.: The quest for efficient boolean satisfiability solvers. In: Brinksma, E., Larsen, K.G. (eds.) CAV 2002. LNCS, vol. 2404, pp. 17–36. Springer, Heidelberg (2002)

Memory-Efficient Tree Size Prediction for Depth-First Search in Graphical Models

Levi H.S. Lelis[1], Lars Otten[2], and Rina Dechter[3]

[1] Departamento de Informática, Universidade Federal de Viçosa, Brazil
[2] Google Inc., USA
[3] Department of Computer Science, University of California, Irvine, USA

Abstract. We address the problem of predicting the size of the search tree explored by Depth-First Branch and Bound (DFBnB) while solving optimization problems over graphical models. Building upon methodology introduced by Knuth and his student Chen, this paper presents a memory-efficient scheme called Retentive Stratified Sampling (`RSS`). Through empirical evaluation on probabilistic graphical models from various problem domains we show impressive prediction power that is far superior to recent competing schemes.

1 Introduction

The most common search scheme for Graphical Models optimization tasks, such as MAP/MPE or Weighted CSP, is Depth-First Branch-and-Bound (DFBnB). Its use for finding both exact and approximate solutions was extensively studied in recent years [1–4]. Our paper addresses the general question of predicting the size of the DFBnB explored search tree, focusing on graphical models optimization tasks.

DFBnB [5] explores the search space in a depth-first manner while keeping track of the current best-known solution cost, denoted $cbound$, which can be initialized with the value of a solution derived by some preprocessing (e.g., local search). DFBnB uses an *admissible* heuristic function $h(\cdot)$, i.e., a function that never overestimates the optimal cost-to-go for every node, and is guided by an evaluation function $f(n) = g(n) + h(n)$, where $g(n)$ is the cost of the path from the root node to node n. Since $f(n)$ is an underestimate of the cost of an optimal solution that goes through n, whenever $f(n) \geq cbound$, n is pruned.

Often the user of search algorithms such as DFBnB does not know a priori how long the search will take to finish solving a problem instance: it could take seconds, hours or years. This is due to a series of factors, including the strength of the heuristic guiding the search. Prediction is particularly elusive for graphical models where solvers originate in diverse communities (e.g., CP, UAI, OR) and employ different principles for (a) traversing the search space, (b) for generating the heuristic lower bound function, and (c) for pruning nodes. In addition to estimating the algorithm's running time, estimates of expanded search tree size could be used to decide which heuristic function to use to solve a particular problem instance: should one use the slow but accurate heuristic or the fast but inaccurate one? (e.g., by controlling the i-bound in the case of the mini-bucket heuristics [1]). Or, in the context of parallelizing search, a prediction scheme

B. O'Sullivan (Ed.): CP 2014, LNCS 8656, pp. 481–496, 2014.

could facilitate load-balancing by partitioning the problem into subproblems of similar EST sizes [6].

Our approach in this paper builds upon the Stratified Sampling (SS) scheme [7, 8]. Knuth [7] proposed a method for estimating the running time of tree search methods by quickly estimating the size of the *expanded search tree* (EST). Under the reasonable assumption that the time required to expand a node is constant throughout the EST, an estimate of the EST's size provides an estimate of the algorithm's running time.

Prediction schemes were investigated in the past primarily in the context of path-finding problems. Specifically, various methods have been developed for estimating EST size of search algorithms such as IDA* [9]. Examples include Partial Backtracking by Purdom [10] and SS by Chen [8], both based on the seminal work of Knuth [7]; other related methods include [11–14]. All of the above work by sampling a small portion of the EST and extrapolating from it. None of those earlier works addressed graphical models tasks which unlike path-finding problems all their solution nodes appear are at a fixed finite depth (i.e., the number of variables). Also, none of these earlier works considered branch and bound search schemes.

Recently Lelis et al. [15] initiated investigating the usage of SS for estimating the DFBnB EST size and evaluated its effectiveness on graphical models. They observed that methods such as SS make the implicit *stable children* assumption, namely that the set of children of node n in an EST can be determined given only the path from the root of the EST to n. Crucially, however, this property does not hold in the context of DFBnB where pruning depends on the upper bound *cbound* that is updated dynamically throughout the search. Lelis et al. thus introduced a new SS scheme called Two-Step Stratified Sampling (TSS), described in more detail later, that mitigates this problem [15]. They also provided an empirical evaluation of their approach by looking at a specific DFBnB solver applied to a collection of typical graphical models benchmarks from the probabilistic domain.

Contributions. TSS presented a substantial advance to the DFBnB search space prediction task, but it was also shown to be limited by its memory requirements. As a result, TSS can produce poor estimates or, in some cases, no estimates at all. In this paper we introduce *Retentive Stratified Sampling* (RSS) that addresses differently the stable children property of DFBnB, resulting in a far more memory-efficient scheme. Namely, instead of memorizing every node expanded during sampling, RSS retains only the encountered solution paths. We show that this scheme is asymptotically unbiased.

We test RSS empirically on optimization benchmarks over probabilistic graphical models [16] using DFBnB guided by the mini-bucket heuristic [1, 17] (BBMB), which has been extended into a competition-winning solver [18, 19]. We compare RSS with TSS and WBE [20], over prediction tasks from 3 problem domains in graphical models. Our empirical results show that RSS overcomes the memory limitation of TSS and yields estimates far superior to any of the currently competing methods of its kind.

2 Background

Given a directed and implicitly defined full search tree representing a state-space problem [21], we are interested in estimating the size of the subtree expanded by a search algorithm seeking an optimal solution. We call the former the *underlying search tree* (UST) and the latter the *Expanded Search Tree* (EST). Let $S(s^*) = (N, E)$ be a tree representing such an EST rooted at s^*. For each $n \in N$ $child(n) = \{n' | (n, n') \in E\}$ defines the node-child relationship in the EST. The prediction task is to estimate the size of N without fully expanding the EST.

2.1 The Knuth-Chen Method

Knuth [7] presented a method to estimate the size of a tree by repeatedly performing a random walk from the root. Under the assumption that all branches have a structure equal to the path visited by the random walk, one branch is enough to estimate the size of the tree. Knuth observed that his method, while guaranteed to converge to the right value, is not effective when the tree is unbalanced. Chen [8] proposed *Stratified Sampling* (SS), which improves upon Knuth's method with a stratification of the tree through a *type system* to reduce the variance of the sampling process.

Definition 1 (Type System). *Let $S(s^*) = (N, E)$ be a tree rooted at s^*, and T a function from N to a finite set of numerical types $\{t_1, \ldots, t_n\}$. We call T a type system, and it yields a partition of N into $T = \{t_1, \ldots, t_n\}$ where $t_i = \{s \in N | T(s) = t_i\}$. We abuse notation: t_i denotes a type and also the set of nodes in N that map to type t_i.*

A type system can be based on any property of the nodes in the search tree. For example, Zahavi et al. [12] used a type system that accounts for the f-value of the nodes to make predictions of the size of the IDA* EST. That is, nodes n and n' have the same type if they have the same f-value. Still in the context of IDA* predictions, Lelis et al. [22] used variations of Zahavi et al.'s type system in which they also account for the f-value of the nodes in the neighborhood of n when computing n's type. In this paper we use the type system introduced by Lelis et al. [15], which is also based on the f-value; we describe such type system in Section 4.1 below.

Chen's Stratified Sampling (SS) is a general method for approximating any function of the form

$$\varphi(s^*) = \sum_{n \in S(s^*)} z(n),$$

where z is any function assigning a numerical value to a node. $\varphi(s^*)$ represents a numerical property of the search tree rooted at s^*. For instance, if $z(n) = 1$ for all $n \in S(s^*)$, then $\varphi(s^*)$ is the size of the tree.

Instead of traversing the entire tree and summing all z-values, SS assumes subtrees rooted at nodes of the same type will have equal values of φ and so only one node of each type, chosen randomly, is expanded. In practice, a type system is good for a function $\varphi(s^*)$ if nodes having identical type root subtrees with similar values of φ. Clearly, we wish to have a good type system with a small number of types. If we have a type for each node, then the type system will be good in the above sense, yet completely ineffective.

SS estimates $\varphi(s^*)$ as follows. First, it samples the tree rooted at s^* and returns a set A of *representative-weight* pairs, with one such pair for every unique type seen during

Algorithm 1. Stratified Sampling, a single probe

Input: tree root s^*, type system T, initial upper bound *cbound*.
Output: a sampled tree ST specified by a set A which is divided into subsets, where $A[i]$ is the set of pairs $\langle s, w\rangle$ for the nodes $s \in ST$ expanded at level i.

1: initialize $A[0] \leftarrow \{\langle s^*, 1\rangle\}$
2: $i \leftarrow 0$
3: **while** i is less then search depth **do**
4:     **for** each element $\langle s, w\rangle$ in $A[i]$ **do**
5:         **for** each child s'' of s **do**
6:             **if** $h(s'') + g(s'') < cbound$ **then**
7:                 **if** $A[i+1]$ contains an element $\langle s', w'\rangle$ with $T(s') = T(s'')$ **then**
8:                     $w' \leftarrow w' + w$
9:                     with probability w/w', replace $\langle s', w'\rangle$ in $A[i+1]$ by $\langle s'', w'\rangle$
10:                 **else**
11:                     insert the new element $\langle s'', w\rangle$ into $A[i+1]$
12:     $i \leftarrow i + 1$

sampling. In the pair $\langle n, w\rangle$ in A for type $t \in T$, n is the unique node of type t that was expanded during search and w is an estimate of the number of nodes of type t in the tree rooted at s^*. $\varphi(s^*)$ is then approximated by $\hat{\varphi}(s^*)$

$$\hat{\varphi}(s^*) = \sum_{\langle n,w\rangle \in A} w \cdot z(n)\,, \tag{1}$$

SS (see Algorithm 1) receives as input a start state s^*, a type system T, and an initial upper bound *cbound* which is derived by some preprocessing (e.g., local search). SS returns a set A which is indexed by depth in the search tree, where $A[i]$ is the set of representative-weight pairs for the types encountered at depth i.

In SS types are required to be partially ordered: a node's type must be strictly greater than the type of its parent. Chen suggests that this can be guaranteed by adding the depth of a node to the type system and then sorting the types lexicographically. In our implementation of SS, types at one level are treated separately from types at another level by the division of A into the $A[i]$. If the same type occurs on different levels the occurrences will be treated as though they were different types – the depth of search is implicitly added to the type system.

The algorithm works as follows: $A[0]$ is initialized to contain only the root of the tree to be probed, with weight 1 (line 1). In each iteration (lines 4 through 11), all nodes in $A[i]$ are expanded. The children of each node in $A[i]$ are considered for inclusion in $A[i+1]$. If a child s'' has a type t that is already represented in $A[i+1]$ by node s' with weight w', then a *merge* of s'' and s' is performed: increase weight w' of s' by the weight w of s'''s parent s (since there were w nodes at level i that are assumed to have children of type t at level $i+1$). With a certain probability (line 9) s'' will replace the s'. Chen [8] proved that this stochastic choice of type representatives reduces the variance of the estimation. Once all the states in $A[i]$ are expanded, we move to the next iteration. In Chen's SS, this process continues until $A[i]$ is empty. The set of nodes in A represents a subtree of the tree SS samples, we call this subtree the *sampled tree*.

A single run of SS is called a *probe*. We denote as $\hat{\varphi}^{(p)}(s^*)$ the estimate produced by SS's p-th probe.

Theorem 1. *[8] Given a set of independent probes $p_1, \cdots, p_m$ produced by* SS *using type system T from tree $S(s^*)$, the average $\frac{1}{m}\sum_{j=1}^{m}\hat{\varphi}^{(p_j)}(s^*)$ converges to $\varphi(s^*)$ as m goes to infinity. Namely,*

$$lim_{m \to \infty} \frac{1}{m} \sum_{j=1}^{m} \hat{\varphi}^{(p_j)}(s^*) = \varphi(s^*)$$

2.2 Two-Step Stratified Sampling (TSS)

Stable Children Property. Lelis et al. [15] observed the implicit assumption in SS that it has access to the generative process of the node-child relationship for every node. In particular, SS prunes child nodes only if their f-value is greater than or equal to the initial upper bound *cbound* (line 6 in Algorithm 1). While such pruning scheme holds for predicting EST size of algorithms such as IDA* [22], it does not hold in the case of DFBnB, where pruning is based on an upper bound that is updated throughout search. As a result, the exact child nodes generated by DFBnB are not available to the sampling algorithm.

Definition 2 (Stable Children Property). *[15] Given an EST $S(s^*) = (N, E)$, the stable children property is satisfied iff for every path π leading from the root s^* to a node n, the set $child(n)$ in EST can be determined based on π alone.*

Lelis et al. [15] overcome the lack of the stable children property in the EST of DFBnB by producing the estimate in two steps. In the first step, their TSS algorithm generates m independent SS probes assuming that the search tree is bounded by the initial upper bound *cbound*. Each SS probe produces a sampled tree, and TSS stores in memory the union of all m sampled trees, denoted $UnionST$. In the second step, TSS emulates DFBnB restricted to the nodes in $UnionST$. $UnionST$ gets larger as we increase the value of m. In particular, as m goes to infinity, $UnionST$ converges to the search tree bounded by *cbound*. In this theoretical scenario, TSS's second step expands exactly the same nodes that DFBnB expands and TSS is able to determine the set $child(n)$ exactly and thus produces perfect estimates.

Although in theory the TSS scheme overcomes the lack of the stable child property, it can have high memory requirement, as it stores every node expanded in the first step of each of the m probes. Therefore, TSS is often limited to only a few probes, frequently producing poor predictions or no predictions at all.

Theorem 2 (TSS's Time and Memory Complexity). *[15] The memory complexity of* TSS *is $O(m \times |T|^2)$, where $|T|$ is the size of the type system being employed and m is the number of* TSS *probes.* TSS *time complexity is $O(m \times |T|^2 \times b)$, where b is the branching factor of $UnionST$.*

2.3 Graphical Models

A *graphical model* is given as a set of variables $X = \{X_1, \ldots, X_n\}$, their respective finite domains $D = \{D_1, \ldots, D_n\}$, a set of functions $F = \{f_1, \ldots, f_m\}$, each defined over a subset of X (the function's *scope*), and a combination operator (typically sum, product, or join) over functions. Together with a marginalization operator such as $\min_X$ and $\max_X$ we obtain a *reasoning problem*. For instance, a *weighted constraint satisfaction* problem is typically expressed through a set of cost functions over the variables, with the goal of finding the minimum of the sum over these costs (i.e., we seek $argmin_X \sum_i f_i)$. In the area of probabilistic reasoning, the *most probable explanation* task over a Bayesian network is defined as maximizing the product of the probabilities $(argmin_X \prod_i f_i)$. The set of function scopes imply a primal graph, an induced width or tree width that is known to control the complexity of variable-elimination and search algorithms for solving a variety of graphical models tasks [23].

The search tree of a graphical model. The most successful schemes for solving optimization tasks over graphical models is by DFBnB search. In its simplest formulation the nodes in UST are consistent partial assignment of values to the variables along a fixed variable ordering $X_1, \cdots, X_n$. The root is the empty assignment, and a node at depth d is $n = (x_1, \cdots, x_d)$ where x_i is a value from the domain of X_i. Child nodes of n extend it by assigning values to the next variable in the ordering. Solutions correspond to full assignments and all appear at depth n. Leaves of the UST correspond either to partial assignments that cannot be extended consistency or to full assignments representing solutions. DFBnB prunes the search tree in the usual manner, comparing its heuristic evaluation function to the current upper-bound.

A popular heuristic function that guides search schemes for graphical models is the mini-bucket heuristic. It is based on mini-bucket elimination, an approximate variant of variable elimination that computes approximations to reasoning problems over graphical models [1]. A control parameter, denoted as i-bound, allows a trade-off between accuracy of the heuristic and its computational requirements: higher values of i yield a more accurate heuristic but take more time and space to compute.

3 Retentive Stratified Sampling

In this section we present Retentive Stratified Sampling (RSS). The central idea is that it is sufficient to have available the full set of solutions subsumed in the DFBnB EST in order to allow SS to determine exactly the set $child(n)$ in the EST.

DFBnB defines a complete ordering on the nodes in the EST, implied by the order in which the child nodes of each parent node are expanded. This expansion ordering also induces an order on the solution leaf nodes.

Definition 3 (Solution Search Tree). *Given that the DFBnB EST is an ordered search tree, the subtree of EST that is restricted to only solution paths is called* Solution Search Tree *(SST). The leaves in the SST are ordered, from left to right, reflecting their discovery order by DFBnB. For each node, its child nodes are ordered from left to right as well. If we have k solutions we assume they are ordered by $s_1, \cdots, s_k$.*

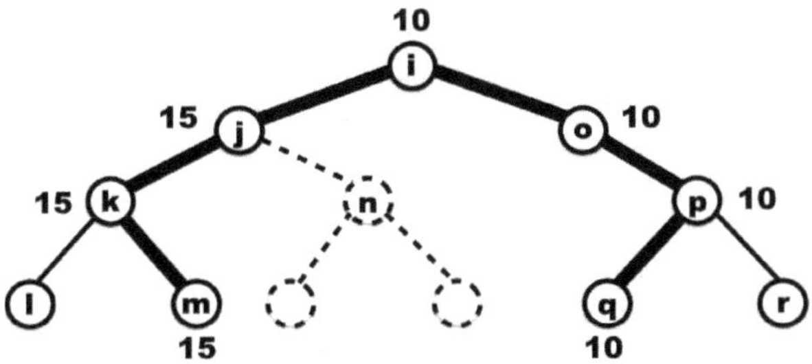

Fig. 1. Example of a UST. The dashed nodes and arcs do not belong to the EST; the arcs in bold represent the SST, with solution nodes m and q, with solution costs of 15 and 10 respectively. The numbers by the nodes a in the SST show the lowest cost solution $l(a)$ going through a.

The assumption that DFBnB has a deterministic ordering of child-node expansion is common since DFBnB usually expands first the subtree rooted at the most promising child (i.e., the child with lowest f-value). By definition the leaves of SST are ordered in decreasing cost from left to right.

Lemma 1. *If the ordered Solution Search Tree SST is available to* SS*, then for every node n in the DFBnB EST,* SS *can determine the set $child(n)$ in the EST.*

Proof (sketch). We prove the theorem constructively, by providing an algorithm (see Algorithm 3) for the task. Given an SST, we will associate each node m in SST with the lowest cost solution in SST that goes through m, denoted $l(m)$. This is easy to compute by a depth-first search traversal on the SST in time linear in $|SST|$. It is also easy to update the $l(\cdot)$-values whenever new solutions are added to SST: after a new solution is added to the SST, its cost is propagated upwards, namely the minimal costs $l(m)$ for all m along the solution path are updated in the obvious way.

Given a partial path $\pi = n_0, n_1, \ldots, n_d$ in UST, from the root n_0 to a node $n = n_d$, we wish to determine the correct upper bound that would be used by DFBnB. This is done as follows: let node n_j be the closest ancestor of $n = n_d$ on the path π going in reverse order from n_d backwards towards n_0 that (1) appears in SST, and (2) has a child node m in SST which is not on π, such that m immediately precedes n_{j+1} (which is the child node of n_j's on π) according to the child-node ordering. Clearly m can be identified in time linear on the depth of SST. It is easy to see that if m exists, then the lowest cost solution encountered by DFBnB prior to n is $l(m)$, which is thus the upper bound available to DFBnB when it visits n.

Example 1. The tree shown in Figure 1 represents a hypothetical UST where the dashed nodes are pruned by DFBnB, and the arcs in bold represent the SST. We are assuming that DFBnB visits the nodes in lexicographical order. If RSS encounters node n during sampling, it identifies $l(k) = 15$ as the relevant upper bound as follows. RSS identifies j as the first ancestor of n along the path π going from n towards the root that appears in SST, and has a child node k in SST which is not on π, which immediately precedes j's child node on π (which in this case is n itself) according to the child-node

Algorithm 2. Retentive Stratified Sampling, a single probe

Input: tree root s^*, type system T, solution branches B, initial upper bound *cbound*.
Output: a sampled tree ST represented by an array of sets A, where $A[i]$ is the set of pairs $\langle s, w \rangle$ for the nodes $s \in ST$ expanded at level i, and solutions B to be reused in next probe.

1: initialize $A[0] \leftarrow \{\langle s^*, 1 \rangle\}$
2: $i \leftarrow 0$
3: **while** i is less then search depth **do**
4:   **for** each element $\langle s, w \rangle$ in $A[i]$ **do**
5:     **if** s is a solution node ending a solution path π_s and π_s is not in B **then**
6:       $B \leftarrow Insert(B, \pi_S)$
7:     **for** each child s'' of s **do**
8:       $curr_b \leftarrow VerifyBound(s'', B, cbound)$ // cf. Algorithm 3
9:       **if** $h(s'') + g(s'') < curr_b$ **then**
10:         **if** $A[i+1]$ contains an element $\langle s', w' \rangle$ with $T(s') = T(s'')$ **then**
11:           $w' \leftarrow w' + w$
12:           with probability w/w', replace $\langle s', w' \rangle$ in $A[i+1]$ by $\langle s'', w' \rangle$
13:         **else**
14:           insert new element $\langle s'', w \rangle$ in $A[i+1]$
15:   $i \leftarrow i + 1$

Algorithm 3. VerifyBound

Input: node $s \in UST$ along path π, ordered tree-structure B whose arcs are labeled $l(n, m)$ denoting the lowest solution cost below m, and initial upper bound *cbound*.
Output: upper bound for s according to B

$(n', m') \leftarrow$ identify $n' \in B$ as the closest ancestor to s along π that has a child node m' on B which immediately precedes m", which is n''s child on π.
If n' exists, then return $l(m')$, return *cbound* otherwise.

ordering. If $f(n) = 16$ (which is greater than $l(k)$), then RSS correctly prunes n. As another example, if RSS encounters node p during sampling, it identifies $l(j) = 15$ as the relevant upper bound as follows. i is the first ancestor of p along the path π going from p towards the root that appears in SST, and has a child node j in SST which is not on π, and which immediately precedes i's child node on π (node o) according to the child-node ordering. In this case, if $f(p) = 10$, then RSS correctly expands p.

Since the SST is generally far smaller than the EST, we are likely to get a memory-efficient algorithm, which is obviously superior to TSS. Algorithm RSS implements the scheme described in Lemma 1 in its pseudo code shown in Algorithms 2 and 3. RSS can be viewed as SS with the following two extensions:

1. Algorithm 2 approximates SST in the initially empty tree structure B, which is updated throughout probes. Specifically, whenever a solution *sol* is generated, it is inserted into B (respecting the parent-child ordering induced by DFBnB). In doing so, pruning within the B structure can be applied by removing from B any solution that succeeds *sol* in B and has a higher cost (note that solutions preceding *sol* in B will always have higher cost due to the pruning in Algorithm 2, line 9). Thus, at all times we maintain in B an ordered tree where leaves have decreasing cost going

from the first to the last solution expanded by DFBnB. The function $Insert(B, \pi_S)$ (line 6 of Algorithm 2) accomplishes this task and can work in time linear in $|B|$ (the function is not formally introduced). The tree B that RSS outputs in the i-th probe is used as input for the $(i+1)$-th probe.

2. RSS does not insert child s'' into $A[i+1]$ if there is a solution in B that appears before s'', and $h(s'') + g(s'') \geq curr_b$ (lines 8 and 9 of Algorithm 2).

3.1 Asymptotically Perfect Predictions

Since every branch in the EST has a non-zero probability of being sampled, it is quite immediate that:

Lemma 2. *The tree structure B converges to SST as the number of probes goes to infinity.*

From Lemma 2 and from Theorem 1 it follows that RSS has the asymptotic guarantee to generate perfect predictions of the size of the DFBnB EST. Formally,

Theorem 3. *Given a set of independent probes $p_1, \cdots, p_m$ produced by* RSS *using type system T from a search tree $S(s^*)$ representing a DFBnB EST, there exists $j_0 \leq m$ such that, the average $\frac{1}{m-j_0}\sum_{j \geq j_0}^{m} \hat{\varphi}^{(p_j)}(S)$ converges to $\varphi(s^*)$ as $m \to \infty$.*

Proof. Eventually, after a finite number of probes j_0, B will be equal (or close) to SST (Lemma 2), allowing RSS to determine the exact set $child(n)$ for any node n in the DFBnB EST (Lemma 1), which in turn allows us to apply Theorem 1.

In practice j_0 is unknown and we use a heuristic estimate as described below.

3.2 Time and Space Complexity of Retentive Stratified Sampling

In each probe, in addition to the B structure, RSS with type system T stores in memory at most $|T|$ nodes. Across multiple probes B converges to SST. Thus,

Theorem 4. RSS*'s time complexity after m probes is $O(m \times |T| \times d)$, where $|T|$ is the size of the type system and d is the EST depth (i.e., the number of variables in the graphical model). The space complexity is $O(|T| + |SST|)$.*

Although RSS's time and space complexities depend on parameters for which we might not know the values in advance, they allow us to contrast, for example, RSS and TSS memory requirements. In particular, we observe that the amount of memory TSS requires is bounded by $|EST|$, while the amount of memory RSS requires is bounded by $|SST|$. In the worst case $|SST|$ is bounded by the number of solutions of the original problem, in the best case SST is a single branch. In practice $|SST|$ tends to be much smaller than the $|EST|$.

4 Experiments

4.1 Empirical Methodology

We evaluate RSS by predicting EST sizes of DFBnB using the mini-bucket heuristic (BBMB) [1, 17]. For a given problem instance we experiment with different mini-bucket i-bounds for producing different heuristic strengths. RSS is currently not able to account for AND/OR search spaces and caching, techniques used in the more advanced solvers such as AOBB [18, 19].

We consider three problem domains, protein side-chain prediction (pdb), computing haplotypes in genetic analysis (pedigree), and randomly generated grid networks, with 14, 4, and 14 problem instances, respectively. We use the following i-bounds values: 3 for pdb; 6, 7, 8, 10, 11, 12, and 13 for pedigree; and 10, 11, 12, and 13 for grids. Thus, we use 14, 28, and 56 prediction tasks (pairs problem instance and i-bound) for pdb, pedigree, and grids, respectively. We remove from our test set the tasks that DFBnB is able to solve in less than a second, and the tasks that DFBnB is not able to solve after several days of running time. After removing such tasks our test set contains 14, 26, and 54 prediction tasks for pdb, pedigree, and grids, respectively. We remove the easy instances from our test set because they are uninteresting and the instances that DFBnB is not able to solve after several days of running time because we are not able to verify the algorithms' prediction accuracy on those instances. The average running time of DFBnB on the tasks of each domain is 89.3 minutes, 72.2 hours, and 10.9 minutes, for pdb, pedigree and grids, respectively, on a 2.6 GHz CPU (10GB RAM).

We compare the performance of RSS against TSS and WBE (described below). We leave SS out of our experiments because Lelis et al. [15] have already shown that SS is not able to produce good predictions of DFBnB EST size even when granted more computation time than the time required by DFBnB to solve the problem. Since TSS and RSS are stochastic algorithms, we consider the average result over five independent runs for each prediction task. In each case we use the following ratio as a measure of accuracy: $\frac{predicted}{actual}$ if $predicted > actual$ and $\frac{actual}{predicted}$, otherwise — this prevents over- and underestimations canceling out when averaging results. Perfect predictions yield a ratio of 1.0.

Weighted Backtrack Estimator. The *Weighted Backtrack Estimator* (WBE) [20] runs alongside DFBnB search with minimal computational overhead. It uses explored branches to predict unvisited ones and thereby the EST size. WBE produces perfect predictions when the search finishes. We implemented WBE in the context of BBMB, yielding an updated prediction every 5 seconds. Kilby et al. presented another prediction algorithm, the Recursive Estimator (RE), whose performance was similar to WBE's in their experiments. Both WBE and RE were developed to predict the size of binary trees, but in contrast to WBE it is not clear how to generalize RE to non-binary search trees.

Type Systems. The use of the f-value to define a node's type has proven effective in other heuristic search tree size estimation problems [22]. In our experiments, in addition to a node's f-value, we also use its depth level (cf. Algorithm 2) for its type, namely nodes n and n' have the same type if they are at the same level of the UST and if

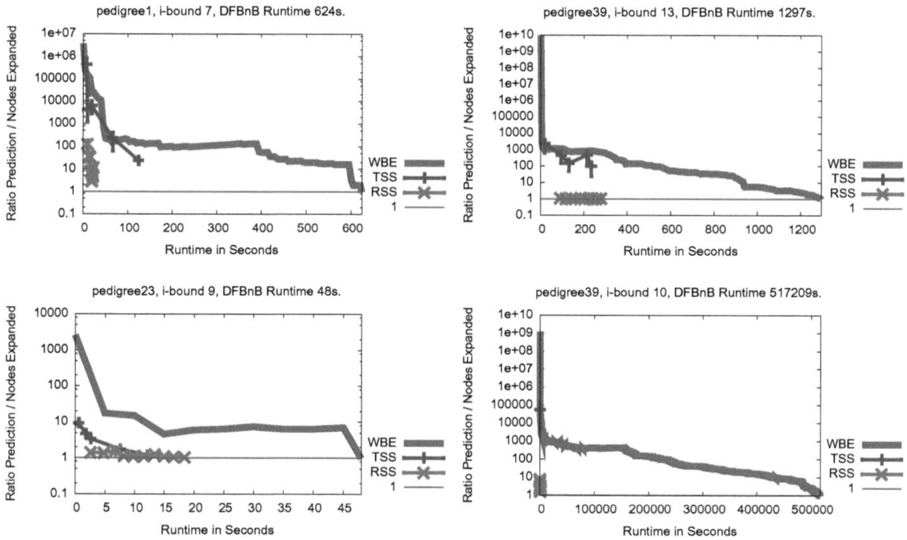

Fig. 2. Prediction accuracy over time of RSS, TSS, and WBE on select representative prediction tasks of the pedigree domain

$f(n) = f(n')$. We note that the cost function and accordingly, the derived heuristic in graphical model problems are often real-valued and a type system based on floating point equality might be far too large. To mitigate this we apply the technique used by Lelis et al. [15], multiplying $f(n)$ by a constant C and truncating to the integer portion. The constant C allows us to control to some extent the size of the type system. That is, larger values of C result in larger type systems, which implies in slower but possibly more accurate predictions. This is because larger range of types yields a larger coverage of the search space.

Warmstarting RSS. Theorem 3 showed that RSS is unbiased in the limit, in particular because RSS is eventually able to determine exactly the sets $child(n)$ in the EST (i.e., $B = SST$) and the number of probes based on this will eventually outweigh the earlier ones. From Theorem 3 we know that the initial set of probes are skewed and should not be included in the estimate. The rule we used is that upon termination of the probes, we compute the RSS estimation by only averaging over probes obtained since the last addition of a new solution to B.

4.2 Select Individual Results

We begin by showing results of RSS, TSS, and WBE on select individual prediction tasks in Figures 2, 3, and 4. These instances are representative in that they highlight different aspects of the prediction methods. For each scheme, we plot in log-scale the ratio of predicted and actual EST size, as defined above, as a function of running time

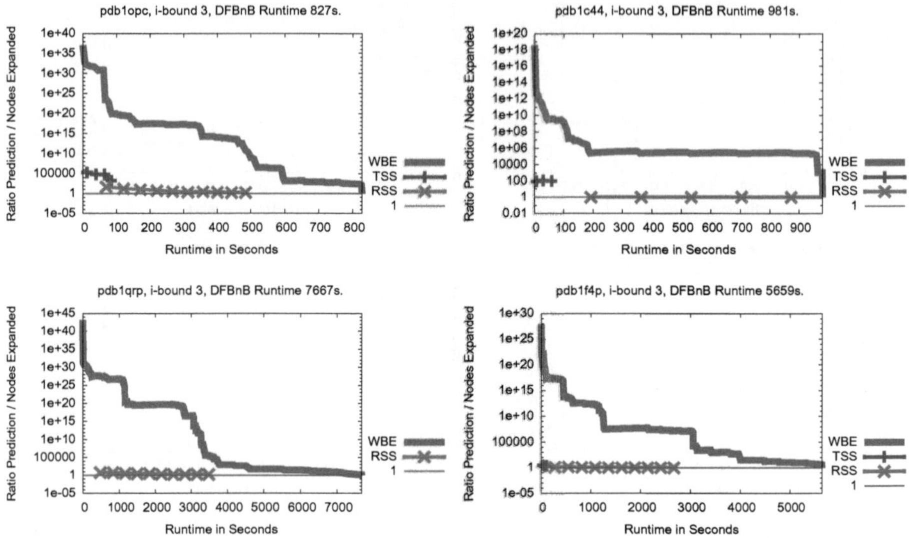

Fig. 3. Prediction accuracy over time of RSS, TSS, and WBE on select representative prediction tasks of the pdb domain

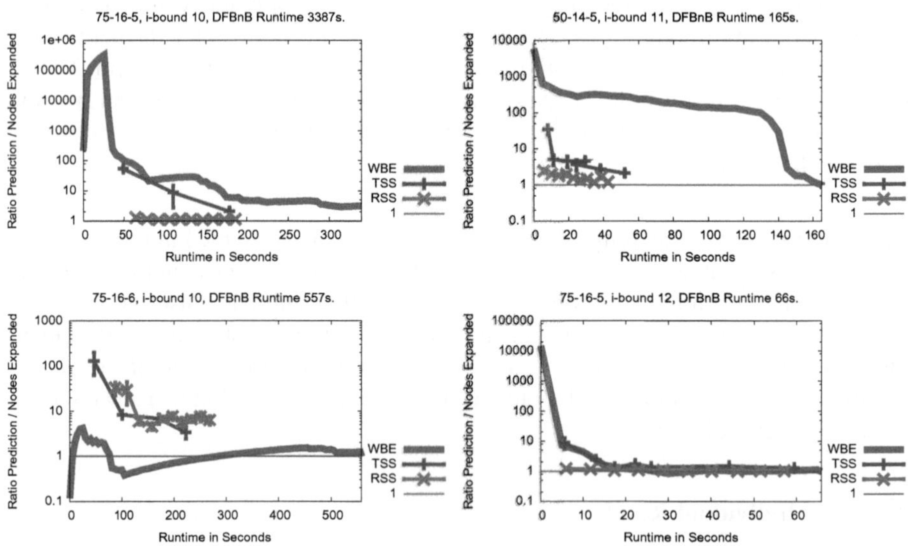

Fig. 4. Prediction accuracy over time of RSS, TSS, and WBE on select representative prediction tasks of the grids domain

in seconds. We run RSS with different number of probes and use the warmstarting strategy to generate predictions with different running time. Results for RSS and TSS are averaged over 5 independent runs. There is very little variance in the prediction accuracy over different runs of RSS as the 95% confidence interval is shown but is hardly noticeable. Note that because we vary the number of probes, RSS is oblivious to the DFBnB total running time. That is why in some plots we do not present the RSS results for larger DFBnB running times. The DFBnB total running time is shown on the top of each plot. For each problem we first run a limited-discrepancy search [24] with a maximum discrepancy of 1 to quickly find an initial bound *cbound* which is provided to both DFBnB and to the prediction algorithms. We use $C = 100$ in this experiment for both RSS and TSS (see Section 4.1 on type systems).

The prediction results in Figures 2, 3, and 4 suggest that RSS is far superior to both TSS and WBE. For instance, RSS quickly produces almost perfect estimates of EST size of pedigree39 with $i = 10$ (Figure 2); TSS is unable to yield good estimates and quickly runs out of memory. WBE is able to produce acceptable predictions only towards the end of DFBnB execution—approximately 6 days on this instance. For pdb1qrp with $i = 3$ (Figure 3) TSS is unable to produce predictions at all—RSS, however, quickly produces near-perfect estimates. Similarly, RSS outperforms the other methods on almost all instances. The pdb1c44 problem instance with i-bound of 3 (Figure 3) shows another situation we would like to highlight. In that instance RSS starts producing estimates after 200 seconds of computation time. This is because in the first 200 seconds RSS is constantly updating B and according to our warmstarting strategy described above RSS does not produce estimates while updating B. 75-16-6 with $i = 10$ (Figure 4) is one of the rare cases in which WBE performs better than both RSS and TSS.

4.3 Comparison with TSS

Table 1 shows a summary of prediction results of TSS and RSS for $C = 10$ and $C = 100$. Here "%" is the average prediction time relative to DFBnB. For instance, a value of 20 means that the prediction was produced in 20% of the DFBnB running time. Given a number of probes for TSS and the resulting %-value, the number of RSS probes is chosen so that it has %-values smaller than TSS, thereby giving the latter a small advantage. In each case we observe that n, the number of instances where TSS does not run out of memory, decreases with the number of probes m. For instance, for pedigrees with $C = 100$ and $m = 50$ probes, TSS is able to produce predictions only for 8 out of 26 prediction tasks (the comparison is performed only on these 8 instances). We also observe in Table 1 the trade-off between prediction accuracy and running time provided by parameter C. RSS using $C = 100$, and thus a larger type system, produces more accurate predictions, but it requires more time to produce such predictions.

Table 1 suggests that RSS produces substantially more accurate predictions than TSS, in less time. In many cases, its average ratio of predicted and actual EST size is orders of magnitude better than TSS. For instance, for grids with $C = 10$ the ratios with $m = 10$ probes for TSS and RSS are over 300,000 and 15.1, respectively, which drops to 416 and 2.17 with $m = 50$. Overall, its accuracy results in Table 1 and its more modest memory requirement suggest that RSS decisively outperforms and thus supersedes TSS.

Table 1. Prediction results of RSS and TSS for $C = 10$ and $C = 100$. For each number of TSS probes m, n is the number of tasks that TSS is able to produce predictions for without running out of memory and which the algorithms are compared on. Ratio of predicted and actual EST size is computed as above, "%" is the average percentage of the full DFBnB search time. Bold results indicate that a scheme produced more accurate predictions in shorter time.

pedigree ($C = 10$)						pedigree ($C = 100$)					
		TSS		RSS				TSS		RSS	
m	n	ratio	%	ratio	%	m	n	ratio	%	ratio	%
1	21	7.45e+06	8.47	**3.5**	**7.04**	1	15	4.2e+06	2.38	**181**	**1.54**
10	18	3.92e+06	5.67	3.2	6.83	10	12	1.07e+03	22.3	**1.6**	**21.4**
50	15	4.37e+06	10.4	**2.21**	**6.62**	50	8	58.8	23.6	**1.71**	**21.5**
100	13	6.46e+04	18.9	**2.3**	**16.7**	100	8	72.4	42.3	**1.7**	**42**
pdb ($C = 10$)						pdb ($C = 100$)					
		TSS		RSS				TSS		RSS	
m	n	ratio	%	ratio	%	m	n	ratio	%	ratio	%
1	14	1.73e+04	0.383	9.83	2.14	1	13	66.4	3.07	**6.66**	2.7
10	14	4.34e+03	1.87	9.83	2.14	10	10	12.1	13.5	**1.2**	**12.3**
50	13	174	4.96	**9.44**	**2.05**	50	2	1.1	12.2	**1.02**	**7.47**
100	11	3.88	8.55	**1.37**	**7.35**	100	0	-	-	-	-
grids ($C = 10$)						grids ($C = 100$)					
		TSS		RSS				TSS		RSS	
m	n	ratio	%	ratio	%	m	n	ratio	%	ratio	%
1	53	2.61e+06	1.61	**1.54e+04**	**1.47**	1	50	373	20.4	**4.18**	**17.5**
10	53	3.02e+05	7.89	**15.1**	**7.49**	10	40	3.47	41.2	**2.61**	**37.3**
50	50	416	17.3	**2.17**	**16.8**	50	20	1.94	61.1	**1.4**	**57.3**
100	48	134	23	**1.87**	**21.8**	100	12	2.22	69.6	**1.24**	**67.3**

4.4 Comparison with WBE

Lastly, we compare the performance of RSS and WBE. Evaluation can occur on the entire set of prediction tasks, since neither of the two schemes had issues running out of memory. The results in Table 2 are again averaged per problem domain, but this time organized by choosing predictions with similar %-value (see table caption for details).

Table 2. Prediction results of WBE and RSS ($C = 100$), averaged per problem domain and arranged by %-value: for RSS we average the results obtained within 0-5%, 5-10%, ..., 20-25% of DFBnB runtime. In each case we then pick the next-highest (in terms of %) WBE result to compare. Bold results indicate that a scheme produced more accurate predictions in shorter time.

	pedigree (26 tasks)				pdb (14 tasks)				grids (54 tasks)			
	WBE		RSS		WBE		RSS		WBE		RSS	
% range	ratio	%	ratio	%	ratio	%	ratio	%	ratio	%	ratio	%
0 - 5	6.92e+06	1.61	**20**	**1.44**	1.7e+25	3.16	**1.78**	**3.14**	8.09e+07	3.48	**142**	**3.27**
5 - 10	6.34e+03	8.82	**12**	**7.16**	2e+26	7.4	**1.97**	**7.36**	2.95e+05	7.65	**85.1**	**7.25**
10 - 15	7.65e+03	15.2	**2.45**	**12.4**	5.74e+25	12.3	**1.86**	**12.2**	2.08e+05	12.6	**1.68**	**11.9**
15 - 20	545	19	**1.06**	**18**	3.11e+19	17.1	**1.72**	**17.1**	1.17e+07	19.5	**5.55**	**17**
20 - 25	111	43.1	**1.41**	**22.4**	6.91e+18	22.8	**1.46**	**22.7**	450	25.4	**1.83**	**22.3**

Note that the %-values for WBE are by design larger than the ones for RSS, thus giving WBE a slight advantage in terms of computation time.

In this prediction setting, RSS performs substantially better than WBE, in all cases being several orders of magnitude more accurate than WBE while taking the same or less amount of time. Most significantly, in case of pdb tasks RSS average ratio in Table 2 never exceeds 2, while WBE overestimates EST size by 18 or more orders of magnitude. Secondly, RSS is able to provide estimations within a factor of 2 (on average) after only 10-15% of DFBnB search time. To the best of our knowledge, this is the first time a sampling algorithm is able to produce such accurate predictions of DFBnB EST size without having memory issues.

5 Related and Future Work

Another approach for DFBnB EST size prediction lies in off-line learning of regression models based on features extracted from the problem instance, the search space, and possibly candidate solvers. These techniques have been applied to satisfiability problems [25], combinatorial auctions [26], and graphical models [6]. These methods generally require collecting a large set of solved training instances, which can take a substantial amount of time. By contrast, RSS does not rely on training data; its output, however, could be used as an input feature for a regression-based approach, an interesting direction we hope to investigate in the future.

RSS has two limitations we hope to address in the future. First, RSS is not able to produce estimates of the size of AND/OR search trees [18]. Second, our sampling scheme does not account for DFBnB implementations that use caching to avoid expanding duplicated nodes.

6 Conclusion

We have introduced *Retentive Stratified Sampling* (RSS), a scheme for estimating the size of DFBnB search trees. RSS repeatedly probes the search tree and remembers solution nodes it encounters in the process, which are used to apply pruning in subsequent probes. We have demonstrated the superiority of RSS over competing schemes like TSS and WBE, namely its ability to produce estimates with high accuracy in relatively little time. In addition, unlike other schemes RSS does not suffer from memory issues, further adding to its attractiveness.

Acknowledgements. This work was sponsored by FAPEMIG, in part by NSF grants IIS-1065618 and IIS-1254071, and in part by the United States Air Force under Contract No. FA8750-14-C-0011 under the DARPA PPAML program.

References

1. Kask, K., Dechter, R.: A general scheme for automatic search heuristics from specification dependencies. Artificial Intelligence, 91–131 (2001)
2. Marinescu, R., Dechter, R.: Memory intensive AND/OR search for combinatorial optimization in graphical models. Artificial Intelligence 173, 1492–1524 (2009)

3. Otten, L., Dechter, R.: Anytime AND/OR depth first search for combinatorial optimization. In: Proceedings of the Symposium on Combinatorial Search, pp. 117–124. AAAI Press (2011)
4. de Givry, S., Schiex, T., Verfaillie, G.: Exploiting tree decomposition and soft local consistency in Weighted CSP. In: Proceedings of the AAAI Conference on Artificial Intelligence, pp. 22–27. AAAI Press (2006)
5. Balas, E., Toth, P.: Branch and bound methods. In: Lawler, E.L., Lenstra, J.K., Kart, A.H.G.R., Shmoys, D.B. (eds.) The Traveling Salesman Problem: A Guided Tour of Combinatorial Optimization. John Wiley & Sons, New York (1985)
6. Otten, L., Dechter, R.: A case study in complexity estimation: Towards parallel branch-and-bound over graphical models. In: Proceedings of the Conference on Uncertainty in Artificial Intelligence, pp. 665–674 (2012)
7. Knuth, D.E.: Estimating the efficiency of backtrack programs. Math. Comp. 29, 121–136 (1975)
8. Chen, P.C.: Heuristic sampling: A method for predicting the performance of tree searching programs. SIAM Journal on Computing 21, 295–315 (1992)
9. Korf, R.E.: Depth-first iterative-deepening: An optimal admissible tree search. Artificial Intelligence 27, 97–109 (1985)
10. Purdom, P.W.: Tree size by partial backtracking. SIAM Journal of Computing 7, 481–491 (1978)
11. Korf, R.E., Reid, M., Edelkamp, S.: Time complexity of Iterative-Deepening-A*. Artificial Intelligence 129, 199–218 (2001)
12. Zahavi, U., Felner, A., Burch, N., Holte, R.C.: Predicting the performance of IDA* using conditional distributions. Journal of Artificial Intelligence Research 37, 41–83 (2010)
13. Burns, E., Ruml, W.: Iterative-deepening search with on-line tree size prediction. In: Proceedings of the International Conference on Learning and Intelligent Optimization, pp. 1–15 (2012)
14. Lelis, L.H.S.: Active stratified sampling with clustering-based type systems for predicting the search tree size of problems with real-valued heuristics. In: Proceedings of the Symposium on Combinatorial Search, pp. 123–131. AAAI Press (2013)
15. Lelis, L.H.S., Otten, L., Dechter, R.: Predicting the size of depth-first branch and bound search trees. In: International Joint Conference on Artificial Intelligence, pp. 594–600 (2013)
16. Pearl, J.: Probabilistic Reasoning in Intelligent Systems. Morgan Kaufmann (1988)
17. Dechter, R., Rish, I.: Mini-buckets: a general scheme for bounded inference. Journal of the ACM 50, 107–153 (2003)
18. Marinescu, R., Dechter, R.: AND/OR Branch-and-Bound search for combinatorial optimization in graphical models. Artificial Intelligence 173, 1457–1491 (2009)
19. Otten, L., Dechter, R.: Anytime AND/OR depth-first search for combinatorial optimization. AI Communications 25, 211–227 (2012)
20. Kilby, P., Slaney, J.K., Thiébaux, S., Walsh, T.: Estimating search tree size. In: Proceedings of the AAAI Conference on Artificial Intelligence, pp. 1014–1019. AAAI Press (2006)
21. Nilsson, N.: Principles of Artificial Intelligence. Morgan Kaufmann (1980)
22. Lelis, L.H.S., Zilles, S., Holte, R.C.: Predicting the Size of IDA*'s Search Tree. Artificial Intelligence, 53–76 (2013)
23. Dechter, R.: Reasoning with Probabilistic and Deterministic Graphical Models: Exact Algorithms. Synthesis Lectures on Artificial Intelligence and Machine Learning. Morgan & Claypool Publishers (2013)
24. Harvey, W.D., Ginsberg, M.L.: Limited discrepancy search. In: Proceedings of the International Joint Conference on Artificial Intelligence, pp. 607–613 (1995)
25. Xu, L., Hutter, F., Hoos, H.H., Leyton-Brown, K.: SATzilla: Portfolio-based Algorithm Selection for SAT. J. Artif. Intell. Res. (JAIR) 32, 565–606 (2008)
26. Leyton-Brown, K., Nudelman, E., Shoham, Y.: Empirical hardness models: Methodology and a case study on combinatorial auctions. Journal of the ACM 56, 1–52 (2009)

Higher-Order Consistencies through GAC on Factor Variables

Chavalit Likitvivatanavong, Wei Xia, and Roland H. C. Yap

School of Computing, National University of Singapore, Singapore
{chavalit,xiawei,ryap}@comp.nus.edu.sg

Abstract. Filtering constraint networks to reduce search space is one of the main cornerstones of Constraint Programming and among them (Generalized) Arc Consistency has been the most fundamental. While stronger consistencies are also the subject of considerable attention, none matches GAC's and for this reason it continues to advance at a steady pace and has become the popular choice of consistency for filtering algorithms. In this paper, we build on the success of GAC by proposing a way to transform a constraint network into another such that enforcing GAC on the latter is equivalent to enforcing a stronger consistency on the former. The key idea is to factor out commonly shared variables from constraints' scopes, form new variables, then re-attach them back to the constraints where they come from. Experiments show that this method is inexpensive and outperforms specialized algorithms and other techniques when it comes to full pair-wise consistency (FPWC).

1 Introduction

Generalized arc consistency (GAC) is one of the most studied filtering algorithms for constraint satisfaction problems (CSPs) due to its simplicity and excellent performance in practice. Domain reduction interspersed with GAC during backtracking search has become the foremost method for solving a general CSP [17]. GAC on positive table constraints, in particular, has received a great deal of attention in recent years [3, 4, 9, 10, 12, 14]. These advances in turn provide a basis for many algorithms that enforce even stronger consistencies than GAC to build on.

In [6] it was shown that a network is pairwise consistent (PWC) iff its dual CSP is arc consistent. PWC is a k-wise consistency [5] for the case where $k = 2$. This is one of the earlier works that demonstrates how (G)AC can be used to achieve other types of consistencies. Consistencies of level/order higher than GAC are also the subject of many recent works [8, 11, 16]. Specifically, maxRPWC, PWC, and FPWC are investigated in [2, 11, 16]. Many of the algorithms that enforce these consistencies are based on well-established GAC algorithms. In [16], the authors extended the GACva algorithm [12] to enforce maxRPWC. Subsequently, STR2 was extended to cope with FPWC, resulting in eSTR2 [11] which gets improvement similar to how STR2 outperforms GACva.

As another area of focus in CSPs, researchers have studied how to transform non-binary constraint networks into equivalent binary constraint networks so that the algorithms and methods from the binary case can be applied [1, 18]. Two techniques emerge as a result: the hidden transformation and the dual transformation. Both rely on the dual

B. O'Sullivan (Ed.): CP 2014, LNCS 8656, pp. 497–513, 2014.

variable associated with each constraint, whose domain values have a one-to-one correspondence with the constraint's tuples. Nevertheless, the benefits of the transformation diminish as filtering algorithms for non-binary CSPs get better.

In this paper, we propose to transform a non-binary constraint network into another non-binary constraint network such that the latter is GAC if and only if the former is full pairwise consistency (FPWC), which means both GAC and PWC. In this respect, our intention is similar to that of the authors of [13] who proposed another kind of transformation. Like the transformation from non-binary to binary network, the one in [13] is based on dual variables. But here the dual variable is included into the scope of the constraint it is associated with. The pruning power comes from the join of tables that uses the dual variables as its scope so that propagation can be transmitted directly to other constraints.

Our transformation works in a fundamentally different way. Instead of forming a dual variable for each constraint, we factor out commonly shared variables among them. These variables form new compound variables that will be augmented to only the constraints that are involved. For FPWC, no new constraints are created. We extend this transformation to cover general k-wise consistency, adding new constraints reduced from the join of k tables. Preliminary experiments show that for FPWC our method is faster than both [13] and a specialized FPWC algorithm. For k-wise consistency where $k \geq 3$, our transformation can lower the number of nodes visited during search but it is more costly and thus of limited application unless the search reduction is large.

2 Preliminaries

A constraint network $\mathcal{P}$ is a $(\mathcal{X}, \mathcal{C})$ where $\mathcal{X}$ is a set of n variables $\{x_1, \ldots, x_n\}$ and $\mathcal{C}$ a set of e constraints $\{c_1, \ldots, c_e\}$. $D(x)$ is the domain of $x \in \mathcal{X}$. During search, $D^c(x)$ denotes the current domain of x. If $a \in D^c(x)$, a is said to be *present* in $D(x)$; otherwise a is *absent* from $D(x)$. We use (x, a) to denote the value $a \in D(x)$ (or simply a when the context is clear). Each $c \in \mathcal{C}$ involves two components: a scope ($scp(c)$) which is an ordered subset of variables of $\mathcal{X}$; and a relation over the scope ($rel(c)$). Given $scp(c) = \{x_{i_1}, \ldots, x_{i_r}\}$, $rel(c) \subseteq \prod_{j=1}^{r} D(x_{i_j})$ represents the set of satisfying combinations of values for the variables in $scp(c)$. We may also refer to c by $c(x_{i_1}, \ldots, x_{i_r})$ to emphasize the scope. A constraint's scope can be made unique by combining the constraint's relation with relations from other constraints with the same scope through intersection. We assume a total ordering for every $rel(c)$ and use $\rho(c, i)$ to denote the ith tuple. The *arity* of c is $|scp(c)|$. Given an ordered set $S \subseteq scp(c)$ and $\tau \in rel(c)$, the projection of τ on S ($\tau[S]$) is the tuple consisting of only the components of τ that correspond to the variables in S. A tuple $\tau = (a_{i_i}, \ldots, a_{i_k})$ where $a_{i_j} \in D(x_{i_j})$ is said to be an tuple over $\{x_{i_1}, \ldots, x_{i_k}\}$. The join of constraints c_i and c_j ($c_i \bowtie c_j$) is a constraint whose scope is $scp(c_i) \cup scp(c_j)$ and whose relation is $\{\tau \mid \tau \text{ is a tuple over } scp(c_i) \cup scp(c_j) \wedge \tau[scp(c_i)] \in rel(c_i) \wedge \tau[scp(c_j)] \in rel(c_j)\}$. The join of tuples $\tau_i \in rel(c_i)$ and $\tau_j \in rel(c_j)$ ($\tau_i \bowtie \tau_j$) is the tuple τ over $scp(c_i) \cup scp(c_j)$ such that $\tau[scp(c_i)] = \tau_i$ and $\tau[scp(c_j)] = \tau_j$. When elements in $rel(c)$ are given explicitly, c is called a *positive table constraint*. A tuple $\tau \in rel(c)$ is *valid* iff $\tau[x] \in D^c(x)$ for each $x \in scp(c)$. Otherwise τ is *invalid*. A tuple $\tau \in rel(c)$ is a *support* of (x, a) in c iff $\tau[x] = a$. A value (x, a) is generalized arc-consistent (GAC) on a constraint c involving x iff there exists a *valid support* τ of (x, a) in c. A value (x, a) is GAC iff it is GAC on every constraint c involving x. A variable x is GAC iff $D^c(x) \neq \emptyset$ and (x, a) is GAC for each $a \in D^c(x)$. $\mathcal{P}$ is GAC iff each of its variables is GAC. A *solution* to $\mathcal{P}$ is a valid tuple over $\mathcal{X}$ such that every constraint is satisfied. $\mathcal{P}$ is *satisfiable* iff one solution exists. The constraint satisfaction problem (CSP) is the NP-hard task of determining whether a given constraint network is satisfiable or not.

A *compound variable* X is a cross-product composition from $\{x_{i_1}, \ldots, x_{i_m}\} \subseteq \mathcal{X}$, called X's *signature* ($\sigma(X)$), where $D(X) \subseteq \prod_{j=1}^{m} D(x_{i_j})$ and its values are sometimes referred to as *compound values*. Given a constraint c and an ordered set $S = \{x_{i_1}, \ldots, x_{i_m}\} \subseteq scp(c)$, we denote $\lambda_c(S)$ to be the compound variable on S with respect to c whose domain $D(\lambda_c(S))$ is $\{\tau[S] \mid \tau \in rel(c)\}$. It follows that $\sigma(\lambda_c(S)) = S$. A value in $D(\lambda_c(S))$ may be written as $\bar{a} = (a_{i_1}, \ldots, a_{i_m})$. We also use $\pi(\lambda_c(S), x_{i_k})$ to denote $\{a_{i_k} \mid \bar{a} \in D(\lambda_c(S))\}$ where $k \in \{1, \ldots, m\}$. Similarly, $\pi^c(\lambda_c(S), x_{i_k}) = \{a_{i_k} \mid \bar{a} \in D^c(\lambda_c(S))\}$. We may drop the subscript and write $\lambda(S)$ if there is no ambiguity. Non-compound variables are called *ordinary* variables. For uniformity, σ is defined for all variables, i.e. $\sigma(x) = \{x\}$ for an ordinary variable x. A value (x, a) is *max-restricted pairwise consistent* (maxRPWC) iff for all $c_i \in \mathcal{C}$ where $x \in scp(c_i)$, (x, a) has a valid support τ_i in $rel(c_i)$ such that for any other $c_j \in \mathcal{C}$ there exists a valid tuple $\tau_j \in rel(c_j)$ and $\tau_i[scp(c_i) \cap scp(c_j)] = \tau_j[scp(c_i) \cap scp(c_j)]$. $\mathcal{P}$ is maxRPWC iff all values are maxRPWC. $\mathcal{P}$ is *k-wise consistent* (kWC) iff given any group of k constraints $\{c_{i_1}, \ldots, c_{i_k}\}$, then for any $\tau \in rel(c_{i_j})$ for some j there exists a valid tuple τ' over $\bigcup_{l=1}^{k} scp(c_{i_l})$ such that $\tau'[scp(c_{i_j})] = \tau$ and $\tau'[scp(c_{i_l})] \in rel(c_{i_l})$ for all $l \in \{1, \ldots, k\}$. If $\mathcal{P}$ is kWC then $\mathcal{P}$ is (k-1)WC. When k is equal to two, it is also called *pairwise consistency* (PWC). $\mathcal{P}$ is *full pairwise consistent* (FPWC) iff it is both GAC and PWC. FPWC is also equivalent to PWC together with maxRPWC [11].

3 Reformulation

First we give a straightforward reformulation of a constraint network that encodes FPWC as follows. Given $\mathcal{P} = (\mathcal{X}, \mathcal{C})$, we construct $\mathcal{P}^+ = (\mathcal{X} \cup \mathcal{W}, \mathcal{C}^+)$ such that $\mathcal{W} = \{\ \lambda_{c_i}(S), \lambda_{c_j}(S) \mid S = scp(c_i) \cap scp(c_j))$ for all $i \neq j \wedge |S| > 1\}$ and $\mathcal{C}^+$ includes constraints of the following three types. The first involves a simple extension of constraints in $\mathcal{C}$. For each $c'_i \in \mathcal{C}^+$, $1 \leq i \leq e$, we have,

- $scp(c'_i) = scp(c_i) \cup \{\lambda_{c_i}(S) \mid \lambda_{c_i}(S) \in \mathcal{W} \wedge S \subseteq scp(c_i)\}\}$
- for any $\tau \in rel(c_i)$, $\tau' \in rel(c'_i)$ is a tuple extended from τ such that
  - $\tau'[x] = \tau[x]$ for any $x \in scp(c_i)$
  - for any $\lambda_{c_i}(S) \in scp(c'_i)$, $\tau'[\lambda_{c_i}(S)] = \tau[S]$

The second type of constraints involves equality between $\lambda_{c_i}(S)$ and $\lambda_{c_j}(S)$ in $\mathcal{W}$ for any i, j, and S. The third involves compatibility constraints between a compound variable and each variable in its signature. That is, given $\lambda_c(S)$ such that $S = \{x_{i_1}, \ldots, x_{i_m}\}$, there is a constraint between $\lambda_c(S)$ and each x_{i_k} that forces $\pi^c(\lambda_c(S), x_{i_k}) = D^c(x_{i_k})$. As a result of this construction, in a generalized arc-consistent $\mathcal{P}^+$ any valid tuple in a constraint c can be extended to a valid tuple over $scp(c) \cup scp(c')$ for any other constraint c' through variables in $\mathcal{W}$. The proof is omitted due to space restrictions.

Theorem 1. *$\mathcal{P}^+$ is GAC if and only if $\mathcal{P}$ is FPWC.*

Next we show how $\mathcal{P}^+$ can be simplified while still preserving Theorem 1. Instead of posting an equality constraint between every pair of compound variables with the same signature, we unify all these compound variables into a single variable. Equality constraints are removed. Given $\mathcal{P}= (\mathcal{X}, \mathcal{C})$ and $\mathcal{P}^+= (\mathcal{X} \cup \mathcal{W}, \mathcal{C}^+)$, the *factor encoding* (FE) of $\mathcal{P}$ is the network $\mathcal{P}^* = (\mathcal{X} \cup \mathcal{W}^*, \mathcal{C}^*)$ where,

$\mathcal{W}^* = \{\lambda(S) \mid D(\lambda(S)) = \bigcup_k D(\lambda_{c_k}(S))$ for all k such that $\lambda_{c_k}(S) \in \mathcal{W}\}$

and for each $c_i^* \in \mathcal{C}^*, 1 \leq i \leq e$,

- $scp(c_i^*) = scp(c_i) \cup \{\lambda(S) \mid \lambda(S) \in \mathcal{W}^* \wedge S \subsetneq scp(c_i)\}\}$
- for any $\tau \in rel(c_i)$, $\tau^* \in rel(c_i^*)$ is a tuple extended from τ such that
  - $\tau^*[x] = \tau[x]$ for any $x \in scp(c_i)$
  - for any $\lambda(S) \in scp(c_i^*)$, $\tau^*[\lambda(S)] = \tau^*[S](= \tau[S])$ (e1)

We call the compound variables in $\mathcal{W}^*$ *factor variables*. $\mathcal{P}^*$ is also referred to as *fe*($\mathcal{P}$). Given $c_k \in \mathcal{C}$, we may denote $c_k^* \in \mathcal{C}^*$ with *fe*(c_k). We observe that the compatibility constraint in $\mathcal{P}^+$ can be decomposed into two conditions. Given $\lambda(S)$ such that $S = \{x_{i_1}, \ldots, x_{i_m}\}$, we have,

(c1) $\bar{a} \in D^c(\lambda(S)) \Rightarrow \forall k \in \{1, \ldots, m\}, a_{i_k} \in D^c(x_{i_k})$
(c2) $a \in D^c(x_{i_k})$ for some $k \in \{1, \ldots, m\} \Rightarrow \exists \bar{a} \in D^c(\lambda(S)), a_{i_k} = a$

We will show that the compatibility constraints in $\mathcal{P}^+$ are actually implied and do not need to be posted explicitly.

Lemma 1. *Enforcing GAC on fe($\mathcal{P}$) imposes the condition* (c1) *between a factor variable and each ordinary variable in its signature.*

Proof. Consider $\lambda(S)$ where $S = \{x_{i_1}, \ldots, x_{i_m}\}$ and $\bar{a} \in D^c(\lambda(S))$. Because *fe*($\mathcal{P}$) is GAC, for any *fe*($c$) such that $\lambda(S) \in scp(fe(c))$, there is a valid support of $\bar{a}$ in $rel(fe(c))$. That is, $\exists \tau \in rel(fe(c))$ such that $\tau[\lambda(S)] = \bar{a}$. Since $\tau[\lambda(S)] = \tau[S]$, $\tau[x_{i_k}] = a_{i_k}$ for $1 \leq k \leq m$, which means a_{i_k} also has a valid support in $rel(fe(c))$. □

Lemma 2. *Enforcing GAC on fe($\mathcal{P}$) imposes the condition* (c2) *between a factor variable and each ordinary variable in its signature.*

Proof. Assume $a_{i_k} \notin \pi^c(\lambda(S), x_{i_k})$ for some a_{i_k}. This indicates that any $\bar{a}$ involving a_{i_k} must be absent from $D(\lambda(S)))$. Due to propagation, every τ in every $rel(fe(c))$ such that $\lambda(S) \in scp(fe(c))$ and $\tau[\lambda(S)] = \bar{a}$ would eventually become invalid. Because $\tau[S] = \tau[\lambda(S)] = \bar{a}$, $\tau[x_{i_k}] = a_{i_k}$. That means such τ is not a valid support of a_{i_k}. Because $D(\lambda(S))$ contains every compound values involving a_{i_k} from all c whose scope subsumes S, there is no other valid tuple τ' such that $\tau'[x_{i_k}] = a_{i_k}$. Hence, $a_{i_k} \notin D^c(x_{i_k})$ after the propagation converges. □

Theorem 2. *fe($\mathcal{P}$) is GAC if and only if $\mathcal{P}$ is FPWC.*

Proof. Follows from Theorem 1, and Lemma 1, and 2. □

Theorem 3. *fe($\mathcal{P}$) is GAC if and only if fe($\mathcal{P}$) is FPWC.*

Proof. As FPWC is both GAC and PWC, ($\Leftarrow$) is immediate. We will prove the ($\Rightarrow$) direction. Assume *fe*($\mathcal{P}$) is GAC. Let $\tau_i \in fe(c_i)$. Now consider another constraint $fe(c_j) \neq fe(c_i)$. If there is no factor variable in $scp(fe(c_i)) \cap scp(fe(c_j))$, then PWC is trivial. Let f be the factor variable[1] in $scp(fe(c_i)) \cap scp(fe(c_j))$ such that $scp(fe(c_i)) \cap scp(fe(c_j)) \setminus \sigma(f) = \{f\}$. Since *fe*($\mathcal{P}$) is GAC, $\tau_i[f]$ must have a valid support in $fe(c_j)$. Call it τ_j. Because τ_i and τ_j agree on f, by definition of factor variable they must agree on $\sigma(f)$ too, which means they agree on $\sigma(f) \cup \{f\} = scp(fe(c_i)) \cap scp(fe(c_j))$. As a result, $\tau_i \bowtie \tau_j$ is well-defined as well as being a tuple extended from τ_i over $scp(fe(c_i)) \cup scp(fe(c_j))$. Hence, *fe*($\mathcal{P}$) is PWC. □

Let $fe^k(\mathcal{P})$ denote $fe(fe(\ldots fe(\mathcal{P})\ldots))$ (the FE is applied k times in a row), then

Corollary 1. *For all $k \geq 1$, $\mathcal{P}$ is FPWC if and only if $fe^k(\mathcal{P})$ is GAC.*

Proof. We consider $k = 2$ as other cases follow from induction. From Theorem 2 and Theorem 3, we have: $\mathcal{P}$ is FPWC iff *fe*($\mathcal{P}$) is FPWC. From this statement and the result of another application of the FE on it we derive: $\mathcal{P}$ is FPWC iff *fe*(*fe*($\mathcal{P}$)) is FPWC. From Theorem 3, *fe*(*fe*($\mathcal{P}$)) is FPWC iff *fe*(*fe*($\mathcal{P}$)) is GAC. □

This shows $fe^k(\mathcal{P})$ for $k \geq 2$ is no different than *fe*($\mathcal{P}$) so applying the FE more than once in succession is pointless. A localized version of this corollary is given as follows.

Corollary 2. *Given any two constraints c_i and c_j, if there exists a factor variable $f \in scp(c_i) \cap scp(c_j)$ such that $scp(c_i) \cap scp(c_j) \setminus \sigma(f) = \{f\}$ then adding the factor variable f' whose signature is $\sigma(f) \cup \{f\}$ to the scopes of both constraints is futile.*

Property 1. *Running GAC on fe($\mathcal{P}$) can be $O(e^2)$ faster and use $O(e^2)$ smaller space than running eSTR2 on $\mathcal{P}$.*

Reasoning: eSTR2 [11] is an extension of STR2 [9] that enforces FPWC. The main difference between enforcing GAC on the FE and enforcing eSTR2 on the original network is the space and time associated with factor variables vs. those associated with the additional data structures for checking PWC in eSTR2. The overhead of running GAC on the FE depends on factor variables, whose number can be lower than the number of intersecting constraints. In eSTR2, the overhead depends on the number of intersecting constraints. If $\mathcal{P}$ consists of only constraints such that a single factor variable is common to all and that no other factor variable exists, the space and time complexity of the GAC on *fe*($\mathcal{P}$) is the same as those on $\mathcal{P}$. By contrast, the space and time of eSTR2 on $\mathcal{P}$ would be at least an order of $O(\binom{e}{2}) = O(e^2)$ larger. □

Property 2. *For any $c \in \mathcal{C}$, $|scp(c)| \leq |scp(fe(c))| \leq |scp(c)| + |\mathcal{C}| - 1$.*

The range is the result of the number of factor variables added. The lower bound is zero, when no other constraint's scope overlaps on more than two variables with $scp(c)$, whereas the upper bound is $|\mathcal{C}| - 1$ when every intersection with another constraint produces a new factor variable.

[1] There may be multiple factor variables if $\mathcal{P}$ itself is the factor encoding of another constraint network, which in turn is the factor encoding of another, and so on (see Corollary 1). The factor variable f is set to be the most recent one.

3.1 Example

We give an example of $\mathcal{P}^*$ and trace some GAC propagation on $\mathcal{P}^*$ in this section. Note that although relations in $\mathcal{P}^*$ are an extension of those in $\mathcal{P}$, the extension to factor variables can be implicit. The expression $\tau[S]$ in (e1) can be given as a function (i.e. the projection) that takes an input S rather than the actual result of the projection of τ on S. Such abstract extension of tuples is demonstrated in this section.

For brevity, compound variables and values are written as a concatenation of ordinary variables and values. Let $\mathcal{P}^* = (\mathcal{X} \cup \mathcal{W}^*, \mathcal{C}^*)$, where $\mathcal{X} = \{x_1, x_2, x_3, x_4, x_5, x_6\}$, $\mathcal{W}^* = \{x_1x_2,\ x_1x_2x_4\}$, $\mathcal{C}^* = \{c_1^*, c_2^*, c_3^*, c_4^*\}$ where $scp(c_1^*) = \{x_1, x_2, x_3,\ x_1x_2\}$, $scp(c_2^*) = \{x_1, x_2, x_4,\ x_1x_2,\ x_1x_2x_4\}$, $scp(c_3^*) = \{x_1, x_2, x_4, x_5,\ x_1x_2,\ x_1x_2x_4\}$, $scp(c_4^*) = \{x_2, x_6\}$. Relations of $\mathcal{P}$ are given as tables for c_i below ($rel(c_i^*)$ will be inferred from $rel(c_i)$). $D^c(x_1) = D^c(x_2) = D^c(x_4) = D^c(x_6) = \{a, b\}$, $D^c(x_3) = D^c(x_5) = \{a, b, c\}$, $D^c(x_1x_2) = \{aa, ab, bb\}$, $D^c(x_1x_2x_4) = \{abb, bba, bbb\}$.

c_1

x_1	x_2	x_3
a	a	a
a	b	a
a	b	c
b	b	b

c_2

x_1	x_2	x_4
a	b	b
b	b	a

c_3

x_1	x_2	x_4	x_5
a	b	b	a
b	b	a	b
b	b	b	c

c_4

x_2	x_6
a	a
b	b

We now look at some GAC propagation on this network. First, we consider whether $(x_1x_2x_4, bbb)$ is GAC. Let $\tau = \rho(c_3, 3) = (b, b, b, c)$. The value $(x_1x_2x_4, bbb)$ is GAC on c_3^* since $\rho(c_3^*, 3) = (\tau[x_1], \tau[x_2], \tau[x_4], \tau[x_5], \tau[x_1x_2], \tau[x_1x_2x_4]) = (b, b, b, c, bb, bbb)$ is found to be a valid support. But $(x_1x_2x_4, bbb)$ is not GAC on c_2^* because no tuple in $rel(c_2^*)$ involves bbb (i.e. $rel(c_2^*) = \{(a, b, b, ab, abb), (b, b, a, bb, bba)\}$), so bbb is removed from $D^c(x_1x_2x_4)$. Propagation leads back to the removal of c from $D^c(x_5)$ as $\rho(c_3^*, 3)$ is no longer valid because $\rho(c_3^*, 3)[x_1x_2x_4] = bbb \notin D^c(x_1x_2x_4)$. Next we look at (x_1x_2, aa). It has no valid support in c_2^* so aa will be removed from the domain of x_1x_2. Because $\rho(c_1^*, 1) = (a, a, a, aa)$, this tuple becomes invalid. Because $\rho(c_1^*, 1)$ is the only tuple involving (x_2, a) in $rel(c_1^*)$, (x_2, a) is no longer GAC on c_1^*. Value a is then removed from $D^c(x_2)$. Further propagation leads to the removal of (x_6, a).

4 The k-Interleaved Encoding

The k-interleaved encoding (kIL) [13] is closely related to the FE as both try to enlarge constraints with auxiliary variables that represent groups of existing variables. Enforcing GAC on the k-interleaved encoding is equivalent to enforcing kWC on the original network in addition to GAC. The following definitions are taken from [13].

Definition 1 (k-Dual Encoding). *Let $\mathcal{P} = (\mathcal{X}, \mathcal{C})$. The k-dual encoding of $\mathcal{P}$ is the constraint network $\mathcal{P}^{kd} = (\mathcal{X}^{kd}, \mathcal{C}^{kd})$ where:*

- *for each $c_i \in \mathcal{C}$, $\mathcal{X}^{kd}$ contains a variable x_i' where $D(x_i') = \{1, \ldots, |rel(c_i)|\}$.*
- *for each subset S of k constraints of $\mathcal{C}$, $\mathcal{C}^{kd}$ contains a constraint c' such that $scp(c') = \{x_i' \mid c_i \in S\}$ and c' is a k-ary table constraint containing the join of all constraints in S (represented with the indexes of the original tuples).*

Definition 2 (Hybrid Constraints). *Let* $\mathcal{P} = (\mathcal{X}, \mathcal{C})$*. The set of hybrid constraints* $\phi(\mathcal{C})$ *of* $\mathcal{P}$ *is the set* $\{\phi(c_i) \mid c_i \in \mathcal{C}\}$ *where:*

- $scp(\phi(c_i)) = scp(c_i) \cup \{x'_i\}$
- *for every* j^{th} *tuple* τ *of* $rel(c_i)$*,* τ' *is a tuple in* $rel(\phi(c_i))$ *such that* $\tau'[x'_i] = j$ *and* $\tau'[x] = \tau[x]$ *for each* $x \in scp(c_i)$

Definition 3 (k-Interleaved Encoding). *Let* $\mathcal{P} = (\mathcal{X}, \mathcal{C})$*. The* k*-interleaved encoding of* $\mathcal{P}$ *is the constraint network* $\mathcal{P}^{ki} = (\mathcal{X}^{ki}, \mathcal{C}^{ki}) = (\mathcal{X} \cup \mathcal{X}^{kd}, \phi(\mathcal{C}) \cup \mathcal{C}^{kd})$ *where* $(\mathcal{X}^{kd}, \mathcal{C}^{kd})$ *is the* k*-dual encoding of* $\mathcal{P}$ *and* $\phi(\mathcal{C})$ *the hybrid constraints of* $\mathcal{P}$*.*

For $k = 2$, enforcing GAC on the 2IL has the same pruning power as enforcing GAC on the FE, but the FE does not add any new constraint. We now look at an example from [13] for a comparison of the 2IL and the FE. Figure 1a shows three constraints from the original network. Figure 1b shows the FE for these constraints. A factor variable's domain of size d is normalized as $\{1, \ldots, d\}$. As a result, $D^c(xy) = D^c(uv) = \{11, 00, 01, 10\} = \{1, 2, 3, 4\}$. After GAC is established, $D^c(y)$ becomes $\{1\}$ and $D^c(v)$ becomes $\{0\}$. Figure 1c shows the 2IL of 1a [13]. This example shows that while enforcing GAC on the kIL gives identical $D^c(y)$ and $D^c(v)$ to Figure 1b, the kIL can take a longer chain of propagation to do so.

x	y	u	v
1	1	1	0
0	0	0	1
0	1	0	0
1	0	1	1

x	y
1	1
0	0
0	1

u	v
1	1
1	0
0	0

(a) Original

x	y	u	v	xy	uv
1	1	1	0	1	4
0	0	0	1	2	3
0	1	0	0	3	2
1	0	1	1	4	1

x	y	xy
1	1	1
0	0	2
0	1	3

u	v	uv
1	1	1
1	0	4
0	0	2

(b) Factor encoding of (a)

x	y	u	v	x'_1
1	1	1	0	1
0	0	0	1	2
0	1	0	0	3
1	0	1	1	4

x	y	x'_2
1	1	1
0	0	2
0	1	3

u	v	x'_3
1	1	1
1	0	2
0	0	3

x'_1	x'_2
1	1
2	2
3	3

x'_1	x'_3
1	2
3	3
4	1

(c) 2-interleaved encoding of (a)

Fig. 1. Comparison of two encodings

We compare the complexity of the FE and the kIL as follows. For simplicity, we assume there are e constraints of arity r, each associated with a table containing t tuples and that every pair of constraints shares at least two variables in their scopes.

Property 3. *The extra cells added to the tables by the FE ranges from* $O(et)$ *to* $O(e^2t)$*.*

Proof. In the best case there is only one factor variable. Each constraint will be extended with an extra column so the total extra space is $O(et)$. In the worst case, every pair of constraint produces one additional factor variable. Each of these factor variables will appear in two different tables. Thus, the total is $O(2t\binom{e}{2}) = O(e^2t)$. □

Because an optimal GAC algorithm traverses every cell of every table in the worst case, the worst-case time complexity of GAC on the FE is thus between $O(ert)$ (i.e. no asymptotic difference) and $O(ert + e^2t) = O(e^2t)$ (i.e. assuming $e > r$).

Property 4. *The extra cells added to the tables by the kIL is* $O(\binom{e}{k}t^k)$

Proof. Each constraint has an extra column for indexing so the space is et. For every subset of $\mathcal{C}$ of size k, a join table of arity k is created. The total space is therefore $O(et + \binom{e}{k}t^k) = O(\binom{e}{k}t^k)$. □

For $k = 2$, this space becomes $O(e^2t^2)$. As far as GAC is concerned, the 2IL is thus a factor of t more expensive in the worst case than the FE.

5 Enforcing k-Wise Consistency through Reduced Join Tables

Given *fe*($\mathcal{P}$), we may post additional constraints so that GAC may also enforce kWC. These new constraints are created from a group of existing constraints and this section studies their effect on the consistency level.

Given $C = \{c_{i_1}, \dots, c_{i_k}\}$ in $\mathcal{P}$ where $k \geq 3$, we define the following notation:

- $mult(C) = \{\lambda(S) \mid \lambda(S) \in \mathcal{W}^* \wedge S = scp(c_{i_j}) \cap scp(c_{i_l})$ for $1 \leq j < l \leq k\}$
- $sing(C) = \{x \mid x \in \mathcal{X} \wedge \{x\} = scp(c_{i_j}) \cap scp(c_{i_l})$ for $1 \leq j < l \leq k\}$
- $join(C) = rel(c_{i_1}) \bowtie \dots \bowtie rel(c_{i_k})$,

Definition 4. *Given a set C of k constraints ($k \geq 3$), the* factor-reduced join *of C (frj(C)) is a constraint constructed as follows. Let $|mult(C)| = o$, and $|sing(C)| = p$,*

- $scp(frj(C)) = mult(C) \cup sing(C) = \{\lambda(S_1), \dots, \lambda(S_o)\} \cup \{x_{j_1}, \dots x_{j_p}\}$
- $rel(frj(C)) = \{(\tau[S_1], \dots, \tau[S_o], \tau[x_{j_1}], \dots, \tau[x_{j_p}]) \mid \tau \in join(C)\}$

The factor-reduced join is not a projection of *join*(C) as its scope may include factor variables. Rather, it can be viewed as a projection of $\bowtie_{c \in C} fe(c)$. In any case, since it is derived from the join of C, its pruning power cannot be greater.

It should be noted that *frj*(C) may end up having the same scope as another existing constraint or another *frj* constraint. For instance, let $C_1 = \{c_1(x_1, x_2, x_3), c_2(x_1, x_2, x_4), c_3(x_1, x_5)\}$ and $C_2 = \{c_4(x_1, x_2, x_6), c_5(x_1, x_2, x_7), c_3(x_1, x_5)\}$. Let $y_1 = x_1x_2$, it follows that $scp(frj(C_1)) = \{x_1, y_1\} = scp(frj(C_2))$. This can also happen in the case where no factor variables are formed by the FE. For instance, let $C_1 = \{c_1(x_1, x_2), c_2(x_2, x_3), c_3(x_3, x_4)\}$. Then $scp(frj(C_1)) = \{x_2, x_3\} = scp(c_2)$. Both cases can be handled by merging constraints with the same scope afterwards.

We assume that every constraint in C must be relevant. Namely, given $c \in C$ there must exist at least one other $c' \in C$ such that $|scp(c) \cap scp(c')| \geq 1$.

Property 5. *The arity of frj(C) ranges from 2 to $\binom{|C|}{2}$.*

The *fe*($\mathcal{P}$) with the additional constraints *frj*(C) for every group C of size k is called the factor encoding of $\mathcal{P}$ for k-wise consistency (FKWC), also denoted by *fkwc*($\mathcal{P}, k$).

Property 6. *Enforcing GAC on fkwc($\mathcal{P}$,k) is strictly weaker than enforcing both FPWC and kWC on $\mathcal{P}$ and strictly stronger than enforcing FPWC on $\mathcal{P}$.*

We show this by an example. Consider the constraints in Figure 2a and their factor encodings in Figure 2b, where $y_1 = x_2x_3$. The networks in both figures are PWC. The join of the three original constraints is given in Figure 2c. The projection of *join*(C) onto each of the original constraint makes the following tuples 3-wise inconsistent: $(1, 1, 0, 0) \in rel(c_1)$, $(1, 0, 1) \in rel(c_2)$, and $(1, 0, 1) \in rel(c_3)$. Now consider the

c_1			
x_1	x_2	x_3	x_4
0	0	0	0
0	0	1	1
1	1	0	0

c_2		
x_2	x_3	x_5
0	0	0
0	1	1
1	0	1

c_3		
x_4	x_5	x_6
0	0	0
1	0	1
1	1	0

(a) Original: $C = \{c_1, c_2, c_3\}$

fe(c_1)				
x_1	x_2	x_3	x_4	y_1
0	0	0	0	0
0	0	1	1	1
1	1	0	0	2

fe(c_2)			
x_2	x_3	x_5	y_1
0	0	0	0
0	1	1	1
1	0	1	2

fe(c_3)		
x_4	x_5	x_6
0	0	0
1	0	1
1	1	0

(b) Factor-encoded constraints

join(C)					
x_1	x_2	x_3	x_4	x_5	x_6
0	0	0	0	0	0
0	0	1	1	1	0

(c) The join of constraints in C

frj(C)		
x_4	y_1	x_5
0	0	0
1	1	1

(d) The factor-reduced join of C

fe(*fe*(c_1))					
x_1	x_2	x_3	x_4	y_1	z_1
0	0	0	0	0	0
0	0	1	1	1	1
1	1	0	0	2	2

fe(*fe*(c_2))				
x_2	x_3	x_5	y_1	z_2
0	0	0	0	0
0	1	1	1	1
1	0	1	2	2

fe(*fe*(c_3))			
x_4	x_5	x_6	z_3
0	0	0	0
1	0	1	1
1	1	0	2

fe(*frj*(C))					
x_4	y_1	x_5	z_1	z_2	z_3
0	0	0	0	0	0
1	1	1	1	1	2

(e) The FE of (b) and (d) where *fe*(*fe*(c_i)) denotes the FE of constraints from (b) with *frj*(C).

Fig. 2. The pruning power of GAC on (b) + (d) lies between FPWC and FPWC + 3-wise consistency on (a), whereas GAC on (e) is equal to FPWC + 3-wise consistency on (a)

frj(C) in Figure 2d. GAC on $\{frj(C), fe(c_1), fe(c_2), fe(c_3)\}$ leads to the inconsistency of $(1,1,0,0,2) \in rel(fe(c_1))$ and $(1,0,1,2) \in rel(fe(c_2))$, but not $(1,0,1) \in rel(fe(c_3))$.

Although the *fkwc*($\mathcal{P}$,k) encoding is only partial kWC, it subsumes *fe*($\mathcal{P}$) so FPWC is guaranteed by GAC. Together with the fact that kWC implies (k-1)WC, we have

Property 7. *Enforcing GAC on fkwc($\mathcal{P}$,k) is strictly weaker than enforcing GAC on the kIL of $\mathcal{P}$ for $k \geq 3$.*

Theorem 4. $\mathcal{Q} = fe(fkwc(\mathcal{P}, k)))$ *is GAC if and only if $\mathcal{P}$ is FPWC and kWC.*

Sketch of Proof: Consider $C = \{c_{i_1}, \ldots, c_{i_k}\}$. It is clear that *join*(C) forces kWC on C through GAC. $fe(frj(C)) \in \mathcal{Q}$ represents all the articulation points of *join*(C) and we will show that both have the same restricting effect on the rest of the network by showing that the "missing columns" can be "rebuilt" via GAC. The proof for ($\Leftarrow$) is omitted for lack of space. Assume $\mathcal{Q}$ is GAC. Let x be a variable in $scp(join(C)) \setminus (\bigcup_{\lambda(S) \in mult(C)} S) \setminus sing(C)$. It follows that there is exactly one constraint $c_{i_j} \in C$ such that $x \in scp(c_{i_j})$. Suppose $a \in D^c(x)$. Because $\mathcal{Q}$ is GAC, so is a. By definition, there exists a valid tuple $\tau_{i_j} \in rel(fe(fe(c_{i_j})))$ such that $\tau_{i_j}[x] = a$. Let $H_{i_j} = scp(fe(c_{i_j})) \cap scp(frj(C))$, $|H_{i_j}| \geq 1$. Because $\mathcal{Q}$ is a factor encoding, there exists a variable $\lambda(H_{i_j})$ in both $scp(fe(fe(c_{i_j})))$ and $scp(fe(frj(C)))$ ($\lambda(H_{i_j})$ is either an ordinary or a factor variable). Since H_{i_j} too is GAC, τ_{i_j} is guaranteed to be extendable to *fe*(*frj*(C)). Let φ be such a tuple in *fe*(*frj*(C)) such that $\tau_{i_j} \bowtie \varphi$ is a tuple over $scp(fe(c_{i_j})) \cup scp(frj(C))$. For each $c_{i_l} \in C \setminus \{c_{i_j}\}$, let $H_{i_l} = scp(fe(c_{i_l})) \cap scp(frj(C))$. By the same argument, there exists a valid tuple τ_{i_l} in *fe*(*fe*(c_{i_l})) such that $\tau_{i_l}[H_{i_l}] = \varphi[H_{i_l}]$. The join of φ, τ_{i_j}, and every such τ_{i_l} would become a valid support of a in $J = (\bowtie_{c \in C} fe(fe(c))) \bowtie$

fe(*frj*(C)). Because the projection of J on *scp*(*join*(C)) is *join*(C) and a is arbitrary, the column x in *scp*(*join*(C)) is thus the same as $D^c(x)$. □

Figure 2e shows another application of the factor encoding on top of *fkwc*($\mathcal{P}, k$), where $z_1 = x_4y_1$, $z_2 = x_5y_1$, and $z_3 = x_4x_5$. GAC on this network would lead to the inconsistency of $(1, 0, 1, 1)$ in the third table. Since $x_2x_3y_1$ is redundant according to Corollary 2 we do not add it to *scp*(*fe*(*fe*(c_1))) and *scp*(*fe*(*fe*(c_2))).

Property 8. *The arity of fe*(*frj*(C)) *ranges from 2 to* $\binom{|\mathcal{C}|}{2} + |\mathcal{C}|$.

Proof The reasoning is similar to the one for Property 2, but here *fe*(*frj*(C)) is not necessarily part of $\mathcal{C}$ so the bound on the number of constraints that it may interact with is $|\mathcal{C}|$ not $|\mathcal{C}| - 1$. Coupled with Property 5, we have,

$$2 \leq |scp(frj(C))| \leq |scp(fe(frj(C)))| \leq |scp(frj(C))| + |\mathcal{C}| \leq \binom{|\mathcal{C}|}{2} + |\mathcal{C}|$$ □

Figure 2e demonstrates: we have $|\mathcal{C}| = |C| = 3$, so the upper bound on the arity of *fe*(*frj*(C)) is $\binom{|\mathcal{C}|}{2} + |\mathcal{C}| = \binom{3}{2} + 3 = 6$, which happens to be the actual arity of *fe*(*frj*(C)).

6 Experiments

In this section, we present experimental results on the effectiveness of the FE and the FKWC in comparison with the kIL and an FPWC algorithm. We will use kFE to denote *fkwc*($\mathcal{P}, k$). Benchmarks are drawn from the CSP solver competition[2] in addition to randomly generated problems. The experiments were conducted on a 2.6GHz quad-core Intel Core i7 on OS X 10.8. The converters take an input in the XCSP format and output the result as another text file. As such, we are not restricted to any particular GAC algorithm and we shall test the encoding on multiple GAC algorithms. Like [7, 11], the search employed the $dom/ddeg$ variable ordering heuristic and the *lex* value ordering. We used AbsCon [15] as the solver. Conversion time is limited to 30 minutes while memory is limited to 8GB for both the converters and the solver.

Because the kIL and the FE augment the original constraints' scope with new variables, the location to which they are inserted has to be considered. Two natural choices are the front and the back. The front leads to slightly faster running time in our experiments, with a big difference on some problem instances, such as when multi-valued decision diagrams (MDDs) are involved. To simplify the presentation, experiments therefore involve only the front insertion. After the conversion the FE and FKWC converters may have to re-sort the tuples since the front insertion may disrupt the ordering of the input that is already sorted. The reason is that AbsCon happens to need sorted relations for some GAC algorithms such as MDDc [3]. The kIL conversion avoids this overhead because it maintains the tuple ordering of the input. For the 2IL any pair of constraints is joined only if their scopes share more than one variable. Unused dual variables are discarded from the output. For instance, consider the 2IL comprising of $c_1(x, y, z, v_1)$, $c_2(x, y, w, v_1)$, $c_3(w, z, v_3)$, where v_1, v_2, v_3 are the dual variables associated with c_1, c_2, c_3 and the rest are ordinary variables. Only v_1 and v_2 are joined to

[2] Available at http://www.cril.univ-artois.fr/CSC09. The modified renault problems with tables all positive are taken from http://becool.info.ucl.ac.be/resources/positive-table-constraints-benchmarks.

form a new constraint $c_4(v_1, v_2)$ as c_1 and c_2 share x and y. Because v_3 is not involved in any constraint it will be removed from $\mathcal{X}$ and c_3.

Table 1 shows the mean results on some series of benchmarks while Table 2 shows the results from selected instances. Five algorithms were tested: GACva [12], MDDc [3], STR3[10], STR2 [9], and an AbsCon's implementation[3] of FPWC based on STR, which we will call Fabs. Fabs is not eSTR2w [11] but can be regarded as a variant of eSTRw (or FPWC-STR^w). It exhibits a profile similar to that of eSTR2w when compared to STR2 on common benchmarks, except in a few cases where the difference in performance with respect to STR2 is noticeably smaller (e.g. *aim*) or larger (e.g. *rand-10-20-10*). Among the four GAC algorithms, GACva and STR3 are generally slower than MDDc and STR2 so for succinctness they do not appear in the main results in Table 1. Their performance on some representative instances can be seen in Table 2. All algorithms were run on the original instance and its two encodings for FPWC: the FE and the 2IL. There are three variants depending on how variables are handled during search. FE-O and 2IL-O (*pref-orig*) force the variable ordering heuristic to choose from the set of original variables until all of them are instantiated before choosing from the set of auxiliary (compound) variables. FE-A and 2IL-A (*pref-aux*) are the opposite, where preference is given to the auxiliary variables. FE-E and 2IL-E (*pref-equal*) give equal treatment to all variables with respect to $dom/ddeg$.

In both tables, tC gives the running time of the converters in seconds, while nV is the number of the variables, e.g. there are 100 ordinary variables in *a2*, 99 extra factor variables in its FE, and 127 extra dual variables in its 2IL. Best times and nodes are set in bold. A node count of zero means unsatisfiability is detected before the search starts. In Table 1, *SS* stands for solving strategy, the combination of GAC algorithm and encoding (if applicable). STR2 is the main GAC algorithm for solving various encodings unless specified otherwise (MDDc is ill-suited to the encodings as will be explained later). (#n) is the number of instances tested in the series. Instances that were not solved within 60 minutes by STR2 (the baseline) or exceeded the memory limit on any solving strategy are excluded, otherwise there is no time limit. Trivial instances (solved within 1 second) of the modified renault benchmark (*modRenault*) are also excluded (17 out of 50). The *mdd-r-n-d* series are randomly generated, based on the RD model [19], by building an MDD in a post-order manner with probability 0.5 that a previously created sub-MDD is reused [3]. The parameters are: arity (r), number of variables (n), and domain size (d). In Table 2, *ENC* is the encoding used. The columns GACva, MDDc, STR3, STR3, and Fabs give their running times. As Fabs reaches the same node count as GAC on the respective FE/2IL encoding, cells on these rows are left blank.

It is worth mentioning that $dom/ddeg$ may not produce the same search tree in both the original network and in its encoding, even when only the ordinary variables are instantiated. Because the dynamic degree counts the number of constraints in which there are at least two uninstantiated variables, the fact that an encoding's scopes have more variables may steer $dom/ddeg$ to pick a different variable than in the original problem. As a result, a weaker consistency may generate lower nodes than a stronger one (e.g. Fabs's node count on *r19* is lower than GAC's on the FE-O and the 2IL-O). Another source of the difference in node count lies in how AbsCon explicitly instantiates all

[3] Available in AbsCon 1.418.

variables, even the ones that are already singletons. For example, both encodings for *a2* are backtrack-free for GAC, but the node count for the FE is 199 because there are 99 more variables, while the 2IL's is 277 because there are 127 additional dual variables.

We make the following observations on these data:

– The FE's conversion time is mostly inconsequential compared to the solver's running time whereas the 2IL's can be a lot more expensive as it is based on the join of tables. The conversion time of the FE can be improved since we have not made an attempt to optimize our converter. For one thing, it always re-sorts relations before writing the output regardless of the solver's requirement.

– The fastest GAC on the original problem is either MDDc or STR2. The best variant of the FE largely improves the running time of STR2 and STR3, while it may help or hurt MDDc and F*abs*. The 2IL has mixed results and the conversion can be very slow due to the join. The FE clearly outperforms the 2IL on the same variant and benchmark, but enforcing MDDc on the original problems is frequently faster than any solving strategy (e.g. *rand-3-20-20* and the *fcd* variation). The implication here is that switching GAC algorithm may improve the running time better than equipping a GAC algorithm with stronger consistency. Experiments in previous works [11, 13, 16] neither considered MDDc nor included more than one GAC algorithm in the same study.

– For MDD compression, a larger scope is associated with lower probability of getting well-compacted MDDs. Any transformation which enlarges the scope may be unfavorable to MDDc. This is especially true with the kIL, which interferes directly with the compression by assigning different index to different tuples. The FE too causes the same problem, but to a lesser extent. However, the pruning from FPWC can more than compensate for this drawback in many cases (e.g. *dubois, dag-rand-1*), although it is not enough to win over STR2 on the same encoding. Since auxiliary variables are put in front of the scope, they will be placed on top of the MDDs by MDDc and this makes the pruning from FPWC more effective. By comparison, putting auxiliary variables in the back of the scope lessens the impact of FPWC to the point where running MDDc on an encoding is always worse off.

– Due to stronger consistency, maintaining F*abs* leads to a lower node count than maintaining GAC during search, but the lower number of nodes does not always translate to faster running time. F*abs* can be faster or slower than STR2. By contrast, all variants of the FE are faster than F*abs* although the node count can be higher.

– When a problem does not present an opportunity for additional pruning beyond GAC, running a stronger algorithm is counterproductive. Given that FPWC is both GAC and PWC, as the FE and the 2IL already builds in PWC propagation into the encoding, the portion of an FPWC algorithm that administers PWC becomes useless and simply incurs overhead when executed. Running an FPWC algorithm on the encoding therefore gets the same number of nodes as running any GAC algorithm on the encoding. It is interesting that the FE can make F*abs* faster in some cases. The reason is that F*abs* enforces only partial FPWC while the encoding provides complete FPWC. When F*abs*'s pruning capability happens to reach the level of complete FPWC on the original problem (i.e. its node count is already the lowest or not too much higher) running it on the FE would be slower (e.g. *r19, dag-rand-1*). Otherwise if F*abs*'s node count is considerably larger, that means there is still room for improvement and running F*abs* on the FE

Table 1. Mean results for selected benchmarks. T/O indicates the converter was timed out. M/O is out-of-memory failure. M stands for millions. For the mdd series, e is the number of constraints while t is the number of tuples in a relation

series	SS	tC	nV	time	nodes
rand-3-20-20-fcd	STR2	–	20	39.61	130,327
(#50)	MDDc	–	20	**21.00**	130,327
	F*abs*	–	20	34.28	37,727
	FE-O	0.21	+45	22.07	40,289
	FE-A	0.21	+45	32.94	71,568
	FE-E	0.21	+45	22.82	**36,195**
	2IL-O	2.83	+55	68.79	40,299
	2IL-A	2.83	+55	105.40	59,885
	2IL-E	2.83	+55	73.09	36,227
rand-3-20-20	STR2	–	20	83.27	256,958
(#50)	MDDc	–	20	**40.94**	256,958
	F*abs*	–	20	74.39	83,529
	FE-O	0.22	+45	41.76	74,825
	FE-A	0.22	+45	54.85	108,696
	FE-E	0.22	+45	42.53	**66,850**
	2IL-O	2.89	+55	130.38	74,830
	2IL-A	2.89	+55	269.64	137,301
	2IL-E	2.89	+55	129.68	66,853
dubois	STR2	–	71	528.45	100.27M
(#8)	MDDc	–	71	541.67	100.27M
	F*abs*	–	71	298.35	75.20M
	FE-O	0.00	+2	92.88	41.78M
STR2	FE-A	0.00	+2	198.42	66.85M
	FE-E	0.00	+2	63.91	16.71M
F*abs* +	FE-E	0.00	+2	**50.90**	16.71M
MDDc +	FE-E	0.00	+2	64.63	16.71M
	2IL-O	0.00	+4	96.74	41.78M
	2IL-A	0.00	+4	196.88	66.85M
	2IL-E	0.00	+4	73.79	16.71M
	4FE-O	0.08	+2	81.88	41.78M
	4FE-A	0.08	+2	194.64	66.85M
	4FE-E	0.08	+2	192.52	66.85M
	4IL-O	0.08	+47	89.80	29.25M
	4IL-A	0.08	+47	76.61	20.89M
	4IL-E	0.08	+47	65.68	**8.36M**
aim-200	STR2	–	200	45.54	637,085
(#6)	MDDc	–	200	32.95	637,085
	F*abs*	–	200	25.42	377,682
	FE-O	0.03	+354	3.85	32,354
	FE-A	0.03	+354	**0.85**	1,767
	FE-E	0.03	+354	2.96	11,397
	2IL-O	0.03	+551	4.36	32,552
	2IL-A	0.03	+551	3.45	22,968
	2IL-E	0.03	+551	3.72	11,594
	3FE-O	79.03	+354	1.39	5,561
	3FE-A	80.52	+354	2.79	19,002
	3FE-E	79.23	+354	1.00	1,434
	3IL-O	31.32	+769	34.73	4,854
	3IL-A	31.28	+769	8.32	690
	3IL-E	31.64	+769	7.73	**683**

instance	SS	tC	nV	time	nodes
rand-8-20-5	STR2	–	20	**12.50**	101,301
(#20)	MDDc	–	20	22.26	101,301
(2IL T/O)	F*abs*	–	20	32.74	18,709
	FE-O	5.87	+130	12.57	5,302
	FE-A	5.87	+130	22.64	**3,111**
	FE-E	5.87	+130	12.73	4,985
mdd-5-15-7	STR2	–	15	18.48	50,402
(#30)	MDDc	–	15	**5.95**	50,402
(e = 42)	F*abs*	–	15	36.26	3,996
(t = 8403)	FE-O	1.05	+175	11.80	1,569
(2IL M/O)	FE-A	1.05	+175	13.63	1,816
	FE-E	1.05	+175	12.33	**1,512**
mdd-7-25-4	STR2	–	25	79.18	231,364
(#10)	MDDc	–	25	**26.19**	231,364
(e = 50)	F*abs*	–	25	287.36	34,636
(t = 8192)	FE-O	1.85	+466	79.95	12,037
(2IL M/O)	FE-A	1.85	+466	71.90	39,366
	FE-E	1.85	+466	73.99	**10,523**
mdd-9-30-3	STR2	–	30	73.16	349,073
(#10)	MDDc	–	30	**39.00**	349,073
(e = 47)	F*abs*	–	30	396.88	66,109
(t = 9,841)	FE-O	2.80	+723	83.68	12,963
(2IL M/O)	FE-A	2.80	+723	79.79	23,578
	FE-E	2.80	+723	84.28	**10,603**
rand-10-20-10	STR2	–	20	0.64	830
(#20)	MDDc	–	20	2.06	830
	F*abs*	–	20	**0.60**	**0**
	FE-O	0.24	+10	0.69	**0**
	FE-A	0.24	+10	0.69	**0**
	FE-E	0.24	+10	0.70	**0**
	2IL-O	4.37	+5	1.41	**0**
	2IL-A	4.37	+5	1.45	**0**
	2IL-E	4.37	+5	1.45	**0**
dag-rand	STR2	–	23	17.48	57,969
(#25)	MDDc	–	23	123.83	57,969
	F*abs*	–	23	12.56	**0**
(2IL T/O)	FE-O	14.07	+120	9.45	**0**
	FE-A	14.07	+120	9.30	**0**
	FE-E	14.07	+120	**9.12**	**0**
modRenault	STR2	–	110	317.18	6.40M
(#12)	MDDc	–	110	295.45	6.40M
	F*abs*	–	110	2.19	**30**
	FE-O	0.71	+102	**1.19**	54
	FE-A	0.71	+102	1.22	58
	FE-E	0.71	+102	1.20	53
	2IL-O	81.16	+148	158.40	66
	2IL-A	81.45	+148	159.93	1023
	2IL-E	80.96	+148	159.14	2411

Table 2. Results from selected instances

instance	ENC	tC	nV	nodes	GAC*va*	MDDc	STR3	STR2	F*abs*	nodes
rand-3-20-20-60-632-19	None	–	20	252,803	49.15	**36.59**	82.39	73.05	64.94	**74,509**
(abbrv. as “r19”)	FE-O	0.22	+47	87,674	119.53	197.73	90.55	49.78	226.84	
	FE-A	0.22	+47	145,296	145.12	296.61	128.35	73.93	301.83	
	FE-E	0.22	+47	77,483	125.08	197.09	97.32	53.58	207.95	
	2IL-O	3.03	+57	87,674	258.08	341.64	222.87	143.26	218.48	
	2IL-A	3.03	+57	147,978	416.86	488.22	340.66	230.14	395.93	
	2IL-E	3.03	+57	77,483	245.40	355.73	222.59	156.22	243.80	
rand-3-20-20-60-632-26	None	–	20	442,871	74.65	67.24	147.71	127.00	35.34	29,765
(abbrv. as “r26”)	FE-O	0.21	+48	34,200	53.73	82.02	32.71	19.54	72.07	
	FE-A	0.21	+48	**23,498**	24.58	52.54	17.64	**10.95**	40.86	
	FE-E	0.21	+48	30,957	53.02	85.44	32.94	19.37	72.55	
	2IL-O	2.73	+57	34,209	117.34	159.62	97.49	62.93	93.67	
	2IL-A	2.73	+57	31,907	117.13	156.49	112.07	66.98	90.20	
	2IL-E	2.73	+57	30,966	117.26	159.61	98.58	63.29	92.36	
dag-rand-1	None	–	23	43,994	74.37	109.04	259.00	15.52	11.68	**0**
(2IL T/O)	FE-O	14.28	+120	**0**	18.44	15.39	12.30	**9.14**	20.57	
	FE-A	14.28	+120	**0**	18.66	15.53	12.54	9.24	21.91	
	FE-E	14.28	+120	**0**	18.74	15.65	13.79	9.91	20.47	
rand-8-20-5-18-800-7	None	–	20	11,063	4.75	4.94	32.95	**4.43**	8.81	980
(2IL T/O)	FE-O	5.71	+128	573	22.77	M/O	10.82	5.98	M/O	
	FE-A	5.71	+128	**177**	6.79	M/O	5.93	4.47	M/O	
	FE-E	5.71	+128	546	23.00	M/O	11.40	6.27	M/O	
aim-100-1-6-sat-2	None	–	100	95.79M	498.67	446.68	1015.90	577.23	163.12	23.11M
(abbrv. as “a2”)	FE-O	0.01	+99	**199**	**0.43**	0.50	0.48	0.45	0.47	
	FE-A	0.01	+99	**199**	0.44	0.50	0.46	**0.43**	0.45	
	FE-E	0.01	+99	**199**	0.44	0.49	0.46	0.46	0.47	
	2IL-O	0.00	+127	227	0.47	0.48	0.52	0.49	0.51	
	2IL-A	0.00	+127	227	0.47	0.48	0.49	0.48	0.52	
	2IL-E	0.00	+127	227	0.46	0.47	0.52	0.49	0.51	
mdd-5-15-7-inst-1	None	–	15	9,975	3.56	**2.13**	10.39	4.33	8.88	694
(2IL M/O)	FE-O	1.05	+190	594	16.58	91.49	5.84	4.11	52.74	
	FE-A	1.05	+190	1,383	35.54	201.29	12.10	9.33	155.91	
	FE-E	1.05	+190	**572**	16.50	88.07	6.01	4.35	52.40	

(or 2IL) could make it faster (e.g. *dubois, a2*). On *dubois*, the combination of F*abs* and FE-E is the fastest, offering an order-of-magnitude improvement over STR2.

– Variable preferences have a strong influence on the performance: the best can be twice as fast and/or halves the node count of the worst. Wide fluctuation also exists within the same series (e.g. in Table 2 FE-A is the best on *r26* but the worst on *r19*). Generally *pref-equal* has an advantage over *pref-orig*, while *pref-aux* is consistently the worst (FE-A on *aim* is the exception). This pattern holds for both the FE and the 2IL.

– As is the case with F*abs*, the node count of various encodings does not correlate well with the running time. However, too many overlapping constraints or factor variables clearly has an adverse effect on the running time. The three *mdd* series illustrate. As arity and number of variable increases, so does the number of overlapping constraints and factor variables. Keep in mind that the latter's number can be lower than the former's. For example, the instance *mdd-9-30-3-inst-1* has 47 relations, so the maximum number of intersecting constraints is $\binom{47}{2} = 1081$, whereas the actual number is 930 and the number of factor variables in the FE is 718. The ratio of the number of factor variables to the number of original variables goes from 11.67 for *mdd-5-15-7* to 18.64 for *mdd-7-25-4* to 24.1 for *mdd-9-30-3*. The ratio of Fabs's running time to STR2's increases accordingly from 1.96 to 3.63 to 5.42. The ratio of the FE's running time increases too, but at a lower pace of 0.54, 0.93, and 1.51 respectively. We also experimented with restricting the number of factor variables allowed in the FE for the mdd series but this does not improve the running time.

We have performed initial experiments with the FKWC and compare it with the kIL. For $k \geq 3$, [13] suggested the cycle heuristic to reduce the number of constraints: each constraint must share at least one variable with the previous and the next constraint in a circular manner. Our converters for the kIL and the FKWC employ this heuristic. Both the kIL and the FKWC are not practical beyond small k (3 or 4) since they are based on join which suffers from exponential growth in computation. The 3IL and 3FE are either timed out or ran out of memory on all the benchmarks in Table 1 except for *dubois* and *aim-200*. On *dubois*, no new constraint is created by the 3IL and the only constraints created by the 3FE are universal (where every combination of value is allowed) so they are useless and ignored. The 4FE does not improve on the FE. The 4IL is better than the 2IL and has the best node count but it is still slower than the FE. Similarly, the 3FE and the 3IL brings down the node count for *aim* but does not improve the running time. We also tried other benchmarks from the solver competition but most exceeded time or memory limit for conversion. Some benchmarks, such as *pret* or *ramsey*, produce only universal constraints for the 3FE. For the benchmarks that can be converted, we found the FKWC to be slower than the FE although the node count is lower.

7 Conclusion

We have introduced a new encoding for non-binary constraint networks that enables stronger consistencies to be acquired through GAC. Thus, this allows stronger consistencies to be incorporated into existing (state-of-the-art) CP solvers. Our experiments suggest FE to be the better method for achieving FPWC than both the 2IL and AbsCon's FPWC algorithm. Unlike specialized FPWC algorithms which are usually slower than GAC when there is little or no overlapping constraint, the preprocessors like the FE or the 2IL converter do not suffer from such computational overhead. Unlike the 2IL which joins constraints to achieve PWC, the FE is more precise and does not require any new constraint to be posted. As a converter, the FE benefits from flexibility: any solver using any GAC algorithm can be used as long as it is able to read the file in the specified format. At the same time, passing information to the solver this way can become a significant expense when large files are involved. Integrating the converter with the solver would eliminate this problem. As for the encodings for general kWC, we found they are not as effective as the FE. Similar to the kIL, the FKWC encoding has limited practical benefits due to the high cost of joins in both time and space and the need for good heuristics that pick only the useful pieces from the large number of possible joins. Success hinges on fine-tuning these heuristics and implementing better join algorithms. Constructing the *frj* constraints directly through search [8] instead of deriving them from join is also less expensive and could be examined in future works.

The MDDc algorithm is faster than STR2 on some sets of benchmarks but its performance on the FE is generally poor due to the factor variable's larger domains and the drop in compression rate as arity increases. Modifying the MDDc algorithm itself to make it aware of factor variables is a promising direction.

Acknowledgments. We thank Christophe Lecoutre for the permission to use AbsCon in our experiment. This work has been supported by grant MOE2012-T2-1-155.

References

1. Bacchus, F., Chen, X., van Beek, P., Walsh, T.: Binary vs. non-binary constraints. AIJ 140(1-2), 1–37 (2002)
2. Bessière, C., Stergiou, K., Walsh, T.: Domain filtering consistencies for non-binary constraints. Artificial Intelligence 172(6-7), 800–822 (2008)
3. Cheng, K.C.K., Yap, R.H.C.: An MDD-based generalized arc consistency algorithm for positive and negative table constraints and some global constraints. Constraints 15(2), 265–304 (2010)
4. Gent, I.P., Jefferson, C., Miguel, I., Nightingale, P.: Data structures for generalised arc consistency for extensional constraints. In: Proceedings of AAAI 2007, Vancouver, Canada, pp. 191–197 (2007)
5. Gyssens, M.: On the complexity of join dependencies. ACM Transactions on Database System 11(1), 81–108 (1986)
6. Janssen, P., Jegou, P., Nouguier, B., Vilarem, M.C.: A filtering process for general constraint-satisfaction problems: Achieving pairwise-consistency using an associated binary representation. In: Proceedings of IEEE Workshop on Tools for Artificial Intelligence, pp. 420–427 (1989)
7. Karakashian, S., Woodward, R., Choueiry, B.Y., Prestwich, S., Freuder, E.C.: A partial taxonomy of substitutability and interchangeability. In: CP 2010 Workshop on Symmetry in Constraint Satisfaction Problems (2010)
8. Karakashian, S., Woodward, R., Reeson, C., Choueiry, B.Y., Bessiere, C.: A first practical algorithm for high levels of relational consistency. In: Proceedings of AAAI 2010, pp. 101–107 (2010)
9. Lecoutre, C.: STR2: Optimized simple tabular reduction for table constraints. Constraints 16(4), 341–371 (2011)
10. Lecoutre, C., Likitvivatanavong, C., Yap, R.H.C.: A path-optimal GAC algorithm for table constraints. In: Proceedings of ECAI 2012, France, pp. 510–515 (2012)
11. Lecoutre, C., Paparrizou, A., Stergiou, K.: Extending STR to a higher-order consistency. In: Proceedings of AAAI 2013, Washington, U.S., pp. 576–582 (2013)
12. Lecoutre, C., Szymanek, R.: Generalized arc consistency for positive table constraints. In: Benhamou, F. (ed.) CP 2006. LNCS, vol. 4204, pp. 284–298. Springer, Heidelberg (2006)
13. Mairy, J.-B., Deville, Y., Lecoutre, C.: Domain k-wise consistency made as simple as generalized arc consistency. In: Simonis, H. (ed.) CPAIOR 2014. LNCS, vol. 8451, pp. 235–250. Springer, Heidelberg (2014)
14. Mairy, J.-B., Van Hentenryck, P., Deville, Y.: Optimal and efficient filtering algorithms for table constraints. Constraints 19(1), 77–120 (2014)
15. Merchez, S., Lecoutre, C., Boussemart, F.: AbsCon: a prototype to solve CSPs with abstraction. In: Walsh, T. (ed.) CP 2001. LNCS, vol. 2239, pp. 730–744. Springer, Heidelberg (2001)

16. Paparrizou, A., Stergiou, K.: An efficient higher-order consistency algorithm for table constraints. In: Proceedings of AAAI 2012, pp. 535–541 (2012)
17. Sabin, D., Freuder, E.C.: Contradicting conventional wisdom in constraint satisfaction. In: Borning, A. (ed.) PPCP 1994. LNCS, vol. 874, pp. 10–20. Springer, Heidelberg (1994)
18. Samaras, N., Stergiou, K.: Binary encoding of non-binary constraint satisfaction problems: Algorithms and experimental results. JAIR 24, 641–684 (2005)
19. Xu, K., Boussemart, F., Hemery, F., Lecoutre, C.: Random constraint satisfaction: easy generation of hard (satisfiable) instances. AIJ 171(8-9), 514–534 (2007)

Incremental QBF Solving*

Florian Lonsing and Uwe Egly

Vienna University of Technology
Institute of Information Systems
Knowledge-Based Systems Group
Vienna, Austria
http://www.kr.tuwien.ac.at/

Abstract. We consider the problem of incrementally solving a sequence of quantified Boolean formulae (QBF). Incremental solving aims at using information learned from one formula in the process of solving the next formulae in the sequence. Based on a general overview of the problem and related challenges, we present an approach to incremental QBF solving which is application-independent and hence applicable to QBF encodings of arbitrary problems. We implemented this approach in our incremental search-based QBF solver DepQBF and report on implementation details. Experimental results illustrate the potential benefits of incremental solving in QBF-based workflows.

1 Introduction

The success of SAT technology in practical applications is largely driven by *incremental solving*. SAT solvers based on conflict-driven clause learning (CDCL) [32] gather information about a formula in terms of learned clauses. When solving a sequence of closely related formulae, it is beneficial to keep clauses learned from one formula in the course of solving the next formulae in the sequence.

The logic of quantified Boolean formulae (QBF) extends propositional logic by universal and existential quantification of variables. QBF potentially allows for more succinct encodings of PSPACE-complete problems than SAT. Motivated by the success of incremental SAT solving, we consider the problem of incrementally solving a sequence of syntactically related QBFs in prenex conjunctive normal form (PCNF). Building on search-based QBF solving with clause and cube learning (QCDCL) [8,13,21,24,36], we present an approach to incremental QBF solving, which we implemented in our solver DepQBF.[1]

Different from many incremental SAT and QBF [27] solvers, DepQBF allows to add clauses to and delete clauses from the input PCNF in a stack-based way by *push* and *pop* operations. A related stack-based framework was implemented in the SAT solver PicoSAT [5]. A solver API with *push* and *pop* increases the usability from the perspective of a user. Moreover, we present an optimization based on this stack-based framework which reduces the size of the learned clauses.

* Supported by the Austrian Science Fund (FWF) under grant S11409-N23. We would like to thank Armin Biere and Paolo Marin for helpful discussions.

[1] DepQBF is free software: http://lonsing.github.io/depqbf/

B. O'Sullivan (Ed.): CP 2014, LNCS 8656, pp. 514–530, 2014.

Incremental QBF solving was introduced for QBF-based bounded model checking (BMC) of partial designs [26,27]. This approach, like ours, relies on selector variables and assumptions to support the deletion of clauses from the current input PCNF [1,11,20,28]. The quantifier prefixes of the incrementally solved PCNFs resulting from the BMC encodings are modified only at the left or right end. In contrast to that, we consider incremental solving of *arbitrary* sequences of PCNFs. For the soundness it is crucial to determine which of the learned clauses and cubes can be kept across different runs of an incremental QBF solver. We aim at a general presentation of incremental QBF solving and illustrate problems related to clause and cube learning. Our approach is *application-independent* and applicable to QBF encodings of *arbitrary* problems.

We report on experiments with constructed benchmarks. In addition to experiments with QBF-based conformant planning using DepQBF [12], our results illustrate the potential benefits of incremental QBF solving in application domains like synthesis [6,33], formal verification [4], testing [17,25,34], planning [9], and model enumeration [3], for example.

2 Preliminaries

We introduce terminology related to QBF and search-based QBF solving necessary to present a general view on incremental solving.

For a propositional variable x, $l := x$ or $l := \neg x$ is a *literal*, where $v(l) = x$ denotes the variable of l. A *clause* (*cube*) is a disjunction (conjunction) of literals. A *constraint* is a clause or a cube. The empty constraint $\emptyset$ does not contain any literals. A clause (cube) C is *tautological* (*contradictory*) if $x \in C$ and $\neg x \in C$.

A propositional formula is in *conjunctive (disjunctive) normal form* if it consists of a conjunction (disjunction) of clauses (cubes), called CNF (DNF). For simplicity, we regard CNFs and DNFs as sets of clauses and cubes, respectively.

A quantified Boolean formula (QBF) $\psi := \hat{Q}.\phi$ is in *prenex CNF (PCNF)* if it consists of a quantifier-free CNF ϕ and a *quantifier prefix* $\hat{Q}$ with $\hat{Q} := Q_1B_1 \ldots Q_nB_n$ where $Q_i \in \{\forall, \exists\}$ are *quantifiers* and B_i are *blocks* (i.e. sets) of variables such that $B_i \neq \emptyset$ and $B_i \cap B_j = \emptyset$ for $i \neq j$, and $Q_i \neq Q_{i+1}$.

The blocks in the quantifier prefix are *linearly ordered* such that $B_i < B_j$ if $i < j$. The linear ordering is extended to variables and literals: $x_i < x_j$ if $x_i \in B_i$, $x_j \in B_j$ and $B_i < B_j$, and $l < l'$ if $v(l) < v(l')$ for literals l and l'.

We consider only *closed* PCNFs, where every variable which occurs in the CNF is quantified in the prefix, and vice versa.

A variable $x \in B_i$ is *universal*, written as $q(x) = \forall$, if $Q_i = \forall$ and *existential*, written as $q(x) = \exists$, if $Q_i = \exists$. A literal l is universal if $q(v(l)) = \forall$ and existential if $q(v(l)) = \exists$, written as $q(l) := \forall$ and $q(l) := \exists$, respectively.

An *assignment* is a mapping from variables to the truth values *true* and *false*. An assignment A is represented as a set of literals $A := \{l_1, \ldots, l_k\}$ such that, for $l_i \in A$, if $v(l_i)$ is assigned to false (true) then $l_i = \neg v(l_i)$ ($l_i = v(l_i)$).

A *PCNF* ψ *under an assignment* A is denoted by $\psi[A]$ and is obtained from ψ as follows: for $l_i \in A$, if $l_i = v(l_i)$ ($l_i = \neg v(l_i)$) then all occurrences of $v(l_i)$ in

ψ are replaced by the syntactic truth constant $\top$ ($\bot$), respectively. All constants are eliminated from $\psi[A]$ by the usual simplifications of Boolean algebra and superfluous quantifiers and blocks are deleted from the quantifier prefix of $\psi[A]$. Given a cube C and a PCNF ψ, $\psi[C] := \psi[A]$ is the formula obtained from ψ under the assignment $A := \{l \mid l \in C\}$ defined by the literals in C.

The *semantics* of closed PCNFs is defined recursively. The QBF $\top$ is satisfiable and the QBF $\bot$ is unsatisfiable. The QBF $\psi = \forall(B_1 \cup \{x\}) \ldots Q_n B_n. \phi$ is satisfiable if $\psi[\neg x]$ and $\psi[x]$ are satisfiable. The QBF $\psi = \exists(B_1 \cup \{x\}) \ldots Q_n B_n. \phi$ is satisfiable if $\psi[\neg x]$ or $\psi[x]$ are satisfiable.

A PCNF ψ is *satisfied under an assignment* A if $\psi[A] = \top$ and *falsified* under A if $\psi[A] = \bot$. Satisfied and falsified clauses are defined analogously.

Given a constraint C, $L_Q(C) := \{l \in C \mid q(l) = Q\}$ for $Q \in \{\forall, \exists\}$ denotes the set of universal and existential literals in C. For a clause C, *universal reduction* produces the clause $UR(C) := C \setminus \{l \mid l \in L_\forall(C) \text{ and } \forall l' \in L_\exists(C) : l' < l\}$.

Q-resolution of clauses is a combination of resolution for propositional logic and universal reduction [7]. Given two non-tautological clauses C_1 and C_2 and a pivot variable p such that $q(p) = \exists$ and $p \in C_1$ and $\neg p \in C_2$. Let $C' := (UR(C_1) \setminus \{p\}) \cup (UR(C_2) \setminus \{\neg p\})$ be the *tentative Q-resolvent* of C_1 and C_2. If C' is non-tautological then it is the *Q-resolvent* of C_1 and C_2 and we write $C' = C_1 \otimes C_2$. Otherwise, C_1 and C_2 do not have a Q-resolvent.

Given a PCNF $\psi := \hat{Q}. \phi$, a *Q-resolution derivation* of a clause C from ψ is the successive application of Q-resolution and universal reduction to clauses in ψ and previously derived clauses resulting in C. We represent a derivation as a directed acyclic graph (DAG) with edges (1) $C'' \rightarrow C'$ if $C' = UR(C'')$ and (2) $C_1 \rightarrow C'$ and $C_2 \rightarrow C'$ if $C' = C_1 \otimes C_2$. We write $\hat{Q}.\phi \vdash C$ if there is a derivation of a clause C from ψ. Otherwise, we write $\hat{Q}.\phi \nvdash C$. Q-resolution is a sound and refutationally-complete proof system for QBFs [7]. A *Q-resolution proof* of an unsatisfiable PCNF ψ is a Q-resolution derivation of the empty clause.

3 Search-Based QBF Solving

We briefly describe search-based QBF solving with conflict-driven clause learning and solution-driven cube learning (QCDCL) [8,13,21,24,36] and related properties. In the context of *incremental* QBF solving, clause and cube learning requires a special treatment, which we address in Section 4.

Given a PCNF ψ, a QCDCL-based QBF solver successively assigns the variables to generate an assignment A. If ψ is falsified under A, i.e. $\psi[A] = \bot$, then a new learned clause C is derived by Q-resolution and added to ψ. If ψ is unsatisfiable, then finally the empty clause will be derived by clause learning. If ψ is satisfied under A, i.e. $\psi[A] = \top$, then a new learned *cube* is constructed based on the following *model generation rule, existential reduction* and *cube resolution.*

Definition 1 (Model Generation rule [13]). *Given a PCNF $\psi := \hat{Q}.\phi$, an assignment A such that $\psi[A] = \top$ is a* model[2] *of ψ. An* initial cube *$C = (\bigwedge_{l_i \in A} l_i)$ is a conjunction over the literals of a model A.*

[2] We adopted this definition of models from [21].

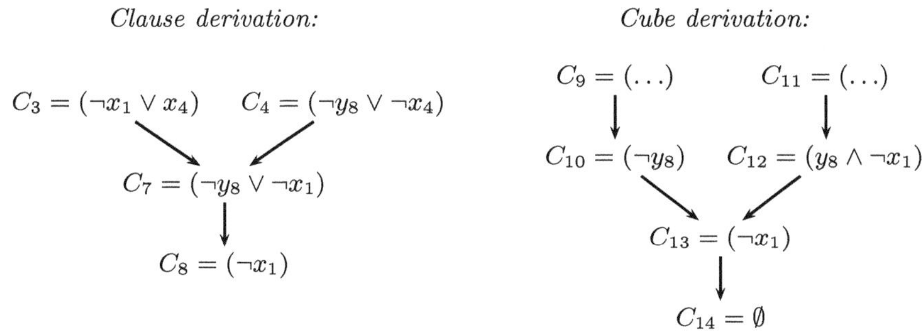

Fig. 1. Derivation DAGs of the clauses and cubes from Example 1. The literals in the initial cubes C_9 and C_{11} have been omitted in the figure to save space.

Definition 2 ([13]). *Given a cube C,* existential reduction *produces the reduced cube $ER(C) := C \setminus \{l \mid l \in L_\exists(C)$ and $\forall l' \in L_\forall(C) : l' < l\}$.*

Definition 3 (Cube Resolution [13,36]). *Given two non-contradictory cubes C_1 and C_2,* cube resolution *is defined analogously to Q-resolution for clauses, except that existential reduction is applied and the pivot variable must be universal. The cube resolvent of C_1 and C_2 (if it exists) is denoted by $C := C_1 \otimes C_2$.*

If ψ is satisfiable, then finally the empty cube will be derived by cube learning (Theorem 5 in [13]). Whereas in clause learning initially clauses of the input PCNF ψ can be resolved, in cube learning first initial cubes have to be generated by the model generation rule, which can then be used to produce cube resolvents. Similar to Q-resolution derivations (DAGs) of clauses and Q-resolution proofs, we define *cube resolution derivations* of cubes and *proofs of satisfiability.*

Example 1. Given the satisfiable PCNF $\psi := \exists x_1 \forall y_8 \exists x_5, x_2, x_6, x_4 . \phi$, where $\phi := \bigwedge_{i:=1,\ldots,6} C_i$ with $C_1 := (y_8 \vee \neg x_5)$, $C_2 := (x_2 \vee \neg x_6)$, $C_3 := (\neg x_1 \vee x_4)$, $C_4 := (\neg y_8 \vee \neg x_4)$, $C_5 := (x_1 \vee x_6)$, and $C_6 := (x_4 \vee x_5)$.

Figure 1 shows the derivation of the clauses $C_7 := C_3 \otimes C_4 = (\neg y_8 \vee \neg x_1)$ and $C_8 := UR(C_7) = (\neg x_1)$ by Q-resolution and universal reduction.

The assignment $A_1 := \{x_6, x_2, \neg y_8, \neg x_5, x_4\}$ is a model of ψ by Definition 1. Hence $C_9 := (x_6 \wedge x_2 \wedge \neg y_8 \wedge \neg x_5 \wedge x_4)$ is an initial cube. Existential reduction of C_9 produces the cube $C_{10} := ER(C_9) = (\neg y_8)$. Similarly, $A_2 := \{y_8, \neg x_4, \neg x_1, x_5, x_6, x_2\}$ is a model of ψ and $C_{11} := (y_8 \wedge \neg x_4 \wedge \neg x_1 \wedge x_5 \wedge x_6 \wedge x_2)$ is an initial cube. Existential reduction of C_{11} produces the cube $C_{12} := ER(C_{11}) = (y_8 \wedge \neg x_1)$. The cube $C_{13} := (\neg x_1)$ is obtained by resolving $C_{10} = (\neg y_8)$ and $C_{12} = (y_8 \wedge \neg x_1)$. Finally, existential reduction of C_{13} produces the empty cube $C_{14} := ER(C_{13}) = \emptyset$, which proves that the PCNF ψ is satisfiable.

A QCDCL-based solver implicitly constructs derivation DAGs in constraint learning. However, typically only selected constraints of these derivations are kept as learned constraints in an *augmented CNF* [36].

Definition 4. *Let $\psi := \hat{Q}.\phi$ be a PCNF. The* augmented CNF (ACNF) *of ψ has the form $\psi' := \hat{Q}.(\phi \wedge \theta \vee \gamma)$, where $\hat{Q}$ is the quantifier prefix, ϕ is the set of original clauses, θ is a CNF containing the learned clauses, and γ is a DNF containing the learned cubes obtained by clause and cube learning in QCDCL.*

Given an ACNF ψ' and an assignment A, the notation $\psi'[A]$ is defined similarly to PCNFs. Analogously to clause derivations, we write $\hat{Q}.\phi \vdash C$ if there is a derivation of a cube C from the PCNF $\hat{Q}.\phi$. During a run of a QCDCL-based solver the learned constraints can be derived from the current PCNF.

Proposition 1. *Let $\psi' := \hat{Q}.(\phi \wedge \theta \vee \gamma)$ be the ACNF obtained by QCDCL from a PCNF $\psi := \hat{Q}.\phi$. It holds that (1) $\forall C \in \theta : \hat{Q}.\phi \vdash C$ and (2) $\forall C \in \gamma : \hat{Q}.\phi \vdash C$.*

Proposition 1 follows from the correctness of constraint learning in *non-incremental* QCDCL. That is, we assume that the PCNF ψ is not modified over time. However, as we point out below, in *incremental* QCDCL the constraints learned previously might no longer be derivable after the PCNF has been modified.

Definition 5. *Given the ACNF $\psi' := \hat{Q}.(\phi \wedge \theta \vee \gamma)$ of the PCNF $\psi := \hat{Q}.\phi$, a clause $C \in \theta$ (cube $C \in \gamma$) is* derivable *with respect to ψ if $\psi \vdash C$. Otherwise, if $\psi \nvdash C$, then C is* non-derivable.

Due to the correctness of model generation, existential/universal reduction, and resolution, constraints which are derivable from the PCNF ψ can be added to the ACNF ψ' of ψ, which results in a satisfiability-equivalent ($\equiv_{sat}$) formula.

Proposition 2 ([13]). *Let $\psi' := \hat{Q}.(\phi \wedge \theta \vee \gamma)$ be the ACNF of the PCNF $\psi := \hat{Q}.\phi$. Then (1) $\hat{Q}.\phi \equiv_{sat} \hat{Q}.(\phi \wedge \theta)$ and (2) $\hat{Q}.\phi \equiv_{sat} \hat{Q}.(\phi \vee \gamma)$.*

4 Incremental Search-Based QBF Solving

We define *incremental QBF solving* as the problem of solving a sequence of PCNFs $\psi_0, \psi_1, \ldots, \psi_n$ using a QCDCL-based solver. Thereby, the goal is to not discard all the learned constraints after the PCNF ψ_i has been solved. Instead, to the largest extent possible we want to re-use the constraints that were learned from ψ_i in the process of solving the next PCNF ψ_{i+1}. To this end, the ACNF $\psi'_{i+1} = \hat{Q}_{i+1}.(\phi_{i+1} \wedge \theta_{i+1} \vee \gamma_{i+1})$ of ψ_{i+1} for $i > 0$, which is maintained by the solver, must be initialized with a set θ_{i+1} of learned clauses and a set γ_{i+1} of learned cubes such that $\theta_{i+1} \subseteq \theta_i$, $\gamma_{i+1} \subseteq \gamma_i$ and Proposition 2 holds with respect to ψ_{i+1}. The sets θ_i and γ_i contain the clauses and cubes that were learned from the previous PCNF ψ_i and potentially can be used to derive further constraints from ψ_{i+1}. If $\theta_{i+1} \neq \emptyset$ and $\gamma_{i+1} \neq \emptyset$ at the beginning, then the solver solves the PCNF ψ_{i+1} *incrementally*. For the first PCNF ψ_0 in the sequence, the solver starts with empty sets of learned constraints in the ACNF $\psi'_0 = \hat{Q}_0.(\phi_0 \wedge \theta_0 \vee \gamma_0)$.

Each PCNF ψ_{i+1} for $0 \leq i < n$ in the sequence $\psi_0, \psi_1, \ldots, \psi_n$ has the form $\psi_{i+1} = \hat{Q}_{i+1}.\phi_{i+1}$. The CNF part ϕ_{i+1} of ψ_{i+1} results from ϕ_i of the previous PCNF $\psi_i = \hat{Q}_i.\phi_i$ in the sequence by addition and deletion of clauses. We write

$\phi_{i+1} = (\phi_i \setminus \phi_{i+1}^{del}) \cup \phi_{i+1}^{add}$, where ϕ_{i+1}^{del} and ϕ_{i+1}^{add} are the sets of deleted and added clauses. The quantifier prefix $\hat{Q}_{i+1}$ of ψ_{i+1} is obtained from $\hat{Q}_i$ of ψ_i by deletion and addition of variables and quantifiers, depending on the clauses in ϕ_{i+1}^{add} and ϕ_{i+1}^{del}. That is, we assume that the PCNF ψ_{i+1} is closed and that its prefix $\hat{Q}_{i+1}$ does not contain superfluous quantifiers and variables.

When solving the PCNF ψ_i using a QCDCL-based QBF solver, learned clauses and cubes accumulate in the corresponding ACNF $\psi'_i = \hat{Q}_i.\,(\phi_i \wedge \theta_i \vee \gamma_i)$. Assume that the learned constraints are derivable with respect to ψ_i. The PCNF ψ_i is modified to obtain the next PCNF ψ_{i+1} to be solved. The learned constraints in θ_i and γ_i might become non-derivable with respect to ψ_{i+1} in the sense of Definition 5. Consequently, Proposition 2 might no longer hold for the ACNF $\psi'_{i+1} = \hat{Q}_{i+1}.\,(\phi_{i+1} \wedge \theta_{i+1} \vee \gamma_{i+1})$ of the new PCNF ψ_{i+1} if previously learned constraints from θ_i and γ_i appear in θ_{i+1} and γ_{i+1}. In this case, the solver might produce a wrong result when solving ψ_{i+1}.

4.1 Clause Learning

Assume that the PCNF $\psi_i = \hat{Q}_i.\,\phi_i$ has been solved and learned constraints have been collected in the ACNF $\psi'_i = \hat{Q}_i.\,(\phi_i \wedge \theta_i \vee \gamma_i)$. The clauses in ϕ_{i+1}^{del} are deleted from ϕ_i to obtain the CNF part $\phi_{i+1} = (\phi_i \setminus \phi_{i+1}^{del}) \cup \phi_{i+1}^{add}$ of the next PCNF $\psi_{i+1} = \hat{Q}_{i+1}.\,\phi_{i+1}$. If the derivation of a learned clause $C \in \theta_i$ depends on deleted clauses in ϕ_{i+1}^{del}, then we might have that $\psi_i \vdash C$ but $\psi_{i+1} \nvdash C$. In this case, C is non-derivable with respect to the next PCNF ψ_{i+1}. Hence C must be discarded before solving ψ_{i+1} starts so that $C \notin \theta_{i+1}$ in the initial ACNF $\psi'_{i+1} = \hat{Q}_{i+1}.\,(\phi_{i+1} \wedge \theta_{i+1} \vee \gamma_{i+1})$. Otherwise, if $C \in \theta_{i+1}$ then the solver might construct a bogus Q-resolution proof for the PCNF ψ_{i+1} and, if ψ_{i+1} is satisfiable, erroneously conclude that ψ_{i+1} is unsatisfiable.

Example 2. Consider the PCNF ψ from Example 1. The derivation of the clause $C_8 = (\neg x_1)$ shown in Fig. 1 depends on the clause $C_4 = (\neg y_8 \vee \neg x_4)$. We have that $\psi \vdash C_8$. Let ψ_1 be the PCNF obtained from ψ by deleting C_4. Then $\psi_1 \nvdash C_8$ because $C_3 = (\neg x_1 \vee x_4)$ is the only clause which contains the literal $\neg x_1$. Hence a possible derivation of the clause $C_8 = (\neg x_1)$ must use C_3. However, no such derivation exists in ψ_1. There is no clause C' containing a literal $\neg x_4$ which can be resolved with C_3 to produce $C_8 = (\neg x_1)$ after a sequence of resolution steps.

Consider the PCNF $\psi_{i+1} = \hat{Q}_{i+1}.\,\phi_{i+1}$ with $\phi_{i+1} = \phi_i \cup \phi_{i+1}^{add}$ which is obtained from $\hat{Q}_i.\,\phi_i$ by *only adding* the clauses ϕ_{i+1}^{add}, but not deleting any clauses. Assuming that $\hat{Q}_i.\phi_i \vdash C$ for all $C \in \theta_i$ in the ACNF $\psi'_i = \hat{Q}_i.\,(\phi_i \wedge \theta_i \vee \gamma_i)$, also $\hat{Q}_{i+1}.\,(\phi_i \cup \phi_{i+1}^{add}) \vdash C$. Hence all the learned clauses in θ_i are derivable with respect to the next PCNF ψ_{i+1} and can be added to the ACNF ψ'_{i+1}.

4.2 Cube Learning

Like above, let $\psi'_i = \hat{Q}_i.\,(\phi_i \wedge \theta_i \vee \gamma_i)$ be the ACNF of the previously solved PCNF $\psi_i = \hat{Q}_i.\,\phi_i$. Dual to clause deletions, the addition of clauses to ϕ_i can make learned cubes in γ_i non-derivable with respect to the next PCNF $\psi_{i+1} = \hat{Q}_{i+1}.\phi_{i+1}$ to be solved. The clauses in ϕ_{i+1}^{add} are added to ϕ_i to obtain the CNF part $\phi_{i+1} = (\phi_i \setminus \phi_{i+1}^{del}) \cup \phi_{i+1}^{add}$ of ψ_{i+1}. An initial cube $C \in \gamma_i$ has been obtained from a model A of the previous PCNF ψ_i, i.e. $\psi_i[A] = \top$. We might have that $\psi_{i+1}[A] \neq \top$ with respect to the next PCNF ψ_{i+1} because of an added clause $C' \in \phi_{i+1}^{add}$ (and hence also $C' \in \phi_{i+1}$) such that $C'[A] \neq \top$. Therefore, A is not a model of ψ_{i+1} and the initial cube C is non-derivable with respect to ψ_{i+1}, i.e. $\hat{Q}_i.\phi_i \vdash C$ but $\hat{Q}_{i+1}.\phi_{i+1} \nvdash C$. Hence C and every cube whose derivation depends on C must be discarded to prevent the solver from generating a bogus cube resolution proof for ψ_{i+1}. If ψ_{i+1} is unsatisfiable, then the solver might erroneously conclude that ψ_{i+1} is satisfiable. That is, Proposition 2 might not hold with respect to non-derivable cubes and the ACNF ψ'_{i+1} of ψ_{i+1}.

Example 3. Consider the PCNF ψ from Example 1. The derivation of the cube $C_{10} = (\neg y_8)$ shown in Fig. 1 depends on the initial cube $C_9 = (x_6 \wedge x_2 \wedge \neg y_8 \wedge \neg x_5 \wedge x_4)$, which has been generated from the model $A_1 = \{x_6, x_2, \neg y_8, \neg x_5, x_4\}$. The cube C_9 is derivable with respect to ψ since $\psi[A_1] = \top$, and hence $\psi \vdash C_9$. The cube C_{10} is also derivable since $C_{10} = ER(C_9)$. Assume that the clause $C_0 := (\neg x_2 \vee \neg x_4)$ is added to ψ resulting in the unsatisfiable PCNF ψ_2. Now C_9 is non-derivable with respect to ψ_2 since $C_0[A_1] = \bot$. Further, $\psi_2 \nvdash C_{10}$.

Consider the PCNF $\psi_{i+1} = \hat{Q}_{i+1}.\phi_{i+1}$ with $\phi_{i+1} = \phi_i \setminus \phi_{i+1}^{del}$ which is obtained from $\hat{Q}_i.\phi_i$ by *only deleting* the clauses ϕ_{i+1}^{del}, but not adding any clauses. If after the clause deletions some variable x does not occur anymore in the resulting PCNF ψ_{i+1}, then x is removed from the quantifier prefix of ψ_{i+1} and from every cube $C \in \gamma_i$ which was learned when solving the previous PCNF ψ_i. Proposition 2 holds for the cleaned up cubes $C' = C \setminus \{l \mid v(l) = x\}$ for all $C \in \gamma_i$ with respect to ψ_{i+1} and hence C' can be added to the ACNF ψ'_{i+1}.

Proposition 3. *Let $\psi'_i := \hat{Q}_i.(\phi_i \wedge \theta_i \vee \gamma_i)$ be the ACNF of the PCNF $\psi_i := \hat{Q}_i.\phi_i$. Let $\psi_{i+1} := \hat{Q}_{i+1}.\phi_{i+1}$ be the PCNF resulting from ψ_i with $\phi_{i+1} = (\phi_i \setminus \phi_{i+1}^{del})$, where the variables V_{i+1}^{del} no longer occur in ϕ_{i+1} and are removed from $\hat{Q}_i$ to obtain $\hat{Q}_{i+1}$. Given a cube $C \in \gamma_i$, let $C' := C \setminus \{l \mid v(l) \in V_{i+1}^{del}\}$. Proposition 2 holds for C' with respect to $\hat{Q}_{i+1}.\phi_{i+1}$: $\hat{Q}_{i+1}.\phi_{i+1} \equiv_{sat} \hat{Q}_{i+1}.(\phi_{i+1} \vee C')$.*

Proof (Sketch). By induction on the structure of the derivations of cubes in γ_i.

Let $C \in \gamma_i$ be an initial cube due to the assignment A with $\psi_i[A] = \top$. For $A' := A \setminus \{l \mid v(l) \in V_{i+1}^{del}\}$, we have $\psi_{i+1}[A'] = \top$ since all the clauses containing the variables in V_{i+1}^{del} were deleted from ψ_i to obtain ψ_{i+1}. Then the claim holds for the initial cube $C' = C \setminus \{l \mid v(l) \in V_{i+1}^{del}\} = (\bigwedge_{l_i \in A'} l_i)$ since $\psi_{i+1} \vdash C'$.

Let $C \in \gamma_i$ be obtained from $C_1 \in \gamma_i$ by existential reduction such that $C = ER(C_1)$. Assuming that the claim holds for $C'_1 = C_1 \setminus \{l \mid v(l) \in V_{i+1}^{del}\}$, it also holds for $C' = C \setminus \{l \mid v(l) \in V_{i+1}^{del}\} = ER(C'_1)$ since existential reduction removes existential literals which are maximal with respect to the prefix ordering.

Let $C \in \gamma_i$ be obtained from $C_1, C_2 \in \gamma_i$ by resolution on variable x with $x \in C_1$, $\neg x \in C_2$. Assume that the claim holds for $C'_1 = C_1 \setminus \{l \mid v(l) \in V_{i+1}^{del}\}$

and $C'_2 = C_2 \setminus \{l \mid v(l) \in V^{del}_{i+1}\}$, i.e. $\hat{Q}_{i+1}.\phi_{i+1} \equiv_{sat} \hat{Q}_{i+1}.(\phi_{i+1} \vee C'_1)$ and $\hat{Q}_{i+1}.\phi_{i+1} \equiv_{sat} \hat{Q}_{i+1}.(\phi_{i+1} \vee C'_2)$. If $x \notin V^{del}_{i+1}$ then the claim also holds for $C' = C \setminus \{l \mid v(l) \in V^{del}_{i+1}\} = C'_1 \otimes C'_2$ with $x \in C'_1$, $\neg x \in C'_2$ due to the correctness of resolution (Proposition 2). If $x \in V^{del}_{i+1}$ then the claim also holds for $C' = C \setminus \{l \mid v(l) \in V^{del}_{i+1}\} = (C'_1 \wedge C'_2)$ since $\{y, \neg y\} \not\subseteq (C'_1 \cup C'_2)$ for all variables y, which can be proved by reasoning with tree-like models of QBFs [30]. □

If a variable x no longer occurs in the formula, then cubes where x has been removed might become non-derivable. However, due to Propositions 2 and 3 it is sound to keep all the cleaned up cubes (resolution is not inferentially-complete). Moreover, due to the correctness of resolution and existential reduction, Proposition 2 also holds for new cubes derived from the cleaned up cubes.

In practice, the goal is to keep as many learned constraints as possible because they prune the search space and can be used to derive further constraints. Therefore, subsets $\theta_{i+1} \subseteq \theta_i$ and $\gamma_{i+1} \subseteq \gamma_i$ of the learned clauses θ_i and cubes γ_i must be selected so that Proposition 2 holds with respect to the initial ACNF $\psi'_{i+1} = \hat{Q}_{i+1}.(\phi_{i+1} \wedge \theta_{i+1} \vee \gamma_{i+1})$ of the PCNF ψ_{i+1} to be solved next.

5 Implementing an Incremental QBF Solver

We describe the implementation of our incremental QCDCL-based solver DepQBF. Our approach is general and fits any QCDCL-based solver. For incremental solving we do not apply a sophisticated analysis of variable dependencies by dependency schemes in DepQBF [22]. Instead, as many other QBF solvers, we use the linear ordering given by the quantifier prefix. We implemented a stack-based representation of the CNF part of PCNFs based on selector variables and assumptions. Assumptions were also used for incremental QBF-based BMC of partial designs [27] and are common in incremental SAT solving [1,11,20,28].

We address the problem of checking which learned constraints can be kept across different solver runs after the current PCNF has been modified. To this end, we present approaches to check if a constraint learned from the previous PCNF is still derivable from the next one, which makes sure that Proposition 2 holds. Similar to incremental SAT solving, selector variables are used to handle the learned clauses. Regarding learned cubes, selector variables can also be used (although in a way asymmetric to clauses), in addition to an alternative approach relying on full derivation DAGs, which have to be kept in memory. Learned cubes might become non-derivable by the deletion of clauses and superfluous variables, but still can be kept due to Proposition 3. We implemented a simple approach which, after clauses have been added to the formula, allows to keep only initial cubes but not cubes obtained by resolution or existential reduction.

5.1 QBF Solving under Assumptions

Let $\psi := Q_1B_1Q_2B_2 \dots Q_nB_n.\phi$ be a PCNF. We define a set $A := \{l_1, \dots, l_k\}$ of *assumptions* as an assignment such that $v(l_i) \in B_1$ for all literals $l_i \in A$. The variables assigned by A are from the first block B_1 of ψ. Solving the PCNF ψ *under the set A of assumptions* amounts to solving the PCNF $\psi[A]$. The definition of assumptions can be applied recursively to the PCNF $\psi[A]$. If A assigns all the variables in B_1, then variables from B_2 can be assigned as assumptions with respect to $\psi[A]$, since B_2 is the first block in the quantifier prefix of $\psi[A]$.

We implemented the handling of assumptions according to the *literal-based single instance (LS)* approach (in the terminology of [28]). Thereby, the assumptions in A are treated in a special way so that the variables in A are never selected as pivots in the resolution derivation of a learned constraint according to QCDCL-based learning. Similar to SAT-solving under assumptions, LS allows to keep all the constraints that were learned from the PCNF $\psi[A]$ under a set A of assumptions when later solving $\psi[A']$ under a different set A' of assumptions.

5.2 Stack-Based CNF Representation

In DepQBF, the CNF part ϕ of an ACNF $\psi'_i = \hat{Q}_i.\,(\phi_i \wedge \theta_i \vee \gamma_i)$ to be solved is represented as a stack of clauses. The clauses on the stack are grouped into *frames*. The solver API provides functions to push new frames onto the stack, pop present frames from the stack, and to add new clauses to the current topmost frame. Each *push* operation opens a new topmost frame f_j. New clauses are always added to the topmost frame f_j. Each new frame f_j opened by a *push* operation is associated with a fresh *frame selector variable* s_j. Frame selector variables are existentially quantified and put into a separate, leftmost quantifier block B_0 i.e. the current ACNF ψ'_i has the form $\psi'_i = \exists B_0 \hat{Q}_i.\,(\phi_i \wedge \theta_i \vee \gamma_i)$. Before a new clause C is added to frame f_j, the frame selector variable s_j of f_j is inserted into C so that in fact the clause $C' = C \cup \{s_j\}$ is added to f_j. If all the selector variables are assigned to *false* then under that assignment every clause $C' = C \cup \{s_j\}$ is syntactically equivalent to C.

The purpose of the frame selector variables is to *enable* or *disable* the clauses in the CNF part ϕ_i with respect to the *push* and *pop* operations applied to the clause stack. If the selector variable s_j of a frame f_j is assigned to *true* then all the clauses of f_j are satisfied under that assignment. In this case, these satisfied clauses are considered disabled because they can not be used to derive new learned clauses in QCDCL. Otherwise, the assignment of *false* to s_j does not satisfy any clauses in f_j. Therefore these clauses are considered enabled.

Before the solving process starts, the clauses of frames popped from the stack are disabled and the clauses of frames still on the stack are enabled by assigning the selector variables to *true* and *false*, respectively. The selector variables are assigned as assumptions. This is possible because these variables are in the leftmost quantifier block B_0 of the ACNF $\psi'_i = \exists B_0 \hat{Q}_i.\,(\phi_i \wedge \theta_i \vee \gamma_i)$ to be solved.

The idea of enabling and disabling clauses by selector variables and assumptions originates from incremental SAT solving [11]. This approach was also applied to bounded model checking of partial designs by incremental QBF solving [27]. In DepQBF, we implemented the *push* and *pop* operations related to the clause stack by selector variables similarly to the SAT solver PicoSAT [5].

In the implementation of DepQBF, frame selector variables are maintained entirely by the solver. Depending on the *push* and *pop* operations, selector variables are automatically inserted into added clauses and assigned as assumptions. This approach saves the user the burden of inserting selector variables manually into the QBF encoding of a problem and assigning them as assumptions via the solver API. Manual insertion is typically applied in incremental SAT solving based on assumptions as pioneered by MiniSAT [10,11]. We argue that the usability of an incremental QBF solver is improved considerably if the selector variables are maintained by the solver. For example, from the perspective of the user, the QBF encoding contains only variables relevant to the encoded problem.

In the following, we consider the problem of maintaining the sets of learned constraints across different solver runs. As pointed out in Section 4, Proposition 2 still holds for learned clauses (cubes) after the addition (deletion) of clauses to (from) the PCNF. Therefore, we present the maintenance of learned constraints separately for clause additions and deletions.

5.3 Handling Clause Deletions

A clause $C \in \theta_i$ in the current ACNF $\psi'_i = \hat{Q}_i. (\phi_i \wedge \theta_i \vee \gamma_i)$ might become non-derivable if its derivation depends on clauses in ϕ^{del}_{i+1} which are deleted to obtain the CNF part $\phi_{i+1} = (\phi_i \setminus \phi^{del}_{i+1}) \cup \phi^{add}_{i+1}$ of the next PCNF ψ_{i+1}.

In DepQBF, learned clauses in θ_i are deleted as follows. As pointed out in the previous section, clauses of popped off frames are disabled by assigning the respective frame selector variables to *true*. Since the formula contains only positive literals of selector variables, these variables cannot be chosen as pivots in derivations. Therefore, learned clauses whose derivations depend on disabled clauses of a popped off frame f_j contain the selector variable s_j of f_j. Hence these learned clauses are also disabled by the assignment of s_j. This approach to handling learned clauses is also applied in incremental SAT solving [11].

The disabled clauses are physically deleted in a garbage collection phase if their number exceeds a certain threshold. Variables which no longer occur in the CNF part of the current PCNF are removed from the quantifier prefix and, by Proposition 3, from learned cubes in γ_i to produce cleaned up cubes. We initialize the set γ_{i+1} of learned cubes in the ACNF $\psi'_{i+1} = \hat{Q}_{i+1}. (\phi_{i+1} \wedge \theta_{i+1} \vee \gamma_{i+1})$ of the next PCNF ψ_{i+1} to be solved to contain the cleaned up cubes.

The deletion of learned clauses based on selector variables is not optimal in the sense of Definition 5. There might be another derivation of a disabled learned clause C which does not depend on the deleted clauses ϕ^{del}_{i+1}. This observation also applies to the use of selector variables in incremental SAT solving.

As illustrated in the context of incremental SAT solving, the size of learned clauses might increase considerably due to the additional selector variables [1,20]. In the stack-based CNF representation of DepQBF, the clauses associated to a frame f_j all contain the selector variable s_j of f_j. Therefore, the maximum number of selector variables in a new clause learned from the current PCNF ψ_i is bounded by the number of currently enabled frames. The sequence of *push* operations introduces a linear ordering $f_0 < f_1 < \ldots < f_k$ on the enabled frames

f_i and their clauses in the CNF with respect to the point of time where that frames and clauses have been added. In DepQBF, we implemented the following optimization based on this temporal ordering. Let C and C' be clauses which are resolved in the course of clause learning. Assume that $s_i \in C$ and $s_j \in C'$ are the only selector variables of currently enabled frames f_i and f_j in C and C'. Instead of computing the usual Q-resolvent $C'' := C \otimes C'$, we compute $C'' := (C \otimes C') \setminus \{l \mid l = s_i \text{ if } f_i < f_j \text{ and } l = s_j \text{ otherwise}\}$. That is, the selector variable of the frame which is smaller in the temporal ordering is discarded from the resolvent. If $f_i < f_j$ then the clauses in f_i were pushed onto the clause stack before the clauses in f_j. The frame f_j will be popped off the stack before f_i. Therefore, in order to properly disable the learned clause C'' after *pop* operations, it is sufficient to keep the selector variable s_j of the frame f_j in C''. With this optimization, *every* learned clause contains *exactly one* selector variable. In the SAT solver PicoSAT, an optimization which has similar effects is implemented.

5.4 Handling Clause Additions

Assume that the PCNF $\psi_i := \hat{Q}_i.\,\phi_i$ has been solved and that all learned constraints in the ACNF $\psi'_i = \hat{Q}_i.\,(\phi_i \wedge \theta_i \vee \gamma_i)$ of ψ_i are derivable with respect to ψ_i. The set ϕ^{add}_{i+1} of clauses is added to ϕ_i to obtain the CNF part $\phi_{i+1} = (\phi_i \setminus \phi^{del}_{i+1}) \cup \phi^{add}_{i+1}$ of the next PCNF $\psi_{i+1} = \hat{Q}_{i+1}.\,\phi_{i+1}$. For learned clauses, we can set $\theta_{i+1} := \theta_i$ in the ACNF $\psi'_{i+1} = \hat{Q}_{i+1}.\,(\phi_{i+1} \wedge \theta_{i+1} \vee \gamma_{i+1})$ of ψ_{i+1}. The following example illustrates the effects of adding ϕ^{add}_{i+1} on the cubes.

Example 4. Consider the cube derivation shown in Fig. 1. As illustrated in Example 3, the cubes $C_9 = (x_6 \wedge x_2 \wedge \neg y_8 \wedge \neg x_5 \wedge x_4)$ and $C_{10} = (\neg y_8)$ are non-derivable with respect to the PCNF ψ_2 obtained from ψ by adding the clause $C_0 := (\neg x_2 \vee \neg x_4)$. The initial cube $C_{11} := (y_8 \wedge \neg x_4 \wedge \neg x_1 \wedge x_5 \wedge x_6 \wedge x_2)$ still is derivable because the underlying model $A_2 := \{y_8, \neg x_4, \neg x_1, x_5, x_6, x_2\}$ of ψ is also a model of ψ_2. Therefore, when solving ψ_2 we can keep the derivable cubes C_{11} and $C_{12} = ER(C_{11})$. The non-derivable cubes C_9 and C_{10} must be discarded. Otherwise, QCDCL might produce the cube resolution proof shown in Fig. 1 when solving the *unsatisfiable* PCNF ψ_2, which is incorrect.

We sketch an approach to identify the cubes in a cube derivation DAG G which are non-derivable with respect to the next PCNF $\psi_{i+1} = \hat{Q}_{i+1}.\,\phi_{i+1}$. Starting at the initial cubes, G is traversed in a topological order. An initial cube C is marked as derivable if $\psi_{i+1}[C] = \top$, otherwise if $\psi_{i+1}[C] \neq \top$ then C is marked as non-derivable. This test can be carried out syntactically by checking whether every clause of ψ_{i+1} is satisfied under the assignment given by C. A cube C obtained by existential reduction or cube resolution is marked as derivable if all its predecessors in G are marked as derivable. Otherwise, C is marked as non-derivable. Finally, all cubes in G marked as non-derivable are deleted.

The above procedure allows to find a subset $\gamma_{i+1} \subseteq \gamma_i$ of the set γ_i of cubes in the solved ACNF $\psi'_i = \hat{Q}_i.\,(\phi_i \wedge \theta_i \vee \gamma_i)$ so that all cubes in γ_{i+1} are derivable and Proposition 2 holds for the next ACNF $\psi'_{i+1} = \hat{Q}_{i+1}.\,(\phi_{i+1} \wedge \theta_{i+1} \vee \gamma_{i+1})$.

However, this procedure is not optimal because it might mark a cube $C \in G$ as non-derivable with respect to the next PCNF ψ_{i+1} although $\psi_{i+1} \vdash C$.

Example 5. Given the satisfiable PCNF $\psi := \exists x_1 \forall y_8 \exists x_5, x_2, x_6, x_4. \phi$, where $\phi := \bigwedge_{i:=1,\ldots,5} C_i$ with the clauses C_i from Example 1 where $C_1 := (y_8 \vee \neg x_5)$, $C_2 := (x_2 \vee \neg x_6)$, $C_3 := (\neg x_1 \vee x_4)$, $C_4 := (\neg y_8 \vee \neg x_4)$, $C_5 := (x_1 \vee x_6)$. Consider the model $A_3 := \{\neg x_1, y_8, \neg x_5, x_2, x_6, \neg x_4\}$ of ψ and the initial cube $C_{15} := (\neg x_1 \wedge y_8 \wedge \neg x_5 \wedge x_2 \wedge x_6 \wedge \neg x_4)$ generated from A_3. Existential reduction of C_{15} produces the cube $C_{16} := ER(C_{15}) = (\neg x_1 \wedge y_8)$. Assume that the clause $C_0 := (x_4 \vee x_5)$ is added to ψ to obtain the PCNF ψ_3. The initial cube C_{15} is non-derivable with respect to ψ_3 since $C_0[A_3] \neq \top$. However, for the cube C_{16} derived from C_{15} it holds that $\psi_3 \vdash C_{16}$. The assignment $A_4 := \{\neg x_1, y_8, x_5, x_2, x_6, \neg x_4\}$ is a model of ψ_3. Let $C_{17} := (\neg x_1 \wedge y_8 \wedge x_5 \wedge x_2 \wedge x_6 \wedge \neg x_4)$ be the initial cube generated from A_4. Then $C_{16} = ER(C_{17})$ is derivable with respect to ψ_3.

In practice, QCDCL-based solvers typically store only the learned cubes, which might be a small part of the derivation DAG G, and no edges. Therefore, checking the cubes in a traversal of G is not feasible. Even if the full DAG G is available, the checking procedure is not optimal as pointed out in Example 5. Furthermore, it cannot be used to check cubes which have become non-derivable after cleaning up by Proposition 3. Hence, it is desirable to have an approach to checking the derivability of *individual* learned cubes which is independent from the derivation DAG G. To this end, we need a condition which is sufficient to conclude that some *arbitrary* cube C is derivable with respect to a PCNF ψ, i.e. to check whether $\psi \vdash C$. However, we are not aware of such a condition.

As an alternative to keeping the full derivation DAG in memory, a *fresh* selector variable can be added to *each* newly learned initial cube. Similar to selector variables in clauses, these variables are transferred to all derived cubes. Potentially non-derivable cubes are then disabled by assigning the selector variables accordingly. However, different from clauses, it must be checked *explicitly* which initial cubes are non-derivable by checking the condition in Definition 1 for all initial cubes in the set γ_i of learned cubes. This amounts to an asymmetric treatment of selector variables in clauses and cubes. Clauses are added to and removed from the CNF part by *push* and *pop* operations provided by the solver API. This way, it is known precisely which clauses are removed. In contrast to that, cubes are added to the set of learned cubes γ_i on the fly during cube learning. Moreover, the optimization based on the temporal ordering of selector variables from the previous section is not applicable to generate shorter cubes since cubes are not associated to stack frames.

Due to the complications illustrated above, we implemented the following simple approach in DepQBF to keep only initial cubes. Every initial cube computed by the solver is stored in a linked list L of bounded capacity, which is increased dynamically. The list L is separate from the set of learned clauses. Assume that a set ϕ_{i+1}^{add} of clauses is added to the CNF part ϕ_i of the current PCNF to obtain the CNF part $\phi_{i+1} = (\phi_i \setminus \phi_{i+1}^{del}) \cup \phi_{i+1}^{add}$ of the next PCNF $\psi_{i+1} = \hat{Q}_{i+1}. \phi_{i+1}$. All the cubes in the current set γ_i of learned cubes are discarded. For every added clause $C \in \phi_{i+1}^{add}$ and for every initial cube $C' \in L$, it is checked whether the assignment A given by C' is a model of the next PCNF ψ_{i+1}. Initial cubes C' for which this check succeeds are added to the set γ_{i+1} of learned cubes in the ACNF ψ'_{i+1} of the next PCNF ψ_{i+1} after existential reduction has been applied to them. If the check fails, then C' is removed from L. It suffices to check the initial cubes in L only with respect to the clauses $C \in \phi_{i+1}^{add}$, and not the full CNF part ϕ_{i+1}, since the assignments given by the cubes in L are models of the *current* PCNF ψ_i. In the end, the set γ_{i+1} contains only initial cubes all of which are derivable with respect to the ACNF ψ'_{i+1}. If clauses are removed from the formula, then by Proposition 3 variables which do not occur anymore in the formula are removed from the initial cubes in L.

In the incremental QBF-based approach to BMC for partial designs [26,27], all cubes are kept across different solver calls under the restriction that the quantifier prefix is modified only at the left end. This restriction does not apply to incremental solving of PCNF where the formula can be modified arbitrarily.

5.5 Incremental QBF Solver API

The API of DepQBF [23] provides functions to manipulate the prefix and the CNF part of the current PCNF. Clauses are added and removed by the *push* and *pop* operations described in Section 5.2. New quantifier blocks can be added at any position in the quantifier prefix. New variables can be added to any quantifier block. Variables which no longer occur in the formula and empty quantifier blocks can be explicitly deleted. The quantifier block B_0 containing the frame selector variables is invisible to the user. The solver maintains the learned constraints as described in Sections 5.3 and 5.4 without any user interaction.

The *push* and *pop* operations are a feature of DepQBF. Additionally, the API supports the manual insertion of selector variables into the clauses by the user. Similar to incremental SAT solving [11], clauses can then be enabled and disabled manually by assigning the selector variables as assumptions via the API. In this case, these variables are part of the QBF encoding and the optimization based on the frame ordering presented in Section 5.3 is not applicable. After a PCNF has been found unsatisfiable (satisfiable) under assumptions where the leftmost quantifier block is existential (universal), the set of relevant assumptions which were used by the solver to determine the result can be extracted.[3]

6 Experimental Results

To demonstrate the basic feasibility of general incremental QBF solving, we evaluated our incremental QBF solver DepQBF based on the instances from *QBFEVAL'12 Second Round (SR)* with and without preprocessing by Bloqqer.[4] We disabled the sophisticated dependency analysis in terms of dependency schemes in DepQBF and instead applied the linear ordering of the quantifier prefix in the

[3] This is similar to the function "analyzeFinal" in MiniSAT, for example.

[4] `http://www.kr.tuwien.ac.at/events/qbfgallery2013/benchmarks/`.

Table 1. Average and median number of assignments ($\overline{a}$ and $\tilde{a}$, respectively), backtracks ($\overline{b}, \tilde{b}$), and wall clock time ($\overline{t}, \tilde{t}$) in seconds on *sequences* $S = \psi_0, \ldots, \psi_{10}$ of PCNFs which were fully solved by DepQBF both if all learned constraints are discarded (*discard LC*) and if constraints which are correct in the sense of Propositions 2 and 3 are kept (*keep LC*). Clauses are *added* to ψ_i to obtain ψ_{i+1} in S.

	QBFEVAL'12-SR		
	discard LC	*keep LC*	*diff.(%)*
$\overline{a}$:	29.37×10^6	26.18×10^6	-10.88
$\tilde{a}$:	3,833,077	2,819,492	-26.44
$\overline{b}$:	139,036	116,792	-16.00
$\tilde{b}$:	8,243	6,360	-22.84
$\overline{t}$:	99.03	90.90	-8.19
$\tilde{t}$:	28.56	15.74	-44.88

	QBFEVAL'12-SR-Bloqqer		
	discard LC	*keep LC*	*diff.(%)*
$\overline{a}$:	39.75×10^6	34.03×10^6	-14.40
$\tilde{a}$:	1.71×10^6	1.65×10^6	-3.62
$\overline{b}$:	117,019	91,737	-21.61
$\tilde{b}$:	10,322	8,959	-13.19
$\overline{t}$:	100.15	95.36	-4.64
$\tilde{t}$:	4.18	2.83	-32.29

Table 2. Like Table 1 but for the reversed sequences $S' = \psi_9, \ldots, \psi_0$ of PCNFs after the original sequence $S = \psi_0, \ldots, \psi_9, \psi_{10}$ has been solved. Clauses are *deleted* from ψ_i to obtain ψ_{i-1} in S'.

	QBFEVAL'12-SR		
	discard LC	*keep LC*	*diff.(%)*
$\overline{a}$:	5.48×10^6	0.73×10^6	-86.62
$\tilde{a}$:	186,237	15,031	-91.92
$\overline{b}$:	36,826	1,228	-96.67
$\tilde{b}$:	424	0	-100.00
$\overline{t}$:	21.94	4.32	-79.43
$\tilde{t}$:	0.75	0.43	-42.66

	QBFEVAL'12-SR-Bloqqer		
	discard LC	*keep LC*	*diff.(%)*
$\overline{a}$:	5.88×10^6	1.29×10^6	-77.94
$\tilde{a}$:	103,330	8,199	-92.06
$\overline{b}$:	31,489	3,350	-89.37
$\tilde{b}$:	827	5	-99.39
$\overline{t}$:	30.29	9.78	-67.40
$\tilde{t}$:	0.50	0.12	-76.00

given PCNFs. For experiments, we constructed a sequence of related PCNFs for *each* PCNF in the benchmark sets as follows. Given a PCNF ψ, we divided the number of clauses in ψ by 10 to obtain the size of a slice of clauses. The first PCNF ψ_0 in the sequence contains the clauses of one slice. The clauses of that slice are removed from ψ. The next PCNF ψ_1 is obtained from ψ_0 by adding another slice of clauses, which is removed from ψ. The other PCNFs in the sequence $S = \psi_0, \psi_1, \ldots, \psi_{10}$ are constructed similarly so that finally the last PCNF ψ_{10} contains all the clauses from the original PCNF ψ. In our tests, we constructed each PCNF ψ_i from the previous one ψ_{i-1} in the sequence by adding a slice of clauses to a new frame after a *push* operation. We ran DepQBF on the sequences of PCNFs constructed this way with a wall clock time limit of 1800 seconds and a memory limit of 7 GB.

Tables 1 and 2 show experimental results[5] on sequences $S = \psi_0, \ldots, \psi_{10}$ of PCNFs and on the reversed ones $S' = \psi_9, \ldots, \psi_0$, respectively. To generate S', we first solved the sequence S and then started to discard clauses by popping the frames from the clause stack of DepQBF via its API. In one run (*discard LC*), we always discarded all the constraints that were learned from the previous

[5] Experiments were run on AMD Opteron 6238, 2.6 GHz, 64-bit Linux.

PCNF ψ_i so that the solver solves the next PCNF ψ_{i+1} (ψ_{i-1} with respect to Table 2) starting with empty sets of learned clauses and cubes. In another run (*keep LC*), we kept learned constraints as described in Sections 5.3 and 5.4. This way, 70 out of 345 total PCNF sequences were fully solved from the set *QBFEVAL'12-SR* by both runs, and 112 out of 276 total sequences were fully solved from the set *QBFEVAL'12-SR-Bloqqer*.

The numbers of assignments, backtracks, and wall clock time indicate that keeping the learned constraints is beneficial in incremental QBF solving despite the additional effort of checking the collected initial cubes. In the experiment reported in Table 1 clauses are always added but never deleted to obtain the next PCNF in the sequence. Thereby, across all incremental calls of the solver in the set *QBFEVAL'12-SR* on average 224 out of 364 (61%) collected initial cubes were identified as derivable and added as learned cubes. For the set *QBFEVAL'12-SR-Bloqqer*, 232 out of 1325 (17%) were added.

Related to Table 2, clauses are always removed but never added to obtain the next PCNF to be solved, which allows to keep learned cubes based on Proposition 3. Across all incremental calls of the solver in the set *QBFEVAL'12-SR* on average 820 out of 1485 (55%) learned clauses were disabled and hence effectively discarded because their Q-resolution derivation depended on removed clauses. For the set *QBFEVAL'12-SR-Bloqqer*, 704 out of 1399 (50%) were disabled.

7 Conclusion

We presented a general approach to incremental QBF solving which integrates ideas from incremental SAT solving and which can be implemented in any QCDCL-based QBF solver. The API of our incremental QBF solver DepQBF provides *push* and *pop* operations to add and remove clauses in a PCNF. This increases the usability of our implementation. Our approach is application-independent and applicable to arbitrary QBF encodings.

We illustrated the problem of keeping the learned constraints across different calls of the solver. To improve cube learning in incremental QBF solving, it might be beneficial to maintain (parts of) the cube derivation in memory. This would allow to check the cubes more precisely than with the simple approach we implemented. Moreover, the generation of proofs and certificates [2,14,29] is supported if the derivations are kept in memory rather than in a trace file.

Dual reasoning [15,16,19,35] and the combination of preprocessing and certificate extraction [18,26,31] are crucial for the performance and applicability of CNF-based QBF solving. The combination of incremental solving with these techniques has the potential to further advance the state of QBF solving.

Our experimental analysis demonstrates the feasibility of incremental QBF solving in a general setting and motivates further applications, along with the study of BMC of partial designs using incremental QBF solving [27]. Related experiments with conformant planning based on incremental solving by DepQBF showed promising results [12]. Further experiments with problems which are inherently incremental can provide more insights and open new research directions.

References

1. Audemard, G., Lagniez, J.M., Simon, L.: Improving Glucose for Incremental SAT Solving with Assumptions: Application to MUS Extraction. In: Järvisalo, M., Van Gelder, A. (eds.) SAT 2013. LNCS, vol. 7962, pp. 309–317. Springer, Heidelberg (2013)
2. Balabanov, V., Jiang, J.R.: Unified QBF certification and its applications. Formal Methods in System Design 41(1), 45–65 (2012)
3. Becker, B., Ehlers, R., Lewis, M.D.T., Marin, P.: ALLQBF Solving by Computational Learning. In: Chakraborty, S., Mukund, M. (eds.) ATVA 2012. LNCS, vol. 7561, pp. 370–384. Springer, Heidelberg (2012)
4. Benedetti, M., Mangassarian, H.: QBF-Based Formal Verification: Experience and Perspectives. JSAT 5, 133–191 (2008)
5. Biere, A.: PicoSAT Essentials. JSAT 4(2-4), 75–97 (2008)
6. Bloem, R., Könighofer, R., Seidl, M.: SAT-Based Synthesis Methods for Safety Specs. In: McMillan, K.L., Rival, X. (eds.) VMCAI 2014. LNCS, vol. 8318, pp. 1–20. Springer, Heidelberg (2014)
7. Büning, H.K., Karpinski, M., Flögel, A.: Resolution for Quantified Boolean Formulas. Inf. Comput. 117(1), 12–18 (1995)
8. Cadoli, M., Schaerf, M., Giovanardi, A., Giovanardi, M.: An Algorithm to Evaluate Quantified Boolean Formulae and Its Experimental Evaluation. J. Autom. Reasoning 28(2), 101–142 (2002)
9. Cashmore, M., Fox, M., Giunchiglia, E.: Planning as Quantified Boolean Formula. In: Raedt, L.D., Bessière, C., Dubois, D., Doherty, P., Frasconi, P., Heintz, F., Lucas, P.J.F. (eds.) ECAI. Frontiers in Artificial Intelligence and Applications, pp. 217–222. IOS Press (2012)
10. Eén, N., Sörensson, N.: An Extensible SAT-Solver. In: Giunchiglia, E., Tacchella, A. (eds.) SAT 2003. LNCS, vol. 2919, pp. 502–518. Springer, Heidelberg (2004)
11. Eén, N., Sörensson, N.: Temporal Induction by Incremental SAT Solving. Electr. Notes Theor. Comput. Sci. 89(4), 543–560 (2003)
12. Egly, U., Kronegger, M., Lonsing, F., Pfandler, A.: Conformant Planning as a Case Study of Incremental QBF Solving. CoRR abs/1405.7253 (2014)
13. Giunchiglia, E., Narizzano, M., Tacchella, A.: Clause/Term Resolution and Learning in the Evaluation of Quantified Boolean Formulas. J. Artif. Intell. Res. (JAIR). 26, 371–416 (2006)
14. Goultiaeva, A., Van Gelder, A., Bacchus, F.: A Uniform Approach for Generating Proofs and Strategies for Both True and False QBF Formulas. In: Walsh, T. (ed.) IJCAI, pp. 546–553. IJCAI/AAAI (2011)
15. Goultiaeva, A., Bacchus, F.: Recovering and Utilizing Partial Duality in QBF. In: Järvisalo, M., Van Gelder, A. (eds.) SAT 2013. LNCS, vol. 7962, pp. 83–99. Springer, Heidelberg (2013)
16. Goultiaeva, A., Seidl, M., Biere, A.: Bridging the Gap between Dual Propagation and CNF-based QBF Solving. In: Macii, E. (ed.) DATE, pp. 811–814. EDA Consortium. ACM DL, San Jose (2013)
17. Hillebrecht, S., Kochte, M.A., Erb, D., Wunderlich, H.J., Becker, B.: Accurate QBF-Based Test Pattern Generation in Presence of Unknown Values. In: Macii, E. (ed.) DATE, pp. 436–441. EDA Consortium, ACM DL, San Jose, CA, USA (2013)
18. Janota, M., Grigore, R., Marques-Silva, J.: On QBF Proofs and Preprocessing. In: McMillan, K., Middeldorp, A., Voronkov, A. (eds.) LPAR-19 2013. LNCS, vol. 8312, pp. 473–489. Springer, Heidelberg (2013)

19. Klieber, W., Sapra, S., Gao, S., Clarke, E.M.: A Non-prenex, Non-clausal QBF Solver with Game-State Learning. In: Strichman, O., Szeider, S. (eds.) SAT 2010. LNCS, vol. 6175, pp. 128–142. Springer, Heidelberg (2010)
20. Lagniez, J.M., Biere, A.: Factoring Out Assumptions to Speed Up MUS Extraction. In: Järvisalo, M., Van Gelder, A. (eds.) SAT 2013. LNCS, vol. 7962, pp. 276–292. Springer, Heidelberg (2013)
21. Letz, R.: Lemma and Model Caching in Decision Procedures for Quantified Boolean Formulas. In: Egly, U., Fermüller, C. (eds.) TABLEAUX 2002. LNCS (LNAI), vol. 2381, pp. 160–175. Springer, Heidelberg (2002)
22. Lonsing, F., Biere, A.: Integrating Dependency Schemes in Search-Based QBF Solvers. In: Strichman, O., Szeider, S. (eds.) SAT 2010. LNCS, vol. 6175, pp. 158–171. Springer, Heidelberg (2010)
23. Lonsing, F., Egly, U.: Incremental QBF Solving by DepQBF (Extended Abstract). In: Hong, H., Yap, C. (eds.) ICMS 2014. LNCS, vol. 8592, pp. 307–314. Springer, Heidelberg (2014)
24. Lonsing, F., Egly, U., Van Gelder, A.: Efficient Clause Learning for Quantified Boolean Formulas via QBF Pseudo Unit Propagation. In: Järvisalo, M., Van Gelder, A. (eds.) SAT 2013. LNCS, vol. 7962, pp. 100–115. Springer, Heidelberg (2013)
25. Mangassarian, H., Veneris, A.G., Benedetti, M.: Robust QBF Encodings for Sequential Circuits with Applications to Verification, Debug, and Test. IEEE Trans. Computers 59(7), 981–994 (2010)
26. Marin, P., Miller, C., Becker, B.: Incremental QBF Preprocessing for Partial Design Verification - (Poster Presentation). In: Cimatti, A., Sebastiani, R. (eds.) SAT 2012. LNCS, vol. 7317, pp. 473–474. Springer, Heidelberg (2012)
27. Marin, P., Miller, C., Lewis, M.D.T., Becker, B.: Verification of Partial Designs using Incremental QBF Solving. In: Rosenstiel, W., Thiele, L. (eds.) DATE, pp. 623–628. IEEE (2012)
28. Nadel, A., Ryvchin, V.: Efficient SAT Solving under Assumptions. In: Cimatti, A., Sebastiani, R. (eds.) SAT 2012. LNCS, vol. 7317, pp. 242–255. Springer, Heidelberg (2012)
29. Niemetz, A., Preiner, M., Lonsing, F., Seidl, M., Biere, A.: Resolution-Based Certificate Extraction for QBF - (Tool Presentation). In: Cimatti, A., Sebastiani, R. (eds.) SAT 2012. LNCS, vol. 7317, pp. 430–435. Springer, Heidelberg (2012)
30. Samulowitz, H., Davies, J., Bacchus, F.: Preprocessing QBF. In: Benhamou, F. (ed.) CP 2006. LNCS, vol. 4204, pp. 514–529. Springer, Heidelberg (2006)
31. Seidl, M., Könighofer, R.: Partial witnesses from preprocessed quantified Boolean formulas. In: DATE, pp. 1–6. IEEE (2014)
32. Silva, J.P.M., Lynce, I., Malik, S.: Conflict-Driven Clause Learning SAT Solvers. In: Biere, A., Heule, M., van Maaren, H., Walsh, T. (eds.) Handbook of Satisfiability, FAIA, vol. 185, pp. 131–153. IOS Press (2009)
33. Staber, S., Bloem, R.: Fault Localization and Correction with QBF. In: Marques-Silva, J., Sakallah, K.A. (eds.) SAT 2007. LNCS, vol. 4501, pp. 355–368. Springer, Heidelberg (2007)
34. Sülflow, A., Fey, G., Drechsler, R.: Using QBF to Increase Accuracy of SAT-Based Debugging. In: ISCAS, pp. 641–644. IEEE (2010)
35. Van Gelder, A.: Primal and Dual Encoding from Applications into Quantified Boolean Formulas. In: Schulte, C. (ed.) CP 2013. LNCS, vol. 8124, pp. 694–707. Springer, Heidelberg (2013)
36. Zhang, L., Malik, S.: Towards a Symmetric Treatment of Satisfaction and Conflicts in Quantified Boolean Formula Evaluation. In: Van Hentenryck, P. (ed.) CP 2002. LNCS, vol. 2470, pp. 200–215. Springer, Heidelberg (2002)

Incremental Cardinality Constraints for MaxSAT

Ruben Martins[1,⋆], Saurabh Joshi[1,⋆], Vasco Manquinho[2,⋆⋆], and Inês Lynce[2,⋆⋆]

[1] University of Oxford, Department of Computer Science, UK
{ruben.martins,saurabh.joshi}@cs.ox.ac.uk
[2] INESC-ID / Instituto Superior Técnico, Universidade de Lisboa, Portugal
{vmm,ines}@sat.inesc-id.pt

Abstract. Maximum Satisfiability (MaxSAT) is an optimization variant of the Boolean Satisfiability (SAT) problem. In general, MaxSAT algorithms perform a succession of SAT solver calls to reach an optimum solution making extensive use of cardinality constraints. Many of these algorithms are non-incremental in nature, i.e. at each iteration the formula is rebuilt and no knowledge is reused from one iteration to another. In this paper, we exploit the knowledge acquired across iterations using novel schemes to use cardinality constraints in an incremental fashion. We integrate these schemes with several MaxSAT algorithms. Our experimental results show a significant performance boost for these algorithms as compared to their non-incremental counterparts. These results suggest that incremental cardinality constraints could be beneficial for other constraint solving domains.

1 Introduction

Plethora of application domains such as software package upgrades [5], error localization in C code [27], debugging of hardware designs [12], haplotyping with pedigrees [24], and course timetabling [6] have benefited from the advancement in MaxSAT solving techniques. Considering such diversity of application domains for MaxSAT algorithms, the continuous improvement of MaxSAT solving techniques is imperative.

Incremental approaches have provided a huge leap in the performance of SAT solvers [47, 22, 45, 8]. However, the notion of incrementality has not yet been fully exploited in MaxSAT solving. Most MaxSAT algorithms perform a succession of SAT solver calls to reach optimality. Incremental approaches allow the constraint solver to retain knowledge from previous iterations that may be used in the upcoming iterations. The goal is to retain the inner state of the constraint solver as well as learned clauses that were discovered during the solving process of previous iterations. At each iteration, most MaxSAT algorithms [23, 38, 25, 43] create a new instance of the constraint solver and rebuild the formula losing most if not all the knowledge that could be derived from previous iterations.

⋆ Supported by the ERC project 280053.

⋆⋆ Partially supported by FCT grants ASPEN (PTDC/EIA-CCO/110921/2009), POLARIS (PTDC/EIA-CCO/123051/2010), and INESC-ID's multiannual PIDDAC funding PEst-OE/EEI/LA0021/2013.

B. O'Sullivan (Ed.): CP 2014, LNCS 8656, pp. 531–548, 2014.

Between the iterations of a MaxSAT algorithm, cardinality constraints are added to the formula [23, 3, 25, 43]. Usually, cardinality constraints are encoded in CNF so that a SAT solver can handle the resulting formula [9, 46, 7]. Otherwise, calls to a SAT solver must be replaced with calls to a pseudo-Boolean solver that natively handles cardinality constraints [38]. This paper discusses the use of cardinality constraints in an incremental manner to enhance MaxSAT algorithms. To achieve this, we propose the following incremental approaches: (i) incremental blocking, (ii) incremental weakening, and (iii) iterative encoding.

The remainder of the paper is organized as follows. Section 2 introduces preliminaries and notations. We describe our proposed techniques in Section 3. In Section 4, we mention prior research work done in relevant areas. We show the superiority of our approaches through experimental results in Section 5. Finally, Section 6 presents concluding remarks.

2 Preliminaries

A Boolean formula in conjunctive normal form (CNF) is a conjunction of clauses, where a clause is a disjunction of literals and a literal is a Boolean variable x_i or its negation $\neg x_i$. A Boolean variable may be assigned truth values $true$ or $false$. A literal x_i ($\neg x_i$) is said to be satisfied if the respective variable is assigned value $true$ ($false$). A literal x_i ($\neg x_i$) is said to be unsatisfied if the respective variable is assigned value $false$ ($true$). A clause is satisfied if and only if at least one of its literals is satisfied. A clause is called a unit clause if it only contains one literal. A formula φ is satisfied if all of its clauses are satisfied. The Boolean Satisfiability (SAT) problem can be defined as finding a satisfying assignment to a propositional formula φ or prove that such an assignment does not exist. Throughout this paper, we will refer to φ as a set of clauses, where each clause ω is a set of literals.

Maximum Satisfiability (MaxSAT) is an optimization version of SAT where the goal is to find an assignment to the input variables such that the number of unsatisfied (satisfied) clauses is minimized (maximized). From now on, it is assumed that MaxSAT is defined as a minimization problem.

MaxSAT has several variants such as partial MaxSAT, weighted MaxSAT and weighted partial MaxSAT [33]. A partial MaxSAT formula φ has the form $\varphi_h \cup \varphi_s$ where φ_h and φ_s denote the set of hard and soft clauses, respectively. The goal in partial MaxSAT is to find an assignment to the input variables such that all hard clauses φ_h are satisfied, while minimizing the number of unsatisfied soft clauses in φ_s. The weighted version of MaxSAT allows soft clauses to have weights greater than or equal to 1 and the objective is to satisfy all hard clauses while minimizing the total weight of unsatisfied soft clauses. In this paper we assume a partial MaxSAT formula. The described algorithms can be generalized to the weighted versions of MaxSAT.

Cardinality constraints are a generalization of propositional clauses. In a cardinality constraint, a sum of n literals must be smaller than or equal to a given value k, i.e. $\sum_{i=1}^{n} l_i \leq k$ where l_i is a literal. As a result, a cardinality constraint over n literals ensures that at most k literals can be satisfied.

Algorithm 1. Linear Search Unsat-Sat Algorithm

Input: $\varphi = \varphi_h \cup \varphi_s$
Output: satisfying assignment to φ
1 $(\varphi_W, V_R, \lambda) \leftarrow (\varphi_h, \emptyset, 0)$
2 **foreach** $\omega \in \varphi_s$ **do**
3 $\quad V_R \leftarrow V_R \cup \{r\}$ // r is a new relaxation variable
4 $\quad \omega_R \leftarrow \omega \cup \{r\}$
5 $\quad \varphi_W \leftarrow \varphi_W \cup \{\omega_R\}$
6 **while true do**
7 $\quad (\mathsf{st}, \nu, \varphi_C) \leftarrow \texttt{SAT}(\varphi_W \cup \{\texttt{CNF}(\sum_{r \in V_R} r \leq \lambda)\}, \emptyset)$
8 $\quad$ **if** st = SAT **then**
9 $\quad\quad$ **return** ν // satisfying assignment to φ
10 $\quad \lambda \leftarrow \lambda + 1$

2.1 MaxSAT Algorithms

Due to the recent developments in SAT solving, different algorithms for solving MaxSAT have been proposed that rely on multiple calls to a SAT solver. A SAT solver call $\texttt{SAT}(\varphi, \mathcal{A})$ receives as input a CNF formula φ and a set of assumptions $\mathcal{A}$. The set of assumptions $\mathcal{A}$ defines a set of literals that must be satisfied in the model of φ returned by the solver call. Assumptions may lead to early termination if the SAT solver learns a clause where at least one of the literals in $\mathcal{A}$ must be unsatisfied. An assumption controls the value of a variable for a given SAT call, whereas a unit clause controls the value of a variable for all the SAT calls after the unit clause has been added.

The SAT call returns a triple (*st*, ν, φ_C), where *st* denotes the status of the solver: satisfiable (SAT) or unsatisfiable (UNSAT). If the solver returns SAT, then the model that satisfies φ is stored in ν. On the other hand, if the solver returns UNSAT, then φ_C contains an unsatisfiable formula that explains the reason of unsatisfiability. Notice that φ may be satisfiable, but the solver returns UNSAT due to the set of assumptions $\mathcal{A}$ (i.e. there are no models of φ where all assumption literals are satisfied). In this case, φ_C contains a subset of clauses from φ and a subset of assumptions from $\mathcal{A}$. Otherwise, if φ is unsatisfiable, then φ_C is a subformula of φ.

The algorithms presented in the paper assume that a SAT solver call is previously performed to check the satisfiability of the set of hard clauses φ_h. If φ_h is not satisfiable, then the MaxSAT instance does not have a solution.

Algorithm 1 performs a linear search on the number of unsatisfied soft clauses. First, a new relaxation variable r is added to each soft clause ω (lines 2-5). The goal is to find an assignment to the input variables that minimizes the number of relaxation variables that are assigned value $true$. If the original clause ω is unsatisfied, then r is assigned to $true$. At each iteration, a cardinality constraint is defined such that at most λ relaxation variables can be assigned to $true$. This cardinality constraint is encoded into CNF and given to the SAT solver (line 7). Algorithm 1 starts with $\lambda = 0$ and in each iteration λ is increased until the SAT solver finds a satisfying assignment. Hence, λ defines a lower bound on the number of unsatisfied soft clauses of φ. At each iteration, the result of the SAT call is UNSAT, except the last one that provides an optimal solution to φ.

Algorithm 2. Fu-Malik Algorithm

Input: $\varphi = \varphi_h \cup \varphi_s$
Output: satisfying assignment to φ
1 $(\varphi_W, \varphi_{W_s}) \leftarrow (\varphi, \varphi_s)$
2 **while** true **do**
3 $\quad (\mathsf{st}, \nu, \varphi_C) \leftarrow \texttt{SAT}(\varphi_W, \emptyset)$
4 $\quad$ **if** st = SAT **then**
5 $\quad\quad$ **return** ν // satisfying assignment to φ
6 $\quad V_R \leftarrow \emptyset$
7 $\quad$ **foreach** $\omega \in (\varphi_C \cap \varphi_{W_s})$ **do**
8 $\quad\quad V_R \leftarrow V_R \cup \{r\}$ // r is a new relaxation variable
9 $\quad\quad \omega_R \leftarrow \omega \cup \{r\}$
10 $\quad\quad \varphi_{W_s} \leftarrow (\varphi_{W_s} \setminus \{\omega\}) \cup \{\omega_R\}$
11 $\quad\quad \varphi_W \leftarrow (\varphi_W \setminus \{\omega\}) \cup \{\omega_R\}$
12 $\quad \varphi_W \leftarrow \varphi_W \cup \{\texttt{CNF}(\sum_{r \in V_R} r \leq 1)\}$

Algorithm 3. MSU3 Algorithm

Input: $\varphi = \varphi_h \cup \varphi_s$
Output: satisfying assignment to φ
$(\varphi_W, V_R, \lambda) \leftarrow (\varphi, \emptyset, 0)$
while true **do**
$\quad (\mathsf{st}, \nu, \varphi_C) \leftarrow \texttt{SAT}(\varphi_W \cup \{\texttt{CNF}(\sum_{r \in V_R} r \leq \lambda)\}, \emptyset)$
$\quad$ **if** st = SAT **then**
$\quad\quad$ **return** ν // satisfying assignment to φ
$\quad$ **foreach** $\omega \in (\varphi_C \cap \varphi_s)$ **do**
$\quad\quad V_R \leftarrow V_R \cup \{r\}$ // r is a new variable
$\quad\quad \omega_R \leftarrow \omega \cup \{r\}$ // ω was not previously relaxed
$\quad\quad \varphi_W \leftarrow (\varphi_W \setminus \{\omega\}) \cup \{\omega_R\}$
$\quad \lambda \leftarrow \lambda + 1$

Algorithm 1 follows an Unsat-Sat linear search. A converse approach is the Sat-Unsat linear search where λ is defined as an upper bound. In that case, λ is initialized with the number of soft clauses. Next, while the SAT call is satisfiable, λ is decreased. The algorithm ends when the SAT call returns UNSAT and the last satisfying assignment found is an optimal solution to φ.

Core-guided algorithms for MaxSAT take advantage of the certificates of unsatisfiability produced by the SAT solver [23]. In Algorithm 2, proposed by Fu and Malik [23], soft clauses are only relaxed when they appear in some unsatisfiable core φ_C returned by the SAT solver. Initially, we consider all hard and soft clauses without relaxation. In each iteration, an unsatisfiable subformula φ_C is identified and relaxed by adding a new relaxation variable to each soft clause in φ_C (lines 7-11). Additionally, a new constraint is added such that at most one of the new relaxation variables can be assigned to *true* (line 12). The algorithm stops when the formula becomes satisfiable.

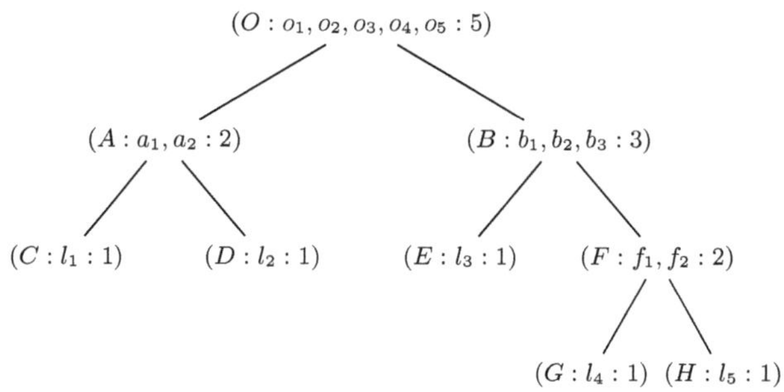

Fig. 1. Totalizer encoding for $l_1 + \cdots + l_5 \leq k$

In Algorithm 2 soft clauses may have to be relaxed several times. As a result, several relaxation variables can be added to the same soft clause. Nevertheless, other core-guided algorithms have already been proposed where at most one relaxation variable is added to each soft clause [3, 40]. Algorithm 3 follows a linear search Unsat-Sat, but soft clauses are only relaxed when they appear in some unsatisfiable core φ_C.

In this section we solely describe MaxSAT algorithms that will be the focus of the enhancements proposed in the paper. We refer to the literature for other approaches such as branch and bound algorithms using MaxSAT inference techniques or procedures to estimate the number of unsatisfied clauses to prune the search [33]. Additionally, there is also an extended overview on core-guided algorithms [43].

2.2 Totalizer Encoding

For the purpose of this paper, we describe the Totalizer encoding [9] for cardinality constraints, as later in the paper we build upon this encoding to present our novel approaches. Totalizer encoding can be better visualized as a tree as shown in Fig. 1. Here, notation for every node is $(node\_name : node\_vars : node\_sum)$. To enforce the cardinality constraint, we need to count how many input literals $(l_1, \ldots, l_n)$ are set to $true$. This counting is done in unary. Therefore, at every node its corresponding $node\_vars$ represents an integer from 1 to $node\_sum$ in the order. For example, at node B, b_2 being set to $true$ means that at least two of the leaves under the tree rooted at B have been set to $true$. The input literals $(l_1, \ldots, l_5)$ are at the leaves where as the root node has the output variables $(o_1, \ldots, o_5)$ giving the finally tally of how many input literals have been set.

Any intermediate node P, counting up to n_1, has two children Q and R counting up to n_2 and n_3 respectively such that $n_2 + n_3 = n_1$. Also, their corresponding $node\_vars$ will be $(p_1, \ldots, p_{n_1})$, $(q_1, \ldots, q_{n_2})$ and $(r_1, \ldots, r_{n_3})$ in that order. In order to ensure that the correct sum is received at P, the following formula is built for P:

$$\bigwedge_{\substack{0 \leq \alpha \leq n_2 \\ 0 \leq \beta \leq n_3 \\ 0 \leq \sigma \leq n_1 \\ \alpha + \beta = \sigma}} \neg q_\alpha \vee \neg r_\beta \vee p_\sigma \quad \text{where, } p_0 = q_0 = r_0 = 1 \tag{1}$$

Essentially, Eq. 1 dictates that if α many leaves have been set to $true$ under the subtree rooted at Q and β many leaves have been set to $true$ under the subtree rooted at R then r_σ must be set to $true$ to indicate that at least $\alpha + \beta$ many leaves have been set to $true$ under P. Eq. 1 only counts the number of input literals set to $true$. In other words, it encodes *cardinality sum* over input literals. To enforce that at most k of the input literals are set to $true$, we conjunct it with the following :

$$\bigwedge_{k+1 \leq i \leq n} \neg o_i \tag{2}$$

Observation 1. *Two disjoint subtrees for the Totalizer encoding are independent of each other. For example, the tree rooted at B counts how many literals have been set from* (l_3, l_4, l_5) *where as, the tree rooted at A counts the set literals from* (l_1, l_2).

Note also that Eq. 1 counts up to n and then Eq. 2 restricts the sum to k. If we only want to enforce the constraint for at most k then we need at most $k+1$ output variables at the root. In turn, we need at most $k + 1$ $node\_vars$ at any intermediate node. Even with this modification, Eq. 1 remains valid. However, the equality $n_2 + n_3 = n_1$ may no longer hold. With this modification, Eq. 2 simplifies to

$$\neg o_{k+1}$$

Without the simplification this encoding requires $O(nlog\ n)$ extra variables and $O(n^2)$ clauses. After the simplification the number of clauses reduces to $O(nk)$ [11, 29]. From here on, we will refer to this simplification as *k-simplification*.

Observation 2. *Let* φ_1 *and* φ_2 *be two formulas, representing cardinality sums* k_1 *and* k_2 *respectively, generated using Eq. 1 and k-simplification. Observe that* $\varphi_1 \subset \varphi_2$, *whenever* $k_1 < k_2$.

3 Incremental Approaches

MaxSAT algorithms that are based on refining unsatisfiable SAT formulas can be enhanced by changing cardinality constraints in an incremental fashion. In this section, we propose the following three techniques to enable incrementality when using cardinality constraints: (i) incremental blocking, (ii) incremental weakening, and (iii) iterative encoding.

3.1 Incremental Blocking

MaxSAT algorithms based on refining unsatisfiable formulas are usually non-incremental. After an unsatisfiable iteration, the formula is refined by removing a certain set of clauses and adding a new set of clauses that imposes a weaker constraint over the relaxation variables. However, SAT solvers do not allow the deletion of clauses that belong to the original formula. Since learned clauses from previous iterations may depend on the clauses that are now being removed, it is not sound to keep all of the learned clauses.

Incremental SAT solving addresses these problems by using assumptions [22]. To the best of our knowledge this approach has not been extended for incremental MaxSAT solving.

We denote b as a *blocking variable* which is used to extend a clause ω to $(\omega \vee b)$. When b is set to $false$ the original clause ω is enforced (enabled). When b is set to $true$ the extended clause $(\omega \vee b)$ is trivially satisfied and ω is no longer enforced (disabled). Thus, adding b (or $\neg b$) as an assumption or unit clause disables (or enables) a clause. Using a blocking variable, we can overcome the limitation of a SAT solver not allowing clause deletions.

MaxSAT Algorithms Based on Cardinality Constraints. Many MaxSAT algorithms are based on refining the formula by encoding and updating cardinality constraints [25, 2, 43]. For these algorithms, the incremental blocking can be done when cardinality constraints are encoded to CNF.

$$\varphi \boxplus b \equiv \{\omega \vee b : \omega \in \varphi\} \tag{3a}$$

$$\Psi(\mathbf{X}, k, b) \equiv \texttt{CNF}_{Tot^k}(\Sigma x_i) \boxplus b \tag{3b}$$

$$\varphi^i \equiv \varphi_W \cup \left(\bigcup_{j=1}^{i} \Psi(\mathbf{X^j}, k^j, b^j) \right) \cup \langle \neg b^i, \neg o_{k^i+1} \rangle \cup \left[b^1, \ldots, b^{i-1} \right] \tag{3c}$$

$$\varphi^{i+1} \equiv \varphi_W \cup \left(\bigcup_{j=1}^{i+1} \Psi(\mathbf{X^j}, k^j, b^j) \right) \cup \langle \neg b^{i+1}, \neg o_{k^{i+1}+1} \rangle \cup \left[b^1, \ldots, b^{i} \right] \tag{3d}$$

Let Eq. 3a define the extension of a CNF formula φ with a blocking variable b. Next, $\Psi(\mathbf{X}, k, b)$ represents a cardinality sum up to $k+1$ over $x_1, \ldots, x_n$ encoded in CNF using Eq. 1 and k-simplification of the Totalizer encoding and extended with a blocking variable b. Then, for incremental blocking, at line 7 in Algorithm 1 and line 3 in Algorithm 3 we call the solver on φ^i as defined in Eq. 3c for the ith iteration. Assumption $\langle \neg b^i \rangle$ enables the cardinality constraint for the current iteration whereas unit clauses $\left[b^1, \ldots, b^{i-1}\right]$ ensure that cardinality constraints from earlier iterations are disabled. In addition, assumption $\langle \neg o_{k^i+1} \rangle$ restricts the sum to k^i. Notice that in the $(i+1)^{th}$ iteration, a new cardinality sum $\Psi(\mathbf{X}^{i+1}, k^{i+1}, b^{i+1})$ is added and earlier constraints are disabled as assumption $\langle \neg b^i \rangle$ moves as unit clause $\left[b^i\right]$.

Assume the MaxSAT formula has a given optimum value k_{opt}. When considering Algorithm 1 and the Totalizer encoding, incremental blocking creates an encoding for each k^i up to k_{opt}. Hence, the overall encoding would have $O(\sum_{i=0}^{k_{opt}} ni) = O(nk_{opt}^2)$ auxiliary clauses. Though incremental blocking creates more clauses as compared to a non-incremental approach ($O(nk_{opt})$), keeping the inner state of the constraint solver across iterations significantly reduces the solving time. A similar reasoning can be made for Algorithm 3 or any other MaxSAT algorithm that uses incremental blocking.

Fu-Malik Algorithm with Incremental Blocking. Incremental blocking can also be used for MaxSAT algorithms that do not update cardinality constraints but modify the

Algorithm 4. Fu-Malik Algorithm with Incremental Blocking

Input: $\varphi = \varphi_h \cup \varphi_s$
Output: satisfying assignment to φ

1 $(\varphi_W, \varphi_{W_s}, \mathcal{A}, \mathcal{B}) \leftarrow (\varphi, \varphi_s, \emptyset, \emptyset)$
2 **while** true **do**
3 $\quad(\mathsf{st}, \nu, \varphi_C) \leftarrow \texttt{SAT}(\varphi, \mathcal{A})$
4 $\quad$**if** st = SAT **then**
5 $\quad\quad$**return** ν // satisfying assignment to φ
6 $\quad V_R \leftarrow \emptyset$
7 $\quad$**foreach** $\omega \in (\varphi_C \cap \varphi_{W_s})$ **do**
8 $\quad\quad V_R \leftarrow V_R \cup \{r\}$ // r is a new relaxation variable
9 $\quad\quad \omega_R \leftarrow (\omega \setminus \mathcal{B}) \cup \{r\} \cup \{b\}$ // b is a new blocking variable
10 $\quad\quad \mathcal{B} \leftarrow \mathcal{B} \cup \{b\}$
11 $\quad\quad \varphi_{W_s} \leftarrow (\varphi_{W_s} \setminus \{\omega\}) \cup \{\omega_R\}$
12 $\quad\quad \mathcal{A} \leftarrow (\mathcal{A} \setminus \{\neg b' : b' \in \mathcal{B} \cap \omega\}) \cup \{\neg b\}$ // enables ω_R
13 $\quad\quad \varphi_W \leftarrow \varphi_W \cup \{\omega_R\} \cup \{b' : b' \in \mathcal{B} \cap \omega\}$ // disables ω
14 $\quad \varphi_W \leftarrow \varphi_W \cup \{\texttt{CNF}(\sum_{r \in V_R} r \leq 1)\}$

formula at each iteration. For example, Fu-Malik algorithm (Algorithm 2, Section 2) can be enhanced with incremental blocking. Algorithm 4 shows the modifications to Fu-Malik algorithm to support incremental blocking. The main differences between the incremental and non-incremental versions of Fu-Malik algorithm are highlighted. For each soft clause ω in φ_C, Algorithm 4 copies ω into ω_R without blocking variables (line 9). Next, it adds a fresh blocking variable b and a fresh relaxation variable r to ω_R (line 9). The current soft clause ω_R is enabled by adding $\langle \neg b \rangle$ to the set of assumptions, where b is the blocking variable that occurs in ω_R (line 12). At the same time, the assumption $\langle \neg b' \rangle$ is removed from the set of assumptions, where b' is the blocking variable that occurs in ω (line 12). Finally, the working formula φ_W is updated with the new clause ω_R, and with the unit clause $[b']$. Note that this unit clause disables ω from the working formula φ_W since ω contains b' and therefore is always satisfied.

The incremental version of Fu-Malik algorithm creates m auxiliary clauses at each iteration, where m is the number of soft clauses in the unsatisfiable subformula. However, the size of unsatisfiable subformulas tends to be small when compared to the total number of soft clauses. Note that the number of auxiliary clauses created by the incremental version of Fu-Malik is not as large as when incremental blocking is directly applied to cardinality encodings.

3.2 Incremental Weakening

Since incremental blocking encodes a new cardinality constraint at each iteration, this results in an increase in formula size at every iteration. To circumvent this increase, one can build the cardinality sum only once, and incrementally weaken the cardinality bound (k).

Incremental weakening is similar to *incremental strengthening* [7], but instead of constraining the output of the cardinality constraint with unit clauses it uses assumptions. Notice that incremental strengthening is used in linear search Sat-Unsat algorithms. In these algorithms, the cardinality bound decreases monotonically at each iteration. Therefore, the unit clauses that constrain the previous cardinality bound remain valid when considering the new bound. On the other hand, incremental weakening is used for MaxSAT algorithms that search on the lower bound of the optimal solution. For these algorithms, the restriction of the cardinality bound is only valid for the current iteration and must be updated for the upcoming iterations.

$$\Gamma(\mathbf{X}, k) \equiv \texttt{CNF}_{Tot^k}(\Sigma x_i) \tag{4a}$$

$$\varphi^i \equiv \varphi_W \cup \Gamma(\mathbf{X}, k_u) \cup \langle \neg o_{k^i+1}, \ldots, \neg o_{k_u} \rangle \tag{4b}$$

$$\varphi^{i+1} \equiv \varphi_W \cup \Gamma(\mathbf{X}, k_u) \cup \langle \neg o_{k^{i+1}+1}, \ldots, \neg o_{k_u} \rangle \tag{4c}$$

Let $\Gamma(\mathbf{X}, k)$ be the cardinality sum over input literals $x_1, \ldots, x_n$ encoded in CNF using Eq. 1 and k-simplification. Then, for incremental weakening, at line 7 in Algorithm 1 and line 3 in Algorithm 3 we call the solver on φ^i as defined in Eq. 4b for the ith iteration. Note that $\Gamma(\mathbf{X}, k_u)$ is encoded only once for a conservative upper bound k_u. For the ith iteration, we restrict the cardinality sum to k^i using assumptions $\langle \neg o_{k^i+1}, \ldots, \neg o_{k_u} \rangle$ (Eq. 2). In the following iteration (Eq. 4c), we only change assumptions to restrict the cardinality sum to k^{i+1}.

To obtain a conservative upper bound k_u, we invoke the SAT solver over φ_h to check if the set of hard clauses itself is satisfiable. If it is not satisfiable, the original MaxSAT formula φ can not be solved. However, if φ_h is satisfiable, one can count the number of soft clauses that remain unsatisfied under the satisfying assignment for φ_h. This number can be used as k_u since we know at least one assignment where k_u many clauses remain unsatisfied. Therefore, the optimum value k_{opt} must be smaller or equal to k_u.

With an upper bound k_u, incremental weakening creates $O(nk_u)$ auxiliary clauses as opposed to $O(nk_{opt})$ of the non-incremental approach. However, a non-incremental approach builds a new formula of size $O(nk_{opt})$ for every iteration, whereas incremental weakening builds the formula only once keeping the internal state and learned clauses across iterations. This results in a significant performance boost for MaxSAT algorithms using incremental weakening.

Incremental weakening does not allow the number of input literals in the cardinality constraint to change. Therefore, it does not directly support the MSU3 Algorithm (Algorithm 3, Section 2). To use incremental weakening with Algorithm 3, we modify the algorithm to relax all soft clauses and build a cardinality constraint over all relaxation variables. The relaxation variables r_i that do not appear in an unsatisfiable subformula φ_C are added as assumptions of the form $\langle \neg r_i \rangle$. This enforces the soft clauses corresponding to the relaxation variables until these clauses occur in φ_C. When they do occur, assumptions $\neg r_i$ are removed and their value is now only restricted by the cardinality constraint. Even though this procedure allows the incremental weakening approach to be used with Algorithm 3, it does not benefit from smaller encodings resulting from having less input literals in the cardinality constraint. Therefore, the non-incremental approach may create a much smaller encoding than the incremental weakening approach for Algorithm 3.

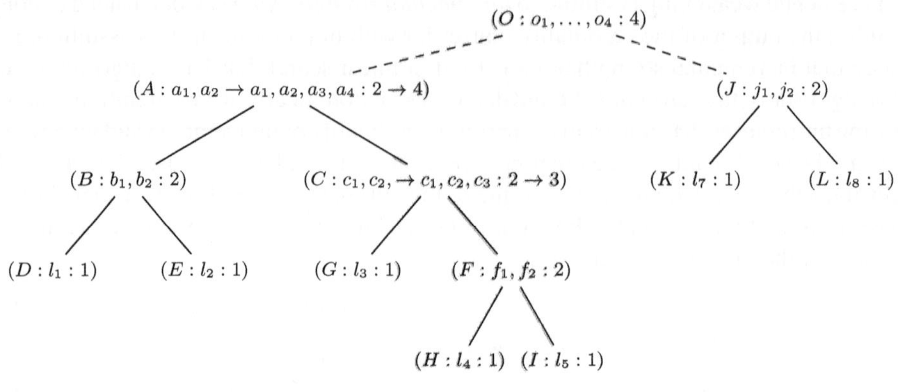

Fig. 2. Transforming $l_1 + \cdots + l_5 \leq 1$ and $l_7 + l_8 \leq 1$ into $l_1 + \ldots + l_5 + l_7 + l_8 \leq 3$

3.3 Iterative Encoding

Incremental weakening uses a conservative upper bound (e.g., k_u) on the number of unsatisfied soft clauses in order to encode the cardinality constraint only once. However, this upper bound may be much larger than the optimum value (e.g. k_{opt}) which may result in a larger encoding than the non-incremental approach. In addition, incremental weakening does not allow the set of input literals in the cardinality constraint to change. Therefore, MaxSAT algorithms that increase the input literals of the cardinality constraint can not take advantage of incremental weakening. To remedy this situation, we propose to encode the cardinality constraint in an iterative fashion. At each iteration of the MaxSAT algorithm, the encoding of the cardinality constraint is augmented with clauses that allow the sum of input literals to go up to k for the current iteration. We call this approach *iterative encoding*.

Let us take a look at Fig. 2 to see how iterative encoding proceeds. Assume that for a particular iteration, we needed to encode $l_1 + \cdots + l_5 \leq 1$. This can be accomplished using the subtree rooted at A. Since the bound for this iteration is $k = 1$, we only need $k + 1 = 2$, $node\_vars$ at every node as described in k-simplification in Section 2.2. In the next iteration, suppose we need to encode $l_1 + \cdots + l_5 + l_7 + l_8 \leq 3$. Observation 2 allows us to augment the formula for subtree rooted at A to allow $l_1 + \cdots + l_5$ to sum up to 4. This is done by increasing the output variables of node A to sum up to 4 and adding the respective clauses that encode sums 3 and 4. Similarly, for node C the output variables are increased to sum up to 3 and the clauses that sum up to 3 are added to the formula. For the additional input literals l_7 and l_8 we encode the subtree rooted at J. Observation 1 allows us to merge trees rooted at A and J by creating a new parent node O which sums up to 4 since A and J have disjoint sets of input literals. To restrict the number of input literals being set to *true* to 3, we only need to add $\neg o_4$ as described in Eq. 2.

In general, if the cardinality constraint changes from $x_1 + \cdots + x_n \leq k_1$ ($k_1 < n$) to $x_1 + \cdots + x_n + y_1 + \cdots + y_m \leq k_2$ where $k_1 \leq k_2$ then we do the following: (1) Remove the assumption over output literal $\neg o_{k_1+1}$ which restricts the sum of $x_1 \ldots, x_n$ to k_1.

(2) Augment the formula for $x_1, \ldots, x_n$ to sum up to $min(k_2 + 1, n)$. (3) Encode the formula over $y_1, \ldots, y_m$ to sum up to $min(k_2+1, m)$. (4) Conjunct these two formulas and augment the resulting formula using Eq. 1 and k-simplification in order to encode $x_1 + \cdots + x_n + y_1 + \cdots + y_m \leq k_2$. Since iterative encoding always adds clauses to the existing formula and changes assumptions, it allows us to retain the internal state of the solver across iterations.

Linear search Unsat-Sat algorithm (Algorithm 1, Section 2) increases the cardinality bound by 1 at each iteration but does not change the set of input literals of the cardinality constraint. Therefore, to apply iterative encoding to this algorithm we only perform steps (1) and (2). On the other hand, MSU3 algorithm (Algorithm 3, Section 2) may change the set of input literals of the cardinality constraint between iterations. Therefore, iterative encoding is applied to MSU3 by performing steps (1) to (4).

Since at every iteration, bare minimum number of clauses necessary to encode the cardinality constraint for that iteration is added, the size of the encoding remains small throughout the run of the MaxSAT algorithm. Iterative encoding is not only faster but allows us to solve more problem instances as compared to non-incremental approaches.

4 Related Work

The first use of incremental SAT solving can be traced back to the 90's with the seminal work of John Hooker [26]. Initially, only a subset of constraints is considered. At each iteration, more constraints are added to the formula. Later, incremental approaches were adopted by constraint solvers in the context of SAT [50, 21] and SAT extensions [29, 8]. Assumptions are widely used for incremental SAT [22, 45]. The minisat solver [21] interface allows the definition of a set of assumptions. Alternatively, the interface of zchaff [36] allows removing groups of clauses.

Although not implemented, the work of Fu and Malik in MaxSAT [23] discusses how learned clauses may be kept from one SAT iteration to the next one. In Pseudo Boolean Optimization (PBO), early implementations include the use of incremental strengthening in minisat+ [20]. Linear search Sat-Unsat algorithms [29, 32] are implemented incrementally. A critical issue is on keeping *safe* learned clauses in successive iterations of a core-guided algorithm [41]. Quantified Boolean Formula (QBF) solving has successfully been made incremental [35] and further applied to verification [39].

In the context of SAT, incremental approaches exist for building encodings and identifying Minimal Unsatisfiable Subformulas (MUSes). For example, an incremental translation to CNF uses unit clauses to simplify the pseudo-Boolean constraint before translating it to CNF [37]. More recent work lazily decomposes complex constraints into a set of clauses [1]. The identification of MUSes has been made incremental by Liffton *et al.* [34]. Later on, the SAT solver Glucose has been made incremental using assumptions and applied to MUS extraction [8].

Incrementality is also present in other SAT-related domains such as Satisfiability Modulo Theories (SMT) and Bounded Model Checking (BMC). The SMT-LIB v2.0 [10] defines the operations *push* and *pop* to work with a stack containing a set of formulas to be jointly solved. The MaxSAT solvers WPM1 and WPM2 [2] use the SMT solver Yices [19] which supports incrementality. Its use resembles the blocking strategy.

Table 1. Number of instances solved by the different incremental approaches and median speedup of solved instances

	None		Blocking		Weakening		Iterative	
	#Inst	Speedup	#Inst	Speedup	#Inst	Speedup	#Inst	Speedup
Fu-Malik	366	1.0	**388**	**2.4**	-	-	-	-
LinearUS	477	1.0	446	1.6	**498**	**2.3**	**509**	**2.4**
MSU3	517	1.0	488	1.6	504	2.0	**541**	**3.6**

The use of SAT solvers in BMC is known to benefit from incrementality, either by implementing incremental SAT solving [47] or by using assumptions [22].

In the context of Constraint Satisfaction Problems (CSPs), incremental formulations, incremental propagation and incremental solving are worth mentioning. Incrementality is naturally present in Dynamic CSPs (DCSPs) [18]. In DCSPs, the formulation of a problem evolves over time by adding and/or removing variables and constraints. *Nogoods* can eventually be carried from one formulation to the next one. DCSPs make use of an incremental arc consistency algorithm [17]. Incremental propagation in CSP [31, 13] makes use of *advisors* which give propagators a detailed view of the dynamic changes between propagator runs. Advisors enable the implementation of optimal algorithms for important constraints. Search in CSP is inherently incremental. From the first implementations, the approach to solve many CSPs is to incrementally build a solution, backtracking when an infeasibility is detected, until a solution is found or the problem is proven to have no solution [48]. More recently, incrementality has been implemented in global constraints mostly due to efficiency reasons [49].

5 Experimental Results

We used all partial MaxSAT instances (627) from the industrial category of the MaxSAT Evaluation 2013[1] as a benchmark for our experiments. The evaluation was performed on two AMD Opteron 6276 processors (2.3 GHz) running Fedora 18 with a timeout of 1,800 seconds and a memory limit of 8 GB. We implemented all algorithms described in section 2 (Linear search Unsat-Sat, Fu-Malik, and MSU3), as well as their incremental counterparts on top of OPEN-WBO [42]. OPEN-WBO is a modular open source MaxSAT solver that is easy to modify and is competitive with state-of-the-art MaxSAT solvers.

Table 1 shows the number of instances solved (*#Inst*) by the described MaxSAT algorithms using the different approaches, namely, non-incremental approach (*none*), incremental blocking (*blocking*), incremental weakening (*weakening*), and iterative encoding (*iterative*). Table 1 also shows the median speedup[2] for instances that have been solved by all incremental approaches for a given algorithm.

Fu-Malik with incremental blocking significantly outperforms the non-incremental algorithm. Incremental blocking not only solves more instances but also is significantly

[1] Benchmarks available at `http://maxsat.ia.udl.cat/13/benchmarks/`

[2] The speedup of an instance is measured as the ratio of the solving time of the non-incremental approach to the solving time of the respective incremental approach.

faster than the non-incremental algorithm. From those instances which were solved by both approaches, 50% of them have a speedup of at least 2.4. Incremental weakening and iterative encoding cannot be used with the Fu-Malik algorithm since it only uses at most one constraints and modifies the formula across iterations of the algorithm.

Linear search Unsat-Sat (LinearUS) with incremental blocking solves less instances than the non-incremental approach. Incremental blocking encodes a new cardinality constraint at each iteration of the MaxSAT algorithm, causing the formula to grow too large resulting in termination due to memory outs. However, for those instances that were solved successfully, incremental blocking was 60% faster than the original LinearUS. Incremental weakening allows MaxSAT algorithms to solve more instances with significant speedup. Since the cardinality constraint is encoded only once, the size of the formula remains almost constant across iterations. The majority of the instances are solved at least 2× faster. Iterative encoding outperforms all other approaches. Smaller formula sizes due to iterative encoding allows it to solve more instances as compared to incremental weakening.

MSU3 with incremental blocking solves less instances as compared to the original MSU3 but it is faster for instances solved by both approaches. Similar results have been observed for the LinearUS algorithm with incremental blocking. Incremental weakening outperforms incremental blocking in the number of solved instances as well as in terms of solving time. However, incremental weakening solves less instances than the non-incremental approach, since incremental weakening is not flexible to directly support the increase in the number of input literals of the cardinality constraint. A non-incremental approach may need to impose the cardinality constraint over a small subset of relaxation variables. Incremental weakening does not enjoy this benefit due to its inflexibility. This may result in incremental weakening producing a larger encoding for certain problem instances. Iterative encoding solves more instances and is significantly faster than the non-incremental approach. Iterative encoding only encodes the clauses that are needed at each iteration of the MaxSAT algorithm, allowing for an encoding with a similar size to the non-incremental approach. Most instances are solved at least 3.6× faster with iterative encoding than without it.

Fig. 3 shows scatter plots that compare the non-incremental and incremental approaches which are highlighted in Table 1. Each point in the plot corresponds to a problem instance, where the x-axis corresponds to the run time required by non-incremental approaches and the y-axis corresponds to the run time required by incremental approaches. Instances that are above the diagonal are solved faster when using a non-incremental approach, whereas instances that are below the diagonal are solved faster when using an incremental approach. Incremental approaches that we propose in this paper clearly assert their dominance over their non-incremental counterparts integrated with all three algorithms as shown in Fig. 3. This is particularly evident in the MSU3 algorithm where the majority of the instances are solved much faster with iterative encoding. For example, for 30% of the instances solved by MSU3 with and without iterative encoding, iterative encoding is at least 6× faster than the non-incremental approach. For 10% of the instances solved by both approaches, iterative encoding boosts MSU3 with at least 14× speedup.

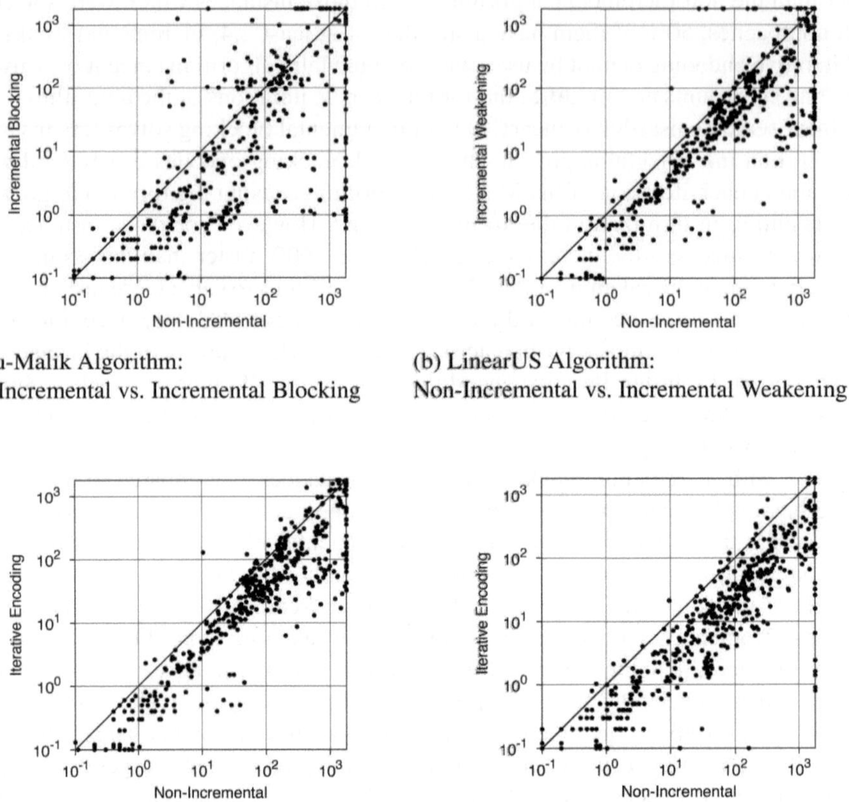

(a) Fu-Malik Algorithm: Non-Incremental vs. Incremental Blocking

(b) LinearUS Algorithm: Non-Incremental vs. Incremental Weakening

(c) LinearUS Algorithm: Non-Incremental vs. Iterative Encoding

(d) MSU3 Algorithm: Non-Incremental vs. Iterative Encoding

Fig. 3. Impact of incremental approaches

Fig. 4 shows a cactus plot with the running times of state-of-the-art MaxSAT solvers used in the MaxSAT Evaluation 2013[3] (WPM1 [3], WPM2 [4, 2], MaxHS [15, 16], BCD2 [44], QMaxSAT2 [29]) and the best incremental algorithms presented in this paper (incremental blocking Fu-Malik, iterative encoding LinearUS and MSU3).

Fu-Malik and WPM1 use similar MaxSAT algorithms. Moreover, WPM1 has a similar incremental strategy due to the incremental SMT solver that is used by WPM1. Since both solvers used similar techniques, it is not surprising that their performance is similar. Even though LinearUS uses a simple MaxSAT algorithm, it is competitive with more complex state-of-the-art MaxSAT algorithms. This is mostly due to the incremental approach that is being used in LinearUS and shows the importance of using an efficient incremental approach. MSU3 and QMaxSAT2 perform complementary

[3] Only single engine solvers have been considered in this evaluation, therefore we did not include ISAC+ (a portfolio MaxSAT solver) [28].

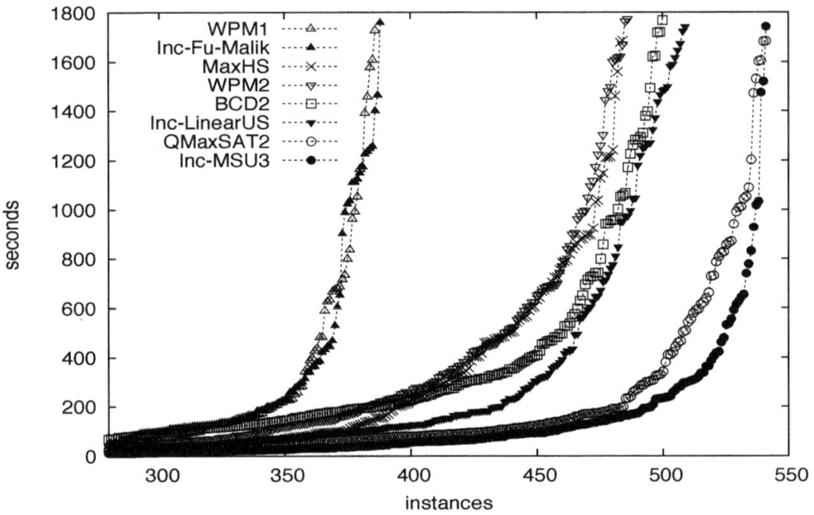

Fig. 4. Running times of state-of-the-art MaxSAT solvers

searches but both use incrementality and have similar performances. Iterative encoding is not restricted to MSU3 and may be used in other MaxSAT algorithms, such as WPM2 and BCD2. It is expected that if those algorithms are enhanced with the incremental iterative encoding, their performance might rise to values similar or higher than those of QMaxSAT2 and MSU3.

6 Conclusions and Future Work

Several state of the art MaxSAT algorithms are based on solving a sequence of closely related SAT formulas. However, although incrementality is not a new technique, it is seldom used in MaxSAT algorithms that search on the lower bound of the optimum solution. In this paper, we describe and propose new techniques to incrementally modify cardinality constraints used in several MaxSAT algorithms, namely in linear Unsat-Sat search, the classic Fu-Malik algorithm and MSU3 core-guided algorithm.

Experimental results show the effectiveness of the techniques proposed in the paper. The incremental versions of the MaxSAT algorithms clearly outperform the non-incremental versions, both in terms of speed and number of solved instances. Furthermore, the proposed techniques can be integrated in other core-guided algorithms such as WPM2 and BCD2, among others.

Finally, the paper also describes that in general it is possible to perform iterative encoding of cardinality constraints using the Totalizer encoding. Therefore, the use of this technique is not limited to the scope of MaxSAT algorithms. As future work, we propose to integrate these techniques in other domains where cardinality constraints are used, and to extend incrementality to other effective cardinality constraints encodings.

References

1. Abío, I., Stuckey, P.J.: Conflict Directed Lazy Decomposition. In: Milano, M. (ed.) CP 2012. LNCS, vol. 7514, pp. 70–85. Springer, Heidelberg (2012)
2. Ansótegui, C., Bonet, M.L., Gabàs, J., Levy, J.: Improving WPM2 for (Weighted) Partial MaxSAT. In: Schulte, C. (ed.) CP 2013. LNCS, vol. 8124, pp. 117–132. Springer, Heidelberg (2013)
3. Ansótegui, C., Bonet, M.L., Levy, J.: Solving (Weighted) Partial MaxSAT through Satisfiability Testing. In: Kullmann (ed.) [30], pp. 427–440
4. Ansótegui, C., Bonet, M.L., Levy, J.: A New Algorithm for Weighted Partial MaxSAT. In: Fox, M., Poole, D. (eds.) AAAI Conference on Artificial Intelligence. AAAI Press (2010)
5. Argelich, J., Berre, D.L., Lynce, I., Marques-Silva, J., Rapicault, P.: Solving Linux Upgradeability Problems Using Boolean Optimization. In: Workshop on Logics for Component Configuration, pp. 11–22 (2010)
6. Asín, R., Nieuwenhuis, R.: Curriculum-based course timetabling with SAT and MaxSAT. Annals of Operations Research, 1–21 (2012)
7. Asín, R., Nieuwenhuis, R., Oliveras, A., Rodríguez-Carbonell, E.: Cardinality Networks: a theoretical and empirical study. Constraints 16(2), 195–221 (2011)
8. Audemard, G., Lagniez, J.M., Simon, L.: Improving Glucose for Incremental SAT Solving with Assumptions: Application to MUS Extraction. In: Järvisalo, M., Van Gelder, A. (eds.) SAT 2013. LNCS, vol. 7962, pp. 309–317. Springer, Heidelberg (2013)
9. Bailleux, O., Boufkhad, Y.: Efficient CNF Encoding of Boolean Cardinality Constraints. In: Rossi, F. (ed.) CP 2003. LNCS, vol. 2833, pp. 108–122. Springer, Heidelberg (2003)
10. Barrett, C., Stump, A., Tinelli, C.: The SMT-LIB Standard: Version 2.0. Tech. rep., Department of Computer Science, The University of Iowa (2010), www.SMT-LIB.org
11. Büttner, M., Rintanen, J.: Satisfiability Planning with Constraints on the Number of Actions. In: Biundo, S., Myers, K.L., Rajan, K. (eds.) International Conference on Automated Planning and Scheduling, pp. 292–299 (2005)
12. Chen, Y., Safarpour, S., Marques-Silva, J., Veneris, A.G.: Automated Design Debugging With Maximum Satisfiability. IEEE Transactions on CAD of Integrated Circuits and Systems 29(11), 1804–1817 (2010)
13. Cheng, K.C.K., Yap, R.H.C.: Maintaining Generalized Arc Consistency on Ad-Hoc n-Ary Boolean Constraints. In: Brewka, G., Coradeschi, S., Perini, A., Traverso, P. (eds.) European Conference on Artificial Intelligence. Frontiers in Artificial Intelligence and Applications, vol. 141, pp. 78–82. IOS Press (2006)
14. Cimatti, A., Sebastiani, R. (eds.): SAT 2012. LNCS, vol. 7317, pp. 2012–2015. Springer, Heidelberg (2012)
15. Davies, J., Bacchus, F.: Exploiting the Power of mip Solvers in maxsat. In: Järvisalo, M., Van Gelder, A. (eds.) SAT 2013. LNCS, vol. 7962, pp. 166–181. Springer, Heidelberg (2013)
16. Davies, J., Bacchus, F.: Postponing Optimization to Speed Up MAXSAT Solving. In: Schulte, C. (ed.) CP 2013. LNCS, vol. 8124, pp. 247–262. Springer, Heidelberg (2013)
17. Debruyne, R.: Arc-Consistency in Dynamic CSPs Is No More Prohibitive. In: International Conference on Tools with Artificial Intelligence, pp. 299–307. IEEE (1996)
18. Dechter, R., Dechter, A.: Belief Maintenance in Dynamic Constraint Networks. In: Shrobe, H.E., Mitchell, T.M., Smith, R.G. (eds.) AAAI Conference on Artificial Intelligence, pp. 37–42. AAAI Press / The MIT Press (1988)
19. Dutertre, B., de Moura, L.M.: A Fast Linear-Arithmetic Solver for DPLL(T). In: Ball, T., Jones, R.B. (eds.) CAV 2006. LNCS, vol. 4144, pp. 81–94. Springer, Heidelberg (2006)
20. Eén, N., Sörensson, N.: Translating Pseudo-Boolean Constraints into SAT. Journal on Satisfiability, Boolean Modeling and Computation 2, 1–26 (2006)

21. Eén, N., Sörensson, N.: An Extensible SAT-solver. In: Giunchiglia, E., Tacchella, A. (eds.) SAT 2003. LNCS, vol. 2919, pp. 502–518. Springer, Heidelberg (2004)
22. Eén, N., Sörensson, N.: Temporal induction by incremental SAT solving. Electronic Notes in Theoretical Computer Science 89(4), 543–560 (2003)
23. Fu, Z., Malik, S.: On Solving the Partial MAX-SAT Problem. In: Biere, A., Gomes, C.P. (eds.) SAT 2006. LNCS, vol. 4121, pp. 252–265. Springer, Heidelberg (2006)
24. Graça, A., Lynce, I., Marques-Silva, J., Oliveira, A.L.: Efficient and Accurate Haplotype Inference by Combining Parsimony and Pedigree Information. In: Horimoto, K., Nakatsui, M., Popov, N. (eds.) ANB 2010. LNCS, vol. 6479, pp. 38–56. Springer, Heidelberg (2012)
25. Heras, F., Morgado, A., Marques-Silva, J.: Core-guided binary search algorithms for maximum satisfiability. In: Burgard, W., Roth, D. (eds.) AAAI Conference on Artificial Intelligence. AAAI Press (2011)
26. Hooker, J.N.: Solving the incremental satisfiability problem. Journal of Logic Programming 15(1&2), 177–186 (1993)
27. Jose, M., Majumdar, R.: Cause clue clauses: error localization using maximum satisfiability. In: Hall, M.W., Padua, D.A. (eds.) Programming Language Design and Implementation, pp. 437–446. ACM (2011)
28. Kadioglu, S., Malitsky, Y., Sellmann, M.: Non-Model-Based Search Guidance for Set Partitioning Problems. In: Hoffmann, J., Selman, B. (eds.) AAAI Conference on Artificial Intelligence. AAAI Press (2012)
29. Koshimura, M., Zhang, T., Fujita, H., Hasegawa, R.: QMaxSAT: A Partial Max-SAT Solver. Journal on Satisfiability, Boolean Modeling and Computation 8, 95–100 (2012)
30. Kullmann, O. (ed.): SAT 2009. LNCS, vol. 5584. Springer, Heidelberg (2009)
31. Lagerkvist, M.Z., Schulte, C.: Advisors for Incremental Propagation. In: Bessière, C. (ed.) CP 2007. LNCS, vol. 4741, pp. 409–422. Springer, Heidelberg (2007)
32. Le Berre, D., Parrain, A.: The Sat4j library, release 2.2. Journal on Satisfiability, Boolean Modeling and Computation 7(2-3), 59–66 (2010)
33. Li, C.M., Manyà, F.: MaxSAT, Hard and Soft Constraints. In: Handbook of Satisfiability, pp. 613–631. IOS Press (2009)
34. Liffiton, M.H., Sakallah, K.A.: Algorithms for Computing Minimal Unsatisfiable Subsets of Constraints. Journal Automated Reasoning 40(1), 1–33 (2008)
35. Lonsing, F., Egly, U.: Incremental QBF Solving. Computing Research Repository - arXiv abs/1402.2410 (2014)
36. Mahajan, Y.S., Fu, Z., Malik, S.: Zchaff2004: An efficient sat solver. In: H. Hoos, H., Mitchell, D.G. (eds.) SAT 2004. LNCS, vol. 3542, pp. 360–375. Springer, Heidelberg (2005)
37. Manolios, P., Papavasileiou, V.: Pseudo-Boolean Solving by incremental translation to SAT. In: Bjesse, P., Slobodová, A. (eds.) International Conference on Formal Methods in Computer-Aided Design, pp. 41–45. FMCAD Inc. (2011)
38. Manquinho, V., Marques-Silva, J., Planes, J.: Algorithms for Weighted Boolean Optimization. In: Kullmann (ed.) [30], pp. 495–508
39. Marin, P., Miller, C., Lewis, M.D.T., Becker, B.: Verification of partial designs using incremental QBF solving. In: Rosenstiel, W., Thiele, L. (eds.) Design, Automation, and Test in Europe Conference, pp. 623–628. IEEE (2012)
40. Marques-Silva, J., Planes, J.: On using unsatisfiability for solving Maximum Satisfiability. Tech. rep., Computing Research Repository, abs/0712.0097 (2007)
41. Martins, R., Manquinho, V., Lynce, I.: Parallel Search for Maximum Satisfiability. AI Communications 25(2), 75–95 (2012)
42. Martins, R., Manquinho, V., Lynce, I.: Open-WBO: a Modular MaxSAT Solver. In: Sinz, C., Egly, U. (eds.) SAT 2014. LNCS, vol. 8561, pp. 438–445. Springer, Heidelberg (2014)
43. Morgado, A., Heras, F., Liffiton, M., Planes, J., Marques-Silva, J.: Iterative and core-guided MaxSAT solving: A survey and assessment. Constraints 18(4), 478–534 (2013)

44. Morgado, A., Heras, F., Marques-Silva, J.: Improvements to Core-Guided Binary Search for MaxSAT. In: Cimatti, Sebastiani (eds.) [14], pp. 284–297
45. Nadel, A., Ryvchin, V.: Efficient SAT Solving under Assumptions. In: Cimatti, Sebastiani (eds.) [14], pp. 242–255
46. Sinz, C.: Towards an Optimal CNF Encoding of Boolean Cardinality Constraints. In: van Beek, P. (ed.) CP 2005. LNCS, vol. 3709, pp. 827–831. Springer, Heidelberg (2005)
47. Shtrichman, O.: Pruning Techniques for the SAT-Based Bounded Model Checking Problem. In: Margaria, T., Melham, T.F. (eds.) CHARME 2001. LNCS, vol. 2144, pp. 58–70. Springer, Heidelberg (2001)
48. van Beek, P.: Backtracking Search Algorithms. In: Rossi, F., van Beek, P., Walsh, T. (eds.) Handbook of Constraint Programming, ch. 4. Elsevier (2006)
49. van Hoeve, W.J., Katriel, I.: Global constraints. In: Rossi, F., van Beek, P., Walsh, T. (eds.) Handbook of Constraint Programming, ch. 6. Elsevier (2006)
50. Whittemore, J., Kim, J., Sakallah, K.A.: SATIRE: A New Incremental Satisfiability Engine. In: Design Automation Conference, pp. 542–545. ACM (2001)

Reducing the Branching in a Branch and Bound Algorithm for the Maximum Clique Problem

Ciaran McCreesh and Patrick Prosser

University of Glasgow, Glasgow, UK
c.mccreesh.1@research.gla.ac.uk,
patrick.prosser@glasgow.ac.uk

Abstract. Finding the largest clique in a given graph is one of the fundamental NP-hard problems. We take a widely used branch and bound algorithm for the maximum clique problem, and discuss an alternative way of understanding the algorithm which closely resembles a constraint model. By using this view, and by taking measurements inside search, we provide a new explanation for the success of the algorithm: one of the intermediate steps, by coincidence, often approximates a "smallest domain first" heuristic. We show that replacing this step with a genuine "smallest domain first" heuristic leads to a reduced branching factor and a smaller search space, but longer runtimes. We then introduce a "domains of size two first" heuristic, which integrates cleanly into the algorithm, and which both reduces the size of the search space and gives a reduction in runtimes.

1 Introduction

A clique in a graph is a subset of vertices, each of which is adjacent to every other vertex in this subset—we illustrate this in Fig. 1. Finding the size of a maximum clique in a given graph is one of Garey and Johnson's fundamental NP-hard problems [1]. The maximum clique problem has been studied in a

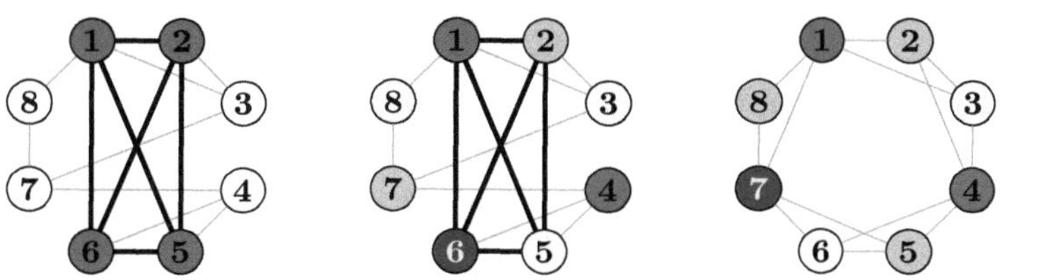

Fig. 1. On the left, a graph with a maximum clique of size four. Next, a greedy four-colouring of this graph: vertices $\{1, 4\}$ have been coloured dark blue, vertices $\{2, 7\}$ are light blue, vertices $\{3, 5, 8\}$ are pale cream and vertex 6 is dark chocolate. On the right, a graph which requires four colours but does not contain a clique of size four.

B. O'Sullivan (Ed.): CP 2014, LNCS 8656, pp. 549–563, 2014.

constraint programming setting by Régin [2], and using MaxSAT and MaxSAT-inspired algorithms by Li et al. [3,4,5,6]. However, we will be looking at a family of dedicated branch and bound algorithms due to Tomita et al. [7,8,9]. These algorithms are widely used on "real" problems in practice [10,11,12,13,14,15,16], and many variations have been proposed [17,18,19,20], notably including bit- and thread-parallel versions [21,22,23,15]; the techniques we will investigate have also been reused to solve other problems, including maximum common subgraph [15] and maximum balanced induced biclique [24]. But despite this wide use, it is not entirely clear *why* these algorithms work so well. In particular, in one part of the algorithm we iterate over a certain array backwards. It is easy to check that this leads to a much smaller search space than iterating over this array in the forwards direction instead, but recent experimental work contradicts the conventional explanation for why this should be the case.

By rephrasing the algorithm using language from constraint programming, we will see that these "forward" and "backward" iterations are effectively two different variable selection heuristics (although the variables only exist implicitly). We take measurements inside search to demonstrate that "backwards", by coincidence, approximates a "smallest domain first" (SDF) heuristic, and that "forwards" is roughly "largest domain first". It is then easy to modify the algorithm to use a genuine SDF heuristic; doing so leads to a smaller search space due to a reduced branching factor, but longer runtimes due to the cost of performing a sort for each recursive call made by the algorithm. Finally, we show how to get both a smaller search space *and* improved runtimes by using a cheaper alternative to SDF, which is effectively "domains of size two first".

2 Algorithms for the Maximum Clique Problem

Throughout, let $G = (V, E)$ be a graph with vertex set V and edge set E. We write $\mathrm{V}(G)$ for V, and $\mathrm{N}(G, v)$ for the neighbourhood of a vertex v (that is, the vertices adjacent to it). The size of a maximum clique is denoted ω. When discussing random graphs, we use $G(n, p)$ to denote an Erdős-Rényi random graph with n vertices, and with edges between each pair of distinct vertices with independent probability p.

We start by describing Algorithm 1, a generic, exact branch and bound algorithm for the maximum clique problem. This algorithm is essentially Tomita et al.'s "MCS" [9] with a simpler initial vertex ordering and the colour repair step omitted (Prosser's computational study [18] calls this combination "MCSa1"), but we describe it in a new, more flexible way that allows us to investigate and explain its behaviour.

The key to this algorithm is a relationship between cliques and colourings. A *colouring* of a graph is an assignment of vertices to colours, where adjacent vertices are given different colours. A set of like-coloured vertices in a given colouring is called a *colour class*. A clique in a coloured graph can contain at most one vertex from each colour class (see Fig. 1), so if we can colour a graph

Algorithm 1. A generic exact algorithm to deliver a maximum clique

1 `maxClique` :: (Graph G) $\rightarrow$ Set
2 **begin**
3   **global** $C_{max} \leftarrow \emptyset$
4   `expand`$(G, \emptyset, \mathrm{V}(G))$
5   **return** C_{max}

6 `expand` :: (Graph G, Set C, Set P)
7 **begin**
8   *colourClasses* $\leftarrow$ `colour`(G, P)
9   **while** *colourClasses* $\neq \emptyset$ **do**
10   *colourClass* $\leftarrow$ `select`(*colourClasses*)
11   **for** $v \in$ *colourClass* **do**
12   **if** $|C| + |colourClasses| \leq |C_{max}|$ **then return**
13             $P' \leftarrow P \cap \mathrm{N}(G, v)$
14             $C \leftarrow C \cup \{v\}$
15   **if** $|C| > |C_{max}|$ **then** $C_{max} \leftarrow C$ **if** $P' \neq \emptyset$ **then** `expand`(G, C, P')
17             $C \leftarrow C \setminus \{v\}$
18   *colourClasses* $\leftarrow$ `remove`(*colourClasses*, *colourClass*)

using k colours, we have shown that $\omega \leq k$. This is not generally an equality: the third graph in Fig. 1 has $\omega = 3$, but cannot be coloured using three colours. Like finding a maximum clique, finding a minimum colouring is NP-hard. However, we may construct a greedy colouring in polynomial time.

These facts are used to grow a maximum clique, as follows. We have a variable C, which contains the current growing clique, and a variable P (the candidate set) which contains vertices which may potentially be added to C. Initially C is empty, and P contains every vertex in the graph (line 4). Now we produce a greedy colouring of the subgraph induced by P (line 8); we describe how this is performed in Algorithm 2. From this colouring, we select a colour class (line 10), and from that colour class we select a vertex v (line 11). We then consider every clique which contains the vertices in C plus v, by filtering from P any vertices not adjacent to v (line 13), and recursing (line 16). Next we consider every clique which contains the vertices in C but not v: we remove v from consideration (line 17), and select a new vertex from the current colour class (line 11). When the current colour class has been explored we remove it (line 18) to consider the possibility of not selecting any vertex at all from this colour class, and then select a new colour class (line 9).

We keep track of the largest clique we have found so far (line 15), which we call the *incumbent*, and store it in C_{max}. At any point during the search, if the number of colour classes remaining plus the number of vertices in C is not strictly greater than $|C_{max}|$, we cannot unseat the incumbent and so we backtrack (line 12).

Algorithm 2. Greedily colour vertices, delivering a list of colour classes

```
colour :: (Graph G, Set P) → List of Set
begin
    colourClasses ← ∅
    uncoloured ← copy(P)
    while uncoloured ≠ ∅ do
        current ← ∅
        for v ∈ uncoloured do
            if current ∩ N(G, v) = ∅ then current ← current ∪ {v}
        uncoloured ← uncoloured \ current
        colourClasses ← append(colourClasses, current)
    return colourClasses
```

Note that we produce a new colouring each time we recurse—we generally get tighter colourings as we consider smaller subproblems. Thus being able to produce a colouring very quickly is important. San Segundo et al. [21,22] observed that using a bitset encoding for the entire algorithm would lead to a performance improvement of between two and twenty times, without changing the steps taken. We will be using (but not explicitly describing) a bitset encoding throughout, and refer the reader to these papers for implementation details.

In constraint programming terms, we can think of colour classes as variables, and vertices within a colour class as values. We are forming a clique by picking a vertex from each colour class. There is also a "nothing from this colour class" value, which we wish to take as infrequently as possible. On line 10 we are selecting a variable, and on line 11 we are trying a value for that variable; the filtering on line 13 is propagation. There is the slight conceptual complication in that we are producing a new set of variables at each recursive call—perhaps this is why this explanation has not been considered previously.

There are three choices to be made when implementing this algorithm. The first is how the colouring is produced (line 8). For this paper, we will use a simple greedy sequential colouring, where colour classes are filled by selecting vertices in order. We show this in Algorithm 2. The order in which vertices are considered has a large effect upon the colourings produced; here we select vertices in a static non-increasing degree order (this is implemented by permuting the graph at the top of search). Other initial vertex orders and more sophisticated (but also more computationally expensive) colouring algorithms sometimes give better results, and sometimes give worse results. A computational study by Prosser [18] examines this issue in depth; our approach is compatible with other colourings and initial vertex orderings, and we are describing the simplest options for clarity.

Note that the choice of vertex ordering here is made to improve the quality of the colouring, in the hope that the greedy colouring will be close to optimal. For computationally challenging graphs, the degree of a vertex is often not an indication of whether it is present in a solution—either by design [25,26], because the degree spread is very narrow (as in Erdős-Rényi random graphs), or because the graph contains many maximum cliques but the difficulty lies in proving optimality [27,28,29].

The second choice to be made is the order in which colour classes are selected (the `select` function used in line 10) and the third is the order in which vertices are selected from within a colour class (iteration over the colour class in line 11). These choices have not been investigated deeply (nor have they been presented explicitly as choices to be made). To emulate Tomita's algorithms we would select (line 10) and then remove (line 18) the last colour class from the list of colour classes constructed by Algorithm 2. Vertices within colour classes would also be selected in reverse order, with the last vertex in that colour class chosen first.

But why select vertices in colour class order, and why select from right to left? This may be implemented very efficiently by using a pair of arrays, as in Fig. 2. The first array (drawn as coloured vertices) controls the iteration order: it contains vertex numbers, in the reverse order that they are to be considered. The second array holds the bound (drawn as a list of numbers): in the ith entry we store the number of colours that were used to colour the induced subgraph which contains only the first i vertices from the first array. Since vertices with the same colour are adjacent in the iteration order, the bound is decreasing when iterating from right to left, which allows Algorithm 1 to be implemented using a single loop—in fact, this algorithm has not previously been described in any other way.

However, recursing from left to right (and thus selecting from the first colour class first, rather than the last colour class first) may be implemented equally efficiently, so why use a reverse order? Tomita claims that vertices in the rightmost colour class are "generally expected [to have a] high probability of belonging to a maximum clique" [8]. This claim was not tested experimentally, beyond verifying that the reverse ordering gives much worse performance, and recent experiments by the authors [29] and by Batsyn et al. [19] suggest that for several families of graphs, these algorithms are not particularly good at finding a maximum clique quickly.

We argue that there is another factor contributing to the success of the reverse selection order. Intuitively, one might think that early colour classes are likely to be larger: colour classes are filled greedily, with vertices being placed in the first available colour class. Selecting from small colour classes first is beneficial: consider Fig. 2, and suppose $C_{max} = 3$. If we select v from the rightmost colour class (which contains only one vertex) first, we make only a single recursive call which cannot be eliminated by the bound. But if we were to select from any other colour class, we would have to make either two or three recursive calls before our bound would decrease. (This also shows why we commit entirely to a selected colour class: we want to eliminate colour classes as quickly as possible.)

In constraint programming terms, selecting from small colour classes first is a "smallest domain first" variable selection heuristic. Such a heuristic tends to give a low branching factor locally (that is, it reduces the number of recursive

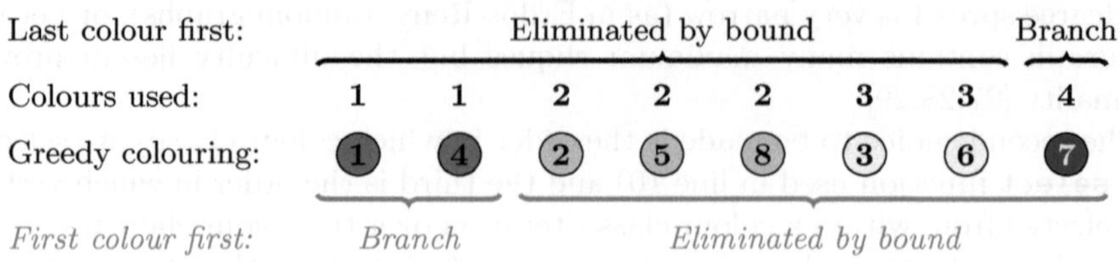

Fig. 2. The colour classes from the third graph of Fig. 1, in colour order. Above the vertices is an array of bounds: the ith entry shows the number of colours used to colour the subgraph containing only the first i vertices from the colouring. Now suppose an incumbent of size three had already been found. If we select from the rightmost colour class first, and discover that there is no clique of size four containing vertex 7, then we may abandon search. But if we select from the leftmost colour class first (as drawn below), we must recurse twice: once to show that there is no clique of size four containing vertex 1, and then again for vertex 4.

calls made) [30]. This does not necessarily produce the best possible search tree globally, but we will demonstrate that it is generally beneficial in this context.

2.1 Are Colour Classes Roughly Sorted by Size?

We will now test our intuition, by augmenting Algorithm 2 to take measurements inside the search. The hypothesis we are testing is as follows: is there a correlation between the position a colour classes is in, and the position it would be in if colour classes were sorted by size (largest first)? To measure this, we use Spearman's rank correlation coefficient (with rank ties) [31]; this will give us a value of 1 if there is a perfect monotonically increasing relationship, -1 if it is perfectly monotonically decreasing, and a value in-between otherwise.

We performed this test for each colouring produced, over 100 samples of random graphs $G(150, 0.9)$. The results are plotted in the top left graph of Fig. 3. For the x-axis, we use the number of colour classes used. For the y-axis, rather than show the average, we show the distribution of the results of the statistical test (so the colours in each column sum to 1). For comparison purposes, the bottom left graph shows what we would see if the colour classes were in no particular order (we shuffle the colour classes before running the test), and the bottom right graph shows the color classes fully sorted (i.e. SDF). These results confirm our suspicions that colour classes are "roughly" sorted by size, as a side effect of the greedy colouring process: the top left graph is much more heavily weighted towards 1 (sorted) than the shuffled graph. In other words, the greedy colouring process and backwards iteration is approximating an SDF heuristic.

The top right graph shows the effects of our "domains of size two first" heuristic, which we describe below. As its name suggests, a partial sort increases the degree to which colour classes are sorted by size, but does not sort them fully—it is a cheap surrogate for SDF.

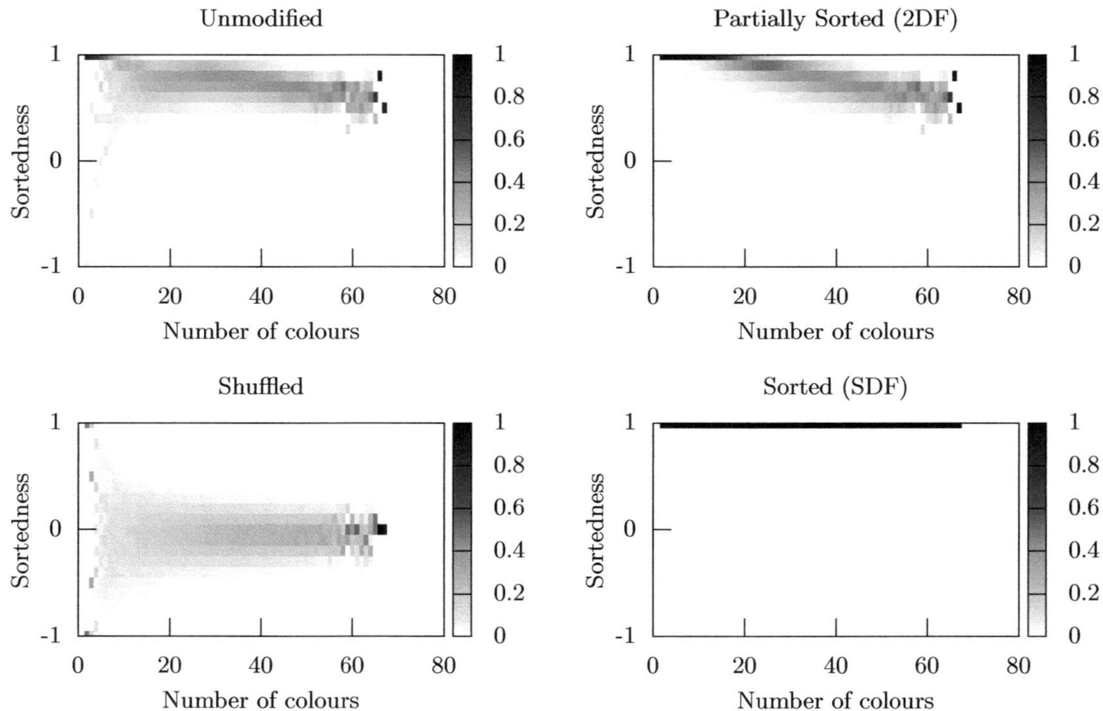

Fig. 3. Are colour classes roughly ordered by size? A value of 1 means "yes, largest first", 0 means "no", and -1 means "yes, smallest first". In the top left graph, the original algorithm, and in the top right, the effects of a partial sort. For comparison, the bottom left graph shows shuffled colour classes, and the bottom right graph shows fully sorted colour classes. Results are from 100 samples of $G(150, 0.9)$.

2.2 Reordering Colour Classes to Reduce the Branching Factor

We have established empirically that smaller colour classes tend to be picked earlier by Tomita's algorithms, and explained theoretically why this is beneficial. Now ask what would happen if we increased this effect. We consider two approaches.

The "sorted", or "smallest domains first" variation. We could explicitly select from the smallest colour class (i.e. the smallest domain) first. This could be implemented directly, via a different `select` function used with Algorithm 1, or we could use Tomita's "two arrays" approach and add a (stable) sort to the end of Algorithm 2.

The "partially sorted", or "domains of size two first" (2DF) variation. We also consider a potentially cheaper alternative: instead of fully sorting colour classes by size, we propose a partial sort that moves colour classes containing only one vertex (which we call *singleton* colour classes) to the end of the list of colour classes, so that they are selected first. In other words, we are picking from domains with two values (a single vertex, plus the "nothing" option) first. We show how to do this in a way which is compatible with a bitset encoding in Algorithm 3: when we produce a colour class containing only a single vertex, we append that colour class onto the list *singletons* (line 12) and when every vertex has been processed we return the concatenated list of colour classes with the singletons appearing at the end (line 15). We then replace the call to `colour` in line 8 of Algorithm 1 with a call to `colourSort2DF`, and implement the `select` step and the bound by using two arrays and selecting the rightmost entry first, as in Tomita's algorithms.

2.3 Tie-Breaking

But why are we preserving the relative order of the partially sorted colour classes—that is, why do we specify a stable sort, or why is it important to put the last singleton colour class at the end of the list of colour classes? Suppose Algorithm 2 produced the colour classes shown in Fig. 4. Due to the greediness of the colouring, vertices 5, 6, 7, 8 and 9 must all be adjacent to vertex 4 (the only member of the dark chocolate colour class), for if one were not, it would also have been coloured dark chocolate. Thus if Algorithm 1 selects colour class $\{4\}$ in preference to the other singleton colour class $\{7\}$, the new candidate set P' will contain *some* of the vertices from the set $\{1, 2, 3\}$, and *all* of the vertices from the sets $\{5, 6\}$, $\{7\}$ and $\{8, 9\}$. However, by the same kind of reasoning, if the colour class $\{7\}$ is selected before $\{4\}$, P' will contain *some* of the vertices from the sets $\{1, 2, 3\}$ and $\{5, 6\}$ and *all* of the vertices in the sets $\{4\}$ and $\{8, 9\}$, so the new candidate set will potentially be smaller. This is why we preserve the order: selecting from the latest-coloured singleton colour class first can increase the amount of filtering done on P, giving a smaller P' in the recursive call (line 16), further reducing the branching in the search process.

Algorithm 3. Greedily colour vertices, delivering a partially sorted list of colour classes, with singleton colour classes deferred to the end.

```
colourSort2DF :: (Graph G, Set P) → List of Set
begin
    colourClasses ← ∅
    singletons ← ∅
    uncoloured ← copy(P)
    while uncoloured ≠ ∅ do
        current ← ∅
        for v ∈ uncoloured do
            if current ∩ N(G, v) = ∅ then current ← current ∪ {v}
        uncoloured ← uncoloured \ current
        if |current| = 1 then
            singletons ← append(singletons, current)
        else
            colourClasses ← append(colourClasses, current)
    return concatenate(colourClasses, singletons)
```

2.4 Compatibility with Other Improvements

Both the "sorted" and "partially sorted" changes are compatible with other recent improvements that have been proposed for this family of algorithms. They do not interfere with priming the search with a heuristic solution [19], they are not sensitive to alternative vertex orderings, and critically, they are compatible with multi-core parallelism [15,23,29].

One improvement with which they are *not* compatible is a relaxed colouring proposed by San Segundo and Tapia [20]. Relaxed colourings also do not interact cleanly with parallel branch and bound or with priming (with relaxed colourings, finding a larger incumbent earlier can be a penalty rather than a benefit), so we do not consider this to be a substantial weakness.

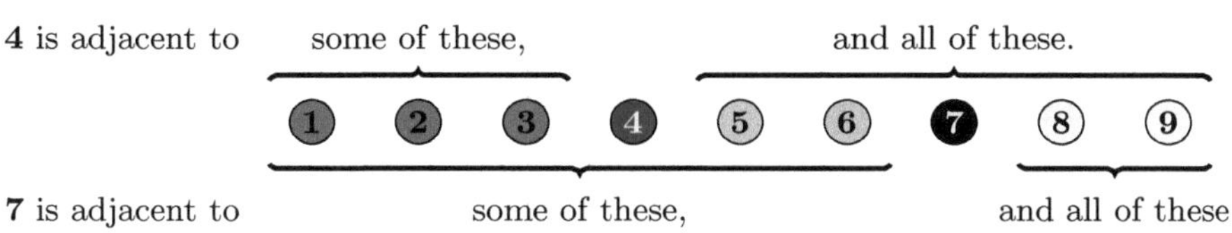

Fig. 4. Due to the greedy colouring, singleton colour classes are not equally powerful from a filtering perspective. For any singleton colour class, its vertex is adjacent to every vertex with a later colour, but only some vertices with an earlier colour. Here, branching on vertex 7 rather than vertex 4 is likely to lead to more filtering, giving a smaller subproblem at the next recursive call.

3 Experimental Results

We now evaluate our "SDF" and "2DF" changes experimentally on a range of standard and random problems. Where runtimes are given, experimental results are produced on a machine with an Intel Xeon E5645 CPU. The time taken to read in a graph from a file is not measured, but preprocessing time is included. The algorithms were implemented in C++. Sometimes we report the number of "nodes" explored by an algorithm, by which we mean the number of recursive calls made to the `expand` function in Algorithm 1.

For our baseline, we use a bitset encoded version of the variant Prosser calls "MCSa1" [18]; our implementation has been shown to perform very competitively with other implementations of the same algorithm [23]. We also have a parallel implementation of these algorithms, although we are reporting sequential results for ease of understanding (parallel branch and bound is speculative, and we often see anomalous speedups [15,23] due to additional diversity from differing search orders [29]). For the "SDF" implementation, we add a stable sort (using the C++ standard library function) at the end of Algorithm 2; for the "2DF" implementation, we replace Algorithm 2 with Algorithm 3. In both cases, we use Tomita's "two arrays" approach, and have `select` pick from right to left.

3.1 Random Graphs

In the top graph in Fig. 5 we show the number of search nodes required for random graphs $G(200, x)$ for values of x between 0.70 and 0.99 (averaged over 100 graph instances for each x). We see that there is a clear benefit to reordering colour classes, either by sorting or by partially sorting. However, sorting and partial sorting give lines which are too close to be easily distinguishable—at least for these graphs, simply deferring singleton colour classes is as effective as a full sort.

But do reductions in the size of the search space help with performance? In the graph below, we present the same data, but measuring runtimes. We see that the improvements to runtimes from a partial sort reflects the improvements to search nodes. On the other hand, it is clear that a full sort is extremely expensive: we get a factor of five slowdown despite the smaller search space.

3.2 Standard Benchmark Problems

Next we consider a range of standard benchmark problems from the Second DIMACS Implementation Challenge[1] and from BHOSLIB ("Benchmarks with Hidden Optimum Solutions for Graph Problems")[2]. From DIMACS, we have omitted graphs where the number of search nodes is below 10^4. Some extremely hard instances, where an optimal solution is either unknown or takes more than

[1] `http://dimacs.rutgers.edu/Challenges/`

[2] `http://www.nlsde.buaa.edu.cn/~kexu/benchmarks/graph-benchmarks.htm`

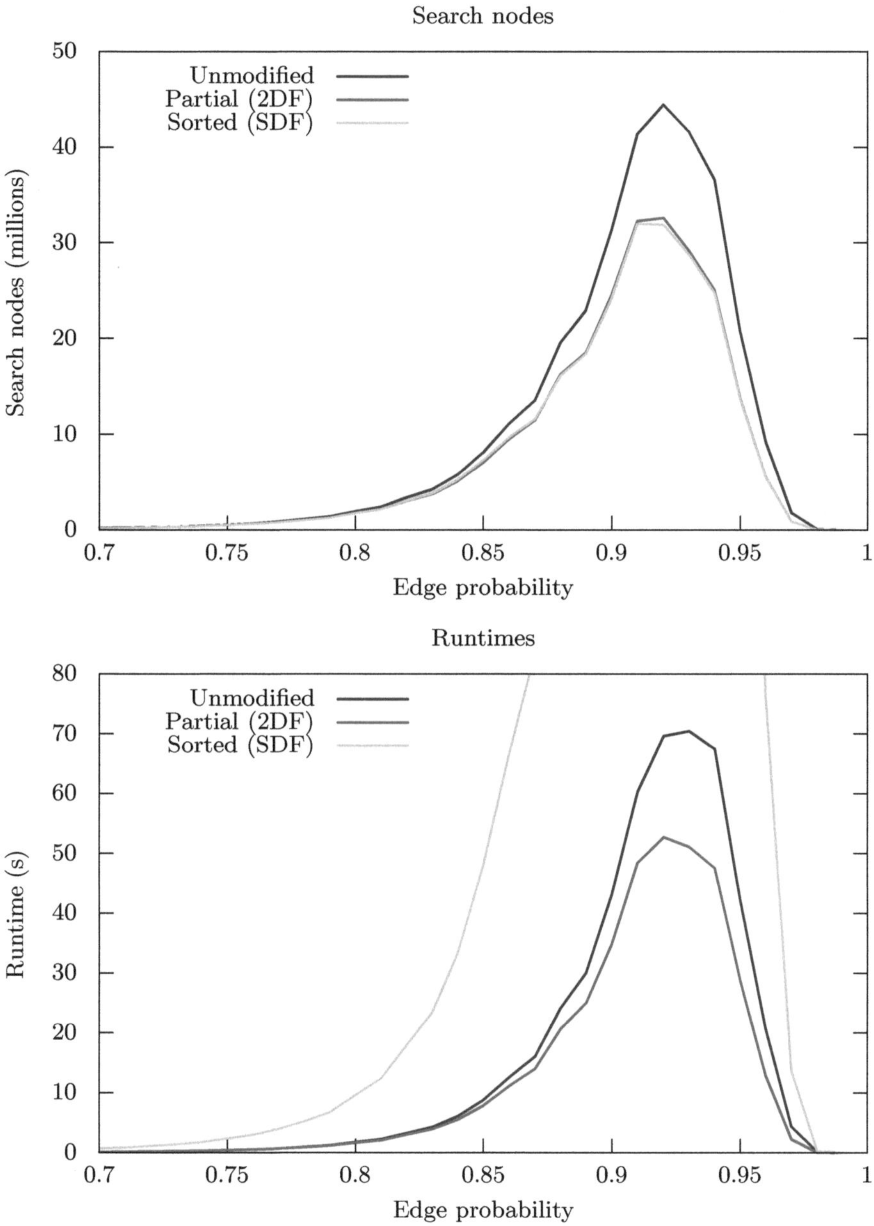

Fig. 5. The number of search nodes (top), and runtimes (bottom) for random graphs $G(200, x)$ with varying edge probabilities (steps of 0.01, 100 samples per probability). In the top graph, the partially sorted and sorted results are nearly indistinguishable, but on the bottom graph we see the high cost to runtime of doing a full sort.

Table 1. Experimental results for medium-sized DIMACS and smaller BHOSLIB graphs. Shown for each instance is the size of a maximum clique, then the number of search nodes (recursive calls) and runtime for the baseline algorithm. We then give the search space size and runtime for partial sorting, as a proportion of the baseline, and then the same for a full sort. Instances in bold are those where a partial sort gives a strict improvement in both nodes and runtimes. The omitted result takes more than two weeks to complete.

Instance	ω	Unmodified		2DF (%)		SDF (%)		Instance	ω	Unmodified		2DF (%)		SDF (%)	
		Nodes	Time	Nodes	Time	Nodes	Time			Nodes	Time	Nodes	Time	Nodes	Time
brock200_1	21	5.25×10^5	411 ms	93.8	96.4	87.5	526.5	**p_hat700-2**	44	7.51×10^5	3.2 s	95.7	95.0	205.8	683.1
brock200_3	15	1.46×10^4	11 ms	97.9	100.0	96.6	509.1	**p_hat700-3**	62	2.82×10^8	1677.9 s	95.3	94.3	324.1	1144.9
brock200_4	17	5.87×10^4	42 ms	96.4	97.6	83.2	476.2	**p_hat1000-1**	10	1.77×10^5	251 ms	99.6	97.6	100.7	278.1
brock400_1	27	1.98×10^8	287.8 s	97.0	99.1	85.3	388.2	**p_hat1000-2**	46	3.45×10^7	164.8 s	94.8	94.2	131.1	444.8
brock400_2	29	1.46×10^8	209.8 s	93.6	94.0	123.0	540.9	**p_hat1000-3**	68	1.30×10^{11}	225.8 h	93.8	98.8		
brock400_3	31	1.20×10^8	166.1 s	94.3	94.6	92.8	423.7	p_hat1500-1	12	1.18×10^6	3.6 s	99.7	102.5	87.4	177.9
brock400_4	33	5.44×10^7	80.6 s	92.2	93.4	193.1	832.8	**p_hat1500-2**	65	2.01×10^9	7.7 h	94.0	96.9	145.7	309.0
brock800_1	23	2.23×10^9	5325.8 s	97.3	95.1	104.6	326.0	san200_0.7_1	30	1.34×10^4	16 ms	100.1	106.2	52.4	243.8
brock800_2	24	2.24×10^9	5313.7 s	97.3	95.0	97.0	305.1	**san200_0.9_1**	70	8.73×10^4	100 ms	76.1	91.0	75.4	516.0
brock800_3	25	2.15×10^9	4924.1 s	97.3	95.5	80.2	255.6	san200_0.9_2	60	2.30×10^5	373 ms	340.9	303.2	316.8	1631.9
brock800_4	26	6.40×10^8	1851.7 s	97.9	95.5	93.8	289.2	**san200_0.9_3**	44	6.82×10^6	9.3 s	72.8	75.1	74.1	453.4
C125.9	34	5.02×10^4	47 ms	71.5	78.7	71.7	510.6	**san400_0.7_1**	40	1.19×10^5	234 ms	94.2	93.2	79.4	435.9
C250.9	44	1.08×10^9	1732.6 s	83.1	83.8	82.8	526.8	**san400_0.7_2**	30	8.89×10^5	2.1 s	89.9	91.1	52.8	186.8
C2000.5	16	1.82×10^{10}	21.4 h	98.9	100.2	95.6	170.4	**san400_0.7_3**	22	5.21×10^5	1.3 s	90.7	91.4	38.6	107.4
DSJC500_5	13	1.15×10^6	1.1 s	98.6	98.4	93.6	372.6	**san400_0.9_1**	100	4.54×10^6	24.2 s	78.8	79.5	73.9	335.5
DSJC1000_5	15	7.70×10^7	145.5 s	98.8	96.3	96.1	274.4	**san1000**	15	1.51×10^5	2.0 s	98.3	98.6	10.9	6.0
gen200_p0.9_44	44	1.77×10^6	2.7 s	80.2	83.2	87.3	536.0	**sanr200_0.7**	18	1.53×10^5	112 ms	95.6	97.3	93.8	546.4
gen200_p0.9_55	55	1.70×10^5	229 ms	86.2	89.1	85.9	533.2	**sanr200_0.9**	42	1.49×10^7	21.4 s	91.5	91.8	87.2	545.3
gen400_p0.9_65	65	1.76×10^{11}	126.5 h	59.5	61.5	58.1	279.5	**sanr400_0.5**	13	3.20×10^5	263 ms	98.5	98.5	93.3	397.7
gen400_p0.9_75	75	1.05×10^{11}	72.2 h	35.7	36.0	34.1	165.8	**sanr400_0.7**	21	6.44×10^7	75.5 s	95.4	95.5	94.2	426.5
hamming8-4	16	3.65×10^4	44 ms	100.9	100.0	60.0	270.5	**frb30-15-1**	30	2.92×10^8	677.8 s	88.6	89.0	86.8	369.4
johnson16-2-4	8	2.56×10^5	49 ms	100.0	106.1	88.8	551.0	**frb30-15-2**	30	5.57×10^8	1228.8 s	81.8	82.8	42.2	183.7
keller4	11	1.37×10^4	8 ms	98.7	125.0	84.1	425.0	**frb30-15-3**	30	1.67×10^8	375.2 s	80.7	82.2	78.5	333.7
keller5	27	5.07×10^{10}	44.1 h	108.8	113.2	69.7	239.5	**frb30-15-4**	30	9.91×10^8	2074.0 s	85.2	86.3	54.0	231.7
MANN_a27	126	3.80×10^4	274 ms	100.0	99.6	100.0	518.6	**frb30-15-5**	30	2.83×10^8	608.9 s	79.8	81.8	80.8	361.5
MANN_a45	345	2.85×10^6	250.0 s	100.0	100.8	100.0	206.9	**frb35-17-1**	35	1.33×10^{10}	14.8 h	82.7	88.4	44.9	154.1
p_hat300-3	36	6.25×10^5	1.1 s	92.9	94.7	109.6	562.3	**frb35-17-2**	35	2.34×10^{10}	26.1 h	82.3	88.8	48.8	171.6
p_hat500-2	36	1.14×10^5	266 ms	95.0	95.5	97.9	423.3	**frb35-17-3**	35	8.25×10^9	9.7 h	80.8	86.5	101.4	336.6
p_hat500-3	50	3.93×10^7	114.1 s	93.6	93.7	211.6	997.8	**frb35-17-4**	35	8.85×10^9	10.7 h	86.6	91.7	95.7	318.9
p_hat700-1	11	2.66×10^4	42 ms	99.4	97.6	106.3	285.7	**frb35-17-5**	35	5.80×10^{10}	58.9 h	76.5	82.2	71.0	249.4

two weeks to produce, are also omitted. However, our results include problems where the baseline runtimes are between less than a second to more than 9 days. For BHOSLIB, we have selected the smaller families, i.e. instances that we can solve within a few days.

Table 1 gives, for each problem instance, the size of the largest clique (ω) and then the performance of the baseline implementation, in number of search nodes and in runtimes. We then give the performance using partial sorting, and then full sorting. Performance is given as a percentage of the baseline.

We see that a full sort (SDF) gives a strict reduction in nodes (in all but 15 cases) but with a substantial increase in runtimes (in all but one instance, "san1000"). Partial sorting (2DF) gives a strict reduction in both search nodes and runtimes in all but 11 instances (those not shown in bold) and generally a reduction in nodes corresponds to a similar reduction in runtimes. The BHOSLIB ("frb" family) problems are worthy of note: these problems take between 5 minutes and 59 hours, and a partial sort gives improvements to both the search space size and runtimes of between ten and twenty five percent. A full sort here would give even more benefit, if it could be done economically, as it roughly halves the size of the search space in four of the instances. However, for some other graphs, a full sort actually gives a larger search space than a partial sort—this should not be a surprise, since a local improvement to the branching factor is not guaranteed to produce the best search tree globally.

4 Conclusion

Previously, in Tomita et al.'s maximum clique algorithms, the vertex selection rules and colourings were tightly coupled and not fully understood. Inspired by constraint programming techniques, we have provided a different way of looking at these algorithms. This allows us to treat the vertex selection and colouring processes separately, and to discuss this process in terms of variable and value ordering heuristics.

By looking inside the search process, we showed that the greedy colouring process produced vertices in an approximation of a "smallest domain first" order, which would reduce the amount of branching done locally at each step. We saw that using that order exactly would reduce the size of the search space in many cases, but that doing a full sort to obtain it had a large impact on runtimes. We introduced a partial sorting technique, and showed that this reduced both the size of the search space and runtimes. (Although not reported, we also investigated exploring colour classes in decreasing size and in the reverse order they were constructed. Both of these resulted in an increase in both nodes and runtimes.)

This is the first investigation into the effects of different vertex selection rules within Tomita's algorithms, and we have shown that genuine improvements can be made. A further benefit is that we now understand more about *why* these algorithms work so well in practice. Our decoupling also gives us more scope for improving the colouring process: previously, a different initial vertex ordering or a more sophisticated colouring algorithm could have undesirable knock-on effects upon which vertex is selected first; we may now ignore these effects, or study them separately. This could make it easier to integrate Li et al.'s MaxSAT-inspired inference into these algorithms.

We stress that more advanced vertex selection rules must not come at the cost of greater runtimes: we saw this issue when a full sort gave a significant slowdown, despite the smaller search space. However, we saw that a very cheap partial sort was an excellent surrogate. We hope that with further investigation into these kinds of techniques, even greater gains can be had.

References

1. Garey, M.R., Johnson, D.S.: Computers and Intractability; A Guide to the Theory of NP-Completeness. W. H. Freeman & Co., New York (1990)
2. Régin, J.-C.: Using constraint programming to solve the maximum clique problem. In: Rossi, F. (ed.) CP 2003. LNCS, vol. 2833, pp. 634–648. Springer, Heidelberg (2003)
3. Li, C.M., Quan, Z.: An efficient branch-and-bound algorithm based on MaxSAT for the maximum clique problem (2010)
4. Li, C.M., Quan, Z.: Combining graph structure exploitation and propositional reasoning for the maximum clique problem. In: 2010 22nd IEEE International Conference on Tools with Artificial Intelligence (ICTAI), vol. 1, pp. 344–351 (October 2010)
5. Li, C.M., Zhu, Z., Manyà, F., Simon, L.: Minimum satisfiability and its applications. In: Proceedings of the Twenty-Second International Joint Conference on Artificial Intelligence - Volume Volume One, IJCAI 2011, pp. 605–610. AAAI Press, Palo Alto (2011)
6. Li, C.M., Fang, Z., Xu, K.: Combining MaxSAT reasoning and incremental upper bound for the maximum clique problem. In: 2013 IEEE 25th International Conference on Tools with Artificial Intelligence (ICTAI), pp. 939–946 (November 2013)
7. Tomita, E., Seki, T.: An efficient branch-and-bound algorithm for finding a maximum clique. In: Calude, C.S., Dinneen, M.J., Vajnovszki, V. (eds.) DMTCS 2003. LNCS, vol. 2731, pp. 278–289. Springer, Heidelberg (2003)
8. Tomita, E., Kameda, T.: An efficient branch-and-bound algorithm for finding a maximum clique with computational experiments. Journal of Global Optimization 37(1), 95–111 (2007)
9. Tomita, E., Sutani, Y., Higashi, T., Takahashi, S., Wakatsuki, M.: A simple and faster branch-and-bound algorithm for finding a maximum clique. In: Rahman, M. S., Fujita, S. (eds.) WALCOM 2010. LNCS, vol. 5942, pp. 191–203. Springer, Heidelberg (2010)
10. Okubo, Y., Haraguchi, M.: Finding conceptual document clusters with improved top-n formal concept search. In: Proceedings of the 2006 IEEE/WIC/ACM International Conference on Web IntelligencE, WI 2006, pp. 347–351. IEEE Computer Society, Washington, DC (2006)
11. Konc, J., Janežič, D.: A branch and bound algorithm for matching protein structures. In: Beliczynski, B., Dzielinski, A., Iwanowski, M., Ribeiro, B. (eds.) ICANNGA 2007. LNCS, vol. 4432, pp. 399–406. Springer, Heidelberg (2007)
12. Yan, B., Gregory, S.: Detecting communities in networks by merging cliques. In: IEEE International Conference on Intelligent Computing and Intelligent Systems, ICIS 2009, vol. 1, pp. 832–836 (November 2009)

13. San Segundo, P., Rodríguez-Losada, D., Matía, F., Galán, R.: Fast exact feature based data correspondence search with an efficient bit-parallel MCP solver. Applied Intelligence 32(3), 311–329 (2010)
14. Fukagawa, D., Tamura, T., Takasu, A., Tomita, E., Akutsu, T.: A clique-based method for the edit distance between unordered trees and its application to analysis of glycan structures. BMC Bioinformatics 12(suppl. 1), S13 (2011)
15. Depolli, M., Konc, J., Rozman, K., Trobec, R., Janežič, D.: Exact parallel maximum clique algorithm for general and protein graphs. Journal of Chemical Information and Modeling 53(9), 2217–2228 (2013)
16. Regula, G., Lantos, B.: Formation control of quadrotor helicopters with guaranteed collision avoidance via safe path. Electrical Engineering and Computer Science 56(4), 113–124 (2013)
17. Konc, J., Janezic, D.: An improved branch and bound algorithm for the maximum clique problem. MATCH Communications in Mathematical and in Computer Chemistry (June 2007)
18. Prosser, P.: Exact algorithms for maximum clique: a computational study. Algorithms 5(4), 545–587 (2012)
19. Batsyn, M., Goldengorin, B., Maslov, E., Pardalos, P.: Improvements to mcs algorithm for the maximum clique problem. Journal of Combinatorial Optimization 27(2), 397–416 (2014)
20. San Segundo, P., Tapia, C.: Relaxed approximate coloring in exact maximum clique search. Computers & Operations Research 44, 185–192 (2014)
21. San Segundo, P., Rodríguez-Losada, D., Jiménez, A.: An exact bit-parallel algorithm for the maximum clique problem. Comput. Oper. Res. 38(2), 571–581 (2011)
22. San Segundo, P., Matia, F., Rodriguez-Losada, D., Hernando, M.: An improved bit parallel exact maximum clique algorithm. Optimization Letters 7(3), 467–479 (2013)
23. McCreesh, C., Prosser, P.: Multi-threading a state-of-the-art maximum clique algorithm. Algorithms 6(4), 618–635 (2013)
24. McCreesh, C., Prosser, P.: An exact branch and bound algorithm with symmetry breaking for the maximum balanced induced biclique problem. In: Simonis, H. (ed.) CPAIOR 2014. LNCS, vol. 8451, pp. 226–234. Springer, Heidelberg (2014)
25. Brockington, M., Culberson, J.C.: Camouflaging independent sets in quasi-random graphs. In: DIMACS Series in Discrete Mathematics and Theoretical Computer Science, vol. 26, pp. 75–88 (1996)
26. Soriano, P., Gendreau, M.: Tabu search algorithms for the maximum clique problem. In: DIMACS Series in Discrete Mathematics and Theoretical Computer Science, vol. 26, pp. 221–242 (1996)
27. Mannino, C., Sassano, A.: Solving hard set covering problems. Operations Research Letters 18(1), 1–5 (1995)
28. Marconi, J., Foster, J.: A hard problem for genetic algorithms: finding cliques in Keller graphs. In: Evolutionary Computation Proceedings of the 1998 IEEE International Conference on IEEE World Congress on Computational Intelligence, pp. 650–655 (May 1998)
29. McCreesh, C., Prosser, P.: The shape of the search tree for the maximum clique problem, and the implications for parallel branch and bound. ArXiv e-prints (January 2014)
30. Haralick, R.M., Elliott, G.L.: Increasing tree search efficiency for constraint satisfaction problems. Artificial Intelligence 14(3), 263–313 (1980)
31. Spearman, C.: The proof and measurement of association between two things. American Journal of Psychology 15, 88–103 (1904)

Core-Guided MaxSAT with Soft Cardinality Constraints

Antonio Morgado[1], Carmine Dodaro[2], and Joao Marques-Silva[1,3,*]

[1] INESC-ID, IST, ULisboa, Portugal
ajrm@sat.inesc-id.pt
[2] Dep. of Mathematics and Computer Science, Unical, Italy
dodaro@mat.unical.it
[3] CASL, University College Dublin, Ireland
jpms@ucd.ie

Abstract. Maximum Satisfiability (MaxSAT) is a well-known optimization variant of propositional Satisfiability (SAT). Motivated by a growing number of practical applications, recent years have seen the development of different MaxSAT algorithms based on iterative SAT solving. Such algorithms perform well on problem instances originating from practical applications. This paper proposes a new core-guided MaxSAT algorithm. This new algorithm builds on the recently proposed unclasp algorithm for ASP optimization problems, but focuses on reusing the encoded cardinality constraints. Moreover, the proposed algorithm also exploits recently proposed weighted optimization techniques. Experimental results obtained on industrial instances from the most recent MaxSAT evaluation, indicate that the proposed algorithm achieves increased robustness and improves overall performance, being capable of solving more instances than state-of-the-art MaxSAT solvers.

1 Introduction

Maximum Satisfiability (MaxSAT) is a well-known optimization version of Propositional Satisfiability (SAT). Recent years have seen a growing number of practical applications of MaxSAT, that include fault localization in C code [12] and design debugging [21], among many others [19]. For practical MaxSAT problem instances, the most effective solutions are based on iterative SAT solving, and a number of alternative approaches exist. One approach iteratively pre-relaxes every clause (by adding to each clause a fresh relaxation variable) and refines bounds on the number of unsatisfied clauses [19]. A recent example of such a MaxSAT solver is QMaxSAT [13]. An alternative approach is based on iterative identification of unsatisfiable cores [10]. Different algorithms based on the identification of unsatisfiable cores have been developed in recent years, e.g. [19]. One additional approach is based on finding minimum hitting sets of a formula representing disallowed sets of clauses [8]. This paper builds on

* This work is partially supported by SFI grant BEACON (09/IN.1/I2618), by FCT grant POLARIS (PTDC/EIA-CCO/123051/2010), by INESC-IDs multiannual PIDDAC funding PEst-OE/EEI/LA0021/2013, and by the European Commission, European Social Fund of Regione Calabria.

B. O'Sullivan (Ed.): CP 2014, LNCS 8656, pp. 564–573, 2014.

recent work on using unsatisfiable cores for solving optimization problems in ASP [1] and shows how the algorithm can be optimized for the case of MaxSAT. Experimental results, obtained on problem instances from the industrial categories of the MaxSAT evaluation, indicate that the new algorithm is more robust in practice than state-of-the-art MaxSAT solvers, being able to solve more problem instances. The paper is organized as follows. Section 1 introduces the paper, followed by the notation and definitions used in the paper in Section 2. The OLL algorithm is presented in Section 3. The experimental results are presented in Section 4 and Section 5 concludes the paper.

2 Preliminaries

This section introduces the notation used throughout the paper. Standard definitions are assumed (e.g. [14,19]). Let $X = \{x_1, x_2 \ldots\}$ be a set of Boolean variables. A *literal* l_i is either a variable x_i or its negation $\neg x_i$. A *clause* c is a disjunction of literals. A *conjunctive normal form* (CNF) formula φ is a conjunction of clauses. An *assignment* $\mathcal{A}$ is a mapping $\mathcal{A} : X \rightarrow \{0, 1\}$, where $\mathcal{A}$ satisfies(falsifies) x_i if $\mathcal{A}(x_i) = 1(\mathcal{A}(x_i) = 0)$. Assignments are extended to literals and clauses in the usual way, that is, $\mathcal{A}(l_i) = \mathcal{A}(x_i)$ if $l_i = x_i$, and $\mathcal{A}(l_i) = 1 - \mathcal{A}(x_i)$ otherwise, while for clauses $\mathcal{A}(c) = \max\{\mathcal{A}(l_i) | l_i \in c\}$. Given a CNF formula φ, a *model* of the formula is an assignment that satisfies all the clauses in φ. The *Propositional Satisfiability* problem (SAT) is the problem of deciding whether there exists a model to a given formula. A subformula of a given unsatisfiable formula which is still unsatisfiable is referred as an *unsatisfiable core* (or simply a core). The calls to the SAT solver are done through function $SATSolver(\varphi)$, that receives a formula and returns a triple (st, $\varphi_C, \mathcal{A}$), where st is either *true* or *false*.If st is true, then $\mathcal{A}$ is a model, otherwise φ_C is a core. Given a clause c and an integer w greater than 0 referred to as *weight*, the pair (c, w) is a *weighted clause*. Weighted clauses may be classified as *hard* or *soft* clauses. Hard clauses have to be satisfied and are associated with the special weight $\top$. Soft clauses may or may not be satisfied, and their weight represents the cost of falsifying the clause. A *weighted CNF formula* (WCNF) is a set of weighted clauses. A *model* of a WCNF formula is an assignment that satisfies all the hard clauses, and the *cost* associated to the model is the sum of the weights of the falsified soft clauses. The *Weighted Partial MaxSAT* problem is the problem of determining the minimum cost of the models of a given WCNF formula.

In the paper, we refer to *Relaxation Variables*, which are fresh Boolean variables. The process of augmenting a clause with a relaxation variable is referred as *relaxing* the clause. We also refer to a special type of constraints called *cardinality constraints*, which have the form $\sum_i x_i \leq k$. The sum $\sum x_i$ is referred to as the *left hand side* (LHS) of the constraint while k is referred to as the *right hand side* (RHS).

3 The OLL Algorithm

Algorithm OLL has been introduced in the unclasp tool for solving ASP optimization problems [1]. Unclasp is the base of the ASP-based Linux configuration system aspuncud [11], which won four tracks of the 2011 Mancoosi International Solver Competition[1]. Algorithm OLL has been reported [1] to be able to solve a higher number of

[1] http://www.mancoosi.org/misc/

MISC optimization instances than clasp (the base solver on which unclasp was built upon).

This section shows how to adapt the OLL algorithm to MaxSAT, but additionally considering the reuse of the cardinality constraints as they are discovered. The idea of the OLL algorithm is to mix the strengths of two MaxSAT algorithms, namely the Fu & Malik algorithm [10] and MSU3 [17,18]. These are core-guided algorithms, which means that the algorithms make use of the unsatisfiable cores in order to relax clauses. Like MSU3 only one relaxation variable is added per clause identified in a core, but similarly to the Fu & Malik algorithm a new cardinality constraint is added for each core as it is found. One of the main difference of the OLL algorithm is that the soft clauses are transformed into hard clauses after relaxing them, while the cardinality constraints are added as soft. Consider the following Example 1.

Example 1. In the example we will abuse the notation and refer to soft constraints as if they were clauses. Consider for example the partial formula $\varphi = \varphi_S \cup \varphi_H$, where φ_S is the set of soft clauses $\varphi_S = \{(x_1, 1), (x_2, 1), (x_3, 1)\}$ and φ_H is the set of hard clauses $\varphi_H = \{(\neg x_1 \vee \neg x_2, \top), (\neg x_1 \vee \neg x_3, \top), (\neg x_2 \vee \neg x_3, \top)\}$. The initial working formula φ_W is $\varphi_S \cup \varphi_H$, which is unsatisfiable. Let the soft clauses in the core returned by the SAT solver be $(x_1, 1)$ and $(x_2, 1)$. The OLL algorithm relaxes both clauses and makes them hard, and adds a cardinality constraint as a soft constraint. As such, the sets of clauses in the working formula are updated as:

$$\begin{aligned} \varphi_S &\leftarrow (\varphi_S \setminus \{(x_1, 1), (x_2, 1)\}) \cup \{(r_1 + r_2 \leq 1, 1)\} \\ \varphi_H &\leftarrow \varphi_H \cup \{(x_1 \vee r_1, \top), (x_2 \vee r_2, \top)\} \end{aligned}$$

where r_1 and r_2 are the new relaxation variables.

The resulting working formula φ_W is again unsatisfiable. Now the unsatisfiable core contains the soft clause $(x_3, 1)$ and the soft constraint $(r_1 + r_2 \leq 1, 1)$. As before the OLL algorithm is going to relax the soft clause and make it hard, that is $(x_3 \vee r_3, \top)$. It will also remove the soft constraint $(r_1 + r_2 \leq 1, 1)$ from the working formula, and add two new constraints $(r_3 + \neg(r_1 + r_2 \leq 1) \leq 1, 1)$ and $(r_1 + r_2 \leq 2, 1)$. The first constraint says that in order to satisfy the constraint, you either have $(r_1 + r_2 \leq 1)$ (the previous constraint) falsified or you are allowed to set r_3 to true.

The second constraint added $(r_1 + r_2 \leq 2, 1)$, if satisfied, allows one more of the previous relaxation variables to be satisfied. The sets in the working formula are then updated as:

$$\begin{aligned} \varphi_S &\leftarrow (\varphi_S \setminus \{(x_3, 1), (r_1 + r_2 \leq 1, 1)\}) \cup \{(r_3 + \neg(r_1 + r_2 \leq 1) \leq 1, 1), (r_1 + r_2 \leq 2, 1)\} \\ \varphi_H &\leftarrow \varphi_H \cup \{(x_3 \vee r_3, \top)\} \end{aligned}$$

Now the resulting working formula is satisfiable and the algorithm returns 2, which is the cost of the satisfying assignment.

As the previous example illustrates, the idea of OLL is to go through unsatisfiable iterations until a satisfiable working formula is obtained. Whenever a new unsatisfiable core is identified, then the working formula is updated such that either all the previous soft constraints in the core are satisfied and allowing a new relaxation variable to be set to true, or one of the soft constraints is allowed to increase its bound by 1.

The example uses soft constraints that correspond to cardinality constraints. On the other hand, SAT solvers only handle clauses. In order to use the OLL algorithm in a MaxSAT solver using a SAT solver, it is necessary to encode the cardinality constraints into CNF each time they are identified. Observe that both cardinality constraints $r_1+r_2 \leq 1$ and $r_1+r_2 \leq 2$, share the same LHS r_1+r_2. In fact, some of the existing encodings of cardinality constraints, encode the sum on the LHS into an array of Boolean variables to represent it as a unary number. In this paper, we propose to use this fact in order to reuse the encodings of the sums of the LHS of the constraints between cardinality constraints that only differ on their RHS. In the previous example, $r_1 + r_2$ would be encoded using an auxiliary function $([s_1, s_2], clauses) \leftarrow createSum(\{r_1, r_2\})$, which receives a set of variables for which we want to encode the sum, and returns a pair containing an array of the Boolean variables that encode the sum in a unary number (the unary number $s_2 s_1$), and a set of clauses that encodes the sum. Whenever a new cardinality constraint is required with the same LHS (sum), then the same variables s_1 and s_2 are set to the appropriate values in order to encode the cardinality constraint.

The pseudo-code of OLL is shown in Algorithm 1. In the following we assume that the input formula is unweighted (weighted case explained later on), partial and the set of hard clauses is satisfiable. Given an input formula φ, OLL maintains three sets of clauses: the current representation of the input formula called the working formula φ_W; the current set of soft clauses φ_S and the set of soft cardinalities φ_{SC} containing (unit) clauses associated to cardinality constraints. Those sets are initialized in line 1. Moreover, function map associates with a literal l (related to one of the cardinality constraints) a pair corresponding to the outputs of the associated sum and a bound (its RHS). OLL starts by calling the SAT solver on the current working formula φ_W. If the formula is satisfiable, then the algorithm terminates and returns the cost of $\mathcal{A}$ (line 5). Otherwise, the working formula is unsatisfiable and an unsatisfiable core is computed. The algorithm proceeds by relaxing all soft clauses of the core that are in φ_S, and making them hard clauses. This is done through function $RelaxAndHarden(\varphi_W, \varphi_C \cap \varphi_S)$, which receives the working formula φ_W, and a set of soft clauses that need to be relaxed. Function $RelaxAndHarden$ returns a pair (L, φ_W), where L is the set of new relaxation variables, and φ_W is updated to the clauses that were in φ_W, but to which the clauses that were in $\varphi_C \cap \varphi_S$, have been relaxed and transformed into hard clauses.

After relaxing soft clauses, the remaining clauses in the core related to cardinality constraints (i.e. clauses in $\varphi_C \cap \varphi_{SC}$) are processed and removed from the working formula φ_W (line 9) and from the set φ_{SC} (line 10). Each of those clauses is a unit clause.The outputs $sumOtps$ of the sum associated with the cardinality constraint corresponding to $\neg s$ are obtained with the function $map(\neg s)$, from which the corresponding bound b (RHS) is also obtained. In fact, the variable s represents the b-th output variable of the sum in $sumOtps$. In line 11, the set L is extended with the variable s. This corresponds to the negation of the previous cardinality constraint with b as the RHS, i.e. if s is true then the sum is greater than b, thus negating the cardinality constraint. The algorithm proceeds by creating a new unit clause $(\neg sumOtps[b + 1])$ that encodes the sum to be less or equal to b. The clause is then added to the working formula φ_W (line 14) and to φ_{SC} (line 15). Moreover, in line 16, the pair (sum, $b + 1$) is added to the map for the $(b + 1)$-th output variable. Note that this is done only if $b + 1$

Algorithm 1. OLL algorithm for (non-weighted) (partial) MaxSAT

Input: A formula φ

1 $(\varphi_W, \varphi_S, \varphi_{SC}) \leftarrow (\varphi, \texttt{Soft}(\varphi_W), \emptyset)$;
2 $\texttt{map} \leftarrow \emptyset$; // $\texttt{map}(lit) = (sumOtps, bound)$
3 **while** true **do**
4 $\quad (\mathsf{st}, \varphi_C, \mathcal{A}) \leftarrow \texttt{SATSolver}(\varphi_W)$;
5 $\quad$ **if** $\mathsf{st} = true$ **then return** $\sum_{(c,1)\in\varphi_S}(1 - \mathcal{A}(c))$
6 $\quad$ **else**
7 $\quad\quad (L, \varphi_W) \leftarrow \texttt{RelaxAndHarden}(\varphi_W, \varphi_C \cap \varphi_S)$;
8 $\quad\quad$ **foreach** $(\neg s, 1) \in \varphi_C \cap \varphi_{SC}$ **do**
9 $\quad\quad\quad \varphi_W \leftarrow \varphi_W \setminus \{(\neg s, 1)\}$;
10 $\quad\quad\quad \varphi_{SC} \leftarrow \varphi_{SC} \setminus \{(\neg s, 1)\}$;
11 $\quad\quad\quad L \leftarrow L \cup \{s\}$;
12 $\quad\quad\quad (sumOtps, b) \leftarrow \texttt{map}(\neg s)$;
13 $\quad\quad\quad$ **if** $b + 1 < |sumOtps|$ **then**
14 $\quad\quad\quad\quad \varphi_W \leftarrow \varphi_W \cup \{(\neg sumOtps[b+1], 1)\}$;
15 $\quad\quad\quad\quad \varphi_{SC} \leftarrow \varphi_{SC} \cup \{(\neg sumOtps[b+1], 1)\}$;
16 $\quad\quad\quad\quad \texttt{map}(\neg sumOtps[b+1]) \leftarrow (sumOtps, b+1)$;
17 $\quad\quad (sumOtps_{New}, sumCls_{New}) \leftarrow createSum(L)$;
18 $\quad\quad \varphi_W \leftarrow \varphi_W \cup \{(c, \top) \mid c \in sumCls_{New}\} \cup \{(\neg sumOtps_{New}[1], 1)\}$;
19 $\quad\quad \varphi_{SC} \leftarrow \varphi_{SC} \cup \{(\neg sumOtps_{New}[1], 1)\}$;
20 $\quad\quad \texttt{map}(\neg sumOtps_{New}[1]) \leftarrow (sumOtps_{New}, 1)$;

is less than the size of the sum, i.e. if incrementing the bound by one does not make the sum trivially satisfied.

When all clauses in the core related to soft cardinality constraints have been processed, a new cardinality constraint, with the corresponding new sum is created, containing all variables in L (line 17). The clauses encoding the sum are added to the working formula as hard clauses while a new unit soft clause $(\neg sumOtpts_{New}[1])$ is added to φ_W and to φ_{SC}. This clause encodes that at most one of the variables in L is true. In addition, the pair $(sumOtps_{New}, 1)$ is associated to the literal $\neg sumOtps_{New}[1]$ by adding a new entry to the map.

Proposition 1. *Given a (partial) MaxSAT formula, Algorithm 1 is correct and returns the optimum MaxSAT solution.*

Proof (sketch). The OLL algorithm goes thought unsatisfiable instances until a satisfiable instance is obtained. Initially the algorithm tries to satisfy all the soft clauses (added to a working formula together with the hard clauses). Whenever the working formula is unsatisfiable, then it is updated such that at most one more of the initial soft clauses is allowed to be falsified (than the previous iteration). When the working formula is satisfiable the algorithm stops and the number of initial soft clauses simultaneously falsified corresponds the optimum MaxSAT solution. This process is similar to other MaxSAT algorithms as MSU3.

Nevertheless, in OLL, the restriction on the number of initial soft clauses that are allowed to be falsified is achieved by adding relaxation variables to soft clauses that belong to a core and have not been relaxed before, and by the addition of soft cardinality constraints (At-Most-K constraints). The soft cardinality constraints (added on line 18, initially with a RHS of 1) allow at most one of the newly relaxed clauses to be falsified or at most one of previous soft cardinality constraints that appeared in the core to be falsified. Falsifying a previous soft cardinality constraint forces the number of associated initial soft clauses that are falsified to increase. The increase is contrained to be at most one by adding a new soft cardinality constraint (added on line 14) equal to the previous soft cardinality but with the RHS increased by 1.

The previous algorithm deals with non-weighted (partial) MaxSAT formulas. In the weighted case the procedure is similar to the MSU1/WPM1 algorithms [15,3], that is, every time a new core is found, the minimum weight of the soft clauses in the core min is computed. Then each clause (c_i, w_i) with a weight greater than the minimum is replaced by two clauses: (c_i, min) and $(c_i, w_i - min)$. Then the algorithm proceeds as in the partial case but as if the core obtained contained only clauses with the same weight min . The result is obtained as in the partial case by considering the cost of the satisfying assignment in the original soft clauses, but considering the original weights.

4 Experimental Results

This section presents the experimental results obtained to validate the performance of the MaxSAT algorithm proposed in Section 3. All experiments were run on an HPC cluster, each node having two processors E5-2620 @2GHz, with each processor having 6 cores, and with a total of 128 GByte of physical memory. Each process was limited to 4GByte of RAM and to a time limit of 1800 seconds. All the industrial instances from the most recent MaxSAT Evaluation[2] 2013 [5] were used, that is the following three categories of benchmarks were considered: (plain) MaxSAT industrial; partial MaxSAT industrial; and weighted partial MaxSAT industrial.

For the experiments, the OLL algorithm proposed in the previous section was implemented in MSUnCore [20][3]. MSUnCore is a state-of-the-art (generic) MaxSAT solver, that won third place in the partial MaxSAT category of the 2013 MaxSAT Evaluation (second place, if portfolio solvers are excluded). The underlying SAT solver in MSUnCore is PicoSAT [7] (version 935). Three different cardinality constraint encodings that are able to encode the sum of all the input variables as a unary number, were considered. Namely Sorting Networks [9], Sequential Counters [22], and Totalizer [6]. In the results the OLL algorithm with the cardinality constraints are referred as *msu-oll-sn*, *msu-oll-sc* and *msu-oll-to* respectively. For weighted instances, we have implemented an OLL solver which includes recent weighted boolean optimizations techniques proposed for MaxSAT solving. When considering the weighted optimizations, and previously to solving, a weighted instance is checked for the BMO condition [16], in which case the instance is solved according to the BMO approach. Otherwise, the stratification technique [4] is considered. The resulting solver is referred in the results as *msu-oll-xx-wo*, where xx corresponds to the cardinality constraint considered. Additionally the experiments include the

[2] `http://www.maxsat.udl.cat`

[3] Logs in `http://sat.inesc-id.pt/~ajrm/oll_statlogs.tgz`

	MSi	PMSi	WPMSi	ALLi
#Instances	55	627	396	1078
msu-oll-sn-wo	25	512	330	867
msu-oll-to-wo	19	517	329	865
msu-oll-to	19	517	315	851
msu-oll-sn	25	512	314	851
msu-oll-sc-wo	18	494	331	843
msu-oll-sc	18	494	289	801
msu-bcd2	22	500	265	787

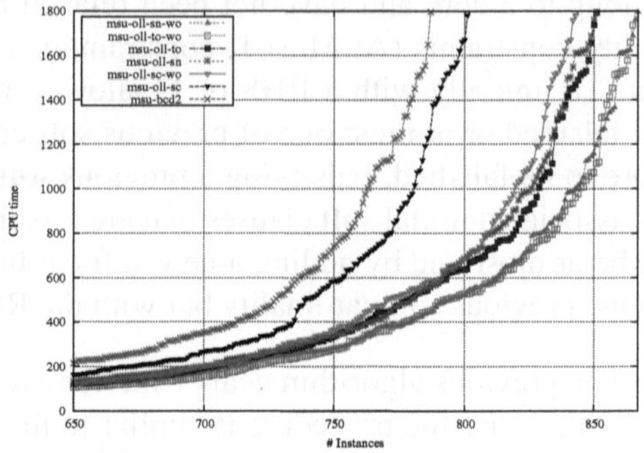

Fig. 1. Cactus plot and statistics for the different configurations of the solvers in MSUnCore

two best solvers for each of the industrial categories from the 2013 MaxSAT Evaluation (among non-portfolio solvers): MiFuMax[4], MSUnCore [20] (BCD2 version), QMaxSAT 0.21 [13], WPM1 [4] (2011 and 2013 versions), and WPM2 [4,2] (2013 version). The solvers are referred in the results as *mifumax*, *msu-bcd2*, *qmaxsat*, *wpm1*, *wpm1-2011*, and *wpm2*.

The table in Figure 1 shows the number of instances solved by each of the algorithms in MSUnCore, that is, the OLL algorithms and msu-bcd2. The first column of the table shows the name of the solver. The second to fourth columns show the number of solved instances by each of the solvers, for the (plain) MaxSAT industrial (MSi), partial MaxSAT industrial (PMSi) and weighted partial MaxSAT industrial instances respectively. The last column shows the total number of solved instances among all of the industrial instances. The first row does not present the number of solved instances, but instead the total number of instances in the category considered in the column. In the table the solvers are ordered according to the number of instances solved in ALLi.

The results in the table show that among the msu-oll-xx solvers, both msu-oll-sn and msu-oll-to have similar performance, being msu-oll-sn slightly better for (plain) MaxSAT instances, while msu-oll-to is slightly better for partial MaxSAT instances. The msu-oll-sc solver performs consistently worse than the other two with a total of 50 less instances solved in ALLi. The table in Figure 1 also allows to conclude that the weighted optimizations included are consistently beneficial for all the msu-oll solvers. This is especially true for msu-oll-sc-wo which solved 32 more instances than msu-oll-sc. Comparing the msu-oll solvers with msu-bcd2 (since they are implemented in the same platform), the results show that for MSi instances, the msu-oll solvers are comparable to msu-bcd2, where msu-oll-sn(-wo) solves 3 more instances than msu-bcd2. In the case of PMSi instances, the difference between msu-bcd2 and the msu-oll solvers is greater, and both the msu-oll-sn and msu-oll-to are able to solve more 12 and 17 instances than msu-bcd2. For weighted instances, all the OLL algorithms outperform msu-bcd2, including the versions that do not make use of weighted optimizations. This can be related to the fact that OLL requires only cardinality constraints to deal with the weights, while msu-bcd2 uses pseudo-Boolean constraints. In fact, the best performing OLL algorithm (msu-oll-sn-wo) is able to solve 80 more instances than msu-bcd2.

[4] `http://sat.inesc-id.pt/~mikolas/sw/mifumax`

	MSi	PMSi	WPMSi	ALLi
#Instances	55	627	396	1078
msu-oll-sn-wo	25	512	330	867
wpm2	12	490	320	822
msu-bcd2	22	500	265	787
wpm1	19	384	342	745
wpm1-2011	37	265	304	606
mifumax	38	273	258	569
qmaxsat	–	540	–	–

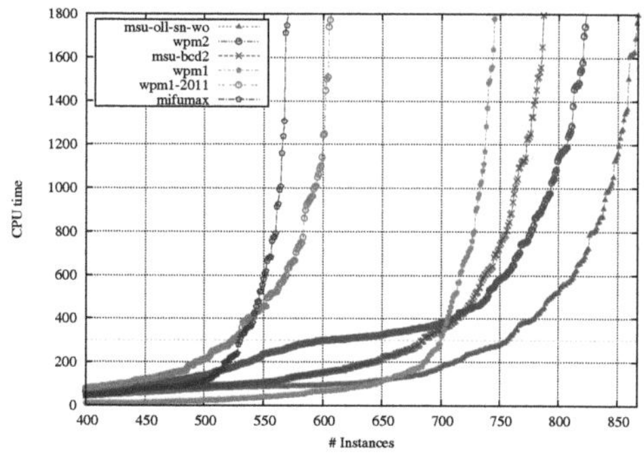

Fig. 2. Cactus plot and statistics for the best OLL algorithm vs non-OLL algorithms

These results are confirmed by the cactus plot in Figure 1, where the msu-bcd2 is the left-most solver, meaning that it solves less instances. On the other end, both msu-oll-sn-wo and msu-oll-to-wo appear close together on the right-most side of the plot.

In order to compare the best performing msu-oll solver (msu-oll-sn-wo) with the remaining solvers, we present in the table of Figure 2 the number of solved instances for the remaining solvers along with msu-oll-sn-wo and msu-bcd2. The table in the Figure 2 has a similar structure to the table in the Figure 1. As before the solvers are ordered according to the number of instances solved in ALLi. The only exception is qmaxsat for which the tested solver only allows to solve partial MaxSAT instances. From the table it is possible to see that for each category, msu-oll-sn-wo is either the third (for MSi) or the second (for PMSi and WPMSi) solver in terms of number of instances solved. Nevertheless, overall msu-oll-sn-wo solves more instances than any of the other solvers (shown in the ALLi column). The closest solver is wpm2 with 822 instances solved, which means a difference of 45 instances to msu-oll-sn-wo. These results are also confirmed by the cactus plot show in Figure 2, where the right-most line corresponds to msu-oll-sn-wo and the gap between the line of msu-oll-sn-wo and next line (wpm2) corresponds to the 45 instances difference. Note that in the figure, qmaxsat is not represented since it only allows solving partial MaxSAT instances.

5 Conclusions

This paper describes how to transform the OLL algorithm, proposed in unclasp for optimization problems in ASP [1], into a core-guided MaxSAT using a modern SAT solver. Additionally, the paper shows how to reuse the encodings of the cardinality constraints as they are added to the working formula. The experimental results indicate that the proposed OLL algorithm represents the currently most robust approach for MaxSAT, being able to solve more instances than state-of-the-art MaxSAT solvers. Despite not being in general the top performer for any specific category of instances, overall the OLL algorithm solves more instances than any of the best performing solvers from the 2013 MaxSAT Evalution, including MiFuMax, MSUnCore (BCD2), WPM1 and WPM2.

Future work will investigate alternative approaches for aggregating soft cardinality constraints, as well as improving the quality of computed unsatisfiable cores.

References

1. Andres, B., Kaufmann, B., Matheis, O., Schaub, T.: Unsatisfiability-based optimization in clasp. In: International Conference on Logic Programming (Technical Communications), pp. 211–221 (2012)
2. Ansótegui, C., Bonet, M.L., Gabàs, J., Levy, J.: Improving WPM2 for (weighted) partial MaxSAT. In: Schulte, C. (ed.) CP 2013. LNCS, vol. 8124, pp. 117–132. Springer, Heidelberg (2013)
3. Ansótegui, C., Bonet, M.L., Levy, J.: Solving (weighted) partial MaxSAT through satisfiability testing. In: Kullmann, O. (ed.) SAT 2009. LNCS, vol. 5584, pp. 427–440. Springer, Heidelberg (2009)
4. Ansótegui, C., Bonet, M.L., Levy, J.: SAT-based MaxSAT algorithms. Artificial Intelligence 196, 77–105 (2013)
5. Argelich, J., Li, C.M., Manyà, F., Planes, J.: The first and second Max-SAT evaluations. Journal on Satisfiability, Boolean Modeling and Computation 4(2-4), 251–278 (2008)
6. Bailleux, O., Boufkhad, Y.: Efficient CNF encoding of boolean cardinality constraints. In: Rossi, F. (ed.) CP 2003. LNCS, vol. 2833, pp. 108–122. Springer, Heidelberg (2003)
7. Biere, A.: Picosat essentials. Journal on Satisfiability, Boolean Modeling and Computation 4(2-4), 75–97 (2008)
8. Davies, J., Bacchus, F.: Solving MAXSAT by solving a sequence of simpler SAT instances. In: Lee, J. (ed.) CP 2011. LNCS, vol. 6876, pp. 225–239. Springer, Heidelberg (2011)
9. Een, N., Sörensson, N.: Translating pseudo-Boolean constraints into SAT. Journal on Satisfiability, Boolean Modeling and Computation 2, 1–26 (2006)
10. Fu, Z., Malik, S.: On solving the partial MAX-SAT problem. In: Biere, A., Gomes, C.P. (eds.) SAT 2006. LNCS, vol. 4121, pp. 252–265. Springer, Heidelberg (2006)
11. Gebser, M., Kaminski, R., Schaub, T.: aspcud: A Linux package configuration tool based on answer set programming. In: International Workshop on Logics for Component Configuration (LoCoCo 2011). Electronic Proceedings in Theoretical Computer Science (EPTCS), vol. 65, pp. 12–25 (2011), `http://www.cs.uni-potsdam.de/wv/aspcud/`
12. Jose, M., Majumdar, R.: Cause clue clauses: error localization using maximum satisfiability. In: International Conference on Programming Language Design and Implementation, pp. 437–446 (2011)
13. Koshimura, M., Zhang, T., Fujita, H., Hasegawa, R.: QMaxSAT: A partial Max-SAT solver. Journal on Satisfiability, Boolean Modeling and Computation 8(1-2), 95–100 (2012)
14. Li, C.M., Manyà, F.: MaxSAT, hard and soft constraints. In: Handbook of Satisfiability, pp. 613–632. IOS Press (2009)
15. Manquinho, V., Marques-Silva, J., Planes, J.: Algorithms for weighted boolean optimization. In: Kullmann, O. (ed.) SAT 2009. LNCS, vol. 5584, pp. 495–508. Springer, Heidelberg (2009)
16. Marques-Silva, J., Argelich, J., Graça, A., Lynce, I.: Boolean lexicographic optimization: algorithms & applications. Annals of Mathematics and Artificial Intelligence 62(3-4), 317–343 (2011)
17. Marques-Silva, J., Planes, J.: On using unsatisfiability for solving maximum satisfiability. Computing Research Repository abs/0712.0097 (December 2007)
18. Marques-Silva, J., Planes, J.: Algorithms for maximum satisfiability using unsatisfiable cores. In: Design, Automation and Testing in Europe Conference, pp. 408–413 (March 2008)

19. Morgado, A., Heras, F., Liffiton, M.H., Planes, J., Marques-Silva, J.: Iterative and core-guided maxsat solving: A survey and assessment. Constraints 18(4), 478–534 (2013)
20. Morgado, A., Heras, F., Marques-Silva, J.: Improvements to core-guided binary search for MaxSAT. In: Cimatti, A., Sebastiani, R. (eds.) SAT 2012. LNCS, vol. 7317, pp. 284–297. Springer, Heidelberg (2012)
21. Safarpour, S., Mangassarian, H., Veneris, A., Liffiton, M.H., Sakallah, K.A.: Improved design debugging using maximum satisfiability. In: International Conference on Formal Methods in Computer-Aided Design (2007)
22. Sinz, C.: Towards an optimal CNF encoding of boolean cardinality constraints. In: van Beek, P. (ed.) CP 2005. LNCS, vol. 3709, pp. 827–831. Springer, Heidelberg (2005)

The IntSat Method for Integer Linear Programming

Robert Nieuwenhuis

Barcelogic.com and Technical University of Catalonia (UPC), Barcelona, Spain

Abstract. Conflict-Driven Clause-Learning (CDCL) SAT solvers can automatically solve very large real-world problems. To go beyond, and in particular in order to solve *and optimize* problems involving linear arithmetic constraints, here we introduce IntSat, a generalization of CDCL to Integer Linear Programming (ILP). Our simple 1400-line C++ prototype IntSat implementation already shows some competitiveness with commercial solvers such as CPLEX or Gurobi. Here we describe this new IntSat ILP solving method, show how it can be implemented efficiently, and discuss a large list of possible enhancements and extensions.

1 Introduction

Conflict-Driven Clause-Learning (CDCL) propositional SAT solving technology can automatically solve hard real-world industrial and scientific problem instances involving large numbers of binary variables and constraints (i.e., clauses). SAT is the particular case of ILP where all variables are binary (0/1) and constraints have the form $x_1 + \ldots + x_m - y_1 \ldots - y_n > -n$, written as *clauses* $x_1 \vee \ldots \vee x_m \vee \overline{y_1} \vee \ldots \vee \overline{y_n}$, i.e., sets (disjunctions) of *literals*. Given a *partial assignment* A, seen as a set of (non-contradictory) literals, a clause C is *true in* A if $A \cap C \neq \emptyset$, it is *false* or a *conflict* if $\overline{l} \in A$ for every literal l in C, and it is *undefined* otherwise. All essential CDCL features are described in the following 14-line algorithm, where A is seen as an (initially empty) stack:

1. *Propagate*: while possible and no conflict appears, if, for some clause $l \vee C$, C is false in A and l is undefined, push l onto A, associating to l the *reason clause* C.
2. **if** there is no conflict
   if all variables are defined in A, output "solution A" and halt.
   else *Decide*: push some undefined literal l, marked as a *decision*, and goto step 1.
3. If A contains no decisions, output "unsatisfiable" and halt.
4. Use a clause data structure C. Initially, let C be any conflict.
   - *Conflict analysis*: Invariant: C is false in A, that is, if $l \in C$ then $\overline{l} \in A$.
     If l is the literal of C whose negation is topmost in A, and D is the reason clause of $\overline{l}$, then replace C by $(C \setminus \{l\}) \cup D$. Repeat this until there is only one literal l_{top} in C such that $\overline{l_{top}}$ is, or is above, A's topmost decision.
   - *Backjump*: pop literals from A until either there are no decisions in A or, for some l in C with $l \neq l_{top}$, there are no decisions above $\overline{l}$ in A.
   - *Learn*: add the final C as a new clause, and go to 1 (where C propagates l_{top}).

Note that replacing C by $(C \setminus \{l\}) \cup D$ is in fact an inference by *resolution* between C and $D \vee \overline{l}$. Essentially all state-of-the-art CDCL SAT solvers use this (so called *1UIP* conflict

B. O'Sullivan (Ed.): CP 2014, LNCS 8656, pp. 574–589, 2014.

analysis-based) algorithm, with efficient data structures for propagation, heuristics that select *recently active* literals as decisions, periodically *forgetting* (removing) the least useful learned clauses, and using clause simplification and other *inprocessing* methods (see, e.g., [16,4] for more details and further references).

Given CDCL's enormous success for SAT, for decades researchers (e.g., from the SAT community, but not only) have tried to produce an effective CDCL-like method for ILP. However, due to a number of obstacles (see, e.g., Example 3 below), the results of such attempts were always orders of magnitude slower than the state-of-the-art commercial MIP/ILP solvers such as CPLEX or Gurobi, based on LP relaxations, simplex, and branch-and-cut (see, e.g., the "mip basics" at `www.gurobi.com`). Typical attempts to generalize CDCL from SAT to ILP are in the following sense, where (possibly subindexed) letters a denote integer coefficients:

SAT		**ILP**	
clause	$l_1 \vee \ldots \vee l_n$	*linear constraint*	$a_1x_1 + \cdots + a_nx_n \leq a_0$
0/1 variable	x	*integer variable*	x
positive literal	x	*lower bound*	$a \leq x$
negative literal	$\overline{x}$	*upper bound*	$x \leq a$
propagation		*bound propagation*	
resolution inference		*cut* inference	

Example 1. By *bound* propagation, from the lower bound $1 \leq x$, the upper bound $y \leq 2$, and the constraint $x - 2y + 5z \leq 5$, we infer that $1 - 4 + 5z \leq 5$, so $5z \leq 8$, and hence $z \leq 8/5$, which is rounded, propagating a new bound $z \leq 1$. Note that any 1-variable constraint propagates a bound by itself, e.g., from $-7x \leq 3$ we have $-3 \leq 7x$, and hence $-3/7 \leq x$, which after rounding propagates the lower bound $0 \leq x$. □

Example 2. From $4x + 4y + 2z \leq 3$ and $-10x + y - z \leq 0$, by multiplying the former by 5 and the latter by 2 and adding them up, we obtain the *cut* $22y + 8z \leq 15$. Here the variable x is *eliminated*, which can always be achieved if in the two premises x has coefficients a and b such that $a \cdot b < 0$. The result $22y + 8z \leq 15$ can be *normalized* dividing by $gcd(22, 8) = 2$, giving $11y + 4z \leq 15/2$ and, by rounding, $11y + 4z \leq 7$. □

Example 3. An important problem for extending CDCL to ILP is the following, which we will call the *rounding problem.* Assume we have the two constraints $x + y + 2z \leq 2$ and $x + y - 2z \leq 0$ and we take the decision $0 \leq x$, which propagates nothing, and later on another decision $1 \leq y$, which due to $x + y + 2z \leq 2$ propagates $z \leq 0$ (since $1 + 2z \leq 2$, and hence $2z \leq 1$ and by rounding $z \leq 1/2$ we get $z \leq 0$). Then $x + y - 2z \leq 0$ becomes a *conflict*: it is false in the current partial assignment $A = \{ 0 \leq x,\ 1 \leq y,\ z \leq 0 \}$.

Now let us attempt a straightforward generalization of the CDCL algorithm: since $z \leq 0$ is the topmost (last propagated) bound, a cut inference eliminating z between both constraints would be needed, generating the new constraint C which is $2x + 2y \leq 2$, or equivalently, $x + y \leq 1$. Then conflict analysis is over because there is only one bound in A at, or above, the last decision relevant for C, namely $1 \leq y$. But unfortunately the new constraint $x + y \leq 1$ is not false in A, breaking (what should be) the invariant. Hence it does not propagate the negation of $1 \leq y$ (which is $y \leq 0$) and is *too weak to force a backjump*. This problem is due to the rounding that takes place when propagating z. □

The rounding problem illustrated in Example 3 was solved in a very ingenious way by Jovanovic and de Moura in their *cutsat* procedure [14], where a decision can only make a variable *equal* to its current upper or lower bound, which permits, at each conflict caused by bound propagations with rounding, to compute *tightly propagating* constraints that justify the same propagations *without* rounding, and to do conflict analysis using tightly propagating constraints only. This makes the learning scheme of [14] similar to the SAT one of doing resolutions until reaching a clause built from decisions only, which is well known to perform significantly worse than 1UIP.

Here we introduce IntSat, a new completely different method for ILP. It admits arbitrary new bounds as decisions and guides the search exactly as with the 1UIP approach in CDCL-based SAT solving, and still it overcomes the rounding problem. The key ideas behind IntSat are as follows. Given a partial assignment A, a set (stack) of bounds, each time a constraint C and a set of bounds R with $R \subseteq A$ propagate a new bound B, this bound is pushed onto A, associating to B not only its *reason constraint* C but also its *reason set* R. Conflict analysis and cuts performed are both guided by successive refinements of a so-called *Conflicting Set* of bounds $CS \subseteq A$ that is infeasible along with the current set of constraints. After each conflict, always a backjump takes place and a new constraint is learned.

This paper is structured as follows. The basic IntSat procedure is introduced in Section 2, its termination, correctness and completeness stated, and extensions are given for, e.g., optimization. In Section 3 we give some details about the efficient implementability of IntSat, and further work on IntSat is discussed in Section 4. Section 5 provides experimental results and in Section 6 we describe related work and conclude.

2 The Basic IntSat Procedure

In a first basic version of IntSat we deal with integer coefficients only and decide the existence of integer solutions, i.e., feasibility problems only, and no optimization yet (extensions are handled later on). Let X be a finite set of *variables* $\{x_1 \ldots x_n\}$. An *(integer linear) constraint* over X is an expression of the form $a_1x_1 + \cdots + a_nx_n \leq a_0$ where, for all i in $0 \ldots n$, the *coefficients* a_i are integers. Below, variables are always denoted by (possibly sub-indexed or primed) lowercase x, y, z and coefficients by a, b, c, respectively. An *Integer Program (IP)* over X is a set S of integer linear constraints over X. A *solution* for an IP S over X is a function $sol\colon X \to \mathbb{Z}$ that *satisfies* every constraint $a_1x_1 + \cdots + a_nx_n \leq a_0$ in S, that is, $a_1 \cdot sol(x_1) + \cdots + a_n \cdot sol(x_n) \leq_{\mathbb{Z}} a_0$. A *bound* is a one-variable constraint $a_1x \leq a_0$. Any bound can equivalently either be written as a *lower bound* $a \leq x$ or as an *upper bound* $x \leq a$. For a constraint $a_1x_1 + \ldots + a_nx_n \leq a_0$, each a_ix_i is called a *monomial* of it, and the monomial is positive (negative) if a_i is.

Bound Propagation. Let C be a constraint of the form $P + ax \leq a_0$, where P is a sum of positive monomials $\{\, b_1y_1, \ldots, b_py_p \,\}$ and negative monomials $\{\, c_1z_1, \ldots, c_qz_q \,\}$. Let R be a set of bounds $\{\, lb_1 \leq y_1, \ldots, lb_p \leq y_p,\ z_1 \leq ub_1, \ldots, z_q \leq ub_q \,\}$. Let E denote the expression $a_0 - b_1lb_1 - \ldots - b_plb_p - c_1ub_1 - \ldots - c_qub_q$. If $a < 0$ then then C and R *propagate* $\lceil E/a \rceil \leq x$. If $a > 0$ then then C and R *propagate* $x \leq \lfloor E/a \rfloor$.

Cuts and Constraint Normalization. In what follows we assume all constraints to be eagerly *normalized:* any constraint $a_1x_1 + \ldots + a_nx_n \leq a_0$ with $d = gcd(a_1, \ldots, a_n) > 1$

is eagerly replaced by $\lceil a_1/d \rceil\ x_1 + \ldots + \lceil a_n/d \rceil\ x_n \leq \lfloor a_0/d \rfloor$. From constraints $a_1x_1 + \cdots + a_nx_n \leq a_0$ and $b_1x_1 + \cdots + b_nx_n \leq b_0$, and natural numbers c and d, a new constraint $c_1x_1 + \cdots + c_nx_n \leq c_0$ called a *cut* can be obtained where $c_i = ca_i + db_i$ for i in $0 \ldots n$. If some $c_i = 0$ then we say that x_i is *eliminated* in this cut. Note that if $a_ib_i < 0$, then one can always choose c and d such that x_i is eliminated. See [13,7,23] for further discussions and references about Chvátal-Gomory cuts and their applications to ILP.

Let A be a set of bounds. A variable x is *defined to a* in A if $\{a \leq x, x \leq a\} \subseteq A$ for some a. We call two bounds $a \leq x$ and $x \leq a'$ *contradictory* if $a > a'$. Note that if all variables of X are defined and there are no contradictory bounds in A then A can be seen as a total assignment $A\colon X \to \mathbb{Z}$. A bound $a \leq x$ is *new* in A if there is no $a' \leq x$ in A with $a' \geq a$. Similarly, $x \leq a$ is new if there is no $x \leq a'$ with $a' \leq a$. A bound is *fresh* in A if it is new in A and contradictory with no bound in A.

False Constraint (Conflict) in A. If C is a constraint $a_1x_1 + \ldots + a_nx_n \leq a_0$ with positive monomials $\{ b_1y_1, \ldots, b_py_p \}$, negative monomials $\{ c_1z_1, \ldots, c_qz_q \}$ and there is a subset of bounds $\{ lb_1 \leq y_1, \ldots, lb_p \leq y_p,\ \ z_1 \leq ub_1, \ldots, z_q \leq ub_q \} \subseteq A$ with $b_1lb_1 + \ldots + b_plb_p + c_1ub_1 + \ldots + c_qub_q > a_0$ then C is *false* or a *conflict* in A.

The Basic IntSat Algorithm. In the following IntSat algorithm, A is seen as an (initially empty) stack of bounds:

1. *Propagate*: while possible and no conflict appears, if C and R propagate some fresh bound B, for some constraint C and set of bounds R with $R \subseteq A$, then push B onto A, associating to B the *reason constraint* C and the *reason set* R.
2. **if** there is no conflict
   if all variables are defined in A, output "solution A" and halt.
   else *Decide*: push some fresh bound B, marked as a *decision*, and go to 1.
3. If A contains no decisions, output "infeasible" and halt.
4. Use data structures C, a constraint, and CS, the *Conflicting Set* of bounds. Initially, C is any conflict and CS is the subset of bounds of A causing the falsehood of C.
   - *Conflict analysis*: Invariants: $CS \subseteq A$ and if S is the current set of constraints, then $S \cup CS$ is infeasible (has no solution). **Repeat** the following three steps:
     - If B is the bound in CS that is topmost in A, and R is the reason set of B, then let CS be $(CS \setminus \{B\}) \cup R$.
     - If a cut eliminating B's variable exists between C and B's reason constraint then replace C by that cut.
     - *Early Backjump*: If for some maximal $k \in \mathbb{N}$, after popping k bounds the last one being a decision, C propagates some new bound in the resulting A, then pop k bounds, learn C as a new constraint, and go to 1.

     until CS contains a single bound B_{top} that is, or is above, A's topmost decision.
   - *Backjump*: Pop bounds from A until either there are no decisions in A or, for some B in CS with $B \neq B_{top}$, there are no decisions above B in A. Then push $\overline{B_{top}}$ with associated reason constraint C and reason set $CS \setminus \{B_{top}\}$.
   - *Learn*: add the final C as a new constraint, and go to 1.

Note that in the 2nd step of conflict analysis indeed sometimes no cut eliminating B's variable exists between C and B's reason constraint; this can be because that variable does not occur in C, or it occurs with the same sign.

Example 4. We apply IntSat to Example 3: there are two constraints $x + y + 2z \leq 2$ and $x + y - 2z \leq 0$ and we take the decision $0 \leq x$, which propagates nothing, and later on another decision $1 \leq y$, which due to $x + y + 2z \leq 2$ propagates $z \leq 0$, which is pushed with associated reason constraint $x + y + 2z \leq 2$ and reason set $\{ 0 \leq x,\ 1 \leq y \}$. *Conflict analysis*: Initially, $x + y - 2z \leq 0$ is the conflict C and CS is the set of bounds $\{ 0 \leq x,\ 1 \leq y,\ z \leq 0 \} \subseteq A$ causing the falsehood C. In the first iteration, in CS we replace $z \leq 0$ by its reason set $\{ 0 \leq x,\ 1 \leq y \}$. The resulting CS is $\{ 0 \leq x,\ 1 \leq y \}$. A cut eliminating z exists (see Example 3) and C becomes $x + y \leq 1$. Then conflict analysis is over because the CS contains exactly one bound B_{top}, which is $1 \leq y$, at or above A's topmost decision. *Backjump*: We pop bounds until for some B in CS with $B \neq B_{top}$, there are no decisions above B in A, in this case, until there are no decisions above $0 \leq x$ in A, and then push $\overline{B_{top}}$, which is $y \leq 0$, with reason set $\{ 0 \leq x \}$, and with reason constraint $x + y \leq 1$. Note that this reason constraint is not a "good" reason, i.e., it does not propagate $y \leq 0$, but still $y \leq 0$ is a valid consequence of the set of constraints together with its reason set $\{ 0 \leq x \}$. *Learn*: The final C, which is $x + y \leq 1$, is learned. □

Example 5. Consider the initial constraints

$$\begin{array}{ll} C_0: & x - 3y - 3z \leq 1 \\ C_1: & -2x + 3y + 2z \leq -2 \\ C_2: & 3x - 3y + 2z \leq -1 \end{array}$$

Below we depict the stack with some initial bounds after doing their propagations and taking and propagating two decisions:

bound	reason set	reason constraint
$2 \leq y$	$\{ 1 \leq x,\ z \leq -2 \}$	C_0
$x \leq 1$	$\{ y \leq 2,\ z \leq -2 \}$	C_0
$z \leq -2$	*decision*	
$z \leq -1$	$\{ x \leq 2,\ 1 \leq y \}$	C_1
$x \leq 2$	*decision*	
$z \leq 0$	$\{ x \leq 3,\ 1 \leq y \}$	C_1
$y \leq 2$	$\{ x \leq 3,\ -2 \leq z \}$	C_1
$1 \leq x$	$\{ 1 \leq y,\ -2 \leq z \}$	C_1
$z \leq 2$	*initial*	
$-2 \leq z$	*initial*	
$y \leq 4$	*initial*	
$1 \leq y$	*initial*	
$x \leq 3$	*initial*	
$-2 \leq x$	*initial*	

Now C_1 is a conflict. The initial CS is $\{ -2 \leq z,\ x \leq 1,\ 2 \leq y \}$, with two bounds above the last decision. In the first conflict analysis step, we replace $2 \leq y$ by its reason set $\{ 1 \leq x,\ z \leq -2 \}$ obtaining the new CS $\{ -2 \leq z,\ 1 \leq x,\ z \leq -2,\ x \leq 1 \}$ which still has two bounds at or above the last decision. Now a cut eliminating y is attempted between the initial C, which is C_1, and the reason constraint of $2 \leq y$, which is C_0. Here this cut exists, with $c = d = 1$, and we obtain and learn the new constraint $C_3 : \ -x - z \leq -1$. It allows us to perform an early backjump to before the second decision, since there it propagates $2 \leq x$ with reason set $\{ z \leq -1 \}$ and reason constraint C_3. Then, after two more propagations, we obtain

$2 \leq y$	$\{ 2 \leq x,\ z \leq -1 \}$	C_0
$-1 \leq z$	$\{ x \leq 2 \}$	C_3
$2 \leq x$	$\{ z \leq -1 \}$	C_3
$z \leq -1$	$\{ x \leq 2,\ 1 \leq y \}$	C_1
$x \leq 2$	*decision*	
$z \leq 0$	$\{ x \leq 3,\ 1 \leq y \}$	C_1
$y \leq 2$	$\{ x \leq 3,\ -2 \leq z \}$	C_1
$1 \leq x$	$\{ 1 \leq y,\ -2 \leq z \}$	C_1
$z \leq 2$	*initial*	
...	...	
$-2 \leq x$	*initial*	

and again C_1 is a conflict, with the initial CS being $\{ x \leq 2,\ -1 \leq z,\ 2 \leq y \}$. After the first conflict analysis step (replacing $2 \leq y$) the CS becomes $\{ x \leq 2,\ z \leq -1,\ 2 \leq x,\ -1 \leq z \}$. As before, the cut eliminates y, between C_1 and C_0 (the initial C and the reason constraint of $2 \leq y$), obtaining $-x - z \leq -1$. After the following step (replacing $-1 \leq z$), the CS becomes $\{ x \leq 2,\ z \leq -1,\ 2 \leq x \}$. The C does not change because no cut eliminating z exists with C_3. In the next step (replacing $2 \leq x$), the CS becomes $\{ x \leq 2,\ z \leq -1 \}$. Again no cut eliminating z exists with C_3. In another step (replacing $z \leq -1$), the CS becomes $\{ 1 \leq y,\ x \leq 2 \}$. Since there is only one bound at or after the last decision, we backjump, in this case to before the first decision, and add there the negation of $x \leq 2$, which is $3 \leq x$.

The result of the cut on C with C_1 eliminating z gives us $-4x + 3y \leq -4$. The backjump with this cut (C_4) can also take us to before the first decision, but propagating $2 \leq x$. Since this is weaker than the bound $3 \leq x$ obtained from the CS, here we choose the CS one. After one further propagation, the procedure returns "infeasible" since the conflict C_2 appears and there are no decisions in the stack:

$-1 \leq z$	$\{ 3 \leq x,\ y \leq 2 \}$	C_0
$3 \leq x$	$\{ 1 \leq y \}$	C_4
$z \leq 0$	$\{ x \leq 3,\ 1 \leq y \}$	C_1
$y \leq 2$	$\{ x \leq 3,\ -2 \leq z \}$	C_1
$1 \leq x$	$\{ 1 \leq y,\ -2 \leq z \}$	C_1
$z \leq 2$	*initial*	
...	...	
$-2 \leq x$	*initial*	

Theorem 1. *The basic IntSat algorithm, when given as input a finite set of constraints S including for each variable x_i a lower bound $lb_i \leq x_i$ and an upper bound $x_i \leq ub_i$, always terminates, finding a solution if, and only if, there exists one, and returning "infeasible" if, and only if, S is infeasible.*

The previous theorem holds even if no cuts are performed and no new constraints are learned (although practical performance depends crucially on these). Its proof follows essentially the same scheme as our termination, soundness and completeness results for SAT and SAT Modulo Theories (SMT) [20]. For termination (from which soundness and completeness are not hard to establish), we define a well-founded ordering $>$ on the states of the stack A, as follows. For a given A, the number of possible values a variable x_i can still take is $v_i(A) = ub_i - lb_i + 1$, where $lb_i \leq x_i$ and $x_i \leq ub_i$ are its topmost lower and upper bounds in A, and the total number of values for all n variables is $v(A) = v_1(A) + \ldots v_n(A)$. Let A_i, for $i \geq 1$ denote the bottom part of A, below (and without) the i-th decision. We define a stack A to be larger (i.e., less advanced, search-wise) than a stack A', written $A > A'$, if $\langle v(A_1), \ldots, v(A_m) \rangle >_{lex} \langle v(A'_1), \ldots, v(A'_m) \rangle$ where m is the maximal number of decisions the stack can contain, at most $n \cdot v(A)$ for the initial A. It is easy to see that this lexicographic ordering $>$ is well-founded and that all steps of the algorithm either halt it or transform A into an A' with $A > A'$.

More General Constraints. It is obvious that a constraint $a_1x_1 + \ldots + a_nx_n \geq a_0$ can be expressed as $-a_1x_1 - \ldots - a_nx_n \leq -a_0$, that $a_1x_1 + \ldots + a_nx_n = a_0$ can be replaced by the two constraints $a_1x_1 + \ldots + a_nx_n \leq a_0$ and $a_1x_1 + \ldots + a_nx_n \geq a_0$, and that rational non-integer coefficients a/b can be removed by multiplying the constraint by b.

Optimization is also possible in a standard way, since, unlike what happens in SAT, linear constraints are first-class citizens (i.e., belong to the core language). For finding a solution that minimizes a linear expression $a_1x_1 + \ldots + a_nx_n$ (or maximizes $-a_1x_1 - \ldots - a_nx_n$), in our current implementation this is done in a completely straightforward way: first an arbitrary solution A is found and then, each time a new solution A is found, it is attempted to improve it by re-running with the additional constraint $a_1x_1 + \ldots + a_nx_n \leq a_0$ where a_0 is $a_1A(x_1) + \ldots + a_nA(x_n) - 1$. This is done until the problem becomes infeasible. Bound propagations from these successively stronger constraints are indeed very effective for pruning (bounding) the resulting branch-and-bound search.

Handling Unbounded Variables. Up to now we have assumed that for each variable there is an initial lower bound and an upper bound, or, equivalently, initial constraints propagating such bounds. Although this is common in practical applications, some problems do have unbounded variables. In theory, any ILP can be converted into an equivalent fully bounded one [23], but these bounds are too large to be useful in practice. One solution is to introduce a fresh auxiliary variable z, with lower bound $0 \leq z$, and for each variable x without lower bound add the constraint $-z \leq x$, and similarly if it has no upper bound add $x \leq z$. Then one can re-run the IntSat procedure with successively larger upper bounds $z \leq ub$ for z, thus guaranteeing completeness for finding (optimal) solutions. Further practical solutions are subject of current work, also for handling the well-known fact that with unbounded variables bound propagation may not terminate in unfeasible problems: consider, e.g., C_1: $x - y \leq 0$ and C_2: $-x + y + 1 \leq 0$ and the bound $0 \leq x$, which makes C_1 propagate $0 \leq y$; then C_2 propagates $1 \leq x$, and so on.

3 Implementation

Here we describe some details of our current prototype IntSat implementation. It currently consists of 1400 lines of simple C++ code that make heavy use of standard STL data structures (this source code can be downloaded from [19]). For instance, a constraint is an STL vector of monomials (pairs of two `int`s: the variable number and the coefficient), sorted by variable number, plus some additional information (independent term, activity). Coefficients are never larger than 2^{30}, and cuts producing any coefficient larger than 2^{30} are simply not performed, which is a straightforward way of guaranteeing that no overflow occurs if bound propagation, cuts, normalization, etc., are done using 64-bit integers for intermediate results. During conflict analysis, the CS is implemented simply as an STL set of `int`s, the heights in the stack of the bounds in the CS. A very large source of inefficiency of conflict analysis is our current implementation of *Early Backjumps*, which, after each cut giving a new C, naively checks, at all (frequently thousands of) prefixes of the stack below a decision, whether C propagates any new bound at that prefix.

The Current Assignment. There is an array, the *Bounds Array*, indexed by variable number, that can return in constant time the current upper and lower bounds for that variable. It always stores, for each variable x_i, the positions pl_i and pu_i in the stack of its current (strongest) upper bound and lower bound, respectively, with $pl_i = 0\,(pu_i = 0)$ if x_i has no current lower (upper) bound. The stack itself is another array containing at each position three data fields: a bound, a natural number pos, and an *info* field containing, among other information, (pointers to) the reason set and the reason constraint. The value pos is always the position in the stack of the previous bound of the same type (lower or upper) for this variable, with $pos = 0$ for initial bounds. When pushing or popping bounds, these properties are easy to maintain in constant time.

Example of bounds array and stack:

	Height in stack of current bound	
	lower:	upper:
x_1	1	2
x_2	0	0
	⋮	⋮
x_7	40	31
	⋮	⋮

		⋮	
40	$5 \le x_7$	23	*info*
		⋮	
31	$x_7 \le 6$	14	*info*
		⋮	
23	$2 \le x_7$	13	*info*
		⋮	
14	$x_7 \le 9$	0	*info*
13	$0 \le x_7$	0	*info*
		⋮	
2	$x_1 \le 8$	0	*info*
1	$0 \le x_1$	0	*info*

Bound Propagation Using Filters. Affordably efficient bound propagation is crucial for performance. In our current implementation, for each variable x, there are two *occurs lists*. The *positive* occurs list for x contains all pairs (I_C, a) s.t. C is a linear constraint where x occurs with positive coefficient a, and the *negative* one contains the same for occurrences with a negative coefficient a. Here I_C is an index to the *constraint header* of C in an array of constraint headers. Each constraint header contains an integer F_C called a *filter*, and (a pointer to) the constraint C itself. The filter F_C is maintained cheaply, and one can guarantee that C *does not propagate anything as long as* $F_C \leq 0$, thus avoiding many useless (cache-) expensive visits to the actual constraint C. This is done as follows.

Let C be a constraint of the form $a_1x_1 + \cdots + a_nx_n \leq a_0$. Let $lb_i \leq x_i$ and $x_i \leq ub_i$ be the current lower and upper bounds (if any) for x_i. Each monomial a_ix_i in C can have a *minimal value* m_i, which is $a_i \cdot lb_i$ if $a_i \geq 0$, and $a_i \cdot ub_i$ otherwise. Here m_i is undefined if there is no such bound lb_i (or ub_i). Initially, if some m_i is undefined, then F_C is set to a special value $\perp$, and otherwise to $-a_0 + m_1 + \cdots + m_n + max_i\{\ |a_i(ub_i - lb_i)|\ \}$. In the latter case, F_C is said to be *precise*: the constraint C propagates *if, and only if*, $\perp \neq F_C > 0$. At all time points, $F_C = \perp$ or F_C is an upper approximation of the precise one, so C can only propagate (or be false) if $F_C > 0$.

To preserve this property, these filters need to be updated when new bounds are pushed onto the stack (and each update needs to be undone when popped, for which other data structures exist). Assume a new lower bound $k \leq x$ is pushed onto the stack. Let the previous lower bound for x (if any) be $k' \leq x$. For each pair (I_C, a) in the positive occurs list of x, using I_C we access the F_C and increase it by $|a(k - k')|$. If there was no previous lower bound, then F_C was $\perp$ and is now set to 1. If F_C becomes positive, the constraint C is visited because it may propagate some new bound. After each time a constraint C is visited, F_C is set to its precise value. If a new upper bound $x \leq k$ is pushed on the stack, exactly the same is done, where $x \leq k'$ is the previous upper bound for x (if any), and using the negative occurs list.

4 Further Work

Both from a theoretical as a practical point of view, a large amount of further ideas around IntSat arise to be explored. From the implementation point of view, aspects such as special treatments for binary variables and for specific types of constraints should be worked out. Our current implementation in fact mimics several ideas from CDCL SAT solving without having tested them thoroughly.

Decision Heuristics. For instance, our current heuristics for selecting the variable of the next decision bound are based on recent activity: the variable with the highest activity score is picked (for this there is a priority queue). The activity score of a variable x is increased each time a bound containing x appears in the CS during conflict analysis, and to reward *recent* activity the amount of increment grows in time. Once a variable is picked, one has to decide the actual decision bound: whether it is lower or upper, and how to divide the interval between the current lower and upper bounds. Another idea is to try to mimic the last-phase polarity heuristic from SAT [22], which would translate into picking some recent upper/lower bound and value for the selected variable.

Restarts and Cleanups. Something similar happens with the periodic *restarts* that SAT solvers apply. We currently follow a rather conservative restart policy based on increasing intervals based on the number of conflicts. Another basically non-tested aspect is the *cleanup* policy for the constraint database; at each cleanup, we remove all non-initial constraints with more than two monomials and activity counter equal to 0. This activity counter is increased each time the constraint is a conflicting or reason constraint at conflict analysis, and is divided by 2 at each cleanup. Cleanups are done periodically, in such a way that the constraint database grows rather slowly over time.

Early Backjump and Conflict Analysis. Performing early backjumps is in fact optional. When omitted (or not done whenever possible), the price to be paid is the loss of an invariant of the stack: it is no longer true that before each decision all bounds are exhaustively propagated. A slight modification of conflict analysis suffices to handle this: before starting conflict analysis, pop bounds from the stack until the initial CS contains at least one bound at or above the topmost decision. We have not done any thorough experiments yet evaluating this option. One could, for example, do, or attempt to do, early backjump with the intermediate conflicting constraint C if it is false in the current stack, or only if it is promising (e.g., short) according to some heuristic. In any case, further work on the implementation should probably cover a much better implementation for early backjumping, which is currently a black hole for efficiency.

Several other improvements exist for conflict analysis. For example, the quality of backjumps and the strength of the reason sets can be improved by doing some more work: the CS can be simplified by removing bounds that are subsumed by stronger ones, and also, instead of using the pre-stored reason sets R, one can re-compute them on the fly during conflict analysis with similar aims. One can also do a bit of search during conflict analysis, e.g., by trying to remove non-topmost bounds and do cuts with these, with the aim of finding good early backjump cuts.

Optimization. For optimization many further ideas exist: heuristics for finding a first solution quickly (which helps bounding the search dramatically), heuristics for choosing the decision bound in a "first-succeed" manner (i.e., steering it towards minimizing the cost function). For problems of a more numerical nature with many solutions, one could also search for the optimal solution with binary search instead of decreasing the objective one by one.

Pre– and in-processing, Arithmetic. For some problems currently too many cuts are discarded because of coefficients larger than 2^{30}. One can look for solutions from the implementation point of view, e.g., by using large integer arithmetic, or using floating point arithmetic, but it might also be the case that more constraint simplification pre- and in-processing techniques can be helpful (and not only for this purpose). For propositional SAT, the so-called lemma shortening techniques introduced in MiniSAT [12] have turned out to be essential for modern SAT solvers and (in fact, several extensions of them) can be applied to IntSat as well. Modern SAT solvers such as lingeling [3] heavily apply different inprocessing techniques to keep the constraint database small but strong.

MIPs. Finally, it needs to be worked out how to apply IntSat in order to solve MIP instances, i.e., where not all variables are subject to integrality. One can decide on the integer variables as it is done now, and at any desired point one can run an LP solver to optimize the values for the rational variables. The inclusion of lower bounding techniques, well-known from modern MIP solvers, also needs to be considered.

5 Experiments

All experiments described in this section were carried out on a standard 2.66GHz 4-core Intel i5 750 desktop. The reader can easily verify these results; in particular our prototype IntSat implementation including source code and all benchmarks can be downloaded from [19].

CPLEX and Gurobi. We compare with the newest versions of the commercial solvers CPLEX (v.12.6) and Gurobi (v.5.6.2). Both use all four processor cores (while IntSat uses only one!). Both are well-known to outperform, in general by far, the existing non-commercial solvers. The technology behind these solvers is extremely mature, after decades of improvements: according to [5], between 1991 and 2012 they have seen a 475000 times speedup from algorithmic improvements only (i.e., not counting another 2000 times from hardware improvements)!.

Rather than using a single method, these solvers apply a large variety of techniques, including, e.g., specialized cuts (Gomory, knapsack, flow and GUB covers, MIR, clique, zerohalf, mod-k, network, submip,etc.), heuristics (rounding, RINS, solution improvement, feasibility pump, diving, etc.) and variable selection techniques (pseudo costs, strong branching, reliability branching, etc.).

They also apply sophisticated *presolve* methods to reduce in advance the size of the problem and to tighten its formulation. Since we have no special-purpose presolve implementation for IntSat (yet), unfortunately here we had to use Gurobi's one, and for fairness, we used Gurobi to presolve and output all instances (which took essentially negligible time) and ran all three solvers on these Gurobi-presolved instances.

Other Classes of Solvers. It is well known that SAT Modulo Theories (SMT) [20] solvers such as Mathsat5 [8], Yices [11], Z3 [10] or our own Barcelogic solver [6] mostly focus on efficiently handling the arbitrary Boolean structure on top of the LIA constraints. Their *Theory Solver* component, the one that handles *conjunctions* of constraints (our aim here), is rather basic, and we do not compere here with SMT solvers since on conjunctive problems they are indeed in general orders of magnitude worse than CPLEX or Gurobi.

Concerning SAT and Lazy Clause Generation (LCG) [21], from our own work (see among many others [1]), we also know too well that solvers that (lazily) encode linear constraints into SAT can be competitive as long as problems are rather Boolean, without a heavy ILP/optimization component. Also CSP solvers such as Sugar [24], which heavily focus on their rich constraint language, are in general very far from the commercial OR solvers on the typical hard pure ILP optimization problems.

Also, most of these SAT/SMT/LCG solvers cannot optimize or are rather bad at it. Cutsat [14] cannot optimize either.

Random Optimization Instances. We used a random generator to create 100 optimization instances with 600 variables (about half of them non-binary) and 750 constraints (instances and generator are available at [19]). Then we discarded the 51 "too easy" instances (for which all three solvers could find an optimal solution and *prove optimality* in less than 2s).

The first columns (I, C, G) in the table below show runtimes in seconds of, respectively, IntSat, CPLEX and Gurobi to prove optimality, and no time indicates timeout after 10s, which happens 17 times for IntSat, 19 times for CPLEX and 9 times for Gurobi.

Since finding good solutions quickly is perhaps as important as proving optimality, the following columns show the cost of the optimal solution ("opt"), and the best solutions found after 10s, only when different from the optimal one. The reader can check that IntSat fails 8 times to find the optimal solution, CPLEX 13 times, and Gurobi 7 times, and that the total sum of distances to the optimal solutions in these cases are 22, 39 and 17, respectively. When given longer runtimes, the commercial solvers tend to behave better on these instances than the current version of IntSat. However, this should of course be re-evaluated after a more mature implementation, heuristics, and pre- and inprocessing, etc., become available for IntSat.

	I	C	G	opt	I	C	G
01.lp				-7	-5	.	-4
03.lp		7.76	5.41	-7	.	.	.
05.lp	1.85	1.08	4.19	-4	.	.	.
06.lp	2.50			-13	.	-7	10
07.lp	1.06	2.98	3.94	-7	.	.	.
10.lp	4.21	0.17	0.07	-11	.	.	.
12.lp			9.80	-9	.	.	.
14.lp	0.77	5.78	3.16	-14	.	.	.
15.lp	1.43	2.56	0.21	-10	.	.	.
16.lp			5.02	-9	.	-8	.
20.lp				-8	-6	-4	-6
21.lp		0.71	0.19	-7	-5	.	.
23.lp				-8	.	.	.
24.lp	2.85	0.08	0.05	-4	.	.	.
26.lp	1.04		2.93	-11	.	-7	.
27.lp		2.71	6.32	-2	.	.	.
28.lp	2.16			-9	.	-7	-8
31.lp				-6	.	-3	-3
33.lp	2.79	6.38	2.92	-6	.	.	.
34.lp	2.62	1.86	0.30	-13	.	.	.
36.lp	6.58	1.53	5.23	-9	.	.	.
40.lp	2.02	3.10	0.05	-18	.	.	.
44.lp	4.76	7.47	8.54	-10	.	.	.
49.lp	2.77	0.47	0.09	-8	.	.	.
50.lp			0.22	-12	-8	-11	.

	I	C	G	opt	I	C	G
53.lp		2.47	3.05	-13	-7	.	.
60.lp	1.69		2.19	-9	.	-8	.
61.lp	1.47	1.14		-16	.	.	-15
62.lp	2.58	0.53	0.02	-3	.	.	.
63.lp			4.55	-12	.	.	.
64.lp	1.56	6.33	2.60	-4	.	.	.
65.lp	3.26	0.80	0.81	-8	.	.	.
66.lp	1.32	9.23	4.47	-5	.	.	.
68.lp				-9	.	.	.
69.lp	5.91	1.20	0.13	-14	.	.	.
70.lp	8.33	0.24	0.09	-6	.	.	.
73.lp			4.75	-11	-9	-9	.
76.lp	0.74	2.89	0.40	-11	.	.	.
78.lp				-8	-5	-2	-4
79.lp	8.26	0.36	0.06	-4	.	.	.
80.lp	7.54	2.59	0.13	-7	.	.	.
81.lp	0.86	4.71	4.78	-9	.	.	.
84.lp	2.93		6.40	-12	.	-5	.
87.lp				-10	.	-8	.
88.lp			9.67	-5	-4	-4	.
91.lp	2.38		2.96	-9	.	.	.
93.lp	2.98	3.03	0.45	-9	.	.	.
95.lp	1.51	3.13	0.14	-10	.	.	.
99.lp	0.62	2.00	4.92	-7	.	.	.

5.1 MIPLIB Instances

From the MIPLIB 2010 Mixed Integer Problem Library, a well-known "standard test set" to compare optimizer performance for the Operations Research (OR) community, cf. `miplib.zib.de`, we considered *all* 30 ILP instances (i.e., with integer and binary variables only) and discarded the 11 instances lacking initial lower and upper bounds for some variable.

For the remaining 19 ones, the next table below indicates runtimes (in s) needed to prove feasibility (as recommended for the commercial solvers, the objective function was replaced by 0). The table also includes some statistics on number of constraints, total number of variables, and among these, the number of binary variables.

We ran IntSat with no presolving, as Gurobi's presolve was harmful for it in some cases. Here we also compare with the Cutsat implementation of [14], which currently can only handle feasibility, and no optimization.

	Feasibility (s)				Problem statistics		
	IntSat	CPLEX	Gurobi	Cutsat	#constr.	total #vars.	#0/1-vars.
30n20b8	20.14	0.83	0.41	>300	666	18380	11036
d10200	10.00	0.20	0.22	>300	1147	2000	733
d20200	0.55	0.34	1.15	>300	1702	4000	3181
lectsched-1	1.11	7.75	39.73	64.84	51608	28718	28236
lectsched-1-obj	1.09	145.25	11.71	45.77	51608	28718	28236
lectsched-2	0.60	3.50	1.14	5.39	31775	17656	17287
lectsched-3	0.99	6.94	6.82	18.43	46615	25776	25319
lectsched-4-obj	0.25	0.15	0.38	0.86	14760	7901	7665
mzzv11	1.27	0.04	1.26	0.08	12871	10240	9989
neos-1224597	0.20	0.11	0.15	>300	3682	3605	3150
neos16	0.09	9.74	16.54	>300	1028	377	336
neos-555424	0.07	0.07	0.07	18.53	2746	3815	3800
neos-686190	1.98	0.44	0.27	>300	3785	3660	3600
ns1854840	3.21	3.31	1.57	3.75	151216	135754	135280
rococoB10-011000	0.06	0.05	0.06	0.06	3063	4456	4320
rococoC10-001000	0.05	0.04	0.05	4.45	2298	3117	2993
rococoC11-011100	0.10	0.16	0.10	0.05	4403	6491	6325
rococoC12-111000	0.16	0.15	0.27	>300	13181	8619	8432
sp98ir	0.16	0.15	0.28	0.92	1531	1680	871

5.2 Optimizing the MIPLIB Instances

We also considered optimizing these same 16 MIPLIB instances (lectsched-1, -2 and -3 are feasibility ones), using all three applicable solvers (IntSat,CPLEX,Gurobi) with a timelimit of 600s.

30n20b8 and Lectsched-4-Obj: For these two instances, all three solvers find the optimal solution and prove optimality in less than 10 seconds. For 30n20b8, in 1.03s, 3.59s and 4.18s, respectively for IntSat, CPLEX, and Gurobi (optimal solution has cost 302) and for lectsched-4-obj, in 0.43s, 3.37s, and 1.61s respectively (optimal is 4). Note that on both instances IntSat is fastest. IntSat is currently not able to prove optimality for any of the other 14 MIPLIB optimization instances in less than 600s.

neos16 and neos-1224597: IntSat does find optimal solutions quickly for these two problems. For neos16, none of the three solvers proves optimality in less than 600s, but IntSat finds the optimal solution (cost 446) in 13.1s, whereas CPLEX needs 39.2s and Gurobi needs 45s. For neos-1224597, all three solvers find the optimal solution (cost -448) in around 1s., but CPLEX and Gurobi moreover prove optimality.

The Other 12 Instances: CPLEX and Gurobi also prove optimality for five other instances, the same ones for both solvers, given in the first table below.

Results and best found solutions for the remaining seven instanes are given in the second table below. Out of these seven, two (d10200, rococoC11) are catalogized in the MIPLIB as "hard" and three further instances (d20200, lectsched-1-obj, and ns1854840) are "open", as their optimal cost is unknown. IntSat is the best solver by far on the open problem ns1854840, even though IntSat's 600s refer to runtime on one core only, whereas the other solvers run 600s on all four cores. In fact, for this instance Gurobi only finds an initial "heuristic" solution that is more than 100 times worse than the one found by IntSat; this happens because Gurobi's root simplex times out after 600s. Of the other "open" problems, IntSat is also better than CPLEX on two other problems: lectsched-1-obj and rococoC11-011100.

In some cases IntSat appears to be quite improvable still, e.g., due to its too naive handling of very large input problems. Sometimes also better optimization heuristics will to be useful on instances with a very numerical nature and slowly decreasing values of the objective function.

	Best solutions found			
	IntSat (600s)	optimal	time CPLEX	time Gurobi
mzzv11	-18368	-21718	16.64s	21.73s
neos-555424	1369300	1324300	4.72s	2.16s
neos-686190	11380	6730	28.95s	24.21s
rococoC10-001000	13402	11460	49.89s	436.31s
sp98ir	279007104	219676790	24.17s	33.07s

	Best solutions found after 600s		
	IntSat	CPLEX	Gurobi
d10200	12809	12441	12438
d20200	13619	12279	12262
lectsched-1-obj	92	93	85
ns1854840	288000	392000	4272000
rococoB10-011000	21462	19449	19810
rococoC11-011100	21427	21800	20957
rococoC12-111000	57118	36988	35845

6 Related Work and Conclusions

We already mentioned the work on Cutsat [14]. The idea of applying conflicting sets is not only reminiscent to the conflict analysis of SAT, but also of *SAT Modulo Theories*

(SMT) [20,2] for the theory of linear arithmetic, with the main difference, among others, that here new ILP constraints are obtained by cut inferences, normalized and learned, and not only new Boolean clauses that are disjunctions of literals representing bounds (usually only those that occur in the input formula). Other SAT/SMT related work, but for rational arithmetic is [17,15,9].

It is also worth mentioning that there may be some possible theoretical and practical consequences of the fact that IntSat's underlying cutting planes proof system is stronger than CDCL's resolution proof system: could IntSat outperform SAT solvers on certain SAT problems for which no short resolution proofs exist? E.g., pigeon-hole-like situations do occur in practical problems (think of timetabling or scheduling). A similar question applies to the current SMT solvers, which are based on resolution as well [18].

It seems unlikely that for ILP or MIP solving one single technique can dominate the others; the best solvers will probably continue combining different methods from a large toolbox, which perhaps will also include IntSat at some point. Still, IntSat by itself already appears to be the first alternative method for ILP that uses no LP relaxations and no simplex that is competitive on certain hard optimization problems, and moreover it still has an enormous potential for enhancement. We expect that this work will trigger quite some further activity on all the improvements mentioned in Section 4.

Acknowledgments. Thanks go to Albert Fiol for his work last year, as an undergraduate student, on a preliminary IntSat implementation, to Enric Rodríguez and Albert Oliveras for always trying to answer my questions, and to Dejan Jovanović for his help with cutsat and to my muse Mariona. Partially supported by the Spanish government under the SweetLogics project (TIN 2010-21062-C02-01).

References

1. Abío, I., Nieuwenhuis, R., Oliveras, A., Rodríguez-Carbonell, E., Stuckey, P.J.: To Encode or to Propagate? The Best Choice for Each Constraint in SAT. In: Schulte, C. (ed.) CP 2013. LNCS, vol. 8124, pp. 97–106. Springer, Heidelberg (2013)
2. Barrett, C.W., Sebastiani, R., Seshia, S.A., Tinelli, C.: Satisfiability modulo theories. In: Handbook of Satisfiability, pp. 825–885. IOS Press (2009)
3. Biere, A. (2010), Lingeling SAT Solver, `http://fmv.jku.at/lingeling/`
4. Biere, A., Heule, M.J.H., van Maaren, H., Walsh, T.: Handbook of Satisfiability. Frontiers in Artificial Intelligence and Applications, vol. 185. IOS Press (February 2009)
5. Bixby, B.: Presentation: 1000X MIP Tricks, Bill Cunninghams 65th (June 12, 2012)
6. Bofill, M., Nieuwenhuis, R., Oliveras, A., Rodríguez-Carbonell, E., Rubio, A.: The Barcelogic SMT Solver. In: Gupta, A., Malik, S. (eds.) CAV 2008. LNCS, vol. 5123, pp. 294–298. Springer, Heidelberg (2008)
7. Chvátal, V.: Edmonds polytopes and a hierarchy of combinatorial problems. Discrete Mathematics 4(4), 305–337 (1973)
8. Cimatti, A., Griggio, A., Schaafsma, B.J., Sebastiani, R.: The MathSAT5 SMT Solver. In: Piterman, N., Smolka, S.A. (eds.) TACAS 2013. LNCS, vol. 7795, pp. 93–107. Springer, Heidelberg (2013)
9. Cotton, S.: Natural domain SMT: A preliminary assessment. In: Chatterjee, K., Henzinger, T.A. (eds.) FORMATS 2010. LNCS, vol. 6246, pp. 77–91. Springer, Heidelberg (2010)

10. de Moura, L., Bjorner, N.: Z3: An Efficient SMT Solver. In: Technical report, Microsoft Research, Redmond (2007), http://research.microsoft.com/projects/z3
11. Dutertre, B., de Moura, L.: The YICES SMT Solver. In: Technical report, Computer Science Laboratory, SRI International (2006), http://yices.csl.sri.com
12. Eén, N., Sörensson, N.: An Extensible SAT-solver. In: Giunchiglia, E., Tacchella, A. (eds.) SAT 2003. LNCS, vol. 2919, pp. 502–518. Springer, Heidelberg (2004)
13. Gomory, R.E.: Outline of an algorithm for integer solutions to linear programs. Bulletin of the American Mathematical Society 64(5), 275–278 (1973)
14. Jovanovic, D., de Moura, L.M.: Cutting to the chase - solving linear integer arithmetic. J. Autom. Reasoning 51(1), 79–108 (2013)
15. Korovin, K., Voronkov, A.: Solving systems of linear inequalities by bound propagation. In: Bjørner, N., Sofronie-Stokkermans, V. (eds.) CADE 2011. LNCS, vol. 6803, pp. 369–383. Springer, Heidelberg (2011)
16. Marques-Silva, J., Sakallah, K.A.: GRASP: A Search Algorithm for Propositional Satisfiability. IEEE Transactions on Computers 48(5), 506–521 (1999)
17. McMillan, K.L., Kuehlmann, A., Sagiv, M.: Generalizing DPLL to Richer Logics. In: Bouajjani, A., Maler, O. (eds.) CAV 2009. LNCS, vol. 5643, pp. 462–476. Springer, Heidelberg (2009)
18. Nieuwenhuis, R.: SAT and SMT Are Still Resolution: Questions and Challenges. In: Gramlich, B., Miller, D., Sattler, U. (eds.) IJCAR 2012. LNCS, vol. 7364, pp. 10–13. Springer, Heidelberg (2012)
19. Nieuwenhuis, R.: Intsat source code, makefile, benchmarks and benchmark generators (2014), http://www.lsi.upc.edu/~roberto/IntSatCP2014.tgz
20. Nieuwenhuis, R., Oliveras, A., Tinelli, C.: Solving SAT and SAT Modulo Theories: From an abstract Davis–Putnam–Logemann–Loveland procedure to DPLL(T). Journal of the ACM, JACM 53(6), 937–977 (2006)
21. Ohrimenko, O., Stuckey, P.J., Codish, M.: Propagation = lazy clause generation. In: Bessière, C. (ed.) CP 2007. LNCS, vol. 4741, pp. 544–558. Springer, Heidelberg (2007)
22. Pipatsrisawat, K., Darwiche, A.: On modern clause-learning satisfiability solvers. Journal of Automated Reasoning 44(3), 277–301 (2010)
23. Schrijver, A.: Theory of Linear and Integer Programming. John Wiley & Sons, Chichester (1986)
24. Tamura, N., Taga, A., Kitagawa, S., Banbara, M.: Compiling finite linear CSP into SAT. Constraints 14(2), 254–272 (2009)

Automatically Improving Constraint Models in Savile Row through Associative-Commutative Common Subexpression Elimination

Peter Nightingale, Özgür Akgün, Ian P. Gent, Christopher Jefferson, and Ian Miguel

School of Computer Science, University of St. Andrews, St. Andrews, Fife KY16 9SX, UK
{pwn1,ozgur.akgun,ian.gent,caj21,ijm}@st-andrews.ac.uk

Abstract. When solving a problem using constraint programming, constraint modelling is widely acknowledged as an important and difficult task. Even a constraint modelling expert may explore many models and spend considerable time modelling a single problem. Therefore any automated assistance in the area of constraint modelling is valuable. Common sub-expression elimination (CSE) is a type of constraint reformulation that has proved to be useful on a range of problems. In this paper we demonstrate the value of an extension of CSE called *Associative-Commutative* CSE (AC-CSE). This technique exploits the properties of associativity and commutativity of binary operators, for example in sum constraints. We present a new algorithm, X-CSE, that is able to choose from a larger palette of common subexpressions than previous approaches. We demonstrate substantial gains in performance using X-CSE. For example on BIBD we observed speed increases of more than 20 times compared to a standard model and that using X-CSE outperforms a sophisticated model from the literature. For Killer Sudoku we found that X-CSE can render some apparently difficult instances almost trivial to solve, and we observe speed increases up to 350 times. For BIBD and Killer Sudoku the common subexpressions are not present in the initial model: an important part of our methodology is reformulations at the preprocessing stage, to create the common subexpressions for X-CSE to exploit. In summary we show that X-CSE, combined with preprocessing and other reformulations, is a powerful technique for automated modelling of problems containing associative and commutative constraints.

1 Introduction

When solving a problem using constraint programming, constraint modelling is widely acknowledged as both important and difficult [13]. A problem may have many models, and it is difficult to know which will be solved most efficiently by a given constraint solver. Even a constraint modelling expert may explore many models and spend considerable time modelling a single problem. Therefore, any automated assistance in constraint modelling is valuable.

We focus on the process we call *tailoring*: given a constraint model in a solver-independent language and a value for each of its parameters, translate it into a form suitable for efficient solving by a given constraint solver. Tailoring must be efficient: it is performed separately for each problem instance, hence any computationally expensive reformulation must pay for itself by saving time during solving.

B. O'Sullivan (Ed.): CP 2014, LNCS 8656, pp. 590–605, 2014.

Common sub-expression elimination (CSE) is a type of constraint reformulation that has proved to be useful on a range of problems [16,15]. Herein we investigate an extension of CSE, *Associative-Commutative* CSE (AC-CSE), which exploits the properties of associativity and commutativity of binary operators (e.g. $+$ and $\times$). Expressions containing these operators can be rearranged to reveal common subexpressions. As an example, take the following two constraints over four variables:

$$w + x + y + z = 6, \quad z + y + w = 5$$

Conventional constraint propagation will not reveal the fact that $x = 1$. AC-CSE could extract $w + y + z$ and replace it with an auxiliary variable a to give the following three constraints. Performing constraint propagation on this set will assign x to 1.

$$x + a = 6, \quad a = 5, \quad a = w + y + z$$

An Associative-Commutative Common Subexpression (AC-CS) of a set of associative and commutative (AC) expressions (e.g. sums) is a set of at least two terms that all appear in each one of the AC expressions (sums). In the example above, the set of three terms $\{w, y, z\}$ appears in both the original sum constraints, hence $\{w, y, z\}$ is an AC-CS of the two sum constraints.

A simple normalisation step, such as sorting the terms in the AC expressions, followed by examining contiguous subsequences of terms within AC expressions, can reveal some but not all of the available AC-CSs. More is necessary to find AC-CSs in general. Consider the example above, with an alphabetical ordering of the terms $w + x + y + z$ and $w + y + z$. The largest contiguous subsequence of both is $y + z$, so this approach would miss the maximal AC-CS $w + y + z$.

In this paper we introduce and describe in detail a new algorithm, X-CSE, to perform AC-CSE in constraint problems. We show that X-CSE is able to find common subexpressions (CSs) automatically in a variety of problems, and that using these subexpressions can greatly reduce search and improve solving time. A particular advantage of X-CSE is that it is able to find and exploit small CSs that occur in many constraints, as well as larger ones that occur in few constraints. This is made possible by finding CSs that contain auxiliary variables introduced at an earlier step of the algorithm. We can thus exploit the occurrence of many small CSs without losing the advantages of finding larger ones. We illustrate this with an example below in Sections 2.3 and 3.3. The reuse of auxiliary variables created by AC-CSE in subsequent common subexpressions is an important advantage of X-CSE.

In addition, we show that X-CSE can be particularly effective in combination with other automated modelling techniques. In this paper we give two examples. First, we show that automated reformulation of an all-different constraint can lead to sum constraints, which can be exploited by X-CSE. Second, we see that by applying Singleton Arc Consistency at the preprocessing stage, we reveal new common subexpressions that can be exploited by X-CSE. These combinations show in particular that X-CSE is a valuable addition to the armoury of automated constraint modelling techniques, both alone and in combination with other techniques.

We evaluate the new algorithm on four problem classes: BIBD, the SONET problem, Killer Sudoku and Molnar's Problem. When applying X-CSE we demonstrate substantial gains in performance. On BIBD we observed speed increases of more than 20 times compared to a standard model. We found that X-CSE outperforms a sophisticated model from the literature with manually derived implied constraints [7]. On the SONET problem we observed speed increases of 5 times on some instances. For Killer Sudoku we found that applying X-CSE can render some apparently difficult instances almost trivial to solve, and we saw more than 300 times speed increases in some cases. Molnar's Problem exhibits more modest gains peaking at 5 times faster for the most difficult instance.

2 Related Work

The context of our work is reformulation of constraint modelling languages such as OPL [19], MiniZinc [18] and ESSENCE′ [16]. These languages have a collection of global constraints, arithmetic and logical operators that act on finite-domain or real interval decision variables. In this paper we consider finite-domain decision variables.

Such languages are not directly accepted by constraint solvers but must be tailored into a form suitable for a constraint solver. During tailoring the model can be *reformulated* to improve the efficiency of the constraint solver. There are many ways of producing better constraint models, some requiring manual interaction [2], and others that are automated [8]. For example these tools can discover global constraints or automatically detect and remove symmetries [11]. These improvements often complement each other, for example Frisch, Jefferson, and Miguel [7] show how breaking symmetries can lead to effective implied constraints for BIBDs among other problems. In this paper we show how to automatically generate a superior model for the BIBD problem.

2.1 Flattening and CSE

Flattening is the process of taking a nested expression and reducing the degree of nesting by replacing a subexpression with a new variable. For example given the product $X \times (Y + Z)$ and a target solver that does not allow sums inside products, the flattening process will add a new variable *aux*, replace the product with the new expression $X \times aux$ and add a new constraint $aux = Y + Z$. We say that $X \times (Y + Z)$ is *flattened* to $X \times aux$ and that $Y + Z$ is *extracted*.

Common sub-expression elimination (CSE) was first applied in the context of finite-domain constraint languages by Andrea Rendl [16,15]. In its simplest form, CSE takes two or more syntactically identical sub-expressions that must be flattened, and flattens them all using the same auxiliary variable. This reduces both the number of constraints and auxiliary variables. Importantly, CSE can reduce the search space dramatically [16,15] by linking different constraints together thus strengthening constraint propagation.

2.2 Normalisation and Active CSE

One way CSE has been fruitfully extended is by matching subexpressions that are not syntactically identical [16]. This is achieved in two ways. The first is *normalisation*, where prior to CSE the expression tree is converted to a normal form (primarily by ordering the arguments of commutative operators and evaluating any constant expressions). This converts some semantically equivalent expressions (such as $C = B + A + 1 - 1$ and $A + B = C$) to syntactically identical expressions. The second is *Active CSE*, where two expressions A and B may be matched if they are identical after some transformation (for example by applying one of De Morgan's laws). For example, Active CSE can match $A < B$ with $A \geq B$ by a simple negation.

The algorithm introduced by Rendl [15] and used by Stuckey and Tack [18] performs CSE during flattening. The algorithm maintains a hash table keyed by the expressions that have been extracted so far, and containing the new auxiliary variable for each. When extracting an expression E, the algorithm looks up E in the hash table, and (if present) uses the auxiliary variable in the hash table rather than creating a new auxiliary. This algorithm has the advantage that is easily extended to active CSE. When looking up E in the hash table, active CSE also looks up each transformation of E. However it is not clear how this algorithm could be extended to AC-CSE. The common subexpressions extracted by AC-CSE would not normally be extracted by flattening.

2.3 Associative-Commutative CSE

Araya, Neveu and Trombettoni [1] exploited common subexpressions among $+$ and $\times$ expressions. Their work is in the context of numerical CSP solved by algorithms such as HC4, but the reformulation is equally applicable to finite-domain CSP. They proposed two algorithms named I-CSE and I-CSE-NC. Both algorithms apply a form of AC-CSE prior to flattening as a separate operation.

The first pass of both algorithms is to transform the abstract syntax tree (AST) into a directed acyclic graph where identical subexpressions are represented once. The second step is to intersect each pair of sums and pair of products to create a set of candidate AC-CSs. As we will see in Section 3.3 other AC-CSs (generated by the intersection of three or more sums) can also be useful, but I-CSE and I-CSE-NC will never generate them. Later passes extract the AC-CSs from the original expressions.

Araya et al. defined two AC-CSs f_1 and f_2 to be *in conflict* if $f_1 \cap f_2 \neq \emptyset$, $f_1 \not\subseteq f_2$ and $f_2 \not\subseteq f_1$. Two AC-CSs in conflict cannot both be extracted from the same expression. When a set of AC-CSs in conflict are subsets of the same original expression s, then I-CSE copies s a sufficient number of times to extract each of the AC-CSs from at least one copy. I-CSE-NC (for No Conflicts) does not copy s, it simply extracts a single maximal subset of the candidate AC-CSs from s. Consider the following example:

$$v + w + x + y = 0, \quad v + w + x + z = 0$$

$$v + w + y + z = 0$$

In this example I-CSE(-NC) would generate three AC-CSs: $v + w + x$, $v + w + y$ and $v + w + z$. I-CSE would duplicate each of the original constraints resulting in six constraints and three further constraints to define the auxiliary variables.

I-CSE-NC can extract only one AC-CS. Suppose it extracts $v + w + x$, then at this point $v + w + y$ and $v + w + z$ cease to be AC-CSs:

$$aux + y = 0, \; aux + z = 0, \; v + w + y + z = 0$$

$$aux = v + w + x$$

I-CSE-NC can only extract CSEs from the original expressions, so fails to exploit the AC-CS $v + w$. Even on this small example I-CSE has increased the size of the model substantially. I-CSE-NC has not, but it has missed a potentially useful AC-CS and has not linked $v + w + y + z$ to the other two sums.

I-CSE and I-CSE-NC are both compared to our algorithm in the experiments below. We implement the algorithms exactly as described in Section 4 of Araya et al. [1]. Both I-CSE and I-CSE-NC only extract AC-CSs from the original expressions, they do not extract AC-CSs from other AC-CSs.

3 The X-CSE Algorithm

The X-CSE algorithm is implemented in Savile Row 1.6 [12]. Savile Row reads the ESSENCE′ language and transforms it in many passes to an output for a constraint solver. X-CSE simply becomes another pass. In this paper we consider the associative and commutative (AC) operators `\/` (or), `/\` (and), `+`, `*`. These are represented as a single AST node with n children in Savile Row 1.6. There are other AC operators in the language, notably `min` and `max`, and `!=` between boolean expressions (exclusive or). We leave these for future work.

Prior to running X-CSE the AST is normalised by sorting the children of all commutative operators. For any AC operator $\diamond$, the goal of X-CSE is to find common sets containing two or more expressions that are contained in more than one $\diamond$ expression. The X-CSE algorithm uses a hash table *map* from *pairs* of expressions $\{a, b\}$ to a list of the $\diamond$ expressions that contain both a and b. Algorithm 2 (populateMap) takes a reference to an AST node and explores the tree, populating *map* for each $\diamond$ expression.

Algorithm 1 (X-CSE) takes a reference to the AST representing all constraints, a reference to the global symbol table, and the AC operator $\diamond$. After initialising data structures it calls populateMap with the entire AST. Following that it enters the main loop on line 4. On line 5 one pair is selected from *map* according to a heuristic. If the pair occurs in more than one $\diamond$ expression then there must exist an AC-CS including that pair. Lines 10-20 find an AC-CS and extract it from all the relevant expressions. The algorithm includes as many $\diamond$ expressions as possible to maximise the effect of extracting the AC-CS. Line 10 intersects all $\diamond$ expressions containing the pair. A new $\diamond$ expression for the AC-CS is constructed, and an auxiliary variable is created. On line 14 a constraint is created to define the auxiliary variable. Each $\diamond$ expression containing the AC-CS is replaced. At this point, lines 19 and 20 update *map* to include all the newly created expressions, allowing X-CSE to extract further AC-CSs from the new expressions. Some references to removed $\diamond$ expressions will remain in *map*; these will be filtered out on line 8.

3.1 Heuristics

X-CSE chooses the next pair to process by calling a heuristic on line 5. We experimented with eight heuristics. There are four basic heuristics: most occurrences (i.e. select the pair that leads to the longest list *ls* after line 8 of X-CSE), fewest occurrences, largest AC-CS and smallest AC-CS. In some cases there exists a pair such that its corresponding AC-CS can be extracted without preventing any other AC-CS. We call these

Algorithm 1. X-CSE(AST, ST, ◇)

```
Require: AST: Abstract syntax tree representing the model
Require: ST: Symbol table containing decision variables
Require: ◇: The associative and commutative operator
newcons ← empty list {Collect new constraints}
map ← empty hash table mapping pairs of expressions to lists
populateMap(AST, map, ◇)
while map not empty do
    pairexp ← heuristic(map)
    ls ← map(pairexp) {ls is a list of ◇ AST nodes}
    delete map(pairexp)
    ls ← filter(isAttached, ls) {Remove ◇ AST nodes no longer contained in AST or newcons}

    if length(ls) > 1 then
        commonset ← ls[1] ∩ ls[2] ∩ · · · ∩ ls[length(ls)]
        e ← fold(◇, commonset)
        bnds ← bounds(e)
        aux ← ST.newAuxVar(bnds)
        newc ← ( e = aux ) {New constraint defining aux}
        newcons.append(newc)
        for all a ∈ ls do
            newe ← fold(◇, (a \ commonset) ∪ {aux})
            Replace a with newe within AST or newcons
            populateMap(newe, map, ◇)
        populateMap(newc, map, ◇)
AST ← AST ∧ fold(∧, newcons)
```

non-blocking pairs and it may be helpful to process them first. We created four more heuristics that select non-blocking pairs first, then fall back to one of the four basic heuristics. We found no clear winner among the eight heuristics. We use the 'most occurrences' heuristic throughout the rest of this paper because it is cheap to compute and often performs well.

3.2 Complexity Analysis

In this analysis we will use n for the number of $\diamond$ expressions, k for the length of the longest $\diamond$ expression, d as the depth of the deepest $\diamond$ expression in the AST, and S as the number of nodes in the AST.

Central to the complexity analysis of X-CSE is the observation that at most $k-1$ AC-CSs may be extracted from one $\diamond$ by X-CSE. Recall (from Section 2.3) that two AC-CSs in conflict cannot both be extracted from the same expression. A pair of AC-CSs may overlap only if one is a subset of the other. Consider an AC-CS f in an expression e. There can be no other AC-CSs involving f in e except possibly some f' where $f \subsetneq f'$. The smallest AC-CS is size two, and extracting this replaces a size two term with a size one term (i.e. the replacement auxiliary variable). If the original expression is size k, we thus find one AC-CS and now have a size $k-1$ expression. Iterating shows that at

Algorithm 2. populateMap(A, *map*, $\diamond$)

Require: A: Reference to an abstract syntax tree
Require: *map*: Hash table mapping pairs of expressions to lists
Require: $\diamond$: The associative and commutative operator
if A is expression of $\diamond$ **then**
    for all $\{e_1, e_2\} \subseteq A$ **do**
        Add A to list *map*$[\{e_1, e_2\}]$
for all *child* $\in$ A.Children() **do**
    populateMap(*child*, *map*, $\diamond$)

most $k - 1$ AC-CSs may be extracted from one $\diamond$ expression by X-CSE. This gives us a global limit of $O(nk)$ AC-CS extractions.

To populate *map*, populateMap traverses the AST with S nodes, and for each $\diamond$ expression e it inserts a reference to e in $O(k^2)$ lists within *map*. Assuming hash table operations are $O(1)$, populateMap takes $O(S + nk^2)$ time.[1]

X-CSE then enters a loop that continues until *map* is empty. Each iteration of the loop is as follows. We assume the heuristic takes $O(1)$ time.[2] For the given pair, its list *ls* has at most n elements. Note that if the pair occurs more than once in an expression it might be entered into *ls* multiple times: to keep the list at size n, when inserting an expression e into *ls* we can check the last element of *ls*: if it is equal to e, we do not insert e for a second time. The list *ls* is filtered in $O(nd)$ time. If the list has length two or greater, then we extract an AC-CS. For the following we assume that an AC expression is represented by a set data structure with $O(1)$ lookup, insertion and removal.[3] Creating *commonset* on line 10 takes $O(nk)$ time. Computing the bounds and creating the auxiliary variable and the new constraint can be done in $O(k)$ time. The algorithm then replaces *commonset* in each *ls* expression in $O(nk)$ time. Re-populating *map* (on lines 19 and 20) takes $O(S+nk^2)$ because the updated AC expressions can contain the entire AST. Therefore the entire cost of extracting one AC-CS is $O(S+nk^2+nd)$, and the total cost of X-CSE is $O(nkS + n^2k^3 + n^2kd)$.

While the complexity may seem high, the algorithm scales with the number of AC-CSs it is able to exploit, therefore it is relatively quick when there are few or no AC-CSs, and it takes more time when there is greater potential benefit.

3.3 Comparison with I-CSE(-NC)

X-CSE differs from the existing algorithms I-CSE(-NC) in that it can extract AC-CSs that are intersections of more than two expressions, and AC-CSs containing auxiliary variables (from earlier steps). Thus it has a larger palette of AC-CSs to choose from. In

[1] This is correct if all expressions to be hashed are size $O(1)$ and computing the hash code is linear. If either assumption is invalid then an additional factor h is necessary, representing the time to hash an expression.

[2] As an example of an $O(1)$ heuristic we could maintain a doubly linked list of keys in *map* and have the heuristic simply remove and return the first element of the list.

[3] Once again we are assuming expressions can be hashed in $O(1)$ time.

the example from Section 2.3, X-CSE would first extract $v + w$ from all three sums as follows.

$$a = v + w,\ a + x + y = 0,\ a + x + z = 0,\ a + y + z = 0$$

Second, X-CSE would extract any one of $a + x$, $a + y$ or $a + z$, as follows. This second step is not possible in I-CSE(-NC).

$$a = v + w,\ b = a + x,\ b + y = 0,\ b + z = 0,\ a + y + z = 0$$

This result is clearly better than I-CSE-NC (Section 2.3) that extracted only $v + w + x$ and thus did not connect the third constraint to the other two. I-CSE produced nine constraints on this example. It is possible that the more compact model produced by X-CSE is better. We investigate this further in Section 5.5.

4 Preprocessing and Reformulation

The number and quality of CSs found can be improved by using MINION to preprocess an initial version of the model then feeding it back into Savile Row for CSE. Our method is as follows. First Savile Row translates the instance to MINION (with or without X-CSE). Then MINION is called to filter domains with SACBounds (no search), which is a variant of SAC [3]. SACBounds applies the SAC test to prune the upper and lower bound of each variable to exhaustion. Savile Row re-starts the translation process with the filtered domains and translates the instance to MINION again (with or without X-CSE). Re-starting translation allows Savile Row to simplify the constraints following domain filtering. For example, on the BIBD problem below, some variables are assigned by SACBounds and this allows constant folding (e.g. $\cdots + a \times x + b \times y + \cdots$ where SACBounds assigns $a = 1$ and $b = 0$ becomes $\cdots + x + \cdots$).

A further step to promote the identification of AC-CSs is in reformulating a model to add implied constraints consisting of AC expressions. Savile Row creates implied sum constraints from all-different and global cardinality constraints. This is done by finding assignments to the all-different (GCC) with the smallest and largest sums (*lb* and *ub* resp.), then adding either $\sum \geq lb$ and $\sum \leq ub$ (when $lb \neq ub$) or $\sum = lb = ub$ where $\sum$ is the sum of the variables in scope of the original constraint (except cardinality variables in GCC). For example, given allDiff(x, y, z) where all variables have domain $\{1 \ldots 4\}$, we add constraints $x + y + z \geq 1 + 2 + 3$ and $x + y + z \leq 2 + 3 + 4$.

5 Case Studies

In this section we study four problems where we found AC-common subexpressions. We use Savile Row 1.6 and the following optimisations are *always* applied: unification of equal variables, domain filtering with SACBounds (as described in the section above), and identical CSE (elimination of identical subtrees in the expression tree). In addition, X-CSE, I-CSE or I-CSE-NC may be applied (before any form of flattening or other CSE) as required for the experiment. Timings include both total time reported by Savile Row (which includes the first preprocessing call to MINION) and total time reported by MINION 1.6.1 64-bit to search for a solution. MINION is given a time limit of 600s to solve the final model. Savile Row is executed in the Java 1.7.0_55 JIT. Each reported timing is a mean of 5 runs. Experiments were performed on a 32-core AMD Opteron 6272 at 2.1 GHz. All model and parameter files are available at `http://pn.host.cs.st-andrews.ac.uk/cp-2013-ac-cse-experiments.tgz`.

5.1 Case Study 1: BIBD

We use Puget's model of the Balanced Incomplete Block Design (BIBD) problem, with Lex^2 symmetry breaking constraints [14]. BIBD is parameterised by (v, k, λ) and has $r = \frac{\lambda(v-1)}{k-1}$ and $b = \frac{\lambda v(v-1)}{k(k-1)}$. The model has a v by b matrix m of boolean variables. Each of the v rows sums to r (row constraints), and each of the b columns sums to k (column constraints). The scalar product of each pair of rows has value λ:

$$\forall i_1, i_2 \in \{1 \dots v\} \,.\, i_1 < i_2 \rightarrow (\sum_{j=1}^{b} m[i_1, j] * m[i_2, j]) = \lambda$$

This model initially has no common subexpressions (identical or AC). As described above the domains are filtered by applying SACBounds. This assigns some of the variables (the entire first two rows and first column, plus some other entries). When translating again with the domains filtered by SACBounds, the scalar product constraints are simplified causing AC-CSs to appear among scalar product constraints, and between scalar product and row sum constraints.

We evaluated X-CSE on the 24 instances in Figure 1 of Puget ([14]). MINION times out for 4 instances without X-CSE. For the remaining 20 instances, X-CSE always decreases the node count. Figure 1 plots the reduction factor for the 20 instances. Harder instances tend to show a greater reduction in node count. For the hardest instance solved within the time limit, the node count is reduced by 78 times.

Figure 1 plots speed-up of total time with X-CSE. For the easiest instances, the reduction in node count does not cause a measurable difference in MINION's run time. The slow down in total time is caused by the up-front cost of X-CSE. On the harder instances, MINION search takes up most of the total time and X-CSE speeds up search substantially by reducing the number of search nodes. Figure 1 (lower) peaks with instance $(10, 3, 6)$, which has a 58-fold reduction in nodes and speed up of 24.5 times. X-CSE typically increases the number of constraints and auxiliary variables, reducing the node rate of MINION. Finally, $(10, 3, 8)$ times out without X-CSE, and takes 138.6 s with X-CSE. Hence it appears on the far right of Figure 1 with a speed up of 4.39.

Implied Constraints for BIBD. Frisch, Jefferson, and Miguel ([7]) derived a set of implied constraints for BIBD that drastically improve the performance of the model. First they observed that the first two rows and first column of the BIBD can be assigned by manually reasoning about the constraints. Second, for each of the remaining rows i, they reformulated the row sum constraint into four sum constraints. For example, for indices where row 1 is set to 0, row i sums to $r - \lambda$. These four constraints are derived

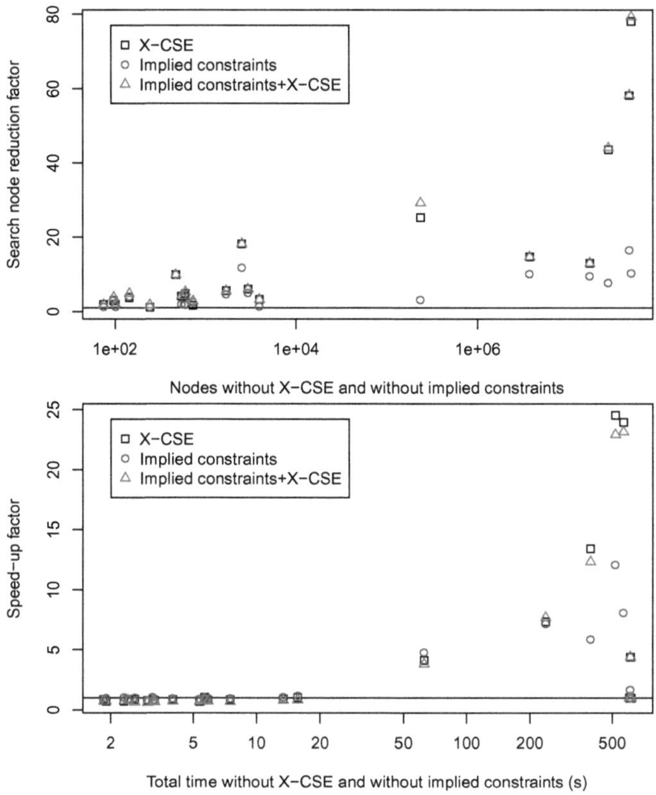

Fig. 1. (Top) BIBD search nodes of instances that do not time out. (Bottom) BIBD total time.

from the row constraint for row i and scalar product with either row 1 or 2 using an approach resembling manual AC-CSE.

The automated approach improves on Frisch et al. in two ways. First SACBounds is able to assign not just the first two rows and first column but also parts of other rows and columns. For example, on the instance $(v = 22, k = 7, \lambda = 2)$ parts of the third and fourth rows and the first eight entries of the second column are assigned. Second, X-CSE is able to link multiple scalar product constraints and a row constraint, whereas the implied constraints are each derived from a single scalar product constraint and a row constraint.

The implied constraints alone do reduce node count (see Figure 1) but are not as effective as X-CSE. For the hardest instances the implied constraints speed up solving but by a smaller degree than X-CSE. Adding the implied constraints then applying X-CSE is slightly more effective than X-CSE alone in reducing node count. However this does not translate to more efficient search. Implied constraints plus X-CSE is slower than X-CSE alone on almost all instances. Remarkably, X-CSE is able to improve the sophisticated model on the hardest four solvable instances.

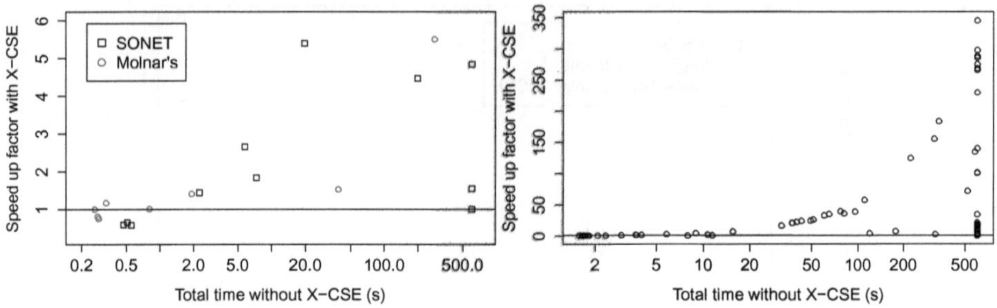

Fig. 2. (Left) Results for SONET and Molnar's Problem total time. (Right) Results for Killer Sudoku total time.

5.2 Case Study 2: The SONET Problem

The SONET problem [17] is a network design problem where each node is connected to a set of *rings* (fibre-optic connections). The simplified SONET problem (Section 3 of [17]) where each ring has unlimited capacity has the following parameters: the number of nodes n, the upper limit on the number of rings m, the maximum number of nodes per ring r, and a set of pairs that must be connected. For each of these pairs there must exist a ring connected to both nodes. The number of node-ring connections is minimised.

The problem is modelled as follows. We have a boolean matrix *rings* indexed by $[1 \ldots m, 1 \ldots n]$. $rings[a, b]$ indicates whether ring a is connected to node b. For each ring a we have the sum constraint $\sum_{b=1}^{n} rings[a, b] \leq r$. The connectedness constraint between two nodes b_1 and b_2 is expressed as a disjunction (refined by Savile Row to a watched-or [9]) of sums:

$$\exists\, i \in \{1 \ldots m\}.\, (rings[a_i, b_1] + rings[a_i, b_2] \geq 2)$$

The minimisation function is simply the sum of *rings*. Rings are indistinguishable so we use lexicographic ordering constraints to order the rows of *rings* in non-decreasing order. The static variable ordering we use is the reading order of *rings* and value order is 0 then 1. This model is very simple and does not include implied or dominance constraints [17]. The problem constraints are already flat and only the minimisation sum needs to be flattened, thus only one auxiliary variable is created by Savile Row without X-CSE. There are AC-CSs between the connectedness constraints, the ring sum constraints and the minimisation sum.

We generated 24 instances with $n \in \{6 \ldots 13\}$, $r \in \{3, 4, 5\}$, and $m = 10$. The demand graph when $n = 13$ is Figure 1 of Smith [17]. For smaller n we take the subgraph with vertices $\{n+1 \ldots 13\}$ and edges adjacent to these vertices removed.

Figure 2 plots the speed-up factor for X-CSE. As before the time limit is 600s. All instances with $n \in \{10 \ldots 13\}$ and also instance $n = 9, r = 4$ timed out both with and without X-CSE. Instances $n = 9, r = 5$ and $n = 8, r = 3$ timed out without X-CSE, and appear on the far right of the plot with a speed-up of 4.84 and 1.55 respectively. X-CSE improves solving speed for all but the most trivial instances.

5.3 Case Study 3: Killer Sudoku

We consider the Killer Sudoku problem. The standard Killer Sudoku has a 9×9 grid where each row and column are all-different, and the nine non-overlapping 3×3 sub-squares are also all-different. Each slot in the grid is initially empty and takes a digit $1 \dots 9$. Clues are sets of squares that sum to a given value (and are also all-different). We found that 9×9 Killer Sudoku instances were very easy. We generalised the puzzle to 16×16 with 16 4×4 subsquares, and each slot takes a number $1 \dots 16$. 100 instances were generated at random. Traditional Killer Sudoku puzzles have exactly one solution. The random 16×16 instances may be unsatisfiable and may have multiple solutions. For brevity we do not describe how these instances are generated. All models and instances are available on the web at the URL given in Section 5 above.

X-CSE alone does nothing because the sums in the clues are the only AC expressions and they do not overlap. However the sums overlap with all-different constraints. Each all-different constraint on a row, column or subsquare represents a permutation of $\{1 \dots 16\}$ which sum to 136. Savile Row automatically adds these implied sum constraints as described in Section 4. X-CSE is able to find common subexpressions among rows, columns, sub-squares and clues.

Figure 2 plots the speed-up quotient for Killer Sudoku. Without X-CSE, 54 instances timed out. With X-CSE, 28 instances timed out. As the instances become more difficult the trend is towards greater speed-up by X-CSE. The plot peaks at 345 times faster. On this instance, without X-CSE Savile Row took 2.26 s and MINION timed out after exploring 2,774,028 nodes. With X-CSE, Savile Row took 1.62 s and MINION took 0.13 s to explore 2 nodes.

5.4 Case Study 4: Molnar's Problem

Molnar's problem [6] (CSPLib problem 035 [5]) is to find a square matrix M of integers. The model has two parameters: the size k (i.e. M has size $k \times k$) and the maximum absolute value of integers in M, named d. The initial domain of each element of M is $\{-d \dots -2\} \cup \{0\} \cup \{2 \dots d\}$. The first constraint is that the determinant of M is 1 or -1 (following the model of Frisch et al. [6]). For the second constraint we construct another matrix S where each entry of S is the square of the corresponding entry of M. The determinant of S must also be 1 or -1.

We used the Leibniz formula for determinants, and expressed a^2 as $a \times a$ to allow more AC-CSs of products. When $k = 3$ we have the following two matrices and two constraints. In addition we break symmetry on M by lexicographically ordering rows and columns.

$$M = \begin{bmatrix} a & b & c \\ d & e & f \\ g & h & i \end{bmatrix}, S = \begin{bmatrix} a^2 & b^2 & c^2 \\ d^2 & e^2 & f^2 \\ g^2 & h^2 & i^2 \end{bmatrix}$$

$$|M| = aei - afh + bfg - bdi + cdh - ceg \in \{-1, 1\}$$

$$|S| = aaeeii - aaffhh + bbffgg - bbddii + ccddhh - cceegg \in \{-1, 1\}$$

There are multiple AC-CSs of products, for example aa and aei. Some connect the two sums, and others connect terms within one sum. X-CSE is able to extract a particular AC-CS from the same product more than once on this problem. Consider aei: extracting it once creates a new constraint $aei = x$ and modifies the expression aei to x, and the expression $aaeeii$ to $x \times aei$. Now X-CSE extracts aei a second time from the new constraint and one of the modified expressions, creating a second auxiliary variable (that will later be unified with x).

Figure 2 plots the speed-up quotient for Molnar's Problem on the eight instances where $k \in \{2, 3\}$ and $d \in \{2 \ldots 5\}$. X-CSE appears to be more useful for the more difficult instances. None of the instances time out. The peak speed-up quotient is 5.5.

5.5 I-CSE and I-CSE-NC

In this section we use all four problem classes to compare X-CSE to I-CSE and I-CSE-NC. Figure 3 plots the speed-up factor for I-CSE and I-CSE-NC compared to X-CSE. It is clear from the lower plot that I-CSE-NC performs much more poorly than X-CSE (since almost all points are below $y = 1$). By comparing the two plots in Figure 3 it is clear that I-CSE outperforms I-CSE-NC on Killer Sudoku, I-CSE-NC is preferable for SONET, and that the two algorithms are very similar for BIBD (without implied constraints) and Molnar's Problem.

X-CSE performs substantially better than I-CSE on BIBD and SONET, and slightly better on Molnar's Problem. For BIBD, both timed out on 3 instances and each solved 21. X-CSE explored fewer search nodes on 16 of the 21 instances, and was much faster overall. The mean time for X-CSE (on the set of 21 instances) was 16.7 s compared to 52.1 s for I-CSE.

For SONET, X-CSE always explores more (or an equal number of) search nodes than I-CSE but total time is lower with X-CSE for all instances taking longer than 1 s. X-CSE was able to solve all instances that I-CSE could within the timeout, and one additional one. Of the 9 that both solved, X-CSE had a mean time of 20.3 s compared to I-CSE's mean of 65.1 s. X-CSE and I-CSE-NC are able to solve the same set of 10 SONET instances. X-CSE had a mean time of 57.1 s while I-CSE-NC had a mean time of 59.7 s.

For Killer Sudoku, the picture is less clear. 70 instances are solved by both I-CSE and X-CSE. I-CSE solves one additional instance in 454 s, and X-CSE solves two additional instances in 2.1 s and 278 s. On the 70 instances solved by both, I-CSE took a mean time of 28.6 s, and X-CSE took a mean time of 35.2 s. I-CSE searches fewer nodes on 16 of these 70 instances and is more than 1.5 times faster than X-CSE on 10 instances. In short, neither X-CSE nor I-CSE is clearly better than the other on Killer Sudoku. The successes of I-CSE show that it can be worthwhile to extract conflicting AC-CSs.

5.6 Other Problems

In this section we investigate the benefit and overhead of X-CSE on a larger set of problems. 47 example ESSENCE′ models were included with Savile Row 1.5 [12]. Four of these are used as case studies above. In this section we use the other 43 problems,

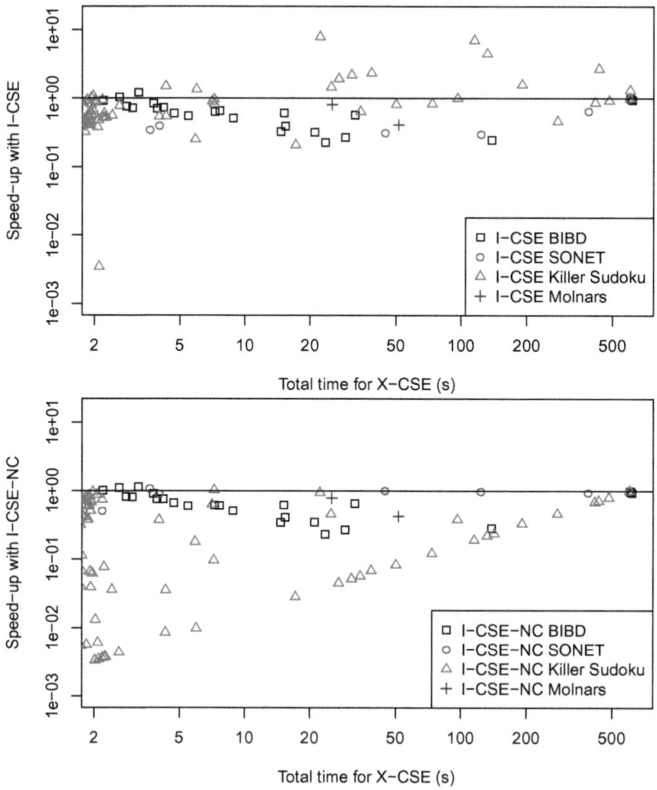

Fig. 3. Comparison of X-CSE with I-CSE (top) and I-CSE-NC (bottom)

almost all of which were written before X-CSE was conceived. Of these 43 problems, 16 have no AC-CSs and 27 have them.

Figure 4 (left) plots the time taken by Savile Row (including running MINION to enforce SACBounds). In some cases applying X-CSE speeds up Savile Row overall. Figure 4 (right) plots total time. Only two problems searched fewer nodes with X-CSE: Plotting (2% reduction) and waterBucket (21% reduction) and for both these problems the search time saved is outweighed by additional time required in Savile Row. For those problems that are sped up overall, there are two reasons: in some cases (e.g. quasiGroup5Idempotent, pegSolitaireState) X-CSE speeds up MINION without reducing the node count; and in other cases X-CSE speeds up Savile Row and not MINION. In summary, X-CSE provides a modest benefit on some of these problems and is a small overhead on others.

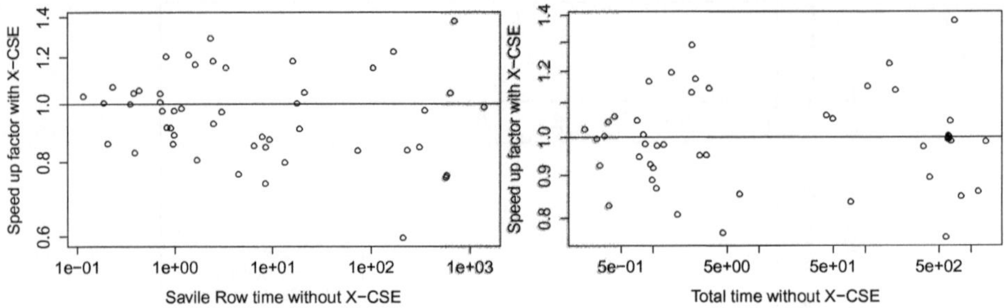

Fig. 4. Savile Row time only (left), and total time (right) on a set of 43 problems

6 Future Work

X-CSE is able to extract sets of *identical* terms shared among a set of AC-expressions. It is unable to match non-identical terms that are equivalent after a simple transformation. On the other hand, Active CSE [16] (described in Section 2.2) can match non-identical expressions that are identical after a simple transformation, but usually cannot extract AC-CSs. Suppose we had expressions $x - y$ and $y - x$. They could in principle be extracted by Active CSE, with one replaced by an auxiliary variable *aux* and the other replaced by $-aux$. However, if we have $x - y + z$ and $y - x + z$, the z term hides the common subexpression and neither X-CSE nor Active CSE can detect it. Exactly this situation arises in a potable water management problem (Choi and Lee [4]). Choi and Lee extracted the common subexpressions manually and proved that constraint propagation is strengthened by doing so.

Our proposed future work is to integrate X-CSE and Active CSE to create a single algorithm that is able to reveal AC-CSs by performing transformations. One (very simple) example of a transformation is multiplying by -1 to reveal the common subexpression in $x - y + z$ and $y - x + z$. A second example is negation (followed by De Morgan's law) to reveal that $\neg A \vee \neg C$ may be extracted from $A \wedge C$ and $\neg A \vee \neg B \vee \neg C$.

7 Conclusions

We have introduced and described a new algorithm, X-CSE, to perform Associative-Commutative Common Subexpression Elimination (AC-CSE) as an automated modelling step for finite domain constraint satisfaction problems. X-CSE is able to find common subexpressions which reduce search in four sample problems: BIBD, SONET, Killer Sudoku and Molnar's Problem. Of particular importance, X-CSE can interact with other automated modelling techniques, thereby magnifying the power of those techniques and X-CSE. We suggest that X-CSE is preferable to an earlier algorithm for AC-CSE, namely I-CSE, because it is able to exploit frequently occurring short common subexpressions. In our experiments X-CSE outperformed I-CSE in most cases. We conclude that X-CSE is a valuable addition to the armoury of automated constraint modelling techniques, both alone and in combination with other techniques.

Acknowledgements. We would like to thank the Royal Society for funding through Dr Jefferson's URF, and the EPSRC for funding this work through grant EP/H004092/1.

References

1. Araya, I., Neveu, B., Trombettoni, G.: Exploiting common subexpressions in numerical CSPs. In: Stuckey, P.J. (ed.) CP 2008. LNCS, vol. 5202, pp. 342–357. Springer, Heidelberg (2008)
2. Beldiceanu, N., Simonis, H.: A constraint seeker: Finding and ranking global constraints from examples. In: Lee (ed.) [10], pp. 12–26
3. Bessiere, C., Cardon, S., Debruyne, R., Lecoutre, C.: Efficient algorithms for singleton arc consistency. Constraints 16(1), 25–53 (2011)
4. Choi, C.W., Lee, J.H.M.: Solving the salinity control problem in a potable water system. In: Bessière, C. (ed.) CP 2007. LNCS, vol. 4741, pp. 33–48. Springer, Heidelberg (2007)
5. Frisch, A., Jefferson, C., Miguel, I.: CSPLib problem 035: Molnar's problem, `http://www.csplib.org/Problems/prob035`
6. Frisch, A.M., Jefferson, C., Miguel, I.: Constraints for breaking more row and column symmetries. In: Rossi, F. (ed.) CP 2003. LNCS, vol. 2833, pp. 318–332. Springer, Heidelberg (2003)
7. Frisch, A.M., Jefferson, C., Miguel, I.: Symmetry-breaking as a prelude to implied constraints: A constraint modelling pattern. In: Proc. 16th European Conference on Artificial Intelligence, ECAI 2004 (2004)
8. Frisch, A.M., Miguel, I., Walsh, T.: CGRASS: A system for transforming constraint satisfaction problems. In: O'Sullivan, B. (ed.) CologNet 2002. LNCS (LNAI), vol. 2627, pp. 15–30. Springer, Heidelberg (2003)
9. Jefferson, C., Moore, N., Nightingale, P., Petrie, K.E.: Implementing logical connectives in constraint programming. Artificial Intelligence 174, 1407–1429 (2010)
10. Lee, J. (ed.): CP 2011. LNCS, vol. 6876. Springer, Heidelberg (2011)
11. Mears, C., Niven, T., Jackson, M., Wallace, M.: Proving symmetries by model transformation. In: Lee (ed.) [10], pp. 591–605
12. Nightingale, P.: Savile Row, a constraint modelling assistant (2014), `http://savilerow.cs.st-andrews.ac.uk/`
13. Puget, J.F.: Constraint programming next challenge: Simplicity of use. In: Wallace, M. (ed.) CP 2004. LNCS, vol. 3258, pp. 5–8. Springer, Heidelberg (2004)
14. Puget, J.F.: Symmetry breaking using stabilizers. In: Rossi, F. (ed.) CP 2003. LNCS, vol. 2833, pp. 585–599. Springer, Heidelberg (2003)
15. Rendl, A.: Effective Compilation of Constraint Models. Ph.D. thesis, University of St Andrews (2010)
16. Rendl, A., Miguel, I., Gent, I.P., Jefferson, C.: Automatically enhancing constraint model instances during tailoring. In: Bulitko, V., Beck, J.C. (eds.) SARA. AAAI (2009)
17. Smith, B.M.: Symmetry and search in a network design problem. In: Barták, R., Milano, M. (eds.) CPAIOR 2005. LNCS, vol. 3524, pp. 336–350. Springer, Heidelberg (2005)
18. Stuckey, P.J., Tack, G.: MiniZinc with functions. In: Gomes, C., Sellmann, M. (eds.) CPAIOR 2013. LNCS, vol. 7874, pp. 268–283. Springer, Heidelberg (2013)
19. Van Hentenryck, P.: The OPL Optimization Programming Language. MIT Press, Cambridge (1999)

Improving GAC-4 for Table and MDD Constraints

Guillaume Perez and Jean-Charles Régin

Université Nice-Sophia Antipolis, I3S UMR 6070, CNRS, France
{guillaume.perez06,jcregin}@gmail.com

Abstract. We introduce GAC-4R, MDD-4, and MDD-4R three new algorithms for maintaining arc consistency for table and MDD constraints. GAC-4R improves the well-known GAC-4 algorithm by managing the internal data structures in a different way. Instead of maintaining the internal data structures only by studying the consequences of deletions, we propose to reset the data structures by recomputing them from scratch whenever it saves time. This idea avoids the major drawback of the GAC-4 algorithm, i.e., its cost at a shallow search-tree depth. We also show that this idea can be exploited in MDD constraints. Experiments show that GAC-4R is competitive with the best arc-consistency algorithms for table constraints, and that MDD-4R clearly outperforms all classical algorithms for table or MDD constraints.

1 Introduction

We consider table and Multi-valued Decision Diagram (MDD) constraints, which list the allowed combinations of values for the variables in the scopes of the constraints. Those constraints are useful for modeling and solving many real-world problems. They can be specified either directly, by input from the user, or indirectly by synthesizing other constraints or subproblems [17,11].

Table constraints are fundamental and implemented in any CP solver. This is the case for the or-tools[1] solver, which won the 2013 MiniZinc Challenge.[2] The or-tools solver does not implement many global constraints, but has an implementation of GAC-4R, which is our efficient algorithm for enforcing arc consistency on table constraints. In this paper, we introduce GAC-4R, and we adapt it to enforce arc consistency on MDD constraints.

Consider an extensional constraint C. Arc-consistency algorithms for C operate as follows: for each value a in the domain of a variable x, they search for a combination of values in the current domains of the other variables in the scope of C that contains (x, a) and satisfies C. A tuple of C is a combination of values in the domain of the variables in the scope of C. We say the the tuple is *allowed*, or a support, when it appears in the constraint's definition. We say that the tuple is *valid* if and only if its values appear in the *current* domains of the respective variables. Note that a valid tuple is not necessarily allowed. Arc-consistency algorithms can be distinguished depending on how they manage allowed and valid tuples [16]. While the allowed tuples do not

[1] http://code.google.com/p/or-tools/
[2] http://www.minizinc.org/challenge2013/call_for_problems.html

B. O'Sullivan (Ed.): CP 2014, LNCS 8656, pp. 606–621, 2014.

change during search because they are listed in the constraint definition, their validity is determined by the current domains.

While some arc-consistency algorithms operate on allowed tuples to check their validity, others first consider the current domains and look for a combination satisfying the constraint.

Existing algorithms mainly differ by how they operate when a value is deleted. Some algorithms are lazy (e.g., GAC-Schema [3], STR-2 [14], or STR-3 [15]). They try to reduce the operations executed at each modification (i.e., deletion of value of a domain) at the cost of increasing the complexity of the implementation. Others, such as GAC-4 [19], operate more systematically, thus keeping the implementation simple.

GAC-4 associates, to each variable-value pair (x, a), the list $S(x, a)$ of valid tuples involving (x, a) that satisfy C. When a value b is deleted from the domain of a variable y, the tuples associated with (y, b) are no longer valid and must be removed. Consequently, for each tuple $t \in S(y, b)$ and for each variable-value pair (z, c) in t, we remove t from $S(z, c)$. If $S(z, c)$ becomes empty, then no valid tuple involving (z, c) and satisfying C exists. Thus, we can safely remove c from $D(z)$.

GAC-4 is efficient when there are only few tuples for each value, which typically occurs at deeper levels of the search tree. However, at shallower levels its performance is qualitatively different in that maintaining the internal data structures is costly. In this paper, we give a solution to this issue.

Indeed, we show that the performance of GAC-4 can be improved by rebuilding, from scratch, the data structures of GAC-4 when the modifications have reached a given threshold. We illustrate such a situation with an example. Consider a table constraint with k tuples and involving a variable x having 10 values in its domain (the arity is not important here). Assume that the tuples are homogeneously distributed among the values of x. In other words, every value of x appears in about $\frac{k}{10}$ tuples. Now, assume that a is assigned to x. Thus, only about $\frac{k}{10}$ tuples remain valid. GAC-4 will consider and propagate deletions of $\frac{9k}{10}$ tuples although only about $\frac{k}{10}$ tuples remain. Thus, it is more effective to reset the constraint with the elements of $S(x, a)$, in other words, to rebuild the constraint from scratch. In this situation, we restart from a tuple set of only $\frac{k}{10}$ tuples and save a factor of 9. We can determine exactly when it is worthwhile to apply such an operation, which is when the sum of the sizes of S lists of the deleted values of x is larger than the sum of the sizes of S lists of the remaining values in the current domain of x.

In this paper, we introduce GAC-4R, which exploits that idea. The challenge is to maintain this mechanism throughout the search, because we need to undo this operation upon backtracking. To this end, we propose to represent the list of supported tuples for each variable-value pair using sparse sets.

Another way for improving the performance of arc-consistency algorithms for table constraints is to reduce the size of the representation of the tuples because the complexity depends on it. Several algorithms for compressing the allowed tuples of a constraint have been proposed, and arc consistency algorithms adapted for dealing with them [12,9,21]. Multi-valued Decision Diagrams (MDDs) are one of the most advanced and powerful representations. Cheng and Yap provide `mddc`, an efficient algorithm for enforcing arc consistency on MDDs based on GAC-3 [8,7]. In this paper, we define MDD-4, which adapts GAC-4 to MDD constraints, and introduce MDD-4R, which implements our improvement in MDD-4.

The paper is organized as follows. First, we recall background information. Then, we describe GAC-4R, our new version of GAC-4 based on resetting data structures. Next, we introduce MDD-4 and MDD-4R, which adapts the idea of reset to MDDs. After reviewing main existing GAC algorithms we discuss experiments that empirically establish the advantages of our algorithms in practice. Finally, we conclude this paper.

2 Preliminaries

2.1 Definitions

Constraint Network. A finite *constraint network* $\mathcal{N} = (X, \mathcal{D}, \mathcal{C})$ is defined as a set of n *variables* $X = \{x_1, \ldots, x_n\}$, a set of *domains* $\mathcal{D} = \{D(x_1), \ldots, D(x_n)\}$ where $D(x_i)$ is the finite set of possible *values* for variable x_i, and a set $\mathcal{C}$ of *constraints* between variables. A value a for a variable x is often denoted by (x, a).

Constraint. A constraint C on the ordered set of variables $X(C) = (x_{i_1}, \ldots, x_{i_r})$ is a subset $T(C)$ of the Cartesian product $D(x_{i_1}) \times \cdots \times D(x_{i_r})$ that specifies the *allowed* combinations of values for the variables $x_{i_1}, \ldots, x_{i_r}$. An element of $D(x_{i_1}) \times \cdots \times D(x_{i_r})$ is called a *tuple on* $X(C)$ and $t[x]$ is the value of the tuple t assigned to x. $|X(C)|$ is the *arity* of C. We will denote by d the size of the largest initial domain and by r the arity.

Arc Consistency. Let C be a constraint. A tuple t on $X(C)$ is *valid* iff $\forall x \in X(C)$, $t[x] \in D(x)$; and t is a *support* for (x, a) iff $t[x] = a$ and $t \in T(C)$. A value $a \in D(x)$ is *consistent with* C iff $x \notin X(C)$ or there exists a valid support for (x, a). C is *arc consistent* iff $\forall x \in X(C), D(x) \neq \emptyset$ and $\forall a \in D(x)$, a is consistent with C.

Table and MDD Constraints. Those constraints are said to be defined in extension. A table constraint is a constraint whose tuples satisfying the constraint are explicitly given in extension. An MDD constraint is a constraint which is defined thanks to a multi-valued decision diagram which is an efficient way to know whether or not a combination of values of the variables involved in the constraints satisfies the constraint.

2.2 GAC-4

The data structures of GAC-4 are quite simple. GAC-4 associates with each value a of each variable x the list $S(x, a)$ of the tuples of $T(C)$ containing (x, a). It maintains the following invariant:

$\forall x \in X(C), \forall a \in D(x)$: *$S(x, a)$ contains the valid tuples $t \in T(C)$ with $t[x] = a$.*
Thus, if a list $S(x, a)$ becomes empty then GAC-4 removes the value a from the domain of x. The initialization is done as follows: for each tuple t and for each value (x, a) belonging to t we add t to $S(x, a)$. Each time there is a deletion of a value of a variable involved in C, this deletion is added to a deletion set and the constraint is pushed in order to be revised later. Function REVISEGAC-4 (See Algorithm 1) is called in order to propagate the consequences of these deletions. It maintains the S lists.

Algorithm 1. REVISEGAC-4

```
REVISEGAC-4(C: constraint; deletionSet: list): Boolean
for each (x, a) ∈ deletionSet do
    for each t ∈ S(x, a) do
        for each (z, c) ∈ t do remove t from S(z, c)
        if S(z, c) = ∅ then remove c from D(z) ; add (z, c) to deletionSet
        if D(z) = ∅ then return False;
return True
```

2.3 Sparse Set

Sparse set is an efficient data structure for manipulating sets with a fixed size universe U [5]. It has been successfully used in CP for representing sets or lists [8,7,14,15]. We will use it in GAC-4R and MDD-4R, so we give details of it.

For convenience, the elements in U are mapped to integers 0 through $|U| - 1$. The representation has three components: two vectors (named dense and sparse), each $|U|$ elements long and a scalar (named members) that records the number of members in the set. The values in the array dense from 0 to members - 1 corresponds to the elements in the set. The array sparse contains indices of the array dense. If a number k is a member of the set, it must satisfy two conditions $0 \leq \texttt{sparse}[k] < \texttt{members}$ and $\texttt{dense}[\texttt{sparse}[k]] = k$. It means that $\texttt{sparse}[k]$ is the index i in the array dense of the value k, that is, we have $\texttt{dense}[i] = k$. Here is a sparse set:

sparse	5	2	-	0	-	1	-	3	4	-
dense	3	5	1	7	8	0				
members	6									

The membership, addition and deletion functions are defined in Algorithm 2. Function DELETE has been modified from its original definition in [5] in order to be able to restore easily the sparse set after some deletions. Consider the sparse set previously defined. Suppose that member 7 is deleted. Before the deletion the scalar members is equal to 6 and after the deletion we have the new sparse set:

sparse	3	2	-	0	-	1	-	5	4	-
dense	3	5	1	0	8	7				
members	5									

When 7 has been deleted, the members 7 and 0 (i.e. the last value of the set) have been exchanged. Precisely, we swap dense[sparse[7]] and dense[sparse[0]] and we swap sparse[7] and sparse[0]. Thanks to these swaps, we can easily restore the sparse set simply by setting the members value to 6. The sparse set contains the same elements but not in the same order.

2.4 Multi-valued Decision Diagram

This presentation is inspired from [22]. Multi-valued decision diagram (MDD) is a method for representing discrete functions. It is a multiple-valued extension of BDD [6].

Algorithm 2. Functions for manipulating a sparse set. k is an element. S is a sparse set with two arrays $S.\texttt{dense}$ and $S.\texttt{sparse}$ and scalar $S.\texttt{members}$.

```
MEMBER(k, S): return S.sparse[k] < S.members and S.dense[S.sparse[k]] = k
ADD(k, S) // assume k is not a member
    S.sparse[k] ← S.members
    S.dense[S.members] ← k
    S.members ← S.members +1
DELETE(k, S) // assume k is a member
    ik ← S.sparse[k]; ie ← S.members −1; e ← S.dense[ie]
    S.sparse[e] ← ik
    S.dense[ik] ← e
    S.sparse[k] ← ie
    S.dense[ie] ← k
    S.members ← S.members −1
```

An MDD is a directed acyclic graph (DAG) used to represent some multi-valued function $f : \{0...d-1\}^r \rightarrow \{0...d-1\}$, based on a given integer d. Given the r input variables, the DAG representation is designed to contain r layers of nodes, such that each variable is represented at a specific layer of the graph. Each node on a given layer has d outgoing arcs to nodes in the next layer of the graph. Each arc is labeled by its corresponding integer. The final layer is represented by the terminal nodes having values in the range $\{0, ...d-1\}$.

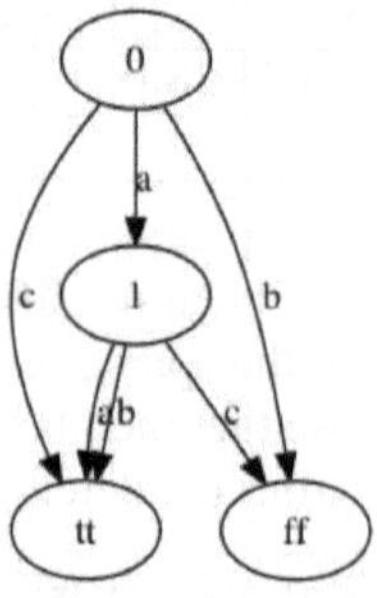

Fig. 1. An MDD representing a function defined on variables x_1 and x_2. Layer 1 (resp. 2) corresponds to x_1 (resp. x_2). Terminal node tt is true while ff is false. The function is true for the tuples (a, a), (a, b), (c, a), (c, b), (c, c).

The function encoded by an MDD may be evaluated on a given set of r input variables by traversing the graph, starting from the root (i.e. the highest-layer variable in the graph), and choosing the outgoing arc at each node corresponding to the input value of the variable represented at the current layer of the graph. This traversal continues until a terminal node is reached. The resulting value of the function is indicated by the value of the terminal node reached along this evaluation path. MDDs have been widely used in CP because there are powerful for modeling problems or for representing some functions or domain store [1,10,11,2]. In CP there are usually two terminal nodes tt, which is true, and ff which is false. Figure 1 gives an example of MDD. We have the path $(0, 1, tt)$ so $f(a, a)$ and $f(a, b)$ are true because tt is true.

An MDD is *ordered* if each variable is encountered at most once on each path from the root to a terminal. An MDD is *fully-reduced* if it does not contain any nodes with all k outgoing arcs pointing to the same node and does not contain duplicate nodes for a given layer in the DAG. Fully-reduced and ordered MDDs are mainly used.

Arc Consistency and MDD. Cheng and Yap provide `mddc`, an algorithm maintaining arc consistency for an MDD constraint based on GAC-3 [8,7]. In an MDD constraint, the MDD models the set of tuples satisfying the constraint. Each variable of the MDD corresponds to a variable of the constraint. An arc associated with an MDD variable corresponds to a value of the corresponding variable of the constraint. It uses a depth first search for traversing the MDD from the root node. Only arcs corresponding to values belonging to the domain of the variables are used. Terminal nodes are either positive or negative. A node x is marked positive if the subgraph traversed from x reaches a positive node. If no terminal node can be reached from x or if only negative nodes can be reached from x then x is marked negative. The algorithm saves the values corresponding to the arcs having a positive terminating node (this means that the value belongs to a valid tuple defined by the path). When the depth first search ends, the values that have not been saved may be safely removed because there is no longer a path from the root to a positive terminal node, which involves an arc corresponding to this value.

In `mddc`, it is important to note that any arc is traversed at most once, because a depth first search is used and because the MDD is not changed during the search by the algorithm. This is only the way the MDD is traversed that depends on the current domains. Note also that it is not straightforward to find an arc corresponding to a value belonging to the current domain and this task is more difficult when the domain size is reduced.

3 GAC-4R

GAC-4 is a simple and easy algorithm to implement. Its worst case complexity is optimal. However it is mainly focused on the study of the consequences of the deletions of values but it could be worthwhile to recompute some data structures instead of maintaining them incrementally. A simple example is the computation of an intersection. Suppose that you want to maintain the intersection C of two sets A and B. You computed this intersection when A and B had 100 values and you determined that C has 50 values. Suppose that a value v of A is deleted. Then, your algorithm will recompute the set C by checking whether C contains v or not, and remove v from C if $v \in C$. Now, suppose that we remove 98 values from A, and we have $A = \{a, b\}$. Then it is faster to check whether a and b belong to B and build C consequently instead of considering the deleted values. Only two operations have to be performed to define the new set C. In this case, we will say that we reset C.

This idea had been applied to define which algorithm should be preferred between AC-2001 and AC-6 [4] or to design an adaptive algorithm [20]. Combining GAC-4 with this idea will save a lot of computations for a shallow depth of the tree search. Such a combination requires to answer two questions:

1. How can we know whether a reset is better or not?
2. How can we perform this reset and the restoration of the previous set efficiently?

The first question answer is simple. Consider a variable x and $\Delta(x)$ the set of values of $D(x)$ that have been deleted and not yet considered by GAC-4 algorithm (i.e. they belong to deletionSet). The number of tuples that are no longer valid, denoted by $\#T\Delta(x)$, is given by the sum of the size of the S lists of the values in $\Delta(x)$, and the number of remaining tuples is the difference between the total number of tuples and $\#T\Delta(x)$ because a tuple contains only one value per variable. So we have:

Property 1. *Let x be a variable, $\Delta(x)$ be the values of x that have been deleted and that must be propagated, $\#T\Delta(x) = \sum_{a \in \Delta(x)} |S(x,a)|$ be the number of tuples that are no longer valid and T be the current number of tuples.*
If $\#T\Delta(x) > \frac{T}{2}$ then a reset operation will consider less tuples than the application of Function REVISEGAC-4

This property is useful only if we can answer the second question. We show that we can use sparse sets for efficiently computing a reset operation in such a way that the restoration is easy.

The question can be reformulated as follows. Consider S a set with two lists of elements R and Q. The lists are disjoint and their union contains exactly all the elements of S. The sets R and Q are not explicitly given, that is, we do not have a set representing them but we can traverse them (in term of programming language, they are given by an iterator) and their size is known. We want to modify S by removing the set of elements $R \subseteq S$ in order to obtain a set containing only the elements of Q. However, instead of performing $|R|$ operations for this task, we want to have a number of operations bounded by $\min(|R|, |Q|)$. In addition, we have to be able to restore the set S after performing the modifications with a similar complexity (or less).

Algorithm 3. Function re-add of a sparse set S. k is an element

```
RE-ADD(k, S) // We assume that S.dense[S.sparse[k]] = k
ik ← S.sparse[k]; e ← S.dense[S.members]
S.sparse[k] ← S.members
S.sparse[e] ← ik
S.dense[S.members] ← k
S.dense[ik] ← e
S.members ← S.members +1
```

Thanks to sparse sets we can give a nice answer to this question. Let S be represented by a sparse set. If $|R| \leq |Q|$ then we delete the elements of R from S as it is explained in Preliminaries section. The restoration of S consists of modifying the scalar `members` of S. Assume that $|Q| < |R|$. S is recomputed as follows. First, we set S.`members` to 0. Then, we traverse Q and for each element $a \in Q$ we add a to S by calling Function RE-ADD (See Algorithm 3) which is a modified version of Function ADD of the sparse set. It exploits the fact that the value which is added was previously in the set. Thus, it proceeds to a swap in a way similar as the one used by Function DELETE in order to be able to restore the set in the future. More precisely, when an element i is re-added to the sparse set, we swap i and the value j at the index defined by `members`. That is, we exchange the value of i and the value of j in the `sparse` array and we exchange i and j in the `dense` array. For instance, consider the left sparse set:

sparse	3	2	-	0	-	1	-	5	4	-
dense	3	5	1	0	8	7				
members	2									

sparse	3	4	-	0	-	1	-	5	2	-
dense	3	5	8	0	1	7				
members	3									

The set contains the values 3 and 5. If we re-add the value 8 then we will exchange the value of `dense[members]`, i.e. 1, with 8. So we will have `dense`$[2] = 8$; `dense`$[4] = 1$; `sparse`$[8] = 2$; `sparse`$[1] = 4$. We obtain the right sparse set.

The advantage of this method is that the restoration of the scalar `members` is enough for restoring the sparse set. Function RE-ADD has a complexity in $O(1)$ per call. Thus, we can re-add $|Q|$ elements in $O(|Q|)$.

Algorithm 4. GAC-4R. T is the current number of tuples

```
REVISEGAC-4R(C: constraint; deletionSet: list, T: number of tuples): Boolean
  for each x ∈ X(C) do #TΔ(x) ← 0
  for each (x, a) ∈ deletionSet do #TΔ(x) ← #TΔ(x) + |S(x, a)|
  #TΔmax ← max_{x∈X(C)}(#TΔ(x))
  if #TΔmax > T/2 then
      // we reset the data structures
      pick a variable x with #TΔ(x) = #TΔmax
      Tset ← ∅; T ← 0
      for each a ∈ D(x) do add S(x, a) in Tset
      for each y ∈ X(C) do
          for each b ∈ D(y) do S(y, b).members ← 0
      // we re-add valid tuples into the S lists
      for each t ∈ Tset do
          if t is valid then
              for each i = 1...n do RE-ADD(t, S(x_i, t[i]))
              T ← T + 1
      // we remove values having an empty S list.
      for each y ∈ X(C) do
          for each b ∈ D(y) do
              if S(y, b) = ∅ then remove b from D(y)
              if D(y) = ∅ then return False;
  else
      // classical GAC-4 deletion process
      for each (x, a) ∈ deletionSet do
          for each t ∈ S(x, a) do
              for each i = 1...n do
                  DELETE(t[i], S(x_i, t[i]))
                  T ← T − 1
                  if S(x_i, t[i]) = ∅ then remove t[i] from D(x_i)
                  if D(x_i) = ∅ then return False;
  return True
```

A possible implementation of GAC-4R is given by Algorithm 4. Each list S is represented by a sparse set with a fixed size universe equal to $|T(C)|$. For convenience, we will consider that t is a tuple and also the index of the tuple in the table of tuples.

The complexity of GAC-4R remains the same as the complexity of GAC-4, because the deletion of a tuple or the re-addition of a tuple have the same complexity which corresponds to the arity of the constraint. In addition traversing the valid tuples costs at least the cost of traversing all the domains of the variables involved in the constraint and since we do this only when there are less valid tuples than non valid tuples, the traversal of all the domains does not impact the complexity.

4 MDD-4R

We propose to adapt the principles of GAC-4R to be able to deal with an MDD instead of a table. First, we will slightly modify the MDD from which the constraint is defined. Then we design the MDD-4 algorithm. Unlike `mddc`, MDD-4 modifies and maintains the MDD during the search for a solution. Next, we will study the reset principle for MDD-4 and explain when the reset should be done.

4.1 MDD Reformulation

ff Removal. The node ff is not useful for maintaining arc consistency. So, we remove it and also all the nodes that do not belong to a path from the root to tt. This can be done by performing a depth first search.

Semi-Reduced MDD. We relax the fully reduced property of the MDD. We accept to have nodes with all k outgoing arcs pointing to the same node. We will say that we have a semi-reduced MDD. Note our algorithm may also work with MDDs having duplicate nodes, i.e., MDDs that are not even semi-reduced. Figure 2 gives an example of a reformulation. Each reduced arc in a fully reduced MDD is replaced by d arcs in the semi-reduced MDD. We do not find any problem for which it really changes the space complexity.

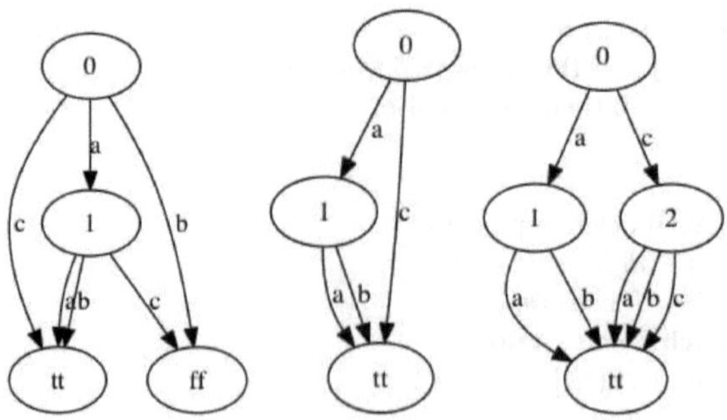

Fig. 2. Reformulation of an MDD. The left graph is the initial MDD. The middle graph represents the deletion of the node ff. The right graph is the semi-reduced MDD that will be used by MDD-4.

4.2 MDD-4 Algorithm

The algorithm MDD-4 is a modification of GAC-4 for dealing with MDDs. We differentiate two parts: the maintenance of the MDD during the search for a solution and the maintenance of the S lists.

In GAC-4 the maintenance of the list of valid tuples is made by managing the S lists. With an MDD this is more complex, because the tuples are not explicitly represented in an MDD. The representation is implicit: a valid tuple corresponds to path from the root to the node tt which traverses only arcs corresponding to valid values. In order to avoid checking all the time the validity of the values, like in `mddc`, and to make sure that we do not uselessly traverse a path we propose to delete the arcs corresponding to values that are no longer valid. Then, we delete the nodes and arcs that do not belong to a path from the root to tt. We call this process the maintenance of the MDD. We do not need to explicitly search for such paths. It is sufficient to delete the nodes from which we can no longer reach the root or the node tt. The idea is to check whether a deleted arc (i, j) is the only outgoing arc from i or the only incoming arc to j, because if a node has no longer any outgoing arc or any incoming arc then we can remove it. We implement this process as follows. When values are removed from domains we delete the corresponding arcs in the MDD and push the extremities into two queues $Q\uparrow$ and $Q\downarrow$ (See Function REMOVEARC). Each queue can contain only once a node. If an arc (i, j) is deleted then i is pushed into $Q\uparrow$ and j is pushed into $Q\downarrow$. Then the queues are proceeded by layer (See Lines 2 and 3). The elements of $Q\uparrow$ are taken from the deepest to the shallowest layer. For each element $i \in Q\uparrow$ we remove i from the queue and we check whether there is an outgoing arc from i. If there is none, then i is removed from the graph, so arcs are deleted and new nodes are added to the queues. The elements of $Q\downarrow$ are taken from the shallowest to the deepest layer. For each element $j \in Q\downarrow$ we remove j from the queue and we check whether there is an incoming arc from j. If there is none, then node j is removed from the graph, so arcs are deleted and new nodes are added to the queues. The process is repeated until the queues are empty.

In order to remove values from domain variables we need to have a relation between the values and the arcs of the MDD. In MDD-4 this relation is defined by the S lists, so the S lists contain arcs instead of tuples as in GAC-4. Precisely, for each value (x, a), the list $S(x, a)$ contains the set of arcs in the MDD labeled with the value a at the layer of x. If there is no more arc for a value (x, a) in the MDD, then $S(x, a)$ becomes empty and we can safely remove a from $D(x)$. Once again, S lists are implemented by sparse sets.

A synopsis of the code of MDD-4 is given by Algorithm 5. In order to restore efficiently the MDD we use sparse sets for representing the neighborhood of the nodes.

We can establish an interesting property:

Property 2. *MDD-4 cannot perform more operations than GAC-4 for establishing arc consistency of a table constraint*

Sketch of Proof. The deletion of a value in GAC-4 implies to consider all the current tuples containing the values. The deletion of a tuple in GAC-4 costs r operations (i.e., the arity of the constraint), because the tuple belongs to r S-list (one for each value of the tuple). When a value(x, a) is deleted in MDD-4, we have to delete all the arcs corresponding to this value. The number of deleted arcs is bounded by the number of tuples

Algorithm 5. MDD-4

REMOVEARC(MDD, $Q\downarrow, Q\uparrow, (i,j)$):Boolean
  delete the arc (i,j) from the MDD
  $y \leftarrow$ variable of the arc (i,j); $b \leftarrow$ value of the arc (i,j)
  remove the arc (i,j) from the $S(y,b)$
  if $S(y,b) = \emptyset$ **then** remove b from $D(y)$
  push nodes i into $Q\downarrow$ and j into $Q\uparrow$ if they are not already in.
  return $(D(y) \neq \emptyset)$

REVISEMDD-4(C: constraint; *deletionSet*: list): Boolean
  $Q\downarrow \leftarrow \emptyset$; $Q\uparrow \leftarrow \emptyset$
1 **for each** $(x,a) \in deletionSet$ **do**
  for each *arc* $(i,j) \in S(x,a)$ **do**
   if $\neg$ REMOVEARC($MDD, Q\downarrow, Q\uparrow, (i,j)$) **then** return False
  while $Q\downarrow \neq \emptyset$ *or* $Q\uparrow \neq \emptyset$ **do**
2 **while** $Q\uparrow \neq \emptyset$ **do**
   pick the node $i \in Q\uparrow$ with the lowest layer
   if *there is no outgoing arc from* i **then**
    for each *arc* (j,i) **do**
     if $\neg$ REMOVEARC($MDD, Q\downarrow, Q\uparrow, (j,i)$) **then** return False
   remove i from $Q\uparrow$
3 **while** $Q\downarrow \neq \emptyset$ **do**
   pick the node $j \in Q\downarrow$ with the highest layer
   if *there is no incoming arc to* j **then**
    for each *arc* (j,i) **do**
     if $\neg$ REMOVEARC($MDD, Q\downarrow, Q\uparrow, (j,i)$) **then** return False
   remove j from $Q\downarrow$
  return True

involving (x,a). The deletion of (x,a) triggers the maintenance of the extremities of the deleted arcs. Since the deleted arcs belong to deleted tuples, the number of arcs that are considered through the nodes maintenance cannot be greater than the number of elements of all the tuples involving (x,a). Thus the property holds.

4.3 MDD-4R

MDD-4 can be improved by integrating the idea of resetting the data structures instead of being focused only on the deletions. MDD-4 works by layer in the MDD, that is, variable per variable. Let $\#A(x)$ be the total number of arcs associated with a value of a variable x. This number can be stored and maintained for each variable. For a given layer corresponding to the variable x, we have to compute $\#DA(x)$ the number of arcs that will be deleted for this layer. By comparing this number to $\#A(x)$ we will know whether it is better to reset the layer or not. Resetting the layer means that we rebuild the layer of the graph from the remaining nodes by adding their remaining arcs instead of deleting the arcs and nodes of the layer. The computation of $\#DA(x)$ depends on the type of modification occurring in the MDD. There are two kinds of modifications:

- First, the arcs corresponding to the deletions of values of a variable x are removed. In this case, we have $\#DA(x) = \sum_{a \in \Delta(x)} |S(x,a)|$. This happens only once.
- Second, the consequences of the deletion of arcs and nodes in the MDD are propagated, in other words the MDD is maintained. MDD-4R proceeds by layer. For a given variable y, MDD-4 has the list $Q(y)$ of nodes that must be deleted, which is for the given layer the content of the queue $Q\uparrow$ or $Q\downarrow$ depending on the sense of propagation. We have $\#DA(y) = \sum_{z \in Q(y)} |N(y,z)|$, where $N(y,z)$ is the list of arcs associated with y having z as extremity.

Property 3. *Let x be any variable, If $\#DA(x) > \frac{\#A(x)}{2}$ then a reset operation for the layer of x will consider less arcs than the application of MDD-4 for this layer.*

Note that this property computes exactly whether it is better to reset or not the data structure.

The reset idea can be implemented in an easy way by using sparse sets for representing the neighborhood of the nodes.

5 Related Work

Allowed of or-Tools. The or-tools solver integrates an efficient algorithm when there are only few tuples. It is based on GAC-3. The algorithm, that we will name `allowed`, indexes the tuples and maintains a bitset of the valid tuples. For each variable x and for each value a, a bitset mask of tuples containing the value (x,a) is maintained. If a value (y,b) is removed then all the tuples of the bitset mask of (y,b) are cleared in the bitset of valid tuples. Then, the algorithm scans the values of all the variables to see if there is an active tuple which supports it thanks to bitwise operators.

STR2. It has been proposed in [14]. It is based on GAC-3. It maintains a sparse set of the valid tuples and improves the test of the validity of a tuple by considering only the variables whose domain recently changed. It also uses the tuples that have been computed valid as support for all the other values involved in the tuple, like in GAC-Schema, and so it decreases the number of values for which we have to search for a new support.

STR3. It has been defined in [15]. It is based on GAC-Schema (or GAC-6). For each variable x and for each value a of x it precomputes the list of tuples involving (x,a). It uses sparse set for maintaining the validity of tuples and to speed-up the test of validity.

We do not include the algorithms of [18] because the authors recognize that when the arity of the table increases, the existing state-of-the-art propagators STR3 and `mddc` are faster than their algorithms.

6 Experiments

Machine: Dell server having four E7- 4870 Intel processors, each having 10 cores with 256 GB of memory and running under Scientific Linux.

Solver: or-tools 3158.

Selected Benchmarks: All problems can be downloaded from the Solver Competition archive [13]. We selected problems having only positive table constraints and at least one variable whose domain is not Boolean. We do not include BDD-based instances and instances involving only binary constraints because we are interested in large arity constraints. rand-8-20-5-18-800 is abbreviated by rand-1 and rand-10-20-10-5-10000 is abbreviated by rand-2. half-n25-d5-e56-r7-1 is abbreviated by half-1 and contains the problems of half-n25-d5-e56-r7 that are solved by mddc in less than 1800s and half-n25-d5-e56-r7-2 is abbreviated by half-2 and contains the problems of half-n25-d5-e56-r7 that are not solved by mddc in less than 1800s.

We also used random problems that we defined. One of the most difficult parameter to define is the tightness of the constraint that is, the ratio between the number of tuples allowed by the constraint and the total number of tuples. We use ratio from 0.00004% to 1%. For a comparison, we can note that an alldiff constraint defined on a set of k variables sharing the same k values is equal to $\frac{k!}{k^k}$ which is 0.6% for $k = 7$; 0.034% for $k = 10$ and 0.0003% for $k = 15$.

Search Strategy: We select the variable that appears in most constraints and its smallest value as proposed for testing mddc [7]. Our algorithm is more robust than the Cheng's one for the strategy, because we maintain the MDD whereas they traverse the initial MDD according to the current domains. Thus mddc algorithm can lose time for finding an arc corresponding to a value belonging to the current domain of the associated variable. This problem does not arise in neither MDD-4 nor MDD-4R.

Results: Times are expressed in seconds. Time Out (T-O) is set at 1800s. All means are geometric.

6.1 General Comparison

Table 1 gives results for the selected benchmarks. MDD-based algorithms perform well in general. Algorithm mddc is clearly improved by MDD-4 and MDD-4R algorithms.

Table 1. Geometric means of the time needed to solve some categories of problems

benchmark	MDD-4	MDD-4R	mddc	GAC-4	GAC-4R	STR2	STR3	allowed
nonograms	0,33	**0,27**	0,8	4,3	3,18	2,77	1,52	1,03
cw-m1c-ogd	3,07	**1,72**	32,69	4,03	2,74	13,21	2,69	3,09
cw-m1c-uk	3,96	1,91	21,14	3,24	2,27	8,32	**1,89**	2,22
rand-1	6,06	2,93	13,71	2,90	1,38	1,56	1,93	**1,09**
rand-2	192,47	**50,51**	186,35	241,56	170,36	141,03	T-O	T-O
half-1	975,49	**471,25**	1438,20	T-O	T-O	T-O	T-O	T-O
half-2	1720	**778,28**	T-O	T-O	T-O	T-O	T-O	T-O

GAC-4R outperforms GAC-4 and is competitive with other GAC algorithms for table constraints. The reset strategy is quite interesting and MDD-4R clearly outperforms all the other algorithms.

We propose to study in detail the behavior of these algorithms according to the tightness and the domain size. For each graph we select randomly 10 problems and run them 30 times and take the mean.

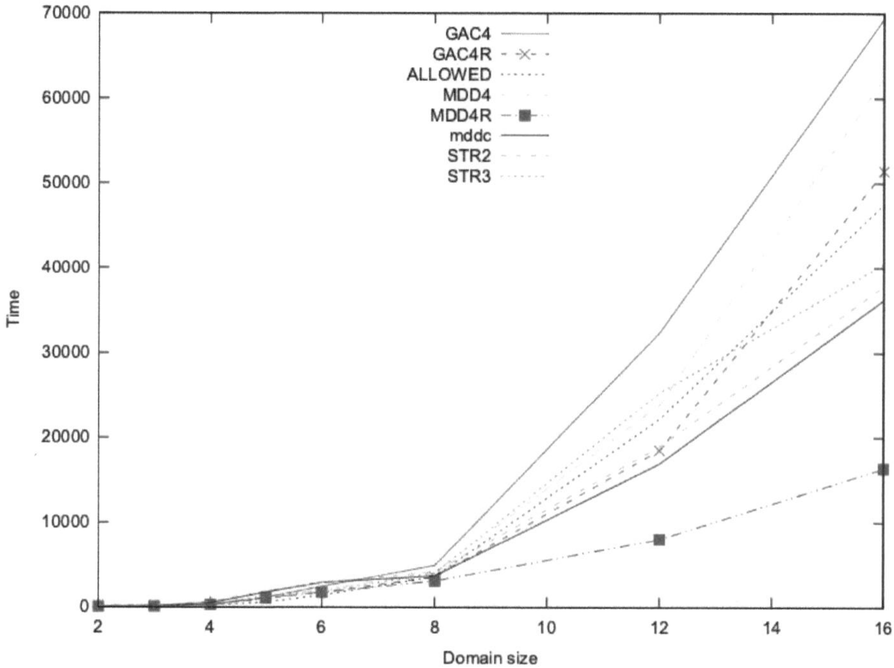

Fig. 3. Time needed to solve problems while increasing the domain size. Tightness is 1%.

Domain Size. We study the behavior of the algorithms while changing the domain size. Each problem involves 100 variables and constraints of arity 7. The results are given in Fig. 3. Modifying the domain may change the tightness so we set it at 1%. Clearly MDD-4R is the best algorithm. The reset idea is also a clear improvement for GAC-4 and MDD-4. GAC-4R is competitive with STR2 and STR3.

Tightness. We set the domain size (8) and the arity (7) and we increase the tightness. Each problem involves 40 variables and 40 constraints.

Fig. 4 presents the results. Once again the reset idea is worthwhile. STR2 outperforms STR3 and GAC-4R.

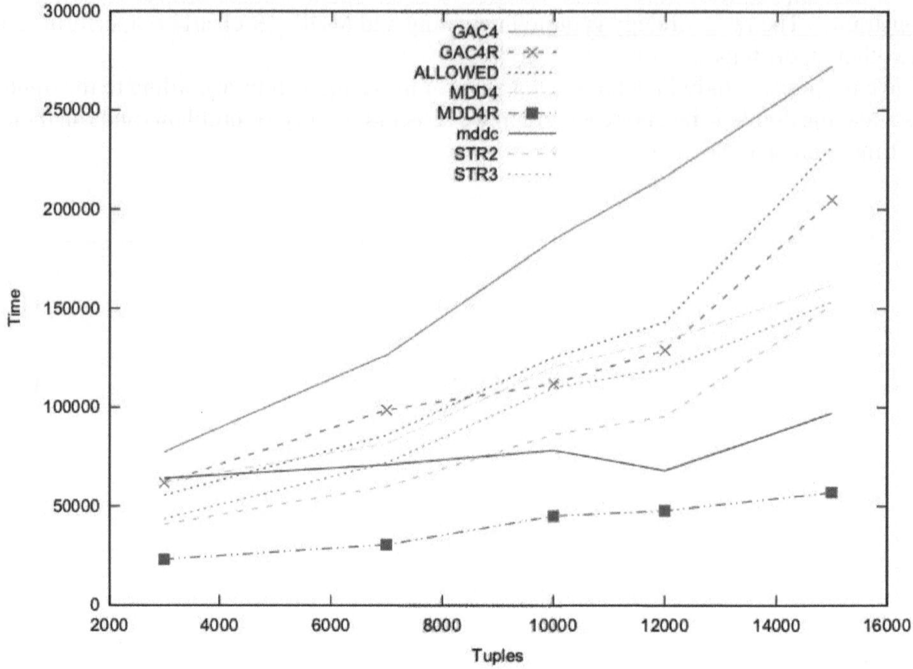

Fig. 4. Time needed to solve problems while increasing the tightness. Domain size is 8.

The conclusion of these experiments is that MDD-based algorithms clearly outperform the algorithms based on table constraints when the tightness and/or the size of the domains grow. The reset idea really improves the algorithms and GAC-4R is a competitive algorithm. The most robust and globally the most efficient algorithm is MDD-4R. This algorithm should be preferred in practice.

7 Conclusion

We have introduced MDD-4 and the two algorithms GAC-4R and MDD-4R which respectively improved the algorithms GAC-4 and `mddc`. These algorithms are mainly based on an adaptive method for maintaining the data structures. If a lot of deletions are made then we reset the data structures from the remaining elements instead of studying the consequences of the deletions. We have defined rules for maintaining the data structures in the best way for GAC-4R and MDD-4R. The experiments show that GAC-4R clearly improves GAC-4 and is competitive with STR2 and STR3. They also show that MDD-4R is a clear improvement of `mddc` and that it outperforms all existing algorithms.

References

1. Andersen, H.R., Hadzic, T., Hooker, J.N., Tiedemann, P.: A Constraint Store Based on Multivalued Decision Diagrams. In: Bessière, C. (ed.) CP 2007. LNCS, vol. 4741, pp. 118–132. Springer, Heidelberg (2007)
2. Bergman, D., van Hoeve, W.-J., Hooker, J.N.: Manipulating MDD Relaxations for Combinatorial Optimization. In: Achterberg, T., Beck, J.C. (eds.) CPAIOR 2011. LNCS, vol. 6697, pp. 20–35. Springer, Heidelberg (2011)
3. Bessière, C., Régin, J.-C.: Arc consistency for general constraint networks: preliminary results. In: Proceedings of IJCAI 1997, Nagoya, pp. 398–404 (1997)
4. Bessière, C., Régin, J.-C.: Refining the basic constraint propagation algorithm. In: Proceedings of IJCAI 2001, Seattle, WA, USA, pp. 309–315 (2001)
5. Briggs, P., Torczon, L.: An efficient representation for sparse sets. ACM Letters on Programming Languages and Systems 2, 59–69 (1993)
6. Bryant, R.E.: Graph-based algorithms for boolean function manipulation. IEEE Transactions on Computers C35(8), 677–691 (1986)
7. Cheng, K., Yap, R.: An mdd-based generalized arc consistency algorithm for positive and negative table constraints and some global constraints. Constraints 15 (2010)
8. Cheng, K.C.K., Yap, R.H.C.: Maintaining generalized arc consistency on ad hoc r-ary constraints. In: Stuckey, P.J. (ed.) CP 2008. LNCS, vol. 5202, pp. 509–523. Springer, Heidelberg (2008)
9. Gent, I., Jefferson, C., Miguel, I., Nightingale, P.: Data structures for generalised arc consistency for extensional constraints. In: Proc. AAAI 2007, Vancouver, Canada, pp. 191–197 (2007)
10. Hadzic, T., Hooker, J.N., O'Sullivan, B., Tiedemann, P.: Approximate Compilation of Constraints into Multivalued Decision Diagrams. In: Stuckey, P.J. (ed.) CP 2008. LNCS, vol. 5202, pp. 448–462. Springer, Heidelberg (2008)
11. Hoda, S., van Hoeve, W.-J., Hooker, J.N.: A systematic approach to MDD-based constraint programming. In: Cohen, D. (ed.) CP 2010. LNCS, vol. 6308, pp. 266–280. Springer, Heidelberg (2010)
12. Katsirelos, G., Walsh, T.: A compression algorithm for large arity extensional constraints. In: Bessière, C. (ed.) CP 2007. LNCS, vol. 4741, pp. 379–393. Springer, Heidelberg (2007)
13. Lecoutre, C.: Csp/maxcsp/wcsp solver competitions (2009), `http://www.cril.univ-artois.fr/~lecoutre/benchmarks.html`
14. Lecoutre, C.: Str2: optimized simple tabular reduction for table constraints. Constraints 16(4), 341–371 (2011)
15. Lecoutre, C., Likitvivatanavong, C., Yap, R.H.C.: A path-optimal gac algorithm for table constraints. In: ECAI, pp. 510–515 (2012)
16. Lhomme, O., Régin, J.-C.: A fast arc consistency algorithm for n-ary constraints. In: Proc. AAAI 2005, Pittsburgh, USA, pp. 405–410 (2005)
17. Lhomme, O.: Practical reformulations with table constraints. In: ECAI, pp. 911–912 (2012)
18. Mairy, J.-B., Van Hentenryck, P., Deville, Y.: Optimal and efficient filtering algorithms for table constraints. Constraints 19(1), 77–120 (2014)
19. Mohr, R., Masini, G.: Good old discrete relaxation. In: Proceedings of ECAI 1988, pp. 651–656 (1988)
20. Régin, J.-C.: AC-*: A configurable, generic and adaptive arc consistency algorithm. In: van Beek, P. (ed.) CP 2005. LNCS, vol. 3709, pp. 505–519. Springer, Heidelberg (2005)
21. Régin, J.-C.: Improving the expressiveness of table constraints. In: CP 2011, Proceedings Workshop ModRef 2011 (2011)
22. Rice, M., Kulhari, S.: A survey of static variable ordering heuristics for efficient bdd/mdd construction. University of California, Tech. Rep. (2008)

Improvement of the Embarrassingly Parallel Search for Data Centers*

Jean-Charles Régin, Mohamed Rezgui, and Arnaud Malapert

Univ. Nice Sophia Antipolis, CNRS, I3S, UMR 7271, 06900 Sophia Antipolis, France

Abstract. We propose an adaptation of the Embarrassingly Parallel Search (EPS) method for data centers. EPS is a simple but efficient method for parallel solving of CSPs. EPS decomposes the problem in many distinct subproblems which are then solved independently by workers. EPS performed well on multi-cores machines (40), but some issues arise when using more cores in a datacenter. Here, we identify the decomposition as the cause of the degradation and propose a parallel decomposition to address this issue. Thanks to it, EPS gives almost linear speedup and outperforms work stealing by orders of magnitude using the Gecode solver.

1 Introduction

Several methods for parallelizing the search in constraint programming (CP) have been proposed. The most famous one is the work stealing [12,14,5,16,3,8]. This method uses the cooperation between computation units (workers) to divide the work dynamically during the resolution. Recently, [13] introduced a new approach named Embarrassingly Parallel Search (EPS), which has been shown competitive with the work stealing method.

The idea of EPS is to decompose statically the initial problem into a huge number of subproblems that are consistent with the propagation (i.e. running the propagation mechanism on them does not detect any inconsistency). These subproblems are added to a queue which is managed by a master. Then, each idle worker takes a subproblem from the queue and solves it. The process is repeated until all the subproblems have been solved. The assignment of the subproblems to workers is dynamic and there is no communication between the workers. EPS is based on the idea that if there is a large number of subproblems to solve then the resolution times of the workers will be balanced even if the resolution times of the subproblems are not.

In other words, load balancing is automatically obtained in a statistical sense. Interestingly, experiments of [13] have shown that the number of subproblems does not depend on the initial problem but rather on the number of workers. Moreover, they have shown that a good decomposition has to generate about 30 subproblems per worker. Experiments have shown good results on a multi-cores machine (40 cores/workers).

* This work was granted access to the HPC and visualization resources of "Centre de Calcul Interactif" hosted by the University of Nice Sophia Antipolis. It was also partially supported by OSEO, with the project ISI "Pajero".

B. O'Sullivan (Ed.): CP 2014, LNCS 8656, pp. 622–635, 2014.

Preliminary experimental results of this method on a data center (512 cores/workers) have shown that the scalability of the overall resolution time without the decomposition is very good. However, the decomposition becomes more difficult and the relative part of the decomposition compared to the overall resolution time grows with the number of workers. There are several reasons for that. First, the number of subproblems that have to be generated grows linearly with the number of cores. Second, the overall resolution time diminishes when the number of workers is increased. At last, the decomposition of the EPS method as proposed in [13] is not efficiently parallelized. In this paper we propose to address this issue by designing an efficient parallel decomposition of the initial problem.

A naive decomposition in parallel has been proposed in [13]. It splits the initial problem into as many subproblems as there are workers and assigns a subproblem to each worker. Then, each worker decomposes its subproblem into 30 subproblems. This gains a factor of 2 or 3 in comparison with a sequential decomposition. This is enough when the number of workers is limited (40 for instance) but it is no longer an efficient method with hundreds of workers. The gain is limited because there is no reason to have equivalent subproblems to decompose. However, from this naive algorithm we learn several things:

1. the difference of total work (i.e. activity time in EPS) made by the workers decreases when the number of subproblems increases. This is not a linear relation. There is a huge difference between the activity time of the workers when there are less than 5 subproblems per worker. These differences decrease when there are more than 5 subproblems per worker.
2. a simple decomposition into subproblems that may be inconsistent causes quickly some issues because inconsistencies are detected very quickly.
3. splitting an initial problem into a small set of subproblems is fast compared to the overall decomposition time and compared to the overall resolution time.

From these observations we understand that we will have to find a compromise and we propose an iterative process decomposing the initial problem in 3 phases. In the first phase, we want to decompose it into only few subproblems because the relative cost is small even if we have an unbalanced workload. However, we should be careful with the first phase (i.e. starting with probably inconsistent subproblems) because it can have an impact on the performance. At last, the most important thing seems to generate 5 subproblems because we could restart from these subproblems to decompose more and such a decomposition should be reasonably well balanced.

Thus, we propose a method which has 3 main phases:

- An initial phase where we generate as quickly as possible one subproblem per worker.
- A main phase which aims at generating 5 subproblems per worker. Each subproblem is consistent with the propagation. This phase can be divided into several steps for reaching that goal while balancing the work among the workers.
- A final phase which consists of generating 30 subproblems per worker from the set of subproblems computed by the main phase.

The paper is organized as follows. First we recall some preliminaries. Next, we describe the existing decomposition and we present an efficient parallelization of the decomposition. Then, we give some experimental results. At last, we conclude.

2 Preliminaries

A worker is a computation unit. Most of the time, it corresponds to a core. We will consider that there are w workers. We present the two methods that we will compare.

2.1 Work Stealing

The work stealing method was originally proposed in [2] and was first implemented in Lisp parallel machines [4]. It splits the problem dynamically during the resolution. The workers solve the subproblems and when a worker finishes a subproblem, it asks the other workers for more work. In general, it is carried out as follows: when a worker W does not have work, it asks another worker V to get some work. If V agrees to give some of its work, then its splits the current subproblem into two subproblems and gives one to W. We say that W "steals" some work of V. If V does not agree to give some work to W, then W asks another worker U for some work until it gets some work or all workers have been solicited.

This method has been implemented in a lot of solvers (Comet [8] or ILOG Solver [12] for instance), and into several manners [14,5,16,3] depending on whether the work to be done is centralized or not, on the way the search tree is split, or on the communication method between workers. Work stealing attempts to solve partially the balancing issue of the workload by decomposing dynamically the subproblems.

When a worker is starving, it should not steal easy problems, because it would ask work again almost instantly. It happens frequently at the end of the search when many workers have no subproblems to solve. Thus, there are a lot of unnecessary communications. [11] proposes to use a threshold to avoid these unnecessary communications but its efficiency depends on the search space. Generally, this method scales well for a few workers but it is difficult to keep a linear speedup with a huge number of workers.

Some methods [15,5] attempt to increase the scalability. In [15], the authors propose a masters/workers approach. Each master has its own workers. The search space is divided between the different masters, then each master puts its attributed sub-trees in a work pool to dispatch to the workers. When a node of the sub-tree is detected that is a root of large sub-tree, the workers generate a large number of sub-trees and put them in a the work pool in order to have a better load balancing.

In [5], the authors experiment with up to 64 cores using a work stealing strategy. A master centralizes all pieces of information (bounds, solutions and requests). The master estimates which worker has the largest amount of work in order to give some work to an idle worker.

Another drawback is that the implementation of the work stealing is intrusive and is strongly dependent of the solver which requires to have a very good knowledge of the CP solver and to access to some internal functions. Some methods try to address this issue [7].

2.2 EPS

The Embarassingly Paralell Search (EPS) method has been defined in [13]. This method splits statically the initial problem into a large number of subproblems that are consistent with the propagation and puts them in a queue. Once this decomposition is over, the workers take dynamically the subproblems from the queue when they are idle. Precisely, EPS relies on the following steps:

- it splits a problem into p subproblems such as $p \geq w$ and pushes them into the queue.
- each worker takes dynamically a subproblem in the queue and solves it.
- a master monitors the concurrent access of the queue.
- the resolution ends when all subproblems are solved.

For optimization problems, the master manages the value of the objective. When a worker takes a subproblem from the queue, it also takes the best objective value computed so far. And when a worker solves a subproblem it communicates to the worker the value of the objective function. Note that there is no other communication, that is when a worker finds a better solution, the other workers that are running cannot use it for improving their current resolution.

The reduction of communication is an advantage over the work stealing. Furthermore, a resolution in parallel can be replayed by saving the order in which the subproblems have been executed. This costs almost nothing and helps a lot the debugging of applications.

2.3 Definitions

A constraint network $\mathcal{CN} = (\mathcal{X}, \mathcal{D}, \mathcal{C})$ is defined by:

- a set of n *variables* $\mathcal{X} = \{x_1, x_2, \ldots, x_n\}$
- a set of n finite *domains* $\mathcal{D} = \{D(x_1), D(x_2), \ldots, D(x_n)\}$ with $D(x_i)$ the set of possible *values* for the variable x_i,
- a set of *constraints* between the variables $\mathcal{C} = \{C_1, C_2, \ldots, C_e\}$. A constraint C_i is defined on a subset of variables $X_{C_i} = \{x_{i_1}, x_{i_2}, \ldots, x_{i_j}\}$ of $\mathcal{X}$ with a subset of Cartesian product $D(x_{i1}) \times D(x_{i2}) \times \ldots \times D(x_{ij})$, that states which combinations of values of variables $\{x_{i_1}, x_{i_2}, \ldots, x_{i_j}\}$ are compatible.

Each constraint C_i is associated with a filtering algorithm that removes values of the domains of its variables that are not consistent with it. The propagation mechanism applies filtering algorithms of $\mathcal{C}$ to reduce the domains of variables in turn until no reduction can be done. For convenience, we will use the word "problem" for designing a constraint network when it is used to represent the constraint network and not the search for a solution. We say that a problem P is consistent with the propagation if and only if running the propagation mechanism on P does not trigger a failure.

Notation 1. *Let Q be a problem, we will denote by $D(Q, x)$ the resulting domain of the variable x when the propagation mechanism has been applied to Q*

3 Decomposition Algorithms

3.1 Sequential Decomposition

EPS method is based on the decomposition of the initial problem into p subproblems consistent with the propagation. It has been shown in [13] that it is essential to generate subproblems consistent with the propagation, because the parallel version of search must not consider problems that would have not been considered by the sequential version of the serach.

If we aim at generating p subproblems then we can apply the simple following algorithm, called SIMPLEDECOMPOSITIONMETHOD.

- we use any variable ordering[1] $x_1, ..., x_n$.
- we compute the value k such that $|D(x_1)| \times \ldots \times |D(x_{k-1})| < p \leq |D(x_1)| \times \ldots \times |D(x_{k-1})| \times |D(x_k)|$.
- we generate all the assignments of the variables from x_1 to x_k and we regroup them if we have too many assignments.

Algorithm 1. Some useful functions

SIMPLEDECOMPOSITION($\mathcal{CN}, p$)
  computes the value k such that $|D(x_1)| \times \ldots \times |D(x_{k-1})| < p \leq |D(x_1)| \times \ldots \times |D(x_{k-1})| \times |D(x_k)|$
  generates all the assignments of the variables from x_1 to x_k
  regroups them and put the resulting subproblems into S
  returns the tuple (S, k)

COMPUTEDEPTH($\mathcal{CN}, cardS, \delta, p$)
  returns d such as $cardS \times |D(x_{\delta+1})| \times \ldots \times |D(x_{d-1})| < p \leq cardS \times |D(x_{\delta+1})| \times \ldots \times |D(x_{d-1})| \times |D(x_d)|$

GETDOMAINS(S)
  returns the set of domains of $\mathcal{D} = \{D(S, x_1), D(S, x_2), \ldots, D(S, x_n)\}$ such that $\forall x \in \mathcal{X}\ D(S, x) = \cup_{P \in S} D(P, x)$

GENERATESUBPROBLEMS($\mathcal{CN}, S, d$)
  runs a search for solution based on a DBDFS with d as depth limit on the constraint network formed by $\mathcal{CN}$ and the table constraint defined from the elements of S.
  returns the set of leaves that are not a failure.

Generating subproblems consistent with the propagation is a more complex task.

In [13], a depth bounded depth first search (DBDFS) is used for computing such problems. More precisely, this decomposition method is defined as follows. First, a static ordering of the variable is considered: $x_1, x_2, \ldots, x_n$. Usually the variables are sorted by non decreasingly domain sizes. Then, the main step of the algorithm is applied: define a depth d and perform a search procedure based on a DBDFS with d as limit. This search triggers the propagation mechanism each time a modification occurs.

[1] In this paper, we do not study the influence of any specific ordering.

For each leaf of this search which is not a failure, the variables $x_1, ..., x_d$ are assigned and so the subproblem defined by this assignment is consistent with the propagation. Thus the set of leaves defines a set S of subproblems. Next, if S is large enough, then the decomposition is finished. Otherwise, we apply again the main step until we reach the expected number of subproblems. However, we do not restart the main step from scratch and we use the previous set for improving the next computations in two ways. We use the cardinal of S for computing the new depth and we define a table constraint from the elements of S to avoid recomputations at the beginning of the search.

Algorithm 2. SEQUENTIALDECOMPOSITION

```
SEQUENTIALDECOMPOSITION(CN, p)
  // CN is a constraint network; p the number of subproblems to be generated
  S ← ∅; d ← 0
  while |S| < p do
    d ←COMPUTEDEPTH(CN, |S|, d, p)
    S ←GENERATESUBPROBLEMS(CN, S, d)
    if S = ∅ then return ∅
    CN ← (X,GETDOMAINS(S),C)
  return S
```

An important part of this method is the computation of the next depth. Currently it is simply estimated from the current number of subproblems that have been computed at the previous depth and the size of the domain. If we computed $|S|$ subproblems at the depth δ and if we want to have p subproblems then we search for the value d such that $|S| \times |D(x_{\delta+1})| \times \ldots \times |D(x_{d-1})| < p \leq |S| \times |D(x_{\delta+1})| \times \ldots \times |D(x_{d-1})| \times |D(x_d)|$

Algorithm 1 gives some useful functions. Algorithm 2 is a possible implementation of the sequential decomposition.

3.2 A Naive Parallel Decomposition

A parallelization of the decomposition is given in [13]. The initial problem is split into w subproblems by domain splitting. Each worker receives one of these subproblems and decomposes it into p/w subproblems consistent with the propagation. The master gathers all computed subproblems. If a worker is not able to generate p/w subproblems because it is not possible, the master asks the other workers to decompose their subproblems into smaller ones until reaching the right number of subproblems.

4 The New Parallel Decomposition

The method we propose has 3 main phases. A fast initial stating the process, a main phase, which is the core of the decomposition and a final phase ensuring that 30 subproblems per worker are generated. In the main phase, we try to progress in the decomposition and to manage the imbalance of the work load between workers, because we

Algorithm 3. EPS: Improved Decomposition in Parallel

```
WORKERDEC(CN, Q, d)
  S ← ∅
  run in parallel
    while Q ≠ ∅ do
      pick P ∈ Q and remove P from Q
      S' ←GENERATESUBPROBLEMS(CN, P, d)
      S ← S ∪ S'
  return S

DECOMPOSE(CN, S, numspb)
  while |S| < numspb do
    d ←COMPUTEDEPTH(CN, |S|, d, numspb)
    S ←WORKERDEC(CN, S, d)
    if S = ∅ then return ∅
    CN ← (X,GETDOMAIN(S),C)
  return S

PARALLELDECOMPOSITION(CN, numPbforStep, numStep, p)
  (S, d) ←SIMPLEDECOMPOSITION(CN, numPbforStep[0])
  S ←WORKERDEC(CN, S, d)
  CN ← (X,GETDOMAIN(S),C)
  for i=0 to numStep-1 do
    S ←DECOMPOSE(CN, S, numPbforStep[i])
    if S = ∅ or |S| ≥ p then return S
  return S
```

cannot avoid this imbalance. We progress by small steps of decomposition that are followed by synchronization of the workers and by merging the set of subproblems computing by each worker in order to correct the imbalance in the future. In other words, we ask the workers to decompose any subproblems into a small number of subproblems, then we merge all these subproblems (the union of subproblems lists computed by each worker) and ask again to decompose each subproblem into a small number of subproblems. When the number of generated subproblems is close to 5 subproblems per worker we know that we will have less problem with the load balancing. Thus, we can move on and trigger the last phase: the decomposition of the subproblems until we reach 30 subproblems per workers. Precisely, the phases are defined as follows:

- An initial phase where we decompose as quickly as possible the problem into as many subproblems as we have workers
- A main phase which aims at generating 5 subproblems per worker. Each subproblem is consistent with the propagation. This phase is divided into several steps in order to balance the work among the workers.
- A final phase which consists of generating 30 subproblems per worker from the set of subproblems computed by the main phase.

Algorithm 3 is a possible implementation of this new parallel decomposition.

The remaining question is the definition of the number of steps and the number of subproblems per worker that have to be generated for each step of the second phase. Clearly the decomposition of the first phase is not good and we have to stop the work for redistributing the subproblems to the worker as quickly as possible. The experiments have shown that we have to stop when 1 subproblem consistent with the propagation per worker have been generated. Then, a stop at 5 subproblems per worker will be enough for the second phase.

5 Experiments

Execution environment. All the experiments have been made on the data center "Centre de Calcul Interactif" hosted by the University of Nice Sophia Antipolis. It has 1152 cores, spread over 144 Intel E5-2670 processors, with a 4,608GB memory and runs under Linux (`http://calculs.unice.fr/fr`). We were allowed to use to up to 512 cores simultaneously for our experiments. The data center uses a scheduler (OAR) that manages jobs (submissions, executions, failures).

Implementation details. EPS is implemented on the top of the solver gecode 4.0.0 [1]. We use MPI (Message Passing Interface), a standardized and portable message-passing system to exchange information between processes. Master and workers are MPI processes. Each process reads a FlatZinc model to init the problem and only jobs are exchanged through messages between master and workers.

Benchmark Instances. We report results for the twenty most significant instances we found. Two types of problems are used: enumeration problems and optimization problems. Some of them are from CSPLib and have been modeled by Håkan Kjellerstrand (see [6]). The others come from the minizinc distribution (see [9]).

To study the decomposition, we select hard instances (i.e. more than 500s and less than 1h) with the Gecode solver.

Tests. It is important to point out that the decomposition must finish in order to begin the resolution of the generated subproblems.
We use the following definitions:

- t_{dec} and t_{res} denote respectively the total decomposition time (by the master and the workers) and the parallel solving time of the subproblems. So, the overall resolution time t is equal to $t_{dec} + t_{res}$
- t_0 is the resolution time of the instance in sequential
- $su = \frac{t_0}{t}$ is the speedup of the overall resolution time compared with the sequential resolution time
- $su_{res} = \frac{t_0}{t_{res}}$ is the speedup of the overall resolution time without taking account the decomposition time compared with the sequential resolution time
- $part_{dec} = \frac{t_{dec}}{t}$ is the ratio in % between the decomposition time and the overall resolution time

5.1 Sensitivity Analysis

Depth of the Decomposition. Figure 1 shows the evolution of the depth depending on the number of workers. As expected, the depth of the decomposition grows with the number of workers. Sometimes, the depth is very high for some instances like ghoulomb_3-7-20 or talent_scheduling_alt_film117 (see table 3). The reason is that the depth estimation we made is based on the Cartesian product of the domains which is sometimes wrong because there are many subproblems that are not consistent with the propagation, so the decomposition goes to a higher depth than the number of considered domains to generate 30 subproblems per worker consistent with the propagation.

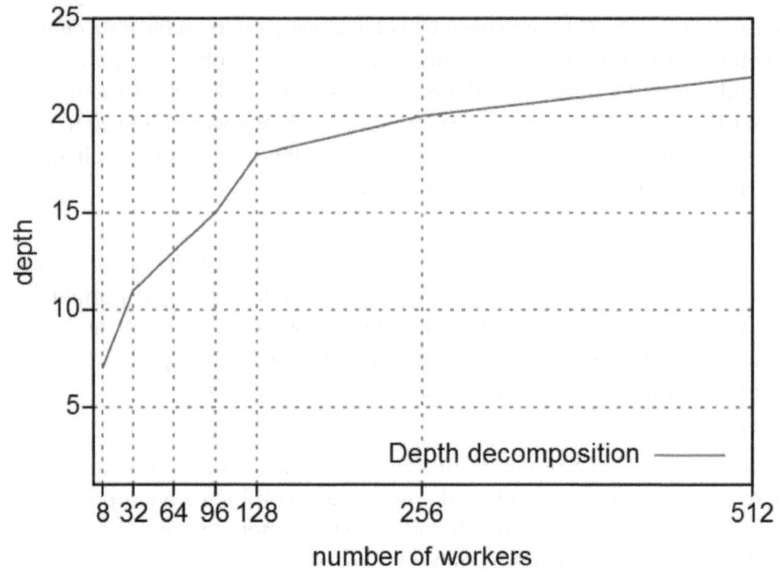

Fig. 1. Depth to reach 30 subproblems per worker related to the number of workers

Sequential Decomposition Issue. Table 1 gives the part of the sequential decomposition according to the overall resolution time with 16 workers and 512 workers. For 16 workers, the sequential decomposition takes a small part of the overall resolution time (an average of 3.5%) because it generates few subproblems ($16 * 30 = 480$ subproblems). Since the number of subproblems to generate is greater with 512 workers than 16 workers (16*30 vs 512*30), the sequential decomposition takes a significant time compared to the overall resolution time (an average of 72.5%). Thus, it takes more time to generate and impacts on the global performances.

Table 1. Part of the sequential decomposition according to the overall resolution time to generate 30 subproblems per worker for 16 workers and 512 workers

Instance	$part_{dec}$ for 16 $workers$ 16 $*$ 30 $=$ 480 $subproblems$ %	$part_{dec}$ for 512 $workers$ 512 $*$ 30 $=$ 15360 $subproblems$ %
market_split_s5-02	0.7%	33.5%
market_split_u5-09	0.6%	29.4%
market_split_s5-06	0.6%	37.6%
prop_stress_0600	4.7%	81.0%
nmseq_400	4.0%	85.3%
prop_stress_0500	3.5%	81.7%
fillomino_18	6.2%	88.7%
steiner-triples_09	3.1%	70.5%
nmseq_300	8.1%	91.2%
golombruler_13	1.7%	78.0%
cc_base_mzn_rnd_test.11	6.1%	83.4%
ghoulomb_3-7-20	6.9%	91.3%
pattern_set_mining_k1_yeast	4.9%	83.6%
still_life_free_8x8	8.8%	90.6%
bacp-6	2.8%	69.7%
depot_placement_st70_6	3.8%	79.5%
open_stacks_01_wbp_20_20_1	7.1%	86.7%
bacp-27	3.8%	76.5%
still_life_still_life_9	6.0%	88.5%
talent_scheduling_alt_film117	5.3%	91.1%
geometric average(%)	3.5%	72.5%

Fixing the Parallel Decomposition Parameters. Now, we must choose the number of subproblems to generate for the main phase. We select some representative instances to fix a good number of subproblems per worker. Table 2 shows the total decomposition time for some instance to choose the number of subproblems for the main phase. We notice that 5 is a good choice. Starting from 7 subproblems per worker, the performances begin to drop.

Table 2. Decomposition time comparison (in seconds) depending on the fixed number of subproblems in the second phase with 512 workers

Instance	**numPbforStep[1]**				
	3	4	5	6	7
prop_stress_0600	10.5	8.6	7.5	9.1	13.0
cc_base_mzn_rnd_test.11	21.5	12.1	10.1	14.5	17.8
ghoulomb_3-7-20	16.4	13.3	12.1	16.5	18.1
pattern_set_mining_k1_yeast	8.5	6.9	5.6	9.2	13.4
still_life_free_8x8	11.5	8.1	8.3	12.7	14.6
total decomposition time(s)	68.4	49.0	**43.6**	62.0	76.9

Table 3. Comparison of the decomposition algorithms with 512 workers

Instance	Seq. t_O	su_{res}	Dec_{seq} t_{dec}	su	$Dec_{//1}$ t_{dec}	su	$Dec_{//2}$ t_{dec}	su
	s	r	s	r	s	r	s	r
market_split_s5-02	3314.4	459.5	3.6	305.5	1.3	388.7	1.0	405.9
market_split_u5-09	3266.6	455.0	3.0	321.2	1.1	394.7	0.8	411.8
market_split_s5-06	3183.9	436.0	4.4	272.0	2.2	334.8	1.0	384.0
prop_stress_0600	2729.2	213.9	54.4	40.7	21.3	80.0	7.5	193.1
nmseq_400	2505.8	429.7	33.7	63.3	14.9	120.9	4.6	240.4
prop_stress_0500	1350.6	265.2	22.7	48.6	9.3	93.7	3.3	161.6
fillomino_18	763.9	301.9	19.8	34.2	6.4	85.7	2.5	150.7
steiner-triples_09	604.9	443.8	3.3	130.8	1.8	191.5	0.5	332.0
nmseq_300	555.3	309.0	18.7	27.1	7.9	57.1	2.4	131.7
golombruler_13	1303.9	492.0	9.4	92.7	1.4	322.9	0.4	427.9
cc_base_mzn_rnd_test.11	3279.5	196.5	83.8	32.6	35.5	62.8	10.1	122.6
ghoulomb_3-7-20	2993.8	279.2	112.6	24.3	50.0	49.3	12.1	131.1
pattern_set_mining_k1_yeast	2871.3	285.5	51.3	46.8	21.0	92.4	5.6	183.2
still_life_free_8x8	2808.9	331.0	82.0	31.1	33.2	67.4	8.3	166.9
bacp-6	2763.3	473.1	13.4	143.5	5.4	245.0	1.5	378.9
depot_placement_st70_6	2665.1	345.6	29.9	70.9	12.5	131.8	3.6	235.1
open_stacks_01_wbp_20_20_1	1523.2	280.7	35.4	37.3	15.6	72.3	4.0	160.8
bacp-27	1499.7	445.3	11.0	104.5	4.4	193.8	1.2	326.5
still_life_still_life_9	1145.1	347.9	25.2	40.1	9.4	90.4	3.0	182.9
talent_scheduling_alt_film117	566.1	386.4	15.0	34.4	6.0	75.8	1.8	175.8
total(s) and geom. average(r)	41694.5	347.1	632.6	66.4	260.8	124.8	**75.0**	**223.9**

Comparison of the Decomposition Algorithms. We denote Dec_{seq}, $Dec_{//1}$ and $Dec_{//2}$ respectively the sequential decomposition, the naive parallel decomposition and the new parallel decomposition.

Table 3 compares the decomposition time between different decomposition methods with 512 workers and the speedup su of each decomposition algorithm. On the first hand, we observe that the average speedup of the resolution time without the decomposition time is closed to a linear factor (347.1). On the other hand, we notice that $Dec_{//1}$ improves Dec_{seq} by a factor of 3.4 and $Dec_{//2}$ improves $Dec_{//1}$ by a factor of 2.4. So, the new parallel decomposition method improves the decomposition. Consequently, the new parallel decomposition $Dec_{//2}$ improves the overall speedup su from 66.4 with the sequential decomposition Dec_{seq} to 223.9.

Scaling Analysis. We test the scalability of EPS for different numbers of workers. Table 4 describes the details of the speedups. We notice that the resolution of Golombruler and the market-split instances scales very well with EPS (speedups su reach around 400 for 512 workers). Instances cc_base_mzn_rnd_test.11 and ghoulomb_3-7-20 give the worst results with all workers. In general, we observe an average speedup near to $w/2$. Figure 2 describes the speedup obtained by EPS for the decomposition phase,

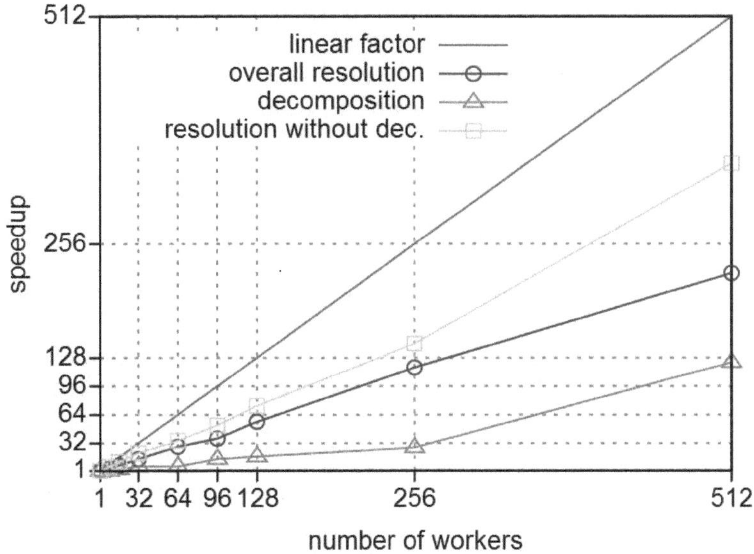

Fig. 2. Geometric Speedup (all instances) for each number of workers with EPS

Table 4. Speedup detailed for each instance and for each number of workers with EPS

Instance	$t_0(s)$ 1w	su 8w	16w	32w	64w	96w	128w	256w	512w
market_split_s5-02	3314.4	7.3	14.2	25.4	50.7	69.7	101.5	201.7	405.9
market_split_u5-09	3266.6	7.3	14.3	25.7	51.5	68.6	103.0	207.4	411.8
market_split_s5-06	3183.9	6.4	12.7	24.0	48.0	64.0	96.0	197.5	384.0
prop_stress_0600	2729.2	3.8	6.7	16.1	24.1	32.2	48.3	104.2	193.1
nmseq_400	2505.8	4.1	7.2	15.0	30.1	40.1	60.1	117.7	240.4
prop_stress_0500	1350.6	2.5	4.4	13.1	20.2	26.9	40.4	81.8	161.6
fillomino_18	763.9	2.4	5.1	11.4	18.8	25.1	37.7	72.4	150.7
steiner-triples_09	604.9	5.7	12.3	21.7	41.5	55.3	83.0	143.2	332.0
nmseq_300	555.3	2.4	5.1	8.2	16.5	21.9	32.9	69.3	131.7
golombruler_13	1303.9	7.3	14.7	27.3	53.7	89.5	117.4	213.1	427.9
cc_base_mzn_rnd_test.11	3279.5	1.7	5.1	8.9	14.3	20.4	30.6	59.7	122.6
ghoulomb_3-7-20	2993.8	2.3	3.9	8.2	17.4	21.8	32.8	76.3	131.1
pattern_set_mining_k1_yeast	2871.3	2.9	5.8	11.5	23.9	30.5	45.8	91.6	183.2
still_life_free_8x8	2808.9	2.6	6.4	10.4	20.9	27.8	41.7	83.5	166.9
bacp-6	2763.3	6.7	12.1	23.7	47.4	63.1	94.7	212.4	378.9
depot_placement_st70_6	2665.1	3.4	7.3	14.7	29.4	39.2	58.8	147.6	235.1
open_stacks_01_wbp_20_20_1	1523.2	3.1	6.5	10.0	23.1	26.8	40.2	95.4	160.8
bacp-27	1499.7	5.6	11.2	20.4	43.8	54.4	81.6	214.3	326.5
still_life_still_life_9	1145.1	3.1	6.1	11.4	22.9	30.5	45.7	89.4	182.9
talent_scheduling_alt_film117	566.1	2.4	4.3	11.0	22.0	31.3	51.7	95.9	175.8
total (t_0) or geometric average (su)	41694.5	3.7	7.5	14.7	28.3	37.9	56.7	117.3	223.9

the resolution phase without taking account the decomposition phase and the overall resolution with all instances (by a geometric average) as a function of the number of workers. EPS scales very well with a near-linear factor of gain for the resolution phase. Thanks to the parallelization of decomposition, EPS obtains good results for the overall resolution.

5.2 Comparison with Work Stealing

Table 5 shows a comparison between EPS and a work stealing implementation with 512 workers. The work stealing used in the datacenter is an MPI implementation based on [10]. In our experiments, the work stealing obtains good speedup until 64 workers but for more workers, the performances drop dramatically. With 512 workers, the average speedup of work stealing is 5.4. Many instances are not solved with 512 workers whereas they are solved by the sequential resolution. However, the average speedup of EPS is 223.9. Note that EPS is better than the work stealing on all selected instances.

Table 5. Comparison between work stealing and EPS with 512 workers

Instance	Seq.	Work stealing		EPS	
	time(s)	*time(s)*	*su*	*time(s)*	*su*
market_split_s5-02	3314.4	-	-	8.2	405.9
market_split_u5-09	3266.6	-	-	7.9	411.8
market_split_s5-06	3183.9	-	-	8.3	384.0
prop_stress_0600	2729.2	1426.4	1.9	14.1	193.1
nmseq_400	2505.8	-	-	10.4	240.4
prop_stress_0500	1350.6	670.0	2.0	8.4	161.6
fillomino_18	763.9	-	-	5.1	150.7
steiner-triples_09	604.9	79.0	7.7	1.8	332.0
nmseq_300	555.3	-	-	4.2	131.7
golombruler_13	1303.9	15.5	83.9	3.0	427.9
cc_base_mzn_rnd_test.11	3279.5	-	-	26.8	122.6
ghoulomb_3-7-20	2993.8	575.4	5.2	22.8	131.1
pattern_set_mining_k1_yeast	2871.3	299.8	9.6	15.7	183.2
still_life_free_8x8	2808.9	1672.8	1.7	16.8	166.9
bacp-6	2763.3	330.1	8.4	7.3	378.9
depot_placement_st70_6	2665.1	1902.9	1.4	11.3	235.1
open_stacks_01_wbp_20_20_1	1523.2	153.9	9.9	9.5	160.8
bacp-27	1499.7	579.6	2.6	4.6	326.5
still_life_still_life_9	1145.1	140.1	8.2	6.3	182.9
talent_scheduling_alt_film117	566.1	95.5	5.9	3.2	175.8
total (time) or geometric average (su)	41694.5	7941	5.4	195.7	**223.9**

6 Conclusion

The previous decomposition algorithms of the Embarrassingly Parallel Search are acceptable methods when there is only a few number of workers. These methods limit the performance of EPS on a data center, that is on a system with hundreds of cores. In this paper, we described an efficient parallel version of the decomposition. With parallelizing the problem decomposition and fixing 2 phases during the process, EPS gets a better workload during the decomposition. Consequently, EPS reaches the scalability with a data center and gives an average speedup at 223.9 with gecode for a set of benchmarks on a machine with 512 cores. This clearly improves the work stealing approach which does not scale well with hundred cores. EPS is more efficient by one or two orders of magnitude.

References

1. Gecode 4.0.0 (2012), http://www.gecode.org/
2. Warren Burton, F., Ronan Sleep, M.: Executing functional programs on a virtual tree of processors. In: Proceedings of the 1981 Conference on Functional Programming Languages and Computer Architecture, FPCA 1981, pp. 187–194. ACM, New York (1981)
3. Chu, G., Schulte, C., Stuckey, P.J.: Confidence-Based Work Stealing in Parallel Constraint Programming. In: Gent, I.P. (ed.) CP 2009. LNCS, vol. 5732, pp. 226–241. Springer, Heidelberg (2009)
4. Halstead Jr., R.H.: Implementation of multilisp: Lisp on a multiprocessor. In: Proceedings of the 1984 ACM Symposium on LISP and Functional Programming, LFP 1984, pp. 9–17. ACM, New York (1984)
5. Jaffar, J., Santosa, A.E., Yap, R.H.C., Zhu, K.Q.: Scalable Distributed Depth-First Search with Greedy Work Stealing. In: ICTAI, pp. 98–103. IEEE Computer Society (2004)
6. Kjellerstrand, H. (2014), http://www.hakank.org/
7. Menouer, T., Le Cun, B., Vander-Swalmen, P.: Partitioning methods to parallelize constraint programming solver using the parallel framework bobpp. In: Nguyen, N.T., van Do, T., Thi, H.A. (eds.) ICCSAMA 2013. SCI, vol. 479, pp. 117–127. Springer, Heidelberg (2013)
8. Michel, L., See, A., Hentenryck, P.V.: Transparent Parallelization of Constraint Programming. INFORMS Journal on Computing 21(3), 363–382 (2009)
9. MiniZinc (2012), http://www.g12.csse.unimelb.edu.au/minizinc/
10. Nielsen, M.: Parallel search in gecode. PhD thesis, Masters thesis, KTH Royal Institute of Technology (2006)
11. Pedro, V., Abreu, S.: Distributed Work Stealing for Constraint Solving. CoRR, abs/1009.3800:1–18 (2010)
12. Perron, L.: Search Procedures and Parallelism in Constraint Programming. In: Jaffar, J. (ed.) CP 1999. LNCS, vol. 1713, pp. 346–361. Springer, Heidelberg (1999)
13. Régin, J.-C., Rezgui, M., Malapert, A.: Embarrassingly parallel search. In: Schulte, C. (ed.) CP 2013. LNCS, vol. 8124, pp. 596–610. Springer, Heidelberg (2013)
14. Schulte, C.: Parallel Search Made Simple. In: Proceedings of TRICS: Techniques for Implementing Constraint programming Systems, a post-conference Workshop of CP 2000, pp. 41–57 (2000)
15. Xie, F., Davenport, A.: Massively parallel constraint programming for supercomputers: Challenges and initial results. In: Lodi, A., Milano, M., Toth, P. (eds.) CPAIOR 2010. LNCS, vol. 6140, pp. 334–338. Springer, Heidelberg (2010)
16. Zoeteweij, P., Arbab, F.: A Component-Based Parallel Constraint Solver. In: De Nicola, R., Ferrari, G.-L., Meredith, G. (eds.) COORDINATION 2004. LNCS, vol. 2949, pp. 307–322. Springer, Heidelberg (2004)

Stochastic MiniZinc*

Andrea Rendl[1], Guido Tack[1], and Peter J. Stuckey[2]

[1] National ICT Australia (NICTA) and Faculty of IT, Monash University, Australia
andrea.rendl@nicta.com.au, guido.tack@monash.edu
[2] National ICT Australia (NICTA) and University of Melbourne, Victoria, Australia
pstuckey@unimelb.edu.au

Abstract. Combinatorial optimisation problems often contain uncertainty that has to be taken into account to produce realistic solutions. However, existing modelling systems either do not support uncertainty, or do not support combinatorial features, such as integer variables and non-linear constraints. This paper presents an extension of the MINIZINC modelling language that supports uncertainty. Stochastic MINIZINC enables modellers to express combinatorial stochastic problems at a high level of abstraction, independent of the stochastic solving approach. These models are translated automatically into different solver-level representations. Stochastic MINIZINC provides the first solving technology agnostic approach to stochastic modelling we are aware of.

1 Introduction

In contrast to deterministic optimisation problems where all problem parameters are known a priori, *stochastic* optimisation problems deal with parameters that are *uncertain*, such as customer demand, resource capacities or travel times. This uncertainty has to be taken into account to provide realistic solutions.

Several stochastic modelling and solving systems have been established in recent years, such as AIMMS [17], AMPL [23,24] or GAMS [11]. These systems provide a strong support for stochastic linear problems on continuous variables, however have only limited or no support for problems with integer variables and non-linear constraints. Moreover, these systems force the modeller to commit to a particular solving approach at the modelling stage. This poses significant limitations to modellers who are interested in formulating stochastic *combinatorial* problems and solving them using different solving techniques.

For combinatorial problems, expressive high-level modelling languages have been developed, such as ESRA [3], Essence [5], Essence' [8], OPL [25], ZINC [12] and MINIZINC [13,14]. The benefit of high-level modelling is that users can focus on the problem formulation without committing to a particular solving approach. Some modelling systems, including MINIZINC, perform automated translations from the high-level model for different backend solvers, such as CP, MIP or SMT solvers. This way problems can be solved using different solving approaches without any background knowledge of the respective technique. Unfortunately, none of the existing combinatorial modelling

* NICTA is funded by the Australian Government through the Department of Communications and the Australian Research Council through the ICT Centre of Excellence Program.

B. O'Sullivan (Ed.): CP 2014, LNCS 8656, pp. 636–645, 2014.

languages provides means for dealing with uncertainty. Stochastic extensions for OPL have been proposed in [27], however, they have never been made available [28].

In this work we present Stochastic MINIZINC, an extension of MINIZINC that supports uncertainty. It allows the user to augment deterministic models to stochastic models without the need to commit to a particular solving strategy. To solve these stochastic models, we present transformations from Stochastic MINIZINC to deterministic MINIZINC for three different stochastic solving approaches: scenario-based [21] and policy-based [26] deterministic equivalents, as well as progressive hedging [16]. These are the only known stochastic solving techniques that can deal with both integer decision variables and non-linear constraints. Our transformations generate standard MINIZINC, enabling the use of any backend solver technique that supports the MINIZINC tool-chain.

2 Background

Stochastic optimisation deals with problems where some parameters are *uncertain*. Uncertain parameters become known at some point in time, which divides the problem into different *stages*: the stage *before* the parameter is known, and the stage *after* it is known. Typically, decisions have to be made *before* the uncertain parameters become known. For instance, a car factory has to decide how many cars to produce before the actual demand is known. These 'beforehand'-decisions are called *first stage* decisions. The *second stage* decisions are made *after* the uncertain parameters are known. In the car factory, after the demand is known (and the production is completed), decisions may have to be made to deal with overproduction or shortage, depending on the actual demand. The aim of second stage decisions is to compensate for 'bad' first stage decisions with respect to the uncertain parameters. This is referred to as *recourse*.

The values of uncertain parameters can often be *estimated*, e.g. from historical data or simulations, resulting in a set of possible outcomes called *scenarios*. All approaches discussed here assume a finite number S of scenarios. Each scenario has a *weight* $w_1, \ldots, w_S$ reflecting its likeliness. In the car factory example, we may have three different scenarios for the demand, $d_1 = 13,000$, $d_2 = 16,000$ and $d_3 = 22,000$, with weights $w_1 = 1$, $w_2 = 6$ and $w_3 = 3$.

Stochastic events may happen repeatedly, resulting in *multiple* stages. For instance, the car factory may have to decide its production every quarter, taking into account the surplus/shortage from the previous quarter. The yearly production plan then contains four stochastic events (one for each quarter), dividing the problem into five stages, where decisions at stage i influence the decisions at stage $i + 1$. Problems with only one stochastic event are called *two-stage* problems, as opposed to *multi-stage* problems with more than one event.

The stages of a stochastic optimisation problem divide its objective into different parts. For instance, in the car factory example, the first stage objective is to minimise production costs, while the second stage objective is to minimise storage costs (in case of overproduction) and unmet demand penalties (in case of underproduction). The first stage objective is *independent* of the stochastic parameters. The objectives in later stages do depend on the stochastic parameters. The overall objective therefore requires a probabilistic interpretation over all scenarios. Here, we focus on the *expected* value of the objective. Given an objective function $f(V)$ of the original problem formulation, the stochastic objective then becomes $\mathbb{E}[f(V)]$. Other interpretations, such as optimising for the worst case, can be realised in a similar manner.

Only three stochastic solving approaches exist for non-linear integer stochastic problems, all of which reformulate the problem into a deterministic model. The *Scenario-based Deterministic Equivalent* [22,21] is a single model that expands the stochastic model for each scenario and stage. All stochastic parameters and each second (and higher) stage variable is copied for each scenario, as well as all constraints involving higher stage parameters or variables. *Policy-based Search* [26] treats stochastic parameters as decision variables and uses *and-or-search* to explore all possible scenarios. Finally, *Progressive Hedging* [16] solves each scenario to optimality in isolation, and then iteratively adapts the objective function to minimise the gap between the first stage variables.

3 Stochastic MiniZinc

The design of Stochastic MiniZinc follows four objectives. (1) The stochastic extension is *conservative*, the stochastic model can be run, debugged, and solved deterministically, without changing the model. (2) The model is *agnostic of the solving approach*, so that the user does not have to commit to a specific stochastic approach during modelling. (3) Basic knowledge of stochastic optimisation should suffice to formulate a stochastic problem. (4) The stochastic extensions are lightweight additions to the language. As a result, Stochastic MiniZinc is a simple extension of standard MiniZinc that includes stochastic annotations.

A stochastic MiniZinc problem specification has three parts: a core problem model, a deterministic and a stochastic data specification. The core problem model is standard MiniZinc, augmented with annotations to mark stochastic parameters and stages. The deterministic data defines deterministic, first stage parameters. The stochastic data is given as a list of deterministic data files, one per scenario. Note that the core model combined with the deterministic data and a single scenario is a valid, deterministic MiniZinc model for that scenario. From the scenario list and a list of scenario weights, a combined stochastic data specification can be generated. Alternatively, the combined representation can be specified directly.

3.1 Stochastic Annotations in MiniZinc

The annotation `::stage(n)` associates a variable or parameter with stage n. A parameter in stage 1 is known from the outset and not stochastic. Variables and parameters without stage annotations belong to the first stage. The objective function is annotated to identify how the probabilistic nature of the scenarios is aggregated. We introduce an annotation `::expected` to optimise the *expected* value over all scenarios. The annotation `::scenario_weights` identifies the weights that reflect the likeliness of each scenario. The scenario weights have to be given as an array of the same length as the number of scenarios.

3.2 An Example: The Stochastic Vehicle Routing Problem

We illustrate Stochastic MINIZINC by formulating a stochastic variant of the vehicle routing problem (VRP) [10]. In the deterministic VRP, the aim is to find tours for m vehicles to serve n customers, minimising travel costs. In the stochastic variant of the VRP, the travel times are uncertain, and the aim is to find a vehicle-customer assignment that minimises the expected travel times. This means that the vehicle-customer assignments are the first stage decisions, and the optimal tours for each vehicle are the second stage decisions.

Fig. 1 shows a stochastic VRP model based on the classical VRP formulation [10], omitting the redundant predecessor variables for brevity. In line 13 we annotate the stochastic parameter, and in lines 16-18 we annotate the stochastic variables with their stages. The objective is annotated with `expected` since we want to find the optimal solution wrt. the expected arrival times of each vehicle. Note that all parameters, variables, constraints and the objective are defined deterministically, and the model can thus be solved as such. For instance, the data sets `d1.dzn` and `d2.dzn` in the bottom left each correspond to a single scenario.

The stochastic data `d_stoch.dzn` is generated from the scenarios `d1.dzn` and `d2.dzn` using the specification in the bottom left. Each parameter has an added dimension for the scenario. The `distance` is now three-dimensional, the first dimension indexing the scenario. The transformations in the next section link the stochastic parameter back to the model using the scenario as an index.

4 Transformations

This section shows how Stochastic MINIZINC can be implemented by transformation into standard MINIZINC. We consider three different formulations: a scenario-based deterministic model, a policy-based search, and a version of progressive hedging. Since the transformations generate standard MINIZINC, all solvers that support MINIZINC can be used. For policy-based search and progressive hedging, the backend solvers need to support search combinators [20]. The results of transforming the VRP from Fig. 1 can be found at [15].

We only present the two-stage version of the transformations, multi-stage problems are a straightforward generalisation [18]. In addition to the first and second stage sets of decision variables V_1 and V_2, we use C_1 and C_2 for the sets of first and second stage constraints, p for the set of stochastic parameters, and o for the original objective function. The transformations rely on a substitution operation $\mathit{substitute}(S, [x_1/y_1, \ldots, x_n/y_n])$, meaning that all occurrences of each x_i in S are simultaneously replaced by y_i.

4.1 The Scenario-Based Deterministic Equivalent

A Stochastic MINIZINC model is transformed into the deterministic equivalent by creating a copy of the second stage variables, with the stochastic parameters substituted by the concrete values for the scenario.

The objective from the stochastic model needs to be modified to represent the expected value over all scenarios. We introduce an array of variables `o` for the contribution of each scenario to the overall objective. The expected value is then computed as

```
% -------------- Stochastic Vehicle Routing Problem ------------ %
include "globals.mzn";
include "stochastic.mzn";

int: nC; int: nV; int: timeBudget;
set of int: VEHICLE = 1..nV;
set of int: CUSTOMER = 1..nC;
set of int: NODES = 1..nC+2*nV;
set of int: START_DEPOT_NODES = nC+1..nC+nV;
set of int: END_DEPOT_NODES = nC+nV+1..nC+2*nV;
set of int: TIME = 0..timeBudget;
array[NODES] of int: serviceTime;
array[NODES, NODES] of int: distance  :: stage(2);

% -------- variables ------------- %
array[NODES] of var VEHICLE: vehicle   :: stage(1);
array[NODES] of var NODES: successor   :: stage(2);
array[NODES] of var TIME: arrivalTime :: stage(2);

% -------- first stage constraints ---------- %
constraint forall (n in START_DEPOT_NODES)       % associate each start
   ( vehicle[n] = n-nC );                        % node with a vehicle
constraint  forall (n in END_DEPOT_NODES)        % associate each end
   ( vehicle[n] = n-nC-nV );                     % node with a vehicle

% -------- second stage constraints ---------- %
constraint forall (n in nC+nV+1..nC+2*nV-1)      % successors of end nodes
   ( successor[n] = n-nV+1 );                    % are start nodes
constraint successor[nC+2*nV] = nC+1;

constraint forall (n in START_DEPOT_NODES)       % vehicles leave the
   ( arrivalTime[n] = 0 );                       % depot at time zero
constraint circuit(successor);                   % hamiltonian circuit

constraint forall (n in CUSTOMER)                % use the same vehicle
   ( vehicle[successor[n]] = vehicle[n] );       % along a subtour
constraint forall (n in 0..nC+nV)
   ( arrivalTime[n] + serviceTime[n] + distance[n,successor[n]]
     <= arrivalTime[successor[n]] );             % time constraints

% -------- objective ------------ %
solve minimize   % expected overall travel time of each vehicle
   (sum (n in END_DEPOT_NODES) (arrivalTime[n])) :: expected;
```

```
% ==================== deterministic data ======================== %
nV = 1;  nC = 3;  timeBudget = 30;
serviceTime = [2,2,2,0,0];
```

```
% ==== d1.dzn ====== %
distance = [| 0, 4, 3, 5, 5
            | 4, 0, 2, 3, 3
            | 3, 2, 0, 2, 2
            | 5, 3, 2, 0, 0
            | 5, 3, 2, 0, 0 |];
```

```
% ==== d2.dzn ====== %
distance = [| 0, 4, 3, 5, 5,
            | 4, 0, 2, 6, 3,
            | 3, 2, 0, 2, 2,
            | 5, 6, 2, 0, 0,
            | 5, 3, 2, 0, 0  |];
```

```
% === implicit stochastic data == %
array[1..2] of string: scenarios =
     ["d1.dzn","d2.dzn"];
array[1..2] of int: weights = [1,1];
```

```
% === d_stoch.dzn === %
distance = array3d(1..2, % scenarios
                   1..5, 1..5,
         [ 0, 4, 3, 5, 5, % scenario 1
           4, 0, 2, 3, 3,
           3, 2, 0, 2, 2,
           5, 3, 2, 0, 0,
           5, 3, 2, 0, 0,

           0, 4, 3, 5, 5, % scenario 2
           4, 0, 2, 6, 3,
           3, 2, 0, 2, 2,
           5, 6, 2, 0, 0,
           5, 3, 2, 0, 0  ]);
array[1..2] of int: weights = [1,1]
           :: scenario_weights;
```

Fig. 1. A stochastic vehicle routing problem

the weighted average using the array of weights w. We use an integer representation for simplicity, but if the target solver supports continuous variables, a version using float variables could be used.

```
function var int: expected(array[int] of int: w, array[int] of var int: o) =
  sum (i in index_set(o)) (w[i]*o[i]) div sum(w);
```

```
V1;
C1;
array[1..S] of var int: o;
constraint forall (s in 1..S) ( let {
     substitute(V_2, [p/p[s], o/o[s]]);
} in substitute(C_2, [p/p[s], o/o[s]]) );
solve minimize expected(w,o);
```

The deterministic equivalent is then constructed by looping over all scenarios and creating a fresh set of second stage variables for each scenario using a let construct. The second stage constraints C_2 are moved into the loop. For both these code sections we add a scenario argument s to each second stage parameter.

4.2 Policy-Based Search

Policy-based search for stochastic constraint programming [26] turns stochastic parameters into decision variables and then uses backtracking search to explore the different scenarios. Instead of copying the second stage model for each scenario as in Sect. 4.1, policy-based search implements the forall over all scenarios using a variant of *and-or search*. Decision variables are searched in the usual *or*-fashion, while stochastic variables represent *and*-nodes.

Our implementation adds decision variables for each stochastic parameter p, and a single integer variable scenario that selects the scenario. Element constraints connect the parameter variables to the actual parameters for the selected scenario. The original objective o is unchanged, but an additional expected objective eo is added as in Sect. 4.1.

```
V1;
C1;
array[1..S] of var int: os;
var 1..S: scenario;
for each array[1..S] of int: p add var int: pV = p[scenario];
substitute(V2, [ p/pV ]);
substitute(C2, [ p/pV ]);
var int: eo = expected(w,os);
solve two_stage(eo,o,os);
```

The *and-or* search can be implemented elegantly using search combinators [20], an expressive domain-specific language for sophisticated search strategies. The combinator two_stage used above can be realised as follows:

```
combinator s_bab(svar int: i, array[int] of svar int: best, var int: o) =
  post(scenario=i /\ o<best[i], and([search_stage_2,assign(best[i],o),prune]));
combinator two_stage(var int: eo, var int: o, array[int] of var int: os) =
  bab(eo,
      let { array[1..S] of svar int: best = [ ∞ | i in 1..S] } in
      and([ search_stage_1,
            portfolio([for (i,1,S, s_bab(i,best,o)),
                        post(os = best)]) ]));
```

The main structure of the combinator is an outer branch-and-bound search (bab in line 4) over the expected objective, combined with an inner branch-and-bound for every scenario (scenario_bab, line 7). The optimum of the second stage for each scenario is collected in an array best, and after all scenarios have been processed, is posted back into the variables os, constraining the expected objective. The search strategies search_stage_1 and search_stage_2 can be user-defined or default searches for the first and second stage variables, respectively.

4.3 Progressive Hedging

Progressive hedging [16] solves each scenario to optimality in isolation, producing different assignments to the first stage variables for each scenario. The objective of each scenario is then augmented with a term to minimise these first stage differences between scenarios. This is iterated until the differences between first stage variables across all scenarios are sufficiently small.

The transformation of a stochastic model using progressive hedging adds weights xw that will be updated after each iteration, and an augmented objective o_hedge that accounts for the differences between the first stage variables:

```
V1;
C1;
var 1..S: scenario;
array[1..|V1|] of int: xw;
for each array[1..S] of int: p add var int: pV = p[scenario];
substitute(V2,[ p/pV ]);
substitute(C2,[ p/pV ]);
var int: o_hedge = o + hedge(xw,V1);
solve progressive_hedging(o_hedge);
```

```
1  combinator progressive_hedging(var int: o) =
2    restart( distance > epsilon,
3      let { array[1..S,1..|V1|] of svar int: V } in
4      portfolio(for (i,1,S,
5                  let { svar int: best = ∞ } in
6                  post(scenario=i /\ o < best,
7                    and([search_stage_1, search_stage_2, assign(best,o),
8                         assign(V[i],V1), prune]))
9        ),
10       update_weights(distance,xw,V)
11   ) );
```

The progressive_hedging combinator iterates over all scenarios (line 4), performing a branch-and-bound search on the modified objective o_hedge similar to scenario_bab. The results of the first stage variables V_1 are stored in the array V (line 8). After search in all scenarios has completed, the distance between the V_1 variables and the weights in the extended objective function are updated (line 10). The update_weights function requires some integration with the underlying solver, since it changes the parameters of an existing constraint. We are currently implementing this feature.

5 Related Work

The closest related approach is Stochastic OPL [27], a proposed extension of the OPL modelling language [25] with support for stochastic variables, chance constraints, and an objective function based on expectation. Similar to Stochastic MINIZINC, the extended language is compiled into the deterministic equivalent in standard OPL. Our approach is more general, with translations into policy-based search and progressive hedging using search combinators. We also took a conservative approach to language extension, where stochastic features are represented using annotations, while the basic model is perfectly valid deterministic MINIZINC. Finally, using MINIZINC as the base language yields a solver agnostic approach, enabling the modeller to experiment with all the different backend solvers that support MINIZINC.

AIMMS [17] is a commercial modelling and solving framework, where models can be formulated using a graphical user-interface or by employing the internal programming language. It provides strong support for stochastic linear, continuous problems, but has very limited support for non-linear integer problems.

AMPL is a commercial algebraic modelling language with a number of stochastic extensions [6,7,4,22,24]. The SAMPL [24] extension, including the Stochastic Programming Integrated Environment (SPiNE) [23], has been integrated into AMPL. It has annotations for stochastic parameters and other stochastic features. However, the modeller has to formulate the deterministic (scenario-based) equivalent and thus has to commit to this approach.

GAMS is a commercial modelling and solving framework that incorporates a high-level modelling language with a stochastic extension [11]. Stochastic models contain annotations that associate parameters to random distributions and assign variables and constraints to stages. The annotation compilation must be explicitly stated by the modeller, and non-linear constraints are not supported.

Lingo [1] is a commercial optimisation framework for Microsoft Excel with support for stochastic programming. It provides a modelling language and strong support for stochastic linear problems, however, integer or non-linear problems can only be translated into their scenario-based deterministic equivalents.

PySP [29] is an open modelling and solving library based on Pyomo [9], an algebraic modelling language extending Python. PySP supports stochastic integer problems that can be solved either as scenario-based deterministic equivalents or by progressive hedging. It has no known support for non-linear constraints.

There is a close connection between stochastic constraint programming and PSPACE-complete formalisms such as QCSP [2] and QBF [19]. It will be interesting future work to explore automatic translations that target solvers for these problem classes.

6 Conclusions

We have presented Stochastic MINIZINC, an extension of the MINIZINC modelling language that introduces syntax for modelling stochastic constraint (optimisation) problems. We have shown how to translate Stochastic MINIZINC automatically into standard MINIZINC, using three different standard approaches for dealing with uncertainty: the scenario-based deterministic equivalent, policy-based search, and progressive hedging. The resulting system enables modellers to express stochastic problems at a high level of abstraction, and to experiment with different solving approaches. Stochastic

MiniZinc is the first solving technology agnostic approach to stochastic modelling we are aware of.

The presented stochastic extensions of MiniZinc are implemented, as well as the automated transformations that take fractions of a second to translate. Stochastic MiniZinc is available at `http://www.minizinc.org`.

References

1. Atlihan, M., Cunningham, K., Laude, G., Schrage, L.: Challenges in adding a stochastic programming/scenario planning capability to a general purpose optimization modeling system. In: Sodhi, M.S., Tang, C.S. (eds.) A Long View of Research and Practice in Operations Research and Management Science. International Series in Operations Research & Management Science, vol. 148, pp. 117–135. Springer, US (2010)
2. Chen, H.M.: The Computational Complexity of Quantified Constraint Satisfaction. Ph.D. thesis, Cornell University, Ithaca, NY, USA (2004)
3. Flener, P., Pearson, J., Ågren, M.: Introducing ESRA, a Relational Language for Modelling Combinatorial Problems (Abstract). In: Rossi, F. (ed.) CP 2003. LNCS, vol. 2833, pp. 971–971. Springer, Heidelberg (2003)
4. Fourer, R., Lopes, L.: StAMPL: A Filtration-Oriented Modeling Tool for Multistage Stochastic Recourse Problems. INFORMS Journal on Computing 21(2), 242–256 (2009)
5. Frisch, A.M., Harvey, W., Jefferson, C., Hernandez, B.M., Miguel, I.: Essence: A Constraint Language for Specifying Combinatorial Problems. Constraints 13(3), 268–306 (2008)
6. Gassmann, H., Ireland, A.: Scenario formulation in an algebraic modelling language. Annals of Operations Research 59, 45–75 (1995)
7. Gassmann, H., Ireland, A.: On the formulation of stochastic linear programs using algebraic modelling languages. Annuals of Operations Research 64, 83–112 (1996), stochastic programming, algorithms and models, Lillehammer (1994)
8. Gent, I.P., Miguel, I., Rendl, A.: Tailoring Solver-Independent Constraint Models: A Case Study with ESSENCE′ and MINION. In: Miguel, I., Ruml, W. (eds.) SARA 2007. LNCS (LNAI), vol. 4612, pp. 184–199. Springer, Heidelberg (2007)
9. Hart, W.E., Watson, J.P., Woodruff, D.L.: Pyomo: modeling and solving mathematical programs in Python. Math. Program. Comput. 3(3), 219–260 (2011)
10. Kilby, P., Shaw, P.: Vehicle routing. In: Rossi, F., van Beek, P., Walsh, T. (eds.) Handbook of Constraint Programming, ch. 23, pp. 799–834. Elsevier Science Inc., New York (2006)
11. Loewe, M., Ferris, M.: Stochastic programming (SP) with EMP (GAMS) (2013), `http://www.gams.com/dd/docs/solvers/empsp.pdf`
12. Marriott, K., Nethercote, N., Rafeh, R., Stuckey, P.J., de la Banda, M.G., Wallace, M.: The design of the zinc modelling language. Constraints 13(3), 229–267 (2008)
13. Nethercote, N., Stuckey, P.J., Becket, R., Brand, S., Duck, G.J., Tack, G.: MiniZinc: Towards a standard CP modelling language. In: Bessière, C. (ed.) CP 2007. LNCS, vol. 4741, pp. 529–543. Springer, Heidelberg (2007)
14. NICTA: MiniZinc (2014), `http://www.minizinc.org`
15. NICTA: Stochastic MiniZinc examples (2014), `http://www.minzinc.org/stochastic/`
16. Rockafellar, R.T., Wets, R.J.B.: Scenarios and policy aggregation in optimization under uncertainty. Mathematics of Operations Research 16(1), 119–147 (1991)
17. Roelofs, M., Bisschop, J.: AIMMS: The language reference 3.9 (2009)
18. Ruszczyński, A., Shapiro, A.: Stochastic Programming. Handbooks in operations research and management science. Elsevier (2003)

19. Samulowitz, H.: Solving Quantified Boolean Formulas. Ph.D. thesis. University of Toronto, Toronto, ON, Canada (2007)
20. Schrijvers, T., Tack, G., Wuille, P., Samulowitz, H., Stuckey, P.J.: Search combinators. Constraints 18(2), 269–305 (2013)
21. Tarim, A., Manandhar, S., Walsh, T.: Stochastic Constraint Programming: A Scenario-Based Approach. Constraints 11(1), 53–80 (2006)
22. Thénié, J., Delft, C., Vial, J.: Automatic formulation of stochastic programs via an algebraic modeling language. Computational Management Science 4(1), 17–40 (2007)
23. Valente, C., Mitra, G., Poojari, C.: A Stochastic Programming Integrated Environement (SPiNE), pp. 115–136. Society for Industrial and Applied Mathematics (SIAM), Philadelphia. MPS, Mathematical Programming Society, Philadelphia (2005)
24. Valente, C., Mitra, G., Sadki, M., Fourer, R.: Extending algebraic modelling languages for stochastic programming. INFORMS Journal on Computing 21(1), 107–122 (2009)
25. Van Hentenryck, P.: The OPL Optimization Programming Language. MIT Press, Cambridge (1999)
26. Walsh, T.: Stochastic Constraint Programming. In: van Harmelen, F. (ed.) ECAI, pp. 111–115. IOS Press (2002)
27. Walsh, T.: Stochastic OPL. In: Proceedings of the Workshop on Modelling and Solving with Constraints (2002)
28. Walsh, T.: Personal Communication (2014)
29. Watson, J.P., Woodruff, D.L., Hart, W.E.: PySP: modeling and solving stochastic programs in Python. Math. Program. Comput. 4(2), 109–149 (2012)

Decomposing Utility Functions in Bounded Max-Sum for Distributed Constraint Optimization

Emma Rollon and Javier Larrosa

Universitat Politècnica de Catalunya, Barcelona, Spain

Abstract. Bounded Max-Sum is a message-passing algorithm for solving Distributed Constraint Optimization Problems (DCOP) able to compute solutions with a guaranteed approximation ratio. In this paper we show that the introduction of an intermediate step that decomposes functions may significantly improve its accuracy. This is especially relevant in critical applications (e.g. automatic surveillance, disaster response scenarios) where the accuracy of solutions is of vital importance.

Introduction

Bounded Max-Sum (BMS) [11] approximately solves Distributed Constraint Optimization Problems (DCOP) with very little computation and communication demands. Arguably, its most interesting feature is that it comes with a *guarantee approximation ratio*, meaning that its approximate solution has a utility which is no more than a factor away from the optimum. The algorithm has been recently revisited and enhanced, producing two improved versions: IBMS [12] and RN-BMS [9], with tighter upper and lower bounds, respectively.

All the BMS algorithms have a relaxation phase in which some functions are replaced by smaller arity functions. In general, such replacement introduces some error, which prevents the algorithms from computing the true optimal solution. In this paper we study the possibility of an exact decomposition in which those binary functions are replaced by pairs of unary functions which faithfully capture the same information. Since, in the general case, functions do not have an exact decomposition, we consider approximate decompositions in which the error introduced is minimized. We theoretically prove that our approach improves the upper bound obtained by IBMS and show its performance in a set of graph coloring problems.

Preliminaries

In this Section we review the main elements to contextualize our work. Definitions and notation are borrowed almost directly from [11]. We urge the reader to visit that reference for more details and examples.

DCOP

A *Distributed Constraint Optimization Problem* (DCOP) is a tuple $P = (\mathbf{A}, \mathbf{X}, \mathbf{D}, \mathbf{F})$, where $\mathbf{A} = \{\mathcal{A}_1, \ldots, \mathcal{A}_r\}$ is a set of agents, and $\mathbf{X} = \{x_1, \ldots, x_n\}$ and $\mathbf{D} =$

B. O'Sullivan (Ed.): CP 2014, LNCS 8656, pp. 646–654, 2014.

$\{\mathbf{d}_1, \ldots, \mathbf{d}_n\}$ are variables and domains. $\mathbf{F} = \{f_1, \ldots, f_e\}$ is a set of cost functions. The objective function is $F(x) = \sum_{j=1}^{e} f_j(x^j)$ where $x^j \subseteq \mathbf{X}$ is the scope of f_j. A *solution* is a complete assignment $\mathbf{x}$. An *optimal solution* is a complete assignment $\mathbf{x}^*$ such that $\forall \mathbf{x},\ F(\mathbf{x}^*) \geq F(\mathbf{x})$. The usual task of interest is to find $\mathbf{x}^*$ through the coordination of the agents.

In the applications under consideration, the agents search for the optimum via decentralized coordination. We assume that each agent can control only its local variable(s) and has knowledge of, and can directly communicate with, a few neighboring agents. Two agents are neighbors if there is a relationship connecting variables and functions that the agents control.

The structure of a DCOP problem $P = (\mathbf{A}, \mathbf{X}, \mathbf{D}, \mathbf{F})$ can be transformed into a factor graph. A *factor graph* is a bipartite graph having a variable node for each variable $x_i \in \mathbf{X}$, a factor node for each local function $f_j \in \mathbf{F}$, and an edge connecting variable node x_i to factor node f_j if and only if x_i is an argument of f_j.

Bounded Max-Sum Algorithms

The *BMS* algorithm [11] and its improved versions IBMS [12] and RN-BMS [9] are approximation methods built on Max-Sum [4,1]. From a possibly cyclic problem P, the idea is to remove cycles in its factor graph by ignoring dependencies between functions and variables, producing a new acyclic problem. Then, Max-Sum is used to optimally solve the acyclic problem while simultaneously computing the approximation ratio.

Here, for simplicity purposes, we restrict ourselves to IBMS, which was proven to be always superior to BMS and usually superior to RN-BMS. For the sake of simplicity, we will restrict ourselves to the case of binary functions $f_j(x_i, x_k)$. The extension to general functions is direct.

IBMS works in three phases:

- **Relaxation Phase:** First, the algorithm assigns a weight w_{ij} to each edge (i, j) of the original factor graph. Then, it finds a maximum spanning tree with respect to the weights. Next, the original problem P is transformed into an acyclic one $\widetilde{P}$ having the spanning tree as factor graph as follows: For each edge (i, j) in the original graph that does not belong to the tree, the cost function $f_j(x_i, x_k)$ is transformed into another function $\widetilde{f}_j(x_k) = \max_{x_i}\{f_j(x_i, x_k)\}$.
  Let T denote the set of functions that have not been simplified. The objective function of $\widetilde{P}$ is
  $$\widetilde{F}(x) = \sum_{j \in T} f_j(x_i, x_k) + \sum_{j \notin T} \widetilde{f}_j(x_k)$$
- **Solving Phase:** IBMS solves $\widetilde{P}$ with Max-Sum. Let $\widetilde{\mathbf{x}}$ be the solution given by IBMS. Since the factor graph of $\widetilde{P}$ is acyclic, $\widetilde{\mathbf{x}}$ is optimal for $\widetilde{P}$. IBMS returns $\widetilde{\mathbf{x}}$ as a sub-optimal solution for P.
- **Bounding Phase:** IBMS computes a guarantee approximation ratio as follows. Note that $F(\widetilde{\mathbf{x}})$ is an obvious lower bound of the the optimal ($F(\widetilde{\mathbf{x}}) \leq F(\mathbf{x}^*)$). Moreover, it can be shown that $\widetilde{F}(\widetilde{\mathbf{x}})$ is an upper bound of the optimal ($\widetilde{F}(\widetilde{\mathbf{x}}) \geq F(\mathbf{x}^*)$). Therefore, $\rho = \frac{\widetilde{F}(\widetilde{\mathbf{x}})}{F(\widetilde{\mathbf{x}})}$ is a guarantee approximation ratio for IBMS.

x_i	x_k	f_j
a	a	15
a	b	30
b	a	10
b	b	25

x_i	g_j
a	10
b	5

x_k	h_j
a	5
b	20

Fig. 1. Example of a binary function f_j that can be exactly decomposed into two unary functions $h_j(x_k)$ and $g_j(x_i)$

Decomposition

The IBMS algorithm relaxes the problem by replacing some binary functions $f_j(x_i, x_k)$ by unary functions $\widetilde{f}_j(x_k)$. Clearly, the relaxed problem is in general not equivalent to the original one because the transformation introduces an error.

Exact Decomposition

The idea of exact decomposition is to replace the binary function $f_j(x_k, x_i)$ by two unary functions $h_j(x_k)$ and $g_j(x_i)$ such that there is no loss of information. Formally, given a binary function $f_j(x_i, x_k)$ we can set a system of linear equations,

$$\forall_{x_i,x_k} \; f_j(x_i, x_k) = h_j(x_k) + g_j(x_i)$$

such that $\forall x_k, h_j(x_k) \geq 0$ and $\forall x_i, g_j(x_i) \geq 0$, where each entry of the unary functions is a variable of the system. If the system has a solution, that solution is an *exact decomposition*. Replacing the binary function by the two unary functions modifies the factor graph without introducing any error. The system of linear equations can be solved very efficiently with one of the many Integer Programming toolkits available.

As an example, consider the binary cost function $f_j(x_i, x_k)$, and the two unary functions $h_j(x_k)$ and $g_j(x_i)$ in Figure 1, respectively. Observe that the combination of the two unary functions is equivalent to the binary one. The reason being that the following equations are satisfied,

$$h_j(a) + g_j(a) = f_j(a, a) \qquad h_j(b) + g_j(a) = f_j(a, b)$$
$$h_j(a) + g_j(b) = f_j(b, a) \qquad h_j(b) + g_j(b) = f_j(b, b)$$

Therefore, in this case, exact decomposition could be achieved.

The natural application of the previous idea constitutes our first algorithm called *exact decomposition-based IBMS* (ED-IBMS). It differs from the previous ones only in the relaxation phase. Before computing each $\widetilde{f}_j(x_k)$, ED-IBMS attempts an exact decomposition. Let D be the set of functions in which exact decomposition was achieved. The resulting objective function is

$$\widetilde{F}_{ED}(x) = \sum_{j \in T} f_j(x_i, x_k) + \sum_{j \in D} (g_j(x_i) + h_j(x_k)) + \sum_{j \notin T \cup D} \widetilde{f}_j(x_k)$$

It is easy to see that the cost of any solution in the relaxed problem with exact decomposition is smaller than or equal to its cost in the relaxed problem without exact decomposition. In other words, ED-IBMS always obtains upper bounds tighter than IBMS. Formally, $F(x) \leq \widetilde{F}_{ED}(x) \leq \widetilde{F}(x)$ holds.

Approximate Decomposition

Note that exact decompositions do not exist in general. In a preliminary set of experiments we observed that exact decomposition occurs very rarely which makes, in practice, IBMS and ED-IBMS behave identically. When we looked into the details, we observed that very often exact decomposition was *almost* achievable, which leaded us to approximate decomposition.

The idea of approximate decomposition is to replace a given binary function $f_j(x_i, x_k)$ by the combination of two unary functions $h_j(x_k)$ and $g_j(x_i)$, and a binary function $r_j(x_i, x_k)$. Formally,

$$\forall_{x_i,x_k}, f_j(x_i, x_k) = h_j(x_k) + g_j(x_i) + r_j(x_i, x_k) \tag{1}$$

such that $\forall_{x_i,x_k}, r_j(x_i, x_k) \geq 0$, $\forall x_k, h_j(x_k) \geq 0$, and $\forall x_i, g_j(x_i) \geq 0$. As before, this expression can be seen as a system of linear equations where each entry in the unary functions and in the binary function $r(x_i, x_k)$ is a variable of the system. Note that $g_j(x_i)$ and $h_j(x_k)$ represent the part of the utility function $f_j(x_i, x_k)$ that has been decomposed, while $r_j(x_i, x_k)$ represents the part that has not been decomposed (the *residual* utility function).

Moreover, we want to ensure that the decomposition improves the upper bound on the original problem. In other words, the optimum of the relaxed problem with approximate decomposition must be tighter than the optimum of the relaxed problem without approximate decomposition. Formally,

$$\forall_{x_i,x_j} \max_{x_i}\{r(x_i, x_k)\} + g(x_i) + h(x_k) \leq \max_{x_i}\{f(x_i, x_k)\} \tag{2}$$

This inequality can be rewritten using Expression 1 as,

$$\forall_{x_i,x_k} \max_{x_i}\{r(x_i, x_k)\} \leq \max_{x_i}\{f(x_i, x_k)\} - f(x_i, x_k) + r(x_i, x_k)$$

Each inequality can be transformed into a set of inequalities without the max operator over function $r(x_i, x_j)$ as follows,

$$\forall_{x_i,x_k} \forall_{a \neq x_i}, r(a, x_k) \leq \max_{x_i}\{f(x_i, x_k)\} - f(x_i, x_k) + r(x_i, x_k) \tag{3}$$

As an example, consider function $f_j(x_i, x_k)$ in Figure 2. The system of linear equations required to approximatelly decompose that function is,

$$\begin{array}{ll}
g_j(a) + h_j(a) + r_j(a,a) = 20 & r_j(b,a) \leq \max\{20,10\} - 20 + r_j(a,a) \\
h_j(a) + g_j(b) + r_j(a,b) = 30 & r_j(a,a) \leq \max\{20,10\} - 10 + r_j(b,a) \\
h_j(b) + g_j(a) + r_j(b,a) = 10 & r_j(b,b) \leq \max\{30,25\} - 30 + r_j(a,b) \\
h_j(b) + g_j(b) + r_j(b,b) = 25 & r_j(a,b) \leq \max\{30,25\} - 25 + r_j(b,b)
\end{array}$$

x_i	x_k	f_j
a	a	20
a	b	30
b	a	10
b	b	25

x_i	g_j
a	5
b	0

x_k	h_j
a	10
b	25

x_i	x_k	r_j
a	a	5
a	b	0
b	a	0
b	b	0

Fig. 2. Example of a binary function f_j that is approximate decomposed into $g_j(x_i)$, $h_j(x_k)$ and $r_j(x_i, x_k)$

subject to,

$$\begin{array}{llll} h_j(a) \geq 0 & g_j(a) \geq 0 & r_j(a,a) \geq 0 & r_j(b,a) \geq 0 \\ h_j(b) \geq 0 & g_j(b) \geq 0 & r_j(a,b) \geq 0 & r_j(b,b) \geq 0 \end{array}$$

where $h_j(\cdot)$, $g_j(\cdot)$, and $r_j(\cdot,\cdot)$ are the variables of the system. Figure 2 shows one solution to the previous system.

Any solution to the system of linear equations from Expression 1 and Expression 3 is an approximate decomposition. Some solutions may be better than other. The worst decomposition would be one in which h_j and g_j are zero because no decomposition would have been achieved. In general, one may prefer those decompositions that minimize in one way or another the residuals (note that exact decomposition coincides with zero residuals). We consider two possibilities:

- Minimize the maximum residual: $\min \max_{x_i, x_k} r_j(x_i, x_k)$.
- Minimize the average residual: $\min \sum_{x_i, x_k} r_j(x_i, x_k)$.

Such objective functions can easily be added to the system of equations and subsequently solved with an Integer Programming toolkit.

Approximate decompositions introduce a new family of IBMS algorithms called AD-IBMS. The idea is to compute an approximate decomposition of function $f_j(x_i, x_k)$ before its relaxation. Then, the relaxation is performed not over the original function f_j, but over the residual r_j (i.e., $\widetilde{r}_j(x_k) = \max_{x_i}\{r_j(x_i, x_k)\}$). Thus, the objective function of the relaxed problem is,

$$\widetilde{F}_{AD}(x) = \sum_{(i,j),(k,j)\in T} f_j(x_i, x_k) + \sum_{(i,j)\notin T} (g_j(x_i) + h_j(x_k) + \widetilde{r}_j(x_k))$$

Since the system of linear equations enforce Expression 2, it is easy to see that AD-IBMS always obtains tighter upper bounds than IBMS. Formally, $F(x) \leq \widetilde{F}_{AD}(x) \leq \widetilde{F}(x)$ holds.

Empirical Evaluation

The purpose of the experiments is to compare IBMS with respect to AD-IBMS. In particular, we want to evaluate the improvement of the bounds and approximation ratio

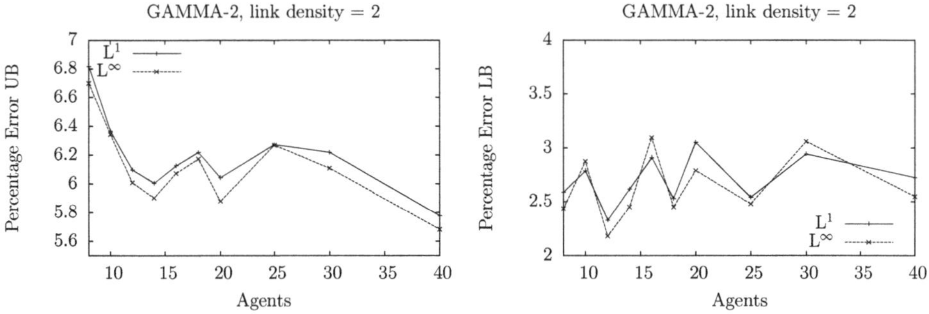

Fig. 3. Percentage error of the upper (left) and lower bound (right) obtained by AD-IBMS minimizing the maximum residual ($L1$) and minimizing the average residual (L^∞)

of the IBMS algorithm using approximate decomposition. For the sake of completeness, we will also report the results for standard BMS and RN-BMS. The percentage error of a value v_{approx} (i.e., upper or lower bound) is computed as $\frac{|v - v_{approx}|}{v} \times 100$ where v is the optimum of the problem.

We consider the same set of problems from the ADOPT repository[1] used in [11]. These problems represent graph coloring problems with two different link densities (i.e., the average connection per agent) and different number of nodes. Each agent controls one node (i.e., variable), with domain $|\mathbf{d}_i| = 3$, and each edge of the graph represents a pairwise constraint between two agents. Each edge is associated with a random payoff matrix, specifying the payoff that both agents will obtain for every possible combination of their variables' assignments. Each entry of the payoff matrix is a real number sampled from two different distributions: a gamma distribution with $\alpha = 9$ and $\beta = 2$, and a uniform distribution with range $(0, 1)$. For each configuration, we report average values over 25 repetitions. For the sake of comparison, we compute the optimal utility with a complete centralized algorithm.

First, we evaluate if the accuracy of AD-IBMS depends on the way the residuals are minimized. Figure 3 shows the percentage relative error of the upper bound (left) and lower bound (right) obtained by AD-IBMS minimizing the maximum residual (L^1) and minimizing the average residual (L^∞). We only report the results on gamma distribution with link density 2 because the behavior pattern is very similar on the other classes of problems. Although theoretically incomparable, L^∞ is always superior to L^1 across all instances. Thus, in the following, AD-IBMS refers to the L^∞ case.

In our second experiment, we compare the bounds obtained by standard BMS, RN-BMS, IBMS, and AD-IBMS. Figure 4 (first two rows) shows the percentage relative error of the upper bound obtained by each algorithm. The behavior of all algorithms is very similar across all link densities and payoff distributions. As theoretically proved, AD-IBMS always computes the tightest upper bound. Figure 4 (last two rows) shows the percentage relative error of the lower bound obtained by the previous set of algorithms. In general, AD-IBMS obtains more accurate lower bounds than BMS and

[1] `http://teamcore.usc.edu/dcop`

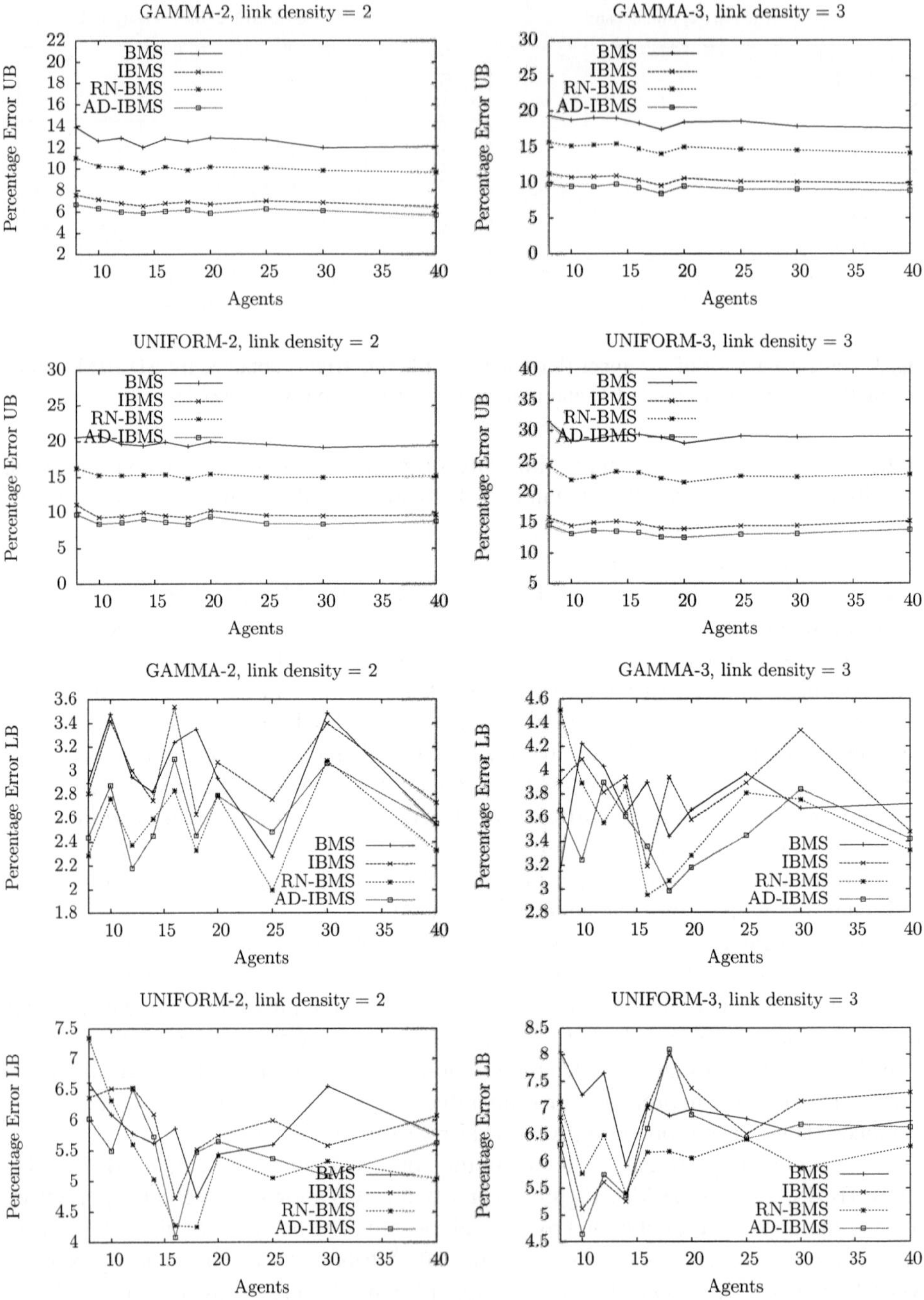

Fig. 4. Percentage error of the upper (first two rows) and lower bound (last two rows) obtained by BMS, IBMS, RN-BMS and AD-IBMS minimizing the maximum residual

IBMS. In some cases, AD-IBMS is also superior to RN-BMS. Note that this improvement is very relevant. On the one hand, recall that this class of algorithms are being developed for applications in which the accuracy of the solution is extremely important. On the other hand, since BMS, IBMS and specially RN-BMS are already very accurate on this type of problems, one cannot expect dramatic improvements. This improvement leads AD-IBMS to obtain very tight approximation ratios. Moreover, the computation time of AD-IBMS is always smaller than twice the computation time of IBMS.

The maximum 95% confidence interval for the gamma and uniform payoff distributions is smaller than 5.5 for the upper bound and smaller than 1 for the lower bound. This small confidence interval shows that 25 repetitions provide, for our experimental setting, a good sample size to assess the statistical significance of the results.

Finally, note that in [12] IBMS was shown to be superior to BMS and to two region-optimal criteria introduced in [13] and [14], which were shown to produce tighter approximation ratios than the approach in [6]. Moreover, in [11] BMS was shown to produce tighter approximation ratios than the approach in [2].

Discussion

The idea of exactly decomposing functions into smaller arity ones is far from new. In the field of probabilistic graphical models [7], where functions are (conditional) probability distributions, exact decomposition is a central issue and can be achieved when the probabilistic variables are (conditionally) independent. In the field of constraint satisfaction, where functions are relations, exact decomposition has been studied at least in [10]. The goal there was to compute the minimal network (i.e, transforming large arity relations into sets of equivalent binary ones). More recently, [5] have studied the power of decomposition in the context of combinatorial optimization graphical models. The goal there was to avoid large arity functions in order to boost local consistency enforcement.

Regarding approximate decomposition, the Mini-Bucket Elimination (MBE) algorithm [3] is the closest to ours. MBE is a dynamic-programming approximation algorithm that decomposes large functions into smaller arity ones in order to keep the space complexity manageable. The algorithm is very general and leaves several aspects undefined. [8] proposed an implementation that decomposes the functions while minimizing the error of the decomposition. One of the main differences wrt our approach is the system of linear equations solved.

Acknowledgment. This work was supported by project TIN2009-13591-C02-01.

References

1. Aji, S.M., McEliece, R.J.: The generalized distributive law. IEEE Trx. on Information Theory 46(2), 325–343 (2000)
2. Bowring, E., Pearce, J.P., Portway, C., Jain, M., Tambe, M.: On *k*-optimal distributed constraint optimization algorithms: new bounds and algorithms. In: Padgham, L., Parkes, D.C., Müller, J.P., Parsons, S. (eds.) AAMAS (2), pp. 607–614. IFAAMAS (2008)

3. Dechter, R., Rish, I.: Mini-buckets: A general scheme for bounded inference. J. of the ACM 50(2), 107–153 (2003)
4. Farinelli, A., Rogers, A., Petcu, A., Jennings, N.R.: Decentralised coordination of low-power embedded devices using the max-sum algorithm. In: AAMAS, pp. 639–646 (2008)
5. Favier, A., de Givry, S., Legarra, A., Schiex, T.: Pairwise decomposition for combinatorial optimization in graphical models. In: IJCAI, pp. 2126–2132 (2011)
6. Kiekintveld, C., Yin, Z., Kumar, A., Tambe, M.: Asynchronous algorithms for approximate distributed constraint optimization with quality bounds. In: AAMAS, pp. 133–140 (2010)
7. Koller, D., Friedman, N.: Probabilistic Graphical Models. The MIT Press (2009)
8. Larkin, D.: Approximate decomposition: A method for bounding and estimating probabilistic and deterministic queries. In: UAI, pp. 346–353 (2003)
9. Larrosa, J., Rollon, E.: Risk-neutral bounded max-sum for distributed constraint optimization. In: SAC, pp. 92–97 (2013)
10. Meiri, I., Pearl, J., Dechter, R.: Tree decomposition with applications to constraint processing. In: AAAI, pp. 10–16 (1990)
11. Rogers, A., Farinelli, A., Stranders, R., Jennings, N.R.: Bounded approximate decentralised coordination via the max-sum algorithm. Artif. Intell. 175(2), 730–759 (2011)
12. Rollon, E., Larrosa, J.: Improved bounded max-sum for distributed constraint optimization. In: Milano, M. (ed.) CP 2012. LNCS, vol. 7514, pp. 624–632. Springer, Heidelberg (2012)
13. Vinyals, M., Shieh, E., Cerquides, J., Rodriguez-Aguilar, J.A., Yin, Z., Tambe, M., Bowring, E.: Quality guarantees for region optimal dcop algorithms. In: AAMAS, pp. 133–140 (2011)
14. Vinyals, M., Shieh, E., Cerquides, J., Rodriguez-Aguilar, J.A., Yin, Z., Tambe, M., Bowring, E.: Reward-based region optimal quality guarantees. In: OPTMAS Workshop (2011)

Insights into Parallelism with Intensive Knowledge Sharing

Ashish Sabharwal[1] and Horst Samulowitz[2]

[1] Allen Institute for Artificial Intelligence (AI2), Seattle, WA 98103, USA
AshishS@allenai.org
[2] IBM Watson Research Center, Yorktown Heights, NY 10598, USA
samulowitz@us.ibm.com

Abstract. Novel search space splitting techniques have recently been successfully exploited to paralleliz Constraint Programming and Mixed Integer Programming solvers. We first show how universal hashing can be used to extend one such interesting approach to a generalized setting that goes beyond discrepancy-based search, while still retaining strong theoretical guarantees. We then explain that such static or explicit splitting approaches are not as effective in the context of parallel combinatorial search with intensive knowledge acquisition and sharing such as parallel SAT, where implicit splitting through clause sharing appears to dominate. Furthermore, we show that in a parallel setting there exists a surprising tradeoff between the well-known communication cost for knowledge sharing across multiple compute nodes and the so far neglected cost incurred by the computational load per node. We provide experimental evidence that one can successfully exploit this tradeoff and achieve reasonable speedups in parallel SAT solving beyond 16 cores.

1 Introduction

There have recently been several successful proposals for parallelizing combinatorial search and optimization, especially in the context of Constraint Programming (CP) and Mixed Integer Programming (MIP), such as by Régin et al. [25], Moisan et al. [22, 23], and Fischetti et al. [13]. A desirable strength of these and prior approaches such as the *guiding path* heuristic [31] is that they achieve parallelization without any communication between the compute cores. In one way or another, they split the underlying search space upfront or *statically* amongst the k available compute cores, which obviates the need for communication. Unlike search schemes based on global load-balancing or work-stealing [10, 21, 26, 27], these communication-less approaches compute a static assignment of subproblems (or of subtrees induced by a static assignment of search tree leaves) to each compute core.

We begin by discussing these static splitting, communication-less approaches and proposing a novel generalized static search space splitting scheme that, unlike some of these recent proposals, is not limited to a particular search strategy (e.g., discrepancy-based search) or a class of problem instances. This general

B. O'Sullivan (Ed.): CP 2014, LNCS 8656, pp. 655–671, 2014.

scheme uses randomly generated XOR or parity constraints and relies on their desirable universal hashing properties in order to achieve a dynamic balanced split of the search space amongst the compute cores. Prior works by Bordeaux et al. [8] and Plaza et al. [24] have alluded to XOR-based splitting, but only from an empirical perspective and without a formal analysis, especially with respect to the balanced effect of pruning *any* large-enough subtree of the whole space.

The formal correctness argument for our XOR-based splitting scheme highlights certain assumptions crucial to the success of some of the recent parallelization proposals. We explain why these assumptions, unfortunately, often fail in the context of search algorithms that dynamically learn from failures, such as CDCL (conflict-directed clause learning) solvers for propositional satisfiability (SAT) and Lazy Clause Generating CP solvers. In combinatorial search algorithms that support knowledge acquisition from failures or conflicts, intensively sharing that learned knowledge can provide an *implicit* way of splitting the search space explored by each core. As long as the solvers running at each core are sufficiently different, one of them will encounter a certain failed state before others and, by informing others of the reason of its failure, will indirectly prevent them from exploring this failed state as well as any search sub-space that fails for the same reason.

When it comes to knowledge sharing, it is common wisdom that communication across a network is costly. When designing distributed constraint solvers running on multiple machines and sharing information, it is deemed desirable to pack as many solvers on compute cores on individual machines as possible, so that inter-machine network latency does not hurt performance. This common wisdom stems from the undisputed fact that communicating across processes (or threads) on a single machine is much faster than communicating across different machines. While true, this reasoning ignores the *memory bandwidth* aspect, which can in principle have a negative impact on solvers. For example, it is folklore knowledge that running multiple copies of the MIP solver CPLEX [18] on the same machine notably degrades performances. The main reason is that more threads running memory intensive applications lead to more *cache misses* and *involuntary context switches*, both of which negatively impact performance. This happens even if one uses fewer threads than the number of available compute cores, because multiple cores tend to share at least the L3 cache. While cache performance of SAT solvers on a single compute node has been analyzed earlier [1, 32], its study in the context of multiple compute nodes and the resulting trade-offs remain unexplored.

As a motivating example, we consider the performance of the state-of-the-art parallel SAT solver Plingeling [6] across $k = 1, 2, 4, \ldots, 32$ cores of a single 32-core machine. Plingeling, like most current parallel SAT solvers, implements *implicit search space partitioning*. The results (geometric average runtime on our dataset, discussed later) are shown in Figure 1. We observe a sharp decline in performance when going from 16 to 32 cores. In this work, we ask: *Is this decline in performance caused mainly by duplication of work by cores given the lack of a static search space splitting mechanism, or by over-utilization of the memory bus?*

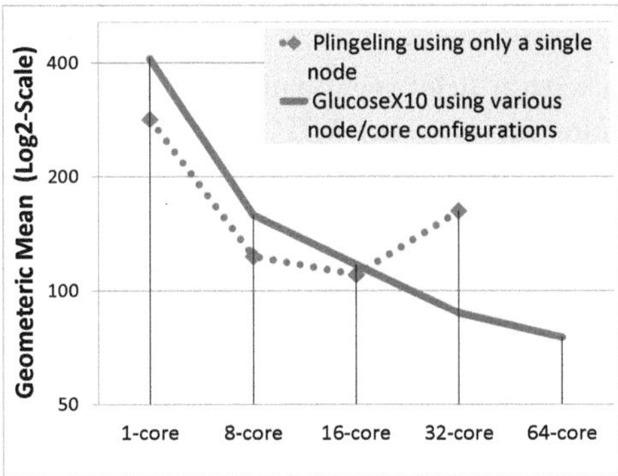

Fig. 1. Performance of Plingeling across $k = 1, 2, 4, \ldots, 32$ core on a single node compared to GlucoseX10 using various node-per-core configurations up to 64 cores

We find that reduced performance can be attributed to the latter and, more surprisingly, to such a large extent that one can, in fact, even successfully trade off the intra-node memory bandwidth bottleneck with the presumably high inter-node network communication cost.

Can this surprising trade off be successfully exploited in practice? To assess this, we first implement as a baseline two *static splitting* strategies that are promising in the context of SAT, one of them based on universal hashing through XOR constraints with strong theoretical guarantees as mentioned earlier. While a careful implementation of these strategies allows full knowledge sharing amongst the compute cores as well as the utilization of the highly effective dynamic search heuristics embedded in SAT solvers, we find that static splitting strategies have limited success in the context of SAT. However, exploiting our findings about the communication vs. node utilization trade-off, we show that a simple distributed variant of the Glucose 3.0 solver [3], created using the SatX10 framework [7] and performing *implicit search space splitting* by sharing the shortest 5% of the learned clauses, can continue to scale (i.e., have increasing speedups) on up to 64 cores when the copies of the solver are split carefully across multiple compute nodes. As shown in Figure 1 on our dataset, this simple distributed solver is more than competitive with Plingeling, the winner of the parallel track in the Application instance category of the 2013 SAT Competition [4]. Even though slower by as much as 1.5x when using 1 core, it clearly outperforms Plingeling as one moves to 32 or more cores, demonstrating that our insights can indeed be successfully exploited in practice.[1]

[1] We emphasize that this comparison is not meant to argue that GlucoseX10 is superior to Plingeling, but rather to illustrate that there exist unusual yet successful ways of making use of available compute resources.

2 Generalizing Static Search Space Splitting

We begin this section with a brief recapitulation of recently proposed successful parallelization strategies for CP and MIP search and optimization using what we will refer to as *static* search space splitting. With k compute cores, the search space is split upfront into k disjoint subspaces and then each core proceeds to search in the subspace it is responsible for. The novelty here is to achieve such a splitting in a way that requires no communication between the compute cores what-so-ever, which means communication cost never becomes a bottleneck as the number of compute cores is increased.

An interesting recent example of explicit search space splitting is the so-called *embarrassingly parallel search* (henceforth referred to as EPS) [25]. Suppose the problem instance has n variables, all of which are binary. The idea is to simply split the entire search space of size 2^n into $2^{\tilde{n}}$ disjoint subspaces, by fixing the value of some $\tilde{n}$ variables, $\tilde{n} < n$, in all possible ways. The resulting $2^{\tilde{n}}$ subproblems are then divided up equally amongst k compute cores. As long as $2^{\tilde{n}} \gg k$, by the law of large numbers, one expects the distribution of overall computation load across the compute cores to be roughly uniform. A related scheme [13] recently proposed in the context of MIP makes each core branch on the top $\tilde{n}$ variables and then choose a $1/k$ fraction of the resulting child nodes in a round robin fashion.

Another explicit parallelization technique—which in fact inspired some of our work—is the recent proposal by Moisan et al. [22] who study parallelization of a particular class of search heuristics in the context of CP, namely limited discrepancy-based search (LDS) [16]. We will refer to this technique as PLDS. It was shown that it is possible to assign the 2^n leaves of the entire search tree to the k available cores such that the leaves are visited by the k cores jointly in roughly the same order as a sequential LDS and the total number of search nodes visited by each core c (in the subtree induced by the leaves assigned to c) is no more than $(2^n/k)\log k$. Further, and more importantly, each of the cores is guaranteed to benefit roughly equally from any dynamic pruning of a subtree T of the entire search space by constraint propagators. All this is achieved by PLDS notably without any communication between the processors. This idea can also be extended to Depth-bounded Discrepancy based search (DDS) [23].

In the context of SAT, earlier work by Bordeaux et al. [8] suggested a completely distributed strategy where each core fixes a small number $\tilde{n}$ of variables (e.g., at random) without coordinating with other cores. Each core is allowed to independently select which $\tilde{n}$ variables it wants to fix and the values it wants to assign to them. To ensure completeness as well as to address the high likelihood of load imbalance in this context, the authors employed an interesting strategy of allowing the solver at each core to backtrack over the top $\tilde{n}$ variables—but only after it had proved its restricted sub-formula to be unsatisfiable. The authors also suggested more traditional search space splitting strategies that added new constraints to split the search space into disjoint subspaces, but, perhaps in part because of no knowledge sharing, they found the distributed variable fixing strategy to be the most effective. More recently, Heule et al. [17] and van der

Tak et al. [30] have proposed the use of more complex inference techniques than unit propagation to split the search space in the first phase of search and then solve the resulting sub-problems in parallel without knowledge sharing.

2.1 Generalized Splitting Using Universal Hashing

The theoretical balancing guarantees provided by the PLDS approach can in fact be extended to a more general setting for dynamic search heuristics that go well beyond LDS and DDS, including the conflict analysis driven heuristics employed by SAT solvers (e.g., VSIDS) as well as impact based search (IBS) in CP. We discuss here a novel way to achieve this in a search-independent and problem-independent manner, using parity or XOR constraints.

XOR constraints of length ℓ over binary variables x_i are constraints of the form $\sum_{i=1}^{\ell} x_i = p \mod 2$, where $p \in \{0, 1\}$ is referred to as the *parity* of the constraint. When generated at random by choosing the set of ℓ variables as well as the parity p uniformly amongst all choices, XOR constraints (of large enough length) act as a family of uniform hash functions, resulting in desirable search space splitting properties that have been exploited in theoretical computer science [5, 29] as well as in the design of practically efficient approaches for approximating the number of solutions of a combinatorial problem and for probabilistic inference [9, 11, 14]. In the interest of space, we refer the reader to any of these other works for formal properties of XOR constraints. In our context, they have precisely the key properties exploited by the PLDS approach. We discuss below how this observation can be exploited.

Consider a sequential search algorithm S. Given a problem instance I on n binary variables, let σ be the ordered sequence of the subset of the 2^n leaves (at depth n) of the underlying search tree $\mathcal{T}$ (of size 2^n) that S visits when operating on I. To create a parallel version S^k of S that utilizes k compute cores, where for simplicity of exposition we assume k is a perfect power of 2, we generate at random $\log k$ sets $X_j, 1 \leq j \leq \log k$, of ℓ variables each and restrict the search space explored by core $i, 1 \leq i \leq k$, to the sub-space determined by XOR constraints C_{ij} defined as $\sum_{x \in X_j} x = b_{ij} \mod 2$, where b_{ij} is the j-th bit of the $\log k$ bit binary representation b_i of the integer i. Let $C_i = \bigwedge_j C_{ij}$. The i-th solver S_i of S^k running on core i operates on the restricted problem instance $I \wedge C_i$. S_i follows the original leaf sequence σ, but simply skips the leaves that do not satisfy its XOR constraints C_i. For efficiency, if there is a subtree T of $\mathcal{T}$ none of whose leaves satisfy C_i, S_i must identify this fact and not waste time exploring T at all. This, in our case, is easily achieved as at least one constraint C_{ij^*} for some j^* must be violated by the partial truth assignment that defines the root node of T, which means that the XOR propagator for C_{ij^*} would make S_i fail immediately as soon as it reaches T.

This restriction scheme ensures that every pair $S_i, S_{i'}$ of solvers operates in disjoint subspaces of $\mathcal{T}$, and that the k cores together cover all of $\mathcal{T}$. Formally:

Proposition 1. *For constraints C_i as defined above and for $i \neq i'$, $C_i \wedge C_{i'} = \bot$ and $\bigvee_i C_i = \top$. Further, $(I \wedge C_i) \wedge (I \wedge C_{i'}) = \bot$ and $\bigvee_i (I \wedge C_i) = I$.*

Proof. Let j^* be a bit in which the $\log k$ bit binary representations of i and i' differ, i.e., $b_{ij^*} \neq b_{i'j^*}$. Then $(C_i \wedge C_{i'}) = (\bigwedge_j C_{ij}) \wedge (\bigwedge_j C_{ij}) \Rightarrow (C_{ij^*} \wedge C_{i'j^*}) \Rightarrow (\sum_{x \in X_{j^*}} x = b_{ij} \mod 2) \wedge (\sum_{x \in X_{j^*}} x = b_{i'j^*} \mod 2) \Rightarrow (b_{ij^*} = b_{i'j^*})$, which, by the choice of j^*, is never the case. Hence, $C_i \wedge C_{i'} = \bot$.

On the other hand, $\bigvee_i C_i = \bigvee_i \bigwedge_j C_{ij} = \bigwedge_j \bigvee_i C_{ij} = \bigwedge_j \bigvee_i (\sum_{x \in X_j} x = b_{ij} \mod 2)$. Since i spans the range $\{0, 1, \ldots, k-1\}$ and we are working with $\log k$ bit representations, for each j there must exist an i such that $b_{ij} = 0$ and an i' such that $b_{i'j} = 1$. Hence, $\bigvee_i (\sum_{x \in X_j} x = b_{ij} \mod 2) = \top$ for every j, implying $\bigwedge_j \bigvee_i (\sum_{x \in X_j} x = b_{ij} \mod 2) = \top$ and finishing the proof that $\bigvee_i C_i = \top$.

The remaining claims now follow from these results. First, $(I \wedge C_i) \wedge (I \wedge C_{i'}) = I \wedge (C_i \wedge C_{i'}) = \bot$. Next, $\bigvee_i (I \wedge C_i) = I \wedge \bigvee_i C_i = I \wedge \top = I$. □

We thus have a static partition of the search space amongst the k solvers. Moreover, the partition is balanced in the sense that each solver S_i gets precisely a $1/k$ fraction of the overall 2^n size search space $\mathcal{T}$ (irrespective of I). Most interestingly, the uniform hashing properties of XORs guarantee that, with large enough ℓ, with high probability, *every large-enough subspace of* $\mathcal{T}$ *has roughly equal representation in each of the* k *compute cores*, which act as k "buckets" for the underlying hash function. This is formalized in the following theorem, which notably is independent of the properties of the search algorithm S (e.g., using LDS or not) or of the problem instance I.

Theorem 1. *Let* S^k *be the parallel constraint solver for* k *cores as described above, operating on a problem instance* I *over* n *binary variables forming the* 2^n *size search space* $\mathcal{T}$. *For* $\ell = n/2, \epsilon \in (0,1), \delta > 0$, *and* $k \leq 2^{n/(2+\delta)}$,

1. *the entire subtree* $\mathcal{T}_i$ *of* $\mathcal{T}$ *induced by the leaves assigned to* S_i *contains no more than* $(2^{n+1}/k) \log k$ *internal and leaf nodes combined; and*
2. *for any arbitrarily chosen subtree* T *of* $\mathcal{T}$ *with* $L \geq k^{2+\delta}/\epsilon$ *leaves, with probability at least* $1 - \epsilon$ *over the choice of the random XOR constraints, the following holds: For any core* i, *the number of leaves of* T *that are assigned to* S_i *lies within* $\mu \cdot (1 \pm k^{-\delta/2})$ *where* $\mu = L/k$ *is the expected value.*

Proof. To argue that the first claim holds, we observe that $\mathcal{T}_i$ has exactly $2^n/k$ leaves, which implies that the number of internal nodes of $\mathcal{T}_i$ with two children must be exactly $2^n/k - 1$. Thus, the total number of nodes in $\mathcal{T}_i$ is higher precisely when it has more internal nodes with only one child. As can be seen from a tree rotation argument, the number of internal nodes with only one child is maximized when the leaves of $\mathcal{T}_i$, all at depth n of $\mathcal{T}$, are uniformly spread apart at distance k from each other. In this case, $\mathcal{T}_i$ contains *all* internal nodes of $\mathcal{T}$ up to depth $n - \log k$, for a total of $2^{n-\log k+1} - 1$ nodes, each of which is extended to depth n by a unique path of length $\log k$ containing nodes with one child. It follows that the number of nodes in $\mathcal{T}_i$ is upper bounded by $(2^{n-\log k+1} - 1) \log k < (2^{n+1}/k) \log k$ as desired.

In order to prove the second claim, we capitalize on the known fact that $\log k$ random XORs of length $n/2$ act as a universal family of hash functions on the 2^n leaves of $\mathcal{T}$, placing the leaves pairwise independently into k different "buckets", which correspond to our k cores. Let L_i be a random variable (with randomness over the choice of XORs) capturing the number of leaves of T assigned to core i. Pairwise independence of the assignment of leaves to cores implies that the variance $\mathrm{Var}(L_i)$ of L_i is no more than its expected value $\mathbb{E}(L_i) = L/k = \mu$. Applying first the Chebychev inequality and then the union bound,

$$\Pr\left[|L_i - \mu| \geq \mu k^{-\delta/2}\right] \leq \frac{\mathrm{Var}(L_i)}{\mu^2 k^{-\delta}} \leq \frac{\mu}{\mu^2 k^{-\delta}} \leq \frac{1}{k}\frac{k^{2+\delta}}{L} \leq \frac{\epsilon}{k}$$

$$\Rightarrow \Pr\left[\exists i.\ |L_i - \mu| \geq \mu k^{-\delta/2}\right] \leq \sum_{i=1}^{k} \Pr\left[|L_i - \mu| \geq \mu k^{-\delta/2}\right] \leq \epsilon$$

Taking the complement of this probability finishes the proof. □

We note that although the theorem is stated for $\ell = n/2$, Ermon et al. [12] have recently shown that certain desirable hashing properties still hold with XORs of length $\ell \ll n/2$, which are often much easier to propagate.

As an illustration of the result, suppose $\epsilon = 1/n, \delta = 1$, and T is *any* subtree of the search space with $L \geq nk^3$ leaves. Then the theorem states that with probability at least $1 - 1/n$ over the choice of random XORs, the number of leaves of T assigned to S_i will be within $L/k \cdot (1 \pm 1/\sqrt{k})$, i.e., very close to the ideal balancing value of L/k. As a consequence, each core will benefit roughly equally if a constraint propagator prunes T.

2.2 Implementing XOR-Based Splitting with Knowledge Sharing

While the above reasoning shows that adding randomly generated XOR constraints can, in principle, qualitatively provide the guarantees of PLDS in a much more generic setting, it is not obvious how best to implement this strategy. One of the parallelization suggestions by Bordeaux et al. [8] was in fact to add random XOR constraints by converting then into a CNF formulation. An XOR constraint with ℓ variables, however, requires adding $2^{\ell-1}$ clauses, which quickly becomes impractical as ℓ grows.[2] Thus, for practical reasons, we limit our evaluation to small values of ℓ. Bordeaux et al. did not find this to be effective, but their tests were performed without communication while we now test the approach in the presence of knowledge sharing, in the context of SAT. This, however, immediately raises an implementation challenge, which we discuss next.

Since each S_i operates on the original instance conjoined with new constraints that differ from core to core, clauses learned by one core may not be valid for other cores. In principle, one can label each learned clause C as sharable or not based on the information used to derive C. However, due to subtleties in the implementation of modern SAT solvers, it is insufficient to simply check whether the conflict analysis that led to the derivation of C involved one of the clauses encoding an XOR constraint. Other operations in the solver, such as propagation

[2] One could alternatively use $O(\ell)$ new variables to encode the XOR constraint, but this is known to slow down the search [14].

of unit literals learnt based on XOR constraints and clause base reductions, must also be appropriately altered to take the effect of the XOR constraints that differ from core to core.

An interesting alternative to explicitly adding new constraints $\mathcal{X}$ to the formula F is to alter the branching heuristic such that the solver *automatically* searches only in the assignment subspace that satisfies the constraints $\mathcal{X}$. The idea is to pre-compute all solutions to $\mathcal{X}$ over the set of variables appearing in $\mathcal{X}$ and add them to a solution pool $\mathcal{S}$. Note that each $\sigma \in \mathcal{S}$ is a partial assignment for F. Now one can iterate through these partial assignments $\sigma_1, \sigma_2, \ldots, \sigma_{|S|} \in \mathcal{S}$, moving from σ_i to σ_{i+1} as soon as the solver refutes the subtree under σ_i. Since we do not add XORs explicitly as constraints and instead just branch in a way that is consistent with XORs, the original formula must logically entail any learnt clause. Furthermore, since we enumerate the solution pool $\mathcal{S}$ exhaustively the approach is both sound and complete.

A notable advantage of this approach is that *all* clauses learnt by any core are valid for all other cores as well, and can thus be freely shared.[3] This obviates the need to implement mechanisms to decide which clauses are safe to share and which aren't. On the other hand, for the approach to be practical, the solution pool $\mathcal{S}$ must have a succinct representation that the solvers can exploit. For example, if $\mathcal{X}$ is the conjunction of $\log k$ XOR constraints of length ℓ each on disjoint sets of variables, then $|\mathcal{S}| = (2^{\ell-1})^{\log k} = k^{\ell-1}$, which can quickly become huge as k grows. To avoid this blow up, we instead use $\log k$ solution sub-pools of size $2^{\ell-1}$, one for each XOR constraint. The branching heuristic first fixes all variables from one sub-pool before moving on to the next sub-pool.

Our implementation includes a range of variations and extensions. For instance, we experimented with the setting where one core remains completely unaltered while all other cores employ XOR based branching. This strategy was motivated by the fact that altering branching decisions can have a tremendous impact on the search, and adding one unchanged solver to the pool of solvers would retain some of the original search pattern and the solver's flexibility. Furthermore, when branching according to the XOR variables at the top of the search tree we take propagation results directly into account so that an implication that falsifies the current XOR assignment causes us to directly move on to the next assignment. Since we have a sequence of XORs to branch on, this often allows us to skip entire sets of assignments.

2.3 Limits of Static Splitting

To our surprise, none of the approaches discussed above was truly effective in parallelizing SAT solvers. We tried several variations and parameter settings and will describe some representative results in Section 3, Table 1. As we discuss next, the reason might lie in the "rigid" static search space splitting interfering with the highly dynamic conflict-directed search performed by SAT solvers.

---

[3] Clauses learnt at core i could be filtered based on the alignment of XORs between cores i and j so that only the ones that have a chance of ever being triggered at core j are shared with it.

In general, the recent static splitting proposals discussed at the beginning of this section and which inspired our XOR-based splitting mechanism, unfortunately, have significant limitations in the context of search strategies that apply aggressive search space pruning through powerful propagators or that use information learned from failures to guide future search.

For example, the EPS approach and its variation for MIP implicitly assume that once the huge pool of $2^{\tilde{n}}$ subproblems has been created, one simply must resolve each subproblem independently and knowledge learned from solving one subproblem cannot significantly help inference on other subproblems. This is clearly not the case for CDCL solvers which are heavily guided by clauses learned from conflicts and are in fact able to quickly prune new subproblems based on experience from previously encountered subproblems. The same applies also to CP solvers that perform conflict analysis and lazy clause generation [28].

The PLDS and XOR-splitting approaches, on the other hand, do not address the rather common case where the number of nodes s visited by a sequential search algorithm is significantly less than 2^n. In fact, s being vastly smaller than 2^n on real-world instances of interest is precisely what allows us to tackle large NP-complete problems in a reasonable amount of time. For example, consider an infeasible instance where fixing well-selected $\tilde{n} \ll (n - \log k)$ variables lets constraint propagators already deduce infeasibility. With $k = 1024$, this condition holds whenever $\tilde{n}$ is much smaller than $n - 10$, which most SAT and CP solvers will guarantee fairly easily. While it does holds that each S_i will not process more than $(2^{n+1}/k) \log k$ nodes and subtree pruning will positively impact each S_i roughly equally, this is not a very useful guarantee as even a well-guided sequential search would take only $2^{\tilde{n}} \ll 2^n/k$ steps to begin with! In general, *uniform splitting of the naïve search space of size* 2^n *across* k *cores does* not *say much about speedups being close to* k *unless the underlying search algorithm works in a rather brute force manner.*

Further, PLDS has so far been demonstrated to be effective in a setting where the leaves of the search tree are much more costly to process than internal nodes. This, however, is usually not the case in most SAT, CP, and MIP applications, where node processing time often *decreases* as one goes deeper in the search tree.

Finally, forcefully fixing variables at the top of the search tree, as is the case in our XOR implementation and the random prefix method [8], seems to often interfere with dynamic branching choices that tend to make search more effective. This is especially true for SAT solvers, which heavily rely on recent conflicts for branching decisions, and variable activities maintained by them often change very rapidly. This behavior is also reflected in improved performance that we observed when using shorter XORs and random prefixes.

3 Implicit Splitting through Intensive Knowledge Sharing

An alternative to static splitting is *implicit* search space splitting, where the k compute cores start exploring the entire search space independently but dynamically communicate to each other—ideally in a succinct fashion—which subspaces

Table 1. Performance of various search space splitting and knowledge sharing approaches for SAT, using $k = 32$ cores

Search Space Splitting Approach	Length	Clause Sharing (%)	Runtime (sec)	#Solved (count)
Implicit	–	2	139	62
Implicit	–	5	119	63
Implicit	–	8	113	64
Static, 5 XORs	2	5	132	60
Static, 5 XORs	3	5	161	59
Static, Random Prefix	5	5	141	61
Static, Random Prefix	10	5	173	59

they have already explored. We argue that in combinatorial search algorithms that learn from failures, such as CDCL SAT solvers and CP solver employing Lazy Clause Generation (LCG), *implicit search space splitting achieved by sharing succinct reasons of failures can be much more effective than explicit search space splitting schemes* such as those discussed above. When such solvers encounter a "conflict" or a failed state, they analyze the reason of the failure in terms of constraint propagations and learn a clause (or a small set of clauses) that would prevent the solver from wasting time in other states that fail for a similar reason. By sharing such learned clauses with other cores, one can implicitly achieve search space splitting—as long as there is enough variation among the solvers executing on different cores that one of them encounters a particular failed state before others do. While simply having different random seeds at each core can make the search sufficiently different, in our experiments we vary also some of the solver's parameters, such as activity decay rate and restart frequency, across various cores.

When viewing knowledge sharing as implicit search space splitting, one must revisit the heuristics commonly used to decide how much to share and when. Currently employed heuristics, again guided by the common wisdom of communication latency being the main hurdle to avoid, tend to share perhaps too little information [19]. As the representative results in Table 1 (to be discussed shortly) demonstrate, it can be better to pay the price of additional communication for implicit search space splitting than use explicit splitting.

The results throughout the paper are for experiments on a cluster of 32-core compute nodes. Each node is a 4x8 3.8 GHz Power7 machine (CHRP IBM 9125-F2C), 4 MB cache per CPU, and 128 GB of RAM. The nodes are connected via a network that supports the PAMI message passing interface [20]. The evaluation is on SAT Race 2010 instances (all industrial/application category) restricted to the 73 that 1-core Glucose 3.0 could not solve within 10 seconds. The timeout (wall-clock) used was 5,000 seconds; our results remain qualitatively unchanged for smaller timeouts as well.

In Table 1, we compare (a) the implicit splitting approach with each core sharing the shortest 2%, 5%, and 8% of its learned clauses;[4,5] (b) the XOR-based explicit splitting approach with XORs of lengths 2 and 3; and (c) the "random prefix" approach of Bordeaux et al. [8] fixing 5 or 10 variables and enhanced with sharing the shortest 5% of the learned clauses. For the XOR-based splitting approach, longer XORs achieved more load balancing as expected but worsened per-core performance as XOR constraints do not unit propagate very well. Shorter XORs led to significantly increased load imbalance, but we could counter this and somewhat improve the overall performance by (i) restarting an early finishing core on the full problem instance (no XORs) or (ii) using one additional core in parallel on this full instance, in both cases continuing to share information. The table reports numbers for the latter, which, as we see, was still insufficient to outperform implicit search space splitting.

Implicit sharing was nearly always the best and sharing 5% yielded good, stable performance for various values of k. This is the setup we use henceforth.

4 The Communication-Utilization Tradeoff

Suppose we have two machines M_1 and M_2 with 32 compute cores each. Let S be a solver that we can run in parallel on one or both of these machines, and we can share information learned from failures across copies of S running on these machines. We are interested in minimizing the time the slowest of the parallel copies of S takes. When running $k \leq 32$ copies of S on a problem instance I, is it better to run k copies on M_1 or $k/2$ copies each on M_1 and M_2?

The answer, it turns out, is not that straightforward. It depends on k (in relation to the number of cores, 32, on each machine), the amount c of communication between each pair of copies of S, and the intensity of memory accesses (and the resulting cache hits and misses) performed by S. As one might expect, when $k \ll 32$ and c is small, either option results in about the same performance. As one might also expect, when $k \ll 32$ but c is substantial, network latency does play a significant role and it is better to run k copies on M_1 rather than communicate across machines. On the other hand, when $k \sim 32$ but the amount c of communication is very small, the latency of inter-machine communication does not play a significant role and it is actually better to run $k/2$ copies on two machines because constraint solvers often require high memory bandwidth and interfere (at the system level) more with each other the more copies of them are run in parallel on a single machine (we quantify this interference later in the

4 Sharing of the shortest x% learned clauses is implemented by maintaining a dynamic cutoff length L such that all learned clauses of length up to L are shared. L is adjusted periodically to achieve the x% sharing target. In principle, x itself could be adjusted based on the total number k of cores or properties of the problem instance.

5 We also experimented with more sophisticated sharing schemes such as those based on the LBD level of the learned clause [3], but our main findings remained unaffected. To avoid unintended consequences of complex sharing mechanisms, we report results on the simplest setting which still achieved state-of-the-art performance on 64 cores.

paper). Remarkably, we find that this trend continues to hold for $k \sim 32$ *even when the amount c of communication is high.* Specifically, even if each copy of solver S shares with every other $k-1$ copy as many as 5% of the clauses that it learns, it is better to run $k/2$ copies each on M_1 and M_2 and pay the price of inter-machine network communication than have k copies compete for memory bandwidth on M_1. E.g., it is often better to run 16 copies on 2 machines, and sometimes even better to run 8 copies on 4 machines, than run 32 copies on a single machine with 32 cores. The best allocation of cores across nodes is clearly dependent on the machine type. However, our empirical observations suggest that the configuration process does not have to be very fine grained.

We note that with $k = 32$ copies, sharing 5% of the learned clauses results in a significant amount of communication as each copy of S, when generating m learned clauses per second on its own, is expected to receive $m*(k-1)*5/100 = 1.55m$ clauses from other copies of S. In other words, each copy listens more to others than spend time doing its own deductions.

4.1 Time Profile of SAT Solvers

When working in a parallel setting where solvers run possibly on different machines and intensively share information, it is important to understand where these solvers spend most of their time and what role does the communication cost play. For CDCL SAT solvers, the total time T can be divided up as follows:

$$\begin{aligned} T &= T_{\text{conf}} + T_{\text{prop}} + T_{\text{comm}} + T_{\text{misc}} \\ &= \frac{N_{\text{conf}}}{R_{\text{conf}}} + \frac{N_{\text{prop}}}{R_{\text{prop}}} + \frac{N_{\text{comm}}}{R_{\text{comm}}} + T_{\text{misc}} \end{aligned}$$

Here T_{conf} represents the total time spent performing conflict analysis, N_{conf} represents the number of times conflict analysis is performed, and $R_{\text{conf}} = N_{\text{conf}}/T_{\text{conf}}$ is the associated rate, i.e., time per conflict analysis. The other symbols similarly correspond to the time, number, and rate of unit propagations, and of communication between cores. Since our main interest is in studying parallelization, we will compute and report all times in terms of wall-clock time.

There have been studies suggesting that SAT solvers spend a substantial amount of time performing unit propagation, and this observation has been used to help understand the behavior and limitations of parallel SAT solvers [15]. To make the numbers concrete in our setting and with our solver, we computed the values of T_{conf} and T_{prop} for Glucose 3.0 on our dataset. Figure 2 shows the result in relative terms as a percentage of T. To make relative numbers meaningful, the dataset was restricted to instances needing at least 30 seconds to solve. Clearly, unit propagation does dominate the time Glucose spends solving most instances, 71.01% on average[6] and never less than 40% on our dataset. Conflict analysis is the next most expensive operation. It can vary from 5% to 45%, with the geometric average being 12.45%. We will return to these observations later.

[6] All results are reported as a geometric mean, which is often more robust to high outliers than arithmetic mean. Results remain qualitatively unchanged either way.

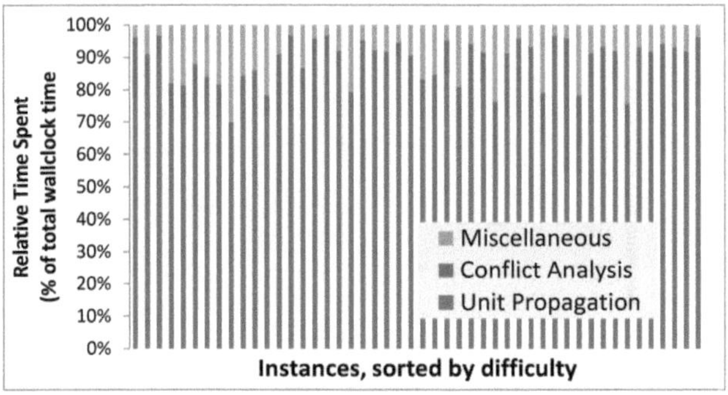

Fig. 2. Relative split of the total time spent by Glucose into T_{conf}, T_{prop}, and T_{misc}

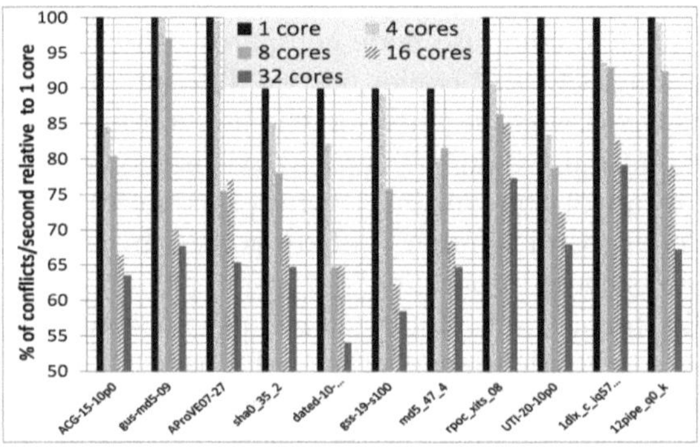

Fig. 3. Impact on R_{conf} when running $k = 4, 8, 16, 32$ independent copies of Glucose. Shown, for a selected set of instances, is R_{conf} as a percentage of the baseline R_{conf} obtained by running only a single copy of Glucose.

While it is folklore knowledge that running k independent processes on a single compute node with m cores can slow down each process as k approaches m, the large extent of slow down is rather surprising for SAT solvers. To quantify this effect, we ran k independent copies of the 1-core solver on a compute node with 32 cores. The plot in Figure 3, which shows the results on individual instances where Glucose on 1 core took between 600 and 1,800 seconds, demonstrates that the rate R_{conf} of conflict analysis is significantly reduced as k increases, by as much as 45% when increasing k from 8 independent copies to 32 copies. We obtained similar results for R_{prop} (omitted due to space limitation).

4.2 Communication Cost vs. Node Utilization

Now suppose we run k cooperating copies of a solver in a parallel setting with information sharing among the copies. How does the slow down seen when running

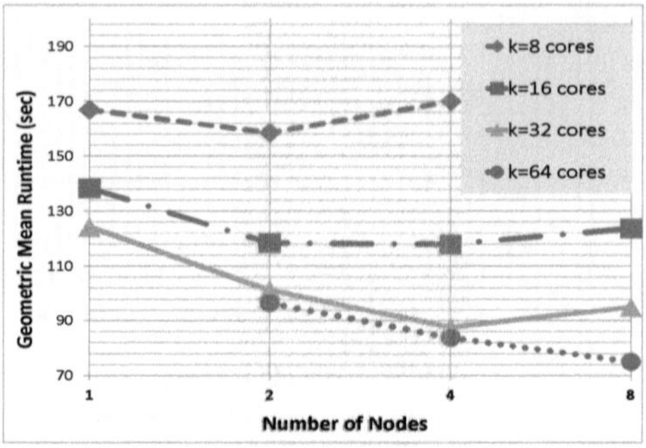

Fig. 4. Trade-off: communication cost vs. node utilization

$k \approx m$ copies on a single compute node compare with the communication cost incurred when running, say, $k/2$ copies on two different compute nodes?

Figure 4 depicts this trade-off, where on the horizontal axis we have the number N of compute nodes used, on the vertical axis is the performance (measured as the geometric mean of the runtimes across instances), and each curve corresponds to a different total number k of cores used in parallel. Each compute node thus runs k/N solvers in parallel.

While for small k (e.g., $k = 8$) performance, as expected, drops when increasing N due to the communication overhead, for larger values of k, increasing N surprisingly leads to significantly better performance. For example, the data shows that when wanting to run $k = 32$ solvers in parallel, it is substantially better to run only $k/N = 16$ or even 8 solvers per compute node than to fully utilize the node by running $k = 32$ solvers on it. Based on these findings, we used the following configurations of N nodes with k/N cores per node (for a total of $N \times k/N = k$ cores) produce data for the GlucoseX10 curve in Figure 1: $1 \times 1 = 1, 2 \times 4 = 8, 2 \times 8 = 16, 4 \times 8 = 32$, and $8 \times 8 = 64$.[7] The figure shows that GlucoseX10, while worse than Plingeling for $k = 1$, continues to scale reasonably well even up to $k = 64$ and clearly outperforms Plingeling for $k > 16$.

To understand this behavior, let us consider the time profile of SAT solvers discussed earlier. By increasing N and keeping everything else unchanged, we must clearly decrease the communication rate R_{comm}. Assuming T_{misc} is not affected significantly and neither are the total numbers N_{conf} and N_{prop} of conflicts and unit propagations, respectively, the only way the overall time T can decrease, is for one or both of the rates R_{conf} and R_{prop} to significantly increase. This, indeed, is the case. Across all problem instances N_{conf} is not systematically altered one way or another by changing N from 1 to 4 with $k = 32$ cores in total. However, as expected, R_{conf} increases quite consistently across all instances and

[7] While the specific numbers reported here are based on evaluation on our compute hardware, our qualitative findings are likely to be applicable to other compute systems as well. Simple experimentation can be used to identify the most effective split of k cores of a parallel solver across multiple compute nodes.

the geometric mean of R_{conf} is roughly 20% larger when using 4 nodes compared to 1 node. A similar trend holds also for N_{prop} and R_{prop}.

These results highlight the rather surprising tradeoff between the utilization of each compute node and the communication cost across multiple nodes. They also show that this tradeoff can be fruitfully exploited.

5 Concluding Remarks

Limited intra-node memory bandwidth has a substantial impact on the performance of today's combinatorial search methods when several such solvers, or their parallel versions, operate on a single machine. One may naïvely consider this impact smaller than the usually high latency of communication across a network. Our results, however, demonstrate that one can significantly gain in performance by distributing a parallel solver across multiple machines *even when the solver employs extensive knowledge acquisition and sharing.* For example, a SAT solver learning around 1,000 clauses per second and sharing 5% of what it learns with other 31 solvers in turn receives 1,000 × 31 × 0.05, or over 1,500 clauses per second. Even then, as our results show, distributing 32 cores across 4 or 8 compute nodes pays off. A testament to the practical importance of this insight is that one is able to significantly outperform the state of the art in parallel SAT solving on 32 or more cores.

This is, of course, not a complete solution to effective parallelization of SAT solvers or CP solvers with lazy clause generation. By using only a subset of the available cores on a machine and letting others idle, we are essentially wasting resources. However, our results suggest that, rather than allocating idle cores to other solvers running in parallel, one should consider other uses of the idle cores. In particular, it may be worthwhile revisiting operations such as unit propagation and conflict analysis, which often take up nearly 90% of the solver's time and must be performed in any case. Our results motivate parallel propagation and conflict analysis schemes as was also suggested earlier by Hamadi and Wintersteiger [15]. In this context, it is important to note that while unit propagation is P-complete [15], the only known theoretical consequence of this completeness observation is that a q-step unit propagation sequence cannot be parallelized to $\log^i q$ parallel steps for any constant i. However, this does not preclude reductions by large constant factors or even asymptotic reductions to, say, $\sqrt{q}$ parallel steps. This may be a more promising use of idle cores than running additional copies of the solver as this is likely to have a more coherent memory footprint across cores and thus be more amenable to better cache performance.

As the total number of compute cores grows, the communication cost must eventually become dominant. It, therefore, remains important to consider more sophisticated knowledge sharing schemes [2] or more parallelizable proofs [19], both of which are promising research directions orthogonal to our findings. Any improvements along these lines will affect our approach positively as well and help it scale to more compute cores.

References

[1] Aigner, M., Biere, A., Kirsch, C.M., Niemetz, A., Preiner, M.: Analysis of portfolio-style parallel SAT solving on current multi-core architectures. In: POS-2013: Intl. Workshop on Pragmatics of SAT, Helsinki, Finland (2013)

[2] Audemard, G., Hoessen, B., Jabbour, S., Lagniez, J.-M., Piette, C.: Revisiting clause exchange in parallel sat solving. In: Cimatti, A., Sebastiani, R. (eds.) SAT 2012. LNCS, vol. 7317, pp. 200–213. Springer, Heidelberg (2012)

[3] Audemard, G., Simon, L.: Predicting learnt clauses quality in modern SAT solvers. In: 21st IJCAI, Pasadena, CA, pp. 399–404 (July 2009)

[4] Balint, A., Belov, A., Heule, M., Järvisalo, M.: SAT competition (2013)

[5] Bellare, M., Goldreich, O., Petrank, E.: Uniform generation of NP-witnesses using an NP-oracle. Information and Computation 163(2), 510–526 (2000)

[6] Biere, A.: Lingeling, Plingeling and Treengeling entering the SAT Competition 2013. In: Proc. of SAT Competition 2013, Univ. of Helsinki. Dept. of Computer Science Series of Publications B, vol. B-2013-1, pp. 51–52 (2013)

[7] Bloom, B., Grove, D., Herta, B., Sabharwal, A., Samulowitz, H., Saraswat, V.: SatX10: A scalable plug&play parallel SAT framework - (tool presentation). In: Cimatti, A., Sebastiani, R. (eds.) SAT 2012. LNCS, vol. 7317, pp. 463–468. Springer, Heidelberg (2012)

[8] Bordeaux, L., Hamadi, Y., Samulowitz, H.: Experiments with massively parallel constraint solving. In: 21st IJCAI, pp. 443–448 (2009)

[9] Chakraborty, S., Meel, K.S., Vardi, M.Y.: A scalable and nearly uniform generator of SAT witnesses. In: Sharygina, N., Veith, H. (eds.) CAV 2013. LNCS, vol. 8044, pp. 608–623. Springer, Heidelberg (2013)

[10] Chu, G., Schulte, C., Stuckey, P.J.: Confidence-based work stealing in parallel constraint programming. In: Gent, I.P. (ed.) CP 2009. LNCS, vol. 5732, pp. 226–241. Springer, Heidelberg (2009)

[11] Ermon, S., Gomes, C., Sabharwal, A., Selman, B.: Taming the curse of dimensionality: Discrete integration by hashing and optimization. In: 30th ICML, pp. 334–342 (June 2013)

[12] Ermon, S., Gomes, C., Sabharwal, A., Selman, B.: Low-density parity constraints for hashing-based discrete integration. In: 31st ICML (2014)

[13] Fischetti, M., Monaci, M., Salvagnin, D.: Self-splitting of workload in parallel computation (May 2014)

[14] Gomes, C.P., Sabharwal, A., Selman, B.: Model counting: A new strategy for obtaining good bounds. In: 21st AAAI, Boston, MA, pp. 54–61 (July 2006)

[15] Hamadi, Y., Wintersteiger, C.M.: Seven challenges in parallel sat solving. In: 26th AAAI (2012)

[16] Harvey, W.D., Ginsberg, M.L.: Limited discrepancy search. In: 14th IJCAI, Montreal, Canada, pp. 607–615 (August 1995)

[17] Heule, M., Kullmann, O., Wieringa, S., Biere, A.: Cube and conquer: Guiding cdcl sat solvers by lookaheads. In: Eder, K., Lourenço, J., Shehory, O. (eds.) HVC 2011. LNCS, vol. 7261, pp. 50–65. Springer, Heidelberg (2012)

[18] IBM ILOG. IBM ILOG CPLEX Optimization Studio 12.6 (2013)

[19] Katsirelos, G., Sabharwal, A., Samulowitz, H., Simon, L.: Resolution and parallelizability: Barriers to the efficient parallelization of sat solvers. In: 27th AAAI (2013)

[20] Kumar, S., Mamidala, A.R., Faraj, D., Smith, B., Blocksome, M., Cernohous, B., Miller, D., Parker, J., Ratterman, J., Heidelberger, P., Chen, D., Steinmacher-Burrow, B.: PAMI: A parallel active message interface for the Blue Gene/Q supercomputer. In: IPDPS-2012: 26th IEEE International Parallel & Distributed Processing Symposium, pp. 763–773 (2012)
[21] Michel, L., See, A., Hentenryck, P.V.: Transparent parallelization of constraint programming. INFORMS Journal on Computing 21(3), 363–382 (2009)
[22] Moisan, T., Gaudreault, J., Quimper, C.-G.: Parallel discrepancy-based search. In: Schulte, C. (ed.) CP 2013. LNCS, vol. 8124, pp. 30–46. Springer, Heidelberg (2013)
[23] Moisan, T., Quimper, C.-G., Gaudreault, J.: Parallel depth-bounded discrepancy search. In: Simonis, H. (ed.) CPAIOR 2014. LNCS, vol. 8451, pp. 377–393. Springer, Heidelberg (2014)
[24] Plaza, S.M., Markov, I.L., Bertacco, V.: Low-latency sat solving on multicore processors with priority scheduling and xor partitioning. In: International Workshop on Logic Synthesis (IWLS) (2008)
[25] Régin, J.-C., Rezgui, M., Malapert, A.: Embarrassingly parallel search. In: Schulte, C. (ed.) CP 2013. LNCS, vol. 8124, pp. 596–610. Springer, Heidelberg (2013)
[26] Rolf, C.C., Kuchcinski, K.: Load-balancing methods for parallel and distributed constraint solving. In: IEEE Conf. on Cluster Computing, pp. 304–309 (September 2008)
[27] Shinano, Y., Achterberg, T., Berthold, T., Heinz, S., Koch, T.: ParaSCIP – a parallel extension of SCIP. In: Competence in High Performance Computing 2010, pp. 135–148. Springer (February 2012)
[28] Stuckey, P.J.: Lazy clause generation: Combining the power of SAT and CP (and MIP?) solving. In: Lodi, A., Milano, M., Toth, P. (eds.) CPAIOR 2010. LNCS, vol. 6140, pp. 5–9. Springer, Heidelberg (2010)
[29] Valiant, L.G., Vazirani, V.V.: NP is as easy as detecting unique solutions. Theoretical Comput. Sci. 47(3), 85–93 (1986)
[30] van der Tak, P., Heule, M., Biere, A.: Concurrent cube-and-conquer - (poster presentation). In: Cimatti, A., Sebastiani, R. (eds.) SAT 2012. LNCS, vol. 7317, pp. 475–476. Springer, Heidelberg (2012)
[31] Zhang, H., Bonacina, M.P., Hsiang, J.: PSATO: a distributed propositional prover and its application to quasigroup problems. J. Symb. Comput. 21(4), 543–560 (1996)
[32] Zhang, L., Malik, S.: Cache performance of SAT solvers: a case study for efficient implementation of algorithms. In: Giunchiglia, E., Tacchella, A. (eds.) SAT 2003. LNCS, vol. 2919, pp. 287–298. Springer, Heidelberg (2004)

The Non-overlapping Constraint between Objects Described by Non-linear Inequalities

Ignacio Salas[1], Gilles Chabert[1], and Alexandre Goldsztejn[2]

[1] Mines de Nantes, LINA UMR 6241, France
{ignacio.salas,gilles.chabert}@mines-nantes.com
[2] CNRS, LINA UMR 6241, France
alexandre.goldsztejn@univ-nantes.fr

Abstract. Packing 2D objects in a limited space is an ubiquitous problem with many academic and industrial variants. In any case, solving this problem requires the ability to determine where a first object can be placed so that it does not intersect a second, previously placed, object. This subproblem is called the non-overlapping constraint. The complexity of this non-overlapping constraint depends on the type of objects considered. It is simple in the case of rectangles. It has also been studied in the case of polygons. This paper proposes a numerical approach for the wide class of objects described by non-linear inequalities. Our goal here is to *calculate* the non-overlapping constraint, that is, to describe the set of all positions and orientations that can be assigned to the first object so that intersection with the second one is empty. This is done using a dedicated branch & prune approach. We first show that the non-overlapping constraint can be cast into a Minkowski sum, even if we take into account orientation. We derive from this an *inner contractor*, that is, an operator that removes from the current domain a subset of positions and orientations that necessarily violate the non-overlapping constraint. This inner contractor is then embedded in a *sweeping* loop, a pruning technique that was only used with discrete domains so far. We finally come up with a branch & prune algorithm that outperforms RSOLVER, a generic state-of-the-art solver for continuous quantified constraints.

1 Introduction

The goal of this article is to calculate the set of all positions and orientations that can be given to an object so that it does not overlap a second one (see the left graphic in Figure 1, which shows the simpler case where no orientation is considered). We address the general case of objects described by nonlinear inequalities. Calculating this set is a key task for solving packing problems, that consists in placing a set of objects in a bounded space so that they do not overlap pairwise.

In this introduction, we first define *objects* in our context and give a precise statement of the non-overlapping constraint. Then, we explicit the type of objects we consider and explain what we mean by *calculating* a set. We finally mention related works.

B. O'Sullivan (Ed.): CP 2014, LNCS 8656, pp. 672–687, 2014.

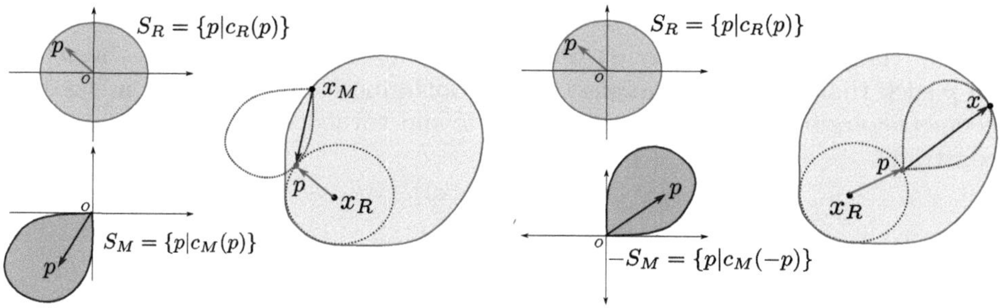

Fig. 1. Non-overlapping constraint. *Left:* two objects S_R and S_M, the region in red represents the set of all positions x_M for S_M that violate the non-overlapping constraint. *Right:* two objects S_R and $-S_M$ and their Minkowski sum, which coincides with the overlapping constraint (the negation of the non-overlapping constraint).

In Section 2, we show that our problem can be cast into the calculation of a Minkowski sum. Based on this observation, we propose a branch & bound algorithm in Section 3. Experimental results are shown in Section 4 and a conclusion follows.

1.1 Object Definition

For clarity, let us first assume that the orientation is fixed, that is, objects can only be translated.

Translating an object means fixing the position of a particular point that we call the *origin*. This origin point can be arbitrarily chosen. For instance, in rectangle packing, the origin of a rectangle can be a vertex or its center point. Once this convention for the origin is made, the shape of the object is just a regular constraint. To illustrate this, let us consider again rectangle packing. If the origin is the lower-left corner, then a rectangle of dimensions l_1 and l_2 is the set of all $p \in \mathbb{R}^2$ satisfying

$$c(p) \iff 0 \leq p_1 \leq l_1 \wedge 0 \leq p_2 \leq l_2. \tag{1}$$

Alternatively,

$$c(p) \iff -\frac{l_1}{2} \leq p_1 \leq \frac{l_1}{2} \wedge -\frac{l_2}{2} \leq p_2 \leq \frac{l_2}{2}. \tag{2}$$

defines a rectangle with the center point as origin, which, of course, is just a shift of the previous constraint. So, the shape of an object can be expressed as a constraint, the latter containing an implicit convention for the origin. As a last example, a circle of radius r is the set of all $p \in \mathbb{R}^2$ such that

$$c(p) \iff \|p\| \leq r \tag{3}$$

the origin being, in this case, the center of the circle.

This definition of $c(p)$ corresponds to an object with no translation nor rotation. The general constraint corresponding to an object translated by x and rotated by α is easily obtained as follows. In the case of translation only, the part of the plane covered by an object placed at some position x is the set of all p such that $c(p-x)$ is satisfied. Let us introduce orientation. By a classical geometric argument, an object placed at x and turned by some angle α is the constraint

$$c(R_{-\alpha}(p-x)) \tag{4}$$

where R_α is the rotation matrix with angle α:

$$R_\alpha = \begin{pmatrix} \cos(\alpha) & -\sin(\alpha) \\ \sin(\alpha) & \cos(\alpha) \end{pmatrix}. \tag{5}$$

Equivalently, we can say that $c(p) \equiv c(R_0(p-0))$ represents the object placed at 0 and rotated by the angle 0. To conclude:

- *an object* is a constraint on the plane (i.e., with two variables),
- *placing* an object means fixing the coordinate of its implicit origin,
- *orienting* an object means fixing the angle of the rotation around its implicit origin.

In this paper we consider objects described by nonlinear inequalities $c(p) \iff f(p) \leq 0$. E.g., a circle of radius 1 which origin is the center point is the set of all $p \in \mathbb{R}^2$ such that $f(p) \leq 0$ with $f : p \mapsto \|p\| - 1$. For the clarity of presentation, we will assume each object to be described by a single inequality, but our results can be easily extended to the more general case of objects described by disjunctions of conjunctions of inequalities. This is possible by introducing min and max operators. For instance, $(f_1(x) \leq 0 \vee f_2(x) \leq 0)$ is equivalent to $\max\{f_1(x), f_2(x)\} \leq 0$. Differentiability is not required by our algorithm. In fact, there is also no assumption on the functions involved, except that they are defined by mathematical expressions based on usual operators ($+$, $\times$, $\sqrt{\cdot}$, exp, etc.). In particular, there is no convexity assumption on the input objects.

1.2 The Non-overlapping Constraint

We can now focus on the non-overlapping constraint. The non-overlapping constraint involves two objects, one being fixed, the other representing the unknowns of the problem. For this reason, we will call "reference object" the first one and denote by c_R its describing constraint. The second one will be called "moving object" and its constraint denoted by c_M.

We will show at the end of this section that the general case, where both objects are translated and/or rotated, can actually be obtained from the simpler case where the reference object has no transformation. Intuitively, the frame where the overlapping constraint is stated can be centered on the reference object and aligned with its orientation, albeit the exact formula is not so trivial.

So, we shall only introduce in our definition below the position x_M and the rotation angle α_M of the moving object. The non-overlapping constraint is the negation of the overlapping constraint that can be stated as follows.

Definition 1 (Overlapping Constraint)
Given two constraints c_R *and* c_M*, a vector* $x_R \in \mathbb{R}^2$ *and* $\alpha_R \in [0, 2\pi]$*:*

$$\text{overlap}_{(c_R,c_M)}(x_M, \alpha_M) \iff \exists p \in \mathbb{R}^2, c_R(p) \wedge c_M\big(R_{-\alpha_M}(p - x_M)\big). \quad (6)$$

In the case of translation only, this simplifies to

$$\text{overlap}_{(c_R,c_M)}(x_M) \iff \exists p \in \mathbb{R}^2 c_R(p) \wedge c_M(p - x_M). \quad (7)$$

Our goal is to *calculate* the overlapping constraint. By *calculating*, we mean here that an explicit (numerical but verified) representation of the solution set $\mathcal{S}' := \{(x_M, \alpha_M),\ \text{overlap}_{(c_R,c_M)}(x_M, \alpha_M)\}$ (or $\mathcal{S} := \{x_M,\ \text{overlap}_{(c_R,c_M)}(x_M)\}$ in the case of translation only) has to be returned by our algorithm.

We show now that the overlapping constraint in the case where the reference object is given a position x_R and an orientation α_R can be tested using $\mathcal{S}'$ (and hence using its explicit representation). More precisely:

Proposition 1. *The moving object with position* x_M *and orientation* α_M *overlaps the reference object with position* x_R *and orientation* α_R *iff*

$$(R_{-\alpha_R}(x_M - x_R), \alpha_M - \alpha_R) \in S' \quad (8)$$

Proof.. By definition, (x_M, α_M) satisfies the overlapping constraint with the reference object at position x_R and orientation α_R iff exists $\exists p \in \mathbb{R}^2$ such that $c_R(R_{-\alpha_R}(p - x_R))$ and $c_M(R_{-\alpha_M}(p - x_M))$. This is equivalent to $\exists q \in \mathbb{R}^2$, namely

$$q = R_{-\alpha_R}(p - x_R) \iff p = R_{\alpha_R} q + x_R, \quad (9)$$

such that $c_R(q)$ and

$$c_M(R_{-\alpha_M}(R_{\alpha_R} q + x_R - x_M)) \iff c_M(R_{-\alpha_M+\alpha_R}(q + R_{-\alpha_R}(x_R - x_M))). \quad (10)$$

This is equivalent to (8). ▲

Remark 1. The non-overlapping constraint is usually considered for packing applications. Its description is easily obtained from the one of the non-overlapping constraint. The former is preferred here because it simplifies the constraints expressions and its link with the Minkowski sum presented in Section 2.

1.3 Intervals, Boxes and Paving

The representation we use is called a *paving*. This representation is a natural choice in the context of constraint programming where the large majority of algorithms dedicated to continuous variables, if not to say all, assume domains of variables to be intervals (see, e.g., [BG06, JKDW01]). In the following definition, we call *box* a Cartesian product of d intervals where d is either 2 in the case of $\mathcal{S}$ or 3 in the case of $\mathcal{S}'$.

Definition 2 (Paving). *A paving of a set* $\mathcal{S} \subset \mathbb{R}^d$ *is a triplet* $(\mathcal{I}, \mathcal{B}, \mathcal{O})$ *where* $\mathcal{I}$ *(for "inside"),* $\mathcal{O}$ *(for "outside") and* $\mathcal{B}$ *(for "boundary") are three sets of boxes verifying*

$$\cup\mathcal{I} \subset \mathcal{S}, \quad (\cup\mathcal{O}) \cap \mathcal{S} = \emptyset \quad \textit{and} \quad \cup(\mathcal{B} \cup \mathcal{I} \cup \mathcal{O}) = \mathbb{R}^d. \quad (11)$$

An example of paving is shown in Figure 2.

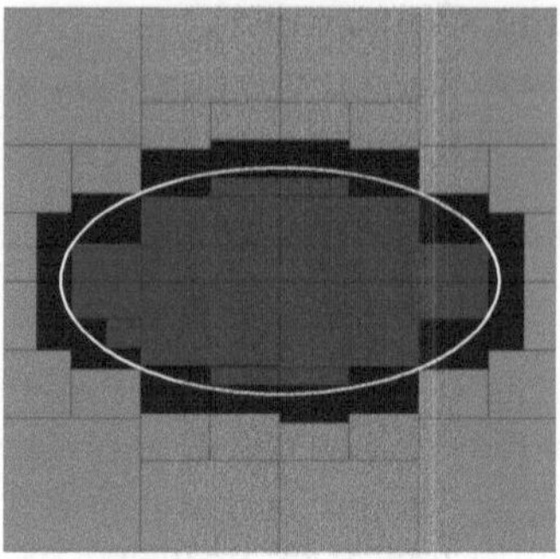

Fig. 2. Paving of an ellipsis. The interior red boxes belongs to $\mathcal{I}$, the unknown blue boxes in the boundary belongs to $\mathcal{B}$ and the green outer boxes belongs to $\mathcal{O}$.

1.4 Contribution and Related Works

This paper proposes an algorithm that computes a paving of the overlapping constraint. One contribution is on the modeling aspect of this problem: we show that the overlapping constraint can be expressed as a Minkowski sum. On the one hand, this generalizes the approach and simplifies the description of the algorithms. On the other hand, it allows handling the simple translation case and the more involved translation and rotation case homogeneously by casting the rotation to a translation into an augmented space (see Proposition 2). The other contribution is algorithmic. We propose an original *inner* contractor for this problem, that is, an operator that identifies parts of the solution set. This operator implements a *sweeping loop* and exploits the properties of the Minkowski sum. The second operator is an outer rejection test based on classical constraint propagation. Both are interleaved in a branch and prune algorithm that computes the desired paving.

From another point of view, definition 1 means that our problem falls into the category of existentially-quantified constraints. A state-of-the-art algorithm for calculating a paving with existentially-quantified inequalities is given in [Rat06] and is implemented in the RSOLVER tool [Rat] (Rsolver implements a general algorithm, somehow a numerical version of CAD, for solving quantified constraints).

General techniques [GJ06, IGJ12] for quantified *equality* constraints could also be used: either by adapting [GJ06] to compute an over approximation of the boundary of the overlapping constraint, or using a necessary condition for the boundary of the overlapping constraint expressed as an under-constrained system of equations and using [IGJ12]. However, the overlapping constraint naturally involves inequality constraints and using costly techniques dedicated to equality constraints turns out to be counterproductive.

Finally, we shall mention that an exact formula for the overlapping set $\mathcal{S}$ has been given in [BGT01] in the case where objects are polytopes. The set, in this case, is the convex hull of the points obtained by summing one vertex of the first polytope to one vertex of the second one.

2 Overlapping as a Minkowski Sum

In this section, we show that the overlapping constraint can be reformulated as a Minkowski sum. This relation underlies our branch & bound solver, that will be presented further. Let us first recall the definition of the Minkowski sum of two sets:

Definition 3 (Minkowski sum)
Given two equi-dimensional sets $S_1, S_2 \subseteq \mathbb{R}^d$, the Minkowski sum is

$$S_1 + S_2 = \{x_1 + x_2 \in \mathbb{R}^d : \ x_1 \in S_1, \ x_2 \in S_2\}. \tag{12}$$

The Minkowski difference is defined accordingly by:

$$S_1 - S_2 = \{x_1 - x_2 \in \mathbb{R}^d : \ x_1 \in S_1, \ x_2 \in S_2\}. \tag{13}$$

The right graphic of Figure 1 (page 673) shows an example of two sets and their Minkowski sum.

Considering S_1 as a constraint c_1 (i.e., $x \in S_1 \iff c_1(x)$) and, similarly, S_2 as c_2, we have, equivalently:

$$S_1 + S_2 = \{x \in \mathbb{R}^d : \ \exists p \in \mathbb{R}^d, \ c_1(p) \wedge c_2(x - p)\}. \tag{14}$$

where d is the number of variables in the constraints. Comparing (14) and (7), we immediately see that $\mathcal{S} = \mathcal{S}_R - \mathcal{S}_M$, i.e. the overlapping constraint can be represented as a Minkowski sum in the case of translation only.

We show now that $\mathcal{S}'$ is also a Minkowski sum, a less trivial result. To this end, we embed the moving object $\mathcal{S}_M \subseteq \mathbb{R}^2$ into $\mathbb{R}^3$ encoding its rotation within the additional dimension:

$$\mathcal{S}'_M := \{(v, \beta) : c_M(R_\beta \, v)\} = \{(v, \beta) : R_\beta \, v \in \mathcal{S}_M\}. \tag{15}$$

Now, the following proposition states that the overlapping constraint with rotation $\mathcal{S}'$ can be written as a Minkowski difference of two such "augmented" sets (see Figure 3).

Proposition 2

$$\mathcal{S}' = \mathcal{S}_R \times \{0\} - \mathcal{S}'_M. \tag{16}$$

Proof. By definition, $(x_M, \alpha_M) \in \mathcal{S}'$ holds if and only if $\exists p \in \mathbb{R}^2$ such that $c_R(p)$ and $c_M(R_{-\alpha_M}(p - x_M))$. Equivalently, there exists $u_R \in \mathcal{S}_R$ and $u_M \in \mathcal{S}_M$ such that $u_R = p$ and $u_M = R_{-\alpha_M}(u_R - x_M) \iff x_M = u_R - R_{\alpha_M} u_M$. Finally, the vector (x_M, α_M) is proved to be the sum of $(u_R, 0) \in \mathcal{S}_R \times \{0\}$ and $(R_{\alpha_M} u_M, \alpha_M) \in \mathcal{S}'_M$. ▲

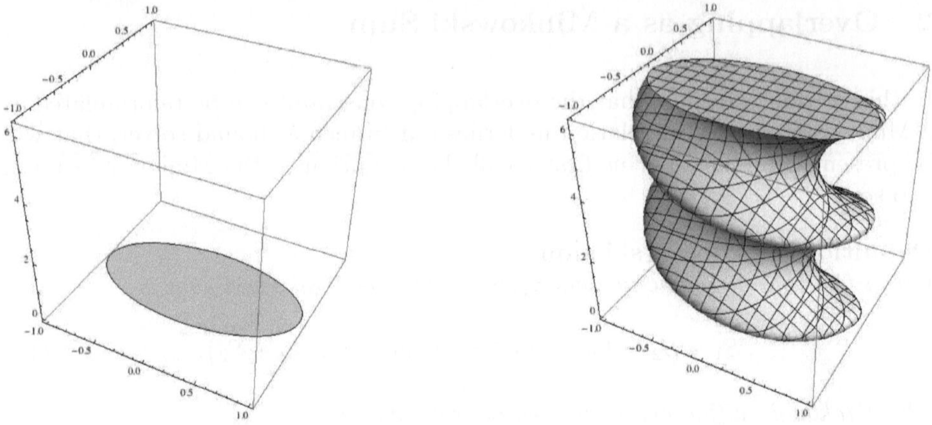

Fig. 3. Sets whose Minkowski difference gives the overlapping constraint of two ellipsis, when rotation is taken into account. *Left:* the augmented reference ellipsis, with rotation coordinate set to 0. *Right*: the augmented moving ellipsis S'_M.

3 Algorithm

From now on, our goal is to calculate a paving $(\mathcal{I}, \mathcal{B}, \mathcal{O})$ of the sum $\mathcal{S}$ of two sets $\mathcal{S}_1$ and $\mathcal{S}_2$. According to the previous section, the link with the non-overlapping constraint is made by setting $\mathcal{S}_1$ to either S_R or S'_R and $\mathcal{S}_2$ to either $-S_M$ or $-S'_M$.

Algorithm 1 below is based on a classical SIVIA-like branch & contract recursive loop [JW93, CJ09]. The core operation made on a box $[x]$ can be broken into three steps. First, $[x]$ is contracted to a box $[x]'$ by an *inner contractor* C_{in}, that is, an operator that guarantees:

$$[x]' \subseteq [x] \ \wedge \ [x] \backslash [x]' \subseteq \mathcal{I}. \tag{17}$$

If $[x]' \neq \emptyset$ then $[x]'$ is contracted to a box $[x]''$ by an *outer contractor* C_{out} that guarantees:

$$[x]'' \subseteq [x]' \ \wedge \ [x]' \backslash [x]'' \subseteq \mathcal{O}. \tag{18}$$

Finally, if $[x]'' \neq \emptyset$ then $[x]''$ is bisected in two new boxes that are pushed in the list of boundary boxes $\mathcal{B}$. The recursion stops when the total surface of boxes in $\mathcal{B}$ is less than ε% of the initial box surface (s_0), ε being a user-defined parameter.

The originality of our approach lies in the inner contractor that we describe now. It is dedicated to the handled problem as it makes use of the specific structure of the quantified constraint (14).

3.1 Inner Contractor

Our inner contractor is based on the *inner arithmetic* and a *sweep* loop: The former builds small inner boxes, and the latter makes the union of these boxes in order to remove slices of the initial box, leading to a so called *inner contraction.*

Algorithm 1. $(\mathcal{I}, \mathcal{B}, \mathcal{O}) = \mathbf{pave}([x], \varepsilon)$

$s_0 \leftarrow$ **surface**$([x])$; // *initial surface*
$\mathcal{I} \leftarrow \emptyset$; $\mathcal{O} \leftarrow \emptyset$; $\mathcal{B} \leftarrow \{[x]\}$;
$s \leftarrow s_0$; // *current surface of* $\mathcal{B}$
while $(s > \varepsilon \times s_0)$ **do**
    $[x] \leftarrow$ box of $\mathcal{B}$ with the largest surface; // *(use a heap structure for that)*
    pop $[x]$ from $\mathcal{B}$;
    $s \leftarrow s-$ **surface**$([x])$; // *update the surface*
    $[x]' \leftarrow C_{in}([x])$; $\mathcal{I} \leftarrow \mathcal{I} \cup ([x] \backslash [x]')$; // *inner contraction*
    $[x]'' \leftarrow C_{out}([x]')$; $\mathcal{O} \leftarrow \mathcal{O} \cup ([x]' \backslash [x]'')$; // *outer contraction*
    if $([x]'' \neq \emptyset)$ **then**
        $([x]_1, [x]_2) \leftarrow$ **bisect**$([x]'')$;
        push $[x]_1$ and $[x]_2$ in $\mathcal{B}$;
        $s \leftarrow s+$**surface**$([x]_1)$+**surface**$([x]_2)$; // *update the surface*
    end
end
return $(\mathcal{I}, \mathcal{B}, \mathcal{O})$;

The inner arithmetic is a variant of the classical interval arithmetic that allows to build a sub-box of a box $[x]$ that is inside a given set $\mathcal{S}$ described by inequalities. This technique was first introduced in §3 of [CB10] and used in [ATNC14] in the context of global optimization. This inner arithmetic can also be used with a initial point (or initial box) that is *inflated*, that is to say, given a box $[x]$ and $\tilde{x} \in [x]$, it produces a box $[\tilde{x}]$ such that

$$\tilde{x} \in [\tilde{x}] \subseteq [x] \ \wedge \ [\tilde{x}] \subseteq \mathcal{S}, \tag{19}$$

or an empty box if $\tilde{x} \notin \mathcal{S}$. This arithmetic has similar properties to its classical counterpart: the time complexity is in the length of the constraint expression and gives an optimal box $[\tilde{x}]$ (i.e., of maximal size in every dimension) if no variable occurs twice in the expression.

Before describing how to contract a box $[x]$ with this new arithmetic, let us first address a simpler question: how to find a subbox of $[x]$ that is inside $\mathcal{S}$?

A possible answer is to look for two boxes $[x]_1$ and $[x]_2$ such that

$$[x]_1 \subseteq \mathcal{S}_1, \quad [x]_2 \subseteq \mathcal{S}_2 \ \text{ and } \ ([x]_1 + [x]_2) \cap [x] \neq \emptyset \tag{20}$$

because, in this case, $([x]_1 + [x]_2) \subseteq [x] \cap \mathcal{S}$. To find such boxes, one can calculate in parallel two pavings, one of $\mathcal{S}_1$ and one of $\mathcal{S}_2$, and stop the process as soon as two boxes satisfying (20) are found. By combining boxes of the first paving with boxes of the second one, this approach amounts to run a branch & bound in a $(2 \times d)$-dimensional space. Note that this branch & bound is a sub-solver embedded in the main one, the one for the x variable. Our goal is to reduce the sub-solver to d dimensions only, which is, by the way, the incompressible price to pay for handling d existentially-quantified parameters.

To this end, we use the same idea as above, but this time based on Relation (14) (see Figure 4). To build an inner box in $[x]$, let us first assume that we

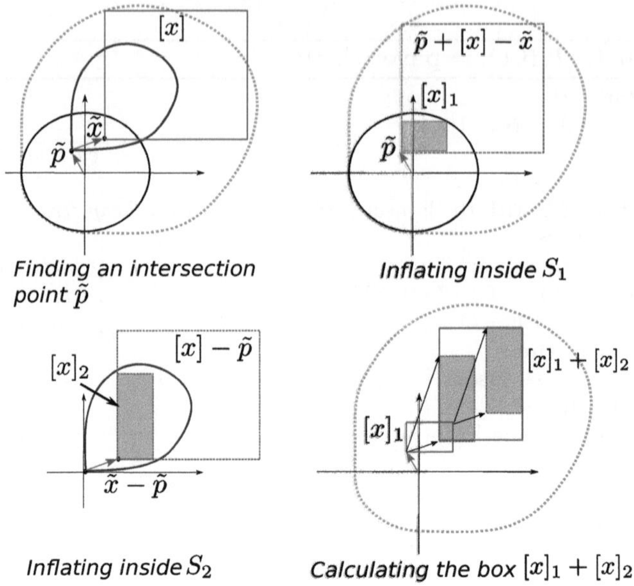

Fig. 4. Steps of the inner Contractor

have picked some point $\tilde{x}$ inside $[x]$ (this point is, in fact, automatically yield by the sweep loop, as this will be explained in Figure 5). Then we look for another point $\tilde{p}$ such that

$$c_1(\tilde{p}) \wedge c_2(\tilde{x} - \tilde{p}). \tag{21}$$

Finding this point is the task of the subsolver.[1]

Once $\tilde{p}$ is found, it is "inflated" to a subbox $[x]_1$ of $(\tilde{p}+[x]-\tilde{x})$ that is inside $\mathcal{S}_1$, which is possible with the inner arithmetic. The point $(\tilde{x}-\tilde{p})$ is also inflated to a sub-box $[x]_2$ of $([x]-\tilde{p})$ that is inside $\mathcal{S}_2$. However, if $\tilde{x}$ turns out to be outside $\mathcal{S}$, the last inflation cannot succeed and the process is interrupted in this case. Otherwise, the resulting box satisfies $([x]_1+[x]_2) \subseteq \mathcal{S}$ and $([x]_1+[x]_2) \cap [x] \neq \emptyset$.

Note that $([x]_1 + [x]_2) \cap [x] \neq \emptyset$ is just a consequence of $\tilde{x} \in [x]$. So the initial boxes used for both inflations are somehow arbitrary, but fixing them as we did is an heuristic that tends to maximize the surface of the final inner box $([x]_1 + [x]_2) \cap [x]$.

Now that we have a technique to build an inner box inside $[x]$ that contains a specific point $\tilde{x}$, we can use this service inside a *sweep* loop. The sweep loop can

[1] This subsolver is implemented with a standard branch & prune based on Hc4 [BG06]. Since only one solution is sought, at each node of the search, we check inequalities with a point $\tilde{p}$ picked randomly in the current domain. If both are satisfied, the search is interrupted and $\tilde{p}$ is returned. The depth of the search is also controlled by a precision on the domain width, which ensures that the subsolver terminates in bounded time. In case of normal termination, no $\tilde{p}$ hence no inner box has been found.

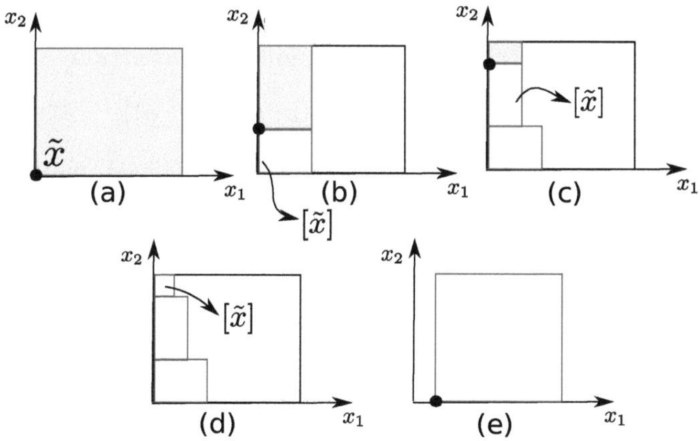

Fig. 5. The sweep loop. The sequence of pictures illustrates a contraction for the lower bound of x_1. At each step, the point $\tilde{x}$ to be inflated is the lower-left corner of the gray box. The inner box $[\tilde{\mathbf{x}}]$ is painted in white. The lower bound of x_1 can be reduced as soon as the projection of the white boxes on x_2 spans the face $[x_2]$, which is the case at step **e**).

be simply viewed as a way to contract a box by "piling up" boxes until some face is entirely covered. This is quickly depicted in Figure 5. The interested reader may refer to [CB10] for further details.

3.2 Outer Contractor

The outer contractor is less sophisticated than the inner one and acts as a simple *rejection test*: the box is either entirely discarded or kept intact.

Rejecting a box $[x]$ means proving $[x] \not\subseteq S_1 + S_2$, that is

$$\forall x \in [x] \quad \forall p \in \mathbb{R}^d, \ \neg(c_1(p) \wedge c_2(x-p)). \tag{22}$$

This assertion can be checked by running the same subsolver we used for the inner contractor, except that the point $\tilde{x}$ is replaced by the current box $[x]$. If the subsolver finds no solution, the previous assertion is proven. Note that only the coordinates of p are bisected, so the subsolver actually proves a stronger assertion if the contraction with respect to c_2 is not optimal (the actual assertion depends on the consistency level enforced by the contraction with c_2). Note also that the precision used in the subsolver is dynamically set to the width of $[x]$ in order to have a somewhat uniform time spent by the subsolver throughout the global search (on x). This dynamic precision also ensures that the outer contractor is *convergent*, that is, it rejects any small enough boxes outside $\mathcal{S}$.

One may be surprised by the simplicity of this rejection test and expect a more elaborated contractor for the outer region, inspired by what we did for the inner region. But the situation could be interpreted in the other way around. Since the overlapping constraint is in two dimensions only, an inner satisfiability test would probably fits, as long as it is fast and convergent. However, such a test amounts to prove for $[x] \subseteq S_1 + S_2$ the following assertion

$$\forall x \in [x] \quad \exists p \in \mathbb{R}^d, \ (c_1(p) \wedge c_2(x-p)).$$

and, contrary to (22), the $\forall$ and $\exists$ quantifiers are involved, which means that the problem is much harder. So, the inner contractor can be seen here as a way to make up for the lack of inner test.

Our previous argument is only based on running time. It is clear that an outer contractor could also lead to a more compact paving, but the size of the paving is anyway conditioned by the representation of the boundary so that a drastic gain on this aspect is not really expectable.

4 Experimental Results

Experimental Setup

The algorithm proposed in this article calculates a paving $(\mathcal{I}, \mathcal{B}, \mathcal{O})$ of the overlapping constraint. The difficulty of this task mainly depends on three criteria:

- *variable occurrences*: the number of times each variable appears in the expressions of the inequalities directly affects the efficiency of the contractors; the more occurrences a variable, the less efficient contractors. This is a well-known drawback of the classical interval arithmetic that carries over the inner arithmetic (used by the inner contractor).
- *convexity*: if objects are non-convex, the boundary of the non-overlapping constraint will be less smooth. So the paving will be more complicated (hence more time consuming), especially near the boundary.
- *degrees of freedom*: that is, whether we take into account rotation or not. Allowing rotation gives a problem of much higher difficulty for multiple reasons. First, the size of the paving is exponential in the number of degrees of freedom so we cannot expect to get a 3D paving within the time scale of a 2D paving. Second, the angle in the inequalities creates a lot of multi-occurrences (see Equations (4) and (5)) and considerably increase non-convexity by the introduction of trigonometric functions.

Our benchmark is based on these criteria. We have made two types of experiments. The first one is with translation only. We have considered three objects of increasing difficulty. Object №1 is a simple ellipsis:

$$\text{Object №1}: \quad (p_1/2)^2 + p_2^2 \leq 1. \tag{23}$$

Object №2 is an ellipsis rotated by some fixed angle. Objects №1 and №2 are obviously of equal complexity if rotation is a degree of freedom, but not if we limit ourselves to translation. This is because the rotated object introduces multi-occurrences for p_1 and p_2:

$$\text{Object №2}: \quad 1.5 \times p_1^2 + 1.5 \times p_2^2 - p_1 \times p_2 - 0.2 \leq 0. \tag{24}$$

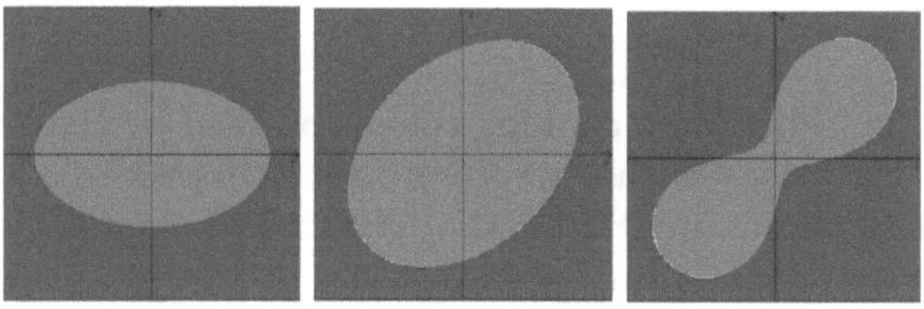

Fig. 6. Objects of increasing complexity. *From left to right*: objects №1, 2 and 3.

Finally, the third object has a "peanut" shape. It cumulates multiple occurrences and non-convexity, as depicted in Figure 6:

$$\text{Object №3}: \quad (p_1^2 + p_2^2)^2 - 2 \times (p_1 \times p_2) - 0.02 \leq 0. \tag{25}$$

The non-overlapping constraint involves two objects: the "reference" one (which coordinates are fixed at the origin of the frame) and the "moving" one that represents the unknowns. We have considered all possible combinations with the three types of objects above, that is, the 6 first cases in Table 1.

Table 1. Cases of study

Case	Reference object	Moving object	Rotation
1	1	1	no
2	1	2	no
3	1	3	no
4	2	2	no
5	2	3	no
6	3	3	no
7	1	1	yes
8	3	3	yes

In the second set of experiments, we have introduced rotation and tested with two ellipsis and two "peanuts" (cases 7 and 8).

In each experiment, the paving process is interrupted when the total surface of the boundary $\mathcal{B}$ is less than $\varepsilon\%$ the surface of the initial box (initial domain for the variables), where ε is a user-defined precision parameter. We have applied the same policy with RSOLVER, the tool we are comparing to.

Since the precision is in proportion of the initial domain surface, it should be noted that the quality of the paving depends also on the width of the initial enclosure. The larger the initial domain, the less precise the resulting paving. For this reason, to give ε a meaningful value, we have set in the experiments the initial box to a fairly accurate enclosure of the overlapping set $\mathcal{S}$ or $\mathcal{S}'$ (as it can be seen in Figure 7 and subsequent).

Results (without Rotation)

We first compare in Table 2 the running times obtained by RSOLVER and our algorithm for the 6 first cases, with a precision ε set each time to 3.25%. This choice for ε corresponds to the minimal value that gives no timeout with RSOLVER. The pavings obtained are depicted in Figure 7.

We then provide in Figures 8 a more detailed analysis for the two extreme cases (case 1 and 6), where ε varies from 10% downto 1%. They show some significant absolute performance gain, as well as some better asymptotic behavior with respect to RSolver.

Table 2. Running time (in s) for the 6 first cases (the precision is set to 3.25%)

Case	Rsolver	Our algorithm
1	4,07	0,37
2	51,55	2,67
3	611,85	7,90
4	132,46	5,58
5	656,11	12,00
6	771,00	26,82

The results first confirm the presumed complexity levels of the different cases, since the "harder" instances indeed require more time to be solved. They also show that our algorithm is more competitive than RSOLVER. But, of course, RSOLVER is a generic solver that does not take advantage of the specific structure of the handled problem. It should also be noted from the graphics that the gap between our approach and RSolver, in a given case, increases as we use smaller values for the precision.

Results (with Rotation)

We only present here preliminary results with rotation.

Figures 9 and 10 show 2D sections of the 3D pavings we have obtained in cases 7 and 8 with a precision set to 3.25%. A 2D section is obtained by fixing the angle to some value and selecting the boxes in the 3D pavings for which the dimension of the angle contains this value. The two other dimensions are plotted.

We have set the domain of the angle to the interval $[0, 0.7]$ for the case 7, and to the interval $[0, 0.3]$ for the case 8. The only paving that was possible to obtain within the time limit was with our algorithm and for the case 7. This paving has been calculated in 9 minutes whereas RSOLVER does not return after 80 minutes. Both algorithm do not terminate after 100 minutes in the case 8.

When a program timeouts, only a partial result is displayed. A partial result means that the surface of $\mathcal{B}$ exceeds ε% the initial width.

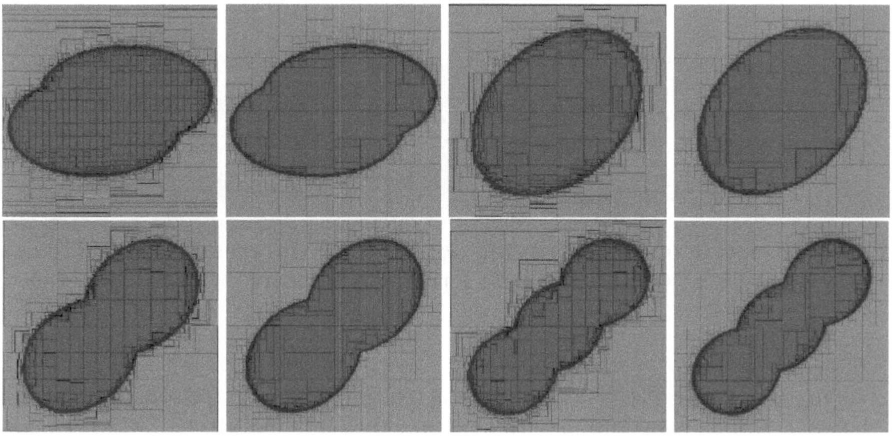

Fig. 7. Pavings obtained for cases 3-6. The precision is set to 3.25%. *Left:* with RSolver. *Right:* with our algorithm.

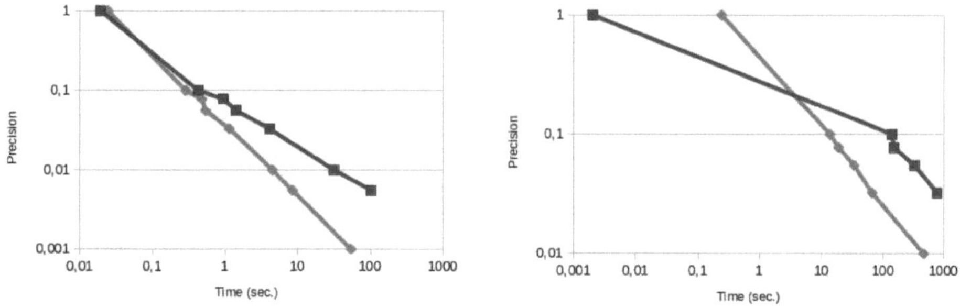

Fig. 8. Time vs Precision (left: case 1; right: case 6). Each curve represents the paving time (vertical axis) with respect to the precision ε (horizontal axis). Both axis are in logarithmic scale. The blue curve is RSolver and the red curve is our method.

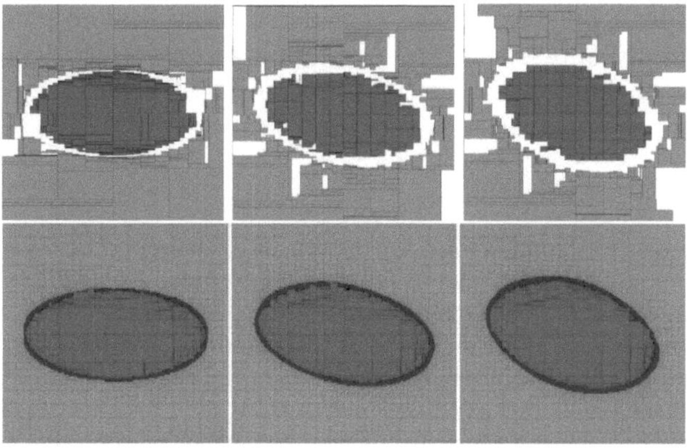

Fig. 9. 2D section of the 3D paving obtained for the case 7. *From left to right:* the rotation angle are 0, 0.4 and 0.7. *Top:* with RSOLVER. *Bottom:* with our algorithm.

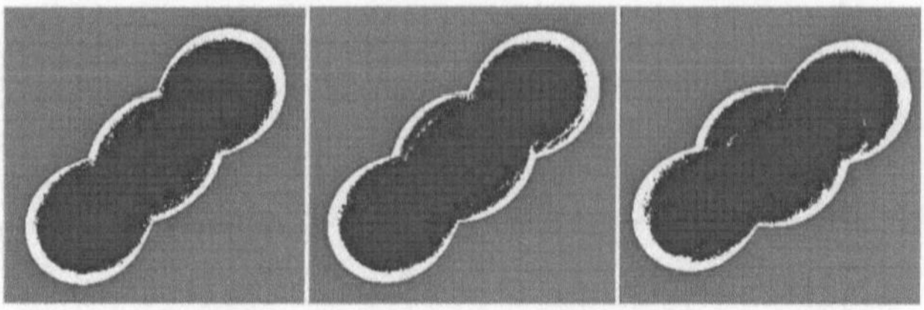

Fig. 10. 2D section of the 3D paving obtained for the case 8. From left to right, the rotation angle are 0, 0.1 and 0.3. This paving is calculated with our algorithm.

The main purpose of this experiment is to show that our approach is still valid in the case where rotations are taken into account. Rotations only transform the expressions that describes the objects, without requiring any change in the algorithm itself. It is clear however that calculating a full 3D paving is a heavy task, whatever the algorithm is.

5 Conclusion

In the case of objects defined by non-linear inequalities, the non-overlapping constraint can only be handled numerically. In this paper, we have given an efficient way to generate verified pavings approximations for this constraint. These pavings represent explicitly the constraint, that is, the set of all acceptable positions and orientations of an object with respect to another. Our preliminary experiments have shown strong efficiency gains with respect to the state of the art solver for quantified numerical constraints RSolver, in particular in the case where the orientation of the objects is taken into account. In our future work, these pre-computed pavings will be incorporated within a packing algorithm, as explicit descriptions of the overlapping constraints. However, with rotation, we have seen that full 3D pavings are clearly too big so that a more adaptative approach will probably have to be considered as well. The idea would be to calculate on-the-fly sub-sets of the overlapping constraint, depending on the actual domains of the objects positions assigned by the packing solver.

References

[ATNC14] Araya, I., Trombettoni, G., Neveu, B., Chabert, G.: Upper Bounding in Inner Regions for Global Optimization under Inequality Constraints. In: Journal of Global Optimization (to appear, 2014)

[BG06] Benhamou, F., Granvilliers, L.: Continuous and interval constraints. In: Handbook of Constraint Programming, ch. 16, pp. 571–604. Elsevier (2006)

[BGT01] Beldiceanu, N., Guo, Q., Thiel, S.: Non-Overlapping Constraints between Convex Polytopes. In: Walsh, T. (ed.) CP 2001. LNCS, vol. 2239, pp. 392–407. Springer, Heidelberg (2001)

[CB10] Chabert, G., Beldiceanu, N.: Sweeping with Continous Domains. In: Cohen, D. (ed.) CP 2010. LNCS, vol. 6308, pp. 137–151. Springer, Heidelberg (2010)

[CJ09] Chabert, G., Jaulin, L.: Contractor Programming. Artificial Intelligence 173(11), 1079–1100 (2009)

[GJ06] Goldsztejn, A., Jaulin, L.: Inner and Outer Approximations of Existentially Quantified Equality Constraints. In: Benhamou, F. (ed.) CP 2006. LNCS, vol. 4204, pp. 198–212. Springer, Heidelberg (2006)

[IGJ12] Ishii, D., Goldsztejn, A., Jermann, C.: Interval-Based Projection Method for Under-Constrained Numerical Systems. Constraints 17(4), 432–460 (2012)

[JKDW01] Jaulin, L., Kieffer, M., Didrit, O., Walter, E.: Applied Interval Analysis. Springer (2001)

[JW93] Jaulin, L., Walter, E.: Set Inversion via Interval Analysis for Nonlinear Bounded-Error Estimation. Automatica 29(4), 1053–1064 (1993)

[Rat] Ratschan, S.: RSolver

[Rat06] Ratschan, S.: Efficient Solving of Quantified Inequality Constraints over the Real Numbers. ACM Transactions on Computational Logic 7(4), 723–748 (2006)

Improving Relational Consistency Algorithms Using Dynamic Relation Partitioning*

Anthony Schneider[1], Robert J. Woodward[1,2], Berthe Y. Choueiry[1], and Christian Bessiere[2]

[1] Constraint Systems Laboratory, University of Nebraska-Lincoln, USA
{aschneid,rwoodwar,choueiry}@cse.unl.edu
[2] LIRMM, CNRS & University of Montpellier, France
bessiere@lirmm.fr

Abstract. Relational consistency algorithms are instrumental for solving difficult instances of Constraint Satisfaction Problems (CSPs), often allowing backtrack-free search. In this paper, we improve an algorithm for enforcing relational consistency by exploiting the property that the constraints of the dual encoding of a CSP are piecewise functional. This property allows us to partition a CSP relation into blocks of equivalent tuples at varying levels of granularity. Our new algorithm dynamically exploits those partitions. Our experiments show a significant improvement of the processing time for enforcing relational consistency.

1 Introduction

Algorithms for enforcing local consistency are a focal point of research in Constraint Programming because they are an efficient means to reduce the size of the search space and effort [1]. In recent years, new techniques for enforcing higher levels of consistency have been proposed. While most consider combinations of two constraints [3,20–22], some operate on combinations of two or more constraints [2, 6, 15, 16, 23, 24]. In this paper, we improve the performance of the algorithm for enforcing the relational consistency property R($*$,m)C [15,16] (originally known as m-wise consistency [10]). This property ensures that any tuple can be consistently extended over every combination of $m-1$ relations.

Samaras and Stergiou showed that the constraints of the dual encoding of a CSP are piecewise functional [11,22]. Given two constraints that are adjacent in the dual graph, this property partitions the tuples of each relation into a set of *blocks*, i.e., equivalence classes of tuples. They exploited those partitions in an algorithm (PW-AC) for enforcing pairwise-consistency (i.e., R($*$,2)C), which is defined on *pairs* of relations. Extending the work of Samaras and Stergiou, we identify as *coarse* blocks those induced on a constraint's relation by one other adjacent constraint and as *fine* blocks those induced by all other adjacent

* This research was supported by NSF Grant No. RI-111795 and EU project ICON (FP7-284715). Woodward was supported by an NSF GRF Grant No. 1041000 and a Chateaubriand Fellowship. Experiments were conducted on the equipment of the Holland Computing Center at the University of Nebraska–Lincoln.

B. O'Sullivan (Ed.): CP 2014, LNCS 8656, pp. 688–704, 2014.

constraints. We modify the PERTUPLE[1] algorithm for enforcing R($*$,m)C into the PERFB algorithm, which exploits not only the fine and coarse blocks but also intermediate ones induced by a subset of the constraint's neighbors.

The contributions of this paper are as follows: *a*) The definitions of levels of relation partitions and the specification of data structures to store and manipulate the coarse and fine blocks; *b*) The design of an algorithm that utilizes those data structures to enforce R($*$,m)C; *c*) A complexity analysis of our data structures and algorithm; and, *d*) An empirical evaluation of PERFB comparing its performance to that of PERTUPLE.

In addition to the contribution of Samaras & Stergiou [22], our approach is related to the following research. Karakashian et al. propose a compact data structure, the index tree, which finds coarse blocks and stores them in the leaves of the tree [16]. However, they fall short of exploiting them to improve constraint propagation. Lecoutre et al. propose the algorithm STR3 to enforce GAC using the size of the blocks induced, on a relation, by a variable in its scope [19]. Further, Lecoutre et al. propose the algorithm eSTR to enforce pairwise-consistency using only the size of the coarse blocks [20]. Our work is also related to the computation of subproblem interchangeability [4, 5, 7, 18], where the variables' domains (instead of constraints' relations) are dynamically partitioned by the constraints of a specified subproblem.

This paper is organized as follows. Section 2 gives some background information. Section 3 discusses relation partitioning. Section 4 describes how to create relation partitions and the data structures for storing them. Section 5 gives the partition-based algorithms for enforcing R($*$,m)C. Section 6 discusses our experiments and results. Finally, Section 7 concludes this paper.

2 Background

A constraint satisfaction problem (CSP) is defined by $\mathcal{P}= (\mathcal{X}, \mathcal{D}, \mathcal{C})$, where $\mathcal{X}$ is a set of variables, $\mathcal{D}$ is a set of domains, and $\mathcal{C}$ is a set of constraints. A variable in $\mathcal{X}$ has a finite domain in $\mathcal{D}$, and is constrained by constraints in $\mathcal{C}$. The constraints restrict the acceptable combinations of values for variables. A solution to the CSP is an assignment to each variable of a value taken from its domain such that all the constraints are satisfied. Deciding the existence of a solution for a CSP is NP-complete.

Each constraint $C_i \in \mathcal{C}$ is defined by a relation R_i specified over the *scope* of the constraint, *scope*(C_i), which is the set of variables to which the constraint applies. The *arity* of a constraint is the cardinality of its scope. In this paper, we study table constraints, where a tuple $t_i \in R_i$ is a combination of allowed values for the variables in *scope*(C_i). We call the *subscope* of a constraint a subset of its scope, and use it to denote the set of variables common to two constraints: *subscope*(C_i, C_j) = *scope*$(C_i) \cap$ *scope*(C_j). We use the relational operator project, π, to restrict a partial assignment (e.g., a tuple) to a particular set of variables.

[1] PERTUPLE was originally called PROCESSQUEUE [16] and later renamed PERTUPLE to contrast it to another algorithm, ALLSOL, that guarantees the same result [8, 14].

Several graphical representations of a CSP exist. In the *hypergraph*, the vertices represent the variables of the CSP, and the hyperedges represent the scopes of the constraints. Figure 1 shows the hypergraph of our running example. In the *dual graph*, the vertices represent the CSP constraints, and the edges connect vertices representing constraints whose scopes overlap (see Figure 2). Thus, two CSP constraints are *adjacent* or *neighbors* in the dual graph when their subscope is not empty. The constraints of the dual graph enforce the equality of the variables in the subscope of the two adjacent CSP constraints.

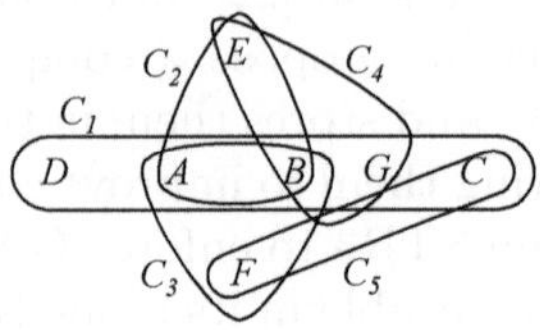

Fig. 1. Hypergraph of a CSP example

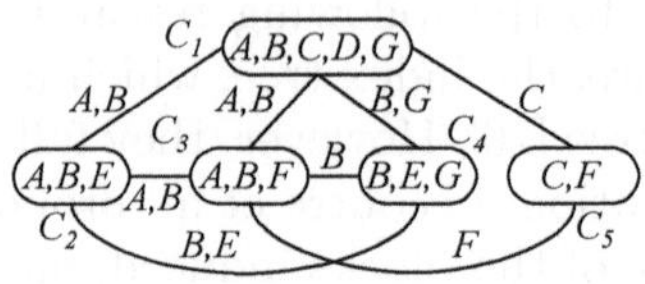

Fig. 2. Dual graph of CSP in Figure 1

Backtrack search is typically used to solve CSPs. To reduce the size of the search tree, CSPs are usually filtered by enforcing a given *local consistency property*. One common property is Generalized Arc Consistency (GAC). A CSP is GAC iff for every constraint, any value in the domain of any variable in the scope of the constraint can be extended to a tuple satisfying the constraint. While GAC is enforced by filtering the domains, other consistency properties are enforced by filtering the relations (which are then typically projected on the domains). Karakashian et al. proposed a relation-filtering algorithm that allows us to control the consistency level enforced while preserving the topology of the constraint network [16]. Their algorithm enforces R($*$,m)C, which guarantees that every relation is *minimal* in every combination of m relations.

Definition 1. *A set of m constraints $\mathcal{C} = \{C_1, C_2, \ldots, C_m\}$ with $m \geq 2$ is said to be R($*$,m)C iff every tuple in the relation of each constraint $C_i \in \mathcal{C}$ can be extended to the variables in $\bigcup_{C_j \in \mathcal{C}} scope(C_j) \setminus scope(C_i)$ in an assignment that satisfies all the constraints in $\mathcal{C}$ simultaneously. A network is R($*$,m)C iff every set of m constraints, $m \geq 2$, is R($*$,m)C.*

PERTUPLE, the algorithm for enforcing R($*$,m)C, ensures that each tuple in a relation appears in a solution of the dual CSP induced by the m relations by conducting a backtrack search on the tuples of the $m-1$ relations (see Figure 3).

Samaras and Stergiou showed that the constraints of the dual graph are piecewise functional [11, 22]. Given two CSP constraints with nonempty overlapping scopes, this property partitions the tuples of each relation into a set of *blocks*, equivalence classes of tuples, where each block is consistent with at most one block in the other relation (see Figure 4). PW-AC, their algorithm for enforcing pairwise-consistency, finds the partition induced on the relation of each constraint by each one of its neighbors in the dual graph. When a block of a

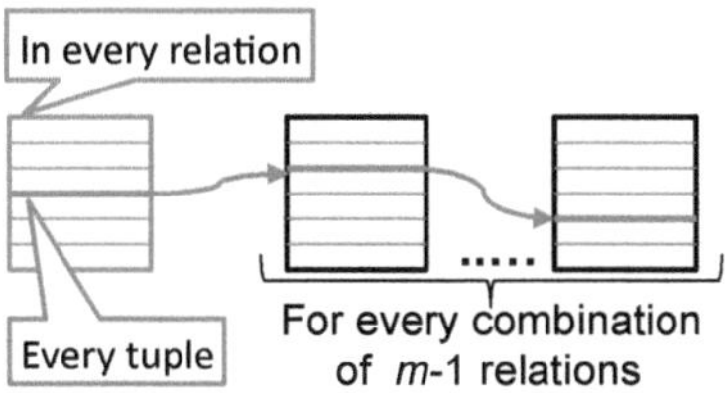

Fig. 3. Illustrating R(∗,m)C

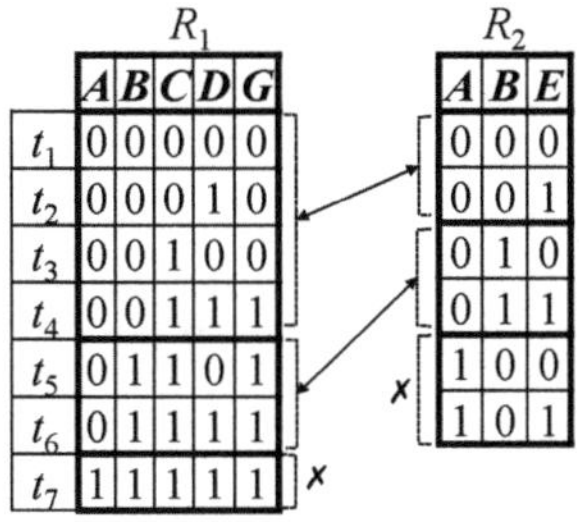

Fig. 4. Piecewise functional constraint

relation is not supported in one of the constraint's neighbors, the block's tuples are deleted. This operation may cause other blocks of the same relation to lose tuples, eventually becoming empty. Propagation stops when the network is pairwise-consistent or when a relation becomes empty. The effectiveness of PW-AC was established on sparse networks and other structured benchmarks [22].[2]

3 Relation Partitioning

We exploit the equivalence classes induced, on the relation R_i of a constraint C_i, by C_i's neighbors in the dual graph. We distinguish three types of such classes depending on the subset of neighbors considered: coarse (cb), fine (fb), and intermediate blocks (ib). Figures 5 and 6 illustrate the above for R_1. The notations and data structures used in the following sections refer to this example.

Coarse blocks: Any single neighbor of C_i in the dual graph partitions R_i into a set of coarse blocks. In Figure 5, *subscope*(C_1,C_2) $= \{A, B, C, D, G\} \cap \{A, B, E\} = \{A, B\} = o_1$. The tuples $t_{i\in[1,4]} \in R_1$ are equivalent for R_1 given o_1=00,[3] and consistent with (0,0,0) and (0,0,1) $\in R_2$. Indeed, $\pi_{o_1}(t_{i\in[1,4]} \in R_1) = (0, 0)$ and $\pi_{o_1}((0, 0, 0) \in R_2) = \pi_{o_1}((0, 0, 1) \in R_2) = (0, 0)$. Further, the above does not hold for any other tuple of R_1. Thus, $cb_1 = \{t_1, t_2, t_3, t_4\}$ is the coarse block of R_1 induced by o_1=00. The other two coarse blocks are $cb_2 = \{t_5, t_6\}$ and $cb_3 = \{t_7\}$. Similarly, *subscope*$(C_1, C_{j\in\{3,4,5\}})$ is $o_1 = \{A, B\}$, $o_2 = \{B, G\}$, and $o_3 = \{C\}$ respectively. Thus, o_1, o_2, and o_3 induce on R_1 the set of coarse blocks $\{cb_1, cb_2, cb_3\}$, $\{cb_4, cb_5, cb_6\}$, and $\{cb_7, cb_8\}$, respectively. Coarse blocks are the partitions identified and exploited by Samaras and Stergiou [22].

Fine blocks: When we consider all the constraints adjacent to C_i in the dual graph, they induce on R_i the finest possible partition, obtained by performing the intersections of all R_i's coarse blocks. As a result, they yield the (unique) set of R_i's fine blocks. In Figure 6, the set of fine blocks of R_1 is $\{fb_1, fb_2, \ldots, fb_5\}$.

[2] We suspect that PW-AC could be shown to be effective on dense networks had the redundant edges of the dual CSPs been removed [12, 16].

[3] Abusing tuple/set assignment notation.

Intermediate blocks: Finally, the partition induced on a relation R_i by a given combination of m constraints depends on the neighboring constraints of C_i that are included in m. The granularity of that partition is intermediate: not finer than R_i's fine partition and not coarser than any of its coarse partitions. For example, $\{C_2, C_5\} \subset neighbors(C_1)$ induce the intermediate blocks $\{ib_1, ib_2, ib_3, ib_4\}$.

	R_1	A	B	C	D	G
fb_1	t_1	0	0	0	0	0
	t_2	0	0	0	1	0
fb_2	t_3	0	0	1	0	0
fb_3	t_4	0	0	1	1	1
fb_4	t_5	0	1	1	0	1
	t_5	0	1	1	1	1
fb_5	t_7	1	1	1	1	1

R_2	A	B	E
fb_6	0	0	0
fb_7	0	0	1
fb_8	0	1	0
fb_9	0	1	1
fb_{10}	1	0	0
fb_{11}	1	0	1

R_3	A	B	F
fb_{12}	0	0	0
fb_{13}	0	0	1
fb_{14}	0	1	1
fb_{15}	1	1	0
fb_{16}	1	1	1

R_4	B	E	G
..	..	..	..

R_5	C	F
..	..	..

Fig. 5. Relations of CSP example

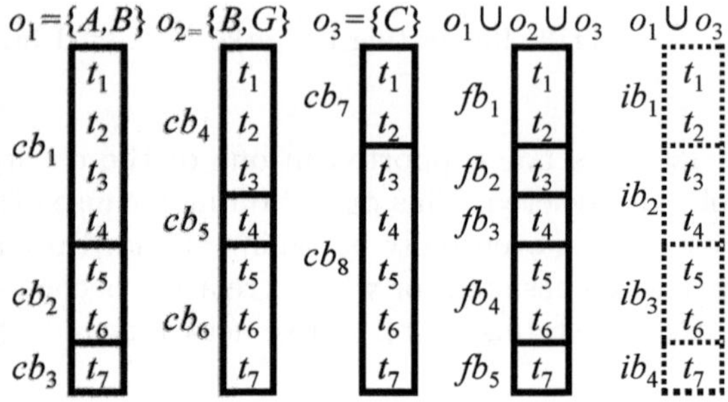

Fig. 6. Coarse, fine, and intermediate blocks of R_1

We compute and store the fine and coarse blocks before preprocessing. PERFB, our new algorithm for enforcing R(∗,m)C, conducts backtrack search over the fine blocks, and uses the coarse blocks during lookahead. It does *not* permanently store any of the intermediate partitions.

4 Generating and Storing Partition Blocks

Because the partitions of a relation R_i are induced by an equivalence relation, any coarse or intermediate block of R_i is made up of a number of R_i's fine blocks (e.g., in Figure 6, $cb_1 = \{fb_1, fb_2, fb_3\}$ and $ib_2 = \{fb_2, fb_3\}$). Also, any fine block appears in exactly one block of a partition of a given granularity. For this reason, we first build the fine blocks of a relation, then we build its coarse blocks as sets of fine blocks. Below, we describe the data structures for storing the blocks (both fine and coarse) of a relation. Then, we describe how to generate them, and discuss their complexity.

4.1 Data Structures

Figure 7 partially depicts the data structures for the example of Figure 5. A coarse block is uniquely determined by three entities: a subscope, values of the subscope, and the relation that is being partitioned. For example, cb_1 of Figure 6 is determined by o_1, o_1=00, and R_1, and stored in $cb_{R1,o1=00}$.

The structure $rel\text{-}cb_{o_1=00}$ stores an entry for all relations R_i (i.e., R_1, R_2, R_3) with at least one tuple t_i where $\pi_{o_1}(t_i) = (0, 0)$. This entry points to the coarse

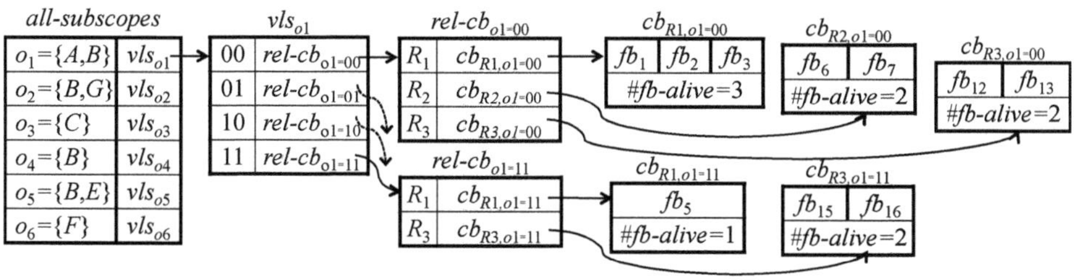

Fig. 7. Data structures for storing coarse and fine blocks for the CSP in Figure 5

blocks of R_i with o_1=00. Further, a back-pointer, not shown in Figure 7 for readability, links each such coarse block back to $rel\text{-}cb_{o_1=00}$. All the coarse blocks accessible from $rel\text{-}cb_{o_1=00}$ are pairwise consistent. If a relation R_i has the subscope o_1 but does not have an entry in $rel\text{-}cb_{o_1=00}$, or if R_i loses during constraint propagation all its tuples t_i where $\pi_{o_1}(t_i) = (0,0)$, then all the coarse blocks accessible from $rel\text{-}cb_{o_1=00}$ are inconsistent. Thus, given any coarse block with o_1=00 (e.g., $cb_{R_2,o_1=00}$), $rel\text{-}cb_{o_1=00}$ gives us constant-time access to the coarse blocks that are pairwise consistent with it in all relations.

The structure vls_{o_1} gives access to the structures storing $rel\text{-}cb_{o_1=v_i}$ for each subscope value v_i of o_1 (i.e., 00, 01, 10, 11). For each v_i, there is one such entry.

The structure *all-subscopes* gives access to the structures storing the subscopes values vls_{o_i} for each subscope o_i in the problem. For example, for $o_1 = \{A, B\}$, *all-subscopes* gives access to vls_{o1}.

In addition to the above structures, we use two constant-time lookup tables. The table *fb-subscope-2-cb* gives the coarse block in which a fine block of a relation appears given a subscope. For example, *fb-subscope-2-cb*$[fb_2, R_1, \{A, B\}]$ points to the coarse block $cb_{R_1,o_1=00}$ shown in Figure 7. Similarly, *tup-2-fb* maps a tuple t_i in a relation to the fine block that contains t_i. In the example in Figures 5 and 6, it maps $t_5 \in R_1$ to the structure storing fb_4.

4.2 Fine Blocks

CreateFineBlocks (Algorithm 1) generates the fine blocks of a relation R_i. We use the following accessors and notations:

- The function FineBlocks(R_i) returns the set of fine blocks of R_i.
- The accessor *#tuples-alive*(fb_i) gives the count of living tuples of fb_i. This count is stored in the structure of fb_i, and updated during search whenever a tuple in fb_i is marked or unmarked as deleted.
- The accessor *#fb-alive*(R_i) gives the number of fine blocks alive in R_i.
- The function FindEquivFB(R_i, *subTuple*) is a relational selection operator: it iterates over the fine blocks in FineBlocks(R_i), and returns the fine block with the values assignment matching *subTuple*.

CreateFineBlocks (Algorithm 1) groups tuples with the same values assignments for the variables in $U_s = \bigcup_{C_j \in neighbors(C_i)} subscope(C_i, C_j)$. In addition to grouping tuples, it keeps a counter storing the total number of living tuples in the given fine partition. When a tuple is deleted during search, we use the array *tup*-2-*fb* to update the count of living tuples in the tuple's fine block.

Algorithm 1. Creating the fine partition of a relation R_i

1: **function** CreateFineBlocks(R_i)
2: FineBlocks(R_i) $\leftarrow \emptyset$
3: $U_s \leftarrow \bigcup_{C_j \in neighbors(C_i)} subscope(C_i, C_j)$
4: **for each** tuple $\tau \in R_i$ **do**
5: $subTuple \leftarrow \pi_{U_s}(\tau)$
6: $fb_{curr} \leftarrow$ FindEquivFB($R_i, subTuple$)
7: **if** $fb_{curr} = nil$ **then**
8: $fb_{curr} \leftarrow$ *create a new fine block*
9: FineBlocks(R_i) $\leftarrow$ FineBlocks(R_i) $\cup \{fb_{curr}\}$
10: *#fb-alive*(R_i) $\leftarrow$ *#fb-alive*(R_i) + 1
11: $fb_{curr} \leftarrow fb_{curr} \cup \{\tau\}$
12: *#tuples-alive*(fb_{curr}) $\leftarrow$ *#tuples-alive*(fb_{curr}) +1
13: *tup*-2-$fb[\tau] \leftarrow fb_{curr}$
14: **return** FineBlocks(R_i)

Complexity: We use the following parameters in the complexity analysis. t is the maximum number of tuples in a relation; k is the maximum constraint arity; $e = |\mathcal{C}|$ is the number of constraints in the CSP; and $|o_i|$ is the size of the largest subscope that a relation shares with a neighbor. The time complexity of each of FineBlocks(R_i) and *#fb-alive*(R_i) is $O(1)$. The creation of U_s on Line 3 of Algorithm 1 is $O(e \cdot |o_i| \cdot log(|o_i|))$ because a constraint may be adjacent to all other constraints in the dual graph, and each edge requires inserting at most $|o_i|$ variables into the set. In the worst case, the function FindEquivFB performs k comparisons on t fine blocks to find an existing equivalent fine block. Its complexity is thus $O(t \cdot k)$. Computing the subTuple in Line 5 is $O(k)$. Thus, CreateFineBlocks is $O(e \cdot |o_i| \cdot log(|o_i|) + (t \cdot (k + k \cdot t))) = O(k \cdot t^2 + e \cdot |o_i| \cdot log(|o_i|))$. When $arity(C_i) = |U_s|$, Lines 5–10 are bypassed. In this case, each fine block of R_i has a single tuple. Thus, while the time complexity is large, the cost of Lines 5–10 is incurred only when a fine block can potentially have more than one tuple.

The space complexity for storing the fine blocks of R_i is $O(t)$, incurred when each fine block in R_i has one tuple. The array mapping tuples to the fine blocks to which they belong is $O(t)$, making the total space complexity $O(t)$ per constraint.

4.3 Coarse Blocks

Below, we describe the creation of the coarse blocks, similar to the ones introduced by Samaras and Stergiou [22]. Our design improves on theirs in that *a*) a coarse block stores fine blocks and not tuples, thus, potentially reducing the size of each coarse block, and *b*) we iterate over subscopes, rather than pairs of relations (e.g., in Figures 5 and 6, R_1 has four neighbors but only three subscopes).

CREATECOARSEBLOCKS (Algorithm 2) generates the coarse blocks of a relation R_i for a subscope o_i out of R_i's existing fine blocks. It first retrieves all the values of o_i (Line 2). Then, it iterates over the fine blocks fb_i of R_i, adding each fb_i to the appropriate coarse partition. When a coarse block does not exist (Line 9), it is created (Line 10). The accessor function *#fb-alive*(cb_i) maintains the number of fine blocks alive in the coarse block cb_i. This function has the same name as *#fb-alive*(R_i) (function-name overload), which operates in a similar manner.

Algorithm 2. Creating the coarse partition of a relation R_i given a subscope o_i

```
1:  function CREATECOARSEBLOCKS(R_i, o_i)
2:      vls_{o_i} ← all-subscopes[o_i]
3:      for each fb_i ∈ FINEBLOCKS(R_i) do
4:          if #tuples-alive(fb_i) = 0 then
5:              continue
6:          v_i ← π_{o_i}(fb_i)
7:          if vls_{o_i}[v_i] does not exist then
8:              vls_{o_i}[v_i] ← create new rel-cb table
9:          curr-rel-cb ← vls_{o_i}[v_i]
10:         curr-cb ← curr-rel-cb[R_i]
11:         if curr-cb does not already exist then
12:             curr-cb ← create new coarse block
13:         curr-cb ← curr-cb ∪ {fb_i}
14:         #fb-alive(curr-cb) ← #fb-alive(curr-cb) + 1
15:         fb-subscope-2-cb[fb_i, R_i, o_i] ← curr-cb
16:     return vls_{o_i}
```

Complexity: The table lookups on Lines 2 and 10 of CREATECOARSEBLOCKS are $O(1)$. The table lookups for vls_{o_i} on Lines 6-9 are $O(\log(t) \cdot |o_i|)$ if vls_{o_i} is represented as a binary search tree that uses v_i as a key. Creating v_i on Line 6 is $O(|o_i|)$. The for-loop is executed $O(t)$ times. The complexity for CREATECOARSEBLOCKS is thus $O(|o_i| \cdot t \cdot \log(t))$.

The time complexity to get the specific coarse block to which a particular fine block belongs provided a subscope is $O(1)$ thanks to the *fb-subscope-2-cb* table. The time complexity to query all fine blocks in a relation R_j that are consistent with a coarse block corresponding to a relation R_i is $O(1)$ thanks to the back-pointer to the *rel-cb* table stored in each coarse block.

The space complexity for a set of coarse blocks for a single subscope and relation is $O(t)$ because each fine block is in exactly one coarse block for a given subscope. The space complexity for all coarse blocks is then $O(k \cdot t \cdot e^2)$ because, in the worst case, each relation is partitioned by every other relation in the problem, and each coarse block is identified by a subtuple of size k. Additionally, the *fb-subscope-2-cb* table requires $O(e^2 \cdot t)$ space, as each tuple is in exactly one coarse block.

Note that the coarse blocks have the same space complexity as the index-tree data structure [16], but can be more efficiently queried.

5 Consistency Algorithm: From PERTUPLE to PERFB

Below, we describe PERFB and FB-SEARCHSUPPORT, which improve PERTUPLE and SEARCHSUPPORT [16], respectively, for enforcing R($*$,m)C on a CSP. Like PERTUPLE, PERFB takes as input $\mathcal{Q}$ and Φ. Φ is the set of all combinations of m relations. The queue $\mathcal{Q}$ is initialized to all the combination-relations pairs $\langle \varphi, R_i \rangle$ such that $\varphi \in \Phi$ and $R_i \in \varphi$. PERFB iterates over all fine blocks of a relation R_i in a combination φ, calling FB-SEARCHSUPPORT to ensure that a fine block can be extended to a solution in the dual CSP induced by φ by conducting a backtrack search that maintains support structures. In addition to the *static* fine and coarse blocks, PERFB and FB-SEARCHSUPPORT make use of intermediate blocks, *dynamically* induced by the relations in φ. In this section, we abuse the notations and use *subscope*(R_i, R_j), *scope*(R_i), *neighbors*(R_i) to refer to *subscope*(C_i, C_j), *scope*(C_i), *neighbors*(C_i), respectively.

5.1 PERFB

PERFB (Algorithm 3) improves PERTUPLE [16] in two ways in order to reduce the number of costly calls to FB-SEARCHSUPPORT:

1. PERFB ensures that all fine blocks, rather than all tuples, in a relation R_i can be extended to a solution over the relations of a combination φ of size m (Line 11). This difference can reduce the number of calls to FB-SEARCHSUPPORT.
2. The number of calls to FB-SEARCHSUPPORT can be further reduced by exploiting the dynamically induced intermediate blocks.

We use the following additional notations to describe how PERFB operates.

- The accessor *CB*((R_i, fb_i)$, o_i$) retrieves the coarse block of R_i containing the fine block fb_i given the subscope o_i. It uses the table *fb-subscope-2-cb*.
- The accessor *Support*(R_j,*CB*(R_i, fb_i, o_{ij})) retrieves the coarse block of R_j containing the fine blocks consistent with fb_i of relation R_i given o_{ij} = *subscope*(C_i, C_j). To this end, it uses the back-pointer to the *rel-cb* structures from the coarse block *CB*(R_i, fb_i, o_{ij}) and accesses *rel-cb*[R_j].
- The structure *shared-fvars*[l] stores, at the search level l in FB-SEARCHSUPPORT where R_i is 'assigned' a fine block, the variables in $\bigcup_{R_j \in \varphi}$ *subscope*(R_i, R_j).

In Figure 6, fb_2 and fb_3 are equivalent in $\varphi = \{R_1, R_2, R_5\}$ yielding $ib_2 = \{fb_2, fb_3\}$ for R_1 by $\{o_1 \cup o_3\}$. PERFB exploits such intermediate blocks. The key

Algorithm 3. Enforces R($*$,m)C using a queue $\mathcal{Q}$ and list Φ of combinations

1: **function** PERFB(Q,Φ)
2:   **while** $Q \neq \emptyset$ **do**
3:     $\langle \varphi, R_i \rangle \leftarrow$ POP(Q)
4:     $deleted \leftarrow false$
5:     $\mathcal{R}_f \leftarrow \varphi \setminus R_i$
6:     **for** $R_j \in \varphi$ **do**
7:       $equiv\text{-}FBs[R_j] \leftarrow \emptyset$
8:     **for** $i = 1$ to m **do**
9:       $shared\text{-}fvars[i] \leftarrow \emptyset$
10:     $shared\text{-}fvars[1] \leftarrow \bigcup_{R_j \in \mathcal{R}_f} subscope(R_i, R_j)$
11:     **for each** living $fb_i \in$ FINEBLOCKS(R_i) **do**
12:       $v_i \leftarrow \pi_{(shared\text{-}fvars[1])}(fb_i)$
13:       **if** $equiv\text{-}FBs[R_i, v_i]$ does not exist **then**
14:         $equiv\text{-}FBs[R_i, v_i] \leftarrow$ FB-SEARCHSUPPORT($fp_i, \langle R_i, \mathcal{R}_f \rangle, equiv\text{-}FBs$)
15:       **if** $equiv\text{-}FBs[R_i, v_i] = false$ **then**
16:         **for each** tuple $\tau \in fb_i$ **do**
17:           DELETE(τ, R_i)
18:         $deleted \leftarrow true$
19:         $\#fb\text{-}alive(R_i) \leftarrow \#fb\text{-}alive(R_i) - 1$
20:         **if** $\#fb\text{-}alive(R_i) = 0$ **then**
21:           **return** *inconsistent*
22:     **if** *deleted* **then**
23:       **for each** $\varphi' \in (\Phi \setminus \{\varphi\}), R_i \in \varphi'$ **do**
24:         **for each** $R' \in (\varphi' \setminus \{R_i\})$ **do**
25:           $Q \leftarrow Q \cup \{\langle \varphi', R' \rangle\}$
26:   **return** *consistent*

to dynamically identifying them is the table $equiv\text{-}FBs[R_i, v_i]$, which is created at each call to PERFB, and returns $true$ or $false$, given a relation $R_i \in \varphi$, and a subset of values v_i from a fine block fb_i. The subset v_i is determined by projecting fb_i over the variables in *shared-fvars*[1] (Lines 10–12 in Algorithm 3). Any other fine block of R_i with the same v_i is necessarily in the same intermediate block. Thus, before executing FB-SEARCHSUPPORT, we check $equiv\text{-}FBs[R_i, v_i]$ to see if a result for this particular v_i was already found (Line 13). If so, the result is reused. Otherwise, FB-SEARCHSUPPORT is called, and its result stored for future use (Line 14). Similar to PERTUPLE, when fb_i has no support, its tuples are marked as deleted, and the count of fine blocks alive in R_i is decremented (Lines 15–19). Inconsistency is detected when all fine blocks in R_i are deleted (Lines 20–21). The updates of $\mathcal{Q}$ are identical to those in PERTUPLE. The only relation used to access $equiv\text{-}FBs$ in PERFB is R_i. Other relations' entries in $equiv\text{-}FBs$ are discussed in Section 5.2.

When $|shared\text{-}fvars[1]| = arity(C_i)$, PERFB reduces to PERTUPLE because no two fine blocks in R_i are equivalent. In this case, the discovery of equivalent fine blocks is bypassed to save on CPU time and memory.

Finally, note that $m = 2$ does not require any calls to FB-SEARCHSUPPORT. For this reason, for $m = 2$, we use PW-AC [22] during preprocessing and PERFB during search. Further, because, when $m = 2$, the intermediate blocks are exactly the stored coarse blocks, checking whether or not a *coarse* block is consistent can be done in constant time by checking the *#fb-alive*(R_j) of the coarse block returned by *Support*(R_j,*CB*(R_i, fb_i, o_{ij})), where R_j is the other relation in the combination. Thus, intermediate blocks are not used.

5.2 FB-SEARCHSUPPORT

FB-SEARCHSUPPORT performs backtrack search with forward checking on the subproblem induced, on the dual of the CSP $\mathcal{P}$, by the relations in the combination φ, denoted as $\mathcal{P}_{D\varphi}$. The variables of $\mathcal{P}_{D\varphi}$ are the relations in the combination $\varphi = \{R_i\} \cup \mathcal{R}_f$. The 'variable' R_i is assigned the 'value' fb_i in the search. FB-SEARCHSUPPORT is called with the argument $(fb_i, \langle R_i, \mathcal{R}_f\rangle, equiv\text{-}FBs)$. The

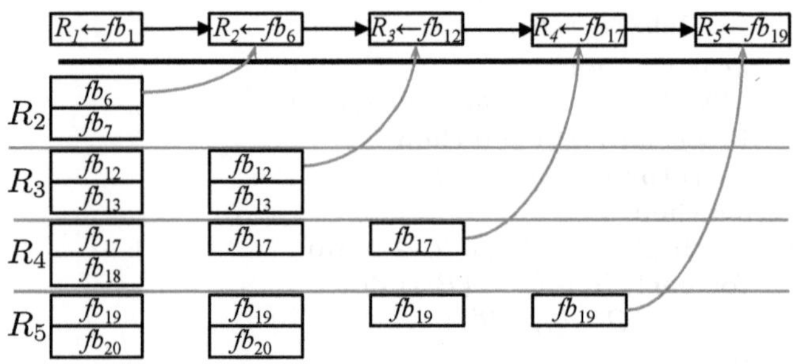

Fig. 8. Backtrack search on fine blocks using coarse and intermediate blocks

domain of a relation $R_j \in \mathcal{R}_f$ is the set of fine blocks (represented by their indices) in the coarse block returned by *Support*(R_j,*CB*($(R_i, fb_i), o_{ij}$)), where o_{ij} is the subscope of R_i and R_j. However, coarse blocks are not defined when R_i and R_j are not neighbors ($subscope(R_i, R_j) = \emptyset$). Thus, the 'domain' of R_j is either *a*) the set of living fine blocks from FINEBLOCKS(R_j) when no relation adjacent to R_j has been instantiated, or *b*) the set of fine blocks in the coarse block *Support*(R_j,*CB*($(R_k, fb_k), o_{jk}$)), where R_k, a relation adjacent to R_j, was 'assigned' fp_k.

Figure 8 illustrates forward checking in FB-SEARCHSUPPORT using the example from Figure 5. Assume R_4 contains only tuples $(0,0,0)$ and $(0,1,0)$, and R_5 only $(0,0)$ and $(0,1)$, denoted $fb_{17}, fb_{18}, fb_{19}, fb_{20}$, respectively. When $R_2 \leftarrow fb_6$, forward checking removes fb_{18} from the domain of the dual variable R_4. As mentioned above, each fine block has an accessor index. The set of fine blocks in a coarse block is represented by a sorted array of indices. Thus, the 'intersection' of the current domain of R_4 and $cb_x(R_4)$, where $cb_x(R_4) =$ *Support*(R_4,*CB*($R_2, fb_6), o_{24}$) is performed by iterating over the index of each fine block in the current domain of R_4, performing a binary search on the fine block indices of $cb_x(R_4)$, and removing, from the current domain of R_4, the indices of the fine blocks not listed in $cb_x(R_4)$.

We further exploit the intermediate partitions in the subproblem $\mathcal{P}_{D\varphi}$ in FB-SEARCHSUPPORT in order to *bypass the exploration of entire redundant subtrees during search.* While this mechanism did not yield significant savings in the number of nodes visited in our experiments for finding one solution, it may prove useful when we search for all solutions (i.e., ALLSOL). Fine blocks are passed over for instantiation by observing the following:

1. When instantiating a relation R_j at level l, we initialize $shared\text{-}fvars[l] \leftarrow \bigcup_{R_k \in \mathcal{R}_f} subscope(R_j, R_k)$.
2. Prior to instantiating $R_j \leftarrow fb_j$ at level l in search, we check in *equiv-FBs* whether or not an equivalent fine block was already instantiated. That is, we check $equiv\text{-}FBs[R_j, v_j]$ where $v_j = \pi_{shared\text{-}fvars[l]}(fb_j)$. If the entry is $false$, fb_j need not be instantiated because an equivalent fine block in the same intermediate partition was already found inconsistent on a previous path in the same search. (Note that the entry cannot be *true* because search terminates after finding the first solution.) When the domain of a future 'variable' is annihilated during forward checking for v_j, $equiv\text{-}FBs[R_j, v_j]$ is marked as $false$.
3. When unlabeling a 'variable' R_j at a level l (upon backtracking), $equiv\text{-}FBs[R_j]$ and $shared\text{-}fvars[l]$ are set to $\emptyset$.

Complexity. When deleting a tuple during search, it is important to maintain the correct counts of fine and coarse blocks. Each tuple deletion costs $O(e^2)$ updates. Updates are constant time thanks to the *fb-subscope-2-cb* and *tup-2-fp* tables. The cost of these updates is, in practice, greatly dwarfed by that of FB-SEARCHSUPPORT. The time complexity of PERFB is identical to that of PERTUPLE, and dominated by the $O(t^{m-1})$ search conducted in FB-SEARCHSUPPORT [16]. Additionally, PERFB performs at most as many calls to FB-SEARCHSUPPORT as PERTUPLE does, because $\bigcup_{R_j \in \varphi \setminus \{R_i\}} subscope(R_i, R_j)$ is the same as $scope(R_i)$ in the worst case, and all fine blocks have a single tuple. Insertion and retrieval of equivalent fine blocks for R_i is done in $O(k \cdot log(t))$ time. Indeed, the entry for $equiv\text{-}FBs[R_i]$ is a binary search tree with sub-tuples of values v_i as its keys, comparing each node in the tree is $O(k)$, and $O(log(t))$ comparisons may be required when each fine block has only one tuple.

At each level of search in FB-SEARCHSUPPORT, *equiv-FBs* holds $O(t)$ fine blocks, each represented by a sub-tuple of size $O(k)$. Thus, an additional $O(m \cdot t \cdot k)$ space is required for PERFB to store the equivalent fine blocks at each level of search in FB-SEARCHSUPPORT.

6 Empirical Evaluations

We compare the performance of PERFB to that of PERTUPLE. We use the latest strategy for enforcing R($*$,m)C obtained after removing redundant edges from the dual graph [12], localizing consistency propagation to the clusters of a tree decomposition of the CSP, and bolstering propagation between adjacent clusters by the addition of constraint projections to the clusters' separators [15]. (The corresponding consistency property is denoted cl+proj-wR($*$,m)C.) Although weakening the dual graph weakens consistency for $m > 2$, it also reduces the number of combinations and, thus, cost. Importantly, localization of the constraints to clusters is an excellent 'set up' for testing intermediate partitions. For both PERFB and PERTUPLE, we used $m = \{2, 3, 4, |\psi(cl)|\}$ where $m = |\psi(cl)|$ is the number of constraints in a cluster in the tree decomposition and corresponds to enforcing the minimality of each cluster .

In our experiments, we find the first solution of an instance by backtrack search, using the dynamic variable ordering dom/deg and doing full lookahead with relational consistency (i.e., cl+proj-wR($*$,m)C for $m = \{2, 3, 4, |\psi(cl)|\}$). For the evaluation, we use benchmarks from the CSP Solver Competition that are either hard to solve, thus requiring high levels of consistency, or are challenging for R($*$,m)C, thus demonstrating the effectiveness of partitioning.[4] We limit maximum processing time to 2 hours, and the maximum memory allocation to 8GB. All CPU times are reported in seconds and include *all* processing operations, including data-structure creation, preprocessing, and search.

Table 1 lists the min, max, and mean values of the fine block sizes averaged over all instances in a benchmark, as well as the size of the largest block in any instance in the benchmark. Table 2 lists similar results for the coarse blocks. Benchmarks not shown in Table 1 all have one tuple per fine block. While the average size tends to be fairly small, some benchmarks show rather large values (e.g., modifiedRenault and tightness0.9). Even though the block sizes may seem small, our technique remains beneficial because cluster-based R($*$,m)C (i.e., cl+proj-wR($*$,m)C) restricts the neighborhood of a relation by localization.

Table 1. Absolute and averaged size of fine blocks

	Absolute	Averages		
Benchmark	Max	Min	Max	Mean
geom	17	1.0	1.2	1.0
graphColoring-hos	3	1.0	2.0	1.0
graphColoring-sgb-book	12	1.0	7.7	1.1
hanoi	2	1.0	2.0	1.0
modifiedRenault	260	1.0	25.6	1.0
rand-10-20-10	2	1.0	1.3	1.0
renault	4	1.0	4.0	1.0
ssa	8	1.0	3.1	1.1
tightness0.9	38	1.0	28.1	1.0
varDimacs	16	1.0	3.4	1.1

[4] Aim-(50, 100, 200), composed-(25-1-2, 25-1-25, 25-1-40, 25-1-80, 25-10-20, 75-1-2, 75-1-25, 75-1-40, 75-1-80), dag-rand, dubois, geom, graphColoring-(hos, mug, register-mulsol, sgb-book, sgb-games, sgb-queen), hanoi, lexVg, modifiedRenault, pret, pseudo-aim, rand-(10-20-10, 3-20-20-fcd), renault, rlfapGraphsMod, rlfapScens-Mod, ssa, super-queens, tightness0.9, varDimacs.

Table 2. Absolute and averaged size of coarse blocks

	Abs	Averages				Abs	Averages		
Benchmark	**Max**	**Min**	**Max**	**Mean**	**Benchmark**	**Max**	**Min**	**Max**	**Mean**
aim-50,100,200,pseudo	4	1.0	4.0	2.1	grCol-sgb-queen	17	10.3	10.3	10.3
cmpsed-25-1-2,25,40,80	10	1.0	10.0	8.0-8.4	hanoi	3	1.0	3.0	2.9
cmpsed-25-10-20	10	1.0	10.0	7.6	lexVg	875	1.0	484.7	3.6
cmpsed-75-1-2,25,40,80	10	1.0	10.0	8.3-8.5	modifiedRenault	48,720	1.0	48,720.0	7.9
dag-rand	108	1.0	91.6	2.9	rand-10-20-10	1,046	1.0	119.2	1.3
dubois,pret	2	1.0	2.0	1.5	rand-3-20-20-fcd	190	1.0	181.5	12.8
geom	20	6.4	20.0	15.0	renault	48,720	1.0	48,720.0	7.7
grCol-hos	6	1.0	3.3	3.3	rlfapGr/ScensMod	44,43	1.0	30.0,35.6	18.5,19.4
grCol-mug	3	1.0	2.5	2.4	ssa	31	1.0	14.7	2.1
grCol-register-mulsol	48	23.2	23.2	23.2	super-queens	49	15.6	17.6	16.4
grCol-sgb-book	12	1.0	7.7	7.5	tightness0.9	40	1.0	36.3	16.9
grCol-sgb-games	8	1.0	6.3	6.1	varDimacs	512	1.0	115.0	5.6

Table 3 summarizes our results. It reports the numbers of instances completed (#Completed) by each algorithm, those completed *only by* one algorithm, and those completed *by both* algorithms. It also reports the average CPU time, the number of calls to SEARCHSUPPORT or FB-SEARCHSUPPORT, and their ratio. For each value of m, the average CPU time is computed over instances completed by both algorithms. The best values are bolded. Note that the entry for SEARCHSUPPORT calls for $m = 2$ is blank because PERFB does not call FB-SEARCHSUPPORT in this case.

PERFB clearly wins across the board. While few instances are solved only by PERTUPLE, many more are solved only by PERFB. The discovery and exploitation of equivalent fine blocks during PERFB clearly greatly reduces the number of calls to find a support, with the poorest reduction ($m = 4$) still reducing the number of searches by over half. This saving is reflected in the reduction of CPU time because the cost of searching for a support in a combination of m relations is much larger than that of identifying and storing equivalent fine blocks (see Section 5). Although not shown here, the average percentages of nodes visited that were skipped in FB-SEARCHSUPPORT thanks to the usage of intermediate partitions are .01%, .04%, and .10% for $m = 3, 4$, and ψ, respectively. Thus, the use of intermediate partitions during FB-SEARCHSUPPORT is largely ineffectual when finding a single solution to the subproblem. (However, it may be useful for improving the performance of ALLSOL.)

The scatter plots in Figure 9a and 9b compare the CPU time of PERFB and PERTUPLE for solving all 853 instances for $m = 2$ and $m = |\psi(cl)|$. Marks below the diagonal line represent instances where PERFB outperformed PERTUPLE. Marks on the right (top) border denote instances that timed out only for PERTUPLE (PERFB). Where PERTUPLE outperforms PERFB, the instances are 'easier' and the time difference is negligible for the majority of these (note the logarithmic scale). On the other hand, for hard instances, PERFB is faster. This difference is likely due to the cost of identifying the intermediate partitions in PERFB; easy instances tend to not make use of the intermediate partitions, but may still incur the cost of identifying them. The cumulative charts in Figures 9c, 9d, 9e, and 9f display the number of instances completed within a given time

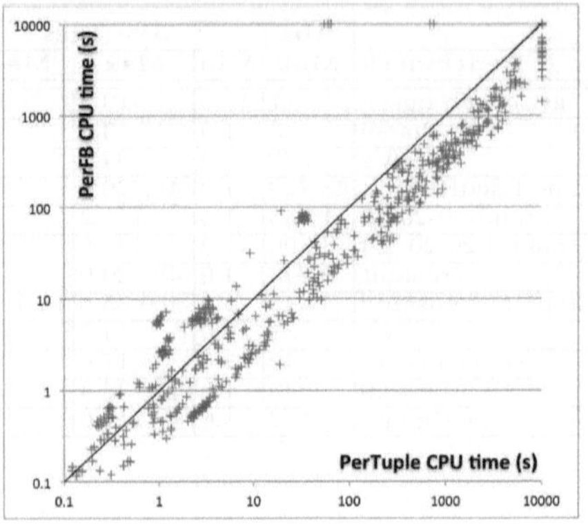

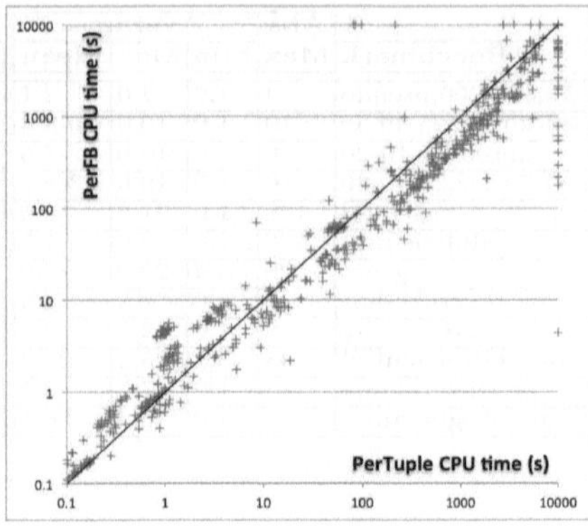

(a) $m = 2$ **(b)** $m = |\psi(cl)|$

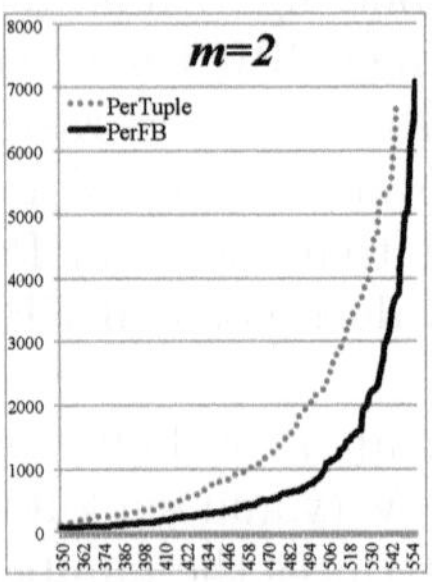

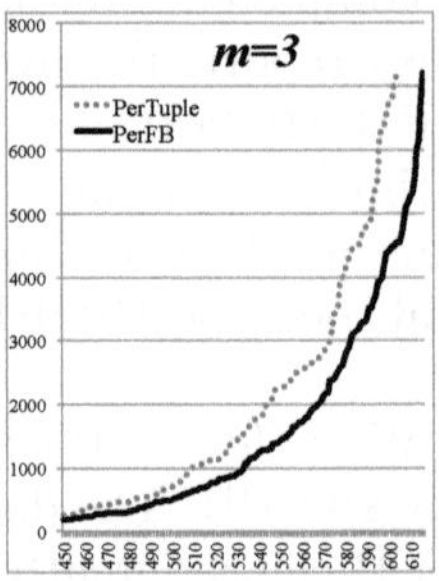

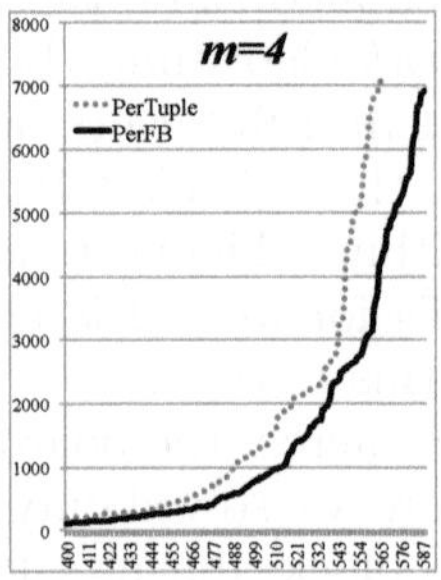

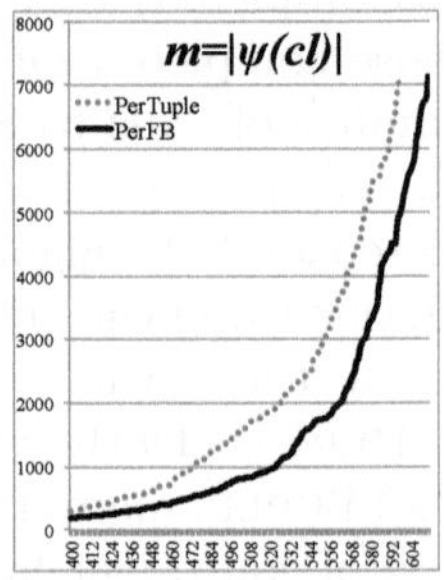

(c) $m = 2$ **(d)** $m = 3$ **(e)** $m = 4$ **(f)** $m = |\psi(cl)|$

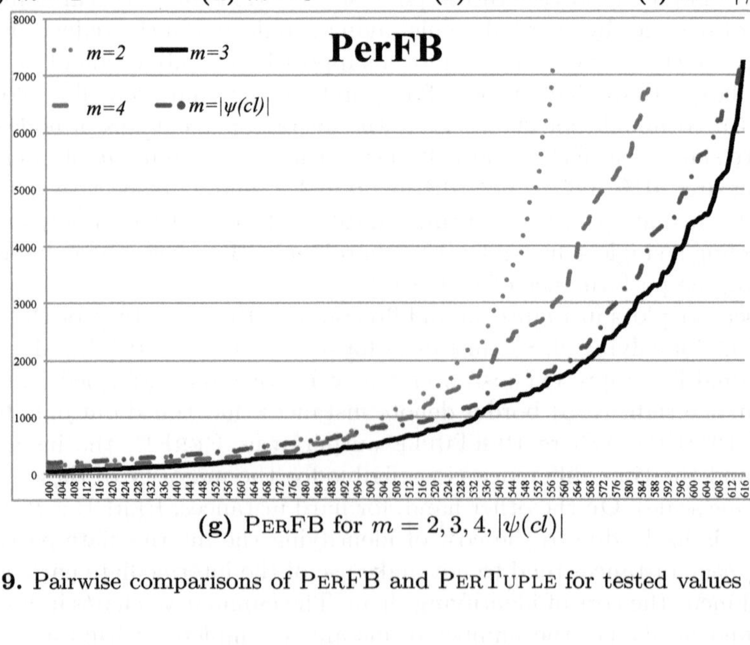

(g) PerFB for $m = 2, 3, 4, |\psi(cl)|$

Fig. 9. Pairwise comparisons of PerFB and PerTuple for tested values of m

Table 3. Summary of results for each tested value of m over 853 instances

	$m = 2$		$m = 3$		$m = 4$		$m = \|\psi(cl)\|$	
	PerTuple	PerFB	PerTuple	PerFB	PerTuple	PerFB	PerTuple	PerFB
#Completed	546	**557**	604	**616**	566	**589**	597	**615**
... only by	5	**16**	1	**13**	2	**25**	8	**26**
... by both	541		603		564		589	
Avg. CPU (sec)	538	**227**	521	**362**	472	**314**	669	**458**
SearchSupport calls (10^9)	86.4	0	88.1	26.1	52.7	19.6	24.7	8.1
ratio	–		3.37		2.69		3.06	

by each algorithm, and show that PerFB outperforms PerTuple for every m. Figure 9g compares PerFB for varying values of m. PerFB with $m = 3, |\psi(cl)|$ are the clear winners on the tested benchmarks.

We establish statistical significance by running a one-tailed paired t-test on instances completed by both PerFB and PerTuple for each value of m. The tests give $p < .01$ for each value of m. Thus, the two algorithms are extremely unlikely to have equivalent performances. This result and those in Table 3 and Figure 9 support our hypothesis that PerFB outperforms PerTuple.

7 Conclusion and Future Work

Given the importance of minimal CSPs in reasoning [9] and higher-level consistencies in solving difficult problems [13], it seems important to improve the performance of the techniques for enforcing them. In this paper, we extend the work of Samaras & Stergiou [22] to improve the initial algorithm of Karakashian et al. [16] for relational consistency by exploiting blocks of equivalent tuples at various levels of granularity, and we empirically validate our approach.

We need to evaluate the effectiveness of the approach on AllSol, the alternative algorithm for minimality [8,14]. We believe that applying the ideas explored in this paper to join computation in relational databases is a promising next step [17], potentially highly rewarding in practice.

References

1. Bessiere, C.: Constraint Propagation. In: Handbook of Constraint Programming, Elsevier (2006)
2. Bessiere, C., Cardon, S., Debruyne, R., Lecoutre, C.: Efficient Algorithms for Singleton Arc Consistency. Constraints 16(1), 25–53 (2011)
3. Bessière, C., Stergiou, K., Walsh, T.: Domain Filtering Consistencies for Non-Binary Constraints. Artificial Intelligence 172, 800–822 (2008)

4. Choueiry, B.Y., Davis, A.M.: Dynamic Bundling: Less Effort for More Solutions. In: Koenig, S., Holte, R. (eds.) SARA 2002. LNCS (LNAI), vol. 2371, pp. 64–82. Springer, Heidelberg (2002)
5. Choueiry, B.Y., Noubir, G.: On the Computation of Local Interchangeability in Discrete Constraint Satisfaction Problems. In: AAAI 1998, pp. 326–333 (1998)
6. Dechter, R., van Beek, P.: Local and Global Relational Consistency. Theor. Comput. Sci. 173(1), 283–308 (1997)
7. Freuder, E.C.: Eliminating Interchangeable Values in Constraint Satisfaction Problems. In: AAAI 1991, pp. 227–233 (1991)
8. Geschwender, D., Karakashian, S., Woodward, R., Choueiry, B.Y., Scott, S.D.: Selecting the Appropriate Consistency Algorithm for CSPs Using Machine Learning Techniques. In: Pre-PhD Student Abstract and Poster Program of AAAI 2013, pp. 1611–1612 (2013)
9. Gottlob, G.: On Minimal Constraint Networks. In: Lee, J. (ed.) CP 2011. LNCS, vol. 6876, pp. 325–339. Springer, Heidelberg (2011)
10. Gyssens, M.: On the Complexity of Join Dependencies. ACM Trans. Database Systems 11(1), 81–108 (1986)
11. Van Hentenryck, P., Deville, Y., Teng, C.-M.: A Generic Arc Consistency Algorithm and its Specializations. Artificial Intelligence 57, 291–321 (1992)
12. Janssen, P., Jégou, P., Nougier, B., Vilarem, M.-C.: A Filtering Process for General Constraint-Satisfaction Problems: Achieving Pairwise-Consistency Using an Associated Binary Representation. In: IEEE Workshop on Tools for AI, pp. 420–427 (1989)
13. Jeavons, P., Petke, J.: Local Consistency and SAT-Solvers. JAIR 43, 329–351 (2012)
14. Karakashian, S.: Practical Tractability of CSPs by Higher Level Consistency and Tree Decomposition. PhD thesis, University of Nebraska-Lincoln (2013)
15. Karakashian, S., Woodward, R., Choueiry, B.Y.: Improving the Performance of Consistency Algorithms by Localizing and Bolstering Propagation in a Tree Decomposition. In: AAAI 2013, pp. 466–473 (2013)
16. Karakashian, S., Woodward, R., Reeson, C., Choueiry, B.Y., Bessiere, C.: A First Practical Algorithm for High Levels of Relational Consistency. In: AAAI 2010, pp. 101–107 (2010)
17. Lal, A., Choueiry, B.Y.: Constraint Processing Techniques for Improving Join Computation: A Proof of Concept. In: Kuijpers, B., Revesz, P.Z. (eds.) CDB 2004. LNCS, vol. 3074, pp. 143–160. Springer, Heidelberg (2004)
18. Lal, A., Choueiry, B.Y., Freuder, E.C.: Neighborhood Interchangeability and Dynamic Bundling for Non-Binary Finite CSPs. In: AAAI 2005, pp. 387–404 (2005)
19. Lecoutre, C., Likitvivatanavong, C., Yap, R.H.C.: A Path-Optimal GAC Algorithm for Table Constraints. In: ECAI 2012, pp. 510–515 (2012)
20. Lecoutre, C., Paparrizou, A., Stergiou, K.: Extending STR to a Higher-Order Consistency. In: AAAI 2013, pp. 576–582 (2013)
21. Paparrizou, A., Stergiou, K.: An Efficient Higher-Order Consistency Algorithm for Table Constraints. In: AAAI 2012 (2012)
22. Samaras, N., Stergiou, K.: Binary Encodings of Non-binary Constraint Satisfaction Problems: Algorithms and Experimental Results. JAIR 24, 641–684 (2005)
23. Woodward, R., Karakashian, S., Choueiry, B.Y., Bessiere, C.: Solving Difficult CSPs with Relational Neighborhood Inverse Consistency. In: AAAI 2011, pp. 112–119 (2011)
24. Woodward, R.J., Karakashian, S., Choueiry, B.Y., Bessiere, C.: Revisiting Neighborhood Inverse Consistency on Binary CSPs. In: Milano, M. (ed.) CP 2012. LNCS, vol. 7514, pp. 688–703. Springer, Heidelberg (2012)

Domain Views for Constraint Programming

Pascal Van Hentenryck[1] and Laurent Michel[2]

[1] NICTA, Australia, RI 02912
[2] University of Connecticut, Storrs, CT 06269-2155, USA

Abstract. Traditional constraint-programming systems provide the concept of *variable views* which implement a view of the type $y = f(x)$ by delegating operations on variable y to variable x. While the traditional support is limited to bound consistency, this paper offers views that support domain consistency without any limitations. This paper proposes the alternative concept of *domain views* which delegate all domain operations. Domain views preserve the benefits of variable views, simplify the implementation of value-based propagation, and also support non-injective views compositionally. Experimental results demonstrate the practical benefits of domain views. The paper also reveals a subtle interaction between views and the exploitation of constraint idempotence, which may lead to incomplete propagation.

1 Introduction

Constraint programming systems provide rich libraries of constraints, each of which models some specific structure useful across a wide range of applications. These constraints are important both from a modeling standpoint, as they make it possible to state problems at a high level of abstraction, and from an efficiency standpoint, as they allow dedicated algorithms to exploit the specific structure. However, each of these constraints potentially come in many different forms as they can be applied, not only on variables, but also on expressions involving variables. This large number of variants presents a challenge for system developers who must produce, validate, optimize, and maintain each version of each constraint. To avoid the proliferation of such variants, system developers often prefer to design a unique variant over variables and introduce new variables and constraints to model the more complex cases. For instance, a constraint $\mathit{alldifferent}(x_1 + 1, \ldots, x_n + n)$ can be modeled by a system of constraints

$$\{\mathit{alldifferent}(y_1, \ldots, y_n), y_1 = x_1 + 1, \ldots, y_n = x_n + n\}$$

where the y_i's are new variables. This approach keeps the system core small but introduces an overhead in time and space. Indeed, the new constraints must be propagated through the constraint engine and the system must maintain additional domains and constraints, increasing space/time costs.

System designers have sought ways to mitigate this difficulty and proposed several solutions. Prolog-style languages offered indexicals [4,2] while C++ libraries like Ilog Solver [5] introduces the concept of variable views. For an injective function f and a variable (or a view) x, a variable view y enforces the

B. O'Sullivan (Ed.): CP 2014, LNCS 8656, pp. 705–720, 2014.

equivalent of the constraint $y = f(x)$ but it does not introduce a new constraint: Instead, it delegates all domain and constraint operations (the ability to wake constraints) on y to x, sometimes after applying f^{-1}. Variable views remove the time and space overhead mentioned above and keep the solver kernel small, thus giving us a valuable abstraction for constraint programming. Recently, [7,8] demonstrated how variable views can be implemented in terms of C++ templates, providing further improvement in speed and memory usage. The idea is to use parametric polymorphism to allow for code reuse and compile-time optimizations based on code expansion and inlining. [8] demonstrates that variable views provide significant software engineering benefits as well as great computational improvements over the standard approach.

This paper considers constraint-programming systems where views are first-class objects and can be composed arbitrarily, which is typically required when implementing high-level modeling languages. It proposes to expand the scope of constraint-programming views with an extremely simple abstraction: The concept of *domain views* which delegate domain operations *only* and keep their own constraint lists. Domain views preserve the benefits of variable views but simplify the implementation of value-based propagation, i.e., the propagation of events of the form $\langle c, x, v\rangle$, meaning that constraint c must be propagated because variable x has lost value v (e.g., [9,3]). Domain views also support *non-injective views* elegantly and compositionally. Extensive experimental results demonstrate the benefits of domain views. They also revealed a subtle interaction between views and idempotence, which is present in a well-known system.

The paper is organized as follows. Sections 2, 3, and 4 present the preliminaries on constraint programming and on views. Section presents the implementation of variable views. Section 5 introduces the concept of domain views. Section 6 demonstrates how to generalize domain views to the case where the function f is not injective. Section 7 briefly discusses how to exploit monotonicity and anti-monotonicity. Section 8 explores the interplay of views and idempotence. Section 9 presents experimental results. Section 10 discusses related work on advisors [6] and Section 11 concludes the paper.

2 Preliminaries

A constraint-programming system is organized around a queue of events $\mathcal{Q}$ and its main component is an engine propagating constraints in the queue, i.e.,

```
while ¬ empty(Q) do propagate(pop(Q));
```

For simplicity, we only consider two types of events: $\langle c, x\rangle$ and $\langle c, x, v\rangle$. An event $\langle c, x\rangle$ means that constraint c must be propagated because the domain of variable x has been shrunk. An event $\langle c, x, v\rangle$ means that constraint c must be propagated because the value v has been removed from $D(x)$. Events of the form $\langle c, x\rangle$ are sometimes called *variable-based propagation*, while those of the form $\langle c, x, v\rangle$ are sometimes called *value-based propagation*. Note that some systems also implement what is called *constraint-based propagation*, where the

```
interface Variable
  bool member(𝒱 v);
  bool remove(𝒱 v);
  void watch(𝒞 c);
  void watchValue(𝒞 c);
  void wake;
  void wakeValue(𝒱 v);
```

Fig. 1. The Variable Interface

```
implementation DomainVariable
  {𝒱} D;
  {𝒞} SC;
  {𝒞} SC_v;
DomainVariable({𝒱} D_o) { D := D_o; SC := ∅; SC_v := ∅; }
bool member(𝒱 v) { return v ∈ D;}
bool remove(𝒱 v) { if v ∈ D { D := D \ {v}; wake();wakeValue(v);}}
void watch(𝒞 c)          { SC := SC ∪ {c}; }
void watchValue(𝒞 c)     { SC_v := SC_v ∪ {c}; }
void wake()              { 𝒬 := 𝒬 ∪ {⟨c, this⟩ | c ∈ SC};}
void wakeValue(𝒱 v)      { 𝒬 := 𝒬 ∪ {⟨c, this, v⟩ | c ∈ SC_v};}
```

Fig. 2. The Implementation of a Domain Variable

event simply consists of constraint to propagate without additional information. We do not discuss constraint-based propagation here since it is easier to handle.

The propagation of a constraint may change the domains of some variables and thus introduce new events in the queue. As a result, a variable x not only maintains its domain $D(x)$ but also keeps track of the constraints it appears in so that the proper events can be inserted in the queue. As a result, a variable x is best viewed as a triple $\langle D, SC, SC_v \rangle$, where D is the domain of the variable, SC is the set of constraints involving x that use variable-based propagation, and SC_v is the set of constraints involving x that use value-based propagation. If x is a variable, we use $D(x)$, $SC(x)$ and $SC_v(x)$ to denote these three components.

For simplicity, a variable implements the interface depicted in Figure 1, where $\mathcal{V}$ denotes the set of values considered (e.g., integers or reals) and $\mathcal{C}$ the set of constraints. For a variable x, method `member(v)` tests $v \in D(x)$, method `remove(v)` implements $D(x) := D(x) \setminus \{v\}$ and returns true if the resulting domain is not empty, method `watch(c)` registers constraint c for variable-propagation, and method `watchValue(c)` registers constraint c for value-propagation. The `wake` methods are used for creating new events in the queue. With our conventions, a variable can be implemented as depicted in Figure 2.

3 Views

The purpose of this paper is to define and implement abstractions for constraints of the form $y = \psi(x)$. In a first step, the paper focuses on injective views, i.e., views in which function ψ is injective, which is the functionality provided by many constraint-programming solvers.

Definition 1 (Injective Function). *A function $\psi : D \to \mathcal{V}$ is injective if*

$$\forall v, v' \in D : \psi(v) = \psi(v') \Rightarrow v = v'.$$

The inverse $\psi^{-1} : \mathcal{V} \to D_\perp$ ($D_\perp = D \cup \{\perp\}$) of injective function ψ is

$$\psi^{-1}(w) = \begin{cases} v & \text{if } v \in D \wedge \psi(v) = w \\ \perp & \text{otherwise} \end{cases}$$

Note that the definition of ψ^{-1} is a specification: An actual implementation uses a dedicated implementation of ψ^{-1} as the following two examples illustrate.

Example 1 (Shift View). *Consider the view $y = x + c$ where c is an integer and x and y are integer variables. $\psi : \mathbb{Z} \to \mathbb{Z}$ can be specified (using lambda calculus notation [1]) as $\lambda k.k+c$. Its inverse $\psi^{-1} : \mathbb{Z} \to \mathbb{Z}$ is defined as $\lambda k.k-c$.*

Example 2 (Affine View). *Consider the view $y = ax + b$ where $a, b \in \mathbb{Z}$ and x, y are integer variables. $\psi : \mathbb{Z} \to \mathbb{Z}$ is $\lambda k.ak + b$. Its inverse $\psi^{-1} : \mathbb{Z} \to \mathbb{Z}$ is*

$$\psi^{-1} = \begin{cases} \lambda k.(k-b)/a & \text{if } (k-b) \bmod a = 0 \\ \lambda k.\perp & \text{otherwise.} \end{cases}$$

In some systems (e.g., `Choco 3.x`), views are first-class objects and *are compositional*: They can be specified by modelers and it is possible to state a view over a view. Compositional views are also important when implementing high-level modeling languages through model compilation and reformulation.

4 Variable Views

The fundamental idea of variable views [7], implemented in many systems [8], is to delegate all domain and constraint operations of variable y to variable x. A variable view thus implements an *adapter pattern* that stores neither domain nor sets of constraints. The variable view simply stores a reference to variable x and delegates all domain and constraint operations to x, possibly after applying function ψ or ψ^{-1} on the arguments. Informally speaking, the membership test $w \in D(y)$ becomes $\psi^{-1}(w) \in D(x)$, the removal operation proceeds similarly and variable x also watches all the constraints of y.

The only difficulty in variable views comes from the fact that variable x now needs to watch constraints on both x and y. For variable-based propagation, it is necessary to remember which variable is being watched for each constraint and

```
1 implementation DomainVariable
2    {V} D;
3    {⟨C,X⟩} SC;
4    {⟨C,X,F⟩} SC_v;
5 DomainVariable({V} D_o) { D := D_o; SC := ∅; SC_v := ∅;}
6 bool member(V v) { return v ∈ D;}
7 bool remove(V v) { if v ∈ D { D := D \ {v}; wake(); wakeValue(v);}}
8 void watch(C c, X y)                { SC := SC ∪ {⟨c,y⟩}; }
9 void watchValue(C c, X y, F ψ)      { SC_v := SC_v ∪ {⟨c,y,ψ⟩};}
10 void watch(C c)                    { watch(c,this); }
11 void watchValue(C c)               { watch(c,this,λk.k); }
12 void wake()          { Q := Q ∪ {⟨c,x⟩ | ⟨c,x⟩ ∈ SC};}
13 void wakeValue(V v){ Q := Q ∪ {⟨c,x,ψ(v)⟩ | ⟨c,x,ψ⟩ ∈ SC_v};}
```

Fig. 3. The Domain Variable for Variable Views

the set SC now consists of pairs $\langle c, z\rangle$ where c is a constraint and z is a variable. For value-based propagation, it is necessary to store the function ψ since it must be applied when method `wakeValue` is applied. Hence the set SC_v now contains triples of the form $\langle c, z, \psi\rangle$. These generalizations are necessary, since when a value v is removed from the domain of x, the value-based events for variable y must be of the form $\langle c, y, \psi(v)\rangle$.

The implementation of variables to support variable views is shown in Figure 3 where $\mathcal{X}$ denotes the set of variables/views and $\mathcal{F}$ the set of first-order functions. Observe the types of SC and SC_v in lines 3–4, the new methods in lines 8–9 allow to watch a constraint c for a view y, the matching redefinition of the `watch` methods, and the `wake` methods that store additional information in the queue by applying the stored function ψ on value v (line 13).

Figure 4 depicts a template for variable views in terms of an injective function ψ. A shift view specialization is shown in Figure 5. Observe that variable views do not store a domain nor constraint sets. Methods `member` and `remove` apply ψ^{-1} as mentioned earlier with only the addition of a test for the $\perp$ case. Methods `watch` and `watchValue` (lines 10–11) state a view on the view itself. In particular, line 11 illustrates the need for function composition in the case of value propagation.

The instantiation for shift views in Figure 5 highlights some interesting points. First, there is no need for a $\perp$ test, since the inverse of ψ is always in the domain of ψ. Second, value-based propagation requires the use of first-order functions (see lines 8 and 10) or objects implementing the same functionalities. In contrast, methods `member` and `remove` "inline" function ϕ^{-1} in the code, which is never stored or passed as a parameter.

Optimization. Variable views now store tuples $\langle c, z, \psi\rangle$ for value-based propagation. Observe however that z is an object so that it is possible to use it to compute function ψ. This only requires the view to provide a method `map` that maps the value v through ψ. Lines 9 and 13 in Figure 3 become

```
implementation VariableView<ψ>
    𝒳 x;
VariableView(𝒳 _x) { x := _x; }
bool member(𝒱 v) {
    if ψ⁻¹(v) ≠ ⊥ return x.member(ψ⁻¹(v)); else return false;
}
bool remove(ℤ v) {
    if ψ⁻¹(v) ≠ ⊥ return x.remove(ψ⁻¹(v)); else return true;
}
void watch(𝒞 c,𝒳 y)                  { x.watch(c,y); }
void watchValue(𝒞 c,𝒳 y,ℱ φ)         { x.watchValue(c,y,φ∘ψ);}
void watch(𝒞 c)                      { x.watch(c,this); }
void watchValue(𝒞 c)                 { x.watchValue(c,this,ψ); }
```

Fig. 4. The Template for Variable Views

```
implementation VariableShiftView
    𝒳 x;
    ℤ c;
VariableShiftView(𝒳 _x,ℤ _c) { x := _x; c := _c; }
bool member(ℤ v) { return x.member(v−c); }
bool remove(ℤ v) { return x.remove(v−c); }
void watch(𝒞 c,𝒳 y) { x.watch(c,y); }
void watchValue(𝒞 c,𝒳 y,ℤ→ℤ φ) {x.watchValue(c,y,φ∘(λk.k+c));}
void watch(𝒞 c)          { x.watch(c,this); }
void watchValue(𝒞 c) { x.watchValue(c,this,λk.k+c); }
```

Fig. 5. A Variable View for Shift Views

```
void watchValue(𝒞 c,𝒳 y)    { SC_v := SC_v ∪ {⟨c,y⟩};}
void wakeValue(𝒱 v)         { 𝒬 := 𝒬 ∪ {⟨c,x,x.map(v)⟩ | ⟨c,x⟩ ∈ SC_v};}
```

The `map` methods on standard variables and on views (defined over variable x with injective function ψ) are respectively defined as

```
𝒱 map(𝒱 v) { return v;}
𝒱 map(𝒱 v) { return ψ(x.map(v));}
```

Observe the recursive call, since views can be posted on views. This optimization clutters a bit the API of variables and views but only minimally.

Variable views are an important concept in constraint programming for injective functions. For constraint-based and variable-based propagation, the implementation is simple and efficient, although it requires to upgrade slightly the data structure to watch constraints. For value-based propagation, the implementation is a bit more cumbersome. It requires a generalization of the constraint queue and the addition of a `map` method on variables and views to avoid manipulating first-order functions. Domain views provide an extremely simple alternative, which also has the benefits of supporting non-injective functions elegantly.

```
1 implementation DomainVariable
2     {V} D;
3     {C} SC;
4     {C} SCv;
5     {X} Views;
6 DomainVariable({V} Do) { D := Do; SC := ∅; SCv := ∅; Views := ∅;}
7 void addView(X x) { Views := Views ∪ {x}; }
8 bool member(V v) { return v ∈ D;}
9 bool remove(V v) {
10     if v ∈ D
11        D := D \ {v};
12        wake();
13        wakeValue(v);
14        forall y ∈ Views { y.wake(); y.wakeValue(v);}
15 }
16 void watch(C c)        { SC := SC ∪ {c};}
17 void watchValue(C c)   { SCv := SCv ∪ {c};}
18 void wake()            { Q := Q ∪ {⟨c, this⟩ | c ∈ SC};}
19 void wakeValue(V v)    { Q := Q ∪ {⟨c, this, v⟩ | c ∈ SCv};}
```

Fig. 6. The Domain Variable for Domain Views

5 Domain Views

The key idea behind domain views is to delegate only domain operations from variable y to variable x: The view for y maintains its own constraints to watch. This removes the need to manipulate first-order functions. To implement domain views, traditional variables (and views) must store which variables are viewing them. When their domains change, they must notify their views.

Figure 6 depicts the revised implementation of domain variables to support domain views. The variable now keeps its views (line 5) and provides a method for adding a view (line 7). The only other change is in method `remove` in line 14: The domain variable calls method `wake` and `wakeValue` on its views to inform them of the loss of value v to let them schedule their own constraints.

Figure 7 shows a template for domain views in terms of an injective function ψ. A specialization for shift views is shown in Figure 8. Observe how the domain view maintains its own set of constraints. It delegates domain operations in methods `member` and `remove` as variable views did, but it does not delegate its `watch` methods, which are similar to those of a traditional domain variable. To implement views on views, the `wake` methods also wake the views (lines 18 and 22), using ψ to send the appropriate value since v is the value removed from $D(x)$. $D(x)$ may be explicit (variable) or implicit (views). The shift view in Figure 8 inlines ψ^{-1} in lines 8 and 9 and ψ in line 18.

Domain views provide an elegant alternative to variable views. They remove the need to modify the data structure for watching constraint and alleviate the need for the `map` function, while preserving the benefits of variable views and enabling more inlining for value-based propagation. They are based on a simple

```
implementation DomainView<ψ>
    𝒳 x;
    {𝒞} SC;
    {𝒞} SC_v;
    {𝒳} Views;
DomainView(𝒳 _x)    { SC := ∅; SC_v := ∅; Views := ∅;}
void addView(𝒳 x) { Views := Views ∪ {x}; }
bool member(𝒱 v) {
   if ψ⁻¹(v) ≠ ⊥ return x.member(ψ⁻¹(v)); else return false;
}
bool remove(ℤ v) {
   if ψ⁻¹(v) ≠ ⊥ return x.remove(ψ⁻¹(v)); else return true;
}
void watch(𝒞 c)          { SC := SC ∪ {c}; }
void watchValue(𝒞 c) { SC_v := SC_v ∪ {c}; }
void wake() {
   𝒬 := 𝒬 ∪ {⟨c, this⟩ | c ∈ SC};
   forall(y ∈ Views) y.wake();
}
void wakeValue(𝒱 v) {
   𝒬 := 𝒬 ∪ {⟨c, this, v⟩ | c ∈ SC_v};
   forall(y ∈ Views) y.wakeValue(ψ(v));
}
```

Fig. 7. The Template for Domain Views

idea: Only delegating the domain operations. Instead of delegating constraint watching, constraints are watched locally. It is interesting to analyze the memory requirements of both approaches. Variable views need to store variables in their constraint lists, which require space proportional to the length of these lists. In contrast, domain views only require a few pointers for their own lists, the constraints themselves being present in both approaches albeit in different lists. The viewed variables must also maintain the list of its views.

6 Non-injective Views

We now generalize domain views to non-injective functions.

Definition 2 (Inverse of a Non-Injective Function). *The inverse* $\psi^{-1} : \mathcal{V} \to 2^D_\perp$ *of non-injective function* $\psi : D \to \mathcal{V}$ *is defined as*

$$\psi^{-1}(w) = \begin{cases} \perp \text{ if } \nexists\, v \in D : \psi(v) = w \\ \{v \in D \mid \psi(v) = w\} \text{ otherwise.} \end{cases}$$

Figure 9 gives the template for non-injective views. There are only a few modifications compared to the template for injective views. The `member` function must now test membership for a set of values (line 9) and the `remove` function must remove a set of values (line 14). Finally, method `wakeValue(w)` must test

```
implementation DomainShiftView
   𝒳 x;
   {𝒞} SC;
   {𝒞} SC_v;
   {𝒳} Views;
   ℤ c;
DomainShiftView(𝒳 _x,ℤ _c) {SC := ∅;SC_v := ∅;Views := ∅;c := _c;}
bool member(ℤ v) { return x.member(v−c); }
bool remove(ℤ v) { return x.remove(v−c); }
void watch(𝒞 c)   { SC := SC ∪ {c}; }
void watchValue(𝒞 c) { SC_v := SC_v ∪ {c}; }
void wake() {
  𝒬 := 𝒬 ∪ {⟨c, this⟩ | c ∈ SC};
  forall(y ∈ Views) y.wake();
}
void wakeValue(𝒱 v) {
  𝒬 := 𝒬 ∪ {⟨c, this, v⟩ | c ∈ SC_v};
  forall(y ∈ Views) y.wakeValue(v + c);
}
```

Fig. 8. A Domain View for Shift Views

membership of $v = \psi(w)$ (line 25), since there may be multiple supports for v in $D(x)$.

The key advantage of domain views is that they own their constraints. In the context of non-injective functions, this is critical since only the view "knows" whether its constraints must be scheduled for propagation. It is more difficult and less elegant, but not impossible, to generalize variable views to support non-injective functions. For a view $y = f(x)$, when a value v is removed from $D(x)$, it is no longer sufficient to just use the `map` function. The view must decide whether $f(v)$ is still supported for y. Moreover, if we have a view $z = g(y)$ and variable x is trying to decide whether to schedule a constraint involving z, it must query z to find out whether $g(f(v))$ is still supported, which depends on whether $f(v)$ is still supported in variable y. Hence, to implement non-injective functions in variable views, waking constraints up must be conditional. It is necessary to implement a method `needToSchedule` on views to determine if the original removal will remove a value on the views. Method `wakeValue` becomes

```
void wakeValue(𝒱 v) {
   𝒬 := 𝒬 ∪ {⟨c, x, x.map(v)⟩ | ⟨c, x⟩ ∈ SC_v & x.needToSchedule(v)};}
```

The implementation of `needToSchedule` must also be recursive (like the `map` function) to handle the case of views on views. The resulting complexity contrasts with the simplicity of domain views.

Literal Views Reified constraints are a fundamental abstraction in constraint programming. For instance, In a magic series s of length n, every s_i must satisfy $s_i = \sum_{j=0}^{n-1}(s_j = i)$, i.e., it states that s_i should be the number of occurrences of value i in s itself. To implement this behavior, one could rely on auxiliary Boolean

```
implementation NonInjectiveDomainView<ψ>
   𝒳 x;
   {𝒞} SC;
   {𝒞} SC_v;
   {𝒳} Views;
NonInjectiveDomainView(𝒳 _x) {SC := ∅; SC_v := ∅; Views := ∅;}
void addView(𝒳 x) { Views := Views ∪ {x}; }
bool member(𝒱 v) {
  if ψ^{-1}(v) ≠ ⊥ return ∃w ∈ ψ^{-1}(v) : x.member(w);
  else return false;
}
bool remove(𝒱 v) {
  if ψ^{-1}(v) ≠ ⊥
     forall(w ∈ ψ^{-1}(v)) if ¬ x.remove(w) return false;
  return true;
}
void watch(𝒞 c)        { SC := SC ∪ {c}; }
void watchValue(𝒞 c) { SC_v := SC_v ∪ {c}; }
void wake() {
  𝒬 := 𝒬 ∪ {⟨c, this⟩ | c ∈ SC};
  forall(y ∈ Views) y.wake();
}
void wakeValue(𝒱 w) {
  v = ψ(w);
  if x.member(v)
     𝒬 := 𝒬 ∪ {⟨c, this, v⟩ | c ∈ SC_v};
     forall(y ∈ Views) y.wakeValue(ψ(v));
}
```

Fig. 9. The Template for Non-Injective Domain Views

variables $b_{ij} \Leftrightarrow s_j = i$. for every i and j in $0..n-1$ leading to a quadratic number of Boolean variables and reified equality constraints. The reification $b \Leftrightarrow x = i$ can be seen as a non-injective view and Figure 10 describes its implementation. The view uses two methods not described before: Method `isBoundTo(i)` on variable x holds if $D(x) = \{i\}$, while method `bind(i)` succeeds if $i \in D(x)$ and reduces the domain $D(x)$ to $\{i\}$. With these two functions, the implementation is direct with the methods `member`, `remove`, and `wakeValue` carried out by case analysis on the value of the "reified variable".

Modulo Views We now show a view for a constraint $y = x \bmod k$ with $k \in \mathbb{Z}$. The view implementation maintains the supports for each value $v \in D(y)$, i.e.,

$$\forall v \in D(y)\ s_v = \{w \mid w \in D(x) \land w \bmod k = v\}$$

Figure 11 depicts a sketch of a simple implementation.

```
implementation ReifedDomainView
   𝒳 x;
   {𝒞} SC;
   {𝒞} SC_v;
   {𝒳} Views;
   ℤ i;
DomainReifiedView(𝒳 _x,ℤ _i) {SC:= ∅;SC_v:= ∅;Views:= ∅; i := _i;}
bool member(ℤ v) {
  if v = 0 return ¬x.isBoundTo(i);
  else return x.member(i);
}
bool remove(ℤ v) {
   if v = 0 return x.bind(i);
   else return x.remove(i);
}
...
void wakeValue(𝒱 v) {
   if v = i
      𝒬 := 𝒬 ∪ {⟨c, this, 1⟩ | c ∈ SC_v};
      forall(y ∈ Views) y.wakeValue(1);
   else
      if ¬member(0)
         𝒬 := 𝒬 ∪ {⟨c, this, 0⟩ | c ∈ SC_v};
         forall(y ∈ Views) y.wakeValue(0);
}
```

Fig. 10. A Domain View for Reified Views

Element Views. Even more strikingly perhaps is the ability to implement an element view[1] replacing the classic constraint $y = c[x]$ where x and y are variables and c is an array of constants. The implementation of the non-injective view (not shown here for brevity's sake) maintains an additional simple data structure tracking the supports for values in c reachable thanks to x and achieves domain-consistency. Awakenings caused by x prompts an update of the data structure and the changes are immediately visible to the view. The benefit is the elimination of the 'micro' constraint and the disappearance of value events related to value losses emanating in $D(x)$ or $D(y)$.

7 Monotone and Anti-Monotone Views

We briefly mention how to exploit monotone and anti-monotone properties to perform additional operations such as `updateMin` and `updateMax`. These techniques are well-known and are only reviewed here for completeness.

Definition 3 (Monotone/AntiMonotone Function). *An injective function ψ is monotone if $\forall v, w : v \leq w \rightarrow \psi(v) \leq \psi(w)$. It is anti-monotone if $\forall v, w : v \leq w \rightarrow \psi(v) \geq \psi(w)$.*

[1] Many thanks to Nicolas Beldiceanu for asking whether this would be possible.

```
implementation ModuloDomainView ...
   int    k;
   {ℤ}[] S;
DomainReifiedView(𝒳 _x,ℤ _k) { ... }
bool member(ℤ v) { return S_v ≠ ∅; }
bool remove(ℤ v) {
  forall(w ∈ S_v) if ¬ x.remove(w) return false;
  return true;
}
void wakeValue(𝒱 w) {
   v := w mod k;
   if ¬ member(v)
       Q := Q ∪ {⟨c, this, v⟩ | c ∈ SC_v};
       forall(y ∈ Views) y.wakeValue(v);
}
```

Fig. 11. A Domain View for a Modulo Function

If $\psi : \mathbb{Z} \rightarrow \mathbb{Z}$ is a monotone function and y is a view on x, then the update operations on bounds becomes

```
bool updateMin(ℤ v) { return x.updateMin(ψ⁻¹(v)); }
bool updateMax(ℤ v) { return x.updateMax(ψ⁻¹(v)); }
```

ignoring the case where $\psi^{-1}(v)$ is not well-defined. When ψ is anti-monotone, they become

```
bool updateMin(ℤ v) { return x.updateMax(ψ⁻¹(v)); }
bool updateMax(ℤ v) { return x.updateMin(ψ⁻¹(v)); }
```

8 Idempotence and Views

Idempotence is a property often exploited in solvers. It enables to terminate the propagation of a constraint even when some of its variables have been modified without affecting the resulting pruning. For instance, the constraint $\sum_{i \in S} a_i \cdot x_i \geq c$ often exploits idempotence and updates its variables at most once: Its propagation is not iterated either internally or through the propagation loop. However, such an implementation is incomplete when using views.

Example 1 (Simple Linear). Consider $x_1 \in \{0,1\}, x_2 = \neg x_1, x_3 \in \{0,1\}$ where x_2 is a negated views on x_1 and both x_1 and x_3 are Boolean variables. Assume the addition of the constraint $5 \cdot x_1 + 3 \cdot x_2 + 3 \cdot x_3 \geq 7$. When posted, the constraint propagation concludes that $x_1 = 1$ must hold in order to reach 7. However, the propagator is unaware of the relation $x_2 = \neg x_1$ and is not iterated. Since the propagator assumes that either x_2 or x_3 could be used to reach the target 7, $D(x_3)$ is left untouched. This is wrong. Indeed, the x_2 view has immediately fixed $x_2 = 0$ and the constraint propagator should have assigned x_3 to 1.

In other words, views create side-channels that can invalidate the propagator assumptions. This behavior is not a pure theoretical musing: Such an incompleteness is in fact present in `Choco 3.1.1`, as reproduced below

```
public void buildModel() {
  BoolVar x1 = VariableFactory.bool("x1",solver);
  BoolVar x2 = VariableFactory.not(x1);
  BoolVar x3 = VariableFactory.bool("x3",solver);
  IntVar[] av = new IntVar[]{x1,x2,x3};
  int[] coef = new int[]{5,3,2};
  IntVar rhs = VariableFactory.fixed(7,solver);
  solver.post(IntConstraintFactory.scalar(av,coef,">=",rhs));
  System.out.format("Before: %s\n",toString(av));
  try { solver.propagate(); } catch(Exception e) {}
  System.out.format("After : %s\n",toString(av));
}
```

which produces the following output:

```
Before: x1 = [0,1] not(x1) x3 = [0,1]
After : x1 = 1 not(x1) x3 = [0,1]
```

The issue can be solved by simplifying the solver and eliminating the notion of idempotence. One could instead analyze the constraint inputs to test for a side-channel through a chain of views and enable idempotence adaptively.

9 Empirical Evaluation

The experiments were run on MacOS X 10.8.3 running on a Core i7 at 2.6Ghz, using the OBJECTIVE-CP optimization system [10]. The complete implementation of the integer and Boolean variables, along with their domain and their views (including literal views) is around 3,200 lines of code, which is similar to the type of code reuse advertised for Gecode [8]. OBJECTIVE-CP pushes the methodology advocated in [8] to the limit, only supporting core constraints and using views to obtain more complex versions. For instance, the CP solver in OBJECTIVE-CP provides $\sum_{i=0}^{n} x_i \leq b$ but not $\sum_{i=0}^{n} a_i \cdot x_i \leq b$. Cost-based propagation for COP would, of course, mandate global constraints retaining the a_i. OBJECTIVE-CP supports value-based propagation and non-injective views, which demonstrates the additional functionalities provided by domain views. The experiments only aim at demonstrating the practicability of domain views: See [8] for the benefits.

Benchmarks. Validation was carried out with classic benchmarks relying on views. The experiments compare implementations with no views, with the optimized variable views (with subtype polymorphism), and domain views. When no-views are used, the implementation uses the constraints and auxiliary variables introduced during the flattening of the model. The implementation uses the same models and the search space and pruning are always identical. For `bibd`, we follow [8] and rewrite Boolean relations like $a \wedge b$ as $\neg(\neg a \vee \neg b)$ to force the use of negation views. **Knapsack** uses linear equations $\sum_{i \in S} x_i = b$ and introduces views for the coefficients. The Steel Mill **Slab** relies on literal views for the color constraint on slabs. **Debruijn** uses linear equations and reifications. **Langford** uses affine views to "shift" indices within element constraints. **Magicseries** relies on reifications. **Sport** is the sport scheduling benchmark and uses globals.

Measurements The benchmarks use a simple first-fail heuristic as decomposition may change the behavior of more advanced heuristics (e.g., WDEG) and these experiments are only interested in assessing view implementations, not inherent speed. Table 1 offers a comparative view of the results. It is based on 50 runs of each benchmark to account for the variability related to modern processors. Columns $\mu(T_{cpu})$ and $\mu(T_{wc})$ give the average user-time or wall-clock times in milliseconds. Columns $\sigma(T_{cpu})$ and $\sigma(T_{wc})$ report the standard deviations for those run times. Column $|M|$ reports the peak memory consumption in kilobytes for the entire process. Measurements were taken at the level of `malloc` and includes all memory allocations done by the executable. Finally, column $P.$ reports the number of propagation events recorded by the engine (in thousands).

Without surprise, a minimalist kernel *must* use views to be competitive as differences in memory consumptions and running times can be quite significant

Table 1. Experimental Results on Variable and Domain Views

Bench	type	$\mu(T_{cpu})$	$\mu(T_{wc})$	$\sigma(T_{cpu})$	$\sigma(T_{wc})$	$\|M\|(KB)$	$P.(\times 1000)$
bibd(6)	No-View	1,088.4	1,130.5	222.1	227.9	44,652	1,984
bibd(6)	Domain-View	729.8	759.2	107.6	111.0	29,197	804
bibd(6)	Var-View	644.7	671.2	59.4	60.2	28,082	805
knapsack(4)	No-View	8,857.5	8,873.9	180.9	184.8	987	33,207
knapsack(4)	Domain-View	6,768.0	6,784.5	166.8	176.1	812	2,949
knapsack(4)	Var-View	6,151.2	6,164.5	109.2	111.0	789	3,062
ais(30)	No-View	1,341.8	1,348.9	52.2	57.6	1,336	2,734
ais(30)	Domain-View	1,355.3	1,361.5	31.8	32.4	1,336	2,734
ais(30)	Var-View	1,354.9	1,362.4	26.1	26.5	1,337	2,734
sport	No-View	4,851.7	4,864.2	111.3	116.0	2,030	3,361
sport	Domain-View	4,850.9	4,864.3	179.4	189.3	2,029	3,361
sport	Var-View	4,936.1	4,949.5	231.8	236.7	2,029	3,361
langford(9/3)	No-View	6,060.5	6,076.1	298.8	306.2	1,375	54,027
langford(9/3)	Domain-View	7,008.3	7,026.6	316.8	323.5	1,368	56,893
langford(9/3)	Var-View	6,859.5	6,877.9	242.7	248.5	1,366	56,887
debruijn(2/12)	No-View	7,665.5	8,437.3	153.4	169.2	624,635	2,558
debruijn(2/12)	Domain-View	7,292.2	8,014.2	111.2	144.1	552,515	946
debruijn(2/12)	Var-View	6,935.2	7,628.2	384.9	422.4	550,792	967
slab	No-View	4,845.2	4,919.4	108.6	115.4	84,092	4,403
slab	Domain-View	2,243.5	2,294.0	128.7	138.8	51,725	909
slab	Var-View	4,509.8	4,578.7	94.7	99.4	73,929	2,968
magicserie(300)	No-View	17,951.2	18,164.7	284.0	300.8	231,622	30,771
magicserie(300)	Domain-View	8,318.3	8,443.6	191.9	201.4	122,026	257
magicserie(300)	Var-View	14,879.9	15,088.0	429.1	441.1	229,288	20,745

when forgoing views. In benchmarks involving only injective views, variable and domain views are essentially similar in time and space efficiency. Given the standard deviations, the differences in efficiency are not statistically significant, although variable views are often slightly more efficient. This is not always the case, as the sport-scheduling problem indicates. Domain views, though, support non-injective views simply and efficiently. This is particularly clear on the benchmarks using reifications, i.e, `slab` and `magicserie`. The benefits are in terms of runtime and memory consumption. The runtime gains are quite substantial, as the running time is halved on the Steel Mill Slab problem. The dramatic drop in the number of propagations is explained by the absence of constraints of the form $b \Leftrightarrow (x = v)$, yet, the same work is still carried out by the view, albeit at a much lower overhead. Finally, the *element view* was also evaluated on the *Steel Mill Slab* problem. The few (about 100) element constraints were replaced by element views. While performance was virtually unchanged, memory usage was reduced a little showing that the approach incurs no penalties. In summary, domain views do not add any measurable overhead on injective views and bring significant benefits on non-injective views, which they support elegantly.

10 Related Work

It is important to contrast the variable and domain view implementations proposed here with another approach using delta-sets and advisors [6,8]. Advisors are another way to "simulate" value-based propagation.[2] An advisor is associated with a variable and a constraint and it modifies the state of the constraint directly upon a domain modification for its variables. Advisors do not go through the propagation queue but modify the state of their constraint directly. Advisors also receive the domain change (called a delta set) which they may query.

Advisors can be associated with variable views. The view must now be upgraded to query, not only the domain, but also the delta sets. In other words, the queries on the delta must transform the domain delta, say $\{v_1, \ldots, v_n\}$, through the view to obtain $\{\phi(v_1), \ldots, \phi(v_n)\}$. Gecode [8] does not compute delta sets exactly but approximates them by intervals instead. A complete implementation of value-based propagation would require the creation of these delta sets. Advisors and delta sets can be used in the case of non-injective functions but that solution would still go through the propagation queue and use a constraint. Domain views in constrast can implement non-injective views without going through the propagation queue and do not need the concept of delta-sets.

11 Conclusion

This paper reconsidered the concept of views and proposed the concept of domain views as an alternative to the concept of variable views. Domain views only

[2] It is only a simulation since an advisor updates the constraint state but does not propagate a constraint itself. They are second-class citizens by choice in Gecode [6].

delegate domain operations and maintain their own set of constraints to watch. As a result, they simplify the implementation of constraint-programming systems featuring value-based propagation as they avoid manipulating first-order functions (or objects implementing a similar functionality). Domain views also make it possible to implement views for non-injective functions, which is particularly convenient for reified constraints. Experimental results demonstrate that domain views introduce a negligible overhead (if any) over variable views and that views over non-injective functions provide significant benefits.

References

1. Barendregt, H.P.: The Lambda Calculus – Its Syntax and Semantics. Studies in Logic and the Foundations of Mathematics, vol. 103. North-Holland (1984)
2. Carlsson, M., Ottosson, G., Carlson, B.: An open-ended finite domain constraint solver. In: Glaser, H., Hartel, P.H., Kuchen, H. (eds.) PLILP 1997. LNCS, vol. 1292, pp. 191–206. Springer, Heidelberg (1997)
3. Dynadec, I.: Comet v2.1 user manual. Technical report, Providence, RI (2009)
4. Hentenryck, P.V., Saraswat, V., Deville, Y.: Constraint processing in cc(fd). Technical report (1992)
5. Ilog Solver 4.4. Reference Manual. Ilog SA, Gentilly, France (1998)
6. Lagerkvist, M.Z., Schulte, C.: Advisors for incremental propagation. In: Bessière, C. (ed.) CP 2007. LNCS, vol. 4741, pp. 409–422. Springer, Heidelberg (2007)
7. Schulte, C., Tack, G.R.: Perfect derived propagators. In: Stuckey, P.J. (ed.) CP 2008. LNCS, vol. 5202, pp. 571–575. Springer, Heidelberg (2008)
8. Schulte, C., Tack, G.: View-based propagator derivation. Constraints 18(1), 75–107 (2013)
9. Van Hentenryck, P., Deville, Y., Teng, C.: A Generic Arc Consistency Algorithm and Its Specializations. Artificial Intelligence 57(2-3) (1992)
10. Van Hentenryck, P., Michel, L.: The Objective-CP Optimization System. In: Schulte, C. (ed.) CP 2013. LNCS, vol. 8124, pp. 8–29. Springer, Heidelberg (2013)

Global Constraints in Distributed CSP: Concurrent GAC and Explanations in ABT

Mohamed Wahbi and Kenneth N. Brown

Insight Centre for Data Analytics, School of Computer Science and IT,
University College Cork, Ireland
{mohamed.wahbi,ken.brown}@insight-centre.org

Abstract. The expressiveness of Distributed CSP has been recently enhanced to include global constraints. Careful reformulation of contractible global constraints has been shown to improve efficiency. In this paper, we first show that explained global constraints further improves the efficiency in distributed problems, sometimes by over two orders of magnitude. We then propose maintaining GAC concurrently for any global constraint, without reformulation. We show empirically that concurrent GAC significantly reduces both message passing and computation time, achieving an order of magnitude improvement on some distributed meeting scheduling problems.

1 Introduction

The *distributed constraint satisfaction problem* models decision problems where physically distributed agents control different decision variables. To solve the problems, the agents must perform local computation and exchange messages, to communicate value choices, inferred no-goods, algorithm control decisions or problem descriptions. One of the main reasons for the success of centralized constraint programming is the use of *global constraints*, in which a relation between a group of variables comes equipped with powerful filtering algorithms, which quickly infer consequences of value assignments and greatly reduce search. Implementing global constraints in Distributed CSP has been less successful, because the distributed control over the variables has led to delayed filtering. If Distributed CSP is to be more widely applied to real problems, more effective filtering for global constraints in distributed problems is required.

Three ways to represent global constraints in DisCSP have been recently proposed [3]. The *nested* representation of *contractible* global constraints offers significant improvements over the other approaches. In addition, propagating unconditional no-goods (that is, values that cannot be in a solution, regardless of other choices) while enforcing generalized arc-consistency (GAC) was also shown to offer an improvement.

Here, we propose two further improvements to the handling of global constraints in Distributed CSP, implemented in ABT.[1] First, we use *explained* global constraints, which allows us to generate more efficient no-goods– that is, given an ordered partial

[1] We focus on ABT, since it is the only DisCSP algorithm (apart from MACA [34]) we know of that has been extended to maintain arc consistency. Our methods are applicable to any algorithm, but without global constraint filtering would provide little benefit.

B. O'Sullivan (Ed.): CP 2014, LNCS 8656, pp. 721–737, 2014.

assignment of values to variables, we identify the earliest inconsistent subset [13, 8]. Each agent evaluating a constraint maintains a copy of the domains of all other variables in the constraint, and whenever it receives either a value assignment or a no-good, it maintains GAC on these domains. Secondly, we introduce a full representation of each global constraint by each agent constrained by it. This allows us to maintain GAC concurrently at each agent. Our aim is to trade off this potentially redundant filtering for faster identification of no-goods, and reduce search and message passing.

We evaluate our algorithms empirically on a set of benchmarks from the literature, including random problems, quasi-groups with holes, and distributed meeting scheduling problems. We demonstrate that the use of explained constraints significantly improves performance over previous methods on the harder problems, sometimes achieving over two orders of magnitude reduction in both non-concurrent computation and messaging. We then show that concurrent filtering offers another significant improvement when added to explained constraints, achieving up to an order of magnitude improvement on the meeting scheduling problems in both non-concurrent computation and messaging, without requiring any reformulation. Finally, we consider the total computation load over all agents, and we show that, surprisingly, concurrent filtering can reduce the total effort – it appears that early identification of relevant no-goods outweighs any redundant filtering.

This paper is structured as follows. Section 2 gives the necessary background on centralized CSP, global constraints, and distributed CSP. It then discusses the use of global constraints in distributed CSP. We present our use of explained global constraints on DisCSP in Section 3, followed by a comparison to previous work. Section 4 introduces concurrent filtering using the full representation of global constraints. We report experimental results in Section 5. Finally, we conclude the paper in Section 6.

2 Background and Related Work

2.1 Centralised CSPs and Global Constraints

The *constraint satisfaction problem* is a triple $(\mathcal{X}, \mathcal{D}, \mathcal{C})$, where $\mathcal{X}$ is a set of variables $\{x_1, \ldots, x_n\}$, $\mathcal{D} = \{D_1, \ldots, D_n\}$ is a set of domains, where D_i is a finite set of values from which one value must be assigned to variable x_i, and $\mathcal{C}$ is a set of constraints. A constraint $C(X) \in \mathcal{C}$, on the ordered subset of variables $X = (x_{j_1}, \ldots, x_{j_k})$, is $C(X) \subseteq D_{j_1} \times \cdots \times D_{j_k}$, and specifies the tuples of values which may be assigned simultaneously to the variables in X. $|X|$ is the *arity* of $C(X)$, and X is its *scope*. A tuple $\tau = (v_{j_1}, \ldots, v_{j_k}) \in C(X)$ is a *support* for $C(X)$, and $\tau[x_i]$ is the value of x_i in τ. We denote by $C_i \subseteq \mathcal{C}$ all constraints that involve x_i. A *solution* is an assignment to each variable of a value from its domain, satisfying all the constraints.

A *global constraint* captures a relation over an arbitrary number of variables. For example, the ALLDIFF constraint states that the values assigned to the variables in its scope must all be different [22]. Filtering algorithms which exploit the specific structure of global constraints are one of the main strengths of constraint programming [23]. A value $v_i \in D_i$, $x_i \in X$ is *generalized arc-consistent (GAC)* with respect to $C(X)$ iff there exists a support τ for $C(X)$ such that $v_i = \tau[x_i]$, and for every $x_j \in X$, $x_i \neq x_j$, $\tau[x_j] \in D_j$. Variable x_i is GAC if all its values are GAC with respect to every constraint in C_i. A CSP is GAC if all its variables are GAC. During search, any value $v \in D_i$ that is not GAC can be removed from D_i.

Global constraint $C(X)$ is *binary-decomposable* without extra variables [4] if it is equivalent to a conjunction of binary constraints involving only variables in X. ALLDIFF is binary-decomposable. For example, ALLDIFF(x_1,x_2,x_3,x_4) is equivalent to $(x_1 \neq x_2) \wedge (x_1 \neq x_3) \wedge (x_1 \neq x_4) \wedge (x_2 \neq x_3) \wedge (x_2 \neq x_4) \wedge (x_3 \neq x_4)$. Global constraint $C(X)$, where $X = (x_{j_1}, \ldots, x_{j_{k+1}})$, is *contractible* [16] iff for any support $(v_{j_1}, \ldots, v_{j_k}, v_{j_{k+1}})$ for $C(X)$, then $(v_{j_1}, \ldots, v_{j_k})$ is a support for $C(x_{j_1}, \ldots, x_{j_k})$. ALLDIFF is a contractible constraint, since the projection of the supports for ALLDIFF(x_1, x_2, x_3, x_4) onto (x_1, x_2, x_3) are also supports for ALLDIFF(x_1, x_2, x_3). However, EXACTLY(k, X, v) that specifies that value v is assigned exactly to k variables in X is not contractible [2].

An *explanation* is a subset of the original constraints of the problem plus a set of decision constraints (variable assignments) made during the search which together are sufficient to justify domain reductions [13, 11]. Explanations were originally introduced [14] to improve intelligent backtracking, allowing the search to jump back to the cause of a failure. Computing an explanation for a reduction caused by binary constraints is relatively simple, but is more complex for a global constraint, where the explanation will depend on the chosen filtering algorithm. For example, for the ALLDIFF constraint [23], the filtering algorithm is based on computing a residual graph constructed from the maximum matching on the variable-value bipartite graph and from the possible values of variables in the constraint. [25] shows how to generate corresponding explanations: given the residual graph, the removal of an arc starting from a vertex belonging to a strongly connected component $\mathcal{S}_1$ to a distinct strongly connected component $\mathcal{S}_2$ is explained by all missing arcs from a descendant component of $\mathcal{S}_2$ to an ancestor component of $\mathcal{S}_1$ (since any one of these arcs would merge $\mathcal{S}_1$ and $\mathcal{S}_2$ into the same strongly connected component). Other global constraints have also been explained [14, 24, 13, 11, 27, 8, 9, 10], with the explanations used to improve the efficiency of backjumping.

2.2 Distributed CSPs

Distributed. CSP (*DisCSP*) [37] models problems where distinct agents control different variables. DisCSP is a 4-tuple $(\mathcal{A}, \mathcal{X}, \mathcal{D}, \mathcal{C})$, where $\mathcal{X}, \mathcal{D}$ and $\mathcal{C}$ are as above, and $\mathcal{A}$ is a set of agents $\{A_1, \ldots, A_a\}$, where each variable $x_i \in \mathcal{X}$ is controlled by a single agent in $\mathcal{A}$. During a solution process, only the agent which controls a variable can assign a value to this variable. Without loss of generality, we assume each agent controls exactly one variable ($a = n$), so we use the terms agent and variable interchangeably and no longer distinguish between A_i and x_i. Each agent A_i knows all constraints relevant to its variable (C_i) and the domains of other variables involved in these constraints (its *neighbours*). A variety of problems have been tackled using DisCSP, including tracking in sensor networks [1], distributed resource allocation [20] and distributed meeting scheduling [18]. Many different algorithms have been proposed, including asynchronous backtracking [36, 5], asynchronous forward checking [19], asynchronous aggregation [29], dynamic programming [21], partially centralised search [17], and dynamic ordering [28, 39, 31].

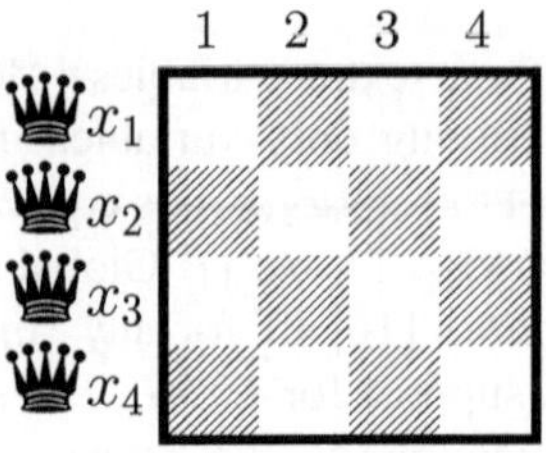

$\mathcal{A} = \{A_1, \ldots, A_4\}$
$\mathcal{X} = \{x_1, \ldots, x_4\}$
$\mathcal{D} = \{D_1, \ldots, D_4\}$,
where $D_i = \{1, 2, 3, 4\}$
$\mathcal{C} = \{c_1, \ldots, c_7\}$

$c_1 : | x_1 - x_2 | \neq 1$
$c_2 : | x_1 - x_3 | \neq 2$
$c_3 : | x_1 - x_4 | \neq 3$
$c_4 : | x_2 - x_3 | \neq 1$
$c_5 : | x_2 - x_4 | \neq 2$
$c_6 : | x_3 - x_4 | \neq 1$
c_7 : ALLDIFF$(\mathcal{X})$

Fig. 1. The distributed n-queens problem (where $n = 4$)

The original algorithm for DisCSP was Asynchronous Backtracking (ABT) [36, 5]. ABT is an asynchronous algorithm executed autonomously by each agent in the problem, and is guaranteed to converge to a global consistent solution (or detect inconsistency) in finite time. Each agent proposes values for its own variable to other agents, and reports no-goods. Agents operate asynchronously, but are subject to a known total priority order, $\prec$. For simplicity, we assume $\prec$ is the lexicographic ordering $(A_1, A_2, \ldots, A_n)$, with A_1 having highest priority. $\prec$ induces a directed acyclic graph, and constraints are directed according to $\prec$.

ABT uses the priority ordering to control the asynchronous search. Each agent selects an assignment for its variable that is consistent with known choices of higher priority neighbours, and then sends its selected value across the directed arcs to its lower priority neighbours. When no value is possible for a variable, the inconsistency is reported to a higher priority agent closest in the ordering, in the form of a no-good. The higher agent then adds the no-good to its constraint store. For a non-binary constraint $C(X)$ (arity > 2), only the lowest agent in the ordering in the scope of the constraint will evaluate $C(X)$, and only when it is totally instantiated [7, 3].

Fig. 1 shows a model of a distributed n-queens problem using an ALLDIFF constraint. The goal is to put 4 queens on the board such that no queen attacks another queen. There is an integer variable x_i for every row i. There are four agents, each of which controls one queen in one row. The domain of each x_i is $D_i = \{1, 2, 3, 4\}$. There exists a global ALLDIFF constraint (i.e., ALLDIFF(x_1, x_2, x_3, x_4)) that forbids the queens being placed in the same column. There is a binary constraint (e.g., c_1) between each pair of queens that forbids those queens being placed on the same diagonal (i.e., $| x_i - x_j | \neq | i - j | \; \forall i, j_{\neq i} \in \{1..4\}$).

2.3 Global Constraints in Distributed CSPs

[3] formulated three different ways to model a global constraint in a DisCSP.

Direct Representation: Applicable to all global constraints. The direct representation of the constraint $C(X)$ is a single copy of $C(X)$ to be evaluated by the lowest agent in its scope, A_i. Directed links from other agents in $C(X)$ to A_i are established.

In the example of Fig. 1, A_1 starts with no constraints. A_2 evaluates constraint c_1, A_3 evaluates c_2 and c_4, while A_4 evaluates all other constraints (i.e., c_3, c_5, c_6 and c_7).

Nested Representation: Restricted to contractible global constraints. The nested representation of the constraint $\{C(x_{j_1},\ldots,x_{j_k})\}$ is the set of constraints $C(x_{j_1},\ldots,x_{j_m}) \mid m \in 2..k$. In the example of Fig. 1, constraint c_7 will be represented by 3 different constraints: (c_7) ALLDIFF(x_1, x_2, x_3, x_4), (c_8^n) ALLDIFF(x_1, x_2, x_3), and (c_9^n) ALLDIFF(x_1, x_2). Constraints c_1 to c_7 will be evaluated by the same agents as in the direct representation. c_8^n is evaluated by agent A_3 and c_9^n by agent A_2.

Binary Representation: Restricted to binary-decomposable global constraints. The binary representation of a constraint $C(X)$ is the set of constraints in its binary decomposition. In the example of Fig. 1, constraint c_7 will be represented by 6 constraints: (c_8^b) ($x_1 \neq x_2$), (c_9^b) ($x_1 \neq x_3$), (c_{10}^b) ($x_1 \neq x_4$), (c_{11}^b) ($x_2 \neq x_3$), (c_{12}^b) ($x_2 \neq x_4$) and (c_{13}^b) ($x_3 \neq x_4$). c_1 to c_6 will be evaluated by the same agents as in the direct representation. Agent A_2 will evaluate c_8^b, agent A_3 will evaluate c_9^b and c_{11}^b and A_3 will evaluate c_{10}^b, c_{12}^b and c_{13}^b.

In addition, in order to take advantage of the standard filtering algorithms for global constraint in DisCSPs, [3] proposed ABT-UGAC (ABT with unconditional GAC). ABT-UGAC maintains a limited form of GAC restricted to unconditional deletions (values removed by a no-good with an empty precondition). In a preprocessing step, the DisCSP is made GAC. Unconditional GAC is then enforced in ABT as follows. When receiving a no-good with an empty precondition justifying the removal of its value, an agent A_i can unconditionally delete its value from D_i. This deletion may then propagate, and cause further deletions (see [3] for details).

[3] showed that the direct representation is the least efficient, while the nested one performs best. ABT-UGAC always improves the performance.

3 Maintaining GAC in ABT

In ABT, the lowest priority agent in the scope X is in charge of evaluating a global constraint $C(X)$ when it is fully instantiated [7, 3]. This method of evaluating global constraints is a major weakness of representing global constraints in ABT. First, it does not take advantage of the global constraints' filtering power. Both nested and direct representations only use global constraints as checkers instead of using their filtering algorithm to prune inconsistent values. Second, it produces chronological backtracks because the deduced no-good will contain all assignments of the agents on the global constraints. Thus, it creates unnecessary work for the agents because it does not address the real reason for the failure.

For example, in the direct representation of the problem of Fig. 1, agent A_4 will not evaluate constraint c_7 until all assignments of variables x_1, x_2 and x_3 are received. Thus, agent A_4 may receive the assignments $x_1 = 1 \wedge x_2 = 1 \wedge x_3 = 1$. In this situation, A_4 will send a no-good $x_1 = 1 \wedge x_2 = 1 \rightarrow x_3 \neq 1$ to A_3. Then, A_3 will change its value to 4 because 2 is removed by c_4 and 3 by c_2. The new assignment ($x_3 = 4$) will lead to another deadend in A_4 with the following no-good ($x_1 = 1 \wedge x_2 = 1 \rightarrow x_3 \neq 4$). Once it receives this no-good, A_3 discovers the real reason for the failure (i.e., $x_1 = 1 \rightarrow x_2 \neq 1$). Thus, unnecessary work is performed when using this evaluation mechanism.

Instead of evaluating a constraint only when it is fully instantiated, we propose maintaining GAC each time an event occurs on a variable involved in a constraint in the agent's constraint store. In order to get more precise no-goods we use explained global constraints, so that all domain reductions are justified [13, 8], and when a dead-end occurs, we will get more informative no-goods. For ALLDIFF, we implement the explained filtering algorithm described previously [23, 25]. For ATMOST(k, X, v) and ATLEAST(k, X, v), we modify the implementations of [14] as below, and then use both methods together for the EXACTLY(k, X, v) constraint.

ATMOST(k, X, v): whenever a constrained variable $x_i \in X$ is assigned the occurrence value (v) or its domain is reduced to the occurrence value ($D_i = \{v\}$), we label it as 'sure' ($sure(i) = true$). A 'sure' variable can be either assigned (i.e, $x_i = v \mid x_i \in X \wedge sure(i)$) or unassigned (i.e, $D_j = \{v\} \mid x_j \in X \wedge sure(j)$). If the number of 'sure' variables equals the number of occurrences (k), the value v will be removed from the domain of all other variables in X. These removals are explained by the union of the set of decision constraints of 'sure' assigned variables x_i (i.e., $x_i = v \mid x_i \in X \wedge sure(i)$) with the union of the explanations that reduces the other sure variable domains to v (i.e., $D_j = \{v\} \mid x_j \in X \wedge sure(j)$).

ATLEAST(k, X, v): all constrained variables $x_i \in X$ that can be assigned v are labelled as 'possible' ($possible(i) = true$). If a variable $x_j \in X$ can not be assigned v its 'possible' flag is false ($possible(j) = false$). Whenever the number of 'possible' variables equals the number of occurrences (k), the domains of all these 'possible' variables are reduced to v (i.e., $D_i = \{v\} \mid x_i \in X \wedge possible(i)$). These reductions are explained by the union of explanations that justify the removal of v from the domain of other variables (i.e., $v \notin D_j \mid x_j \in X \wedge \neg possible(j)$).

To maintain GAC, some minor changes to ABT are required. First, when an agent A_i receives a message containing an assignment or a no-good, it updates its AgentView with the given assignment. The AgentView of an agent stores the most up-to-date assignments of its higher neighbours [5]. Then, A_i adds the received constraint (assignment or no-good) to its constraint store (no-goods inconsistent with the AgentView are removed to keep the space complexity polynomial [5]). Next, A_i maintains GAC. Second, whenever an agent A_i assigns a value to its variable ($x_i = v_i$) it adds its assignment to its constraint store (C_i). Next, it maintains GAC. When agent A_i maintains GAC using explained constraints each domain reduction is justified by a precise explanation. Thus, whenever a dead-end occurs, the no-good that will be generated will be more precise (a subset of the AgentView). However, if non-explained constraints are used, whenever a dead-end occurs, the generated no-good contains all assignments in the AgentView.[2] If the domain of a variable is emptied while maintaining GAC, agent A_i generates a new no-good from the explanations stored for value removals of that variable. If the generated no-good contains the assignment of agent A_i (i.e., v_i is not consistent after maintaining GAC), A_i tries to assign another value to x_i. If no value is possible for x_i or if the generated no-good doesn't contain its assignment, A_i backtracks by sending the no-good justifying the failure to the closest agent in

[2] Note that replacing the global constraint checker with the global constraint filter does not change the operation of ABT, apart from detecting earlier no-goods and generating more informed messages.

the no-good. If C_i is GAC, A_i sends its new assignment ($x_i = v_i$) to all its lower priority neighbours.

Now, in the direct representation example, when agent A_4 maintains GAC on its constraints (i.e., c_3, c_5, c_6 and c_7) it can directly discover the no-good (i.e., $x_1 = 1 \rightarrow x_2 \neq 1$) without the assignment of x_3.

3.1 Evaluation of Explained Global Constraints in ABT

We experimentally compare ABT(direct), ABT(nested), ABT-UGAC(direct) and ABT-UGAC(nested) as presented in [3] (evaluating constraints when they are totally instantiated) to ABT-GAC(direct) and ABT-GAC(nested) that maintain GAC thanks to explained constraints as presented above. Algorithms are tested using the static *max-degree* agent ordering. For all ABT versions we implemented an improved version of Silaghi's solution detection [30] and counters for tagging assignments. All experiments were performed on the DisChoco 2.0 platform [33],[3] in which agents are simulated by Java threads that communicate only through message passing.

We reproduce the benchmark problems of [3]. Algorithms are evaluated on uniform random binary DisCSP where some global constraints are injected. We evaluate the performance of the algorithms by the total number of exchanged messages among agents during algorithm execution ($\#msg$) [15] and non-concurrent computation. The non-concurrent computation is measured by the number of non-concurrent constraint checks ($\#ncccs$) [38], as a proxy for elapsed time. All GAC effort is counted in $\#ncccs$, as in [3], where for each call of the ALLDIFF propagator, which computes a maximum matching in a graph, we increase $\#ncccs$ by the degree of that graph. For other constraints we increase $\#ncccs$ by the size of the data structure used in that constraint.

Uniform binary random DisCSPs are characterized by $\langle n, d, p_1, p_2 \rangle$, where n is the number of agents/variables, d is the number of values in each domain, p_1 is the network connectivity defined as the ratio of existing binary constraints to possible binary constraints, and p_2 is the constraint tightness defined as the ratio of forbidden value pairs to all possible pairs. We solved instances of two classes of constraint graphs: sparse graphs $\langle 20, 5, 0.2, p_2 \rangle$ and dense graphs $\langle 20, 5, 0.7, p_2 \rangle$. We varied the tightness from 0.1 to 0.9 by steps of 0.1. For each pair of fixed density and tightness (p_1, p_2) we report the average over 100 instances.

We generate two types of benchmarks: the ALLDIFF benchmark and the ATMOST benchmark. An ATMOST(k, X, v) constraint specifies that at most k variables in X are assigned value v. In the ALLDIFF benchmark, each binary instance includes 2 ALLDIFF constraints, each involving 5 randomly chosen variables. In the ATMOST benchmark, each binary instance includes 10 ATMOST(k, X, v) constraints, each involving from 3 to 10 randomly chosen variables (i.e., $3 \leq |X| \leq 10$). The value v is randomly chosen in the set of values in domains and its number of occurrences $k = 2$.[4]

[3] http://dischoco.sourceforge.net/

[4] ATMOST is not a contractible constraint [2]. To use nested representation, we changed the number of occurrences on nested constraints.

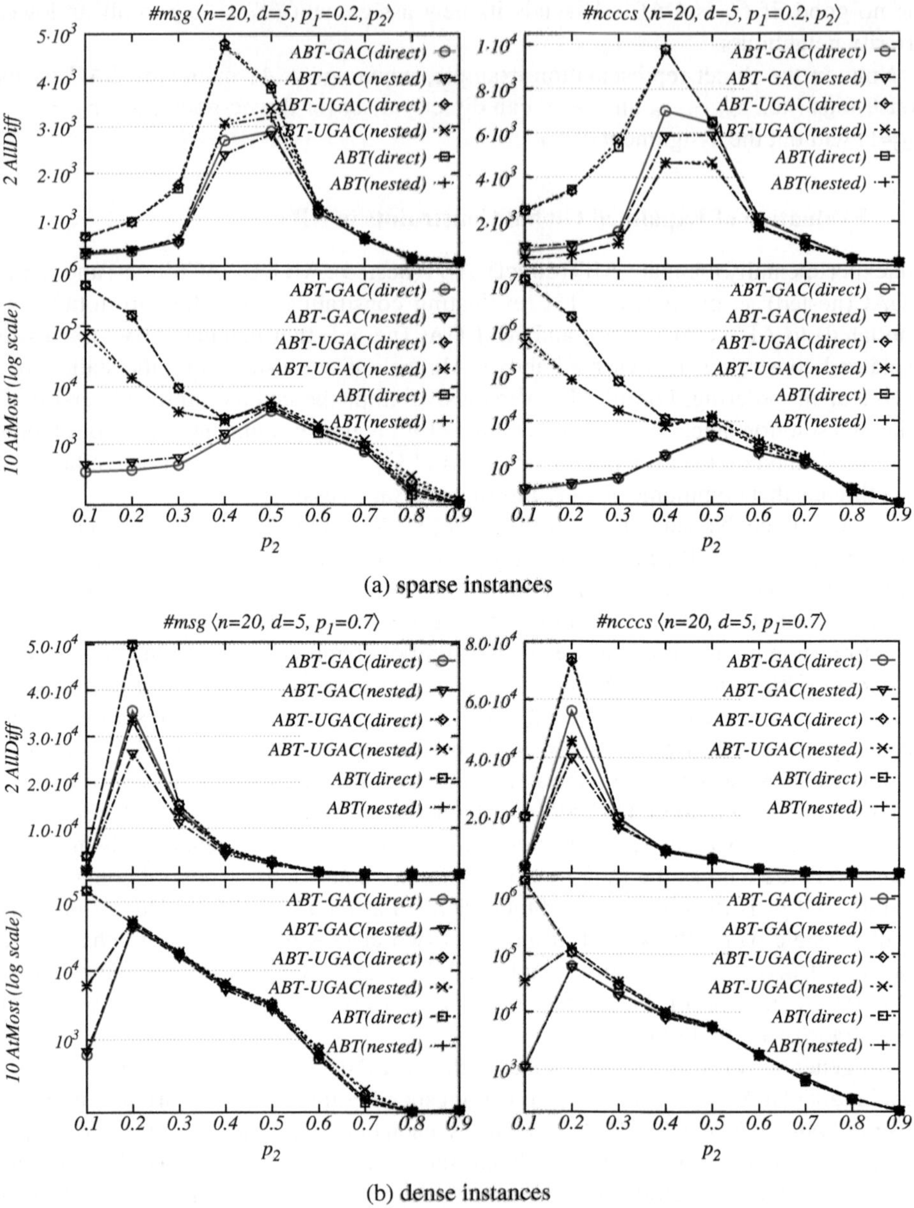

Fig. 2. Performance on the ALLDIFF benchmark and ATMOST benchmark (in logarithmic scale)

Results are shown in Fig. 2. ABT-GAC(direct) and ABT-GAC(nested) always require fewer messages than other algorithms, achieving up to two orders of magnitude improvement in the harder region for the non-contractible ATMOST constraint. For $\#ncccs$, the results are similar, except for the sparse instances of ALLDIFF, where

	Constraints	Explanations	Chessboard
direct	$c_1 : \mid x_1 - x_2 \mid \neq 1$ $c_{x_1} : x_1 = 1$	$c_{x_1} \rightarrow D_1 = \{1\}$ $c_{x_1} \wedge c_1 \rightarrow x_2 \neq 2$	1 2 3 4 x_1 x_2 
nested	$c_1 : \mid x_1 - x_2 \mid \neq 1$ c_9^n : ALLDIFF(x_1, x_2) $c_{x_1} : x_1 = 1$	$c_{x_1} \rightarrow D_1 = \{1\}$ $c_{x_1} \wedge c_9^n \rightarrow x_2 \neq 1$ $c_{x_1} \wedge c_1 \rightarrow x_2 \neq 2$	1 2 3 4 x_1 x_2 
binary	$c_1 : \mid x_1 - x_2 \mid \neq 1$ $c_8^b : x_1 \neq x_2$ $c_{x_1} : x_1 = 1$	$c_{x_1} \rightarrow D_1 = \{1\}$ $c_{x_1} \wedge c_8^b \rightarrow x_2 \neq 1$ $c_{x_1} \wedge c_1 \rightarrow x_2 \neq 2$	1 2 3 4 x_1 x_2 
full	$c_1 : \mid x_1 - x_2 \mid \neq 1$ $c_4 : \mid x_2 - x_3 \mid \neq 1$ $c_5 : \mid x_2 - x_4 \mid \neq 2$ $c_{x_1} : x_1 = 1$ c_7 : ALLDIFF(x_1, x_2, x_3, x_4)	$c_{x_1} \rightarrow D_1 = \{1\}$ $c_{x_1} \wedge c_7 \rightarrow x_2 \neq 1$ $c_{x_1} \wedge c_7 \rightarrow x_3 \neq 1$ $c_{x_1} \wedge c_7 \rightarrow x_4 \neq 1$ $c_{x_1} \wedge c_1 \rightarrow x_2 \neq 2$	1 2 3 4 x_1 x_2 x_3 x_4 

Fig. 3. Agent A_2 running ABT with different representation after receiving the assignment of $x_1 = 1$ when solving the distributed 4-queen (Fig. 1)

ABT-GAC(direct) and ABT-GAC(nested) are improved by ABT(nested) and ABT-UGAC(nested), and for dense instances of ALLDIFF, where ABT(nested) and ABT-UGAC(nested) improve ABT-GAC(direct).

4 Maintaining GAC Concurrently Using a Full Representation of Global Constraints

In the direct representation each global constraint will be represented by a single constraint. In the nested representation a contractible global constraint will be represented by a set of constraints, which allows concurrent processing and earlier identification of some inconsistencies. However, the filtering in these new constraints is weaker than that of the original constraint (ALLDIFF for example). More importantly, not all global constraints are contractible.

We now present a new way to represent a global constraint in DisCSPs, which we call the *full* representation. In the full representation, the original constraint is evaluated in the DisCSP by all agents that are involved in it, using their own copies of other agents' domains. As with direct, the full representation is applicable to all global constraints. The motivation is to benefit from the strength of filtering algorithms for global constraints and to do more concurrent computation, to detect unfruitful decisions earlier, and thus decrease the number of messages and $\#ncccs$. Much of this concurrent pruning may be redundant, though, and so we may increase the total amount of work averaged over all agents. We note that the full representation weakens the domain privacy of lower priority agents, but is fair for all agents involved in the constraint.

In the example of Fig. 1, A_1 evaluates the constraints c_1, c_2, c_3 and c_7. A_2 evaluates the constraints c_1, c_4, c_5 and c_7. A_3 evaluates the constraints c_2, c_4, c_6 and c_7. A_4 evaluates the constraints c_3, c_5, c_6 and c_7. Fig. 3 shows the reasoning by agent A_2 after receiving the assignment of A_1 (i.e., $x_1 = 1$) when running ABT with direct, nested, binary and full representation. Once it receives this assignment, A_2 adds the constraint $c_{x_1} : (x_1 = 1)$ to its constraint store then maintains GAC. In direct representation only value 2 is removed from D_2. Thus, A_2 assigns 1 to its variable. In nested and binary representation, two values (1 and 2) are removed from D_2. Thus, A_2 assigns 3 to its variable. In full representation, two values (1 and 2) are removed from D_2, and value 1 from D_3 and D_4. Thus, A_2 tries to assign value 3 to its variable. A_2 again maintains GAC and removes value 3 from D_2 because it has no support in D_3 (Fig. 3 (full), circles). It then assigns 4 to its variable, i.e., $x_2 = 4$.

We also extend ABT-GAC with the propagation of unconditional value deletions from [3], to get ABT-GAC+U. ABT-GAC(full) and ABT-GAC+U(full) inherit the correctness, completeness and termination of ABT(direct) and ABT-UGAC(direct) [3]. The only changes we make are adding redundant copies of constraints and allowing agents to do more powerful correct filtering.

4.1 Theoretical Analysis

We demonstrate that ABT-GAC(full) is sound, complete and terminates.

Theorem 1. *ABT-GAC(full) is sound.*

Proof. (Sketch) When the state of quiescence is reached, all agents know the assignments of all their higher priority neighbours. Thus, any constraint has been successfully checked by the lowest priority agent in its scope when it is fully instantiated. Otherwise, that agent would have tried to change its value and would have either sent an message containing its new value or a no-good, breaking the quiescence.

Theorem 2. *ABT-GAC(full) is complete.*

Proof. All explanation and no-goods are generated by logical inferences from existing constraints. Therefore, an empty no-good cannot be inferred if a solution exists.

Theorem 3. *ABT-GAC(full) terminates.*

Proof. ABT-GAC(full) inherits the termination of ABT [37]. We can prove by induction on the agent ordering that agents can never fall into an infinite loop . First, we can show that agent A_1, never falls into an infinite loop. Then, assuming that all agents higher that an agent A_i $(i > 2)$ are in a stable state, we can show that agent A_i never falls into an infinite loop.

Table 1. Performance on hard region of sparse 2 ALLDIFF benchmark ($p_1 = 0.2, p_2 = 0.7$)

$p_2 = 0.7$	$\#msg$		$\#ncccs$		$\#ccs$	
	ABT-GAC	ABT-UGAC	ABT-GAC	ABT-UGAC	ABT-GAC	ABT-UGAC
full	**6 984**	**7 086**	30 439	28 012	203 934	193 403
nested	8 173	8 605	26 099	26 859	170 798	179 706
direct	7 774	8 253	**25 875**	**26 560**	**164 106**	**173 683**
binary	7 857	8 358	28 491	29 499	188 500	198 213

Table 2. Performance on hard regions of instances with 5 ATMOST benchmark

	$(p_1 = 0.2, p_2 = 0.7)$				$(p_1 = 0.7, p_2 = 0.3)$			
	$\#msg$		$\#ncccs$		$\#msg$		$\#ncccs$	
	GAC	GAC +U	GAC	GAC +U	GAC	GAC +U	GAC	GAC +U
full	7 352	**7 755**	24 847	23 927	**737 854**	**737 590**	2 534 734	2 530 072
nested	8 509	9 153	21 756	22 709	747 080	745 303	1 745 611	1 739 681
direct	**7 288**	7 897	**21 157**	**22 061**	752 748	752 669	**1 737 450**	**1 733 302**

5 Experimental Results

We empirically compare the full representation to direct, nested and binary representations, all implemented within ABT-GAC and ABT-GAC+U. In our experiments, we imposed a cut-off on $\#ncccs$ of 10^9 and a cut-off on $\#msg$ of 10^9. In almost all instances, ABT without explanations (ABT(direct), ABT(nested) and ABT(binary)) exceeds this limit. Moreover, it runs out of memory in many instances. Thus, here we only present results on explained versions (ABT-GAC). For the same reason, we only show results for ABT-GAC+U.

5.1 Uniform Binary Random DisCSPs with Global Constraints

We solved instances of two classes of constraint graphs: sparse graphs $\langle 20, \mathbf{10}, 0.2, p_2\rangle$ and dense graphs $\langle 20, \mathbf{10}, 0.7, p_2\rangle$. We varied the tightness from 0.1 to 0.9 by steps of 0.1. For each pair of fixed density and tightness (p_1, p_2) we report averages over 100 instances. From a binary instance, we generate 4 types of benchmarks: the ALLDIFF, ATMOST, ATLEAST and EXACTLY benchmarks. In the ALLDIFF benchmark, each binary instance includes 2 ALLDIFF constraints, each involving 5 randomly chosen variables. In the ATMOST benchmark, each binary instance includes 5 ATMOST$(3, X, v)$ constraints, each involving from 5 to 7 randomly chosen variables. The value v is randomly chosen from the set of values in domains. In the ATLEAST benchmark, each binary instance includes 10 ATLEAST$(3, X, v)$ constraints, each involving from 5 to 7 randomly chosen variables. The value v is randomly chosen from the set of values in domains. In the EXACTLY benchmark, each binary instance includes 5 EXACTLY$(3, X, v)$ constraints, each involving from 5 to 7 randomly chosen variables. The value v is randomly chosen from the set of values in domains.

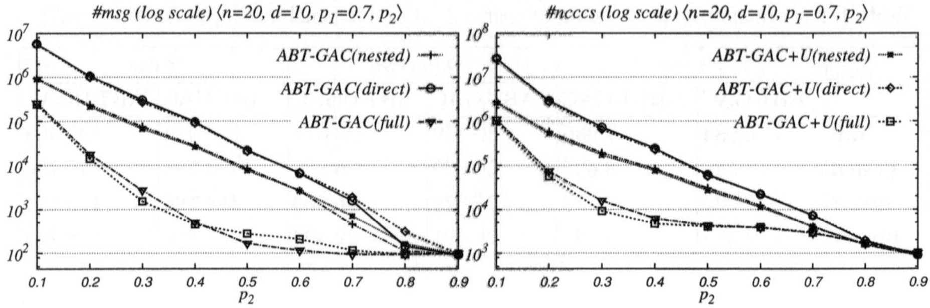

Fig. 4. Performance on dense instances with 10 ATLEAST benchmark

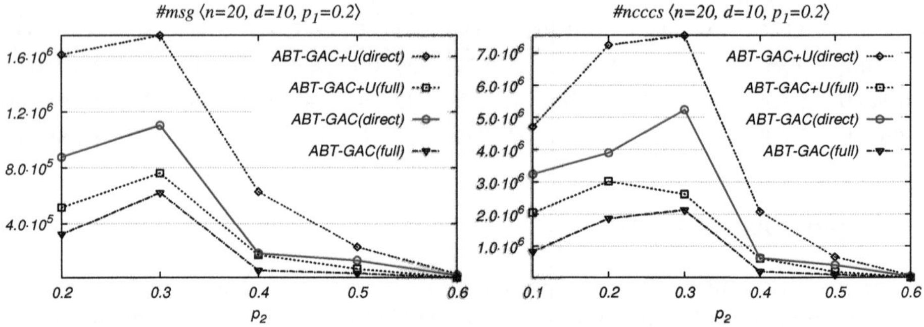

Fig. 5. Performance on sparse instances with 5 EXACTLY benchmark

For space reasons, we only show a selection of results. In Table 1, for sparse ALLDIFF problems, in the harder region ($p_2 = 0.7$), we see a small improvement in $\#msg$ and a small deterioration in $\#ncccs$ compared to direct, for the full representation over the explained versions of the other representations. In Table 1, we also show the total computational effort (total number of constraint checks by all agents $\#ccs$), and as expected, we show a small increase. For ATMOST problems (Table 2), the three representations are similar in $\#msg$, but full is almost 50% poorer on $\#ncccs$ for dense problems. On dense ATLEAST problems (Fig. 4), we see at least an order of magnitude improvement for the full representation over the nested for both $\#msg$ and $\#ncccs$, while direct is consistently worst. On sparse EXACTLY problems Fig. 5, we gain a two-fold improvement compared to direct representation. ABT-GAC+U deteriorates compared to ABT-GAC. It seems that extra messages needed to propagate the unconditional removals slows the search for both representations.

Table 3. Performance on Quasi-Groups With Holes problems

	#instances solved		$\#msg$		$\#ncccs$	
	ABT-GAC	ABT-GAC+U	ABT-GAC	ABT-GAC+U	ABT-GAC	ABT-GAC+U
full	**98**	**99**	**5 882 353**	**5 190 937**	**737 296**	**782 601**
nested	95	**99**	10 495 843	6 852 796	1 413 990	1 087 459
direct	96	**99**	10 044 646	7 060 771	1 971 401	1 380 700
binary	87†	84‡	11 765 454	11 910 842	3 930 123	3 884 403

† ABT-GAC(binary) runs out of memory for 8 instances
‡ ABT-GAC+U(binary) runs out of memory for 11 instances

5.2 Quasi-Groups with Holes

We also evaluate ABT-GAC and ABT-GAC+U on a set of satisfiable balanced quasi-groups with holes (QGWH) instances [26, 6].[5] The set contains 100 different instances. Each instance contains 106 variables and 30 ALLDIFF constraints, and as before, each variable is controlled by a different agent. Only 71 instances were solved by all algorithms, and the average performance over these instances is presented in Table 3. We see that the binary representation in QGWH performs relatively poorly. The concurrent filtering (i.e., *full*) outperforms all other strategies.

5.3 Distributed Meeting Scheduling Problem

The *distributed meeting scheduling problem* (DMSP) [18, 35] consists of a set of n agents having a personal private calendar and a set of m meetings each taking place in a specified location. Each agent knows the set of the k among m meetings she must attend, and knows the traveling time between the locations where her meetings will be held. The traveling time between two meetings m_i and m_j is denoted by $TT(m_i, m_j)$. The following constraints apply: (i) all agents attending a meeting must agree on when it will occur, (ii) an agent cannot attend two meetings at the same time, (iii) an agent must have enough time to travel from one meeting to the next.

We encode the DMSP in DisCSP as follows. Each DisCSP agent represents a real agent and contains k variables representing the k meetings in which the agent participates. These k meetings are selected randomly among the m meetings. The domain of each variable contains the $d \times h$ slots when a meeting can be scheduled. A slot is one hour long, and there are h slots per day and d days. There is an ALLEQUAL constraint for all variables corresponding to the same meeting in different agents (constraint (i)). There is an *arrival-time* constraint between all variables/meetings belonging to the same agent. The arrival-time constraint between two variables m_i and m_j is $\mid m_i - m_j \mid - duration > TT(m_i, m_j)$, where *duration* is the duration of

[5] `http://www.cril.univ-artois.fr/~lecoutre/benchmarks.html`

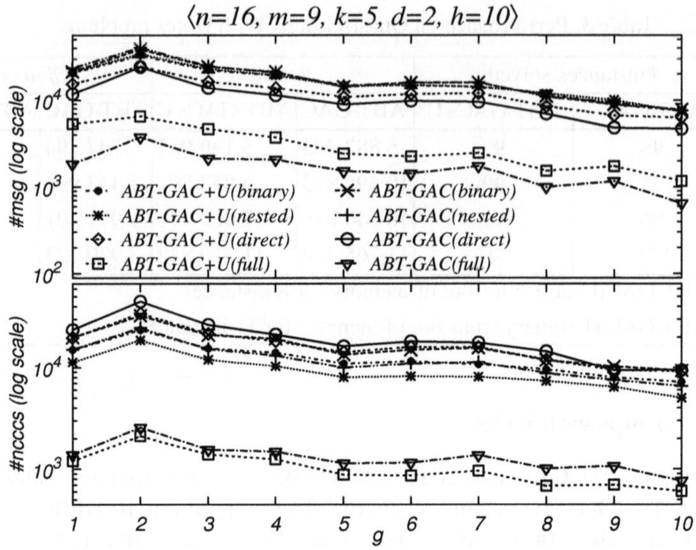

Fig. 6. Performance (log scale) on 16-agent DMSP

every meeting. This arrival-time constraint allows us to express both constraints (ii) and (iii). We place meetings randomly on the nodes of a uniform grid of size $g \times g$ and the traveling time between two adjacent nodes is 1 hour. The traveling time between two meetings equals the Euclidean distance between their locations. To vary the tightness of the arrival-time constraint we vary the size of the spatial grid.

Problems are characterized by $\langle n, m, k, d, h, g \rangle$, where n is the number of agents, m is the number meetings, k is the number of meetings/variables per agent, d is the number of days and h is the number of hours per day, and g is the grid size. The duration of each meeting is one hour. In our implementation this encoding is translated into an equivalent formulation where we have k (number of meetings per agent) virtual agents for each real agent. Each virtual agent handles a single variable but $\#msg$ does not take into account messages exchanged between virtual agents belonging to the same real agent [35]. We solved instances for two classes $\langle 20, 11, 3, 2, 10, g \rangle$ and $\langle 16, 9, 5, 2, 10, g \rangle$ where we vary g from 1 to 10 by steps of 1. For each g we generated 100 instances.

In Fig. 6, for the 16-agent problems, we see a consistent order of magnitude improvement for the full representation over the other approaches for both $\#msg$ and $\#ncccs$ across all grid sizes. In Fig. 7, for the larger 20 agent problems, but with fewer meetings, we see an improvement in both measures between a factor of 5 and a factor of 10. We also plot in the same figure the total number of constraint checks over all agents. Surprisingly, we see a reduction in total computational effort up to a factor of 5. It appears that the gains from more powerful filtering, more efficient no-goods and the reduced number of messages outweighs the extra effort of the redundant filtering.

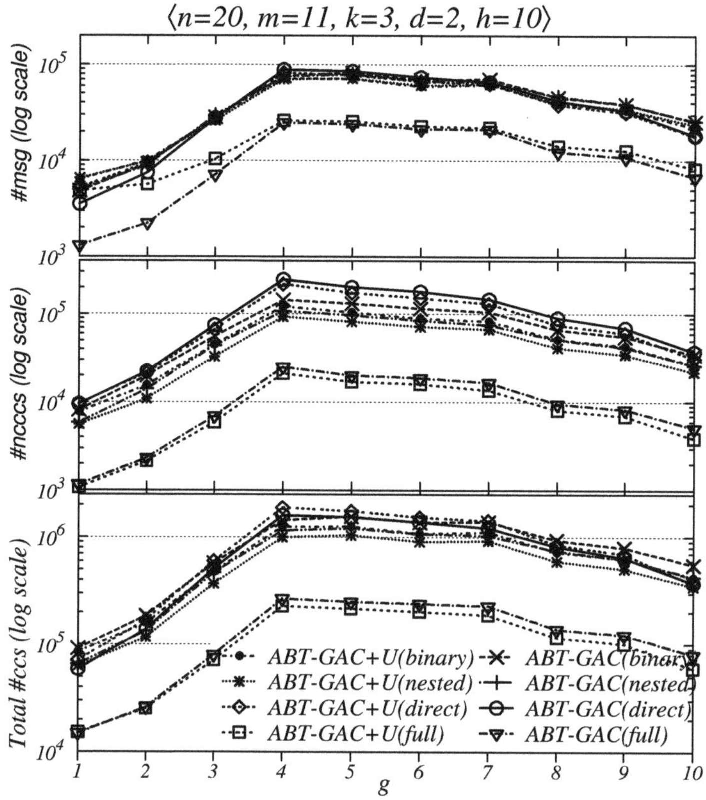

Fig. 7. Performance (log scale) on 20-agent DMSP

6 Conclusion

The power of filtering algorithms for global constraints is one of the main features of the success of constraint programming. In Distributed CSPs, however, global constraints have received little attention, because of the distributed control of the variables and their domains. Recently, a nested representation of contractible global constraints was shown to reduce computational effort and communication. In this paper, we make two contributions to the handling of global constraints in Distributed CSP. First, we show that maintaining GAC using explained constraints significantly improves the performance over the previous approaches. Secondly, we introduce a full representation for any global constraint, allowing every agent to evaluate any constraint it is involved in, and we use this to implement concurrent maintenance of GAC. We demonstrate empirically that concurrent GAC on the full representation offers a further significant improvement in both non-concurrent computation and messaging. This appears to contradict recent results that suggest reducing redundancy in Distributed CSP always improves performance [12, 32]. We also show that for some problems, despite the redundant filtering, we reduce the total computation cost over all the agents.

Future work will focus on extending other DisCSP algorithms to include MAC and then exploiting full concurrent GAC, and on implementing global constraints and full concurrent GAC with distributed dynamic ordering algorithms.

Acknowledgements. This work is funded by Science Foundation Ireland (SFI) under Grant Number SFI/12/RC/2289.

References

[1] Béjar, R., Domshlak, C., Fernández, C., Gomes, C., Krishnamachari, B., Selman, B., Valls, M.: Sensor networks and distributed csp: communication, computation and complexity. Artif. Intel. 161, 117–147 (2005)
[2] Beldiceanu, N., Carlsson, M., Rampon, J.X.: Global constraint catalog. SICS Research Report (2005)
[3] Bessiere, C., Brito, I., Gutierrez, P., Meseguer, P.: Global constraints in distributed constraint satisfaction and optimization. The Computer Journal (2013)
[4] Bessière, C., Van Hentenryck, P.: To be or not to be... a global constraint. In: Rossi, F. (ed.) CP 2003. LNCS, vol. 2833, pp. 789–794. Springer, Heidelberg (2003)
[5] Bessiere, C., Maestre, A., Brito, I., Meseguer, P.: Asynchronous backtracking without adding links: a new member in the ABT family. Artif. Intel. 161, 7–24 (2005)
[6] Boussemart, F., Hemery, F., Lecoutre, C., Sais, L.: Boosting Systematic Search by Weighting Constraints. In: Proceedings of ECAI 2004, pp. 146–150 (2004)
[7] Brito, I., Meseguer, P.: Asynchronous backtracking for non-binary discsp. In: DCR Workshop at ECAI-2006, DCR 2006, Riva di Garda, Italia (2006)
[8] Downing, N., Feydy, T., Stuckey, P.J.: Explaining alldifferent. In: Proceedings of ACSC 2012, Darlinghurst, Australia, Australia, pp. 115–124 (2012)
[9] Downing, N., Feydy, T., Stuckey, P.J.: Explaining flow-based propagation. In: Beldiceanu, N., Jussien, N., Pinson, É. (eds.) CPAIOR 2012. LNCS, vol. 7298, pp. 146–162. Springer, Heidelberg (2012)
[10] Francis, K., Stuckey, P.: Explaining circuit propagation. Constraints 19(1), 1–29 (2014)
[11] Gaudin, E., Jussien, N., Rochart, G.: Implementing explained global constraints. In: Proceedings of the CP 2004 Workshop on Constraint Propagation and Implementation (CPAI 2004), Toronto, Canada, Canada, pp. 61–76 (2004)
[12] Gutierrez, P., Meseguer, P.: Saving redundant messages in bnb-adopt. In: AAAI 2010 (2010)
[13] Jussien, N.: The versatility of using explanations within constraint programming. HDR, Université de Nantes (September 2003)
[14] Jussien, N., Barichard, V.: The palm system: explanation-based constraint programming. In: Dechter, R. (ed.) CP 2000. LNCS, vol. 1894, pp. 118–133. Springer, Heidelberg (2000)
[15] Lynch, N.A.: Distributed Algorithms. Morgan Kaufmann Series (1997)
[16] Maher, M.J.: Open contractible global constraints. In: Proceedings of IJCAI 2009, pp. 578–583. Morgan Kaufmann Publishers Inc., San Francisco (2009)
[17] Mailler, R., Lesser, V.R.: Asynchronous partial overlay: A new algorithm for solving distributed constraint satisfaction problems. JAIR 25(1), 529–576 (2006)
[18] Meisels, A., Lavee, O.: Using additional information in DisCSP search. In: Proceedings of DCR 2004 (2004)
[19] Meisels, A., Zivan, R.: Asynchronous Forward-checking for DisCSPs. Constraints 12(1), 131–150 (2007)

[20] Petcu, A., Faltings, B.V.: A Value Ordering Heuristic for Distributed Resource Allocation. In: Faltings, B.V., Petcu, A., Fages, F., Rossi, F. (eds.) CSCLP 2004. LNCS (LNAI), vol. 3419, pp. 86–97. Springer, Heidelberg (2005)
[21] Petcu, A., Faltings, B.: DPOP: A Scalable Method for Multiagent Constraint Optimization. In: Proceedings of IJCAI 2005, pp. 266–271 (2005)
[22] Régin, J.C.: A filtering algorithm for constraints of difference in csps. In: Proceedings of AAAI 1994, pp. 362–367 (1994)
[23] Régin, J.C.: Global constraints: A survey. In: van Hentenryck, P., Milano, M. (eds.) Hybrid Optimization, pp. 63–134. Springer, New York (2011)
[24] Rochart, G.: Explanations for global constraints. In: Rossi, F. (ed.) CP 2003. LNCS, vol. 2833, pp. 993–993. Springer, Heidelberg (2003)
[25] Rochart, G.: Explications et programmation par contraintes avancée. Ph.D. thesis, Nantes University, France (2005)
[26] Roussel, O., Lecoutre, C.: Xml representation of constraint networks: Format xcsp 2.1. CoRR (2009)
[27] Schutt, A., Feydy, T., Stuckey, P.J., Wallace, M.G.: Explaining the cumulative propagator. Constraints 16(3), 250–282 (2011)
[28] Silaghi, M.C.: Generalized Dynamic Ordering for Asynchronous Backtracking on DisCSPs. In: Proceedings of DCR 2006 (2006)
[29] Silaghi, M.C., Faltings, B.: Asynchronous aggregation and consistency in distributed constraint satisfaction. Artif. Intel. 161, 25–53 (2005)
[30] Silaghi, M.C., Sam-Haroud, D., Faltings, B.: Asynchronous Search With Aggregations. In: Proceedings of AAAI 2000/IAAI 2000, pp. 917–922 (2000)
[31] Wahbi, M.: Algorithms and Ordering Heuristics for Distributed Constraint Satisfaction Problems. John Wiley & Sons, Inc. (2013)
[32] Wahbi, M., Ezzahir, R., Bessiere, C.: Asynchronous Forward Bounding Revisited. In: Schulte, C. (ed.) CP 2013. LNCS, vol. 8124, pp. 708–723. Springer, Heidelberg (2013)
[33] Wahbi, M., Ezzahir, R., Bessiere, C., Bouyakhf, E.H.: DisChoco 2: A Platform for Distributed Constraint Reasoning. In: Proceedings of workshop on DCR 2011, pp. 112–121 (2011), `http://dischoco.sourceforge.net/`
[34] Wahbi, M., Ezzahir, R., Bessiere, C., Bouyakhf, E.H.: Maintaining Arc Consistency Asynchronously in Synchronous Distributed Search. In: Proceedings of ICTAI 2012, Athens, Greece, pp. 33–40 (November 2012)
[35] Wahbi, M., Ezzahir, R., Bessiere, C., Bouyakhf, E.H.: Nogood-Based Asynchronous Forward-Checking Algorithms. Constraints 18(3), 404–433 (2013)
[36] Yokoo, M., Durfee, E.H., Ishida, T., Kuwabara, K.: Distributed constraint satisfaction for formalizing distributed problem solving. In: Proceedings of 12th IEEE Int'l Conf. Distributed Computing Systems, pp. 614–621 (1992)
[37] Yokoo, M., Durfee, E.H., Ishida, T., Kuwabara, K.: The Distributed Constraint Satisfaction Problem: Formalization and Algorithms. IEEE Trans. on Knowledge and Data Engineering 10, 673–685 (1998)
[38] Zivan, R., Meisels, A.: Message delay and DisCSP search algorithms. Annals of Mathematics and Artificial Intelligence 46(4), 415–439 (2006)
[39] Zivan, R., Zazone, M., Meisels, A.: Min-Domain Retroactive Ordering for Asynchronous Backtracking. Constraints 14(2), 177–198 (2009)

The Impact of Wireless Communication on Distributed Constraint Satisfaction

Mohamed Wahbi and Kenneth N. Brown

Insight Centre for Data Analytics, School of Computer Science and IT,
University College Cork, Ireland
{mohamed.wahbi,ken.brown}@insight-centre.org

Abstract. Distributed constraint satisfaction (*DisCSP*) models decision problems where physically distributed agents control different decision variables, but must communicate with each other to agree on a global solution. Most DisCSP research assumes an abstract communication layer based on a peer-to-peer wired network. However, many practical applications of distributed reasoning require to be implemented over wireless networks, which impose different communication costs, and may affect the performance of DisCSP algorithms. We study the impact of wireless network topology and routing on two leading DisCSP algorithms – ABT and AFC-ng. We introduce a new framework for experiments which models different communication layers. We show that the communication layer has a significant impact on the messaging costs, which can vary by over an order of magnitude. We also show the impact on computation time, where the equivalent non-concurrent constraint checks can vary by a factor of 6. Finally, we show that given a fixed agent ordering, changing the communications topology can increase the number of messages by up to 50%.

1 Introduction

Distributed Constraint Satisfaction (*DisCSP*) models constrained decision problems where different variables are controlled by physically distributed agents. The agents must communicate with each other to reach a global assignment which satisfies all constraints. The main algorithms and protocols include synchronous [37, 23, 33] and asynchronous [38, 2] search, backtracking, local search [13, 39] and dynamic programming methods [26], and algorithms which respect privacy and autonomy [34, 4, 18] versus those which partially centralise the decision making [20]. The main metrics are non-concurrent constraint checks ($\#ncccs$) [21], which measures the longest sequence of computation among the agents as a proxy for elapsed time, and messages ($\#msg$)[19], which counts the total number of messages exchanged between agents. Communication delays can be modelled by adding extra increments to get $\#encccs$, the equivalent non-concurrent constraint checks [5]. Most research assumes an abstract model of the underlying communication network [41, 30, 17, 32, 11], equivalent to essentially a peer-to-peer wired network. This model works well for applications operating over standard internet architectures.

There is a rich domain of application problems for DisCSP which assume wireless communication, including, for example, dynamic coordination of missions in unmanned

B. O'Sullivan (Ed.): CP 2014, LNCS 8656, pp. 738–754, 2014.

aerial vehicles (*UAVs*) [28], mobile robot coordination [6], decision making in wireless sensor networks (*WSNs*) [1, 15, 36, 3], and dynamic channel selection [25] and time-division scheduling for the self-configuration of ad hoc wireless networks. However, the standard communication layer for DisCSP does not account for all the relevant details of wireless communication, and so algorithms may behave radically differently when implemented on physical devices. For example, in wireless communication, each radio transmission consumes energy, and so the number and size of individual transmissions required to deliver a message is significant, as opposed to the number of end-to-end messages. This is particularly important in remote operation (e.g., UAVs or WSNs) where the agents have limited battery power. In addition, there are many different protocols for exchanging information in a wireless network, ranging from peer-to-peer unicast to one-to-many local broadcast, and from static routing to dynamic routing based on flooding the network. Thus, the communications layer may significantly change the number of individual message transmissions required. More message transmissions means longer delays, and it is known [5, 7] that delays in exchanging messages adversely affects the $\#encccs$ metric in some DisCSP algorithms. To establish DisCSP for wireless applications, we require a better understanding of how the underlying communication layer affects the performance of different algorithms.

In this paper, we propose a new framework for analysing *wireless* DisCSP, based on wireless communication networks. The framework distinguishes between the *constraint graph* and the *communications graph*, which may be different. Direct communication is only possible between adjacent nodes in the communication graph. Exchanging messages between neighbours in the constraint graph thus may require multiple transmissions in the communications graph. We use the framework to re-evaluate the asynchronous ABT algorithm [38, 2] and the partially synchronised AFC-ng [7, 33] algorithm. We consider a range of different communication network topologies, from linear chain trees to complete graphs. We consider the abstract source routing protocols N-way unicast, multicast and multicast* (multicast with local broadcasts), which vary in the number of individual transmissions required to send messages to a set of recipients. We show that changing the topology has a significant impact on the number of message transmissions, sometimes causing an order of magnitude increase for the same algorithm and routing protocol. Similarly, we show that the topology and routing protocol also combine to affect messaging, with N-way unicast on linear topologies requiring significantly more messages than the standard DisCSP model assumes, while multicast* on complete communication networks reduces the number of messages compared to the standard model. Varying the communications layer also has an impact on $\#encccs$, increasing it in the worst case by a factor of six. The performance of static DisCSP ordering heuristics is also influenced by the communication layer, with different topologies increasing the message count by up to 50%. Finally, we propose future directions for wireless DisCSP research, with the eventual aim of modifying the algorithms and ordering heuristics to adapt to different communication layers.

2 Background

The Distributed Constraint Satisfaction Problem (DisCSP) is a 5-tuple $(\mathcal{A}, \mathcal{X}, \mathcal{D}, \mathcal{C}, \phi)$, where $\mathcal{X}$ is a set of variables $\{x_1, \ldots, x_n\}$, $\mathcal{D} = \{D_1, \ldots, D_n\}$ is a set of domains, where D_i is a finite set of values from which one value must be assigned to variable x_i, $\mathcal{C}$ is a set of constraints, $\mathcal{A}$ is a set of agents $\{A_1, \ldots, A_a\}$, and $\phi : \mathcal{X} \rightarrow \mathcal{A}$ is a function specifying an agent to control each variable. During a solution process, only the agent which controls a variable can assign it a value. A constraint $C(X) \in \mathcal{C}$, on the ordered subset of variables $X = (x_{j_1}, \ldots, x_{j_k})$, is $C(X) \subseteq D_{j_1} \times \cdots \times D_{j_k}$, and specifies the tuples of values which may be assigned simultaneously to the variables in X. For this paper, we restrict attention to binary constraints. We denote by $C_i \subseteq \mathcal{C}$ all constraints that involve x_i. A *solution* is an assignment to each variable of a value from its domain, satisfying all constraints. Each agent A_i knows all constraints relevant to its variables (C_i) and the other variables involved in its constraints (its *neighbours* in the constraint graph). Without loss of generality, we assume each agent controls exactly one variable ($a = n$), so we use the terms agent and variable interchangeably and do not distinguish between A_i and x_i. A variety of problems have been tackled using DisCSP, including tracking in sensor networks [1], resource allocation [27] and meeting scheduling [22].

Asynchronous Backtracking (ABT) [38, 2] is an asynchronous algorithm executed autonomously by each agent in the problem, and is guaranteed to compute a global consistent solution (or detect inconsistency) in finite time. Each agent proposes values for its own variable to other agents, and reports no-goods. Agents operate asynchronously, but are subject to a known total priority order, o.

Nogood-Based Asynchronous Forward Checking (AFC-ng) [33] is a partially synchronised algorithm that uses no-goods as justification for value removals. Following a total priority agent ordering o, agents assign their variables one by one, recording assignments in a data structure called the Current Partial Assignment (CPA). Once an agent adds its variable assignment to the CPA, it sends the CPA to its unassigned neighbours to perform forward checking (FC [12]) asynchronously. AFC-ng allows different agents to perform backtracks concurrently to the same or different destinations. As a result, several CPAs can be generated simultaneously by the destination agents. Due to the timestamps integrated in the CPAs, the CPA coming from the highest level in the agent ordering will eventually dominate.

In the standard DisCSP communication model [41, 30, 17, 32] each exchange of information from one agent to another (e.g., value choice, acknowledgement, no-good, constraint description) is represented as a *message*. This is an abstraction of the communication layer, based on the assumption that the number of source-to-destination messages is the main factor in communication cost. Zivan and Meisels (2006) proposed an Asynchronous Message Delay Simulator (AMDS) [41] for distributed constraint reasoning algorithms. AMDS is a Mailer thread through which all messages are passed to simulate message delays. The mailer holds a counter of non-concurrent computation steps (LTC, Logical Time Counter) performed by agents, represented as the number of non-concurrent constraint checks ($\#ncccs$). Communication delays can be modelled by adding extra increments to get $\#encccs$. Each agent maintains its own LTC, and attaches it to each message she sends. An agent that receives a message updates her counter to the maximum value between the received LTC and her own counter. Next, she performs a computation step and sends her outgoing messages. Immediately prior sending a message, the agent increments her LTC by the number of constraints checks performed during that step. When an agent desires to send a message she passes it to

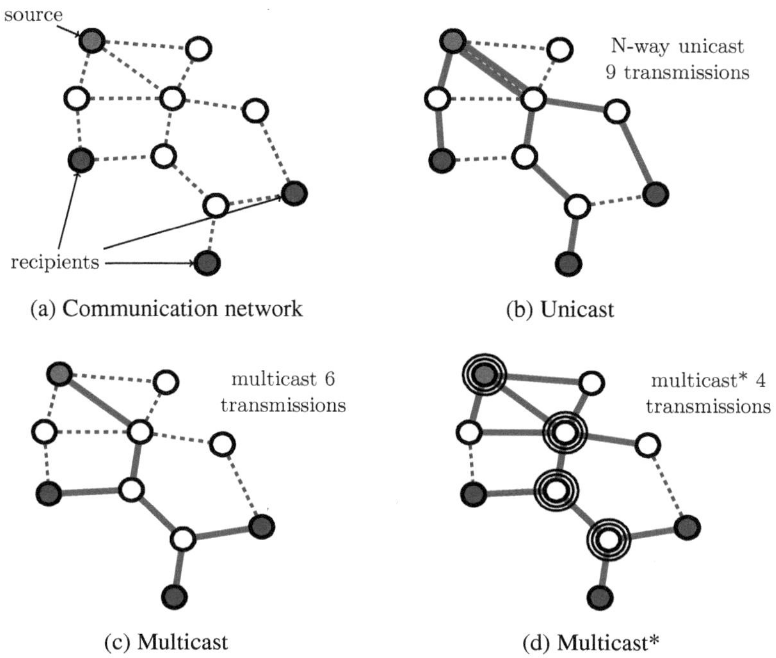

(a) Communication network (b) Unicast

(c) Multicast (d) Multicast*

Fig. 1. Different communication protocols

the dedicated mailer thread. Upon receiving that message, the mailer updates its LTC by the value of the LTC carried by the message if its value is larger than that held in the mailer. A delay for the message is then chosen and the message is added to the outgoing queue. The queue is ordered by increasing LTC. When a message reaches the front of the queue, it is removed, and delivered to the incoming queue of the receiving agent.

Each agent maintains a count of the messages it sends ($\#msg$), incrementing it by 1 for each message sent to the mailer. Note that each message counts 1 regardless of size, and is thus recording the instance of a single source-to-destination communication. Most experimental comparisons of DisCSP algorithms record the maximum $\#ncccs$ value over all agents at termination, and the sum of the message counts ($\#msg$) over all agents. Occasionally, the total number of constraints checks (summed over all agents) or the number of communication cycles is also reported. The most important metric is usually considered to be $\#ncccs$, consistent with work in general distributed algorithms [19, p. 22]. This model works well with applications implemented over standard wired TCP/IP networks, where variation in routing and packet retransmissions can be averaged over all agents and messages.

There are many practical application problems which rely on wireless networking. For example, in dynamic mission scheduling in unmanned aerial vehicles (*UAVs*), clusters of UAVs are remote from the base and must coordinate their actions and revise their plans through negotiation with each other using wireless communication [28]. In wireless sensor networks, small sensor nodes must relay sensed data to a base, and may be required to coordinate their sensing in order to ensure the target phenomena is adequately covered [1, 15, 36, 3]. This is further extended to wireless sensor actuator networks, in which some of the nodes can control aspects of the physical environment. In addition, the operation and configuration of ad hoc and mesh wireless networks requires individual radio nodes to sense the topology of the network and to agree spectrum use, time schedules for communication, and routing paths [25], all of which involve distributed combinatorial problems.

In wireless communication [24], each individual transmission of data consumes energy in order to radiate and receive the signal. The energy consumed depends on the distance over which the data is transmitted, and on the amount of data. In wireless sensor networks, the cost of transmitting one bit of information is estimated to be equal to the cost of executing 1000 to 2000 logical instructions on the sensor node [14, p. 104], and so computation is considered much cheaper than communication. For battery powered nodes, limiting communication cost is the key to maintaining node, and thus network, lifetime. In multi-hop networks (e.g., wireless sensor networks or ad hoc networks), some nodes are not in range of each other, and so intermediate nodes must receive and re-transmit the data in order for it to be delivered. Thus, the delivery of a single message between nodes A_i and A_j may require significantly more transmissions than the same message delivered from A_i to A_k. Also, wireless communication is subject to radio interference, and so messages may have to be retransmitted in order to be received successfully; alternatively, to avoid interference, nodes may have to wait until the radio spectrum is free before they transmit. Thus, as well as incurring additional energy costs, longer transmission paths impose additional delays on message delivery.

Within a multi-hop network, there are many different approaches for routing the data [16]. Methods include flooding the network with messages, decentralised routing with each node maintaining a routing table, and source-level routing, in which each node decides upon the end-to-end route it will use for each recipient. Within source-level routing, options include *N-way unicast* , *multicast* , and multicast using the broadcast medium (which we refer to as *multicast**)(Fig. 1). For N-way unicasting, a source sending to N recipients creates N copies of the message, and then initiates each message along its chosen route. In multicasting, the source constructs a rooted tree with itself as root and containing all the intended recipients. Messages are then sent down the tree, with multiple copies only created when multiple branches leave from a single node. Multicast* takes advantage of the fact that multiple nodes can receive a single transmission, and thus each node in the multicast tree only needs to transmit a single copy of the message (assuming an omnidirectional antenna).

The differences between the standard DisCSP communication model and the wireless communication model raises the question of how our algorithms perform when deployed on wireless networks. The standard DisCSP model is essentially N-way unicasting, either on a complete communication network or on the constraint graph. Every edge in the graph requires the same energy for a transmission – any agent can communicate directly with any neighbour, for a cost of 1 message (plus a delay increment of δ) for each communication. In a problem instance with n agents, sending a variable assignment to m (constraint graph) neighbours takes m transmissions each with delay δ. Even if we assume that each transmission does consume the same energy, implementing the same instance on a wireless network may incur different costs. If the communication topology is a linear chain, with the source at one end and the recipients at the other, the same variable assignment would require $n*m-(m(m-1)/2)$ transmissions, with the longest delay $n*\delta$. Multicast on the same topology would require just n transmissions, but with the longest delay still $n*\delta$. Finally, if the topology is a complete graph, using multicast*, then we require just 1 transmission, with delay δ. Since increasing delays in messages are known to adversely affect $\#encccs$ for DisCSP algorithms, we may also see variation in the $\#encccs$ metric as we vary the communications layer. Therefore, if we are to deploy DisCSP algorithms on wireless networks, we need to revisit the algorithms, and assess their performance under different communication assumptions.

3 Network Communication Simulator Framework (NeCoS)

The standard DisCSP model views agents as distributed autonomous entities. Almost all distributed constraint reasoning simulators implement agents as Java Threads [40, 30, 17, 32, 11]. Zivan and Meisels (2006) proposed AMDS counting non-concurrent constraints-checks ($\#ncccs$) for systems with message delays. In AMDS, agents run concurrently, exchanging messages using a common mailer. In this section we generalize AMDS to simulate different communication topologies and routing protocols (unicast, multicast, and multicast*). We call the new simulator Network Communication Simulator (*NeCoS*). NeCoS is a Thread to which all messages are passed to simulate delays in communication networks using different communication protocols. We assume two graphs: (i) the standard DisCSP constraint graph, where two agents are neighbours if and only if they share a constraint, and (ii) the communications graph, where two nodes are neighbours if and only if they can communicate directly with each other in a single transmission. There is a function from the constraint graph vertex set (agents) to the communications graph vertex set (nodes), but the edge sets may be arbitrarily different. NeCoS requires as input the communications graph and the function. When A_i sends a message to A_j, the message must traverse a path in the communications graph, which may require multiple retransmissions of the message.

As in AMDS, NeCoS maintains a Logical Time Counter (LTC), which measures the longest sequence of computation and communication between agents. Each agent maintains its own counter. To simulate delays on message transmissions, each message in the system carries the LTC value of its sender. Whenever an agent receives a message, it updates its counter to the maximum value of the received LTC and its own counter. It then performs its computation step and sends messages with the value of its counter incremented by the amount of computation required during this step.

The NeCoS pseudo-code is presented in Figs. 2 and 3. In initialisation, NeCoS stores the communications graph in $network$, and then computes all shortest paths using [8]. When an agent desires to send a message msg to a set of recipients, it emits the message to NeCoS by calling `sendMessage(`$msg, recipients$`)`. Depending on the routing protocol used, NeCoS runs different procedures (lines 15-17):

unicast: for each recipient, NeCoS creates a copy (m) of the original message and computes the routing tree for that copy. The routing tree is the shortest path in the communications graph from the sender to the recipient (line 23). Then, NeCoS calls procedure `chooseDelay(`m`)` (line 25) to select a random delay needed to

```
procedure NeCoS()
    outQueue ← ∅; LTC ← 0; end ← false;
    network ← getCommunicationGraph();
    network.computeAllShortestPaths();            /* Use Floyd-Warshall [8, 35] */
    while ( ¬end ) do
        if ( all agents are terminated ) then end ← true;
        else if ( all agents are idle ) then LTC ← outQueue.first().getLTC();
        deliverMessages();
procedure deliverMessages()
    foreach ( msg ∈ outQueue s.t. msg.getLTC() < LTC ) do
        tree ← msg.gettree();
        A_s ← tree.getRoot();
        if ( A_s ∈ msg.getRecipients() ) then deliver(msg) to A_s;
        if ( routingProtocol ≠ multicast* ) then routingMessage(msg, tree, A_s);
        else routingMulticast*(msg, tree, A_s);
procedure sendMessage(msg, recipients)
    switch ( routingProtocol ) do // routingProtocol ∈{unicast,multicast,multicast* }
        unicast     : sendUnicast(msg, recipients);
        multicast   : sendMulticast(msg, recipients);
        multicast*  : sendMulticast*(msg, recipients);
procedure sendUnicast(msg, recipients)
    A_s ← msg.getSender();
    foreach ( A_i ∈ recipients ) do
        A_s.nbMsgSent ← A_s.nbMsgSent + 1;
        m ← msg;
        m.setRecipients(A_i);
        tree ← network.shortestPath(A_s, A_i);
        m.setRoutingTree(tree);
        chooseDelay(m);
        outQueue.add(m);
procedure sendMulticast(msg, recipients)
    A_s ← msg.getSender();
    tree ← network.steinerTree(A_s, recipients);
    routingMessage(msg, tree, A_s);
procedure sendMulticast*(msg, recipients)
    A_s ← msg.getSender();
    tree ← network.steinerTree(A_s, recipients);
    routingMulticast*(msg, tree, A_s);
procedure routingMessage(msg, tree, node)
    foreach ( subtree ∈ tree.getSubtreesOf(node) ) do
        node.nbMsgSent ← node.nbMsgSent + 1;
        m ← msg;
        m.setRecipients(recipients ∩ subtree);
        m.setRoutingTree(subtree);
        chooseDelay(m);
        outQueue.add(m);
```

Fig. 2. Network Communication Simulator (Part 1)

procedure `routingMulticast*` $(msg, tree, node)$
40. $node.nbMsgSent \leftarrow node.nbMsgSent + 1$;
41. `chooseDelay` (msg) ;
42. **foreach** ($subtree \in tree.$`getSubtreesOf` $(node)$) **do**
43. $m \leftarrow msg$;
44. $m.setRecipients(recipients \cap subtree)$;
45. $m.setRoutingTree(subtree)$;
46. $outQueue.add(m)$;

procedure `chooseDelay` (msg)
47. $LTC \leftarrow$ `max` $(LTC, msg.getLTC())$;
48. $msg.LTC \leftarrow msg.getLTC() + \delta$;

Fig. 3. Network Communication Simulator (Part 2)

transmit the message to the next node in the routing tree (lines 47-48). The copy of the message is then added to the outgoing message queue (i.e., *outQueue*, line 26).

multicast: NeCoS constructs a rooted Steiner tree with the source (A_s) as root and containing all recipients, line 28.[1] Then, NeCoS mimics the multicast routing protocol (`routingMessage` call, line 29). In `routingMessage`, a message is transmitted from the root *node* to each of its children (roots of its sub-trees, *subtree*, line 33) in the routing tree, *tree*, and each of these in turn queues the message for retransmission to its children (lines 38-39). We increment the number of messages transmitted by *node* by the number of its children in routing tree (line 34).

multicast*: NeCoS constructs a rooted Steiner tree with the source (A_s) as root and containing all recipients, line 31. Then, NeCoS mimics the multicast* routing protocol (`routingMulticast*` call, line 32). In `routingMessage`, a message is transmitted from the root *node* to each of its children (roots of its sub-trees, *subtree*, line 42) in the routing tree, *tree* requiring only one transmission (line 40) from *node* with the same delay (line 41). However, to simulate this, NeCoS creates a copy m of msg for all children of *node* in the routing tree, *tree* (lines 43) and each of these in turn will queue the copy for retransmission to its children (line 46).

In the three communication protocols, the LTC of each transmitted message is updated to the sum of the value of the message LTC and a random selected delay (line 48). Then, the message is added to the outgoing queue (*outQueue*). The outgoing queue is a priority queue in which messages are sorted by their LTC, so that the first message is the message with the lowest LTC.

When there are no incoming messages, and all agents are idle, NeCoS increases the value of its LTC to the LTC value of the first message in the outgoing queue (line 6) and calls procedure `deliverMessages` (line 7). When `deliverMessages` is invoked by NeCoS all messages carrying an LTC smaller than the counter of the simulator are transmitted line 4. Thus, this message is delivered to the root of the routing tree if it is one of the final recipients of that message, line 11. Next, we simulate a routing node for that agent depending on the routing protocol used (lines 12-13).

[1] In our experiments, we use a heuristic algorithm to compute good Steiner trees.

4 Experiments

In this section we evaluate ABT and AFC-ng under different network conditions. Algorithms are tested on the same static agent ordering using the *max-degree* heuristic. For ABT we implemented an improved version of Silaghi's solution detection [29] and counters for tagging assignments. All experiments were performed on the DisChoco 2.0 platform [32],[2] in which agents are simulated by Java threads that communicate only through message passing.

The algorithms are tested on uniform binary random DisCSPs which are characterized by $\langle n, d, p_1, p_2\rangle$, where n is the number of agents/variables, d the number of values per variable, p_1 is the constraint graph connectivity defined as the ratio of existing binary constraints to possible binary constraints, and p_2 is the constraint tightness defined as the ratio of forbidden value pairs to all possible pairs. We solved instances of two classes of constraint graphs: sparse constraint graphs $\langle 20, 5, 0.2, p_2\rangle$ and dense ones $\langle 20, 5, 0.7, p_2\rangle$. We varied the tightness from 0.1 to 0.9 by steps of 0.1. For each pair of fixed density and tightness (p_1, p_2) we report the average over 20 instances.

We evaluate the performance of the algorithms by communication load and computation effort. Communication load is measured by the total number of transmission messages in the communication network during algorithm execution ($\#transmission$). Computation effort is measured by the average of the equivalent non-concurrent constraint checks ($\#encccs$) [5] that extends the non-concurrent constraint checks ($\#ncccs$) [9]. The $\#encccs$ are a weighted sum of processing and communication time. We simulate uniform random message delays on the communication network. For each message a delay is randomly chosen between 10 and 100 constraint checks. We simulated three communication protocols: unicast, multicast, and multicast*.

To assess the behaviour of ABT and AFC-ng on different communication layers we generate 8 different connected network topologies, using only agents in the problem:
complete: the communication network is a complete graph, where all agents are connected to each other – the maximum number of hops between any two agents is one;
constraint: the communication network matches the constraint graph exactly;
star: the communication network has a star topology where one randomly selected agent is directly connected to all other agents, and there are no other connections;
random (0.7): dense random communication networks, where exactly $0.35*(n(n-1))$ randomly selected binary connections are created;
random (0.2): sparse random communication networks, where exactly $0.1*(n(n-1))$ randomly selected binary connections are created;
spanning: the communication network is a random spanning tree that spans the agents of the problem;
ring: the communication network is a random ring, and so the maximum distance between any agent pair is $n/2$ hops;
chain: the communication network is a linear chain tree, randomly selected; agents at the ends of the chain are $(n-1)$ hops from each other.

[2] `http://dischoco.sourceforge.net/`

Table 1. Performance on hard region when simulating unicast communication protocol

Communication graph	$(p_1 = 0.2, p_2 = 0.7)$				$(p_1 = 0.7, p_2 = 0.3)$			
	$\#transmission$		$\#encccs$		$\#transmission$		$\#encccs$	
Algorithm	ABT	AFC-ng	ABT	AFC-ng	ABT	AFC-ng	ABT	AFC-ng
complete	5 769	**2 724**	26 143	**25 968**	627 742	**188 934**	1 784 567	**1 223 455**
constraint	7 834	**2 886**	**27 807**	30 115	772 906	**182 744**	1 836 074	**1 292 729**
star	12 061	**5 076**	**39 587**	48 301	1 283 398	**361 580**	2 286 323	**2 181 065**
random (0.7)	7 846	**3 518**	**29 640**	32 344	846 561	**242 956**	1 929 618	**1 505 303**
random (0.2)	15 299	**6 147**	**43 790**	55 555	1 609 192	**431 490**	**2 476 220**	2 541 963
spanning	28 236	**10 936**	**67 390**	97 036	3 083 883	**739 513**	**3 673 841**	4 158 617
ring	40 274	**13 654**	**86 113**	120 862	4 602 738	**972 699**	**4 564 197**	5 528 475
chain	51 075	**19 872**	**99 331**	164 056	6 183 110	**1 321 164**	**5 262 304**	6 680 264

Table 2. Performance on hard region when simulating multicast communication protocol

Communication graph	$(p_1 = 0.2, p_2 = 0.7)$				$(p_1 = 0.7, p_2 = 0.3)$			
	$\#transmission$		$\#encccs$		$\#transmission$		$\#encccs$	
Algorithm	ABT	AFC-ng	ABT	AFC-ng	ABT	AFC-ng	ABT	AFC-ng
complete	5 604	**2 774**	**26 187**	26 758	624 311	**188 967**	1 776 765	**1 241 699**
constraint	6 629	**2 764**	**28 361**	29 793	678 895	**183 789**	1 832 203	**1 279 464**
star	8 657	**3 483**	**38 680**	46 047	881 339	**215 746**	2 276 400	**2 051 918**
random (0.7)	6 976	**3 103**	**29 708**	32 157	736 382	**197 971**	1 929 399	**1 527 361**
random (0.2)	12 022	**4 442**	**45 536**	55 573	1 153 096	**242 460**	2 625 804	**2 588 631**
spanning	19 066	**6 497**	**66 879**	91 341	1 741 666	**304 504**	**3 610 608**	3 823 730
ring	28 641	**8 086**	**90 159**	121 029	2 593 760	**367 807**	**4 843 591**	5 465 870
chain	32 553	**10 917**	**97 718**	153 332	3 055 454	**430 339**	**5 182 070**	5 984 795

Table 3. Performance on hard region when simulating multicast* communication protocol

Communication graph	$(p_1 = 0.2, p_2 = 0.7)$				$(p_1 = 0.7, p_2 = 0.3)$			
	$\#transmission$		$\#encccs$		$\#transmission$		$\#encccs$	
Algorithm	ABT	AFC-ng	ABT	AFC-ng	ABT	AFC-ng	ABT	AFC-ng
complete	2 643	**993**	**25 493**	25 959	230 833	**35 028**	1 764 498	**1 113 742**
constraint	4 167	**1 273**	**27 977**	30 186	313 675	**37 869**	1 817 239	**1 181 323**
star	5 416	**1 805**	**37 564**	44 317	487 496	**67 289**	2 242 006	**1 905 299**
random (0.7)	4 013	**1 554**	**29 361**	32 211	366 740	**62 809**	1 913 308	**1 407 144**
random (0.2)	9 004	**3 257**	**44 894**	56 056	874 364	**151 131**	2 613 529	**2 505 835**
spanning	15 522	**5 382**	**65 817**	91 692	1 442 528	**213 293**	**3 600 303**	3 728 196
ring	25 648	**7 741**	**88 114**	122 376	2 517 252	**351 788**	**4 826 414**	5 452 225
chain	31 882	**10 580**	**97 689**	153 234	2 972 596	**415 085**	**5 161 530**	5 977 598

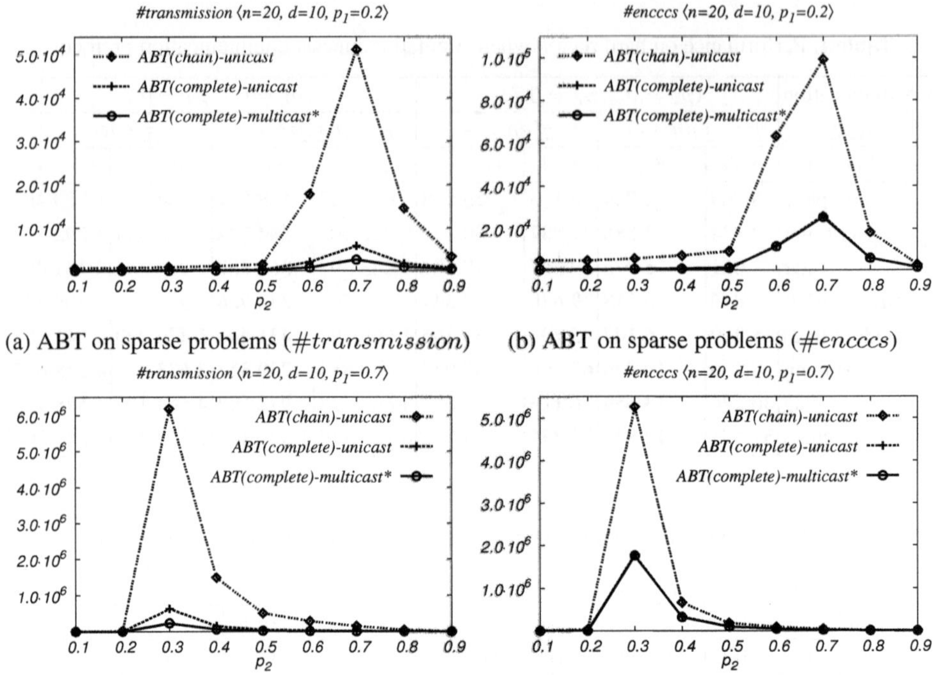

(a) ABT on sparse problems ($\#transmission$) (b) ABT on sparse problems ($\#encccs$)

(c) ABT on dense problems ($\#transmission$) (d) ABT on dense problems ($\#encccs$)

Fig. 4. Comparison of the performance of ABT using unicast on chain communication tree and complete graph, and using multicast* on complete graph

For each topology (except complete and constraint), we generated 5 different random communication networks, and solve each problem instance on each one. The results presented are averaged over 20 problem instances.

For space reason, we show only a subset of the obtained results. Tables 1, 2, 3 present the obtained results on the hard regions ($p_1 = 0.2, p_2 = 0.7$) and ($p_1 = 0.7, p_2 = 0.3$). On dense problems on complete communications graphs with multicast* (Table 3), ABT can use 7 (resp. 1.7) times more $\#transmission$ (resp. $\#encccs$) than AFC-ng. On sparse problems on complete graphs when simulating multicast*, ABT requires 2.5 times more $\#transmission$ than AFC-ng while the $\#encccs$ value for both algorithms is similar. On dense problems on chain communications network with multicast, (Table 2), ABT requires 7 times more $\#transmission$ than AFC-ng, but ABT requires slightly fewer $\#encccs$. On sparse problems on chain communications network with multicast, ABT requires 3 times more $\#transmission$ than AFC-ng, but AFC-ng records 1.5 times more $\#encccs$. For unicast (Table 1) we see a similar pattern, but smaller differences. For sparse problems on the chain communication networks ABT requires 2.5 times more $\#transmission$ than AFC-ng, but AFC-ng performs 1.6 times more $\#encccs$. Thus, when we use unicasting on sparse communications networks, ABT is better on $\#encccs$, otherwise, AFC-ng is better, and particularly for

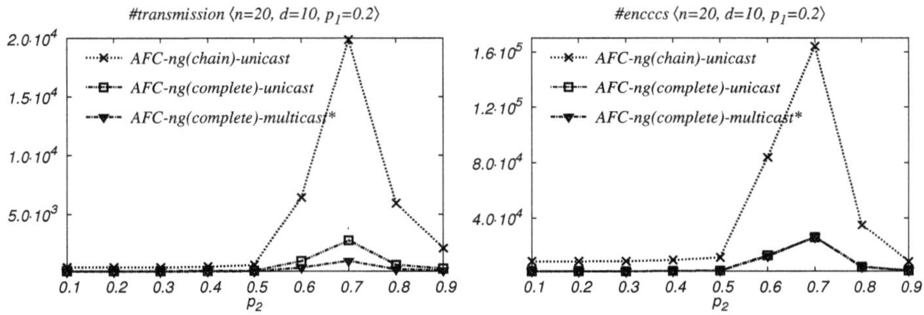

(a) AFC-ng on sparse problems ($\#transmission$) (b) AFC-ng on sparse problems ($\#encccs$)

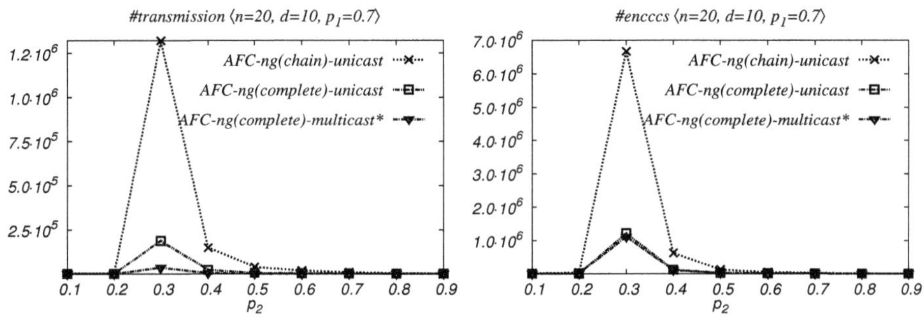

(c) AFC-ng on dense problems ($\#transmission$) (d) AFC-ng on dense problems ($\#encccs$)

Fig. 5. Comparison of the performance of AFC-ng using unicast on chain communication tree and complete graph, and using multicast* on complete graph

dense communication networks with multicast*, where ABT requires 7 times more $\#transmission$ than AFC-ng.

We note that, for the two algorithms we studied, changing just the routing protocol rarely changes the ranking, but it does affect the margin of improvement. For example, for multicast on dense networks that match the constraint graph, the ratio of $\#transmission$ for ABT against AFC-ng is 3.7, but for multicast* on the same problems, the ratio increases to 8.2. A change in the topology of the communications graph does affect the ranking for $\#encccs$. For example, for unicast on dense problems, on complete networks AFC-ng offers an improvement over ABT of 45%, but for chain tree networks, ABT is better by a factor of 27%.

Looking at just ABT, for sparse problems, for different communication layers, the $\#transmission$ can vary by a factor of 20 (multicast* on complete, vs unicast on chain), Fig. 4. If we assume unicast on a complete network is the standard DisCSP model, then varying the communication layer could drop $\#transmission$ by a factor of 2.5, or raise it by a factor of 8. For dense problems, the biggest factor is 25 for $\#transmission$ (this factor can drop by 1/3 or rise by 9), and 2.7 for $\#encccs$.

Looking at just AFC-ng on sparse problems (Fig. 5), again the range in $\#transmission$ varies by a factor of 20, and for $\#encccs$ by a factor of 6.5. For

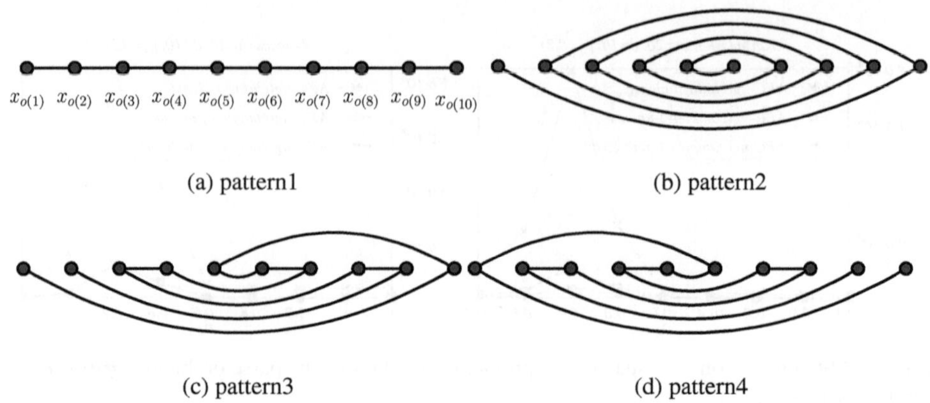

Fig. 6. Patterns used to generate communication chain trees from the max-degree ordering *o*

dense problems, the variation factor for $\#transmission$ is 37 when comparing the multicast* protocol on complete communications graph to the unicast protocol on chain communications network. The variation is 6 for $\#encccs$. These results show that the $\#transmission$ for AFC-ng using the standard model used so far to compare DisCSP algorithms (unicast on complete communications graph) could be factor of 6 too high or a factor of 6 too low.

Changing just the routing protocol has only a small effect for ABT, varying by a factor of approximately 2, but a larger effect for AFC-ng, of up to 5.5. We believe this is because of the nature of the algorithms. Multicast* does not change the number of no-goods, and ABT sends significantly more no-goods than AFC-ng. In general, multicasting (multicast and multicast*) offers an improvement over unicast when the communications graph is sparse, while multicast* improves over multicast when the communications graph is dense.

On the chain communications network, AFC-ng appears to require more $\#encccs$ than ABT, while ABT requires more $\#transmission$. This is investigated more closely in the next section. On chain communications, multicast and multicast* are similar, apart from random variation in message delays.

4.1 Communication Chain Trees

In the following we evaluate the performance of ABT and AFC-ng on different chain communication networks. Based on the max-degree ordering (o) (computed from the constraint graph) 5 patterns are used to generate chain communication trees (T). In the following we denote by $x_{o(k)}$ the kth agent in o. This patterns are presented on Fig. 6.

p1: every pair of adjacent agents in o are connected in the chain communication tree ($T_E = \{\{x_{o(k)}, x_{o(k+1)}\} \mid k \in 1..n-1\}$.

p2: for each pair of adjacent agents on the resulting communication tree T, we try to maximise their separation in the ordering o. Thus, the first agent ($x_{o(1)}$) is connected to the last ($x_{o(n)}$), the last the second, and so on: $T_E = \{\{x_{o(1)}, x_{o(n)}\}; \{x_{o(n)}, x_{o(2)}\}; \{x_{o(2)}, x_{o(n-1)}\}; \ldots\}$.

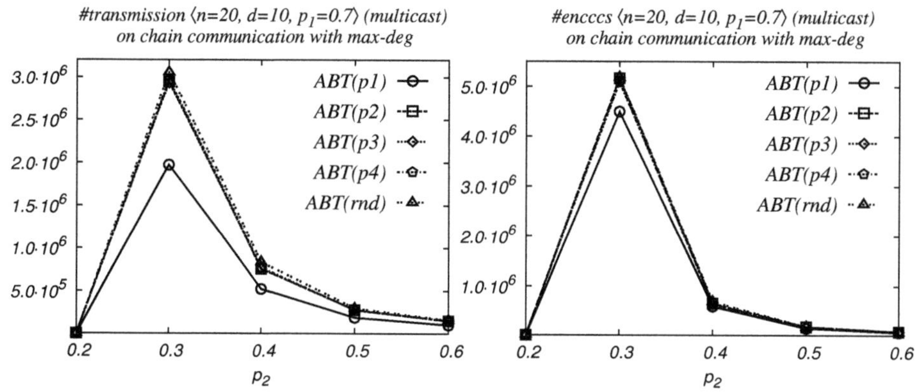

(a) $\#transmission$ (multicast) (b) $\#encccs$ (multicast)

Fig. 7. Performance of ABT on dense problems, multicast routing protocol, different communication chain trees

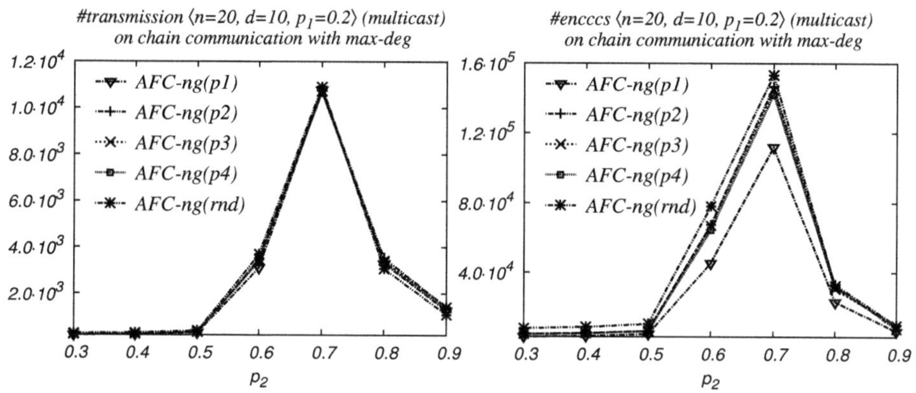

(a) $\#transmission$ (b) $\#encccs$

Fig. 8. Performance of AFC-ng on sparse problems, multicast routing protocol different communication chain trees

p3: we try to maximise the distance between the first and the second agent on o. Thus, $x_{o(1)}$ and $x_{o(2)}$ are the extremities of the resulting chain communication tree T, $T_E = \{\{x_{o(2)}, x_{o(n-1)}\}; \{x_{o(n-1)}, x_{o(n-2)}\}; \{x_{o(n-2)}, x_{o(3)}\}; \ldots; \{x_{o(\frac{n}{2})}, x_{o(n)}\}; \{x_{o(n)}, x_{o(1)}\}\}$.

p4: we try to maximise the distance between the last and the second last agent on o. Thus, $x_{o(n-1)}$ and $x_{o(n)}$ are the extremities of the resulting chain communication tree T, $T_E = \{\{x_{o(n-1)}, x_{o(2)}\}; \{x_{o(2)}, x_{o(3)}\}; \{x_{o(3)}, x_{o(n-2)}\}; \ldots; \{x_{o(\frac{n}{2}+1)}, x_{o(1)}\}; \{x_{o(1)}, x_{o(n)}\}\}$.

rnd: we generate random chain trees.

Again we show only a subset of results due to space reasons. Looking at ABT performance on dense problems (Fig. 7), when the chain communication tree does not match the max-degree agent ordering (pattern 1) the $\#transmission$ required is increased by 50%. For $\#encccs$ a small improvement can be seen for pattern 1 compared to other patterns. For AFC-ng on sparse problems (Fig. 8), the pattern used doesn't really matter for the $\#transmission$. For $\#encccs$, if the chain communication fits the max-degree we get an improvement of 30%.

5 Conclusion

There are many potential applications of Distributed Constraint Satisfaction that rely on wireless networking for the agent to communicate. The standard DisCSP communication model does not represent important features of wireless communication, particularly the topology of the communications graph and the routing protocols. We have introduced a new simulator for modelling wireless communication in DisCSP, NeCoS, which allows different topologies and routing protocols to be modelled. We have shown that varying the communications layer can have a significant effect on the performance metrics for existing DisCSP algorithms, sometimes varying the number of messages by a factor of over 30. The topology of the network also has an impact on the performance of the algorithms, causing a variation of up to 50% in the number of transmissions for ABT and almost 30% in the number of $\#encccs$ for AFC-ng.

These results indicate that, if DisCSP is to be applied to wireless network applications, further research is required on the interaction between the algorithms and the communications layer. In particular, we will investigate ordering heuristics that adapt to the wireless network structure. We will explore algorithm variants that exploit the communication structure; for example, in distributed optimisation, we believe the broadcast mechanism of the AFB family [10, 31] will work well with multicast*. Finally, we will extend these ideas to explore more features of wireless networking, including cases where the communication network is dynamic or agents are mobile, and so areas of the network may become temporarily disconnected [28].

Acknowledgements. This work is funded by Science Foundation Ireland (SFI) under Grant Number SFI/12/RC/2289. We are grateful to Cormac Sreenan for many conversations about wireless networking, and to Pedro Meseguer and the anonymous reviewers for their helpful comments on this submission.

References

[1] Béjar, R., Domshlak, C., Fernández, C., Gomes, C., Krishnamachari, B., Selman, B., Valls, M.: Sensor networks and distributed csp: communication, computation and complexity. Artif. Intel. 161, 117–147 (2005)

[2] Bessiere, C., Maestre, A., Brito, I., Meseguer, P.: Asynchronous backtracking without adding links: a new member in the ABT family. Artif. Intel. 161, 7–24 (2005)

[3] Bijarbooneh, F.H., Flener, P., Ngai, E., Pearson, J.: Optimising quality of information in data collection for mobile sensor networks. In: 2013 IEEE/ACM 21st International Symposium on Quality of Service (IWQoS), pp. 1–10. IEEE (2013)

[4] Brito, I., Meisels, A., Meseguer, P., Zivan, R.: Distributed Constraint Satisfaction with Partially Known Constraints. Constraints 14, 199–234 (2009)

[5] Chechetka, A., Sycara, K.: No-commitment Branch and Bound Search for Distributed Constraint Optimization. In: Proc. of the Fifth International Joint Conference on Autonomous Agents and Multiagent Systems, AAMAS 2006, pp. 1427–1429. ACM, New York (2006)

[6] Doniec, A., Bouraqadi, N., Defoort, M., Le, V.T., Stinckwich, S.: Distributed Constraint Reasoning Applied to Multi-robot Exploration. In: Proc. of the 21st IEEE International Conference on Tools with Artificial Intelligence, ICTAI 2009, pp. 159–166. IEEE Computer Society, Washington, DC (2009)

[7] Ezzahir, R., Bessiere, C., Wahbi, M., Benelallam, I., Bouyakhf, E.H.: Asynchronous Inter-level Forward-checking for DisCSPs. In: Gent, I.P. (ed.) CP 2009. LNCS, vol. 5732, pp. 304–318. Springer, Heidelberg (2009)

[8] Floyd, R.W.: Algorithm 97: Shortest path. Commun. ACM 5(6), 345 (1962)

[9] Gershman, A., Meisels, A., Zivan, R.: Asynchronous Forward-Bounding for Distributed Constraints Optimization. In: Proc. of the 2006 Conference on ECAI 2006: 17th European Conference on Artificial Intelligence, pp. 103–107. IOS Press, Amsterdam (2006)

[10] Gershman, A., Meisels, A., Zivan, R.: Asynchronous Forward Bounding for Distributed COPs. JAIR 34, 61–88 (2009)

[11] Grubshtein, A., Herschorn, N., Netzer, A., Rapaport, G., Yaffe, G., Meisels, A.: The Distributed Constraints (DisCo) Simulation Tool. In: Proc. of the IJCAI workshop on DCR 211, Barcelona, Catalonia, Spain, pp. 30–42 (2011)

[12] Haralick, R.M., Elliott, G.L.: Increasing tree search efficiency for constraint satisfaction problems. Artif. Intel. 14(3), 263–313 (1980)

[13] Hirayama, K., Yokoo, M.: The Distributed Breakout Algorithms. Artif. Intel. 161, 89–116 (2005)

[14] Karl, H., Willig, A.: Protocols and Architectures for Wireless Sensor Networks. Wiley (2005)

[15] Kho, J., Rogers, A., Jennings, N.R.: Decentralized control of adaptive sampling in wireless sensor networks. ACM Trans. Sen. Netw. 5(3), 19:1–19:35 (2009)

[16] Kurose, J.F., Ross, K.W.: Computer Networking, p. 63. Addison Wesley (2013)

[17] Léauté, T., Ottens, B., Szymanek, R.: FRODO 2.0: An Open-Source Framework for Distributed Constraint Optimization. In: Proceedings of the IJCAI 2009 Workshop on Distributed Constraint Reasoning, Pasadena, California, USA, pp. 160–164 (2009)

[18] Léauté, T., Faltings, B.: Protecting privacy through distributed computation in multi-agent decision making. J. Artif. Int. Res. 47(1), 649–695 (2013)

[19] Lynch, N.A.: Distributed Algorithms. Morgan Kaufmann Series (1997)

[20] Mailler, R., Lesser, V.R.: Asynchronous partial overlay: A new algorithm for solving distributed constraint satisfaction problems. JAIR 25(1), 529–576 (2006)

[21] Meisels, A., Kaplansky, E., Razgon, I., Zivan, R.: Comparing Performance of Distributed Constraints Processing Algorithms. In: Proc. of DCR 2002, pp. 86–93 (2002)

[22] Meisels, A., Lavee, O.: Using additional information in DisCSP search. In: Proc. of DCR 2004 (2004)

[23] Meisels, A., Zivan, R.: Asynchronous Forward-checking for DisCSPs. Constraints 12(1), 131–150 (2007)

[24] Molisch, A.F.: Wireless Communications, 2e. Wiley-Blackwell (2010)

[25] Newton, M., Pham, D., Tan, W., Portmann, M., Sattar, A.: Stochastic Local Search Based Channel Assignment in Wireless Mesh Networks. In: Schulte, C. (ed.) CP 2013. LNCS, vol. 8124, pp. 832–847. Springer, Heidelberg (2013)

[26] Petcu, A., Faltings, B.: DPOP: A Scalable Method for Multiagent Constraint Optimization. In: Proc. of IJCAI 2005, pp. 266–271 (2005a)

[27] Petcu, A., Faltings, B.: A value ordering heuristic for local search in distributed resource allocation. In: Faltings, B.V., Petcu, A., Fages, F., Rossi, F. (eds.) CSCLP 2004. LNCS (LNAI), vol. 3419, pp. 86–97. Springer, Heidelberg (2005)

[28] Pujol-Gonzalez, M., Cerquides, J., Meseguer, P., Rodríguez-Aguilar, J., Tambe, M.: Engineering the Decentralized Coordination of UAVs with Limited Communication Range. In: Bielza, C., Salmerón, A., Alonso-Betanzos, A., Hidalgo, J.I., Martínez, L., Troncoso, A., Corchado, E., Corchado, J.M. (eds.) CAEPIA 2013. LNCS, vol. 8109, pp. 199–208. Springer, Heidelberg (2013)

[29] Silaghi, M.C., Sam-Haroud, D., Faltings, B.: Asynchronous Search With Aggregations. In: Proc. of AAAI 2000/IAAI 2000, pp. 917–922 (2000)

[30] Sultanik, E.A., Lass, R.N., Regli, W.C.: Dcopolis: a framework for simulating and deploying distributed constraint reasoning algorithms. In: Proc. of AAMAS 2008, pp. 1667–1668 (2008)

[31] Wahbi, M., Ezzahir, R., Bessiere, C.: Asynchronous Forward Bounding Revisited. In: Schulte, C. (ed.) CP 2013. LNCS, vol. 8124, pp. 708–723. Springer, Heidelberg (2013)

[32] Wahbi, M., Ezzahir, R., Bessiere, C., Bouyakhf, E.H.: DisChoco 2: A Platform for Distributed Constraint Reasoning. In: Proceedings of workshop on DCR 2011, pp. 112–121 (2011), `http://dischoco.sourceforge.net/`

[33] Wahbi, M., Ezzahir, R., Bessiere, C., Bouyakhf, E.H.: Nogood-Based Asynchronous Forward-Checking Algorithms. Constraints 18(3), 404–433 (2013)

[34] Wallace, R.J., Freuder, E.C.: Constraint-based reasoning and privacy/efficiency tradeoffs in multi-agent problem solving. Artif. Intel. 161, 209–228 (2005)

[35] Warshall, S.: A theorem on boolean matrices. J. ACM 9(1), 11–12 (1962)

[36] Wu, X., Brown, K.N., Sreenan, C.J.: Data pre-forwarding for opportunistic data collection in wireless sensor networks. In: 2012 Ninth International Conference on Networked Sensing Systems (INSS), pp. 1–8 (2012)

[37] Yokoo, M.: Algorithms for distributed constraint satisfaction problems: A review. Journal of AAMAS 3(2), 185–207 (2000)

[38] Yokoo, M., Durfee, E.H., Ishida, T., Kuwabara, K.: Distributed Constraint Satisfaction for Formalizing Distributed Problem Solving. In: Proc. of 12th IEEE International Conference on Distributed Computing Systems, pp. 614–621 (1992)

[39] Zhang, W., Wang, G., Xing, Z., Wittenburg, L.: Distributed stochastic search and distributed breakout: properties, comparison and applications to constraint optimization problems in sensor networks. Artif. Intel. 161, 55–87 (2005)

[40] Zivan, R., Meisels, A.: Dynamic Ordering for Asynchronous Backtracking on DisCSPs. Constraints 11(2-3), 179–197 (2006)

[41] Zivan, R., Meisels, A.: Message delay and DisCSP search algorithms. Annals of Mathematics and Artificial Intelligence 46(4), 415–439 (2006)

Adaptive Parameterized Consistency for Non-binary CSPs by Counting Supports*

Robert J. Woodward[1,2], Anthony Schneider[1], Berthe Y. Choueiry[1], and Christian Bessiere[2]

[1] Constraint Systems Laboratory, University of Nebraska-Lincoln, USA
{rwoodwar,aschneid,choueiry}@cse.unl.edu
[2] CNRS, University of Montpellier, France
bessiere@lirmm.fr

Abstract. Determining the appropriate level of local consistency to enforce on a given instance of a Constraint Satisfaction Problem (CSP) is not an easy task. However, selecting the right level may determine our ability to solve the problem. Adaptive parameterized consistency was recently proposed for binary CSPs as a strategy to dynamically select one of two local consistencies (i.e., AC and maxRPC). In this paper, we propose a similar strategy for non-binary table constraints to select between enforcing GAC and pairwise consistency. While the former strategy approximates the supports by their rank and requires that the variables domains be ordered, our technique removes those limitations. We empirically evaluate our approach on benchmark problems to establish its advantages.

1 Introduction

There is an abundance of local consistency techniques of varying cost and pruning power to apply to a Constraint Satisfaction Problem (CSP), but choosing the right one for a given instance remains an open question. In a portfolio approach [22,11,7], we typically choose a single consistency level and enforce it on the entire problem (or a subproblem). Heuristic-based methods have been proposed to dynamically switch, at various stages of search and depending on the constraint, between a weak and a strong level of consistency, AC and maxRPC for binary CSPs [20] and GAC and maxRPWC for non-binary CSPs [18]. The above-mentioned approaches do not allow us to enforce different levels of consistency on the values in the domain of the same variable. To this end, Balafrej et al. introduced *adaptive parameterized consistency*, which selects, for each value in the domain of a variable, one of two consistency levels based on the value of a parameter [1]. That parameter is determined by the *rank* of the support of the value in a constraint (assuming a fixed total ordering of the variables' domains),

* This research was supported by NSF Grant No. RI-111795 and EU project ICON (FP7-284715). Woodward was supported by an NSF GRF Grant No. 1041000 and a Chateaubriand Fellowship. Experiments were conducted on the equipment of the Holland Computing Center at the University of Nebraska–Lincoln.

B. O'Sullivan (Ed.): CP 2014, LNCS 8656, pp. 755–764, 2014.

and updated depending on the weight of the constraint [5]. Their study targeted enforcing AC and maxRPC on binary CSPs.

In this paper, we extend their mechanism to enforcing GAC and pairwise-consistency on non-binary CSPs with table constraints. Our approach is based on *counting* the number of supporting tuples, which is automatically provided by the algorithms that we use. Thus, we remove the restriction on maintaining ordered domains and the approximation of a support's count by its rank. We establish empirically the advantages of our approach.

The paper is structured as follows. Section 2 provides background information. Section 3 describes our approach, and Section 4 discusses our empirically evaluation on benchmark problems. Finally, Section 5 concludes the paper.

2 Background

We first summarize the main concepts and definitions used.

2.1 Constraint Satisfaction Problem

A *Constraint Satisfaction Problem* (CSP) is defined by a tuple $(\mathcal{X}, \mathcal{D}, \mathcal{C})$, where $\mathcal{X}$ is a set of variables, $\mathcal{D}$ is a set of domains, and $\mathcal{C}$ is a set of constraints. Each variable $x_i \in \mathcal{X}$ is associated a finite domain $dom(x_i) \in \mathcal{D}$. We denote a variable-value pair as $\langle x_i, v_i \rangle$, where $v_i \in dom(x_i)$. Each constraint $c_j \in \mathcal{C}$ is defined in extension by a relation R_j specified over the scope of the constraint, $scope(c_j)$, which is the set of variables to which the constraint applies. For readability, we refer to the scope of a relation $scope(R_j) = scope(c_j)$. A tuple $\tau \in R_j$ is a combination of allowed values for the variables in $scope(R_j)$. $\tau[x_i]$ is the value that the variable x_i takes in τ. We denote $cons(x_i)$ as the set of constraints that apply to variable x_i, and $neigh(c_j)$ the set of constraints whose scopes overlap with c_j. When $|scope(c_j)| = 2$, c_j is said to be a binary constraint, otherwise, it is non-binary. A solution to the CSP assigns, to each variable, a value taken from its domain such that all the constraints are satisfied. Deciding the existence of a solution for a CSP is NP-complete.

2.2 Local Consistency Properties

CSPs are typically solved with backtrack search. To reduce the severity of the combinatorial explosion, CSPs are usually filtered by enforcing a given *local consistency property* [2].

A variable-value pair $\langle x_i, v_i \rangle$ has an arc-consistent support (AC-support) $\langle x_j, v_j \rangle$ if the tuple $(v_i, v_j) \in R_{ij}$ where $scope(R_{ij}) = \{x_i, x_j\}$ [16,3]. A CSP is arc consistent if every variable-value pair has an AC-support in every constraint. Generalized Arc Consistency (GAC) generalizes arc consistency to non-binary CSPs [16]. $\langle x_i, v_i \rangle$ has a GAC-support in constraint c_j if $\exists \tau \in R_j$ such that $\tau[x_i] = v_i$. A CSP is GAC if every $\langle x_i, v_i \rangle$ has a GAC-support in every constraint in $cons(x_i)$. GAC can be enforced by removing domain values that have no GAC-support, leaving the relations unchanged. Simple Tabular Reduction (STR) algorithms not only enforce GAC on the domains, but also remove all tuples $\tau \in R_j$ where $\exists x_i \in scope(R_j)$ such that $\tau[x_i] \notin dom(x_i)$ [21,13,14].

A CSP is m-wise consistent if, every tuple in a relation can be extended to every combination of $m-1$ other relations in a consistent manner [8,10]. Keeping with relational-consistency notations, Karakashian et al. denoted m-wise consistency by R($*$,m)C, and proposed a first algorithm for enforcing it [12]. Their implementation finds an extension (i.e., support) for a tuple by conducting a backtrack search on the other $m-1$ relations, and removes the tuples that have no support. After all relations are filtered, they are projected onto the domains of the variables. Pairwise consistency (PWC) corresponds to m=2, R($*$,2)C≡PWC. Lecoutre et al. introduced the algorithm extended STR (eSTR) [15], which enforces PWC on a CSP using the STR mechanism [21]. eSTR maintains counters on the intersections of two constraints to determine if a tuple is pairwise consistent or not. In this paper, we enforce PWC using the algorithm for R($*$,2)C [12], and not eSTR, because it is prohibitively expensive to continuously maintain the counters of eSTR in a strategy where PWC is only selectively enforced.

2.3 Adaptive Parameterized Consistency

Balafrej et al. introduced *the distance to the end of value v_i for variable x_i* as:

$$\Delta(x_i, v_i) = \frac{|dom^o(x_i)| - rank(v_i, dom^o(x_i))}{|dom^o(x_i)|}$$

where $dom^o(x_i)$ is the original, unfiltered domain of x_i, and $rank(v_i, dom^o(x_i))$ is the position of v_i in the ordered set $dom^o(x_i)$ [1]. In Figure 1, borrowed from [1], $\Delta(x_2, 1) = 0.75$, $\Delta(x_2, 2) = 0.50$, $\Delta(x_2, 3) = 0.25$, and $\Delta(x_2, 4) = 0.00$.

Further, for a given parameter p, they defined $\langle x_i, v_i\rangle$ to be *p-stable for AC* for c_{ij} where $scope(c_{ij}) = \{x_i, x_j\}$ if there exists an AC-support $\langle x_j, v_j\rangle$ with $\Delta(x_j, v_j) \geq p$ for c_{ij}. Figure 1 illustrates an example for the constraint $x_1 \leq x_2$ with $p = 0.25$. $\langle x_1, 1\rangle$,$\langle x_1, 2\rangle$,$\langle x_1, 3\rangle$ are all 0.25-stable for AC for the constraint, but $\langle x_1, 4\rangle$ is not, because its only AC-support, $\langle x_2, 4\rangle$, has distance 0.

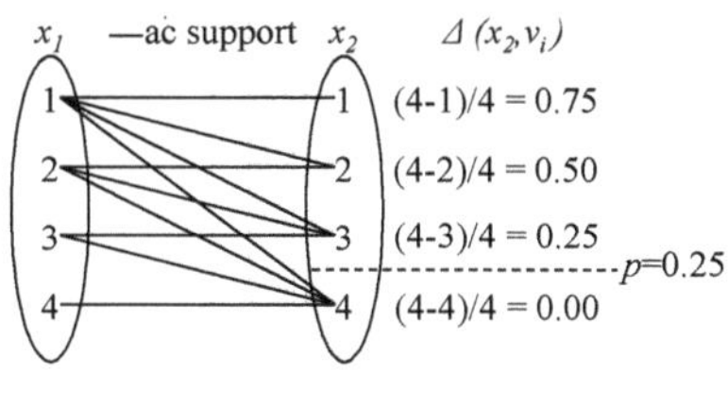

Fig. 1. The constraint $x_1 \leq x_2$. $\langle x_1, 4\rangle$ is not 0.25-stable for AC [1]

The *parameterized* strategy p-LC [1] enforces, on each variable-value pair, either AC or some local consistency (LC) property strictly stronger than AC depending on the value of the parameter p. The idea is to enforce LC only on the variable-value pairs with few supports, approximated with the rank ($< p$) of the first found AC-support. We focus on the constraint-based version, *pc*-LC, where $\langle x_i, v_i \rangle$ is *pc*-LC if for *every* constraint $c_j \in cons(x_i)$, $\langle x_i, v_i \rangle$ is p-stable for AC on c_j or $\langle x_i, v_i \rangle$ is LC on c_j. In *pc*-LC, the value of p is given as input. In the *adaptive* version, *apc*-LC, it is dynamically determined for each constraint c_j using the *weight* of c_j, $w(c_j)$, which is the number of times c_j caused a domain wipe-out like in the variable-ordering heuristic dom/wdeg [5]:

$$p(c_j) = \frac{w(c_j) - \min_{c_k \in \mathcal{C}}(w(c_k))}{\max_{c_k \in \mathcal{C}}(w(c_k)) - \min_{c_k \in \mathcal{C}}(w(c_k)) + 1}. \tag{1}$$

In [1], *apc*-maxRPC was experimentally shown to outperform AC and maxRPC [6].

3 Modifying *apc*-LC for Non-binary CSPs

For binary CSPs, p-stability for AC of $\langle x_i, v_i \rangle$ estimates how many supports are left for $\langle x_i, v_i \rangle$ in other constraints using the rank of the AC-support in the corresponding domain. This estimate should not directly applied to non-binary table constraints because the GAC-support of $\langle x_i, v_i \rangle$ is a tuple in a relation that is unsorted, which would make the estimate way too imprecise. Consider the example with $\langle x_i, v_i \rangle$ and a relation R_j of 100 tuples. Assume that the only tuple $\tau \in R_j$ supporting $\langle x_i, v_i \rangle$ appears at the top of the table of R_j. The estimate would indicate that there are many supports for $\langle x_i, v_i \rangle$ because there are 99 tuples that appear after it. However, in reality, $\langle x_i, v_i \rangle$ has a unique support. Below, we introduce p-stability for GAC, which counts the number of supports for each variable-value pair. Then, we introduce a mechanism to compute p-stability for GAC, and finally give an algorithm for enforcing *apc*-LC, which adaptively enforces STR or LC. In this paper, we study R(∗,2)C as LC, and discuss the implementation of *apc*-R(∗,2)C.

3.1 *p*-Stability for GAC

We say that $\langle x_i, v_i \rangle$ is *p-stable for GAC* if for every constraint $c_j \in cons(x_i)$,

$$\frac{|\sigma_{x_i=v_i}(R_j)|}{|R_j^o|} \geq p(c_j),$$

where $\sigma_{x_i=v_i}(R_j)$ selects the tuples in R_j where $\langle x_i, v_i \rangle$ appears, and R_j^o is the original, unfiltered relation. A CSP is p-stable for GAC if every variable-value pair is p-stable for GAC for every constraint that applies to it.

Figure 2 gives the relation for the constraint $x_1 \leq x_2$. $\langle x_1, 1 \rangle$ and $\langle x_1, 2 \rangle$ are 0.25-stable for GAC. Indeed, $\sigma_{x_1=1}$ returns four rows $\{0, 1, 2, 3\}$ in the table, and $\langle x_1, 1 \rangle$ is 0.25-stable: $\frac{4}{10} \geq 0.25$. Similarly, $\langle x_1, 2 \rangle$ also is 0.25-stable: $\frac{3}{10} \geq 0.25$. $\langle x_1, 3 \rangle$ and $\langle x_1, 4 \rangle$ are not 0.25-stable, because $\frac{2}{10} \ngeq 0.25$ and $\frac{1}{10} \ngeq 0.25$. This example illustrates how, on binary constraints, and for a given p, p-stable for AC does not guarantee p-stable for GAC. (Recall that $\langle x_1, 3 \rangle$ is 0.25-stable for AC in Figure 1).

	x_1	x_2	
0	1	1	$gacSupports[R_j](\langle x_1,1\rangle)=\{0,1,2,3\}$
1	1	2	$gacSupports[R_j](\langle x_1,2\rangle)=\{4,5,6\}$
2	1	3	$gacSupports[R_j](\langle x_1,3\rangle)=\{7,8\}$
3	1	4	$gacSupports[R_j](\langle x_1,4\rangle)=\{9\}$
4	2	2	
5	2	3	
6	2	4	$gacSupports[R_j](\langle x_2,1\rangle)=\{0\}$
7	3	3	$gacSupports[R_j](\langle x_2,2\rangle)=\{1,4\}$
8	3	4	$gacSupports[R_j](\langle x_2,3\rangle)=\{2,5,7\}$
9	4	4	$gacSupports[R_j](\langle x_2,4\rangle)=\{3,6,8,9\}$

Fig. 2. The relation of $x_1 \leq x_2$. $\langle x_1, 3\rangle$ and $\langle x_1, 4\rangle$ are not 0.25-stable for GAC.

3.2 Computing *p*-Stability for GAC

For each constraint c_j, we introduce for every $\langle x_i, v_i\rangle$ a set of integers indicating the position of the tuples returned by $\sigma_{x_i=v_i}(R_j)$, which is similar to the data structure in GAC4 [17]. We denote this table $gacSupports[R_j][\langle x_i, v_i\rangle]$. The check for *p*-stable can be verified by using $|gacSupports[R_j][\langle x_i, v_i\rangle]|$. Figure 2, shows the $gacSupports[R_j]$ for the constraint $x_1 \leq x_2$. For each relation, the space complexity to store each $gacSupports[R_j]$ is $\mathcal{O}(k \cdot t)$, where k is the maximum constraint arity and t is the maximum number of tuples in a relation. The time complexity to generate $gacSupports[R_j]$ is $\mathcal{O}(k \cdot t)$, by iterating through every tuple.

3.3 Algorithm for Enforcing *apc*-LC

With the *gacSupports* data-structure, we can apply STR by verifying, for each constraint c_j, that every variable $x_i \in scope(c_j)$ and $v_i \in dom(x_i)$ has a nonzero $|gacSupports[R_j][\langle x_i, v_i\rangle]|$. LIVING-STR (Algorithm 1) does precisely this operation (ignoring Lines 1 and 1, which apply to the *apc*-LC operation introduced next). $past(\mathcal{P})$ denotes the variables of the CSP $\mathcal{P}$ already instantiated by search, and delTuples$(R_k, S, level)$ deletes all the tuples in the subset $S \subseteq R_k$, and marks their removal level at the level of search *level*. When deleting a tuple from the relation R_k, c_k's neighboring constraints, $neigh(c_k)$, should be re-queued to be processed with LIVING-STR. Initially, all constraints are in the queue. LIVING-STR is similar to STR3 in that it iterates over variable-value pairs rather than over tuples. However, it does not use as much book-keeping for optimizing the number of STR checks as STR3 [14]. Instead, LIVING-STR uses the same data structures as STR and STR2(+) to manage tuple deletions in a relation [13,21].

Including Lines 1 and 1 in Algorithm 1 yields *apc*-LC, which adaptively applies LC. The adaptive level $p(c_j)$ is defined by Balafrej et al. [1] and recalled in Equation (1). The local consistency technique used here is the implementation of R(∗,2)C [12], *apc*-R(∗,2)C. APPLY-R(∗,2)C (Algorithm 2) takes as input the list of tuples of a constraint on which R(∗,2)C must be enforced.

Algorithm 1. LIVING-STR(c_i): set of variables

Input: c_j: a constraint of $\mathcal{P}$
Output: Set of variables in $scope(c_j)$ whose domains have been modified

$X_{modified} \leftarrow \emptyset$
foreach $x_i \in scope(c_j) \mid x_i \notin past(\mathcal{P})$ **do**
  foreach $v_i \in dom(x_i)$ **do**
    if $|gacSupports[R_j](\langle x_i, v_i \rangle)| \neq 0$ ***and*** $\frac{|gacSupports[R_j](\langle x_i, v_i \rangle)|}{|R_j^o|} \ngtr p(c_j)$ **then**
      APPLY-LC($R_j, gacSupports[R_j](\langle x_i, v_i \rangle)$)
    if $|gacSupports[R_j](\langle x_i, v_i \rangle)| = 0$ **then**
      foreach $c_k \in cons(x_i)$ **do**
        delTuples($c_k, gacSupports[R_k](\langle x_i, v_i \rangle), |past(\mathcal{P})|$)
      $dom(x_i) \leftarrow dom(x_i) \setminus \{v_i\}$
      if $dom(x_i) = \emptyset$ **then throw** INCONSISTENCY
      $X_{modified} \leftarrow X_{modified} \cup \{x_i\}$
return $X_{modified}$

Algorithm 2. APPLY-R(*,2)C(c_i, *tuples*)

Input: c_i: a constraint; *tuples*: a set of tuples from the constraint c_i
Output: The *tuples* are either R(*,2)C or deleted

1 **foreach** $\tau \in tuples$ **do**
2   **foreach** $c_j \in neigh(c_i)$ **do**
3   **if** SEARCHSUPPORT($R_i, \tau, \{R_j\}$) *returns inconsistent* **then**
4   delTuples($c_i, \{\tau\}, |past(\mathcal{P})|$)

SEARCHSUPPORT($R_i, \tau, \{R_j\}$) on Line 2 of Algorithm 2 searches for a support for the tuple $\tau \in R_i$, the pairwise check [12].

Theoretical analysis: Let k be the maximum constraint arity, d the maximum domain size, and δ the maximum number of neighbors of a constraint. The time complexity of Algorithm 1 is $\mathcal{O}(k \cdot d)$. Algorithm 2 is $\mathcal{O}(\delta \cdot t^2)$ because it makes $\mathcal{O}(\delta \cdot t)$ calls to SEARCHSUPPORT, which is $\mathcal{O}(t)$ in our context. The correctness of Algorithms 1 and 2 can be shown in straightforward manner by contradiction.

4 Empirical Evaluations

The goal of our experimental analysis is to assess if *apc*-R(*,2)C effectively selects when to apply STR and R(*,2)C when used in a pre-processing step and in a real full lookahead strategy [9] during backtrack search to find the first solution to a CSP. In our experiments, we use the variable ordering dom/wdeg [5]. The experiments are conducted on the benchmarks of the CSP Solver Competition[1]

[1] http://www.cril.univ-artois.fr/CPAI08/

with a time limit of two hours per instance and 8 GB of memory. Because STR and R(∗,2)C enforce the same level of consistency on binary CSPs [4], we focus our experiments on 21 non-binary benchmarks[2] consisting of 623 CSP instances. We chose these benchmarks because they are given in extension and at least one algorithm completed 5% of the instances in the benchmark.

Table 1 summarizes the results in terms of number of instances solved. Importantly, *apc*-R(∗,2)C completes the largest number of instances (552). Considering the instances solved by all algorithms (485 instances), *apc*-R(∗,2)C has the smallest average and median CPU time. Row 3 indicates the number of instances STR solved but R(∗,2)C and *apc*-R(∗,2)C did not solve (18 and 11 instances, respectively), thus showing that *apc*-R(∗,2)C, although it may have enforced R(∗,2)C too often, outperformed R(∗,2)C and missed fewer instances than it (11 vs. 18). Row 4 exhibits similar results showing the number of instances that R(∗,2)C could solve, but that were missed by STR and *apc*-R(∗,2)C (64 and 6 instances, respectively). Here, *apc*-R(∗,2)C did not enforce R(∗,2)C often enough, but managed to outperform STR missing significantly fewer instances than STR (6 vs. 64).

Table 1. Number of instances completed by the tested algorithms

		STR	R(∗,2)C	*apc*-R(∗,2)C
1	#instances completed by	504	550	**552**
2	#instances completed only by	10	5	0
3	#instances solved by STR, but missed by	0	18	11
4	#instances solved by R(∗,2)C, but missed by	64	0	6
5	#instances solved by *apc*-R(∗,2)C, but missed by	59	8	0
	Average CPU time (sec.) over 458 instances	328.41	378.12	**313.31**
	Median CPU time (sec.) over 458 instances	7.23	17.35	**7.21**

Table 2 gives a finer analysis of the data, showing the number of completions and average and median CPU time per benchmark. Averages computed over only the instances completed by all techniques are shown in the column *All*. We split the table into four categories based on the *average* CPU time of *apc*-R(∗,2)C: *a)* *apc*-R(∗,2)C performs the best (5 benchmarks); *b)* *apc*-R(∗,2)C is competitive, performing between STR and R(∗,2)C (13 benchmarks); *c)* *apc*-R(∗,2)C performs the worst (2 benchmarks); and *d)* STR does not solve the benchmark but R(∗,2)C and *apc*-R(∗,2)C do (1 benchmark). The best average CPU time appears in bold face in the corresponding column. The median CPU time of *apc*-R(∗,2)C is bold faced when its rank differs from that of the average CPU time (on which the four categorized are based). On TSP-20, *apc*-R(∗,2)C ranks bottom on average CPU time but between STR and R(∗,2)C on median CPU time. On aim-100, jnhUnsat, rand-8-20-5, and ukVg, *apc*-R(∗,2)C is between STR and R(∗,2)C for average CPU time, but best for median CPU time.

[2] Aim-(50,100,200), allIntervalSeries, dag-rand, dubois, jnh(Sat/Unsat), lexVg, modifiedRenault, pret, rand-10-20-10, rand-3-20-20(-fcd), rand-8-20-5, ssa, travellingSalesman-20, travellingSalesman-25, ukVg, varDimacs, wordsVg.

Table 2. Results of the experiments per benchmark, organized in four categories

Benchmark	#Instances	#Completed STR	#Completed R(∗,2)C	#Completed *apc*-R(∗,2)C	#Completed All	Average CPU time (sec) STR	Average CPU time (sec) R(∗,2)C	Average CPU time (sec) *apc*-R(∗,2)C	Median CPU time (sec) STR	Median CPU time (sec) R(∗,2)C	Median CPU time (sec) *apc*-R(∗,2)C
a) apc-R(∗,2)C is the best											
aim-50	24	24	24	24	24	**0.04**	0.07	**0.04**	0.02	0.04	**0.03**
allIntervalSeries	25	22	22	22	22	7.09	141.85	**6.00**	0.13	0.31	0.12
jnhSat	16	16	16	16	16	13.07	357.66	**11.74**	8.15	142.24	7.21
modifiedRenault	50	50	50	50	50	6.39	11.17	**6.29**	7.24	8.79	6.98
rand-3-20-20	50	31	43	41	31	1,666.10	939.88	**932.77**	1,211.50	822.54	811.74
b) apc-R(∗,2)C is competitive											
aim-100	24	24	24	24	24	0.38	**0.26**	0.41	0.18	0.25	**0.16**
aim-200	24	22	24	24	22	414.48	**6.52**	286.27	2.39	1.37	**2.60**
jnhUnsat	34	34	34	34	34	**13.61**	294.77	13.95	10.74	153.50	**9.78**
lexVg	63	63	63	63	63	**69.81**	341.87	338.74	0.50	1.38	0.89
pret	8	4	4	4	4	**117.89**	347.03	136.04	115.81	354.82	145.70
rand-3-20-20-fcd	50	39	48	47	39	928.06	**546.84**	615.23	501.30	422.24	464.00
rand-8-20-5	20	9	20	20	9	2,564.94	**355.57**	372.76	1,987.35	314.26	**261.68**
rand-10-20-10	20	12	12	12	12	6.72	**1.67**	2.76	6.40	1.66	2.75
ssa	8	6	5	6	5	**64.60**	100.64	69.59	1.51	1.60	1.58
TSP-25	15	13	10	13	10	**232.38**	1,072.72	743.33	69.00	211.41	131.69
ukVg	65	37	31	34	31	**166.82**	796.90	421.35	36.29	54.65	**30.39**
varDimacs	9	6	6	6	6	**89.23**	587.55	319.20	1.56	6.43	2.94
wordsVg	65	65	58	58	58	**119.76**	532.05	400.22	0.39	0.95	0.59
c) apc-R(∗,2)C is the worst											
dubois	13	7	8	6	6	1,000.54	**451.91**	1,456.01	552.13	255.25	779.57
TSP-20	15	15	15	15	15	**101.20**	318.37	335.13	23.32	61.55	**46.34**
d) Not solved by STR											
dag-rand	25	0	25	25	0	-	**123.70**	149.64	-	124.47	151.33

Table 3 shows the average number of STR and R(∗,2)C checks that *apc*-R(∗,2)C performs per benchmark. In allIntervalSeries, *no* calls are made to R(∗,2)C because the instance is solved backtrack free with STR alone. For *apc*-LC, *no call to LC is done during pre-processing* because the weights of all the

Table 3. Number of calls to STR and R(∗,2)C by benchmark

Benchmark	STR checks	R(∗,2)C checks	Benchmark	STR checks	R(∗,2)C checks
a) apc-R(∗,2)C is the best			*b) apc*-R(∗,2)C is competitive		
aim-50	456,823	39,491	aim-100	7,731,585	894,353
allIntervalSeries	38,281,694	0	aim-200	1,160,334,482	163,177,907
jnhSat	22,119,135	599,080	jnhUnsat	51,688,166	1,918,781
modifiedRenault	4,618,778	601,641	lexVg	564,010,457	2,180,503,026
rand-3-20-20	489,441,126	3,480,216,943	pret	422,987,946	13,973,748
			rand-3-20-20-fcd	455,664,100	2,956,467,994
c) apc-R(∗,2)C is the worst			rand-8-20-5	77,470,561	184,764,543
dubois	3,343,830,604	4,668,288	rand-10-20-10	72,608	3,972
TSP-20	622,949,698	991,590,957	ssa	156,631,370	11,689,961
			TSP-25	2,903,953,315	3,947,391,769
			ukVg	341,565,892	1,002,334,753
d) Not solved by STR			varDimacs	720,843,958	84,123,204
dag-rand	359,248	21,870	wordsVg	514,840,737	2,052,367,934

constraints are set to 1 (giving $p(c_j) = 0$ for all $c_j \in \mathcal{C}$) and updated only during search. For dag-rand, there is a smaller number of R(∗,2)C calls than STR calls (21,870 vs. 359,248). However, those few calls allow us to solve *all* the instances of this benchmark whereas STR alone could not solve *any* instance. This result is a glowing testimony of the ability of *apc*-R(∗,2)C to apply the appropriate level of consistency where needed.

5 Conclusions

In this paper, we extend the notion of *p*-stability for AC to GAC, and provide a mechanism for computing it. We give an algorithm for enforcing *apc*-R(∗,2)C on non-binary table constraints, which adaptively enforces GAC and R(∗,2)C. We validate our approach on benchmark problems. Future work is to investigate other adaptive criteria for selecting the level of consistency to apply, in particular one that operates during both pre-processing and search. To apply our approach to constraints defined in intension and other global constraints, we could use techniques that approximate the number of solutions in those constraints [19].

References

1. Balafrej, A., Bessiere, C., Coletta, R., Bouyakhf, E.H.: Adaptive Parameterized Consistency. In: Schulte, C. (ed.) CP 2013. LNCS, vol. 8124, pp. 143–158. Springer, Heidelberg (2013)
2. Bessiere, C.: Constraint Propagation. In: Handbook of Constraint Programming, pp. 29–83. Elsevier (2006)
3. Bessière, C., Régin, J.C., Yap, R.H., Zhang, Y.: An Optimal Coarse-Grained Arc Consistency Algorithm. Artificial Intelligence 165(2), 165–185 (2005)
4. Bessière, C., Stergiou, K., Walsh, T.: Domain Filtering Consistencies for Non-Binary Constraints. Artificial Intelligence 172, 800–822 (2008)
5. Boussemart, F., Hemery, F., Lecoutre, C., Sais, L.: Boosting Systematic Search by Weighting Constraints. In: Proc. ECAI 2004, pp. 146–150 (2004)
6. Debruyne, R., Bessière, C.: From Restricted Path Consistency to Max-Restricted Path Consistency. In: Smolka, G. (ed.) CP 1997. LNCS, vol. 1330, pp. 312–326. Springer, Heidelberg (1997)
7. Geschwender, D., Karakashian, S., Woodward, R., Choueiry, B.Y., Scott, S.D.: Selecting the Appropriate Consistency Algorithm for CSPs Using Machine Learning Techniques. In: Proc. of AAAI 2013, pp. 1611–1612 (2013)
8. Gyssens, M.: On the Complexity of Join Dependencies. ACM Trans. Database Systems 11(1), 81–108 (1986)
9. Haralick, R.M., Elliott, G.L.: Increasing Tree Search Efficiency for Constraint Satisfaction Problems. Artificial Intelligence 14, 263–313 (1980)
10. Janssen, P., Jégou, P., Nougier, B., Vilarem, M.C.: A Filtering Process for General Constraint-Satisfaction Problems: Achieving Pairwise-Consistency Using an Associated Binary Representation. In: IEEE Workshop on Tools for AI, pp. 420–427 (1989)
11. Kadioglu, S., Malitsky, Y., Sabharwal, A., Samulowitz, H., Sellmann, M.: Algorithm Selection and Scheduling. In: Lee, J. (ed.) CP 2011. LNCS, vol. 6876, pp. 454–469. Springer, Heidelberg (2011)

12. Karakashian, S., Woodward, R., Reeson, C., Choueiry, B.Y., Bessiere, C.: A First Practical Algorithm for High Levels of Relational Consistency. In: Proc. AAAI 2010, pp. 101–107 (2010)
13. Lecoutre, C.: STR2: Optimized Simple Tabular Reduction for Table Constraints. Constraints 16(4), 341–371 (2011)
14. Lecoutre, C., Likitvivatanavong, C., Yap, R.H.C.: A Path-Optimal GAC Algorithm for Table Constraints. In: Proc. of ECAI 2012, pp. 510–515 (2012)
15. Lecoutre, C., Paparrizou, A., Stergiou, K.: Extending STR to a Higher-Order Consistency. In: Proc. AAAI 2013, Bellevue, WA, pp. 576–582 (2013)
16. Mackworth, A.K.: Consistency in Networks of Relations. AI 8, 99–118 (1977)
17. Mohr, R., Masini, G.: Good Old Discrete Relaxation. In: European Conference on Artificial Intelligence (ECAI 1988), pp. 651–656. W. Germany, Munich (1988)
18. Paparrizou, A., Stergiou, K.: Evaluating Simple Fully Automated Heuristics for Adaptive Constraint Propagation. In: Proc. of ICTAI 2012, pp. 880–885 (2012)
19. Pesant, G., Quimper, C.G., Zanarini, A.: Counting-Based Search: Branching Heuristics for Constraint Satisfaction Problems. JAIR 43, 173–210 (2012)
20. Stergiou, K.: Heuristics for Dynamically Adapting Propagation. In: Proc. of ECAI 2008, pp. 485–489 (2008)
21. Ullmann, J.R.: Partition Search for Non-binary Constraint Satisfaction. Information Sciences 177(18), 3639–3678 (2007)
22. Xu, L., Hutter, F., Hoos, H.H., Leyton-Brown, K.: SATzilla: Portfolio-Based Algorithm Selection for SAT. JAIR 32, 565–606 (2008)

Proactive Workload Dispatching on the EURORA Supercomputer

Andrea Bartolini, Andrea Borghesi, Thomas Bridi, Michele Lombardi, and Michela Milano

DISI, University of Bologna, Italy
{a.bartolini,andrea.borghesi3,michele.lombardi2,michela.milano}@unibo.it
thomas.bridi@gmail.com

Abstract. In the era of Cloud Computing, Big Data, and Quantum Physics Simulations, data centers play in the world ICT infrastructure a role as big as (sadly) their power consumption. In many cases, a surprising amount of such consumption is due to idle resources, either introduced to face workload peaks or leftovers of workload fragmentation. In this context, proactive workload dispatching has the chance to improve the utilization of computing resources, thus reducing the idle time *and* improving the ability to handle peaks. In this paper, we devise a CP based approach for proactive workload dispatching on the EURORA supercomputer placed at the CINECA computing center in Bologna. The new system is evaluated on simulated job traces, where it leads to remarkable improvements in terms of both machine utilization and waiting times for queued jobs with respect to the currently used dispatcher, i.e., Portable Batch System (PBS).

1 Introduction

Computing centers play a key role in modern ICT architectures: they run our internet services, keep track of our savings, make our research possible. They are also well known to be power hungry: in Italy, data centers make for ~2% of the national energy consumption, for a total of 6.6 TWh (roughly that of the Calabria region, according to data by Fondazione Politecnico di Milano, 2010).

The mainstream solution to reduce such a gigantic consumption is to employ efficient hardware or efficient design. By doing so, it is possible to obtain remarkable reductions of the PUE index (Power Usage Effectiveness), i.e. the ratio between the power consumption of the whole data center and the power consumption of the IT equipment alone. Recently, a joint effort by the CINECA inter-university consortium [1] in Italy and the Eurotech group [2] has led to the design of the EURORA system. Thanks to an innovative liquid based cooling system and carefully chosen hardware components, this new machine has a PUE of just 1.05 and managed to reach the top of the Green 500 ranking in the first half of 2013, effectively becoming for a time the most efficient supercomputer on earth. As a comparison, PUE values of around 3 were still common in 2009.

B. O'Sullivan (Ed.): CP 2014, LNCS 8656, pp. 765–780, 2014.

However, reducing the PUE is just a half of the problem. Data by McKinsey [3] for US data centers reveals that on average only 6-12% of the power is employed for actual computation. The reason for this dramatically low value lies in *how efficiently the existing IT resources are used.* In particular, redundant resources are usually employed to maintain the quality of service under workload peaks. More redundant resources are also needed to compensate for the fragmentation resulting from suboptimal dispatching choices. As a consequence, a typical data center ends up packing a lot of idle muscles. Unfortunately, idle resources still consume energy: for a 1MW center with a 1.5 PUE, a 30% utilization means a 1M€ annual cost and 3,500 tons of CO_2. In this context, optimization techniques can enable dramatic improvements in the resource management, leading to lower costs, better response times, and fewer emissions.

In this paper, we tackle the problem of performing workload dispatching over the EURORA supercomputer, operating at the CINECA computing center in Bologna. The machine is employed for High Performance Computing (HPC) applications and has a job submission system currently managed by a PBS Dispatcher (Portable Batch System [9]). The dispatcher relies on a number of heuristic techniques to tentatively maintain a high machine utilization and keep the waiting times as small as possible. The CINECA staff has hints that the current system operation could be improved, but finding a more effective PBS configuration is a cumbersome and error-prone task: hence there is interest in alternative approaches. We propose to tackle workload dispatching via proactive scheduling using Constraint Programming. We adopt a rolling horizon approach, where our scheduler is awakened at certain events. At each of such activations, we build a full schedule and resource assignment for all the waiting jobs, but then we dispatch only those jobs that are scheduled for immediate execution. By taking into account forthcoming jobs, we avoid making dispatching decisions with undesirable consequences; by starting only the ones scheduled for immediate execution, the system can manage uncertain execution times.

Our long-term goal is the development of a state of the art workload dispatching approach to replace the current PBS logic. However, at this stage, our main objective is just is to assess the degree of improvement (in terms of waiting times and reduced idleness) that can be obtained by acting on the dispatching decisions. Since our focus is on investigating the solution quality, we do not enforce tight restrictions on the approach run-time (over-exploiting a bit the fact that HPC jobs tend to have large durations). We evaluated our approach by simulating its behavior on real workload traces from the EURORA machine. We compare the results of our approach with those of the currently operating PBS system, demonstrating that substantial improvements are indeed possible.

2 System Description and Motivations for Using CP

This section contains a brief presentation of the architecture of the EURORA supercomputer, a discussion about the current dispatching system, and a review to the motivations behind our choice of CP for building an alternative dispatcher.

The EURORA Supercomputer: As described in [5] EURORA has a modular architecture based on nodes (blades). In its the current state, the system counts 64 nodes, each one comprising 2 octa-core CPUs and 2 expansion cards configured to host an accelerator module: currently, 32 nodes host 2 powerful NVidia GPUs, while the remaining ones are equipped with 2 Intel MIC accelerators. Each node has 16GB of installed RAM memory. EURORA is interfaced with the outside world through a few dedicated computing nodes, physically positioned outside the rack: in particular, a designated login node connects EURORA to the users and runs the job dispatcher (PBS). One of the main boosting factors for the energy efficiency of the supercomputer is the adoption of a hot liquid cooling technology, i.e. the water inside the system can reach up to 50°C. This strongly reduces the energy required for operating the system, since no power is used for actively cooling down the water, and the waste-heat can be recovered as an energy source for other applications.

The PBS Dispatcher: The tool currently used to manage the workload on EURORA system is PBS (Portable Batch System), a proprietary job scheduler by Altair PBS Works with the primary duty of allocating computational tasks, i.e. batch jobs, among available computing resources. The main components of PBS are a server (which manages the jobs) and several daemons running on the execution hosts (i.e. the 64 nodes of EURORA), which track the resource usage and answer to polling request about the host state issued by the server component.

Jobs are submitted by the users into one of multiple queues, each one characterized by different access requirements and by a different approximate waiting time. Users submit their jobs by specifying 1) the number of required nodes; 2) the number of required cores per node; 3) the number of required GPUs and MICs per node (never both of them at the same time); 4) the amount of required memory per node; 5) the maximum execution time. All processes that exceed their maximum execution time are killed. The main available queues on the EURORA system are called *debug*, *parallel*, and *longpar*, and are described in Table 1 - for each of those queues we report the maximum number resources that a job could ask if it desires to belong to that queue, i.e. maximum number of nodes, maximum number of cores and GPUs (second column), maximum execution time, and also the approximate time it might wait before starting its execution.

Cyclically, PBS selects a job for execution by polling the state of one or more nodes, trying to find enough available resources to actually start the job execution. If the attempt is unsuccessful, the job is sent back to its queue and PBS proceeds to consider the following candidate. The choices are guided by priority values and hard-coded constraints defined by the EURORA administrators with the aim to have a good machine utilization and small waiting times. For example, the administrators decided to reserve some nodes to the debug queue and to force jobs in the *longpar* queue to start at night.

Why CP? In its current state, the PBS system works mostly as an on-line heuristic, incurring the risk to make poor resource assignments due to the lack

Table 1. Access requirements and waiting times for the PBS queues in EURORA

Queue	*Max Nodes*	*Max Cores/GPUs*	*Max Time*	*Approx. Wait*
debug	2	32/4	00:30:00	seconds
parallel	32	512/64	06:00:00	minutes
longpar	16	256/32	24:00:00	hours

of an overall plan. Also the hard-coded mapping constraints, designed as a way to ensure low waiting times for specific job classes (e.g. the *debug* queue), may easily cause resource under-utilization, and long waiting times for the remaining jobs (e.g. those in the *longpar* queue). A proactive dispatching approach should intuitively be able to improve the resource utilization and reduce the waiting times without the need of devising such hard-coded restrictions. The task of obtaining a proactive dispatching plan on EURORA can be naturally framed as a resource allocation and scheduling problem, for which CP as a long track of success stories.

3 Design of a CP Approach

We adopt a rolling horizon approach, in which our scheduler is awakened whenever a job 1) enters the system or 2) ends its execution. At each iteration, we build a full schedule and mapping for all the jobs in the input queues, taking into account resource capacity limitations. We consider different performance metrics, which we treat either as objective functions or as soft-constraint. Then we dispatch only those jobs that are scheduled for immediate execution.

The schedule is computed based on the worst-case durations (as provided by the users), but the dispatcher reactivation is triggered by the job *actual* terminations (besides of course by their arrivals). Whenever this occurs, the jobs currently in execution cannot be migrated, but all the waiting ones can be re-scheduled to take advantage of the released resources.

3.1 Formal Problem Definition

We can now provide a precise definition of the scheduling problem solved at each activation of the dispatcher. Each job i enters the system at a certain arrival time q_i, by being submitted to a specific queue (depending on the user choices and on the job characteristics). By analyzing existing execution traces coming from PBS, we have determined an estimated waiting time for each queue, which applies to each job it contains: we refer to this value as ewt_i.

When submitting the job, the user has to specify several pieces of information, including the maximum allowed execution time D_i, the maximum number of nodes to be used rn_i, and the required resources (cores, memory, GPUs, MICs). By convention, the PBS systems consider each job as if it was divided into a set of exactly rn_i identical "job units", to be mapped each on a single node.

It is therefore convenient to specify the resource requirements on a job-unit basis. Formally, let R be a set of indexes corresponding to the resource types (cores, memory, GPUs, MICs), and let the capacity of a node k for resource $r \in R$ be denoted as $cap_{k,r}$. We recall that the system has $m = 64$ nodes, each with 16 cores and 16 GB of RAM memory; 32 nodes have 2 GPUs each (and 0 MICs), and the remaining 32 nodes have 2 MICs each (and 0 GPUs). Finally, let $rq_{i,r}$ be the requirement of a unit of job i for resource r. The dispatching problem at time t consists in assigning a start time $s_i \geq t$ to each waiting job i and a node to each of its units. All the resource capacity limits should be respected, taking into account the presence of jobs already in execution. Once the problem is solved, only the jobs having $s_i = t$ are actually dispatched.

Informally speaking, in the big picture, the goal is to increase the resource utilization and reduce the waiting times, but those metrics can be meaningfully evaluated only once the actual job durations become known. Hence we formulate the problem in terms of several objective functions that are intuitively correlated with the metrics we are interested in. After extensive preliminary experimentations, we settled for the following possible problem objectives:

$$\max_{i=0..n-1} (s_i + D_i) \qquad \text{(makespan)} \qquad (1)$$

$$\sum_{i=0..n-1} \max\left(0, \frac{s_i - q_i - ewt_i}{ewt_i}\right) \qquad \text{(weighted tardiness)} \qquad (2)$$

$$\sum_{i=0..n-1} [[s_i - q_i > ewt_i]] \qquad \text{(num of late jobs)} \qquad (3)$$

where n is the number of jobs and the notation $[[-]]$ stands for the reification of the constraint between brackets. The makespan has been chosen because compressing the schedule length tends to increase the resource utilization. For the tardiness and the number of late jobs, we consider a job to be late if it stays queued for a time larger than ewt_i. The tardiness is weighted, because we assume that users that are already expecting to wait more (i.e. jobs with higher ewt_i) should adjust better to prolonged queue times. Both the tardiness based objectives are chosen to improve the perceived response time, in one case by avoiding (proportionally) long waiting times, in the second by reducing the number of jobs in the queues.

3.2 CP Model

Employed CP Techniques: We defined for the described scheduling problem a CP model that is based on Conditional Interval Variables (CVI, see [8]). A CVI τ represents an interval of time: the start of the interval is referred to as $s(\tau)$ and its end as $e(\tau)$; the duration is $d(\tau)$. The interval may or may not be present, depending on the value of its existence expression $x(\tau)$. In particular, if $x(\tau) = 0$ the interval is not present and does not affect the model: for this situation we also use the notation $\tau = \perp$.

CVIs can be subject to a number of constraints, including the classical *cumulative* [4] to model finite capacity resources, and the more specific *alternative* constraint [8]. This last global constraint has the following signature:

$$alternative(\tau_0, [\tau_1, .., \tau_{n_\tau}], m_\tau) \tag{4}$$

The constraint forces all the interval variables $\tau_1, \tau_2, \ldots$ to have the same start and end time as τ_0. Moreover, exactly m_τ of $\tau_1, \tau_2, \ldots$ will be actually present if τ_0 is present. Formally, the constraint enforces:

$$s(\tau_0) = s(\tau_i),\ e(\tau_0) = e(\tau_i) \quad \forall i = 1..n_\tau \qquad \sum_{i=1}^{n_t} x(\tau_i) = m_\tau\, x(\tau_0) \tag{5}$$

Modeling Decisions and Constraints: In our model, we use a CVIs to model the scheduling decisions. In particular, we introduce an interval variable τ_i with duration D_i for each job waiting in the input queues or already in execution. Then, we fix the start of all τ_i corresponding to running jobs to their real value (which is known at this point). For the waiting jobs we have $s(\tau_i) \in t..eoh$, where t is the time instant for which the model is built and eoh can be given for example by t plus the sum of the maximum duration of all jobs[1]. All the τ_i variables are mandatory, i.e. $x(\tau_i) = 1$.

Mapping decisions should be taken at the level of single job-units. The modeling style we adopt for them is best explained by temporarily introducing a simplifying assumption, namely that no two units of the same job can be mapped on a single node. With this assumption, the mapping decisions can be modeled by introducing a second set of *optional* interval variables $\upsilon_{i,k}$ such that $x(\upsilon_{i,k}) = 1$ iff a unit of job i is mapped to node k.

However, mapping multiple units of the same job on the same node is possible and can be beneficial. To account for this possibility, we have to introduce for each job i multiple sets of υ variables. Specifically, we add one more index and we maintain the semantic, so that we have variables $\upsilon_{i,j,k}$ such that $x(\upsilon_{i,j,k}) = 1$ iff a unit of job i is mapped to node k. The j index is only used to control the number of job units that can be mapped to the same node. Finding a suitable range for the index is a critical step: on the one hand, allowing j to range on $0..rn_i - 1$ (i.e. one set of υ variables for each requested node) is a safe choice. On the other hand, it is impossible to map multiple units of the same job on the same node if doing so would exceed the availability of some resource. Hence, a valid upper bound on the number of υ variable sets for a single job i is given by:

$$p_i = \min\left(rn_i, \min_{r \in R} \left\lfloor \frac{cap_{k,r}}{r_{i,r}} \right\rfloor\right) \tag{6}$$

[1] Note that it is possible to shift all the domains by subtracting the smallest s_i to all values, so that at least one $s(\tau_i)$ has a minimum of 0.

and for each job i, the index j can range in $0..p_i - 1$. Then we have to specify that exactly rn_i job-units should be mapped, i.e. that exactly such number of $v_{i,j,k}$ intervals should be present. This can be done by using an *alternative* constraint:

$$alternative(\tau_i, [v_{i,j,k}], rn_i) \qquad \forall i = 0..n-1 \tag{7}$$

Additionally, the alternative constraint forces all the job-units to start at the same time instant as τ_i. Now, the resource capacity restrictions can be modeled via a set of cumulative constraints:

$$cumulative([v_{i,j,k}], [D_i^{(p_i)}], [r_{i,r}^{(p_i)}], cap_{i,r}) \qquad \forall k = 0..m-1, \forall r \in R \tag{8}$$

where m is the number of nodes and the notation $D_i^{(p_i)}$ stands for a vector containing D_0 repeated p_0 times, then D_1 repeated p_1 times, and so on. As mentioned in Section 2 we disregard all the hard-coded constraints introduced by the PBS administrator and we trust the decision making capabilities of our optimization system with providing waiting times as low as possible.

Handling the Objective Function: We consider several variants of our dispatching problem, differing one from each other for the considered objective and for the possible presence of soft constraints. First, we have three "pure" models, obtained by adding on top of the presented formulation one of the problem objectives that we have discussed in Section 3.1:

$$\min \max_{i=0..n-1} e(\tau_i) \qquad \text{(makespan)} \tag{9}$$

$$\min \sum_{i=0..n-1} \max\left(0, \frac{s(\tau_i) - q_i - ewt_i}{ewt_i}\right) \qquad \text{(weighted tardiness)} \tag{10}$$

$$\min \sum_{i=0..n-1} [[s(\tau_i) - q_i - ewt_i > 0]] \qquad \text{(num. of late jobs)} \tag{11}$$

Then we consider three "composite" formulations obtained by choosing as a main cost function one of Equations (9)-(11), and then by posting a constraint on the value of the remaining ones. For example, assuming the makespan is the main objective, we get:

$$\min \max_{i=0..n-1} e(\tau_i) \tag{12}$$

$$\text{s.t.} \sum_{i=0..n-1} \max\left(0, \frac{s(\tau_i) - q_i - ewt_i}{ewt_i}\right) \leq \delta_0\, \theta_0 \tag{13}$$

$$\sum_{i=0..n-1} [[s(\tau_i) - q_i - ewt_i > 0]] \leq \delta_1\, \theta_1 \tag{14}$$

The values θ_0 and θ_1 are obtained by solving the pure models corresponding to the constrained functions. The parameters δ_0, δ_1 allow to tune the tightness of the constraints. The three new composite formulations are loosely inspired by multi-objective optimization approaches and aim at obtaining good solutions according to one global metric (say, resource utilization), while keeping acceptable levels for the other (say, waiting times).

Table 2. An example of problem instance

i	rn_i	$rq_{i,core}$	$rq_{i,gpu}$	$rq_{i,mic}$	$rq_{i,mem}$	D_i
000	32	4	1	0	1000000	14000
001	1	14	1	0	400000	600
002	2	4	1	0	200000	14400
003	32	16	0	0	400000	800
004	32	3	0	2	800000	400

Table 3. A feasible solution for the instance from Table 2

i	$s(\tau_i)$	$v_{i,0,0}$	$v_{i,0,1}$	$v_{i,0,2..31}$	$v_{i,0,32..63}$	$v_{i,1,0}$	$v_{i,1,1}$	$v_{i,1,2..31}$	$v_{i,1,32..63}$
000	0	$\bot$	0	0	$\bot$	$\bot$	0	$\bot$	$\bot$
001	0	1	$\bot$	$\bot$	$\bot$	$\bot$	$\bot$	$\bot$	$\bot$
002	600	600	$\bot$	$\bot$	$\bot$	600	$\bot$	$\bot$	$\bot$
003	0	$\bot$	$\bot$	$\bot$	0	$\bot$	$\bot$	$\bot$	$\bot$
004	800	$\bot$	$\bot$	$\bot$	800	$\bot$	$\bot$	$\bot$	$\bot$

Example of a solution: Let us suppose we have the set of waiting jobs described in Table 2, then a feasible solution to this instance is described in Table 3. As reported in the table, jobs 000, 001 and 002 can execute only on the nodes equipped with GPUs (i.e. node 0 to 31), job 004 can execute only in nodes with MICs (i.e. node 32 to 63). Tow units of job 000 are allocated on node 1, the other 30 units of job 000 are allocated in nodes from 2 to 31; node 0 is completely free and can run job 001 while job 000 is executing; job 003 can execute on nodes from 32 to 63; after the termination of job 001, job 002 can start its execution with two units on node 0 and after the termination of job 003, job 004 can start in nodes from 32 to 63.

4 Added Value of CP

The scheduler we realized is currently a prototype: it will eventually be deployed on the EURORA supercomputer, but this requires still considerable development and research effort. At this stage we (and the CINECA consortium) are interested in investigating the kind of improvements that could be obtained by changing the dispatcher behavior. On this purpose, we have compared the results we obtained with our dispatcher and the ones achieved by PBS as it is currently configured on EURORA.

We performed the comparison on real PBS execution traces, which contain all the information that is usually available at the job arrival times (i.e. the chosen queue, the resource requirements, the maximum execution time). Additionally, the traces report for each job two important pieces of information, namely the *actual* duration (which we use together with the arrival time to simulate the scheduler activation events) and the start time assigned by PBS.

Our approach was implemented using IBM ILOG CP Optimizer [6] using its default search strategy, which is based on Self-adapting Large Neighborhood

Table 4. Models comparison, queue times

	Average Queue Time			
Model	*all*	*debug*	*parallel*	*longpar*
MKS	187.14	4.77	161.81	0.01
MKS WT/NL	165.98	0.10	160.04	0.01
NL	722.04	2.30	316.92	369.14
NL MKS/WT	201.32	0.31	145.59	18.99
WT	662.18	2.16	203.50	446.34
WT MKS/NL	861.81	0.76	278.60	572.29
PBS	6840.81	17.34	2825.05	3600.40

Table 5. Models comparison, system load

	Average Resource Utilization				
Model	*cores*	*GPUs*	*MICs*	*cores (%)*	*Avg jobs*
MKS	678.81	45.21	3.99	66%	121.68
MKS WT/NL	701.92	45.61	3.99	68%	121.92
NL	614.75	45.89	3.99	60%	116.58
NL MKS/WT	670.75	45.00	3.99	65%	121.21
WT	671.41	47.67	3.98	66%	120.50
WT MKS/NL	620.45	41.72	3.99	61%	119.07
PBS	447.98	29.16	0.33	46%	63.04

Search [7] guided by an Linear Programming relaxation. At each scheduler activation we use the best solution found within a time limit to decide the jobs that should start. To allow a fair comparison, all traces were pre-processed to reset the waiting time of all jobs that are in queue at the beginning of the trace, so that this is not taken into account. Additionally, we have subtracted from the PBS waiting times the overhead required for implementing the dispatching decision. This was experimentally identified by analyzing the traces themselves.

4.1 Evaluation of Our Models

We performed an evaluation of all our models on a PBS execution trace containing data for a batch of jobs that was considered for dispatching in a 2-hour long interval. The main performance metrics considered are (1) the time spent by the jobs in the queues while waiting their execution to begin (ideally as low as possible), and (2) the overall utilization of the system (ideally as large as possible). Waiting times are measure of the perceived quality of services, while a high utilization directly translates to a low number of idle (but still power consuming) resources.

The results for the first batch (BATCH1) are presented in Table 4 and Table 5; the models evaluated are the three "pure" ones (Makespan [MKS], Weighted Tardiness [WT] and Num. of late jobs [NL]) plus the three composite ones (i.e. with Makespan as main objective and constraints on Weighted tardiness and

Table 6. Job traces composition

	BATCH1	*BATCH2*	*BATCH3*
#jobs	437	434	619
#jobs DEBUG	237	133	127
#jobs PAR	130	240	415
#jobs LONGPAR	62	25	12
#jobs req. GPUs	85	203	224
#jobs req. MICs	3	1	1
#jobs req. 1 core	298	197	258
#jobs req. 2 cores	2	73	38
#jobs req. 4 cores	1	4	7
#jobs req. 5 cores	1	1	0
#jobs req. 6 cores	6	2	3
#jobs req. 8 cores	59	56	187
#jobs req. 8+ cores	70	101	126

Num. of late jobs [MKS WT/NL], and similarly for the others). In the table we can see the average waiting time per job (both total and per-queue). There is a remarkable improvement w.r.t PBS for all the models, and those using the Makespan as main objective (MKS and MKS WT/NL). All the composite models perform better than their pure counterparts when dealing with the jobs from *debug* queue (short and with relatively low requirements). The models with Makespan as primary objective do their best when dealing with the long jobs from the *longpar* queue.

The corresponding resource utilization statistics are reported in Table 5, showing for each model and PBS the average number of used cores, GPUs and MICs over time. Again, we can see a significant improvement in comparison to PBS performance, but in this case the differences between our models are less clear. In particular, the average numbers of used GPUs and MICs is very similar – probably because not every job needs an accelerator –, but we can notice that MKS WT/NL is the model which performs a bit better in terms of the average number of active cores. In the fifth column of the table we see the average number of jobs that are in execution at each time instant: more running jobs usually correspond to a higher utilization and a smaller time to complete the execution of the batch. Finally in the last column we report the average percentage of active cores on EURORA, which is a good index for the utilization of the whole system. As one can see, our best results (coming from the MKS WT/NL) are around 20% better than those of PBS. No approach was able to reach a 100% utilization: to a large extent, this appears to be due to the presence of bottleneck resources (e.g. GPUs) and to their allocation.

4.2 Comparison with PBS

The previous results show that our best model is a composite one, namely MKS WT/NL, thus such mode was chosen for a more detailed comparison with PBS

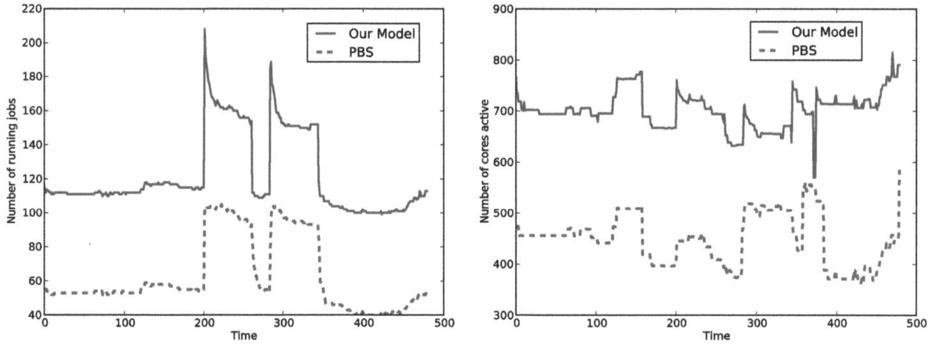

(a) Running Jobs (b) Active cores

Fig. 1. EURORA utilization on the first trace (BATCH1)

on three PBS execution traces, each one corresponding once again to a two-hour time frame of the EURORA activity. The features of the job batches considered in each trace (i.e. BATCH1, BATCH2, BATCH3) are summarized in Table 6, which reports the total number of jobs, the number of jobs in each queue[2], the number of jobs requiring at least one GPU or MIC and the number of jobs requiring a certain number of cores.

We start by presenting the results for BATCH1, which is the same we used for evaluating the model. The jobs considered in this trace belong to a wide range of classes, with different resource requirements and different execution times. In Fig. 1a we can observe the number of active jobs in the considered time frame, for both our approach (solid line) and PBS (dashed line). Fig. 1b reports instead the number of active cores. Our approach significantly outperforms PBS, being able to execute more jobs concurrently and to use a larger fraction of the available cores. Neither approach managed to reach the optimal system usage: this could be due to (a combination of) the presence of bottleneck resources, to suboptimal allocation choices, or simply to the lack of more workload to be dispatched. Fig. 2a shows the number of waiting jobs at each time step for our approach and PBS. From the data in the figure, we can deduce that our approach managed to dispatch most of the incoming jobs immediately, suggesting that the machine underutilization is at least in part to blame on the lack of more jobs. Still, suboptimal choices and resource bottlenecks cause some jobs to wait (a relatively high number of them, in the case of PBS).

Fig. 2b contains a histogram with the waiting times for our model, weighted by the (inverse of) the Estimated Waiting Time of the queue they belong to. The histogram shows how many jobs (y-axis) wait for a certain amount of times their ewt_i (y-axis). The majority of the waiting jobs with our approach stay in their queue for a very short time, unlike in the case of PBS, where especially the jobs

[2] The sum of those values may be lower than the total, because we do not report detailed statistics for some minor queues.

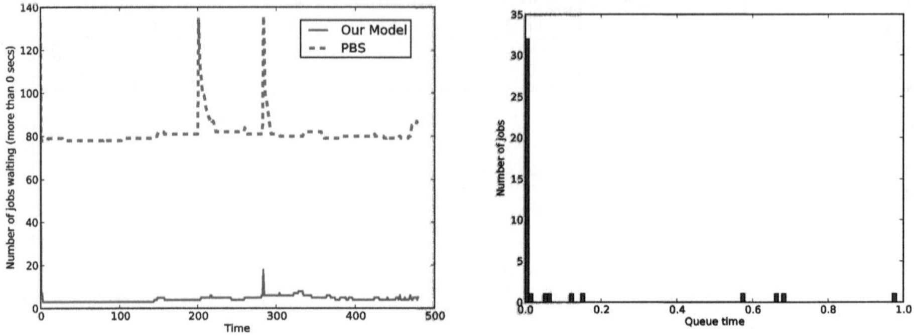

(a) Jobs in Queue (b) Times in Queue

Fig. 2. Waiting jobs and queue time for BATCH1

in the *longpar* queue tend to be considerably delayed. We recall that currently these jobs (which are characterized by longer durations than the remaining ones) are forced to execute only at night, for fear or delaying jobs in the *debug* or *parallel* queue. The evidence we provide here leads us to believe that such a strong constraint is in fact not needed when using a proactive approach, and its removal could provide benefits in terms of both queue time and average utilization of the supercomputer resources.

Fig. 3 and Fig. 4 refer instead to our second trace, i.e. to the jobs in BATCH2. This is another mixed group of jobs in terms of computational and resource requirements, but in this case we have many more GPU requests, putting a great strain on the dispatcher since GPUs in EURORA are a much fewer than cores. The consequences of this situation can be observed in Fig. 3a and Fig. 3a, respectively showing the number of running jobs and active cores over time. For both PBS and our model we notice that the number of jobs in execution, after an initial spike, reaches a cap in the middle section of the trace, although the percentage of actives cores is not even close to 100%. This cap occurs because in many cases, basically all waiting jobs are requiring a GPU and hence, even if there are have available cores, they cannot be used. Despite that, we still manage to achieve a largely improved schedule than the one of PBS in term of number of running jobs. In particular, the average number of active GPUs with our dispatcher is higher than 63: given that the whole supercomputer counts only 64 GPUs, this means that the performance obtained by our approach for the GPU-requiring jobs is very close to the theoretical limit.

We owe this result to the proactive nature of our scheduler, which allow us to more efficiently use constrained resources. For example, suppose we have node A and B, where A has n cores and 1 GPU while B has only n cores, and suppose that A and B are fully occupied by a previous job. We also have *job1* and *job2* waiting to start their execution: *job1* needs n cores and a GPU, whereas *job2* requires only the cores and has higher priority (for PBS). When the job currently

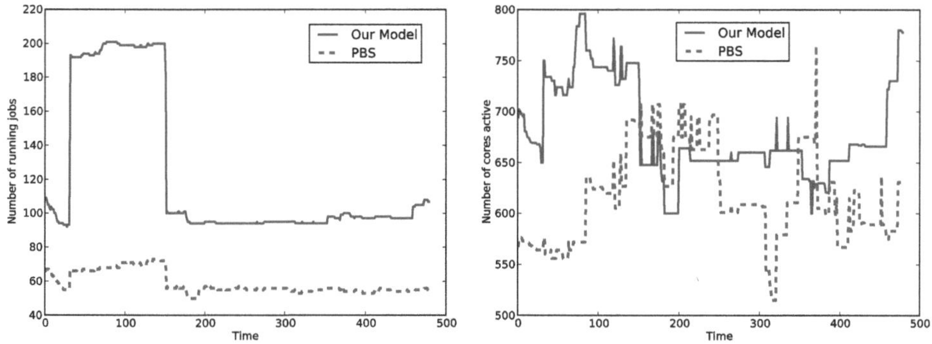

(a) Running Jobs (b) Active cores

Fig. 3. EURORA utilization on the second trace (BATCH2)

occupying nodes A and B terminates, PBS selects *job2*, then it checks if on A there are enough cores to satisfy the requirements. Since this is true in our example, PBS dispatches *job2* on node A, using up all node cores and leaving the GPU idle. In this scenario, *job2* cannot start executing until the other job has terminated. Conversely our dispatcher would have made a smarter - and in this particular case obvious - decision, that is putting *job1* on B, since it only needs cores, and *job2* on A, without further delay. In Figure 4 we can see our performance in terms of queue times for BATCH2. We outperform PBS again but at the same time we notice how the number of jobs in queue (Fig 4a) follow a similar pattern in both systems, with a distinctive spike after a relatively low initial value: this happens because of the congestion on the GPUs resources we mentioned earlier – after all, optimization can provide improvement only as long as spare resources are available.

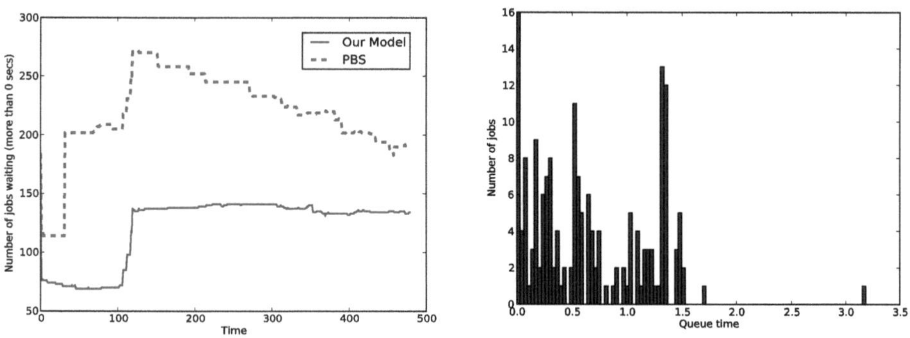

(a) Jobs in Queue (b) Times in Queue

Fig. 4. Waiting jobs and queue time for BATCH2

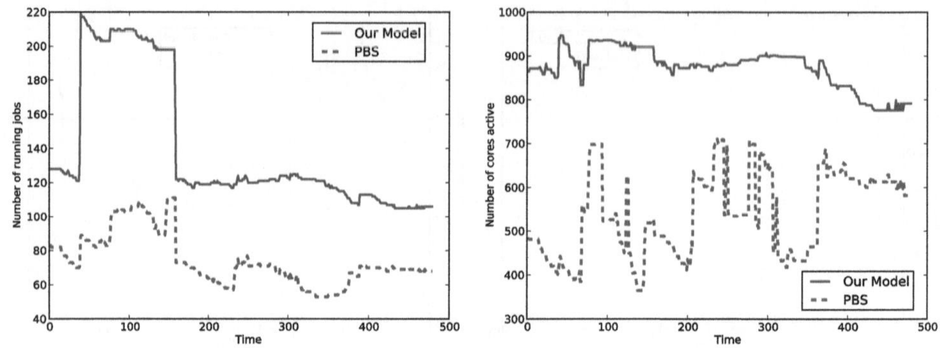

(a) Running Jobs (b) Active cores

Fig. 5. EURORA utilization on the third trace (BATCH3)

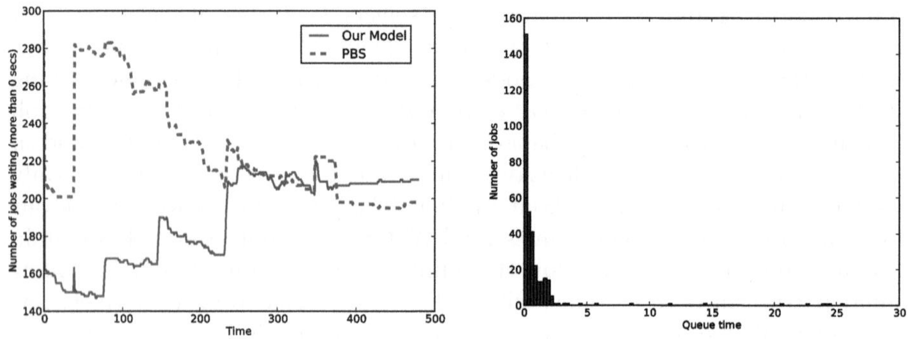

(a) Jobs in Queue (b) Times in Queue

Fig. 6. Waiting jobs and queue time for BATCH3

Finally, we can eventually consider BATCH3 and the results are displayed in Fig. 5 and Fig. 6. The jobs considered in this trace require, on average, a higher number of cores than all other traces and, for a large part, were submitted to the *parallel* queue. They require proportionally fewer GPUs than the jobs in BATCH2, but still more than BATCH1. We manage again to obtain a better usage of computational resources on EURORA, as revealed in Fig. 5b and from the average percentage of actives cores (85% in our model versus 55% with PBS). One more time, these results are due to a smarter management of the different types of resources, although the limitations imposed by the relatively low number of available GPUs still has an impact on the number of running jobs (Fig. 5a). In Figure 6a we can see our model is able not to force to wait as many jobs as PBS, but only during the first half of the trace, while after that point the number of jobs in queue is comparable between the two dispatchers. One possible explanation for this is again the limit imposed by the GPUs availability, given that not all the cores are occupied, which forces more jobs to wait when a certain threshold for the number of GPUs required is reached.

5 Conclusions

In this paper have presented a CP based proactive workload dispatcher for the HPC EURORA supercomputer and compared its performance with those of the system currently in use (PBS). Our goal is to manage the computational resources on the platform so as to achieve a twofold result: increase the machine utilization and then reduce the job waiting times. A higher machine utilization translates into a lower consumption from idle resources and a large number of accepted jobs, with benefits for the supercomputer owner and on the environmental side. Short waiting times correspond to a higher quality of service for the system users.

The problem we tackled was not an easy one, owing to the need to manage multiple objectives and to the limited availability of multiple, heterogeneous, resources. In both the considered metrics (machine utilization and waiting times) we considerably outperformed the current scheduler, showing that there are great margins for improvement when a proactive approach is used. The current, fundamentally reactive approach currently in use proved to have particular difficulties with the simultaneous management of different classes of resources (e.g. cores and GPUs). As a future long-term goal, we plan to further develop our model to replace (or at least complement) the scheduler currently in use on EURORA, with focus on improving its energetic behavior. To achieve this result, we will need to research and develop techniques to allow our approach to operate quickly enough to match the frequency of job arrivals. Moreover, we will need to make some adjustments to take into account the complex policies which regulate exactly the services provided by the supercomputer to its users.

Acknowledgement. This work was partially supported by the FP7 ERC Advance project MULTITHERMAN (g.a. 291125). We also want to thank CINECA and Eurotech for granting us the access to their systems.

References

1. Cineca inter-university consortium web site, `http://www.cineca.it//en` (accessed: April 14, 2014)
2. Eurotech group web site, `http://www.eurotech.com/en/` (accessed: April 14, 2014)
3. Ny times article about a survey by mc kinsey & co., `http://www.nytimes.com/2012/09/23/technology/data-centers-waste-vast-amounts-of-energy-belying-industry-image.html` (accessed: April 14, 2014)
4. Baptiste, P., Laborie, P., Le Pape, C., Nuijten, W.: Constraint-based scheduling and planning. Foundations of Artificial Intelligence 2, 761–799 (2006)

5. Bartolini, A., Cacciari, M., Cavazzoni, C., Tecchiolli, G., Benini, L.: Unveiling eurora - thermal and power characterization of the most energy-efficient supercomputer in the world. In: Design, Automation Test in Europe Conference Exhibition (DATE) (March 2014)
6. Laborie, P.: IBM ILOG CP Optimizer for detailed scheduling illustrated on three problems. In: van Hoeve, W.-J., Hooker, J.N. (eds.) CPAIOR 2009. LNCS, vol. 5547, pp. 148–162. Springer, Heidelberg (2009)
7. Laborie, P., Godard, D.: Self-adapting large neighborhood search: Application to single-mode scheduling problems. In: Proc. of MISTA (2007)
8. Laborie, P., Rogerie, J.: Reasoning with conditional time-intervals. In: Proc. of FLAIRS, pp. 555–560 (2008)
9. Altair PBS Works. Pbs professional®12.2 administrator's guide (2013), `http://resources.altair.com/pbs/documentation/support/PBSProAdminGuide12.2.pdf`

Scheduling B2B Meetings

Miquel Bofill*, Joan Espasa*,**, Marc Garcia, Miquel Palahí*,***, Josep Suy*, and Mateu Villaret*

Departament d'Informàtica, Matemàtica Aplicada i Estadística
Universitat de Girona, Spain
{mbofill,jespasa,mgarciao,mpalahi,suy,villaret}@imae.udg.edu

Abstract. In this work we deal with the problem of scheduling meetings between research groups, companies and investors in a scientific and technological forum. We provide a CP formulation and a Pseudo-Boolean formulation of the problem, and empirically test the performance of different solving techniques, such as CP, lazy clause generation, SMT, and ILP, on industrial and crafted instances of the problem. The solutions obtained clearly improve expert handmade solutions with respect to the number of idle time slots and other quality parameters.

1 Introduction

Business-to-business (B2B) events typically provide bilateral meeting sessions between participants with affine interests. These B2B events occur in several fields like sports, social life, research, etc. In this paper we describe the application developed to generate the timetable of such meetings in the *4th Forum of the Scientific and Technological Park of the University of Girona.*[1] The goal of this forum is to be a technological marketplace in Girona by bringing the opportunity to companies, research groups, investors, etc., to find future business partnerships.

The scheduling of the meetings is a tough task and requires expertise at distinct levels: on the one side, the human matchmaker should choose the appropriate matches maximizing somehow the potential results of the meetings according to the interests of the participants. On the other hand side, the timetable generation must satisfy several constraints, e.g., avoid meeting collisions, avoid unnecessary idle time between meetings for each participant or too many meeting location changes, etc.

In previous editions of the forum, and in other events with B2B meetings held in the park of the University of Girona, the human matchmaker made both tasks by hand. This process showed to be highly unsatisfactory with respect to

* Supported by the Spanish Ministry of Science and Innovation (project TIN2012-33042).

** Supported by UdG grant (BR 2013).

*** Supported by UdG grant (BR 2010).

[1] `http://www.forumparcudg.com`

B. O'Sullivan (Ed.): CP 2014, LNCS 8656, pp. 781–796, 2014.

the human effort required and also with respect to the final result obtained. Moreover, the growing participation makes the work much harder.

As far as we know, there are no many works dealing with this problem. On the one hand, in [12] we can find a system that is used by the company *piranha womex AG* for computing matchmaking schedules in several fairs. This system differs from ours in some aspects. For instance, it does not consider forbidden time slots but unpreferred ones, and it allows meeting collisions under the assumption that the companies can send another participant. Moreover, it is based on answer set programming, whereas we follow a different model-and-solve approach. On the other hand, the system B2Match [1] is a commercial system which does not support location change minimization.

Our work focuses on the generation of the meetings' timetable. That is, the human matchmaker provide us with the meetings that she wants to be scheduled and we generate the timetable. We provide both a CP model and a Pseudo-Boolean model for the problem, and compare the efficiency of different solving techniques such as CP, lazy clause generation, SMT, and ILP, on several real instances.

The obtained results for this year's edition of the Forum of the Scientific and Technological Park of the University of Girona have been very satisfactory. Moreover, using instances of previous years we have observed that the number of idle time slots could be dramatically improved.

The rest of the paper is structured as follows. In Section 2 we define the problem at hand. In Section 3 we present the different models considered, and in Section 4 we check the efficiency of distinct solvers on these models, using real instances. Conclusions are given in Section 5.

2 The B2B Problem

In the Forum of the Scientific and Technological Park of the University of Girona, the participants answer some questions about their interests and expertise, in the event registration phase. This information is made public to the participants, who may ask for bilateral meetings with other participants with a (normal or high) priority. They also indicate their time availability for the meetings (notice that in the forum there is also a conference with talks given by the participants and therefore they should reserve some time slots for their own talks). Moreover, the participants may ask for meetings in the morning or the afternoon session. Then, according to this information (participants availability and priorities for the desired meetings), the human matchmaker proposes a set of matches (meetings) to be scheduled. Once this schedule is obtained, the matchmaker asks the participants for confirmation of the meetings timetable. Then, the confirmed ones are fixed and the rejected ones are retracted. With this partial timetable, and considering the late arrivals of participants interested in having some meeting, the expert tries to add other meetings in the timetable by individually contacting with the participants. Finally, the distribution of tables is made. Below we formally define the basic problem.

Definition 1. *Let P be a set of participants, T a list of time slots and L a set of available locations (tables). Let M be a set of unordered pairs of participants in P (meetings to be scheduled). Additionally, for each participant $p \in P$, let $f(p) \subset T$ be a set of forbidden time slots.*

A feasible B2B schedule *S is a total mapping from M to $T \times L$ such that the following constraints are satisfied:*

- *Each participant has at most one meeting scheduled in each time slot.*

$$\forall m_1, m_2 \in M \text{ such that } m_1 \neq m_2 : \quad \pi_1(S(m_1)) = \pi_1(S(m_2)) \implies m_1 \cap m_2 = \emptyset \quad (1)$$

- *No meeting of a participant is scheduled in one of her forbidden time slots.*

$$\forall p \in P, m \in M : \quad p \in m \implies \pi_1(S(m)) \notin f(p) \quad (2)$$

- *At most one meeting is scheduled in a given time slot and location.*

$$\forall m_1, m_2 \in M \text{ such that } m_1 \neq m_2 : \quad \pi_1(S(m_1)) = \pi_1(S(m_2)) \implies \pi_2(S(m_1)) \neq \pi_2(S(m_2)) \quad (3)$$

The B2B scheduling problem *(B2BSP) is the problem of finding a feasible B2B schedule.*

The B2BSP is clearly in NP, and can be easily proved to be NP-complete by reduction from the restricted timetable problem (RTT) in [11].

Typically, we are interested in schedules that minimize the number of idle time periods. By an *idle time period* we refer to a group of idle time slots between a meeting of a participant and her next meeting. Before formally defining this optimization version of the B2BSP, we need to introduce some auxiliary definitions.

Definition 2. *Given a B2B schedule S for a set of meetings M, and a participant $p \in P$, we define $L_S(p)$ as the list of meetings in M involving p, ordered by its scheduled time according to S:*

$$L_S(p) = [m_1, \ldots, m_k], \text{ with } \begin{array}{l} \forall i \in 1..k : p \in m_i \\ \forall m \in M : p \in m \Rightarrow \exists! i \in 1..k : m_i = m \\ \forall i \in 1..(k-1) : \pi_1(S(m_i)) < \pi_1(S(m_{i+1})) \end{array}$$

By $L_S(p)[i]$ we refer to the i-th element of $L_S(p)$, i.e., m_i.

Definition 3. *We define the* B2B Scheduling Optimization Problem (B2BSOP) *as the problem of finding a feasible B2B schedule S, where the total number of idle time periods of the participants is minimal, i.e., minimizes*

$$\sum_{p \in P} \#\{L_S(p)[i] \mid i \in 1..|L_S(p)| - 1, \pi_1(S(m_i)) + 1 \neq \pi_1(S(m_{i+1}))\} \quad (4)$$

It is also quite common to ask for the minimization of location changes between consecutive meetings of the same participants. Due to the requirements of the event organization (there is a phase where the human matchmaker manually arranges new meetings by reusing time slots and participants' availabilities), we do this minimization with the time slots of the meetings already fixed.

Definition 4. *We define the* B2B location scheduling optimization problem (B2BLOP) *as the problem of, given a B2B schedule S, re-assign to each meeting a location in a way such that the total number of location changes for consecutive meetings of the same participant is minimal, i.e., minimizes*

$$\sum_{p \in P} \#\{L_S(p)[i] \mid$$
$$i \in 1..|L_S(p)| - 1, \pi_1(S(m_i)) + 1 = \pi_1(S(m_{i+1})), \pi_2(S(m_i)) \neq \pi_2(S(m_{i+1}))\}$$

As an additional constraint, we consider the case where meetings may have a morning/afternoon requirement, i.e., that some meetings must necessarily be celebrated in the morning or in the afternoon. Let's then consider that the set of time slots T is divided into two disjoint sets T_1 and T_2 and, moreover, that we have a mapping t from meetings m in M to $\{1, 2, 3\}$, where 1 means that the meeting m must take place at some time slot in T_1, 2 means that it must take place at some time slot in T_2, and 3 means that it does not matter. Then the schedule should also satisfy the following requirement:

$$\forall m \in M :$$
$$(t(m) = 1 \Longrightarrow \pi_1(S(m)) \in T_1) \ \wedge \ (t(m) = 2 \Longrightarrow \pi_1(S(m)) \in T_2) \quad (5)$$

Summing up, the scheduling process goes as follows:

1. The human matchmaker arranges the meetings taking into account the participants' requirements and preferences (who they want to meet, with which priority, and a possible restriction to the morning or afternoon frame for the meeting) and their forbidden hours (for example, the hours where the participant is giving a talk at the event).
2. We solve the B2BSOP with the chosen meetings.
3. The human matchmaker removes the revoked meetings (for instance, one participant may not be interested in a meeting requested by another participant) from the solution found in Step 2, and manually adds new last-arrival meetings.
4. We solve the B2BLOP with the meetings proposed in Step 3.

3 Models

It is our aim to model the aforementioned problem by means of a declarative CP language and evaluate the performance of several solvers taking this formulation as input. To this purpose, MiniZinc [15] is almost a perfect target, as it is a medium-level solver-independent CP modelling language, which is supported by an increasing number of backend solvers, constituting a de facto standard.

For the sake of completeness, we also consider WSimply [6], a system with a similar language to that of MiniZinc, but supporting weighted constraints and several predefined meta-constraints which allow to express, e.g., homogeneity in violations, in order to enforce fairness of solutions. A proposal for incorporating similar ideas into MiniZinc has been presented in [5].

Moreover, since a priori it is not clear if a high-level or a low-level model will be better in terms of performance, we have considered two different models: a CP model (with finite domain variables like t_i, denoting the time at which meeting i takes place) and a Pseudo-Boolean model (with 0/1 variables like $x_{i,j}$, stating that meeting i is celebrated at time j).

3.1 A WCSP Model for the B2BSOP

Here we give the WSimply version of the CP model for the B2BSOP, expressed as a weighted CSP.

Parameters

```
int nParticipants;                       % # of participants
int nMeetings;                           % # of meetings
int nTables;                             % # of locations
int nTimeSlots;                          % # of time slots
int nMorningSlots;                       % # of morning time slots
int meetings[nMeetings,3];               % Meetings to schedule
int tnForbidden;                         % Total # of forbidden slots
int forbidden[tnForbidden];              % Forbidden slots
int indexForbidden[nParticipants+1];     % Start indices in forbidden
int nMeetingsParticipant[nParticipants]; % # of meetings of each part.
```

The number of afternoon time slots is `nTimeSlots` − `nMorningSlots`, and for this reason it is not explicitly defined.

The `meetings` matrix has three columns, denoting the number of the first participant, the number of the second participant and the type of session required (1: morning, 2: afternoon, 3: don't care) for each meeting to be scheduled.

The `indexForbidden` array contains the indices where the forbidden time slots of each participant do begin in the `forbidden` array. This array has an extra position in order to ease the modelling of the constraints (see the `Forall` statements below), with `indexForbidden[nParticipants+1] = tnForbidden+1`.

Variables and Domains

```
Dom dTimeSlots  = [1..nTimeSlots];
Dom dUsedSlots  = [0..1];
Dom dTables     = [0..nTables];
```

```
IntVar schedule[nMeetings]::dTimeSlots;
IntVar tablesSlot[nTimeSlots]::dTables;
IntVar usedSlots[nParticipants,nTimeSlots]::dUsedSlots;
IntVar fromSlots[nParticipants,nTimeSlots]::dUsedSlots;
```

The array variable `schedule` shall contain the time slot assigned to each meeting.

The array variable `tablesSlot` will indicate the number of meetings to be celebrated at each time slot (and hence the number of tables needed). Note that by setting the domain to `[0..nTables]` we are already restricting the number of tables available.

The variable `usedSlots` is a two dimensional 0/1 array representing, for each participant i and time slot j, if i has a meeting scheduled at time j.

Finally, the variable `fromSlots` is a two dimensional 0/1 array such that, for each participant i, `fromSlots`$[i, j] = 1$ for all j from the first time slot at which a meeting for i is scheduled on.

These two last variables are used for optimization, as shown below.

Constraints

- *Each participant has at most one meeting scheduled at each time slot*, i.e., constraint (1). We force two meetings sharing one member to be scheduled at a different time slot:

```
Forall (n in [1..nMeetings]) {
  Forall (m in [n+1..nMeetings]) {
    If (meetings[n,1] = meetings[m,1] Or
        meetings[n,2] = meetings[m,1] Or
        meetings[n,1] = meetings[m,2] Or
        meetings[n,2] = meetings[m,2])
    Then { schedule[n] <> schedule[m]; };
  };
};
```

- *No meeting of a participant is scheduled in one of her forbidden time slots*, i.e., constraint (2). We force, for each meeting, to be scheduled in a time slot distinct from all forbidden time slots for both members of the meeting:

```
Forall(j in [1..nMeetings]) {
  Forall(k in [ indexForbidden[meetings[j,1]]..
                indexForbidden[meetings[j,1]+1]-1 ]) {
    schedule[j] <> forbidden[k];
  };
  Forall(k in [ indexForbidden[meetings[j,2]]..
                indexForbidden[meetings[j,2]+1]-1 ]) {
    schedule[j] <> forbidden[k];
  };
};
```

- *At most one meeting is scheduled in a given time slot and location*, i.e., constraint (3). This constraint is ensured by matching the number of meetings scheduled in time slot i with tablesSlot[i], whose value is bounded by the number of tables available:

```
Forall(i in [1..nTimeSlots]) {
 Sum([ If_Then_Else(schedule[n] = i)(1)(0) | n in [1..nMeetings] ],
     tablesSlot[i]);
};
```

Note that we are not assigning a particular table to each meeting, but just forcing that there are enough tables available for the meetings taking place at the same time.
- *Each meeting must be scheduled in a required (if any) set of time slots*, i.e., constraint (5). Since we know the number of morning time slots, we can easily enforce this constraint:

```
Forall (n in [1..nMeetings]) {
  If (meetings[n,3] = 1) Then {schedule[n] =< nMorningSlots;}
  Else {If (meetings[n,3] = 2) Then {schedule[n] > nMorningSlots;};
  };
};
```

- *Channeling constraints.* In order to be able to minimize the number of idle time periods, i.e., objective function (4), we introduce channeling constraints mapping the variable schedule to the variable usedSlots. That is, if meeting k is scheduled at time slot j, we need to state that time j is used by both members of meeting k, by setting accordingly the corresponding value in usedSlots:

```
Forall(j in [1..nTimeSlots]) {
  Forall(k in [1..nMeetings]) {
    (schedule[k] = j) Implies
       (usedSlots[meetings[k,1],j] = 1 And
        usedSlots[meetings[k,2],j] = 1);
  };
};
```

In the reverse direction, for each participant e, her number of meetings as derived from usedSlots must match her known total number of meetings nMeetingsParticipant[e]:

```
Forall(e in [1..nParticipants]) {
  Sum([ usedSlots[e,f] | f in [1..nTimeSlots] ],
      nMeetingsParticipant[e]);
};
```

Next, we impose the required constraints on the variable fromSlots:

```
Forall(e in [1..nParticipants]) {
  Forall(f in [1..nTimeSlots]) {
    usedSlots[e,f] = 1 Implies fromSlots[e,f] = 1;
  };

  Forall(f in [1..nTimeSlots-1]) {
    fromSlots[e,f] = 1 Implies fromSlots[e,f+1] = 1;
  };
};
```

It is worth noting that, for any participant e, having fromSlots$[e, f] = 1$ for all f is possible, even if e has no meeting at time slot 1. However, the soft constraints used for optimization will prevent this from happening, as commented below.

Optimization. Minimization of the objective function (4) is achieved by means of soft constraints, where a group of contiguous idle time slots between two meetings of the same participant is given a cost of 1.

Soft constraints are labeled with Holes$[e, f]$, where e denotes a participant and f denotes a time slot, and state that if e does not have any meeting in time slot f, but it has some meeting before, then she does not have any meeting in the following time slot:

```
Forall(e in [1..nParticipants], f in [1..nTimeSlots-1]) {
  #Holes[e,f]:((usedSlots[e,f] = 0 And fromSlots[e,f] = 1) Implies
                usedSlots[e,f+1] = 0)@{1};
};
```

We claim that, with these constraints, an optimal solution will be one having the least number of groups of contiguous idle time slots between meetings of the same participant. Note that, for each participant, we increase the cost by 1 for each meeting following some idle period. Moreover, it is not difficult to see that the (possibly null) period of time preceding the first meeting of a participant e will have no cost, since fromSlots$[e, f]$ can freely take value 0 for all f prior to the time of the first meeting.

Finally, since we are not only interested in optimal solutions, but in fair ones, we can add the following meta-constraint, stating that the difference between the number of idle periods of any two distinct participants is, e.g., at most 2:

```
homogeneousAbsoluteNumber([ [ Holes[e,f] | f in [1..nTimeSlots-1] ] |
                            e in [1..nParticipants] ], 2);
```

Note that this meta-constraint makes use of the labels introduced in the soft constraints. It has as first argument a list of lists ll of soft constraint labels, and as second argument a natural number n. It ensures that, for each pair of lists in ll, the difference between the number of violated constraints in the two lists is at most n. The meta-constraints supported by WSimply are described in [6]. The precise syntax and semantics of the language can be found in a technical report.[2]

The WSimply model presented above has been translated to MiniZinc in order to test the performance of a greater number of different solvers (see Section 4). The translation of hard constraints is straightforward, due to the similarities between the two languages. Soft constraints have been translated into a linear objective function, as they are not directly supported in MiniZinc. The meta-constraint used in the WSimply model has been translated into the MiniZinc constraints that would result from the translation process implemented in the WSimply compiler.

3.2 A PB Model for the B2BSOP

Here we give the WSimply version of the Pseudo-Boolean model for the B2BSOP. The parameters are the same as in the CP model but, in this case, we only use 0/1 variables.

Variables and Domains

```
Dom pseudo = [0..1];

IntVar schedule[nMeetings,nTimeSlots]::pseudo;
IntVar usedSlots[nParticipants,nTimeSlots]::pseudo;
IntVar fromSlots[nParticipants,nTimeSlots]::pseudo;
IntVar holes[nParticipants,nTimeSlots-1]::pseudo;
IntVar nHoles[nParticipants,5]::pseudo;
IntVar max[5]::pseudo;
IntVar min[5]::pseudo;
```

Here, the array variable `schedule` shall contain a 1 at position i, j if and only if meeting i is celebrated at time slot j.

The variables `usedSlots` and `fromSlots` are the same as in the CP case.

The variable `holes` will indicate, for each participant, when a time slot is the last of an idle period of time.

The variable `nHoles` will hold the 5-bit binary representation of the number of idle time periods of each participant, i.e., the number of groups of contiguous idle time slots between meetings of each participant.

The variables `max` and `min` will hold the 5-bit binary representation of an upper bound and a lower bound of the maximum and minimum values in `nHoles`, respectively. As we will see, these variables will be used to enforce fairness of solutions, by restricting their difference to be less than a certain value.

All variables but `schedule` are only necessary for optimization.

[2] http://imae.udg.edu/recerca/lap/simply/docs/technical-report.pdf

Constraints

- *Each participant has at most one meeting scheduled at each time slot*:

```
Forall (f in [1..nTimeSlots]) {
  Forall (n in [1..nParticipants]) {
    AtMost([ schedule[m,f] |
               m in [1..nMeetings],
               (meetings[m,1] = n Or meetings[m,2] = n) ], 1, 1);
  };
};
```

The atMost(l, e, n) global constraint (where l is a list, e is an expression of the same type of the elements in l, and n is an integer arithmetic expression) forces the number of elements in l that match e to be at most n.

- *No meeting of a participant is scheduled in one of her forbidden time slots*:

```
Forall (r in [1..nMeetings]) {
  Forall (p in [indexForbidden[meetings[r,1]]..
                indexForbidden[meetings[r,1]+1]-1]) {
      schedule[r,forbidden[p]] = 0;
  };
  Forall (p in [indexForbidden[meetings[r,2]]..
                indexForbidden[meetings[r,2]+1]-1]) {
      schedule[r,forbidden[p]] = 0;
  };
};
```

- *At most one meeting is scheduled in a given time slot and location.* We ensure this by forcing that no more than nTables meetings are scheduled in the same time slot:

```
Forall(f in [1..nTimeSlots]) {
  AtMost([schedule[n,f] | n in [1..nMeetings]],1,nTables);
};
```

- *Each meeting must be scheduled in the required (if any) set of time slots.* By means of sums, we force that each meeting is scheduled exactly once in its required set of time slots:

```
Forall (n in [1..nMeetings]) {
  If (meetings[n,3] = 1) Then {
    Sum([ schedule[n,f] | f in [1..nMorningSlots] ], 1);
    [ schedule[n,f] = 0 | f in [nMorningSlots+1..nTimeSlots] ];
  } Else {
  If (meetings[n,3] = 2 ) Then {
    [ schedule[n,f] = 0 | f in [1..nMorningSlots] ];
    Sum([ schedule[n,f] | f in [nMorningSlots+1..nTimeSlots] ], 1);
  } Else {
    Sum([ schedule[n,f] | f in [1..nTimeSlots] ], 1);
  }};
};
```

Note that list comprehensions can be used to post constraints.
– *Channeling constraints.* The channeling constraints are analogous to before:

```
Forall(j in [1..nTimeSlots]) {
  Forall(k in [1..nMeetings]) {
    (schedule[k,j] = 1) Implies
       (usedSlots[meetings[k,1],j] = 1) And
       (usedSlots[meetings[k,2],j] = 1));
  };
};

Forall(e in [1..nParticipants]) {
  Sum([ usedSlots[e,f] | f in [1..nTimeSlots] ],
      nMeetingsParticipant[e]);
};

Forall(e in [1..nParticipants]) {
  Forall(f in [1..nTimeSlots]) {
    usedSlots[e,f] = 1 Implies fromSlots[e,f] = 1;
  };

  Forall(f in [1..nTimeSlots-1]) {
    fromSlots[e,f] = 1 Implies fromSlots[e,f+1] = 1;
  };
};
```

Optimization. The soft constraints are the same as in the CP model:

```
Forall(e in [1..nParticipants], f in [1..nTimeSlots-1]) {
  ((usedSlots[e,f] = 0 And fromSlots[e,f] = 1) Implies
  usedSlots[e,f+1] = 0)@{1};
};
```

The homogeneity meta-constraint `homogeneousAbsoluteNumber` used in the CP model cannot be used in the Pseudo-Boolean model since, in its current implementation, WSimply will translate this meta-constraint into a set of non (Pseudo-)Boolean constraints. For this reason, this meta-constraint needs to be simulated here. We proceed as follows.

On the one hand, we post the constraints for the array variable `holes` which contains, for each participant, the last time slot of each idle period of time:

```
Forall(e in [1..nParticipants], f in [1..nTimeSlots-1]) {
  (usedSlots[e,f] = 0 And fromSlots[e,f] = 1 And usedSlots[e,f+1] = 1)
     Implies holes[e,f]=1;
  holes[e,f]=1 Implies
  (usedSlots[e,f] = 0 And fromSlots[e,f] = 1 And usedSlots[e,f+1] = 1);
};
```

On the other hand, we post the constraints defining the 5-bit binary representation of the number of idle time periods of each participant (the `sum1` function returns the sum of the elements in the list it receives as argument):

```
Forall(e in [1..nParticipants]){
  sum1([ holes[e,f] | f in [nTimeSlots-1] ]) =
  16*nHoles[e,1]+8*nHoles[e,2]+4*nHoles[e,3]+2*nHoles[e,4]+nHoles[e,5];
};
```

Finally, we constrain the difference between the number of idle time periods of any two distinct participants to be at most 2. We do this by constraining the difference between an upper bound and a lower bound of the maximal and minimal number, respectively, of idle time periods of all participants, to be at most 2:

```
Forall(e in [1..nParticipants]){
  16*max[1]+8*max[2]+4*max[3]+2*max[4]+max[5] >=
  16*nHoles[e,1]+8*nHoles[e,2]+4*nHoles[e,3]+2*nHoles[e,4]+nHoles[e,5];

  16*nHoles[e,1]+8*nHoles[e,2]+4*nHoles[e,3]+2*nHoles[e,4]+nHoles[e,5]
  >= 16*min[1]+8*min[2]+4*min[3]+2*min[4]+min[5];
};

2 >= 16*max[1]+8*max[2]+4*max[3]+2*max[4]+max[5] -
     16*min[1]+8*min[2]+4*min[3]+2*min[4]+min[5];
```

4 Experiments and Comparisons with Manually Generated Solutions

In this section we compare the performance of several state-of-the-art CSP, WCSP, ILP and PB solvers with the proposed models. We also analyse which is the improvement of our solution over some handmade solutions from past editions of the forum.

As said, apart from the two (CP and PB) WSimply models presented, we have also considered a MiniZinc model in order to test the performance of a greater number of different solvers. The MiniZinc model considered is an accurate translation of the first WSimply (CP) model. All models and data used in this paper can be found in `http://imae.udg.edu/recerca/lap/simply/`.

For solving the WSimply instances we have used the SMT solver Yices 1.0.33 [10] through its API, with two different solving methods:

- WPM1, as described in [6]. This is an adaptation of the algorithms introduced in [7,14] for (weighted) MaxSAT to the context of weighted MaxSMT.
- SBDD, as described in [9]. In this approach, the weighted SMT constraints are replaced by a linear objective function, which is encoded as a shared BDD using the compact and generalized arc-consistent encoding of [3].

We have also considered the translation of the CP model written in WSimply into ILP, and used IBM ILOG CPLEX 12.6 for solving the resulting instances. The translation of high-level CP models into ILP is a new feature supported by the WSimply system, using similar transformations to that applied for the PB case, as described in [8].

For solving the MiniZinc instances we have used Gecode 4.2.1 [17], a state-of-the-art CP solver, and Opturion 1.0.2 [2], winner of the 2013 MiniZinc Challenge[3] in the free search category, using lazy clause generation [16].

For solving the PB WSimply instances we have used CPLEX 12.6, SCIP 3.0.1 [4] and clasp 3.1.0 [13]. The two former obtained the best results in the last PB competition[4] in the category "optimisation, small integers, linear constraints", which fits our problem. However, the latter has shown a much better performance on this problem. The transformations used to obtain plain PB instances from WSimply are described in [8].

All experiments have been run on a cluster of Intel® Xeon™CPU@3.1GHz machines, with 8GB of RAM, under 64-bit CentOS release 6.3, kernel 2.6.32, except for the experiments with Opturion, where we have used a slightly different computer (Intel® Core™CPU@2.8GHz, with 12GB of RAM, under 64-bit Ubuntu 12.04.3, kernel 3.2.0) due to some library dependence problems. We have run each instance with a cutoff of 2 hours.

We have considered four industrial instances provided by the human expert: instances tic-2012a and tic-2013a, from past editions of a technology and health forum, and instances forum-2013a and forum-2014a, from the previous year and this year's editions of the scientific and technological forum.

Taking as basis these four instances, we have crafted five more instances of distinct hardness by increasing the number of meetings, reducing the number of locations and changing the morning/afternoon preferences of some meetings. Table 1 summarizes the results of the experiments.

Looking at the results, it can be said that clasp is the best solver on the easier instances (only beaten by CPLEX in two cases), and SBDD is the best on the harder instances, both of them using conflict driven solvers. The latter is also the most robust method when considering all instances, with a good compromise between solving time and quality of the solutions. Due to the fact that SBDD is based on representing the objective function as a BDD, and iteratively calling the decision procedure (an SMT solver) with successively tighter bounds, we can obtain a feasible solution at each stage of the optimization process. The WPM1 method also exhibits good performance on many instances, but it cannot provide suboptimal solutions as, roughly, it approaches to solutions from the unsatisfiable side.

An interesting aspect is that, in all real instances from past editions (tic-2012a, tic-2013a and forum-2013a) we obtained an optimum of 0 idle time periods between meetings of the same participant, whereas the expert handmade solutions included 20, 39 and 99 idle time periods respectively, as shown in Table 2. Note also that only for some crafted instances, and the bigger real instance from the last forum, we could not certify the optimality of the solutions found within two hours.

In Section 3 we have not included the model used for the B2BLOP due to lack of space. However, one aspect of the model that deserves a comment is the

---

[3] `http://www.minizinc.org/challenge2013/results2013.html`

[4] `http://www.cril.univ-artois.fr/PB12/`

Table 1. Solving time (in seconds) and optimum found (number of idle time periods) per instance, model and solver. The instances are tagged with (#meetings, #participants, #locations, #time slots, #morning time slots). TO stands for time out and MO for memory out. The cutoff is 2 hours. For aborted executions we report the (sub)optimum found if the solver reported any. Best running times and upper bounds of the objective function are indicated in boldface.

Instance	CP Model										PB Model					
	WSimply						MiniZinc				WSimply					
	WPM1		SBDD		CPLEX		Gecode		Opturion		CPLEX		SCIP		clasp	
tic-2012a (125,42,21,8,0)	2.0	0	2.7	0	3375.6	0	1.4	0	8.1	0	7.3	0	126.1	0	**0.1**	**0**
tic-2012c (125,42,16,8,0)	51.1	0	235.4	0	TO	4	TO	-	2206.2	0	**18.0**	**0**	1692.7	0	977.9	0
tic-2013a (180,47,21,10,0)	25.0	0	65.2	0	TO	9	2923.1	0	394.5	0	1345.3	0	TO	13	**2.1**	**0**
tic-2013b (184,46,21,10,0)	5.6	0	24.1	0	TO	8	3.4	0	108.0	0	121.0	0	4975.4	0	**2.1**	**0**
tic-2013c (180,47,19,10,0)	TO	-	TO	8	TO	18	TO	-	TO	8	**TO**	**4**	TO	31	TO	-
forum-2013a (154,70,14,21,13)	5542.3	0	3128.1	0	MO	83	TO	-	1142.8	0	TO	20	TO	-	**52.4**	**0**
forum-2013b (195,76,14,21,13)	TO	-	**TO**	**12**	TO	-	TO	-	TO	50	MO	-	TO	-	TO	24
forum-2013c (154,70,12,21,13)	TO	-	**TO**	**20**	MO	50	TO	-	TO	23	TO	30	TO	-	TO	-
forum-2014a (302,78,22,22,12)	TO	-	**TO**	**7**	TO	-	TO	-	TO	90	MO	-	TO	-	TO	-

use of meta-constraints, to ensure fairness in the number of location changes of the participants. With the meta-constraint

```
homogeneousAbsoluteNumber([[Changes[e,f] | f in [1..nTimeSlots-1]]
                          | e in [1..nParticipants]], Hfactor);
```

the user can look for solutions where the difference on the number of location changes between participants is at most `HFactor`, and with the meta-constraint

```
maxCost([[Changes[e,f] | f in [1..nTimeSlots]]
        | e in [1..nParticipants]], MaxChanges);
```

the user can look also for solutions where the number of location changes per participant is at most `MaxChanges`.

We have solved the B2BLOP with the schedules obtained from real instances of previous editions (tic-2012a, tic-2013a and forum-2013a), in order to compare the obtained results to the handmade ones, and with the schedule proposed for this year's edition of the forum. This last schedule, which we refer to as forum-2014a-t, has been obtained by the human expert by retracting 6 cancelled meetings, and by adding 15 last arrival meetings, to the schedule obtained for forum-2014a with the WPM1 method, in approximately 2.5 hours. The resulting B2BLOP instances have been solved with the WPM1 method in approximately 3, 120, 420 and 480 seconds respectively.

In Table 2 we provide a comparison on the quality of the solutions with respect to the number of idle time periods and location changes, when solved by hand and with WSimply.

Table 2. Number of idle time periods and location changes for the real instances when solved by hand and with WSimply. The maximum difference (homogeneity) between those numbers for distinct participants is given between parentheses.

Instance	# idle periods		# location changes	
	Handmade	WSimply	Handmade	WSimply
tic-2012a	20 (4)	0 (0)	112 (7)	103 (3)
tic-2013a	39 (4)	0 (0)	191 (9)	156 (4)
forum-2013a	99 (5)	0 (0)	27 (5)	105 (4)
forum-2014a-t		22 (2)		249 (6)

Note that the number of idle time periods is reduced to 0 when solving the problem with WSimply in almost all cases. In spite of this, we are still able to reduce the number of location changes with respect to the handmade solutions, except for the case of forum-2013a. But the solution obtained with WSimply in this case is still significantly better than the handmade one, which implied 99 idle time periods and was far less homogeneous. In any case, we are prioritizing the minimization of idle time periods of participants and the fairness of solutions, and leave location change minimization as a secondary desirable property.

5 Conclusion

In this work we have provided two distinct models for the B2BSOP and compared the efficiency of several types of solvers when dealing with some industrial and crafted instances of this problem, in a 'model-and-solve' approach. We also provide several new nontrivial industrial timetabling instances to the community.

The solutions found have been highly satisfactory for the human matchmaker, as they dramatically improve the handmade ones with respect to the number of idle time periods as well as location changes for the participants and, obviously, with quite less effort. Another aspect that the human matchmaker has really appreciated is the facility of adding meta-constraints to the model, like the ones we have used for achieving some level of fairness in the solutions. Fairness is crucial since participants may complain if they have the feeling of being discriminated, either with respect to idle time periods or with respect to location changes. The possibility of being able to fix partial solutions and adding new meetings to schedule has also been appreciated, since this is a hard task to do typically the day before the event due to last meeting request arrivals.

With respect to performance, we have noted that clasp is especially good on small instances, while WSimply (with the SBDD approach) appears to be better on bigger ones. However, it is not easy to draw conclusions, as we have not tuned the model for any solver in particular. The good results in the PB approach encourage us to develop a plain SAT model in the future, and to use MaxSAT solvers on this problem. Finally, it would be interesting to consider a handcrafted MIP model, and compare it against the MIP models obtained automatically from our models.

References

1. http://www.b2match.com (accessed April 11, 2014)
2. http://www.opturion.com (accessed April 11, 2014)
3. Abío, I., Nieuwenhuis, R., Oliveras, A., Rodríguez-Carbonell, E., Mayer-Eichberger, V.: A New Look at BDDs for Pseudo-Boolean Constraints. Journal of Artificial Intelligence Research (JAIR) 45, 443–480 (2012)
4. Achterberg, T.: SCIP: solving constraint integer programs. Mathematical Programming Computation 1(1), 1–41 (2009)
5. Ansótegui, C., Bofill, M., Palahí, M., Suy, J., Villaret, M.: W-MiniZinc: A Proposal for Modeling Weighted CSPs with MiniZinc. In: Proceedings of the 1st International Workshop on MiniZinc (MZN 2011) (2011)
6. Ansótegui, C., Bofill, M., Palahí, M., Suy, J., Villaret, M.: Solving weighted CSPs with meta-constraints by reformulation into Satisfiability Modulo Theories. Constraints 18(2), 236–268 (2013)
7. Ansótegui, C., Bonet, M.L., Levy, J.: Solving (Weighted) Partial MaxSAT through Satisfiability Testing. In: Kullmann, O. (ed.) SAT 2009. LNCS, vol. 5584, pp. 427–440. Springer, Heidelberg (2009)
8. Bofill, M., Espasa, J., Palahí, M., Villaret, M.: An extension to Simply for solving Weighted Constraint Satisfaction Problems with Pseudo-Boolean Constraints. In: XII Spanish Conference on Programming and Computer Languages (PROLE 2012), Almería, Spain, pp. 141–155 (September 2012)
9. Bofill, M., Palahí, M., Suy, J., Villaret, M.: Boosting Weighted CSP Resolution with Shared BDDs. In: Proceedings of the 12th International Workshop on Constraint Modelling and Reformulation (ModRef 2013), Uppsala, Sweden, pp. 57–73 (September 2013)
10. Dutertre, B., de Moura, L.: The Yices SMT solver (August 2006) Tool paper available at, http://yices.csl.sri.com/tool-paper.pdf (accessed April 11, 2014)
11. Even, S., Itai, A., Shamir, A.: On the complexity of time table and multi-commodity flow problems. In: Foundations of Computer Science, 16th Annual Symposium, pp. 184–193. IEEE (1975)
12. Gebser, M., Glase, T., Sabuncu, O., Schaub, T.: Matchmaking with Answer Set Programming. In: Cabalar, P., Son, T.C. (eds.) LPNMR 2013. LNCS, vol. 8148, pp. 342–347. Springer, Heidelberg (2013)
13. Gebser, M., Kaufmann, B., Neumann, A., Schaub, T.: *clasp*: A Conflict-Driven Answer Set Solver. In: Baral, C., Brewka, G., Schlipf, J. (eds.) LPNMR 2007. LNCS (LNAI), vol. 4483, pp. 260–265. Springer, Heidelberg (2007)
14. Manquinho, V., Marques-Silva, J., Planes, J.: Algorithms for Weighted Boolean Optimization. In: Kullmann, O. (ed.) SAT 2009. LNCS, vol. 5584, pp. 495–508. Springer, Heidelberg (2009)
15. Nethercote, N., Stuckey, P.J., Becket, R., Brand, S., Duck, G.J., Tack, G.R.: MiniZinc: Towards a Standard CP Modelling Language. In: Bessière, C. (ed.) CP 2007. LNCS, vol. 4741, pp. 529–543. Springer, Heidelberg (2007)
16. Ohrimenko, O., Stuckey, P.J., Codish, M.: Propagation via lazy clause generation. Constraints 14(3), 357–391 (2009)
17. Schulte, C., Lagerkvist, M., Tack, G.: Gecode. Software download and online material at the website (2006), http://www.gecode.org (accessed April 11, 2014)

Solving a Judge Assignment Problem Using Conjunctions of Global Cost Functions

Simon de Givry[1], Jimmy H.M. Lee[2], Ka Lun Leung[2], and Yu Wai Shum[2]

[1] MIA-T, UR 875, INRA, 31320 Castanet Tolosan, France
Simon.Degivry@toulouse.inra.fr
[2] Department of Computer Science and Engineering,
The Chinese University of Hong Kong, Shatin, N.T., Hong Kong
{jlee,klleung,ywshum}@cse.cuhk.edu.hk

Abstract. The Asia Pacific Information and Communication Technology Alliance (APICTA) Awards has been held for 12 years, rewarding the most innovative solutions in different categories of Information and Communication Technology (ICT). To maintain professionalism, judges are nominated from each economy, and appointed to panels of different categories. Judge assignment is a difficult task, since it has to optimize between expertise, distribution of workloads, fairness and sometimes even political correctness. In this paper, we describe our experience in analyzing and automating the APICTA judge assignment process using Soft Constraint Programming for the 13th APICTA hosted in Hong Kong on November, 2013. We chose the *weighted constraint satisfaction* (WCSP) framework since both hard constraints and preferences can be modeled by cost functions. Consistency algorithms can effect strong propagation by redistributing costs among cost functions. We observe that a number of restrictions in the judge assignment problem involves counting. In our first attempt, we utilized the SOFT_AMONG^{var} global cost function for these counting conditions but we could not solve the problem within a day. SOFT_GCC^{val} is another possible global cost function to model counting, which is what we used in the second attempt. We can compute the optimum in a few hours, which is far from practical.

We apply similar techniques as Régin to show that the combination of SOFT_GCC^{val} and SOFT_AMONG^{var} is flow-based. We further prove that the combination results in a *flow-based projection-safe* cost function, meaning that soft arc consistencies can be enforced efficiently. By using this combination in our final model, we can solve the judge assignment problem within a few minutes. We consider this a success story where theory and practice meet.

1 Introduction

The Asia Pacific Information and Communication Technology Alliance (APICTA) Awards[1] is an international Awards Program. It aims at increasing the awareness

[1] http://www.apicta.org/

B. O'Sullivan (Ed.): CP 2014, LNCS 8656, pp. 797–812, 2014.

of Information and Communication Technology (ICT) in the community, facilitating technology transfers, and offering business matching opportunities, by providing networking and product benchmarking opportunities to ICT innovators and entrepreneurs. Up to 2013, 13 economies have joined the APICTA Awards, including China, Malaysia, Thailand, to name a few. In 2013, the APICTA Awards was hosted in Hong Kong[2]. Preparation becomes a hard work due to limited resources. The APICTA Awards organizers require planning for two different scenarios.

1. Scheduling the presentations for nominated candidates, and;
2. Assigning judges into panels for each award category representing different aspects of the current ICT fields.

The former subjects to venue and time constraints, which has been completed beforehand. The latter involves various logistical and political factors, which are optimization in nature. Judges are nominated by different economies. After nomination, judges of each category are selected by a standard procedure to ensure professionalism and fairness. However, the manual procedure is tedious and inefficient, resulting in large number of complaints from judges and economies. As the resources are limited, the organizers seek for automation that can produce high-quality assignments under a tight schedule.

To eliminate manual and inefficient processes and improve the quality of assignments, we introduce an automated solution to generate judge assignments using weighted constraint satisfaction [9]. We identify a number of *global cost functions* [1,5] in the problem model, which help capture all the restriction and preference requirements much more succinctly. More importantly, global cost functions provide much stronger propagation. We also benefited further by conjoining global cost functions into a single global cost function, which utilizes the flow algorithm from Régin [8]. During the assignment process, we can deliver a new assignment within a few minutes every time requirements are changed. With our automation, judge assignments can be finalized in two weeks.

The rest of the paper is arranged as follows. Section 2 further describes the scenario. We give the current practice in Section 3, which also explains our choice of solving techniques. Section 4 analyzes the problem and lists all the constraints and preferences. Section 5 gives the corresponding WCSP model, and shows how global cost functions can be conjoined in this problem. Section 6 shows that conjoining global cost functions can give optimal solutions in a few minutes. Section 7 discusses the consequence after introducing automated approaches. We conclude the paper in Section 8.

2 Problem Description

The APICTA Awards is an international awards program organized by APICTA. The competition is divided into 16 categories, ranging from School Project to Industry Application. Each category opens for candidates in 13 economies to

[2] `http://www.apicta2013.com/`

join. Each economy nominates at most three entries in each category. Nominated candidates travel to the economy hosting the Awards, and present their ICT solutions to a panel of judges specialized in the corresponding category. The judge panel picks the award winners of each category from all candidates according to their innovation and quality of work. The assessment of each category usually takes one day, but some categories require two days due to a large number of entries.

The hosting economy of APICTA Awards needs to plan for the presentation schedule for each nominated candidate and the assignments of categories for each nominated judge. The presentation schedule is constrained by the availability of presentation rooms and the number of entries in each category. In APICTA 2013, the presentation schedule had been planned by the organizers and given as parts of the input to the judge assignment process.

To ensure a fair and equitable judging process, the organizers follow a set of procedures to assign judges. Each economy has nominated at most five specialists or experts in ICT fields as judges of the APICTA Awards. The chief judge panel examines their qualifications and finalizes a list of eligible judges. In the APICTA Awards 2013, 61 judges from 13 economies are eligible. Each judge has a set of declared specialties, which corresponds to the categories in the APICTA Awards. Some judges have experience in judging the APICTA Awards before but some are new. The judge assignment process begins by assigning judges into different categories according to their specialties and experiences. Judges comment on the assignments and the organizers make modifications accordingly. This process iterates until all judges are satisfied. After the judge assignments have been confirmed, the chief judge and advisory judge panels choose a head judge for each category. The whole process is completed after all head judges are chosen. The assignments will then be announced to the public along with the presentation schedule. Therefore, the process must be completed within one month before the APICTA Events. However, the assignments usually go through a series of modifications due to negotiations and political arguments. An automated solution is desirable to ensure the judge panels can be formed on time.

The assignment of judge panels must satisfy the following criteria.

Size Restriction. Each panel must be approximately the same size, consisting of 3 to 5 judges. Larger judge panels are preferred.

Maintain Fairness. To ensure no domination of an economy, judges in the same panel must come from different economies. The assignment must also minimize chances that judges assess entries from their own economies, which are unavoidable but give an impression of conflicts during evaluation.

Balanced Workload. As most judges are sponsored by APICTA, the assignment must ensure a reasonable amount of workloads to every judge. No judges should be left behind, and no judges should be in more than one panel in a day of the APICTA Events. To avoid work overload, if judges are in a category spanning two days, they should not serve in other categories throughout the Events.

Respect Expertise. To ensure professional judgments, no judges should be assigned to categories outside of their declared expertises.

Ensure Experienced Leaders. Each judge panel must have at least one judge that had been a head judge before, so that their experience can be passed to inexperienced judges. More experienced judges in each panel are preferred.

Political Correctness. Due to political issues, judges from some economies should not be placed together in the same categories.

Allow Partial Assignments. To increase flexibility, the organizers can force or avoid certain judges to be in a particular category.

3 Current Practice versus Constraint Programming

In the past 12 years, the judge assignments were done purely by hand following hunches. Since the criteria and objectives are not laid down explicitly, the process was far from transparent and the initial results were always complained by judges and economy leaders. The results were revised purely by hand and iterated for numerous times until (forced) consensus were reached. The processes were tedious and inefficient, especially when there were always many modifications due to negotiations and political arguments, but the resources were limited.

We propose a CP approach to develop the optimization engine in solving the judge assignment problem. A key advantage of CP is the separation of concerns in modeling and solving. Modeling involves determination of variables, domains, constraints, and objective functions. The rich constraint language allows for the model being relatively close to problem statements, making the model easy to verify and adapt. Indeed, after we delivered the first prototype, the hosting organization proposed various changes in problem statements before the assignment was finalized, and we could deliver a new solution by simply changing the model but not the implementation of solvers.

A number of attempts were tried before we arrived at the final approach and model. Instead of traditional constraint optimization, we chose the *weighted constraint satisfaction problem* (WCSP) framework after we analyzed the problem. Due to its optimization nature, the problem contains not just hard constraints but also a number of preferences. The "soft-as-hard" approach [7] in traditional constraint optimization is weak when compared with WCSP framework, which specialized in modeling preferences. As shown by Lee and Leung [5], the strong $\varnothing$-inverse consistency [5], which is a weak consistency in WCSP, is stronger in propagation than the "soft-as-hard" approach [7] in constraint optimization. We also identified global cost functions in the problem. Global cost functions in WCSP provide not just a simple language to express complex ideas, but also help increase propagation power. However, the native WCSP model failed to solve the problem within a day. We re-modeled the problem by grouping multiple global cost functions as a single one. The result could be found in a few hours, still far from being practical. We further conjoined more global cost functions as a single one. We show that the conjoined global cost function can utilize the flow algorithm from Régin [8]. Eventually, we manage to deliver a new solution within a few minutes even if the problem statements are changed. The assignment could be finalized after two weeks of blood and sweat.

4 Domain Analysis

In the following, we formally analyze the judge assignment problem and identify the corresponding constraints and objectives. The APICTA Events are hosted during $DAY = \{day_1, \ldots, day_P\}$. In APICTA 2013, $P = 2$. In day_j, a set of categories $CAT_j \subseteq CAT$, where $CAT = \{cat_1, \ldots, cat_M\}$ and $M = 16$ in APICTA 2013, are judged. Each category cat_k will have a set of entries, denoted by $entry(cat_k)$, for judges to assess. We represent a judge as $jud_i \in JUD$, where $JUD \in \{jud_1, \ldots, jud_N\}$ and $N = 61$ in APICTA 2013. Each judge jud_i is nominated from the economy $From(jud_i) \in ECO$, where $ECO = \{eco_1, \ldots, eco_Q\}$ and $Q = 13$ in APICTA 2013. Each judge $jud_i \in JUD$ had also declared the specialties defined as $SP_i \subseteq CAT$. The task at hand is to find out a time table TT, where $TT_{i,j}$ represents the set of categories assigned to the judge $jud_i \in JUD$ at $day_j \in DAY$, subject to a set of hard constraints and preferences.

4.1 Hard Constraints

The judge assignment must obey the following constraints.

Constraints on judge panel sizes

1. The size of each judge panel is at most 5 and at least 3.

Constraints on judge attendance

2. Each judge jud_i can only be in at most 1 judge panel in each day.
3. Each judge jud_i must be in at least 1 judge panel throughout the event.
4. To ensure fairness of judging, if jud_i and jud_j are in the same judge panel, they cannot be from the same economy, *i.e.* $From(jud_i) \neq From(jud_j)$.

Constraints on cross-day categories A *cross-day category* $cat_i \in CAT_{cross}$ is one spanning across two days due to a large number of entries. In APICTA 2013, 3 out of 17 are cross-day categories.

5. Judges in a cross-day category cannot be in another category, and;
6. Each judge can only be in at most 1 judge panel of a cross-day category.

Constraints on experienced judges

7. A judge jud_i is a *previous head judge*, *i.e.* $jud_i \in JUD_{head} \subseteq JUD$ iff jud_i was a head judge in APICTA before APICTA 2013. Each judge panel of a category must have at least 1 previous head judge.

Constraints on judge placement

8. Given two specific economies $eco_X \in ECO$ and $eco_Y \in ECO$, where $eco_X \neq eco_Y$. Judges from eco_X and eco_Y cannot be placed in the same panel.

Constraints on specialties
9. A judge jud_i is in category cat_k iff cat_k is a specialty of jud_i, *i.e.* $cat_k \in SP_i$.

Constraints on pre-setting We define, for each judge $jud_i \in JUD$, $Assign_i \subseteq SP_i$ to be the categories that jud_i must be in, and $Avoid_i \subseteq SP_i$ to be the categories that jud_i must not be in. In APICTA 2013, $|Assign_i| \leq 1$ and $|Avoid_i| \leq 1$ for every judge $jud_i \in JUD$.
10. A judge jud_i is in category $cat_k \in Assign_i$ iff $Assign_i \neq \varnothing$, and;
11. A judge jud_i is not in category $cat_k \in Avoid_i$ iff $Avoid_i \neq \varnothing$.

Constraint on judge workload
12. The *workload* of a judge jud_i in a category cat_k is the number of entries in cat_k. All judges evaluate at least 7 entries in total.

4.2 Preferences

The objective is to minimize the weighted sum of the following preference functions. The weights determine the importance of each preference, *i.e.* the most important one will have the highest weight. We adjusted the weights through experiments.

Preference on conflicts Define a *conflict* as a function $conf : JUD \times CAT \mapsto \mathbb{N}$, which returns the number of entries in cat_k from $From(jud_i)$, *i.e.* the same economy as jud_i.
13. Minimize the total number of conflicts, *i.e.*

$$\min \sum_{jud_i \in JUD} \sum_{day_j \in DAY} \sum_{cat_k \in TT_{i,j}} conf(jud_i, cat_k)$$

Preference on maximizing judge panel sizes The assignment prefers larger judge panels,and penalize the judge panels with size less than 5.
14. Minimize the total penalties due to small panel sizes, *i.e.*

$$\min \sum_{day_j \in DAY} \sum_{cat_k \in CAT_j} (5 - |\{jud_i \mid cat_k \in TT_{i,j}\}|)$$

Preference on maximizing experienced share A judge $jud_i \in JUD_{exp}$ is *experienced* iff jud_i has judged in APICTA before. Note that $JUD_{head} \subseteq JUD_{exp}$. More experienced judges in each panel are preferred, and penalize the judge panels with number of experienced judges less than 5.
15. Minimize the total penalties due to less experienced judges in panels, *i.e.*

$$\min \sum_{day_j \in DAY} \sum_{cat_k \in CAT_j} (5 - |\{jud_i \mid cat_k \in TT_{i,j} \wedge jud_i \in JUD_{exp}|)$$

5 Problem Modeling

We first give a background on weighted constraint satisfaction problems (WCSP) and global cost functions. While there are many ways of formulating the judge assignment problem into WCSP, we give the one that allows natural expression of cost functions and utilizes global cost functions. Based on the model, we further propose how the global cost functions can be combined to give stronger propagation power.

5.1 Weighted Constraint Satisfaction and Global Cost Functions

A *WCSP* [9] is a tuple $(\mathcal{X}, \mathcal{D}, \mathcal{C}, \top)$. $\mathcal{X}$ is a set of variables $\{x_1, x_2, \ldots, x_n\}$. Each variable has its finite domain $D(x_i) \in \mathcal{D}$ containing possible values for x_i. A tuple $\ell \in \mathcal{L}(S) = D(x_{s_1}) \times \ldots \times D(x_{s_n})$ is used to represent an assignment on $S = \{x_{s_1}, \ldots, x_{s_n}\} \subseteq \mathcal{X}$. The notation $\ell[x_i]$ denotes the value assigned to x_i in ℓ, and $\ell[S']$ denotes the tuple formed from projecting ℓ onto $S' \subseteq S$. $\mathcal{C}$ is a set of cost functions. Each cost function $W_S \in \mathcal{C}$ has its scope $S \subseteq \mathcal{X}$, and maps $\ell \in \mathcal{L}(S)$ to a cost in the valuation structure $V(\top) = ([0 \ldots \top], \oplus, \leq)$. $V(\top)$ contains a set of integers $[0 \ldots \top]$ with standard integer ordering $\leq$. $\top$ is a finite or infinite integer corresponding to forbidden assignments. Addition $\oplus$ is defined by $a \oplus b = \min(\top, a+b)$. Subtraction $\ominus$ is defined only for $a \geq b$, $a \ominus b = a - b$ if $a \neq \top$ and $\top \ominus a = \top$ for any a. The *cost* of a tuple $\ell \in \mathcal{L}(\mathcal{X})$ in a WCSP is defined as $cost(\ell) = \bigoplus_{W_S \in \mathcal{C}} W_S(\ell[S])$. A tuple ℓ is an optimal valid *solution* of a WCSP if $cost(\ell)$ is minimum among all tuples in $\mathcal{L}(\mathcal{X})$ and $cost(l) < \top$.

A *global cost function* [1,5] is a cost function with special semantics, based on which efficient algorithms can be designed for consistency enforcements. In particular, we denote a global cost function as $\text{SOFT\_GC}^{\mu}_{m}(S)$ if it is derived from the corresponding hard global constraint GC with variable scope S, a violation measure μ, and a weight constant m. The cost function $\text{SOFT\_GC}^{\mu}_{m}(S)$ returns $m \cdot \mu(\ell)$ to indicate how much a tuple $\ell \in \mathcal{L}(S)$ has violated GC, or 0 if the tuple satisfies GC. Two examples of violation measures for global constraints involves counting are μ_{var} and μ_{val}: $\text{SOFT\_GC}^{var}_{1}(S)$ returns the minimum number of assignments modified to satisfy GC [7]; while $\text{SOFT\_GC}^{val}_{1}(S)$ returns the number of values exceeding the boundaries allowed by GC [11]. We assume $m = 1$ if m is not specified. If $m = \top$, a global cost function represents a **hard** constraint.

5.2 Problem Formulation

Define $P = (\mathcal{X}, \mathcal{D}, \mathcal{C}, \top)$ to be the WCSP model for the judge assignment problems. The variable $x_{i,j} \in \mathcal{X}$ gives the category that jud_i assesses on day_j. The domain $D(x_{i,j})$ of each variable $x_{i,j}$ is the set of categories $CAT_j \subseteq CAT$ judged on day_j, with a dummy category cat_0 to indicate no judging on a day. Each constraint and preference are enforced as follows.

Constraints on judge panel sizes. Constraint 1 can be enforced by the global cost function $\text{SOFT\_AMONG}^{var}(S, lb, ub, V)$, which returns $max(0, lb - t(\ell), t(\ell) - ub)$

for each tuple $\ell \in \mathcal{L}(S)$, where $t(\ell) = |\{x_i \in S \mid \ell[x_i] \in V\}|$ [10]. For each $day_j \in DAY$, we place one $\textsc{Soft\_Among}^{var}_{\top}(\mathcal{X}_j, 3, 5, \{cat_i\})$ for each $cat_i \in CAT_j$, where $\mathcal{X}_j = \{x_{i,j} \mid jud_i \in JUD\}$.

Constraints on judge attendance. Constraint 2 is always satisfied for all valid assignments. Constraint 3 can be enforced by placing one $\textsc{Soft\_Among}^{var}_{\top}(\{x_{i,j} \mid day_j \in DAY\}, 1, |DAY|, CAT)$ for each jud_i. As $|DAY| = 2$, binary table cost functions were used instead. Constraint 4 can be enforced by the global cost functions $\textsc{Soft\_GCC}^{val}(S, LB, UB)$ [11]. Given $\Sigma = \bigcup_{x_i \in S} D(x_i)$. Define the number of occurrences of a value $v \in \Sigma$ in ℓ by $\#(\ell, v)$ and two functions $s(\ell, v)$ and $e(\ell, v)$ as follows.

$$s(\ell, v) = \begin{cases} LB(v) \ominus \#(\ell, v), & \text{if } \#(\ell, v) \leq LB(v) \\ 0, & \text{otherwise} \end{cases} \quad e(\ell, v) = \begin{cases} \#(\ell, v) \ominus UB(v), & \text{if } \#(\ell, v) \geq UB(v) \\ 0, & \text{otherwise} \end{cases}$$

The global cost function $\textsc{Soft\_GCC}^{val}$ is defined for each tuple $\ell \in \mathcal{L}(S)$ as follows [11].

$$\textsc{Soft\_GCC}^{val}(S, LB, UB)(\ell) = \bigoplus_{v \in \Sigma} s(\ell, v) \oplus \bigoplus_{v \in \Sigma} e(\ell, v)$$

Define $\{\mathcal{X}_{j,eco_t}\}$ to be the partition of $\mathcal{X}$ according to the day of the events and the economies the judges represent, *i.e.* $x_{i,j} \in \mathcal{X}_{j,eco_t}$ iff $From(jud_i) = eco_k$. For each day_j and economy eco_t, we place one $\textsc{Soft\_GCC}^{val}_{\top}(\mathcal{X}_{j,eco_t}, LB, UB)$, which $LB(cat_k) = 0$ for all $cat_k \in CAT_j \cup \{cat_0\}$ and UB are defined as follows.

$$UB(cat_k) = \begin{cases} 1, & k \neq 0, \\ |\mathcal{X}|, & \text{otherwise.} \end{cases} \tag{1}$$

Constraints on cross-day categories. Constraints 5 and 6 can be enforced by the binary cost function $JoinOnlyCrossCat$ placed on each pair of variables $\{(x_{i,h}, x_{i,k}) \mid h \neq k \wedge day_h, day_k \in DAY\}$. The cost function is defined for each pair of values (v_h, v_k), where $v_h \in D(x_{i,h})$ and $v_k \in D(x_{i,k})$ as follows.

$$JoinOnlyCrossCat_{\{x_{i,h}, x_{i,k}\}}(v_h, v_k) = \begin{cases} 0, & \text{if } v_h \notin CAT_{cross} \text{ and } v_k \notin CAT_{cross} \\ 0, & \text{if } v_h \in CAT_{cross} \text{ and } v_k = v_h \\ \top, & \text{otherwise} \end{cases}$$

Constraints on experienced judges. Constraint 7 enforced by placing one $\textsc{Soft\_GCC}^{val}_{\top}(\{x_{i,j} \mid jud_i \in JUD_{head}\}, LB, UB)$ for each $day_j \in DAY$, where LB and UB are defined as follows.

$$LB(cat_j) = \begin{cases} 1, & j \neq 0, \\ 0, & \text{otherwise.} \end{cases} \quad UB(cat_j) = \begin{cases} 1, & j \neq 0, \\ |\mathcal{X}|, & \text{otherwise.} \end{cases}$$

Constraints on judge placement. Constraint 8 can be fused in constraint 4 by considering eco_X and eco_Y as a single economy.

Constraints on specialties. Constraint 9 can be enforced by placing the unary cost function *Specialty* on each variable $x_{i,j} \in \mathcal{X}$. The function *Specialty* is defined for each value $v \in D(x_{i,j})$ as follows.

$$Specialty_{\{x_{i,j}\}}(v) = \begin{cases} 0, & \text{if } v = cat_0 \text{ or } v \in SP_i \\ \top, & \text{otherwise} \end{cases}$$

Constraints on pre-setting. Constraints 10 and 11 can be enforced by the unary cost functions *Set* and *Unset* placed on each variable $x_{i,j} \in \mathcal{X}$, defined for each value $v \in D(x_{i,j})$ as follows.

$$Set_{\{x_{i,j}\}}(v) = \begin{cases} 0, & \text{if } v \in Assign_i \\ & \text{or } Assign_i = \varnothing; \\ \top, & \text{otherwise.} \end{cases} \quad Unset_{\{x_{i,j}\}}(v) = \begin{cases} \top, & \text{if } v \in Avoid_i \\ & \text{and } Avoid_i = \varnothing; \\ 0, & \text{otherwise.} \end{cases}$$

Constraint on judge workload. Define $entry(cat_0) = 0$. Since $|DAY| = 2$, constraint 12 can be simply enforced by the binary cost function *WLLimit* placed on each pair of variables $\{(x_{i,h}, x_{i,k}) \mid h > k \wedge day_h, day_k \in DAY\}$. The cost function is defined for each pair of values (v_h, v_k), where $v_h \in D(x_{i,h})$ and $v_k \in D(x_{i,k})$, as follows.

$$WLLimit_{\{x_{i,h}, x_{i,k}\}}(v_h, v_k) = \begin{cases} \top, & \text{if } entry(v_h) + entry(v_k) < 7 \text{ if } v_h \neq v_k, \text{ or} \\ & entry(v_h) < 7 \text{ if } v_h = v_k; \\ 0, & \text{otherwise} \end{cases}$$

Preference on conflicts. We define the weight of preference 13 as $\epsilon_{conflict}$. Preference 13 can be enforced by the unary cost function *Conflict* placed on each variable $x_{i,j} \in \mathcal{X}$, which $Conflict_{\{x_{i,j}\}}(v) = \epsilon_{conflict} \cdot conf(jud_i, v)$ for every $v \in D(x_{i,j})$.

Preference on maximizing judge panel size. We observe that preference 14 is a special case of SOFT_AMONG^{var}. We define the weight of preference 14 as $\epsilon_{maxsize}$. One $\text{SOFT\_AMONG}^{var}_{\epsilon_{maxsize}}(\mathcal{X}_j, 5, 5, \{cat_k\})$ is posted for each day_j and each category cat_k.

Preference on maximizing experienced share. Define the weight of preference 15 as ϵ_{exp}. One $\text{SOFT\_AMONG}^{var}_{\epsilon_{exp}}(\{x_{i,j} \mid jud_i \in JUD_{exp}\}, 5, 5, \{cat_k\})$ is posted for each day_j and each category cat_k.

5.3 Conjoining Cost Functions

As the original WCSP model cannot be solved within a day, we consider conjoining global cost functions to further reduce the search space. Lee *et al.* [6] give a general technique on conjoining global cost functions using linear programs, and show that enforcing consistencies on a conjoined cost function is stronger than enforcing the same consistencies on separate cost functions. In this section, we give a special case on conjoining global cost functions using *flow networks*.

A *flow network* $G = (V, E, w, c, d)$ is a connected directed graph (V, E), in which each edge $e \in E$ has a weight w_e, a capacity c_e, and a demand $d_e \leq c_e$. An (s,t)-*flow* f from a source $s \in V$ to a sink $t \in V$ of a value $value(f)$ in G is defined as a mapping from E to real numbers such that:

- $\sum_{(s,u)\in E} f_{(s,u)} = \sum_{(u,t)\in E} f_{(u,t)} = value(f)$;
- $\sum_{(u,v)\in E} f_{(u,v)} = \sum_{(v,u)\in E} f_{(v,u)} \; \forall\, v \in V \setminus \{s,t\}$;
- $d_e \leq f_e \leq c_e \; \forall\, e \in E$.

For simplicity, we call an (s,t)-flow as a flow if s and t have been specified. The *cost* of a flow f is defined as $cost(f) = \sum_{e\in E} w_e f_e$. A *minimum cost flow* problem of a value α is to find the flow f of $value(f) = \alpha$ such that its cost is minimum. If α is not given, it is assumed to be the maximum value among all flows.

A global cost function W_S is *flow-based* if W_S can be represented as a flow network $G = (V, E, w, c, d)$ such that $min\{W_S(\ell) \mid \ell \in \mathcal{L}(S)\} = min\{cost(f) \mid f$ is the maximum (s,t)-flow of $G\}$, where $s \in V$ is the fixed source and $t \in V$ is the fixed destination. One example of flow-based global cost function is $\textsc{Soft\_GCC}^{val}$ [11].

The global cost functions $\{\textsc{Soft\_Among}^{var}(\mathcal{X}_j, lb_i, ub_i, \{cat_k\}) \mid cat_k \in CAT\}$ in constraint 1, preferences 14 and 15 can be conjoined respectively as a single $\textsc{Soft\_GCC}^{val}$ for each $day_j \in DAY$. We show as follows.

Proposition 1. *Given a set* $\{\textsc{Soft\_Among}^{var}(S, lb_i, ub_i, \Omega_i) \mid i = 1 \ldots h\}$. *If* $\Omega_i \cup \Omega_j = \varnothing$ *for* $i \neq j$ *and* $|\Omega_i| = 1$ *for every* i, *the following holds for every tuple* $\ell \in \mathcal{L}(S)$:

$$\textsc{Soft\_GCC}^{val}(S, LB, UB)(\ell) = \bigoplus_{i=0}^{m} \textsc{Soft\_Among}^{var}(S, lb_i, ub_i, \Omega_i)(\ell)$$

The lower bound LB *is defined as* $LB(v) = lb_i$ *iff* $v \in \Omega_i$, *and the upper bound* UB *is defined as* $UB(v) = ub_i$ *iff* $v \in \Omega_i$.

Proof. Define $v_i \in \Omega_i$. By definitions, for every tuple $\ell \in \mathcal{L}(S)$, $max(0, lb_i - t(\ell), t(\ell) - ub_i) = s(\ell, v_i) \oplus e(\ell, v_i)$. Results follow. □

By conjoining $\textsc{Soft\_Among}^{var}$ into $\textsc{Soft\_GCC}^{val}$, the problem instance can be solved within a few hours, but is still far from being practical. We further conjoin $\textsc{Soft\_Among}^{var}$ in constraint 1 and preference 14 and $\textsc{Soft\_GCC}^{val}$ in constraints 4 and 8 into a single global cost function $\textsc{Soft\_GCC\_Among}^{val+bvar}$. We found that the $\textsc{Soft\_GCC\_Among}^{val+bvar}$ is *flow-based*, allowing consistencies in WCSP to be enforced by flow networks.

The cost functions $\{\textsc{Soft\_Among}^{var}_{\top}(\mathcal{X}_j, 3, 5, \{cat_k\}) \mid cat_k \in CAT\}$ in constraint 1 and $\{\textsc{Soft\_Among}^{var}(\mathcal{X}_j, 5, 5, \{cat_k\}) \mid cat_k \in CAT\}$ in preference 14 can be conjoined into a single $\textsc{Soft\_Among}^{bvar}(\mathcal{X}_j, 3, 5, CAT)$ for each $day_i \in DAY$. The new violation measure μ_{bvar} forbids the number of specified values exceeding the boundaries given by cost functions, and favor the tuple containing more specified values. The cost function $\textsc{Soft\_Among}^{bvar}(S, lb, ub, \Omega)$ is defined for each $\ell \in \mathcal{L}(S)$ as follows.

$$\textsc{Soft\_Among}^{bvar}(S, lb, ub, \Omega) = \begin{cases} ub \ominus t(\ell), & \text{if } t(\ell) \geq lb \text{ and } t(\ell) \leq ub \\ \top, & \text{otherwise} \end{cases}$$

We further conjoin $\textsc{Soft\_Among}^{bvar}$ and $\textsc{Soft\_GCC}^{val}$ into $\textsc{Soft\_GCC\_Among}^{val+bvar}$. The violation measure $\mu_{val+bvar}$ is a conjoined violation measure from μ_{bvar} and μ_{val} used by $\textsc{Soft\_Among}^{bvar}$ and $\textsc{Soft\_GCC}^{val}$ respectively. Given a set of global cost functions $\mathcal{C}_{GCC} = \{\textsc{Soft\_GCC}^{val}(S_i, LB_i, UB_i) \mid i = 1, \ldots, m\}$ and $\mathcal{C}_{Among} = \{\textsc{Soft\_Among}^{bvar}(K_j, lb_j, ub_j, \Omega_j) \mid j = 1, \ldots, h\}$, the cost function $\textsc{Soft\_GCC\_Among}^{val+bvar}(\{(S_i, UB_i, LB_i)\}\{(K_j, lb_j, ub_j, \Omega_j)\})$ is defined as a global cost function formed by conjoining $\mathcal{C}_{GCC}$ and $\mathcal{C}_{Among}$, *i.e.* for every tuple $\ell \in \mathcal{L}(\bigcup_{i=1}^{m} S_i \cup \bigcup_{j=1}^{h} K_j)$:

$$\textsc{Soft\_GCC\_Among}^{val+bvar}(\ell) = \bigoplus_{i=1}^{m} \textsc{Soft\_GCC}^{val}(S_i, LB_i, UB_i)(\ell[S_i]) \oplus \bigoplus_{j=1}^{h} \textsc{Soft\_Among}^{bvar}(K_j, lb_j, ub_j, \Omega_j)(\ell[K_j])$$

We show $\textsc{Soft\_GCC\_Among}^{val+bvar}$ is flow-based as follows.

Theorem 1. *The cost function* $\textsc{Soft\_GCC\_Among}^{val+bvar}(\{(S_i, LB_i, UB_i)\}, \{(K, lb_j, ub_j, \Omega_j)\})$ *is flow-based if the following condition holds.*

- $S_i \cap S_j = \varnothing$ *if* $i \neq j$*;*
- $LB(v) = 0$ *for* $v \in \Sigma_i$*, where* $\Sigma_i = \bigcup_{x_j \in S_i} D(x_j)$*, for each* $i = 1, \ldots, m$*;*
- $\Omega_i \cap \Omega_j = \varnothing$ *if* $i \neq j$*, and;*
- $K = \bigcup_{i=1}^{m} S_i$ *for each* $j = 1, \ldots, h$*.*

Proof. Without loss of generality, we assume $\Sigma = \bigcup_{j=1}^{h} \Omega_j$. If there exists a value $u \in \bigcup_{j=1}^{h} \Omega_j$ that does not exist in Σ, we can add a dummy $\textsc{Soft\_Among}^{var}(K, 0, |K|, \{u\})$ into the set of cost functions.

The flow network can be constructed using the method suggested by Régin [8]. We construct a single flow network $G = (V, E, w, c, d)$ representing a set of $\textsc{Soft\_GCC}^{val}$ and $\textsc{Soft\_Among}^{var}$ as follows.

- $V = K \cup \{v_{iv} \mid v \in \Sigma_i\} \cup \{\mu_j \mid j = 1, \ldots, h\} \cup \{s, t\}$;
- $E = A_s \cup A_K \cup A_v \cup A_t \cup A_{gcc-ex} \cup A_{among-short} \cup A_{among-vio} \cup A_{among-ex}$, where:
  - $A_s = \{(s, x_j) \mid x_j \in \mathcal{X}\}$;
  - $A_K = \{(x_j, v_{iu}) \mid x_j \in \mathcal{X}_i \wedge u \in D(x_j)\}$;
  - $A_v = \{(v_{iu}, \mu_j) \mid u \in V_j \wedge j = 1, \ldots, h\}$;
  - $A_t = \{(\mu_j, t) \mid j = 1, \ldots, h\}$;
  - $A_{gcc-ex} = \{(v_{iu}, \mu_j) \mid u \in \Sigma_i \wedge u \in \Omega_j\}$;

- $A_{among-vio} = \{(s, \mu_j) \mid j = 1, \ldots, h\}$;
- $A_{among-short} = \{(s, \mu_j) \mid j = 1, \ldots, h\}$, and;
- $A_{among-ex} = \{(\mu_j, t) \mid j = 1, \ldots, h\}$.

– $c_e = \begin{cases} UB_i(u), & \text{if } e = (v_{iu}, \mu_j) \in A_v \\ ub_j, & \text{if } e = (\mu_j, t) \in A_t \\ ub_j \ominus lb_j, & \text{if } e = (s, \mu_j) \in A_{among-short} \\ |K|, & \text{if } e \in A_{gcc-ex} \cup A_{among-ex} \\ 1, & \text{otherwise} \end{cases}$

– $d_e = \begin{cases} ub_j, & \text{if } e = (\mu_j, t) \in A_t \\ 0, & \text{otherwise} \end{cases}$

– $w_e = \begin{cases} \top, & \text{if } e \in A_{among-vio} \cup A_{among-ex} \\ 1, & \text{if } e \in A_{gcc-ex} \cup A_{among-short} \\ 0, & \text{otherwise} \end{cases}$

In the flow network, there may exist multiple edges between two nodes. The edges A_v enforce SOFT_GCC^{val} while A_{gcc-ex} give the corresponding violation cost. The edges A_t enforce SOFT_AMONG^{bvar} while $A_{among-short}$ give the corresponding violation cost. The edges $A_{among-ex}$ and $A_{among-vio}$ ensure all tuples have corresponding maximum flows.

Using the similar reasoning as in Proposition 7 given by Régin [8], a maximum (s,t)-flow in G corresponds to an assignment to K. With the similar reasoning by van Hoeve [11], the minimum cost of maximum flows corresponds to the minimum of SOFT_GCC_AMONG$^{val+bvar}$. Results follow. □

An example of the flow network G is shown in Figure 1, based on a set of variables $S = \{x_i \mid i = 1, \ldots 5\}$ and $D(x_i) = \{cat_1, cat_0\} \subseteq CAT$. The SOFT_GCC_AMONG$^{val+bvar}$ consists of the following cost functions.

- SOFT_AMONG$^{bvar}(\{x_1, x_2, x_3, x_4, x_5\}, 2, 5, \{cat_1\})$
- SOFT_AMONG$^{bvar}(\{x_1, x_2, x_3, x_4, x_5\}, 0, 5, \{cat_0\})$
- SOFT_GCC$^{val}(\{x_1, x_2\}, LB, UB)$, and;
- SOFT_GCC$^{val}(\{x_3, x_4, x_5\}, LB, UB)$, where LB and UB are defined as in constraint 4.

The pair of numbers on the edges represent the demands and capacities of the edges. If no numbers are on the edge, the edge has zero demand and unit capacity. If the edge is dotted, the edge has unit weight. Otherwise, the edge has zero weight. For simplicity, we omit the edges with weight equal to $\top$. The thick lines show the flow corresponding to the tuple $\ell = (cat_1, cat_1, cat_0, cat_0, cat_0)$ having a cost of 4: 1 from SOFT_GCC^{val} and 3 from SOFT_AMONG^{bvar}.

We further show that the conjoined cost function is *flow-based projection-safe* [5]. If the cost function is flow-based projection-safe, stronger consistencies like GAC* [2,5], FDGAC* [5], and weak EDGAC* [5] can be enforced in polynomial time throughout the search. Stronger consistencies help remove more search places, and reduce the runtime if the reduction in search space can compensate the time on enforcing consistencies.

A global cost function W_S is *flow-based projection-safe* [5] iff W_S is flow-based, and for all W'_S derived from W_S by a series of projections and extensions, W'_S is

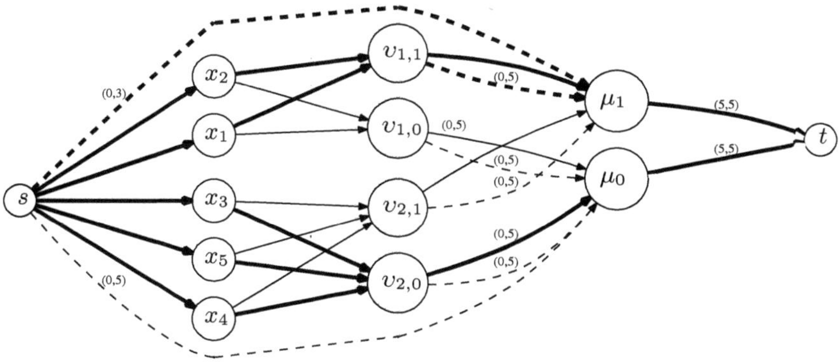

Fig. 1. An example of the flow network

flow-based. Lee and Leung [5] give sufficient conditions on flow-based projection-safety.

1. W_S is flow-based, with the corresponding network $G = (V, E, w, c, d)$ with a fixed source $s \in V$ and a fixed destination $t \in V$;
2. there exists a subjective function mapping each maximum flow f in G to each tuple $\ell_f \in \mathcal{L}(S)$, and;
3. there exists an injection mapping from an assignment (x_i, v) that set the variable x_i to $v \in D(x_i)$ to a subset of edges $\bar{E} \subseteq E$ such that for all maximum flow f and the corresponding tuple ℓ_f, $\sum_{e \in \bar{E}} f_e = 1$ whenever $\ell_f[x_i] = v$, and $\sum_{e \in \bar{E}} f_e = 0$ whenever $\ell_f[x_i] \neq v$

We show that SOFT_GCC_AMONG$^{val+bvar}$ is flow-based projection-safe as follows.

Theorem 2. *If the cost function* SOFT_GCC_AMONG$^{val+bvar}(\{(S_i, UB_i, LB_i)\}, \{(K_j, lb_j, ub_j, \Omega_j)\})$ *satisfies the conditions stated in Theorem 1, it is flow-based projection-safe.*

Proof. Theorem 1 already shows the SOFT_GCC_AMONG$^{val+bvar}$ satisfies conditions 1 and 2. In addition, the edge (x_j, v_{iu}) corresponds to assigning u to x_j. Results follow. □

The cost functions in constraints 1, 4, 8, and preference 14 satisfies the conditions given in Theorem 1 if they are grouped by $day_j \in DAY$. The conjoined cost function shown below, defined for $day_j \in DAY$, is flow-based projection-safe.

$$\text{SOFT\_GCC\_AMONG}^{val+bvar}\ (\ \{(X_{j,eco_t}, LB, UB) \mid eco_t \in ECO\},$$
$$\{(X_j, 3, 5, \{cat_i\}) \mid cat_i \in CAT\})$$

By applying the above conjoined cost functions, the assignments can be found within a few minutes, even if the requirements are changed frequently. In the following, we show the robustness of our solution by experiments.

6 Experiments

As the data from previous years are not available, the experiments are purely based on the data from APICTA 2013 and conducted on Toulbar2 [3], an open-source WCSP solver. In the experiments, variables are assigned in lexicographic order. Value assignments start with the values having the minimum unary costs. Weak EDGAC* [5] is enforced during search with no initial upper bound. Each test is conducted on a Linux Cluster (4 × 2GHz CPU) machine with 3GB memory. In all experiments, we set $\epsilon_{conflict} = 1$, $\epsilon_{maxsize} = 200$, and $\epsilon_{exp} = 50$.

We first compare the runtime in solving the APICTA 2013 instance using different models. The solver cannot give the optimal for the native model as stated in Section 5.2 after 3 days of execution. The model applying Proposition 1 can be solved after 5.3 hours. With SOFT_GCC_AMONG$^{val+bvar}$, the solver gives the optimal solution in 107.6 seconds.

We further show the robustness of our solution by simulating the judge assignment process using the conjoined model. We mark the instance from APICTA 2013 as instance 0. Each instance i, where $i > 0$, is modified from instance 0. We keep around 10% of judge assignments in the solution of instance 0 as the preset assignments of instance i. We randomly modify the requirements of the remaining 90% of judges on top of instance i as follows.

- Remove a judge and add a new judge from a different economy;
- Modify the specialties of a judge;

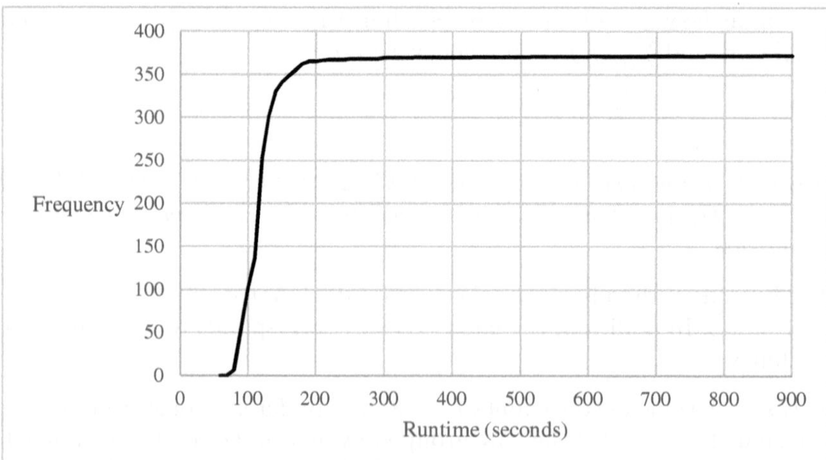

Fig. 2. Frequency distribution according to the runtime

[3] `https://mulcyber.toulouse.inra.fr/projects/toulbar2/`

- Avoid a judge to be in a specific category, and;
- Withdraw some entries so that the conflict is changed.

We generate 400 instances, each of which is allowed to run for 15 minutes.[4]

We present the results as the frequency distribution as shown in Figure 2. The x-axis is the runtime and the y-axis gives the number of instances able to be solved within the runtime. We observe that 93% of the instances can be solved within 200 seconds, while only 29 out of 400 instances cannot be solved within 15 minutes.

7 Discussion

After we gave an initial solution to the organizers, we were asked to do modifications, as simulated in the previous section, before finalization. The modifications were unpredictable, but we could cope with the modifications and gave an updated optimal solution within a few minutes.

The organizers also received far less requests on modifications from the judges on the assignments. Throughout the process, only one judge complained the assignments, and it was resolved by correcting the specialties.

Besides, our solution speeded up the process. The judge assignment needs to be completed one month before the APICTA event. In previous years, the assignment was completed only a few days before the deadline. With automation, the process was completed in two weeks, including endorsement from advisory judges. Compared with previous years, the time required was greatly reduced.

8 Conclusion

Judge assignment problems have been studied in the field of sport tournaments. Lamghari and Ferland [4] formulates the problems into linear programs and solves by tabu search. Fernando *et al.* [3] also use integer linear programming to compute referee assignments in football matches.

Our contributions are three-fold. First, we implement an automated solution for APICTA 2013 to generate the most preferred assignments of each judge to each category. Our solution helps lay down all restrictions and preference explicitly, and generate a new assignment within a few minutes every time the requirements and preferences are changed, shortening the process to two weeks. Second, we give a real-life example on global cost functions [1, 5] in WCSP. By using SOFT_GCC^{val} [11] and SOFT_AMONG^{var} [10], WCSP can model restrictions that involves several variables. Third, we further give special cases of conjoining a group of SOFT_GCC^{val} and SOFT_AMONG^{var} via flow networks. The original model cannot be solved in a day. We refine the model by grouping a set of SOFT_AMONG^{var} into SOFT_GCC^{val}, but still not practical. We further conjoin SOFT_AMONG^{var} and SOFT_GCC^{val} in the model as a single

[4] The solver and instances can be found online:
`http://www.cse.cuhk.edu.hk/~klleung/cp14/JudgeAssign.tar.gz`

SOFT_GCC_AMONG$^{val+bvar}$, which is flow-based projection-safe [5]. We show by experiments that conjoining global cost functions can solve an instance within a few minutes after requirements are modified.

We plan to refine our solution for judge assignments in APICTA 2014[5] to produce solutions in a shorter time. Eventually, we hope our technique can be applied to other similar scenarios such as banquet seating planning.

Acknowledgment. We are grateful to the anonymous referees for their constructive comments. The work described in this paper was substantially supported by grant CUHK413710 from the Research Grants Council of Hong Kong SAR and grant F-HK019/12T from the Consulate General of France of Hong Kong and the Research Grants Council of Hong Kong SAR.

References

1. Cooper, M., de Givry, S., Sanchez, M., Schiex, T., Zytnicki, M., Werner, T.: Soft Arc Consistency Revisited. Artificial Intelligence 174, 449–478 (2010)
2. Cooper, M., Schiex, T.: Arc Consistency for Soft Constraints. Artifical Intelligence 154, 199–227 (2004)
3. Fernando, A., Durán, G., Guajardo, M.: Referee Assignment in the Chilean Football League Using Integer Programming and Patterns. International Transactions in Operational Research 21(3), 415–438 (2014)
4. Lamghari, A., Ferland, J.A.: Assigning Judges to Competitions of Several Rounds Using Tabu Search. European Journal of Operational Research 210(3), 694–705 (2011)
5. Lee, J.H.M., Leung, K.L.: Consistency Techniques for Global Cost Functions in Weighted Constraint Satisfaction. Journal of Artificial Intelligence Research 43, 257–292 (2012)
6. Lee, J.H.M., Leung, K.L., Shum, Y.M.: Consistency Techniques for Polytime Linear Global Cost Functions in Weighted Constraint Satisfaction. CONSTRAINTS 19(3), 270–308 (2014)
7. Petit, T., Régin, J.-C., Bessière, C.: Specific Filtering Algorithm for Over-Constrained Problems. In: Walsh, T. (ed.) CP 2001. LNCS, vol. 2239, pp. 451–463. Springer, Heidelberg (2001)
8. Régin, J.-C.: Combination of among and cardinality constraints. In: Barták, R., Milano, M. (eds.) CPAIOR 2005. LNCS, vol. 3524, pp. 288–303. Springer, Heidelberg (2005)
9. Schiex, T., Fargier, H., Verfaillie, G.: Valued Constraint Satisfaction Problems: Hard and Easy Problems. In: Proceedings of IJCAI 1995, pp. 631–637 (1995)
10. Solnon, C., Cung, V., Nguyen, A., Artigues, C.: The Car Sequencing Problem: Overview of State-of-the-Art Methods and Industrial Case-Study of the ROADDEF 2005 Challege Problem. European Journal of Operational Research 191(3), 912–927 (2008)
11. van Hoeve, W.-J., Pesant, G., Rousseau, L.-M.: On Global Warming: Flow-based Soft Global Constraints. J. Heuristics 12(4-5), 347–373 (2006)

[5] https://www.facebook.com/APICTA2014

Worst-Case Scheduling of Software Tasks
A Constraint Optimization Model to Support Performance Testing

Stefano Di Alesio[1,2], Shiva Nejati[2], Lionel Briand[2], and Arnaud Gotlieb[1]

[1] Certus Centre for Software Verification and Validation, Simula Research Laboratory, Norway
{stefano,arnaud}@simula.no

[2] Interdisciplinary Centre for Reliability, Security and Trust (SnT), University of Luxembourg, Luxembourg
{shiva.nejati,lionel.briand}@uni.lu

Abstract. Real-Time Embedded Systems (RTES) in safety-critical domains, such as maritime and energy, must satisfy strict performance requirements to be deemed safe. Therefore, such systems have to be thoroughly tested to ensure their correct behavior even under the worst operating conditions. In this paper, we address the need of deriving worst case scenarios with respect to three common performance requirements, namely task deadlines, response time, and CPU usage. Specifically, we investigate whether this worst-case analysis can be effectively re-expressed as a Constrained Optimization Problem (COP) over the space of possible inputs to the system. Solving this problem means finding the sets of inputs that maximize the chance to violate performance requirements at runtime. Such inputs can in turn be used to test if the target RTES meets the expected performance even in the worst case. We develop an OPL model for IBM ILOG CP OPTIMIZER that implements a task priority-based preemptive scheduling, and apply it to a case study from the maritime and energy domain. Our validation shows that (1) the input to our model can be provided with reasonable effort in an industrial setting, and (2) the COP effectively identifies test cases that maximize deadline misses, response time, and CPU usage.

1 Introduction: Performance Testing in Safety-Critical Systems

Systems in domains such as avionics, automotive, and maritime are often safety-critical, implying that their failure could result in catastrophic consequences. For this reason, their safety-related software components are usually subject to third-party certification to be deemed operationally safe. In particular, software certification has to take into account performance requirements specifying how the system should execute on its hardware platform, and how it should react to its environment [15]. Such requirements often specify constraints on response time, jitter, task deadlines, and computational resources utilization [13, 20]. Widely used safety standards, such as IEC 61508 and IEC 26262, state that performance testing is highly recommended to provide an evidence that the system is safe [6]. However, safety-critical systems are progressively relying on real-time embedded software that features multi-threaded application design, highly configurable operating systems, and multi-core architectures for computing platforms [17]. The concurrent nature of embedded software also entails that the order of

B. O'Sullivan (Ed.): CP 2014, LNCS 8656, pp. 813–830, 2014.

external events triggering the systems tasks is often unpredictable [12]. Such increasing software complexity renders performance testing more and more challenging. This aspect is reflected by the fact that most existing software testing approaches target only the system functionality, even though the degradation in performance can have more severe consequences than mere incorrect behavior [27].

In this paper, we consider three common classes of performance requirements, concerning respectively hard real-time, soft real-time, and resource constraints. Specifically, (1) we relate the hard real-time constraints to *task deadlines* requirements, stating that the system tasks should always terminate before a given completion time. Such strict requirements entail that even a single deadline miss severely compromises the system operational safety. (2) For the soft real-time constraints, we consider *response time* requirements, stating that the system should respond to external inputs within a specified time. Failure to do so poses negative consequences over the Quality of Service (QoS). (3) Finally, we consider *CPU usage* requirements, stating that the system should always keep a certain percentage of free CPU. Limiting the CPU usage is a necessary safety precaution. Indeed, if the CPU usage trespasses a certain threshold, the system may fail to timely respond to safety-critical alarms.

To check whether the system satisfies these requirements, we need to identify worst-case scenarios with respect to deadline misses, response time, and CPU usage. Such scenarios are determined by the way tasks are scheduled to execute at runtime by the Real-Time Operating System (RTOS). The schedules in turn depend on real-time, unpredictable events, on constraints deriving from software design, and on the system execution platform. For instance, critical tasks in RTES are usually ready to be executed upon triggers that depend on the external environment. Furthermore, the system design constrains the way tasks interact with each other, specifying temporal relationships, and communication through shared resources with exclusive access. Finally, RTOS are in general configured with a priority-based scheduling policy, which entails that the lowest priority task must be preempted when a higher priority task is ready for execution. However, this preemption can only occur when the running task is not locking a shared resource, and is only necessary when there is no available processor core. This mix of real-time, software design, and execution platform constraints on task scheduling renders the analysis of runtime scenarios challenging in the context of RTES.

Our work focuses on performance testing, whose goal is to identify scenarios that exercise a system in a way to either violate performance requirements, or be as close as possible to doing so. Consistent with widely used terminology [5], we refer to this activity as *stress testing*. We propose a strategy to find combinations of system inputs that maximize the likelihood of violating performance requirements. Such input combinations are characterized by sequences of *arrival times* for aperiodic tasks in the target software system, and we refer to each sequence as a *stress test case*. Finding these test cases is not trivial, since the set of all possible arrival times for aperiodic tasks quickly grows as the system size increases. Therefore, it is practically impossible to investigate all the potential ways in which the arrival times can determine task schedules at runtime. This reason motivates the need of a systematic search that effectively finds stress test cases likely to reveal deadline misses, long response time, and high CPU usage.

Contributions of This Paper. The main contribution of this paper is to address the systematic generation of stress test cases by applying Constraint Programming. Specifically, we present a Constraint Optimization Model (COP) to automate the generation of stress test cases, which is inspired by the work done in this field to solve traditional scheduling problems [4]. We evaluate our model using a practical application on a RTES from the maritime and energy domain. We cast the problem of generating stress test cases as an OPL model designed for the IBM ILOG CP OPTIMIZER, that models the system design, executing platform, and performance requirements. The OPL model in this paper builds on our previous work [10, 11, 22]. Specifically, we started [22] with a COMET optimization model, where we addressed the validation of CPU Usage and response time. Then [10], we devised an OPL version of the model where we focused on task deadlines. Finally [11], we included a dedicated search procedure for a smarter labeling of variables, and compared our constraint-based approach with metaheuristic search techniques. Our earlier work included a variable boolean matrix showing tasks execution over time, that proved to severely limit the efficiency of our model. In this work, we significantly improve the data structures representing task executions, and demonstrate the applicability of our new COP to an industrial case study. Specifically,

1. We provide a detailed OPL model which implements a task priority-based scheduling process by considering a discretized matrix, as opposed to a boolean one, which represents task executions over time. In addition, we address the search for solutions adapting the dedicated heuristic proposed in previous work [11].
2. We demonstrate the efficiency of our OPL model by applying it to an industrial case study representing a multi-threaded I/O driver with several instances running concurrently on a multi-core platform. Our approach successfully found scenarios violating the three performance requirements in a few minutes.

2 The Fire and Gas Monitoring System

The main motivation behind our work comes from a case study of a Fire and gas Monitoring System (FMS) in the maritime and energy domain. The goal of the system is to monitor potential gas leaks in oversea oil extraction platforms, and trigger an alarm in case a fire is detected. The system displays to human operators data coming from smoke and heat detectors, and gas flow sensors. When the system receives critical data from the hardware sensors, it automatically triggers actuators, such as fire sprinklers and audio/visual alarms. An older version of this case study was presented in our previous work [22], where we discussed the detailed design and modeling of the I/O drivers. The FMS software architecture is shown in Figure 1a.

Drivers implement I/O communication between the control modules of the system, and the external environment, such as hardware sensors, actuators, and human operators. In the FMS, thousands of instances of I/O drivers run concurrently interacting with several hundreds sensors. The software components of the FMS are executed on a Real-Time Operating System that runs on a tri-core computing platform.

Drivers in the FMS share the same design pattern, featuring six tasks that communicate through three buffers which have fixed capacity and cannot be simultaneously accessed by different tasks. Figure 1b shows the typical operational scenario, that is a

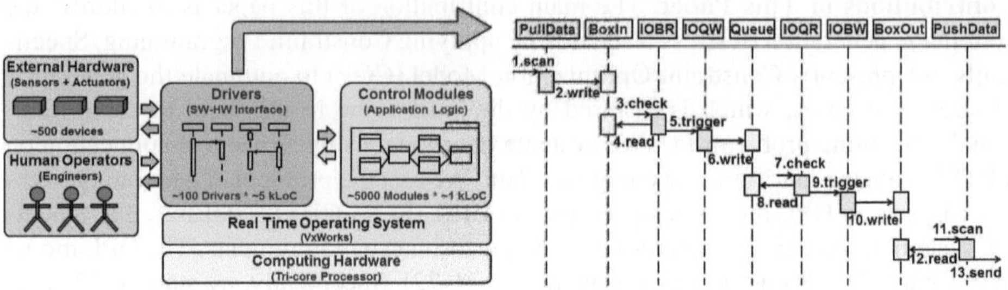

(a) Software architecture of the FMS (b) Typical drivers data transfer scenario

Fig. 1. Description of the Fire and gas Monitoring System

unidirectional data transfer between hardware sensors and control modules. (1) *PullData* periodically receives data from sensors or human operators, formats the data in an appropriate command form, and (2) writes it in the buffer *BoxIn*. (3) When *BoxIn* is full, the *check* signal activates *IOBoxRead* that (4) reads the data from the buffer and (5) triggers *IOQueueWrite*. *IOQueueWrite* extracts the commands from the data, and (6) stores them in the priority *Queue*. When *Queue* reaches a critical capacity, (7) the *check* signal activates *IOQueueRead* that (8) reads the highest priority command and (9) triggers *IOBoxWrite* which in turn (10) writes the command to *BoxOut*. When the periodic *scan* signal (11) activates *PushData*, the task (12) reads the commands from *BoxOut* and finally (13) sends them to the control modules for processing.

The data transfer functionality is subject to strict performance requirements. Specifically, in each FMS driver, (1) no task should miss its deadline, (2) the response time should be less than one second, and (3) the average CPU usage should be below 20%. The main variables determining whether or not these requirements will be satisfied at runtime are the arrival times of the *check* signal. These arrival times depend on the external environment, in the sense that depend on the data sent by the hardware sensors via *PullData*. The arrival times also vary across different system executions, as a consequence of the impossibility to predict the data coming from the sensors. Therefore, in order to evaluate task deadlines, response time, and CPU usage, we need a strategy to search all the possible task arrival times. This search has the objective of finding scenarios that are predicted to violate the requirements, or be close to violating them. Indeed, the more likely a scenario is predicted to violate a performance requirement, the higher the chances that the test case characterized by such scenario will stress the system.

3 Related Work

Testing multi-threaded concurrent software has largely focused on functional properties, rather then system performance [27]. Specific methods [12] for design-time performance analysis have been proposed to estimate the schedulability of a set of tasks through formulas and theorems from the Real-Time Schedulability Theory [26], or with model checking techniques [2]. In our experience, performance analysis is addressed in industry mainly with Performance Engineering, which extensively relies on profiling and benchmarking tools to dynamically analyze performance properties [16]. Such tools, however, are limited to producing a small number of system executions, and require their manual inspection. Performance analysis can check the overall sanity of the system performance, but cannot replace systematic stress testing.

For performance testing, search-based approaches have extensively been used [1], especially in the domain of distributed systems. Specifically, Genetic Algorithms (GA) have successfully been used to support performance testing, in particular with respect to QoS constraints [24] or computational resources consumption [7]. GA have also been used to generate test cases for testing tasks timeliness [23]. As for hard real-time properties such as deadline misses, the state-of-the-art is represented by the work of Briand et al. [8], that we used as a baseline for comparison in our previous work [11].

For schedulability analysis, CP approaches [4] have been studied for long time, especially in the domain of job-shop scheduling problems [19]. Among those, several approaches target task real-time constraints such as task deadlines [14], or timeliness [21]. Preemptive scheduling problems have also been solved both with pure CP [9], and with hybrid approaches featuring combinations with GA [28]. Furthermore, recent implementations [18] have successfully used IBM ILOG CP OPTIMIZER and OPL for scheduling problems, albeit not addressing task preemption.

Despite the extensive literature for constraint based scheduling, we are unaware of CP approaches targeted to test case generation, such as the generation of worst case scenarios, and of approaches addressing all the complexities of RETS such as multi-core architectures, task dependencies, aperiodic tasks, and preemptive scheduling policies.

4 Supporting Performance Testing: A New Application of COPs

We address the problem of determining worst-case schedules of tasks with an approach inspired by the work done in Constraint Programming to solve traditional scheduling problems [4]. Specifically, we cast the search for real-time properties that characterize the worst-case schedules, namely arrival times for aperiodic tasks, as a Constraint Optimization Problem (COP). The key idea behind our formulation relies on five main points. (1) First, we model the system design, which is static and known prior to the analysis, as a set of constants. The system design mainly consists of the tasks of the real-time application, their dependencies, period, duration, deadline, and priority. Constants of our model are described in Section 4.3. (2) Then, we model the system properties that depend on runtime behavior as a set of variables. The main real-time properties are the number of task executions, the arrival times of aperiodic tasks, and the specific runtime schedule of the tasks. (3) We model the RTOS scheduler as a set of constraints among such constants and variables. Indeed, the real-time scheduler periodically checks for triggering signals of tasks and determines whether tasks are ready to be executed or need to be preempted. (4) We model the performance requirement to be tested (i.e., task deadlines, response time, or CPU usage) as an objective function to be maximized. (5) Finally, we encapsulate the logic behind the RTOS scheduler in an effective labeling strategy over the variables of the model. By design, the scheduler tries to execute high priority tasks as soon as possible, potentially preempting tasks with lower priority.

We exploit this behavior by proposing a labeling strategy for the variables related to tasks execution. Our analysis is subject to two main assumptions:

1. The RTOS scheduler checks the running tasks for potential preemptions at regular and fixed intervals of time, called *time quanta*. Therefore, each time value in our problem is expressed as a multiple of a time quantum. Accordingly to the specification of the RTOS executing the FMS, we will consider the length of ten milliseconds for time quanta.
2. The interval of time in which the scheduler switches context between tasks is negligible compared to a time quantum.

These two assumptions are reasonable in the context of RTES, as the scheduling rate of operating systems varies in the ranges of few milliseconds, while the time needed for context switching is usually in the order of nanoseconds [25]. These assumptions allow us to consider time as discrete, and model the COP as an Integer Program (IP) over finite domains. We implemented the COP in OPL, and solved it with IBM ILOG CPLEX CP OPTIMIZER. This choice was motivated by practical reasons, such as extensive documentation, strong supporting community, and its acknowledged efficiency to solve optimization problems. Despite the scheduling nature of our problem, we implemented our model as a traditional IP as opposed to using the scheduling features of OPL and CP OPTIMIZER. This is because we could not express a preemptive priority-driven scheduling behavior in an effective way that exploited the capabilities of the solver.

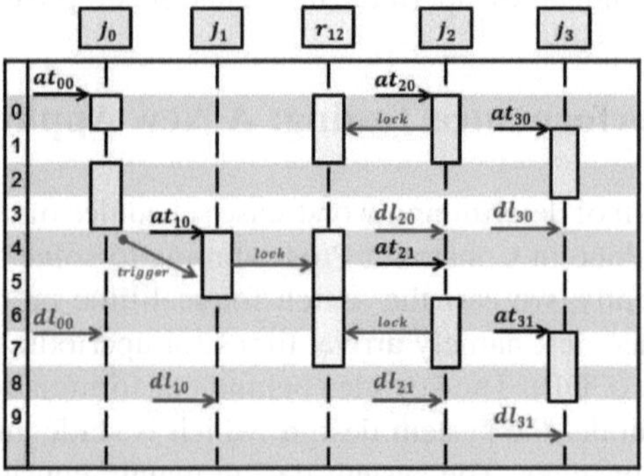

Fig. 2. Real-time scheduling example of four tasks on a dual-core platform

The rest of this section details our constraint model using the example shown in Figure 2. This system features four tasks in increasing priority order, j_0 to j_3, running on a dual-core platform for 10 time units. j_0 and j_1 are executed once, while j_2 and j_3 are executed twice. The figure reports the arrival times and deadlines of the tasks, respectively labeled by at and dl, where the first index represents the task, and the second the task execution. In this example, j_0 is aperiodic, while j_2 and j_3 are periodic.

Note that task j_1 is triggered by j_0 upon termination, and that j_1 and j_2 share the resource r_{12} with exclusive access.

4.1 Constants

Constants are implemented as integers (*int*), integer ranges (*range*), tuples (*tuple*), sets of tuples (*setOf*) and integer expressions. Integers values are defined as external data.

Observation Interval. Let T be an integer interval of length tq, i.e., $T \stackrel{\text{def}}{=} [0, tq - 1]$, representing the time interval during which we observe the system behavior. T is an integer interval, as a consequence that time is discretized in our analysis. Therefore, each time value $t \in T$ is a time quantum. In Figure 2, $tq = 10$ and $T = [0, 9]$.

Number of Platform Cores. Let c be the number of cores in the execution platform. Therefore, c represents the maximum number of tasks that can be executed in parallel. In Figure 2 $c = 2$, as at most two tasks are allowed to run in parallel.

Set J of Tasks. Let J be the set of tasks of the system. Each task $j \in J$ has a set of static properties, defined as constants, and a set of dynamic properties, defined as variables. Let J_p, J_a, and J_g respectively be the set of periodic, aperiodic, and triggered tasks of the system. J_p, J_a, and J_g define a partition over J. We assume that OS tasks have lower priority than system tasks and can be preempted at any time, and hence, can be abstracted away in our analysis. Each task j is implemented as an OPL tuple named *Task*, whose fields are the following non-relation constants. J is implemented as an OPL tuple set, while J_p, J_a, and J_g are OPL generic sets derived from J. In Figure 2, $J = \{j_0, j_1, j_2, j_3\}$, $J_a = \{j_0\}$, $J_p = \{j_2, j_3\}$, and $J_g = \{j_1\}$.

Priority of Tasks. Let $pr(j)$ be the priority of task j. For simplicity, we define the set HP_j of tasks having higher or equal priority than j: $HP_j \stackrel{\text{def}}{=} \{j_1 \in J \mid j \neq j_1 \wedge pr(j_1) \geq pr(j)\}$. In Figure 2, $pr(j_0) = 0, pr(j_1) = 1, pr(j_2) = 2$, and $pr(j_3) = 3$.

Period of Tasks. Let $pe(j)$ be the period of the task j, only defined if j is periodic. In Figure 2, $pe(j_2) = 5$ and $pe(j_3) = 6$.

Offset of Tasks. Let $of(j)$ be the offset of the task j, i.e., the time, counted from the beginning of T, after which the first period of task j begins. of is only defined if j is periodic. In Figure 2, $of(j_2) = 0$ and $of(j_3) = 1$.

Minimum and Maximum Inter-Arrival Times of Tasks. Let $mn(j)$ and $mx(j)$ respectively be the minimum and maximum inter-arrival times of task j, i.e., the minimum and maximum time separating two consecutive arrival times of j. $mn(j)$ and $mx(j)$ are only defined if j is aperiodic since for all periodic tasks j, $mn(j) = mx(j) = pe(j)$ trivially holds. In Figure 2, we assumed $mn(j_0) = 5$ and $mx(j_0) = 10$.

Duration of Tasks. Let $dr(j)$ be the estimated Worst Case Execution Time (WCET) of task j. For simplicity, we define the integer interval P_j of *execution slots* as $P_j \stackrel{\text{def}}{=} \big[0,\ dr(j) - 1\big]$. Since OPL does not support indexed ranges, P_j is implemented as a single range $P \stackrel{\text{def}}{=} [0, \max_{j \in J} dr(j) - 1]$. This definition entails that $\forall j \in J \cdot P_j \subseteq P$.

The iteration through values in P_j is emulated with a logic implication. Indeed, the following properties hold for every logic predicate C and arithmetic expression E:

$$\forall p \in P_j \ \cdot \ C(p) \iff \forall p \in P \ \cdot \ p < dr(j) \implies C(p) \tag{1}$$

$$\sum_{p \in P_j} E(p) = \sum_{p \in P} (p < dr(j)) \cdot E(p) \tag{2}$$

Note that in Equation 2 $(p < dr(j))$ is a boolean expression that is true if $p < dr(j)$, and false otherwise. For the rest of this paper, equalities and inequalities written within parentheses represent boolean expressions that evaluate to the integer 1 if true, and to the integer 0 if false. This is also the default behavior in CP OPTIMIZER. In Figure 2, $dr(j_0) = 3$ and $P_{j_0} = [0, 1, 2]$.

Deadline of Tasks. Let $dl(j)$ be the time, with respect to its arrival time, before which task j should terminate. In Figure 2, $dl(j_0) = 7$, $dl(j_1) = 6$, $dl(j_2) = 4$, and $dl(j_3)$=3.

Triggering Relation between Tasks. Let $tg(j_1, j_2)$ be a binary relation between tasks j_1 and j_2 that holds if the event triggering j_2 occurs when j_1 finishes its execution. The relation *triggers* is defined as irreflexive and antisymmetric. For simplicity, we respectively define the sets TS_j and ST_j of tasks triggered by and triggering j: $TS_j \stackrel{\text{def}}{=} \{j_1 \in J \mid tg(j, j_1)\}$ and $ST_j \stackrel{\text{def}}{=} \{j_1 \in J \mid tg(j_1, j)\}$. tg is implemented as an OPL tuple with two fields, the first being the task triggering, and the second being the task triggered. In Figure 2, $tg(j_0, j_1)$ holds.

Dependency Relation between Tasks. Let $de(j_1, j_2)$ be a binary relation between tasks j_1 and j_2 that holds if there exists a computational resource r such that j_1 and j_2 access r during their execution in an exclusive way. This implies that j_1 and j_2 cannot be executed in parallel nor can preempt each other, but one can execute only after the other has released the lock on the resource. The relation *dependent* is defined as reflexive and symmetric. For simplicity, we define the set DS_j of tasks depending on j: $DS_j \stackrel{\text{def}}{=} \{j_1 \in J \mid j \neq j_1 \wedge de(j_1, j)\}$. de is implemented as an OPL tuple with two fields, each being one of the task depending on the other. In Figure 2, $de(j_1, j_2)$ holds.

4.2 Variables

Independent variables in our model are implemented as OPL finite domain variables (*dvar int*). Dependent variables are implemented as OPL variable expressions (*dexpr int*) defined through equality constraints. The first three variables described hereafter, namely the number of task executions, their arrival times, and active sets, are independent variables. The remaining variables described in this section are all dependent.

Number of Task Executions. Let $te(j)$ be the number of times task j is executed within T. For simplicity, we define the integer interval K_j of task executions for the task j as $K_j \stackrel{\text{def}}{=} [0,\ te(j) - 1]$. Furthermore, we refer to the kth execution of task j as the couple (j, k). We assume the minimum and maximum inter-arrival times bound the number of executions of an aperiodic task. This means that, for aperiodic tasks,

$te(j)$ is defined as a variable with domain $\left[\left\lfloor \frac{tq}{mx(j)} \right\rfloor, \left\lfloor \frac{tq}{mn(j)} \right\rfloor\right]$. Similarly, we assume that offset and period statically determine the number of executions of periodic tasks so that $te(j) = \left\lfloor \frac{tq - of(j)}{pe(j)} \right\rfloor$. Therefore, the number of task executions of periodic tasks is constant, rather than variable. However, we do not formally distinguish it from the number of task execution for aperiodic tasks. te is implemented as an integer array ranging over J_p if the task is periodic (or ranging over J_g if triggered by a periodic task), and as an integer variables array ranging over J_a if the task is aperiodic (or ranging over J_g if triggered by an aperiodic task). Since OPL does not support ranges with variable bounds, K_j is implemented as a single constant range K:

$$K \stackrel{\text{def}}{=} \left[0, \max\left(\max_{j \in J_p} \left\lfloor \frac{tq - of(j)}{pe(j)} \right\rfloor, \max_{j \in J_a} \left\lfloor \frac{tq}{mn(j)} \right\rfloor\right)\right]$$

Note that K is defined as a range from 0 to the largest upperbound for task executions of periodic and aperiodic tasks. This definition entails that $\forall j \in J \cdot K_j \subseteq K$. The iteration through values in K_j is performed in a similar way as the case of P_j, thanks to the following properties for each logic predicate C and arithmetic expression E:

$$\forall k \in K_j \cdot C(k) \iff \forall k \in K \cdot k < te(j) \implies C(k) \tag{3}$$

$$\sum_{k \in K_j} E(k) = \sum_{k \in K} (k < te(j)) \cdot E(k) \tag{4}$$

In Figure 2 $te(j_0) = 1$, $te(j_3) = 2$, $K_{j_1} = [0]$, and $K_{j_2} = [0, 1]$.

Arrival Time of Task Executions. Let $at(j, k)$ be the time when an event notifies the RTOS that task j should be executed for the kth time. We say that j *arrives* for the kth time at time t iff $at(j, k) = t$. When the specific execution k of j is understandable from the context, we will simply say that j arrives at time t. In our analysis, we assume that the arrival time of periodic tasks is constant: $\forall j \in J_p,\ k \in K_j \cdot at(j, k) = of(j) + k \cdot pe(j)$. Similarly to the case of te, we do not formally distinguish the arrival times of periodic and aperiodic tasks. at has domain T for aperiodic tasks. In Figure 2, $at(j_0, 0) = 0$ and $at(j_2, 1) = 5$.

Active Set of Task Executions. Let $ac(j, k, p)$ be the pth time quantum in T in which task j is running for the kth execution. We refer to the set of all ac variables as the *schedule* produced by the arrival times of the tasks in J. ac variables have domain T. In Figure 2, $ac(j_0, 0, 0) = 0$, $ac(j_0, 0, 1) = 2$.

Preempted Set of Task Executions. Let $pm(j, k, p)$ be the number of time quanta for which the kth execution of task j is preempted for the pth time: $pm(j, k, p) \stackrel{\text{def}}{=} ac(j, k, p) - ac(j, k, p-1) - 1$. pm is only defined for $p > 0$. In Figure 2, $pm(j_0, 0, 1) = 1$, and $pm(j_0, 0, 2) = 0$.

Start and End Times of Task Executions. Let $st(j, k)$ and $en(j, k)$, respectively be the first and the one after the last time quantum in which task j is executing for the kth time. We say that j *starts* or *ends* for the kth time at time t iff respectively $st(j, k) = t$ or

$en(j,k) = t-1$. By definition, $st(j,k) \stackrel{\text{def}}{=} ac(j,k,0)$ and $en(j,k) \stackrel{\text{def}}{=} ac(j,k,dr(j)-1)+1$. In Figure 2, $st(j_0,0)=0$ and $en(j_1,0)=6$.

Waiting Time of Task Executions. Let $wt(j,k)$ be the time that j has to wait after its arrival time before starting its kth execution. By definition, $wt(j,k) \stackrel{\text{def}}{=} st(j,k) - at(j,k)$. In Figure 2, $wt(j_0,0)=0$, and $wt(j_2,1)=1$.

Deadline of Task Execution. Let $ed(j,k)$ be the absolute deadline of the kth execution of j, i.e., the time, with respect to the beginning of T, before which j should terminate to meet its deadline. By definition, $ed(j,k) \stackrel{\text{def}}{=} at(j,k) + dl(j) - 1$. ed is implemented as two-dimensional array of integer variable expressions ranging over the set J and the range K. In Figure 2, $ed(j_0,0)=6$, and $ed(j_1,0)=8$.

Deadline Miss of Task Execution. Let $dm(j,k)$ be the amount of time by which j missed its deadline during its kth execution. By definition, $dm(j,k) \stackrel{\text{def}}{=} en(j,k) - ed(j,k) - 1$. dm is implemented as two-dimensional array of integer variable expressions ranging over the set J and the range K. In Figure 2, $dm(j_0,0) = -3$.

Blocking Task Execution Time Quantum. Let $bl(j,k,j_1,k_1,p_1)$ be a boolean variable that is true if in the interval $\left[at(j,k), st(j,k)\right)$ the task execution (j_1,k_1) is active at the time slot p_1:

$$bl(j,k,j_1,k_1,p_1) \stackrel{\text{def}}{=} at(j,k) \leq ac(j_1,k_1,p_1) < st(j,k)$$

In Figure 2, $bl(j_2,1,j_1,0,1) = true$, since $(j_2,0)$ waits at $t=5$ for the last time quantum of $(j_1,0)$ before starting.

Higher Priority Active Tasks. Let $ha(j,k)$ be the number of time quanta in the interval $\left[at(j,k), st(j,k)\right)$ where exactly c tasks having higher priority of j and not depending on j are active. Consider the summation indexes j_1, k_1, p_1 respectively defined in the sets $HP_j \setminus DS_j$, K_{j_1}, and P_{j_1}, and the summation indexes j_2, k_2, p_2 respectively defined in the sets $HP_j \setminus DS_j$, K_{j_2}, and P_{j_2}. We define:

$$ha(j,k) \stackrel{\text{def}}{=} \sum_{j_1,k_1,p_1} \left(bl(j,k,j_1,k_1,p_1) \wedge \left(\Big(\sum_{j_2,k_2,p_2} bl(j,k,j_2,k_2,p_2) \Big) = c \right) \right)$$

Note that for the definition of $ha(j,k)$, it is important that HP_j also includes tasks with equal priority than j. This is because, in the RTOS scheduling policy we consider, tasks can only preempt tasks with strictly lower priority. In Figure 2, $ha(j,k)=0$ for all task executions (j,k), since in no case there are 2 tasks active when a task is waiting.

Dependent Active Tasks. Let $da(j,k)$ be the number of time quanta in the interval $\left[at(j,k), st(j,k)\right)$ where task executions depending on j is active. Consider the summation indexes j_1, k_1, p_1 respectively defined in the sets DS_j, K_{j_1}, and P_{j_1}:

$$da(j,k) \stackrel{\text{def}}{=} \sum_{j_1,k_1,p_1} bl(j,k,j_1,k_1,p_1)$$

In Figure 2, $da(j_2,1)=1$, because j_1 is active for the time quantum $t=5$ between the arrival and the start of j_2.

Dependent Preempted Tasks. Let $dp(j,k)$ be the number of time quanta in the interval $\big[at(j,k), st(j,k)\big)$ where task executions depending on j have been preempted. Consider the summation indexes j_1, k_1, p_1 respectively defined in the sets DS_j, K_{j_1}, and P_{j_1}. Then, we define

$$dp(j,k) \stackrel{\text{def}}{=} \sum_{j_1,k_1,p_1} pm(j_1,k_1,p_1) \cdot bl(j,k,j_1,k_1,p_1)$$

In Figure 2, $dp(j,k) = 0$ for all task executions (j,k), since there are no dependent task preempted that block the execution of any task.

System Load. Let $ld(t)$ be the load of the system at time t, i.e., the number of tasks active at time t. Consider the summation indexes j, k, p respectively defined in the sets J, K_j, and P_j. Then we define

$$ld(t) \stackrel{\text{def}}{=} \sum_{j,k,p} \big(ac(j,k,p) = t\big)$$

In Figure 2, $ld(0) = 2$, and $ld(3) = 1$.

4.3 Constraints

We define five groups constraints related to different aspects of the RTOS.

Well-formedness Constraints, specifying relations among variables that directly follow from their definition in the schedulability theory.

Each task execution starts after its arrival time, and ends after the task duration.

$$\forall j \in J,\ k \in K_j \ \cdot \ at(j,k) \le st(j,k) \le end(j,k) - dr(j) \qquad \text{(WF1)}$$

Arrival times of aperiodic tasks are separated by their minimum and maximum interarrival times.

$$\forall j \in J_a,\ k \in K_j \setminus \{0\} \ \cdot \ at(j,k-1) + mn(j) \le at(j,k) \le at(j,k-1) + mx(j) \qquad \text{(WF2)}$$

The time indexes $p \in P_j$ define an order over the active time quanta of tasks.

$$\forall j \in J,\ k \in K_j,\ p \in P_j \setminus \{0\} \ \cdot \ ac(j,k,p-1) < ac(j,k,p) \qquad \text{(WF3)}$$

Temporal Ordering Constraints, specifying the relative ordering of tasks basing on their dependency and triggering relations.

Each triggered task is executed the same number of times of its triggering task.

$$\forall j_1 \in J,\ j_2 \in TS_j \ \cdot \ te(j_1) = te(j_2) \qquad \text{(TO1)}$$

Each triggered task execution arrives when its triggering task execution ends.

$$\forall j_1 \in J,\ k \in K_{j_1}\ j_2 \in TS_j \ \cdot \ end(j_1,k) = at(j_2,k) \qquad \text{(TO2)}$$

Executions of dependent tasks cannot overlap, i.e., one task can only start after the one it depends on has ended.

$$\forall j_1 \in J,\ k_1 \in K_{j_1}\ j_2 \in DS_j,\ k_2 \in K_{j_2} \cdot en(j_1, k_1) \leq st(j_2, k_2) \vee en(j_2, k_2) \leq st(j_1, k_1) \tag{TO3}$$

If two tasks that depend on each other arrive at the same time, the higher priority task executes first.

$$\forall j_1 \in J,\ k_1 \in K_{j_1},\ j_2 \in \big(DS_j \cap (J \setminus HP_j)\big),\ k_2 \in K_{j_2} \cdot at(j_1, k_1) = at(j_2, k_2) \implies st(j_1, k_1) < st(j_2, k_2) \tag{TO4}$$

Multi-core Constraint, capturing the specification of the number c of cores of the computing platform, and stating that no more than c tasks are allowed to be active in parallel at any time.

The system load should be less than the number of cores at any time.

$$\forall t \in T \cdot ld(t) \leq c \tag{MC}$$

Note that, when $c = 1$, MC is equivalent to an *alldifferent* constraint over ac.

Preemption Constraint, capturing the priority-driven preemptive scheduling of the RTOS, and stating that each task should be preempted when a higher priority task is ready to be executed and no cores are available.

The number of time quanta where a task execution is preempted times c is equal to the number of time quanta where higher priority tasks are active. Considering the summation indexes j_1, k_1, p_1 respectively defined in the sets HP_j, K_{j_1}, and P_{j_1},

$$\forall j \in J,\ k \in K_j,\ p \in P_j \cdot pm(j, k, p) \cdot c = \sum_{j_1, k_1, p_1} \big(ac(j, k, p-1) < ac(j_1, k_1, p_1) < ac(j, k, p)\big) \tag{PC}$$

Scheduling Efficiency Constraint, ensuring that there is no unnecessary task preemption, and that tasks are executed as soon as possible.

For each time quanta in which a task execution (j, k) is waiting, there should be either (1) exactly c tasks with higher priority that do not depend on j active, or (2) one task execution dependent on j that is active, or (3) one task execution dependent on j that is preempted.

$$\forall j \in J,\ k \in K_j \cdot wt(j, k) = ha(j, k) + da(j, k) + dp(j, k) \tag{SE}$$

4.4 Objective Functions

We formalized three objective functions, each modeling one performance requirement, and each meant to be maximized in a separate constraint model having the same constants, variables, and constraints. In this way, solutions to each of the three constraint models characterize worst-case scenarios for the requirement modeled by the function.

Task Deadline Misses Function, modeling the performance requirement involving task deadlines with the function F_{DM}:

$$F_{DM} = \sum_{j \in J,\ k \in K_j} 2^{\,dm(j,k)}$$

To properly reward scenarios with deadline misses, F_{DM} assigns an exponential contribution to deadline misses towards the sum [11]. Recall from Section 4.2 that $dm(j,k)$ is positive if the task execution (j,k) misses its deadline, and negative otherwise.

Response Time Function, modeling the system response time with the function F_{RT}:

$$F_{RT} = \left(\max_{j \in J,\ k \in K_j} en(j,k)\right) - \left(\min_{j \in J,\ k \in K_j} at(j,k)\right)$$

F_{RT} measures the total length in time quanta of the schedule, starting from when the first task arrives, up to when the last ends. This function is also known in traditional scheduling as *makespan*.

CPU Usage Function, modeling the system CPU usage with the function F_{CU}:

$$F_{CU} = \frac{\sum_{t \in T} \big(ld(t) > 0\big)}{tq}$$

F_{CU} measures the average CPU usage of the system over T, by counting all the time quanta where at least one task is active, i.e., where the system load is greater than 0.

4.5 Search Heuristic

We defined a search heuristic that refines the branching process of the CP OPTIMIZER solving algorithm. The heuristic specifies that the solver should mimic the behavior of a RTOS by first trying to schedule tasks with higher priority. This is done by choosing the ac variables to branch on by decreasing priority, and then by assigning their time values in increasing order. For example, consider a system where $c = 1,\ j_0, j_1 \in J,\ pr(j_1) > pr(j_0)$. Suppose that, for given k_0, p_0, k_1, j_1 the filtering algorithm reduced the domains of the ac variables to the set $[0,1]$. Figure 3a shows the branching tree in case the solver runs with default settings.

In the root node, the ac variables have domain $[0,1]$. The solver then tries the first variable assignment in the branch b_1, stating that j_0 is executing at time 0. Then, the solver tries the second assignment in the branch b_2, stating that j_1 is executing at time 0. This variable assignment violates the multi-core constraint MC since both j_0 and j_1 are executing at the same time. Therefore, the solver prunes the node, backtracks to the father node, and tries the assignment in b_3 where j_1 is executing at time 1. This assignment violates the preemptive scheduling constraint PC, since j_1 has higher priority, but j_0 is running instead. Only after backtracking up to the root node, the solver tries the

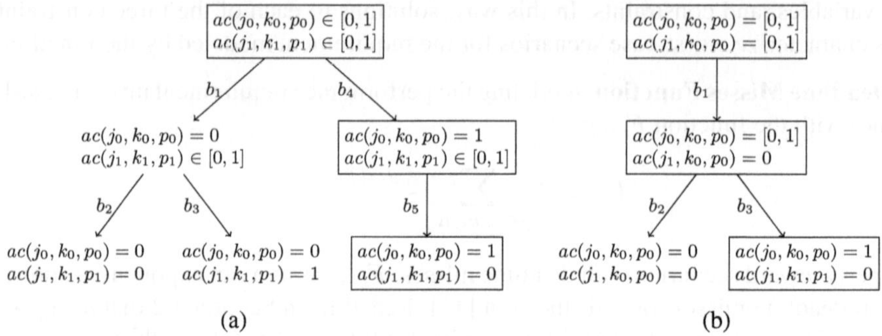

Fig. 3. Branch and bound backtracking without (a) and with (b) our search heuristic

assignments in b_4 and b_5 which do not violate any constraint. Note that several other branching steps might have been necessary if $ac(j_1, k_1, p_1)$ had a larger domain.

Consider Figure 3b, where the solver has been instructed to first branch by assigning the smallest value in its domain to the ac variable associated with the highest priority task. In this case, the solver tries the first assignment $ac(j_1, k_1, p_1) = 0$ in the branch b_1. Then, it tries the second assignment in the branch b_2, that violates MC. However, the third assignment in b_3 does not violate any constraint, making the solver perform only one backtracking step.

The semantics of this heuristic, i.e., highest priority tasks should be scheduled first, is the same as the semantics of the RTOS scheduler, which in turn is captured by preemptive scheduling constraint. By using this concept in the branching process, the solver will be less likely to assign values for ac that violate the preemptive scheduling constraint, and thus will find solutions faster. We implemented the search heuristic within a stand-alone application that solves the OPL model using the .NET CONCERT library to interface with the CP OPTIMIZER. Experimentation with our search heuristic showed a significant decrease in the time needed by the solver to find solutions [11].

5 Industrial Experience

Context and Process. The work reported in this paper originates from the interaction over the years with Kongsberg Maritime (KM)[1], a leading company in the maritime and energy field. KM has pressing needs to improve its practices related to safety certification, and this involves improving the validation of performance requirements. Therefore, we proposed the work reported in this paper to provide support for systematic performance testing.

Through regular meetings with KM engineers, we first identified the need for a model-based testing approach defining the abstractions required for performance analysis [22]. Then, we casted such analysis as an optimization problem over a mathematical model of the tasks preemptive scheduling policy. To prepare for industrial adoption,

[1] http://www.km.kongsberg.com

we initially evaluated our methodology in five publicly available case studies of several RTES domains [11]. This preliminary evaluation showed encouraging results when comparing our approach with a state-of-the-art optimization strategy based on Genetic Algorithms [8]. In this paper, we provide an improved constraint model and evaluate it in the context of the KM Fire and gas Monitoring System.

Results. The main goal of our evaluation is to investigate whether CP can effectively be used for performance testing in an industrial context. For our approach to be used in practice, we need to discuss (1) whether the input data to our approach, i.e., the values for the constants in the constraint model, can be provided with reasonable effort, and (2) whether engineers can use the output of our analysis, i.e., the values for the variables in the constraint model, to derive stress test cases for different performance requirements. We discussed the first question in our previous work [22]. Specifically, we demonstrated that the effort to capture the input data for our approach was approximately 25 man-hours of effort. This was considered worthwhile as such drivers have a long lifetime and are certified regularly. We discuss the second question below.

Recall from Section 2 that we characterize stress test cases by arrival times of aperiodic tasks in the FMS drivers. Therefore, such arrival times are the main variables in our constraint model. We performed an experiment with the FMS drivers with an observation interval T of five seconds, assuming time quanta of 10 ms. We run our OPL model for three times on a single Amazon EC2 m2.xlarge instance [1]. Each run maximized one objective function defined in Section 4.4, and had a duration of five hours. Figure 4 shows the feasible solutions with the best objective value that were found within five hours. Consistent with the terminology used in Integer Programming, we refer to these solutions as *incumbents* [3]. In each graph, the x-axis reports the incumbent computation times in the format *hh:mm:ss*, and the y-axis reports the corresponding objective value. The constraint problems had almost 600 variables and more than one million constraints, and used up to 10 GB RAM during resolution.

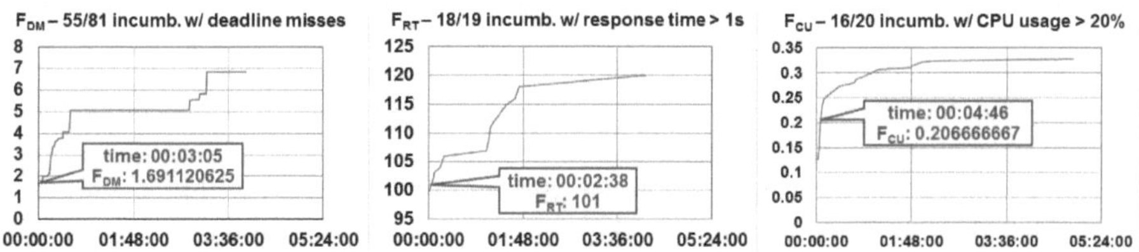

(a) F_{DM} value over time (b) F_{RT} value over time (c) F_{CU} value over time

Fig. 4. Objective values over time, where we highlighted the time when the first incumbent predicted to violate a performance requirement was found

Note that, for practical use, software testing has to accommodate time and budget constraints. It is then essential to investigate the trade-off between the time needed to

[1] `http://aws.amazon.com`

generate test cases, and their revealing power for violations of performance requirements. For this reason, we also recorded the computation times of the first incumbents predicted to violate the three performance requirements as expressed in Section 2.

The run optimizing F_{DM} is shown in Figure 4a. The solver found 55 out of a total of 81 incumbents with at least one deadline miss in their schedule; the first of such solutions was found after three minutes. The solution yielding the best value for F_{DM} produced a schedule where the *PushData* task missed its deadline by 10 ms in three executions over T. Figure 4b shows the results for the run optimizing F_{RT}. The solver found 18 out of 19 incumbents with response time higher than one second; the first of such solutions was found after two minutes. The best solution with respect to F_{RT} produced a schedule where the response time of the system was 1.2 seconds. Finally, the solutions found by optimizing F_{CU} are shown in Figure 4c. The solver found 16 out of 20 incumbents with CPU usage above 20%; the first of such solutions was found after four minutes. The solution with the highest value for F_{CU} produced a schedule where the CPU usage of the system was 32%. In all of the three runs the solver terminated after five hours, our time budget, without completing the search with proof of optimality. However, for each objective function, the solver was able to find, within few minutes, solutions that are candidates to stress test the system as they may lead to requirements violations. These solutions can be used to start testing the system while the search continues, because the highest the objective value, the more likely the solutions are to push the system to violating its performance requirements.

6 Conclusions and Future Work

Currently, KM engineers spend several days simulating the behavior of the FMS and monitoring its performance requirements. We expect that, by following the systematic strategy proposed in this paper, they can effectively derive stress test cases to produce satisfactory evidence that no safety risks are posed by violating performance requirements at runtime. We note that our methodology draws on context factors (Section 4) that need to be ascertained prior to successful application. While the generalizability of these factors needs to be further studied, we have found the factors to be commonplace in many industry sectors relying on RTES. Furthermore, we note how casting the worst-case scenario analysis as a search problem relies on modeling the property to stress test as an objective function to be maximized. This is a flexible design when it comes to adapting the constraint model to test different performance requirements. In such cases, it is only needed to substitute the objective function with one modeling a difference performance requirement. Moreover, the final users of our approach, i.e., software testers and engineers, do not need to be aware of the mathematical details of the constraint model, as they can simply use our methodology as a black box test cases generator. Finally, note that there is no randomization process in the search: this means that solving a model multiple times will always find the same set of solutions in a given time budget. To achieve diversity among the test cases, we plan to consider hybrid approaches combining CP with meta-heuristic search strategies. As future work, we also plan to further investigate the scheduling capabilities of CP OPTIMIZER and OPL.

Acknowledgments. We gratefully thank Arnaud Malapert and Jean-Charles Régin for their invaluable input in refining the constraint model. The first and fourth authors acknowledge funding from the Research Council of Norway (ModelFusion project and Certus). The second and the third authors are supported by the Luxembourg National Research Fund (FNR/P10/03 Validation and Verification Laboratory).

References

1. Afzal, W., Torkar, R., Feldt, R.: A systematic review of search-based testing for non-functional system properties. Information and Software Technology 51(6), 957–976 (2009)
2. Alur, R., Courcoubetis, C., Dill, D.: Model-checking for real-time systems. In: Proceedings of the Fifth Annual IEEE Symposium on Logic in Computer Science, LICS 1990, pp. 414–425. IEEE (1990)
3. Atamtürk, A., Savelsbergh, M.W.: Integer-programming software systems. Annals of Operations Research 140(1), 67–124 (2005)
4. Baptiste, P., Le Pape, C., Nuijten, W.: Constraint-based scheduling: applying constraint programming to scheduling problems, vol. 39. Springer (2001)
5. Beizer, B.: Software testing techniques. Dreamtech Press (2002)
6. Bell, R.: Introduction to IEC 61508. In: Proceedings of the 10th Australian Workshop on Safety Critical Systems and Software, vol. 55, pp. 3–12. Australian Computer Society, Inc. (2006)
7. Berndt, D.J., Watkins, A.: High volume software testing using genetic algorithms. In: Proceedings of the 38th Annual Hawaii International Conference on System Sciences 2005. IEEE (2005)
8. Briand, L.C., Labiche, Y., Shousha, M.: Using genetic algorithms for early schedulability analysis and stress testing in real-time systems. Genetic Programming and Evolvable Machines 7(2), 145–170 (2006)
9. Cambazard, H., Hladik, P.E., Déplanche, A.M., Jussien, N., Trinquet, Y.: Decomposition and learning for a hard real time task allocation problem. In: Wallace, M. (ed.) CP 2004. LNCS, vol. 3258, pp. 153–167. Springer, Heidelberg (2004)
10. Di Alesio, S., Gotlieb, A., Nejati, S., Briand, L.: Testing deadline misses for real-time systems using constraint optimization techniques. In: 2012 IEEE Fifth International Conference on Software Testing, Verification and Validation (ICST), pp. 764–769. IEEE (2012)
11. Di Alesio, S., Nejati, S., Briand, L., Gotlieb, A.: Stress testing of task deadlines: A constraint programming approach. In: 2013 IEEE 24th International Symposium on Software Reliability Engineering (ISSRE), pp. 158–167. IEEE (2013)
12. Gomaa, H.: Designing concurrent, distributed, and real-time applications with UML. In: Proceedings of the 28th International Conference on Software Engineering, pp. 1059–1060. ACM (2006)
13. Henzinger, T.A., Sifakis, J.: The embedded systems design challenge. In: Misra, J., Nipkow, T., Sekerinski, E. (eds.) FM 2006. LNCS, vol. 4085, pp. 1–15. Springer, Heidelberg (2006)
14. Hladik, P.E., Cambazard, H., Déplanche, A.M., Jussien, N.: Solving a real-time allocation problem with constraint programming. Journal of Systems and Software 81(1), 132–149 (2008)
15. Jackson, D., Thomas, M., Millett, L.I., et al.: Software for Dependable Systems: Sufficient Evidence? National Academies Press (2007)
16. Jain, R.: The art of computer systems performance analysis. John Wiley & Sons (2008)
17. Kopetz, H.: Real-time systems: design principles for distributed embedded applications. Springer (2011)

18. Laborie, P.: IBM ILOG CP Optimizer for detailed scheduling illustrated on three problems. In: van Hoeve, W.-J., Hooker, J.N. (eds.) CPAIOR 2009. LNCS, vol. 5547, pp. 148–162. Springer, Heidelberg (2009)
19. Le Pape, C., Baptiste, P.: An experimental comparison of constraint-based algorithms for the preemptive job shop scheduling problem. In: Smolka, G. (ed.) CP 1997. LNCS, vol. 1330, Springer, Heidelberg (1997)
20. Lee, E.A., Seshia, S.A.: Introduction to embedded systems: A cyber-physical systems approach. Lee & Seshia (2011)
21. Malapert, A., Cambazard, H., Guéret, C., Jussien, N., Langevin, A., Rousseau, L.M.: An optimal constraint programming approach to the open-shop problem. INFORMS Journal on Computing 24(2), 228–244 (2012)
22. Nejati, S., Di Alesio, S., Sabetzadeh, M., Briand, L.: Modeling and analysis of CPU usage in safety-critical embedded systems to support stress testing. In: France, R.B., Kazmeier, J., Breu, R., Atkinson, C. (eds.) MODELS 2012. LNCS, vol. 7590, pp. 759–775. Springer, Heidelberg (2012)
23. Nilsson, R., Offutt, J., Mellin, J.: Test case generation for mutation-based testing of timeliness. Electronic Notes in Theoretical Computer Science 164(4), 97–114 (2006)
24. Shams, M., Krishnamurthy, D., Far, B.: A model-based approach for testing the performance of web applications. In: Proceedings of the 3rd International Workshop on Software Quality Assurance, pp. 54–61. ACM (2006)
25. Singh, A.: Identifying Malicious Code Through Reverse Engineering. Springer (2009)
26. Tindell, K., Clark, J.: Holistic schedulability analysis for distributed hard real-time systems. Microprocessing and Microprogramming 40(2), 117–134 (1994)
27. Weyuker, E.J., Vokolos, F.I.: Experience with performance testing of software systems: issues, an approach, and case study. IEEE Transactions on Software Engineering 26(12), 1147–1156 (2000)
28. Yun, Y.S., Gen, M.: Advanced scheduling problem using constraint programming techniques in SCM environment. Computers & Industrial Engineering 43(1), 213–229 (2002)

Continuous Casting Scheduling with Constraint Programming

Steven Gay[1], Pierre Schaus[1], and Vivian De Smedt[2]

[1] Université Catholique de Louvain, Belgium
[2] PSI Metals, Belgium

Abstract. Although the Steel Mill Slab problem (prob 38 of CSPLib) has already been studied by the CP community, this approach is unfortunately not used anymore by steel producers since last century. Continuous casting is preferred instead, allowing higher throughput and better steel quality. This paper presents a CP model related to scheduling of operations for steel making with continuous casting. Activities considered range from the extraction of iron in the furnace to its casting in continuous casters. We describe the problem, detail a CP scheduling model that is finally used to solve real-life instances of some of the PSI Metals' customers.

Keywords: Continuous Casting, Steel Production, Scheduling, Constraint Programming.

1 Introduction

Steel Production problems have already been tackled with CP. In particular the Steel Mill Slab problem (prob 38 of CSPLib) has been studied in [4,5,14,7]. Although interesting from a theoretical point of view, this problem is of limited practical interest since this technique has been replaced since 1950's by continuous casting.

A schematic representation of continuous casting is given on Fig. 1. A ladle is poured into the tundish, a reservoir of hot metal, to feed the casting machine. The strand (solidifying metal, output of the casting machine) passes through straightening rolls before being cut into predetermined lengths by mechanical shears. We refer to [16] for more information related to steel production in general.

The continuous casting depicted on Fig. 1 is the last of a three step process, the first two steps being the carbon removal and the steel refining. The goal of the continuous casting scheduling problem (CCSP) is to determine the timing of operations on molten steel at the three steps of production and the device/facility where they happen, when some flexibility is allowed.

There are two reasons for using computer optimization for scheduling operations related to continuous casting. The first one is the cost of the machines involved in the process. They are so costly that they should effectively be used 24 hours per day. The cost of a modern steel making facility is on the order of 10 billion euros. A single continuous caster costs on the order of half a billion euros.

B. O'Sullivan (Ed.): CP 2014, LNCS 8656, pp. 831–845, 2014.

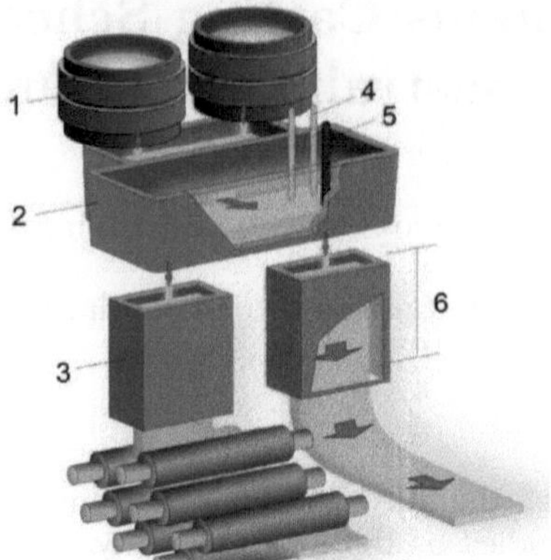

Fig. 1. 1: Ladle. 2: Tundish. 3: Mold. 4: Plasma torch. 5: Stopper. 6: Straight zone (Image from [19]).

Since the lifetime of a caster is around 30 years of lifetime, which is around 45 thousand euros per day. Thus, one of the main objective is to ensure continuity of the production process at the caster. The second reason is satisfiability. Many instances are very constrained, some constraints compete and make it difficult to solve the problem manually or using heuristics. The problem requires typically to plan operations up to 36 hours which means about 500 activities to schedule.

PSI Metals develops software for decision making in steel production. This work aims at developping flexible and maintainable tools to deal with continuous casting in the software suite of PSI Metals.

Outline. Section 2 describes the CCSP problem in detail. Section 3 motivates the usage of CP for solving the problem. Section 4 details the CP model, before experimenting the model in Section 5 on real-life instances.

2 Description of the Problem

The problem is a scheduling problem presenting some similarities with the job-shop problem in the sense that some activities must be processed sequentially by different *machines*. Each job is a *heat*, which is a pocket of molten metal, that undergoes several transformations ; each transformation may be modeled as an activity with start date, duration and end. However, application-specific requirements makes this problem different from a pure job-shop problem.

Continuous Casting as a Scheduling Problem. A *heat job* is composed of activities that must take place in a fixed order, as represented in Fig. 2.

The first activity of a heat is its creation in a *converter*, either a blast oxygen furnace that processes iron ore or an electric furnace that melts recycled scrap. At this end of this activity, metal is tapped in liquid state into a ladle. The metal

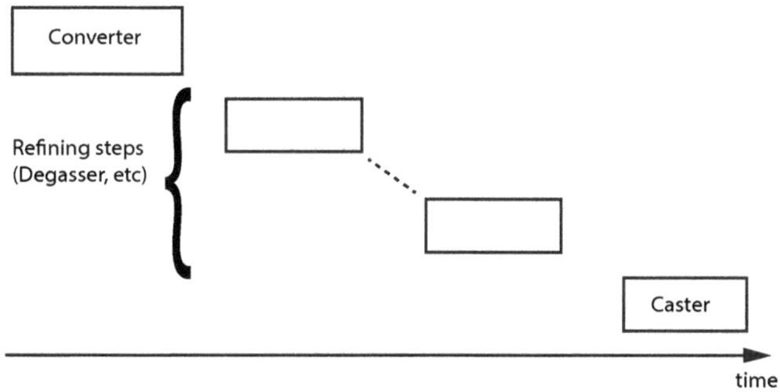

Fig. 2. A heat job: the converter activity must happen first, then the refining activities (degasser, stew, transports, etc), and finally the caster activity

contents of a ladle is a *heat*, it embodies a discrete unit of molten metal. Note that this discretization into heat/ladles comes from a physical reality.

Then the heat is treated in various facilities, to undergo treatments that will modify the metal's final properties. These treatments and the transports between the facilities can be described as activities of the heat, with start date, duration, end date, and associated resource. For the plant considered in our experiments, there are two intermediate treatments, alloying (addition of external components) and vacuum degassing (removal of excess oxygen blown by the conversion process).

Finally a heat is tapped into a continuous caster, to be transformed into solid metal. This is the last activity of a heat and the end of the problem under consideration.

Application-Specific requirements. Although the problem presents some similarities with the job-shop problem, there are some constraints specific to the CCSP. The first is the temperature requirement, depicted in Fig. 3. From its generation in the converter facility to its final transformation in the caster, a heat *has* to remain in liquid state. Even though there are no release times nor deadlines specified, there is a maximum time a heat can spend in the relatively cold atmosphere before solidifying. To avoid this situation, a maximum duration is specified between the exit of the converter and the entry in the caster. Note that it is not possible to raise the initial temperature at the converter artificially to ensure arrival at the caster in liquid state, since the metal could boil away and the containing equipment could be damaged. Moreover, heating metal has an energy cost.

The second is the *continuity* requirement, also depicted in Fig. 3. Continuous casting is a technique where to make n shapes that are effectively truncated cylinders (such as I-beams, sheets ...), a single cylinder is cast in a continuous process in a first stage, and cut into the required lengths in the following stage, on-the-fly. The advantages are better metal quality and cheaper processing costs,

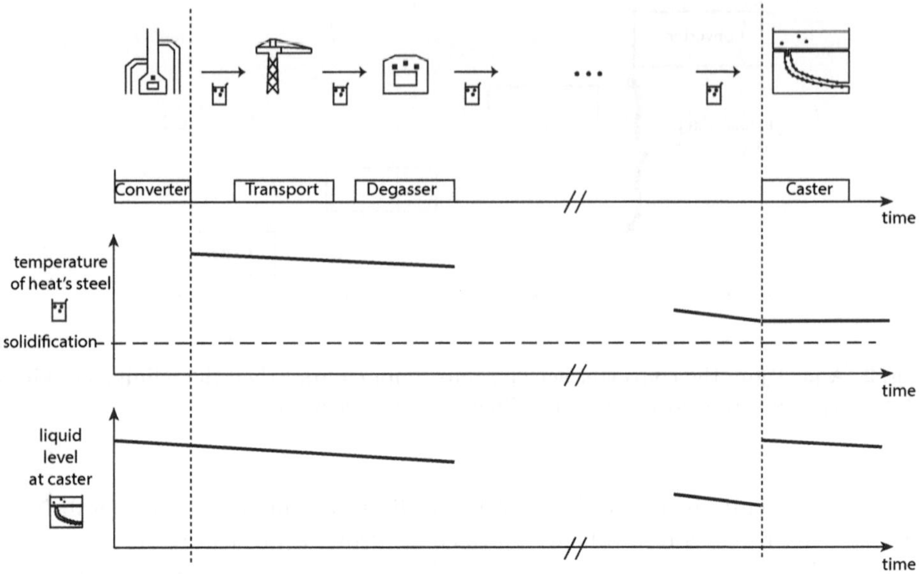

Fig. 3. As a heat undergoes treatments in different facilities, it cools down and the level of the molten metal in the caster waiting for it decreases

but the drawback is that during the first stage, the caster can not be stopped. This translates into additional constraints for the scheduling: a *program* is a set of heats that will be used for the continuous casting of one object. The heats of a program have thus to be treated consecutively at the caster (without any gap between corresponding activities). This is shown in the bottom part of Fig. 3 Adjacent heats may arrive on-time, early, but not late. Typically, the allocation of programs to the casters and the order of programs at each caster is received from a previous computation in the planning and must remain unchanged.

Example 1. A heat has to undergo several transformations before being cast in a solid shape, as shown in Fig. 3. The top part of this diagram shows a heat going through different facilities along the timeline, from conversion to casting. The middle part shows the temperature of the metal inside the heat, which starts decreasing as soon as the heat leaves the converter. Then, the temperature starts decreasing, and the heat must undergo all its treatments and arrive at the caster before it turns solid. Temperature is only relevant between these two events. The bottom part shows the level of the liquid metal at the caster as it is waiting for the heat to arrive: if the caster is in the middle of a program, then the heat must arrive before the liquid level reaches 0.

The last typical variation on job-shop is the running time production, where some activities have already been scheduled. Keeping in mind that steel making is a 24 hours per day operation, in practice, there are always fixed activities in the schedule: when adding activities required by a command from a customer, the schedule is not remade from scratch ; instead, some activities that may be happening while the algorithm is re-computing a new schedule are fixed.

Example 2. This example is depicted on Fig. 4. There are two programs represented with gray and white activities. There are two casters, the gray program must be scheduled on caster 1 and the white program on caster 2. The converter and the degasser can each process two heats in parallel. Each activity has duration 1 to simplify the example. There is a maximum delay of 2 from exit of converter to entry into caster to avoid solidifying metal. Notice how heat number 3 of the gray program is scheduled at the last possible moment to avoid solidifying. On the right, an unexpected event occurred: heat 2 of the white program took more time than expected at caster. Unfortunately, heat 3 of the white program was started at the converter and heat 1 is already being treated at caster 1, so no matter the schedule, heat 3 of the gray program can never arrive on time at the caster.

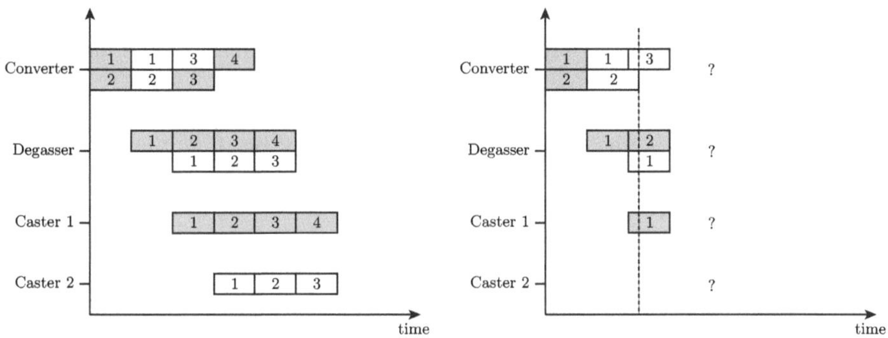

Fig. 4. Representation of Example 2

Instance-Specific requirements There may be additional constraints depending on the plant and its casting facilities (converter, degasser, etc).

Although facilities may be cumulative (able to treat several heats in parallel) with fixed capacity, some might require that sub-parts of the activities do not overlap. In particular, referring to Fig. 5, the converter activity of a heat is made of three internal continuous parts: Loading, Converting and Tapping. The converter has two chambers and can be seen as a resource with capacity 2. However, there is only one oxygen blower, so that the actual conversion part can not overlap the conversion part of another. While the loading and tapping parts may still overlap with the same conditions as a resource of capacity 2, the non overlapping constraint induced by the oxygen blower prevents the two activities of chamber 2 to be scheduled consecutively: some gap must be introduced to prevent conversion parts from overlapping.

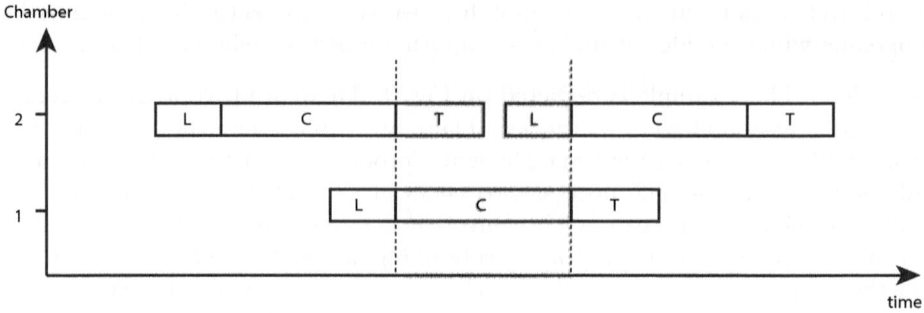

Fig. 5. Illustration of the disjunctive aspect for sub-part of the converter activities

Generally a program can be treated only in a specific caster, but some activities might have alternatives, i.e. they can be treated indifferently in one facility or another (say they might use vacuum degasser number 1 or 2 indifferently). However, another activity may require vacuum degasser number 1 only, distinguishing the two facilities and preventing us from using a single cumulative constraint for both degassers.

Some facilities might be unavailable for a fixed interval of time for maintenance; in flexible cases, the maintenance activity may be schedulable. Furthermore, casters require a setup time in-between programs.

An incident may force an activity to take longer than scheduled, or may break down some facility. In such an event, activities that have not yet happened have to be re-scheduled just for feasibility, since temperature and continuous casting constraints may become violated, see Example 2. In the event where a schedule is infeasible due to a continuity violation at a caster, a program may be split in two parts, even though it will probably interfere with the subsequent operations on the slab (typically the subsequent cutting will not be able to stop at a whole number of smaller cylinders). If the temperature constraint of a heat will be violated, the liquid metal can be re-heated in some facilities or even recycled in the converter. In these exceptional cases, the schedule has to be heavily penalized; we do not take the possibility of splitting programs into account here.

3 Using CP for Caster Scheduling

In order to satisfy clients at reasonable development costs, PSI needs a framework that enables the coding of algorithms that yield good feasible solutions at industrial scales while requiring little tuning to minimize the development time. The framework should also be expressive enough to allow direct encoding of side constraints. We believe CP offers these advantages thanks to its high level declarative modeling. Furthermore, CP has proved competitive for scheduling applications, mainly due to effective filtering algorithms for global scheduling constraints [18,10] ; Large

Neighborhood Search (LNS) [15] makes it scalable [8,12,3,11]. Consequently, we believe CP is a good candidate for solving the CCSP.

3.1 Two Sources of Difficulty

Two other problem-specific arguments support the choice for CP on this problem:

1. The continuous casting problem is very constrained, making it sometimes difficult to find a feasible solution.
2. Bottleneck resources are instance dependent, making it difficult to build generic heuristics.

These two points are discussed next.

A very constrained problem. There is a basic strategy to ensure continuity of heats of a same program at the caster: make heats ahead of time, then deliver stored heats at the caster. The major obstacle to this strategy is the temperature constraint, that prevents us from creating heats too long in advance. Recall that the temperature constraint is ensured by allowing a maximum delay from the exit of the converter to the entry at the caster, representing a maximum initial temperature. This tension between the desire to produce in advance to ensure continuity and the just-in time aspect induced by the temperature constraint may cause some instances to be unsatisfiable or at best difficult to solve.

Instance-dependent bottlenecks. As in most scheduling problems, some resources behave more or less like a bottleneck in the schedule. An interesting aspect of the continuous casting scheduling problem is that the bottleneck resource(s) are very instance-dependent. Let us approximate the metal as a continuous flow going through facilities instead of being discretized in heats. Then one can think of facilities as having a continuous throughput, say in tonne/minute; each step of the transformation process is taken care of by one or several facilities, amounting to a total throughput of the step. The step with the smallest throughput is a bottleneck that limits the maximum throughput of the whole factory. From experience, this approximation still holds when the liquid metal is discretized as heats (remind that this is a physical constraint due to the metal being contained, moved and treated in ladles). Identifying the bottleneck facility/resource is a valuable indicator in order to construct good feasible solutions. Indeed, those resources will be used at maximal throughput in a good solution, and a partial instantiation that satisfies the resource constraint of the bottleneck facility can most likely be extended into a complete feasible schedule. The key factors impacting bottleneck resources are the wide range of parameters possible in each step of the process. For instance, a customer may want a cylinder of small diameter, and thus the throughput at the corresponding caster will be decreased enough to move the bottleneck from another step to the casting step. If the customer wants a higher quality metal, a step corresponding to some secondary treatment will take longer and have a smaller throughput, making it the new bottleneck.

3.2 Alternative/Existing Approaches

Although we were not able to find exactly the same problem description in the literature, the problem described in [1] presents many similarities with our CCSP: although the central problem is similar, the objective function and the modeling of facilities are different.

The authors decompose the resolution in two phases. First an Ant Colony Optimization (ACO) metaheuristic is used to sequence the jobs using MATLAB. Then a second phase assigns starting time with CONOPT non-linear optimization solver. In [17], the authors describe a decomposition approach for solving a similar problem combining Lagrangian relaxation, dynamic programming and heuristics.

Although similar local search, decomposition and heuristics have been engineered in the past for other problems, PSI finds it requires much manpower for tuning heuristics and designing moves. Recall that the CCSP problem is strongly constrained, it is thus difficult to design efficient feasible moves manually. LS was also evaluated as a difficult approach to maintain on this CCSP because requirements change according to both the specificities of each production plant and the typical set of programs to schedule.

MILP requires time discretization, and the logical constraints used lead to rather weak linear relaxations. The current solution developed by PSI uses MIP but it required a lot of simplifications (such as time bucketing) and tuning, just to come up with solvable instances. Moreover, these simplifications generally lead to sub-optimal final solutions, when compared to solutions found by CP.

4 Modeling Using CP

Our model for the CCSP uses standard resource constraints for scheduling with additional side constraints where needed mainly to impose offsets, set-up times and precedence constraints between activities.

4.1 Model Overview

The activities to schedule are structured hierarchically in heats, programs and casters: an activity belongs to a heat, a heat to a program, and a program to a caster.

- $\forall c$ caster, the activities of programs $p_1, \ldots, p_{n_c}$ at the caster must be scheduled in a predefined order, separated by a setup time of the caster.
- $\forall p$ program, the activities of heats $h_1, \ldots, h_{n_p}$ of p happening at the caster must be scheduled in listed order, and be contiguous.
- $\forall h$ heat, activities $a_1, \ldots, a_{n_h}$ of the heat must be scheduled in listed order.

Every activity a has a fixed duration d_a, a set of resources R_a it can be scheduled on, and a delay t_a that modelizes the temperature constraint. Every resource has an associated capacity, ranging from 0 to $+\infty$. The demand of our activities is always 1.

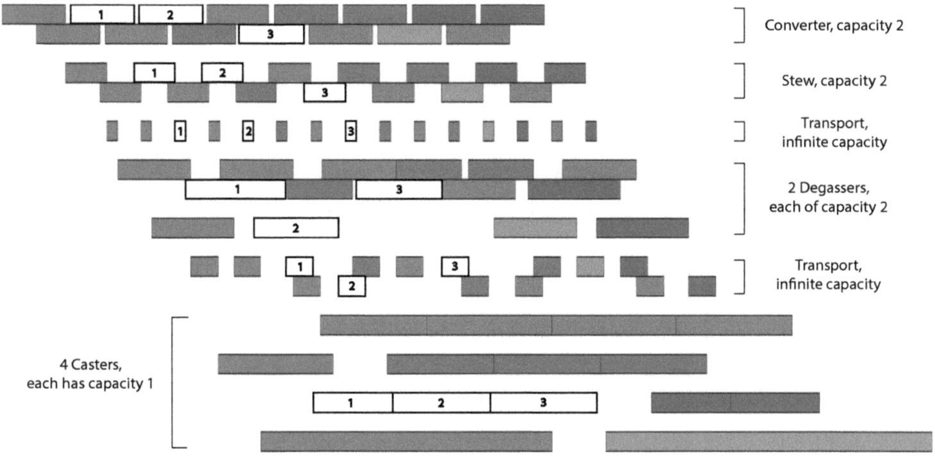

Fig. 6. A small example of continuous casting schedule

Example 3. Fig. 6 represents a schedule respecting all given constraints. Every resource has a capacity limit, e.g. here 1 for the casters, 2 for the converter, $+\infty$ for the transports, etc. Activities of the same program have the same color, we single out the white program in this example. Caster order constrains the white program to be scheduled before the blue one. Program order forces the activities of the white program at the caster to be consecutive. Heat order constrains the order of activities of the same heat, here the converter activity of white-1 must happen before its stew activity, which must happen before its first transport, etc. Temperature constrains the gap between the converter and caster activities to be smaller than some value, given as an input.

4.2 CP Model

The pure model uses conventional scheduling constraints for disjunctive and cumulative resources (see [2] for more information on scheduling constraints). The temperature constraint is modeled with a fixed time delay depending on the heat.

Variables. We write the decision variables in boldface. They are, for each activity a, start $\mathbf{s}_a$, duration $\mathbf{d}_a$, end $\mathbf{e}_a$, and resource $\mathbf{r}_a$. The set of all activities equipped with their decision variables is written $\mathbf{A}$. Durations are fixed, the initial domain for starts and ends is $\mathbb{N}$, and resources are initialized according to user specifications, for instance, it is expected, for a resource variable of activity a at caster c, that the initial domain of $\mathbf{r}_a$ is $\{c\}$.

Constraints. There are two types of constraints in this problem, resource limitations and time dependencies:

- Resource constraints. For every resource r, depending on its capacity $C(r)$, we add a constraint :
  - $C(r) = 0$: the resource is not available, add the constraints $\forall a \in A, \mathbf{r}_a \neq r$
  - $C(r) = 1$: **disjunctive**$(\mathbf{A})$
  - $C(r) > 1$: **cumulative**$(\mathbf{A}, C(r))$
  - $C(r) = +\infty$: the resource does not constrain the problem, no constraint added
- Time Dependencies. We write $h(c)$ the activity of heat h at caster c, and $\mathtt{setup}_c$ the setup time that must separate programs at caster c.
  - Caster order. $\forall c$ caster, $\forall p_i, p_{i+1}$ consecutive programs of c with $p_i = h_1 \ldots h_n$ and $p_{i+1} = h'_1 \ldots h'_{n'}$, $\mathbf{e}_{h_n(c)} + \mathtt{setup}_c \leq \mathbf{s}_{h'_1(c)}$.
  - Program order. $\forall p$ program, $\forall h_j, h_{j+1}$ consecutive heats of p, $\mathbf{e}_{h_j(c)} = \mathbf{s}_{h_{j+1}(c)}$.
  - Heat order. $\forall h$ heat, $\forall a_k, a_{k+1}$ consecutive activities of h, $\mathbf{e}_{a_k} \leq \mathbf{s}_{a_{k+1}}$.
  - Temperature. $\forall h = a_1 \ldots a_n$, $\mathbf{e}_{a_1} + delay_h \geq \mathbf{s}_{a_n}$

Objective. The goal of the optimization here is to minimize the sum of the completion times at casters, so if $\mathtt{last}(c)$ is the last activity at the caster c, the objective value to minimize is $\sum_c \mathbf{e}_{\mathtt{last}(c)}$. This forces the resources to be free earlier, to deal with new customer demands that are unknown at scheduling time: a makespan objective would only be relevant if all jobs were known in advance.

Search Heuristic. The problem sizes are so large (more than 500 activities) that there is little hope to complete the search on real instances. Designing custom heuristic and search strategies is thus crucial for CCSP. An important aspect for PSI customers is the anytime behavior of the application. The operator should have a solution available quickly, and the application can not afford to return no solution. To this end, we designed a search heuristic behaving similarly to a greedy manual schedule construction minimizing the risk of any backtrack before reaching the first feasible solution. Recall that one important constraint is that activities of a same program at a caster should be contiguous. We order the heats (jobs) in a static order to ensure the feasibility of the continuity property. This order on the jobs is obtained as follows (see Fig. 7):

1. Caster activities are placed without any gap between them (unfortunately this schedule for caster activities is surely not feasible because of capacity limitations at the converter, degasser, etc).
2. Programs are numbered according to their earliest starting time (see numbers 1 to 7 on top of the programs on Fig. 7)).
3. Heats are numbered increasingly inside a program by numbering first programs with lowest starting time (see numbers 1 to 18 in each heat on Fig. 7)).

Then for a given heat (job) activities are scheduled facility by facility, starting at the converter and finishing with the caster activity, in their program order. Because the domains are huge (duration are specified in seconds) a binary split

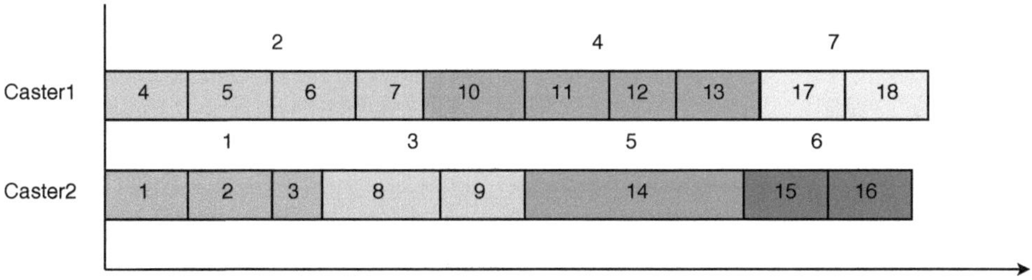

Fig. 7. Job ordering strategy obtained by first ordering programs, then heats inside each programs. Heats in a same program have the same color.

search is used instead of a labeling search. This static search is randomized by applying some random modifications on the activity durations when creating the program ordering.

Although the previous strategy is very good to obtain quickly feasible solutions, it is too conservative to obtains high quality values for the sum of completion times. Indeed, this strategy may introduce large gaps between the programs at the casters.

We use a Large Neighborhood Search (LNS) to improve the incumbent solution. Some structure of the current best solution is kept by restricting a randomized partial order schedule between the programs at the caster (inspired from [6]). At each LNS restart we randomly choose between two searches:

- The safe search with static (randomized) ordering, or
- A schedule or postpone search (also called SetTimes) described in [9] on all the activities of the problem.

SetTimes is a search that tries to assign early activities at their earliest starting time (EST), postponing on backtrack the scheduling of activities that failed until their EST changes: it only considers them again when their EST change from propagation. This search exploits dominance properties that do not hold for our problem. Indeed, it may be interesting in our case to postpone activities on the converter to satisfy the itemperature constraints. Thus SetTimes is not necessarily complete on our problem. However, this search proves very useful in practice to minimize the sum of completion times when it finds a feasible solution.

5 Experiments

Table 1 shows the results obtained for 48 CCSP instances. Those instances were generated as variants of 2 real-life instances coming from a customer of PSI. The horizon varies from 12 to 48 hours of operations to schedule, where 36 hours is a good enough time horizon for the industrial setting.

The number of activities given in the table does not take into account subpart activities, in this particular instance 3 resources need subparts, our modeling

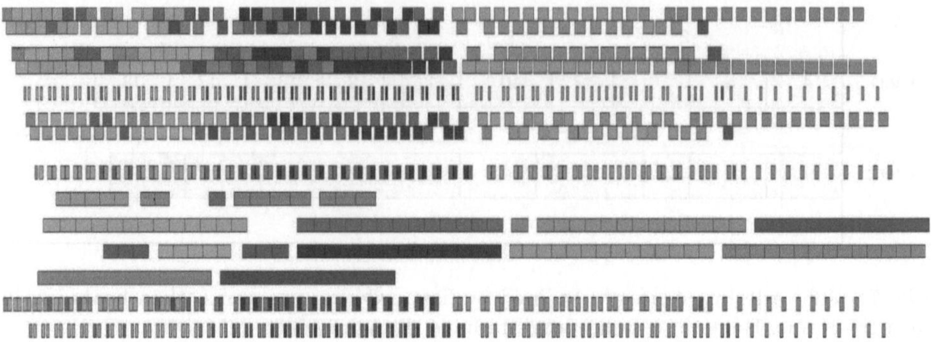

Fig. 8. A real-life schedule

codes them as additional activities ; the number of activities the solver actually has to handle is roughly the number in the table times 1.5.

Durations of activities were artificially changed to make the bottleneck resource change, allowing us to test the robustness of the search. In table 1, "step" is the bottleneck resource (Converter or STEW), and "extension" is a fixed number of seconds added to the duration of each activity on this resource, making the step slower and more bottleneck-like.

Example 4. Fig. 8 shows an example of a real-life schedule produced by our model. The two bottom lines are subpart activities (as explained on Fig. 5). This schedule is well optimized since activities are very well packed on the bottleneck resources. For the first half of the schedule, the second facility is the bottleneck, then the casters become the bottleneck.

A time-out of 30 seconds is given to the algorithm[1] and we compare the final objective function (sum of completion times) using different searches:

- A "Safe" search as the one described above diving quickly toward feasible solution.
- A SetTimes search on the whole set of activities.
- A LNS search using a partial order schedule relaxation and alternating between the two previous searches on each LNS restart.

The LNS search obtains the best objective most of the time on 34/48 instances. The SetTimes search obtains the best solutions on 25/48 instances but is not able to find any feasible solution for 10/48 instances. As expected the "safe" search is always able to find a feasible solution, but the quality of the objective is generally worse compared to the two previous searches. When the LNS search is not the optimal one, it is reasonably close to the SetTimes results and it has the main advantage to always produce a feasible solution. This search is thus the most robust and should be preferred for the final deployment of the application.

[1] 30s was chosen as a realistic wait an operator can allow, 300s would be unrealistic.

Table 1. Results on 48 CCSP instances testing 3 different search strategies with a timeout of 30 seconds

					Safe	SetTimes	LNS
instance	limit	activities	step	extension	objective	objective	objective
Instance1	12	234	Converter	600	274620	**250740**	**250740**
				900	308280	270060	**265200**
				1200	342720	**298140**	**298140**
			STEW	600	253980	222600	**221640**
				900	255180	223800	**221640**
				1200	282300		**221640**
	24	436	Converter	600	493560	440760	**437760**
				900	570720	**474480**	**474480**
				1200	630420	515940	**513900**
			STEW	600	448440	**378060**	382441
				900	449640	**379260**	383344
				1200	482100		**424500**
	36	598	Converter	600	588660	540720	**536641**
				900	718560	**610440**	**610440**
				1200	760920	658860	**652920**
			STEW	600	539820	**469440**	475800
				900	541020	**470640**	477181
				1200	**573480**		574140
	48	736	Converter	600	658260	610680	**610441**
				900	776940	**676380**	**676380**
				1200	861900	759660	**757981**
			STEW	600	614580	**553020**	559080
				900	615780	**554160**	559980
				1200	649200		**589140**
Instance2	12	240	Converter	600	294258	**268228**	268583
				900	318596	284783	**279656**
				1200	358828	296848	**290128**
			STEW	600	266398	**243448**	**243448**
				900	271672	**251478**	**251478**
				1200	282533	**264348**	**264348**
	24	420	Converter	600	427218	**421996**	**421996**
				900	547138		**462369**
				1200	609786	**482013**	492857
			STEW	600	388098	**344727**	347307
				900	386397	**352757**	**352757**
				1200	414408		**374121**
	36	576	Converter	600	514408	509266	**475486**
				900	641972		**579460**
				1200	740546	**687380**	691115
			STEW	600	475288	**431917**	436177
				900	473587	**439947**	**439947**
				1200	501598		**461311**
	48	666	Converter	600	595276	590134	**557004**
				900	723262		**580207**
				1200	863598	**782701**	873943
			STEW	600	556156	**512876**	516145
				900	554455	**520815**	**520815**
				1200	582466		**542179**

6 Conclusion and Future Work

We have described the continuous casting scheduling problem (CCSP) for steel production and introduced a complete CP model and search strategy using LNS for solving this problem with good any-time behavior. Although the application is still in the prototyping phase, the high quality results obtained using CP and the ease of maintaining and modifying the CP model should allow to deploy it in the near future at some of PSI's customers.

As a future work, we would like to experiment the Variable Objective Large Neighborhood Search (VO-LNS) introduced in [13]. We believe VO-LNS might be a good strategy to focus on a different restricted number of casters at each LNS restart. We may also try to optimize activities according to a rolling horizon scheduling strategy. This should allow us to consider a limited number of activities, fix some of the earliest scheduled ones before sliding the horizon and considering a few more activities. Finally we believe dominances and/or redundant constraints may be inferred to improve the filtering on this problem.

References

1. Atighehchian, A., Bijari, M., Tarkesh, H.: A novel hybrid algorithm for scheduling steel-making continuous casting production. Computers & Operations Research 36(8), 2450–2461 (2009)
2. Baptiste, P., Pape, C.L., Nuijten, W.: Constraint-based scheduling: applying constraint programming to scheduling problems, vol. 39. Springer (2001)
3. Carchrae, T., Beck, C.: Principles for the design of large neighborhood search. Journal of Mathematical Modelling and Algorithms 8(3), 245–270 (2009)
4. Frisch, A.M., Miguel, I., Walsh, T.: Modelling a steel mill slab design problem. In: Proceedings of the IJCAI 2001 Workshop on Modelling and Solving Problems with Constraints, Citeseer (2001)
5. Gargani, A., Refalo, P.: An efficient model and strategy for the steel mill slab design problem. In: Bessière, C. (ed.) CP 2007. LNCS, vol. 4741, pp. 77–89. Springer, Heidelberg (2007)
6. Godard, D., Laborie, P., Nuijten, W.: Randomized large neighborhood search for cumulative scheduling. In: ICAPS, vol. 5, pp. 81–89 (2005)
7. Heinz, S., Schlechte, T., Stephan, R., Winkler, M.: Solving steel mill slab design problems. Constraints 17(1), 39–50 (2012)
8. Laborie, P., Godard, D.: Self-adapting large neighborhood search: Application to single-mode scheduling problems. In: Proceedings MISTA 2007, Paris, pp. 276–284 (2007)
9. Pape, C.L., Couronné, P., Vergamini, D., Gosselin, V.: Time-versus-capacity compromises in project scheduling. In: Proceedings of the Thirteenth Workshop of the UK Planning Special Interest Group (1994)
10. Letort, A., Carlsson, M., Beldiceanu, N.: A synchronized sweep algorithm for the k-dimensional cumulative constraint. In: Gomes, C., Sellmann, M. (eds.) CPAIOR 2013. LNCS, vol. 7874, pp. 144–159. Springer, Heidelberg (2013)
11. Monette, J.-N., Deville, Y., Hentenryck, P.V.: Just-in-time scheduling with constraint programming. In: ICAPS (2009)

12. Pacino, D., Van Hentenryck, P.: Large neighborhood search and adaptive randomized decompositions for flexible jobshop scheduling. In: Proceedings of the Twenty-Second International Joint Conference on Artificial Intelligence, vol. 3, pp. 1997–2002. AAAI Press (2011)
13. Schaus, P.: Variable objective large neighborhood search: A practical approach to solve over-constrained problems. In: IEEE International Conference on Tools with Artificial Intelligence (ICTAI 2013) (2013)
14. Schaus, P., Hentenryck, P.V., Monette, J.-N., Coffrin, C., Michel, L., Deville, Y.: Solving steel mill slab problems with constraint-based techniques: Cp, lns, and cbls. Constraints 16(2), 125–147 (2011)
15. Shaw, P.: Using constraint programming and local search methods to solve vehicle routing problems. In: Maher, M.J., Puget, J.-F. (eds.) CP 1998. LNCS, vol. 1520, pp. 417–431. Springer, Heidelberg (1998)
16. steel.org. The online resource for steel (accessed: April 20, 2014)
17. Tang, L., Luh, P.B., Liu, J., Fang, L.: Steel-making process scheduling using lagrangian relaxation. International Journal of Production Research 40(1), 55–70 (2002)
18. Vilım, P.: Global constraints in scheduling. PhD thesis, PhD thesis, Charles University in Prague, Faculty of Mathematics and Physics, Department of Theoretical Computer Science and Mathematical Logic, KTIML MFF, Universita Karlova, Malostranské námestı 2/25, 118 00 Praha 1, Czech Republic (2007)
19. Wikipedia. Continuous casting — Wikipedia, the free encyclopedia (2013) (accessed: April 20, 2014)

Case Study: Constraint Programming in a System Level Synthesis Framework

Shuo Li and Ahmed Hemani

Department of Electronic Systems
School of Information and Communication Technology
Royal Institute of Technology
Isafjördsgatan 39, 16440, Stockholm, Sweden
{shuol,hemani}@kth.se

Abstract. This article presents a case study of using a constraint programming solver in a system level synthesis framework called SYLVA. The solver is used to find the repetition vector of a synchronous data flow graph and serving as the design space exploration engine, which rapidly finds qualified system implementations by solving a constraint satisfaction optimization problem. Each system implementation is a combination of a number of function implementation instances and their cycle accurate execution schedules. The problem to be solved is automatically generated based on the user inputs: 1) a system model to be synthesized, 2) a library containing all the usable function implementations, 3) the performance/cost constraints, and 4) the optimization objectives. Use of constraints programming technique enabled a low cost development of design space exploration engine in addition to gaining ease of use.

Keywords: System Level Synthesis, Design Space Exploration, Constraint Programming.

1 System Level Architectural Synthesis (SYLVA)

System level hardware synthesis is an evolutionary next step after the high-level synthesis. It synthesizes abstract Digital Signal Processing (DSP) sub-systems like modems and codecs, e.g. WLAN, LTE mode, etc. modeled as Synchronous Data Flow (SDF) graphs [1] in terms of pre-characterized Function Implementations (FIMPs). SYLVA [2] is a system level hardware synthesis framework under development in our group. Currently, SYLVA only supports acyclic SDF. It explores the design space in terms of a) number and type of FIMPs, b) number of buffers for each FIMP, and c) pipeline parallelism among them. It also automatically generates the global interconnect and control logic to glue the FIMPs and buffers together into a working system. The design flow based on SYLVA has four steps of which the first two are the focus of this paper. 1) SDF to HSDF conversion 2) Design Space Exploration (DSE) 3) Global interconnect and control synthesis, and 4) Code generation

The key difference with the existing work on CP methods for scheduling tasks on processors, e.g. [3], is that of hardware synthesis vs. software compilation.

B. O'Sullivan (Ed.): CP 2014, LNCS 8656, pp. 846–861, 2014.

Processors can host multiple tasks that are scheduled. In the hardware synthesis case (SYLVA), a FIMP (corresponding to a processor) is a dedicated hardware implementation for a specific function like FFT and only one instance of it can be executed at a time. The scheduling in the context of SYLVA is the relative ordering of the HSDF nodes that are mapped to FIMP instances. The relative order decides the number of FIMPs but not the type. The type selection has been formulated as a CP problem.

Another key difference is that in software compilation, the critical path is known because the arithmetic parallelism of each task is fixed because the processor hardware is fixed. Whereas in case of SYLVA, critical path is not fixed. It depends on the FIMP types that the CP tool evaluates as part of the design space exploration. FIMP types vary in the degree of arithmetic parallelism and can change the total number of cycles that a particular path takes.

1.1 SDF to HSDF Conversion

In this step, the system modeled as SDF graph is converted into a Homogeneous SDF (HSDF) graph [1].

An SDF graph is a directed graph $S = \{A, E\}$. Each vertex is called an actor $a \in A$ that represents a DSP function. The number of data tokens produced or consumed by each actor on each invocation is specified a priori. Each communicational edge $e \in E$ represents a data dependency between two actors. By this definition, two communicating actors a_s (the source) and a_d (the destination) may have different producing and consuming data rates (data token per invocation). Therefore, a_s and a_d have to be invoked at different frequencies to match the data rate of each other. An example SDF graph with four actors is shown in Fig. 1(a), where a, b, c, and d are names of four DSP functions and the data rates are not matched at all.

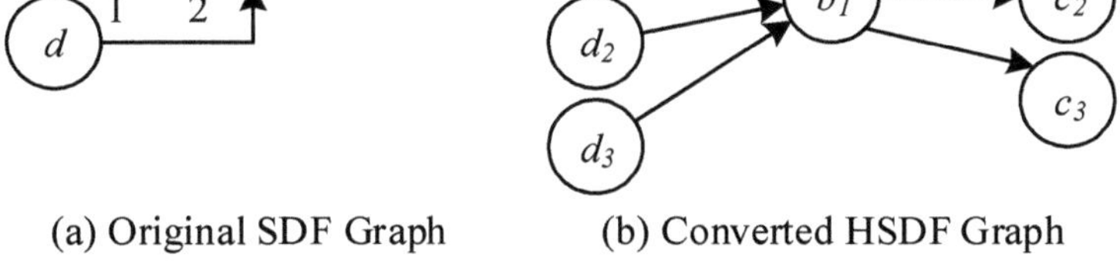

Fig. 1. SDF to HSDF Conversion

An HSDF graph in this article is an SDF graph that all the data rates for all data producer and consumer actor pairs are matched as formulated in 1, where s_e and d_e are the source and destination actors for edge e, respectively.

$$s_e = d_e \quad \forall e \in E \tag{1}$$

Each HSDF actor represents an execution of a DSP function, while each SDF actor represents a DSP function. And the SDF to HSDF conversion will increase the number of nodes. For example, the SDF graph shown in Fig. 1(a) can be converted into the HSDF graph with 11 actors shown in Fig. 1(b), where a_0 is an execution of a, $\{b_0, b_1\}$ are two executions of b, $\{c_0, c_1, c_2, c_3\}$ are four executions of c and, $\{d_0, d_1, d_2, d_3\}$ are four executions of d.

SDF to HSDF Conversion can be done by finding the null space of the topology matrix [1]. In SYLVA, we compute the null space by solving a simple Constraint Satisfaction Problem (CSP). The number of variables equals to the number of actors ($|A|$) in the original SDF graph. The possible values of each variable are set to integers from 1 to $2^{32} - 1$. Although the number of possible values is large, experience shows that only a few branches are enough to get the result. For all the experiments in this article, one branch is enough to get the number of SDF actor repetitions and the solver runtime is similar to computing the null space using C# program. The details of the CSP modeling and solving can be found in section 3.

1.2 Design Space Exploration (DSE)

After the first step, we have a number of function executions (HSDF actors) to be implemented by hardware (Function Implementations - FIMPs, which will be explained later in this subsection). The second step is to find the optimal system implementation (solution) in a reasonable time. Each solution specifies the type and number of FIMPs, the number of buffers for each FIMP, and the cycle accurate schedules for them.

The design space can be quite large, since for each HSDF actor, we need to determine following parameters: 1) the FIMP instance to execute it (multiple HSDF actors may share the same FIMP instance in a time-multiplexing fashion) 2) the buffer usage (if each function execution has its own output buffer or not), and 3) the cycle accurate execution schedule (when to start function execution to achieve the desired parallelism) The total number of solutions is the size of the design space $|D|$ and it can be calculated by 2.

$$|D| = \prod_{i=0}^{|A|-1} (2M_i \cdot (T_{max} - t_i \cdot f_i)). \tag{2}$$

$|A|$ is the number of SDF actor. M_i is the number of possible FIMPs for the ith SDF actor (a_i). The term $2M_i$ represents the decision of having output buffer for each HSDF actor or each FIMP instance. T_{max} is the maximum system latency constraint and t_i is a horizontal vector that contains the execution times for all the FIMPs for a_i. The jth element in t_i is the execution time of the jth FIMP for a_i. f_i is a vertical vector that represents the FIMP assignment decision. The values of each element in f_i can be only 1 or 0 and only one element equals 1 ($\forall i \quad f_i \in \{0, 1\}$ and $\sum f_i = 1$). The jth element represents the decision of using the jth FIMP (1) or not (0) to implement a_i. The term $T_{max} - t_i \cdot f_i$ represents the number of possible schedules. In most of the cases, we have more freedom on the schedule side and less freedom on the FIMP type selection side.

If $|A| = 10$, $M_i = 5$ for all actors, $T_{max} = 100$ and all FIMPs have the same execution time of 10 clock cycles ($\forall i \quad t_i \cdot f_i \equiv 10$), the total number of solutions $|D|$ will be $(2 * 5 * (100 - 10))^{10} = 900^{10} = 3.49 \times 10^{29}$. In this example, the schedule term $100 - 10$ contributes much more than the FIMP type selection term $2 * 5$. The Constraint Satisfaction Optimization Problem (CSOP) model will be explained in detail in section 3.

FIMP represents an implementation of a DSP function in terms of three views: *interface view*, *execution view* and *implementation view*. In the rest of this article, a FIMP in a library is called a *FIMP* while a FIMP instantiated for executing one or more HSDF actors is called a *FIMP instance*.

The interface view provides the unique function name, and the input/output data structure of the FIMP. The unique function name defines the name of the DSP function that can be executed by the FIMP. For example, if a FIMP has name = FFT_64, this FIMP can only be used to execute 64-point FFT function, i.e. the HSDF actors derived from the SDF actor whose name = FFT_64. The input and output data structure provides the number of producing/consuming data tokens on each invocation. Three data structure examples with different degrees of parallelism are shown in Fig. 2. All of them has 64 data tokens. Each data token is a 32-bit complex numbers (16-bit fixed point for real and imaginary components). In Fig. 2(a), all 64 numbers are input simultaneously (fully parallel). In Fig. 2(b), all 64 numbers are input one by one (fully sequential). In Fig. 2(c), all 64 numbers are input four by four (partially parallel). At present, two communicating FIMP instances, source and destination, should share same data structure or an explicit data structure conversion should be applied.

The execution view provides a) the function execution timing model (Fig. 3), b) the energy consumption in nJ (nanojoule), and c) the area usage in equivalent gate count. The timing model and the energy consumption are for single execution. In Fig. 3, t_s is the FIMP start time. t_{ie} is the input end time. All input data should be read up by the FIMP till t_{ie}. t_{os} is the output start time. From t_{os}, the FIMP starts sending the output data. t_e is the FIMP end time. Everything is done for one execution till t_e. t_{ie}, t_{os} and t_e are relative times referring to t_s. All of these times are in number of clock cycles and used in the CSOP model.

The implementation view provides a) the implementation style (e.g. ASIC, FPGA, or a specified CGRA) and b) the implementation template (e.g. VHDL code, CGRA configware) for the computation and the output buffer.

(a)	(b)	(c)
{16, 16} ⋮ {16, 16} {16, 16}	{16, 16} {16, 16} ... {16, 16}	{16, 16} {16, 16} ... {16, 16} {16, 16} {16, 16} ... {16, 16} {16, 16} {16, 16} ... {16, 16} {16, 16} {16, 16} ... {16, 16}

Fig. 2. Data Structure Examples

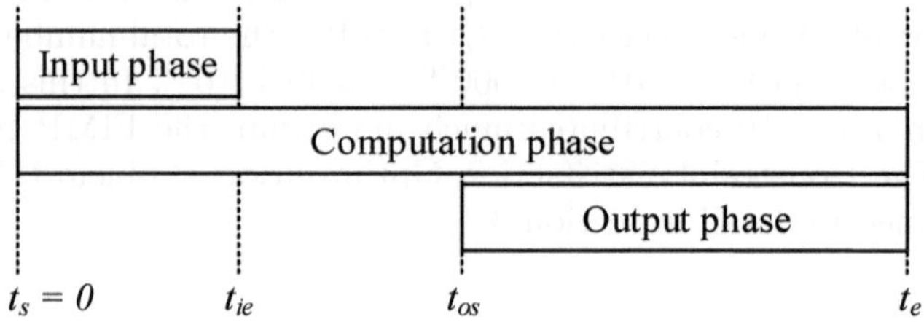

Fig. 3. Function Execution Timing Model

For example, the implementation view of an 64-point FFT can be either a computation core with SRAM in VHDL for ASIC or a MATLAB function for a CGRA to be compiled by the CGRA specific compiler to a configware. This view is not involved in DSE.

Cycle accurate schedule defines the execution schedule of an HSDF actor on a FIMP instance. It consists of a function execution timing model (in terms of t_s, t_{ie}, t_{os} and t_e) and an output buffer end time t_{be} (when the output buffer is released by the data consumer HSDF actor). For example, the HSDF in Fig. 1(b) can have the cycle accurate schedule shown in Fig. 4(a) and (b).

Fig. 4 also illustrates the influence of the output buffer on the FIMP execution schedule. In Fig. 4(a), the FIMP instance B_0 has only one data buffer. Therefore, the second execution of B_0, which is HSDF actor b_1, cannot start until the output buffer of B_0 is released by FIMP instance C_0 and C_1 (HSDF actors c_0 and c_1). The overall system latency is 22 clock cycles. In Fig. 4(b), each of the HSDF actors that are executed by FIMP instance B_0 has its own data buffer. In this case, b_1 can start directly after b_0. The overall system latency becomes 20 clock cycles. The decision of having output buffer for each FIMP instance or each HSDF actor is made be solving the DSE CSOP problem.

1.3 Other Steps

In these steps, the FIMP interconnection and control logics are automatically generated. and the final system implementation is generated. The details can be found in [4] and [5].

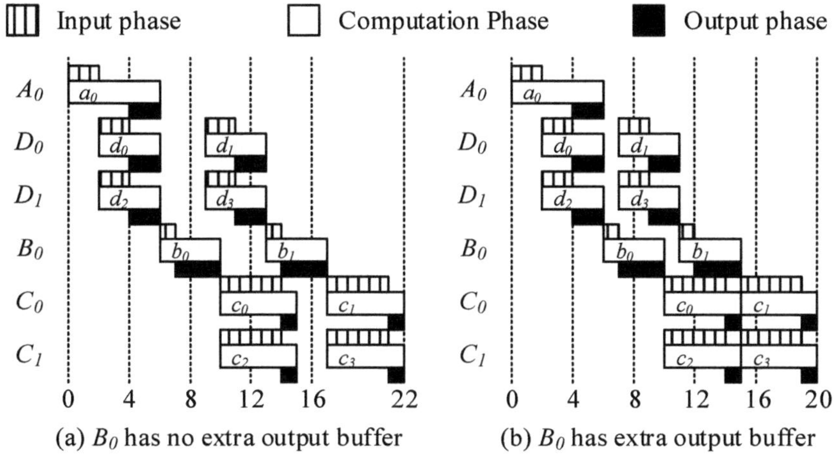

Fig. 4. Cycle Accurate Schedule Example

2 Why Choose Constraint Programming

SYLVA is a research project in which new ideas are constantly emerging and we need a system which enables a high level formal modeling of the problem with minimal effort. This led us to choose the Constraints Programming (CP) framework. Specifically, we list three main motivations:

1) CP has much lower modeling complexity compared with other approaches.
2) CP is suitable for solving scheduling problems, which is critical for us.
3) CP is easy to integrate with other components (e.g. code generator).

2.1 Low Modeling Complexity

CP model supports logical constraints (e.g. logical AND and logical OR) and a full range of arithmetic expressions such as minimum, maximum, or an expression which indexes an array of values by a decision variable. By using CP, constraints can be easily ported from the specification in human language to the solver supported format. In contrast, if we use another approach (e.g. integer linear programming, simulated annealing, genetic algorithm or any other evolutionary algorithms) we need to model the problem in a more complex manner and also define the searching strategy in detail. For example, if integer linear programming is used, we need to formulate the constraints by using linear equations or inequalities, and we cannot use a variable to index another variable or perform logical operations. If an evolutionary algorithm is used, we need to model the problem as well as define the searching strategy by assigning parameters for searching (e.g. randomness simulated annealing or mutation function in genetic algorithm). These parameters have strong impact on the quality of the result and the time required for searching. Finding a good search parameter is in itself a complex problem and there are no *default* values to be used. If CP is used, the modeling is much more flexible than other approaches and we can benefit from the default constraint propagator and searching strategies.

The low modeling complexity of CP gives us the opportunity to have very fast prototyping and high maintainability. We could try out our new ideas and set up new experiments just by adding/deleting/modifying a few constraints or selecting another search strategy among the built ones. For example, finding an ASAP (As Soon As Possible) schedule can be achieved by searching from the lowest value, and finding an ALAP (As Late As Possible) schedule can be achieved by searching from the largest value.

2.2 Suitable for Scheduling Problem

As stated in section 1, the schedule term ($T_{max} - t_i \cdot f_i$ in 2) contributes much more than the FIMP type selection term ($2M_i$ in 2) to the number of solutions ($|D|$). Since CP is suitable for finding solutions to scheduling problems, it is suitable in our case.

2.3 Easy to be Integrated

Since CP is a programming paradigm and it is rooted in computer science, a CP solver usually has interfaces to a number of programming languages. For example, in the SYLVA project, we choose the CP solver from Google's or-tools [6]. It has interfaces to C++, JAVA, Python and C#. Another popular solver GECODE [7] has interfaces to ECLiPSe, AMPL, YAP Prolog, Python, Haskell, Ruby and Common Lisp.

In our case, most of the SYLVA components are implemented in C#. If the DSE problem modeling and solving is also implemented in C#, model translation (converting CSOP model in C#to the format the solver supports) and the resulting deserialization (converting CSOP solution to an C#object) can be eliminated.

3 Constraint Satisfaction Optimization Problem Model

In SYLVA, we have two Constraint Satisfaction Optimization Problem (CSOP) models. The first one (named as P_1) is for the SDF to HSDF conversion. The other one (named as P_2) is for the Design Space Exploration (DSE). In the rest of this section, we use the following annotations. The number of actors in the SDF graph is denoted as $|A|$ and the number of edges in the SDF graph is denoted as $|E|$. The ith SDF actor is denoted as a_i and the jth SDF edge is denoted as e_j.

3.1 P_1, SDF to HSDF Conversion

Variables in P_1 are the set of function execution counts X. The ith variable $x_i \in X$ represents the number of executions of the ith SDF actor a_i. $\forall i \quad x_i \in N_{\geq 0}$

($N_{\geq 0}$ stands for Natural number that greater or equal to zero). The number of variables $|X|$ equals to the number of SDF actors $|A|$. In our model, we constraint the possible values to be less than $2^{32}-1$. Although the number of possible values is large, the constraints will result in a small amount of branches when solving P_1, since we only need the simplest valid solution. It can be found by taking the smallest value in the domain of the variable associated with the root node, propagating this choice and iterating. For example, the conversion in Fig. 1 only has one branch.

Constraints in P_1 are such that all edges have matched data rate. Denoting the set of constraints of P_1 as C_1, the number of constraints $|C_1|$ equals to the number of SDF edges $|E|$.

Before explaining the implementation of the constraints, we need to introduce topology matrix T. T is a $|E| \times |A|$ matrix. Each row of T represents one edge and each column of T represents one SDF actor. The ith row of T is denoted as T_i and the jth element in T_i is denoted as $T_{i,j}$. $T_{i,j}$ is the number of data tokens produced or consumed by the jth SDF actor a_j on the ith edge e_i. Also denoting $a_{s,i}$ as the source actor in e_i and a_d as the destination actor in e_i, the value of $T_{i,j}$ is defined in 3.

$$T_{i,j} = \begin{cases} s_i & a_j = a_{s,i} \\ -d_i & n_j = a_{d,i} \\ 0 & n_j \neq a_{s,i} \text{ and } n_j \neq a_{d,i} \end{cases} \tag{3}$$

The SDF to HSDF conversion constraint can be expressed by 4.

$$X \cdot T_i = 0 \qquad \forall \quad 0 \leq i < |A| \tag{4}$$

For example, the topology matrix T of the SDF graph illustrated in Fig. 1(a) is listed in 5. The first row is for the edge from a to b. The second row is for the edge from d to b. The last row is for the edge from b to c. The solution for the SDF graph shown in Fig. 1(a) is $v_1 = [1, 2, 4, 4]$.

$$P = \begin{bmatrix} 2 & -1 & 0 & 0 \\ 0 & -2 & 0 & 1 \\ 0 & 2 & 1 & 0 \end{bmatrix} \tag{5}$$

HSDF Construction can be done by the pseudo-code shown in Algorithm 1. It has two steps. The first step (line 6 to 8) is to create HSDF actors based on the value of X. The second step (line 9 to 31) is to create edges based on the original SDF graph. Note that in this article, we assume that we only have multiple source actors to single destination actor or single source actor to multiple destination actors data communications. It is formulated by 6, where $x_{s,i}$ and $x_{d,i}$ are the number of executions of the source and destination actors for e_i.

$$(x_{s,i} - \lfloor x_{d,i}/x_{s,i} \rfloor \cdot x_{s,i}) \cdot (x_{d,i} - \lfloor x_{s,i}/x_{d,i} \rfloor \cdot x_{d,i}) \equiv 0 \qquad \forall \quad 0 \geq i < |E| \tag{6}$$

3.2 P_2, Design Space Exploration

In the current development stage, the number of used FIMP instances is not decided using CP but an exhaustive search. All the *load balanced* SDF schedules (which FIMP for which HSDF actors) are checked from the most parallel one (each FIMP instance executes one HSDF actor) to the most serial one (each FIMP instance executes all HSDF actors for the same SDF actor).

The *load balance* is defined as that HSDF actor executions should be distributed equally or approximately equally on proper FIMPs. It is formulated by 7, where $A_{i,f}$ is the set of HSDF actors to be executed by the fth FIMP for SDF actor a_i and F_i is the set of FIMP instances for executing all HSDF actors derived from a_i.

$$0 \leq |A_{i,f}| - \lfloor x_i/|F_i| \rfloor \leq 1 \qquad \forall \;\; 0 \leq i < |A| \text{ and } 0 \leq j < |F_i| \tag{7}$$

The algorithm to generate the load balanced SDF schedules is shown in Algorithm 2 line 1 to 10. The used functions: **PossibleNumbers**, **Modify** and **Generate** are defined in line 12 to 23. Each SDF schedule produced by the **Generate** function is for further design space exploration.

Variables in P_2 are denoted as V and it has four parts, which are FIMP type selection matrix FT, buffer usage indicator vector BU, execution start time vector T_s and buffer end time vector T_{be}.

The first part FT represents the FIMP types. FT is a $M_{max} \times |A|$ matrix, where M_{max} is the maximum number of FIMPs for implementing one function. The ith row ft_i represents the FIMP type selection vector for SDF actor a_i. The possible values of the elements in FT are 0 (the FIMP is not used) or 1 (the FIMP is used). All the FIMP instances for the same SDF actor are in the same FIMP type. For the SDF graph example in Fig. 1(a), if the numbers of possible FIMPs to implement actors [a, b, c, d] are [4, 3, 2, 5]. Thus, $M_{max} = 5$.

The second part BU represents the output buffer usage. The ith element is bu_i and it represents the buffer usage for SDF actor a_i. BU is a vector with $|A|$ elements. The value of each element in BU can be 0 (one FIMP has one output buffer) or 1 (one HSDF actor has one buffer).

The third part T_s represents the execution start time for all HSDF actors. It is a vector and $|T_s = |A_H$. For the ith element $t_{s,i}$ in T_s, its value is set to integers from 0 to $T_{max} - t_{i,max}$, where $t_{i,max}$ is the maximum execution time of a_i and it is determined by the FIMP types for implementing a_i. Since other time values (t_{ie}, t_{os} and t_e in Fig. 3) are in constant relation to t_s, we do not need to search them as well.

The fourth part T_{be} represents the time to free the output buffer of each HSDF actor. It is a vector and $|T_{be} = |A_H$. The ith element in T_{be} is denoted as $t_{be,i}$ and its value can be integers from 0 to $T_{max} - 1$.

Therefore, $|V| = M_{max} \cdot |A| + |A| + |A_H| + |A_H|$. In the example shown in Fig. 1, $|V| = 5 * 4 + 4 + 10 + 10 = 44$.

Algorithm 1. HSDF Construction

```
A is the actor set of the SDF graph
E is the edge set of the SDF graph
A_H is the actor set of the HSDF graph
E_H is the edge set of the HSDF graph
initialize A_H = E_H = empty array
for all actor a_i in A do
    for j in 0 to x_i - 1 do
        A_{H i,j} = a_i
for all edge e_i in E do
    p is the index of the source actor a_s in e_i
    q is the index of the destination actor a_d in e_i
    s_i is the number of source data token of e_i
    d_i is the number of destination data token of e_i
    if x_s > x_d then
        for j in 0 to x_d - 1 do
            for k in 0 to x_s/x_d - 1 do
                add a new edge e_n to E_H
                s_n is the number of source data token of e_n
                d_n is the number of destination data token of e_n
                a_s of e_n is A_{H p,j*x_s/x_d+k]}
                a_d of e_n is A_{H q,j}
                s_n = d_n = s_i
    else
        for j in 0 to x_s - 1 do
            for k in 0 to x_d/x_s - 1 do
                add a new edge e_n to E
                s_n is the number of source data token of e_n
                d_n is the number of destination data token of e_n
                a_s of e_n is A_{H p,j}
                a_d of e_n is A_{H q,j*x_d/x_s+k]}
                s_n = d_n = d_i
```

Constraints in P_2 can be sorted into six categories. They are automatically generated based on the user input.

1. The first category is data dependency for execution. A function can only start its execution after all the dependent functions are complete. For example in Fig. 4, the actor b_0 should start execution after actor a_0 is complete. This category of constraints is formulated in 8, where $t_{s,d}$ is the execution start time of the destination HSDF actor and $t_{e,s}$ is execution end time of the source HSDF actor. For the example, it is expressed as $t_{s,b_0} > t_{e,a_0}$.

$$t_{s,d} > t_{e,s} \qquad \forall \text{ edges in } E_H. \tag{8}$$

There are $|E_H|$ number of constraints in this category. $|E_H|$ is the number of edges in the HSDF graph.

Algorithm 2. HSDF Schedule

```
A is the actor set of the SDF graph
F is the FIMP instance set for the SDF graph
∀ 0 ≤ i < |A|, F_i is the set of FIMP instances used for a_i
F_max is the set of maximum numbers of FIMP instances used for A
for all actor a_i in A do
    F_max,i = x_i
for all actor a_i in A do
    |F_i| = F_max,i
    for all n in PossibleNumbers(F_max,i) do
        Modify(i, n)
        Generate()

PossibleNumbers(u):
output an array of numbers {n | 0 < n ≤ u and ⌊u/n⌋ * n = u}

Modify(i, n):
if |F_i| > n then
    ratio r = F_i/n
    for all actor a_j in A do
        F_i = ⌈F_i/r⌉

Generate():
for all actor a_i in actor set of the SDF graph do
    equally and linearly assign HSDF actors that are derived from a_i to F_i FIMPs
```

2. The second category is data dependency for output buffer. The producing and consuming actors cannot output and input at the same time to/from the output buffer. For example in Fig. 4, the actor b_0 should complete its input phase before the buffer end time of actor d_0. This category of constraints is formulated in 9, where $t_{ie,d}$ is the input end time of the destination HSDF actor and $t_{be,s}$ is output buffer end time of the source HSDF actor. For the example, it is expressed as $t_{ie,b_0} \leq t_{be,d_0}$.

$$t_{ie,d} \leq t_{be,s} \qquad \forall \text{ edges in } E_H. \tag{9}$$

There are $|E_H|$ number of constraints in this category.

3. The third category is resource dependency for execution. Only one actor can be executed on a FIMP instance at a time. For example in Fig. 4, the actor b_1 should start execution after actor b_0 is complete. This category of constraints is formulated in 10. $A_{i,j}$ is the HSDF actors that are executed on the jth FIMP instance for executing SDF actor a_i. $t_{s,p+1}$ is the execution start time for the actor a_{p+1}, which is the $p+1$th actor in $A_{i,j}$. $t_{e,p}$ is the execution end time for actor a_p, which is the pth actor in $A_{i,j}$. For the example in Fig. 4, $A_{d,1}$ has d_2 and d_3, and both of them are executed on FIMP instance D_1.

$$t_{s,p+1} > t_{e,p} \qquad \forall\, a_p \in A_{i,j}. \tag{10}$$

There are $|A_H|-|F|$ number of constraints in this category. $|A_H|$ is the number of HSDF actors and $|F| = \sum F_i$ $(0 \leq i < |A|)$ is the total number of used FIMP instances and it is determined in the HSDF scheduling step.

4. The fourth category is resource dependency for output buffer. One output buffer can only be written by one HSDF actor at a time. For example in Fig. 4(a), the actor b_1 cannot start writing data into the output buffer before it is freed by c_0 and c_2. This category of constraints is formulated in 11. $t_{os,p+1}$ is the output start time for the actor a_{p+1}, which is the $p+1$th actor in $A_{i,j}$. $t_{be,p}$ is the output buffer end time for actor a_p, which is the pth actor in $A_{i,j}$. $A_{i,j}$ is the HSDF actor set that is derived from the SDF actor a_i and executed on the jth FIMP instances a_i. For the example in Fig. 4(a), t_{os,b_1} should be larger than t_{be,b_0} since no extra buffer is used ($bu_b = 0$). While in Fig. 4(b), t_{os,b_1} can be smaller than t_{be,b_0} since extra buffer is used ($bu_b = 1$). There are $|A_H| - |F|$ number of constraints in this category.

$$(t_{os,p+1} > t_{be,p} \text{ or } bu_i) = True \qquad \forall\, a_p \in A_{i,j}. \tag{11}$$

5. The fifth category is cost constraints. There are area, energy and timing cost constraints.

The total *area* usage $AREA$ can be computed by 12, where $|F_i|$ is the number of FIMP instances for executing SDF actor a_i, $area_i$ is a vector that represents the area usage for all the FIMPs for executing SDF actor a_i and ft_i is the ith row of FT (FIMP type variables for a_i). The term $area_i \cdot ft_i$ is the area usage for the used FIMP instance for a_i.

$$AREA = \sum_{i=0}^{|A|-1} (|F_i| \cdot area_i \cdot ft_i). \tag{12}$$

The total *energy* cost $ENERGY$ can be computed by 13, where $energy_i$ is a vector that represents the energy cost for all the FIMPs for executing SDF actor a_i. The term $energy_i \cdot ft_i$ is the energy cost for one execution of a_i.

$$ENERGY = \sum_{i=0}^{|A|} (x_i \cdot energy_i \cdot ft_i). \tag{13}$$

The *timing* costs consist of system latency $LATENCY$ (the time between the first input data token is consumed till the last output data token is produced) and system sample interval $INTERVAL$ (the time for one system iteration). They can be computed by 14 and 15, respectively. A_H is the actor set of the HSDF graph. a_q and a_0 are the last and the first HSDF actor, respectively, in $A_{i,j}$, which is a actor set containing all HSDF actors that are derived from SDF actor a_i and executed on the jth FIMP instance for a_i.

$$LATENCY = max(t_{e,p}) \qquad \forall\, a_p \in A_H \tag{14}$$

$$INTERVAL = max(t_{be,q} - t_{os,0}, t_{e,q} - t_{s,0}) \qquad \forall\, a_q, a_0 \in A_{i,j} \tag{15}$$

In the example shown in Fig. 4(a), $LATENCY = 22$ and $INTERVAL = 14$.

$$\begin{aligned} INTERVAL &= max(t_{e,a_0} - t_{s,a_0}, t_{be,a_0} - t_{os,a_0}, ...) \\ &= max(6, 7-4, 13-2, 14-4, 17-6, 21-7) \\ &= max(6, 3, 11, 10, 11, 14) = 14 \end{aligned}$$

The area, energy and timing cost constraints are shown in 16.

$$\begin{array}{ll} AREA & \leq A_{max}, \\ ENERGY & \leq E_{max}, \\ LATENCY & \leq T_{max}, \\ INTERVAL & \leq R_{max}. \end{array} \tag{16}$$

In SYLVA, providing values to A_{max}, E_{max}, T_{max} and R_{max} is optional. If any of them is not provided by user, the default value $2^{64}-1$ will be used. There are 4 constraints in this category as shown in 16.

6. The sixth category is the FIMP type constraint. Only one FIMP can be used to implement an SDF actor a_i. This constraint is shown in 17, where $ft_{i,j}$ is usage of the jth FIMP for implementing a_i

$$\sum_{j=0}^{M_{max}} ft_{i,j} = 1 \qquad \forall\, 0 \leq i < |A| \tag{17}$$

There are $|A|$ number of constraints in this category.

The number of constraints in P_2 is $|C_2| = 2 \cdot |E_H| + 2 \cdot (|A_H| - |F|) + 4 + |A|$. In the example shown in Fig. 1, $|C_2| = 2 * 10 + 2 * 6 + 4 + 4 = 40$.

Optimization Objective is to minimize the total cost C, which is calculated by 18. K_A, K_E, K_T and K_R are four predefined constants for optimizing the solution in terms of area usage, energy cost, system latency and system sample interval, respectively. They can be provided by user or be the default value ([0, 1, 0, 0] for [K_A, K_E, K_T, K_R]).

$$C = K_A \cdot AREA + K_E \cdot ENERGY + K_T \cdot LATENCY + K_R \cdot INTERVAL \tag{18}$$

3.3 Solving CSOPs

In SYLVA, P_1 and P_2 are modeled and solved in C# by using the CP solver in or-tools from Google. For branching strategy, we set the solver to always select the first unbound variable (INT_VAR_DEFAULT) and assign the minimum possible value first (ASSIGN_MIN_VALUE).

The efficiency and efficacy of SYLVA are evaluated by synthesizing five examples (Fig. 5): 1) a sub-system composed of two FIR filters feeding an FFT, 2) a correlation pool (part of UMTS rake receiver), 3) a sigma delta demodulator, 4) a JPEG Encoder for 1920 × 1080, and 5) a simplified MPEG2 Encoder for a 720 × 480 @25 frames per second.

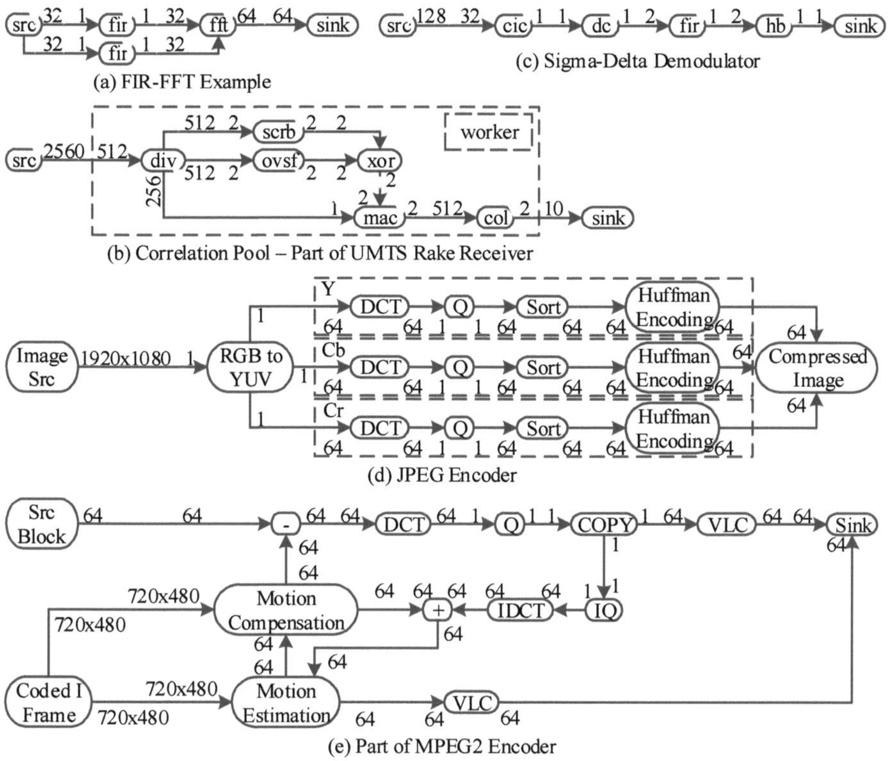

Fig. 5. Examples for Experiments

The individual components of the SYLVA flow have been implemented in C# and integrated at script level. This script was invoked for each of the five examples (in the form of SDF graphs) with the maximum sampling interval (R_{max}) and the maximum total latency (T_{max}) as command line parameters. The experimental result shows that on average, the SYLVA runtime for the five examples are 15s, 13s, 20s, 74s, and 97s, respectively, and for all the examples, only one branch is required to find the SDF repetition vectors. They are much faster than the commercial high level synthesis tool we compared with. SYLVA gets speed advantage compared to commercial synthesis tools because of its use of very large grain design objects (FIMPs) that are 2-3 orders smaller than the objects used by commercial tools. This dramatically reduces the design space for SYLVA. The details of the quality of result comparison can be found in [2].

4 Conclusion and Future Work

4.1 Development Cost

Learning Cost: Before coding SYLVA, one of the authors has taken a CP course, which lasts for two months. The author also spent an additional month to get familiar with the C# port of the CP solver in Google's or-tools. **Time Cost:** By using CP, the first SYLVA release came out after one month of intensive coding in C#. The time spent on the Design Space Exploration (DSE) CSOP model is only one week and it includes coding, debugging and experimenting. By using CP, modifying and maintaining the DSE model are quite straightforward and simple. We only need to add/remove/modify the concerned constraints. **Software Cost:** The used software copies are all free. The text editor and or-tools from Google are free software and the Windows SDK is also free to use.

4.2 Usage Difficulty

Using SYLVA to synthesize a system SDF graph into a hardware description is quit simple and does not require any knowledge of CP. The user is required to provide an SDF graph. Providing A_{max}, E_{max}, T_{max}, R_{max}, K_A, K_E, K_T and K_R are all optional. In most of the cases of hardware design, the final system implementations should be optimized to have minimal area or energy. In this case, K_A and K_E should be 1, while K_T and K_R are 0's.

4.3 Conclusion

By using CP, we saved a lot of time on implementing the model of the problem. Compared to other approaches, e.g. integer linear programming or evolutionary algorithms, CP provides a straightforward framework for modeling and solving problems. After three months of non-intensive learning of CP fundamentals and a CP solver, we were able to write a complex CSOP model, which has four categories of variables and six categories of constraints in a short time. We hope the work in this article could serve as a good application example of CP.

4.4 Future Work

Currently, we are updating SYLVA to support more features, such as scenario-aware SDF that supports dynamic streaming and signal processing applications, and I/O data structure matching that match not only the number of data tokens but the data structure of the data tokens to improve the timing performance of the system.

References

1. Lee, E., Messerschmitt, D.: Synchronous Data Flow. Proceedings of the IEEE 75(9) (September 1987)
2. Li, S., Farahini, N., Hemani, A., Rosvall, K., Sander, I.: System Level Synthesis of Hardware for DSP Applications Using Pre-Characterized Function Implementations. In: International Conference on Hardware/Software Codesign and System Synthesis (CODES+ISSS), pp. 1–10 (September 2013)

3. Bonfietti, A., Benini, L., Lombardi, M., Milano, M.: An efficient and complete approach for throughput-maximal sdf allocation and scheduling on multi-core platforms. In: Design, Automation Test in Europe Conference Exhibition (DATE), pp. 897–902 (March 2010)
4. Li, S., Hemani, A.: Global Interconnect and Control Synthesis in System Level Architectural Synthesis Framework. In: Euromicro Conference on Digital System Design (DSD), pp. 11–17 (September 2013)
5. Li, S., Malik, J., Liu, S., Hemani, A.: A Code Generation Method for System-Level Synthesis on ASIC, FPGA and Manycore CGRA. In: Proceedings of the First International Workshop on Many-core Embedded Systems (2013)
6. Operations Research Tools from Google, `https://code.google.com/p/or-tools/`
7. Gecode: generic constraint development environment, `http://www.gecode.org/`

Scheduling Agents Using Forecast Call Arrivals at Hydro-Québec's Call Centers

Marie Pelleau[1], Louis-Martin Rousseau[2], Pierre L'Ecuyer[1], Walid Zegal[3], and Louis Delorme[3]

[1] Université de Montréal, Montreal, Canada
me.pelleau@umontreal.ca,lecuyer@iro.umontreal.ca
[2] Polytechnique Montréal, Montreal, Canada
louis-martin.rousseau@polymtl.ca
[3] Institut de recherche d'Hydro-Québec, Montreal, Canada
{zegal.walid,delorme.louis}@ireq.ca

Abstract. The call center managers at Hydro-Québec (HQ) need to deliver both low operating costs and high service quality. Their task is especially difficult because they need to handle a large workforce (more than 500 employees) while satisfying an incoming demand that is typically both time-varying and uncertain. The current techniques for determining the schedule of each employee according to the forecast call volumes at HQ are often unreliable, and there is a need for more accurate methods. In this paper, we address the concerns of the call center managers at HQ by modeling the problem of multi-activity shift scheduling. This model has been implemented and tested using real-life call center data provided by HQ. The main contribution of this paper is to demonstrate that a constraint programming (CP) model with regular language encoding can solve large problems in an industrial context. Furthermore, we show that our CP-based formulation has considerably better performance than a well-known commercial software package.

1 Introduction

The management of call center operations at Hydro-Québec (HQ) is a highly complicated task that often involves balancing contradictory objectives. Managers need to achieve simultaneously high levels of service quality and operational efficiency. The service quality is typically measured by key target performance metrics such as the average caller waiting time, i.e., the *délai moyen de réponse* (DMR). The daily target DMR is 120 seconds. The operational efficiency is typically measured by the proportion of time that agents are busy handling calls.

It can readily be seen that high levels of service quality are associated with low levels of operational efficiency, and vice versa. It is challenging to achieve the right balance. First, there is the problem of determining the appropriate *staffing level*, weeks or even months in advance, based on long-term forecasts of future incoming demand (i.e., future call volumes); this demand is typically both time-varying and random. Second, there is the problem of scheduling (and

B. O'Sullivan (Ed.): CP 2014, LNCS 8656, pp. 862–869, 2014.

rescheduling) the available pool of agents based on updated forecasts, typically made several days or weeks in advance, which is a problem of *resource deployment.* Finally, short-term decisions must be made, such as the routing of incoming calls in real-time to available agents or the mobilizing of agents at short notice because of unforeseen fluctuations in the incoming demand. At HQ, additional real-time control involves moving agents between the *front office* (where they answer calls) and the *back office* (where they perform other tasks such as paperwork). These decisions are based on short-term forecasts, updated one day or a few hours in advance.

In this paper, we focus on the *resource deployment* problem. Given a staffing level, we wish to specify which activity (such as taking calls, responding to emails, or working in the back office) each worker should perform in each period of the day. The creation of such a detailed schedule is generally referred to as the *multi-activity assignment problem* and is much more difficult than the traditional mono-activity version. The need to specify the occupation of an employee in each time period drastically increases the number of possible work shifts, which makes classical formulations (such as set covering [4]) intractable in this context.

In recent years, researchers have investigated using formal languages to solve complex scheduling problems where employees need to perform several tasks during a shift. These approaches can be combined with MIP (mixed integer programming) formulations and then solved directly, through column generation [5,11,2], via metaheuristics such as large neighborhood search [10] or tabu search [3], via a hybrid MIP/CP approach [12], or directly in CP through lazy clause generation [7].

As in [8,6] we solve the activity assignment problem, which involves shift scheduling where the shift positions and breaks are fixed. The main contribution of this paper is to demonstrate that a constraint programming (CP) model with regular language encoding can solve large problems in an industrial context. In the data provided by HQ, there are 40 periods of 15 minutes covering the period from 8 a.m. to 6 p.m. There are between 129 and 142 activities, and from 369 to 544 employees to schedule. The problem is significantly larger than the test cases used in the existing literature on multi-activity scheduling. Furthermore, we show that our CP-based formulation has considerably better performance than a well-known commercial software package.

This paper is organized as follows. Section 2 introduces the problem and the main limitations of the current tool used by HQ. In Section 3, we present the model we used to address these limitations. Section 4 provides concluding remarks.

2 Problem Description

We consider the problem of designing work schedules for the agents. Given a day divided into periods, a set of employees, and a staffing requirement, we must select which activities should be performed by each employee during each period of the day, in order to satisfy the demand.

2.1 Problem Characteristics

We now detail the different characteristics of the problem and give some of the notation.

Activities. In HQ, there are more than 100 different activities. Let the set of all the activities be A. There are two types of activities. The *nonproductive* activities are the breaks. We denote by A_{rest} the set of nonproductive activities. The *productive* activities are all the activities that are not in A_{rest}. The set of such activities is denoted by A_{prod}. And may be split into three disjoints sets: the phone activities A_{phone} (taking the calls), the back-office activities A_{office} (doing paperwork), and the web activities A_{web} (responding to emails). We must distinguish between the different activities to formalize some of the concerns of the call center managers; see Section 2.2.

Periods. The call arrival process is uncertain and time-varying. In order to approximate it, the day is divided into m periods of 15 minutes. We retained this time division in the scheduling process. We define $T = \{1, \ldots, m\}$ to be the set of periods in a day.

Employees. Let E be the set of employees. Each employee has a set of skills $S_e \subseteq A$. For each employee $e \in E$ in each period $t \in T$, we must determine which activity should be performed. Let $x_{e,t} \in S_e$ be the activity performed by employee e in period t.

Staffing and Deviation. The staffing requirement indicates the desired number of staff for each activity in each period of the day; it is calculated beforehand. We denote by $d_{a,t}$ the staffing requirement for activity $a \in A$ in period $t \in T$. The goal is to ensure that the number of employees for each activity in each period of the day is greater than or equal to the requirement. However, it may not be possible to provide a schedule that fully satisfies the demand. In this context, the aim is to minimize the sum of the weighted staff shortages over all activities and periods, with weights that depend on both activities and time. This is referred to as *deviation* or *undercover* and denoted by $u_{a,t}$ with $a \in A$ and $t \in T$.

Thus, given a set of skills and a staffing requirement, building an optimal schedule corresponds to assigning activities to each employee for every period of the day. Each assigned employee must have the necessary skills, and the goal is to minimize the total deviation.

2.2 HQ's Solution

HQ currently applies a widely used commercial scheduling software. This system is able to compute the staffing requirement, and given a staffing requirement it

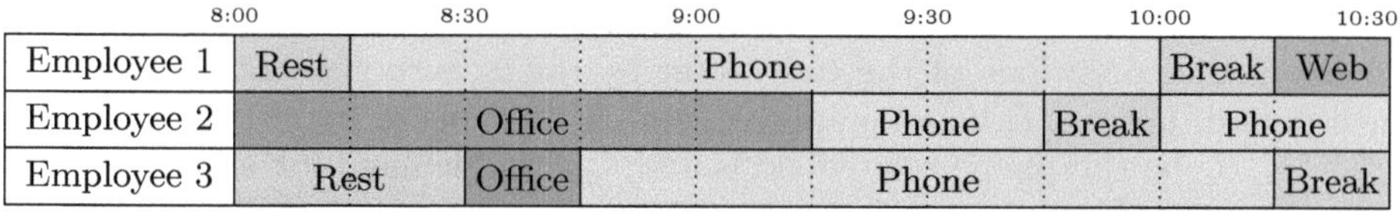

Fig. 1. Example of HQ schedule for three employees between 8 and 10:30

designs a schedule that satisfies the demand. Figure 1 shows an example of a schedule for three employees between 8:00 and 10:30. However, the current tool does not consider some of the following critical aspects of the management of HQ's call centers.

Priority Management. Some activities have priority over others. For instance the activity of answering *failure* call type has priority over other call types. Moreover, any phone activity has priority over office and web activities. One may also want to prioritize certain times of the day, such as lunchtime when many employees are unavailable. The tool currently used at HQ does not allow the prioritization of activities, resources, and calls.

Activity Transitions. The current tool is unable to efficiently manage sequences of different activities. It computes solutions in which employees must switch from one activity to another after only a short interval. For example, in Figure 1, Employee 3 has just 15 minutes of office work before switching to phone activity. However, HQ's managers wish to establish a minimum duration of one hour for each activity type. The current tool cannot enforce this rule, so the schedules must be corrected manually.

Multi-skill Management. Multi-skill management is necessary only for the phone activities. Given the stochastic context of call arrivals, it is desirable to assign an agent to a set of phone activities during the same period. However, the current tool is unable to select a subset of activities and assigns an employee to all the phone activities in his or her set of skills. In practice, the multi-skill assignment is performed using an internal simulator. This simulator is part of the software and is not documented. We do not know if it is stochastic or deterministic, or how the call arrival process is modeled. Furthermore, we do not have access to the results of the simulator. It generates a distribution of skills per agent per period. We define the *phone activity distribution ratio* to be the percentage of time allocated to a specific call type for an agent who is assigned to phone activities for a given period. This distribution is important because the good coverage of calls by agents relies on it. HQ thus uses the ratio below that is proportional to each activity's demand to evaluate the quality of a schedule.

Proportional Ratio for Multi-skill Management. This ratio is designed to satisfy the conditions of the call center by taking into account the demand for a period. Let $e \in E$ be an employee, and $S_e \subseteq A$ his or her set of skills. Let $S_e^{phone} \subseteq S_e$ be the phone-related skills and $d_{a,t}$ the demand for activity $a \in A$ in period $t \in T$. If employee e is assigned to phone activities during period t, the ratio for each phone activity $a \in S_e^{phone}$ is

$$r_{e,a,t} = \frac{d_{a,t}}{\sum_{a' \in S_e^{phone}} d_{a',t}}.$$

For all the other activities, this ratio is equal to 1. This ratio can be seen as a realistic assessment of the current conditions in HQ's call centers.

In previous work, HQ has implemented a MIP model using CPLEX. However, for large instances, this method sometimes returns an out-of-memory error. To address the limitations introduced above, and to reduce the large number of variables, we decided to use CP.

3 Constraint Programming Formulation

3.1 Proposed Solutions to Address the Limitations

We address the issues of priority management and sequence limitations. For multi-skill management we currently use the method applied in HQ's existing tool: we assign an employee to all the phone activities in his or her set of skills, and we compute the deviation using the ratio introduced in the previous section.

Priority Management. Assigning a cost $C_{a,t}$ to each activity $a \in A$ and each period $t \in T$ allows us to prioritize some activities and periods over others.

Activity Transition. As part of the scheduling process, we propose to use a regular language to model the transition rule between the different types of activities. This rule states that an employee must perform productive activities of the same type (phone, office, or web) for a fixed duration (e.g., 1 hour) before switching to another activity. The rule is based not on activities but rather on families of activities. In addition, the rule must be validated for each employee $e \in E$. To model this sequence rule, we use an automaton and state that, for an employee, the word formed by the activities performed during the day must be recognized by the automaton.

Let $\Sigma = \{n, p, o, w\}$ be the alphabet of the automaton, where $n \in A_{\text{rest}}$, $p \in A_{\text{phone}}$, $o \in A_{\text{office}}$, and $w \in A_{\text{web}}$ belong respectively to the set of nonproductive, phone, office, and web activities. Let $Q = \{0, ..., 9\}$ be the set of states. The automaton defined on (Σ, Q) is given in Figure 2. Starting from the initial state 0, the automaton ensures that if an employee is assigned to a phone activity (state 1), an office activity (state 4), or a web activity (state 7) for a period of

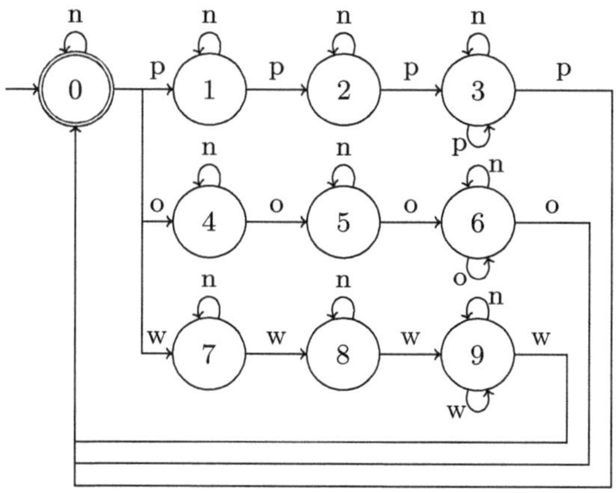

Fig. 2. Automaton for the activity transition rule

15 minutes, then they must perform an activity from that category for at least one hour. Breaks can occur during or after this block of activities. Moreover, the fact that state 0 is accepting ensures that even the last block of activities lasts at least one hour.

This automaton is represented by its transition table and modeled using a *Table* constraint. For each employee e and each period t, $q_{e,t} \in Q$ gives the state in the automaton.

3.2 Model

This model is a simplified version because all the breaks are fixed. We used the schedules designed by HQ's software and fixed the breaks at the same periods.

Variables

$x_{e,t} \in S_e$	Activity performed by employee e during period t
$u_{a,t} \in \mathbb{N}$	Shortage of employees performing activity a during period t
$q_{e,t} \in Q$	State in the automaton

Constraints

$$\min \sum_{a \in A_{\text{prod}}} \sum_{t \in T} C_{a,t} \times u_{a,t} \tag{1}$$

s.t.

$$\sum_{e \in E} ((x_{e,t} == a) \times r_{e,t,a}) + u_{a,t} \geq d_{a,t}, \quad \forall a \in A_{\text{prod}}, t \in T \tag{2}$$

$$\text{Table}((q_{e,t}, x_{e,t}, q_{e,t+1}), \text{automaton}) \quad \forall e \in E, t \in \{1, \ldots, m-1\} \tag{3}$$

$$q_{e,1} = 0 \quad \forall e \in E \tag{4}$$

$$q_{e,m} = 0 \quad \forall e \in E \tag{5}$$

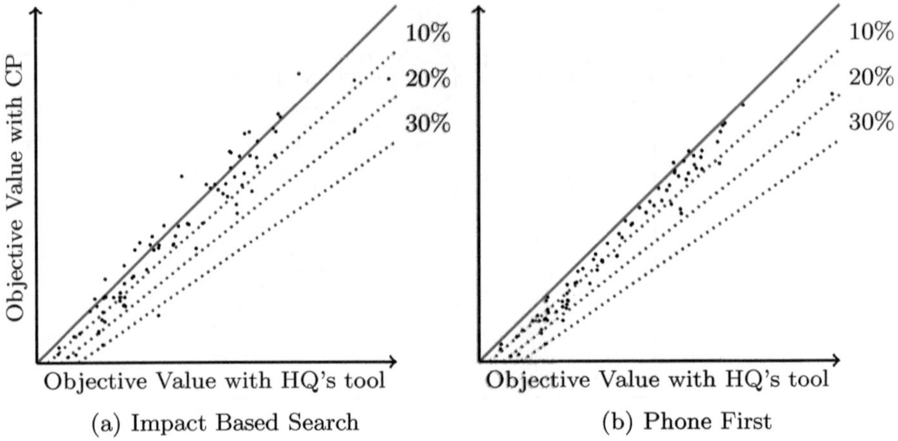

Fig. 3. Comparison of the CP results and the HQ schedules

3.3 Implementation

We implemented this model using the Or-Tools[9] library for Java. The main reason for using Java is to be able in future work to communicate directly with the call-center simulator [1]. In this version we used pure CP, with a timeout of 5 minutes. We try several strategies to select the next branching variable during the search. We report here the results for the classical *Impact Based Search* and a dedicated search strategy we call *Phone First*, which consist of first assigning employees that can perform phone activities and focus on office, web, and rest activities afterwards.

We tested our implementation on daily data for February to May 2011. In these instances, there are 40 periods of 15 minutes corresponding to 8 a.m. to 6 p.m. There are between 129 and 142 activities, of which 40 are phone activities, and from 369 to 544 employees to schedule. The experiments were run on a 1.7-GHz Intel Core i7.

A comparison of our solution and the HQ solution on 82 instances shows that we reduced the understaffing by an average of 5% with the *Impact Based Search*, and 9% with the *Phone First* strategy. Figure 3 compares the HQ's currently used tool with CP for both search strategies. Points below the bisector are instances on which CP outperforms the current solution, we can thus see that both search strategies perform quite well. *Phone First*, however, seems to be more robust as it degrades the solution in only two cases.

4 Conclusion

In this paper, we have investigated a large industrial multi-activity assignment problem. We took into account the concerns of the call center managers and proposed a solution that considers issues not handled by their current tool. We implemented this model with Or-Tools and tested it on a significant number of real instances. Future work will involve solving the full multi-activity shift scheduling problem, where the shift positions and breaks are not fixed.

References

1. Buist, E., L'Ecuyer, P.: A java library for simulating contact centers. In: Proceedings of the 37th Conference on Winter Simulation, pp. 556–565 (2005)
2. Côté, M.-C., Gendron, B., Rousseau, L.-M.: Grammar-based column generation for personalized multi-activity shift scheduling. INFORMS Journal on Computing 25(4), 461–474 (2013)
3. Dahmen, S., Rekik, M.: Solving multi-activity personalized shift scheduling problems with a hybrid heuristic. Technical report, Faculté des sciences de l'administration, Université Laval (2012)
4. Dantzig, G.B.: A comment on edie's "traffic delays at toll booths". Journal of the Operations Research Society of America 2(3), 339–341 (1954)
5. Demassey, S., Pesant, G., Rousseau, L.-M.: A cost-regular based hybrid column generation approach. Constraints 11(41), 315–333 (2006)
6. Elahipanah, M., Desaulniers, G., Lacasse-Guay, È.: A two-phase mathematical-programming heuristic for flexible assignment of activities and tasks to work shifts. Journal of Scheduling 16(5), 443–460 (2013)
7. Gange, G., Stuckey, P.J., Van Hentenryck, P.: Explaining propagators for edge-valued decision diagrams. In: Schulte, C. (ed.) CP 2013. LNCS, vol. 8124, pp. 340–355. Springer, Heidelberg (2013)
8. Lequy, Q., Bouchard, M., Desaulniers, G., Soumis, F., Tachefine, B.: Assigning multiple activities to work shifts. Journal of Scheduling 15(2), 239–251 (2012)
9. Google OR-Tools, `https://code.google.com/p/or-tools/`
10. Quimper, C.-G., Rousseau, L.-M.: A large neighbourhood search approach to the multi-activity shift scheduling problem. Journal of Heuristics 16(3), 373–392 (2010)
11. Restrepo, M.I., Lozano, L., Medaglia, A.L.: Constrained network-based column generation for the multi-activity shift scheduling problem. International Journal of Production Economics 140(1), 466–472 (2012)
12. Salvagnin, D., Walsh, T.: A hybrid mip/cp approach for multi-activity shift scheduling. In: Milano, M. (ed.) CP 2012. LNCS, vol. 7514, pp. 633–646. Springer, Heidelberg (2012)

Deployment of Mobile Wireless Sensor Networks for Crisis Management: A Constraint-Based Local Search Approach

Cédric Pralet and Charles Lesire

ONERA – The French Aerospace Lab, 31055, Toulouse, France
{firstname.lastname}@onera.fr

Abstract. In this paper, we consider a problem of management of crisis situations (incidents on nuclear or chemical plants, natural disasters...) that require remote sensing, using a set of ground and aerial robots. In this problem, sensed data must be transmitted in real-time to an operation center even in case of unavailability of traditional communication infrastructures. This implies that an ad hoc wireless communication network must be deployed, for instance using a fleet of UAVs acting as communication relays. From a technical point of view, we tackle a scheduling problem in which activities of mobile sensing robots and mobile relays must be synchronized both in time and space. Schedules produced must also be flexible and robust to the uncertainty about the duration of robot moves at execution time. The problem is modeled and solved using constraint-based local search, with some calls to graph algorithms that help defining good communication networks.

1 Problem Description

The first step in crisis management consists in performing sensing operations, to assess the situation before choosing appropriate measures. In many cases, sensing cannot be directly performed by humans, due either to the difficulty to reach some areas, *e.g.* in case of natural disasters, or to the dangerousness to do so, *e.g.* in case of incidents on nuclear or chemical plants. As a result, the use of Unmanned Ground Vehicles (UGVs) and Unmanned Aerial Vehicles (UAVs) is more and more considered to help rescue forces in such situations. These vehicles can be equipped with various sensors, allowing to measure radioactivity, to measure the concentration of a chemical element, to take pictures of damaged buildings, or to capture audio/video streams on critical areas.

Data collected may have to be transmitted to an operation center in real-time (or at least quite fast), first because the amount of memory available on-board each vehicle may be limited, and second because providing immediate feedback allows the operators to instantaneously analyze the situation. However, traditional communication infrastructures may be unavailable during crises. A solution for maintaining communication links is to deploy an ad hoc wireless communication network, using a fleet of UAVs acting as communication relays. These relays also allow operators to take the remote control of a vehicle.

B. O'Sullivan (Ed.): CP 2014, LNCS 8656, pp. 870–885, 2014.

In this paper, we gather all these features and consider a system composed of mobile sensors and mobile relays, respectively in charge of realizing acquisitions and transmitting acquisition data in real-time. Some vehicles may change roles dynamically during the mission, such as UAVs equipped with both a camera and a wireless router. The goal is to decide on the sequence of activities of each vehicle so that a set of requested acquisitions is performed as fast as possible, and sensors-operators communication links are maintained. The main constraint is that activities of UAVs and UGVs must be synchronized both in time and space, since communication between two vehicles is possible only at a certain distance. We must also manage cumulative resource consumptions, since each relay can transmit data coming from several sensors only if the total amount of data to be transmitted simultaneously does not exceed a given relay capacity. The problem obtained is a combinatorial scheduling problem, potentially hard to solve for human operators, and we propose to use automatic optimization tools [1].

Fig. 1 illustrates the kind of deployment strategies which we obtain, on a mission involving seven acquisitions (a1 to a7) and five UAVs (r1 to r5). Fig. 1(a) shows the trajectories of vehicles. Four specific relay positions are used (positions p1 to p4). For each relay position p, the large dotted circle around p represents the communication range of a relay placed at p. Fig. 1(b) gives the schedule, which involves two kinds of activities: acquisitions and communications. The setup durations between activities are induced by the durations of moves between locations. Acquisitions such as a4 do not require any communication relay, because they are near enough from the operation center. Acquisitions such as a1 require the simultaneous use of several relays (r2 at position p3, r4 at position p2, r5 at position p1). The communication network is dynamic because relays are mobile, such as robot r2 which moves from position p3 to position p4 as soon as it is no more needed in p3. Last, a communication relay can receive data from several robots simultaneously: see the example of robot r5, which receives data from both r3 and r4 when acquisitions a6 and a8 are performed.

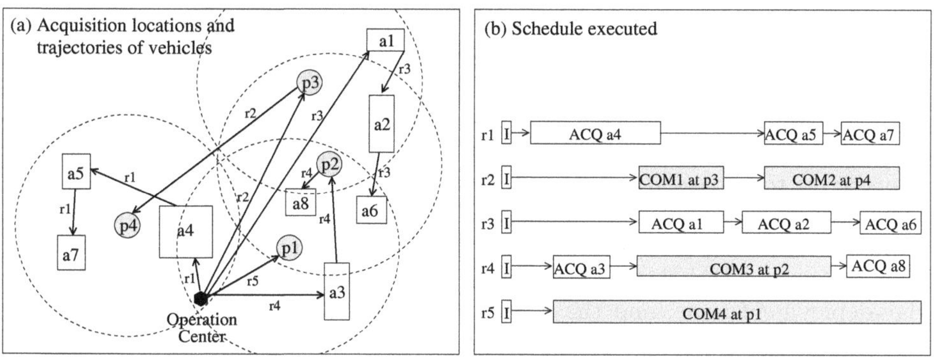

Fig. 1. Example of deployment of sensors and communication relays

The paper is organized as follows. We first describe some related work (Section 2). We then introduce a constraint-based model (Section 3). Next, a local search procedure is defined (Section 4). Last, experimental results are presented (Section 5). This work was performed during the French-German ANR ANCHORS project, whose goal is the definition of UAV-assisted ad hoc networks for crisis management and hostile environment sensing.

2 Related Work

The mobile sensor and relay deployment problem considered can be decomposed into two subproblems: (1) *exploration*, consisting in allocating acquisition activities to sensors and in ordering these acquisition activities; (2) *communication*, consisting in maintaining a communication network for transmitting acquisition data to the operation center.

The exploration subproblem can be seen as a kind of *multiple Traveling Salesman Problem* (mTSP [2]). This problem involves a set of salesmen and a set of cities, and the goal is to find minimum cost tours for the salesmen so that each city is visited exactly once. In our case, cities correspond to acquisitions and salesmen correspond to mobile sensors. In the robotics field, viewing the realization of a set of tasks by a set of robots as an mTSP is not new [3]. Constraint programming approaches for solving mTSPs also exist, with an emphasis on the flexibility of constraints for modeling additional specifications [4].

The communication subproblem is related to the literature on *wireless sensor networks* (WSNs [5,6]). WSNs are composed of two types of nodes: sensor nodes, which collect data, and relay nodes, which transmit data. Each node is usually placed at a static position, and two nodes can communicate when the distance between them is within range. A major problem on WSNs is to place relay nodes so that in the network, there is a path between any two sensor nodes. When the number of relays must be minimized, the problem obtained is called the *Steiner Minimum Tree with Minimum Steiner Points*, which is NP-hard [7]. Another problem on WSNs is to build networks robust to relay failures. This has already been tackled in [8] using a constraint programming approach. Deployment aspects can also be considered, *e.g.* when minimizing the length of a tour which deploys new relays to repair a broken WSN [9].

The two above subproblems have been combined in the robotics field, where strategies were defined to explore an unknown environment while maintaining connectivity to a base station [10]. These strategies consist in maintaining an exploration frontier and in extending this frontier progressively, with some exploration robots taking the role of relays when the frontier to base station distance becomes too big [11]. In other contexts, static relays are deployed using one relay-deployment node, and the objective is both to place relays and minimize the length of the path for the relay-deploying robot [12,13].

Contribution. In previous approaches mixing exploration and communication maintenance, sensors and relays are deployed during successive rounds. At each round, the planning process decides on how to extend the exploration frontier and on how to place relays. Once relays are placed, sensing and data transmission occur. When planning acquisitions for the current round, future rounds are not considered. Such a greedy deployment strategy can lead to suboptimal plans, therefore we propose to reason over a larger horizon by viewing the problem as a scheduling problem. As scheduling is one of the most successful application area of constraint programming, we explore the use of a constraint-based approach for deploying the team of cooperative vehicles. To the best of our knowledge, no prior constraint programming or scheduling approach has been proposed for solving the mixed exploration/communication problem.

Another drawback of existing exploration strategies with communication maintenance is that they synchronize actions of all mobile robots at each exploration round. This induces executions in which at each round, each vehicle waits for the placement of all relays, even if it could start its exploration task earlier, and then each vehicle waits for the end of the exploration of all other vehicles, even if it could perform another exploration task. One benefit offered by the scheduling approach we propose is that it has the capacity to avoid synchronizing activities which do not need to be synchronized. Schedules produced only require an ordering between conflicting activities, and not between all activities. This may improve reactivity during crises.

Operationally speaking, we consider that schedules are produced at the operation center, in a centralized way, before being dispatched and executed on-board each vehicle, in a distributed way. For this reason, schedules produced must also be flexible and robust to the uncertainty about the durations of robot moves. These durations may be shorter or longer than expected, especially for ground robots which may encounter unforeseen terrain conditions.

3 Modeling

Good deployment strategies must be generated quickly, in order to be reactive during the crisis. As problems considered may involve numerous acquisitions and possible positions for communication relays, we define a model in the framework of Constraint-Based Local Search [14].

3.1 Constraint-Based Local Search (CBLS)

In CBLS, models are defined by decision variables, constraints, and criteria, as in classical constraint programming. One specificity of CBLS models is the use of so-called *invariants*, which correspond to one-way constraints $x \leftarrow exp$, where x is a variable and exp is a functional expression of other variables of the problem, such as $x \leftarrow sum(i \in [1..N])\, y_i$. The set of invariants in a model must be acyclic, so that a variable is not a function of itself. Fig. 2 shows a CBLS model together with the Directed Acyclic Graph (DAG) of invariants associated with it.

In CBLS, the search space is explored more freely than with standard tree search with backtrack. When searching for a solution to a given CBLS model, all

Decision variables:	*Invariants:*
$\mathbf{var}\{bool\}\, b$	$\mathbf{var}\{int\}\; z \leftarrow ite(b, x, y)$
$\mathbf{var}\{int\}\, x \in [0..10]$	$\mathbf{var}\{int\}\; t \leftarrow (x - y)$
$\mathbf{var}\{int\}\, y \in [2..5]$	$\mathbf{var}\{bool\}\; u \leftarrow (z < t)$
	$\mathbf{var}\{int\}\; v \leftarrow (t + 2)$

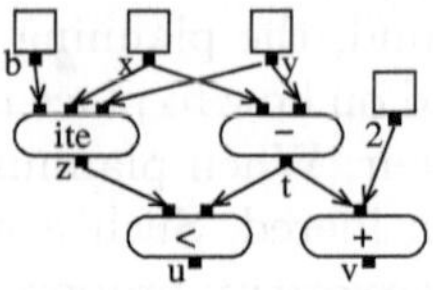

Fig. 2. Example of a CBLS model ($ite(b, x, y)$ stands for "if b then x else y")

decision variables are always assigned, *i.e.* the approach manipulates complete variable assignments. At each step during search, a local move is performed by reassigning some decision variables. Afterwards, all invariants impacted by the local move are reevaluated, following a topological order of the DAG of invariants. A specific procedure is attached to each type of invariant, so that the reevaluation is performed as fast as possible. On previous example $x \leftarrow sum(i \in [1..N])\, y_i$, in case of change of y_k for some $k \in [1..N]$, x can be incrementally reevaluated by adding to it the difference between the current and previous values of y_k, instead of recomputing the sum from scratch (reevaluation in constant time). More generally, invariants allow combinatorial constraints, temporal constraints, resource constraints, and criteria to be very quickly evaluated from a variable assignment and reevaluated from a small change in this assignment.

Several CBLS solvers were developed in the past few years, from the seminal work on Localizer [15] to solvers like COMET [14], iOpt [16], LocalSolver [17], Kangaroo [18], OscaR.cbls [19], or InCELL [20]. In this paper, we use InCELL, which offers flexibility for modeling complex scheduling problems, *e.g.* involving time-dependent scheduling aspects or continuously evolving states. In InCELL, invariants are formally defined as triples (I, O, f) with I and O sequences of variables called the input and output variables respectively, and f a function mapping assignments of I to assignments of O.

3.2 Data of the Mobile Sensor and Relay Deployment Problem

In the following, **R** denotes the number of robots involved in the mission, **A** denotes the number of acquisitions to be performed, and **P** denotes the number of 3D-positions (x, y, z) used in the modeling. $[\mathbf{Hs}, \mathbf{He}]$ denotes the scheduling horizon: every activity must start after Hs and end before He.

Each robot $r \in [1..R]$ is available from time $\mathbf{TimeIni}[r]$, and located at position $\mathbf{PosIni}[r]$ at that time. The duration required by r to move between two positions $p, p' \in [1..P]$ is given by $\mathbf{DuTrans}[r](p, p')$. This duration depends on the robot, because robots may have different motion capabilities. $DuTrans[r]$ is defined implicitly by a specific code: for a UAV, $DuTrans[r](p, p')$ can return the Euclidean distance between p and p' divided by the speed of robot r; for a UGV, $DuTrans[r](p, p')$ can be computed by a path-planning algorithm.

Each acquisition $a \in [1..A]$ can be realized following a certain number of acquisition modes $\mathbf{Nmodes}[a]$. The latter correspond to different ways of scanning the acquisition area. For instance, an acquisition between two points p and p' can be performed from p to p' or from p' to p. With each acquisition mode m are associated positions **AcqPosSta**$[a, m]$ and **AcqPosEnd**$[a, m]$ at the start and end of a, and an acquisition duration **AcqDu**$[a, m]$. Each acquisition generates a data flow with rate **Qos**$[a]$ (quality of service requested for a, in Mb/s).

For each acquisition $a \in [1..A]$ and each robot $r \in [1..R]$, boolean data **AcqFeas**$[a, r]$ takes value *true* iff robot r is equipped with the instrument required for realizing a. Boolean data **IsRelay**$[r]$ takes value *true* iff robot r is equipped with a wireless router and can serve as a relay. The maximum capacity of each relay in terms of data transmission (in Mb/s) is denoted by **RelCap**. We assume that $Qos[a] \leq RelCap$ holds for every acquisition a.

As for communications, we consider that relays can be placed only at predefined locations called *candidate communication nodes*. To define these nodes, one can discretize the environment into a certain number of cells and put one candidate communication node at the center of each cell. In the following, **N** denotes the number of candidate communication nodes and **NodePos**$[n]$ denotes the position of a node n. We assume that one particular node denoted by **OpNode** is associated with the operation center. We also associate with each acquisition $a \in [1..A]$ a node **AcqNode**$[a]$ such that a relay placed at this node is able to receive data sensed during the whole realization of a. If an acquisition is too wide to be covered by a unique node, it is always possible to split it into smaller acquisitions. Boolean function **Linked**(n, n') returns *true* iff a relay placed at node n can communicate with a relay placed at node n'. For communication between UAVs, this function checks whether the distance between the two nodes is not greater than the communication range. Last, the length of each communication path from a sensor node to the base station (number of relays on this path) must not be greater than a given limit, denoted by **NhopsMax**. We assume that for every acquisition a considered individually, a valid communication path from $AcqNode[a]$ to $OpNode$ can be built.

3.3 Decision Variables

Decision variables are given in Eq. 1 to 6. Similarly to IBM ILOG CpOptimizer or to the CAIP framework [21], InCELL represents activities based on the notion of *interval*. An interval itv is defined by a boolean presence variable **pres**(itv), indicating whether the activity is present, and two time-points denoted by **start**(itv) and **end**(itv), representing respectively the start and the end of the activity.

Eq. 1 defines one acquisition interval $acqItv[a]$ per acquisition $a \in [1..A]$, and decision variable $acqMode[a]$ introduced in Eq. 2 represents the realization mode chosen for a. Next, intervals $comItv[k]$ are introduced in Eq. 3 for representing communication activities. Performing a communication activity consists in placing a robot at one of the N candidate communication nodes and in using this robot as a data transmission relay. The communication node in which the kth communication interval is placed corresponds to decision variable $comNode[k]$ given in Eq. 4. Through this choice of variables, we impose that during a single communication activity, the relay robot used must stay at the chosen node. As they are at most A acquisitions and as each communication path can contain at most *NhopsMax* relays, we bound the number of possible communication activities by $\mathbf{K} = A \cdot NhopsMax$. The schedule provided in Fig. 1(b) involves eight acquisition activities (ACQ(a1) to ACQ(a8)) and four communication activities (activities "COM1 at p3" and "COM2 at p4" for robot r2, activity "COM3 at p2" for robot r4, and activity "COM4 at p1" for robot r5).

In order to synchronize acquisition and communication intervals, we introduce, for each acquisition $a \in [1..A]$ and for each communication interval index $k \in [1..K]$, one integer decision variable $useCom[a,k]$ representing the transmission rate (in Mb/s) reserved by a in communication interval k (Eq. 5).

The last set of decision variables (Eq. 6) represents the choice in the sequences of activities *activitySeqs* performed by robots. We use here a type of InCELL called *DisjointIntSequences*(M, T). A variable of this type allows to compactly represent M sequences $s_m = [i_{m,1}, \ldots, i_{m,k_m}]$ composed of integers belonging to $[1..T]$, and such that any integer appears at most once over all sequences. By viewing integers as task indices and sequences as machines, a variable of type *DisjointIntSequences*(M, T) represents, for each of the M machines, the sequence of indices of tasks which are successively performed on this machine. For the mobile sensor and relay deployment problem, we need to consider R machines (one per robot) and $A{+}K$ tasks (one task per acquisition and communication interval), hence we use a variable of type *DisjointIntSequences*$(R, A{+}K)$. An integer $i \in [1..A]$ corresponds to the ith acquisition, and an integer $i \in [A{+}1..A{+}K]$ corresponds to the $(i-A)$-th communication interval. The ordering of activities in Fig. 1 would be represented by the five sequences of integers $r_1 = [4,5,7]$, $r_2 = [9,10]$, $r_3 = [1,2,6]$, $r_4 = [3,11,8]$, $r_5 = [12]$. Variables of type *DisjointIntSequences*(M, T) support several local moves, including the insertion of an integer $i \in [1..T]$ in the sequence of a machine $m \in [1..M]$, or the removal of an integer $i \in [1..T]$ from sequences. Internally, as shown in Eq. 6, they are implemented using three variables per integer $i \in [1..T]$: two integer variables $prev[i]$ and $next[i]$ representing the integer preceding and following integer i in some sequence, and one integer variable $seq[i]$ representing the sequence in which integer i appears, with value 0 when i does not appear in any sequence. Specific indices and values not detailed here are also added to represent the start and end of each sequence.

$$\forall a \in [1..A],\ \textbf{Interval}\ acqItv[a] \in [Hs, He]\quad \texttt{//acquisition intervals} \tag{1}$$

$$\forall a \in [1..A],\ \textbf{var}\{int\}\ acqMode[a] \in [1..Nmodes[a]]\quad \texttt{//acquisition modes} \tag{2}$$

$$\forall k \in [1..K],\ \textbf{Interval}\ comItv[k] \in [Hs, He]\quad \texttt{//communication intervals} \tag{3}$$

$$\forall k \in [1..K],\ \textbf{var}\{int\}\ comNode[k] \in [1..N]\quad \texttt{//communication node} \tag{4}$$

$$\forall a \in [1..A], \forall k \in [1..K],\ \textbf{var}\{int\}\ useCom[a,k] \in [0, RelCap]\quad \texttt{//relay use} \tag{5}$$

$$\textbf{DisjointIntSequences}(R, A{+}K)\ activitySeqs\quad \texttt{//sequences of activities} \tag{6}$$

$$\left[\begin{array}{l} \forall i \in [1..A{+}K{+}R],\ \textbf{var}\{int\}\ prev[i] \in [1..A{+}K{+}R] \\ \forall i \in [1..A{+}K{+}R],\ \textbf{var}\{int\}\ next[i] \in [1..A{+}K{+}R] \\ \forall i \in [1..A{+}K],\ \textbf{var}\{int\}\ seq[i] \in [0..R] \end{array}\right.$$

3.4 Invariants, Constraints, and Criterion

Invariants and constraints, as well as the criterion, are defined in Eq. 7 to 20.

Constraint 7 expresses that every acquisition a must be performed and assigned to a robot. Constraint 8, expresses that a communication interval is present iff it appears in one of the sequences of activities. Constraint 9 enforces that each acquisition is performed by an appropriate robot (we assume that $AcqFeas[a, 0] = true$). Similarly, Constraint 10 enforces that the sequence which contains the kth communication interval must correspond to a robot capable of being a relay (we assume that $IsRelay[0] = true$). Constraint 11 expresses that relay capacities can only be reserved on communication intervals which are present. Invariants given in Eq. 12 to 14 define the start and end positions of each acquisition, function of the acquisition mode, as well as the position of each communication interval, function of the communication node chosen.

Next, Eq. 15 to 17 specify the temporal constraints of the model. These constraints are all simple precedence constraints between start/end time-points of activities. Constraint 15 expresses that the duration of an acquisition interval must be equal to the duration associated with the chosen realization mode. Constraint 16 expresses that if acquisition a uses communication interval k for transmitting data (boolean condition $useCom[a, k] > 0$), then acquisition interval $acqItv[a]$ must be included in communication interval $comItv[k]$. Constraint 17 imposes that for each robot r, there is no overlap between activities assigned to r when these activities are sequenced as specified in *activitySeqs*. Constraint **noOverlap**($TimeIni, PosIni, DuTrans, Itvs, PosSta, PosEnd, activitySeqs$) used in Eq. 17 is a generic temporal constraint of InCELL. It has seven inputs: (1) a table *TimeIni* defining the initial availability time of each machine usable for realizing tasks, (2) a table *PosIni* defining the configuration of each machine at that time, (3) a table *DuTrans* of functions giving the setup time required from one configuration to another, (4) a table of intervals *Itvs* which may be placed on machines, (5) a table *PosSta* such that $PosSta[i]$ defines the configuration required at the start of interval $Itvs[i]$, (6) a table *PosEnd* such that $PosEnd[i]$ gives the configuration obtained at the end of interval $Itvs[i]$, (7) an element *activitySeqs*, of type *DisjointIntSequences*, which defines the successive indices of intervals to be realized on each machine. In InCELL, the *noOverlap* constraint is implemented using several invariants, and formally it ensures that:

- for an activity j placed just after activity i on machine r, $\mathbf{start}(Itvs[j]) \geq \mathbf{end}(Itvs[i]) + DuTrans[r](PosEnd[i], PosSta[j])$;
- for the first activity j on a machine r, $\mathbf{start}(Itvs[j]) \geq TimeIni[r] + DuTrans[r](PosIni[r], PosSta[j])$.

Constraint 18 enforces that for every acquisition a, there must exist a communication path, *i.e.* a sequence of communication nodes $[n_1, \ldots, n_l]$, such that: (1) n_1 is node $AcqNode[a]$ associated with a, (2) n_l is node *OpNode* associated with the operation center, (3) the length l of the path is not greater than *NhopsMax*, (4) every two successive nodes are such that $Linked(n_i, n_{i+1})$ holds. Also, it imposes that the total capacity reserved by a on a node n, defined by $\sum_{k\in[1..K]\,|\,comNode[k]=n} useCom[a,k]$, is equal to $Qos[a]$ if n belongs to the path, and to 0 otherwise. The constraint takes as an input the nodes in which communication intervals are placed as well as all capacities reserved by a on communication intervals. It is expressed in a rather global form because it is handled using specific graph algorithms (see Sect. 4.3). Note that we do not forbid several relays to be placed side by side at the same node. More generally, we consider that possible conflicts on trajectories of robots are handled at execution time, using online collision avoidance techniques.

Constraint 19 imposes that for every communication interval k, the sum of the maximum resource usages of robots on k does not exceed the relay capacity. For example, in the figure on the right, if m_1, m_2, m_3 are the maximum resource usages in interval k by robots r_1, r_2, r_3 resp., then $m_1+m_2+m_3 \leq RelCap$ must hold. This guarantees that relay capacity is not exceeded whatever the real activity dates are at execution time. The approach may be suboptimal, but it is robust and does not require any synchronization between users of a communication interval.

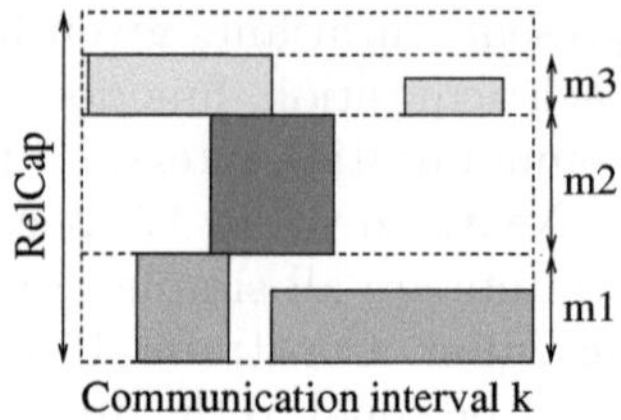

The criterion given in Eq. 20 corresponds to the makespan, defined as the earliest end time of the last acquisition in the schedule.

$$\forall a \in [1..A], \mathbf{pres}(acqItv[a]) \wedge (seq[a] \neq 0) \quad (7)$$

$$\forall k \in [1..K], \mathbf{pres}(comItv[k]) \leftrightarrow (seq[k+A] \neq 0) \quad (8)$$

$$\forall a \in [1..A], AcqFeas[a, seq[a]] \quad (9)$$

$$\forall k \in [1..K], IsRelay[seq[k+A]] \quad (10)$$

$$\forall a \in [1..A], \forall k \in [1..K], (useCom[a,k] > 0) \rightarrow \mathbf{pres}(comItv[k]) \quad (11)$$

$$\forall a \in [1..A], \mathbf{var}\{int\}\ spos[a] \leftarrow AcqPosSta[a, acqMode[a]] \quad (12)$$

$$\forall a \in [1..A], \mathbf{var}\{int\}\ epos[a] \leftarrow AcqPosEnd[a, acqMode[a]] \quad (13)$$

$$\forall k \in [1..K], \mathbf{var}\{int\}\ comPos[k] \leftarrow NodePos[comNode[k]] \quad (14)$$

$$\forall a \in [1..A], \mathbf{durationEq}(acqItv[a], AcqDu[a, acqMode[a]]) \quad (15)$$

$$\forall a \in [1..A], \forall k \in [1..K], \mathbf{during}(useCom[a,k] > 0, acqItv[a], comItv[k]) \quad (16)$$

$$\mathbf{noOverlap}(TimeIni, PosIni, DuTrans, Itvs, PosSta, PosEnd, activitySeqs) \quad (17)$$

$$\begin{aligned} with:\ & Itvs = (\mathbf{all}(a \in [1..A])\ acqItv[a]) \cdot (\mathbf{all}(k \in [1..K])\, comItv[k]) \\ & PosSta = (\mathbf{all}(a \in [1..A])\ spos[a]) \cdot (\mathbf{all}(k \in [1..K])\, comPos[k]) \\ & PosEnd = (\mathbf{all}(a \in [1..A])\ epos[a]) \cdot (\mathbf{all}(k \in [1..K])\, comPos[k]) \end{aligned}$$

$$\begin{aligned} \forall a \in [1..A],\ & \mathbf{ComPathConstraint}(AcqNode[a], OpNode, Linked, \\ & NhopsMax, Qos[a], comNode, \mathbf{all}(k \in [1..K])\, useCom[a,k]) \end{aligned} \quad (18)$$

$$\forall k \in [1..K], \Big(\sum_{r\in[1..R]} \max_{a\in[1..A]\,|\,seq[a]=r} useCom[a,k]\Big) \leq RelCap \quad (19)$$

$$\mathbf{minimize} \max_{a\in[1..A]} \mathbf{earliestTime}(\mathbf{end}(acqItv[a])) \quad (20)$$

4 Local Search Algorithm

We now describe a local search procedure which produces schedules satisfying all constraints of the model, as the schedule given in Fig. 1.

A possible strategy could be to solve first the exploration problem and then the communication maintenance problem. Solving the exploration problem would consist in choosing sequences of acquisition activities, while solving the communication maintenance problem would consist in adding communication relay activities in the sequences found at the first step. The drawback of such a decomposition approach is that synchronization constraints between acquisition and communication activities are taken into account too late, and poor quality schedules may be produced. See Fig. 3 for an example.

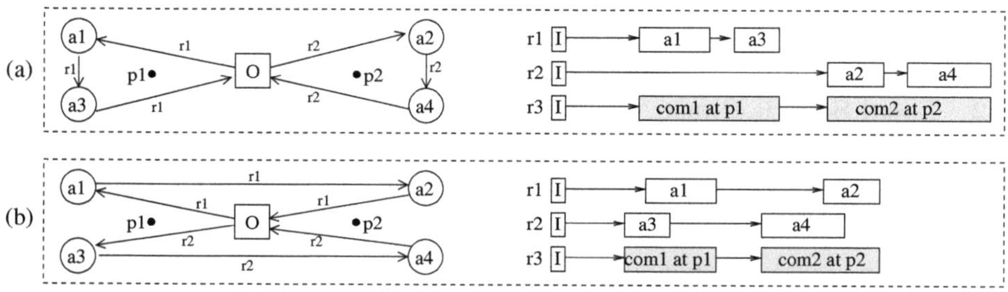

Fig. 3. An example involving two robots r1, r2 capable of making acquisitions and one relay robot r3: (a) schedule obtained by first computing optimal exploration tours, and then adding communication relay activities; (b) a better schedule, which uses longer exploration tours but synchronizes the accesses to relay r3 by robots r1 and r2

The strategy we propose uses two phases: (1) *a constructive phase*, which produces an initial schedule containing all acquisitions; this phase iteratively adds activities at the end of the robot schedules, using a greedy randomized heuristics; (2) *a local search phase*, during which the makespan of the schedule found at the previous step is improved; unlike the constructive phase, the local search phase can modify the schedule in a non chronological way. The two phases are iterated: when the local search phase does not create any new improvement, a restart from an empty schedule occurs, as in the GRASP metaheuristics [22].

4.1 Constructive Phase

The constructive phase starts from an empty schedule. While there exist acquisitions not scheduled yet, we select a pair (r, a) composed of a robot $r \in [1..R]$ and an acquisition $a \in [1..A]$. Robot r is selected randomly among robots which are capable of realizing an acquisition not performed yet, with a probability function of the current end time of the schedule of r (earliest idle robot heuristics). Acquisition a is an acquisition which is (1) feasible by r, (2) not performed yet, and (3) as near as possible from the position of r at the end of its current schedule (nearest neighbor heuristics). These selection operations are implemented with the help of *set invariants*, such as $candidateRobots \leftarrow \{r \in [1..R] \mid \exists a \in [1..A] \neg \mathbf{pres}(acqItv[a]) \wedge AcqFeas[a, r]\}$. The latter CBLS invariant efficiently maintains the set of robots which are candidates for selection.

Acquisition a is then inserted at the end of the schedule of robot r. To do this, interval $acqItv[a]$ is marked as present, and integer a is added at the end of the rth sequence in *activitySeqs*. Next, the communication network allowing a to be covered is built following the graph-based procedure described in Section 4.3. This procedure computes a communication path from a to the operation center, and adds a set of communication intervals $comItv[k]$ at the end of plans of some robots $r' \neq r$. The procedure also chooses the resource usage of a in each communication interval (decision variables $useCom[a, k]$).

The schedule obtained after the constructive phase satisfies all constraints and contains all acquisitions (provided that the horizon end He is large enough).

4.2 Local Search Phase

To improve the schedule generated by the constructive phase, we use local moves that try to relocate acquisitions belonging to the critical path, that is to the list of successive activities justifying the value of the makespan.

The main issue is to avoid considering local moves which create cycles in the temporal precedence graph, and which are therefore trivially inconsistent. To solve this issue, we maintain an ordered list containing all acquisitions. This ordered list is denoted by **NetAccess**, and the meaning of this list is that if an acquisition a_1 appears before an acquisition a_2 in **NetAccess**, then it is guaranteed that no communication interval reserved for a_2 on a relay r is placed strictly before a communication interval reserved for a_1 on r.

Then, each step of the local search works as follows:

1. we randomly select an acquisition a on the critical path; a is removed from the schedule by removing interval $acqItv[a]$, as well as all capacity usages reserved by a; if a was the only user of a communication interval $comItv[k]$, the latter is also removed from the schedule;
2. we then choose a permutation of robots capable of realizing a; as long as a better insertion position has not been found for a, we select the next robot r in the permutation and go to point 3 below; if all robots of the permutation have already been considered, the schedule before the local move is restored and a is marked as currently not relocatable;
3. we try to perform a local move of addition of a into the schedule of r, for each position in the **NetAccess** order; assume that **NetAccess** corresponds to list $[a_1, \ldots, a_n]$; the insertion of a in the schedule of r, between a_i and a_{i+1} in **NetAccess**, is tested as follows; first, we determine, for each robot r', the ongoing activity $lastItv[r']$ on r' after the realization of $[a_1, \ldots a_i]$; if r' is free after the realization of $[a_1, \ldots a_i]$, new activities can be added to the plan of r' just after $lastItv[r']$ without creating precedence cycles; the only case

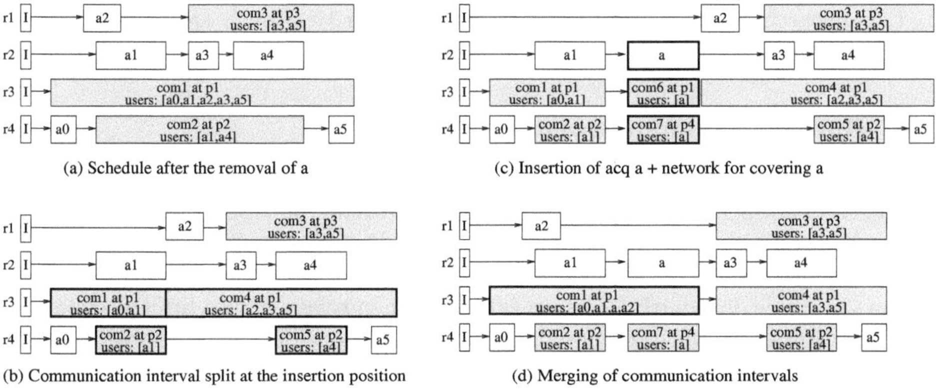

Fig. 4. Example of a local move: addition of acquisition a in the schedule of robot $r2$, between $a1$ and $a2$ in the **NetAccess** order given by $[a1, a2, a3, a4, a5]$

in which r' may not be free is when activity $lastItv[r']$ is a communication interval k, and this communication interval is also used by an acquisition in $[a_{i+1}, \ldots, a_n]$; in this case, inserting new activities for r' between a_i and a_{i+1} may create cycles; to avoid this, we create a new communication interval k' just after k in the schedule of r', and all acquisitions in $[a_{i+1}, \ldots, a_n]$ that use capacity on k are redirected to k'; an example is given in Fig. 4(b), which explicitly mentions the list of acquisitions which use a given communication interval; in this figure, the insertion of acquisition a between a_1 and a_2 induces a split of communication intervals com1 and com2 compared to the schedule given in Fig. 4(a); the two new intervals created, com4 and com5, contain all users of com1 and com2 placed after a_2 in **NetAccess**; note that splitting communication intervals may postpone acquisition tasks;

4. acquisition a is added to the schedule of r at the chosen position; from the state of the communication network after the realization of $[a_1, \ldots a_i]$, a communication network is built for covering a, using the procedure described in Sect. 4.3; communication activities corresponding to this new network are also added to the schedules, as in Fig. 4(c);
5. to reduce the makespan, we try to merge communication intervals that were split at step 3 for avoiding the creation of cycles, or that have become contiguous following the removal of a at step 1; merging two intervals k, k' means transferring from k' to k as many relay capacity usages as possible; transfers are performed following the **NetAccess** order and they stop as soon as one transfer fails; for instance, in Fig. 4(d), intervals com1 and com6 are merged, and capacity usage of a_2 over com4 is transferred to com1;
6. insertions are tested at all positions in the **NetAccess** order, with the best possible acquisition mode; the best insertion on robot r is kept provided that it improves the makespan; ties are broken by keeping the option that minimizes the sum of the durations of the robot schedules, so as to occupy resources as least as possible; if the relocation of a succeeds, all acquisitions marked as non-relocatable are marked as relocatable again.

Local search ends when all acquisitions of the critical path are marked as not relocatable. A restart is performed if there is still some computing time left.

4.3 Building Communication Paths

To build a communication path for covering an acquisition a, we first build a sequence of connected communication nodes starting at $AcqNode[a]$ and ending at $OpNode$. This point is tackled using EQAR [23], an algorithm capable of building wireless sensor networks with quality of service requirements, when sensors produce data with a certain rate and when relays have a limited capacity. In short, EQAR is an iterative algorithm which considers each sensor in turn. For covering a sensor, it adds a set of relays at some nodes of the communication grid. At each step of the algorithm, each node n of the grid may already contain some relays and have a so-called *residual capacity* $resCap[n]$, which corresponds to the amount of Mb/s that are still available on relays placed at n. EQAR then computes a good communication path by solving a shortest-path problem (using Dijkstra's algorithm) in the graph where there are arcs between any two connected communication nodes, and each arc pointing to node n is weighted by 0 if the residual capacity $resCap[n]$ of n is greater than the quality of service qos required for the sensor (traversing n is free in this case), and by $1-resCap[n]/qos$ otherwise. An illustration of EQAR is given in Fig. 5.

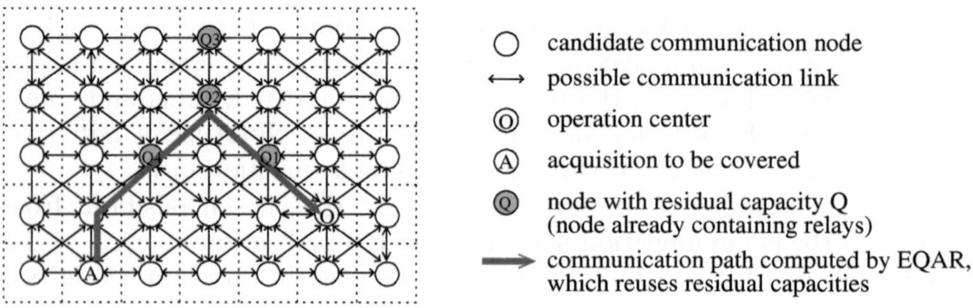

Fig. 5. Illustration of the EQAR algorithm [23]

Using EQAR for covering an acquisition a is quite straightforward. When a must be added to the schedule of robot r, the ongoing activity on each robot $r' \neq r$ is first determined. If the ongoing activity on r' is a communication activity in node n, then we compute the residual capacity offered by r' in n. Using all residual capacities of all robots, except for robot r, we compute a good communication path following the EQAR procedure. This communication path may require the use of new relays at some communication nodes. Deciding on which robot to send to which node in order to build the communication path as fast as possible can be seen as a *Linear Bottleneck Assignment Problem* [24]. Polynomial algorithms do exist for such problems, but we use here a simple greedy procedure which successively sends to each node n a relay robot r able to reach n as fast as possible. An associated communication interval k is added to the schedule of r, and the capacity consumed by acquisition a on k is chosen as high as possible, in order to use the minimum number of relays for covering $Qos[a]$. The communication intervals introduced are merged with previous communication intervals when possible. These steps guarantee the satisfaction of all constraints given in Eq. 18.

5 Experiments

We consider here a 1km×1km crisis area. A communication grid containing 100 cells of size 100m×100m is built, and candidate communication nodes are placed at the center of these cells. The range of a communication relay is 150m, hence each cell can communicate with its eight neighbors. The operation center is placed at a corner of the environment. Two third of the robots can relay communications, the capacity of a relay is 45Mb/s, and we do not limit the length of communication paths. Acquisitions are performed using two types of instruments. Instruments of the first (resp. second) type generate 3Mb (resp. 6Mb) of data per second. Each robot has zero, one or two instrument(s).

As we do not dispose of real data, we build instances by defining sets of acquisition strips of a certain length. These strips are specified by end points chosen randomly, and each strip is split into acquisitions which can be covered by a unique communication node. Table 1 reports some statistics on such instances, concerning the CBLS model and the search phases. Results are obtained on an Intel i5-520 1.2GHz, 4GBRAM. They show that the approach scales quite well when the number of tasks involved in schedules increases (column *nItvs*).

Fig. 6 details results for two specific scenarios involving 14 UAVs. The first one involves 20 acquisitions generated as previously described. The second one involves 100 acquisitions positioned regularly at the center of the 100 cells of the communication grid, which simulates an exploration of the whole environment. Fig. 6 shows the best schedules generated after 50 restarts, as well as the evolution of the makespan during the first 10 restarts. It appears that restarts help in escaping local minima, and that local search quickly improves the value of the makespan given by the greedy constructive phase.

Table 1. Statistics on the CBLS approach; column *nMovesPerSec* counts one local move each time an acquisition is added to a robot using a particular position in the **NetAccess** order (one local move corresponds to the set of operations given in Fig. 4)

A	*R*	*nVars*	*nInvariants*	*nItvs*	*tCreateModel* (sec.)	*tGreedySol* (sec.)	*nMovesPerSec*
20	14	69550	12343	200	0.26	0.002	8295
50	14	238778	30433	500	0.45	0.03	1626
100	14	700827	60583	1000	1.09	0.093	637
20	30	161178	33465	357	0.49	0.008	1532
50	30	632050	97287	1050	1.21	0.06	573
100	30	1760500	193987	2100	2.21	0.197	351

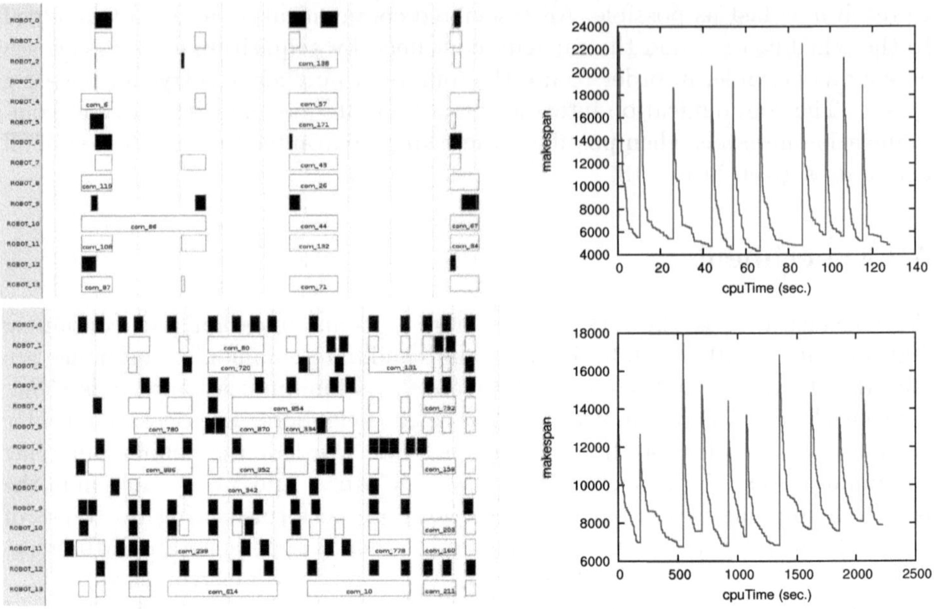

Fig. 6. Schedules produced by the local search and evolutions of the makespan; upper part: scenario involving 20 acquisitions; lower part: scenario involving 100 acquisitions; acquisitions (resp. communications) are depicted in black (resp. white)

6 Conclusion

This paper introduced a CBLS approach for deploying mobile wireless sensor networks. This approach computes efficient schedules which remain executable despite the uncertainty about the duration of robot moves. In terms of modeling, additional resource constraints such as energy limitations could be taken into account, and we could build a CP model in which relays can move during communications. The latter point requires to interleave more finely scheduling with the planning of robot paths, which raises new modeling issues. The next step will be to tackle real scenarios and perform real demonstrations. On this point, we are currently developing a supervision layer able to manage plan execution and to request plan repairs/optimizations when robot moves are longer/shorter than expected, or when operators request new acquisitions during the mission.

References

1. Hentenryck, P.V.: Computational disaster management. In: Proc. of IJCAI 2013, pp. 12–18 (2013)
2. Bektas, T.: The multiple traveling salesman problem: an overview of formulations and solution procedures. OMEGA: The International Journal of Management Science 34(3), 209–219 (2006)

3. Sariel-Talay, S., Balch, T., Erdogan, N.: Multiple traveling robot problem: A solution based on dynamic task selection and robust execution. IEEE/ASME Transactions on Mechatronics, Special Issue on Mechatronics in Multirobot Systems 14(2), 198–206 (2009)
4. Pesant, G., Gendreau, M., Potvin, J.Y., Rousseau, J.M.: On the flexibility of constraint programming models: From single to multiple time windows for the traveling salesman problem. European Journal of Operational Research 117, 253–263 (1999)
5. Akyildiz, I., Su, W., Sankarasubramaniam, Y., Cayirci, E.: Wireless sensor networks: a survey. Computer Networks 38, 393–422 (2002)
6. Younis, M., Akkaya, K.: Strategies and techniques for node placement in wireless sensor networks: A survey. Ad Hoc Networks 6(4), 621–655 (2008)
7. Hwang, F., Richards, D., Winter, P.: The Steiner tree problem. Annals of Discrete Mathematics, vol. 53. Elsevier (1992)
8. Quesada, L., Brown, K., O'Sullivan, B., Sitanayah, L., Sreenan, C.: A constraint programming approach to the additional relay placement problem in wireless sensor networks. In: Proc. of ICTAI 2013, pp. 1052–1059 (2013)
9. Truong, T.T., Brown, K.N., Sreenan, C.J.: Repairing wireless sensor network connectivity with mobility and hop-count constraints. In: Cichoń, J., Gębala, M., Klonowski, M. (eds.) ADHOC-NOW 2013. LNCS, vol. 7960, pp. 75–86. Springer, Heidelberg (2013)
10. Pal, A., Tiwari, R., Shukla, A.: Communication constraints multi-agent territory exploration task. Applied Intelligence 38(3), 357–383 (2013)
11. Mukhija, P., Krishna, K.M., Krishna, V.: A two phase recursive tree propagation based multi-robotic exploration framework with fixed base station constraint. In: Proc. of IROS 2010 (2010)
12. Pei, Y., Mutka, M.W.: Steiner traveler: Relay deployment for remote sensing in heterogeneous multi-robot exploration. In: Proc. of ICRA 2012, pp. 1551–1556 (2012)
13. Pei, Y., Mutka, M.W., Xi, N.: Connectivity and bandwidth-aware real-time exploration in mobile robot. Wireless Communications and Mobile Computing 13(9), 847–863 (2013)
14. Hentenryck, P.V., Michel, L.: Constraint-based local search. The MIT Press (2005)
15. Michel, L., Hentenryck, P.V.: Localizer. Constraints 5(1-2), 43–84 (2000)
16. Voudouris, C., Dorne, R., Lesaint, D., Liret, A.: iOpt: A software toolkit for heuristic search methods. In: Walsh, T. (ed.) CP 2001. LNCS, vol. 2239, pp. 716–719. Springer, Heidelberg (2001)
17. Benoist, T., Estellon, B., Gardi, F., Megel, R., Nouioua, K.: Localsolver 1.x: a black-box local-search solver for 0-1 programming. 4OR: A Quarterly Journal of Operations Research 9(3), 299–316 (2011)
18. Newton, M.H., Pham, D., Sattar, A., Maher, M.: Kangaroo: An efficient constraint-based local search system using lazy propagation. In: Lee, J. (ed.) CP 2011. LNCS, vol. 6876, pp. 645–659. Springer, Heidelberg (2011)
19. Landtsheer, R.D.: OscaR.cbls: a Constraint-Based Local Search Engine (2012)
20. Pralet, C., Verfaillie, G.: Dynamic online planning and scheduling using a static invariant-based evaluation model. In: Proc. of ICAPS 2013 (2013)
21. Frank, J., Jónsson, A.: Constraint-based attribute and interval planning. Constraints 8(4), 339–364 (2003)
22. Feo, T., Resende, M.: Greedy randomized adaptive search procedures. Journal of Global Optimization 6, 109–133 (1995)
23. Lee, S., Younis, M.F.: EQAR: Effective QoS-aware relay node placement algorithm for connecting disjoint wireless sensor subnetworks. IEEE Transactions on Computers 60(12), 1772–1787 (2011)
24. Burkard, R.E., Dell'Amico, M., Martello, S.: Assignment Problems. SIAM (2009)

Air Traffic Controller Shift Scheduling by Reduction to CSP, SAT and SAT-Related Problems*

Mirko Stojadinović

Faculty of Mathematics
University of Belgrade, Serbia
mirkos@matf.bg.ac.rs

Abstract. In this paper we present our experience in solving Air Traffic Controller Shift Scheduling Problem. We give a formal definition of this optimization problem and introduce three encodings. The encodings make possible to formulate a very wide set of different scheduling requirements. The problem is solved by using SAT, MaxSAT, PB, SMT, CSP and ILP solvers. In combination with these solvers, three different optimization techniques are presented, a basic technique and its two modifications. The modifications use local search to modify some parts of the initial solution. Results indicate that SAT-related approaches outperform other solving methods used and that one of the introduced techniques which uses local search can significantly outperform the basic technique. We have successfully used these approaches to make shift schedules for one air traffic control center.

1 Introduction

In the last few decades, personnel scheduling problems have been extensively studied (e.g., nurse scheduling problem [7], course timetabling [9]). Given the input parameters (e.g., the number of available workers, their skills and skills needed for working positions) and constraints (e.g., maximum number of consecutive working days for each worker), a schedule satisfying specified constraints needs to be generated.

In this paper we consider a type of scheduling problem called *Air Traffic Controller (ATCo) Shift Scheduling Problem* (*ATCoSSP*). The objective is to make a shift schedule, so that at each working hour every position is filled by a sufficient number of controllers with adequate skills. Because of the nature and the importance of their job, controllers need to be fully concentrated while they are on position. Therefore, their schedule must satisfy a stricter set of constraints compared to other personnel scheduling problems. To the best of our knowledge, there are no papers which describe encodings or instances of the problem, nor its solving methods in detail. We focus on solving this problem by using different exact methods: *CSP (COP)*, *SAT*, *Partial MaxSAT*, *SMT*, *ILP* and *PB*.

* This work was partially supported by the Serbian Ministry of Science grant 174021.

B. O'Sullivan (Ed.): CP 2014, LNCS 8656, pp. 886–902, 2014.

Constraint satisfaction problems (CSP) and *constraint optimization problems (COP)* [1] are wide classes of problems that include many problems relevant for real world applications (e.g., scheduling, timetabling, sequencing, routing, rostering, planning). Many different approaches for solving CSPs and COPs exist (e.g., constraint programming, mathematical programming, systematic search algorithms, forward checking, answer set programming). *Global constraints* [3] describe relations between a non-fixed number of variables and their purpose is to improve the readability and the efficiency of CSP solving.

Propositional satisfiability problem (SAT) [6] is the problem of deciding if there is a truth assignment under which a given propositional formula (in conjunctive normal form) evaluates to true. It is a canonical NP-complete problem [10] and it holds a central position in the field of computational complexity. *Partial MaxSAT problem* [6] is an optimization version of SAT which consists of finding an assignment that satisfies all *hard clauses* and maximizes the number of satisfied *soft clauses*. *Satisfiability modulo theories (SMT)* [6] is a research field concerned with the satisfiability of formulae with respect to some decidable background theory (or combination of them). Some of these theories are *Linear Integer Arithmetic*, *Integer Difference Logic*, *Linear Real Arithmetic*, etc. *Integer linear programming (ILP) problem* deals with minimizing a linear function while satisfying a set of linear constraints over integer variables. *Pseudo-boolean (PB) problem* [6] is a restriction of ILP problem where the domains of variables are restricted to $\{0, 1\}$. It can be considered as a generalization of SAT problem.

Contributions of this work are the following.

- A formal definition of *ATCoSSP* is given (to the best of our knowledge only informal descriptions are available in literature).
- Three encodings of the problem are introduced. The first two encodings formulate problem as a COP: the first encoding uses linear arithmetic constraints and the well-known global constraint *count*, whereas the second uses linear arithmetic constraints only. The third encoding formulates the problem as a PB problem.
- Different solving methods are compared and a variety of solvers are used. Experimental results indicate that, due to the nature of the constraints, non-SAT-related solvers cannot be on a par with SAT-related solvers when solving this problem is in question. Results show that solver Sugar [29] using reduction to SAT outperforms all other solvers. Sugar efficiently handles arithmetic constraints and can process large number of problem constraints.
- Three optimization techniques are defined and applied to *ATCoSSP*. In each of them we run the solver on instances that differ only in values of optimization variable. The first technique uses a variant of binary search to determine the next value of this variable. The second is intended to improve the solutions of the first approach by using local search specifically adapted to this problem. The third is even more adapted to this problem. It uses a two-step approach that can significantly speed up the solution process by overcoming the main difficulty: a great number of variables and constraints. It finds an

initial solution and iteratively searches for a better solution by fixing some parts of the initial solution and then optimizing its other parts.
- As stated by Burke et al. [7], the drawback of many developed scheduling algorithms is that they were not applied in practice. We have applied the solving methods described in this paper to make shift schedules for one air traffic control center. Instances of the problem are made available online.

Related work. The overview of available results related to *ATCoSSP* is presented by Arnvig et al. [2]. Some software tools for generating *ATCo* schedules already exist [16]. The advantages and the disadvantages of using these tools have already been recognized [31]. Some of them are in-house tools, and the details about them are not available. The others are more general tools (e.g., *Shift Scheduler Continuous*[1]) that can be used only with restricted versions of *ATCoSSP* (e.g., controllers are divided into teams and this is usually the case at big airports).

Scheduling problems have been extensively studied in literature. One of these problems is *Nurse Scheduling Problem* (NSP) [7]. Although there are many similarities between this problem and the problem we discuss, the important difference is that (in its most frequent form) NSP does not include scheduling on an hourly basis. *Course Timetabling* (CTT) [9] is well studied problem that does scheduling on an hourly basis. *ATCoSSP* differs from both cited problems as schedule requirements are much stricter. This implies a great number of hard and a small number of soft constraints, and finding any solution is often very difficult. Most often, known heuristic methods which successfully solve NSP and CTT assume easy way of finding initial solution. The described differences make it very hard to adapt known heuristic methods to *ATCoSSP*.

Overview of the paper. In Section 2 we give the problem description. In Section 3 we describe encodings of the problem and the optimization techniques used. We present our implementation and experimental evaluation in Section 4. Section 5 contains details about the real world application of our implementation. In Section 6 we draw some final conclusions and present ideas for further work.

2 Problem Description

ATCoSSP is a problem of assigning shifts to controllers in a considered period (usually a month or a year) with respect to some requirements. There are many documents that describe these requirements (e.g., [2], [16], [17]). The period consists of a number of days and the days consist of time slots. Each day a controller takes exactly one of three possible types of shifts. During *working shifts* a controller works in an ATCo facility on a given day from the first until the last time slot of that shift (including both these time slots) and rests in the remaining time slots. Working shifts may have different lengths and depending on the first time slot, we distinguish between morning, day, afternoon and night shifts. It is assumed that the time slots of working shifts are known in advance.

[1] `http://www.bizpeponline.com/Helpssce.html`

If a controller does not take working shift on some day, we say the controller takes a *rest shift* on that day (it is equivalent to weekend day for the majority of professions as teachers, lawyers, etc.). Each controller is allowed a number of paid vacation days and we say this person takes a *vacation shift* on these days. Vacations for the period are approved or disapproved in advance by the officials. For each controller, a number of *working hours* in the considered period must be greater than some value min (in order to get a full wage) and smaller than some value max (to avoid fatigue). Each working shift implies a number of working hours equal to the duration of that shift. A rest shift is not counted as working hours. A vacation shift implies some predetermined number of working hours.

Each controller must not take more than a specified number of consecutive working/rest shifts (usually 2 or 3). Only some controllers have the licence to be the heads of the working shift. On each day in each time slot when a facility works, at least one of these controllers has to be in the facility[2]. Each controller must take at least a minimum number of rest shifts per month. All controllers need rest between working shifts, and regulations usually specify a minimum number of rest time slots between working shifts (e.g., 12 hours).

Assigning controllers to positions within their working shifts is also a part of the problem. There are different types of positions in ATCo facilities (e.g., tower, terminal, en route) and depending on a facility size some or all of the positions are present. In any time slot of a working shift a controller can either be on position or can have a break. In any time slot a controller can be assigned to maximum one position. In two consecutive time slots a controller can be on two different positions. A controller must not be on position longer than the specified number m of consecutive time slots. Based on expected air traffic intensity, for each day in each time slot of working hours of a facility (some facilities work 24 hours a day while others do not) a number of controllers is needed for each position. A controller needs certain skills in order to obtain a licence to work on some position. It is assumed that the licences of controllers and the number of needed controllers for each time slot are known in advance[3].

The description so far has been focused only on *hard constraints* which are essential for shift schedule correctness and thus have to be satisfied. *Soft constraints* represent staff wishes (or preferences). Controllers may prefer different working shifts (e.g., they may prefer morning shifts), they may prefer to take consecutive working shifts as rarely as possible, etc. The reasons for including the staff to make schedules and some of most usual preferences are described by Arnvig et al. [2], subsection 6.5.

3 Encodings of the Problem

Time slots can be of any fixed specified length but we assume the length of time slot is 1 hour. Shift schedules are generated for a period of one month and

[2] This requirement can be expressed in terms of shifts instead of time slots.

[3] Actually, determining the number of controllers for each position is a problem itself, and it has been extensively studied in literature (e.g., [17], [31]).

for each month, a new shift schedule needs to be generated. There are many reasons for this: expected monthly traffic intensity changes, different controllers take vacation days in different months, etc. We assume that there are only two types of positions: tower and terminal, and that there are no night shifts. These assumptions are not a limitation since the encodings can be easily extended to support more types of positions and night shifts.

Let us assume that the days are $1, \ldots, n_d$, the controllers are $1, \ldots, n_c$ and the shifts are $1, \ldots, n_s$. In order to make the encodings compact and more efficient, we assume that working shifts are $1, \ldots, n_s - 2$, that $res = n_s - 1$ is rest shift and $vac = n_s$ is vacation shift. Time slots take values $0, \ldots, 23$ and for each shift s, the first (s_f) and the last (s_l) working hour of that shift are fixed.

We experimented with different encodings and constraints. In the following 3 subsections, we describe 3 encodings which showed good results. Only the values of variables that determine controllers shifts for each day and controllers positions for each time slot are used when making tabular schedule for employees. Other variables are auxiliary and they are used to improve the readability and the efficiency of the encodings. The fact that vacation shifts and vacation working hours are fixed for the period is used to make the encodings more compact. Due to space limit, we omit the descriptions of some constraints[4]. In subsection 3.4 we describe how optimization instances are solved.

3.1 The First Encoding

Linear arithmetic constraints and global constraint *count* [3] are imposed on integer variables. The *count* constraint requires that the number of occurrences of the value of the expression e in the set of expressions $e_1, \ldots, e_k$ is in some arithmetic relation ($=, \neq, \leq, <, \geq, >$) with the expression n. E.g., $count(\{x_1, x_2, x_3, x_4\}, 5) > 3$ (where $e = 5$, $e_i = x_i$, the relation is $>$, and $n = 3$) specifies that the value 5 occurs more than 3 times in the set of variables $\{x_1, x_2, x_3, x_4\}$.

Integer variables.

- $dc_{d,c}$: on the day d the controller[5] c can be assigned any from the possible shifts $1, \ldots, n_s$. Note that the fact that the working shifts are $1, \ldots, n_s - 2$ allows us to state that the controller c is working in a facility on the day d by imposing constraint $dc_{d,c} \leq n_s - 2$.
- $dhc_{d,h,c}$: in the hour h of the day d the controller c can be assigned different tasks: c can be on position on tower ($TOW = 0$) or terminal ($TER = 1$), c can have break hour in a facility ($B = 2$), vacation hour ($V = 3$) or a rest time ($R = 4$). Note that this allows us to state that c is having working hour in the facility in the hour h of the day d by imposing constraint $dhc_{d,h,c} \leq B$.
- $h_{d,c}$: on the day d the controller c is counted a certain number of working hours $(0, \ldots, 12)$.

[4] The omitted variable relationships and the constraints of the encodings are available online from: `http://jason.matf.bg.ac.rs/~mirkos/Atco.html`

[5] Most of the constraints have to be true for all controllers, but we use some fixed controller c in the descriptions. Similarly for days, hours and shifts.

Variable Relationships

- If the controller c takes the working shift s on the day d, then c works on working hours of that shift and rests during other hours of the day. So, for any $j \in \{s_f, \ldots, s_l\}$: $dc_{d,c} = s \rightarrow dhc_{d,j,c} \leq B$. For any $j \in \{0, \ldots, 23\} \setminus \{s_f, \ldots, s_l\}$: $dc_{d,c} = s \rightarrow dhc_{d,j,c} = R$.
- If the controller c takes the non-rest shift s with working hours $s_f, \ldots, s_l$ on the day d, then this implies c is working $s_l - s_f + 1$ hours on that day: $dc_{d,c} = s \rightarrow h_{d,c} = s_l - s_f + 1$. In case of the rest shift: $dc_{d,c} = res \rightarrow h_{d,c} = 0$.

Hard Constraints

- *Assigning shifts:* On the day d the controller c takes one of the possible shifts (already imposed as variable $dc_{d,c}$ takes one from the values $1, \ldots, n_s$).
- *Consecutive working shifts:* The controller c must take at least one rest shift in cws days in a row starting from the day d, if c does not take vacation shift on any of these days: $count(\{dc_{d,c}, \ldots, dc_{d+cws,c}\}, res) \geq 1$.
- *Maximum working hours:* The controller c must not work more than the specified max working hours in a month: $\sum_{d=1}^{n_d} h_{d,c} \leq max$.
- *Positions filled:* If in the hour h of the day d at least k controllers are needed for tower position, then the following constraint is imposed: $count(\{dhc_{d,h,1}, \ldots, dhc_{d,h,n_c}\}, TOW) \geq k$. Analogously for terminal position[6].
- *Consecutive time slots on position:* On the day d starting from the hour h the controller c is on position not more than the specified number m of hours in a row: $dhc_{d,h,c} \geq B \vee \ldots \vee dhc_{d,h+m,c} \geq B$.

Soft Constraints. Each wish of a controller is expressed as a constraint that is true iff the wish is not satisfied. Each of these constraints is made equivalent to a fresh integer variable with the domain $\{0, 1\}$. If all these variables take value 0, then all wishes are satisfied. For the fixed controller c a new fresh variable is denoted by $x_{c,i}$, where index i takes the smallest unused non-negative number.

- *Shift preferences:* If the controller c prefers working shifts $s_1, \ldots, s_z$, then each shift s that is different from these shifts, rest or vacation shift is considered undesirable on any day d: $x_{c,i} \leftrightarrow dc_{d,c} = s$. E.g., if the month has 30 days and if the smallest unused non-negative index of the variables associated with the controller c is j, then for undesired shift s variables $x_{c,j}, \ldots, x_{c,j+29}$ are introduced.
- *Minimize consecutive working shifts:* The controller c prefers to take consecutive working shifts as rarely as possible. For each day d: $x_{c,i} \leftrightarrow dc_{d,c} \leq n_s - 2 \wedge dc_{d+1,c} \leq n_s - 2$.

3.2 The Second Encoding

As the syntax of some solvers does not allow the usage of global constraints, we adapt the first encoding not to use these constraints. *Integer variables*, *variable relationships* and *soft constraints* are the same as in the first encoding.

---

6 Actually, several *count* constraints are replaced by one *global cardinality* constraint [3] in order to obtain stronger filtering, but we omit the details.

Hard Constraints. As most of the hard constraints are the same as in the first encoding, we only describe differently encoded constraints. New variables and constraints specifying their relationships are introduced for if-then-else expressions, so this encoding can be of much greater size than the first one.

- *Consecutive working shifts:* The controller c must take at least one rest shift in cws days in a row starting from the day d, if c does not take vacation shift on any of these days: $dc_{d,c} = res \vee \ldots \vee dc_{d+cws,c} = res$.
- *Positions filled:* If in the hour h of the day d at least k controllers are needed for tower position, then the following constraint is imposed: $\sum_{c=1}^{n_c}$(if $(dhc_{d,h,c} = TOW)$ then 1 else 0) $\geq k$. Analogously for terminal position.

3.3 The Third Encoding

If $l_1, \ldots, l_n$ are Boolean literals, then the formula $l_1 + \ldots + l_n \# k, \quad k \in \mathbb{N}, \quad \# \in \{\leq, <, \geq, >, =\}$ is called *Boolean cardinality constraint* (BCC) [27]. In our presentation of the constraints we use equivalences, implications and clauses as often as possible in order to improve the readability of the paper, but the third encoding actually uses BCCs only. Each equivalence can be converted to 2 implications (from left to right and vice versa). The implication $a \rightarrow b$ can be directly translated to a clause $\neg a \vee b$ and more complicated implications can be translated to clauses by using De Morgans laws and distributivity rules[7]. Note that each clause $l_1 \vee \ldots \vee l_n$ is actually BCC $l_1 + \ldots + l_n \geq 1$.

Propositional Variables

- $dcs_{d,c,s}$: on the day d the controller c takes the shift s.
- $dc_{d,c}$: on the day d the controller c takes a working shift.
- $dhc_{d,h,c}$: the hour h of the day d for the controller c is a working hour (in a facility or on vacation). If false, c is having rest hour.
- $tow_{d,h,c}/ter_{d,h,c}/pos_{d,h,c}$: in the hour h of the day d the controller c is on position on tower/on position on terminal/on any position.
- $b_{d,h,c}/v_{d,h,c}$: in the hour h of the day d the controller c has break hour in a facility/has vacation working hour.

Variable Relationships

- If the controller c takes the working shift s on the day d, then c works on working hours of that shift and does not work during other hours of the day. So, for any $j \in \{s_f, \ldots, s_l\}$: $dcs_{d,c,s} \rightarrow dhc_{d,j,c}$. For any $j \in \{0, \ldots, 23\} \setminus \{s_f, \ldots, s_l\}$: $dcs_{d,c,s} \rightarrow \neg dhc_{d,j,c}$.
- The controller c works in the hour h of the day d iff c is on position on tower or on position on terminal or has break hour in a facility or has vacation working hour: $dhc_{d,h,c} \leftrightarrow tow_{d,h,c} \vee ter_{d,h,c} \vee b_{d,h,c} \vee v_{d,h,c}$.
- The controller c is on position in the hour h of the day d iff c is on position on tower or on terminal: $pos_{d,h,c} \leftrightarrow tow_{d,h,c} \vee ter_{d,h,c}$.

[7] There is no risk of exponential blow-up as implications in this encoding have small number of literals.

Hard Constraints

- *Assigning shifts:* On the day d the controller c takes one of the possible shifts (any of working shifts, rest or vacation shift): $dcs_{d,c,1} + \ldots + dcs_{d,c,n_s} = 1$.
- *Consecutive working shifts:* The controller c must not work in a facility more than cws days in a row starting from the day d, if c does not take vacation shift on any of these days: $\neg dc_{d,c} \vee \ldots \vee \neg dc_{d+cws,c}$.
- *Maximum working hours:* The controller c must not work more than the specified max working hours in a month: $\sum_{d=1}^{n_d} \sum_{h=0}^{23} dhc_{d,h,c} \leq max$.
- *Positions filled:* If in the hour h of the day d at least k controllers are needed for tower position, then the following constraint is imposed: $\sum_{c=1}^{n_c} tow_{d,h,c} \geq k$. Analogously for terminal position.
- *Consecutive time slots on position:* On the day d starting from the hour h the controller c is on position not more than the specified number m of hours in a row: $\neg pos_{d,h,c} \vee \ldots \vee \neg pos_{d,h+m,c}$.

Soft Constraints. Fresh Boolean variables are introduced in the same way as integer variables with the domain $\{0,1\}$ in the first encoding.

- *Shift preferences:* If the controller c prefers working shifts $s_1, \ldots, s_z$, then any other shift s that is different from these shifts, rest or vacation shift is considered undesirable on any day d: $x_{c,i} \leftrightarrow dcs_{d,c,s} = 1$.
- *Minimize consecutive working shifts:* The controller c prefers to take consecutive working shifts as rarely as possible. For each day d: $x_{c,i} \leftrightarrow dc_{d,c} \wedge dc_{d+1,c}$.

3.4 Search for Optimum

Controllers indicate importance for their wishes and this is expressed by associating integer weight with each wish they have. In order to make the schedule fair, the weights are scaled so that for each controller the sum of weights of all wishes is equal to some fixed value. If the controller c specifies m_c wishes expressed by the Boolean variables $x_{c,i}$ (introduced in the description of soft constraints), and if associated weights are scaled to the values $w_{c,i}$, then *controllers penalty* is defined: $c_{penalty} = \sum_{i=1}^{m_c} w_{c,i} \cdot x_{c,i}$. For each controller c, the constraint of the form $c_{penalty} \leq cost$ is imposed (for the fixed value of some integer variable $cost$). The goal is to find a minimum non-negative value for this variable (the maximum of all controllers penalties is to be minimized). Note that all $x_{c,i}$ have the domain $\{0,1\}$, so the upper constraint can be encoded either as a linear expression (for the first two encodings) or BCC (for the third encoding).

Three optimization techniques are used. For all of them, instances for different values of $cost$ (with bounds $cost_l$ and $cost_r$) are generated and solved by new runs of the associated solver. The solving process starts from the beginning for each new value of $cost$ on the instance which differs from the previous instance only in this value. We are aware that there are approaches that can improve efficiency by using incremental solving [14,30], but the improvements should not be significant due to results of conducted experiments (Subsection 4.2). Linear search, binary search and many other algorithms [6] can be used to find an optimal value of $cost$ variable. In all three techniques we use (asymmetric) binary search algorithm combined with some additional techniques.

The first technique (*bsBasic*). In this approach pure (asymmetric) binary search is used. If a solution is found and the maximum controllers penalty is some value sol ($sol = \max_{c=1}^{n_c} c_{penalty}$, where $c_{penalty}$ is calculated for the found values of $x_{c,i}$), then $cost_r = sol - 1$. If no solution exists, then $cost_l = cost + 1$. Next instance considers the value $cost = cost_l + k \cdot (cost_r - cost_l)$. For $k = 1/2$ this is symmetric binary search, and for $k \geq 1/2$ the satisfiable instances are favored (they are usually easier). The search is ended and an optimum is found when $cost_l$ becomes greater than $cost_r$.

The second technique (*bsShExc*). Having found a solution with the maximum penalty sol, a local search is used in order to improve the solution (to reduce sol). The local search is based on shift exchanges. Two shifts can be exchanged between two controllers if they are of the same length and if the exchange does not violate any of the hard constraints. In the example schedule given in Table 1, if Alice and Charlie have licences for the same positions and Alice prefers working in the morning, then their working shifts on the day 4 are promising candidates for exchange. The shifts are exchanged whenever the greater of two controllers penalties is not increased. This assures that sol cannot be increased. In order to escape from local minimum, after a number of iterations a certain number of random shift exchanges is performed (thus maybe increasing sol). After a number of local search iterations, the binary search continues.

The third technique (*bsNoPos*). The third and the second technique are similar. The difference is in the way the local search is performed. This technique assumes that all controllers have the licence to work on all positions. We aim to get much smaller encodings by replacing position requirements with working shift requirements. A number of controllers is needed for each position in each time slot, as noted in Section 3. In the initially found solution, a sufficient number of controllers is assigned to each position in each time slot and at the same time a number of controllers is assigned to each working shift on each day (e.g., on the day 2 in the example given in Table 1, 2 controllers are assigned to the shift 3 and no controllers are assigned to the shift 1).

The goal is to reduce the number of assigned controllers to each working shift and to get less filled shift schedule than the one initially found. Let us first assume that for some days d_1 and d_2 the same number of controllers is needed for each position in each time slot of these days (we assume this is the case with days 1 and 3 in the example). The second assumption is that for each working shift, the

Table 1. Small example of schedule for only 4 days (no position schedule presented). Shifts are 1 (04-12), 2 (08-16), 3 (12-20), 4 (rest), 5 (vacation).

Name	Day 1	Day 2	Day 3	Day 4
Alice	2 (08-16)	4 (rest)	2 (08-16)	3 (12-20)
Bob	4 (rest)	3 (12-20)	3 (12-20)	4 (rest)
Charlie	1 (04-12)	4 (rest)	1 (04-12)	2 (08-16)
Dave	4 (rest)	3 (12-20)	5 (vac.)	5 (vac.)

number of assigned controllers to that shift on day d_1 is less or equal than the one on day d_2. The third assumption is that for at least one working shift, the number of assigned controllers to that shift on day d_1 is strictly less than the one on day d_2 (days 1 and 3 fulfill both the second and the third requirement). If these three assumptions are met, then the number of needed controllers for each working shift of d_2 is made equal to the corresponding number of d_1, thus decreasing the number of needed controllers for working shifts of the day d_2 (no controller is needed for the shift 3 on day 3). The positions assignment for d_1 is copied to position assignment for d_2. We repeat this until there is not a day for which the number of needed controllers for any working shift can be reduced.

The search is now continued in the encodings where instead of positions variables and constraints, the constraints specifying the number of needed controllers for each working shift on each day are imposed (e.g., in the first encoding for day 2: $count(\{dc_{2,1}, \dots, dc_{2,4}\},\ 3) \geq 2$). Each time slot of these shifts is already associated with a position. This significantly reduces the encodings size (we denote the original encoding as *complete* and this encoding as *reduced* encoding). When an optimum on the *reduced* encoding is found, it is not necessary the optimum of the initial problem. The binary search then continues on *complete* encoding.

4 Experimental Evaluation

All the tests were performed on a multiprocessor machine with AMD Opteron(tm) CPU 6168 on 1.9Ghz with 2GB of RAM per CPU, running Linux. Value $k = 4/5$ is used for optimization as it showed good results in initial experiments. The timeout is 600 minutes (10 hours) for each instance, including both encoding and solving time (the first is negligible in comparison to the second).

Instances. We experimented with making monthly shift schedules for 2 different months for an airport in Vršac, Serbia, that employs 12 controllers. Officials specified different input parameters as they wanted to choose among several schedules. Parameters differed in number of working shifts (from 3 to 5) and their working hours (from 8 to 12), in number of controllers allowed to take vacation, and in few other parameters. For each month 9 instances were generated. Officials selected one of the solutions to be an official schedule[8]. In this way, 18 real-world instances were generated[9]. Additionally, we generated 4 harder instances in order to estimate the scalability of different solving methods. These instances include more controllers (17-25), longer periods (60-120 days) and more controllers are needed for certain positions in certain time slots.

Table 2 shows the average number of variables, constraints and average domain size for each of the encodings. The numbers are presented both for the

[8] The most important parameter in selection was the number of controllers allowed to go on vacation - the intention was to maximize this number.

[9] The source code of our implementation and the instances used in experiments (but without third-party solvers, due to specific licensing) are available online from: `http://jason.matf.bg.ac.rs/~mirkos/Atco.html`

Table 2. Average numbers of variables and constraints, and average domain size on all instances

	Variables		Constraints		Domain	
Encoding	*complete*	*reduced*	*complete*	*reduced*	*complete*	*reduced*
1	10215	3951.7	81295.3	13400	4.79	2.79
2	35222.2	4467	131840	14430	2.82	2.70
3	55447.9	12964	160314	65293.1	2	2

complete and *reduced* encodings used during local search in *bsNoPos* technique. The numbers indicate that the size of *reduced* encoding is significantly smaller compared to *complete* encoding. This is due to a large number of variables and constraints directly connected to position requirements. Both these variables and constraints need to be introduced for each controller, day and time slot, so we advocate that the sizes of *complete* encodings can not be significantly reduced. Although the first two encodings differ only in the way some of the constraints are encoded, the first encoding is much more compact as it uses global constraints.

4.1 Solving Methods and Preliminary Experiments

Solving methods. Four exact solving methods were used to make schedules.

The first method uses state-of-the-art non-SAT-oriented CSP solvers for solving generated CSP and COP instances in two formats. The first two encodings specifications can be directly translated to these formats. XCSP [25] format is used by solvers *Mistral* 1.545 [19] and *Abscon* 112v4 [23]. MiniZinc [24] input format is used by solvers mzn-g12cpx, mzn-g12fd, mzn-g12lazy, mzn-g12mip included in MiniZinc 1.6 distribution and mzn-gecode from Gecode-4.2.0 [26] distribution.

The second method reduces CSP and COP instances in the described formats to satisfiability problems SAT, Partial MaxSAT and SMT (in Linear Integer Arithmetic theory). In this approach, input instances are translated to instances of satisfiability problems in standardized input formats (e.g., DIMACS[10] for SAT, WCNF[11] for Partial MaxSAT, SMT-LIB[12] for SMT). Modern efficient satisfiability solvers are used for finding solutions that are then converted back to the solutions of the original CSPs and COPs. Systems Sugar v2-0-3 [29] and Sat4j 2.3.4 [4] are used for reduction to SAT. We performed the reduction to SMT and solvers Z3 v4.2 [11] and Yices 2.2 [12] are used for solving generated instances. Partial MaxSAT solvers QMaxSAT 0.4 [21] and MSUNCORE-20130422 [22] are used for solving (slightly modified) instances generated by Sugar.

The third method solves the problem instances of the third encoding directly encoded in satisfiability input formats (PBS[13] for PB and already mentioned DIMACS format for SAT) by corresponding satisfiability solvers. We used SAT

[10] ftp://dimacs.rutgers.edu/pub/challenge/satisfiability/doc/satformat.dvi
[11] http://maxsat.ia.udl.cat/requirements
[12] http://www.smt-lib.org
[13] http://www.cril.univ-artois.fr/PB12/format.pdf

Table 3. Comparison of techniques on interesting instances (timeout 600 minutes): cells contain average objective value achieved, the smaller the number, the better the result is; rows represent techniques; *clasp* is used for solving PB instances

Encoding	1	2	2	3	3
Method	Sugar	Sugar	Yices	*clasp*	MINISAT
bsBasic	31.5	29.3	50.8	53.2	66.3
bsNoPos	23.1	22.5	33.2	72.7	73.7

solvers *clasp* 2.1.3 [18], MINISAT 2.2 [13] and Lingeling aqw-27d9fd4-130429 [5], and PB solvers *clasp* 2.1.3 and MINISAT+ 1.0 [15]. In all further experiments in case of SAT and Partial MaxSAT solvers reduction to clauses is done using sequential counters (as they outperformed cardinality networks, that we used in experiments) implemented in system *meSAT* [28]. This system implements 5 encodings of CSP problems to SAT and it uses different encodings of BCCs to SAT. When number of variables is greater than 20, at-most-one constraint (special type of BCC where # is $\leq$ and k is 1) is encoded in a way described by Chen [8]. Otherwise, it is encoded in a way described by Klieber [20].

The fourth method uses ILP solver IBM ILOG CPLEX Optimization Studio[14] with the second encoding specification translated to its input format.

Preliminary experiments. These were conducted on 5 randomly selected instances with the goal to eliminate less efficient solvers. All solvers except Partial MaxSAT solvers were used with the described optimization techniques, as they outperformed the built-in optimization algorithms. All solvers were used in their default configurations.

4.2 Experimental Results

In this subsection we present the results only for the solvers that achieved the best results in preliminary experiments and for the *interesting instances*, the ones for which not all of the best solvers found optimum within given timeout.

Comparison of techniques. In *bsShExc* approach we used 10^6 iterations. After each 10^4 iterations random shift exchanges were performed. However, there was no improvement in the value of *sol*, so we excluded this approach from the experimental results. Table 3 summarizes the results of comparison between the two remaining techniques using the best solvers. The average objective value achieved on interesting instances is given. The results show that in case of the best solvers (Sugar and Yices) *bsNoPos* technique outperforms *bsBasic* technique.

Detailed results. Table 4 shows the results of the best solver/technique combinations. Instances 19-22 are the additionally generated harder instances. If the best achieved objective value from all the methods used (column *Best*) equals the optimum, the content of the cell is printed in italic font. For cells containing 2 numbers, the first number is the value of objective variable achieved by

[14] `http://www-01.ibm.com/software/commerce/optimization/cplex-optimizer/`

Table 4. Results on interesting instances (timeout 600 minutes): each cell contains objective value; the time needed to achieve this value is given in parenthesis; rows represent different instances; *clasp* stands for using *clasp* on PB instances

	Encoding	1	2			3	
Instance	Best	Sugar *bsNoPos*	Sugar *bsNoPos*	Yices *bsNoPos*	QMaxSAT built-in opt.	*clasp* *bsBasic*	MINISAT *bsBasic*
2	52	61 (1)	55 (1)	**52 (1)**	160 (0)	54 (51)	52 (35)
3	52	55 (3)	60 (1)	54 (8)	160 (0)	58 (0)	**52 (129)**
4	28	34 (8)	**28 (41)**	36 (413)	160 (0)	36 (99)	38 (538)
5	*6*	**6 (54)**	10 (455)	12 (457)	160 (0)	20 (95)	10 (587)
6	*20*	**20 (28)**	28 (272)	20 (77)	20 (45)	20 (70)	20 (342)
10	*16*	16 (137)	20 (6)	22 (15)	26 (11)	22 (2)	**16 (105)**
14	*0*	6 (78)	0 (36)	0 (37)	0 (223)	**0 (21)**	0 (447)
15	*4*	20 (587)	18 (125)	**4 (418)**	160 (0)	42 (109)	66 (577)
18	*12*	13 (88)	20 (18)	14 (414)	**12 (77)**	15 (242)	18 (411)
19	*8*	**8 (25)**	8 (40)	8 (329)	160 (0)	10 (522)	110 (563)
20	8	14 (220)	15 (67)	**8 (329)**	160 (0)	95 (109)	160 (0)
21	12	13 (573)	**12 (300)**	42 (145)	160 (0)	160 (0)	160 (0)
22	18	34 (89)	**18 (125)**	160 (0)	160 (0)	160 (0)	160 (0)
average	18.2	23.1	22.5	33.2	115.2	53.2	66.3

the method in the given timeout. Number 160 denotes that the solver did not find any solution on a given instance (160 is used as it is greater than all the objective values obtained). The number in parenthesis is the time (in minutes) needed to achieve the objective value. The cells which represent the best method for each instance are printed bold. In the last row we present the arithmetic mean (average objective value) from the obtained objective values.

Results show that among the best solvers there were no non-SAT-oriented solvers (mzn-g12cpx and *Abscon* were the only ones that managed to solve some of the instances but could not refine any of the found solutions). Interestingly, CPLEX could not solve any instance. By removing constraint *Positions filled*, CPLEX managed to solve some of the instances so probably these constraints are the reason for inefficiency. We attribute success of SAT-oriented solvers to small domains of variables and large number of connection constraints (they represent about 75% of the generated constraints). The best results in average are achieved by reduction to SAT using Sugar. However, there are cases when other solvers achieved better performance. When we look at the hard instances (19-22), we can see that Sugar significantly outperforms other solvers. As BCC encodings of hard instances for the third encoding are very large, MINISAT and *clasp* are less successful with these instances. Sugar efficiently handles arithmetic constraints by using order encoding. This solver can process large number of constraints of the problem as it uses efficient built-in propagation. These are the reasons why it outperforms other solvers. Its success cannot be attributed to the underlying SAT solver, as MINISAT achieved only slightly better performance than Lingeling.

Figure 1 shows the change of average objective value achieved on the interesting instances during time (as if tests were run in parallel) for each of the solving methods presented in Table 4. It is taken that objective value achieved in the

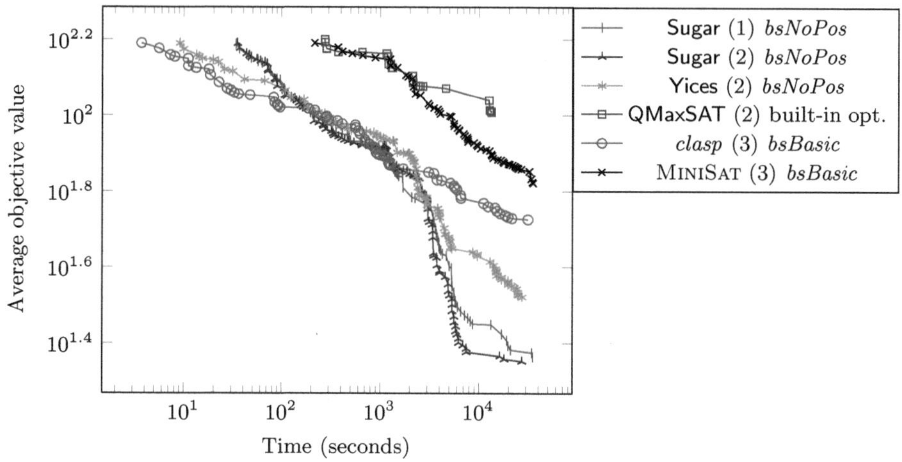

Fig. 1. Average objective value achieved in time - each mark on the curves represents one decrease in value; the encoding for each solver is given in parentheses

beginning is 160 for all solving methods and all instances. Solver Sugar shows the best performance and it achieves similar performance on the first and the second encoding. This can be attributed to similarity of these encodings, with the difference in encoding some of the constraints only. Reduction to PB achieves good results in the beginning and is the best solving method for finding quick solutions. However, it does not scale well in time. The average number of solver runs is 19 when an optimum was found, otherwise it is 3. In presentation of Sugar++ [30] (a version of Sugar using incremental solving) in Third International CSP Solver Competition[15], the authors indicate that the solving time of incremental search can be significantly increased when the number of solver runs is small (less than 6). For these reasons, we advocate that incremental search could not significantly improve the objective values found, although it could reduce the time in cases when optima were found.

5 Real World Applications

The schedules for the airport in Vršac were generated manually prior to using techniques described in this paper. The need for automated generation of schedules is seasonal. The number of needed working hours in summer months increases and the number of available controllers decreases, as this is the time when most of employees go on vacation. Therefore, it becomes too hard to generate schedules manually for these months, and the airport staff is trying to find the way to automate this process. Last year scheduling was offered as a service, not as a tool, as we have not yet developed GUI-based tool to enter input (but we plan to address this issue in our further work). We generated schedules for

[15] `http://www.cril.univ-artois.fr/CPAI08/`

summer months of 2013. It took us about a month to develop the application that reads the input, automatically generates instances, solves them and outputs tabular schedules (their correctness is automatically checked). The automated generation of schedules is also planned for summer months of 2014.

We generated two schedules manually in order to compare them with the automatically generated ones. On average, we achieved objective value 50 in 3 hours by manually generating and objective value 28 in 11 minutes by automated generation (Sugar (2) *bsNoPos*). Many wishes in soft constraints became possible to satisfy by using the automation. This lead to higher satisfaction of the staff. Since automatically generated schedules reduce the overall controllers workload, the controllers are less subject to fatigue, so the overall safety is improved. Also, the manual scheduling was very error-prone. No improvement of solution could be made by using simple shift exchanges, as suggested by the results of *bsShExc* technique. The solution can be improved if we consider the exchange of a larger number of controllers and shifts of different lengths. But in our experience, this is almost impossible to achieve manually.

6 Conclusions and Further Work

In this paper, we have presented the air traffic controller shift scheduling problem. We have described three encodings of the problem in detail and presented three optimization techniques for solving the problem. A variety of solvers have been used for this problem. We have used the described solving methods to design shift schedules for one air traffic control center.

To our knowledge, the presented encodings are the most compact way to introduce position variables and constraints as they are needed for each controller, for each day and for each time slot in all three encodings. These variables and constraints represent the largest part of generated instances. Non-SAT-related approaches are inefficient in processing these instances. SAT-related approaches are significantly more efficient as they compactly encode the variables with small domains and directly encode connection constraints. Experimental results show that the technique *bsNoPos* that fixes assigned positions in local search improves efficiency of the best solver (Sugar). Generally, main lessons learned from the use of CP are that different CP techniques should be tried and that it is beneficial to hybridize CP with other techniques (e.g., local search).

Our experience shows that stated requirements can be very diverse and can change over time. Existing software cannot express these requirements. The encodings we have developed offer a rich modeling language and a possibility to formulate a very diverse set of requirements.

We plan to address unexpected condition changes (e.g., sick leave) in future. Also, the promising directions for further work are improvements of *bsShExc* technique and the usage of other types of local search.

Acknowledgments. We thank Filip Marić, Milan Banković, Mladen Nikolić and anonymous reviewers for valuable comments on the first versions of this manuscript.

References

1. Apt, K.R.: Principles of constraint programming. Cambridge University Press (2003)
2. Arnvig, M., Beermann, B., Köper, B., Maziul, M., Mellett, U., Niesing, C., Vogt, J.: Managing shiftwork in european atm. Literature Review. European Organisation for the Safety of Air Navigation (2006)
3. Beldiceanu, N., Carlsson, M., Rampon, J.-X.: Global constraint catalog. Technical report, SICS (2005)
4. Berre, D.L., Parrain, A.: The sat4j library, release 2.2. JSAT 7(2-3), 59–64 (2010)
5. Biere, A.: Lingeling, plingeling, picosat and precosat at sat race 2010. FMV Report Series Technical Report 10(1) (2010)
6. Biere, A., Heule, M.J.H., van Maaren, H., Walsh, T. (eds.): Handbook of Satisfiability, February 2009. Frontiers in Artificial Intelligence and Applications, vol. 185. IOS Press, Amsterdam (2009)
7. Burke, E.K., De Causmaecker, P., Berghe, G.V., Van Landeghem, H.: The state of the art of nurse rostering. J. Scheduling 7(6), 441–499 (2004)
8. Chen, J.: A new sat encoding of the at-most-one constraint. In: Proceedings of the 9th International Workshop on Constraint Modelling and Reformulation (2010)
9. Chiarandini, M., Birattari, M., Socha, K., Rossi-Doria, O.: An effective hybrid algorithm for university course timetabling. J. Scheduling 9(5), 403–432 (2006)
10. Cook, S.A.: The complexity of theorem-proving procedures. In: Harrison, M.A., Banerji, R.B., Ullman, J.D. (eds.) STOC, pp. 151–158. ACM (1971)
11. de Moura, L., Bjørner, N.S.: Z3: An efficient smt solver. In: Ramakrishnan, C.R., Rehof, J. (eds.) TACAS 2008. LNCS, vol. 4963, pp. 337–340. Springer, Heidelberg (2008)
12. Dutertre, B., De Moura, L.: The yices smt solver, vol. 2, p. 2 (2006), Tool paper at, `http://yices.csl.sri.com/tool-paper.pdf`
13. Eén, N., Sörensson, N.: An extensible sat-solver. In: Giunchiglia, E., Tacchella, A. (eds.) SAT 2003. LNCS, vol. 2919, pp. 502–518. Springer, Heidelberg (2004)
14. Eén, N., Sörensson, N.: Temporal induction by incremental sat solving. Electr. Notes Theor. Comput. Sci. 89(4), 543–560 (2003)
15. Eén, N., Sörensson, N.: Translating pseudo-boolean constraints into sat. JSAT 2(1-4), 1–26 (2006)
16. EUROCONTROL. Shiftwork practices study - atm and related industries. DAP/SAF-2006/56 Brussels: EUROCONTROL (2006)
17. Committee for a Review of the En Route Air Traffic Control Complexity and Workload Model. Air traffic controller staffing in the en route domain: A review of the federal aviation administration's task load model (2010)
18. Gebser, M., Kaufmann, B., Neumann, A., Schaub, T.: *clasp*: A Conflict-Driven Answer Set Solver. In: Baral, C., Brewka, G., Schlipf, J. (eds.) LPNMR 2007. LNCS (LNAI), vol. 4483, pp. 260–265. Springer, Heidelberg (2007)
19. Hebrard, E.: Mistral, a constraint satisfaction library. In: Proceedings of the 3rd International CSP Solver Competition, pp. 31–39
20. Klieber, W., Kwon, G.: Efficient cnf encoding for selecting 1 from n objects. In: Proc. International Workshop on Constraints in Formal Verification (2007)
21. Koshimura, M., Zhang, T., Fujita, H., Hasegawa, R.: Qmaxsat: A partial max-sat solver. JSAT 8(1/2), 95–100 (2012)
22. Marques-Silva, J.: The msuncore maxsat solver. In: SAT 2009 competitive events booklet: preliminary version, p. 151 (2009)

23. Merchez, S., Lecoutre, C., Boussemart, F.: Abscon: A prototype to solve csps with abstraction. In: Walsh, T. (ed.) CP 2001. LNCS, vol. 2239, pp. 730–744. Springer, Heidelberg (2001)
24. Nethercote, N., Stuckey, P.J., Becket, R., Brand, S., Duck, G.J., Tack, G.R.: Minizinc: Towards a standard cp modelling language. In: Bessière, C. (ed.) CP 2007. LNCS, vol. 4741, pp. 529–543. Springer, Heidelberg (2007)
25. Roussel, O., Lecoutre, C.: Xml representation of constraint networks: Format xcsp 2.1. CoRR, abs/0902.2362 (2009)
26. Schulte, C., Lagerkvist, M., Tack, G.: Gecode (2006), Software download and online material at the website, http://www.gecode.org
27. Sinz, C.: Towards an optimal cnf encoding of boolean cardinality constraints. In: van Beek, P. (ed.) CP 2005. LNCS, vol. 3709, pp. 827–831. Springer, Heidelberg (2005)
28. Stojadinović, M., Marić, F.: mesat: Multiple encodings of csp to sat. Constraints (2014), doi:10.1007/s10601-014-9165-7
29. Tamura, N., Banbara, M.: Sugar: A csp to sat translator based on order encoding. In: Proceedings of the Third Constraint Solver Competition, pp. 65–69 (2008)
30. Tanjo, T., Tamura, N., Banbara, M.: Sugar++: a sat-based max-csp/cop solver. In: Proc. the Third International CSP Solver Competition, pp. 144–151 (2008)
31. EATCHIP Human Resources Team. Ats manpower planning in practice: Introduction to a qualitative and quantitative staffing methodology. HUM.ET1.ST02.2000-REP-01 Brussels: EUROCONTROL (1998)

Optimization Bounds from Binary Decision Diagrams*
(Extended Abstract)

David Bergman[1], Andre A. Ciré[2], Willem-Jan van Hoeve[2], and John N. Hooker[2]

[1] University of Connecticut
david.bergman@business.uconn.edu
[2] Carnegie Mellon University
andrecire@cmu.edu, {vanhoeve,jh38}@andrew.cmu.edu

1 Introduction

Bounds on the optimal value are often indispensable for the practical solution of discrete optimization problems, particularly in the branching procedures used by constraint programming (CP) and integer programming solvers. Such bounds are frequently obtained by solving a continuous relaxation of the problem, perhaps a linear programming (LP) relaxation. In this paper, we explore an alternative strategy of obtaining bounds from a *discrete* relaxation, namely a binary decision diagram (BDD). Such a strategy is particularly suitable for CP, because BDDs provide enhanced propagation as well [2–5].

Binary decision diagrams are compact graphical representations of Boolean functions [6–8]. They were originally introduced for applications in circuit design and formal verification [9, 7] but have since been used for a variety of other purposes, such as sequential pattern mining and genetic programming [10, 11].

A BDD can represent the feasible set of a 0-1 optimization problem, because the constraints can be viewed as defining a Boolean function $f(x)$ that is 1 when x is a feasible solution. Unfortunately, a BDD that exactly represents the feasible set can grow exponentially in size. We circumvent this difficulty by creating a *relaxed* BDD of limited size that represents a superset of the feasible set. The relaxation is created by merging nodes of the BDD in such a way that no feasible solutions are excluded. A bound on any additively separable objective function can now be obtained by solving a longest (or shortest) path problem on the relaxed BDD. The idea is readily extended to general discrete (as opposed to 0-1) optimization problems by using *multivalued decision diagrams* (MDDs).

As a test case, we apply the proposed method to the maximum independent set problem on a graph. We find that BDDs can deliver tighter bounds than those obtained by a strong LP formulation, even when the LP is augmented by cutting planes generated at the root node by a state-of-the-art mixed integer solver. In most instances, the BDD bounds are obtained in less computation time, even though we used a non-default barrier LP solver that is faster for these instances.

* This is a summary of the paper: D. Bergman, A.A. Ciré, W.-J. van Hoeve, J.N. Hooker, Optimization bounds from binary decision diagrams *INFORMS Journal on Computing* **26** (2014) 253–268.

B. O'Sullivan (Ed.): CP 2014, LNCS 8656, pp. 903–907, 2014.

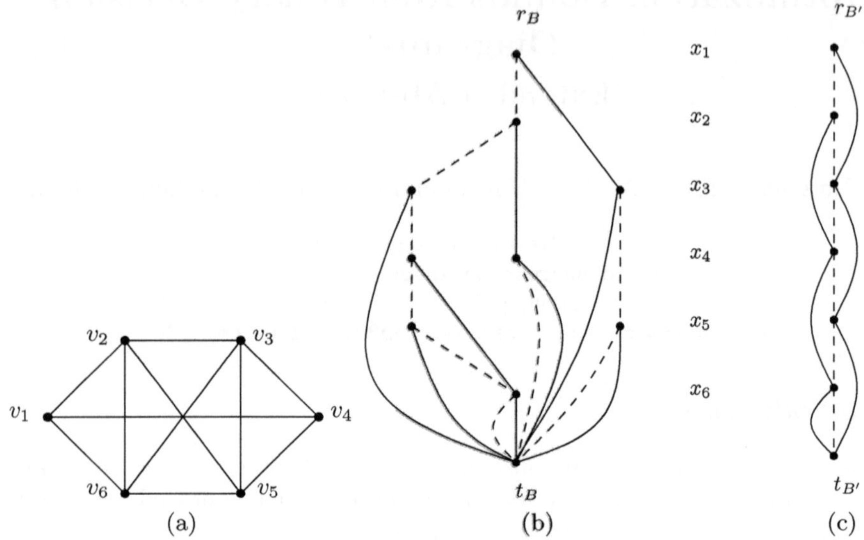

Fig. 1. (a) Instance of the independent set problem. (b) Exact BDD for the instance. (c) Relaxed BDD for the instance.

2 Exact and Relaxed BDDs

An *exact* BDD for an optimization problem is one that represents precisely the feasible solutions of the problem. A *relaxed* BDD, introduced in [2], represents a superset of the feasible solutions.

Figure 1(b) illustrates an exact *zero-suppressed* BDD for the maximum indepedent set problem on the graph in Fig. 1(a). The problem is to find a largest subset of nonadjacent vertices. Binary variable x_i is 1 when vertex v_i is selected (solid arc) and 0 otherwise (dashed arc). The 10 paths from top to bottom correspond to the 10 possible indepedent sets. If a length of 1 is assigned to solid arcs and 0 to dashed arcs, any longest path is an optimal solution.

Figure 1(c) illustrates a relaxed BDD, which encodes a superset of the independent sets. The width (maximum number of nodes per layer) of a relaxed BDD can be restricted while it is generated. In this case, the width is 1.

3 BDD Compilation

We exhibit an efficient top-down compilation algorithm that generates exact reduced BDDs for the independent set problem, and prove its correctness. We then modify the algorithm to generate a limited-size relaxed BDD, prove its correctness, and show that it has polynomial time complexity.

We also discuss variable ordering for exact and relaxed BDD compilation, as this can have a significant impact on the size of the exact BDD and the bound

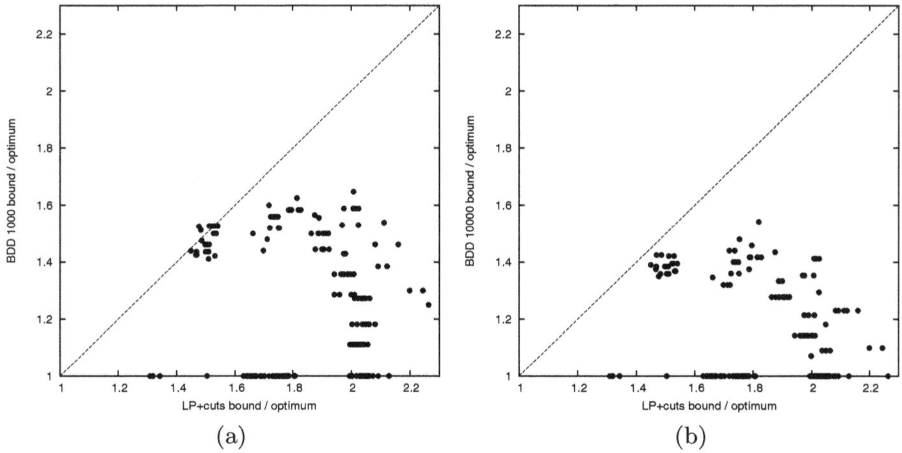

Fig. 2. (a) Bound quality for an LP relaxation with cuts vs. width 1000 BDDs for random instances. The BDD bounds are obtained in about 5% of the time required for the LP bounds. (b) Bound quality comparison for width 10,000 BDDs. The BDD bounds are obtained in less time overall than the LP bounds, but somewhat more time for sparse instances.

provided by relaxed BDDs. We prove that for the independent set problem on n vertices, there is a variable ordering such that the width of an exact BDD is bounded by the nth Fibonacci number. In addition, we describe heuristics for deciding which nodes to merge while building a relaxed BDD and investigate their effectiveness experimentally. This is further explored in [1].

4 Computational Results

We provide here a sampling of computational results on two sets of instances of the maximum independent set problem. One set consists of 180 randomly generated graphs on 200 vertices in which each pair of vertices is joined by an edge with probability $p \in \{0.1, 0.2, \ldots, 0.9\}$. The second set consists of complement graphs of the well-known DIMACS benchmark for the maximum clique problem, obtained from `http://cs.hbg.psu.edu/txn131/clique.html`. The tests ran on an Intel Xeon E5345 with 8 GB RAM in single core mode.

We compared the bound obtained from the BDD relaxation with that obtained from a traditional LP relaxation and cutting planes. To obtain a tight initial LP relaxation, we used a *clique cover* model [12] of the maximum independent set problem, which requires computing a clique cover before the model can be formulated. We then augmented the LP relaxation with all cutting planes generated at the root node by the CPLEX 12.4 MILP solver.

Scatterplots comparing bound quality appear in Fig. 2 for random instances and in Fig. 3 for DIMACS instances. BDD bounds are shown for BDDs of

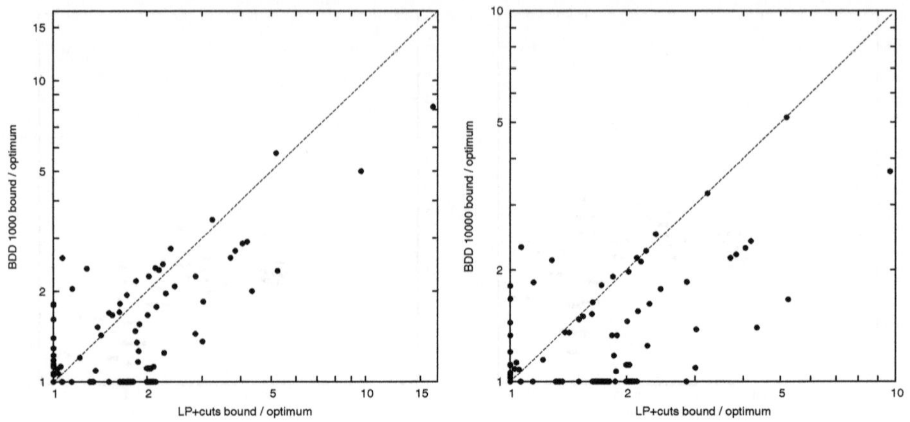

Fig. 3. (a) Bound quality for an LP relaxation with cuts vs. width 1000 BDDs for DIMACS instances. The BDD bounds are obtained in about 15% as much time overall as the LP bounds. (b) Bound quality comparison for width 10,000 BDDs. The BDD bounds are generally obtained in less time for all but the sparsest instances.

maximum width 1000 and 10,000. The fact that almost all points lie below the diagonal indicates the superior quality of BDD bounds.

5 Conclusion

We found that when applied to the maximum independent set problem, BDD-based bounding usually yields significantly better bounds, in less time, than cutting plane technology obtains at the root node in a state-of-the-art mixed-integer solver. This suggests that BDD-based relaxations may have promise as a general technique for bounding the optimal value of discrete problems. They have particular relavance to CP, because BDDs provide enhanced propagation as well.

In addition, BDD-based bounds can be obtained for combinatorial problems that are not formulated as mixed integer models. Unlike LP relaxations, BDD relaxations do not presuppose that the constraints take the form of linear inequalities. Finally, The BDD algorithms presented here are relatively simple, compared with the highly developed technology of LP and mixed-integer solvers. Future research may yield improvements in BDD-based bounding and extend its usefulness to a broader range of discrete optimization problems.

Future research will include the use of BDD-based bounding within a general-purpose BDD-based solver for discrete optimization. Testing of such a solver is now underway for the maximum stable set problem, the maximum cut problem, and the maximum 2-satisfiability problem.

References

1. Bergman, D., Cire, A.A., van Hoeve, W.J., Hooker, J.N.: Variable ordering for the application of BDDs to the maximum independent set problem. In: Beldiceanu, N., Jussien, N., Pinson, É. (eds.) CPAIOR 2012. LNCS, vol. 7298, pp. 34–49. Springer, Heidelberg (2012)
2. Andersen, H.R., Hadzic, T., Hooker, J.N., Tiedemann, P.: A constraint store based on multivalued decision diagrams. In: Bessière, C. (ed.) CP 2007. LNCS, vol. 4741, pp. 118–132. Springer, Heidelberg (2007)
3. Hadzic, T., Hooker, J.N., O'Sullivan, B., Tiedemann, P.: Approximate compilation of constraints into multivalued decision diagrams. In: Stuckey, P.J. (ed.) CP 2008. LNCS, vol. 5202, pp. 448–462. Springer, Heidelberg (2008)
4. Hoda, S., van Hoeve, W.J., Hooker, J.N.: A systematic approach to MDD-based constraint programming. In: Cohen, D. (ed.) CP 2010. LNCS, vol. 6308, pp. 266–280. Springer, Heidelberg (2010)
5. Bergman, D., van Hoeve, W.-J., Hooker, J.N.: Manipulating MDD relaxations for combinatorial optimization. In: Achterberg, T., Beck, J.C. (eds.) CPAIOR 2011. LNCS, vol. 6697, pp. 20–35. Springer, Heidelberg (2011)
6. Akers, S.B.: Binary decision diagrams. IEEE Transactions on Computers C-27, 509–516 (1978)
7. Lee, C.Y.: Representation of switching circuits by binary-decision programs. Bell Systems Technical Journal 38, 985–999 (1959)
8. Bryant, R.E.: Graph-based algorithms for boolean function manipulation. IEEE Transactions on Computers C-35, 677–691 (1986)
9. Hu, A.J.: Techniques for efficient formal verification using binary decision diagrams. Thesis CS-TR-95-1561, Stanford University, Department of Computer Science (December 1995)
10. Loekito, E., Bailey, J., Pei, J.: A binary decision diagram based approach for mining frequent subsequences. Knowl. Inf. Syst. 24(2), 235–268 (2010)
11. Wegener, I.: Branching programs and binary decision diagrams: theory and applications. SIAM monographs on discrete mathematics and applications. Society for Industrial and Applied Mathematics (2000)
12. Grötschel, M., Lovász, L., Schrijver, A.: Geometric Algorithms and Combinatorial Optimization, vol. 2. Springer (1993)

Reformulation Based MaxSAT Robustness* (Extended Abstract)

Miquel Bofill[1,**], Dídac Busquets[2,***], and Mateu Villaret[1,**]

[1] Departament d'Informàtica, Matemàtica Aplicada i Estadística,
Universitat de Girona, Spain
{mbofill,villaret}@imae.udg.edu

[2] Department of Electrical and Electronic Engineering,
Imperial College London, UK
didac.busquets@imperial.ac.uk

Abstract. The presence of uncertainty in the real world makes robustness a desirable property of solutions to Constraint Satisfaction Problems (CSP). A solution is said to be robust if it can be easily repaired when unexpected events happen. This has already been addressed in the frameworks of Boolean satisfiability (SAT) and Constraint Programming (CP). In this paper we consider the unaddressed problem of robustness in weighted MaxSAT, by showing how robust solutions to weighted MaxSAT instances can be effectively obtained via reformulation into pseudo-Boolean formulae. Our encoding provides a reasonable balance between increase in size and performance. We also consider flexible robustness for problems having some unrepairable breakage, in other words, problems for which there does not exist a robust solution.

1 Introduction

Uncertainty is inherent to most real world problems. For instance, in job-shop scheduling, if a machine breaks down, a new solution must be computed. Such a new solution should be fast to compute and, ideally, should also be close to the initial one (e.g., in the job-shop problem, it is not desirable to reassign a large number of tasks). Thus, instead of looking for an optimal solution, which may be brittle and not comply with these two requirements, one could directly look for a solution that can be *easily* repaired (*easy* referring both to time and

* This is a summary of the paper: Miquel Bofill, Dídac Busquets, Víctor Muñoz, Mateu Villaret, "Reformulation based MaxSAT robustness", *Constraints*, April 2013, Volume 18, Issue 2, pp 202-235, doi link: 10.1007/s10601-012-9130-2. A preliminary version of this work was published in: Miquel Bofill, Dídac Busquets, Mateu Villaret, "A declarative approach to robust weighted Max-SAT", *Proceedings of the 12th international ACM SIGPLAN symposium on Principles and practice of declarative programming (PPDP 2010).*

** Supported by the Spanish Ministry of Science and Innovation (project TIN2012-33042).

*** Supported by the UK EPSRC Grand Challenge Project Autonomic Power System.

B. O'Sullivan (Ed.): CP 2014, LNCS 8656, pp. 908–912, 2014.

number of repairs). Such a solution is said to be *robust*. Obviously, this robust solution may be suboptimal, compared to a non-robust one. This fact has been sometimes called *the price of robustness* [2] but, in many real world situations, it is worth sacrificing some optimality for a stronger solution.

In this paper we consider the unaddressed problem of seeking robust solutions in weighted MaxSAT. MaxSAT is becoming a competitive approach for solving combinatorial optimization problems [11] in the CP framework, as well as to deal with Max-Constraint Satisfaction Problems (Max-CSP) [1].

Existing works on robust solutions for (plain) SAT [4,13] and CP [6,7,8,9,10] are based on two main approaches: reformulation and search-based algorithms. The idea of reformulation is to extend the initial instance so that a solution to the extended instance is a robust solution to the initial one [4]. On the other hand, search-based algorithms look for robust solutions with backtracking, propagation and consistency techniques [6,7,8,9,10]. Previous works [5] claim that, at least for CP, the reformulation approach results in prohibitively large formulas.

In this paper we show that reformulation is still feasible in the setting of weighted MaxSAT. This has several advantages. First, the notion of robustness can be directly expressed in the original formulation of the problem with no need of changing the underlying solving method. Additionally, the reformulation can be easily adapted to interesting extensions of the notion of robustness such as adding dependencies between breakable and repairable variables, or introducing failure probabilities, among others. Contrarily, in search-based approaches, if the notion of robustness is modified, the algorithm must probably be modified too.

2 Weighted MaxSAT Robustness

The following definition generalizes the one of Ginsberg et al. [4] to partial weighted MaxSAT.

Definition 1. *Let F be a partial weighted MaxSAT formula and S_1, S_2 and S_3 be sets of variables occurring in F, such that $(S_1 \cup S_2) \cap S_3 = \emptyset$. A $(S_1^a, S_2^b, S_3, \beta)$-* supermodel *of F is a (minimal cost) model of F such that if we modify the values taken by the variables in a subset of S_1 of size at most* **a** *(breakage), then another model can be obtained by modifying the values of the variables in a disjoint subset of S_2 of size at most* ***b*** *(repair) and the values of any number of variables in S_3 (don't-care variables), and moreover the solution and all possible repaired solutions have a cost of at most β. When the set of don't-care variables S_3 is empty we simply talk of (S_1^a, S_2^b, β)-supermodels. Also, if the sets S_1, S_2 and S_3 are unrestricted, we talk of an (a, b, β)-supermodel.*

The idea behind S_3 is that sometimes a formula contains auxiliary or redundant variables, whose values are implied by others, and a change in their values should not be counted neither as a break nor as a repair.

In the following we assume that $F = C \wedge W$ is a partial weighted MaxSAT formula, where C denotes the set of mandatory clauses and W denotes the set of weighted, non-mandatory clauses. W.l.o.g., we assume that W consists only of unary clauses, i.e., $W = (l_1, w_1) \wedge \cdots \wedge (l_k, w_k)$, where l_i is a literal and w_i is a weight for all i in $1..k$. We define $B = \sum_{j \in 1..k} \overline{l_j} \cdot w_j$, which amounts to the cost of the unsatisfied clauses in W.

We now show how we can reformulate an initial formula F to an extended formula F_{SM}, whose solution is a robust solution to F. We achieve this by means of Boolean *cardinality constraints*, which allow us to state, for a given set of Boolean variables E and a given number p, that at most p variables of E can be true [12]. A formula which is only $\mathcal{O}(n^a)$ larger than F is obtained (where n is the number of variables). This is especially important, since it means that the complexity (in size) of our approach does not depend on the number of repairs, but only on the number of breakages, which is usually assumed to be low. This is an important improvement from the encoding in [4], whose size is $\mathcal{O}(n^{a+b})$.

The key idea of the encoding is the following: instead of encoding the different repairs by explicitly flipping (i.e., negating) the variables (as done in [4]), simply rename the variables and restrict the number of variables that can change their value by means of cardinality constraints. As we need a different repair for each possible breakage, a different renaming of the repair (and don't-care) variables is necessary for each possible breakage. To this end, each variable of a repair set R is tagged with the name of the breakage set S which is repairing.

Definition 2. (Variable renaming). *Let R and S be sets of variables. The function $\mathsf{ren}_{R,S} : \mathcal{X} \rightarrow \mathcal{X}$ is defined as $\mathsf{ren}_{R,S}(x) = x^S$ for every variable $x \in R$, where x^S is a new atom, and $\mathsf{ren}_{R,S}(x) = x$ if $x \notin R$.*

Definition 3. (Formula renaming). *Let F be a Boolean (or pseudo-Boolean) formula, and R and S be sets of variables. Then $F^{\mathsf{ren}_{R,S}}$ denotes the formula F where all occurrences of each variable x have been replaced by $\mathsf{ren}_{R,S}(x)$.*

Definition 4. (Difference cardinality). *Let R and S be sets of variables, and $\mathsf{ren}_{R,S}$ be a variable renaming function. Then we define the* difference cardinality formula *as $\nabla^{\mathsf{ren}_{R,S}} = \sum_{x \in R}(x \neq \mathsf{ren}_{R,S}(x)) = \sum_{x \in R}(x \neq x^S)$.*

The encoding of F_{SM}, which will give us a $(S_1^a, S_2^b, S_3, \beta)$-supermodel of F, is the following pseudo-Boolean Optimisation (PBO) instance:

$$
\begin{aligned}
F_{SM}^{\nabla} = \textbf{Minimize} \quad & B \\
\textbf{Subject To} \quad & C \wedge (B \leq \beta) \wedge \\
\bigwedge_{S \subseteq S_1,\, 1 \leq |S| \leq a} \quad & \left(C_{\overline{S}}^{\mathsf{ren}_{(S_2 \setminus S) \cup S_3,\, S}} \wedge B_{\overline{S}}^{\mathsf{ren}_{(S_2 \setminus S) \cup S_3,\, S}} \leq \beta \wedge \nabla^{\mathsf{ren}_{S_2 \setminus S,\, S}} \leq b \right)
\end{aligned}
$$

Roughly speaking, the meaning of the PBO instance F_{SM} is the following: we have first replaced the weighted clauses W of the original formula F by the objective function to be minimized, which corresponds to B, the sum of the weights of the falsified clauses. Then we have C, that is, the mandatory clauses of the original formula, and we add $B \leq \beta$ to bound the cost. Next, we need to be able to repair all possible breakages. The big *and* accounts for all possible (*breakage*) sets S of size smaller than or equal to a. For each of these breakages, we flip the broken variables in the original mandatory clauses and rename those allowed to change in $C_{\overline{S}}^{\mathsf{ren}_{(S_2\setminus S)\cup S_3,\, S}}$, we bound the cost of the new solution with $B_{\overline{S}}^{\mathsf{ren}_{(S_2\setminus S)\cup S_3,\, S}} \leq \beta$, and limit the number of repairs with $\nabla^{\mathsf{ren}_{S_2\setminus S,\, S}} \leq b$. Note that $\mathsf{ren}_{(S_2\setminus S)\cup S_3,\, S}$ is a renaming of the variables in $(S_2 \setminus S) \cup S_3$, by labeling them with S. Since a different renaming is needed for every considered subset S of S_1, we just choose that set S for the renaming, as it improves readability.

In [3] we also provide: the correctness proofs of the F_{SM}^{∇} encoding for finding supermodels of a partial weighted MaxSAT formula F; the generalizations of our definitions and encodings so that partially robust solutions can be obtained for problem instances lacking totally robust solutions, and an extensive benchmarking section considering resource allocation problem instances.

3 Conclusion

In this paper we have proposed a mechanism for finding robust solutions to weighted MaxSAT problems. We have extended the approach of Ginsberg et al. [4] to deal with cost constraints and don't-care variables. By using cardinality constraints, the reformulation results in a much smaller problem in the pseudo-Boolean framework. Moreover, with our approach, the solution to the extended instance provides not only the supermodel for the initial problem, but also a possible repair for each of the potential breakages.

(a, b, β)-super solutions do not exactly match (a, b)-super solutions of [7] because the cost β is not explicitly considered there. Notice that imposing a constraint on the cost of (a, b)-super solutions would only guarantee it for the solution found, but not on the possible repairs. Therefore, if we wanted such restriction to hold also for the repairs, we should modify the algorithm to find (a, b)-super solutions, whilst our approach guarantees a cost of β both for the initial solution and for every possible repair. Using state-of-the-art pseudo-Boolean solvers we have been able to find (a, b, β)-super solutions for combinations of a ranging from 1 to 2 breakages, b ranging from 1 to 8 repairs, and β ranging from a 60% to a 90% of optimality, for several resource allocation problems, most of the times in far less than 1000 seconds. This is quite successful, especially compared to previous works on robustness, which were restricted to (1,0)- and (1,1)-supermodels [4] for reformulation approaches, and to at most (1,3)-super solutions to CSP problems with search-based approaches [5].

Our approach can be seen as a generic framework for robustness through reformulation, since slight changes in the encoding allow to model other notions of robustness. For example, a variant could be to directly designate the potential breaks to handle: instead of using S_1^a, we could decide what (combinations of) breaks deserve being repaired, which would be always a subset of 2^{S_1}. This would be useful if we only want to consider those breaks having a non negligible probability of occurring, particularly in very large problems. We could also think of a robustness notion where each breakable variable has a corresponding set of associated repairable variables.

References

1. Argelich, J., Cabiscol, A., Lynce, I., Manyà, F.: Modelling Max-CSP as Partial Max-SAT. In: Kleine Büning, H., Zhao, X. (eds.) SAT 2008. LNCS, vol. 4996, pp. 1–14. Springer, Heidelberg (2008)
2. Bertsimas, D., Sim, M.: The Price of Robustness. Operations Research 52(1), 35–53 (2004)
3. Bofill, M., Busquets, D., Muñoz, V., Villaret, M.: Reformulation based MaxSAT robustness. Constraints 18(2), 202–235 (2013)
4. Ginsberg, M.L., Parkes, A.J., Roy, A.: Supermodels and robustness. In: 15th National Conference on Artificial Intelligence and 10th Innovative Applications of Artificial Intelligence Conference, AAAI/IAAI 1998, pp. 334–339. AAAI Press / The MIT Press (1998)
5. Hebrard, E.: Robust Solutions for Constraint Satisfaction and Optimisation under Uncertainty. PhD thesis, University of New South Wales (2006)
6. Hebrard, E., Hnich, B., O'Sullivan, B., Walsh, T.: Finding Diverse and Similar Solutions in Constraint Programming. In: 20th National Conference on Artificial Intelligence and 17th Innovative Applications of Artificial Intelligence Conference, AAAI/IAAI 2005, pp. 372–377. AAAI Press / The MIT Press (2005)
7. Hebrard, E., Hnich, B., Walsh, T.: Robust Solutions for Constraint Satisfaction and Optimization. In: 16th Eureopean Conference on Artificial Intelligence, ECAI 2004, pp. 186–190. IOS Press (2004)
8. Hebrard, E., Hnich, B., Walsh, T.: Super Solutions in Constraint Programming. In: Régin, J.-C., Rueher, M. (eds.) CPAIOR 2004. LNCS, vol. 3011, pp. 157–172. Springer, Heidelberg (2004)
9. Hebrard, E., Hnich, B., Walsh, T.: Improved Algorithm for Finding (a,b)-Super Solutions. In: Workshop on Constraint Programming for Planning and Scheduling, pp. 236–248 (2005)
10. Holland, A., O'Sullivan, B.: Weighted Super Solutions for Constraint Programs. In: 20th National Conference on Artificial Intelligence and 17th Innovative Applications of Artificial Intelligence Conference, AAAI/IAAI 2005, pp. 378–383. AAAI Press / The MIT Press (2005)
11. Li, C.M., Manyà, F.: MaxSAT, Hard and Soft Constraints. In: Handbook of Satisfiability, pp. 613–631. IOS Press (2009)
12. Roussel, O., Manquinho, V.: Pseudo-Boolean and Cardinality Constraints. In: Handbook of Satisfiability, pp. 695–734. IOS Press (2009)
13. Roy, A.: Fault Tolerant Boolean Satisfiability. Journal of Artificial Intelligence Research 25, 503–527 (2006)

Probabilistic Constraints for Nonlinear Inverse Problems*

(Extended Abstract)

Elsa Carvalho, Jorge Cruz, and Pedro Barahona

Centro de Inteligência Artificial, Universidade Nova de Lisboa, Portugal
elsac@uma.pt, {jcrc,pb}@fct.unl.pt

The probabilistic continuous constraint (PC) framework complements the representation of uncertainty by means of intervals with a probabilistic distribution of values within such intervals. This paper, published in Constraints [8], describes how nonlinear inverse problems can be cast into this framework, highlighting its ability to deal with all the uncertainty aspects of such problems, and illustrates this new methodology in Ocean Color (OC), a research area widely used in climate change studies with significant applications in water quality monitoring.

Many problems of practical interest can be formulated as nonlinear inverse problems [20]. Such problems aim to estimate parameters from observed data based on a model of the system behavior. The model variables are divided into model parameters, $\boldsymbol{m} = (m_1, \ldots, m_n)$, whose values completely characterize the system and observable parameters, $\boldsymbol{o} = (o_1, \ldots, o_k)$, that can be measured. The model, $\boldsymbol{o} = \boldsymbol{g}(\boldsymbol{m})$, is typically a forward mapping $\boldsymbol{g}$ from the model parameters to the observable parameters. It allows predicting the results of measurements based on the model parameters.

Uncertainty arises from measurement errors on the observed data or approximations in the model specification. When the model equations $\boldsymbol{g}$ are nonlinear, the problem is a nonlinear inverse problem. Nonlinearity and uncertainty play a major role in modeling the behavior of most real systems.

Nonlinear inverse problems are typically ill-posed problems: they may have no exact solutions (no combination of parameter values are capable of predicting exactly all the observed data), solutions are not necessarily unique (different combinations of parameter values may induce the same observable values) and the stability of solutions is not guaranteed (small change in the observed data may induce arbitrarily large changes in the model parameters).

Classical approaches for these problems are based on nonlinear regression methods [2] which search for the model parameter values that best-fit a given criterion. Best-fit approaches, often based on local search methods, provide a non reliable single scenario which may be inadequate to characterize the parameters.

In contrast continuous constraint programming [14,4,19], provides a framework to characterize the set of all scenarios consistent with the constraints of a problem given the uncertainty on its parameters modeled by intervals including all their possibilities. This is achieved through constraint reasoning, which

* This is a summary of the paper: E. Carvalho, J. Cruz, and Pedro Barahona. Probabilistic constraints for nonlinear inverse problems. Constraints, 18(3):344-376, 2013.

B. O'Sullivan (Ed.): CP 2014, LNCS 8656, pp. 913–917, 2014.

relies on branch-and-prune algorithms to obtain sets of boxes that cover exact solutions for the constraints (the feasible space). These algorithms begin with an initial crude cover of the feasible space which is recursively refined by interleaving pruning and branching until a stopping criterion is satisfied. Branching splits a box from the covering into sub-boxes. Pruning either eliminates a box from the covering or reduces it into a smaller (or equal) box maintaining all the exact solutions. Safe pruning is based on safe methods from interval analysis [16] combined within a constraint propagation algorithm [3].

Nevertheless, the application of classical constraint approaches to nonlinear inverse problems [12,10] suffers from the major pitfall of considering the same likelihood for all values in the intervals. To account for different likelihoods, stochastic Monte Carlo techniques [11,1] use extensive random sampling over the different scenarios to characterize the distribution of the model parameter values given the forward model and the observations. However, even after intensive computations, such characterization may be inaccurate, because a significant subset of the probabilistic space may have been missed.

This is not the case of our previous work [6] where we developed an extension to the continuous constraint framework that complements the interval bounded representation of uncertainty with a probabilistic characterization of values distribution. Such information makes it possible to characterize scenarios with a likelihood value, allowing their comparison. Our main emphasis was on the formalization of the framework, which relied on a simplified integration method for computing probability distributions. In [7] we applied this previous approach to two types of simple applications (inverse and reliability problems).

In the present paper we a) provide a validated integration method supported on constraint-based algorithms to compute these distributions, b) study approximations obtained by their hybridization with Monte-Carlo methods, and c) obtain a better uncertainty characterization, by including methods to compute expected values and standard deviations.

The validated integration method relies on the efficiency of constraint reasoning to get a tight box covering of the region of integration, and on the efficacy of interval Taylor methods [5,9] to obtain sharp enclosures for the integrals over the obtained boxes (see figure 1).

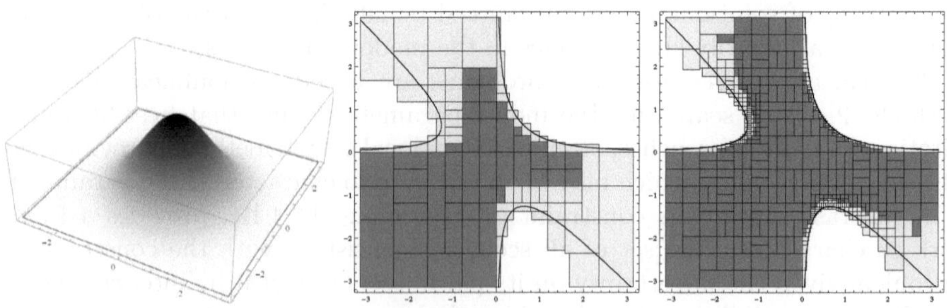
(a) (b) (c)

Fig. 1. Given a probability distribution a), the probability of an event is computed through interval Taylor quadrature over tighter box coverings b) and c)

Although this alternative outputs guaranteed results it is computationally demanding. This justified an hybrid approach which relies on constraint programming to obtain the feasible space and then uses Monte Carlo Integration to sample on this reduced space. We show that this technique, although outputting approximate values, achieves quite accurate results even with small sampling rates and it is much faster than the previous one. Its success relies on the hybridization with constraint programming, since a pure non-naive Monte Carlo method is not only hard to tune but also impractical in small error settings.

Both alternatives allow the computation of probability distributions (both joint and marginal, conditional or unconditional) and extra information, not previously addressed in [6], such as expected values and variances. All these features are illustrated in the OC application (see figure 3 and tables 1 and 2).

OC satellite missions can provide cost-effective environmental indicators at large spatial scales by deriving optically active seawater compounds (OC products) through remote sensing measurements of the sea-surface reflectance [18]. Semi-analytical approaches [15,21,13] handle this problem as a nonlinear inverse problem where field data are used to configure a forward model [17,22] that expresses sea-surface reflectance as a function of the OC products (see figure 2).

In the paper we show how to apply the PC framework to invert the forward model and compute all OC product scenarios consistent with the model, characterized by a probability distribution conditioned by the measurement error. Such information is of extreme importance to understand the impact of measurement uncertainties on the derived OC products, providing support to: a) investigate the applicability of ocean color inversion schemes in different water types; and b) define accuracy requirements for the radiometric sensors to guarantee specified levels of uncertainty for the estimated concentrations. This is an innovative and remarkable contribution to the OC community and can be extended to different parameterizations of the semi-analytical model.

To assess the approach we studied a set of 12 simulated cases representative of different seawater types found in nature. Figure 3 shows the results obtained for the uncertainty on the model parameters given the measurements. Table 1 shows the results obtained to compute the Chla expected value and standard deviation, where the enclosures get sharper as time proceeds.

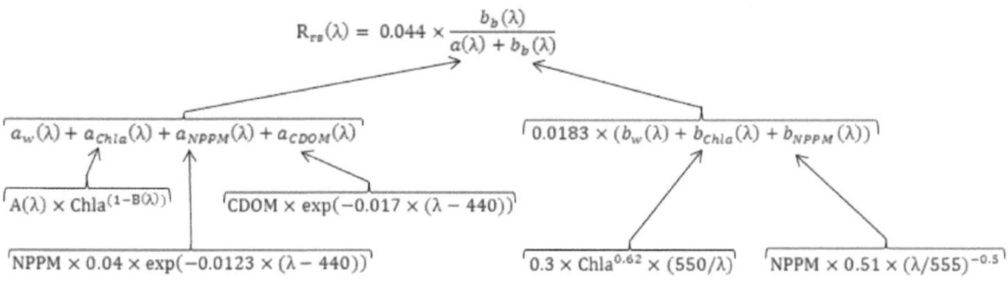

Fig. 2. The forward model is a function from the OC products (*Chla*, *NPPM* and *CDOM*) to the remote sensing reflectance (R_{rs}) at a given wavelength (λ)

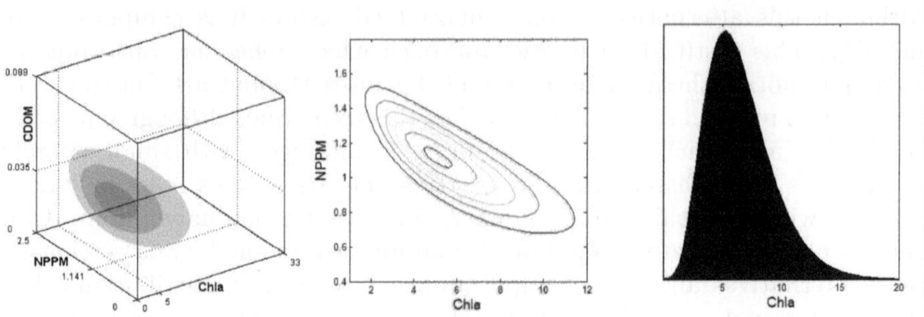

(a) (b) (c)

Fig. 3. Joint and marginal uncertainty distributions computed by the PC framework

Table 1. Interval enclosures computed for $E[Chla]$ and $STD[Chla]$

	$E[Chla]$			$STD[Chla]$		
	enclosure	midpoint	error	enclosure	midpoint	error
10 min	[6.5366, 6.7694]	6.6530	0.1164	[2.6983, 2.9418]	2.8201	0.1218
20 min	[6.5924, 6.7092]	6.6508	0.0584	[2.7716, 2.9011]	2.8364	0.0648
60 min	[6.6260, 6.6742]	6.6501	0.0241	[2.8193, 2.8753]	2.8473	0.0280
300 min	[6.6393, 6.6609]	6.6501	0.0108	[2.8400, 2.8644]	2.8522	0.0122

The hybrid approach, that combines constraint reasoning and Monte Carlo integration, is shown to provide very accurate results in a fraction of the computation time (see table 2). It was also demonstrated that this technique clearly benefits from the contribution of constraint programming to reduce the sample space into a sharp enclosure of the feasible space, combined with the efficiency of Monte Carlo integration.

Table 2. Approximations computed for $E[Chla]$ and $STD[Chla]$

	N	$E[Chla]$	$STD[Chla]$
2 min	5	6.6543	2.8627
3 min	10	6.6545	2.8631
4 min	20	6.6553	2.8629
7 min	50	6.6547	2.8629

In summary the paper overviews the Ocean Color inversion problem and discusses the preliminary results obtained with the PC framework, confirming the relevance of improving methods to control error propagation in the semi-analytical models, an important issue for decisions about the sensors used in satellite-based studies.

References

1. Alrefaei, M.H., Abdul-Rahman, H.M.: An adaptive Monte Carlo integration algorithm with general division approach. Math. Comput. Simul. 79, 49–59 (2008)
2. Bates, D.M., Watts, D.G.: Nonlinear Regression Analysis and Its Applications. Wiley Series in Probability and Mathematical Statistics. John Willey & Sons (1988)
3. Benhamou, F., Goualard, F., Granvilliers, L., Puget, J.-F.: Revising hull and box consistency. In: Procs. of ICLP, Cambridge, MA, USA, pp. 230–244. MIT (1999)
4. Benhamou, F., McAllester, D., van Hentenryck, P.: CLP(intervals) revisited. In: ISLP, pp. 124–138. MIT Press (1994)
5. Berz, M., Makino, K.: New methods for high-dimensional verified quadrature. Reliable Computing 5, 13–22 (1999)
6. Carvalho, E., Cruz, J., Barahona, P.: Probabilistic continuous constraint satisfaction problems. In: ICTAI (2), pp. 155–162 (2008)
7. Carvalho, E., Cruz, J., Barahona, P.: Reasoning with uncertainty in continuous domains. In: Huynh, V.-N., Nakamori, Y., Lawry, J., Inuiguchi, M. (eds.) Integrated Uncertainty Management and Applications. AISC, vol. 68, pp. 357–369. Springer, Heidelberg (2010)
8. Carvalho, E., Cruz, J., Barahona, P.: Probabilistic constraints for nonlinear inverse problems. Constraints 18(3), 344–376 (2013), `http://dx.doi.org/10.1007/s10601-012-9139-6`
9. Goldsztejn, A., Cruz, J., Carvalho, E.: Convergence analysis and adaptive strategy for the certified quadrature over a set defined by inequalities. J. of Comp. and Applied Math. 260, 543–560 (2014)
10. Granvilliers, L., Cruz, J., Barahona, P.: Parameter estimation using interval computations. SIAM J. Scientific Computing 26(2), 591–612 (2004)
11. Hammersley, J.M., Handscomb, D.C.: Monte Carlo Methods. Methuen, London (1964)
12. Jaulin, L., Walter, E.: Set inversion via interval analysis for nonlinear bounded-error estimation. Automatica 29(4), 1053–1064 (1993)
13. Lee, Z., Arnone, R., Hu, C., Werdell, P.J., Lubac, B.: Uncertainties of optical parameters and their propagations in an analytical ocean color inversion algorithm. Appl. Opt. 49(3), 369–381 (2010)
14. Lhomme, O.: Consistency techniques for numeric CSPs. In: Proc. of the 13th IJCAI, pp. 232–238 (1993)
15. Maritorena, S., Siegel, D.A.: Consistent Merging of Satellite OC Data Sets Using a Bio-optical Model. Remote Sensing of Environment 94(4), 429–440 (2005)
16. Moore, R.: Interval Analysis. Prentice-Hall, Englewood Cliffs (1966)
17. Morel, A., Prieur, L.: Analysis of Variation in Ocean Colour. Limnol. Oceanogr. 22, 709–722 (1977)
18. Robinson, I.S.: Discovering the Ocean from Space. Springer (2010)
19. Sam-Haroud, D., Faltings, B.: Consistency techniques for continuous constraints. Constraints 1(1/2), 85–118 (1996)
20. Tarantola, A.: Inverse Problem Theory and Methods for Model Parameter Estimation. SIAM, Philadelphia (2004)
21. Wang, P., Boss, E.S., Roesler, C.: Uncertainties of inherent optical properties obtained from semianalytical inversions of ocean color. Applied Optics 44 (2005)
22. Zibordi, G., Voss, K.: Field radiometry and ocean colour remote sensing. In: Barale, V., Gower, J., Alberotanza, L. (eds.) Oceanography from Space, ch. 18, pp. 307–334. Springer (2010)

Multivalued Decision Diagrams for Sequencing Problems*
(Extended Abstract)

Andre A. Ciré and Willem-Jan van Hoeve

Tepper School of Business, Carnegie Mellon University
5000 Forbes Avenue, Pittsburgh, PA 15213, USA
{acire,vanhoeve}@andrew.cmu.edu

1 Introduction

Sequencing problems are among the most widely studied problems in operations research. Specific variations of sequencing problems include single machine scheduling, the traveling salesman problem with time windows, and precedence-constrained machine scheduling. In this work we propose a new approach for solving sequencing problems based on *multivalued decision diagrams* (MDDs). Decision diagrams are compact graphical representations of Boolean functions, originally introduced for applications in circuit design by Lee [7], and widely studied and applied in computer science. They have been recently used to represent the feasible set of discrete optimization problems, as demonstrated in [2] and [3, 4]. This is done by perceiving the constraints of a problem as a Boolean function $f(x)$ representing whether a solution x is feasible. Nonetheless, such MDDs can grow exponentially large, which makes any practical computation prohibitive in general.

To circumvent this issue, Andersen et al. [1] introduced the concept of a *relaxed MDD*, which is a diagram of limited size that represents an over-approximation of the feasible solution set of a problem. We argue in this paper that such MDDs can be particularly useful as a discrete relaxation of the feasible set of sequencing problems. In particular, we embed relaxed MDDs within a state-of-the-art constraint-based scheduling system, and show that the resulting MDD propagation can reduce the solving time by several orders of magnitude.

2 Problem Definition

Let $\mathcal{J} = \{j_1, \ldots, j_n\}$ be a set of n jobs to be processed on a machine that can perform at most one job at a time. Each job $j \in \mathcal{J}$ has an associated *processing time* p_j, which is the number of time units the job requires from the machine, and a *release date* r_j, the time from which job j is available to be processed. For each pair

* This is a summary of the paper "A. A. Cire and W.-J. van Hoeve. Multivalued Decision Diagrams for Sequencing Problems. *Operations Research*, 61(6):1411–1428, 2013".

B. O'Sullivan (Ed.): CP 2014, LNCS 8656, pp. 918–922, 2014.

of distinct jobs $j, j' \in \mathcal{J}$ a *setup time* $t_{j,j'}$ is defined, which indicates the minimum time that must elapse between the end of j and the beginning of j' if j' is the first job processed after j finishes. We assume that jobs are *non-preemptive*, i.e., we cannot interrupt a job while it is being processed on the machine.

We are interested in assigning a *start time* $s_j \geq r_j$ for each job $j \in \mathcal{J}$ such that job processing intervals do not overlap, the resulting schedule observes a number of constraints, and an objective function f is minimized. Two types of constraints are considered in this work: *precedence constraints*, requiring that $s_j \leq s_{j'}$ for certain pairs of jobs $(j, j') \in \mathcal{J} \times \mathcal{J}$, which we equivalently write $j \ll j'$; and *time window constraints*, where the *completion time* $c_j = s_j + p_j$ of each job $j \in \mathcal{J}$ must be such that $c_j \leq d_j$ for some *deadline* d_j. Furthermore, we study three representative objective functions in scheduling: the *makespan*, where we minimize the completion time of the schedule, or $\max_{j \in \mathcal{J}} c_j$; the *total tardiness*, where we minimize $\sum_{j \in \mathcal{J}} (\max\{0, c_j - \delta_j\})$ for given *due dates* δ_j; and the *sum of setup times*, where we minimize the value obtained by accumulating the setup times $t_{j,j'}$ for all consecutive jobs j, j' in a schedule. Note that for these objective functions we can assume that jobs should always be processed as early as possible (i.e., idle times do not decrease the value of the objective function).

Since jobs are processed one at a time, any solution to such scheduling problem can be equivalently represented by a total ordering $\pi = (\pi_1, \pi_2, \ldots, \pi_n)$ of $\mathcal{J}$. The start time of the job j implied by π is given by $s_j = r_j$ if $j = \pi_1$, and $s_j = \max\{r_j,\ s_{\pi_{i-1}} + p_{\pi_{i-1}} + t_{\pi_{i-1},j}\}$ if $j = \pi_i$ for some $i \in \{2, \ldots, n\}$. We say that an ordering π of $\mathcal{J}$ is *feasible* if the implied job times observe the precedence and time window constraints, and *optimal* if it is feasible and minimizes f.

3 MDD Representation

For the purpose of this work, an (exact) MDD $\mathcal{M}$ is a directed acyclic graph whose paths represent the feasible orderings of $\mathcal{J}$. The set of nodes of $\mathcal{M}$ are partitioned into $n+1$ layers $L_1, \ldots, L_{n+1}$, where layer L_i corresponds to the i-th position π_i of the feasible orderings encoded by $\mathcal{M}$, for $i = 1, \ldots, n$. Layers L_1 and L_{n+1} are singletons representing the root $\mathbf{r}$ and the terminal $\mathbf{t}$, respectively. An arc $a = (u, v)$ of $\mathcal{M}$ is always directed from a *source node* u in some layer L_i to a *target node* v in the subsequent layer L_{i+1}, $i \in \{1, \ldots, n\}$. We write $\ell(a)$ to indicate the layer of the source node u of the arc a (i.e., $u \in L_{\ell(a)}$).

With each arc a of $\mathcal{M}$ we associate a label $val(a) \in \mathcal{J}$ that represents the assignment of the job $val(a)$ to the $\ell(a)$-th position of the orderings identified by the paths traversing a. Hence, an arc-specified path $(a_1, \ldots, a_n)$ from $\mathbf{r}$ to $\mathbf{t}$ identifies the ordering $\pi = (\pi_1, \ldots, \pi_n)$, where $\pi_i = val(a_i)$ for $i = 1, \ldots, n$. Every feasible ordering is identified by some path from $\mathbf{r}$ to $\mathbf{t}$ in $\mathcal{M}$, and conversely every path from $\mathbf{r}$ to $\mathbf{t}$ identifies a feasible ordering.

A *relaxed MDD* is an MDD $\mathcal{M}$ that represents a superset of the feasible orderings of $\mathcal{J}$; i.e., every feasible ordering is identified by some path in $\mathcal{M}$, but not necessarily all paths in $\mathcal{M}$ identify a feasible ordering. We construct relaxed

Job Parameters

Job	r_j	d_j	p_j
j_1	2	20	3
j_2	0	14	4
j_3	1	14	2

Setup Times

	j_1	j_2	j_3
j_1	-	3	2
j_2	3	-	1
j_3	1	2	-

a. Instance data

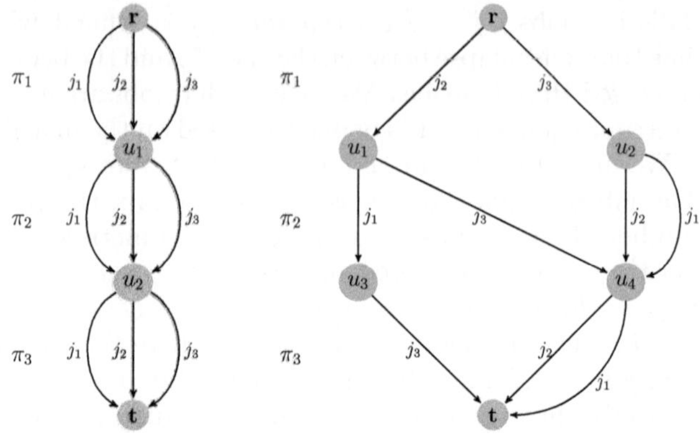

b. MDD relaxation (width 1)

c. MDD relaxation (width 2)

Fig. 1. Example MDD relaxations for a scheduling problem

MDDs by limiting the size to a fixed maximum allowed width W. Thus, the strength of the relaxed MDD can be controlled by increasing W; we obtain an exact MDD by setting W to infinity.

Figures 1.b and 1.c present two examples of a relaxed MDD with maximum width $W = 1$ and $W = 2$, respectively, for the sequencing problem given in Figure 1.a. In particular, the MDD in Figure 1.b encodes all the orderings represented by permutations of $\mathcal{J}$ with repetition, hence it trivially contains the feasible orderings of any sequencing problem.

4 Filtering and Refinement

Considering that a relaxed MDD $\mathcal{M}$ can be easily constructed for any sequencing problem (e.g., the width-1 relaxation of Figure 1.b), we modify $\mathcal{M}$ in order to strengthen the relaxation it provides while observing the maximum width W. These are based on the compilation procedures developed by Hadzic et al. [5] and Hoda et al. [6] for general constraint satisfaction systems, which apply MDD *filtering* and *refinement*.

MDD *filtering* consists in identifying infeasible arcs and removing them from $\mathcal{M}$, which hence eliminates one or more infeasible orderings that are encoded in $\mathcal{M}$. We provide such filtering rules for all constraints and objective functions given in Section 2. Of particular note is the following result regarding MDD filtering for precedence constraints, which shows that MDDs can provide stronger inference than existing scheduling rules based on domain propagation, such as edge finding or not-first/not-last:

Theorem 1. *Let $\mathcal{M}$ be an exact MDD representing a sequencing problem. The set of all precedence relations that must hold in any feasible ordering can be extracted from $\mathcal{M}$ in $O(n^2 |\mathcal{M}|)$ time.*

MDD *refinement* consists in identifying nodes in $\mathcal{M}$ that are encompassing multiple equivalence classes, and *splitting* them into two or more new nodes to represent such classes more accurately (as long as the maximum width W is not violated). Our refinement strategies aim at exactly representing the most important jobs in $\mathcal{J}$.

5 Constraint-Based Scheduling with MDD Propagation

We next provide a brief summary of the application of MDD propagation to the constraint-based scheduling system of IBM ILOG CP Optimizer. We compare the performance of CP Optimizer with (`CPO+MDD`) and without (`CPO`) the additional inference that the MDD relaxations provide.

As illustration, we present results for the TSP with time windows (TSPTW) in Figure 2. Figure 2.a shows that MDD propagation can improve the performance of CPO by several orders of magnitude, which translates in being able to optimally solve more than twice as many problems (Figure 2.b).

Similar results were obtained for other problem classes, including the TSP with precedence constraints (Sequential Ordering Problem), and objective functions, including makespan and total tardiness minimization. We note that using our approach we were able to close three open instances of the Sequential Ordering Problem from TSPLIB.

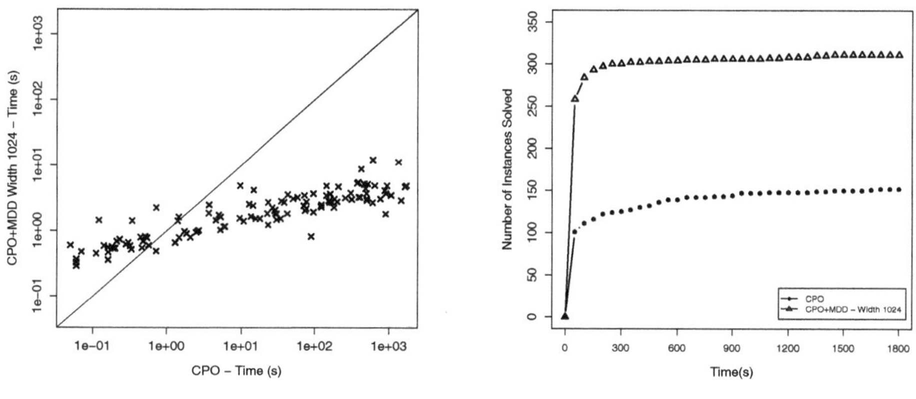

a. Scatter plot of solution time

b. Performance plot

Fig. 2. Performance comparison between `CPO` and `CPO+MDD` for minimizing sum of setup times on Dumas, Gendreau, and Ascheuer TSPTW classes using default depth-first CPO search. The horizontal and vertical axes in (a) are in logarithmic scale.

References

[1] Andersen, H.R., Hadzic, T., Hooker, J.N., Tiedemann, P.: A constraint store based on multivalued decision diagrams. In: Bessière, C. (ed.) CP 2007. LNCS, vol. 4741, pp. 118–132. Springer, Heidelberg (2007)

[2] Becker, B., Behle, M., Eisenbrand, F., Wimmer, R.: BDDs in a branch and cut framework. In: Nikoletseas, S.E. (ed.) WEA 2005. LNCS, vol. 3503, pp. 452–463. Springer, Heidelberg (2005)

[3] Bergman, D., van Hoeve, W.-J., Hooker, J.N.: Manipulating MDD relaxations for combinatorial optimization. In: Achterberg, T., Beck, J.C. (eds.) CPAIOR 2011. LNCS, vol. 6697, pp. 20–35. Springer, Heidelberg (2011)

[4] Bergman, D., Cire, A.A., van Hoeve, W.-J., Hooker, J.N.: Variable ordering for the application of bdds to the maximum independent set problem. In: Beldiceanu, N., Jussien, N., Pinson, É. (eds.) CPAIOR 2012. LNCS, vol. 7298, pp. 34–49. Springer, Heidelberg (2012)

[5] Hadzic, T., Hooker, J.N., O'Sullivan, B., Tiedemann, P.: Approximate compilation of constraints into multivalued decision diagrams. In: Stuckey, P.J. (ed.) CP 2008. LNCS, vol. 5202, pp. 448–462. Springer, Heidelberg (2008)

[6] Hoda, S., van Hoeve, W.-J., Hooker, J.N.: A systematic approach to MDD-based constraint programming. In: Cohen, D. (ed.) CP 2010. LNCS, vol. 6308, pp. 266–280. Springer, Heidelberg (2010)

[7] Lee, C.Y.: Representation of switching circuits by binary-decision programs. Bell Systems Technical Journal 38, 985–999 (1959)

Robustness and Stability in Constraint Programming under Dynamism and Uncertainty*
(Extended Abstract)

Laura Climent[1], Richard J. Wallace[1], Miguel A. Salido[2], and Federico Barber[2]

[1] Insight Center for Data Analytics, University College Cork, Ireland
[2] Instituto de Automática e Informática Industrial, Universitat Politècnica de València, Spain
{laura.climent,richard.wallace}@insight-centre.org,
{msalido,fbarber}@dsic.upv.es

1 Introduction

Because of the dynamism and uncertainty associated with many real life problems, these problems and their associated Constraint Satisfaction Problem (CSP) models may change over time; thus an earlier solution found for the latter may become invalid. Moreover, many approaches proposed in the literature cannot be applied when the required information about dynamism is unknown ([9], [4], [5], [11], [10], etc.). This fact has motivated us to consider dynamic situations where, in addition, only limited assumptions about changes can be made. Our analysis focuses on CSPs with ordered and discrete domains that model problems for which the order over the elements of the domain is significant. In these cases, a common type of change that problems may undergo is restrictive modifications over the bounds of the solution space. A discussion of these assumptions, their motivation and real life examples can be found in [3].

In this paper, we present an algorithm that meets the goal of combining solution *stability* (meaning that solutions can often be repaired using other similar values if they undergo a value loss) and robustness (meaning that solutions have a high likelihood of remaining solutions after changes). The desireability of this combination of features was noted in the survey [8]. Furthermore, in this work we have extended both concepts to apply to the type of CSP analyzed. The paper is organized as follows. Section 2 presents the new conceptions of robustness and stability. Section 3 describes our approach for finding solutions that simultaneously meet both these criteria. In Section 4 we present some experimental results. Section 5 gives conclusions.

2 Extending Robustness and Stability Concepts

Given CSPs with ordered domains, where only limited assumptions are made about changes in the problem that are related to their inherent structure, it is reasonable to

* This is a summary of the paper: L. Climent, R. J. Wallace, M. A. Salido, and F. Barber. Robustness and Stability in Constraint Programming under Dynamism and Uncertainty. Journal of Artificial Intelligence Research, 49:49-78, 2014.
http://www.jair.org/media/4126/live-4126-7626-jair.pdf

B. O'Sullivan (Ed.): CP 2014, LNCS 8656, pp. 923–927, 2014.

assume that the original *bounds* of the solution space (delimited by the domains and constraints of the CSP) can only be restricted or relaxed, even if this does not cover all possible changes. Note that the possibility of solution loss only exists in the restrictive case. For this reason, we specialize the definition of robustness as follows.

Definition 1. *The most robust solution of a CSP with ordered domains without detailed dynamism data is the solution that maximizes the distance from all the dynamic bounds of the solution space.*

In addition, we can define the notion of stability more precisely in this framework because it is possible to define a more specific notion of closeness between two solutions thanks to the existent order over domain values than the one introduced in [6]. Here, we use the Manhattan distance ($\sum_{i=1}^{n} |s1_i - s2_i|$, where $s1$ and $s2$ are solutions).

Definition 2. *Given an order relationship over the values of a set of solutions, a solution $s1$ is more stable than another solution $s2$ iff, in the event of a change that invalidates them, there exists an alternative solution to $s1$ with lower Manhattan distance than the Manhattan distance of any alternative solution to $s2$.*

3 Searching for Robust and Stable Solutions

In this section we explain our strategy for searching for solutions that combine robustness and stability according to the definitions of Section 2. The measure of the distance from the dynamic bounds of the solution space (required for the robustness measurement) is not always obvious or easy to derive, since the constraints of the CSP may be extensionally expressed. However, some deductions about minimum distances to the bounds can be made based on the feasibility of the neighbours of a solution. This idea is first motivated with a very simple example and then is formalized.

Example 1. Figure 1 shows two solution spaces (composed by the variables x and y) whose dynamic bounds are marked by contiguous lines. The most robust solutions according to Definition 1 are highlighted. Note that there are two contiguous feasible neighbours on both sides of each assignment (discontinuous lines).

From Example 1, we conclude that *we can only ensure that a solution s is located at least at a distance d from a bound in a certain direction of the n-dimensional space if all the tuples at distances lower or equal to d from s in this direction are feasible.* Therefore, the number of feasible contiguous surrounding neighbours of the solution is a measure of its robustness and also of its stability. Because if the value assigned to a variable has at least one feasible neighbour value, then this variable is repairable.

Let $\mathcal{N}_k$ denote the neighbourhood of feasible contiguous surrounding values for a given value (assuming a specific variable and a feasible partial or complete assignment) at distance not greater than k in increasing, or decreasing, or both directions with respect to the order relationship. For the general case of CSPs with ordered domains, the desirable goal is to find contiguous surrounding feasible neighbours on both sides. For instance, in Figure 1(b), considering the partial assignment $\{x = 2\}$, $\mathcal{N}_k = \{1, 3\}$ for

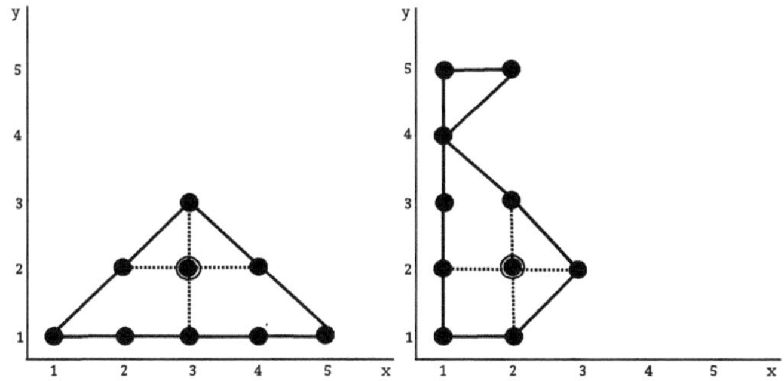

(a) Convex Space (b) Non-convex Space

Fig. 1. Most robust solutions for different solution spaces

the value 2 in the y axis (for any k value). However, for some problems it is important to consider the feasibility of neighbours only in an increasing/decreasing order [2].

In order to find solutions with the maximum number of contiguous feasible neighbours, we implemented a Branch & Bound algorithm that maximizes the sum of the sizes of $\mathcal{N}_k$ for all the variables of the assignment s. If s is an incomplete assignment, we calculate the maximum size of $\mathcal{N}_k$ of the analyzed variable, for each value of its feasible domain with respect s. Note that this objective function is an upper bound on the final total number of feasible contiguous neighbours of the solution. For the inference process, we developed an extension of the well-known Generalized Arc Consistency (GAC) that checks the feasiblity of both the analyzed value and its $\mathcal{N}_k$. An earlier description of this search algorithm can be found in [1]. For the highlighted solutions of both Figures 1(a) and 1(b) the objective function is equal to four for $k \geq 1$ (every value assigned to each solution has two contiguous neighbours on both sides).

4 Experimental Results

In this section, we describe a very limited part of the evaluation that was carried out. Experiments were run on an Intel Core i5-650 Processor (3.20 Ghz) with a time cutoff of 100 seconds. Figure 2 shows, for a fixed $k = 1$, the solutions obtained by our search algorithm ("neighbour solutions" and "(R)" is a variant). We also evaluated an ordinary CSP solver ("simple solutions") and two other methods: a WCSP modeling technique [3] ("WCSP-mod solutions") and the $(1, 0)$-super-solutions [6].

Figure 2(a) shows an analysis of robustness as a function of the tightness of the constraints (ratio of the number of forbidden assignments to the total number possible) of random CSPs with 25 variables, domain size 30 and 200 binary constraints. Here we made 500 random changes to each solution by increasing or decreasing two of their values and then checking if they were still solutions. The more neighbours that are not solutions of the CSP, the higher the likelihood of the solution becoming infeasible after changes over the bound/s. Note that our search algorithm outperformed the other approaches, specially for lower tightness. At higher tightness values, there is a lower probability of neighbour solutions (i.e. *all* solutions are located close to the bounds).

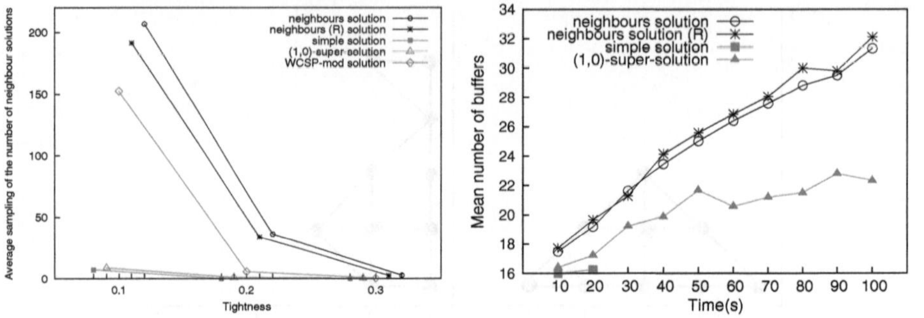

(a) Robustness analysis based on the tightness (b) Stability analysis based on computing time

Fig. 2. Robustness and stability analysis for random CSPs and scheduling problems

Figure 2(b) shows a stability analysis based on the computing time of the algorithms (discretization of 10 seconds) of the CSPs of the "e0ddr1" scheduling benchmark [7]. The mean number of buffer times is a measure of the stability because the start time of a task with an associated buffer can be delayed, for instance when there are delays in previous tasks. The most noteworthy aspect is that our search algorithm clearly outperformed the other approaches, specially when the computation time cutoff was higher.

5 Conclusions

In this paper we extend the concept of robustness and stability to deal with CSPs with discrete and ordered domains where only limited assumptions can be made about changes in these problems due to a lack of detailed dynamism information. Furthermore, we present a new search algorithm that combines criteria for both robustness and stability in this framework by searching for a solution that maximizes the sum of contiguous feasible surrounding neighbours at distances of k or less from the values of the solution. The obtained solutions have a higher probability of remaining valid after possible future restrictive changes over the constraints and domains of the original problem (robustness criterion), and they also have a high number of variables that can be easily repaired with a value at a distance lower or equal to k if they undergo a value loss (stability criterion).

Our experiments showed that our search algorithm outperformed other approaches that need only limited information about dynamism, with respect to robustness and stability as we have defined them, in cases where there were real differences in the robustness of solutions that could be obtained. The latter occurs when the problem is not so constrained that there are only a few valid solutions.

Acknowledgements. The first author is supported by the research grant from Science Foundation Ireland (SFI) under Grant Number SFI/12/RC/2289. The work was also supported by the FPU fellowship (Min. de Ciencia e Innovación, Spain) and the project TIN2013-46511-C2-1 (pending).

References

1. Climent, L., Wallace, R., Salido, M., Barber, F.: An algorithm for finding robust and stable solutions for constraint satisfaction problems with discrete and ordered domains. In: 24th IEEE International Conference on Tools with Artificial Intelligence (ICTAI 2012), pp. 874–879 (2012)
2. Climent, L., Wallace, R.J., Salido, M.A., Barber, F.: A constraint programming approach to solve scheduling problems under uncertainty. In: Workshop on Constraint Satisfaction Techniques for Planning and Scheduling Problems (COPLAS 2013) in ICAPS 2013, pp. 28–37 (2013)
3. Climent, L., Wallace, R.J., Salido, M.A., Barber, F.: Finding robust solutions for constraint satisfaction problems with discrete and ordered domains by coverings. In: Artificial Intelligence Review (AIRE) (2013), doi:10.1007/s10462-013-9420-0
4. Fargier, H., Lang, J.: Uncertainty in constraint satisfaction problems: A probabilistic approach. In: Moral, S., Kruse, R., Clarke, E. (eds.) ECSQARU 1993. LNCS, vol. 747, pp. 97–104. Springer, Heidelberg (1993)
5. Fargier, H., Lang, J., Schiex, T.: Mixed constraint satisfaction: A framework for decision problems under incomplete knowledge. In: Proceedings of the 13th National Conference on Artificial Intelligence (AAAI 1996), pp. 175–180 (1996)
6. Hebrard, E.: Robust Solutions for Constraint Satisfaction and Optimisation under Uncertainty. PhD thesis, University of New South Wales (2006)
7. Sadeh, N., Fox, M.: Variable and value ordering heuristics for the job shop scheduling constraint satisfaction problem. Artificial Intelligence 86(1), 1–41 (1996)
8. Verfaillie, G., Jussien, N.: Constraint solving in uncertain and dynamic environments: A survey. Constraints 10(3), 253–281 (2005)
9. Wallace, R., Freuder, E.: Stable solutions for dynamic constraint satisfaction problems. In: Maher, M.J., Puget, J.-F. (eds.) CP 1998. LNCS, vol. 1520, pp. 447–461. Springer, Heidelberg (1998)
10. Walsh, T.: Stochastic constraint programming. In: Proceedings of the 15th European Conference on Artificial Intelligence (ECAI 2002), pp. 111–115 (2002)
11. Yorke-Smith, N., Gervet, C.: Certainty closure: Reliable constraint reasoning with incomplete or erroneous data. Journal of ACM Transactions on Computational Logic (TOCL) 10(1), 3 (2009)

Monotone Temporal Planning: Tractability, Extensions and Applications*
(Extended Abstract)

Martin C. Cooper, Frédéric Maris, and Pierre Régnier

IRIT, University of Toulouse III, 31062 Toulouse, France
{cooper,maris,regnier}@irit.fr

Abstract. We describe a polynomially-solvable class of temporal planning problems. Polynomiality follows from two assumptions. Firstly, by supposing that each fluent (fact) can be established by at most one action, we can quickly determine which actions are necessary in any plan. Secondly, the monotonicity of fluents allows us to express planning as an instance of $STP^{\neq}$ (Simple Temporal Problem with difference constraints). This class includes temporally-expressive problems requiring the concurrent execution of actions, with potential applications in the chemical, pharmaceutical and construction industries. Any (temporal) planning problem has a monotone relaxation, which can lead to the polynomial-time detection of its unsolvability in certain cases. Indeed our relaxation is orthogonal to the relaxation based on ignoring deletes used in classical planning since it preserves deletes and can also exploit temporal information.

1 Temporal Planning

Temporal planning is an important extension of classical planning in which actions are durative and may overlap. Classical propositional planning is already PSPACE-Complete [1], and temporal planning is EXPSPACE-complete [8]. An important aspect of temporal planning is that, unlike classical planning, it permits us to model so-called temporally-expressive problems in which the execution of two or more actions in parallel is essential in order to solve the problem [5]. We define the first polytime-solvable class of temporal planning. This class includes temporally-expressive problems. It also leads to a novel relaxation of arbitrary temporal planning problems which provides a polytime sufficient condition for the detection of certain properties of actions, fluents and instances. Preliminary (and weaker) versions of this tractable class and temporal relaxation appeared in conference proceedings [2,3] before being improved in the journal paper [4] corresponding to this extended abstract.

A *fluent* is an atomic proposition (such as `door-open`). Changes to the value of a fluent are instantaneous, but conditions on the value of a fluent may be

* This is a summary of the paper Cooper M.C., Maris F., Régnier P., Monotone Temporal Planning: Tractability, Extensions and Applications, *JAIR* 50, 447-485, 2014. This research is supported by ANR Project ANR-10-BLAN-0210.

B. O'Sullivan (Ed.): CP 2014, LNCS 8656, pp. 928–932, 2014.

imposed over an interval. An *action* $a = \langle \text{Cond}(a), \text{Add}(a), \text{Del}(a), \text{Constr}(a) \rangle$ consists of a set Cond(a) of fluents which are required to be true for a to be executed, a set Add(a) of fluents which are established by a, a set Del(a) of fluents which are destroyed by a, and a set Constr(a) of interval constraints between the relative times of events which occur during the execution of a. An *event* corresponds to one of four possibilities: the establishment or destruction of a fluent by an action a, or the beginning or end of an interval over which a fluent is required by an action a. We represent an action by a rectangle whose length corresponds to its duration. Conditions are written above an action, and effects (adds or deletes) below. For example, consider the two actions shown below: LIGHT-MATCH and LIGHT-CANDLE. The action LIGHT-MATCH requires that the match be live, in order to light it. The match remains lit until it is blown out at the end of the action. A constraint in Constr(LIGHT-MATCH) imposes that the duration of the action is at most 10 seconds (at which moment the whole match has burnt). The second action LIGHT-CANDLE requires that the match be lit during two seconds for the candle to be lit.

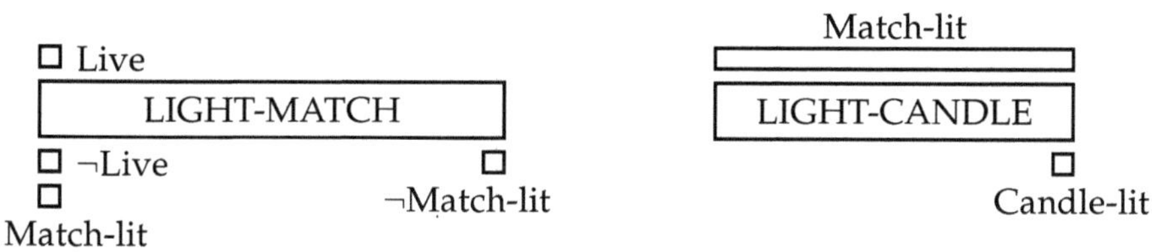

A *temporal planning problem* $\langle I, A, G \rangle$ consists of a set of actions A, an initial state I and a goal G, where I and G are sets of fluents. In a *positive* problem all fluents in G and Cond(a) (for all actions a) are positive. A *temporal plan P* for the problem $\langle I, A, G \rangle$ is a mapping τ from the events in a set of instances of actions from A to the time dimension such that all conditions of actions are true when required, all goal fluents $g \in G$ are true at the end of the execution of P and the constraints Constr(a) of each action a are satisfied (together with a technical condition ensuring that P is robust under infinitesimal shifts in the starting times of actions). Thus a temporal plan does not schedule its action-instances directly but schedules all the events in its action-instances. A plan is *minimal* if removing any non-empty subset of action-instances produces an invalid plan. For an initial state I = {live, ¬Match-lit} and a set of goals G = {Candle-lit}, it is clear that all minimal temporal plans for our example problem involve executing the two actions in parallel with the start (respectively, end) of LIGHT-MATCH being strictly before (after) the start (end) of LIGHT-CANDLE.

2 Monotonicity and Establisher-Uniqueness Imply Tractability

A set of actions A is *establisher-unique (EU)* if no fluent can be established by two distinct actions of A.

A fluent f is *−monotone** (relative to a positive temporal planning problem $\langle I, A, G \rangle$) if, after being destroyed f is never re-established in any minimal temporal plan for $\langle I, A, G \rangle$. Similarly, a fluent f is *+monotone** if, after having been established f is never destroyed in any minimal temporal plan. A fluent is monotone* if it is either + or −monotone*.

An action $a \in A$ is *unitary* for a temporal planning problem $\langle I, A, G \rangle$ if each minimal temporal plan for the $\langle I, A, G \rangle$ contains at most one instance of a. An *action landmark* is an action which occurs in each temporal plan [7].

In our example problem, both actions are clearly essential and hence landmarks. There is only one match available, which means that LIGHT-MATCH can be executed at most once (and is hence unitary). This means that the fluent Match-lit is −monotone* since it cannot be established after being destroyed. This same fluent Match-lit is not +monotone* since it is destroyed after being established. If $\nexists a_i, a_j \in A$ such that $f \in \mathrm{Add}(a_i) \cap \mathrm{Del}(a_j)$, then f is both +monotone* and −monotone*. This is the case for f = Candle-lit in our example. In certain IPC benchmark domains (**parcprinter, crewplanning, tms**), we found that many fluents were monotone* (respectively, 100%, 95% and 50% of those fluents that are either goals or liable to be established in minimal plans).

The following theorem follows from a reduction to $\mathrm{STP}^{\neq}$ [6]. The constraints created by this reduction are given in Section 3. The proof of this and all other results are given in the journal version [4] of this extended abstract.

Theorem 1. *The class of positive temporal planning problems $\langle I, A, G \rangle$ in which A is establisher-unique, all fluents are monotone* and all fluents in I are −monotone* can be solved in $O(n^3)$ time and $O(n^2)$ space, where n is the total number of events in the actions in A. Indeed, we can even find a temporal plan with the minimum number of action-instances or of minimal cost in the same complexity. Furthermore, if all actions in A are rigid (i.e. intervals between different events in the action are fixed) then the problem of finding a plan with minimum makespan is polytime approximable.*

3 Temporal Relaxation

Relaxation is ubiquitous in Artificial Intelligence. A valid relaxation of an instance I has a solution if I has a solution. Hence when the relaxation has no solution, this implies the unsolvability of the original instance I. A tractable relaxation can be built and solved in polynomial time. Our tractable class of EU monotone planning allows us to define a relaxation TR (temporal Relaxation) which is an alternative to the traditional relaxation of propositional non-temporal planning problems consisting of simply ignoring deletes. In fact, TR is a solution procedure for the class described in Theorem 1 (see [4] for a proof).

We use the notation $a \to f$ (resp., $a \to \neg f$) to denote the event that action a establishes (destroys) fluent f, and $f|\to a$ and $f \to |a$, respectively, to denote the beginning and end of the interval over which action a requires condition f. We use the notation $\tau_{\text{first}}(e)$ (respectively, $\tau_{\text{last}}(e)$) to represent the time in a plan at which an event e occurs first (resp., last).

By applying the following simple rule until convergence we can transform (in polynomial time) any temporal planning problem into a relaxed version which is EU: if a fluent f is established by two distinct actions, then delete f from the goal G and from Cond(a) for all actions a. From now on we assume the temporal planning problem is EU. We denote by A^{LM} the set of action landmarks that have been detected. Establisher-uniqueness implies that we can easily identify many such actions. The constraints of TR are as follows:

intrinsic constraints: $\forall a \in A^{\text{LM}}$, for all events e of a, $\tau_{\text{first}}(e) \leq \tau_{\text{last}}(e)$.

inherent constraints: $\forall a \in A^{\text{LM}}$, τ_{first} and τ_{last} both satisfy the interval constraints in Constr(a).

contradictory-effects constraints: no fluent is simultaneously established and destroyed by two actions.

$-$**authorisation constraints:** For each positive fluent f which is known to be $-$monotone*, $\forall a_i, a_j \in A^{\text{LM}}$, if $f \in \text{Del}(a_j) \cap \text{Cond}(a_i)$, then $\tau_{\text{last}}(f \rightarrow\mid a_i) < \tau_{\text{first}}(a_j \rightarrow \neg f)$. (If $i = j$ then the inequality is not strict [4]).

$+$**authorisation constraints:** For each positive fluent f which is known to be $+$monotone*, $\forall a_i, a_j \in A^{\text{LM}}$, if $f \in \text{Del}(a_j) \cap \text{Add}(a_i)$, then $\tau_{\text{last}}(a_j \rightarrow \neg f) < \tau_{\text{first}}(a_i \rightarrow f)$.

causality constraints: For each positive fluent f, $\forall a_i, a_j \in A^{\text{LM}}$, if $f \in (\text{Cond}(a_j) \cap \text{Add}(a_i)) \setminus I$ then $\tau_{\text{first}}(a_i \rightarrow f) < \tau_{\text{first}}(f\mid\rightarrow a_j)$. (If $i = j$ then the inequality is not strict).

goal constraints: $\text{Cond}(A^{\text{LM}}) \subseteq I \cup \text{Add}(A)$, $G \subseteq (I \setminus \text{Del}(A^{\text{LM}})) \cup \text{Add}(A)$, and for each $g \in G$, $\forall a_i, a_j \in A^{\text{LM}}$, if $g \in \text{Del}(a_j) \cap \text{Add}(a_i)$, then $\tau_{\text{last}}(a_j \rightarrow \neg g) < \tau_{\text{last}}(a_i \rightarrow g)$.

unitary constraint: For each action a which is known to be unitary (see [4] for rules for the polytime detection of unitary actions), for all events e in a, $\tau_{\text{first}}(e) = \tau_{\text{last}}(e)$.

Theorem 2. *A temporal planning problem in the tractable class described in Theorem 1 has a solution if and only if TR has a solution.*

We can use TR to detect certain properties of actions, fluents and problems.

Lemma 1. *If the temporal relaxation TR(I, A, G) of a positive temporal planning problem $\langle I, A, G\rangle$ has no solution, then $\langle I, A, G\rangle$ has no solution. If TR($I, A \setminus \{a\}, G$) has no solution, then a is a landmark action in $\langle I, A, G\rangle$.*

The detection of monotonicity* is theoretically as difficult as temporal planning, since it is EXPSPACE-complete [4]. However, TR together with extra constraints provides a powerful polytime method for detecting monotonicity*.

Lemma 2. *If $\forall a, b \in A$ s.t. $f \in \text{Add}(a) \cap \text{Del}(b)$, TR together with the constraint $\tau_{\text{first}}(a \rightarrow f) < \tau_{\text{last}}(b \rightarrow \neg f)$ is inconsistent, then f is $+$monotone*. If $\forall a, b \in A$ s.t. $f \in \text{Add}(a) \cap \text{Del}(b)$, TR together with the constraint $\tau_{\text{first}}(b \rightarrow \neg f) < \tau_{\text{last}}(a \rightarrow f)$ is inconsistent, then f is $-$monotone*.*

The class of temporal planning problems described in Theorem 1 which also have the property that all fluents can be detected as monotone* by Lemma 2 constitutes a tractable class that can be detected and solved in polynomial time.

Further research is required to determine if interesting tractable classes can be defined without the restrictive assumption of establisher-uniqueness.

References

1. Bylander, T.: The Computational Complexity of Propositional STRIPS Planning. Artificial Intelligence 69(1-2), 165–204 (1994)
2. Cooper, M.C., Maris, F., Régnier, P.: Tractable monotone temporal planning. In: Proceedings ICAPS 2012, pp. 20–28 (2012)
3. Cooper, M.C., Maris, F., Régnier, P.: Relaxation of Temporal Planning Problems. In: International Symposium on Temporal Representation and Reasoning (TIME), pp. 37–44 (2013)
4. Cooper, M.C., Maris, F., Régnier, P.: Monotone Temporal Planning: Tractability, Extensions and Applications. JAIR 50, 447–485 (2014), `http://www.irit.fr/publis/ADRIA/MonotoneJAIR.pdf`
5. Cushing, W., Kambhampati, S., Mausam, W.D.S.: When is Temporal Planning Really Temporal? In: Proc. International Joint Conference on Artificial Intelligence (IJCAI 2007), pp. 1852–1859 (2007)
6. Gerevini, A., Cristani, M.: On Finding a Solution in Temporal Constraint Satisfaction Problems. In: Proc. International Joint Conference on Artificial Intelligence (IJCAI 1997), pp. 1460–1465 (1997)
7. Karpas, E., Domshlak, C.: Cost-optimal planning with landmarks. In: Proc. International Joint Conference on Artificial Intelligence (IJCAI 2009), pp. 1728–1733 (2009)
8. Rintanen, J.: Complexity of Concurrent Temporal Planning. In: Proc. 17th International Conference on Automated Planning and Scheduling (ICAPS 2007), pp. 280–287 (2007)

Anytime AND/OR Depth-First Search for Combinatorial Optimization*
(Extended Abstract)

Lars Otten[1,**] and Rina Dechter[2]

[1] University of California, Irvine, USA
lotten@uci.edu
[2] University of California, Irvine, USA
dechter@ics.uci.edu

Abstract. One popular and efficient scheme for solving combinatorial optimization problems over graphical models *exactly* is depth-first Branch and Bound. However, when the algorithm exploits problem decomposition using AND/OR search spaces, its anytime behavior breaks down. This article 1) analyzes and demonstrates this inherent conflict between effective exploitation of problem decomposition (through AND/OR search spaces) and the anytime behavior of depth-first search (DFS), 2) presents a new search scheme to address this issue while maintaining desirable DFS memory properties, and 3) analyzes and demonstrates its effectiveness through comprehensive empirical evaluation. Our work is applicable to *any* problem that can be cast as search over an AND/OR search space.

1 Introduction

Max-product problems over graphical models, generally known as MPE (most probable explanation) or MAP (maximum a posteriori) inference, have many applications with practical significance, ranging from computational biology and genetics to scheduling tasks and coding networks. One established and efficient class of algorithms for solving these problems exactly is depth-first Branch and Bound over AND/OR search spaces. Developed in the past decade within the probabilistic reasoning and constraint communities, these methods are effective because they use sophisticated lower bound schemes such as soft arc-consistency [1] or the mini-bucket heuristic [2,3], because they avoid redundant computation using caching schemes, and most significantly, because they take advantage of problem decomposition by exploring an AND/OR search space [4] or an equivalent representation. The efficiency of these algorithms was established in several evaluations, including recent UAI competitions [5], and their properties when used for exact computation are well documented [6,3,7].

A principled alternative is presented by best-first schemes, but while provably superior in terms of number of node expansions, these often fail when a problem has large induced width due to the generally exponential size of the algorithm's OPEN list; moreover, they can only provide a solution at termination [7]. Depth-first search is therefore

* This is a summary of the full article published in AI Communications, Volume 25(3) [13].
** Now at Google Inc. (ottenl@google.com).

B. O'Sullivan (Ed.): CP 2014, LNCS 8656, pp. 933–937, 2014.

often preferred because of its flexibility in working with bounded memory – the OPEN list of nodes grows linearly – and because of its *anytime behavior*. Namely, when finding a feasible solution is easy but an optimal one is hard, depth-first Branch and Bound generates solutions that get better and better over time, until it eventually discovers an optimal one. Thus it can function also as an approximation scheme for otherwise infeasible problems or when time is limited [8].

Indeed, in the 2010 UAI Approximate Inference Challenge participating Branch and Bound solvers performed competitively with respect to approximation (placing 1st and 3rd in some categories). But we also observed an inability to produce even a single solution on some instances, especially when the time bound was small. Thus motivated, this article demonstrates that the issue is rooted in the underlying AND/OR search space.

These search spaces were originally introduced to graphical models to facilitate problem decomposition during search (e.g. [9]) and can be explored by any search strategy. When traversed depth-first, however, all but one decomposed subproblem will be *fully solved* before a single overall solution can be composed, voiding the algorithm's anytime characteristics.

This article's main contribution is a new Branch and Bound scheme over AND/OR search spaces, called *Breadth-Rotating AND/OR Branch and Bound (BRAOBB)* that addresses the anytime issue in a principled way, while maintaining the favorable complexity guarantees of depth-first search. The algorithm combines depth-first and breadth-first exploration by periodically rotating over the different subproblems, each of which is processed depth-first.

Experimental evaluation is conducted on a variety of benchmark domains, including haplotype computation problems in genetic pedigrees, random grid networks, and protein side-chain prediction instances. We compare BRAOBB against one of the best variants of AND/OR branch and Bound search, AOBB [3], and against an "ad hoc" fix that we suggest – the latter algorithm relies on a heuristic to quickly find a solution to each subproblem before reverting to depth-first search. We furthermore compare against a state-of-the-art stochastic local search solver, which is specifically targeted at anytime performance but cannot provide any proof of optimality [10].

The empirical results demonstrate superior anytime behavior of BRAOBB, especially over problematic cases where standard AOBB and its ad hoc fix fail, including several very hard instances from the 2010 UAI Approximate Inference Challenge that were made available and three weighted constraint satisfaction problem instances that are known to be very complex. We also show how combining local search and exhaustive AND/OR search lets us enjoy the benefits of both approaches. Notably, a solver based on this concept recently won all three categories (20 seconds, 20 minutes, and 1 hour) in the MPE track of the PASCAL 2011 Inference Challenge [11,12], the successor to the 2010 UAI Challenge.

2 Brief Overview of Results

As noted above, in AND/OR search spaces depth-first traversal of a set of independent subproblems will solve to completion all but one subproblem before the last one is even considered. As a consequence, the first generated overall non-optimal solution contains

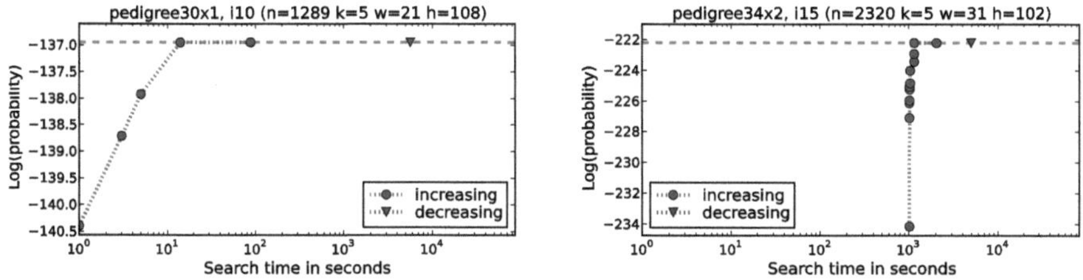

Fig. 1. Anytime performance of AOBB for differing subproblem oderings. Specified for each network: number of variables n, max. domain size k, induced width w along the chosen ordering, height of the corresponding pseudo tree h. The dashed gray line indicates the optimal solution.

conditionally optimal solutions to all subproblems but the last one. Furthermore, depending on the problem structure and the complexity of the independent subproblems, the time to return even this first non-optimal overall solution can be significant, practically negating the anytime behavior of depth-first search (DFS).

To illustrate, consider Figure 1, which depicts the anytime performance (best-known solution cost over time) of AOBB on two example problem instances. For demonstration purposes we apply a simple heuristic which has AOBB consider independent subproblems by increasing or decreasing hardness, based on the subproblem induced width. If decomposition yields only one large subproblem and several smaller ones, the latter can be solved depth-first in relatively little time, to be then combined with the incrementally improving solutions of the larger subproblem. This is exemplified by applying the "increasing" order to pedigree30x1, which has one hard subproblem and several other, simple ones: we see a suboptimal overall solution right away which is gradually improved upon; using the "decreasing" order AOBB spends a long time on solving the hardest subproblem to completion before returning any overall solution.

In case of pedigree34x2, however, decomposition yields two complex subproblems: the increasing subproblem order still outperforms its inverse, yet it returns the initial solution only after about 1,000 seconds. In fact, no possible subproblem ordering can lead to acceptable anytime behavior in this case due to the structure of subproblems, clearly highlighting the limits of this approach.

2.1 Breadth-Rotating AND/OR Branch and Bound (BRAOBB)

To remedy this issue, BRAOBB combines depth-first exploration with the notion of "rotating" through different subproblems in a breadth-first manner. Namely, node expansion still occurs depth-first as in standard AOBB, but the algorithm takes turns in processing subproblems, each up to a given number of operations at a time, round-robin style. This lets us develop all branches of the solution tree "simultaneously".

More systematically, the algorithm maintains a list of currently open subproblems and repeats the following high-level steps until completion:

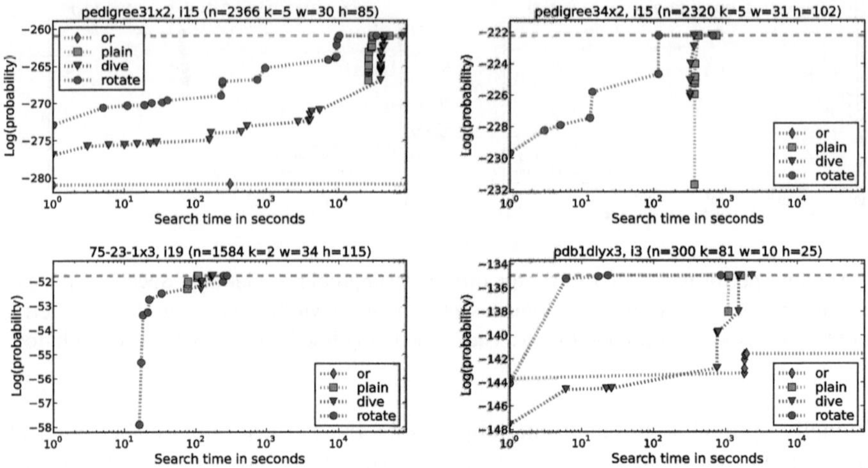

Fig. 2. Anytime performance of BRAOBB ("rotate") compared against "plain" AOBB and two other schemes (OR search and AOBB with "dive" extension, as outlined in the full article)

1. Move to next open subproblem P in a breadth-first fashion.
2. Process P depth-first, until either:
   (a) P is solved optimally,
   (b) P decomposes into child subproblems, or
   (c) a predefined threshold number of operations is reached.

The threshold in (c) is needed to ensure the algorithm does not get stuck in one large subproblem where the other two conditions, (a) and (b), do not occur for a long time. Furthermore, in order to focus on a single solution tree at a time, a subproblem is only considered "open" if it does not currently have any child subproblems. More details, algorithm pseudo code, and theoretical analysis are given in the full article [13].

Figure 2 shows four examples for the anytime perfomance of BRAOBB. For reference the plots also include AOBB and plain OR search, as well as AOBB with a "dive" extension (which performs an initial greedy dive into each subproblem – details in the full article). From the results it is clear that BRAOBB holds a decisive advantage over the other schemes evaluated here. It generally returns a first solution quickly and is consistently the first scheme, or one of the first, to do so. Furthermore, in almost all cases it proceeds to improve upon the initial solution quickly, again outperforming other schemes in the evaluation.

The full article also contains a number of summary statistics, for instance showing that after 5 seconds BRAOBB has found an initial solution for 510 out of 543 problem instances, compared to 269 for plain AOBB. And after 1 minute, BRAOBB has found the optimal solution for 321 instances compared to 274 for plain AOBB – again, refer to the full article for more details [13].

Moreover, the article also provides analysis of BRAOBB from several angles, including complexity analysis that shows that BRAOOB retains the favorable asymptotic guarantees of "plain" AND/OR search.

References

1. Larrosa, J., Schiex, T.: Solving weighted CSP by maintaining arc consistency. Artif. Intell. 159(1-2), 1–26 (2004)
2. Dechter, R., Rish, I.: Mini-buckets: A general scheme for bounded inference. Journal of the ACM 50(2), 107–153 (2003)
3. Marinescu, R., Dechter, R.: AND/OR Branch-and-Bound search for combinatorial optimization in graphical models. Artif. Intell. 173(16-17), 1457–1491 (2009)
4. Nilsson, N.: Artificial Intelligence: A New Synthesis. Morgan Kaufmann, San Francisco (1998)
5. Elidan, G., Globerson, A.: UAI 2010 approximate inference challenge, `http://www.cs.huji.ac.il/project/UAI10/`
6. Jégou, P., Terrioux, C.: Decomposition and good recording for solving max-CSPs. In: Proc. of the 16th Eureopean Conference on Artificial Intelligence (ECAI 2004), pp. 196–200. IOS Press, Amsterdam (2004)
7. Marinescu, R., Dechter, R.: Memory intensive AND/OR search for combinatorial optimization in graphical models. Artif. Intell. 173(16-17), 1492–1524 (2009)
8. Zilberstein, S.: Using anytime algorithms in intelligent systems. AI Magazine 17(3), 73–83 (1996)
9. Dechter, R., Mateescu, R.: AND/OR search spaces for graphical models. Artif. Intell. 171(2-3), 73–106 (2007)
10. Hutter, F., Hoos, H.H., Stützle, T.: Efficient stochastic local search for MPE solving. In: Proc. of the 19th International Joint Conference on Artificial Intelligence (IJCAI 2005), pp. 169–174. Professional Book Center, Denver (2005)
11. Elidan, G., Globerson, A., Heinemann, U.: PASCAL 2011 probabilistic inference challenge, `http://www.cs.huji.ac.il/project/PASCAL/`
12. Otten, L., Ihler, A., Kask, K., Dechter, R.: Winning the PASCAL 2011 MAP Challenge with Enhanced AND/OR Branch-and-Bound.
13. Otten, L., Dechter, R.: Anytime AND/OR depth-first search for combinatorial optimization. AI Communications 25(3) (2012)

View-Based Propagator Derivation*
(Extended Abstract)

Christian Schulte[1] and Guido Tack[2]

[1] SCALE, KTH Royal Institute of Technology, Sweden
cschulte@kth.se
[2] National ICT Australia (NICTA) and Monash University, Australia
guido.tack@monash.edu

Abstract. When implementing a propagator for a constraint, one must decide about variants: When implementing min, should one also implement max? Should one implement linear equations both with and without coefficients? Constraint variants are ubiquitous: implementing them requires considerable effort, but yields better performance. This abstract shows how to use *views* to *derive* propagator variants where derived propagators are *perfect* in that they inherit essential properties such as correctness and domain and bounds consistency. Techniques for systematically deriving propagators are developed, and the abstract sketches an implementation architecture for views that is independent of the underlying constraint programming system. Evaluation of views implemented in Gecode shows that derived propagators are efficient and that views often incur no overhead. Views have proven essential for implementing Gecode, substantially reducing the amount of code that needs to be written and maintained.

1 Introduction

When implementing a propagator for a constraint, one typically must also decide whether to implement some of its variants. When implementing a propagator for the constraint $\max\{x_1,\ldots,x_n\} = y$, should one also implement $\min\{x_1,\ldots,x_n\} = y$? The latter can be implemented using the former as $\max\{-x_1,\ldots,-x_n\} = -y$. When implementing a propagator for the reified linear equation $(\sum_{i=1}^n x_i = c) \Leftrightarrow b$, should one also implement $(\sum_{i=1}^n x_i \neq c) \Leftrightarrow b$? These two constraints only differ by the sign of b, as the latter is equivalent to $(\sum_{i=1}^n x_i = c) \Leftrightarrow \neg b$.

The two straightforward approaches for implementing constraint variants are to either implement dedicated propagators for the variants, or to decompose. In the last example, for instance, the reified constraint could be decomposed into two propagators, one for $(\sum_{i=1}^n x_i = c) \Leftrightarrow b'$, and one for $b \leftrightarrow \neg b'$, introducing an additional variable b'.

Implementing the variants inflates code and documentation and is error prone. Given the potential code explosion, one may be able to only implement some variants (say, min and max). Other variants important for performance (say, ternary min and max) may be infeasible due to excessive programming and maintenance effort. Decomposing, on the other hand, massively increases memory consumption and runtime.

* This is a summary of the paper *View-based propagator derivation* by Christian Schulte and Guido Tack, Constraints 18(1), 75–107 (2013).

B. O'Sullivan (Ed.): CP 2014, LNCS 8656, pp. 938–942, 2014.

Our paper *View-based Propagator Derivation* [9] presents a third approach, introducing *views* to *derive* propagators, which combines the efficiency of dedicated propagator implementations with the simplicity and effortlessness of decomposition. Some preliminary results from this paper were presented as a poster at CP 2008 [8] which contains a formal model of views and derived propagators, which was used to prove that derived propagators inherit properties such as correctness and propagation strength.

This extended abstract focuses on the majority of the material in the journal paper that has not been presented at a conference. In particular, it describes techniques for *systematically* deriving propagators such as transformation, generalization, specialization, and type conversion; it presents an *implementation architecture* for views based on parametric polymorphism; and it summarizes the experimental evaluation that shows that derived propagators are efficient and that views often incur no runtime overhead.

2 Deriving Propagators Using Views

We call the basic building block for propagator derivation a *view*. A view can be regarded as a restricted form of a bi-directional *indexical* [2,10], or an expression such as those supported by IBM ILOG CP Optimizer [4], where the restrictions have been chosen carefully such that the resulting derived propagator satisfies important properties concerning correctness and effectiveness, and such that the implementation does not incur any overhead.

Consider a propagator for the constraint $\max(x,y) = z$. Given three additional propagators for $x' = -x$, $y' = -y$, and $z' = -z$, we could propagate the constraint $\min(x',y') = z'$ using the propagator for $\max(x,y) = z$. Instead of these three additional propagators, we will derive a propagator for max from the propagator for min using views that perform the simple negation transformations.

Views transform input and output of a propagator. For example, a minus view on a variable x transforms the variable domain of x by negating each element. When a propagator *reads* the domain of the minus view, the view returns this transformed domain of x. When the propagator *updates* the domain (e.g. changing a bound or removing a value), the view performs the inverse transformation before updating the domain of x. With views, the implementation of the maximum propagator can be reused: a propagator for the minimum constraint can be derived from a propagator for the maximum constraint and a minus view for each variable.

3 Propagator Derivation Techniques

This section introduces systematic techniques for deriving propagators using views. The techniques capture the transformation, generalization, specialization, and type conversion of propagators. Each technique is illustrated with an example.

Transformation. A transformation view performs a simple operation such as a negation or inversion of a variable domain. The most basic examples are *negation views* for Boolean variables, which correspond to *literals* in SAT solvers. From disjunction $x \vee y = z$ one can then derive conjunction $x \wedge y = z$ with negation views on x, y, z, and implication $x \rightarrow y = z$ with a negation view on x. From equivalence $x \leftrightarrow y = z$ one can derive exclusive or $x \oplus y = z$ with a negation view on z.

Scheduling propagators. Many propagation algorithms for constraint-based scheduling [1] are based on tasks t characterized by start time, processing time, and end time. These algorithms are typically expressed in terms of earliest start time ($\mathrm{est}(t)$), latest start time ($\mathrm{lst}(t)$), earliest completion time ($\mathrm{ect}(t)$), and latest completion time ($\mathrm{lct}(t)$).

Another particular aspect of scheduling algorithms is that they are often required in two dual variants. Let us consider not-first/not-last propagation as an example. Assume a set of tasks T and a task $t \notin T$ to be scheduled on the same resource. Then t cannot be scheduled before the tasks in T, if $\mathrm{ect}(t) > \mathrm{lst}(T)$ (where $\mathrm{lst}(T)$ is a conservative estimate of the latest start time of all tasks in T) and hence $\mathrm{est}(t)$ can be adjusted. The dual variant is that t is not-last: if $\mathrm{ect}(T) > \mathrm{lst}(t)$ then $\mathrm{lct}(t)$ can be adjusted. Running the dual variant of a scheduling algorithm on tasks $t \in T$ is the same as running the original algorithm on the *dual tasks* $t' \in T'$, which are simply mirrored at the 0-origin of the time scale: $\mathrm{est}(t') = -\mathrm{lct}(t)$, $\mathrm{ect}(t') = -\mathrm{lst}(t)$, $\mathrm{lst}(t') = -\mathrm{ect}(t)$, and $\mathrm{lct}(t') = -\mathrm{est}(t)$. The dual variant of a scheduling propagator can be automatically derived using a minus view that transforms the time values. In our example, only a propagator for not-first needs to be implemented and the propagator for not-last can be derived (or vice versa).

Generalization. Common views for integer variables capture linear transformations of the integer values: an *offset view* for a variable x and a constant offset o behaves like $x + o$, while a *scale view* for $a \in \mathbb{Z}$ on x behaves like $a \times x$.

Offset and scale views are useful for generalizing propagators. Generalization has two key advantages: simplicity and efficiency. A more specialized propagator is often simpler to implement (and simpler to implement *correctly*) than a generalized version. The specialized version can save memory and runtime during execution.

We can devise an efficient propagation algorithm for the common case of a linear equality constraint with unit coefficients $\sum_{i=1}^{n} x_i = c$. The more general case $\sum_{i=1}^{n} a_i x_i = c$ can be derived by using scale views for a_i on x_i.

Specialization. We employ *constant views* to specialize propagators. A constant view behaves like an assigned variable. In practice, specialization has two advantages. Fewer variables require less memory. And specialized propagators can be compiled to more efficient code (constants are known at compile time). Few examples for specialization are: a propagator for binary linear inequality $x + y \leq c$ derived from a propagator for $x + y + z \leq c$ by using a constant 0 for z; propagators for the counting constraints $|\{i \mid x_i = c\}| = z$ and $|\{i \mid x_i = y\}| = c$ from a propagator for $|\{i \mid x_i = y\}| = z$; a propagator for set disjointness from a propagator for $x \cap y = z$ and a constant empty set for z.

Type conversion. A type conversion view lets propagators for one type of variable work with a different type, by translating the underlying representation. For example, Boolean variables are essentially integer variables restricted to the values $\{0, 1\}$. CP systems may choose a more efficient implementation for Boolean variables and hence the types for integer and Boolean variables differ. By wrapping an efficient Boolean variable in an *integer view*, all integer propagators can be directly reused.

4 Implementation

The implementation architecture for views and derived propagators is based on making propagators *parametric*. Deriving a propagator then means *instantiating* a parametric propagator with views. The architecture is an orthogonal layer of abstraction on top of any solver implementation.

In an object-oriented implementation of this model, a propagator is an object with a `propagate` method that *accesses* and *modifies* a domain through the methods of variable objects. Such an object-oriented model is used for example by ILOG Solver [7] and Choco [5], and is the basis of most of the current propagation-based constraint solvers.

In order to derive a propagator using view objects, we use *parametricity*, a mechanism provided by the implementation language that supports the instantiation of the same code (the propagator) with different parameters (the views). In C++, for example, a propagator is based on C++ templates, it is *parametric* over the types of the views it uses and can be *instantiated* with any view that provides the necessary operations. This type of parametricity is called *parametric polymorphism*, and is available in other programming languages for example in the form of Java generics [3] or Standard ML functors [6].

In Gecode (version 3.7.2), views are used to derive 616 propagators from 154 propagator implementations. On average, every propagator implementation therefore results in four derived propagators. Propagator implementations in Gecode account for almost 40 000 lines of code and 21 000 lines of documentation. As a rough estimate, deriving propagators using views thus saves around 120 000 lines of code and 60 000 lines documentation to be written, tested, and maintained. Views make up less than 8 000 lines of code, yielding a 1500% return on investment.

5 Experimental Evaluation

This section is a summary of our experimental evaluation of views in Gecode (version 3.7.2).

Our experiments showed that generalization and specialization can be implemented without any performance overhead compared to handwritten propagators. Transformation views on Boolean and integer variables (negation) showed negligible overhead. For set constraints, the basic view operations cannot be fully optimized by the compiler, resulting in an overhead of 32% (geometric mean) across a number of experiments. We also evaluated a decomposition of a constraint compared to a propagator derived using views. The geometric mean of the runtime overhead in this case was 175%, with some examples running almost four times slower. Finally, we evaluated the effect of using parametric polymorphism in C++, comparing it to virtual function calls, which we found to be 28% slower in the geometric mean.

The experiments show that deriving propagators using C++ templates yields competitive (in many cases optimal) performance compared to dedicated handwritten propagators. The results also clarify that deriving propagators is vastly superior to decomposing the constraints into additional variables and simple propagators.

Acknowledgements. Christian Schulte has been partially funded by the Swedish Research Council (VR) under grant 621-2004-4953. NICTA is funded by the Australian Government through the Department of Communications and the Australian Research Council through the ICT Centre of Excellence Program.

References

1. Baptiste, P., Le Pape, C., Nuijten, W.: Constraint-based Scheduling. International Series in Operations Research & Management Science. Kluwer Academic Publishers (2001)
2. Carlsson, M., Ottosson, G., Carlson, B.: An open-ended finite domain constraint solver. In: Hartel, P.H., Kuchen, H. (eds.) PLILP 1997. LNCS, vol. 1292, pp. 191–206. Springer, Heidelberg (1997)
3. Gosling, J., Joy, B., Steele, G., Bracha, G.: The Java Language Specification, 3rd edn. Addison-Wesley Professional (2005)
4. IBM Corporation: IBM ILOG CP Optimizer V2.3 User's Manual (2009)
5. Laburthe, F.: Choco: Implementing a CP kernel. In: Beldiceanu, N., Harvey, W., Henz, M., Laburthe, F., Monfroy, E., Müller, T., Perron, L., Schulte, C. (eds.) Proceedings of TRICS: Techniques foR Implementing Constraint Programming Systems, a Post-Conference Workshop of CP 2000, pp. 71–85 (2000)
6. Milner, R., Tofte, M., MacQueen, D.: The Definition of Standard ML. MIT Press, Cambridge (1997)
7. Puget, J.F.: A C++ implementation of CLP. In: Proceedings of the Second Singapore International Conference on Intelligent Systems (SPICIS), Singapore, pp. B256–B261 (November 1994)
8. Schulte, C., Tack, G.: Perfect derived propagators. In: Stuckey, P.J. (ed.) CP 2008. LNCS, vol. 5202, pp. 571–575. Springer, Heidelberg (2008)
9. Schulte, C., Tack, G.: View-based propagator derivation. Constraints 18(1), 75–107 (2013), `http://dx.doi.org/10.1007/s10601-012-9133-z`
10. Van Hentenryck, P., Saraswat, V.A., Deville, Y.: Design, implementation, and evaluation of the constraint language cc(FD). Journal of Logic Programming 37(1-3), 293–316 (1998)

Author Index

MIX
Papier aus verantwortungsvollen Quellen
Paper from responsible sources
FSC® C105338

If you have any concerns about our products,
you can contact us on
ProductSafety@springernature.com

In case Publisher is established outside the EU,
the EU authorized representative is:
Springer Nature Customer Service Center GmbH
Europaplatz 3, 69115 Heidelberg, Germany

Printed by Libri Plureos GmbH
in Hamburg, Germany